DISEASES OF TREES AND SHRUBS

DISEASES OF TREES AND SHRUBS

By **WAYNE A. SINCLAIR,**

Professor of Plant Pathology,
Cornell University

HOWARD H. LYON,

Photographer, Department of Plant Pathology,
Cornell University

and **WARREN T. JOHNSON,**

Professor of Entomology,
Cornell University

COMSTOCK PUBLISHING ASSOCIATES, a division of
CORNELL UNIVERSITY PRESS | ITHACA AND LONDON

First published 1987 by Cornell University Press.

Library of Congress Cataloging-in-Publication Data

Sinclair, Wayne A., 1936–
    Diseases of trees and shrubs.

    Bibliography: p. 512.
    Includes index.
    1. Trees—Diseases and pests—United States. 2. Shrubs—Diseases and pests—United States. 3. Trees—Diseases and pests—Canada. 4. Shrubs—Diseases and pests—Canada. 5. Trees—Wounds and injuries—United States. 6. Shrubs—Wounds and injuries—United States. 7. Trees—Wounds and injuries—Canada. 8. Shrubs—Wounds and injuries—Canada. I. Lyon, Howard H. II. Johnson, Warren T. III. Title.
    SB762.S56 1987   635.9'77   86-29984
    ISBN 0-8014-1517-9

Printed in Japan

The paper in this book is acid-free and meets the guidelines for permanence and durability of the Committee on Production Guidelines for Book Longevity of the Council on Library Resources.

Contents

About This Book   7
Acknowledgments   9

Downy Mildews, Plate 1   12
Powdery Mildews, Plates 2–4   14
Witches'-broom of Hackberry and Black Witches'-broom of
  Serviceberry, Plate 5   20
Taphrina Diseases: Leaf Blisters and Leaf Curls, Bladder
  Plums, and Witches'-brooms, Plates 6–7   22
Exobasidium Galls and Blisters, Plate 8   26
Algal Leaf Spot, or Green Scurf, Plate 9   28
Sooty Molds, Black Mildews, and Other Dark Superficial
  Fungi, Plate 10   30
Lophodermium Needle Casts, Plate 11   32
Ploioderma Needle Casts of Pine, Plate 12   34
Elytroderma Needle Cast of Pine, Plate 13   36
Lophodermella and Cyclaneusma (Naemacyclus) Needle
  Casts of Pine, Plate 14   38
Rhabdocline and Phaeocryptopus Needle Casts of
  Douglas-fir, Plate 15   40
Rhizosphaera and Isthmiella Needle Casts of Spruce,
  Plate 16   42
Meria Needle Blight of Larch and Didymascella Leaf Blight
  of Cedar, Plate 17   44
Brown Spot Needle Blight of Pine, Plate 18   46
Dothistroma Needle Blight of Pine, Plate 19   48
Brown Felt Blights and Snow Blights of Conifers,
  Plates 20–21   50
Tar Spots of Maple, Plate 22   54
Ink Spot Leaf Blight of Aspen and Poplar, Plate 23   56
Ovulinia Petal Blight of Azalea and Rhododendron and
  Ciborinia Petal Blight of Camellia, Plate 24   58
Botrytis Blight, Plate 25   60
Cristulariella Leaf Spots, Plate 26   62
Entomosporium Leaf Spot of Pomoidae, Plate 27   64
Black Spot of Rose and Coccomyces Leaf Spot of Cherry
  and Plum, Plate 28   66
Spot Anthracnoses and Scabs Caused by Elsinoë and
  Sphaceloma, Plate 29   68
Mycosphaerella Leaf Spots of Ash and Yucca, Plate 30   70
Greasy Spot of Citrus and Black Rot and Cladode Spot of
  Prickly Pear, Plate 31   72
Septoria Leaf Spots and Septoria Canker of Poplar, Plate 32   74
Phyllosticta Leaf Spots, Plates 33–34   76
Guignardia Blotch of Horse-chestnut and Buckeye and
  Guignardia Leaf Spot of Boston Ivy, Plate 35   80
Coniothyrium and Hendersonia Leaf Spots, Plate 36   82
Linospora Leaf Blight of Balsam Poplar and Fly Speck Leaf
  Spot of Huckleberry, Plate 37   84
Cercospora Blights and Leaf Spots, Plates 38–39   86
Alternaria Blights and Leaf Spots, Plate 40   90
Stigmina Leaf Spot and False Smut of Palm, Plate 41   92
Insolibasidium Blight of Honeysuckle and Rhizoctonia
  Blight of Natal Plum, Plate 42   94
Apple Scab and Venturia Leaf Blight of Maple, Plate 43   96
Scabs of Firethorn, Loquat, Toyon, and Pear, Plate 44   98

Venturia Leaf and Shoot Blights of Aspen and Poplar,
  Plate 45   100
Scab and Black Canker of Willow, Plate 46   102
Anthracnoses and Didymosporina Leaf Spot of Maple,
  Plate 47   104
Ash Anthracnose, Plate 48   106
Black Spot of Elm and Gnomonia Leaf Spot of Hickory,
  Plate 49   108
Oak Anthracnose, Plate 50   110
Sycamore Anthracnose, Plate 51   112
Anthracnoses of Filbert, Redbud, and Birch, Plate 52   114
Anthracnose of Hornbeam and Hop Hornbeam, Plate 53   116
Walnut Anthracnose and Downy Leaf Spot of Walnut and
  Hickory, Plate 54   118
Marssonina Spots and Blights of Poplar, Birch, and
  Bittersweet, Plate 55   120
Apioplagiostoma Blight of Aspen, Plate 56   122
Anthracnoses and Diebacks Caused by Glomerella
  cingulata and Colletotrichum gloeosporioides,
  Plates 56–57   122
Brown Rot of Stone Fruit Trees, Plate 58   126
Pestalotiopsis Spots, Blights, and Diebacks, Plate 59   128
Ascochyta Blight of Lilac, Plate 60   130
Sirococcus Canker of Butternut, Plate 61   132
Sirococcus Blight of Conifers, Plate 62   134
Sphaeropsis Blight of Pine and Other Conifers, Plate 63   136
Phomopsis, Kabatina, and Sclerophoma Blights of Juniper
  and Other Gymnosperms, Plate 64   138
Phomopsis Canker of Russian Olive, Plate 65   140
Diaporthe and Phomopsis Blights, Cankers, and Diebacks,
  Plates 66–67   142
Phomopsis Galls, Plate 68   146
Phomopsis Canker of Gardenia and Nectriella Gall of
  Ornamental Plants, Plate 69   148
Sphaeropsis Knot, Plate 70   150
Black Knot of Prunus, Plate 71   152
Bacterial Galls of Olive, Oleander, and Douglas-fir, Plate 72   154
Crown Gall, Plate 73   156
Bacterial Spots of English Ivy and California Laurel and
  Bacterial Rots of Cacti, Plate 74   158
Bacterial Blights and Cankers Caused by Pseudomonas
  syringae pv. syringae and Xanthomonas campestris pv.
  juglandis, Plate 75   160
Fire Blight, Plates 76–77   162
Bacterial Cankers, Plate 78   166
Smooth Patch and Bark Rot and Hendersonula Dieback,
  Plate 79   168
Seiridium Canker of Cypress, Plate 80   170
Botryosphaeria and Botryodiplodia Cankers and Diebacks,
  Plates 81–86   172
Cryptodiaporthe Cankers, Plates 87–88   184
Cryphonectria Cankers, Plates 88–90   186
Endothia Canker, Plate 91   192
Leucostoma and Valsa Cankers, Plates 92–96   194
Cryptosporiopsis Canker of Maple, Plate 96   202

Thyronectria Canker of Honey Locust, Plate 97 — 204

Volutella Blight of Pachysandra, Plate 98 — 206

Tubercularia and Nectria Cankers and Diebacks, Plates 99–101 — 208

Twig Blight of Mulberry and Fusarium Cankers of Golden-chain and Other Trees, Plate 102 — 214

Pitch Canker of Pine, Plate 103 — 216

Cryptosphaeria Canker of Aspen, Plate 104 — 218

Eutypella Canker of Maple, Plate 105 — 220

Hypoxylon Cankers and Diebacks, Plates 106–108 — 222

Sooty-bark Canker of Aspen, Plate 109 — 228

Cenangium Dieback of Pine, Plate 110 — 230

Scleroderris (Ascocalyx) Canker of Conifers, Plate 111 — 232

Atropellis Cankers of Pine, Plate 112 — 234

Strumella Canker of Oak and Lachnellula Cankers of Conifers, Plate 113 — 236

Rose Rust and Fig Rust, Plate 114 — 238

Gymnosporangium Rusts, Plates 115–119 — 240

Puccinia Rusts of Currant and Buckthorn and Cumminsiella Rust of Oregon Grape, Plate 120 — 250

Ash Rust and Birch Rust, Plates 121–122 — 252

Melampsora Rusts, Plates 123–124 — 256

Pucciniastrum Rusts, Plates 125–126 — 260

Broom Rust of Fir and Fir-Fern Rusts, Plate 127 — 264

Chrysomyxa Rusts of Spruce, Plate 128 — 266

Needle Rusts of Pine, Plate 129 — 268

Stem and Cone Rusts of Pine, Plates 130–136 — 270

Diseases Caused by *Phytophthora* Species, Plates 137–140 — 284

Phymatotrichum and Charcoal Root Rots, Plate 141 — 292

Cylindrocladium Root Rot and Blight and Thielaviopsis Root Rot, Plate 142 — 294

Southern Blight and Fusarium Root Rot, Plate 143 — 296

Nematode Diseases of Roots, Plates 144–145 — 298

Root and Butt Rot and Basal Canker Caused by *Hypoxylon deustum*, Plate 146 — 304

Dematophora (Rosellinia) and Xylaria Root Rots, Plate 147 — 306

Armillaria Root Rot, Plates 148–150 — 308

Annosum Root and Butt Rot, Plates 151–152 — 314

Red Root and Butt Rot of Conifers, Caused by *Inonotus circinatus* and *I. tomentosus*, Plates 153–154 — 318

Brown Root and Butt Rot of Conifers, Caused by *Phaeolus schweinitzii*, Plate 155 — 322

Laminated Root Rot of Conifers, Caused by *Phellinus weirii*, Plates 156–157 — 324

Root and Butt Rots Caused by *Inonotus dryadeus* and *Oxyporus latemarginatus*, Plate 158 — 328

Ganoderma Root and Butt Rots and Trunk Decay, Plates 159–161 — 330

Compartmentalization of Wound-associated Discolored and Decayed Wood, Plates 162–163 — 336

Drought Cracks and Frost Cracks in Relation to Internal Defect, Plate 164 — 340

Trunk and Limb Rots of Hardwoods, Plates 165–167 — 342

Trunk Rots of Conifers, Plates 168–169 — 348

Canker-rots of Hardwoods, Plates 170–173 — 352

Ceratocystis Cankers and Sapstreak Disease, Plates 174–175 — 360

Oak Wilt, Plate 176 — 364

Dutch Elm Disease, Plate 177 — 366

Black Stain Root Disease of Conifers, Plate 178 — 368

Procera Root Disease of Conifers, Plate 179 — 370

Persimmon Wilt, Plate 180 — 372

Verticillium Wilt, Plates 181–182 — 374

Mimosa Wilt, Plate 183 — 378

Wilt of Conifers Caused by the Pine Wood Nematode, Plate 184 — 380

Bacterial Wetwood and Slime Flux, Plate 185 — 382

Bacterial Scorch Diseases, Plates 186–187 — 384

Elm Yellows, Plate 188 — 388

Lethal Yellowing of Palm, Plate 189 — 390

X-disease of Cherry and Peach, Plate 190 — 392

Ash Yellows, Plate 191 — 394

Witches'-brooms Caused by Mycoplasmalike Organisms, Plate 192 — 396

Pecan Bunch, Pear Decline, and Citrus Stubborn, Plate 193 — 398

Viral Diseases of Aspen and Poplar, Plates 194–195 — 400

Cherry Leafroll Virus in Walnut and Dogwood, Plate 196 — 404

Viruslike Symptoms Associated with Decline of Redbud, and Elm Mosaic, Plate 197 — 406

Some Viral Diseases of Cherry, Plate 198 — 408

Rose Mosaic Complex and Viral Diseases of Hibiscus, Plate 199 — 410

Diseases Caused by Apple Mosaic Virus, and Apple Flatlimb Disease, Plate 200 — 412

Tobacco Mosaic, Tobacco Ringspot, and Tomato Ringspot Viruses in Ash and Other Woody Plants, Plate 201 — 414

Viral Diseases of Azalea, Rhododendron, and Forsythia, Plate 202 — 416

Viral Diseases of Cactus, Camellia, Magnolia, and Oak, Plate 203 — 418

American Mistletoes, Plates 204–206 — 420

Dwarf Mistletoes, Plates 207–211 — 426

Cassytha and Dodder, Plate 212 — 436

Vines That Damage Trees, Plate 213 — 438

Beech Bark Disease, Plates 214–215 — 440

Decline Caused by Multiple or Unknown Factors, Plates 216–219 — 444

Damage Caused by Salt Spray and by Salt and Other Inorganic Poisons in Soil, Plates 220–221 — 452

Damage Caused by Misapplied Pesticides, Plates 222–224 — 456

Diseases and Injuries Caused by Air Pollutants, Plates 225–229 — 462

Nutrient Deficiencies, Plates 230–231 — 472

Damage Caused by Drought, Heat, and Freezing, Plates 232–236 — 476

Damage Caused by Excess Water, Plate 237 — 486

Damage Caused by Girdling Roots, Hail, and Sheet Ice, Plate 238 — 488

Lightning Injury, Plate 239 — 490

Noninfectious and Unexplained Growth Abnormalities, Plates 240–241 — 492

Bark Formation and Restoration, Plate 242 — 496

Color Changes Associated with Foliar Shedding, Plate 243 — 498

Symbiotic Relationships of Roots, Plates 244–245 — 500

Epiphytes and Lichens, Plates 246–247 — 504

Glossary — 509

References — 512

Index — 547

Diseases of Trees and Shrubs is a pictorial survey of diseases of, as well as environmental damage to, forest and shade trees and woody ornamental plants in the United States and Canada. Disorders caused by more than 350 pathogens and environmental factors to plants of more than 260 species are portrayed, and the text treats hundreds of other diseases that are related to those shown. The book is designed to serve as a diagnostic aid, as an authoritative reference to the diseases and pathogens that are illustrated, and as a guide to further information. The text emphasizes description, biology, and ecology, including host and geographic ranges of pathogens. Information about biological and cultural (but usually not chemical) control is included if it was available and if space permitted.

Since 1971, when the most recent previous general reference about tree diseases in North America was published, many new diseases have been found, and some that were insignificant have become important. Much has been learned about biology, taxonomy, and host-parasite relations of plant-pathogenic bacteria, fungi, mollicutes, nematodes, plants, and viruses. Subtle interactions of pathogens and stress-inducing environmental factors are now recognized. This book assembles information about new or newly known disorders and new understanding of old ones in a context of general knowledge to which thousands of previous authors have contributed.

Background and scope. The project that resulted in this volume and its predecessor, *Insects That Feed on Trees and Shrubs,* began in the mid-1960s with the objective of producing a few dozen copies of a loose-leaf pictorial handbook to be used by extension agents for identification of common insect pests and disorders of shade trees and woody ornamental plants in New York State. The scope of the project soon grew, however, because we perceived that photographic aids for diagnosis and instruction about disorders of trees and shrubs were needed nationally and in Canada. The volume about insects appeared in 1976. The present book, to be complemented by a revised, expanded edition of *Insects That Feed on Trees and Shrubs* (reference 965), completes the project.

The photographs in this volume, except those noted in the Acknowledgments, are a partial record of our observations in more than 50 states and provinces. Some significant diseases that we had planned to treat are not represented, but we have included many minor diseases and some major ones that were not mentioned in previous references. Plants grown in subtropical areas of southern states receive less attention than those in cooler regions.

Many more fungal diseases of trees and shrubs are known than diseases caused by other biotic agents, and the book reflects this imbalance: there are 176 plates for diseases caused by fungi versus 49 for all other biotic agents. Our pictorial survey is less extensive for disorders caused by nonliving agents than for those caused by pathogenic organisms. The number of plates (22) reserved for diseases and injuries caused by abiotic agents does not reflect the frequency or magnitude of damage by those agents.

Organization. No classification of plant diseases is satisfactory for multiple purposes. Similar or even single pathogens may cause diverse symptoms, and diverse pathogens or environmental agents may cause similar symptoms. We have blended three schemes in organizing this book (see the table on the next page). First, we separate diseases caused by biotic agents from those caused by other environmental stimuli. Second, most diseases are arrayed according to the plant part affected: leaves, twigs, limbs, roots, trunks, or the entire plant. This scheme allowed us to bring together similar diseases of diverse plants. Third, diseases are grouped according to taxonomic relationships among pathogens. Rusts and parasitic seed plants are treated as discrete groups regardless of the plant part parasitized. General information about a group of diseases or pathogens is presented with the first plate in a series.

The plates show symptoms and signs visible with the unaided eye or with a hand lens. Microscopic details are omitted from most plates and are described in the text only to the extent deemed necessary for practical understanding. Such information is available in the references. For each photo we indicate the place and month it was made, or for specimens photographed some time after collection, the place and month of collection. We did so partly to indicate where our observations were made and partly because woody plants and their diseases and pathogens vary in appearance from one region and season to another. Black-and-white illustrations have been inserted where we judged them useful adjuncts to the text and color plates and where space permitted. Sources of black-and-white illustrations are indicated in the Acknowledgments.

The text is designed to accommodate readers whose technical training does not extend beyond general biology and to be informative as well to advanced students and plant pathologists. Scientific terms other than the Latin names of pathogens are used only as necessary for precise communication, and a glossary of terms is provided. Cross-references to the volume on insects appear where arthropods are mentioned as vectors, predisposing agents, or contributors to plant damage.

Names of pathogens and plants. Scientific names are the only ones available for most microorganisms. The names used here reflect the surge of activity in nomenclature and taxonomy that has occurred since approximately 1955. Because many contemporary fungal names are unfamiliar even to plant pathologists, we give important synonyms (abbreviated *syn.*). We do not cite authorities for names of organisms, but they are available in listed references. For most fungal pathogens we indicate the division, class, or order and usually the family within which the pathogen is classified. This information makes plain the relationships among pathogens.

Binomial Latin names are spelled out when they are first used in a text article or in the index. Thereafter, the genus name is usually shortened to its first letter until another genus name beginning with the same letter appears. *Plasmopara viburni,* for example, might be abbreviated *P. viburni,* but this name would again be spelled in full after the appearance of a name such as *Peronospora sparsa* later in the article. A reader seeking the full spelling of a genus name should scan back toward the beginning of the article.

Only common names are available for plant viruses and only coined words for groups of related viruses. Latin nomenclature for viruses has not been adopted. The name of a virus usually reflects the common name of the plant in which a disease caused by the virus was first described, even though the virus may be unimportant in that plant. Virus group names usually reflect the name of the first virus that was characterized in the group; for example, tobacco mosaic virus (TMV) is a member of the tobamovirus group. Virus names are spelled out at first use and are abbreviated thereafter in the same article.

Common names of plants are used in text and captions wherever it is possible to do so without ambiguity. The index contains the Latin equivalents of common names. Latin names of plants are as used in *Hortus Third* (reference 1137) or, if not listed there, as in *Index Kewensis* (874).

Geographic and host ranges of pathogens and lists of plants that resist pathogens or tolerate environmental stress. We present more information about geographic and host ranges of pathogens than is available elsewhere, but this information is incomplete because we lack the resources and the space to provide comprehensive lists. Similarly, our lists of plants that resist pathogens or tolerate adverse environments could not be comprehensive. In any event, such lists should be used with caution, since plant response varies with circumstance.

References. Because space is limited, and in the interest of readable discussions, we list references after each discussion rather than cite all sources of information in the text. The references contain further illustrations of the diseases and injuries shown, descriptions and microscopic illustrations of pathogens, and additional information about pathogen biology, ecology and classification, disease physiology, and epidemiology. The references also deal with many diseases related to those illustrated. References were accumulated and text articles updated to at least January 1986, but for many topics, updating continued until mid-1986. To conserve space, we did not usually list a reference published as part of a series or as a book chapter apart from the entry for the entire series or book. There is, for example, only one entry for the *CMI Descriptions of Pathogenic Fungi and Bacteria,* published by Commonwealth Mycological Institute.

The distribution of technical publications about tree diseases is skewed toward a small number of important diseases, while little information is available about many others. For most diseases in the latter group, we usually write little rather than belabor the obvious,

extrapolate from other diseases, or speculate much about the unknown. Any extrapolation or speculation is identified as such.

Index. Illustrations of subjects listed are indicated by boldface page numbers. Because some plants have many common names and the same common name may apply to more than one plant, information is arranged under the Latin name of the plant genus or species.

We have seen and learned much while preparing this volume, and we have been stimulated by the thought of sharing some of the images and knowledge with you, the reader. We hope that you will find the book useful and that its usefulness will persist.

Wayne A. Sinclair
Howard H. Lyon
Warren T. Johnson

Ithaca, New York

Reader's guide to plant diseases and injuries

Plant part affected and type of disorder or association	Causal agent (numbers refer to plates)						
	Fungi	Bacteria and mollicutes	Viruses	Nematodes	Plants	Abiotic factors	Unknown or autogenous
Foliage, flowers, and fruit							
Anthracnoses	47–57					237	
Blisters, galls	6–8					222–233, 236, 237, 240	240
Chlorosis, nonnecrotic foliar markings	22, 37, 119–125	188–191, 193	194–203	145			
Flower blights, fruit rots and spots	24, 25, 31, 58, 66, 137	75					
Mildews, algae	1–5, 10				9		
Necrotic spots, blights, scorch	22, 23, 25–42, 45–60, 66	74, 186, 187	197, 198			220–229, 232–236	
Needle blights, needle casts	11–21, 38					226, 228, 229, 233, 234	
Normal foliar shedding							243
Rusts	114–130						
Scabs	29, 43–46						
Twigs, branches, and trunk							
Cankers and diebacks	32, 45, 46, 50, 51, 53, 56–67, 69, 71, 79–113, 117, 119, 121, 131–138, 165, 169–175, 181, 214–219	74–78, 186, 192	196, 197, 200, 202		204, 205, 211	220–221, 223, 232–236	
Cracks and wounds, reparative processes						162–164, 236, 238, 239, 242	
Galls, stem distortion	68–71, 115–117, 126, 132, 135, 136	72, 73	200		204–205, 213		240, 241
Mistletoes, cassytha, dodder, vines					204–213		
Rusts	114–117, 119, 120, 125–128, 131–136						
Shoot blights	25, 42, 45, 46, 50, 51, 62–64, 137	75–77				233, 234	
Trunk and limb rots	104, 161–173						
Wetwood and slime flux		185					
Witches'-brooms	4, 5, 7, 13, 70, 116, 117, 126–128	191–193			207–211		241
Roots and butt							
Basal cankers, collar rots	138, 139, 146, 179					236, 237	
Cankers on roots, necrosis of feeder roots	138–143, 179			145		237	
Galls		73		144			
Root and butt rots	146–161						
Entire plant							
Declines	137–140, 146–160, 214–217	190, 191, 193	196, 197, 202	145	204–211	216–224, 233	219
Scorch	181	186, 187				220, 221, 224, 228, 229, 232	
Wilts	175–183	188, 189		184		233, 234	
Nonpathogenic associations	79, 108, 244, 245	245			246, 247		

Acknowledgments

Preparation of this book was sponsored by the New York State College of Agriculture and Life Sciences at Cornell University through its Department of Plant Pathology. We are grateful to two former deans of the college, Charles E. Palm and W. Keith Kennedy, and to two chairmen of the department, Durward F. Bateman and William E. Fry, for their encouragement and support and for fostering an academic environment that made this project feasible.

We gratefully acknowledge the following organizations and individuals, whose financial contributions partially defrayed the cost of printing.

American Association of Nurserymen
Bartlett Tree Experts
Larry Borger
Philip Brogan
Kenneth Brownell, Sr.
George Callaway
Richard J. Campana
Chemlawn Corporation
Ted Collins Associates, Ltd.
Davey Tree Expert Company
Frank Donovan
Greymont Tree Specialists, Inc.
Grace Griswold Fund
Horticultural Research Institute, Inc.
Stanley L. Hunt
International Society of Arboriculture
Long Island Arborist Association
Maryland Arborist Association
J. M. McDonald Foundation
Fred Micha
National Arborist Association
New England Chapter, International Society of Arboriculture
New York State Arborists ISA Chapter, Inc.
Pennsylvania-Delaware Chapter, International Society of Arboriculture
Kenneth Post Foundation
Reader's Digest Association, Inc.
Rosedale Nurseries
Richard Sears
Carmine Serpe
James Sheehan
Gregory R. Smith
Southeastern New York Nurserymen's Association
Richard Synnestvedt
James and Dorothy Taylor in honor of James W. Taylor, Sr.
Western Chapter, International Society of Arboriculture
Weyerhaeuser Company Foundation
Whitmore Worsley, Inc.
Wisconsin Arborist Association
Barney Zipkin

Our thanks go also to arborists Peter Poltrak and Robert Mullane for their assistance in fund raising, to L. W. Schneski for technical help, and to Linda Glovack, Cheryl Hurwitz, Janet Kazimir, and Janet Steele for help with library work.

The following colleagues contributed time, knowledge, and materials that aided us in field work and photography. In Canada: H. L. Gross, D. T. Myren, and R. D. Whitney of the Canadian Forestry Service, Sault Ste. Marie, Ont.; G. B. Ouellette of the Canadian Forestry Service, St. Foy, Que.; and S. Navratil, formerly of Lakehead University, Thunder Bay, Ont. In Arizona: S. M. Alcorn, H. E. Bloss, and R. L. Gilbertson of the University of Arizona, Tucson; and F. H. Mahr, formerly of the University of Arizona, Mesa. In California: F. W. Cobb, Jr., and R. A. Raabe of the University of California, Berkeley, and J. W. Kloepper, formerly of that university; G. Nyland of the University of California, Davis; J. Ryder, apple grower, of Watsonville; and R. Sanborn of Cooperative Extension, Contra Costa County. In

Colorado: F. G. Hawksworth and T. E. Hinds of the U.S. Forest Service, Fort Collins; J. G. Laut of the Colorado Forest Service, Fort Collins; and J. M. Staley, formerly of the U.S. Forest Service, Fort Collins. In Connecticut and New York: D. R. Houston and P. M. Wargo of the U.S. Forest Service, Hamden, CT. In the District of Columbia: R. Hammerschlag and H. V. Wester of the U.S. National Parks Service. In Florida: S. A. Alfieri, J. C. Denmark, and R. P. Esser of the Department of Agriculture and Consumer Services, Division of Plant Industry, Gainesville; E. W. Barnard of the Division of Forestry, Gainesville; and G. M. Blakeslee and R. E. McCoy of the University of Florida, Gainesville and Fort Lauderdale, respectively. In Georgia: W. C. Bryan and W. A. Campbell, formerly of the U.S. Forest Service, Athens; and F. F. Hendrix, Jr., and W. Wynn of the University of Georgia, Athens. In Idaho: A. D. Partridge of the University of Idaho, Moscow. In Illinois: E. B. Himelick of the Illinois Natural History Survey, Urbana; and D. Stenger of the Sinnissippi Forest, Oregon, IL. In Louisiana: G. E. Holcomb of Louisiana State University, Baton Rouge; and A. L. Welden of Tulane University, New Orleans. In Michigan: J. H. Hart of Michigan State University, East Lansing. In Minnesota: R. A. Blanchette and D. W. French of the University of Minnesota, St. Paul. In Mississippi: D. J. Blasingame of Mississippi State University, State College; T. H. Filer, Jr., R. Lewis, Jr., and F. I. McCracken of the U.S. Forest Service, Stoneville; and A. G. Kais of the U.S. Forest Service, Gulfport. In Missouri: V. Dropkin of the University of Missouri, Columbia; and A. Foudin of the USDA Animal and Plant Health Inspection Service, Columbia. In Nebraska: G. W. Peterson and J. W. Riffle of the U.S. Forest Service, Lincoln. In New Hampshire: A. L. Shigo of the U.S. Forest Service, Durham. In New York: G. Callaway, arborist, of Argyle; J. D. Castello, P. D. Manion, and R. A. Zabel of the State University College of Environmental Science and Forestry, Syracuse; G. W. Hudler of Cornell University, Ithaca; and C. F. Scheer, formerly of Suffolk County Cooperative Extension Service. In North Carolina: R. I. Bruck of North Carolina State University, Raleigh. In North Dakota and northeastern states: R. W. Stack of North Dakota State University, Fargo. In Oklahoma: G. Simmons of the Samuel Roberts Noble Foundation of Ardmore. In Oregon: H. R. Cameron, E. M. Hansen, and L. F. Roth of Oregon State University, Corvallis. In Pennsylvania: W. Merrill of The Pennsylvania State University, University Park, and F. A. Wood, formerly of that university; and B. Towers of the Department of Environmental Resources, Middletown. In South Carolina: O. W. Barnett of Clemson University, Clemson. In Tennessee: C. Hadden of the University of Tennessee, Knoxville. In Texas: J. Amador of Texas A&M University, Weslaco; and C. W. Horne and D. Rosburgh of Texas A&M University, College Station. In Virginia: S. A. Alexander, R. J. Stipes, and R. L. Wick of Virginia Polytechnic Institute and State University, Blacksburg. In Washington: K. W. Russell of the Department of Natural Resources, Olympia; L. F. Roth of Oregon State University, Corvallis; and J. M. Staley, formerly of the U.S. Forest Service, Fort Collins, CO.

We are grateful to the following colleagues for certain photographs that are used in the figures and plates: E. J. Braun of Iowa State University, Ames, for Figure 18; M. F. Brown of the University of Missouri, Columbia, for Figure 1; M. N. Cline of Mallinkrodt, Inc., St. Louis, for Figure 7; P. Fenn of the University of Arkansas, Fayetteville, for Plate 107B, C; D. W. French of the University of Minnesota, St. Paul, for Plate 176I, J; J. Hopkins of the Canadian Forestry Service, Victoria, for Plate 112A, C; G. W. Hudler of Cornell University, Ithaca, NY, for Figure 5 and Plates 175A and 242E; S. N. Jeffers, formerly of Cornell University, for Plate 138F; K. J. Kessler of the U.S. Forest Service, Carbondale, IL, for Plate 54F; J. Laurence of Boyce Thompson Institute, Ithaca, NY, for Plate 229E; M. Linit of the University of Missouri, Columbia, for Plate 184E; J. A. Matteoni of Agriculture Canada, Vineland, Ont., for Plates 54D, E and 172B; M. E. Ostry of the

U.S. Forest Service, St. Paul, MN, for Plate 32F, I; B. C. Raju of Yoder Brothers, Inc., Alva, FL, for Plate 193C, E; C. Schoulties of the Florida Department of Agriculture and Consumer Services, Gainesville, for Plate 143A, C, F; A. L. Shigo of the U.S. Forest Service, Durham, NH, for Plate 113E; J. M. Staley, formerly of the U.S. Forest Service, Fort Collins, CO, for Plate 152G; D. A. Stark of the Maine Department of Conservation, Augusta, for Plate 113F; and L. C. Weir of the Oregon Department of Forestry, Salem, for Plate 112B.

The following black-and-white figures are reprinted or adapted by permission of the publishers. Figure 4 is from CMI Descriptions of Pathogenic Fungi and Bacteria, Nos. 568 and 570 (reference 387), by permission of CAB International. Figure 8 is from R. Schneider (1710) by permission of the author. Figure 9 is from M. B. Ellis (543) by permission of CAB International. Figures 10–12 are from J. A. von Arx (59) by permission of the author. The remaining black-and-white illustrations and color plates belong to the collection of the Department of Plant Pathology, Cornell University, and are reproduced by permission.

We extend special thanks to colleagues who provided critical review of the manuscript and suggestions for its improvement. D. W. French, G. W. Hudler, and P. D. Manion served as principal reviewers. S. A. Alfieri, E. L. Barnard, G. W. Blakeslee, J. D. Castello, F. W. Cobb, Jr., T. H. Filer, Jr., R. L. Gilbertson, F. G. Hawksworth, T. E. Hinds, R. D. Raabe, and T. S. Schubert examined particular sections.

Special thanks also go to those whose work began where ours ceased: Mary Brodie of the Department of Plant Pathology, Cornell University, for superb typing and word processing; Marcia Brubeck of The Guilford Group for editing the manuscript; and Hélène Maddux, Robb Reavill, and Richard Rosenbaum of Cornell University Press for editorial and production services.

W.A.S.
H.H.L.
W.T.J.

DISEASES OF TREES AND SHRUBS

Downy Mildews (Plate 1)

We begin with downy mildews, a group of diseases named for the fungi that cause them. These fungi produce delicate, soft ("downy") white tufts or clusters of spore-bearing structures (sporangia) on surfaces of infected plant parts in humid environments. Downy mildew fungi are members of the family Peronosporaceae of the class Oomycetes, division Mastigomycotina. They are destructive to many field and vegetable crops, but on woody plants only the downy mildews of grape and rose are economically significant. They occur wherever their host plants grow. Downy mildew of viburnum is often conspicuous east of the Great Plains, but it has no economic impact. Other members of the group are infrequent to rare. The downy mildew fungi are obligate parasites and thus in nature derive sustenance only from living plants. Like most obligately parasitic fungi, they display considerable host specificity. That is, each species attacks only one or a few closely related kinds of plants.

Downy mildew fungus	Host plants
Peronospora rubi	Rubus. Blackberry and other brambles.
P. sparsa	Rosa. Roses, both cultivated and wild.
Plasmopara cercidis	Cercis. Redbud.
P. ribicola	Ribes. Currants and gooseberries.
P. viburni	Viburnum. Arrowwood, cranberry bush, and other viburnums.
P. viticola	Ampelopsis, Parthenocissus, and Vitis. Ampelopsis, Virginia creeper, grapevines, and Boston ivy.
Pseudoperonospora celtidis	Celtis. Hackberry and sugarberry.

All downy mildews on woody plants develop and affect their hosts similarly. Attack is most conspicuous on leaves, although all succulent aerial parts are susceptible. Symptoms vary according to plant type and weather conditions. The first symptoms are off-color (usually light green) spots on upper surfaces of leaves (D), especially near ground level. The spots darken to shades of reddish brown, then brown (B, G). On the lower side of the leaf beneath each spot while the tissue is still alive, the fungus produces minute, colorless spore-bearing structures (sporophores) that appear grayish to white and often downy in mass (E, F). Leaves and other organs become distorted if they are infected before reaching full size. Numerous infections result in large dead areas and premature shedding of leaves.

Downy mildew fungi have two developmental phases. One consists of mycelium within plant tissues and white clusters of sporophores on surfaces. In this phase the fungus reproduces rapidly and causes a buildup of disease. The other phase consists of microscopic spores (oospores) that form within killed plant tissue, ensure winter survival of the fungus, and cause the first or primary infections annually. Oospores withstand environmental extremes and microbial attack. As dead, previously infected leaves decay on the ground during winter and early spring, quiescent oospores are released to the soil. During wet weather early in the growing season, some of them are splashed by rain to the lower leaves of the host, where they germinate if leaves remain wet for several hours. The stage is thus set for a rapid succession of generations. The fungus penetrates the leaf, grows as mycelium there, and produces sporophores that emerge through stomata. Details of spore production and germination vary according to the type of downy mildew fungus and the temperature. Some of the fungi produce motile spores (zoospores); others do not. In each species, however, spore production requires high humidity, spore dispersal is by air and by running or splashing water, and spores germinate only in a film of water. Hyphae from germinated spores on a leaf surface may penetrate and start new infections.

Downy mildews develop under moist conditions in cool or warm but not hot weather. Measures that promote dry foliage suppress these diseases. In warm regions, infections may be continuous throughout the year, and the pathogens of rose and grape may survive as mycelium in winter buds. In temperate regions, however, oospores in dead plant parts are the means of survival.

Downy mildew of viburnum. Lesions on leaves of viburnums are angular and are often bounded by veins. They usually coalesce, forming marginal or interveinal blotches that die and shrivel (B). Severely infected leaves fall. The browning and defoliation make shrubs or hedges unsightly. White sporophores form a sparse covering on the lower surfaces of lesions. Currants and gooseberries show similar symptoms, and in addition the flowers and young fruit die and shrivel. Lesions on redbud leaves also resemble those on viburnum.

Downy mildew of rose. The sporadic epidemics typical of this disease occur most often in greenhouses but also in nurseries and outdoor plantings in warm, humid regions. Leaves, stems, buds, and flower parts are attacked. The lesions in leaves and stems are reddish purple to brown, irregular, and up to 2.5 cm long on stems. Severe disease causes defoliation and sometimes death of shoots. Bud infection leads to retardation and malformation of flowers and flower buds, sometimes without leaf damage. The pathogen can be transported as dormant mycelium in cuttings and plants. Rose cultivars vary in susceptibility, but where disease builds up on highly susceptible types, most others become infected to some extent.

Downy mildew of grapevine. Infections on leaves of Virginia creeper, grapevine, or Boston ivy first appear as round, light green areas with indefinite margins, up to 2 cm across (G). These areas turn yellowish and then darken, forming irregular patches that are often bounded by the large veins of the leaf. Severe infection defoliates vines. Shoots, tendrils, and grape berry clusters are also attacked while young. A downy mat of sporophores develops on all diseased parts in humid air. It is white initially and becomes gray with age. The berries become resistant when partly grown, but if infection is severe, berries that escape infection are likely to be small and of poor quality. In humid regions grape yield can be reduced by half or more if vines are not protected. The pathogen is native to North America. Cultivated grapes derived from North American species are less susceptible than European types.

References: 89, 291, 387, 577, 646, 682, 883, 1757, 1758, 1868

A–D, F. Downy mildew of viburnum, caused by *Plasmopara viburni*. A. Browning and partial defoliation of a hedge of *Viburnum opulus* 'Nanum' (NY, Sep). B. Symptoms on leaves of *V. opulus* (NY, Aug). C, D. Lower and upper surfaces of *V. opulus* 'Nanum', showing white areas of spore production on the lower surface and inconspicuous chlorotic spots on the upper surface. The leaf in (C) is distorted because of infection that began before it expanded fully (NY, Aug). F. Sporophores of *P. viburni* on the lower surface of a leaf of *V. opulus* 'Nanum' as they would be seen with a hand lens (NY, Sep).

E. Downy mildew on rose cultivar 'Samantha'. Grayish aerial mycelium on a spreading lesion is typical (from a greenhouse, CA, Nov).

G. Signs and symptoms of downy mildew on grape leaves. The lower surface has white clumps of sporophores; the upper surface has yellowish spots (NY, Aug).

Plate 1

Powdery Mildews (Plates 2–4)

The powdery mildew fungi are a family of obligately parasitic Ascomycotina, the Erysiphaceae, that grow mainly on the surfaces of leaves and on other tender aerial plant parts and there produce spores on wefts or mats of white or light-colored mycelium. This superficial growth often appears powdery. Most of these fungi penetrate and parasitize only the epidermal cells of their hosts, producing within the living host cell an absorbing structure (haustorium) that diverts water and food materials to the fungus. The result is slow debilitation of the infected plant part; symptoms include dwarfing, distortion, chlorosis, premature senescence and browning of leaves, subnormal growth rate, blemished or aborted fruits, and depressed yields.

Experts place the number of powdery mildew species at between 125 and more than 300, in 7–19 genera, on more than 7000 host plants worldwide. Taxonomy of this group is evolving rapidly, with trends toward recognition of more genera and the description of multiple species in place of single broad species such as *Microsphaera penicillata*. Unfortunately, most host records are based on old taxonomic concepts; contemporary names of powdery mildews on woody plants have not been applied in host indexes. Several mildew-host combinations that we mention belong to this category. A few of the most common or conspicuous powdery mildew species or groups that infect woody plants in North America are

Erysiphe polygoni on gardenia, hydrangea, and black locust.
Microsphaera penicillata (syn. *M. alni*) complex on many hosts.
Phyllactinia guttata (syn. *P. corylea*) on many hosts.
Podosphaera clandestina and *P. tridactyla* (both formerly called *P. oxycanthae*) on hawthorn, stone fruit trees, and their ornamental relatives.
P. leucotricha on apples and crabapples.
Sphaerotheca lanestris on oaks in California (reported also on northern red oak in North Carolina).
S. pannosa on roses, peach, and photinia.
Uncinula necator on ampelopsis, Virginia creeper, and grapevine.
U. adunca (syn. *U. salicis*) on poplars and willows.

Gymnosperms are not attacked, and powdery mildews are unimportant in forests.

The characteristic fruiting structure of powdery mildew fungi is the cleistothecium, a tiny sphere usually 0.1–0.2 mm in diameter that forms on or in the mycelial mat on the plant surface (Plate 2E, F) and at maturity liberates ascospores from microscopic sacs (asci) that develop within the sphere. Cleistothecia are at first colorless, then yellow, brown, and finally black in most species. They have microscopic appendages of various types that diagnosticians use together with internal features for identification. In a few species, cleistothecia are so distinctive as to be identifiable with just a hand lens. Those of *Brasiliomyces trina* (syn. *Erysiphe trina*) on oaks in California, for example, are yellow-brown, relatively transparent, and consistently abundant. Those of *Pleochaeta* (*Uncinula*) *polychaeta* on sugarberry are two to three times as large as most others, reaching more than 0.3 mm in diameter, and are easily visible to the unaided eye.

Cleistothecia are absent from young powdery mildew colonies, and many powdery mildew fungi in warm regions do not produce cleistothecia. Diagnosis must then be based upon the identity of the host plant and upon the microscopic characteristics of the colorless sporophores and asexual spores (conidiophores and conidia) on the mycelial mat. Many plants—roses, for example—are attacked by only one species of powdery mildew, and practical diagnosis can be based on plant identification. In contrast, several species attack oaks, and microscopic study is required for inexperienced observers to identify them. Powdery mildews that are known only in their mycelial or conidial states belong to the form-genera *Oidiopsis*, *Oidium*, *Ovulariopsis*, and *Streptopodium*.

The annual cycle of a powdery mildew fits one of three patterns. (1) The fungus may continually produce mycelium and conidia on plant surfaces. This behavior is common in warm areas and in greenhouses. *Microsphaera euonymi-japonici* on *Euonymus japonica* (Plate 2D) persists this way. (2) The fungus may survive winter or hot dry summer as cleistothecia, then start new infections by means of ascospores released from these fruit bodies. *Pleochaeta polychaeta* on sugarberry and *Brasiliomyces trina* on oaks do this. Cleistothecia do not function for survival in all species, however. Cleistothecia of *Sphaerotheca pannosa* on rose and *Podosphaera clandestina* on hawthorn, for example, degenerate during winter in some areas. (3) The fungus may enter buds, infect the primordial leaves, shoots, or flower parts, and spend the winter there. Plant parts that develop from infected buds become stunted, distorted, and covered by mildew. Severe disease on some young shoots while adjacent shoots show no mildew indicates that the pathogen overwintered in buds. Powdery mildews of apple, cotoneaster, crape myrtle, currant and gooseberry, firethorn, grape, hawthorn, peach, photinia, plum, and rose overwinter in this way. So also does *Sphaerotheca lanestris* on California live oak (Plate 4A, B). Some powdery mildews have more than one mode of survival, but in a given situation, one predominates.

Within a growing season, multiple overlapping disease cycles (secondary cycles) may develop on plants that continue to produce succulent shoots. Infection is caused by conidia that are dispersed by air currents in greatest numbers near midday. The conidia germinate on dry surfaces even when atmospheric humidity is low, and germination is inhibited by free water. Most other fungi, in contrast, require free water for germination. Spore germination and penetration of the plant surface usually occur in 6 hours or less except in cold weather, and many powdery mildews can produce a new generation of spores in 4–6 days under favorable conditions.

The mycelium of most powdery mildews is exclusively superficial, but a few species send hyphae into leaves. *Leveillula taurica* on mesquite and many herbaceous hosts, *Phyllactinia guttata* on many hosts, *Pleochaeta polychaeta* on sugarberry, and *Sphaerotheca lanestris* on oaks are examples. Infection soon causes a large reduction in the rate of photosynthesis and sometimes an increase in transpiration of the affected part. The abnormal water loss is mainly through mycelium of the parasite.

Some plants are most susceptible while new shoots are developing and leaves are expanding; mature leaves and fruits become resistant. In other plant-mildew interactions, mature foliage is most susceptible. Cultural measures, such as trimming, fertilizing, and cultivating, that stimulate and prolong succulent plant growth encourage many powdery mildews. These diseases are also favored by dry weather with warm days and cool nights. They tend to build up most conspicuously in areas with a definite dry season. Although rain inhibits them, high atmospheric humidity does not. In fact, fog and dew occur frequently in West Coast areas where powdery mildews are perennially severe.

A. *Microsphaera syringae* on common lilac (NY, Sep).
B. Ash from Mt. St. Helens volcano on honeysuckle (*Lonicera utahensis*) (ID, Jul). Even experienced observers must look closely to distinguish powdery mildew from dust.
C. Secondary infections by *Brasiliomyces trina* on California live oak. Each mildew colony may have arisen from a single spore (CA, Mar).
D. *Microsphaera euonymi-japonici* on upper and lower leaf surfaces of *Euonymus japonica* (CA, Apr).
E. *Phyllactinia guttata* on the undersurface of a leaf of California black oak. Black dots are cleistothecia (CA, Sep).
F. Cleistothecia of *P. guttata* in various stages of development on a sparse mycelial mat on a lower leaf surface of *Calycanthus occidentalis* (CA, Nov).

Plate 2

Host specialization is prominent in the powdery mildews. Some species, such as *Microsphaera euonymi-japonici,* are restricted to one plant species. Others, such as *Phyllactinia guttata,* that attack many kinds of plants are actually species complexes consisting of subgroups that each infect only certain hosts. Mildew races capable of attacking resistant cultivars have been detected on plant groups such as rose that have been bred or selected for resistance.

Microsphaera penicillata complex. Species of this complex affect a broad array of plants, mainly trees and shrubs. Fungi in the group include *M. abbreviata* on oak; *M. aceris* on maple; *M. americana* on chestnut and chinkapin; *M. azaleae* on azalea; *M. calocladophora* on oak; *M. caprifoliacearum* on honeysuckle; *M. caryae* on hickory and pecan; *M. ceanothi* on ceanothus; *M. cinnamomi* on cinnamon; *M. elevata* on catalpa; *M. ellisii* on American filbert, hornbeam, and hop hornbeam; *M. erineophila* on beech; *M. extensa* on oak; *M. fraxini* on ash and privet; *M. hommae* on beaked filbert; *M. juglandis-nigrae* on walnut; *M. lonicerae* on honeysuckle; *M. magnifica* on magnolia; *M. neglecta* on elm; *M. nemopanthis* on holly and mountain holly; *M. neomexicana* on forestiera; *M. ornata* on birch; *M. penicillata* (narrow sense) on alder; *M. platani* on plane tree and sycamore; *M. pulchra* on dogwood; *M. pusilla* on euonymus; *M. ravenelii* on honey locust; *M. semitosta* on buttonbush; *M. sparsa* on viburnum; *M. sydowiana* on ceanothus; *M. symphoricarpi* on snowberry; *M. syringae* on forestiera, fringe tree, lilac, and privet; and *M. vaccinii* on azalea, blueberry, *Gaultheria,* American laurel, rhododendron, and other ericaceous plants.

Other *Microsphaera* species, not part of the *M. penicillata* complex, on woody plants in North America include *M. alphitoides* on oak and *M. grossulariae* and *M. vanbruntiana* on gooseberry and currant.

Most *Microsphaera* species become noticeable in summer and increase during the remainder of the growing season, especially on sprout growth. Some species produce prominent white mats. In most areas the first infections each year are probably initiated on leaves by spores from overwintered cleistothecia. It has been suggested that *M. syringae* overwinters in lilac buds (Plate 2), but if this mode of survival were common, it would be reasonable to expect mildewed shoots early in the season. Instead, mildew appears on scattered leaves and usually on oldest leaves first. Powdery mildew causes little long-term damage to most lilacs, but it blemishes the leaves of susceptible types annually. This disease is so prevalent that plantings of mixed cultivars can readily be evaluated for resistant and susceptible types. Cultivars of the common lilac, *Syringa vulgaris,* are more severely affected than those of most other *Syringa* species.

Microsphaera euonymi-japonici. Although this powdery mildew (Plate 2) occurs only on *Euonymus japonica,* it is found wherever its host plant grows in southern and western states. It forms a thick felt with abundant conidia on leaves, causes some yellowing and defoliation, and produces successive infections throughout the year. Cleistothecia of this fungus have been found only in Europe.

Phyllactinia guttata (syn. P. corylea). This broadly adapted species, or perhaps species-group (Plate 2), attacks many trees and shrubs, occurs worldwide, builds up during the last half of the growing season, rarely causes important damage, and includes specialized subgroups (formae speciales). Its formae speciales include *alni, betulae, carpini, coryli,* and *fagi,* indicating subgroups specialized for parasitism of alder, birch, hornbeam, hazelnut, and beech, respectively. In addition to these host groups, *P. guttata* attacks ash, mountain ash, barberry, boxwood, California buckeye, buttonbush, catalpa, bitter cherry, chestnut, Chinaberry, coffeeberry, crabapple, currant, dogwood, elder, elm, fringe tree, gooseberry, hawthorn, hickory, holly, linden, magnolia, maple, mulberry, crape myrtle, ninebark, oak, mock orange, *Paulownia,* pear, plane tree, *Prunus,* quince, sassafras, sumac, sycamore, tulip tree, walnut, and willow. Species and cultivars within these plant groups vary in susceptibility.

P. guttata commonly appears as a mat, often with prominent cleistothecia (Plate 2E, F) on the lower surfaces of mature leaves. On deciduous hosts, *P. guttata* may cause distortion of succulent late-season shoots. On evergreen hosts it sometimes causes premature yellowing of leaves 1 year after the onset of infection. It probably survives winter as cleistothecia in temperature areas.

Podosphaera clandestina (syn. P. oxycanthae). Susceptible members of the rose family include apple, mountain ash, sour cherry, sweet cherry, chokecherry, hawthorn, medlar, peach, pear, plum, quince, and spirea. Nonrosaceous hosts include common persimmon, snowberry, and *Vaccinium uliginosum. P. clandestina* f. sp. *crataegi* attacks hawthorn and related plants, causing much disfigurement of hedges in England. Another forma specialis, *cydonia,* is specialized for infection of pear and quince. In North America, *P. clandestina* occurs coast to coast and is most important on stone fruit trees in western fruit-growing regions. It overwinters in buds; thus the first infections each year appear early in the growing season on distorted shoots. Secondary infections begin a few weeks later and make the mildew conspicuous on fruits, leaves, and succulent shoots. At first it appears as a delicate weblike mat, but this is soon replaced by a white mass of hyphae and spores. Diseased parts may remain dwarfed or may become distorted (Plate 3A, D) or chlorotic before premature death. Powdery mildew mats persist on twigs even into winter. *P. clandestina* produces cleistothecia at the end of the growing season, but these are apparently unimportant for winter survival.

Erysiphe polygoni. Many kinds of plants, most of which are herbaceous, are infected. Woody hosts include acacia, amelanchier, gardenia, hydrangea, and black locust. This powdery mildew is most important on hydrangea because the environmental conditions under which pot plants grow, especially in greenhouses, promote infection. The fungus attacks leaves, causing abnormal purple color, yellowing, browning (Plate 3C), and premature leaf fall. Some hydrangea cultivars are highly resistant to common strains of *E. polygoni. E. liriodendri,* which forms prominent white mats on leaves of tulip tree, was formerly included in *E. polygoni.*

Sphaerotheca pannosa. This is one of the best-known and most studied powdery mildews of woody plants. Its variety *rosae* is a pathogen of roses and photinia, and the variety *persicae* attacks apricot, peach, and plum. A strain in New Zealand infects apricot, pomegranate, rose, and some eucalyptus species. *S. humuli,* the powdery mildew of the hop plant, was once blamed for some infections on roses, but specialists now recognize only *S. pannosa* var. *rosae* on rose in North America. This pathogen survives winter in buds or, in warm areas and greenhouses, produces mycelium and conidia continually on succulent parts. During the growing season, it produces successive generations of mycelium and spores on foliage and green stems of peach and rose. On green plum fruits (*Prunus salicina*), *S. pannosa* causes chlorotic areas that later turn into scabby lesions lacking signs of the fungus.

A, B. *Podosphaera clandestina* on fruits of snowberry (*Symphoricarpos albus*). Berries attacked while young ceased growth. Black specks are cleistothecia (CA, Sep). Another powdery mildew, *Microsphaera symphoricarpi* (not shown but similar in appearance), is more common than *P. clandestina* on snowberry.

C. *Erysiphe polygoni* causing premature senescence and browning of hydrangea leaves (CA, Nov).

D. *P. clandestina* on holly-leaved cherry. Infection causes distortion and reddening of leaf tissue beneath the white mat of mildew (CA, Jul).

E–G. *Sphaerotheca pannosa* var. *rosae* on rose. Uniformly mildewed leaves (E) grew from an infected bud. Discrete colonies on leaves (G) represent secondary infections. Powdery mildew on canes (F) typically appears as white, feltlike patches. Cleistothecia are most common on canes and thorns (CA; Aug, Sep, and Jul).

Plate 3

On susceptible roses, white mycelium and conidiophores of *S. pannosa* var. *rosae* first appear in discrete patches on unfolding leaves. The patches enlarge, covering the young leaves and stems and causing distortion, curled leaves, premature leaf fall, and even death of the shoot. Mats of powdery mildew persist in patches on stems, particularly near bases of thorns. The fungus does not usually attack petals but does grow abundantly on the sepals, pedicels, and receptacles of blossoms and on unopened buds. Blossoms from such buds are of low quality. Rose leaves are highly susceptible just as they emerge from buds, but they become less susceptible beginning about 4 days later and are highly resistant after 2 weeks. Pedicels lose susceptibility more slowly than leaves, perhaps because the cuticle forms more slowly. For both pedicels and leaves, the thicker the cuticle, the slower the mildew develops.

Conditions highly favorable for powdery mildew of roses are temperature near 15°C at 90–99% relative humidity at night and 26–27°C at 40–70% relative humidity during the day.

Rose species and cultivars vary from highly susceptible to apparently immune to powdery mildew, and the mildew fungus varies in pathogenic ability. *S. pannosa* includes races with the ability to attack certain rose cultivars but not others. Unfortunately, resistant rose types are eventually attacked by new races.

Cleistothecia of *S. pannosa* form erratically on rose, seldom on the leaves but often in the mycelium around thorns and receptacles of blossoms. In California, where *S. pannosa* is a significant pathogen of both peach and rose, cleistothecia are rare on rose and have never been found on peach. In England the spores in cleistothecia on rose plants degenerate during winter. There, as in North America, diseased buds are responsible for the first infected leaves and stems each year.

Sphaerotheca lanestris. This pathogen infects several species of oaks in California and in some cases causes the shoots to develop into witches'-brooms (Plate 4A, B). Brooms are most common on California live oak and occur also on canyon, holly, Kellogg, laurel, tanbark, water, and willow oaks. The brooms on water oak may be leafless.

Broom induction by *S. lanestris* results from infection of buds. When diseased buds open in spring, they produce abnormally thick and frequently branched stems and dwarfed, pale yellow leaves. The brooms may develop an open form or may consist of tight clusters of dwarfed shoots. Diseased parts turn white while the fungus is sporulating. Brooms often die during winter. Those that survive may produce either more infected shoots or normal ones the next spring. In southern California, colonies of *S. lanestris* survive the winter on leaves of live oak. Farther north, the fungus overwinters in buds that give rise to infected shoots. Spores from either source cause discrete colonies on leaves (Plate 2C). Most colonies arise on the lower leaf surface, but the upper surface also becomes infected both by direct penetration and by growth of the fungus from the lower surface. With age, the superficial mycelium turns tan, then brown, and cleistothecia develop on it. Pruning and fertilization of live oak trees at the wrong time of year may promote growth of succulent shoots that, if infected, develop into witches'-brooms.

Two other powdery mildews, *S. phytoptophila* on hackberry and sugarberry (Plate 5) and *Sphaerotheca* sp. on ninebark (Plate 4E), are also believed to cause witches'-brooms, but experimental evidence is lacking. Sometimes a witches'-broom caused by another agent may favor development of powdery mildew by providing abundant succulent plant tissue and a favorable environment among the tightly clustered leaves and twigs. Examples include *Microsphaera loniceae* on brooms caused by the aphid *Hyadaphis tataricae* on honeysuckle (see Johnson and Lyon, 2nd ed., Plate 149) and *Podosphaera leuco-*

tricha associated with apple proliferation, which is caused by a mycoplasmalike organism.

Powdery mildew of apple. *Podosphaera leucotricha* is the ubiquitous and economically important powdery mildew of apple and crabapple (Plate 4C, D). This fungus infects several members of the rose family, including pear and quince, and it is blamed for rusty spots on peach fruits. It has been studied mainly on apple trees. Leaves, flower parts, green shoots, and fruits are attacked. The fungus overwinters in leaf buds and sometimes in flower buds. After these open, mildew develops rapidly on unfolding leaves and on the elongating stem, causing both to be dwarfed, distorted, and white in contrast to normal shoots. Spores from the white mat cause secondary infections that continue throughout the summer until plant growth stops. Leaves that become infected while expanding are abnormally narrow, folded longitudinally or curled, crinkled, and brittle. Infection of blossom buds leads to shriveling of flower parts soon after growth begins. Fruits infected while young become stunted and have brown blemishes (russeting) on the surface. Severe disease causes defoliation and sometimes death of shoots during the growing season. The mycelial mat on leaves and shoots, at first a loose weft, then thick and white during spore production, eventually turns tan or brown. Cleistothecia develop in the mat after midseason in some areas, but these fruit bodies are apparently unimportant for survival and propagation of the pathogen.

Mildew on apple leaves spreads down the petioles and infects some axillary buds soon after they form. Diseased buds and shoots not killed by the fungus are abnormally susceptible to death by freezing in winter. Death of diseased buds is the reason why, in northern orchards after an abnormally cold winter, powdery mildew is less severe than usual. In experiments, the death rate among diseased apple buds frozen at −22°C was as great as that among uninfected buds frozen at −26°C.

Apple and crabapple varieties vary in susceptibility to powdery mildew. A list of some resistant crabapples appears opposite Plate 43.

Erysiphe lagerstroemiae. This powdery mildew (Plate 4F, G) attacks crape myrtle and a related plant, *Lagerstroemia parviflora*, in eastern, southern, and western states. It is often severe on shaded plants and on those in hedges. The pathogen overwinters in buds and develops quickly in spring on shoots from diseased buds. These shoots serve as foci from which, under favorable conditions, the fungus may spread all over the plant in less than 2 weeks. Infection on young shoots causes stunting and floral abortion. Reddish pigment may develop beneath the white mat on leaves. Diseased leaves and buds usually drop within a few weeks, but the tips of diseased shoots often outgrow infection. The disease subsides during hot summer weather and develops again in autumn. The mycelial mat becomes tan or brown toward the end of the growing season, and cleistothecia rarely develop in it.

References for Plate 2: 6, 245, 541, 617, 618, 672, 833, 1017, 1026, 1441, 1455, 1867, 2255, 2257, 2278
References for Plate 3: 142, 246, 247, 808, 1017, 1920, 2211
References for Plate 4: 143, 391, 541, 618, 1276, 1403, 1554, 1877, 2153, 2211, 2257

A, B. Witches'-brooms caused by *Sphaerotheca lanestris* on California live oak. Broomed shoots covered with white mycelium and spores grew from infected buds. One-year-old leaves are unaffected (CA, Jul).

C, D. *Podosphaera leucotricha* on an apple shoot that grew from an infected bud. Such severe infection can kill shoots and blossoms. Mildew at leaf axils enters buds such as the one that was exposed (arrow in D) by removing a leaf near the base of the shoot (NY, Jun).

E. Witches'-broom believed to be caused by *Sphaerotheca* sp. on ninebark. Dark color of the broom is caused by large numbers of black cleistothecia on leaves (NY, Oct).

F, G. *Erysiphe lagerstroemiae* on crape myrtle. F. A shoot from a diseased bud is white with powdery mildew; green leaves at its base are from a different bud (FL, Apr). G. Secondary infection has caused distortion and death of leaves and blossoms (CA, Oct).

Plate 4

Witches'-broom of Hackberry and Black Witches'-broom of Serviceberry (Plate 5)

Witches'-broom of hackberry. Throughout much of its range hackberry (*Celtis occidentalis*) is disfigured by brooms attributed to two agents acting in concert: a powdery mildew, *Sphaerotheca phytoptophila,* and an eriophyid mite, *Eriophyes celtis* (syn. *Aceria snetsingeri*) (see Johnson and Lyon, 2nd ed., Plate 230). The two agents can readily be found attacking buds and shoots in brooms, but neither has been tested for ability to induce broom formation. Observations suggest that the mites may induce brooms and that the mildew invades the habitat thus created. Large trees may bear hundreds of brooms without obvious loss of vitality. Trees growing in the open are more frequently and severely affected than those in woodlands. Sugarberry (*C. laevigata*) is also affected.

Each broom consists of numerous short twigs that arise together, often at a conspicuous swelling or knot on the branch. Many twigs in the broom die back during the dormant season. Buds on surviving twigs are abnormally numerous, unusually grayish, broader, and with looser scales than those on normal twigs. Loose bud scales are the result of swollen internal parts.

Brooms arise initially from single infested buds, each of which produces shoots with more infested buds. On a vigorously growing branch, a loose broom may form, often with subsidiary tight clusters of twigs along its axis (B). Slowly growing branches lose the apically dominant habit and develop tight brooms centered on knots.

During spring and early summer, a sparse mat of mycelium and conidiophores of the powdery mildew fungus grows on succulent stems, petioles, buds, and sometimes the lower surfaces of leaves. Fruit bodies (cleistothecia) (C) soon form on this mat and may be found throughout the year. Ascospores mature beginning in autumn. Whether the first infection each year is caused by spores from cleistothecia or mycelium within buds is unknown. Whether conidia cause secondary infections is also unknown.

Mites in all stages of development can be found throughout the year and are most numerous in late summer. They occur and overwinter beneath bud scales and on the primordial shoots within buds (D, E). Populations as large as 2000 per bud have been reported. In the central United States, females begin to produce eggs in May, and new generations develop throughout the growing season. The severity of attack varies greatly among trees in close proximity to one another, but reasons for this variation are unknown.

Black witches'-broom of serviceberry. Various species of serviceberry (*Amelanchier*) from coast to coast in Canada and as far south as the Carolinas and New Mexico are affected by black witches'-broom.

The disease is most common in woodland or stream bottom habitats. Landscape specimens, although susceptible, are seldom affected. The pathogen, *Apiosporina collinsii* (Pleosporales, Venturiaceae), produces perennial mycelium in branches and dark mycelium, conidiophores, and fruit bodies (pseudothecia) on leaf surfaces. This fungus can be isolated in pure culture.

Either conidia or ascospores can cause infection. They germinate in bark fissures or axils of leaves or buds, and the fungus then invades twigs. Mycelium grows mainly toward the tip of the twig and enters buds, petioles, leaf blades, and flowers. Soon after leaves have unfolded from diseased buds, hyphae emerge from stomata or from between epidermal cells and form a mycelial mat on which conidiophores develop. The conidial state is a *Cladosporium*. The mycelial mat, initially olive colored, thickens, expands, and becomes dark brown to black (H), eventually covering the entire undersurface of the leaf except for the midrib and major veins. Pseudothecia begin developing in early summer. When fully formed, these are black spheres 150–250 μm across (G), each with a tiny pore through which spores will be released the following spring. Pseudothecia may cover the entire lower leaf surface, and the leaves die soon afterward.

As additional shoots develop in the axils of infected leaves, the fungus grows into the new tissues and causes development of abnormally short, thick, and numerous twigs. On shaded branches, the brooms thus formed are compact, and the branch at the point of infection is swollen and bent toward the ground. Diseased branches on trees in open areas develop loose brooms with more regular branching. If the broom maintains a distinct axis, this may become S-shaped as twig tips turn upward. Many twigs in brooms die back during winter. Leaves on a broom may be dwarfed and yellowish, depending on the host species and on the age and vigor of the broom. Where summers are dry, as in the Northwest, the first crop of diseased leaves may be cast and replaced at midsummer by a second crop more sickly than the first. The second leaves are likely to be killed by autumn frost.

Broomed branches become less vigorous as years pass, but the parasite does not kill them directly, and the disease has little net effect on the plant unless brooms are very numerous. Brooms may snap off under the weight of snow. Pruning for control of the disease on the occasional infected ornamental plant is feasible because the pathogen does not grow toward mainstems.

References: 996, 999, 1006, 1377, 1683, 1853, 1878

A–E. Witches'-broom of hackberry. A, B. Brooms on a mature tree (NY, May). C. A diseased bud with black fruit bodies of the powdery mildew *Sphaerotheca phytoptophila* (NY, Sep). D, E. A bud dissected (D) to detect eriophyid mites. Mites are visible in magnified view (arrows in E) (NY, Sep).

F–H. Black witches'-broom of serviceberry (*Amelanchier*). F. A broom on *A. canadensis* (NY, Jul). G. Fruit bodies of *Apiosporina collinsii* on the lower surface of a leaf (NY, Aug). H. Shoots of serviceberry in the 1st year of infection by *A. collinsii*. Leaves have a dark mycelial mat on their undersides except on veins. Infected axillary buds will give rise to a witches'-broom (NY, Jul).

Plate 5

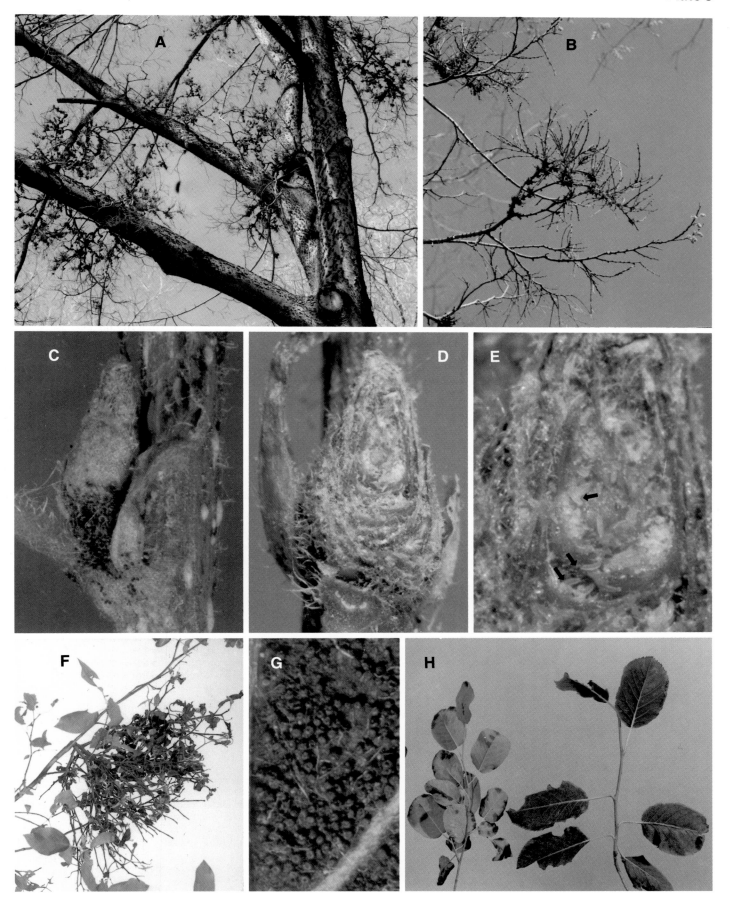

Taphrina Diseases: Leaf Blisters and Leaf Curls, Bladder Plums, and Witches'-brooms (Plates 6–7)

Blisterlike distortions on leaves are the hallmark of parasitism by a group of 90–100 fungal species composing the genus *Taphrina* (Ascomycotina, Taphrinales). The best-known diseases attributed to members of this group in North America are peach leaf curl, caused by *T. deformans* (Plate 6A, B), and oak leaf blister, caused by *T. caerulescens* (Plate 6D–F). Peach leaf curl is the only economically significant disease. Most species of *Taphrina* that have been studied infect leaves annually in spring and produce spores in late spring or early summer. These species survive during the rest of the year by growing as saprophytes on plant surfaces during moist weather at moderate temperatures and by persisting as quiescent spores on twigs and among bud scales during dry periods and in winter. Some species attack flowers or fruit rather than leaves. *T. communis* (Plate 6C) and *T. pruni*, for example, each induce grotesquely swollen plum fruits, called plum pockets or bladder plums. A few species grow as perennial mycelium in host branches and induce the formation of brooms. *T. wiesneri* (syn. *T. cerasi*) (Plate 7) on various cherries is the most common broom inducer.

Some additional, relatively common foliar parasites are *T. populina* (syn. *T. aurea*) and *T. aesculi,* which cause yellow leaf blisters on poplars and California buckeye, respectively; *T. carnea* on birches; *T. coryli* on hazelnut; *T. dearnessii* and *T. sacchari* on maples; *T. virginica* on hop hornbeam; and *T. ulmi* on elm. Fruit and flower pathogens include *T. farlowii* on fruit, leaves, and shoots of black cherry; *T. johansonii* on poplar catkins; and *T. occidentalis* and *T. robinsoniana* on alder catkins. Other broom inducers are *T. amelanchieri* on serviceberry in California and *T. flectans* on cherries in the Northwest.

Most species of *Taphrina* infect only one or a few closely related host plants and can be identified quickly if one recognizes the type of disease and the host. *T. deformans* on peach is an example. Leaf blisters of oaks and maples, however, illustrate a need for refinement in naming species of *Taphrina*. Only one name, *T. caerulescens,* is given to a somewhat diverse group of fungi that cause leaf blisters on about 50 species of oaks. At the opposite extreme are 11 differently named but virtually indistinguishable leaf blister fungi on maples. Perhaps these are specialized variants of one or a few biological species.

On leaves, *Taphrina* species often cause unthickened spots that turn brown and generally go unnoticed. Deformities, on the other hand, attract attention. These result from abnormal cell division (hyperplasia) and enlargement (hypertrophy) that the parasites induce in part by secreting growth-regulating chemicals. Several species of *Taphrina,* including *T. deformans* and *T. wiesneri,* produce cytokinins (cell-division factors) and auxins (promoters of cell enlargement). Bulging spots on leaves must be examined closely to distinguish between *Taphrina* leaf blisters and symptoms caused by eriophyid mites or midges (see Johnson and Lyon, 2nd ed., Plates 223, 233, and 234).

The life cycles of *Taphrina* species have sharply delimited parasitic and saprophytic phases in which the fungi assume different forms. In the parasitic phase they infect leaves or flower parts at about the time these emerge from buds. This phase consists of intercellular or subcuticular mycelium that may stimulate host overgrowth and pigment formation and that produces a single layer of fertile cells (asci) on the leaf or fruit surface. Asci of *Taphrina* are microscopic sacs in each of which many spores develop. When fresh the layer of asci appears as a granular, sometimes glistening deposit, usually white or translucent.

The spores are forcibly discharged in late spring to midsummer, depending on the species and locality, and the infected plant tissue dies soon afterward.

Spores from asci give rise to the saprophytic phase, in which the fungus produces successive crops of spores on plant surfaces by budding during warm, moist weather. Some of these spores survive winter on twigs or among bud scales and are thus on hand to cause new infections or to produce more spores that in turn infect young plant tissue in spring. *Taphrina* species can be isolated from noninfected plant surfaces throughout most of the year. In cultures on laboratory media, these fungi resemble yeasts. Species with perennial mycelium have the same type of annual cycle outlined above, and in addition, an occasional infection progresses into stem tissue and becomes perennial, causing distortion or broom formation.

Peach leaf curl. *Taphrina deformans* causes leaf curl and deformity of succulent shoots of peach and nectarine in orchards worldwide. Almond in Europe and New Zealand and apricot in New Zealand are also affected, but the latter two hosts apparently escape infection in North America. Symptoms appear annually as leaves unfold. Leaves become puckered and curled during the development of thickened bulging areas that change from green to yellow and then become reddish (Plate 6A). Succulent shoots may also be deformed. Within a few weeks a white bloom appears on the deformed part of the leaf, usually on the upper surface (Plate 6B). This is the layer of asci, similar in magnified view to that of *T. wiesneri* (Plate 7I). Infected parts of leaves soon degenerate, and the leaves fall prematurely. Severe infection can cause such defoliation that a second flush of shoot growth occurs in early summer. Blossoms sometimes become infected and then shrivel and fall.

Only undifferentiated tissues are susceptible. Thus most infections occur on leaves during or just after emergence from winter buds. Mycelium of the fungus grows intercellularly in and beneath either the upper or lower epidermis and develops extensively beneath the cuticle, giving rise there to the layer of asci. Spores from asci on deformed leaves splash about on the plant during rains and multiply by budding. The fungus is thereby distributed all over the plant and keeps pace with the growth of new shoots but seldom infects them. The spores can tolerate hot dry summers and freezing winters. Spores lodged in winter buds start primary infections annually. Secondary cycles possibly occur but are never important.

Peach and nectarine cultivars vary in resistance to leaf curl, and resistance has been shown to be heritable. The peach cultivar 'Redhaven' and others derived from it are said to be tolerant of infection.

Periods of cool wet weather during leaf emergence favor leaf curl. *T. deformans* in its saprophytic phase grows most rapidly at temperatures near 20°C.

Plum pocket, or bladder plum. Swollen plum fruits infected by *Taphrina communis* (Plate 6C) or occasionally by *T. pruni* most often appear on native species, especially Canada plum and wild, or American, plum. These diseases are less common on domestic (European) plum and are uncommon on Japanese plum. Other suscepts are beach, Chickasaw, and wild-goose plums, the last-named including both *Prunus hortulana* and *P. munsoniana.* Plum pocket occurs all across North America, caused usually by *T. communis.* In Eurasia, *T. pruni* is common and *T. communis* absent.

A, B. Peach leaf curl, caused by *Taphrina deformans*. A. Infected leaves are puckered and develop red pigment (OR, Jul). B. A superficial layer of asci (white area) produces spores. Then leaves die or fall (CA, Jun).

C. Bladder plums, or plum pockets, caused by *T. communis* on Canada plum. Enlarged fruits darken and fall after spore production. Dark spots on green fruits are injuries made by an insect, the plum curculio (NH, Jun).

D–F. Oak leaf blister, caused by *T. caerulescens*. D. On willow oak (MS, May). E. On black oak (NY, Jul). F. Asci on the lower surface of a northern red oak leaf. These areas turn brown after spore production (NY, Jun).

Plate 6

Infection by *T. communis* or *T. pruni* apparently begins soon after blossoms open and is first evident on fruits 6–12 mm in diameter as elevated, spongy, greenish white spots that enlarge rapidly until, in most cases, the entire plum is involved. The plum grows quickly to three or four times normal size, becomes hollow with pocketlike invaginations, lacks a seed, and develops a leathery surface. A white layer of asci develops on the epidermis and breaks through the cuticle. Sporulation occurs in late March to early July, depending on latitude. After spore dispersal the affected fruit darkens, shrivels, and falls. *T. communis* and *T. pruni* also cause thickening and deformity of succulent stems, and *T. pruni* produces perennial mycelium in them. In other respects, however, the disease cycles resemble that of peach leaf curl.

Oak leaf blister. *Taphrina caerulescens* (Plate 6D–F) occurs around the globe on about 50 species in both the red oak (subgenus *Erythrobalanus*) and white oak (*Leucobalanus*) groups. The disease is more common and severe in the southeastern and Gulf states than in the North, perhaps because the pathogen population remains high on plant surfaces throughout the winter. Leaf blister is favored by mild moist conditions during the early phases of leaf growth.

Leaves become infected just as the buds open; expanded leaves are not susceptible. Symptoms are bulges, or depressions as viewed from the reverse side, usually 3–20 mm across on either leaf surface. These may cause severe curling on narrow-leaved species such as willow oak. Numerous blisters sometimes coalesce and involve entire leaves. Severe disease may cause premature defoliation. Heavily infected water oaks in Mississippi, for example, have been observed to lose half their leaves by the end of May.

The annual cycle of *T. caerulescens* is as outlined for the genus as a whole. In addition, however, the pathogen occasionally causes secondary cycles of disease on leaves from buds that open unseasonably in summer. The mycelium develops subcuticularly and intercellularly in the epidermis. Sporulation occurs between late March and mid-August, depending on latitude and elevation. The layer of asci, usually on the concave surface of the blister, is at first colorless, then darkens after spores have been released (Plate 6F).

Asci of *T. caerulescens* vary in form and in relationship to epidermal cells on different oaks, and there is circumstantial evidence of pathogenic specialization within the species. More than one observer has noted a severely diseased oak of one species growing amid healthy oaks of a second species known to be susceptible. Such observations have led to speculation that *T. caerulescens* is actually a group of biologically distinct organisms.

Leaf curl and witches'-broom of cherry. *Taphrina wiesneri* causes leaf curl and witches'-broom on wild and domestic cherries across North America and around the world in both temperate zones. Hosts include bitter (Oregon) cherry, European dwarf cherry, Japanese flowering cherries (*Prunus serrulata* and *P. yedoensis*), pin cherry, *P. pseudocerasus*, sour cherry, and sweet cherry. Apricot is also susceptible. The disease achieved minor notoriety in the 1920s by occurring in the plantings of Japanese cherries in Washington, D.C., from which it was quickly eradicated by pruning. It is seen yearly in cherry orchards of the Pacific Northwest and on wild pin cherry in eastern states and provinces.

Leaf curl apparently precedes broom formation and may occur independent of brooms (Plate 7C). Infected parts of leaves become slightly thickened, discolored yellowish to reddish or reddish brown (Plate 7B), and curled, crinkled, or droopy. Asci develop on the lower leaf surface (Plate 7E, I), rarely on both surfaces, about a month after leaves begin growth. Spores are dispersed from a given leaf during a 3- to 4-day period, both night and day. Infected parts of leaves darken and die soon after sporulation. This foliar phase of the disease apparently has an annual cycle like that of peach leaf curl.

Brooms are induced by perennial intercellular mycelium in twigs. Whether the fungus grows through petioles to stems or infects them directly is unclear. Once in a twig, however, *T. wiesneri* causes localized swelling and twig proliferation that may continue for several years, with the result that brooms range in in length from a few

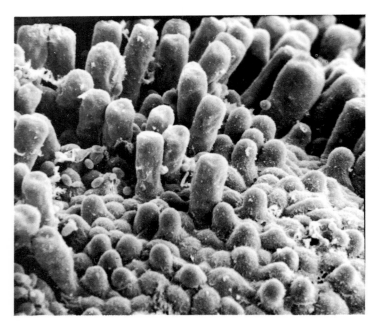

Figure 1. Asci of *Taphrina deformans* on a peach leaf as seen in a scanning electron micrograph.

centimeters to over a meter. The knotlike swellings at the bases of old brooms sometimes exude gum (Plate 7D). Twigs in brooms tend to die back in winter. Brooms produce foliage earlier than normal and bear no flowers or fruit. Leaves in brooms are heavily infected early in the season and soon die or fall. New, normal leaves then develop, making brooms inconspicuous during much of the growing season.

Yellow blisters on poplar leaves. *Taphrina populina* is globally distributed and causes yellow leaf blister of various poplars coast to coast in North America as far south as South Carolina. *T. populi-salicis* causes yellow leaf blister in the Pacific Northwest on black cottonwood (Plate 7F–H), Fremont cottonwood, and red willow (*Salix laevigata*). The yellow color is imparted by pigment in oil droplets within asci during the period of sporulation. More than one unwary observer seeing yellow patches on leaves has confused these diseases with Melampsora rust (Plate 123).

Infected areas on leaves vary from round depressions or mounds a few millimeters across to large, irregularly shaped, thickened areas on distorted leaves. The upper surfaces turn light green or greenish yellow and the lower ones bright yellow when asci develop. The annual cycle is as generalized for the genus *Taphrina*. Extensive defoliation is unreported; apparently these diseases are insignificant to both humans and trees.

References: 6, 61, 70, 301, 387, 389, 574, 778, 966, 1010, 1045, 1166, 1339, 1583, 1624, 1708, 1709, 1951, 1970, 1998, 2211

A–E, I. Taphrina broom and leaf curl of cherry, caused by *Taphrina wiesneri*. A. A broom several years old at the beginning of the growing season. Winter-killed twigs are evident (NY, May). B, C. Diseased foliage on a small broom and on a nonbroomed branch (NY, Jun). D. A swollen knot identifies the initial site of perennial infection at the base of a broom (NY, Jun). E. A diseased leaf. The white area on the lower surface is a layer of asci (NY, Jun). I. Magnified view of asci. The midvein of the leaf is at the left side of the picture (NY, Jun).

F–H. Yellow leaf blister of black cottonwood caused by *T. populi-salicis*. F. Small blisters viewed on the upper surface of a leaf (ID, Jul). G, H. Upper and lower surfaces of leaves distorted by *T. populi-salicis*. Yellow patches on lower surfaces are large groups of asci (ID, Jul).

Plate 7

Exobasidium Galls and Blisters (Plate 8)

Fungi of the genus *Exobasidium* (Basidiomycotina, Exobasidiales) attack plants in the families Empetraceae, Lauraceae, Symplocaceae, Theaceae, and especially Ericaceae. Symptoms are of four general kinds: annual galls involving buds, leaves, floral parts, or shoot tips (A–C, E, G); yellow or red annual leaf spots, thickened or not (D); enlarged reddish leaves (F) resulting from perennial systemic infections of stems and buds; and perennial stem galls, either spindle shaped or cylindrical, some of which give rise to witches'-brooms. The best known of these disorders in North America are leaf and flower galls of azalea, red leaf of lowbush blueberry, and rose bloom of bog cranberry plants, all caused by strains of *E. vaccinii;* and leaf and shoot galls of camellia, caused by *E. camelliae.* Economic damage is slight.

Red color in infected leaves results from secretion of anthocyanin pigments in palisade cells. Swelling results from hypertrophy and hyperplasia. Some species of *Exobasidium* produce indoleacetic acid, which is presumed to play a role in gall formation. Photosynthesis in diseased parts is depressed, respiration increases, and phenolic substances accumulate.

About 50 species of *Exobasidium* are known worldwide, including 10–20 in North America, mainly on woody hosts. The total depends upon whether *E. vaccinii* is considered as one species or as several. The name *Exobasidium* refers to the habit of producing spore-bearing structures (basidia, Figure 2) on plant surfaces without any fruit body or covering. Basidia develop from mycelium in a layer that also includes conidiophores and conidia. This gives one surface of the diseased organ a white, velvety appearance.

E. vaccinii (broad sense) occurs throughout the United States and Canada on many ericaceous plants, causing systemic or localized symptoms that vary with the host plant and strain of the fungus. Symptoms include swollen shoots, stem galls, brooms, and red-leaf disorders that represent perennial infections; red or yellow leaf spots; and annual galls of buds, leaves, and flower parts. *E. vaccinii* includes varieties and host-specialized strains that look nearly alike but cause distinct diseases in various host plants. In some references, infraspecific groups are given the following separate names: *Exobasidium andromedae, E. angustisporum, E. arctostaphyli, E. azaleae, E. japonicum, E. ledi, E. oxycocci, E. parvifolii, E. rhododendri,* and *E. uviursae.* Hosts include azalea, mock azalea, bearberry, blueberry, cassiope, cranberry, farkleberry, gaultheria, huckleberry, kalmia, leatherleaf, leucothoe, madrone, manzanita, rhododendron, bog rosemary, box sandmyrtle, and Labrador tea.

E. vaccinii-uliginosi, also transcontinental in the North, causes perennial systemic infections that result in swollen shoots, red swollen leaves, and sometimes brooms. This species does not cause localized fleshy galls or leaf spots. Hosts are bearberry, bog and dwarf bilberries, mountain cranberry, evergreen and thinleaf huckleberries, mountain heather, West Coast rhododendron, and Labrador tea. *E. dimorphosporum* and *E. empetri* cause swollen shoots and leaves of dwarf bilberry and black crowberry, respectively, in the Northwest. *E. camelliae* causes leaf and shoot galls on common and sasanqua camellias in the South. *E. symploci* causes bud galls of sweetleaf in southern and southwestern areas. Species that cause yellow leaf spots without swelling include *E. burtii* and *E. decolorans* on rhododendrons and azaleas in the West (*E. burtii* occurs occasionally on cultivated rhododendrons in the East also), *E. canadense* on rhodora in the Northeast, and *E. cordilleranum* on huckleberries in the Northwest.

Life cycles and pathogenic activities of *Exobasidium* fungi fit within two general schemes. In one scheme, fungi that cause leaf spots or galls develop annually in buds, leaves, or flower parts while the host is growing. These fungi produce localized intercellular mycelium that draws nourishment from lobed haustoria (absorbing structures) formed within host cells. In late spring to midsummer, depending upon latitude and elevation, basidia emerge from between epidermal cells, rupture the cuticle, and produce basidiospores that are dispersed by air currents. In a variation of this scheme, *E. camelliae* forms a layer of basidia at a depth of five or six plant cells inward from the lower epidermis and exposes its basidia by causing the entire lower surface of the infected part of the leaf to slough off as a sheet four or five cell layers thick. After sporulation, diseased leaves or spots slowly degenerate. Basidiospores germinate on wet surfaces and may cause new infections by direct hyphal penetration of young, late-developing stems and leaves. Conidia, dispersed by water, are also infectious.

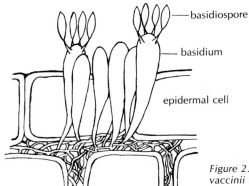

Figure 2. Basidia of *Exobasidium vaccinii* emerging from between epidermal cells on a leaf surface.

Basidiospores often germinate by budding and thus give rise to more spores on uninfected plant surfaces. Whether this reproduction continues is unknown, but saprophytic growth is possible, as judged from the ability of species of *Exobasidium* to grow and sporulate readily on laboratory media. Survival during winter occurs in buds from which infected plant parts develop. Whether the pathogens infect undifferentiated tissues in buds before winter or in spring after survival as spores among bud scales is not clear.

The second scheme is like the first except that mycelium becomes perennial. It may be localized, inducing stem galls and brooms, or systemic, as in red-leaf disease of lowbush blueberry and rose bloom of cranberry. Perennial mycelium infects buds and thus infects all organs that grow from them. The spore-producing layer arises on the lower surfaces of diseased leaves or on hornlike projections that develop annually from stem galls.

Secondary cycles of some Exobasidium diseases probably occur occasionally on late-developing shoots, but they are not significant.

Control by hand picking galls or pruning localized infections before annual sporulation has often been suggested. Whether these measures accomplish more than cosmetic improvement is uncertain.

References: 6, 287, 387, 389, 425, 668, 885, 1018, 1157, 1584, 1687, 2233

A. Leaf galls on sasanqua camellia, caused by *Exobasidium camelliae.* Sporulation occurs on white areas on the lower surfaces of leaves. Dark spots on the diseased leaf at left are mycelium and spores of a secondary fungus (MS, May).

B, C. Galls caused by *E. vaccinii* on azalea cultivar 'Anchorite'. Upper surfaces of transformed leaves are light green; lower surfaces are white during sporulation (WA, Jun).

D. Spots caused by *E. vaccinii* on evergreen huckleberry (OR, Jul).

E. A sporulating gall caused by *E. vaccinii* on an azalea, *Rhododendron periclymenoides* (NY, Jun).

F. Leaves of manzanita infected by *E. vaccinii-uliginosi* are reddish brown and shriveled at the end of winter. Infected buds will soon give rise to new leaves with red upper surfaces (CA, Mar).

G. Galls caused by *E. vaccinii* on Hiryu azalea. Such galls have fragile surfaces that discolor darkly when bruised (NY, May).

Plate 8

Algal Leaf Spot, or Green Scurf (Plate 9)

Algae that parasitize higher plants, although uncommon in comparison to free-living algae, are widespread in warm, humid regions. At least 15 species of these parasites belonging to three families are known. As a practical matter, however, only a few species in the genus *Cephaleuros* (Chlorophyta, Chroolepidaceae) are significant pathogens. *C. virescens* is the only species common in the United States. It affects plants of more than 200 species in over 60 families in a region extending from North Carolina to Florida and Texas. *C. parasiticus,* a significant pathogen of tea abroad, causes a leaf spot of southern magnolia in Louisiana but is otherwise unknown in the United States. Plants that have leaves with somewhat leathery surfaces are most conspicuously attacked by parasitic algae.

C. virescens grows on leaves, twigs, branches, and sometimes fruits, producing superficial colonies that appear as raised spots or blotches (A, B). Host tissue beneath the colony slowly dies, and a brown spot appears. Numerous colonies cause premature yellowing and defoliation. During a period of algal reproduction in summer, the colonies on leaves appear as reddish brown pads with velvety surfaces (C). At other times the colonies are grayish green to greenish brown with green margins (D). Colonies are usually less than 1 cm across but may coalesce and form a continuous sheet (E).

Twigs and branches are the principal sites of damage. Infections begin in bark crevices or (rarely) on green twigs. The first symptom is thickening of bark around an algal colony. The raised bark eventually cracks superficially into small, irregularly shaped plates or becomes somewhat shredded. This indicates a shallow canker that in time can encircle the branch and cause stunting or premature yellowing of leaves and death of twigs. Citrus branches as large as 5 cm in diameter may be killed. The depth of infections in bark and the mechanism by which the alga invades bark are unreported. Mechanical rupture by the pressure of expanding algal cells in bark fissues has been suggested.

The alga may cause conspicuous damage to leaves or bark of acacia (*Acacia auriculiformis*), anatto, azalea, banana shrub, common camellia, citrus of several types, cotoneaster, franklinia, sea grape, guava, holly of several species, longan, southern magnolia, candleberry myrtle, Jambolan plum, pigeon plum, and other species.

The causal organism is conspicuous before symptoms arise. Colonies begin development after rainy periods in summer and are evident by early autumn. Colonies on leaves survive only if they arise in minute wounds or somehow breach the leaf surface. Those that remain entirely superficial are either washed away by rains or shrivel and blow off. Developing colonies become disc shaped, several cell layers thick at their centers and one layer thick at their margins. The margins often have spokelike branching radii visible with a hand lens. The radii are chains of algal cells. The surface appears velvety to roughened and bears numerous sterile hairs (trichomes) that are present throughout the year on all but the youngest colonies. Initially greenish, and often remaining so at margins (B, D), the colonies turn orange-brown to rusty red as a pigment called haematochrome forms in sterile hairs and in the stalks of hairlike asexual spore-producing structures (sporangiophores). Several microscopic sporangia develop at the tip of each sporangiophore during wet weather in summer. Sporangia, which look and function like spores, are dispersed by splashing or running water or by air. Some sporangia develop without stalks. In a film of water on the plant surface, they open and release motile spores (zoospores) that swarm about for a few minutes before settling. These give rise to new algal colonies that produce another generation of sporangia in late spring or early summer of the next year. Thus the asexual reproductive cycle of *C. virescens* is 8–9 months long. Successive crops of sporangia develop as long as leaves remain on the tree. Sexual reproduction involves fusion of zoospores to form a dwarf sporophyte that in turn produces microzoospores. Whether or not these infect the plant is unknown.

Colonies on leaves usually develop between the cuticle and epidermis and rarely beneath the epidermis or among mesophyll cells. Just how the alga causes separation of cells or of cuticle from epidermis is unknown. Plant tissue beneath old colonies sometimes shrivels, leaving an air space between the leaf and the center of the algal disc. Columns of algal cells extend across the space, presumably providing anchorage and allowing conduction of water and nutrients to the disc. Whether the alga derives organic nutrients in addition to water and minerals from its hosts is unknown. The alga can be cultivated readily on artificial media.

In the leaves of citrus and probably in those of other plants with relatively thick leaves (but not loquat, magnolia, or tea), and also in bark, a layer of cork cells often develops beneath the algal disc, separating healthy tissue from diseased. The cork may impede growth of the colony and cause the disc and killed tissue to project cushionlike above the normal plant surface.

The life span of an individual colony is determined mainly by the fate of the infected plant part. Colonies on leaves survive until sometime after the leaves drop. The centers of old colonies often support fruiting structures of fungi, the roles of which are unclear. Some are fungal symbionts in lichenized colonies (B). Two lichens, *Strigula complanata* and *S. elegans,* are considered to be parasitic. See the discussion accompanying Plate 247 for more information about lichens. Other fungi in old algal colonies may be parasites of weakened leaf tissue or of degenerating parts of the algal disc, or they may simply be saprophytic colonists of dead cells.

Abundant rainfall, high temperature, and direct sunlight favor infection. *C. virescens* is most aggressive and damaging on slowly growing, weakened plants. Such plants are also most subject to secondary attack by opportunistic fungi, but the possible role of these fungi in damage attributed to *C. virescens* has not been reported.

References: 343, 344, 862, 975, 1038, 1219, 1682, 2226

Algal spots caused by *Cephaleuros virescens*.
A. Numerous colonies apparently induce premature senescence of leaves of southern magnolia (FL, Mar).
B. Lichenized colonies on southern magnolia display green margins, grayish brown centers, and black fruit bodies (perithecia) of a fungal partner. The lichen is a species of *Strigula* (FL, Mar).
C. The rust-colored appearance of algal colonies on common gardenia indicates haematochrome pigment (FL, Mar).
D. Young colonies on red bay (FL, Mar).
E. Numerous small colonies on a southern magnolia leaf arose after spore production by the large colony at lower left. Arrangement of colonies indicates that spores were carried in a film of water (FL, Apr).

Plate 9

Sooty Molds, Black Mildews, and Other Dark Superficial Fungi (Plate 10)

In a plant pathologist's perspective, three kinds of dark colored fungi blacken the surfaces of plants. Sooty molds are entirely superficial saprophytes that derive nourishment from insect and plant secretions. Black mildews are obligately parasitic fungi that grow on the surfaces of leaves or green twigs and derive nourishment by penetrating epidermal cells and producing haustoria in them. A third group, which lacks a unifying common name and has been lumped with both of the above groups, includes minor pathogens that grow both within and on the surfaces of leaves, cause premature senescence or localized necrosis, and produce dark superficial masses of hyphae, fruiting structures, and spores. Although widespread and conspicuous, these groups of fungi cause little damage to plants and have therefore received attention only from mycologists. Diagnosticians rarely attempt to identify such fungi to the species level, and standard diagnostic references promote confusion by unnaturally categorizing various dark fungi as sooty molds or black mildews. In warm regions, black mildews and sooty molds may occur on the same plants, which explains some of the confusion.

Sooty molds. Several species of this diverse group of dark-walled, epiphytic, saprophytic Ascomycotina (sac fungi) and Deuteromycotina (asexual fungi) often intermingle on a leaf or stem. They reproduce by ascospores, by conidia, and sometimes by fragmentation of hyphae. Species that have been studied grow readily in pure cultures on artificial media. Many belong to the Capnodiaceae and to allied families of the order Dothideales. Common genera are *Aethaloderma, Capnodium, Euantennaria, Morfea, Phragmocapnias, Scorias,* and *Trichomerium.* The most common nutritional substrate for these fungi is honeydew, a liquid secretion released from the anus of aphids, soft scales, mealy bugs, and some species of leafhoppers. Honeydews are complex mixtures of sugars, amino acids, proteins, other organic substances, and minerals. Droplets of honeydew often fall from insect-infested leaves or branches and stick to other plant parts or to objects such as automobiles or outdoor furniture. Both honeydew and spores of sooty mold fungi are dispersed in water during rain. Thus dark fungal deposits often occur on plant parts and on other surfaces not infested by insects. In addition, some sooty molds apparently grow on organic substances secreted by normal leaves and green twigs. On catalpa in eastern states, for example, sooty molds commonly associate with the glandular trichomes on leaves. Most sooty molds display no host preferences, but there are exceptions; for example, *Metacapnodium juniperi* occurs only on *Juniperus,* and *Antennatula pinophila* only on true firs.

Sooty molds vary in gross appearance from thin dark patches or coverings to irregular masses measuring several centimeters in each dimension. Species of *Capnodium* on citrus leaves may resemble a layer of black tissue paper that tears after aging and drying (E). On lindens and other trees, a dark superficial mold long known as *Fumago vagans* was eventually shown to be a mixture of two fungi, *Aureobasidium pullulans* and *Cladosporium herbarum,* which are ubiquitous on plant surfaces and in dead plant tissue. *A. pullulans* is also notorious for its ability to discolor wood and painted surfaces.

One of the most conspicuous and widespread sooty molds is *Scorias spongiosa* (Figure 3), which occurs on twigs and branches of alder, beech, birch, pines, and other trees in eastern states from Florida to Maine. It forms spongelike masses of mycelium that have been recorded as large as 5 cm high and 11 cm long. With moisture, these masses, called stromata, change from brittle to soft. On their surfaces in autumn, long-necked asexual fruit bodies that produce conidia arise. Ascocarps develop in spring and appear when viewed with a hand lens as swellings about 0.1 mm in diameter at the tips of short stalks that extend from the stroma. Hyphae in the stroma of *S. spongiosa,* massive by common standards, are composed of short cells with walls up to 16 μm thick.

Black mildews. Perhaps 1800 species of obligately parasitic Ascomycotina of the order Meliolales develop as dark patches or mats, often with prominent setae (sterile hairs), on leaves and sometimes on green twigs. They obtain nourishment by sending haustoria into epidermal plant cells, and they reproduce only by means of ascospores. They occur mainly in tropical and nonarid warm-temperate regions around the world. Many black mildews greatly resemble one another, and each species parasitizes only a small number of plant species, usually within a single family. To identify black mildews, therefore,

Figure 3. A sooty mold, *Scorias spongiosa,* growing on the remains of a colony of woolly alder aphids (see Johnson and Lyon, 2nd ed., Plate 144) on an alder twig.

one must first identify the host plant. *Meliola,* the largest genus, includes more than 1000 species. Other principal genera are *Amazonia, Asteridiella, Asterina,* and *Irenopsis. Meliola palmicola,* which infects scrub palmetto (F, G), is typical of the group. It appears as circular black patches that on close examination are seen to consist of branched, radiating hyphae (G) that have many short, lateral, microscopic appendages. The appendages, called hyphopodia, provide anchorage and points from which the parasite sends haustoria into epidermal cells.

Other dark superficial fungi. A third group includes but is not restricted to members of the Dimeriaceae, order Pleosporales, and of the Asterinaceae, order Hemisphaeriales. These fungi are apparently host-specialized, although this specialization has not been studied critically. Unlike the black mildews, they reproduce by both conidia and ascospores, and those studied grow readily on artificial media. Several species of *Asterina* and *Epipolaeum* (syn. *Dimerosporium*) form dark patches on leaves. *E. abietis,* for example, forms dark patches on the lower surfaces of fir needles in the Northwest (B). Its parasitism and pathogenicity are inferred from observations that it hastens the senescence of old needles. On oaks in several southeastern states, *Lembosia* (*Morenoella*) *quercina* forms distinct black patches of hyphae that are initially superficial and that by late summer penetrate and develop between the cuticle and epidermis, causing browning and disorganization of epidermal cells.

References: 6, 119, 600, 612, 659, 734, 735, 789, 912, 1176, 1320, 1530, 1597, 1598, 1791

A. Unidentified sooty mold on 1-year-old needles of ponderosa pine. New needles are not colonized (OR, Jul).

B. *Epipolaeum abietis* on lower surfaces of needles of grand fir, associated with premature senescence (ID, Jul).

C, D. Unidentified sooty mold on common gardenia. Sterile hairs on the torn superficial fungal layer are visible in (D) (FL, Apr).

E. A sooty mold, *Capnodium* sp., on a citrus leaf (FL, Nov).

F, G. A black mildew, *Meliola palmicola,* on scrub palmetto. Close view shows branched, radiating dark hyphae (FL, Mar).

Plate 10

Lophodermium Needle Casts (Plate 11)

More than 20 species of *Lophodermium* (Rhytismatales, Rhytismataceae) colonize the needles of coniferous trees and shrubs, but only one of these fungi, *Lophodermium seditiosum,* is a major pathogen. Several weakly pathogenic species invade living needles 1 or more years old and cause premature senescence and casting. Still others, such as *L. juniperi* (F, G), are always saprophytic, invading needles that die naturally or are killed by other agents. Several additional species of *Lophodermium* are minor pathogens or saprophytic colonists of broadleaf plants, but none of these is well known or important. Some of the lophodermia of conifers, their host plants, attack habits, and geographic ranges in North America are listed below.

On pines:

L. australe, on two- and three-needle pines; saprophyte or weak pathogen; Southeast and Gulf region.

L. durilabrum, on western white pine; weak pathogen; causes needle browning; Northwest.

L. nitens, on five-needle pines; weak pathogen; causes spots and yellowing and casting of old needles; transcontinental.

L. pinastri, on two-, three-, and occasionally five-needle pines; weak pathogen; invades green needles, causing yellowing and casting of the oldest ones; also invades dying needles as a saprophyte; transcontinental.

L. ponderosae, on ponderosa pine; pathogen; causes browning and casting of old needles; Rocky Mountains.

L. seditiosum, on two-, three-, and some five-needle pines; virulent pathogen; infects new needles and kills them before the 2nd growing season; transcontinental.

On spruces and firs:

L. crassum, on Brewer spruce; weak pathogen; California.

L. decorum, on grand fir; weak pathogen; West.

L. lacerum, on firs; weak pathogen on shaded foliage; East and West.

L. piceae, primarily on spruces, occasionally on firs; pathogen of spruces; causes browning and casting of old needles; East and West.

On cedars, incense cedar, and junipers:

L. juniperinum; saprophyte; transcontinental.

L. seditiosum causes needle cast of pines in forest nurseries and plantations, especially Christmas tree plantations. Austrian, red, and Scots pines sustain greatest damage. Other hosts include Aleppo, rough-barked Mexican, and Virginia pines. Additional pine species are undoubtedly susceptible, but records are confused with those for the similar species, *L. pinastri.*

In the mid-1960s, epidemic needle cast in nurseries in the Great Lakes region killed or so damaged red and Scots pine seedlings that millions could not be shipped for planting. The damage was attributed to *L. pinastri.* Not long thereafter, because diseased seedlings had been shipped from nurseries, epidemics developed in Christmas tree plantations across North America. Short-needle strains of Scots pine with seed origins in France and Spain were damaged most. The epidemics gave impetus to research that led to recognition of the real pathogen, *L. seditiosum,* as a species distinct from *L. pinastri.*

Lophodermium species produce elliptical black fruit bodies (apothecia, also called hysterothecia) that develop in or just beneath the epidermal layer, causing the needle surface to bulge (D). Ripe apothecia of *Lophodermium* open widely by a medial longitudinal split. Apothecia of *L. seditiosum* develop beneath all needle surfaces, are 0.8–1.5 mm long, and lie entirely beneath the epidermis (Figure 4), which causes them to appear gray unless they are wet. The apothecia of *L. pinastri* (D) are so similar to those of *L. seditiosum* that only specialists can distinguish between the two species.

Lophodermium species also produce inconspicuous pycnidia, often called spermagonia. These develop before the apothecia and remain visible as tiny elongate marks among the apothecia (D). They also develop on laboratory media. The spores produced in them are not infectious and perhaps act as male gametes in the life cycle of the fungus.

L. seditiosum, by attacking current season's foliage and killing it before the next growing season (A, C) reduces photosynthetic capability and causes growth loss in small trees. Killed needles either droop and remain attached for a time or are cast soon after death. Twigs that bear only diseased needles often wither during late winter or early spring, and buds that survive produce abnormally small shoots and leaves (E). Thus needle cast stunts and disfigures trees and may kill young seedlings in nurseries.

Pathogenic species of *Lophodermium* undergo one generation and cause one cycle of needle cast per year. The time of fruit body maturation and spore release varies among species. *L. seditiosum* fruits on dead 1-year-old needles on the ground or lodged among branches, sometimes on needles still attached to twigs, and even on cones. The apothecia mature in late summer and, in response to rainfall during August–November, release long, thin, colorless ascospores that have sticky sheaths. Spores germinate on wet needles, and penetration occurs directly through the cuticle of current-season needles. Most infections occur during August and September. Yellow spots develop in late autumn to early spring and become brown with yellow margins. Infected parts of needles turn yellow and then reddish brown as the spots enlarge and merge. Browning is most conspicuous in April and May, after which new foliage partially hides discolored needles and bare twigs. Some diseased needles remain partly green and have dead spots and tips during the 2nd growing season (B). Foliage on low branches is usually most damaged, although entire trees may be involved (A). Dead needles, which gradually turn from reddish brown to straw colored, fall in June–July.

Resistance to *L. seditiosum* varies within and among pine species. Trial plantings of Scots pine seedlings from many seed origins have shown various combinations of growth rate, color, needle characteristics, and resistance. Red pines seem highly susceptible only in the seedling stage. Jack pine and eastern white pine remained unaffected in nurseries where Scots and red pine seedlings were destroyed by needle cast.

Abundant ascospores and a cool, moist environment are required for a destructive outbreak of needle cast. In nurseries, diseased needles in windbreaks of susceptible pines are sources of spores. This threat can be avoided by growing immune trees such as cedars for windbreaks. In dense or weedy plantations, poor air circulation promotes needle cast by prolonging periods when foliage and needle litter are wet. The result is enhanced spore release, germination, and penetration because the pathogen requires free moisture for these activities. Effective weed control enhances air circulation near ground level and suppresses *L. seditiosum.*

References: 387, 444, 445, 612, 921, 1106, 1283, 1285, 1303, 1323–26, 1400, 1822, 1883, 1893

A, C, E. Scots pine Christmas trees infected by *Lophodermium seditiosum.* One-year-old needles are dead, and casting has begun. New shoots are stunted, and many buds failed to open. Small tree size is the result of repeated loss of year-old needles (WA and PA, Jun).

B. Symptoms caused by *L. seditiosum* on 1-year-old needles of Scots pine: brown spots with yellow borders, dead needle tips. Gaps along the twig show where needles were cast (IL, Jun).

D. Apothecia (large gray-brown structures) and pycnidia (small brown marks) of *L. pinastri* on a needle of Japanese black pine (NY, Sep).

F, G. Apothecia of *L. juniperi* on foliage of a juniper branch that dried following mechanical damage in winter (NY, May).

Plate 11

Ploioderma Needle Casts of Pine (Plate 12)

Hard pines, those with two or three needles per fascicle, from Maine to Florida and across the Southeast are subject to needle browning and defoliation caused by *Ploioderma* (*Hypoderma*) *lethale*. The fungus infects newly formed needles in summer. It causes no symptoms during that season, but it kills the needles during late winter and early spring. Severe disease results in loss of 1-year-old needles and lost growth. Susceptible species are Austrian, Japanese black, Cuban, loblolly, pitch, pond, sand, shortleaf, slash, spruce, table-mountain, and Virginia pines. Native species are affected in forests and landscapes. Exotic species such as Austrian and Japanese black pines are damaged in nurseries, landscapes, and Christmas tree plantations.

P. lethale and three related species, *P. hedgcockii*, *P. lowei*, and *P. pedatum* (Rhytismatales, Rhytismataceae) cause similar diseases, but only *P. lethale* is regularly destructive over a wide area. *P. hedgcockii* affects several hard pines in the Southeast and Gulf region, *P. lowei* slash pine in Louisiana, and *P. pedatum* Monterey pine in California. We deal only with the disease caused by *P. lethale*.

Symptoms caused by *P. lethale* arise during winter as yellow spots and bands that soon turn brown (B), giving the year-old needles a mottled appearance. In late spring the bands turn straw color to grayish brown, and the intervening tissue and needle tips turn brown. Needle bases may remain green, however (C, E). Severe disease often causes death of all 1-year-old needles on a twig. Needles with various amounts of tip necrosis tend to fall prematurely, or dead parts break off before the whole needle is dead. Needle drop during spring and summer may leave just the new foliage on the tree. This foliage grows in tufts at the branch tips because shoots from defoliated twigs do not attain normal length and needles remain close together. Damage is usually most severe in the low parts of tree crowns but may involve entire trees of any size.

In early spring, inconspicuous pycnidia develop in infected spots on needles. Pycnidia are colorless and embedded in the needles and appear as minute longitudinal wrinkles in the epidermis. The pycnidia, like those of other needle-cast pathogens, have no known role in the disease cycle.

In May in the South and June in the Northeast, apothecia (also called hysterothecia) develop. They appear as black lines 0.4–1.4 mm long, covered by the epidermis (Figure 4), in all surfaces of straw-colored needles or in dead bands on green needles (D–H). Sometimes they aggregate into continuous lines up to 5 mm long. During wet weather in late spring to early summer, the time depending on locality, mature apothecia open by a medial lengthwise split and disperse ascospores. The spores are colorless, one-celled, and equipped with a sticky

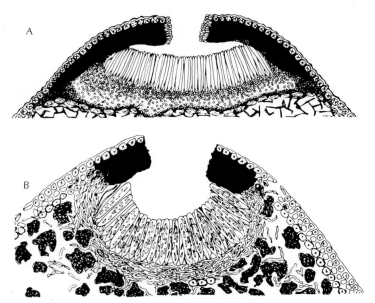

Figure 4. Apothecia of *Lophodermium seditiosum* (A) and *Ploioderma lethale* (B) as seen in cross-sections. Both fungi form apothecia entirely beneath the epidermis. A black roof of fungal tissue covers the entire fruit body of *L. seditiosum*, which therefore appears elliptical as viewed from above. Black tissue in *P. lethale* is not as wide as the apothecium, which therefore appears as a dark line on the needle surface. Erect parallel elements in the apothecia are asci, from which spores are ejected.

sheath that allows them to stick to foliage and other surfaces. They are presumed to germinate and penetrate new needles immediately, but no symptoms develop until after the growing season.

Individual trees of slash and loblolly pine vary markedly in apparent resistance to needle cast (A). Such differences are strongly heritable and not correlated with growth rate.

Longleaf pine is unaffected by *P. lethale* even when needle cast is severe on associated southern pines. Red pine seedlings, after inoculation with ascospores under controlled conditions, remained healthy, while southern pines similarly treated developed needle cast.
References: 229, 387, 438, 444, 445, 921, 1056, 1324, 1352

Needle cast caused by *Ploioderma lethale*.

A. Resistant (left) and susceptible slash pines. Sparsely foliated appearance of the susceptible tree is caused by absence of 2-year-old needles and tufting of 1-year-old blighted needles on short twigs. New needles are beginning to elongate (GA, May).

B. Early symptoms on Austrian pine: yellow to brown spots and bands (NY, Feb).

C. Repeated severe infection of Austrian pine caused death of 1-year-old foliage except green needle bases and caused the absence of needles on 2-year-old twig segments (NY, Jun).

D, E. Austrian pine needles less severely diseased than those in (C) show dead bands and killed tips on which apothecia are forming (NY, Jun).

F. Tip necrosis, a secondary symptom, results from withering of green tips beyond fungus-killed bands (Austrian pine, NY, Jun).

G, H. Mature apothecia on Austrian pine needles before wetting (F) and after opening in response to moisture (NY, Jun).

Plate 12

Elytroderma Needle Cast of Pine (Plate 13)

Elytroderma needle cast is among the most conspicuous, damaging, and least understood foliar diseases of North American conifers. It is the only one characterized by perennial growth of the parasite in twigs and branches. The causal fungus, *Elytroderma deformans* (Rhytismatales, Rhytismataceae), occurs only in North America. A related fungus, *E. torres-juanii*, causes needle blight of pines in southern Europe.

E. deformans causes needle cast of jack, lodgepole, pinyon, and shortleaf pines; and witches'-brooms and needle cast of Coulter, Jeffrey, knobcone, and ponderosa pines. Severe damage is confined to the last two species. The disease is common in the northern Sierra and Cascade mountain ranges and is widespread but infrequent eastward to Colorado. Single records in Ontario and Georgia account for the listing of jack and shortleaf pines, respectively.

Incidence of this disease is low in most areas, but local epidemics develop occasionally and last for several years because the parasite grows perennially in twigs and branches. Major waves of infection began in some northwestern ponderosa pine forests early in the 20th century and again during the 1930s and 1940s. One particularly destructive outbreak in Oregon resulted in premature harvest to salvage over 100 million board feet of sawtimber. Repeated outbreaks in Jeffrey pine have been noted around Lake Tahoe and in other localities in the Sierra Nevada.

E. deformans attacks trees of all ages and sizes. Severe infection in 2–3 consecutive years can cause nearly complete defoliation and is sometimes lethal. Seedlings in their first few years of growth are seldom killed but may become permanently stunted and deformed. Mortality begins among saplings. Severe needle blight results in growth loss, death of small trees in dense stands, and increased susceptibility of trees of all sizes to killing by bark beetles and root-infecting fungi.

Symptoms appear in spring when clusters of 1-year-old needles, often all needles on a twig, simultaneously turn red-brown except for the basal 6–12 mm. These "flags" are conspicuous from long distances and indicate infections that began at least a year earlier. Flags gradually fade (A) and become partly obscured by new foliage until by late summer they are inconspicuous among normally senescing old foliage. Most diseased needles fall during autumn and winter, but a few remain attached, take on a weathered gray appearance, and slowly disintegrate.

Both needles and twigs in the flags are infected by *E. deformans*. Twigs flagged once usually flag again each spring, slowly lose vigor, and die. Initially vigorous branches may grow slowly for some years and curve upward (B). Twigs flagged more than 1 year lack old needles. This characteristic and lost vigor may make tree crowns appear thin.

E. deformans grows in the inner bark of twigs and branches of ponderosa pine. Mainstems of saplings sometimes become infected via needles. In bark the fungus often but not always causes small, elongate, resinous lesions that appear as reddish brown streaks in longitudinal view (E) or as a ring of dots in living bark near the cambium in cross-sections. The lesions, sometimes called resin cysts, are visible in living bark of ponderosa pine branch segments 2–12 years old and also occur in knobcone pine.

Infection of vigorous branches often results in broom formation in Jeffrey, knobcone, and ponderosa pines. Small trees in dense stands are sometimes completely converted into loose flat-topped brooms.

The cycle of the causal fungus is incompletely known. Inconspicuous pycnidia of *E. deformans*, up to 1.2 mm long and the same color as the leaf surface, appear as minute blisters in newly reddened needles in spring. Translucent beads or tendrils of spores emerge from these in moist weather. The function of these spores is unknown. In May–June, dark lines aligned with stomata on needles signal the development of apothecia. These arise beneath the epidermis of all surfaces of dead parts of 1-year-old needles and become mature from midsummer until spring. They are 1–7 mm long by 0.35–0.45 mm wide and open by a longitudinal fissure between the lines of stomata.

Apothecia liberate sticky-sheathed ascospores that become airborne during wet weather, mainly in late summer and autumn but also in spring. New infections begin as spores germinate and the fungus penetrates either the current season's needles or older ones. Needles infected in autumn remain asymptomatic until at least the next spring.

The interval between infection and display of flags is unknown and perhaps variable. During this time, the fungus grows in phloem of needles, into twigs, and thence into buds, new shoots, and new needles. Perennial infection of twigs perpetuates needle disease in the absence of new infections from spores. Hyphae of *E. deformans* grow with twigs and occur in buds as much as 1 year before the development of symptoms in needles grown from those buds. The fungus probably does not grow far in the proximal direction from the point of entry in a twig.

E. deformans does not grow on media commonly used for culturing fungi.

Outbreaks of Elytroderma needle cast begin almost simultaneously in localities hundreds of miles apart, indicating regional occurrence of conditions favorable for new infections by spores, but the specific conditions have not been described. The infrequency of destructive epidemics indicates that spring and summer conditions favoring fungal development, and autumn conditions favoring spore production and discharge, germination, and penetration of needles, seldom all occur in a single year. Local outbreaks start in sheltered places such as narrow valley bottoms, edges of meadows, thickets of saplings, and on low branches on north and northwest sides of pole-sized and larger trees. The most destructive epidemics in ponderosa pine have occurred in localities at about 1500 meters elevation in Oregon and about 1000 meters in northern Washington.

Hot dry weather in spring, while allowing development of foliar symptoms, suppresses spore production by pycnidia and is followed by suppressed apothecial development or failure of spore production in apothecia. This relationship is a basis for speculation that the pycnidia and spores from them have a sexual function in the life cycle of the fungus.

In areas where susceptible pines are subject to severe infection, damage can be minimized by logging selectively to eliminate severely diseased trees and those with infections near the top of the crown and accelerating logging schedules in mature stands. The latter practice reduces the impact of secondary pests and pathogens.

Individual ponderosa pines practically free from symptoms have been noted in areas of severe disease.

References: 353, 356, 387, 444, 445, 1324, 1650, 1797, 2116

Symptoms and signs of needle cast caused by *Elytroderma deformans* on ponderosa pine.

A. Blighted foliage on flagged branches, reddish brown in spring, fades through shades of brown during the summer (ID, Jul).
B. Twig and branch dieback and witches'-brooms. Diseased twigs turn upward (ID, Jul).
C. A perennially infected branch has stunted twigs and needles, dieback, and needle blight. Usually all needles in a fascicle die simultaneously (CO, Jun).
D. Brooms, usually less than 60 cm but occasionally up to 2 meters broad, are tight clusters of twigs on which only needles of the current season are green. Dead needles remain attached until autumn or winter (ID, Jul).
E. Brown lesions in inner bark of infected twigs are diagnostic but not always present (CO, Jun).
F–H. Apothecia in 1-year-old needles from infected twigs. Needles in (G) still have green bases (ID, Jul).

Plate 13

Lophodermella and Cyclaneusma (Naemacyclus) Needle Casts of Pine (Plate 14)

Lophodermella needle cast. Five species of *Lophodermella* (Rhytismatales, Rhytismataceae) kill and cause premature casting of the needles of pines in North America. *L. arcuata,* the only one that attacks five-needle pines, infects sugar, western white, and whitebark pines in the West. *L. cerina* attacks loblolly, Scots, and slash pines in southeastern states and lodgepole and ponderosa pines in Colorado and the Southwest. *L. concolor,* the most widespread and damaging species, and *L. montivaga* cause needle cast of jack and lodgepole pines from Ontario westward to the Pacific coast. *L. concolor* also ranges southward in the Rocky Mountains into Colorado. The severity of needle cast caused by this fungus decreases with elevation below 2000 meters. *L. morbida* affects ponderosa and knobcone pines west of the Cascade crest in California, Oregon, and Washington. In Europe, *L. sulcigena* and *L. conjuncta* are noteworthy defoliators of two- and three-needle pines.

All of these fungi are native to the respective continents and typically cause sporadic needle-cast epidemics in natural forests in single years. If weather or local site conditions permit severe infections in successive years, disease removes needles of all but the youngest age class and retards tree growth.

The biology of *L. concolor* is typical of the group. Its life cycle is 1 year long. Fruit bodies (apothecia, also called hysterothecia) in dead, 2nd-year needles open by means of a longitudinal fissure during wet weather in summer and release elongate, colorless, unicellular spores that have sticky, gelatinous sheaths. After the airborne spores settle on young, developing needles, the fungus penetrates them but causes no symptoms until the next spring. Diseased needles then turn reddish brown (C). By July, they are straw colored (B). Apothecia mature during June and July. Those of *L. concolor* appear as inconspicuous, shallow elliptical depressions 0.4–0.8 mm long in all surfaces of the straw-colored needle (D). Apothecia are colorless and thus seem to be the same color as the pine needle. Needles begin to fall as apothecia mature, and casting is complete about 14 months after the time of infection. During outbreaks of disease caused by *L. concolor,* nearly all of the 2nd-year foliage in the lower third or half of the crown may die and fall.

Other fungi sometimes parasitize or compete with *Lophodermella* species in recently killed needles. *Hendersonia* species parasitize *Lophodermella* in both Europe and North America. In the Northwest, *Hemiphacidium longisporum* invades needles infected by *L. concolor.* These depredations suppress sporulation of the needle-cast fungi, but whether this suppression influences the amount of disease is unknown.

Cyclaneusma needle cast. *Cyclaneusma minus* (Rhytismatales, Rhytismataceae) causes needle cast of many pine species in plantations and nurseries throughout the world. In North America, damage to Scots pine grown for Christmas trees is noteworthy because the premature yellowing and casting of 1- and 2-year-old needles impairs or destroys the economic value of the trees. Pines recorded as hosts include but are not restricted to Austrian, Cuban, digger, Jeffrey, limber, lodgepole, Monterey, mugo, ponderosa, Virginia, and eastern white. The pathogen was long known as *Naemacyclus niveus,* but taxonomic and biologic studies showed the existence of two nearly identical species, one a strong parasite and the other inconsequential. The innocuous fungus, which apparently does not infect Scots pine, retained the familiar name *N. niveus,* and the pathogen was named *N. minor.* Both fungi were later transferred to a new genus, *Cyclaneusma. C. minus* causes much needle discoloration that was once attributed not only to *C. niveus* but also to *Lophodermium pinastri* (Plate 11) and to the aphid *Eulachnus agilis* (see Johnson and Lyon, 2nd ed., Plate 35).

The dominant symptom is yellowing and casting of 2nd-year foliage in autumn. Symptoms appear as spots that are light green, then yellowish on needles that became infected during the previous year or earlier. The spots enlarge to yellow bands, and the needles eventually become entirely yellow. Prominent transverse brown bars appear on yellow needles of Scots pine (F, G), but this symptom is uncommon or indistinct on other hosts. On Monterey pine, whole needles or bands on them may turn reddish brown—symptoms similar to those of Dothistroma blight (Plate 19). Symptoms usually arise in late summer and autumn, but some infected needles remain green until spring. Severely diseased trees appear distinctly yellow before needles fall (E). After several years of severe disease, trees retain only the current season's needles, which although infected have no external symptoms.

Fruit bodies (apothecia) of *C. minus* develop in the brown bars and then beneath the entire surface of the needle, appearing at first as elliptical swellings 0.2–0.6 mm long, the same color as the rest of the needle. At maturity they open by a longitudinal fissure (H) and when wet expand, pushing flaps of host epidermis upward like tiny covers with hinges at their sides. This exposes the dull cream white to yellow spore-producing surface. In some regions the apothecia are preceded and accompanied by inconspicuous pycnidia, also colorless, which are deeply embedded in the needles. The spores produced in these are not infectious. In Pennsylvania, where Cyclaneusma needle cast has been studied intensively, pycnidia have not been observed in nature, although they form readily on laboratory media.

C. minus propels ascospores into the air when mature apothecia are moistened by rainfall or by dew at temperatures from near freezing to near 30°C (optimum 22°C). Spores in air reach peak concentrations 4–6 hours after the onset of rain.

Seasonal events in Cyclaneusma needle cast are complicated because *C. minus* liberates ascospores throughout the year, so that there are overlapping generations. In Pennsylvania, most infections begin in mid-April to late August, and a secondary wave of infection begins in late autumn. Needles of all ages are susceptible. The fungus enters them through stomata. Diseased needles turn yellow after 10–15 months, and apothecia form before or after needles drop. The length of the incubation period, before needles turn yellow, may depend in part on host vigor. The pathogen's cycle on Monterey pine is just 1 year long, which may reflect the fact that this tree species grows in climates milder than that of Pennsylvania.

References for Lophodermella: 228, 385, 439, 750, 853, 921, 1303, 1335, 1760, 1885, 2282
References for Cyclaneusma: 289, 315, 387, 444, 445, 487, 1033, 1286, 1324, 1614

A–D. Needle cast of shore pine (*Pinus contorta*) caused by *Lophodermella concolor* (WA, Jun). A. Diseased 1-year-old needles in the lower half of the crown are brown to straw colored. B. A twig with healthy developing needles, dead 2nd-year needles, spaces where some diseased needles were cast previously, and green 3rd-year needles that escaped infection while young. C. Diseased needles turn brown, then straw colored. D. Apothecia become visible as inconspicuous depressions in all needle surfaces.

E–I. Cyclaneusma (Naemacyclus) needle cast of Scots pine. E. A Christmas tree with yellow to straw-colored 2nd-year needles (MI, Aug). F. Brown bars on yellow needles are useful for diagnosis. Green needles that escaped infection during their 1st growing season remain susceptible. Spaces on the twig indicate that needle casting has begun (MI, Aug). G. Apothecia the same color as their substrate are prominent on dead foliage (MI, Aug). H, I. At maturity apothecia open by median fissures (NY, Oct).

Plate 14

Rhabdocline and Phaeocryptopus Needle Casts of Douglas-fir (Plate 15)

Rhabdocline and Phaeocryptopus needle casts, caused by native fungi endemic in the Douglas-fir region of western North America, came to prominence first in Europe and later in eastern North America on account of damage in young plantations. The pathogens, *Rhabdocline* species (Rhytismatales, Rhytismataceae) and *Phaeocryptopus gaeumannii* (Pleosporales, Venturiaceae), are restricted to Douglas-fir and have been transported with their host plants around the globe.

Rhabdocline needle cast. *Rhabdocline pseudotsugae*, including the subspecies *epiphylla* and *pseudotsugae*, and *R. weirii*, including subspecies *oblonga*, *obovata*, and *weirii*, cause similar symptoms. Precise identification is therefore seldom important. *R. weirii* subsp. *oblonga* and *R. pseudotsugae* subsp. *pseudotsugae* are the predominant forms in plantations in the Great Lakes region and the Northeast, while various subspecies are intermingled throughout the range of Douglas-fir in the West. Big-cone Douglas-fir in southern California is also affected. The disease in Europe is caused by *R. pseudotsugae* subsp. *pseudotsugae*.

Defoliation by Rhabdocline needle cast leads to suppressed growth and to value loss in Christmas trees and ornamental nursery stock. Young seedlings are seldom severely damaged, but those infected in nurseries can become foci of subsequent damage in plantations.

Symptoms arise in late summer or early autumn as tiny chlorotic spots on one or both surfaces of 1st-year and occasionally older needles. These lesions, initiated by ascospores during the preceding May–July, enlarge and darken to purplish brown spots and bands that are conspicuous by late autumn to early spring. When they are numerous, lesions coalesce and involve the entire needle except for a short basal portion. Such needles begin casting during winter. Those less severely affected remain until the next summer. Discolored foliage is most conspicuous in early spring.

Rhabdocline species produce apothecia in May–June in lesions in needles still attached to the twigs. The apothecia develop beneath the epidermis and appear as swellings up to several millimeters long and the same color as the lesion. They open by splitting the needle surface, usually near the midrib, and pushing a flap of epidermis aside. Except for *R. pseudotsugae* subsp. *epiphylla*, most apothecia arise along the stomatal lines on the lower needle surface.

Ascospores are released in May–July, and the fungus penetrates needles directly through the cuticle. Only one infection period occurs per year, but it may last several weeks. Colonization and cellular disruption in needles proceed slowly during summer but accelerate during autumn as symptoms become apparent. During July–September, the only conspicuous symptom may be missing foliage because most diseased needles fall during the summer 1 year after infection. Lesions found on needles of various age classes indicate the possibility of fungal development during more than 1 year in some cases.

No asexual state of *Rhabdocline* is known except perhaps for *R. weirii* subsp. *weirii*. Apothecia of this fungus are regularly associated with acervuli of *Rhabdogloeum pseudotsugae* in the same lesions on Douglas-fir needles. Whether the conidia of *Rhabdogloeum* are infectious is unknown.

Rhabdocline needle cast is favored by cool, moist environments. In British Columbia, *Rhabdocline pseudotsugae* subsp. *pseudotsugae* requires about 3 days of cool (e.g., 13°C), wet conditions for maximum infection from ascospores. Shorter wet periods or higher temperatures inhibit infection. Spores of *R. pseudotsugae* germinate best near 10°C.

Susceptibility of Douglas-fir to Rhabdocline needle cast varies with geographic origin of seeds. Trees of northwest coastal origin are least affected by *R. pseudotsugae* subsp. *pseudotsugae*; those from southern Rocky Mountain sources tend to be severely damaged, and those from northern interior regions are intermediate. The relation of seed origin to susceptibility to other *Rhabdocline* forms is less clear. Within even highly susceptible populations under epidemic conditions, however, some trees remain nearly free from infection, indicating the possibility of selection for resistance.

Rhabdocline needle cast is most damaging in plantations where weed growth, close spacing of trees, or dense foliage induced by shearing Christmas trees impedes air circulation and prolongs wetness on low branches. *Rhabdocline* species do not mature in fallen needles or in foliage on cut branches. Therefore the disease is often adequately suppressed by weed control, removal of low branches bearing diseased needles, and destruction of severely diseased trees. Susceptible trees that reach a height of 6–8 meters in affected plantations seem to outgrow the disease, perhaps because microclimatic conditions in upper, exposed parts of tree crowns do not favor the pathogen. Rhabdocline needle cast seldom causes significant damage to ornamental trees in landscapes, presumably because open environments are unfavorable for infection.

Phaeocryptopus needle cast. Once called Adelopus needle cast or Swiss needle cast, this disease can also be destructive in regions where the climate is moist. Christmas tree plantations coast to coast have sustained significant defoliation and loss of value. The disease develops insidiously because the causal fungus can produce fruit bodies (pseudothecia) and spores inconspicuously on green needles. Most infections occur on 1st-year needles. One to three years later, these become yellow-green, mottled with brown (E), or entirely brown before casting. When disease becomes noticeable, much green foliage is already infected. Defoliation begins with the oldest needles (D), and in severe attacks, all but the youngest needles may fall. After harvest, Christmas trees with green, infected needles dry out and lose needles more rapidly than normal.

Phaeocryptopus gaeumannii produces black pseudothecia 40–100 μm in diameter that grow from the stomata in lines on the lower needle surface (G, H). Pseudothecia arise throughout the year on needles 1 or more years old and are sometimes visible on 1st-year needles as early as August. Diseased needles may produce pseudothecia for 1, 2, or 3 seasons before falling. Most spores are dispersed during early summer, and most new infections occur during shoot elongation. No asexual spore-producing state of *P. gaeumannii* is known.

Phaeocryptopus needle cast is favored by rainy summer weather. Individual trees vary greatly in apparent susceptibility, but differences related to regional seed origins have not been identified.

Additional species of *Phaeocryptopus* that cause needle casts are *P. nudus* on true firs and hemlock and *P. pinastri* on pines. These are inconsequential compared with *P. gaeumannii* on Douglas-fir.

References for Rhabdocline: 238, 387, 612, 1262, 1448, 1449, 1891
References for Phaeocryptopus: 226, 346, 347, 526, 612, 1179, 1289, 1290, 1356

A–C, F, H, I. Rhabdocline needle cast. A. Brown to purplish brown diseased foliage on low branches (NY, May). B. A tree defoliated by severe infection for several years (NY, Apr). C. Apothecia of *Rhabdocline weirii* subsp. *oblonga* on lower surfaces of Douglas-fir needles (NY, May). F. Twigs with severely diseased 1-year-old needles. Alternating green and brown foliage on the twig at right resulted from infection of needles produced in spring, and escape of those produced during a second flush of growth in midsummer (NY, Mar). H. See below. I. Lesions, purplish brown in spring, give diseased needles a mottled appearance (NY, Mar).

D, E, G, H. Phaeocryptopus needle cast. D. An affected branch from which most 2nd- and 3rd-year needles have fallen. The current season's growth is stunted because of defoliation (NH, Jul). E. Mottled, chlorotic 1- and 2-year-old needles infected by *Phaeocryptopus gaeumannii* will soon fall. Yellow spots and deformity of 1st-year needles were caused by the Cooley spruce gall adelgid (see Johnson and Lyon, 2nd ed., Plate 49) (NH, Jul). G, H. Pseudothecia of *P. gaeumannii* on lines of stomata on lower surfaces of Douglas-fir needles. Fruiting on the needle in (H) is most prolific in a lesion caused by *R. weirii* (NY, Nov).

Plate 15

Rhizosphaera and Isthmiella Needle Casts of Spruce (Plate 16)

Rhizosphaera needle cast. Five species of *Rhizosphaera* (Deuteromycotina, Coelomycetes) infect needles of various conifers, but only *R. kalkhoffii* is a significant pathogen in North America. This fungus causes premature death and casting of needles of conifers throughout the Northern Hemisphere. It causes significant defoliation of spruces in nurseries, plantations, and landscapes outside the native range of the trees, but damage in natural forests is negligible. Colorado blue spruce sustains the greatest damage, especially in Christmas tree plantations in the East. Engelmann spruce is highly susceptible also. White spruce is intermediate, and Norway spruce is relatively resistant. Other spruces affected are black, Serbian, Sitka, and two Asian species, *Picea orientalis* and *P. schrenkiana*. *R. kalkhoffii* causes a significant needle blight of Japanese red pine in Japan. Other pine hosts are Austrian, Japanese black, mugo, eastern white, and Himalayan white. Additional coniferous hosts include Douglas-fir, true firs (noble, Siberian, and silver), and western hemlock.

Infection usually begins in spring in needles on low branches. Under favorable conditions during a period of years, it spreads gradually upward and around the tree. Sometimes infection starts 1–2 meters or more above ground and then spreads mainly downward and outward in a cone pattern, creating defoliated "holes" among normal branches. Branches die if they are defoliated in 3–4 successive years. Trees of any size may be infected, but serious defoliation leading to branch death is confined mainly to trees less than 7 meters tall. In most areas, symptoms arise in late summer as yellow mottling of 1st-year needles, some of which soon turn brown or, in blue spruce, purplish brown. Browning becomes general during late winter and early spring, and infected needles fall during summer and autumn, 12–15 months after the onset of infection. *R. kalkhoffii* overwinters in infected needles on the tree and also in fallen needles.

Rhizosphaera species produce fruit bodies (pycnidia) on stalks that emerge from stomata. Thus pycnidia of *R. kalkhoffii* develop in rows on all faces of spruce needles. These pycnidia usually appear on discolored needles in spring but often arise at other times and often on needles that are still green. At full size, pycnidia appear as minute black spheres less than 0.1 mm in diameter. Each pycnidium is capped by a bit of white wax that was pushed outward from the stoma by the hyphae that gave rise to the fruit body.

Spores (conidia) are dispersed from pycnidia by splashing and dripping water beginning in spring and continuing into autumn. The spores infect either newly grown needles of the current season or needles of any age that are senescent or stressed by other pathogens or environmental insults. Pycnidia of *R. kalkhoffii* are sometimes conspicuous on needles of blue spruce twigs dying of Sirococcus shoot blight (Plate 62), and in Europe *R. kalkhoffii* was once suspected of a causal role in "top dying" of Norway spruce. Experiments in Japan showed that development of Rhizosphaera needle blight in pine was accelerated in foliage previously stressed by exposure to sulfur dioxide.

In highly susceptible blue spruce Christmas trees, symptoms sometimes develop only 3–4 months after needles first become infected. In parts of Pennsylvania, for example, some 1st-year needles infected in spring turn yellow and then purple-red as early as August. Gradually they become purple-brown as the fungus sporulates on them from August to November. Such needles fall during autumn and winter.

R. kalkhoffii grows readily on laboratory media, producing mycelium, yeastlike bud cells, and pycnidia. Isolates usually grow most rapidly at 24–25°C, but some from Great Britain grow best at 18°C.

Conidia of *R. kalkhoffii* germinate relatively slowly. On wet foliage at about 25°C, the fungus may require about 48 hours to infect spruce needles. More time is required at higher and lower temperatures. The need for prolonged wetness may explain why *R. kalkhoffii*, although widespread, causes significant damage mainly in scattered northern localities. Christmas tree culture favors disease because waterborne spores may be splashed or carried inadvertently by people among closely spaced trees and because the dense foliage of sheared trees retains moisture longer than the foliage of unsheared trees.

Isthmiella needle cast. *Isthmiella crepidiformis* (Rhytismatales, Rhytismataceae) is one of three so-called needle-cast fungi that commonly attack spruces in natural forests across North America. The other two fungi are *Lirula macrospora*, which is more virulent, and *Lophodermium piceae*, which is less so, than *I. crepidiformis*. Hosts of *I. crepidiformis* are black, Engelmann, and white spruces. The fungus is most common in moist areas, especially in western Canada. Its pycnidia and apothecia (hysterothecia) are always found on dead needles 2 or more years old. It does no economic damage, although it often causes conspicuous foliar browning of black spruce in dense stands.

Although we use the term *needle cast* here, needles killed by *I. crepidiformis* or *Lirula macrospora* remain attached until they have weathered off after several years. As a result branch segments with dead needles sometimes alternate with segments lacking needles if severe infections have not occurred in successive years. The normal needles fall off first, leaving only dead ones on segments several years old (C).

Apothecia of *I. crepidiformis* (D) are black, are 0.3–0.6 mm wide and up to 3 mm long, and develop on lower needle surfaces. Externally they resemble those of *L. macrospora*, but the two species produce distinctly different elongate ascospores. Those of *I. crepidiformis* are constricted in the center, while those of *L. macrospora* are club shaped. *Lophodermium piceae* fruits only on fallen needles.

The life cycle of *I. crepidiformis* is incompletely known but seems to be similar to those of the better known *I. faullii* on balsam fir and *I. quadrispora* on alpine fir. The fungi on fir have 2-year cycles characterized by ascospore dispersal and infection of 1st-year needles in summer, death of diseased needles the next spring, production of pycnidia one year after infection, and maturation of apothecia after 1 additional year.

References for Rhizosphaera: 351, 358, 387, 485, 667, 1077, 1284, 1377, 1399, 1614, 2111

References for Isthmiella: 444, 445, 612, 921, 1886, 2053, 2280

A, B. Rhizosphaera needle cast of spruce. A. A landscape specimen of white spruce with needles browned by *Rhizosphaera kalkhoffii* (NY, Jul). B. Close view of needles of Colorado blue spruce with pycnidia of *R. kalkhoffii* along the rows of stomata (NY, Sep).

C–E. Isthmiella needle cast of black spruce. C. A black spruce branch with 3rd-year and older needles browned by *Isthmiella crepidiformis* (Ont, Jun). D. Apothecia of *I. crepidiformis* on a black spruce needle (Ont, Aug). E. A 3rd-year segment of a black spruce twig with needles killed by *I. crepidiformis* (Ont, Aug).

Plate 16

Meria Needle Blight of Larch and Didymascella Leaf Blight of Cedar (Plate 17)

Meria needle blight. *Meria laricis*, a Hyphomycete, causes conspicuous browning of needles of western larch in the northern Rocky Mountain region. It also affects western larch seedlings in nurseries in the Pacific Northwest and various larch species in Great Britain and Europe. European larch has been found infected in Ontario. Western and European larches are highly susceptible, Japanese and Siberian larches much less so, and Dahurian larch least so among species tested. Japanese-European larch hybrids (Dunkeld larch) display intermediate susceptibility. Tamarack (eastern larch) is not a reported host, but one instance of infection of Douglas-fir is on record.

Infection by *M. laricis* does not kill large trees, but extensive loss of foliage slows growth. It weakens and sometimes kills seedlings. Needle blight in nurseries increases the number of larch seedlings culled before shipment.

Symptoms appear rather suddenly during or after wet weather while shoots are still growing. Yellow to brown spots arise on any needle surface and enlarge to involve the whole needle. Brown needles wither and soon fall from the tree. If moist conditions continue, the pathogen produces colorless spores (conidia) in tiny clusters along the lines of stomata, mainly on the lower leaf surface. Viewed with a hand lens, the spore clusters appear as white dots more prominent than the wax deposits normally associated with stomata. Viewed microscopically, the spores develop on conidiophores that emerge from stomata.

Although *M. laricis* is capable of producing many generations of spores per year, infection apparently occurs only during the early part of the annual growth period while succulent foliage is present. Needles that escape early infection remain green.

Damp weather favors the disease, but a dry season checks it almost entirely. If dry conditions prevail during the first part of the growth period, and wet weather follows, the first-formed foliage may remain green while that toward the shoot apex becomes blighted.

M. laricis overwinters in fallen needles and in the dead terminal tufts of needles that adhere to seedlings. This pathogen has caused conspicuous damage in British and European nurseries but only among seedlings growing in one place for 2 years or longer. This illustrates the importance of fallen needles as the source of spores for infection in spring. How the fungus is dispersed through the crowns of large trees in the forest is unknown.

Experiments in Great Britain indicated that, under moist conditions, temperatures of 10–25°C favor infection. Slight infection occurred at 0–5° and none at 30°C. Optimum temperature for both vegetative growth and sporulation was 17–20°C. Inoculation of various larch species with isolates from the same or other species showed no pathogenic specialization.

At times the symptoms of Meria needle blight arise so quickly as to suggest frost damage. They can be distinguished because lesions caused by the fungus arise as discrete, enlarging spots, whereas frost quickly kills entire needles. Also, *M. laricis* kills only needles, whereas frost kills succulent young stem tissue as well. Needles infected by *M. laricis* usually fall after becoming totally browned, whereas frost-killed needles tend to hang on for several weeks.

Gross symptoms of Meria needle blight can also be confused with needle cast caused by *Hypodermella laricis* (Rhytismatales, Rhytismataceae) in the Rocky Mountain region. Needles on dwarf shoots become infected by *H. laricis* as they emerge in April, and they turn brown around June 1. Needles on terminal shoots usually escape browning because the terminals elongate after spores of the pathogen have dispersed. Killed needles tend to remain on the short shoots. In some of these needles, hysterothecia develop and produce spores the

next spring. Hypodermella needle cast and Meria needle blight sometimes occur together.

Didymascella leaf blight. *Didymascella thujina* (Helotiales, Hemiphacidiaceae) causes leaf blight of western red cedar. The fungus, which was once known as *Keithia thujina*, causes significant damage on moist sites in western North America and in Great Britain and Europe, where it was introduced with North American seedlings. *D. thujina* also infects northern white cedar from the Great Lakes region eastward but causes negligible harm to this host. Similarly, Port Orford cedar in the West, although susceptible, is seldom affected. In Europe, Oriental arborvitae has been damaged in nurseries.

The disease is most common and severe in damp places such as ravines where snow lingers and on shaded foliage in dense stands. Damage tends to occur as the result of localized epidemics. Western red cedar trees of all sizes are susceptible, but seedlings and saplings are damaged most. Seedlings less than 4 years old may be killed within a single season, and mortality rates up to 97% have been recorded. Severe disease on trees beyond the seedling stage retards growth and may make low branches of trees in dense stands appear scorched. Foliage on upper branches may also become generally infected but never as intensely as that on lower branches.

Symptoms appear first on 1-year-old foliage in spring as small bleached areas, one or more per leaf, usually on the upper surface. These become confluent, turn brown, and involve the entire leaf. On seedlings, foliage of the juvenile type is most commonly infected. On older plants, foliage of all ages seems equally susceptible. The fungus sometimes spreads from one scale-leaf to others at the same level on a twig, but it does not move to leaves above or below.

Fruit bodies (apothecia) become conspicuous by late June. Young apothecia appear as slightly swollen spots, usually on the upper leaf surface, red-brown to olive brown against the dull brown background of the killed tissue. Usually one apothecium, but occasionally as many as three, form per leaf and become nearly black at maturity. Under moist conditions they swell until the epidermis splits, exposing the olive-colored spore-producing layer.

Ascospores are expelled from June to October, with a peak in late summer and early autumn. These can be dispersed many kilometers in air without loss of viability. They are thick-walled and equipped with a sticky sheath by which they may adhere to foliage for long periods before germination. No other spore type is known.

Twigs with severely diseased scale-leaves fall during autumn. Killed leaves that remain on the tree weather to a gray color and often contain dark cavities where spent apothecia either shriveled or fell out.

Although some infections may begin in autumn, most start in spring when ascospores that have overwintered on green foliage germinate and the fungus penetrates the leaves.

Western red cedar trees vary in susceptibility, resistant individuals having been detected in the midst of populations affected by natural epidemics. In Europe, western red cedar and Japanese arborvitae have been hybridized with the aim of combining the size and rapid growth of the American species with the resistance of the Asiatic one. Resistance to *D. thujina* appears to be simply inherited.

Three species of *Didymascella* other than *D. thujina* are known, and all are inconsequential pathogens of conifer needles. *D. chamaecyparisii* infects Atlantic white cedar in New Jersey, *D. tetraspora* infects junipers in Europe and western North America, and *D. tsugae* occurs on eastern hemlock in the Great Lakes region and the Northeast.

References for needle blight of larch: 120, 315, 374, 397, 1469, 1470
References for needle blight of cedar: 677, 1446, 1466, 2141

A–D. Meria needle blight of western larch (ID, Jul). A, B. Blighted foliage extending high into tree crowns. C. A branch with blighted foliage and twig dieback. Cause of the dieback is unknown. D. Close view of diseased leaves on dwarf shoots.

E, F. Didymascella needle blight of western red cedar. Brown to black apothecia are visible (F) in lesions that enlarge to involve entire scale leaves (OR, Jun).

Plate 17

Brown Spot Needle Blight of Pine (Plate 18)

Brown spot needle blight kills foliage and retards growth of many species of pine, mainly in North America. Long known for causing significant growth loss in longleaf pine in the Gulf states, the disease gained added notoriety in recent decades because of needle browning and defoliation of landscape and Christmas tree plantings of ponderosa and Scots pines in the central plains and Great Lakes regions. It occurs from Central America northward to Oregon, Manitoba, Ontario, and New York and in scattered South American and European localities.

The pathogen, *Mycosphaerella dearnessii* (syn. *Scirrhia acicola*) (Dothideales, Dothideaceae), attacks only pines. Hosts in addition to those named above include Aleppo, Austrian, Japanese black, cluster, Cuban, jack, knobcone, loblolly, lodgepole, Monterey, mugo, pitch, pond, red, shortleaf, slash, spruce, Italian stone, Virginia, eastern white, and western white pines, and several Mexican and Central American species.

Damage to longleaf pine is favored by its peculiar growth habit. Young seedlings grow only slightly in height until about the age of 5 years and then rise rapidly. Before the onset of height growth, seedlings are said to be in the grass stage (B). Foliage of these plants is subject to severe infection because moisture conditions are favorable and spores of the pathogen are numerous close to the ground. Severe blight 3 years in succession can kill seedlings. Less severe attack may delay the onset of height growth 2–10 years.

Brown spot affects trees of all sizes but is most damaging on small ones. Symptoms appear as spots that often enlarge to bands that encircle needles and cause death of parts beyond the bands (C, I). In the South, spots appear at any time of year but mostly from May to October. Symptoms are delayed until June–July in the central plains and until August in the Great Lakes region. The lesions are of two kinds, each about 3 mm wide. The more common type is initially straw yellow, becoming light brown with a dark border. Lesions of this type on longleaf pine needles eventually appear raised because uninfected tissue between lesions shrinks as the needles die. The second type of lesion consists of a brown spot on an amber yellow band. The yellow tissue is infiltrated with resin. Diseased needles often have dead tips, central zones with several spots in green tissue, and green bases. Needles with many lesions appear mottled.

Killed needles of longleaf pine gradually droop, turn reddish brown, and slowly weather to drab hues (B). After successive years of severe infection, longleaf pine seedlings beyond the grass stage may display a needle-free stem surmounted by a tuft of diseased foliage on the terminal shoot.

On Christmas trees and ornamental pines with dense foliage, infection is most common on low branches. Killed needles drop during October–November, leaving bare branches if infection was severe. Defoliated branches sometimes die, but usually buds remain alive and, during the next year, produce new foliage that in turn becomes infected. In lightly infected Scots and mugo pines, brown spot may merely accelerate the loss of 2nd- and 3rd-year needles (D).

In the South, *M. dearnessii* produces two kinds of fruiting structures on killed needle tissue. Conidial stromata (F–H) arise throughout the year as small dark masses of mycelium that break through the epidermis 2 weeks or more after spots form. Masses of sticky conidia produced by them are dispersed by splashing rain, wind-driven water, or contaminated people and equipment. Conidia are most abundant from May to August. They cause the first infections in each growing season, and they cause disease intensification throughout the season. Sexual fruit bodies (pseudothecia) differentiate within some stromata 6–8 weeks after conidial formation. Ascospores, discharged from pseudothecia during rain, dew formation, or fog, are windblown within and beyond the immediate locality. Ascospores can be found throughout the year in pseudothecia on necrotic tips of longleaf pine needles but not in brown spots on green needles. Numbers of ascospores increase from early spring to a peak in late summer. Showers of ascospores are blamed for occasional rapid development of brown spot needle blight in nursery beds or plantations previously free from the disease.

In central and northern areas, all infections are apparently due to conidia; pseudothecia and ascospores are not found. Conidia infect new needles in June–September, mainly in June–July, and the fungus produces another generation of conidial stromata beginning in August. Although these release many spores in late summer and early autumn, little infection seems to occur then. The major period of infection is late spring, when spores from overwintered stromata are dispersed to developing needles. After spore germination, the fungus enters needles through stomata.

The period from inoculation to display of symptoms varies with temperature, time of year, and species of pine from about a month on young, succulent foliage of longleaf or Scots pine to 6 months or more on old foliage of loblolly pine. In general, succulent needle tissue is more susceptible than mature tissue.

Warm, wet weather favors brown spot needle blight. Fruit bodies must be moist to produce and expel spores. Spores germinate and penetrate needles only when they are wet. Although infection can occur over a wide range of temperature, it is most rapid in longleaf pine if day and night temperatures are about 30°C and 21°C, respectively.

Brown spot can easily be confused with Dothistroma needle blight (Plate 19). Where both diseases occur, diagnosis often depends upon microscopic identification of the conidial state of the causal fungus. The conidial state of *M. dearnessii*, known as *Lecanosticta acicola*, is characterized by multicellular, olive-colored spores 15–35 × 3–4 μm in size. Conidia of *Dothistroma septospora* are similar in size, shape, and septation but are colorless.

Susceptibility of pines to brown spot needle blight varies both within and among populations of different seed origins. Some individual longleaf pines in natural populations remain free from brown spot year after year in spite of high infection levels in the populations as a whole. Such trees grow more rapidly than diseased neighbors, and their resistance is heritable. In Scots pine, short-needle varieties from southern France and Spain are especially susceptible, and long-needle varieties from Germany and Austria are resistant.

Generally 2–3 years are required for brown spot needle blight to reach epidemic status in a susceptible plantation. Many infections of Christmas trees result from transmission of spores while trees are being sheared. Damage by brown spot is minimized in both forest and Christmas tree plantations by the use of seedlings derived from resistant trees or populations, the planting of disease-free seedlings on sites free from *M. dearnessii*, adequate space between seedlings in nurseries and plantations, and the avoidance of shearing and other cultural operations that could spread spores while foliage is wet.

In longleaf pine stands, prescribed ground fires suppress brown spot. Young trees in the grass stage can survive these low-intensity fires, which are used in winter to destroy dead foliage that harbors the pathogen.

References: 106, 387, 552, 983–85, 1493, 1519, 1614, 1795, 1821, 1855, 2229

A. Needle browning caused by *Mycosphaerella dearnessii* on large slash pines (MS, Apr).

B, C. Brown spot needle blight on longleaf pine seedlings in the grass stage. One-year-old needles are brown, and elongating green needles of the current season already show dead tips bearing conidial stromata (MS, May).

D. Brown spot needle blight on mugo pine. First-year needles have blighted tips and many 2nd-year needles are missing (NY, Dec).

E–I. Symptoms and conidial stromata (black spots) on 1st-year needles of mugo pine, typical of brown spot needle blight on many pine species (NY, Sep).

Plate 18

47

Dothistroma Needle Blight of Pine (Plate 19)

Dothistroma needle blight, also called red band needle blight, was discovered in 1917–1919 in Idaho and in the Georgian S.S.R. Within the next 60 years the disease earned notoriety for retarding growth and sometimes killing pines in plantations around the world. It became the most damaging disease of pines in the Southern Hemisphere; on Monterey pine in New Zealand it earned further fame as the first forest tree disease to be controlled on a practical basis by application of fungicides from aircraft. The causal fungus, *Mycosphaerella pini* (syn. *Scirrhia pini*) (Dothideales, Dothideaceae), was perhaps distributed widely with pine planting stock. The pathogen is usually encountered in its conidial state, *Dothistroma septospora* (syn. *D. pini*).

More than 30 pine species and varieties are subject to infection. Austrian, Monterey, and ponderosa pines are most severely damaged. Other susceptible pines include but are not limited to Aleppo, Japanese black, Canary Island, cluster, Cuban, Jeffrey, knobcone, loblolly, lodgepole, mugo, Japanese red, slash, and western white. Bishop, red, Scots, and Mexican yellow pines are usually resistant. In rare cases, Douglas-fir, European larch, and Sitka spruce have been reported to be mildly infected where they were growing adjacent to severely diseased pines.

In North America the disease occurs in Newfoundland and from Ontario to Virginia and westward to the Pacific. Its distribution in the East is spotty, however, and it has not been reported from the western Great Plains or central Rocky Mountain regions. It is widespread in natural forests in western North America but causes significant damage mainly in parks, shelterbelts, Christmas tree plantations, and forests where pines are planted outside their natural ranges. Young trees are likely to be damaged more than older ones, and small seedlings can be killed within 1 year if disease is severe. In Monterey pine, growth loss becomes significant if more than 25% of a tree's foliage is killed, and growth is virtually halted if defoliation reaches 75%.

On the Great Plains, infection occurs from spring into autumn. Water-soaked or translucent spots or bands appear, and their centers turn yellow to brown or reddish brown. The transition from lesions to normal green tissue is abrupt. Along the Pacific coast, infection occurs throughout the year, and symptoms appear 4–12 weeks later. The incubation period is shortest in spring to early summer and longest in winter. Lesions develop into tan to brown or, in the West, reddish brown bands, and needle tips beyond lesions die. Needle bases sometimes remain green. When disease is severe, dying needle tips become prominent 2–3 weeks after spots appear. Infection is most severe within 2 meters of the ground and in most areas is slight on foliage more than 6 meters above ground.

The red color in lesions is due to accumulation of dothistromin, a toxin that *M. pini* produces in pure cultures and in pine tissue. Dothistromin apparently kills leaf tissue in advance of the fungus. Strong light enhances symptoms caused by the toxin, and shade suppresses them.

Within a few weeks after death of diseased tissues, or after winter in inland areas, small black fruit bodies (stromata) of the fungus break through the epidermis singly or in groups. Conidia form in the stromata and push to the surface in masses during wet weather. Conidia are dispersed by splashing or wind-driven rain or mist. Cloud mist was found to carry spores from diseased pine plantations at high elevations in East Africa. In California, conidia have been trapped from air apparently independent of water droplets. How they become airborne is unknown. Spore production begins in late summer in western areas of mild climate and may continue all year there, but it occurs only from spring to late autumn in the Great Plains.

In mild Pacific areas, infection can begin at any time of year, and more than one cycle of disease may occur annually. In the Great Plains, there is one cycle per year. Infection of 2nd-year and older needles occurs from May into October, and symptoms develop and intensify from September onward. First-year needles of Austrian and ponderosa pine become infected beginning in midsummer. Thus lesions in all stages of symptom development and sporulation can be found throughout the growing season.

Infection occurs by growth of conidial germ tubes or hyphae through stomata. This process takes at least 3 days but often much longer when temperature and moisture level are suboptimal. In general, several days of cool, rainy, or overcast weather after a period of conidial dispersal are needed. For most strains of the fungus, temperatures of 5–25°C permit conidial germination and mycelial growth. These processes are most rapid at 17–20°C. The period from infection until appearance of symptoms varies from 5 weeks to 6 months, depending upon environment and host factors.

Diseased needles drop prematurely, the older ones first. On ponderosa pine, needles infected during their 1st growing season usually have no symptoms until autumn and often persist until late summer of the following year. Newly infected 2nd-year needles may be shed during autumn or as late as early summer of the next year.

The pathogen survives winter or dry summer conditions mainly in foliage on the tree. Survival time in killed needles on the ground is usually less than 2 months under moist conditions, although survival for up to 6 months has been recorded. Needles on the ground are much less important than those on the tree as a source of infectious spores.

Three varieties of *D. septospora* have been proposed on the basis of conidial size and shape. *D. septospora* var. *linearis*, which has long spores (23–42 × 1.8–2.9 μm), occurs in western North America, and variety *pini*, with spores 15–28 × 2.6–4 μm, is found in the Great Plains region. *D. septospora* var. *keniensis*, with conidia of intermediate length, occurs in Africa.

The sexual or ascospore-producing state of the fungus occurs in scattered localities throughout the world and is associated with all three conidial varieties. In North America it has been found only in California and the Pacific Northwest, associated with the conidial variety *linearis*. Stromata that give rise to pseudothecia of the sexual state begin to form in autumn of the year of infection and may rupture the epidermis, but ascospores do not develop until spring. Ascospores are dispersed mainly in late spring, although they have been trapped all year in California. The role of ascospores in the development of epidemics is unknown.

Epidemics of Dothistroma needle blight develop most quickly in areas of mild climate with high rainfall or frequent fog or mist. Disease increases logarithmically and thus often seems explosive. Plantations of highly susceptible trees can be destroyed in 5 years or less. Observations to detect or assess the disease are best made in late autumn or early spring except along the Pacific coast, where symptoms are visible throughout the year.

Resistance to *M. pini* varies widely within and among pine species and may depend in part on the variety of the fungus in a given locality. In the Great Plains, ponderosa pines from certain sources in Arizona, New Mexico, and Nebraska and Austrian pine from Yugoslavia have shown high resistance. Blight-resistant individuals of Monterey pine have been shown capable of passing resistance to their seedling offspring. The susceptibility of Monterey pines also diminishes with age, beginning when the trees reach age 3, but Austrian and ponderosa pines remain fully susceptible throughout their lives.

References: 387, 552, 630, 631, 938, 1486, 1493–95, 1497, 1499, 1500, 1614, 1754

Dothistroma needle blight of Austrian pine.
A. A mature landscape tree with foliage on low branches browned by disease (NE, Jun).
B. Blighted tips on 2nd-year needles. First-year needles become susceptible shortly after the stage of development shown here (IL, Jun).
C, D. Reddish pigment is sometimes prominent in tan or brown spots and bands (IL, Jun).
E, F. Conidial stromata breaking through the surfaces of lesions on 2nd-year needles (IL, Jun).

Plate 19

Brown Felt Blights and Snow Blights of Conifers (Plates 20–21)

Several fungi, collectively called snow molds, kill the foliage of conifers beneath snow. Those fungi with brown mycelium that grows perennially on plant surfaces are sometimes called brown or black snow molds, and they cause brown felt blights. Those producing annual mycelium that disappears from plant surfaces soon after snow melts cause snow blights.

Brown felt blights. Feltlike mats of dark brown mycelium overgrow needles and twigs, binding them together and killing them. The felts develop while branches are beneath snow. They become visible and cease growth in spring and early summer as the snow melts. On sites exposed to harsh weather, brown felts are most common on the leeward sides of trees. Felt-covered twigs and branches often die, but they may survive and resume growth if their apical buds are not covered. Any foliage deeply buried in snow is subject to attack, and conversely, exposed parts escape.

Two fungi cause brown felt blights. *Herpotrichia juniperi* (syn. *H. nigra*) (Pleosporales, Massarinaceae) attacks plants in 10 genera of gymnosperms throughout the snowy parts of the Northern Hemisphere. It is most common on various firs and spruces. *Neopeckia coulteri* (Pleosporales, Coccoideaceae), which usually attacks only pines, occurs in North America and Europe. It is the more specialized as regards host plants, and studies in Europe have shown it to be restricted to a narrower high-elevation zone than is *H. juniperi*. Any pine species growing in the snowy habitat of the fungus may be attacked. Neither fungus causes economic damage in North America, but both occasionally do so in Europe, killing spruces and other conifers in northern nurseries and on mountains where seedlings have been planted for avalanche control. In North America, although both fungi have been found in the Northeast, they are common only in the mountains of the West.

The life cycles and related pathogenic behavior of brown felt blight fungi are incompletely known. Apparently both fungi can be dispersed either as ascospores or as fragments of felts that fall with dead twigs and needles from diseased trees. Litter infested with mycelium and ascospores falls onto snow and is deposited on branches below as the snow melts. This mechanism provides for eventual infection. Another possible means of infection is contact between healthy foliage and litter on the ground as branches are forced down by the weight of snow. The pathogens are quiescent during summer.

Both *H. juniperi* and *N. coulteri* can grow at temperatures of 0° to −3°C. Temperatures in this range may occur near ground level in deep snow even when the air above is much colder. Most felt development, however, occurs at temperatures slightly above 0°C in early spring when melting snow provides both free water and a water-saturated atmosphere around buried branches. The mycelium grows not only from needle to needle on an affected branch but also from one plant to another, even among different kinds of plants where their foliage intermingles. Growth from coniferous hosts may explain the reports of brown felt blight on fir dwarf mistletoe and on mountain heather, a tiny evergreen plant in the heath family.

As killed foliage is released from snow, the mycelial mat at first appears dull grayish and fuzzy (Plate 21F). Within days it turns dark and often somewhat shiny as the hyphae become thick-walled and brown. Felts in this condition can withstand sunlight and drying without rapid loss of viability.

Close inspection of felts during summer may show nearly globose fruit bodies (pseudothecia), 0.3–0.5 mm in diameter, embedded in the mycelium with their tips protruding slightly (Plate 20E). These become conspicuous after some of the felt weathers away (Plate 20F). The pseudothecia mature and produce ascospores during summer and early autumn, but the role of these spores in disease cycles is unclear.

Both *H. juniperi* and *N. coulteri* cover host needles with mycelium, and *N. coulteri* penetrates the cuticles of living needles, producing haustoria in cells beneath. *H. juniperi*, on the other hand, seems to remain superficial while the needles are alive but penetrates both needles and twigs after they die. During felt development, both fungi produce compact masses of cells, called either stromata or microsclerotia, that plug stomata. These plugs have been credited with a "smothering" action, but the mechanisms by which brown felts actually kill needles have not been studied.

Microsclerotia serve as survival structures for *N. coulteri* during summer. This fungus also survives as mycelium at the bases of needles and among bud scales. In early spring, hyphae capable of starting new infections grow from the survival structures.

The life cycle of *H. juniperi* requires 2 years. Needles and twigs are killed by mycelial felts during the first winter, and the fungus completes its sexual development, producing pseudothecia and ascospores, during the 2nd summer after felt formation.

The two brown felt blight pathogens can be distinguished on the basis of host preferences, minor differences in temperature response (both fungi grow most rapidly at 10–15°C in pure cultures, but *H. juniperi* grows well over a broader temperature range than does *N. coulteri*), and by the microscopic appearance of ascospores. Those of *H. nigra* are four-celled and somewhat olive colored at maturity, while those of *N. coulteri* are two-celled and dark brown. This distinction is important for specific diagnosis of brown felt blight on pines.

Some host plants of brown felt blight fungi

Herpotrichia juniperi
Alaska cedar
incense cedar
western red cedar
Douglas-fir
firs: alpine, balsam, grand, noble, red, silver, Pacific silver, white
mountain heather
mountain hemlock
common juniper
fir dwarf mistletoe
pines: lodgepole, mugo, western white, whitebark
spruces: black, Colorado blue, Engelmann, Norway, Sitka
western yew

Neopeckia coulteri
pines: Apache, Austrian, foxtail, jack, Jeffrey, limber, lodgepole, mugo, ponderosa, shore, sugar, eastern white, western white, whitebark, and unspecified "hard pines"
spruce: unspecified (uncommon)
western yew

Herpotrichia juniperi causing brown felt blight on Engelmann spruce (CO, Jun).
A, B. Young trees shortly after snow melt. New felts are nearly black; old, weathered ones are gray-brown.
C. A felted twig with green tips that indicate capability for renewed growth.
D. A 1-year-old felt with pseudothecia on a dead twig.
E. Pseudothecia embedded in a felt, visible as pimplelike protrusions.
F. Pseudothecia exposed after brown mycelium has weathered away.

Plate 20

Snow blights. Widespread in natural and planted forests and nurseries where deep snow persists, these diseases may cause severe damage. In an outbreak, hundreds of thousands of spruce and fir seedlings may be killed. Small trees beyond the seedling stage, although often severely defoliated, usually retain some green foliage and survive.

Gross aspects of snow blights are similar to brown felt blights in that the pathogens kill coniferous foliage beneath snow, spread from one needle to another by means of mycelium, and cease spreading after the snow melts. The mycelium is ephemeral and disappears soon after snow melts away from blighted parts.

For many years nearly all snow blight was attributed to one fungus, *Phacidium infestans* (Phacidiales, Phacidiaceae), but research in Europe and North America showed that several fungi, somewhat specialized in host preferences, are involved. *P. infestans,* important in Europe on Scots pine, has been reported to occur on red pine and white spruce in Quebec, but most records of its occurrence in North America are erroneous. This is the case for eastern arborvitae and for jack, Scots, and eastern white pines.

Two pathogens account for most of the snow blight records in North America. *P. abietis* is the common snow mold on Douglas-fir and on alpine, balsam, grand, Pacific silver, and white firs. It is pathogenic also to eastern white pine and eastern hemlock as well as to Colorado blue, Norway, red, and white (but not black) spruces. It produces fruit bodies (apothecia) only on firs. *Lophophacidium hyperboreum* (Phacidiales, Phacidiaceae), which is common in North America and Europe, attacks mainly spruces: black, Colorado blue, Engelmann, Norway, red, and white. It is also pathogenic to eastern white pine and balsam fir but does not sporulate on the latter.

A dozen additional fungi related to these first two are sometimes listed as snow molds or are blamed for snow blight, but most of them require more study before their roles are known with certainty. *P. dearnessii* on western yew, *P. pini-cembrae* on Douglas-fir, and *P. sherwoodiae* on western red cedar are examples. *Sarcotrichila balsameae* and *Nothophacidium phyllophilum* on firs and *S. macrospora* and *Hemiphacidium longisporum* on pines occur on foliage killed by needle-cast fungi and inhibit their fruiting. *S. piniperda* is associated with snow blight on black, Norway, red, and white spruces, but its pathogenicity is unproved. *H. planum* fruits on limber, pinyon, ponderosa, and eastern white pines. *H. planum* apparently follows a needle-cast fungus on ponderosa pine, but this possible sequence has not been studied for other hosts. *H. convexum* on pitch pine is rare and its pathogenicity unproved. *Korfia tsugae* on eastern hemlock is little known and also of unproved pathogenicity. *P. taxicola* on American

and western yews is a proven pathogen but not a typical snow mold. Apparently it infects wounded twigs and then grows into needles, on which it eventually sporulates. Typical snow molds do not infect stems.

Snow blight, regardless of the specific causal fungus, can be recognized by death and discoloration of patches of foliage up to 60–70 cm in diameter in spring at about the time that snow melts away from nursery beds or from low branches of trees. Blighted foliage appears ashy or waxy (glaucous) brown (Plate 21E, G). During the ensuing summer it becomes somewhat bleached, and it slowly weathers away during the next 1–2 years.

During the first summer after blight development, brownish black fruit bodies (apothecia) develop, usually on the lower sides of needles if on fir. The apothecia of *P. abietis* are oval to round, 0.3–0.5 mm wide, and arise in rows on each side of the midrib. Apothecia of *L. hyperboreum* are elongate, appearing as a dark line that commonly extends half the length of the needle and sometimes the entire length. Both species mature in autumn, open by irregular fissures, and discharge ascospores into the air during wet weather from late September until early November. These spores infect 1st-year foliage after snow covers it. The next spring during snow melt, mycelium emerges from stomata of diseased needles and spreads to adjacent healthy ones of the same or neighboring plants, including plants of other species. Thus green needles are infrequent within a patch of snow blight.

Some snow blight fungi have been reported to produce minute black structures, termed microsclerotia, on recently killed needles during summer. The potential role of these in survival of the fungi or initiation of infection is unknown. Indeed, their identity as structures of snow molds remains to be verified.

The European snow blight fungus, *P. infestans,* has been studied more than any other. It grows on laboratory media at temperatures from −3° to over 20°C and grows most rapidly at about 15°C. *P. abietis* and *L. hyperboreum* also grow on laboratory media and are assumed to resemble their European relative in environmental preferences. Moist autumn weather favors severe snow blight in Europe, presumably by enhancing ascospore production and dispersal. Conversely, dry weather before snowfall leads to a lesser amount of the disease. In Finland, snow blight is most severe on sites deficient in potassium, and a dense ground cover of ericaceous plants inhibits spread of the snow blight fungus. Such relationships have not been studied in North America.

References for brown felt blights: 108, 132, 387, 612, 1800
References for snow blights: 182, 387, 488, 554, 611, 612, 783, 1092, 1499, 1591, 1828, 1832, 2282

A, B, D, F. *Herpotrichia juniperi* causing brown felt blight on alpine fir. A, B. A branch infected for several years shows new felts (dark), grayish tan felts formed the previous year, and bare twigs where old felts with dead needles weathered off (WA, Jun). D, F. A new felt just after release from snow. Close inspection of its edge reveals the fuzzy gray-brown appearance of freshly exposed mycelium (CO, Jun).

C. Brown felt blight on whitebark pine. This felt, in its 1st year, lacked fruit bodies and spores necessary for identification of the fungus. Some needles are detached from the twig but are bound by mycelium (OR, Sep).

E, G. Snow blight on alpine fir. Mottled grayish brown color is typical of symptoms caused by *Phacidium abietis,* which would have produced apothecia several weeks after the photos were made (CO, Jun).

Plate 21

Tar Spots of Maple (Plate 22)

Tar spots are among the most showy and least damaging foliar diseases. Caused by species of *Rhytisma* (Rhytismatales, Rhytismataceae), they occur on several kinds of woody plants and on a few herbaceous ones. They are best known on maples, on which *R. acerinum* causes large tar spots (0.5–2 cm across) and *R. punctatum* causes small ones, about 1 mm in diameter, sometimes called speckled tar spots. *R. acerinum* occurs wherever maples grow in moist environments in North America, while *R. punctatum* is found coast to coast from the latitude of North Carolina northward. *R. acerinum* is the more common in the East, *R. punctatum* in the West. Both species also occur in Europe.

Maples reported susceptible to both tar spot fungi are bigleaf, mountain, red, Rocky Mountain, silver, sugar, and sycamore. Box elder is also susceptible to both fungi. Only *R. acerinum* is recorded on Amur, hedge, and Norway maples. *R. punctatum* is the only form listed on striped and vine maples. Probably the exclusive listings are more the result of incomplete observation than of strict host specialization by the causal fungi. Some other tar spot fungi and their woody plant hosts are *R. arbuti* on madrone and menziesia, *R. liriodendri* on tulip tree, *R. prini* on hollies and related plants, and *R. salicinum* on willows.

Spots on maples arise in late spring or early summer after leaves attain full size. At first the infected tissue turns light green or yellowish green. Then, during mid to late summer, raised, black tarlike stromata develop within yellow spots on the upper leaf surfaces. In most circumstances the large tar spots of *R. acerinum* are readily distinguished from the small ones of *R. punctatum*. On Norway maple and sometimes on red maple, however, *R. acerinum* produces large stromata by coalescence of small ones within one yellow spot. Some small stromata at the edges of large tar spots may remain separate from the main stroma, as shown in Figure 5A. In early stages of development, such stromata closely resemble those of *R. punctatum*, shown in panels D and F. Stromata arise within the upper epidermis and rupture the vertical walls of epidermal cells during development. Large stromata of *R. acerinum* become about 0.5 mm thick. The lower surface of a leaf beneath a large tar spot turns brown, but the surface beneath speckled tar spots remains yellow. Leaves with multiple spots may wither and drop prematurely but seldom so early or in such numbers as to threaten the general health of the tree. In autumn as leaves turn yellow or red, glossy tar spots present a striking contrast (E, G). Leaves infected by *R. punctatum* often retain chlorophyll longer within the infected area than in surrounding normal tissue. This retention has been termed the green island effect.

Soon after stromata of *R. acerinum* form, their surfaces become convoluted, the ridges showing locations of fertile zones where asexual spores and eventually apothecia are produced. Masses of the asexual spores break through the surface of the stroma during summer, but they are not infectious. It has been suggested that they function as gametes in the sexual development of the fungus. During autumn and early spring, apothecia develop in stromata on fallen leaves that remain moist. Apothecia ripen during spring, split the surface of the stroma open, and eject sticky ascospores as much as 1 mm into the air. Some spores alight on young maple leaves, thereby starting a new annual cycle. Symptoms appear 1–2 months after infection. *R. punctatum* develops similarly.

R. acerinum will grow and produce germinable conidia in laboratory cultures, but these spores, like those formed in stromata during summer, are not infectious.

Strains of *R. acerinum* differ in host preference. In experiments, ascospores from apothecia on hedge, Norway, and sycamore maple leaves caused abundant infections on source species but few or no spots on the other two maples. In laboratory studies, cuticular wax from maple leaves stimulated conidial germination of the fungal strain specialized for parasitism of the species from which the wax was obtained. Spores of an isolate from Norway maple, for example, germinated best on a medium containing cuticular wax of that species and germinated least well in the presence of wax from sycamore maple. It was suggested that the relative susceptibility or resistance of maple hosts depends upon the composition of cuticular wax, but the idea has not been well tested.

Early research on *R. acerinum* indicated infection only through the lower leaf surface, but a contemporary study of *R. punctatum* showed

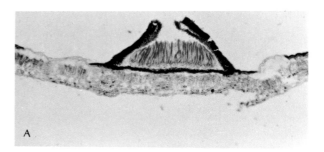

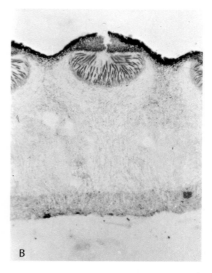

Figure 5. Apothecia of *Rhytisma acerinum* as seen in sections through stromata on dead maple leaves collected from the ground at the end of winter. Each apothecium, less than 0.3 mm wide, appears as a cavity with a thick black roof split to allow discharge of ascospores. Vertically oriented elements within the cavity are asci. A. A small, thin stroma containing only one apothecium that rests directly on dead Norway maple leaf tissue. Small stromata on Norway maple coalesce to form large ones (NY, May). B. Part of a thick stroma of the sort shown in panel G. The dark band across the bottom of the photo is leaf tissue of red maple; the lighter tissue above is all composed of fungal hyphae.

just the opposite. This ordinarily inconsequential matter would be of interest in relation to questions about cuticular wax, which on most plants is more abundant on the upper leaf surface than on the lower.

Conspicuous outbreaks of tar spots are infrequent except in moist, sheltered locations favorable for winter survival of the causal fungi. Tar spot caused by *R. acerinum* is scarce in urban and industrial areas. One reason is the sensitivity of the fungus to polluted air. Low concentrations of sulfur dioxide prevent new infections, although they do not influence spot development after infection.

References: 387, 389, 680, 783, 972, 1324, 1729, 2236

A. *Rhytisma acerinum* causing tar spot on red maple (NY, Sep).

B, E, G. *R. acerinum* on sycamore maple, causing abnormal color around stromata. The convoluted upper surface of a stroma shows the arrangement of numerous immature apothecia (England, Sep).

C, D. *R. punctatum* causing small tar spot, or speckled leaf spot, on bigleaf maple (OR, Sep).

F. *R. punctatum* on striped maple (NY, Sep).

Plate 22

Ink Spot Leaf Blight of Aspen and Poplar (Plate 23)

Trembling aspen, which grows all across North America, is accompanied in northern areas and in the central Rocky Mountains by *Ciborinia whetzelii* (Helotiales, Sclerotiniaceae), one of two or possibly three fungi that cause ink spot leaf blight. These fungi kill leaves in late spring and quickly produce in them prominent dark brown to black sclerotia, or ink spots. The diseases are sporadic in young stands of aspen, killing 25–100% of the foliage in localized outbreaks but not causing long-term or consistent suppression of tree growth. Saplings in dense thickets can be killed by intense, repeated defoliation, but this mortality is properly viewed as a natural thinning required for normal development of the survivors.

C. whetzelii occasionally infects but never causes severe leaf blight on other aspens and poplars including bigtooth aspen, eastern cottonwood, balsam poplar, and black poplar. Trembling and bigtooth aspens sometimes hybridize, and hybrid families commonly display the susceptibility of the trembling aspen parent.

The disease is usually first noticed in summer after many leaves are dead, but events of the annual cycle begin in spring as aspen leaves are expanding. At that time, sclerotia that have wintered in contact with the soil produce stalked, cuplike sporocarps (apothecia, Figure 6) from which ascospores are discharged into the air. If the spores are deposited in water on developing leaves, they germinate immediately and initiate infection. After 2–3 weeks, reddish brown blotches develop and usually expand until the entire leaf blade is dead. During expansion the blotches sometimes display alternating concentric zones of light and dark color. Dead leaves soon become tan.

Within 2–3 weeks after blight appears, one to four discrete, dark, circular to oval waferlike masses of mycelium form in the dead tissue of each diseased leaf. These are the sclerotia. They measure 2–4 × 3–6 mm and are about 0.5 mm thick. They include all leaf tissues at the site of formation (D). Initially light brown, they slowly darken to nearly black. Dead leaf tissue breaks at the edges of fully formed sclerotia, releasing them to the ground during July or August, the time depending on the locality. The dead leaves, then conspicuously perforated, hang on for a while.

Sclerotia on the ground ripen during autumn and early spring, and each is capable of producing one to three apothecia. The apothecia,

Figure 6. Apothecia of *Ciborinia whetzelii*, approximately natural size.

easily overlooked by the untrained observer, are usually 4–6 mm across, with stalks up to 1.5 cm long or occasionally longer.

C. whetzelii, like many other members of the Sclerotiniaceae, has a confusing taxonomic history. Synonymous names common in reference manuals are *C. bifrons*, *Sclerotinia bifrons*, and *S. whetzelii*. In western North America, ink spot leaf blight was formerly attributed to *C. seaveri* (syn. *Sclerotinia confundens*), but it has never been clearly demonstrated that *C. seaveri* is distinct from *C. whetzelii*. On the other hand, *C. pseudobifrons*, a distinct although less common species, occurs on trembling aspen from Quebec to Alberta.

Other tree pathogens that cause diseases similar to the ink spot leaf blights of aspen are *C. candolleana* on oaks (red oak in North America) and chestnuts, *C. foliicola* and *C. wisconsinensis* on willows, and *C. gracilipes* on sweet bay.

Ink spot fungi grow readily and produce sclerotia in laboratory cultures. They do not produce infectious asexual spores.

Severe outbreaks of ink spot leaf blight are favored by dense stands of young trees, low temperature and high humidity during the period of spore dispersal and leaf expansion, and the survival of a large number of sclerotia on the ground throughout winter. These conditions occur erratically, and thus outbreaks are unpredictable.
References: 95, 121, 389, 612, 700, 1046, 1540, 1942

A. Moderately intense ink spot leaf blight on trembling aspen (Ont, Jul).

B, C. Symptoms and signs on trembling aspen. Usually the entire leaf but not the petiole is blighted. Brown stromata form in blighted leaves, turn nearly black, and drop out, leaving holes with smooth edges (NY & Ont, Jul).

D. Magnified view of a young sclerotium in a blighted leaf of hybrid (bigtooth × trembling) aspen (Ont, Jul).

Plate 23

Ovulinia Petal Blight of Azalea and Rhododendron and Ciborinia Petal Blight of Camellia (Plate 24)

Azaleas, rhododendrons, and camellias throughout the regions of relatively mild climate in the United States are subject to disfiguring flower blights. Ovulinia petal blight of azalea and rhododendron is caused by *Ovulinia azaleae* (syn. *Sclerotinia azaleae*). Camellia blossoms are blighted by *Ciborinia camelliae* (syn. *Sclerotinia camelliae*). Both fungi belong to the Sclerotiniaceae, order Helotiales, and infect only the blossoms of their hosts. They were unknown in the United States until the 1930s and may have been introduced from the Orient.

Ovulinia petal blight. Many kinds of native and introduced azaleas and rhododendrons as well as mountain laurel are affected from Connecticut southward to Florida and west to California. The disease is most destructive in the humid South. It may also occur on greenhouse-grown azaleas, but this is usually prevented by rigorous exclusionary measures. Indian azaleas are notably susceptible, and Kurume types are somewhat less so. Petal blight can be confused with frost damage on the latter types. Carolina and catawba rhododendrons and related hybrids as well as mountain laurel are most likely to become infected when they are growing near azaleas.

When warm, moist weather favors the disease, nearly all the azalea blossoms in an entire locality may seem to collapse simultaneously. Blight begins as small spots about 1 mm across, whitish on colored petals or brownish on white petals. Within 24 hours the spots are 5–10 mm across, and if numerous, they coalesce and cause the petal or the whole flower to collapse. Limp, moist petal tissue, degraded by fungal enzymes, disintegrates to mush if it is touched lightly. Many blighted flowers remain attached or collapse against nearby leaves, dry out, and remain hanging on the plants.

Multiple cycles of azalea petal blight develop each year, beginning when azaleas start to bloom—in January–February near the Gulf Coast but much later in the North. Resting bodies (sclerotia) of *O. azaleae,* having wintered on or just below the soil surface beneath azalea branches, produce tiny, stalked fruiting structures (apothecia) from which ascospores are shot into the air during moist weather. Sclerotia are variously shaped, flattened black objects 3–12 mm long and up to 8 mm broad. Apothecia are shaped like wine glasses. They are 1.5–2.5 mm across, dull reddish brown, with light brown to buff-colored stalks 3–10 mm long. As many as three apothecia may grow from one sclerotium. Considered collectively, the apothecia in a locality develop and discharge spores during a period of 4–5 weeks, but an individual fruit body requires only 5–8 days for these functions.

Ascospores that alight on moist petals germinate, and the fungus penetrates immediately. If mild, moist weather continues, colorless, infectious spores (conidia) are produced asexually within a few days on the surfaces of blighted blossoms. New cycles of disease initiated by successive crops of conidia then occur at intervals as short as 3–4 days under favorable conditions. Conidia are dispersed by wind, splashing rain, and pollinating insects. Bumblebees and carpenter bees are the most important vectors.

If moist conditions continue, sclerotia begin to form in blighted flowers 2–3 days after petals collapse. The sclerotia darken and remain conspicuous for weeks. Most sclerotia eventually fall out or are carried with falling flower parts to the ground, where the fungus survives until the next flowering period. Sclerotia can survive for 2 years or longer, which explains the failure of attempts to control flower blight by removing all flower buds from azalea plants for a year.

Weather conditions most favorable for infection are fog, dew, or rain and continual high humidity at temperatures near 18°C. Disease can become severe quickly in the temperature range of 10–22°C. Losses in greenhouses have been greatest at 10–15°C, probably because this temperature range promotes condensation or retention of water on plants. The disease ordinarily develops from first spots to full blight in about 3 days, but development can be delayed by dry, cold, or very hot weather. Hot dry weather may arrest symptom development and may inhibit sclerotial development.

In southern areas, early-blooming azaleas are most likely to be disfigured, and late ones may escape because of diminished conidial production by the causal fungus. In the North, however, the pattern is reversed because ascospores are not produced until after the blooming period of early varieties.

Ciborinia petal blight. This disease of camellia is similar to Ovulinia petal blight of azalea in most major respects. The differences are that, since *Ciborinia* produces only ascospores, there are no secondary disease cycles and no significant vector activities of insects. *C. camelliae* occurs wherever camellias grow outdoors in North America, from Virginia southward to Florida and the Gulf Coast, westward to California, and north into the Pacific Northwest. It attacks all varieties of common camellia and sasanqua camellia but is unknown on plants in other genera. It causes rapidly developing flower blight and produces sclerotia in the bases of killed flowers. Sclerotia overwinter on or just beneath the soil surface, and apothecia develop in late winter or spring as camellias come into bloom. Ascospores infect the blossoms. The disease is favored by wet conditions and temperatures of 15–21°C during the period of bloom.

Infection takes place any time after the tips of the petals emerge from buds and begin to show color. Symptoms, beginning as brownish specks on expanding petals, are visible as soon as 24 hours after spore deposition. The lesions enlarge until the entire flower is dead and dull brown. Killed tissues are dry or leathery except during wet weather. In early stages of disease, veins of infected petals are darker than surrounding tissue. The resulting netted appearance distinguishes flower blight from injury by frost or wind. Blighted flowers tend to retain their shape and soon drop to the ground where, under moist conditions, dark, shiny streaks or masses of noninfectious microconidia form on them. Sclerotia usually form within petal bases and frequently unite into compound structures up to 25 mm or more across, mimicking the original arrangement of the sepals. They are initially buff-olive and become nearly black with age. In spring some sclerotia produce apothecia, while others remain dormant for another year or more. Some sclerotia produce apothecia annually for several years. The apothecia are buff-olive and similar in shape to those of *C. whetzelii* (Figure 6). They measure 5–20 mm across, with stalks 3–40 mm long. The length of a stalk depends on how far the sclerotium rests below the soil surface.

Most outbreaks of camellia flower blight can be traced to shipments of plants in containers that presumably carried sclerotia of *C. camelliae*. Once established in a locality, the pathogen cannot be eradicated by practical means, but the chance of introduction to noninfested areas can be minimized by shipping only bare-rooted plants on which flower buds show no color.

References: 643, 702, 733, 882, 1046, 2144

A, B. Early stages of Ovulinia petal blight on two azalea cultivars. Small lesions progress quickly to the limp blight stage (FL, Mar–Apr).

C, E, G, H. Late stages of symptoms and associated sclerotial development on rhododendron blossoms blighted by *Ovulinia azaleae*. Sclerotia form on and within blossom parts (PA, Jun).

D, F. Ciborinia petal blight on two camellia cultivars. Darkened veins are diagnostic (MS, May; and CA, Feb).

Plate 24

Botrytis Blight (Plate 25)

Botrytis cinerea (Deuteromycotina, Hyphomycetes) causes blight of flowers, leaves, and shoots and decay of fruit of hundreds of woody and herbaceous plant species worldwide. The disease is often called gray mold blight or gray mold rot. *B. cinerea* is most prevalent in humid areas, colonizing dead vegetable matter and attacking living tissues predisposed by poor nutrition, low light intensity, low temperature, prolonged succulence, senescence, or toxic chemicals, including air pollutants. The fungus often colonizes dead plant parts first and then spreads into living ones.

Strains of *B. cinerea* that attack woody plants show no host specificity. They cause diseases including but not restricted to twig blight of arborvitae; flower and bud blight of camellia; twig blight, gummosis, and fruit rot of citrus (especially of lemon); flower and leaf blight of dogwood; blight of flowers, peduncles, and twigs of male American holly plants; leaf blight of poinsettia and Indian rubber tree; blossom and spur blight of pear; flower, twig, and seedling blight of rhododendron; bud or flower blight and stem canker of rose; flower blight of fuchsia, hydrangea, lilac, rose-of-Sharon, and viburnum; blight of red alder, blueberry, Japanese cedar, cherry, false cypress, eucalyptus, filbert, hawthorn, juniper, lavender, osage orange, and pistachio. Coniferous seedlings in dense seedbeds or young natural stands are subject to leaf and shoot blight. Dark lesions on succulent shoots commonly develop at the bases of killed needles and may then girdle the shoots. This pattern is most common on giant Italian and Monterey cypresses, redwood, and giant sequoia. Leaf blight occurs on Douglas-fir, true firs, hemlocks, larches, and spruces. *B. cinerea* also causes storage mold and rot of many kinds of woody horticultural plants and forest tree seedlings; snow mold of Japanese cedar and Douglas-fir; damping-off of coniferous seedlings including jack, ponderosa, and red pines; and germination failure in Scots pine seeds.

Latent infections—those not immediately causing external symptoms—are common in many plants, including flowers of grape, hydrangea, pear, and rose and the vascular system of alpine currant. Latent infection of hydrangea blossoms may cause heightened color.

Symptoms are similar on petals or leaves of many plants and are not diagnostic. Expanding lesions that are tan to brown and eventually grayish are common. Lesions in delicate leaves or blossoms become wrinkled as they dry. Those in thick leaves such as those of rhododendron or Indian rubber tree may have concentric zones of lighter and darker color. In seedbeds or gardens, shaded leaves and shoots near the ground are most likely to become blighted. When succulent shoots are attacked, they quickly collapse and wither. On dogwood, leaf blight often results when blighted bracts fall on developing leaves. On dormant nursery stock in storage, the pathogen kills bark and cambium, causing them to become brown and mushy.

Field diagnosis requires recognition of the typical structures of *B. cinerea*. In humid air this fungus produces a sparse web of gray-brown mycelium on diseased parts. Conidiophores bearing conidia in clusters like bunches of grapes arise on the web, and these clusters are readily visible with a hand lens (C, E). The spores are dry, and clouds of them may rise when killed leaves or blossoms or moldy bundles of nursery stock are disturbed.

Most conidia are released naturally during periods of rapidly rising or falling humidity. They also are dispersed by rain. Impacting drops dislodge dry spores by shock waves of air, and conidia may also be carried 1–2 meters on the surfaces of splashing drops. Conidia may go for long periods without germinating, held in check by lack of moisture or nutrients or by microbial antagonism on plant surfaces.

B. cinerea enters living plant tissue through wounds or dead plant parts or by direct penetration of healthy leaves, blossoms, or fruits. The usual mode of penetration of intact surfaces involves germination of conidia and differentiation of the germ tube into a holdfast structure, the appressorium. A penetration hypha from the appressorium then breaches the cuticle and epidermal wall. Apparently *B. cinerea* kills plant tissues by a combination of enzymatic and toxic action. It produces cutinolytic, pectolytic, and cellulolytic enzymes by which it can penetrate the cuticle and degrade cell walls. It also produces a toxic polysaccharide and potentially toxic amounts of citric and oxalic acids.

Many strains of *B. cinerea* produce small, dark resting bodies (sclerotia) in moist blighted tissue. These function as survival structures during winter in temperate areas. In spring sclerotia germinate to produce either conidia or stalked apothecia of the sexual state, *Botryotinia*, from which ascospores are discharged to the air. Although the ascospores may cause some infections, they are not considered important because the fungus also overwinters as mycelium in decaying vegetation and produces conidia as temperature and moisture conditions allow. In fact, *B. cinerea* is such a competent saprophyte that it can reproduce indefinitely by colonizing dead and dying plant materials. Parasitism seems to be an unnecessary and incidental capability.

Strains of *B. cinerea* from apple, grape, and other hosts have been shown to be the conidial state of *Botryotinia fuckeliana* (syn. *Sclerotinia fuckeliana*), of the order Helotiales, family Sclerotiniaceae. By custom, these strains should be called *Botryotina fuckeliana*, but most plant pathologists prefer to use the familiar name of the conidial state.

Botrytis cinerea flourishes where air is moist and stagnant. It is common on low leaves of plants in dense seedbeds, notably on conifers in forest nurseries. The climatic events likely to promote blight are a warm period in early spring, causing growth to begin, then cool humid weather, which prolongs succulence in developing leaves and shoots. Frost or freeze damage, whether it kills or only weakens plant tissues, also sets the stage for attack, especially if followed by damp or wet weather. Sublethal injury causes abnormal leakage of nutrients from leaf cells onto plant surfaces. The nutrients stimulate spore germination of *B. cinerea*. Volatile substances from leaves of several conifers can apparently also stimulate spore germination, even at low temperatures such as might occur in storage.

Given high humidity, *B. cinerea* grows at temperatures from 0° to over 25°C. The optimum for growth and conidial production is 20–22°C, but spore germination, growth, and infection can occur, albeit slowly, at 0–5°C. For this reason *B. cinerea* is important in storage and is occasionally encountered as a snow mold. The minimum time required for infection is about 20 hours when plant surfaces are wet, humidity high, and temperature near 20°C. A few days of warm dry weather will prevent or check the disease.

Although plants vary in susceptibility to *B. cinerea*, no major genes controlling resistance are known in either woody or herbaceous plants.

Botrytis blight can be prevented, or losses can be minimized, by any measure that prevents plant stress or promotes air circulation and helps keep plant surfaces dry.

References: 214, 383, 387, 783, 882, 883, 950, 1046, 1469, 1499, 1730, 1936

A–C. Botrytis blight on rose. Early stage (A) is followed quickly by collapse of blossoms (B) and sporulation of the causal fungus as a gray mold on dead parts (C) (NY, Jul).

D. Early stage of Botrytis blight on blossoms of star magnolia (CA, Apr).

E. Sporulation of *Botrytis cinerea* on a dead blossom of bauhinia (FL, Mar).

F. A spreading lesion on a rhododendron leaf. *B. cinerea* is often a secondary invader of lesions like this one that are caused by other agents (NY, Jul).

G. Botrytis blight of a redbud leaf. This succulent sprout leaf was growing in humid air in deep shade (NY, Aug).

Plate 25

Cristulariella Leaf Spots (Plate 26)

Cristulariella leaf spots, also known as bull's-eye, target, or zonate leaf spots, are caused by *Cristulariella depraedens* and *C. moricola* (syn. *C. pyramidalis*) (Deuteromycotina, Hyphomycetes). Both fungi were first found on maples, and *C. depraedens* is still known mainly as a maple pathogen. *C. moricola*, however, infects and may defoliate many herbaceous and woody plants. It occurs throughout the eastern half of the United States and in eastern Asia. It causes economically significant defoliation of pecan and black walnut trees. *C. depraedens* occurs in northeastern and northwestern North America, Europe, and Great Britain.

The two fungi cause indistinguishable symptoms on maples. Lesions begin as water-soaked greenish gray spots and when fully developed are of all sizes up to about 3 cm, round or irregular, and grayish brown with dark margins. If irregular, the lesions are pointed along leaf veins. When weather is cool and moist, lesions may coalesce quickly and give a scalded appearance to large sections of leaf blades. Dead tissue soon begins to weather away, leaving holes in leaves, and severely damaged leaves drop.

On its many other host plants, *C. moricola* causes lesions that vary from pinpoints to irregular or target-shaped grayish brown blotches, usually on leaves but sometimes on other succulent parts such as the pods of orchid tree (E). The target spots have light centers and alternating light and dark concentric rings. On most hosts, spots that remain small have light centers and dark margins. Severe infection causes premature leaf senescence and defoliation as shown on pawpaw (A) and box elder (G). White crystalline deposits often appear in concentric rings (B) or as a film on the lesion (I).

Beginning a few days after lesion formation, distinctive reproductive structures (called either conidia or propagules) may develop. They are found mostly on lower surfaces of lesions and sometimes on green tissue at their edges and are visible with a hand lens. The propagules of *C. depraedens* are scattered, colorless, multicellular globes, about 0.15 mm across, on slender stalks. *C. moricola* produces conical structures that average about 0.4 mm long and 0.15 mm broad at the base. These also are stalked, multicellular, and colorless (Figure 7 and Plate 26G, I). Sometimes they are branched.

The propagules function as multicelled spores and are infectious. Their abundance varies with host and weather conditions. Ordinarily they are numerous on lesions on butternut, magnolia, maples, and sassafras. They eventually become detached from the stalks and turn light tan late in the season.

Propagules are dispersed by splashing water. Under conducive conditions (wet leaves and temperature near 21°C), propagules of *C. moricola* on black walnut germinate within about 6 hours by multiple germ tubes. These penetrate host tissue and allow establishment of the parasite within 12–24 hours after inoculation. After germination, propagules remain in place for a time; thus a propagule can often be found in the center of the upper surface of a small lesion. A new generation of propagules of *C. depraedens* may develop as soon as 5 days after inoculation of maples.

On some hosts, including white ash, mountain magnolia, maples, and probably many others, black resting bodies (sclerotia) form within large lesions, close against the veins. Sclerotia sometimes begin to develop while leaves are still attached, but they are usually found near the end of the growing season on fallen leaves. They form also in laboratory cultures after 2–3 weeks of mycelial growth. Those in culture are round in outline, 2–4 mm wide, 1–2 mm thick, and arrayed in concentric rings or clusters. *C. moricola* also produces distinctive eight-sided crystals in culture media.

The sclerotia presumably play a role in winter survival and development of the sexual (apothecial) states of *Cristulariella* species, but facts are not yet available. Likewise, the form and source of inoculum for initiation of disease in spring are unknown. The apothecial state of *C. moricola*, named *Grovesinia pyramidalis* (Helotiales, Sclerotiniaceae), has developed from sclerotia in laboratory cultures but has not been found in nature. No apothecial state of *C. depraedens* is known.

C. moricola apparently kills plant tissues by a combination of toxic and enzymatic activity. It produces toxic amounts of oxalic acid in infected tissue, and it degrades cell walls by means of pectic enzymes that are synergized by the acid.

Epidemics of Cristulariella leaf spots are favored by cool wet weather in midsummer. In temperate regions the disease becomes

Figure 7. Propagules of *Cristulariella moricola* on a leaf as viewed with a scanning electron microscope.

noticeable on low branches by the middle of July and can completely defoliate susceptible plants within 1 month. Normally, however, it is retarded by weather fluctuations. When relative humidity is less than about 94%, for example, lesions stop expanding, and no propagules form. The pathogen remains alive in lesions, however, and may produce propagules again if, after wetting, lesions are kept moist for 4 days or longer.

C. depraedens is known to infect only Japanese, mountain, Norway, red, Rocky Mountain, striped, sugar, sycamore, and vine maples and the rosaceous plant goatsbeard. *C. moricola*, by contrast, infects many diverse woody plants. Alder (two Oriental species), apple, avocado, basswood, two species of blueberry, butternut, western catalpa, box elder, Chinese elm, dogwoods (flowering and silky), three species of grapevine, shagbark hickory, magnolias (*Magnolia fraseri* and *M. tripetala*), maples (black, Norway, red, silver, sugar, and sycamore), poison oak, black olive, orchid tree, pawpaw, pecan, sassafras, serviceberry (*Amelanchier arborea*), sourwood, spicebush, sugarberry, sycamore and London plane tree, tree-of-heaven, tulip tree, tung, two species of *Viburnum*, Virginia creeper, and black walnut are reported hosts.

References: 224, 364, 365, 612, 673, 1078, 1111, 1112, 1407, 1589, 2037, 2115

A, B. Small spots and large lesions with concentric markings (bull's-eye spots) caused by *Cristulariella moricola* on leaves of pawpaw (NY, Sep).

C, D. Lesions on white ash. Purple borders are typical of small lesions on this host (NY, Aug).

E. Lesions on a pod of orchid tree (FL, Mar).

F. An irregular, expanding necrotic blotch on a leaflet of shagbark hickory (NY, Aug).

G, I. Propagules of *C. moricola*. G. Magnified view showing conical form of a propagule on a leaflet of box elder (NY, Aug). I. The lower surface of a lesion on silver maple, showing abundant propagules and a whitish crystalline deposit (WV, Aug).

H. Spots, blotches, and shriveling of box elder leaflets caused by *C. moricola*. Such leaves soon drop (NY, Aug).

Plate 26

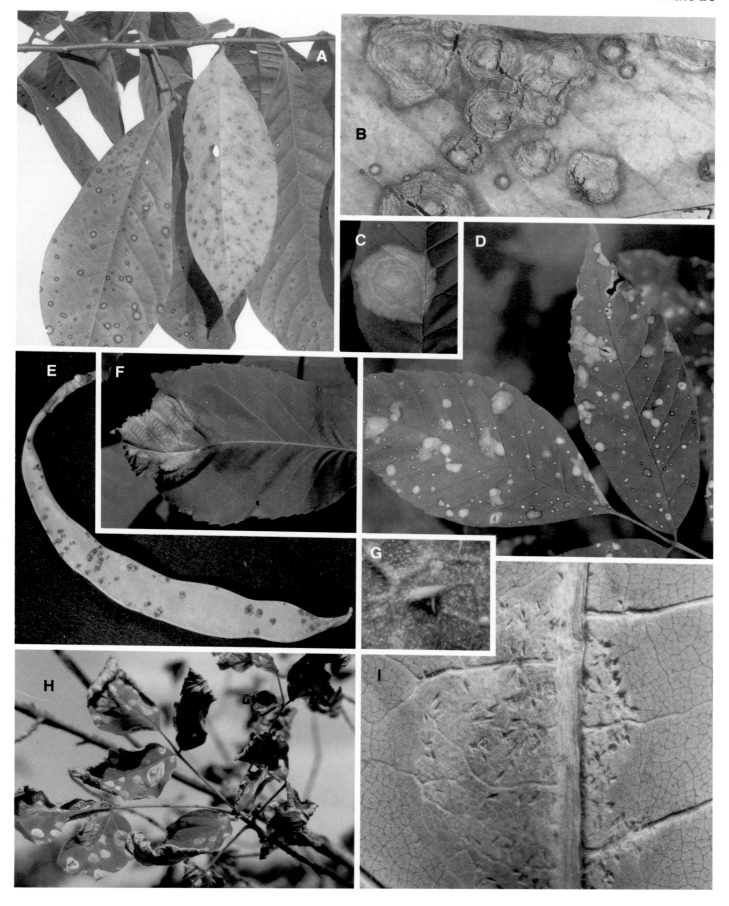

Entomosporium Leaf Spot of Pomoidae (Plate 27)

Diplocarpon mespili (Helotiales, Dermateaceae) in its conidial state, *Entomosporium mespili* (Coelomycetes), causes necrotic spots on leaves, on fruits, and sometimes on succulent stems of more than 60 species of plants, all in the subfamily Pomoidae of the rose family. The disease is known as Entomosporium leaf spot and also as Diplocarpon or Fabraea scald, leaf spot, leaf blight, or fruit spot. The alternative names call attention to the severity of the disease on some plants and to other names of the pathogen. *D. mespili* was formerly known as *D. maculatum* and before that as *Fabraea maculata*. Synonyms of the conidial state are *E. thuemenii* and *E. maculatum*.

Entomosporium leaf spot occurs worldwide. Susceptible plants or plant groups include apple and crabapple, mountain ash, chokeberry, cotoneaster, firethorn, hawthorn, Indian hawthorn, Yedda hawthorn, loquat, medlar, pear, evergreen pear, photinia, common and flowering quinces, serviceberry or Juneberry, stranvaesia, and toyon. Peach is sometimes listed as susceptible also, but this may be an error.

Leaf and fruit spot caused by *D. mespili* is said to be the most important fungal disease of pear trees in some parts of North America. Trees can be defoliated and fruit spotted, deformed, and cracked by July. Yield is reduced accordingly. Defoliation of quince understocks can cause them to harden prematurely and thus to be unsuitable for grafting of pear scions. Loquat seedlings in their 1st year have been killed by multiple lesions on succulent stems. Highly susceptible species of serviceberry develop twig and branch dieback if they are defoliated in several successive years. Hawthorn plants in landscapes may be defoliated before midsummer (A).

Lesions arise as minute dots and when fully developed appear as brown to gray irregular spots 2–5 mm in diameter. Those on leaf veins or succulent stems are elongate. On highly susceptible plants such as Yedda hawthorn, spots may be as large as 1 cm in diameter. Lesions may be scattered or so numerous that they coalesce to form large dead blotches. If plants defoliate before midsummer, new leaves may grow, and these soon also become infected.

On photinia and stranvaesia, dark brown to black spots form on upper and lower leaf surfaces, and tissue around the spots becomes reddish or purplish (G). Spots on loquat develop purplish margins with indistinct yellowish green halos. On Indian hawthorn, spots are gray-brown with a dark red or brown border, and diseased leaves frequently drop. Diseased leaves of crabapple gradually turn yellow and drop. Lesions on leaves of common quince and hawthorn cause relatively rapid yellowing and casting. On these plants, green color remains in tissues around lesions. This is the so-called green island effect.

Dark, blisterlike, irregularly shaped fruiting structures (acervuli) form beneath the cuticle in the centers of lesions soon after death of the plant tissue (F, H). The sizes of acervuli vary with hosts and environment. They may be up to 1 mm or more in diameter, and they often coalesce. Black at first, they and the surrounding leaf surface become gray with the accumulation of large numbers of colorless conidia that appear white in mass (H). The conidia are five-celled, with four cells normally visible and seemingly arranged in a cross. Each cell has a thin appendage. As the name of the conidial state, *Entomosporium*, suggests, the spores remind some people of microscopic insects. Conidia are dispersed by splashing water, and each conidial cell is capable of germinating.

In early stages of infection, hyphae grow intercellularly in living tissue and send absorptive structures (haustoria) into host cells, especially cells with chloroplasts.

Formation of conidia begins on diseased leaves in early summer in northern areas but can continue all year in regions with a mild winter climate on evergreen hosts such as Indian and Yedda hawthorns, loquat, photinia, stranvaesia, and toyon. On deciduous hosts, the fungus may persist through winter in lesions on young, green twigs. From either year-old leaves or twigs, conidia are splashed to developing leaves in spring. Conidia can also survive winter in acervuli on fallen leaves, and some new conidia are also produced there in spring. Primary infection begins when spores are splashed to new leaves. Thereafter many repeating, or secondary, cycles of disease start as successive generations of conidia cause new lesions.

In autumn in northern areas, the fungus shifts from conidial production to development of its sexual phase. Structures that look like abnormally small acervuli appear, especially on lower surfaces of lesions, and produce tiny, noninfectious one-celled spores that may function as gametes. Apothecia then begin to form in fallen leaves. They break through the leaf surface and complete their development in spring. At maturity they are black, cup shaped, and only about 150 μm wide and 100 μm deep. They produce colorless, two-celled ascospores that at maturity are ejected into the air. The role of ascospores in initiating disease in spring is unknown but is generally considered to be unimportant because of continual conidial production in warm regions and the apparent absence of apothecia there. During prolonged wet periods in spring, conidia sometimes begin to form in place of asci in apothecia on fallen leaves.

D. mespili will grow and produce *Entomosporium* conidia on laboratory media. Laboratory conditions favorable for the fungus are a guide to natural conditions favorable for disease. It grows most rapidly at about 20°C, and both spore germination and germ tube elongation are most rapid at 22–26°C. Like most fungi, *D. mespili* is inhibited by temperatures near or above 30°C. This information lends credence to the common observation that cool, wet summer weather promotes epidemics.

Although *D. mespili* affects many kinds of plants, strains of the fungus may be specialized to attack particular plant types. In one study, isolates from pear and quince were highly virulent on pear and loquat but not on *Photinia glabra* or *P. serrulata*, while isolates from the two *Photinia* species were virulent on their original hosts and also on pear and loquat.

Host plants of *D. mespili* vary widely in susceptibility. Crabapples and flowering quinces usually develop only small lesions and may not defoliate prematurely. Among pear species, *Pyrus amygdaliformis*, *P. betulifolia*, and *P. calleryana* are highly resistant. The resistance of callery pear has been transmitted to hybrid progeny resulting from crosses with common pear. Therefore it may be possible to breed commercial pear cultivars resistant to *D. mespili*. Among hawthorns, the Washington thorn has remained only lightly spotted while nearby 'Paulii' English hawthorns were completely defoliated. Cockspur thorn is also resistant.

Several cultural methods have been suggested but not critically tested for suppression of Entomosporium leaf spot. Removal of fallen leaves might be beneficial but only for plants on which the fungus does not overwinter on twigs or persistent leaves. In areas with dry summer weather, avoidance of sprinkler irrigation may help, but in the humid eastern part of North America, rainfall is usually adequate for severe disease on susceptible plants. Summer pruning should be avoided because it tends to promote a flush of highly susceptible, succulent new growth.

References: 123, 124, 387, 881, 1888, 1918, 1941

A. *Crataegus laevigata* 'Paulii' defoliated before midseason (NY, Jul).

B. Lesions on leaves and fruit of quince. Leaves showing yellow soon drop (NY, Sep).

C, D. Spots on upper and lower leaf surfaces and on a stem of *C. laevigata* 'Paulii'. The leaf at left shows yellowing and green island effect around lesions (NY, Jul).

E, H. Lesions on leaves of Indian hawthorn. Black acervuli form in the centers of spots on upper or lower leaf surfaces. Whitish deposits on brown background around acervuli are masses of colorless conidia (CA, Nov).

F. Lesions on 1-year-old leaves of toyon (CA, May).

G. Lesions and associated reddening on photinia leaves (MS, May).

Plate 27

Black Spot of Rose and Coccomyces Leaf Spot of Cherry and Plum (Plate 28)

Black spot. Caused by *Diplocarpon rosae* (Helotiales, Dermateaceae), which infects only roses, black spot occurs wherever roses are grown and is their most important disease. Rose species and cultivars vary from highly susceptible to apparently immune. Yellow or gold flowering types tend to be more susceptible than red or pink ones. Austrian Brier roses, Pernetians, Polyanthas, Tea roses, and Hybrid Teas are notably susceptible.

The damage is of two types, disfigurement and general weakening. Black lesions disfigure leaves, which turn yellow and drop. Repeated severe defoliation weakens the plants so that they produce fewer and poorer blooms, are difficult to graft, are abnormally sensitive to winter cold, survive poorly in storage, and fail to thrive after planting.

More or less circular black spots or blotches up to 15 mm across and sometimes larger develop on leaves throughout the season. They usually arise on the upper surface and have fimbriate, or fringed, margins, which are diagnostic. Minute black blisters (acervuli), visible with a hand lens and glistening when wet, arise beneath the cuticle in the central parts of the spots. The acervuli are 0.1–0.4 mm in diameter and produce colorless, two-celled spores (conidia). The conidial state is named *Marssonina rosae*. Petioles, fruit, and flower parts except petals develop symptoms similar to those on leaves. Petal infections, relatively unimportant, appear as red dots with some distortion of surrounding tissues.

Rose canes are susceptible during their 1st season of growth. Lesions develop as purplish red, raised, irregular blotches that usually remain small and become black and dotted with acervuli but lack fimbriate margins.

Multiple cycles of black spot develop each year. The first, or primary, cycle is usually initiated by conidia that are dispersed with splashing water from overwintered fruit bodies (modified acervuli) on fallen leaves or on diseased canes. The first infections give rise to successive generations of spots and conidia. Both severity of the disease and susceptibility of the plants seem to reach annual peaks in late summer. Infection occurs through either leaf surface as spores germinate and produce appressoria from which penetration pegs pierce the cuticle. Hyphae then form a radiating network of subcuticular and intercellular mycelium, sending tiny absorptive structures (haustoria) into living epidermal and palisade cells. With time, the infected cells die and darken, causing the characteristic black spot. Hyphae near the margins of the spot, organized into strands radiating from the point of infection, give the spot its fringed appearance.

Environmental conditions most conducive to black spot are wet plant surfaces, high humidity, and temperatures of 20–24°C for several days. Under these conditions, conidial germination and infection occur within 1 day, symptoms are apparent after 4–5 days, and acervuli with a new generation of conidia form within 10–11 days.

D. rosae grows saprophytically in fallen rose leaves and in spring disperses either conidia from acervuli or ascospores from apothecia. The apothecia, found in only a few localities in the northern United States, Canada, and Europe, are black, globose to disc shaped, and less than 0.3 mm in diameter. The ascospores, although infectious, are considered to be relatively unimportant inoculum in comparison with conidia.

D. rosae exists as numerous pathogenic races. Some are able to attack many rose types, others few. Some, while able to disfigure roses that are known for resistance, are unable to cause spots on cultivars that are normally susceptible. Thus roses resistant in one locality may become spotted if grown in a different area inhabited by different races of the pathogen.

Several rose species and cultivars resist a wide array of isolates and are thus useful either for ornament or as breeding stock. In general, resistance levels seem higher in natural rose species than in cultivars derived by extensive selection and breeding for horticultural characters, but cultivars selected expressly for resistance to black spot are an exception to this rule. Several types of resistance are known. Some cultivars prevent cuticular penetration by the fungus. Others limit lesion size after penetration. Resistance is often expressed as unusually slow symptom development and sporulation, which retards the rate of disease increase.

Black spot spreads and intensifies rapidly where many plants of susceptible cultivars grow together. Conversely, development of the disease is retarded in plantings where susceptible and resistant cultivars are artfully mixed. Good ventilation also suppresses black spot by promoting rapid drying of rose leaves.

Black spot infection is suppressed or prevented by low concentrations of sulfur dioxide. Thus the disease is scarce in areas where air is chronically polluted by this gas.

Coccomyces leaf spot. The causal fungus occurs across North America on cherries and plums and is best known to plant pathologists as a species of *Coccomyces* or *Higginsia*. These names are invalid, however, and the fungi once known as *Coccomyces hiemalis*, *C. lutescens*, and *C. prunophorae* are now grouped within one species, *Blumeriella jaapii* (Helotiales, Dermateaceae). We retain the old genus name in the name of the disease because it is specific and familiar. Few would recognize the name Blumeriella leaf spot.

The disease is economically important on cherry trees, especially sour cherry, because severe infection causes defoliation and reduces yields and winter hardiness. Other plants affected include almond and dwarf Russian almond; bitter, black, Japanese flowering, mahaleb, pin, sand, and sweet cherries; chokecherry; cherry laurel; and Canada, common, Japanese, and wild plums. Strains of *B. jaapii* that infect cherries are generally unable to attack plums, and vice versa.

Each year the first spots appear about the time leaves reach full size, and new spots appear until late summer. Initially they are dark purple and scattered or aggregated on one part of the leaf blade. Within about a week they turn reddish brown. Mature spots measure up to 3 mm across. Many spots close together cause large sections of the leaf blade to turn brown. Soon afterward, either the affected leaf turns yellow and falls, or the spots drop out, leaving shot holes. Shot holes, while indicative, are not diagnostic because many different agents induce them in leaves of cherries and related plants. Spots sometimes develop on fruits and pedicels but are not important there.

The first infections each spring are caused by airborne ascospores from inconspicuous apothecia (about 0.2 mm across) that develop on dead leaves on the ground. The fungus enters leaves via stomata, grows intercellularly, and sends haustoria into living cells, which soon die. During wet weather, white masses of conidia similar to those shown in Plate 30D extrude from acervuli, usually on the lower leaf surface. The conidia cause repeating cycles of disease.

Mild, wet summer weather promotes Coccomyces leaf spot. Conidial germination and infection occur only on wet foliage and are most rapid at 20–28°C. Symptoms appear as soon as 5–8 days and acervuli with spores as soon as 6 days after infection. Diagnosis is usually based on recognition of the conidial state of the pathogen, known as *Phloeosporella* (syn. *Cylindrosporium*) *padi*.

References for black spot of rose: 90, 337, 387, 393, 882, 883, 1744
References for Coccomyces leaf spot of Prunus: 6, 389, 538, 812, 997

A–C. Black spot symptoms on rose leaflets. Lesions in (B) and (C) show typical radiate, or fringed, appearance. Leaflets showing yellow will soon drop (NY, Jun–Aug).

D. Magnified view of part of a black spot lesion showing acervuli, which appear as minute black blisters (CA, Jul).

E, F. Coccomyces leaf spot on sour cherry. Leaves showing yellow will soon drop (NY, Jul).

G. Coccomyces leaf spot on chokecherry. Many lesions have dropped out, leaving shot holes (NY, Aug).

Plate 28

Spot Anthracnoses and Scabs Caused by *Elsinoë* and *Sphaceloma* (Plate 29)

More than 40 host-specialized species of *Elsinoë* (Myriangiales, Myriangiaceae) cause small lesions and scabby eruptions on leaves, fruits, green shoots, and other succulent parts of trees and shrubs. *Elsinoë* produces minute (less than 100 μm) black fruit bodies (ascostromata) in killed tissue on living plant parts. The ascostromata are preceded by inconspicuous acervuli in which conidia are produced. The acervuli are visible with a hand lens as near-microscopic cuticular blisters. The conidial state, *Sphaceloma*, is often the only one found, and some pathogens in this group lack a recognized ascostromal state. If the ascostromal state is known, we use the name *Elsinoë*; if not, *Sphaceloma*.

The name *spot anthracnose* is widely recognized as meaning a disease characterized by tiny, sunken, necrotic lesions caused by a species of *Elsinoë* or *Sphaceloma*. *Scab*, on the other hand, may mean a disease caused by a fungus in this group or by the conidial state of a species of *Venturia* (Plates 43–46). Scab diseases caused by *Elsinoë* or *Sphaceloma* are characterized by localized scabby overgrowth of host tissues. The overgrowth results in part from the action of hormones secreted by the pathogen. Fungal stimulation of plant growth reaches an extreme in the superelongation disease of cassava, caused by *E. brasiliensis*. This pathogen secretes a gibberellic acid that causes supernormal stem growth.

Spot anthracnose of snowberry. *Sphaceloma symphoricarpi* causes this disease, which disfigures leaves and fruits of snowberry across North America. Spots ranging from dark purple to black appear on leaves in spring and enlarge slowly, their centers becoming grayish. Spots sometimes coalesce into large, irregular blotches of dead, cracked tissue. Multiple lesions near the margins of young, expanding leaves may cause marked leaf deformity, and severe infections partially defoliate the plants. Small lesions similar to those on leaves also form on flower and leaf buds and on green shoots. On fruits, circular, somewhat sunken, pink-brown lesions, most common at the calyx end, may cause lopsided growth. This damage is of little direct consequence, but the diseased berries are often invaded by secondary fungi, particularly a species of *Alternaria*, that cause berry rot. Killed berries shrivel and remain in place for a time as dry brown mummies.

Euonymus scab. This disease, caused by *Elsinoë euonymi-japonici*, disfigures Japanese euonymus. Discovered first in Japan, the causal fungus perhaps came to the southern United States with its host plant. Lesions develop on both surfaces of leaves but are most common on the upper one. Lesions are up to 2 mm across, grayish white with a raised orange-cinnamon, waxy-appearing margin and, in the larger spots, a raised, dark center. When they are numerous, spots may coalesce to form extensive lesions. Minute white dots can be seen on individual spots and on coalesced lesions. The centers of leaf spots sometimes fall out. Lesions coalesce more frequently on stems than on leaves and are usually darker, with raised, wrinkled, or fissured surfaces. Japanese euonymus is apparently the only plant affected.

Spot anthracnose of flowering dogwood. Perhaps the most widely known of the spot anthracnoses of woody plants is that caused in flowering dogwood by *E. corni*. It occurs from Florida into the Northeast and also in California. The pathogen attacks bracts, leaves, petioles, fruit, peduncles, and green stems, resulting in malformed leaves and bracts and dead spots on shoots and fruits. White and pink flowering dogwoods are both susceptible, but the latter is less often disfigured. Other types of dogwoods are not affected. Like most diseases of leaves and shoots, spot anthracnose is favored by wet weather. If it is uncontrolled for several years under these conditions, it can weaken trees so that floral display and foliage become sparse.

The spots on leaves and bracts are reddish purple with tan centers and are 2 mm or less in diameter. Spots on young leaves usually have yellow borders, but spots on mature leaves may lack these. Young leaves with numerous lesions sometimes become grossly deformed. Grayish tissues in the lesions tend to fall out, giving severely diseased leaves a ragged appearance. Lesions on stems and fruits appear similar to those on leaves.

Acervuli form in lesions until late in the growing season, but they are inconspicuous and sometimes so sparse that even skilled observers cannot find them. Conidia produced in the acervuli are presumed responsible for secondary cyles and for some increase of the disease, but this matter has not been adequately studied. Black ascostromata mature in leaf and twig lesions in August in the mid-Appalachian region and are postmature by September. The role of ascospores in causing spot anthracnose is unknown. Primary infection in spring may be caused by conidia from overwintered lesions on twigs.

Poinsettia scab. *Sphaceloma poinsettiae* causes this disease. It occurs in Florida, Hawaii, the Caribbean region, and Central and South America. The fungus attacks leaves and stems and under conditions of severe infection causes wrinkling and distortion of leaves, premature leaf drop, and dieback of young stems. Lesions on stems are raised, pale buff, often surrounded by a reddish purple margin, as much as 1 cm long, and usually numerous. Coalescing lesions cause dieback. Lesions on petioles and midribs of leaves are similar to those on stems, and petiole infection is primarily responsible for leaf drop. On leaf blades, however, the fungus causes brown spots as much as 3 mm or more across. These are typically concave on the upper surface and convex below, causing a puckered appearance. Old stem lesions become grayish brown and covered with a microscopically velvety layer of condiophores and conidia. Conidia, dispersed by splashing water, cause new infections. Symptoms appear within 7 days, and a new generation of spores is produced within 14–30 days after infection.

Other *Elsinoë* and *Sphaceloma* pathogens and their woody host plants include:

E. ampelina, grapevine	*E. randii*, pecan
E. cinnamomi, camphor tree	*E. rosarum*, rose
E. diospyri, persimmon	*E. tiliae*, linden
E. eucalypti, eucalyptus	*E. venezuelensis*, croton
E. fawcetti, citrus	*S. araliae*, Hercules club
E. ilicis, Chinese holly	*S. cassiae*, golden-wonder
E. jasminae, jasmine	*S. catalpae*, western catalpa
E. ledi, Labrador tea and salal	*S. cerocarpi*, birch-leaf mahogany
E. leucospila, sasanqua camellia	*S. hederae*, English ivy
E. magnoliae, southern magnolia	*S. meliae*, chinaberry
E. mangiferae, mango	*S. murrayi*, willow
E. mattriolianum, madrone and strawberry tree	*S. oleanderi*, oleander
E. oleae, olive	*S. perseae*, avocado
E. piri, apple, pear, quince	*S. psidii*, guava
E. populi, poplar	*S. punicae*, pomegranate
E. pruni, black cherry	*S. ribis*, gooseberry
E. quercicola, water oak	*S. viburni*, viburnum
E. quercus-falcatae, southern red oak	

References: 115, 577, 882, 954, 955, 957, 1659

A, B. Spot anthracnose of snowberry, caused by *Sphaceloma symphoricarpi* (NY, Sep).

C, D. Scab of *Euonymus japonica*, caused by *S. euonymi-japonici*. Coalesced lesions become large dead blotches that may be extended by secondary fungi (FL, Apr).

E–H. Spot anthracnose of flowering dogwood. Tiny purplish red spots with yellow halos (G) (MS, May) cause leaf distortion if they are numerous (E, F) (NY, Jul). Lesions on bracts have grayish centers with reddish purple margins (FL, Mar).

I. Scab, caused by *S. poinsettiae*, on a poinsettia stem (FL, Apr).

Plate 29

Mycosphaerella Leaf Spots of Ash and Yucca (Plate 30)

Mycosphaerella species cause a wide array of plant diseases, including several dozen on trees and shrubs. We sample the array, the diversity of symptoms and signs, and some of the mycological confusion in this group of fungi by presenting diseases of pines on Plates 18 and 19, ash and yucca here, citrus and prickly pear on Plate 31, and poplars on Plate 32.

All species of *Mycosphaerella* (Dothideales, Dothideaceae) produce small black pseudothecia, usually in dead fallen leaves but sometimes in lesions on living leaves. In the latter, microscopic characters of pseudothecia and ascospores can be used for diagnosis, but usually only a conidial or spermagonial state of the pathogen develops while diseased leaves are attached. The three spore-producing states of a fungus in the same family are shown in Figure 8 (with Plate 35). Conidial and spermagonial states of *Mycosphaerella* species that parasitize woody plants are assigned to various form-genera. These and some major differences among them are as follows. *Cercospora* (Plates 38–39), *Cercoseptoria, Isariopsis,* and *Stigmina* produce dark conidiophores and conidia in minute clumps on the surface of diseased tissue. *Dothistroma* (Plate 19) and *Lecanosticta* (Plate 18) produce colorless conidia and dark conidia, respectively, within erumpent stromata. *Cylindrosporium* and *Marssonina* (Plates 30, 54, 55) produce colorless conidia in acervuli. *Cercosporella* produces colorless conidiophores and conidia on the surface of diseased tissue without fruit bodies or stromata; and *Asteromella, Phyllosticta* (Plates 33–34), and *Septoria* (Plate 32) produce conidia (or in some cases spermatia) in black, flasklike pycnidia that may be nearly superficial on host tissue or immersed in it. Conidial shape and septation vary widely among the above form-genera. One must understand, moreover, that *Mycosphaerella* is not the only genus of Ascomycotina whose member species may have conidial or spermagonial states of the types named.

Leaf spots of ash. Two or more species of *Mycosphaerella* cause necrotic spots on ash leaves. *M. fraxinicola,* which causes the disease illustrated here, occurs across North America on various species: Biltmore, black, blue, green, Oregon, red, and white ash. It causes brown spots with yellow borders, usually 5–15 mm in diameter, which develop most conspicuously in late summer. When they are very numerous, the lesions coalesce to form large brown blotches. As a result, leaflets may drop 4–6 weeks early, but apparently no significant harm is done to the trees. In fact, premature defoliation of ash seedlings by leaf-spotting fungi in early autumn is welcomed in some nurseries because it accelerates the development of seasonal dormancy and thus allows early lifting from the soil.

M. fraxinicola has two asexual fruiting stages. Acervuli of a *Cylindrosporium* state produce colorless, cylindrical conidia in masses on upper surfaces of the lesions (D), and tiny black spermagonia develop on the lower sides. The latter impart a distinct dark hue to the lesion (B, E). The conidial state, sometimes listed in references as the cause of a separate disease, is perhaps *C. fraxini* and presumably causes repeating cycles of disease. The spermagonial state has been called *Phyllosticta viridis,* but *Asteromella* would be a more appropriate generic name. The names *C. fraxini* and *P. viridis* have also been associated with corresponding life stages of *Mycosphaerella effigurata* (see below), but this seems to have been an error.

Pseudothecia of *M. fraxinicola* mature in fallen leaves during host dormancy. Ascospores are discharged from pseudothecia from late spring to early summer and presumably initiate the first infections each year.

M. effigurata also infects leaves of several ash species across the continent and can sometimes be found on leaves infected by *M. fraxinicola.* Dual infections have led to confusion among mycologists, diagnosticians, and writers of reference manuals. Lesions caused by *M. effigurata* are usually numerous and less than 3 mm across. This fungus produces colorless acervuli of a conidial state, *Marssonina fraxini,* followed by black spermagonia (*Asteromella fraxini*), both on the lower leaf surface. Its pseudothecia, found on fallen, overwintered leaves, produce ascospores that differ microscopically from those of *M. fraxinicola.*

Leaf spots of yucca. *Mycosphaerella yuccae* causes brown lesions with tan centers, usually 15 mm or less in extent, on the leaves of *Yucca filamentosa, Y. glauca,* and *Y. gloriosa.* The disease is noticed mainly in warm, humid regions. If they are numerous, the spots cause premature yellowing and death of leaves. Foliage on an affected plant may slowly become unsightly as the frequency of lesions increases. In spring the fungus produces a conidial state that fits the description of *Stigmina concentrica* (syn. *Cercospora concentrica*). This state appears as tiny dark eruptions that create starlike patterns as they break through the cuticle on both surfaces of the lesion (F). Each projection consists of a stroma bearing a dense tuft of olive-brown conidiophores and conidia. In late spring, spermagonia differentiate in stromata that may still be producing conidia. After conidial production ceases, pseudothecia develop. These are brown at first and black at maturity.

The disease cycle has not been studied in detail. From observations, we infer that inoculum in the form of conidia or ascospores is available at least during spring and early summer. Conidia of the *Stigmina* state can be dispersed by either wind or water, and ascospores are dispersed by wind. The random distribution of lesions on leaves indicates that leaves are susceptible during the entire period of elongation and perhaps much longer.

References: 56, 389, 425, 471, 552, 783, 882, 1464, 1614, 2228, 2231

A, B, D, E. Leaf spot caused by *Mycosphaerella fraxinicola* on Oregon ash. A, B. Symptoms. D. Acervuli, seen in magnified view, produce masses of colorless conidia on upper surfaces of spots throughout summer. E. Spermagonia, magnified here, darken the lower surfaces of spots in late summer, as shown in (B) (OR, Sep).

C, F, G. Leaf spot caused by *M. yuccae* on *Yucca gloriosa.* In spring, pseudothecia may be found in dark brown to black erumpent stromata such as those in (F) and at right in (G). While pseudothecia are immature, the conidial state of the pathogen, *Stigmina* sp., may be found on the same stromata. Tan stromata, as at left in (G), bear only the conidial state (MS, Apr).

Plate 30

Greasy Spot of Citrus and Black Rot and Cladode Spot of Prickly Pear (Plate 31)

Greasy spot of citrus. Caused by *Mycosphaerella citri*, this disease disfigures fruit, on which it is called greasy spot rind blotch, and it causes premature leaf drop beginning in autumn and continuing into spring. This leaf loss, if severe and early, reduces fruit yield the next season. Leaf infections occur on all commercial citrus and kumquat cultivars, orange jasmine, and *Aeglopsis chevalieri*.

Symptoms arise as tiny, raised, blisterlike eruptions on the lower surfaces of leaves and inconspicuous yellow mottling on upper surfaces. With time the blisters darken through shades of orange and brown to black, and both leaf surfaces appear as though spattered with a black greasy substance (D). The greasy spots consist of groups of mesophyll cells that become swollen, infused with gum, and necrotic. Severely affected leaves turn yellow except where chlorophyll is retained close to the greasy spots (B, left). On fruit, pinpoint black specks appear between the oil glands, and development of yellow or orange color in the vicinity of the specks is delayed.

M. citri commonly produces conidiophores and conidia on living leaves, but these usually arise sparsely on superficial hyphae without showing a relationship to the greasy spot lesions. If infected tissues are damaged further by a secondary agent such as frost, however, the parasite produces fruiting structures on necrotic parts of living leaves. Clumps of conidiophores with conidia develop and are followed by pseudothecia. The conidial state of *M. citri* is *Cercospora citri-grisea*. After diseased leaves fall, pseudothecia form abundantly on both leaf surfaces.

Although infectious conidia are available all year, most infections are started by ascospores. These can be found in pseudothecia on fallen leaves throughout the year. The peak period of their dispersal and of new infection occurs from June to early September, depending upon weather and locality.

M. citri requires rain or sprinkler irrigation to trigger the release of ascospores, but liquid water is not essential for germination and penetration. Relative humidity near 100% and temperature near 30°C allow spores to germinate and produce a sparse, superficial mycelium. The amount of infection that follows is determined in part by organic nutrients on the plant surface. Greasy spot often becomes severe, for example, if leaves are coated with honeydew secreted by insects. This coating promotes the growth of superficial mycelium and increases the frequency with which hyphae penetrate the plant surface. The fungus penetrates only through stomata. Greasy spot lesions visible to the unaided eye arise only through the mass action of many individual infections close together. A period of 1–8 months, depending upon weather and type of citrus host, elapses between infection and the appearance of macroscopic symptoms.

Greasy spot occurs also in Japan and Australia, but apparently two species of *Mycosphaerella* different from *M. citri* are the causal agents there.

Black rot and cladode spot of prickly pear. *Mycosphaerella opuntiae* and the diseases called "anthracnose" and black rot of various cacti and cladode spot of prickly pear constitute a puzzle that illustrates the general problem of insufficient information about identity and biology of pathogens of noncommercial plant species. "Anthracnose," the best known and most damaging fungal disease of prickly pear, is commonly attributed to *M. opuntiae* and more specifically to its putative conidial state, *Gloeosporium lunatum*. We use quotation marks because the term *anthracnose* is inappropriate here. The name *Gloeosporium* is also inappropriate. *G. lunatum* is identical with *Fusarium dimerum* var. *violaceum*. *M. opuntiae*, if really the sexual state of *F. dimerum* var. *violaceum*, may be not a *Mycosphaerella* at all

but rather a member of the unrelated genus *Plectosphaerella*. Furthermore as species of *Fusarium* do not cause anthracnoses, the disease must be called by another name. *Black rot* is a name sometimes applied, so it is used here even though the name *Fusarium rot* would be more informative. Unfortunately, the connection between *M. opuntiae* and *G. lunatum* is not consistent and was never confirmed by experimentation, and so it remains possible that *M. opuntiae* is a legitimate *Mycosphaerella* that causes a disease quite distinct from black rot. Or *M. opuntiae* may be only a secondary invader of lesions caused by *F. dimerum* var. *violaceum*.

Some of the symptoms attributed to *M. opuntiae* on prickly pears are the same as those of cladode spot, which is putatively caused by *Phyllosticta concava*. The tiny spores, only $1.0–1.5 \times 4$ μm, make *P. concava* seem likely to be the spermagonial state of some other fungus.

Black rot (Fusarium rot) is a common and destructive disease of prickly pears, affecting several species of these as well as cacti in the genera *Cereus*, *Echinocactus*, and *Mammillaria*. Most infections begin after rainy periods, usually on newly formed plant parts. Lesions arise as soft, sunken brownish areas on which after several days abundant flesh-colored tufts (sporodochia) of conidiophores and conidia usually arise. Progress of the disease thereafter varies with environment, species of host plant, and age of tissue. Under favorable conditions, rot develops rapidly in young stems or segments and may spread through an entire prickly pear segment within a few days. Often, however, lesions cease enlarging while less than 25 mm in diameter and become separated from healthy tissue by a layer of cork cells. Infection of mature tissue results in mostly small lesions that may dry out before conidia form. In the southwestern United States, black fruit bodies about 0.1 mm in diameter develop about a month after lesion formation. Typical lesions several weeks old are nearly circular, grayish depressed areas, sometimes with brown borders, dotted in the center with black fruit bodies. On prickly pears with thin segments, the lesions commonly extend from one surface to the other, and the dead tissue sometimes falls out, leaving a hole.

Wounds made by midges and other insects are important sites of infection. In Australia, where *F. dimerum* var. *violaceum* was introduced under the name *G. lunatum* for possible biological control of prickly pear, the fungus became established and assisted in the destruction of stands of prickly pear attacked by larvae of the cactus moth, *Cactoblastis cactorum*. There the fungus caused a spreading decay of prickly pear segments. Black fruit bodies of *Mycosphaerella* did not form in lesions initiated after introduction of conidia of *G. lunatum* into wounds. The difference in symptoms and signs between Australia and the southwestern United States indicates that two or more fungi may be involved in the disease attributed to *M. opuntiae* in the latter area.

Phyllosticta concava, according to its discoverer, causes large dead spots similar to those attributed to *M. opuntiae*. These lesions, sometimes called cladode spot, are brownish to blackish, concave, often with an elevated rim, and extend from one face to the other of thin prickly pear segments. The dead tissue becomes thin and often breaks away, leaving a hole. Black pycnidia, scattered over the lesion, are 0.15–0.20 mm in diameter. *P. concava* is not typical of *Phyllosticta* as construed in contemporary taxonomy. It may be the spermagonial state of a species of *Mycospherella*, possibly *M. opuntiae*, but no specific relationship has been reported.

References: 15, 59, 387, 425, 500, 504, 1394, 1734, 2174, 2176–78, 2222

A, B, D. Greasy spot, caused by *Mycosphaerella citri*, on leaves of grapefruit (A) and 'Valencia' orange (B, D). In magnified view (D), lesions appear as black flecks or small, black, blisterlike eruptions (FL, Mar–Apr).

C. Lesions attributed to *M. opuntiae* on prickly pear (TX, Apr).

E, F. Lesions attributed to *Phyllosticta concava* on prickly pear (AZ, Apr).

Plate 31

Septoria Leaf Spots and Septoria Canker of Poplar (Plate 32)

Septoria leaf spots. Several dozen species of *Septoria* (Deuteromycotina, Coelomycetes) parasitize the leaves of trees and shrubs in North America. Most of these fungi are inconsequential pathogens that cause brown spots and sometimes premature leaf fall but are otherwise little known. They produce inconspicuous black fruit bodies (pycnidia) in lesions on green leaves (D). Some species of *Septoria* are the conidial states of species of *Mycosphaerella*.

S. exotica causes necrotic spots on leaves of shrubs in the genus *Hebe* and on herbaceous plants in the related genus *Veronica* around the world. The spots are round, mostly 1–3 mm across, and brown at first, becoming grayish brown and finally gray to white with dark borders. When they are numerous, they cause premature yellowing and abscission. Green color remains for a time around spots on yellowing leaves.

S. azaleae causes brown, usually angular spots with yellow halos on leaves of evergreen azaleas outdoors in areas of mild, humid climate and occasionally in greenhouses. The disease is called angular leaf spot or scorch, the latter name referring to the general browning of tips and margins that sometimes develops when lesions are numerous. The spots, usually 1.5–4 mm across, are yellow to reddish yellow at first and become rusty brown with age. The beaks of pycnidia protrude from both surfaces. The disease usually appears late in the growing season and intensifies slowly during cool weather in regions with a mild climate. Water from rain and overhead irrigation disperses conidia. Temperatures of 16–28°C favor spore germination and growth of the fungus, but up to 2 months may elapse between infection and appearance of symptoms. This lag may explain the late-season development of the disease. *S. azaleae* occasionally causes economic damage by inducing defoliation that makes plants unsalable and that, if repeated, leads to death of terminal buds. Significant outbreaks have been noted on many azalea cultivars in Japan, Europe, and North America. Hiryu, macranthum, and snow azaleas are notably susceptible.

S. cornicola, *S. corni-maris*, and *S. floridae* cause colorful but generally inconsequential spots on the leaves of flowering dogwood. *S. cornicola* also affects red-osier and Pacific dogwoods. Spots caused by each pathogen are more or less angular, often limited by veins, at first brown and later tan to grayish or nearly white in the center with a purplish brown margin. Spots caused by *S. cornicola* on flowering dogwood may have a two-toned halo, reddish next to the lesion and yellow beyond. The spots, 1–6 mm across, appear in early July in the South but later in northern areas. All three diseases intensify during the latter part of the growing season. They can be distinguished reliably only by microscopic examination and identification of the associated fungus.

Septoria diseases of poplar. *Mycosphaerella populicola* and *M. populorum* cause leaf spots and canker on various poplars across North America. The diseases are innocuous in natural stands of native poplars, but *M. populorum* can devastate young plantings of susceptible types. Hybrids and species from abroad sustain the most damage. Diagnosis requires microscopic identification of the conidial state of each pathogen. The conidial state of *M. populorum* is *Septoria musiva*, and that of *M. populicola* is *S. populicola*. Diagnosis of Septoria canker, which is caused by *M. populorum*, often requires isolation and identification of the pathogen in pure culture because secondary fungi such as *Cytospora chrysosperma* (Plate 95) invade cankers and may prevent the sporulation of *Septoria*. Apart from occasional development in cankers, pseudothecia of *M. populicola* and *M. populorum* mature in fallen overwintered leaves and are therefore not useful for diagnosis.

All native North American species of poplar and aspen are susceptible to *M. populorum*. This fungus causes only leaf spots on most species but also causes cankers on eastern cottonwood, natural hybrids of North American poplars, and many introduced poplars. *M. populicola* affects only a few species and is less virulent than *M. populorum*, usually causing only leaf spots. Hosts of *M. populicola* include black and eastern cottonwoods, balsam poplar, and *Populus angustifolia*.

Leaf infections by *M. populorum* usually precede stem infections and are initiated annually either by airborne ascospores from pseudothecia or by conidia from overwintered pycnidia in cankers. Ascospores are discharged in greatest numbers at 22–26°C during moist weather. Leaves become infected soon after they unfold, and lesions develop 1–2 weeks later. They are most numerous on the foliage of low branches.

The appearance of lesions varies considerably within and between host species. Usually they first appear as sunken black flecks that enlarge to more or less round spots 1–15 (mostly 2–5) mm across, coalescing and forming dead blotches where they are numerous. The dead tissue fades to shades of brown, tan, or white with a brown or black margin (see such spots on balsam poplar in Plate 45). On plants that are somewhat resistant, the spots may remain small (1–2 mm) and may appear silvery.

Pycnidia develop 3–4 weeks after infection and appear as black specks on one or both surfaces of the dead tissue. Pink masses or tendrils of conidia exude from the pycnidia under moist conditions, and the spores, dispersed by splashing water, infect leaves and stems that have not yet formed corky bark. Leaf lesions increase rapidly in size and number under favorable conditions.

Cankers develop only on trees with leaf infections, and canker severity is correlated with leaf spot severity. Most cankers originate within 1.5 meters of the ground at wounds, lenticels, stipules, or leaf bases. Diseased bark is initially black and often becomes tan in the center of the lesion as pycnidia develop. Pycnidia are common in young cankers but are difficult to find in older ones. Depending upon the level of susceptibility of the host plant, the fungus may girdle shoots during the 1st season or may be halted by host defenses. Callus rolls develop at the margins of arrested lesions, and they eventually grow together, restoring the symmetry of the stem. Cankers on side shoots of young plants commonly extend to the mainstem during the 2nd season of infection. Mainstems of highly susceptible trees thus attacked may be girdled within 1–2 additional years. On eastern cottonwood, however, cankers usually fail to girdle mainstems except as the result of secondary infection by opportunistic pathogens.

Because poplars vary greatly in susceptibility to *Mycosphaerella* species, diseases caused by these fungi can be controlled through selection and breeding of resistant trees.

References: 133, 170, 389, 776, 783, 957, 1430, 1499, 1614, 1675, 1978, 2000, 2112, 2274

A, C. Leaf spot caused by *Septoria exotica* on *Hebe speciosa* (CA, Mar).

B. Spots caused by *S. cornicola* on leaves of flowering dogwood (GA, Aug).

D. Magnified view of an angular lesion with pycnidia of *S. corni-maris* in a leaf of flowering dogwood. Minute buff-colored specks with black edges are masses of conidia atop the protruding beaks of black pycnidia. The short white spines are leaf hairs (PA, Aug).

E. Angular leaf spots and associated yellowing caused by *S. azaleae* on macranthum azalea (FL, Apr).

F, G. Spots caused by *Mycosphaerella populorum* on hybrid poplar (MN, Jul) and on balsam poplar (Ont, Jul).

H. A canker caused by *M. populorum* on an artificially inoculated shoot of eastern cottonwood (MS, Apr).

I. A canker caused by *M. populorum* on a naturally infected hybrid poplar (MN, May).

Plate 32

Phyllosticta Leaf Spots (Plates 33–34)

Leaf spots—discrete lesions caused by fungi or bacteria—represent a finely tuned balance between host and parasite. The host leaf is initially defenseless but soon localizes the infection within a small area. Thus, even if infections are numerous, they usually cause no great loss of photosynthetic surface unless they trigger abscission. On the other hand, the localized lesion is a sufficient substrate for reproduction of the parasite.

Hundreds of fungi that produce black pycnidia in lesions on leaves were described by early mycologists and were assigned to the form-genus *Phyllosticta* (Deuteromycotina, Coelomycetes). These fungi are similar in form and production of colorless, unicellular spores, but they are not necessarily related. Because few species of *Phyllosticta* that infect trees are considered to be important, few have been studied beyond the point of description and naming. Given conducive conditions—wet weather in spring and early summer in successive years—populations of some *Phyllosticta* species may build up enough to kill a significant proportion of leaf area or to cause premature leaf shedding. The usual situation, however, is inconsequential spotting, mainly on the leaves of low branches.

Many, perhaps most, fungi called *Phyllosticta* are developmental states of Ascomycetes whose sexual fruiting states are perithecia or pseudothecia. If the name of the sexual state is known, it is usually used in preference to the name of the *Phyllosticta* state. Leaf blotch caused by *Guignardia aesculi* on buckeye and horse chestnut (Plate 35), for example, is diagnosed by identifying the conidial state of the pathogen, *P. sphaeropsoidea* (Figure 8, with Plate 35), but the latter name is seldom used.

Species of *Phyllosticta* that infect deciduous plants are generally assumed to survive winter in fallen leaves and, in spring, to produce spores that are dispersed to new foliage and start another cycle of disease. The conidia of *Phyllosticta* are adapted for dispersal by water rather than by air. This explains the increase of disease during rainy seasons, but it does not help us understand dispersal from the ground to new leaves. Common species such as *P. minima* and *P. hamamelidis* for which ascigerous states have not been reported perhaps survive on or in buds or twigs on the host plants and produce conidia in spring. Or perhaps they have ascigerous states that mature in fallen leaves in spring and liberate airborne ascospores that initiate the first infections annually. These remain matters for speculation because the diseases have not stimulated the research necessary to describe annual cycles.

Many fungi originally described as species of *Phyllosticta* produce only tiny spores (microconidia, spermatia), on the order of 1×3 μm, that are presumed to be noninfectious and to function as gametes in the sexual cycle. *P. rosicola* on rose and *P. platani* on sycamore (Plate 34) are examples. The sexual state of the former is *Mycosphaerella rosicola* and that of the latter may be *M. platanifolia*. Both species of *Mycosphaerella* also produce infectious conidia. The conidial states are *Cercospora rosicola* and *C. platanicola*, respectively. Diagnosticians may see the spermatial or conidial state but not the ascigerous state (*Mycosphaerella*), because the latter develops on fallen leaves in spring.

In modern fungal classification, spermatial states are no longer considered to be members of *Phyllosticta*. The spermatial states of species of *Guignardia*, for example, are assigned to the form-genus *Leptodothiorella* (Figure 8). True species of *Phyllosticta* have spores larger than spermatia, often with a gelatinous sheath and an appendage (Figure 8). We present *Phyllosticta* in the original broad sense because spermagonia can be distinguished from true pycnidia only by microscopic examination. The diagnostician must approach either type of structure as a potential *Phyllosticta*. Accordingly, Plates 33 and 34 present some of the leaf spots caused by fungi with conidial or microconidial states currently or previously assigned to *Phyllosticta*.

Phyllosticta spots of maple. *Phyllosticta negundinis* (A, B), a true *Phyllosticta* in the modern sense, affects box elder (a member of the maple genus) throughout central and eastern North America from Texas and Manitoba eastward. It causes lesions that are round or nearly so, up to 8 mm across, yellowish brown at the edges, and pale yellow in the center, where they become thin and translucent. Each is bounded by a very narrow ridge. The fragile central tissue often breaks out, leaving a ragged hole (B). Severe outbreaks may impart a distinctly yellowish color and sparseness to foliage in tree crowns viewed from a distance (A). Pycnidia are usually numerous in lesions, more so on the upper side than on the lower, scattered or arranged in a circle. Their small size—100–125 μm across—makes them barely visible to the unaided eye. No ascigerous state is known.

P. minima (C–H), also a true *Phyllosticta* in the modern sense, causes eye spot, or purple-bordered leaf spot, of maples. The disease affects several species: Amur, hedge, Japanese, mountain, red, silver, sugar, sycamore, and tatarian maple. It occurs nearly everywhere that maples grow, from the Great Plains eastward, and occasionally becomes so severe as to cause partial defoliation of red maple. The spots are irregularly round and usually less than 5 mm across, at first brown and later with a tan central portion and dark border. The border is often reddish or purplish, and its contrast with the tan center is the basis of the name *eye spot*. Pycnidia, as much as 150 or rarely 200 μm in diameter, are visible from both sides of the lesion, but most of them open to the upper surface. Often they are arranged in a circle. If infection is severe, lesions may coalesce to form large, irregular dead areas. The conidia of this fungus each have an apical appendage similar to the conidial states of *Guignardia* species, but as mentioned, no ascigerous state has been reported.

Leaf blight of witch hazel. *Phyllosticta hamamelidis* (I–L), another true *Phyllosticta*, causes necrotic spots and sometimes blight on leaves of witch hazel. The disease is widespread in the United States east of the Great Plains. The spots are irregular in outline, brown, and darker on the upper surface than the lower. Each has a narrow dark purple margin (I). Lesions often coalesce to involve large areas or even whole leaves (J, K). Black pycnidia about 100 μm in diameter break through both surfaces of the lesions (I, L). When severe, this disease severely disfigures witch hazel shrubs. No ascigerous state of the causal fungus is known.

A, B. Leaf spot caused by *Phyllosticta negundinis* on box elder. A. Trees partially defoliated by severe infection. B. Typical lesions, many with central tissues weathered away (NY, Aug).

C–H. Leaf spots caused by *P. minima* on maples. C, D. Purple-leaved Japanese maple (NY, Jun). E, F. Silver maple (NY, Jul). G, H. Sugar maple (NY, Jun).

I–L. Leaf spot and leaf blight of witch hazel caused by *P. hamamelidis*. J, K. Typical symptoms. I, L. Close views of lesions, showing pycnidia (PA, Jun).

Plate 33

Leaf spot of rose. *Phyllosticta rosicola* (A, D) is the spermatial state of *Mycosphaerella rosicola*, which occurs coast to coast wherever roses are grown. Lesions caused by *M. rosicola* are round, usually 2–4 mm but occasionally as much as 10 mm across, and may coalesce to kill large irregular areas. The spots are light brown to tan or gray with a narrow dark border, and some are surrounded by diffuse purple halos (D). In comparison with the species of *Phyllosticta* considered with Plate 33, the pycnidia of *P. rosicola* are small (50–70 μm in diameter), and its spores are rod-shaped cells measuring only 4 × 1 μm. Such small dimensions of fruit bodies and spores typify spermagonia and spermatia. These spermagonia are usually preceded by microscopic stromata of *Cercospora rosicola*, which is the conidial state of *M. rosicola*. At summer's end, however, the spermagonia may be the only signs available for diagnosis. Fruit bodies of the sexual state, *Mycosphaerella*, mature in fallen leaves and discharge ascospores in spring. These spores are presumed to cause the first infections each year. The disease increases during the growing season as conidia of the *Cercospora* state are dispersed by air and splashing water. Severe infection causes defoliation during summer.

Leaf spot of mountain ash. *Phyllosticta sorbi* (B, C) is found on brown lesions on leaves of mountain ash from Maine to Illinois and Texas. Severe infection induces premature yellowing and casting of leaflets, but the disease is usually inconsequential. *P. sorbi* has been suggested to be a spermagonial state of *Mycosphaerella aucupariae*, but this is still a matter for research. In the Pacific Northwest, *P. globigera* also spots the leaves of mountain ash.

Leaf spot of filbert. *Phyllosticta coryli* occurs from coast to coast on filberts: American, beaked, and European. It causes lesions that are round to irregular with dark margins, light brown centers, and sometimes yellow halos (E). When the lesions are numerous, they often give rise to irregular dead patches at the edges of leaves. Pycnidia, often in groups of five to seven, break through the upper leaf surface in lesions. Viewed with a hand lens, the pycnidia appear much like those illustrated in Plate 33. The conidia of *P. coryli*, unlike those of *P. rosicola*, are probably infectious. No sexual state has been reported.

Large leaf spot of magnolia. *Phyllosticta magnoliae* infects sweet bay and Fraser, saucer, and southern magnolias and is found from New York to the Gulf states. The lesions first appear as minute purplish black spots on the upper leaf surface. They enlarge to 2 cm or more across and vary from nearly circular to irregular. The center takes on a dirty white color, and the edge remains purplish black. A faint chlorotic halo (F) often surrounds the spot but tends to fade as the spot matures. The lower surface of the spot is yellowish. Black pycnidia, 0.3–0.4 mm in diameter, become numerous on the upper surface. The conidia are probably infectious. No sexual state of *P. magnoliae* has been reported. Two other species of *Phyllosticta* also spot the leaves of magnolias. *P. cookei* occurs coast to coast on at least three magnolia species. *P. glauca* infects leaves of sweet bay in the Southeast.

Leaf blotch of American linden. This leaf blotch (G) is caused by a fungus long known as *Asteroma tiliae*, which occurs in North America and Europe on several species of linden. It produces minute, very inconspicuous acervuli with microconidia on the upper surfaces of lesions. This blotch disease is presented here with Phyllosticta leaf spots because often the only fungal signs are minute pycnidia that are dark brown to black and open to the lower leaf surface. These resemble *Phyllosticta* but produce microconidia. Thus they may be the spermagonia of an undetermined Ascomycete. Some early herbarium records associate the name *Phyllosticta tiliae* with leaf blotch, but *P. tiliae* is clearly a separate fungus that causes light brown spots 3–4 mm across with abrupt dark brownish or purplish borders.

Blotches on American linden develop after midsummer, mainly on the upper surface, appearing dark brown with feathery margins. Wide chlorotic halos develop around the margins as leaves become senescent. The lower surface of a lesion is tan with an indistinct margin. Until the disease cycle is studied, the reason for the late-season appearance of symptoms will be unknown.

Leaf spot of dogwood. *Phyllosticta cornicola* (H) occurs on leaves of flowering dogwood and kousa dogwood in the Appalachian region and west to Kansas. It is one of several species of *Phyllosticta* reported to spot dogwood leaves, but this group of fungi has not been studied critically to determine how many names should be synonymized with *P. cornicola*. In addition, it has been suggested that *P. cornicola* is merely a variant of *Ascochyta cornicola*. The main difference between these fungi is that spores of *A. cornicola* are one-septate, while those of *C. cornicola* are unicellular. *P. cornicola* and *A. cornicola* do cause similar symptoms—round to irregular spots 1–6 mm across, with tan to gray centers and brown or reddish brown borders. Pycnidia arise in the tan portion, which becomes fragile and tends to break out of mature lesions (H). Outbreaks of *A. cornicola* in the Appalachian mountains sometimes result in blight, that is, in blackening and shriveling of leaves on low branches in mid-June. *P. cornicola* has not been associated with blight.

Leaf spot of sycamore. *Phyllosticta platani* is a microconidial fungus found on the lower surfaces of round to irregular lesions on leaves of sycamore in late summer. It occurs from Massachusetts to Alabama and Kansas and may, as has been suggested, be the spermatial state of *Mycosphaerella platanifolia*, which occurs across the Southeast. Leaf spot and blight caused by *M. platanifolia* are more common and conspicuous than anthracnose of sycamore in parts of the South.

Lesions bearing *P. platani* are generally 1–2 cm across but may become larger as the result of coalescence. They are tan to grayish in the center and dark brown near the edge, with indefinite margins. When numerous, they induce generalized yellowing and necrosis of affected leaves (I). Pycnidia (spermagonia) break through the lower leaf surface in the lesion and appear as inconspicuous dark brown specks barely visible with a hand lens. They are up to 90 μm in diameter.

The possible connection between *P. platani* and *M. platanifolia* is suspect because symptoms attributed to *M. platanifolia* are tiny (1 mm) irregular brown spots that become noticeable in early summer and fuse when numerous. The conidial state of *M. platanifolia*, *Cercospora platanicola*, sporulates on both surfaces of the lesions, producing conidia on tiny stromata that emerge from stomata. In North Carolina, spermagonia arise in lesions with *Cercospora*, but it has not been established that these spermagonia are identical with *P. platanicola*.

References for Plate 33: 1, 425, 562, 1734
References for Plate 34: 360, 425, 453, 577, 883, 957, 1734, 1941, 2227

A, D. Leaf spot of rose (*Rosa* sp.) caused by *Mycosphaerella rosicola*. Minute spermagonia (arrow), designated *Phyllosticta rosicola*, are barely visible with a hand lens in the centers of the lesions in (D) (NH, Sep).

B, C. Leaf spot of American mountain ash. Pycnidia of *P. sorbi* (arrow) may be the spermagonial state of *M. sorbi* (NH, Sep).

E. Leaf spot of *Corylus* sp. caused by *Phyllosticta* sp. (NJ, Jun).

F. Large leaf spot of southern magnolia, caused by *P. magnoliae* (FL, Apr).

G. Leaf blotch of American linden. Spermagonia (*Phyllosticta* sp. in its original sense) are the only fungal signs present. The disease is attributed to *Asteroma tiliae* (NY, Aug).

H. Leaf spot of kousa dogwood caused by *P. cornicola* (NJ, Jun).

I, J. Leaf spot of sycamore. *P. platani*, associated with these lesions, is perhaps a spermagonial state of *M. platanifolia*. Its tiny fruit bodies appear as dark brown dots in (J) (MD, Sep).

Plate 34

Guignardia Blotch of Horse-chestnut and Buckeye and Guignardia Leaf Spot of Boston Ivy (Plate 35)

Several species of *Guignardia* (Dothideales, Dothideaceae) cause leaf spots, blotches, and fruit rots of woody plants. Two of these pathogens that attack woody ornamentals are well known. *G. aesculi* (syn. *Botryosphaeria aesculi*) causes blotch of horse-chestnut and buckeye. *G. bidwellii* (syn. *B. bidwellii*) causes black rot of grape and leaf spot of ampelopsis, Virginia creeper, and Boston ivy.

Guignardia blotch. This disease disfigures the foliage of species of *Aesculus* (horse-chestnut and buckeye) wherever they grow east of the Great Plains. The disease occurs in Europe also but apparently not in the American West, perhaps because of the dry summer climate there. Susceptible plants include California, Ohio, red, and yellow buckeyes; several less-known buckeye species; and common, Japanese, and red horse-chestnuts. Observations of *Aesculus* species in Illinois revealed probable resistance only in certain varieties of Ohio buckeye (*A. glabra* varieties *arguta, monticola,* and *sargentii*) and in bottlebrush buckeye (*A. parviflora*) and its variety *serotina.*

Blotch lesions first appear as water-soaked irregular areas that enlarge rapidly. Within a few days they turn reddish brown to brown, often bordered by a yellow band that merges gradually with normal green tissue. Lesions vary greatly in size. Often the small ones are limited by veins. Large lesions frequently coalesce and cause grotesque curling and distortion of leaflets (A). Petioles and immature fruits also become infected occasionally. Lesions on these parts are small reddish brown spots, somewhat elongate when on petioles.

Black pycnidia of the conidial state of *G. aesculi, Phyllosticta sphaeropsoidea,* appear soon after lesions form. The pycnidia are 90–175 μm in diameter and nearly globose and develop mainly but not exclusively beneath the upper leaf surface (Figure 8A). When moistened, they extrude masses of colorless, one-celled conidia that are dispersed by splashing rain or in moving water on wet plant surfaces. In late summer, black spermagonia and immature pseudothecial stromata appear on both surfaces of lesions. The spermagonia (in the form-genus *Leptodothiorella*) are only 40–110 μm in diameter (Figure 8B). The pseudothecial stromata appear similar to pycnidia but produce no spores initially. They mature in fallen leaves (Figure 8C) during early spring and liberate ascospores to the air during wet weather while new foliage is developing. The ascospores, if deposited on leaves that remain wet for several hours, initiate the first infections of the new season. Blotches appear within 10–20 days after inoculation. Pycnidia develop beginning early in June, and conidia initiate secondary cycles throughout the summer as permitted by wet weather.

Guignardia blotch typically becomes severe in plantings where tree crowns are close together because dense foliage retards drying of leaves after rainfall. The disease tends to develop after most of the annual growth of host plants is complete. Thus it does not greatly influence growth.

G. aesculi grows readily but slowly on common laboratory media.

Black rot of grapevine and Boston ivy. *Guignardia bidwellii,* the cause of black rot, was once confined to North America, but the pathogen was apparently carried with propagating material to Europe. Now it occurs on every continent except Australia, and it is a significant threat to grape crops in the humid parts of eastern North America.

G. bidwellii consists of host-specialized strains called formae speciales. *G. bidwelli* f. sp. *parthenocissi* infects ampelopsis, Virginia creeper, and Boston ivy. This strain is not pathogenic to grapevines. The strains that do infect grapes are *G. bidwellii* f. sp. *euvitis,* which infects fox and European grapes, and *G. bidwellii* f. sp. *rotundifolia,* which infects muscadine and European grapes.

Lesions on leaves of Boston ivy or Virginia creeper are small (to 6–8 mm across), tan to brown, and often somewhat angular where limited by veins of the leaf. They have dark brown margins with an abrupt transition to normal green tissue. With time, the dead tissue breaks away, leaving small ragged holes. If numerous spots develop before leaf expansion is complete, the leaves become more or less distorted as the result of continued expansion of green tissues around the dead spots.

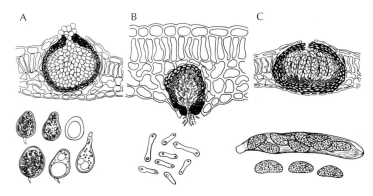

Figure 8. Microscopic details of fruiting states of *Guignardia aesculi* in leaves of horse-chestnut. A–C: pycnidium with conidia, spermagonium with spermatia, and pseudothecium with asci and ascospores.

As soon as lesions attain full size, black pycnidia protrude through the upper surfaces, either scattered or arranged in a ring. These fruit bodies are 120–230 μm across and extrude colorless, one-celled conidia when moistened. White masses of these spores sometimes adhere to the tips of pycnidia. The pycnidial state of *G. bidwellii* is named *Phyllosticta ampelicida.* Spermagonia (of the *Leptodothiorella* type) and immature pseudothecia develop in late summer, and the pseudothecia mature in fallen leaves the next spring.

On grapevines, *G. bidwellii* causes leaf spots much like those on Boston ivy (sometimes much larger on certain varieties of European grapes). It also causes small lesions on shoots, fruit stalks, and tendrils, but all these symptoms are inconsequential compared with the black rot of the fruit. This typically appears when berries are about half grown. The first symptom is a pale spot with a dark ring, but the entire fruit becomes black, shriveled, and hard (mummified) within 7–10 days. Pycnidia develop in the mummified fruit and overwinter there as well as on canes, tendrils, and fallen leaves. Pseudothecia also form in fallen leaves. Infections in spring are initiated by both ascospores from pseudothecia and conidia from pycnidia in dead tissues on the vines or on the ground.

Epidemiological studies of black rot have been conducted in the north-central United States. There the annual peak of ascospore dispersal occurs in May–June. Conidia are liberated throughout the season but in greatest numbers in July–August, after pycnidia develop in new lesions on leaves. Rains lasting 1–3 hours are optimal for release of both types of spores. Release begins within an hour after the onset of rain and may continue for several hours afterward. If leaf or fruit surfaces remain wet after spore dispersal, infection occurs within 6–24 hours, depending upon temperature. Infection by spores of either type is most rapid at at 26–27°C. The incubation period (interval from infection to appearance of symptoms) is about 1 week under the most conducive conditions but often lasts 2–3 weeks. Pycnidia usually appear within an additional 3–6 days. Spores of both types are susceptible to desiccation, surviving 48 hours or less on dry leaves.

References: 1, 106, 387, 425, 564, 565, 1177, 1377, 1385, 1388, 1588, 1710, 1876, 1902

A–C. Leaf blotch of horse-chestnut caused by *Guignardia aesculi.* A. Foliage distorted by severe infection. B. Typical lesions. C. Pycnidia in the upper surface of a lesion as viewed with a hand lens. The light dot in the center of each pycnidium is the pore through which a colorless mass of conidia emerges when the leaf is moist (NY, Aug–Sep).

D–F. Leaf spot of Boston ivy caused by *G. bidwellii* f. sp. *parthenocissi.* D, E. Lesions viewed on upper and lower leaf surfaces; dead tissue has broken out of the centers of old lesions (NY, Sep). F. Pycnidia in upper surfaces of lesions as viewed with a hand lens (NY, Jun).

Plate 35

Coniothyrium and Hendersonia Leaf Spots (Plate 36)

Fungi in the form-genera *Coniothyrium* and *Hendersonia* (Deuteromycotina, Coelomycetes) inhabit leaves and the cortical tissues of stems. They often cause leaf spots, twig dieback, or cankers, but some act as secondary invaders of dying or dead plant parts. Records of their occurrence on various plants frequently do not distinguish between parasitic and saprophytic occurrence, and very few of the diseases have been studied.

Diseases caused by *Coniothyrium*. In its original sense, the name *Coniothyrium* connoted small dark pycnidia in which tiny unicellular, olive to brown conidia usually less than 10 μm long were produced. In modern taxonomy, many of the fungi once placed in *Coniothyrium* would be assigned to the form-genus *Microsphaeropsis* on the basis of the manner of conidial production and characteristics of the conidia. Two of the best-known species, *C. fuckelii* and *C. olivaceum,* would thus be transferred, and this change is already in effect for the latter fungus. Many species remaining in *Coniothyrium* have two-celled conidia. As a practical matter, however, few of these fungi have been studied enough to permit us to understand their biology and taxonomic relationships. Some species of *Coniothyrium* are the conidial states of Ascomycetes in the genus *Leptosphaeria*.

Three pathogens in this group cause cankers on rose canes. *C. fuckelii* (the conidial state of *Leptosphaeria coniothyrium* [Pleosporales, Pleosporaceae]) causes common canker, *C. rosarum* causes graft canker, and *C. wernsdorffiae* causes brand canker. *C. fuckelii* is cosmopolitan and attacks many plants, causing diseases as diverse as leaf spot of magnolia (A, C) and, in Europe, shoot blight of juniper. Whether a strain from magnolia could infect rose canes or vice versa is unknown because host specialization has not been studied. On leaves of southern magnolia, *C. fuckelii* causes round lesions up to 1 cm across, at first brown with a nearly black outer ring and sometimes with a diffuse brown border that merges gradually with the surrounding green tissue (A, C). The center of the lesion becomes light grayish brown (C), and numerous pycnidia develop beneath its upper surface. This fungus also causes cane blight of currant and stem cankers of apple and Virginia creeper.

Both *C. fuckelii* and *Microsphaeropsis olivacea* (syn. *C. olivaceum*) are ubiquitous secondary invaders of senescent or dead tissues, and they possibly suppress such common foliar and stem pathogens as *Lophodermium* species on pines (Plate 11) and *Leucostoma* species on stone fruit trees (Plate 94). The many substrates of *M. olivacea* include apple, bean tree, broom, citrus, eucalyptus, fig, grapevine, juniper, macadamia, magnolia, mulberry, mock orange, persimmon, pine, and quince.

The Coniothyrium leaf spot of salal (B) has apparently not been described, and the pathogenic species is undetermined. The disease is characterized by small brown spots (up to 5–6 mm) with dark purplish brown borders. The relationship, if any, of this *Coniothyrium* to *Leptosphaeria gaultheriae*, which occurs on stems of salal, should be studied.

On *Ixora*, a species of *Coniothyrium* is sometimes associated with irregularly spreading lesions, especially along margins of old leaves (D, E) but not in young leaves of the same plant. The fungus in this case may be a saprophyte or a weak pathogen that invades only senescent leaves or those previously damaged by environmental stress.

C. concentrica, the conidial state of *Leptosphaeria obtusispora,* causes leaf spots of yucca and century plant. The lesions are round to oval and 10–25 mm long, dark brown initially, and later tan in the center, with concentric zones of black pycnidia. Severely affected leaves turn yellow and are shed prematurely. The disease apparently occurs wherever the host plants grow in North America.

Several diseases caused by species of *Coniothyrium* have been described in Europe. *C. juniperi* causes canker and twig blight of various junipers, and *C. fuckelii* causes shoot blight of the 'Compressa' cultivar of common juniper. *C. clematidis-rectae* causes wilting disease of clematis, and *C. viburni* causes necrotic blotches on leaves of *Viburnum × burkwoodii.*

Many other species of *Coniothyrium* are associated with symptoms on leaves or stems of woody plants in North America, but details and confirmation of pathogenicity are lacking. Leaf spots occur on apple, bearberry, beech, buttonbush, elder, elm, hickory, holly, madrone, manzanita, and pear. Cankered or killed twigs with pycnidia of *Coniothyrium* are recorded for acacia, beautyberry, box elder, grapevine, mallow, mimosa, and tree-of-heaven.

Diseases caused by *Hendersonia*. Species of *Hendersonia* are characterized by black, globose pycnidia that are at first immersed in bark or leaf tissue and then become erumpent. The conidia are elongate, usually brown, and multicellular. The name *Hendersonia* has been rejected by mycologists, and the many fungi heretofore classified in this genus must be assigned to other form-genera such as *Stagonospora*. Some of these fungi, like members of *Coniothyrium*, have ascigerous states in the genus *Leptosphaeria*. We use the name *Hendersonia* because most of the species have not yet been reassigned to other genera.

H. eucalypticola is one of several related species that cause leaf spots on eucalypts. In Australia, at least three species of *Hendersonia* are specialized for attack of particular eucalypt hosts. On blue gum (*Eucalyptus globulus*) in California, *H. eucalypticola* causes small, brown to tan lesions with dark purplish borders, visible on both sides of the leaf (F, G). Apparently only the juvenile foliage is susceptible.

H. rubi causes leaf blight of almond in California. The fungus infects petioles and, beginning in June, causes sudden drying of scattered leaves. The petioles remain attached to the twigs, and pycnidia of the pathogen develop on light tan areas near the bases of petioles during winter. Infection may progress into twigs and lead to death of flower buds in autumn or during the next spring.

H. desmazierii, the conidial state of *Massaria platani,* causes a minor twig dieback and canker of sycamore and plane trees from New Jersey to California. *H. cerei* reportedly causes a cortical rot of saguaro in the Southwest. *H. opuntiae* is associated with symptoms on prickly pear cactus that have been called scorch or sunscald. Infected segments turn reddish brown and die. Lesions in these segments become grayish brown and cracked. The fungus may be an opportunistic invader of heat-stressed tissue.

H. pinicola in North America and *H. acicola* in Europe invade pine needles already infected by *Lophodermella* species (Plate 14). The secondary fungi suppress fruiting of the pathogens and thereby exert some natural control of needle cast. *H. thyoides* is a secondary invader of twigs of western red cedar.

At least two dozen other species of *Hendersonia* have been reported as occurring on woody plants in the United States and Canada, but as we noted, details are usually lacking. Leaf spots occur on apple, hawthorn, hickory, magnolia, pear, quince, rhododendron, and viburnum. Twig cankers occur on box elder and Russian olive. In addition there are reports of unspecified associations of *Hendersonia* species with ash, bladdernut, camellia, Atlantic white cedar, clematis, currant, alternate-leaved dogwood, grapevine, hackberry, witch hazel, hydrangea, mountain laurel, honey locust, maple, Siberian pea tree, rose, serviceberry, spicebush, and willow.

References for Coniothyrium: 389, 425, 439, 883, 914, 915, 1334, 1347, 1728, 1858, 1889, 1939, 1941

References for Hendersonia: 281, 315, 389, 425, 783, 1336, 1940, 1941, 2211

A, C. Leaf spot caused by *Coniothyrium fuckelii* on southern magnolia. Tiny pycnidia dot the center of a lesion as viewed with a hand lens (MS, Aug).

B. Leaf spot apparently caused by *Coniothyrium* sp. on salal (WA, Jun).

D, E. *Coniothyrium* sp. associated with necrotic blotches on old leaves of *Ixora* sp. Pycnidia are prominent in some lesions (FL, Mar).

F, G. Leaf spot caused by *Hendersonia eucalypticola* on *Eucalyptus globulus* (CA, Apr).

Plate 36

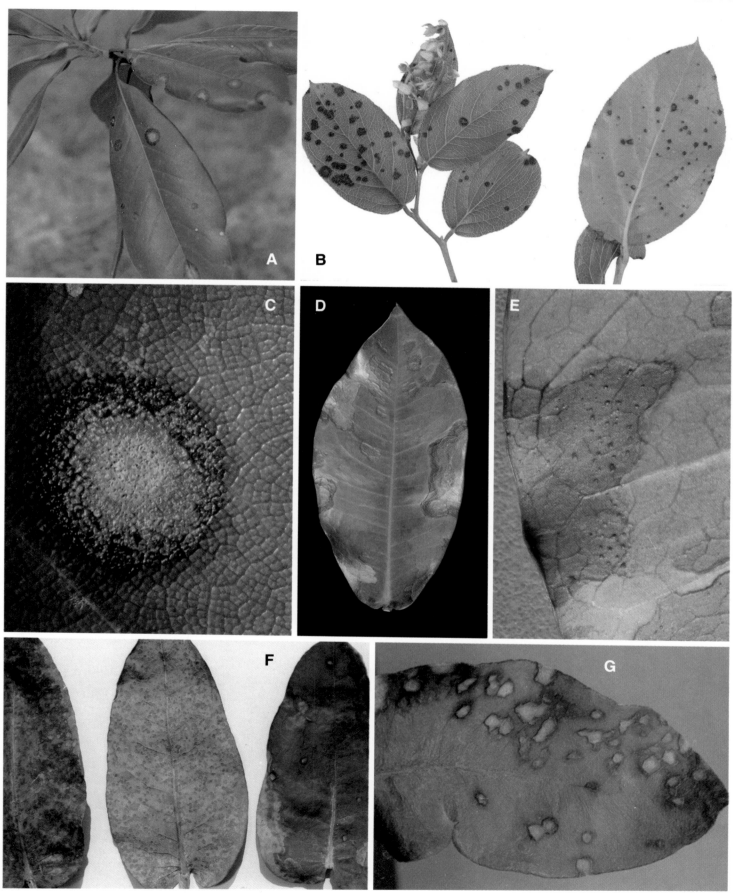

Linospora Leaf Blight of Balsam Poplar and Fly Speck Leaf Spot of Huckleberry (Plate 37)

Linospora leaf blight. *Linospora tetraspora* (Diaporthales, Valsaceae) causes a distinctive leaf blight of balsam poplar and black cottonwood. The disease occurs from Quebec and New England to the Pacific Northwest. Severe outbreaks are confined to small localities and usually to pure stands of susceptible trees on which 80–100% of the foliage may be blighted.

The lesions vary in shape and size and often involve whole leaves. The upper surfaces of lesions are dark brown near the margins, becoming grayish brown to ashen in older portions. The lower surface is reddish brown. Enlarging lesions follow leaf veins and thus have irregular outlines (B). Infections usually spread from leaf blades into petioles, which then shrivel.

The diagnostic signs are black stromata about 0.5 mm across, which develop in midsummer in the ashen upper surfaces of lesions. In late summer these structures produce microconidia that are apparently noninfectious. Perithecia develop beneath the black stromata in fallen leaves and mature in late spring. Ascospores dispersed from these fruit bodies are presumed responsible for starting the single annual cycle of infection. The perithecia are unusual in that their bodies are aligned in the plane of the leaf blade, while their long necks bend upward at right angles and project above the surface of the black stroma.

Unlike many other little-known leaf-blighting fungi, *L. tetraspora* is a proven pathogen; artificial inoculations with asci plus ascospores dissected out of perithecia have resulted in leaf blight. The fungus grows slowly in pure cultures on laboratory media.

The fungus *Plagiosphaeria gleditschiae* (syn. *Linospora gleditsiae*) causes a disease similar to Linospora leaf blight but on honey locust and water locust. *P. gleditschiae* occurs from New England to Kansas and Texas. The disease is called tar spot. Fruit bodies of the pathogen open to the lower surfaces of leaflets.

Fly speck leaf spot of huckleberry. This inconsequential disease, which affects various species of farkleberry, huckleberry, and blueberry, is characterized by colorful and distinctive lesions. The causal fungus, *Ophiodothella vaccinii* (Sphaeriales, Polystigmataceae), occurs from Georgia and Florida to Texas and Illinois. Symptoms first appear as nearly round yellow spots 1–5 mm across on upper leaf surfaces. The spots enlarge to 10–13 mm, sometimes have reddish purple borders, and become visible on the lower surface as yellowish brown areas with indistinct margins. Numerous small black stromata, the "fly specks," develop beneath both surfaces of each lesion, and colorless linear conidia are produced in acervuli just beneath the blackened epidermis in each stroma during summer and autumn. During wet weather these spores extrude in masses through the ruptured epidermis. The conidia are apparently dispersed by water and cause numerous secondary infections.

Perithecia begin to develop between the black surfaces during early autumn and mature in fallen leaves in spring. These perithecia are unusual in that each opens simultaneously to both sides of the leaf.

Other species of *Ophiodothella* cause leaf spots of various plants. *O. fici* and *O. floridanum* cause tar spot, also known as black leaf spot, and premature leaf drop of strangler fig and other species of *Ficus* in Florida and elsewhere in the Caribbean region. The spots consist of raised, uneven, shiny black stromata measuring 1–10 mm across, solitary or in groups, visible on both leaf surfaces and often surrounded on the lower surface by a yellow halo.

References for Linospora blight: 315, 389, 425, 1308, 1343, 1526, 1759, 1999

References for fly speck leaf spot: 232, 425, 577

A–C. Linospora leaf blight of balsam poplar. A. Leaves distorted by lesions. B. A typical lesion with a dark brown irregular margin and an ash gray center with stromata of *Linospora tetraspora*. C. Close view of stromata on an upper leaf surface (VT, Aug).

D–F. Fly speck leaf spot of farkleberry. D. Spotted leaves. E, F. Close views of upper and lower leaf surfaces bearing black stromata of *Ophiodothella vaccinii* (GA, Aug).

Plate 37

Cercospora Blights and Leaf Spots (Plates 38–39)

Cercospora is a large genus of plant-parasitic Hyphomycetes (more than 1200 species were recognized in 1953), with many members that attack leaves or in some cases green shoots of trees and shrubs. Most diseases caused by *Cercospora* species are characterized by chlorotic to necrotic localized lesions or by diffuse infected areas that remain alive and only slightly discolored while the pathogen sporulates on them. These diseases occur in all temperate habitats and abound in warm, humid regions such as the southeastern United States.

Cercospora species produce long (to 1000 μm or more), slender (1–7 μm), often dark-colored, multicelled conidia on conidiophores that arise singly or in groups. Many species produce their conidiophores and conidia on olive-brown to black, cushionlike stromata that break through plant surfaces (Plates 38B, 39C). Species are differentiated on the basis of morphological characters, host plants, and the symptoms caused. Most species of *Cercospora* seem to be specialized for attack of just one or a few closely related plant species, but this matter has not been investigated by means of extensive cross-inoculations. Some species are conidial states of ascomycetes in the genus *Mycosphaerella* (Plates 30, 34). The genus *Cercospora* has long been recognized as heterogeneous, and its taxonomy is changing with the transfer of many species to such genera as *Cercosporidium*, *Pseudocercospora*, *Stenella*, *Stigmina* (Plate 41), and others. As of 1986, most of the species discussed here had not been reassigned.

Cercospora blight of Cupressaceae. The most destructive of the Cercospora diseases of trees in North America are the needle blights of evergreens in the cypress family, caused by *Cercospora sequoiae* and its variety *juniperi* (Plate 38A–C). These fungi cause progressive browning and loss of foliage beginning on low branches close to the mainstem and moving upward and outward until the plant is dead or until green shoots remain only on the topmost branches. Although described in 1887 from Pennsylvania and Wisconsin, respectively, *C. sequoiae* and *C. sequoiae* var. *juniperi* (which cause identical symptoms) were not recognized as economically significant pathogens until after 1945. In that year *C. sequoiae* was reported to be the cause of extensive damage to plantings of Arizona cypress in the Gulf states. Within the next 20 years it became destructive in Christmas tree and ornamental plantings of this species throughout the Southeast. In the 1960s *C. sequoiae* var. *juniperi* was recognized as one of the major pathogens of eastern red cedar and Rocky Mountain juniper in windbreaks and other plantings in the Great Plains. Cercospora blight occurs now in western Canada and in the eastern half of the United States from North Dakota and Texas to Connecticut and Florida. It is relatively uncommon, however, in the northern part of this area. Both *C. sequoiae* and its variety *juniperi* occur in the Southeast. In the Great Plains north of Texas and in most northeastern areas, however, only the variety has been found. *C. sequoiae* also occurs on various species of Cupressaceae in South America and Japan.

Plants susceptible to *C. sequoiae* and/or *C. sequoiae* var. *juniperi* include Oriental arborvitae; eastern and southern red cedars and Japanese cedar; Arizona, bald, Italian, MacNab, Monterey, Portuguese, and Sawara cypresses; cherrystone and Rocky Mountain junipers; coast redwood; and giant sequoia.

Symptoms develop during summer in temperate regions and are similar on the various hosts. Leaves (scale leaves or needles) on small shoots near branch bases become bronze to tan or light brown and eventually grayish. Often all leaves on a shoot are affected. Soon after the leaves die, dark, cushionlike stromata of the causal fungus break through the epidermis (Plate 38B). Microscopic examination of the surface of a stroma reveals brownish conidiophores and dry, elongate, multicelled, brownish conidia. The stromata of *C. sequoiae* are 50–115 μm across, while those of the variety *juniperi* are conspicuously larger (125–300 μm). The variety also differs in having shorter conidiophores and narrower conidia.

Within several weeks to a few months after infection (during October–November in the Great Plains), the diseased shoots drop. As infection on a given plant spreads and intensifies during 2 years or longer, foliage and shoots progressively nearer the branch tips die and fall. As a result the inner, lower part of the crown becomes open or devoid of green foliage, while the extremities appear normal (Plate 38C). On junipers so affected, juvenile foliage often develops from dormant or adventitious buds on the denuded low branches. Highly susceptible plants may be killed within 1–3 years.

In North America, Cercospora blight seldom affects young plants in nurseries, but it may develop on plants that are held in the nursery for several years. On Japanese cedar in Japan, *C. sequoiae* causes not only leaf and twig blight but also cankers that may encircle young stems or deform older ones in plantations. The canker phase is unknown in North America.

Green branch tips and juvenile foliage on low branches, where it ordinarily does not occur, serve to distinguish Cercospora blight from Phomopsis, Kabatina, and Sclerophoma blights of eastern red cedar and other members of the cypress family (Plate 64). The latter diseases are characterized by tip blight.

In temperate regions, conidia of *C. sequoiae*, including the variety *juniperi*, are produced throughout the growing season, most abundantly during wet weather in late spring and summer. They are dispersed short distances by splashing or wind-driven water. Long-distance movement by air is possible but unproved. Shipment of diseased plants is important for long-distance movement.

Infections occur throughout the summer. Conidia germinate on wet new or 1-year-old leaves, and the fungus enters leaves through stomata or by direct penetration of the cuticle. Symptoms appear beginning 2–3 weeks after infection from midsummer to late autumn. The pathogen overwinters in diseased leaves on living trees. Neither *C. sequoiae* nor its variety *juniperi* has a known sexual state.

Wet weather and moderate temperature favor Cercospora blight. Water is necessary for dispersal and germination of spores and for infection. Growth of the fungus in culture is most rapid at about 24°C. At this temperature, conidial germination begins within 6 hours after wetting and is nearly complete within 16 hours. In one test, spores of *C. sequoiae* var. *juniperi* germinated at temperatures of 18–28°C but failed to germinate at 8° or 32°C.

Little information is available about resistance to Cercospora blight. In the Great Plains, the disease develops more rapidly and to greater intensity on Rocky Mountain juniper than on eastern red cedar. Also, cherrystone juniper was more severely damaged than eastern red cedar in one comparison.

Cercospora leaf spots. *Cercospora epicoccoides* causes irregularly shaped, somewhat angular chestnut brown lesions on leaves of several species of *Eucalyptus*, especially silver dollar tree (Plate 38D) and Tasmanian blue gum. Originally known only in Australia, this leaf spot now occurs in South America and Florida. A similar disease affects Tasmanian blue gum in California. Lesions caused by *C. epicoccoides* develop on either leaf surface, vary in size from pinpoints to more than 1 cm across, often coalesce, and may then involve entire leaves. The centers of large lesions become grayish tan. The causal fungus produces black stromata on the upper surfaces of lesions.

Privet leaves are spotted by several species of *Cercospora* and related fungi. These include *C. adusta*, *C. lilacis*, *Pseudocercospora ligustri*, and one or more others that have not been named. *P. ligustri* occurs on wax-leaf privet (Plate 38E) and also on Amur, California, common, and glossy privets in the Gulf states and as far north as Kansas. It also occurs in France and Japan. A similar fungus spots the leaves of privet in California. *P. ligustri* causes circular lesions that appear as chlorotic spots 10–12 days after inoculation. Lesions enlarge to 5–15 mm across and become depressed and tan to brown in

---

A. Cercospora blight of Italian cypress (FL, Apr).

B, C. Cercospora blight of eastern red cedar, caused by *Cercospora sequoiae* var. *juniperi* (NE, Jun). B. Dark brown stromata on killed leaves. C. Only the terminal shoots on a diseased branch usually remain green.

D. Leaf spot of silver dollar eucalyptus caused by *C. epicoccoides* (FL, Apr).

E. Leaf spot of wax-leaf privet caused by *Pseudocercospora ligustri*. These lesions, atypically small, were either immature or arrested in development. Yellow halos around spots are typical (FL, Apr).

F. Leaf spot caused by *C. liquidambaris* on sweet gum (GA, Aug).

G. Diffuse leaf spot of nandina caused by *C. nandinae* (FL, Apr).

Plate 38

the center with wide reddish purple margins. A chlorotic halo surrounds most lesions. Small brown stromata (to about 40 μm across) arise on the upper surfaces of lesions 2–3 weeks after infection. In pathogenicity tests on wax-leaf and glossy privets, *P. ligustri, C. adusta,* and *C. lilacis* each caused distinct symptoms. Nevertheless, precise diagnosis of Cercospora leaf spots on privets requires mycological expertise.

C. liquidambaris spots the leaves of sweet gum (Plate 38F) and Formosan gum. The fungus is found throughout the Southeast and as far north as Delaware and Maryland. It also occurs in Taiwan. The lesions are angular to nearly round, 2–10 mm across, and dark brown with a purplish black border and a diffuse purplish halo. The pathogen sporulates on both surfaces of lesions, producing conidia on dark brown stromata 20–40 μm in diameter. No biological studies have been reported.

C. nandinae apparently came to North America with its host plant, *Nandina domestica* (heavenly bamboo), from Japan. The fungus is widely distributed on this plant in the southeastern United States. It causes irregular reddish to rust-colored blotches, some with much darker centers, on the upper surfaces of leaves (Plate 38G). Stromata seldom form, but the lower surface of each lesion is slightly darkened by the effuse development of olivaceous conidiophores and conidia.

C. kalmiae (Plate 39A–C) causes irregular to circular necrotic spots on the leaves of mountain laurel throughout the Appalachian region. Severe infection seems to retard plant growth and suppress flowering, particularly on plants growing in moist, shady places. Mature lesions are 5–10 mm across, at first medium to dark brown on both surfaces but fading to grayish brown in the center of the upper surface. The margin remains dark brown to purplish brown. Black stromata 50–150 μm in diameter dot the upper surface. Most infections begin on newly expanded leaves shortly before and during bloom, and lesions appear in late summer. Most leaves, even if heavily infected, remain on the shrubs until the next spring. Thus lesions on 1-year-old leaves seem to be the source of inoculum for new infections.

Rhododendrons and azaleas of several types are attacked by *C. handelii,* which causes irregular to circular lesions on both sides of the leaves (Plate 39D) and occasionally on petioles. The pathogen occurs across the United States and also in Europe, Japan, and New Zealand.

Common host plants include Hiryu and macranthum azaleas, catawba rhododendron, and *Rhododendron ponticum.* The lesions are various shades of brown or, after weathering, grayish brown above and brown below and 2–10 mm across with an orange to black line margin. Under moist conditions, a thin, olive-colored mat of hyphae with conidiophores and conidia may develop on the lesions. Dark brown stromata 15–70 μm in diameter form on both surfaces but chiefly on the upper one. Immature pseudothecia of an unidentified sexual stage sometimes show on the lower surface. On azaleas, severe disease causes chlorosis and premature defoliation.

Six species of *Cercospora* infect the leaves of persimmon. Criteria for distinguishing these fungi include the type of lesions they cause (distinct versus indistinct), effuse sporulation versus conidial production on tufts of conidiophores, and the color and dimensions of conidiophores and conidia. *C. fuliginosa* causes distinct purple-black, usually angular spots, 0.4–4 mm across, on both leaf surfaces of common persimmon in most places where the plant grows naturally (Plate 39E, F). The lesions have minute reddish brown centers and may have indistinct chlorotic halos. The pathogen usually sporulates on the lower surface. Stromata, formed inconsistently, are up to 50 μm in diameter. Severely spotted leaves turn yellow and drop prematurely.

C. pittospori causes angular leaf spot of pittosporum (Plate 39G). Chlorotic to yellowish brown or dull brown angular spots 1–5 mm (occasionally up to 12 mm) in diameter form on the leaves of both variegated and green varieties, appearing first on the upper leaf surface. Infection occurs during the warm season. In early stages the spots are pale green, but during a period of months they slowly pass through shades of yellow to brown and remain bounded by veinlets of the leaves. The causal fungus grows out through stomata and, in humid air, produces an olive-colored felt of hyphae, conidiophores, and conidia on the lower surfaces of old spots. Many diseased leaves remain attached; thus symptoms are visible all year. Severe infection of young leaves may cause distortion or premature leaf drop.

References for Cercospora blight of Cupressaceae: 387, 612, 846, 936, 1490, 1493, 1499, 1533, 1614
References for other Cercospora diseases: 360, 471, 472, 540, 649, 1243, 1502, 1531, 1857

A–C. Leaf spot of mountain laurel caused by *Cercospora kalmiae.* A. Spots are conspicuous on 1-year-old leaves. B. Lower (left) and upper surfaces of leaves with typical lesions. C. Close view of the upper surface of a lesion showing dark line margin, purplish brown halo, and stromata of *C. kalmiae* (NY, Jun–Jul).

D. Necrotic spots caused by *C. handelii* on leaves of *Rhododendron* sp. (CA, Oct).

E, F. Leaf spots caused by *C. fuliginosa* on common persimmon. E. Severely spotted leaves are prematurely senescent. F. Close view of typical black lesions (GA, Aug).

G. Angular leaf spot of pittosporum, caused by *C. pittospori.* Infected spots remain chlorotic but alive for months (FL, Apr).

Plate 39

Alternaria Blights and Leaf Spots (Plate 40)

Fungi in the form-genus *Alternaria* (Deuteromycotina, Hyphomycetes) are common saprophytic colonists of plant surfaces and decaying leaves and fruits. Of about 10 species that are pathogenic to plants, at least four cause diseases of trees and shrubs. These species are *A. alternata* (syn. *A. tenuis*), *A. citri*, *A. panax*, and *A. tenuissima*. In most instances, fungi in this group cause negligible damage. Often they act as secondary invaders of foliage or fruit killed or weakened by other agents. When acting as primary pathogens, they cause lesions ranging from tiny spots to large necrotic blotches. They sporulate on dead plant surfaces, producing brown conidiophores and conidia singly or in chains. These structures may be visible with a hand lens as a brownish mold (E). Viewed microscopically, the conidia are large (e.g., 50–125 μm long), dry, and multicellular, often with both transverse and longitudinal or oblique septa and often with a long apical cell. Even within a species, these spores vary considerably in shape, septation, and surface ornamentation (Figure 9).

Relatively few associations of *Alternaria* species with tree diseases have been studied enough for us to learn whether the fungi are primary or secondary pathogens or only saprophytes. An incomplete list of reported disease associations in the United States and Canada follows.

Leaf spots on mountain ash, aucuba, catalpa, stinking cedar, currant, Chinese elm, forsythia, gardenia, sea grape, hibiscus, holly, English ivy, ixora, lilac, magnolia, mahogany, maple, myrtle, oleander, palms of several species, Sago palm, pittosporum, poinsettia, wax-leaf privet, rose, rubber plant, Chinese tallow tree, wisteria, and yucca.

Leaf blight of black locust and hybrid poplar, also of schefflera, dwarf schefflera, and other members of the aralia family.

Leaf spot and fruit rot of apple, cherry, grape, papaya, pear, and plum.

Fruit rot of apricot, avocado, citrus, guava, kumquat, black persimmon, and snowberry.

Leaf and twig blight of blueberry.

Needle blight of juniper and twig dieback of arborvitae.

Petal blight of rose, flower spot of rhododendron, and damping-off of catawba rhododendron.

A. alternata is the species most commonly associated with disease. This fungus is a cosmopolitan saprobe and a weak, opportunistic pathogen. When found on necrotic spots or blight lesions it is often considered to be a secondary invader. As a primary pathogen, however, it does cause a petal blight of rose with symptoms similar to those of Botrytis blight (Plate 25). This disease has been noted in the South on field-grown Floribunda and Hybrid Tea roses during wet summer weather. Spores produced on dead blossoms or rose hips are responsible for infection. When this fungus infects stored citrus seeds, the result is diminished germinability and an unusually high proportion of albino seedlings among the germinants.

A. tenuissima infects leaves of pittosporum, causing tiny necrotic lesions with yellow halos.

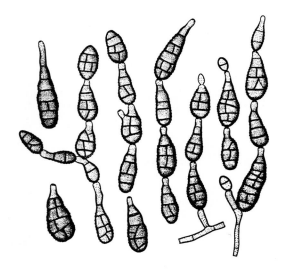

Figure 9. Conidia of *Alternaria alternata*. The longest spores here are about 60 μm long.

A. citri causes necrotic spots and blight of tangerine leaves, leaf spot of rough lemon, and fruit rot of several kinds of citrus.

Perhaps the most important of the Alternaria diseases of woody plants is the leaf blight of members of the aralia family, caused by *A. panax*. This fungus causes discrete necrotic spots or spreading dark lesions, often surrounded by yellow zones, in leaves of false aralia, geranium-leaf aralia (G), ming aralia, schefflera (A, B), dwarf schefflera, *Acanthopanax sieboldianus*, *Tupidanthus calyptratus*, and several herbaceous members of the family. The disease is most important in seedlings and young nursery plants with succulent foliage. It develops in nurseries and landscapes in the South and occasionally in greenhouses in northern areas. Lesions in leaves of schefflera are tan to dark brown, circular to irregular in outline, and surrounded by chlorotic zones. They either remain as discrete dead spots or expand into large blighted areas. Severe disease causes defoliation. In humid air, the fungus soon sporulates on lesions. Dispersal of the conidia to nearby plants by air or splashing rain leads to secondary cycles of disease, with new symptoms appearing 4–5 days after inoculation. Severe disease is favored by temperatures in the range 24–27°C and intermittent wetting of foliage. The wetness is necessary for spore germination and penetration. *A. panax* attacks only members of the aralia family, but isolates have shown no host preference within the family.

References: 20, 67–69, 99, 376, 387, 425, 543, 577, 672, 1798, 1799, 2051

A, B. Necrotic spots and blight of schefflera leaves caused by *Alternaria panax* (FL, Mar–Apr).

C. Lesions on a leaf of western catalpa. The pathogenicity of *Alternaria* sp. found sporulating on such lesions has not been proved (NY, Jul).

D. Close view of a lesion in a leaf of *Hibiscus* sp. Pathogenicity of the associated *Alternaria* sp. has not been reported.

E, F. Leaf blight of hybrid poplar. *Alternaria* sp. that sporulates on such lesions (magnified in E) under moist conditions is presumed to cause them (MI, Aug).

G. Necrotic spots and blight of leaves of aralia caused by *A. panax* (Fl, Mar).

Plate 40

Stigmina Leaf Spot and False Smut of Palm (Plate 41)

Palms growing in humid regions are subject to a few disfiguring foliar diseases among which Stigmina leaf spot and false smut are examples.

Stigmina leaf spot. This disease is caused by *Stigmina palmivora* (Deuteromycotina, Hyphomycetes). The fungus, formerly known as *Exosporium palmivorum,* is common in many regions where palms grow outdoors, and it also occurs in temperate areas on palms grown under glass. It was described originally from a greenhouse in Nebraska.

Palms of many types are susceptible. Those reported to be affected in the United States include cabbage, saw cabbage, Christmas, date, Canary Island date, cliff date, pygmy date, Senegal date, fishtail, key, lady, pindo, queen, Cuban royal, and Florida royal palms; palmetto; *Borassus aethiopum; Chrysalidocarpus* sp.; and *Phoenix loureirii.*

The disease attracts attention because of prominent brown dead spots or tips on leaflets. These lesions arise as tiny, circular, translucent tan spots that become depressed and often somewhat irregular or elongate with a dark brown to black spot in the center, an indefinite brown margin, and a diffuse yellow halo. Mature lesions are elongate, $2-3 \times 1-4$ mm or larger, and may coalesce to form irregular dead blotches that in time turn grayish brown. These often cause or contribute to dieback of the tips of leaflets. Lesions also occur on the rachises of fronds.

The fungus sporulates on both surfaces of lesions. First, however, it forms minute stromata that fill substomatal cavities beneath the leaf surfaces and expand outward through the stomata. On the surface the stroma becomes visible as a minute dark brown to black dome about 150 μm wide (B, C) on which dark-pigmented conidiophores and elongate, warty, multicelled conidia are produced.

Epidemiological aspects of Stigmina leaf spot have not been reported beyond the observation that the disease is common on plants grown under glass or with insufficient light.

Four other species of *Stigmina* are confirmed or possible pathogens of trees. *S. carpophila* causes necrotic spots and shot holes in peach leaves. *S. platani,* the conidial state of *Mycosphaerella stigminaplatani,* causes a leaf spot of sycamore. *S. juniperina* and *S. glomerulosa* are associated with needle blight on junipers.

False smut. First described from greenhouse specimens in the United States, this disease affects palms of many species around the globe and causes economic damage to date palms in Africa and Asia. It suppresses photosynthesis and fruit production and shortens the life span of leaves. It occurs across the southern United States and is most noticeable in humid areas from Florida to Texas. The causal fungus, *Graphiola phoenicis,* belongs to a distinctive group of about six species that compose the order Graphiolales of the Heterobasidiomycetes.

G. phoenicis is restricted to palms. Those affected in the United States include areca, arenga, cabbage, coconut, date, Canary Island date, pygmy date, Senegal date, pindo, queen, Florida royal, silver, and Washingtonia palms (including thread palm) and dwarf palmetto. Cabbage palm is also host to *G. congesta* and *G. thaxteri,* but these species are poorly known in comparison with *G. phoenicis.*

The symptoms of false smut are premature senescence and tiny yellow to brown spots that are much less noticeable than the causal fungus. *G. phoenicis* develops on both leaf surfaces as numerous black wartlike projections (sori) $1-3$ mm in diameter, each with a central crater from which light yellowish filaments extrude (E, F). The sori appear first near the tips of leaves and subsequently arise on subapical leaflets and sometimes on rachises. They are generally absent from the youngest leaves but increase progressively on leaves $2-3$ years old, eventually numbering 20 or more per square centimeter where infection is severe. On date palms, leaves that normally live $6-8$ years may die after 3 years.

The false smut sorus consists of a hard, dark, persistent layer that encloses a yellowish fertile region where powdery spores are produced among the conspicuous filaments. The filaments extend as much as 2.5 mm beyond the rim of the crater and are often found coated with spores. Filaments are thought to play a role in dispersal of the fungus, perhaps dislodging spores when disturbed by wind or water drops. The spores are small (e.g., 5×7 μm), two-celled, and yellowish in mass. They can germinate either by germ tubes or by budding in the manner of yeasts. When initial growth is yeastlike, colonies convert to mycelial growth after $18-24$ hours.

The disease cycle has not been studied extensively, but apparently its duration is 10 months or more. In one experiment, the sowing of spores of *G. phoenicis* on young palm fronds in December resulted in characteristic sori the following October. Hyphae of the pathogen are believed to enter leaves through stomata. The fungus grows among and within leaf cells, eventually producing a compact mass of mycelium beneath the epidermis. Emerging sori rupture the leaf surface, but the pathogen causes no other derangement of host tissues. Rough, dark craters of old sori remain after spores have dispersed.

Diseased leaves of date palm, in comparison to healthy ones, were found to contain smaller amounts of chlorophyll, soluble sugars, and major nutrient elements (nitrogen, phosphorus, and potassium) and larger amounts of phenolic compounds and micronutrients (especially iron and manganese). Activities of the enzymes IAA oxidase and amylase were greater than normal.

Date palms in Africa are often pruned for control of false smut, but whether this reduction of photosynthetic surface aids productivity is unreported.

References for Stigmina diseases: 20, 542, 577, 846, 910, 2227
References for false smut: 20, 378, 989, 1772, 1814, 1837, 1975

A–C. Stigmina leaf spot of Canary Island date palm. A. Dieback of leaflets, the result of severe infection and coalesced lesions. B, C. Successively closer views of killed portions of a leaflet, showing sporulation of *Stigmina palmivora.* Dark brown spots in the centers of lesions are characteristic (FL, Apr).

D–F. False smut of Canary Island date palm. D, E. Sori projecting from leaflets. F. Sori and associated chlorotic to necrotic spots. Twisted yellowish filaments extrude from a central cavity in each wartlike sorus (FL, Apr).

Plate 41

Insolibasidium Blight of Honeysuckle and Rhizoctonia Blight of Natal Plum (Plate 42)

Insolibasidium blight. Many species of honeysuckle are subject to leaf blight caused by *Insolibasidium deformans* (syn. *Herpobasidium deformans*) (Auriculariales, Auriculariaceae). This fungus, found only on honeysuckle, is common in the Great Plains and eastward in the United States and Canada and occurs also in the Pacific Northwest. Epidemics in nurseries may cause defoliation, dieback, and lost growth. Susceptible species include *Lonicera* ×*amoena*, ×*bella*, *caerulea*, *canadensis* (fly honeysuckle), *dioica*, *discolor*, *gracilipes*, *korolkowii*, *maackii*, ×*minutiflora*, *involucrata* (twinberry), *japonica* (Japanese honeysuckle), *morrowii* (Morrow's honeysuckle), *muendeniensis*, *nervosa*, ×*notha*, *oblongifolia* (swamp fly honeysuckle), *orientalis*, *prolifera* (grape honeysuckle), *prostrata*, *quinquelocularis*, *ruprechtiana*, *sempervirens* (trumpet honeysuckle), *tatarica* (tatarian honeysuckle), *tatsiensis*, and *vilmorinii*. Snowberry, also in the honeysuckle family, is susceptible by artificial inoculation but has not been found infected in nature.

Blight appears beginning in spring, often on some of the first leaves on new shoots. Symptoms arise as slight crinkling or rolling of the infected areas. These turn yellowish green and then tan within several days. The veins tend to remain green for a time after interveinal tissues turn brown. Diffuse yellow zones develop around young lesions (C) and become more conspicuous with time. As lesions expand and coalesce, diseased leaves curl and twist, and entire leaves or large parts of them become distorted, dry, and brown (A, B). Although stem tissues are not attacked, an appearance of shoot blight may result from infection of all leaves on a shoot. If conducive weather prevails, blight appears on new leaves throughout the season. Symptoms appear 8 or more days after each infection period, and severely diseased leaves fall prematurely.

I. deformans produces both basidiospores and conidia. The basidial state develops first, during humid weather at moderate temperatures, as a continuous thin white layer on the lower surface of a lesion. This layer develops from hyphae that emerge through stomata. The basidia are distinctive in that each curves in a semicircle and produces four basidiospores along the arc. Basidiospores are discharged in humid air at temperatures of 5–28°C. Discharge is most abundant at 14–21°C, and germination is maximal at 14–26°C. This sporulation occurs from spring through early autumn during and after periods of general rainfall.

The conidial state appears in summer as a white powdery mass (F) on or at the margins of the same lesions where basidia previously formed. This state is most common on severely blighted, shaded leaves. Conidiophores emerge singly or in clusters from the stomata and are more numerous on the lower leaf surface than on the upper. Conidial production and germination are favored at temperatures similar to those for basidiospores. Conidia are common on fallen diseased leaves from summer until early winter. Apparently the conidia, which form in groups of six, are not dispersed but serve as resting spores. The conidial state is named *Glomopsis lonicerae*.

In experiments, basidiospores but not conidia have caused blight. Moreover, basidiospores are liberated from overwintered leaves and are responsible for the first infections in spring. Perhaps the conidial state functions only for survival and gives rise to basidia in spring.

Honeysuckle species found to be resistant in Iowa were *L. dioica*, *L. gracilipes*, and *L. sempervirens*. *L. japonica* cv. 'Halliana' was immune in Iowa, but the parent species was found naturally infected in New Zealand.

Rhizoctonia blight. Leaves and shoots of several species of woody shrubs in warm humid regions are subject to blight caused by another basidiomycete, *Thanatephorus cucumeris* (Tulasnellales, Ceratobasidiaceae). This cosmopolitan fungus is usually encountered in and called by the name of its mycelial, sclerotium-forming state, *Rhizoctonia solani*. *R. solani* and related fungi are best known as pathogens of roots and other subterranean organs, but this description greatly understates their capability. They parasitize hundreds of species of plants, causing seed rot, damping-off, root rot, fruit decay, foliar lesions, and, on some herbaceous plants, stem cankers. These fungi reside in the soil and, in the absence of living hosts, grow freely as saprophytic mycelium or persist as quiescent sclerotia.

R. solani is characterized by pale to dark brown mycelium with hyphae of large diameter that branch frequently at nearly right angles. The fungus also produces black sclerotia that vary in size and shape from spheres less than 1 mm across to crusts several millimeters in extent. This fungus penetrates plant surfaces by hyphal growth into lenticels on stems or stomata in leaves. Or it may penetrate intact roots by means of narrow infection pegs produced from appressoria or cellular aggregates called infection cushions. Hyphae then grow among and within host cells, and the cells die after contact or penetration. The precise mode of killing remains undescribed. Sclerotia often form in killed tissues.

On woody plants, *R. solani* causes damping-off, root rot of seedlings and cuttings, and occasional top rot of small coniferous seedlings. Exudates from roots and seeds stimulate the fungus. It invades succulent tissues and, under humid conditions, grows as weblike mycelium on the soil surface and on plant bases. This behavior led to the name *web blight*. Occasionally basidia of the sexual state arise on such mycelium. The killing of basal parts close to infested soil is illustrated on cuttings of peperomia (G).

Foliar blight caused by *R. solani* affects Natal plum (D, E) and azalea. This disease differs from the root or cutting rot syndrome in that symptoms arise without any obvious relationship to the soil. Blight usually begins in summer on low branches. On Natal plum, lesions are initially dark green, water-soaked areas at margins and bases of leaves. The lesions eventually become irregularly shaped sunken brown spots. Some remain discrete, but others enlarge, display concentric zones of varying brownish hues, and involve entire leaves and often the succulent tips of branches (D, E). If the blight is promoted by high temperature and humidity, it spreads upward. Killed leaves, which tend to curl or roll on drying, either drop or are held by strands of mycelium. Severely affected plants often also have decayed roots.

Rhizoctonia blight of azalea may result in severe defoliation if hot wet weather persists after an outbreak begins. Conversely, this disease subsides under dry conditions or with seasonal cooling in autumn. Symptoms are similar to those on Natal plum. Under continuously moist conditions, mycelium can be seen extending from soil to leaves, and matted dead leaves may hang from the mycelium. Tiny azalea plants may be killed outright, but the woody parts of larger ones survive and eventually produce normal shoots and leaves.

R. ramicola causes a similar disease, called silky thread blight, in coral bean, thorny elaeagnus, Japanese holly, pineapple guava, crape myrtle, and Japanese pittosporum in Florida.

References for Insolibasidium blight: 665, 777, 1417, 1612, 1614
References for Rhizoctonia blight: 22, 286, 387, 577, 1456, 1673, 2131, 2133

A–C, F. Insolibasidium blight of tatarian honeysuckle. A, B. Mature lesions and blighted leaves. C. Young lesions viewed from the upper surface. F. Conidial (resting spore) state on lower surfaces of lesions (NY, Aug–Sep).

D, E. Rhizoctonia blight of Natal plum. Foliar lesions (E) are succeeded by death of whole leaves and shoots (FL, Mar).

G. Rhizoctonia rot of peperomia cuttings. Grayish mycelium is visible on the soil and on the base of a diseased plant (FL, Apr).

Plate 42

Apple Scab and Venturia Leaf Blight of Maple (Plate 43)

This plate begins a series on scab and blight diseases caused by species of *Venturia* (Pleosporales, Venturiaceae) and related conidial fungi. Of more than 30 species of *Venturia* recognized in North America, about 12 are pathogens of woody plants and perhaps 6 are economically or aesthetically important. Parasitic species of *Venturia* infect succulent, aboveground parts of their host plants, causing localized swelling with distortion and necrosis (scab), necrotic spots on leaves, or blight of leaves and shoots. They produce olive-brown masses of conidiophores and conidia on infected or recently killed parts during the growing season, and they overwinter as mycelium or developing pseudothecia in dead, previously diseased leaves or twigs. Primary infections are initiated by ascospores, conidia, or both, depending upon the disease. If weather is conducive during the growing season, conidia cause secondary cycles of disease as long as immature host tissue is available. Pathogens in this group display considerable host specificity, each fungal species restricted to one plant genus or a few closely related genera. One species, *Venturia rhamni* on *Rhamnus californica* in California, completes its life cycle in living leaves during the growing season, but the common scheme is an alternation of conidial parasitic and ascigerous saprophytic states. The conidial states are in the form-genera *Cladosporium*, *Fusicladium*, *Pollaccia*, and *Spilocaea*. Some conidial scab fungi such as *Spilocaea pyracanthae* (Plate 44) have not been associated with pseudothecial states.

Apple scab. The most famous and economically important pathogen in this group is *Venturia inaequalis,* the cause of apple scab. In the absence of costly control measures, scab reduces both quality and yield of apples in all but a few of the warmest or most arid areas of apple production. The disease is most severe where humidity is high and temperature moderate during spring and early summer. Host plants include nearly all commercial cultivars of apple, most crabapples and many other species of *Malus*, and various other members of the tribe Pomoidiae of the rose family. The list includes but is not restricted to cultivars of the following species: *M. angustifolia*, ×*arnoldiana*, *baccata*, *brevipes*, *coronaria*, *florentina*, *glaucesens*, *ioensis*, ×*micromalus*, ×*platycarpa*, *pumila* (common apple), ×*purpurea*, ×*scheideckeri*, *sieboldii*, and *sylvestris*. Other hosts include several species of mountain ash (*Sorbus aria*, *aucuparia*, *domestica*, and *torminalis*), *Cotoneaster affinis* and *C. integerrima*, firethorn (*Pyracantha* sp.), and common pear.

The apple scab fungus infects leaves (both surfaces), flower parts, fruit, and succulent twigs but penetrates only the cuticle while the plant part is alive. It forms colonies between the cuticle and epidermis and draws nutrients from living tissues beneath. Symptoms on leaves and fruit arise in spring as small olive green spots that enlarge and darken to become more or less circular superficial lesions with radiate margins. Almost as soon as lesions appear, they become velvety as masses of olivaceous brown conidiophores and conidia break through the cuticle. Chlorosis and death of tissues beneath the colonies follow during the next several weeks. The conidial state, named *Spilocaea pomi*, causes secondary infections on all succulent parts throughout the period of host growth. Secondary lesions on leaves are often diffuse and tend to be prominent along veins (E, F). On fruit, secondary lesions sometimes appear as small spots around large primary scab lesions. The same is true for pear scab (Plate 44G).

Tissues that become infected while enlarging are at first stimulated to overgrowth that results in raised lesions (scabs) on fruit, bumps on succulent twigs, and curled or puckered leaves. Such tissues cease growth prematurely, however, resulting in further deformity that is most noticeable on fruit. Photosynthesis is suppressed in infected leaves. Severely infected leaves and fruit fall prematurely. Lesions on remaining fruit and on twigs become delimited by a corky layer, and the surfaces become cracked and rough. Old lesions turn grayish as tissues die and sporulation ceases. Mature tissues resist new infection but may continue to support colonies of the parasite that were previously established. Symptoms on mountain ash and other hosts are similar to those on apple, with emphasis on foliar lesions and premature leaf fall.

On some cultivars of apple and crabapple, scab lesions on twigs become numerous and may remain active for 2 seasons. Thus conidia from overwintered lesions on twigs are sometimes responsible for primary infection in spring, particularly in years following severe scab epidemics. Most primary infections, however, are initiated by airborne ascospores dispersed from pseudothecia in previously diseased leaves on the ground.

In fallen leaves, *V. inaequalis* colonizes tissues beneath former scab lesions and produces pseudothecia during the dormant season of the host. Ascospores mature in spring and are expelled into the air beginning at about the time of renewed host growth and continuing for 1–3 months, depending upon local climate. Ascospores are liberated within 1 hour after the onset of rain, and peak numbers are found in the air after 3–6 hours. Small amounts of rain, for example 0.2 mm, may induce the expulsion of many ascospores. Total ascospore production from apple or crabapple leaves reflects the severity of scab during the previous season and varies with the species or cultivar and with local environmental conditions. The peak of ascospore dispersal often occurs near the end of the period of bloom.

Infections by ascospores and conidia may occur simultaneously and are indistinguishable. Conidia are dispersed primarily with water that splashes and runs along plant surfaces, but some are airborne during dry weather.

Given wet plant surfaces and available ascospores or conidia, scab infections occur at temperatures from 2° to about 26°C. The severity of resulting disease increases with the duration of wetting. The minimum wet period needed for infection is about 6 hours if temperature is near the optimum of 20°C. Symptoms appear and conidial production begins usually 8–18 days after infection, but the incubation period is longer under very cool conditions or if dry weather intervenes during incubation.

V. inaequalis includes subpopulations specialized for parasitism of particular host plants. Several races that have been identified can cause scab on species and cultivars of *Malus* that are resistant to the general population of the fungus. Subpopulations specialized for attack of mountain ash, cotoneaster, or apple but incapable of cross-infecting these hosts have been designated formae speciales: *V. inaequalis* f. sp. *aucupariae*, f. sp. *cotoneasteris*, and f. sp. *mali*. These designations are not widely used, however, because of insufficient information about specialization of *V. inaequalis* on other hosts. It has been reported to occur on common pear and firethorn but has not been evaluated for possible specialization on these hosts. A second apple scab fungus, *V. asperata*, has been reported to occur on 'Almey' crabapple in Ontario. Its North American distribution and host range are unknown.

Natural controls on apple scab, aside from resistance or dry weather, are few. In moist areas, earthworms and litter-decomposing microorganisms degrade fallen leaves and reduce the population of the scab fungus. Natural decomposition can be hastened by composting leaves and applying nitrogenous fertilizer to the compost pile.

Many species and cultivars of *Malus* are resistant to *V. inaequalis*. Resistance has been bred into new commercial apple cultivars (e.g., 'Freedom' and 'Liberty'), and resistant types have been used directly or as breeding stocks for ornamental crabapples. Some crabapple types sufficiently resistant to apple scab and also to powdery mildew (Plate 4) and cedar-apple rust (Plate 115) to be useful in ornamental horticulture include *Malus* cv. 'Adams'; *M. baccata* cv. 'Jackii'; *M.* cv. 'Baskatong', 'Beverly', 'Bob White', 'David', 'Dolgo', and 'Donald Wyman'; *M. floribunda*; *M.* cv. 'Henry Kohankie', 'Liset', 'Ormiston Roy', 'Professor Sprenger', and 'Red Jewel'; *M. sargentii*; *M. sargentii* cv. 'Tina'; *M.* cv. 'Sugartyme'; *M.* '*tschonoskii*'; and *M.* ×*zumi* var. *calocarpa*.

A, C. Scab caused by *Venturia inaequalis* on European mountain ash. Infections, often concentrated along midveins of leaflets, cause premature yellowing and death or casting (NY, Aug).

B, D–G. Apple scab. B. Lesions and premature yellowing of leaves of an unidentified crabapple. Chlorophyll is retained (green island effect) beneath lesions. D. Scab on a leaf and a fruit of apple. E. Chlorosis and distortion of an infected leaf of *Malus* sp. F. Close view of olive-brown colonies of *V. inaequalis* on a leaf; radiate margins of colonies are characteristic. G. Crabapple shoots defoliated by scab (NY, Jun–Jul).

H. Lesions caused by *V. acerina* on red maple; upper and lower leaf surfaces are shown (NY, Aug).

Plate 43

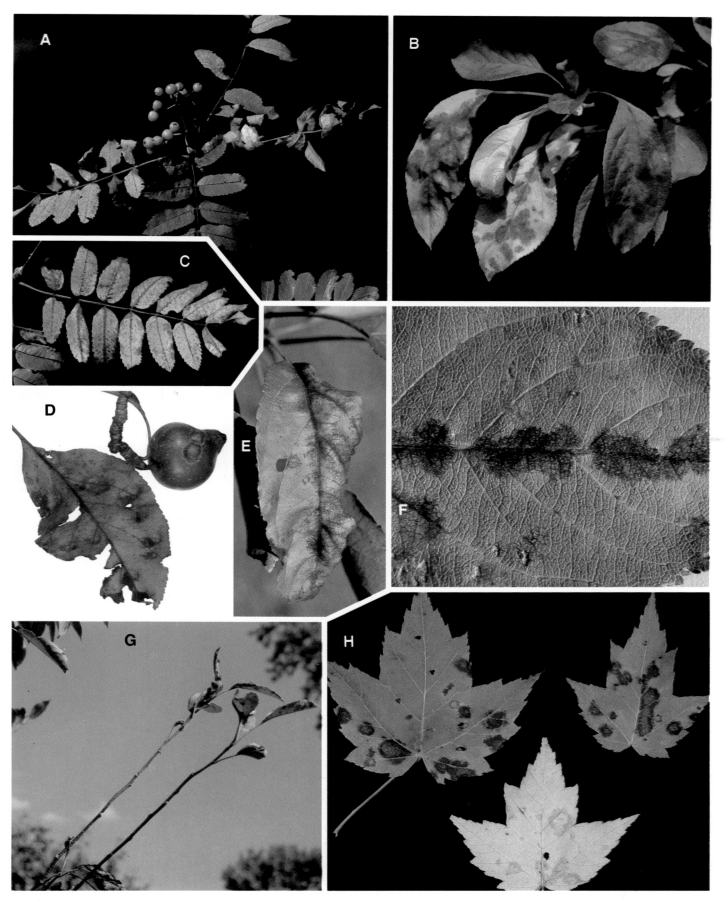

Venturia leaf blight of maple. *Venturia acerina* causes an inconsequential leaf blotch of maples (Plate 43H) in the Great Lakes region and eastward to North Carolina and New Brunswick. Mountain, red, silver, and sugar maples are affected. The disease is best known on red maple, on which necrotic lesions up to 2 cm in diameter develop in midsummer. The lesions are more or less round except where they are bounded by major veins or the leaf edge. On the upper surface they have deep reddish brown centers and dark brown edges with diffuse margins. The lower surface is grayish green to grayish tan. Coalescing lesions kill large areas of leaf blades. Severe infection leads to premature reddening and casting of red maple leaves.

Diagnosis of the maple pathogen requires recognition of its conidial state, *Cladosporium humile*, which arises in humid air on a network of light brown mycelium on both surfaces of lesions. The olive-brown, one- or two-celled conidia form in chains on short conidiophores that occur singly or in clusters.

V. acerina overwinters in fallen leaves as mycelium and as developing pseudothecia. Ascospores mature in spring. Germinable conidia have been found on overwintered leaves, but their role in the disease is unknown.

References for apple scab: 82, 83, 105, 387, 821, 841, 953, 998, 1278, 1311, 1312, 1341, 1377, 1403–5, 1762, 1816, 1839, 1944, 2019, 2201, 2206, 2211

References for Venturia leaf blight of maple: 105, 425, 1532, 1816

Scabs of Firethorn, Loquat, Toyon, and Pear (Plate 44)

Scab of firethorn and other plants. Scab is perhaps the most important disease of firethorns (*Pyracantha* species). It is caused by the conidial fungus *Spilocaea pyracanthae*, which also infects *Kageneckia oblonga*, loquat, and toyon. The symptoms are similar to those of apple scab and pear scab: olive green to black velvety spots (B, F) on leaves, fruit, and young twigs; premature yellowing or reddening of leaves; and premature casting of leaves and fruit. Severe disease on highly susceptible firethorns such as *P. coccinea* cv. 'Lalandei' may cause nearly all the leaves and fruit to drop, thus greatly reducing the ornamental value of the shrub. Severe infection of young leaves and shoots of toyon may kill those parts (E) even though the pathogen is confined to surface cells. After leaves of this host have expanded fully, however, infection results in discrete spots and gradual reddening and browning of leaves. Scab causes green toyon fruit to shrivel, and it may cause severe defoliation or death of *K. oblonga*.

The host plants of *S. pyracanthae* are evergreen. Thus infected leaves as well as twigs on the plants are sites where the pathogen survives winter. Conidia dispersed from old leaves and twigs by water and perhaps also by air cause the first infections in spring. There would seem to be little need for airborne spores in such a system, and in fact, no ascigerous state has been found.

The incubation period after conidial inoculation in California was about 2 weeks at 16°C and 3–5 weeks under natural conditions at temperatures of 6–17°C.

S. pyracanthae occurs across the United States and Canada. This pathogen was formerly known by several names that connoted distinct host-specific fungi or, incorrectly, identity with the pear scab fungus. In California, however, *S. pyracanthae* was shown by cross-inoculations to be capable of infecting several species of firethorn, loquat, and toyon as well as *K. oblonga* but incapable of causing scab on apple or pear. Furthermore, conidia of a *Spilocaea* on mountain ash (presumably a strain of the apple scab pathogen) did not infect firethorn, loquat, or toyon. Thus *S. pyracanthae* on these hosts was recognized as an entity distinct from the better-known pathogens of pear and apple.

Firethorn species and cultivars vary greatly in susceptibility. Species affected to various degrees include *P. angustifolia*, *P. atalantioides*, *P. coccinea*, *P. crenulata*, *P. fortuneana*, *P. koidzumii*, and *P. rogersiana*. Of approximately 30 cultivars exposed to natural inoculation in Washington State, the following seemed resistant: *P. coccinea* 'Government Red' and 'Prostrata'; *P. koidzumii* 'Bella', 'Duval', and 'Santa Cruz Prostrata'; *P. rogersiana* 'Flava'; and the hybrid 'Shawnee'. In eastern trials, *Pyracantha* cv. 'Firey Cascade' and *P. coccinea* 'Rutgers' were resistant.

Pear scab. This disease occurs around the globe and is economically significant in the Pacific Northwest. It is similar to apple scab (Plate 43) in modes of infection, symptoms, annual cycle, and epidemiology. The most significant difference is that twig infections are more numerous and persistent. These cause roughening and distinct lesions in the bark of twigs, and the fungus remains active in twig lesions throughout 2 or more seasons. There it produces conidia that complement and in some areas supplant ascospores as inoculum for primary infections. *Venturia pirina* also differs from the apple scab pathogen in characteristics of ascospores, in its conidial state (*Fusicladium pyrorum*), and in host specificity.

Host plants of *V. pirina* include most cultivars of common pear, other *Pyrus* species such as *P. sativa* and *P. syriaca*, and, in England, loquat. Several races have been designated on the basis of differential pathogenicity to pear cultivars and species. Pear species used for ornament are apparently resistant.

References for scab of firethorn and other plants: 58, 105, 1247, 1377, 2075

References for pear scab: 105, 387, 1019, 1020, 1104, 1750, 1816, 2211

A–C. Scab caused by *Spilocaea pyracanthae* on fruit of *Pyracantha coccinea* 'Lalandei' (NY & MD, Sep–Oct).

D. Scab on a yellow-fruited firethorn (CA, Sep).

E. Discoloration and death of leaves and shoots of toyon resulting from infection of immature tissues by *S. pyracanthae* (CA, Sep).

F. Velvety scab lesions caused by *S. pyracanthae* on a leaf of loquat (CA, Jun).

G. A 'Bartlett' pear fruit with scab caused by *Venturia pirina*. A large primary lesion is surrounded by small lesions resulting from secondary infection (OR, Jul).

Plate 44

Venturia Leaf and Shoot Blights of Aspen and Poplar (Plate 45)

Leaves and shoots of aspens and poplars across Canada and the northern United States are subject to blight caused by *Venturia tremulae* and *V. populina*. Bigtooth and trembling aspens, white poplar, various poplar hybrids, and occasionally eastern cottonwood are hosts of *V. tremulae*, while *V. populina* infects black cottonwood and balsam poplar. These diseases are significant mainly in young forest stands.

In wet seasons, shoot blight caused by *V. tremulae* may kill virtually all terminal shoots in aspen stands that are regenerating by sprouts after the harvest of mature trees. This damage reduces height growth and deforms trees by causing a bend in the stem at the point where, after death of a terminal shoot, a lateral shoot becomes the new leader (G). Successive leaders may be killed during seasons of severe disease. Plants less than 3 meters tall are at greatest risk, and damage becomes negligible as trees attain heights greater than 5 meters. *V. populina* attacks plants of all ages and sizes but has less impact than does *V. tremulae* in North American forests.

Both pathogens infect only succulent leaves and shoots, but the patterns and seasonal development of symptoms differ. The first infections by *V. tremulae* on aspen occur in spring on leaf blades or petioles and on young stem tissue. Dark brown to black lesions expand rapidly, causing leaves and shoots to droop, wither, and become brittle. Primary infections may be caused by either conidia or ascospores. The conidial state of *V. tremulae* is *Pollaccia radiosa*. Conidia are dispersed by splashing rain from the stubs of shoots blighted the previous season, where the pathogen overwintered as mycelium. Thus primary infections are often found adjacent to old blighted twigs. Lesions unrelated to dead twigs may be caused by ascospores that are airborne from pseudothecia in previously blighted, overwintered leaves on the ground.

Within a few days after the onset of lesion formation, the blackened surface turns olive green with the formation of a layer of conidiophores and conidia (C). Secondary cycles, promoted by wet weather, are initiated throughout the period of shoot elongation by conidia from newly blighted shoots. New shoots frequently grow adjacent to blighted ones and are blighted in turn, but lesions do not extend into woody twigs.

Most infections of aspen by *V. tremulae* occur on terminal shoots because these continue to grow and produce susceptible tissue throughout the season. Lateral shoots, however, cease elongation early in the season and become resistant. Thus most of the foliage of diseased aspens remains intact even though much damage is caused by repeated deformity of the stem.

V. populina overwinters as developing pseudothecia in blighted shoots of balsam poplar or black cottonwood. In spring its ascospores cause primary infections on the tips of leaves emerging from buds adjacent to the old dead, previously infected shoots. The killing of leaf tips results in distortion, but the leaf bases typically stay alive. Conidia develop within a few days on the dead leaf tips, are dispersed to shoots that are by then lengthening, and initiate secondary cycles characterized by shoot blight and leaf spot. In eastern Ontario the primary cycle typically begins near the end of May and secondary cycles early in June.

As in shoot blight of aspen, diseased tissues of black cottonwood and balsam poplar turn black, and killed shoots droop and become brittle. The mature parts of the current year's shoots and all parts formed in previous years escape infection. Unlike the disease of aspens, blight caused by *V. populina* in black cottonwood and balsam poplar may destroy both terminal and lateral shoots throughout the crown. During midsummer *V. populina* ceases to sporulate on twigs that became infected in June. The cessation of conidial production apparently signals the onset of sexual reproductive processes, for immature perithecia can be found in blighted shoots beginning in August. They remain immersed in cortical tissues and at maturity are black objects 160–220 μm in diameter. Mature ascospores are ready for discharge in early spring.

Balsam poplars severely damaged by shoot blight can be recognized at any time of year. When leafless, their branches appear to be crowded together, stunted, and decadent. The crowding results from development of adventitious shoots and multiple leaders after death of terminals. Canker-causing fungi may enter through blighted terminals and contribute to the appearance of decadence.

Given wet surfaces, ascospores and conidia of both pathogens germinate over a wide range of temperatures. Germination is most rapid at 15–25°C, depending upon the isolate. In experiments with *V. populina*, most ascospores germinated within about 40 hours after expulsion from pseudothecia.

In experiments wherein ascospores of *V. populina* were allowed to germinate on leaves of balsam poplar, the fungus produced appressoria and penetrated epidermal cells directly, causing conspicuous lesions within 4 days. Conidial production began within an additional 7 days. When ascospores were applied to leaves of Lombardy poplar, they caused lesions, but no sporulation followed. Thus Lombardy poplar was judged resistant. In parts of Europe, however, *V. populina* causes severe damage to Lombardy poplar and to its parent species, black poplar. A related plant, *Populus ×berolinensis,* and European aspen are also susceptible.

The conidial state of *V. populina* is *Pollaccia elegans,* which forms an olive to olive-brown mat or sometimes tufts of conidiophores and conidia on recently killed parts (D). The elongate conidia are usually three-celled, with the central cell larger than the others. This species was formerly confused with *V. tremulae,* but the latter has smaller conidia, smaller pseudothecia (40–180 μm in diameter), and smaller ascospores that are septate just above the middle rather than in the lower third like those of *V. populina.*

The aspen pathogen, *V. tremulae,* includes three varieties that differ in conidial morphology, geographic distribution, and host preference. *V. tremulae* var. *grandidentatae* is the principal North American pathogen. Its conidial state, *Pollaccia radiosa* var. *lethifera,* is the form once designated as *P. americana. V. tremulae* var. *populi-albae* is a little-known and apparently uncommon form that causes necrotic spots on aspen and white poplar leaves in North America and Europe. Its conidial state, *P. tremulae* var. *populi-albae,* was formerly designated *P. ramulosa. V. tremulae* var. *tremulae* occurs in Europe on European aspen, gray poplar, white poplar, and various hybrids. Its conidial state, *P. radiosa* var. *radiosa,* differs from the North American pathogen in having longer, narrower conidia. *V. tremulae* var. *tremulae* was formerly called *V. macularis,* but the latter name is now applied to be a different fungus that inhabits dead aspen leaves. Its conidial state and pathological role, if any, are unknown.

Clones of European aspen resistant to *V. tremulae* have been identified, but similar observations for North American poplars and aspens are lacking.

References: 105, 387, 441, 442, 695, 1351, 1424, 1816

A, E. Leaf and shoot blight of balsam poplar caused by *Venturia populina.* A. Severe blight on saplings (Alta, Jun). E. A blighted shoot and foliar lesions (Que, Aug).

B, C, G. Leaf and shoot blight of trembling aspen caused by *V. tremulae.* B. A diseased branch (Alta, Jun). C. A young lesion on a leaf; conidia are produced on the olive green areas (VT, May). G. Foliar lesions, apical crook in a dead shoot, and a bend in the twig where a lateral shoot assumed apical dominance after the death of the terminal (Ont, Jul).

D, F. Leaf blight of black cottonwood caused by *V. populina.* D. Tufts of conidiophores and conidia of the asexual state, *Pollaccia elegans,* on lesions. F. Diseased leaves. (WA, Jun).

Plate 45

Scab and Black Canker of Willow (Plate 46)

Scab and black canker of willow are distinct diseases with essentially the same gross symptoms—rapid blighting of leaves and shoots. Scab is caused by *Venturia saliciperda* (Pleosporales, Venturiaceae) and black canker by *Glomerella miyabeana* (Phyllachorales, Melogrammataceae). The two pathogens often occur together and may kill highly susceptible trees by repeated defoliation and destruction of shoots. The name *willow blight* is often used for these simultaneous infections.

Most published information about the relative susceptibilities of various willows applies to trees naturally attacked by both pathogens. Many species are susceptible to some degree. Those most damaged in North America are black, goat, heart-leaved, Niobe, and white (especially its golden variety) willows. Others affected include bebb, crack, peach-leaved, pussy, shining, and silky willows, also *Salix aurita*, *S.* ×*mollissima*, and *S. nigricans*. Species considered to be more or less resistant are bay-leaved, osier, purple, and weeping willows, also *S. alba* var. *tristis* and *S. triandra*. The cricket-bat willow is reportedly immune. In one report from British Columbia, Sitka willow remained healthy while growing adjacent to blighted golden willow.

The pathogens that cause willow blight apparently came to North America from Europe. Blight appeared in the Maritime Provinces in the 1920s and soon became widespread in eastern Canada and the northeastern United States, decimating populations of golden willow and other highly susceptible types. Subsequent reports extended its known range into the Appalachian and Great Lakes regions and the Pacific Northwest. Early workers disagreed about the pathogenicity of the two associated fungi because in several instances experimental inoculations of *V. saliciperda* to willow shoots or leaves failed to cause disease. Eventually, however, both fungi were shown to be aggressive parasites independently capable of causing leaf and shoot blight.

The name *willow scab* indicates the relationship of *V. saliciperda* to the well-known pathogens of apple and pear (Plates 43–44). The symptoms it causes, however, are best described by the word *blight*. Attacks begin in spring on the youngest leaves, which are often killed just as they emerge from buds. Killed leaves shrivel and gradually drop. Leaves that initially escape infection may be attacked later or may die of attrition because of death of a portion of the stem below. The fungus grows through petioles and into stems, causing cankers that kill small shoots. Olive-brown, velvety masses of conidiophores and two- to four-celled conidia develop within a few days as tiny mounds or as a continuous layer on diseased leaves and shoots (C, D). This sporulation is most common along major veins on the lower sides of leaves. The conidial state is *Pollaccia saliciperda*.

During wet weather, conidia cause repeating cycles of scab, leading not only to death of the first complement of leaves and shoots but also to infection of adventitious shoots that arise after defoliation. Only leaves and stem tissues produced in the current year are attacked, however, and tissues become less susceptible with age during the first season. The fungus remains alive in killed twigs, overwinters as mycelium, and in spring produces conidia that cause primary infections. Wet seasons promote epidemics, and dry weather halts them.

The ascigerous state of *V. saliciperda* has been produced in laboratory cultures but has not been reported to occur in nature. Thus ascospores probably have no role in the disease cycle.

Four other species of *Venturia* occur on willows but not so as to cause diagnostic confusion. Three have no known conidial states, while the fourth, *V. chlorospora*, causes inconsequential leaf spots. Its conidial state has been described from laboratory cultures but not found in nature. *V. chlorospora* was long thought to be the ascigerous state of the willow scab fungus because its pseudothecia were common in overwintered leaves where scab occurred, but careful study showed no relationship.

Black canker was described first in Japan, and the causal fungus, thought to be new to science, was named *Physalospora miyabeana*. Early researchers noticed a close resemblance of the willow pathogen to *G. cingulata*, which causes many plant diseases, including bitter rot of apple (see Plates 56–57). Inoculations of the willow pathogen to apple fruits, however, caused no disease in most cases. We treat the willow pathogen as a distinct species, *G. miyabeana*, but a strong argument could be made for considering it a host-specialized strain of *G. cingulata*. These fungi are indistinguishable in the conidial state.

Minor differences exist between scab and black canker in the timing of symptoms and in host parts attacked. The black canker fungus tends to attack leaves and twigs relatively later in the season, and it causes cankers on larger woody twigs (F) than does scab. Conidia or ascospores may cause primary infections. The first symptoms of black canker are blackened areas on leaf blades, extending quickly toward the leaf base. Lesions may remain discrete, but usually entire leaves die, droop, and shrivel (F, G). Lesions on twigs most often arise at nodes where the pathogen has grown down petioles. Cankers that fail to encircle twigs are more or less elliptical and usually 2–3 cm long but sometimes extend 5–8 cm. They become depressed, and the dead bark splits as twigs increase in girth.

Fruit bodies (acervuli) of the conidial state of *G. miyabeana*, *Colletotrichum gloeosporioides*, develop beginning early in summer on cankers and on killed twigs. Acervuli rarely develop on blighted leaves or shoot tips unless wet weather prevents them from drying. The acervuli produce conidia that are light pinkish in mass (E). Individual conidia are elliptical, unicellular, and nearly colorless and measure $12–24 \times 3–6$ μm. Dispersed by water, the conidia germinate on wet leaves or succulent stems and produce appressoria from which the fungus penetrates the host. This initiates secondary cycles of disease. Conidial germination and mycelial growth are most rapid at temperatures near 25°C. In experiments, lesions became visible within 40 hours when wet leaves were inoculated with conidia and incubated at 25°C.

In autumn, perithecia follow or accompany acervuli in the same lesions. Perithecia are nearly globose and 140–200 μm in diameter at maturity, and they remain immersed in plant tissues. They liberate one-celled colorless ascospores in spring.

In recent years, although damage by scab and black canker persist in wild willow populations, these diseases have become uncommon in managed landscapes. Perhaps this scarcity indicates depleted populations of highly susceptible types of willows in areas where climatic conditions favor these diseases.

References: 57, 105, 228, 367, 387, 604, 611, 744, 1377, 1378, 1415, 1521, 1816

A–D. Willow scab caused by *Venturia saliciperda*. A, B. An unidentified willow defoliated by the disease (NY, Jul). C, D. Masses of olive-brown conidia on a killed stem tip and on leaves of purple willow. Conidia on leaves are most abundant on veins (NY, Jun).

E–G. Black canker caused by *Glomerella miyabeana*. E. Acervuli with pinkish masses of conidia on killed buds within a twig canker on an unidentified willow (NY, Sep). F, G. Leaf blight, shoot blight, and twig cankers on unidentified willows (WA, May).

Plate 46

Anthracnoses and Didymosporina Leaf Spot of Maple (Plate 47)

This plate leads a series about anthracnoses—diseases caused by fungi that produce conidia in acervuli (Figures 10–11) in necrotic lesions on leaves, flower parts, fruit, and stems. Symptoms vary with the host-pathogen combination, ranging from innocuous leaf spots through blight of leaves and shoots to cankers and dieback of twigs and branches. The pathogens are conidial fungi (Deuteromycotina, Coelomycetes), and many are known to be the asexual states of ascomycetes. Under conducive conditions of wet weather at moderate temperatures, they may cause several cycles of infection annually. Conidia, dispersed by splashing and running water, start each cycle, and more conidia are produced in anthracnose lesions. In warm regions on evergreen hosts, this pattern continues indefinitely, interrupted only by dry weather or periods when host plants between growth flushes are relatively resistant. In temperate regions, however, the pathogens overwinter as mycelium in lesions, most often on twigs, or as immature perithecia in twig cankers or fallen leaves. New symptoms in spring may result from expansion of twig lesions initiated the previous season or from primary infections initiated by conidia or ascospores on succulent shoots. Details vary with particular host-pathogen combinations and with spring weather.

Anthracnoses of maple. Two or more different fungi cause anthracnoses of maples. These fungi, in common with hundreds of others that produce one-celled, colorless conidia in acervuli, were once classified in the genus *Gloeosporium*. Old records in herbaria, host indexes, and plant disease manuals include many references to *Gloeosporium* species associated with leaf spots or leaf blight of maples. These fungi would now be referred to genera such as *Discella*, *Discula*, and *Monostichella* or to *Kabatiella apocrypta* (syn. *Aureobasidium apocryptum*). Three types of foliar symptoms are induced by these pathogens: vein-associated lesions (A) that extend secondarily into interveinal areas; discrete necrotic spots (B); and irregular, spreading necrotic blotches and blight (E).

Two forms of *Discula* that are possibly distinct species cause the first two types of symptoms. One form attacks the veins of sugar and striped maples in the northeastern United States, causing elongate lesions that are brown above and tan below. Leaf sectors bounded by dead veins become yellow or red, then brown, and extend to the tips and margins. These symptoms become conspicuous in late summer. Assorted secondary fungi invade and sporulate in the dead sectors. The causal fungus produces inconspicuous light brown acervuli on the lower surfaces of lesions, mainly on and immediately adjacent to the veins (C). The other form, which resembles the *Discula* state of *Apiognomonia errabunda* (Figure 11; also Plates 50–51) causes necrotic spots that often but not invariably center on veins (B). On sugar maple the mature lesions are reddish brown with tan centers. Brown acervuli become prominent on the upper surfaces of the lesions, especially along veins (D).

The *Discula* state of *A. errabunda* includes many anthracnose fungi that are morphologically indistinguishable but are thought to be specialized for attack of particular host plants. Host specificities of *Discula* species found on maples have not been studied.

K. apocrypta is the best known anthracnose pathogen of maples. It occurs across North America, causing necrotic spots or scorchlike blight on leaves and occasional blight of succulent sprouts. Hosts include Japanese, mountain, Norway, red, silver, sugar, and tatarian maples as well as box elder. Symptoms develop during or shortly after wet weather from late spring until late summer as discrete, irregularly

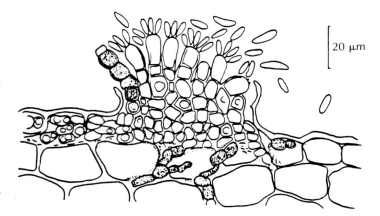

Figure 10. An acervulus of *Kabatiella apocrypta* on a leaf, microscopic representation.

shaped, randomly arrayed lesions that merge and kill large areas of a leaf. The color of lesions varies with the host, from reddish brown on Norway and sugar maples to light tan on Japanese maple (E). Very young leaves attacked by *K. apocrypta* or other anthracnose fungi may become blackened and shriveled, but leaves infected after reaching mature size retain nearly normal form. The disease is economically inconsequential but in wet years may cause locally conspicuous foliar browning. Severe infection leads to premature defoliation. Inconspicuous light brown acervuli of *K. apocrypta* form on the lower surfaces of lesions, especially along veins. This fungus is distinguished from other anthracnose pathogens on maples by the formation of its one-celled, colorless conidia in groups of four to eight on the tips of short, broad conidiophores (Figure 10). In pure cultures, *Kabatiella* species produce conidia from hyphal cells without any distinct conidiophores.

Annual cycles of the various maple anthracnose diseases have not been studied. Thus the nature and source of inoculum for primary infections is unknown. Also, the ascigerous states of these fungi, if any, are generally unknown. *Discella acerina*, however, is the conidial state of *Cryptodiaporthe hystrix*, the perithecia of which occur on dead twigs.

Didymosporina leaf spot. *Didymosporina aceris* (syn. *Marssonina truncatula*) causes large necrotic spots on the leaves and samaras of bigleaf, hedge, and Norway maples and box elder. It occurs in the northeastern and northwestern United States. On Norway maple the spots are distinctly two-toned with reddish brown centers, grayish green outer portions, and an abrupt transition to normal green tissue. When they are numerous they coalesce, making leaves appear blighted. Tiny dark acervuli (about 100 μm in diameter) develop between the epidermis and cuticle of the lower surfaces of lesions and produce two-celled, yellow-brown conidia in masses that rupture the cuticle. The disease appears and intensifies at the end of summer. The disease cycle has not been studied, and no ascigerous state of the pathogen is known. This disease could be called an anthracnose, but the more specific name, Didymosporina leaf spot, is appropriate because the symptoms and the causal fungus are distinctive.

References: 59, 389, 395, 425, 789, 857, 1239

A, C. Anthracnose of sugar maple caused by *Discula* sp. The lower surface of a vein-associated lesion (C) shows acervuli as brown flecks or short brown lines on the vein (NH, Sep).

B, D. Another anthracnose of sugar maple caused by a fungus resembling the *Discula* state of *Apiognomonia errabunda*. Brown acervuli are prominent on the upper surface of a lesion (D) (NY, Aug).

E. Anthracnose caused by *Kabatiella apocrypta* on Japanese maple (NY, Jun).

F, G. Didymosporina leaf spot of Norway maple. A magnified view of the lower surface of a lesion (G) discloses minute acervuli as dark brown stippling (NY, Sep).

Plate 47

Ash Anthracnose (Plate 48)

Ash anthracnose is common wherever ash trees grow in the relatively cool regions of eastern North America. It occurs also in parts of Arizona, California, British Columbia, and Alberta. Hosts include black, green, red, velvet, and white ash. Of these, the Modesto variety of velvet ash seems most susceptible; anthracnose is its most important disease. Green ash is relatively resistant.

The disease is characterized by blight of very young leaves and shoots or by irregular necrotic blotches on expanded leaflets. In years of prolonged wet spring weather with moderate temperatures, the leaf and shoot blight phases of anthracnose may kill almost the entire first flush of shoots on highly susceptible trees (A). Severe defoliation in several successive years may lead to dieback as has been noted in California. Usually, however, the damage is minor, and defoliation is restricted to low branches.

The pathogen, long known as *Gloeosporium aridum,* is one of several morphologically indistinguishable conidial fungi now grouped artificially within *Discula* sp., the conidial state of *Apiognomonia errabunda* (Diaporthales, Gnomoniaceae). This arrangement indicates the similarity in appearance of the pathogens but not their probable host specificity. The conidial states of the fungi that cause sycamore anthracnose and oak anthracnose, for example, are in this group, but their host specialization has been established by experiments, and they are regarded as distinct species. The life cycle and host specificity of the ash pathogen have not been studied enough to provide a basis for designation of a full species name for the conidial state or to indicate whether the conidial state is really related to *A. errabunda* or to any ascomycete.

Symptoms of ash anthracnose arise in spring on succulent, expanding shoots and leaves as water-soaked spots that may enlarge and coalesce rapidly. Young lesions are greenish brown to dark brown. Lesions that girdle leaf rachises or the bases of young shoots cause the distal parts to droop and shrivel. In timing and appearance this damage is similar to frost injury. Leaves that are approaching full size and twigs that are becoming woody are somewhat resistant, and lesions on these cease to enlarge. The result is discrete brown or, more commonly, tan blotches in distorted leaflets. Lesions on nongirdled twigs remain as tiny elliptical cankers. Defoliated branches produce new shoots by midsummer.

Within a few days after infection, if moist weather continues, disc-shaped acervuli form on killed parts and produce tiny masses of conidia. At first the acervuli are the same color as the plant tissue and because of small size (100–250 μm in diameter) can be discerned only by careful examination with a magnifying lens (D). They become somewhat darker and more prominent with age. Conidia cause repeating cycles of disease as wet weather permits.

In California, conidia of the ash pathogen germinated well at 15–27°C but poorly at 9°C. Germination began 8–12 hours after wetting

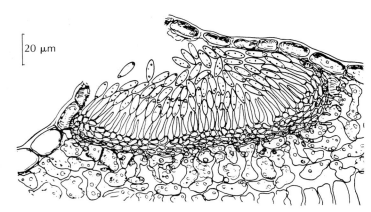

20 μm

Figure 11. An acervulus of the *Discula* state of *Apiognomonia errabunda,* microscopic representation.

and was most rapid at 24–27°C. Thus the relatively cool conditions that seem to favor anthracnose are not those that promote the most rapid development of the causal fungus. Cool temperatures may promote disease primarily by retarding host development and prolonging susceptibility. Dry weather halts epidemics.

The anthracnose pathogen overwinters on blighted twigs and petioles and in spring liberates conidia that cause primary infections on succulent parts. Whether the conidia come from newly formed acervuli or from old ones that overwintered is not known.

As mentioned above, *A. errabunda* is a cluster of host-specialized entities, and it is not established that any of them produces ascospores capable of infecting ash or gives rise to conidia that can do so. The known extent of the relationship between the ash pathogen and *A. errabunda* is simply that their conidial states look alike. The *Discula* conidial state of *A. errabunda* (Figure 11) is characterized on its various hosts by subcuticular or subepidermal disc-shaped acervuli and ellipsoidal, one-celled colorless conidia that measure 6–16 × 2.5–5.5 μm. Acervuli on leaves may consist mainly of a spore mass and a colorless, spore-producing layer of basal cells and conidiophores. Overwintered acervuli or those on twigs may develop a cupped and pigmented layer of basal cells that in extreme cases is nearly closed and resembles the wall of a pycnidium. Such variation in form may occur within a single host-specific anthracnose fungus and is greater than the variation between fungi that infect different hosts. Thus several fungi that infect different host plants came to be grouped under one name.

References: 59, 107, 372, 389, 425, 1343, 1420

A, E–G. Anthracnose on Modesto ash. A. A tree defoliated by early-season attack. E. Pinpoint lesions and spreading necrosis. Lesions are arrested when hot dry weather follows infection, but wet weather favors spreading necrosis and blight. F, G. Mature lesions (CA, Apr).

B–D. Anthracnose on white ash. B, C. Young and mature lesions (NY, May–Jun). D. Inconspicuous acervuli (arrows) and masses of conidia on the lower surface of a young lesion (NY, May).

Plate 48

Black Spot of Elm and Gnomonia Leaf Spot of Hickory (Plate 49)

Black spot of elm. In North America before the advent of Dutch elm disease, the best known disease of elms was black spot, caused by the native fungus *Stegophora ulmea* (syn. *Gnomonia ulmea*) (Diaporthales, Gnomoniaceae). The disease was rarely fatal, even to very small elms defoliated repeatedly, but it could be observed in most years throughout the natural range of elms. It is widespread from the Great Plains to the Atlantic Ocean. The usual symptoms are black spots and the shedding of diseased leaves somewhat before the normal time, but severe infection in wet seasons may cause blight of young leaves and succulent shoots or complete defoliation by early August. Susceptible species include American, cedar, Chinese, Dutch, English, Japanese, rock, Scotch, September, Siberian, slippery, smooth-leaf, European white, and winged elms and *Ulmus laciniata*. Japanese zelkova is also susceptible. In California, black spot or a similar anthracnose (attributed to "*Gloeosporium* sp.") results in leaf spots and leaf and shoot blight of Chinese elm (A) and Siberian elm.

Symptoms arise as yellow spots about 1 mm across on upper surfaces of leaves beginning while they are unfolding from buds and continuing throughout much of the season. After several days, an acervulus forms in the center of the lesion and a black stroma forms beneath the acervulus. The stroma is visible as a black dot about 0.5 mm in diameter. Usually several acervuli and stromata develop very close together and may coalesce to irregular black masses up to 5 mm wide, surrounded by a narrow band of whitish dead tissue. Sometimes the stroma covers the entire lesion, leaving no necrotic border. The cuticle over each acervulus splits irregularly, exposing a white mass of conidia. Conidia mature 10–20 days after infection, are dispersed by splashing rain, and cause secondary cycles of disease. Symptoms progress from low branches to higher ones.

Green fruit of some elm species are susceptible and develop a crumpled appearance as the result of infection. During the early part of the growing season, lesions may form on petioles and succulent stems. These infections when numerous cause blight of leaves and shoot tips. The disease is arrested during dry summers, and even severely blighted parts seem to recover as the result of growth from buds that would normally remain dormant until the next spring.

Two conidial states develop successively in acervuli in black spot lesions. The first produces colorless, unicellular spores 8–10 × 3–3.5 μm during spring and early summer. This conidial state, responsible for secondary cycles and long known as *Gloeosporium ulmicolum*, lacks a proper contemporary name. For 50 years it was thought to be the cause of a separate disease because it was first noticed on elongate lesions associated with the veins of elm leaves and had no evident relation to typical black stromata or perithecia. Later research revealed that its spores were consistently present in black spot acervuli during the first half of the growing season. In midsummer a transition to a microconidial form, *Cylindrosporella ulmea* (syn. *Asteroma ulmeum*, *Gloeosporium ulmeum*), occurs in black spot lesions. Microconidia typically measure 4–6 × 1–2 μm, are not infectious, and may serve as spermatia in the sexual cycle of the parasite. Perithecia begin to develop in lesions while microconidia are present. The beaks of immature perithecia may be found protruding through the lower surfaces of black spot lesions during late summer and autumn.

Perithecia continue development in fallen leaves during autumn and late winter and at maturity are flask shaped with bodies 200–385 μm wide and 150–230 μm deep, and beaks 80–100 μm long. In northern areas the two-celled, colorless ascospores are expelled to the air in spring under conditions of alternating wetness and drying after several days with temperatures of at least 7°C. Expulsion of ascospores is usually synchronized with foliar development of elms. The lowermost leaves and shoots are most likely to be infected by airborne ascospores.

S. ulmea may also overwinter in persistent leaves and in buds. Dormant buds have been shown to contain the fungus, and black spot lesions often appear on the youngest expanding leaves in spring. Foliar disease continues all year in parts of California where Chinese and Siberian elms are evergreen.

S. ulmea is adapted to activity during both cool spring weather and warm conditions of early summer. The optimum temperature for germination of its ascospores is about 8°C. In experiments, germination was reduced at temperatures above 16°C and did not occur at 24°C. Conidia, by contrast, germinated at 4–28°C with no particular optimum. Mycelium in pure cultures grew at 8–24°C.

Resistance of elms to *S. ulmea* varies greatly within and between species and is strongly heritable. One expression of resistance is nonsporulating lesions. In Wisconsin, species ranked in order of increasing susceptibility were rock elm, *U. laciniata*, and Chinese, Siberian, smooth-leaf, Japanese, American, Scotch, and European white elms.

Gnomonia leaf spots of hickory and pecan. Hickory trees throughout the eastern United States and southern Ontario are host to *Gnomonia caryae* (Diaporthales, Gnomoniaceae), which causes brown spots with yellowish indefinite edges. Susceptible species include bitternut, mockernut, pignut, and shagbark hickories. Individual spots are nearly circular and vary from 5 to 20 mm in diameter. Where confluent, they cause browning and curling of large areas of leaflets, premature yellowing, and leaf cast. A conidial state, *Cylindrosporella caryae* (syn. *Asteroma caryae*, *Gloeosporium caryae*), becomes prominent on the lower surface of each lesion. Acervuli 70–150 μm in diameter, at first pinkish, later brown, develop in abundance (H). The conidia are unicellular, colorless, and 7–10 × 1–2 μm. No test of their pathogenicity has been reported.

Perithecia of *Gnomonia caryae* form in fallen leaves during winter. Ascospores presumably function as primary inoculum, but this disease cycle has not been studied. Symptoms, however, are invariably delayed until late summer and early autumn. One possible explanation is latent infection. Young leaves perhaps become infected in spring but resist colonization and fail to show symptoms until late summer. This is a matter for research. For practical purposes, the disease is inconsequential. It disfigures hickory foliage briefly and causes leaves to fall only a short time before they would normally do so.

Pecan in the Gulf states is subject to two or three Gnomonia leaf spots, depending upon the authority one follows. *G. nerviseda* causes elongate lesions along the veins (vein spot), and *G. pecanae* causes brown lesions bounded by major lateral veins in leaflets (liver spot). Their conidial states are similar to that of *G. caryae* and appear late in the season. In Florida, *G. dispora*, named for the unusual habit of producing just two ascospores per ascus, causes necrotic spots on pecan leaves. Its perithecia mature in leaves on the tree during late summer. This fungus has no known conidial state.

References for black spot of elm: 59, 107, 1263, 1264, 1301, 1343, 1539, 1909, 1941

References for leaf spots of hickory and pecan: 59, 107, 379, 380, 389, 425, 473, 1343, 1941, 2223

A. Leaf and shoot blight of Chinese elm caused by "*Gloeosporium* sp." (CA, Jun).

B, C, F. Symptoms and signs of black spot, caused by *Stegophora ulmea*, on European white elm. B. A leaf is blighted where lesions are most numerous. C. Close view of lesions, each with a prominent black subepidermal stroma. F. Stromata with whitish borders remain prominent on killed parts of leaves (NY, Aug).

D, E. Black spot on American elm. D. A senescent leaf with unusually large black stromata. E. Close view of stromata on the leaf shown in (D) (NY, Sep).

G, H. Leaf spot of shagbark hickory caused by *Gnomonia caryae*. H. Magnified view of brown acervuli of the conidial state of the pathogen on the lower surface of a lesion (NY, Sep).

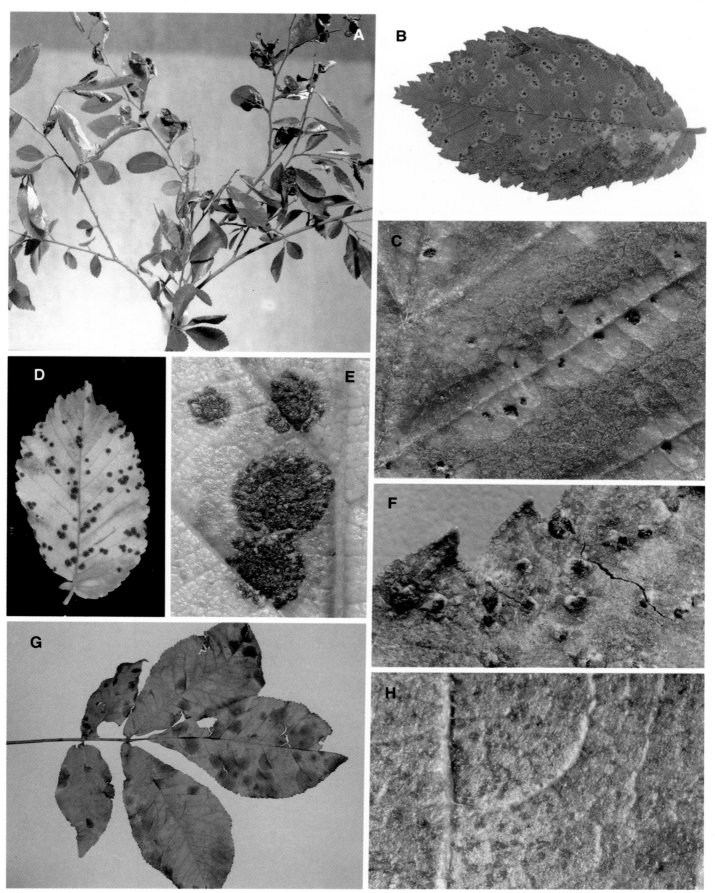

Plate 49

Oak Anthracnose (Plate 50)

Oak anthracnose occurs from the eastern provinces of Canada to the Gulf of Mexico, in Pacific coast states, and also in Europe. In a given locality the disease is innocuous in most years, but during outbreaks it can kill virtually all the foliage and many twigs on highly susceptible trees. Severe leaf blight on California live oak first drew attention to the disease in Pacific coast states. In the Northeast, anthracnose outbreaks on trees already infested by the golden oak scale (see Johnson and Lyon, 2nd ed., Plate 168) seem to trigger more dieback than would be caused by either agent alone.

The disease is caused, at least in part, by *Apiognomonia quercina* (Diaporthales, Gnomoniaceae) in its conidial state, *Discula quercina*. We qualify this statement because of the possibility that additional fungi related to *D. quercina* also cause anthracnose on oaks.

Host plants in North America include at least the following oaks: black, California black, bur, cherrybark, chestnut, English, laurel, live, California live, Oregon, pin, northern pin, post, northern red, southern red, scarlet, Shumard, water, white, swamp white, and willow.

Symptoms vary with the host, weather, and stage of plant development at the time of infection. On white oak, which is perhaps most susceptible, symptoms tend toward one of three patterns: rapidly developing blight of leaves and shoots, characterized by browning and shriveling of young leaves during the period of leaf expansion (A); large, irregular dead areas on distorted leaves that otherwise remain green (B); and small necrotic spots on leaves that have reached mature size. When outbreaks of anthracnose are promoted by rainy weather in spring, leaf and shoot blight become prominent on low branches, then spread upward. Enlarging lesions on leaves tend to follow the veins or midrib and to be bounded by them, often killing all the leaf tissue on one side of a midrib or major vein and thus causing distortion (B). After drying, lesions are papery in texture and turn tan to nearly white before weathering to grayish white (B, C). At the edges of lesions there is an abrupt transition from brown to normal green tissue. Leaves approaching full size become quite resistant, which may account for the association of the anthracnose fungus with only small necrotic spots on mature leaves.

The usual foliar symptoms on black oak are brown spots up to 2 cm in diameter. On this and most other hosts, large irregular lesions and blight are infrequent. On live oak, the tips of the narrow leaves are frequently killed.

Acervuli of the pathogen become visible, even to the unaided eye, as raised brown flecks on lower surfaces of foliar lesions (F) or as pustules on dead twigs. The acervuli on leaves are circular to elliptical in outline, often about 250 μm in diameter, and more common on or adjacent to major veins than in interveinal areas. On twigs in spring, they develop as orange-brown blisters (E) that later darken and may resemble pycnidia. The colorless, unicellular conidia, similar to those illustrated in Figure 11 (with Plate 48), are usually in the size range of 9–14 × 3.5–5 μm. The fungus also produces microconidia that average about half this size. Conidia from foliar lesions cause secondary infections.

Twig infections and dieback are recorded for black, red, scarlet, and white oaks and undoubtedly occur on other species. The twigs die before buds open in spring; thus this aspect of the disease is inconspicuous. It would be unimportant except for the fact that acervuli on dead twigs produce conidia that cause foliar infection. Presumably twigs become infected as the result of hyphal growth down petioles as in

sycamore anthracnose, but this aspect has not been carefully studied. *D. quercina* causes damage to acorns in the Soviet Union, but this has not been noticed in North America.

Anthracnose outbreaks usually subside before midsummer, but succulent shoots such as those developing on trees defoliated by insects may be affected any time during the season. The main environmental requirement is wet weather.

The foregoing description characterizes oak anthracnose as it is usually encountered, but a different syndrome has been reported from Mississippi. There a fungus similar to the *Discula* state of *A. quercina* has caused epidemics of tiny necrotic spots similar to those in (G) on leaves of laurel, pin, Shumard, water, and willow oaks. When very numerous the spots cause leaf tips to wither and become ragged. Spots vary from pinpoint size to 2–3 mm in diameter. They are dark brown to black on willow and laurel oaks and brown with yellow halos on cherrybark oak. The diseased tissues fall out of pin oak and Shumard oak leaves, leaving more or less ragged holes surrounded by pale green or yellowish halos. Cherrybark, Durand, live, post, and southern red oaks are resistant. The pathogen in this disease has been neither associated with a perithecial state nor compared critically with isolates of *A. quercina*.

Where perithecia of *Apiognomonia* do occur, they form in fallen leaves and mature in spring. They are black, 130–440 μm in diameter, and distinctive because their beaks are at least as long as the perithecium is wide. The ascospores, as in all representatives of *Apiognomonia*, are colorless and unequally two-celled. They are expelled to the air and are presumed to initiate primary infection. The relative importance of ascospores versus conidia from killed twigs as primary inoculum has not been studied.

A. quercina belongs to a group of closely related fungi that can scarcely be distinguished from one another except on the basis of host specificity. Synonyms for the perithecial state include *A. errabunda, A. veneta, Gnomonia errabunda, G. quercina,* and *G. veneta*. Synonyms for the conidial state, *Discula quercina*, include *D. umbrinella, Gleosporium nervisequum, G. quercinum, G. quercuum,* and many others. Host specificity is indicated by the fact that, when an array of isolates of *A. veneta* from sycamore and *A. quercina* from oak were used in cross-inoculations, nearly every isolate caused disease only in its original host. In contrast, certain isolates of *Discula* from American elm, sycamore, and black walnut have been reported capable of causing disease in oak, and isolates from oaks have caused infection in sycamore and black walnut. The relationship of these isolates to host-specific *A. quercina* is unknown.

Frequent spring rainfall and moderate temperatures are conducive to outbreaks of oak anthracnose. Infection occurs over a wide temperature range if plant surfaces remain wet for several hours. Symptoms develop rapidly at temperatures of 20–28°C, but lower temperatures (16–20°C) promote the greatest eventual severity of symptoms, presumably by prolonging the period of high susceptibility of the host plant.

Oak species and individual oak trees vary in apparent susceptibility to the oak anthracnose pathogen, but no systematic observations or selections for resistance have been reported.

References: 24, 59, 107, 389, 425, 535, 772, 1343, 1391, 1458, 1726, 1759, 1941

A–D, F. Anthracnose on white oak (NY, Jun). A. Adjacent trees, one with severe leaf blight, the other unaffected. B. Lesions ranging from large irregular dead areas on expanding leaves to spots a few millimeters across on mature leaves. C. Leaves partially killed while immature are distorted, their dead parts shriveled. D. Dieback and tufted sprout growth in the crown of a tree perennially damaged by anthracnose and the golden oak scale. F. Magnified view of brown acervuli of *Discula quercina* on and beside leaf veins on the lower surface of an anthracnose lesion.

E. Close view of young acervuli of *D. quercina* on a twig of red oak (NY, May).

G. Unidentified leaf spot on blackjack oak. Small spots similar to these, with or without yellow halos, depending upon oak species, are caused by *Discula* sp. on several oak species in the South (GA, Aug).

Plate 50

Sycamore Anthracnose (Plate 51)

Anthracnose of plane trees and sycamores is the best known and most widely distributed disease of these trees. It is characterized by blight of leaves and shoots, cankers and dieback of twigs, and deformation of branches. It occurs in North and South America and in Eurasia and Australasia, and it is an important factor limiting the selection of sycamores and plane trees for landscape plantings. Large trees that repeatedly sustain severe damage by anthracnose are weakened, as evidenced by loss of vigor, dieback of large branches, and apparent increased susceptibility to borers. The pathogen, *Apiognomonia veneta* (Diaporthales, Gnomoniaceae), closely resembles anthracnose pathogens of ash, maple, oak, and other trees (see discussion with Plates 47–48) but is distinct where host specialization is concerned. Host plants are London plane, Oriental plane, and American, Arizona, and California sycamores.

Sycamore anthracnose occurs in three phases corresponding to the plant parts attacked. In order of appearance each season, the phases are canker formation (including bud and twig mortality), shoot blight, and leaf blight. Year-to-year weather patterns influence the severity of each phase relative to the others.

Cankers form and buds and twigs die during dormancy because the pathogen overwinters at nodes in twigs and is active whenever temperature permits its growth during autumn, winter, and early spring. It enters twigs via petioles during the growing season, remains more or less quiescent until host dormancy, and then may begin to colonize and kill bark and cambium. Sycamores effectively resist twig colonization during the growing season. Buds, because of their location immediately above leaf scars, are often invaded and killed (C). Many twig lesions are restricted to the immediate vicinity of buds, but others expand and encircle twigs, causing death of parts beyond the lesions. This type of damage tends to intensify over a period of years in a given tree as the population of the pathogen increases. Twig death is most conspicuous on large old sycamores. In years of severe bud and twig infection, fewer than 5% of buds on these trees may survive to produce new shoots. In early summer the trees stand leafless except for a few tufts of foliage at branch tips (A).

By midsummer, regrowth is under way. New shoots arise from buds that would have remained dormant. Leaves on these shoots may become infected but often escape significant disease because summer heat and dryness suppress the pathogen. The new twigs will be subject to canker formation and dieback during the dormant season following the next major attack of leaf blight.

Repeated twig dieback alters the form of sycamores in two characteristic ways. First, when the terminal twig on a branch is killed, a lateral takes over as the new leader. The branch axis thus changes direction repeatedly, and crooked branches result (A, B). The second alteration is the development of a cluster of twigs around a common point on a branch because of the repeated killing of terminals.

Most anthracnose cankers enlarge during only 1 dormant season, but cankers that arise near twig bases may extend into and girdle branches 2 or more years old. If a diseased twig or branch is not girdled, the lesions are eventually enclosed beneath callus rolls. Some cankers expand during 2 or more seasons. In these cases the host produces a callus roll at the edge of the lesion during the growing season. Then during the ensuing dormant season the pathogen kills the callus plus some surrounding bark and cambium. The anthracnose canker in (E), for example, was active for 2 years.

In early spring *A. veneta* produces black pycnidia in the bark of new cankers and newly killed twigs. Conidia from these fruit bodies are dispersed by rain splash beginning about the time sycamore buds open. The conidia infect new shoots and leaves.

Occasionally anthracnose cankers form on the mainstems of young, highly susceptible trees. These cankers tend to involve long strips of bark and seem to develop in plants stressed by drought or untimely pruning.

Shoot blight, the second phase of anthracnose, involves the rapid death of expanding shoots and leaves. It occurs when the pathogen in twig cankers either kills twigs on which shoot growth has begun or enters succulent new shoots soon after they emerge from buds. The stems of young shoots may also become infected by growth of hyphae from petioles, but this aspect is less important. Shoot blight tends to develop suddenly during or immediately after a period of cold spring weather. It has occasionally been confused with injury by late spring frost.

Leaf blight results from direct infection of leaves. Most severe on low branches, leaf blight intensifies and spreads upward during wet seasons, causing premature leaf drop. Leaves are most susceptible during the first few weeks of growth. Conidia on wet leaves at temperatures near 20°C germinate within as short a time as 6 hours and produce appressoria, which generate infection pegs that penetrate the cuticle. The fungus may grow beneath the cuticle for a time or remain quiescent before invading tissues beneath. Conditions conducive to lesion formation as opposed to latent infection have not been defined.

Foliar lesions characteristically extend along the veins (D) and involve interveinal tissue secondarily. Large irregular marginal lesions develop occasionally (F). Infection may occur on any part of the leaf blade or on the petiole. Acervuli form on the lower surfaces of leaf lesions, especially along veins, and conidia from these initiate secondary cycles as wet conditions permit. The acervuli are intraepidermal, brownish, and 100–300 μm in diameter, and they produce colorless, unicellular conidia that measure 9–15 × 4–6 μm. Hyphae of the fungus often grow down petioles into the twigs, setting the stage for the next year's damage.

Conidia are the most common and important inoculum, but *A. veneta* does produce ascospores in black perithecia in fallen leaves. The perithecia are about 150–300 μm in diameter with beaks protruding 50–100 μm from the leaves. Perithecia mature during winter and expel unequally two-celled, colorless ascospores into the air in spring.

A. veneta has received much attention from mycologists, and the result is an array of names for the various forms of the fungus. In some references the group-species *A. errabunda* includes the sycamore pathogen. We prefer to emphasize host specificity, and if this is done, *A. veneta* is correct. Synonyms are *Gnomonia veneta* and *G. platani*. The conidial state, *Discula platani*, varies greatly in the form of its fruit bodies according to the substrate and time of year. Conidia indistinguishable from those in acervuli on leaves on the tree in summer may be found in spring in cup-shaped or nearly covered, walled structures in fallen, overwintered leaves and in black pycnidia in the bark of twigs. The pycnidia on twigs are relatively large (500–900 μm), easily visible to the unaided eye, and arise beneath the cork layer of the outer bark. They are preceded by and develop beneath small masses of fungal tissue (pressure cushions) that rupture the bark surface. This provides for the escape of conidia. All of the conidial stages are now united under the name *Discula platani*, synonyms of which include *Gloeosporium platani*, *G. nervisequum*, and many others.

Temperature is the main environmental determinant of the canker

---

Anthracnose, caused by *Apiognomonia veneta*, on eastern sycamore and London plane.

A. A large sycamore with twig blight, nearly devoid of foliage at the time leaves would normally be approaching full size (NY, Jun).

B. A sycamore branch, crooked because of repeated twig blight; the branch axis changed direction each time a lateral twig became the leader after death of a terminal (NY, Jun).

C. Twig blight in close view. Bark has been shaved to reveal a girdling canker at a node. The lowermost shoot in the photo was killed by a separate infection at its node of origin (NY, Jun).

D. Typical vein-associated leaf blight lesions (NY, Jul).

E. A 2-year-old canker on a small branch. The healthy callus roll beginning to cover bare wood indicates that the pathogen is no longer active (NY, Aug).

F. Twigs killed by anthracnose, one before and the other after shoot growth; also atypical scorch-type leaf blight (NY, Aug).

G. Shoot blight on London plane (NY, Jun).

H. An anthracnose canker on the main stem of a young London plane. Bark cracks at canker edges were caused by callus formation (NY, Aug).

I. Close view of a twig killed by anthracnose. Black dots are old fruit bodies that discharged spores at the time nearby buds opened (NY, Aug).

Plate 51

phase of anthracnose, including bud and twig mortality, and of shoot blight. Mild weather during host dormancy promotes fungal activity and lesion formation in twigs and branches. Beginning at the time of bud break, temperatures above 15–16°C favor quick shoot growth and cessation of twig killing. On the other hand, temperatures averaging below 12–13°C for the first 2 weeks after buds open retard host development, prolong the susceptibility of twig tissues and are conducive to severe shoot blight. Rainfall at that time is relatively unimportant because the pathogen already resides within twigs. Wet conditions are important mainly for leaf blight. Foliar infection, given wet conditions, is highly favored at 16–20°C.

Because of the high susceptibility of native sycamores, plane trees are preferred for landscape plantings. Oriental plane, resistant to anthracnose, gives significant resistance to its progeny when crossed with susceptible species. This was the basis of the well-known resistance of the original London plane tree, which resulted from an accidental cross between eastern sycamore and Oriental plane in England before 1700. Unfortunately, such hybrids themselves hybridize with susceptible sycamores, and resistance in the seedling offspring is diluted. Seedling origin is the reason why London plane trees in North America vary considerably in resistance. Resistant clonal lines can be maintained, however, by vegetative propagation. Examples are the 'Bloodgood' clone of London plane and the new hybrid clones 'Columbia' and 'Liberty'.

References: 59, 107, 535, 629, 822, 1343, 1384, 1389, 1390, 1677–79, 1726, 1740

Anthracnoses of Filbert, Redbud, and Birch (Plate 52)

Monostichella coryli (syn. Gloeosporium coryli) (Deuteromycotina, Coelomycetes) causes brown spots on leaves of filberts across North America and in Europe. Depending upon host and weather conditions, the spots range from small with rather abrupt edges (A, B) to large with indefinite edges. The causal fungus forms subcuticular acervuli that are more numerous on the upper surface of lesions than on the lower. The acervuli are 60–150 μm in diameter, and the overlying cuticle is stained dark brown. The conidia are single-celled, colorless, and oblong and measure 12–16 × 5–7 μm. Host plants include American, beaked, and European filberts. Severe infection leads to marginal scorch, premature yellowing, and leaf cast, but significant defoliation before late summer has not been reported. The sexual state of the fungus, if any, is unknown, and the disease cycle has not been studied. The name Piggotia coryli has been suggested for this pathogen but is inappropriate because it was first proposed for a different fungus also originally called Gloeosporium coryli. In Europe another fungus in the "Gloeosporium" group causes bud rot, twig canker, and death of male catkins of European filbert.

Redbud has no important foliar diseases. The anthracnose shown here is uncommon and apparently undescribed. It is characterized by necrotic lesions that at first follow veins and then expand to become large, irregular interveinal blotches with abrupt transition between dead and green or chlorotic tissue. A species of Kabatiella (Deuteromycotina, Coelomycetes) fruits on the undersides of the lesions, especially along veins, and produces colorless, unicellular conidia that appear white in mass (F).

Birches in eastern Canada, the eastern United States, and Europe are subject in midsummer to partial defoliation caused by Discula betulina (syn. Gloeosporium betulinum, G. betulicola) (Deuteromycotina, Coelomycetes). Gray, paper, and river birches are attacked. The causal fungus was first described as causing spots 5–10 mm across, but the prevailing symptoms are large brown blotches that have indefinite margins and become surrounded by yellow tissue. Affected leaves often fall while part of the blade is still green. The disease is most prominent on low branches, and defoliation progresses up the tree. Landscape specimens as well as forest trees are affected. Brown acervuli form on the undersides of lesions, either along veins or scattered (H, I). The colorless, ovoid, unicellular conidia measure 7–17 × 2.0–4.5 μm. No sexual state is known, and the disease cycle is undescribed.

Four other anthracnose fungi cause leaf spots on birches, but the fungi are microscopically distinct and the symptoms, mainly small spots, differ from those caused by D. betulina. An anthracnose caused by Marssonina betulae is shown on Plate 55.

References: 59, 389, 390, 672, 691, 1345, 1437, 1521, 1941, 2066

A–D. Leaf spots of filberts caused by Monostichella coryli. A. Lesions on European filbert (OR, Aug). B. Numerous lesions associated with minor veins on American filbert, causing yellowing and marginal scorch (WA, Jun). C, D. Successively magnified views of the undersurface of a lesion, showing tiny dark brown acervuli (OR, Aug).

E, F. An undescribed anthracnose of redbud (NY, Jun). E. Lesions that form initially along veins expand to involve interveinal and marginal areas. F. Close view of the undersurface of a lesion. White masses of conidia of the putative causal fungus, Kabatiella sp., are associated with veins.

G–I. Anthracnose of paper birch caused by Discula betulina. G. Large brown blotches cause yellowing and leaf cast (NY, Jul). H, I. Brown acervuli on undersurfaces occur either along veins (H) or scattered over the surface (I) (Que & NY, Jul).

Plate 52

Anthracnose of Hornbeam and Hop Hornbeam (Plate 53)

Gnomoniella carpinea (syn. *Sphaerognomonia carpinea*) (Diaporthales, Gnomoniaceae) in its conidial state, *Monostichella robergei*, causes anthracnose of American hornbeam and hop hornbeam in landscapes and forests in a large region of eastern North America. Ontario, Georgia, Oklahoma, and Wisconsin are the reported limits of its distribution. The same disease is widespread on European hornbeam in Europe and affects Japanese hornbeam in Japan, but infection of these species in North America has not been reported.

The usual symptoms on leaves are irregular necrotic spots of various sizes from pinpoints to 5–6 mm in diameter, increasing in number and becoming confluent during the season, causing marginal and apical browning, curling, and leaf cast. Lesions are reddish brown on hop hornbeam and brown to gray-brown on American hornbeam. Subcuticular acervuli develop on both surfaces of lesions, mainly the upper surface. These acervuli are visible with a magnifying lens as dark gray to nearly black dots. They measure 60–120 μm in diameter and produce colorless, unicellular conidia that measure 12–16 × 5–7 μm (Figure 12). Diseased leaves fall prematurely, and low branches of hop hornbeam may be defoliated by August in some seasons.

A twig canker phase of the disease occurs on hop hornbeam and becomes prominent if twig death is delayed until after shoots develop. Reddish brown cankers, dramatic in contrast to greenish healthy bark, are centered on leaf scars. This indicates that the pathogen either infected fresh abscission wounds or entered the twigs through petioles. Axillary buds die (B, C), and the cankers often encircle 1-year-old twigs, causing dieback. These events usually occur during the dormant season and thus are inconspicuous. Occasionally the canker phase becomes severe, killing both 1- and 2-year-old twigs not only during the dormant season but also after leaves expand. Pathogenicity tests have confirmed that *G. carpinea* can cause cankers during the period of host growth in spring. Canker expansion during the growing season results in prominent small "flags" consisting of wilted and dead shoots with curled, somewhat bleached leaves. A twig canker epidemic in western New York in the mid-1970s was associated with summer drought, which was presumed to have impaired the ability of hop hornbeam to restrict the pathogen in twigs.

Since the twigs of hop hornbeam are delicate, those killed in the dormant season tend to dry, shrivel, and fail to support fruiting of the pathogen. Acervuli do form, however, in cankers and on some of the largest twigs killed by the fungus. These fruit bodies arise in spring and summer in cortical tissue, break through the periderm (C), and become

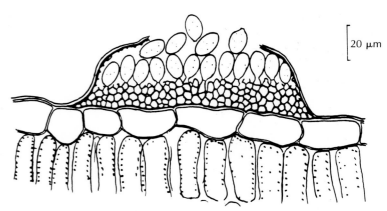

Figure 12. An acervulus of *Monostichella robergei* on a leaf, microscopic representation.

much larger and more conspicuous than those in leaves. They are black, persistent (D, E), and contain conidia for up to 1 year. Cankers do not enlarge after the 1st season, but their locations may become conspicuous as twigs swell when callus develops around cankers.

The ascigerous state of *G. carpinea* develops in fallen leaves and matures in spring. The black perithecia appear as somewhat flattened spheres 110–250 × 80–170 μm in size. The ascospores are colorless and unicellular. Perithecia of *G. carpinea* have been collected from fallen leaves of numerous trees—alder, ash, birch, chestnut, filbert, maple, and oak in addition to hornbeam and hop hornbeam. Whether the fungus is pathogenic to plants other than those in the last two groups is unknown.

From observations of the seasonal development of symptoms and signs, we infer that the disease cycle of anthracnose on hop hornbeam is like that of sycamore anthracnose (Plate 51). The pathogen is presumed to overwinter on twigs of American hornbeam.

G. carpinea grows readily in pure culture, producing both conidia and, if pieces of host tissue are present, fruit bodies. It also causes amber-brown crystals to form in the medium, a characteristic useful for diagnosis.

References: 59, 107, 425, 1343, 1546, 1712, 1806

A–E, G, H. Anthracnose caused by *Gnomoniella carpinea* on hop hornbeam. A. Flags consisting of dead shoots with bleached leaves identify twigs girdled during the growing season. B. Twig dieback and cankers centered on leaf scars. C. Enlarged view of a canker with young acervuli of the conidial state of the pathogen on a 1-year-old twig. D, E. One-year-old cankers in 2-year-old twigs. Swelling is caused by callus formation at canker margins. Black acervuli are still prominent in the cankers (A–E: NY, Jul). G. Necrotic spots and a blighted tip on a diseased leaf. H. Magnified view of acervuli on the undersurface of a foliar lesion, most numerous along the vein (G, H: NY, Aug).

F, I. Anthracnose caused by *G. carpinea* on American hornbeam. F. Necrotic spots and marginal scorch on leaves. I. Enlarged view of the upper surface of a foliar lesion. Acervuli appear as dark dots on the grayish brown dead tissue, most numerous along a lateral vein (NY, Aug).

Plate 53

Walnut Anthracnose and Downy Leaf Spot of Walnut and Hickory (Plate 54)

Walnut anthracnose. This disease occurs worldwide where walnuts and butternuts grow in humid climates. It is common throughout the eastern United States and adjacent Canada and also in the Pacific Northwest. Persian walnut in California, however, escapes significant infection because of the hot, dry summer there.

The main effects of the disease are defoliation and diminished nut quality. Anthracnose builds up during wet seasons, causing highly susceptible trees to be nearly leafless by late July or early August. Before that time, however, seasonal growth of the trees is nearly complete; thus anthracnose has little effect on growth unless trees are defoliated repeatedly. Nuts from severely defoliated trees, however, are apt to have dark, somewhat shriveled kernels.

Anthracnose is caused by *Gnomonia leptostyla* (Diaporthales, Gnomoniaceae) in its conidial state, *Marssoniella juglandis* (syn. *Marssonina juglandis*). Hosts found naturally infected in North America include butternut; black, California black, Hinds, and Persian walnuts; and assorted hybrid walnuts. Black walnut, although not the most highly susceptible host, is the most severely affected under natural conditions. Arizona and Hinds walnuts are more susceptible than black walnut, while butternut; heart nut; Japanese, little, and Persian walnuts; and *Juglans sinensis* are less susceptible. Clones of Persian walnut range from highly susceptible to quite resistant.

Symptoms on leaves are more or less circular brown lesions that appear first on the undersurfaces about the time leaves reach full size. Lesions soon become equally apparent from both sides. They range in size from pinpoints to about 5 mm in diameter and are typically surrounded by a yellow zone. When they are numerous, they cause yellowing, marginal browning, curling, and casting of leaflets. Defoliation, although a valid measure of anthracnose severity, is not closely correlated with the number of lesions per leaflet.

Brown acervuli of the *Marssoniella* state form abundantly on the surfaces of lesions, particularly along the veins (D). Viewed with magnification when moist, the acervuli are tiny blisters 100–200 μm in diameter. They form beneath the epidermis and produce crescent-shaped, colorless, two-celled conidia 14–30 × 3–4 μm in size. Conidia are dispersed mainly by splashing and dripping water.

Lesions also arise on petioles, rachises, fruit husks, and occasionally twigs of the current season. The lesions on husks are tiny dark depressions that if numerous cause premature fruit drop. Lesions on twigs are circular to oval, gray-brown spots up to 2 mm across, with dark margins.

Beginning in late summer as night temperatures drop below about 10°C, microconidia may be found in some acervuli. The microconidia are rod shaped, about 10 × 1 μm in size, and noninfectious. They function as gametes in the sexual cycle of the fungus.

Black perithecia form in fallen leaflets, rachises, and fruit husks during host dormancy. These fruit bodies mature in spring and expel ascospores in response to rain during a period of several weeks. Airborne ascospores cause most primary infections. In some regions, however, the pathogen may also survive winter in twig lesions, from which conidia are dispersed to initiate primary infections in spring.

Wet plant surfaces are required for spore germination and infection. After germination, the parasite produces appressoria from which penetration pegs breach intact leaf or fruit surfaces. Increases in lesion numbers can be perceived 15–17 days after periods when leaves are wet for 12 hours or longer, and acervuli are visible beginning 3 days after lesions appear. Thereafter successive cycles of infection are caused by conidia.

Environmental conditions that favor anthracnose are relatively well understood. Perithecial formation is most favored at temperatures of 7–10°C and ascospore production at about 10°C. Expulsion of ascospores begins about 1 hour after the onset of rain and reaches a peak after 5–6 hours. Conidia are dispersed whenever precipitation is sufficient for water to run or splash on plant surfaces. Temperatures of 18–26°C are most favorable for production of acervuli and conidia, and both ascospores and conidia germinate most rapidly at 24–26°C.

Susceptibility of black walnut to anthracnose is influenced by tree nutrition. Application of nitrogenous fertilizers to the root zones of trees has resulted in more growth and less severe disease than occurred in untreated trees.

Downy leaf spot. Also called white mold or white leaf spot, this disease occurs on various species of walnut and hickory. Host plants include butternut; bitternut, mockernut, pignut, sand, shagbark, and water hickories; pecan; and Arizona, black, California black, Hinds, little, and Persian walnuts. The pathogen, *Microstroma juglandis*, is thought by some mycologists to be an asexual fungus and by others to be a basidiomycete. It produces clusters of fertile cells that resemble basidia in producing spores on pointed projections (sterigmata) atop each fertile cell. White mold occurs in all regions of the world where hickories and walnuts grow. Unlike anthracnose, it is economically insignificant.

Microstroma species sporulate on living tissues. Hyphae emerge from stomata, especially along veins, and produce white or silvery masses of fertile cells and spores either in discrete spots or in a continuous mass over the lower leaf surface. The spores of *M. juglandis* are unicellular, colorless, and 6–8 × 3–5 μm in size.

The most common symptoms are leaf spots of various sizes, pale yellow on the upper surface and downy or powdery white below. Lesions often coalesce to large angular areas and eventually become brown and dry. *M. juglandis* var. *robustum* causes blight of pecan catkins and leaflets.

On several hosts, both hickory and walnut, *M. juglandis* induces the formation of brooms that arise near branch tips and attain sizes up to about 2 meters in diameter. Apparently the fungus grows perennially within twigs and buds, for white mold appears beginning in early spring on the undersurfaces of nearly all leaves in a broom. Viewed from above, leaves in brooms are somewhat dwarfed and yellowish. They tend to shrivel and drop beginning in midsummer. *M. album* (syn. *Articularia quercina*) causes similar brooms on several species of oak in the southwestern United States.

The annual cycles of these diseases have not been worked out for North American conditions. In eastern Europe, *M. juglandis* is said to overwinter as stromata in fallen leaves, from which infectious spores are liberated to the air in spring.

References for anthracnose: 59, 107, 159, 183, 184, 366, 389, 425, 558, 1236, 1386, 2018
References for downy leaf spot and related diseases: 61, 389, 392, 425, 813, 1138, 1759, 1901, 2203, 2225

A–D. Walnut anthracnose, caused by *Gnomonia leptostyla*. A. Yellowing, browning, and curling of diseased leaflets (NY, Sep). B. Close view of lesions that, although individually small, cause marginal browning where numerous (NY, Sep). C. Close view of one lesion; dark dots are acervuli; yellow halo around the lesion is typical (NY, Sep). D. Magnified view of a portion of a lesion bearing blisterlike acervuli (IL, Jul).

E, F. White mold of black walnut, caused by *Microstroma juglandis*. E. White mold, associated with veins, prominent on lower surfaces of leaves (IL, Jul). F. Chlorotic spots on upper surfaces correspond to sites of infection on lower surfaces (IL, Jul).

Plate 54

Marssonina Spots and Blights of Poplar, Birch, and Bittersweet (Plate 55)

Leaf spots and blights of aspen and poplar. Marssonina leaf spots and blights of poplars and aspens are among the important forest diseases because they occur worldwide and cause defoliation, dieback, and lost growth of trees that are used in intensive forestry programs. Three fungi are involved primarily: *Marssonina brunnea*, *M. castagnei*, and *M. populi* (Deuteromycotina, Coelomycetes). Each is the conidial state of an ascomycete: *Drepanopeziza punctiformis*, *D. populi-albae*, and *D. populorum* (Helotiales, Dermateaceae), respectively. Since the ascigerous states (brown apothecia only 150–350 μm across, on fallen, overwintered leaves) have not been reported to occur in North America, we follow the common custom of using the names of the conidial forms.

These pathogens occur across North America and cause damage when wet weather favors epidemics on groups of highly susceptible trees such as natural clones of aspen, plantations of poplars, or seedlings in nursery beds. The symptoms are brown spots or blotches on leaves and pustules on green twigs. Acervuli up to 400 μm in diameter with unequally two-celled, colorless conidia form in the epidermal cell layer of the lesions and appear initially as tiny blisters. Severe outbreaks of disease may cause foliar browning in midsummer and nearly complete defoliation by early August. Regrowth follows in late summer and early autumn, and twig dieback may follow in winter because late-season shoots lack normal cold hardiness. Seedlings in nurseries may be killed. Defoliated trees produce less wood for 1 or more years after an outbreak.

Annual cycles of the three pathogens are similar. Primary infection, occurring soon after leaves emerge in spring, is caused by water-dispersed conidia from acervuli in overwintered twig lesions or in fallen leaves. In Europe, ascospores also initiate early infections. Symptoms commonly appear after 14–16 days. Conidia from acervuli in primary lesions initiate repeating cycles of disease that continue as wet weather permits until leaves fall. Symptoms intensify and seem to ascend trees as the season advances. In late summer and early autumn the pathogens produce microconidia (less than 6 μm long) in the same lesions. These fungi survive winter as tiny stromata in fallen leaves and in twig lesions.

M. brunnea is the most important of these pathogens. Originally described from New Jersey in the late 1800s, it remained obscure until the 1960s. Typical hosts of *M. brunnea* are eastern cottonwood; balsam, black, Carolina, Fremont, and Lombardy poplars; and various Euroamerican poplar hybrids. We mention "typical" hosts because strains of this pathogen and also of *M. castagnei* and *M. populina* vary in pathogenic capability. Some infect diverse poplar species, while others are specialized on clusters of closely related species or hybrids.

M. brunnea causes dark brown, discrete spots that usually measure about 1 mm in diameter and are visible beginning soon after leaves expand. Numerous spots may cause general yellowing or browning, especially along leaf margins (C). Leaves with numerous spots drop. An acervulus in the center of each lesion appears at first faintly orange, then grayish white as conidia are exposed (B). The conidia are mostly in the size range 13–18 × 4.5–6 μm. On green twigs this fungus causes tiny brown cankers that appear as pustules. Conidia are produced in these as environmental conditions permit until early the next summer. *M. brunnea* also infests poplar seed and may infect seedlings grown from infested seed. These capabilities help explain its global distribution.

M. castagnei typically attacks white and gray poplars and sometimes bigtooth aspen. It causes dark brown to reddish brown spots, often with lighter centers, mostly about 3 mm in diameter (A). The lesions usually coalesce into irregular blotches that may cover much of the leaf surface. Acervuli develop primarily on upper surfaces and produce conidia that for the most part measure 16–21 × 6–7.5 μm.

M. populi typically attacks trembling aspen, eastern and black cottonwoods, and balsam poplar. It causes irregular bronze to chestnut brown spots 2–5 mm in diameter that usually coalesce into vein-limited blotches. These may involve nearly the entire leaf. The upper surfaces of lesions are dotted with grayish white acervuli. Conidia of this species are mostly in the size range 18–25 × 8.5–10.5 μm, much wider in the upper cell than the basal one, and usually curved. This fungus is responsible for occasional spectacular outbreaks of leaf blight and defoliation of trembling aspen in the central Rocky Mountain region.

A fourth species, *M. balsamiferae*, spots the leaves of balsam poplar in Manitoba and Ontario. Neither its importance nor its ascigerous state is known.

Aspen and poplar clones vary widely in susceptibility to a particular strain of *Marssonina*. In the Rocky Mountain region during outbreaks of leaf blight, natural aspen clones, each consisting of many stems, may differentiate into a mosaic of green (resistant), brown (highly susceptible), and intermediate foliar hues. Similarly, in any large array of clonal poplar hybrids challenged by one of the *Marssonina* species, some clones are highly resistant or immune.

Anthracnose of birch. *Marssonina betulae* causes a striking but little-noticed anthracnose of paper birch and European white birch in eastern Canada, the northeastern United States, and Europe. The lesions are large dark reddish brown spots with radiate or stellate margins. Subcuticular acervuli form on the upper surfaces. The acervuli are flat, dark brown, and difficult to discern. They produce colorless oblong spores that at maturity are two-celled and measure 17–22 × 5–10 μm. Leaves with multiple lesions turn yellow and drop prematurely. The ascigerous state of *M. betulae*, if any, is unknown, and the disease cycle has not been studied.

Anthracnose of climbing bittersweet. This anthracnose is caused by an apparently host-specific strain of *Marssonina thomasiana* that infects leaves, green stems, and fruit. On leaves it causes dark lesions 1–2 mm across that extend through the leaf and are sometimes surrounded by a yellow zone. Lesions on fruit are shallow, yellowish to light tan cankers about 1 mm across with abrupt dark edges (G). Lesions on green stems are elliptical, up to 3 mm long, and flat to slightly raised. Where they are numerous, lesions may coalesce and kill stems, fruit, or leaves. The surfaces of young lesions are covered by raised yellowish acervuli that contain masses of two-celled conidia. The spores are colorless individually but appear yellow in mass. *M. thomasiana* occurs also on burning bush in the midwestern United States. In New York State, however, burning bush adjacent to heavily infected bittersweet remained free from disease. Thus *M. thomasiana* may consist of host-specialized strains, one on burning bush and another on bittersweet. No information is available on the disease cycle or the life history of the pathogen.

References for leaf blights of aspen and poplar: 234, 683, 684, 737, 837, 968, 986, 1294, 1377, 1430, 1499, 1527, 1871–74
References for other Marssonina diseases: 217, 389, 699, 956, 1521

A. Leaf spot caused by *Marssonina castagnei* on white poplar (NY, Sep).

B, C. Leaf spots caused by *M. brunnea* on eastern cottonwood. B. Magnified view of the upper surface of a diseased leaf showing tiny dark brown spots with grayish centers where masses of conidia in acervuli lift the host cuticle. C. Typical punctate leaf spots (NY, Aug).

D, G–I. Anthracnose of climbing bittersweet caused by *M. thomasiana*. D. Lesions on stems. Blight results from coalescence of lesions. G. Lesions on fruit. H. Leaf and shoot blight. I. Close view of lesions on leaves. Each lesion is covered by a prominent acervulus with conidia that are yellowish in mass (NY, Aug).

E, F. Leaf blotch of European white birch caused by *M. betulae*. E. Close view of the upper surface of one lesion showing the typical radiate margin. F. Diseased leaves, some prematurely yellow (NY, Aug).

Plate 55

Apioplagiostoma Blight of Aspen (Plate 56)

Reddish brown to chocolate brown leaves on scattered branches of trembling aspen or aspen hybrids in August and September typify the leaf and shoot blight caused by *Apioplagiostoma populi* (Diaporthales, Gnomoniaceae). This disease, known only in North America, is common in Ontario, Quebec, and the northeastern United States. Neither the disease cycle nor the life cycle of the pathogen is fully known.

Symptoms arise in midsummer as yellowish to reddish regions near leaf margins. These lesions expand rapidly between the veins toward leaf bases but leave the veins green for a time. All leaves on a shoot tend to be involved, or all remain green. Both short shoots and terminals are affected and often die back. Affected branches become prominent as foliage turns orange-brown and then reddish brown while leaves are still moist and at least partly alive. Dead leaves turn chocolate brown and may adhere to dead twigs throughout the winter.

Whether twigs are invaded by *A. populi* or die because of infection by opportunistic fungi after leaf death is unknown. The relative uniformity of symptoms on a given shoot, however, suggests the possibility that the shoot axis is infected by *A. populi*, perhaps before or during bud opening. This is a matter for research.

Diagnosis requires recognition of the *Discula* (syn. *Gloeosporium*) conidial state of *A. populi*. Minute, exceedingly numerous brown acervuli of this state develop beneath the cuticle all over the upper surfaces of reddish brown leaves (E). These structures produce unicellular colorless conidia that average 7 × 3.3 μm. In one study, these spores did not germinate; they may function as spermatia.

Perithecia begin to develop in brown leaves in early autumn before leaf fall and mature in fallen leaves in spring. The perithecia are flasklike and nearly black when mature. They lie parallel to the leaf surface, and each has a beak that extends through the lower leaf surface at a right angle to the axis of the fruit body. The ascospores are colorless and unequally two-celled. They are presumably infectious, but we do not know this from research.

Suscepts in addition to trembling aspen include hybrids between trembling and bigtooth aspen and between bigtooth aspen and white poplar. Aspen clones vary in apparent susceptibility as indicated by severity of symptoms.

Symptoms of Apioplagiostoma blight are often confounded by other diseases that build up late in the season. Leaf spots or blotches caused by *Marssonina* (Plate 55) or *Melampsora* (Plate 123), or leaf blight or dieback caused by *Colletotrichum gloeosporioides* (Plate 57) may occur on the same branches affected by *A. populi*. Symptoms on the leaf at lower left in (D), for example, were caused mainly by Melampsora rust. Only *A. populi* causes leaves to turn reddish brown and then chocolate brown.

References: 107, 335, 440, 1343

Anthracnoses and Diebacks Caused by *Glomerella cingulata* and *Colletotrichum gloeosporioides* (Plates 56–57)

Glomerella cingulata (Phyllachorales, Melogrammataceae) is one of the most common plant pathogenic fungi in tropical and temperate regions around the world. This highly variable species consists of many strains that differ in physiological characteristics, mating reactions, or host specificity but cannot be distinguished reliably by differences in form. Hundreds of annual and perennial plant species as diverse as century plant, palm, pear, and pine are attacked. Symptoms include seedling blight, blossom blight, leaf spots, leaf and shoot blight, fruit rot, cankers, dieback of twigs and branches, and death of whole plants. *G. cingulata* is one of the few parasitic fungi to have been recorded on the ginkgo tree. A partial list of woody hosts appears below.

Some woody hosts of *Glomerella cingulata*

acacia	crabapple	oleander
apple	cranberry	sweet olive
mountain ash	croton	palms
trembling aspen	currant	sago palm
Japanese aucuba	flowering dogwood	papaya
avocado	euonymus	peach
azalea	fig	pear
heavenly bamboo	China fir	prickly pear
bauhinia	firethorn	pecan
European birch	ginkgo	persimmon
climbing bittersweet	grapefruit	Cuban pine
blueberry	grapevine	screw pine
bougainvillea	greenbrier	Chickasaw plum
bunya-bunya	guava	privet
camellia	hawthorn	quince
camphor tree	Chinese holly	rhododendron
cassava	horse-chestnut	rose
catalpa	French hydrangea	India rubber plant
century plant	English ivy	Para rubber tree
sour cherry	lemon	sassafras
Chinese chestnut	common lilac	snowberry
Chinaberry	lime	soapberry
cinnamon	linden	spicebush
citrus	honey locust	tea
clematis	loquat	tulip tree
coffee	southern magnolia	tung-oil tree
eastern cottonwood	mango	orange wattle

G. cingulata is particularly aggressive toward plant parts weakened by environmental stress, improper nutrition, or natural senescence. The susceptibility of fruit, for example, rises markedly as it begins to ripen. Many tropical fruits are subject to rot by *G. cingulata*, but in the United States the fungus is best known as the cause of bitter rot of apples, one of the major diseases of this crop.

G. cingulata enters host tissues through wounds or by penetration of intact surfaces after spore germination. This fungus has the ability to infect at an early stage of plant development and then remain quiescent near the point of penetration for several weeks or months (latent infection) until the plant organ becomes more susceptible. Then the pathogen resumes activity and rapidly colonizes host tissues, causing death and decay. Latent infection ensures success of the pathogen even though environmental conditions during periods of high susceptibility of the plant may not favor new infections.

G. cingulata is usually encountered in its conidial state, *Colletotrichum gloeosporioides*, which is characterized by subepidermal to subcuticular acervuli that develop in recently killed tissue on any aboveground organ. Acervuli vary in appearance from pinkish eruptions (B) to small dark blisters up to 0.5 mm in diameter (I, J). The pinkish masses are actually composed of conidia in a mucilaginous matrix, produced so abundantly that they hide the underlying fruiting structure. Acervuli often contain dark setae (sterile hairs) interspersed with conidiophores. The conidia of typical forms of *C. gloeosporioides* are cylindrical with rounded ends, unicellular, colorless individually, and 12–24 × 3–6 μm. They become two-celled shortly before germination. Conidia are dispersed by rain splash and by insects that become contaminated during visits to necrotic tissues.

A–E. Apioplagiostoma blight of aspens. A, C. Reddish brown blighted leaves and dead, leafless twig tips of trembling aspen in contrast with normal foliage (NY, Sep). B, D. Leaf blight and twig dieback of hybrid aspen (bigtooth aspen × white poplar). E. Brown acervuli of the *Discula* conidial state of *Apioplagiostoma populi* on the upper surface of a lesion in a hybrid aspen leaf (NY, Aug).

F–H. Canker and dieback of camellias, caused by *Glomerella cingulata*. Description appears with Plate 57. F. A canker that originated at the base of a killed twig on sasanqua camellia. G. Dried leaves cling to a killed twig of sasanqua camellia. H. Glomerella dieback of common camellia (FL, Apr).

Plate 56

Many isolates of this pathogen readily form perithecia in culture, either autonomously or when crossed with a different sexually compatible strain. For some conidial strains, however, no ascigerous state has been found. Perithecia-forming strains produce these fruit bodies following the acervuli. Perithecia develop singly or in clusters and, on woody plants, break through the bark surface of cankers and dead twigs in spring. They are globose to somewhat pear shaped and 85–300 μm in diameter. The ascospores are colorless and unicellular, average around 12 × 4 μm in diameter, are released in response to rainfall, and are dispersed by air. They occasionally become one-septate and faintly brown before germination.

Conidia or ascospores can act as inoculum for primary infection. Both spore types germinate within 5–12 hours under warm (24–29°C) wet conditions and form dark appressoria from which plant surfaces are penetrated directly. Penetration may occur soon after germination or after a delay of weeks, with the appressoria acting as resting structures in the interim. If penetration is immediate, lesions may enlarge quickly, or the fungus may establish itself in microlesions and remain quiescent for a period before resuming host colonization. Colonization is most rapid at temperatures of 24–29°C.

G. cingulata commonly overwinters as mycelium and developing fruit bodies in killed bark and twigs. In warm regions on evergreen hosts such as citrus, however, conidia may be present all year on foliage, and their numbers are augmented by new arrivals from acervuli during rains. In this sense, *G. cingulata* is part of the normal surface microflora of its host plants.

As evidence of the broad host range and variability of *G. cingulata*, about 120 synonyms exist for the perithecial state and more than 600 for the conidial state. Most of these reflect an early mycological custom of assigning different names to similar microfungi found on different host plants on the supposition that the fungi were more or less host-specific. The asexual state was frequently described as a *Gloeosporium,*—for example, *G. acaciae* on acacia, *G. elasticae* on India rubber plant, and *G. fructigenum* on apple. Now all these are classified as *C. gloeosporioides* regardless of host specificity or the lack of it. Strains capable of causing bitter rot in apple fruits tend to be unspecialized, causing disease in such other hosts as birch, blueberry, grape, and privet. Reciprocal pathogenicity of isolates from apple and other hosts has often been a criterion for identifying conidial strains of *G. cingulata* found causing "new" anthracnose diseases.

The diseases shown here represent the diversity of symptoms caused by *G. cingulata*. Anthracnose on India rubber plant, characterized by unsightly brown blotches on foliage and cankers on stems, is widespread. On acacia seedlings of several species, coalescing anthracnose lesions may cause dieback or death. Beach acacia is thus affected in Florida. The fungus is transmitted in acacia seeds. Evergreen euonymus species are subject to leaf spots and stem cankers. The fungus on Japanese euonymus was long known as *C. griseum*. Twig infection by *G. cingulata* causes dieback of Japanese aucuba. Symptoms on citrus include necrotic blotches on leaves, dieback, tear-staining or spotting of fruits, and fruit rot. Anthracnose on citrus is generally considered to be a disease of weak or poorly tended trees.

Key lime, however, is subject to tip blight caused by a strain of the fungus (once called *C. limetticolum*) that attacks vigorous, well-tended plants. Similarly, a blossom blight and postbloom fruit drop of citrus in Belize is significant in well-tended groves.

Other noteworthy diseases caused by *G. gingulata* include canker and dieback of camellia (Plate 56) and privet; leaf blight, canker, and dieback of horse-chestnut; seedling and cutting blight of flowering dogwood; leaf spots of European birch and southern magnolia; leaf spot and defoliation of azalea; blossom end rot of Chinese chestnut; and anthracnose of cultivated blueberry, characterized by leaf spot, canker, dieback, and fruit rot.

Glomerella canker and dieback of camellia. Dieback and canker affect camellias in most areas where these plants grow in the United States, but the trouble is most common and severe in the Southeast because of warm, wet weather favorable for dispersal of the fungus and for infection. Susceptible species of *Camellia* include common camellia, sasanqua camellia, *C. oleifera*, *C. reticulata*, *C. saluensis*, and *C. sinensis* (tea). Cultivars within susceptible species vary in susceptibility. 'Governor Mouton', 'Professor Sargent', 'Rose Emery', and 'Woodville Red', all cultivars of common camellia, are somewhat resistant. So also are the sasanqua cultivars 'Daydream' and 'Setsugekka'.

Dieback begins in early spring. Young shoots wilt and turn dull green and then brown. At the junction of dead and living stem tissue, often at the base of the new shoot, is a band about 20–25 mm long in which the bark and wood are discolored dark brown. Death of tissues in this zone, from which the pathogen can be isolated readily, causes the shoots beyond to die from lack of water and nutrients. Twigs and small branches may be girdled similarly.

Lesions that fail to encircle twigs persist as elliptic cankers. Usually they are less than 25 mm long, at first slightly sunken, and eventually surrounded by a prominent ridge of callus (F). The fungus remains active in cankers, slowly killing sapwood until symptoms of stress (sparse foliage and dieback) develop in the parts beyond the canker.

Most infections occur through leaf scars in early spring just after old leaves drop. Freeze injuries, graft unions, and wounds made by horticultural tools are also common sites of infection. Fresh leaf scars become available as the annual production of conidia of *G. cingulata* peaks in spring. Pinkish orange masses of conidia are produced in acervuli on cankers and fallen leaves and branches. The spores are dispersed by water, and under warm (18–27°C) wet conditions germinate and allow the fungus to penetrate into leaf scars or wounds within 16 hours. Dieback may develop within 7 days on new shoots or after months to years as cankers affect branches and main stems.

Sometimes *G. cingulata* causes necrotic spots on young leaves of camellia, but this symptom is seldom prominent because affected leaves drop.

References for Glomerella dieback of camellia: 129–31, 1397
References for other Glomerella diseases: 57, 59, 60, 102, 387, 425, 474, 577, 584, 585, 672, 749, 762, 935, 1038, 1178, 1196, 1216, 1338, 1553, 1887, 1945, 1981, 1982

A, B. Anthracnose of India rubber plant. A. Lesions on leaves, each lesion surrounded by a yellowish zone. B. A canker on a stem, with pinkish masses of conidia of the *Colletotrichum* state of *Glomerella cingulata* (FL, Apr).

C–E. Anthracnose of beach acacia (orange wattle). C, D. Lesions on leaves and green twigs are small (1–2 mm) spots, tan with a black border on leaves, purplish black surrounded by a raised purplish zone on twigs. E. Minute, dark acervuli occur in foliar lesions (FL, Apr).

F. Orange-brown, roughened bark indicates superficial cankers on a twig of Japanese euonymus (TX, Apr).

G, I. Dieback of Japanese aucuba. G. Leaves, isolated from their source of water and nutrients by a lethal stem infection, droop and turn brown. I. Magnified view of acervuli breaking the surface of the bark just above the leaf scar in G (TN, Apr).

H, J. Anthracnose on a lemon leaf. Prominent acervuli are arrayed concentrically (FL, Mar).

Plate 57

Brown Rot of Stone Fruit Trees (Plate 58)

Stone fruit trees in home and commercial orchards around the world are subject to brown rot. This disease is characterized by blight of blossoms and shoots, decay of fruit, and formation of cankers on twigs. Common hosts are almond, apricot, sweet and sour cherries, nectarine, peach, plums, and prune. Ornamental and wild members of the genus *Prunus* are damaged also. Brown rot of stone fruits is usually caused by either *Monilinia fructicola* or *M. laxa* (Helotiales, Sclerotiniaceae). The former is native to North America and the latter to Eurasia. *M. fructicola* predominates in eastern North America, and *M. laxa* occurs there in scattered northern localities. *M. laxa* predominates in California, where *M. fructicola* is sporadically important. Both species are important in the northwestern states and in British Columbia. The two fungi may cause nearly identical symptoms, depending upon host and environmental conditions. Normally, however, *M. laxa* is the more important cause of cankers and dieback, while *M. fructicola* predominates in fruit rot. Both cause conspicuous blossom blight that precedes other phases of the disease. We deal first with generalized symptoms and pathogen biology and then with aspects that vary according to the fungus involved.

The first symptom annually is blossom blight—the sudden collapse and browning of blossoms. All parts of blossoms are susceptible and are penetrated directly after germination of conidia or ascospores of the pathogen. Blight appears 3–6 days after infection. Shoot and twig blight may follow during the next 3–4 weeks as the result of fungal growth from blighted blossoms into spurs and twigs. Often, however, primary infection is restricted to blossom parts.

Many blossom clusters and leafy shoots wilt and wither not because of direct infection but because of spur, twig, or branch lesions that cut off the supply of water and nutrients. On plums and sometimes on cherries, brown rot fungi may infect young leaves and spread thence into shoots, resulting in the symptom called wither-tip.

Nongirdling cankers on twigs become apparent as small darkly discolored areas from which gum often exudes. A dead twig or fruit spur typically extends from the center of the lesion. Cankers usually remain inconspicuous except for gummosis, but some enlarge during more than 1 season. This is often the result of invasion of brown rot cankers by secondary fungi, especially species of *Leucostoma* (Plate 94). Gum production at cankers varies with host and size of infected branch. Small twigs may be girdled and die back without gummosis except perhaps at the junction with the parent branch. Apricot trees produce gum most profusely.

The final phase of symptoms is fruit rot. This is usually initiated by conidia from blossoms or peduncles and is characterized by rapidly enlarging brown lesions that spread from one fruit into another if they are in contact. Ripe or ripening fruits are most susceptible. Some decaying fruits fall, but many remain attached and shrivel. Dried, shriveled fruits are called mummies.

Diagnostic signs are powdery buff to gray tufts (sporodochia) of conidiophores and conidia that break through the surfaces of blighted blossoms, peduncles, twigs, and decaying fruit. Viewed microscopically, the conidia are colorless, unicellular, ovoid to lemon shaped, and borne dry in chains. The conidial stage is classified in the form-genus *Monilia*.

Conidia from blighted blossoms may infect additional blossoms and occasionally leaves and may thus contribute to intensification of disease. More important, blighted blossoms and peduncles are sources of conidia that infect fruit. Wild hosts such as the chickasaw plum are also sources of spores that infect domesticated species. Brown rot fungi can penetrate intact leaves and fruits by way of stomata and hair sockets, but wounds are the usual route of entry.

Conidia are dispersed by splashing water, air, or insects. The most important insect vectors are those that make wounds or visit wounds in fruit. These include the plum curculio, the oriental fruit moth, and certain sap beetles (dried fruit beetles) of the family Nitidulidae.

Given free moisture and moderate temperature, most conidia germinate within 2–4 hours, whereas ascospores require at least 6 hours. The optimum temperature for spore germination, infection, and disease development is near 24°C. At temperatures above and below the optimum in the range 4–30°C, spore germination and infection are delayed but not prevented.

In most areas where stone fruit trees grow, brown rot fungi survive in blighted blossoms and peduncles, twig cankers, and mummified fruit on the tree, producing conidia during the growing season and again in late winter and early spring as temperature and moisture permit. These spores cause blossom blight and thus start the annual cycle.

Where winters are cold, brown rot fungi survive poorly on trees, but fallen mummified fruits are important sites of survival. In spring, the fungi may produce conidia on surfaces of fallen mummies. More important in the case of *M. fructicola*, fruit bodies (apothecia) of the sexual state grow from mummified fruit that has wintered in contact with moist soil. Apothecia of this species are light brown and are shaped somewhat like champagne glasses. Most are 10–15 mm across, with stems rising from buried or partly buried mummies. The interior surface becomes light buff color and at maturity consists of a layer of asci from which ascospores are ejected into the air during the period of host bloom. One mummy typically produces several apothecia. Apothecia of *M. laxa*, rarely found, are smaller than those of *M. fructicola*.

M. fructicola and *M. laxa* differ in several other features. *M. laxa* kills many twigs, and its smoke gray sporodochia may be prominent on those newly killed. *M. fructicola*, by contrast, causes much fruit rot but relatively little twig blight and produces dusty buff to fawn-colored sporodochia. In pure cultures, *M. fructicola* forms colonies with regular margins, whereas *M. laxa* produces lobed colonies.

Both fungi occasionally infect rosaceous plants other than stone fruits. *M. fructicola* causes fruit rot of apples, and *M. laxa* causes blossom wilt and rarely fruit rot of pear. Both species infect quinces.

Several other species of *Monilinia* cause diseases similar to the two we have described. Species on stone fruit trees include *M. demissa* on chokecherry, *M. padi* on sour cherry, and *M. seaveri* on black cherry and sour cherry. *M. amelanchieris* infects serviceberry, and *M. johnsonii*, hawthorn. In Europe, *M. fructigena* and *M. laxa* forma *mali* infect pome fruits (apple, medlar, pear, quince). *M. fructigena* occurs on pear in at least one North American location. The heath family is another major habitat for *Monilinia* species. *M. azaleae* causes blossom and shoot blight of azaleas, and *M. vacinii-corymbosae* causes mummy berry disease of cultivated blueberries.

References: 122, 294, 871, 1046, 1101, 1582, 1627, 2160, 2211, 2242

A. Shoot blight and branch dieback of apricot caused by *Monilina laxa* (CA, Jun).

B. Twig dieback caused by *M. laxa* on peach. The twig was girdled by an infection two nodes above the fruit (OR, Jul).

C. Shoot and blossom clusters that wilted and died when the parent twig of Japanese flowering cherry was girdled by a brown rot canker (WA, Jun).

D. Twig blight and gummosis caused by *M. laxa* on apricot (CA, Jun).

E. *M. laxa* sporulating on blighted blossoms of Japanese flowering cherry (WA, Jun).

F. *M. laxa* causing rot of sweet cherries. Powdery buff-gray sporodochia cover the decaying parts of the fruits (OR, Jul).

G. *M. fructicola* sporulating on a fallen plum (NY, Sep).

H. Apothecia of *M. fructicola* (NY, May).

Plate 58

Pestalotiopsis Spots, Blights, and Diebacks (Plate 59)

Diagnosticians commonly encounter species of *Pestalotiopsis* (syn. *Pestalotia*) and related fungi associated with dead or dying spots, blotches, tips or margins of leaves, twig dieback, or cankers on many different plants. The fungi may be found fruiting on dead tissues or may develop in cultures from diseased plant parts. Often the symptoms or the circumstances of their occurrence suggest that tissues colonized by the fungi were first weakened or killed by environmental injuries such as freezing or sunscald. On evergreen plants, these fungi are commonly found on the oldest leaves. Often they are clearly secondary invaders of plant parts killed or damaged by primary pathogens or insects.

A few species of *Pestalotiopsis* can behave as primary pathogens, invading very young parts of healthy plants, but most fungi in this group are nonpathogenic, and all of the pathogens are opportunistic invaders of tissues predisposed or damaged by other agents. Thus although these fungi may increase the damage done by primary agents, they are usually regarded as insignificant to plant health.

Pestalotiopsis species (Deuteromycotina, Coelomycetes) produce conidia in acervuli that develop beneath the epidermis of leaves, stems, fruit, and some flower parts. While the colonized plant tissue is moist, masses of conidia in a sticky matrix rupture the epidermis and push upward as glistening black droplets or, under drier conditions, as tiny black tongue- or hornlike projections (cirrhi) that may attain lengths of 2 mm (H). When wet by dew or rain, the matrix dissolves, and a sooty film of spores spreads out around the acervulus. The spores are dispersed by splashing or running water and presumably also by vagrant insects.

Conidia of *Pestalotiopsis* (Figure 13) are five-celled and pointed at the ends. They appear black in mass because the three central cells of each spore have darkly pigmented walls. The spores have several colorless appendages (setulae) at the apex and a basal pedicel that provides attachment during spore formation.

Fungi now in the genus *Pestalotiopsis* were formerly grouped with others in *Pestalotia*. In contemporary taxonomy, fungi once called *Pestalotia* are arrayed in several genera. Most of those with four-celled conidia are assigned to *Truncatella* and those with five-celled conidia (the majority) to *Pestalotiopsis*. The few with six-celled spores are placed in *Labridella*, *Seiridium*, and, for just one species, *Pestalotia*.

Ascigerous states of only a few of these fungi are known. That of *Pestalotiopsis sydowiana* on rhododendron is *Pestalopezia rhododendri* (Helotiales, Leotiaceae). That of *Pestalotiopsis gibbosa*, which is found in lesions in living leaves of salal in the Northwest, is *Pestalopezia brunneo-pruinosa*. *Pestalotiopsis palmarum* is the conidial state of a species of *Leptosphaeria* (Pleosporales, Pleosporaceae).

Pestalotiopsis funerea, one of the most common fungi in this group, occurs worldwide. Although reported to colonize diverse plants, it inhabits gymnosperms primarily, fruiting on dead or dying foliage and stems. It has been associated with damping-off, root and collar rot of seedlings, needle blight, tip blight, twig dieback, and stem cankers, especially of members of the cypress family. Its pathogenicity to several coniferous hosts has been proved by inoculations. Host genera include *Araucaria*, *Calocedrus*, *Cedrus*, *Chamaecyparis*, *Cryptomeria*, ×*Cupressocyparis*, *Cupressus*, *Dacrydium*, *Gingko*, *Juniperus*, *Picea*, *Pinus*, *Pseudotsuga*, *Sequoia*, *Taxus*, *Thuja*, and *Tsuga*. This fungus is prominent on junipers and arborvitae damaged by freezing or predisposed by unfavorable cultural conditions.

The needle blight with which *Pestalotiopsis funerea* is associated on conifers is characterized by yellowing and browning that progress from tips toward the bases of leaves. Shoot blight, less common, occurs when lesions form on the bases of succulent new shoots. The shoots wilt and turn brown. Browning of foliage on entire branches is usually the result of primary damage by an agent other than *Pestalotiopsis*. *P. funerea* grows rapidly on laboratory media, forming white colonies in which acervuli develop and produce greenish black spore masses.

After *P. funerea*, the most widely known fungus in this group is *P. maculans* (syn. *P. guepinii*), which causes gray blight of common camellia and tea plants around the world. Lesions on leaves, brown at first, turn gray to white, may coalesce, and eventually involve large areas of leaves. Lesions are bordered by a raised, narrow dark line. *P. maculans* also kills camellia buds and causes fruit rot, dieback, and cankers. This fungus is considered to be a primary pathogen because it was shown capable of infecting intact lower surfaces of tea leaves. It

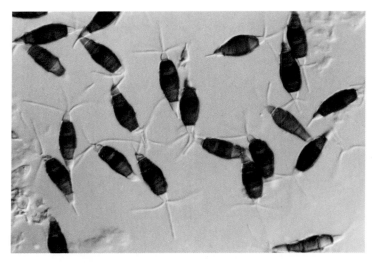

Figure 13. Conidia of *Pestalotiopsis funerea*.

infects twigs or upper surfaces of leaves only through various injuries. *P. maculans* has been reported to occur on a wide variety of plants in addition to *Camellia* species, but experts differ on the validity of these records. We follow the interpretation that *P. maculans* is restricted to species of *Camellia*. A second species resembling *P. macrochaeta* causes blossom blight and twig dieback of common camellia in the southern United States.

Pestalotiopsis sydowiana (syn. *Pestalotia macrotricha*) and *Pestalotiopsis rhododendri* attack leaves and stems of andromeda, various azaleas, mountain laurel, and rhododendrons across North America and in Europe. Necrotic spots or marginal areas on mature leaves are initially brown to reddish brown, then turn silver gray on the upper surface. Acervuli and spore masses dot the upper surface. Brown lesions on stems also develop silvery surfaces. *P. sydowiana* commonly follows winter drying, sunscald, or insect injury or enters lesions caused by other fungi on rhododendrons. In pathogenicity tests, it has infected leaves only through wounds or areas artificially scalded. Once established, however, it causes large lesions. It is also capable of infecting undamaged leaf axils, petioles, and soft green twigs while these are very young.

The biology and pathogenicity of *Pestalotiopsis* species on privets and *Ixora* have not been studied. *P. ixorae* was described from lesions on living leaves of *Ixora* sp. in Brazil.

Other noteworthy associations include the following. *Pestalotiopsis palmarum* enlarges lesions caused by other agents on various palms around the world but is incapable of infecting healthy, intact leaves. *Pestalotiopsis foedans* (syn. *Pestalotia peregrina*) is associated with needle blight of various pines, seedling blight of Hinoki cypress, and dieback of western red cedar planted abroad. The fungus is pathogenic when inoculated to wounds. *Truncatella hartigii* (syn. *Pestalotia hartigii*) is associated with strangling disease (basal constriction followed by death; see Plate 236) of young coniferous seedlings. The primary cause of this malady is considered to be heat at the soil surface.

References: 387, 701, 933, 1577, 1903, 1927, 1938, 1941, 2150, 2170, 2171

A, D. Tip blight of American arborvitae. A. Tan to brown twig tips colonized by *Pestalotiopsis funerea*. D. Black acervuli of *P. funerea* breaking through the epidermis of dead scale leaves (NY, Aug).

B, C. *Pestalotiopsis* sp. fruiting on a lesion caused by *Phytophthora* sp. (see Plate 137) on a branch of rhododendron. B. Overview of lesion. C. Close view of young acervuli beginning to break through the epidermis (NY, Oct).

E, H. *Pestalotiopsis* sp. in lesions in 1-year-old leaves of *Ixora* sp. Magnified view (H) shows how acervuli break the epidermis and produce black spore horns (arrow) (FL, Mar).

F, G. *Pestalotiopsis* sp. in lesions on leaves of wax-leaf privet. Viewed under magnification (G), acervuli of this fungus are minute black dots quite inconspicuous in comparison with the fungus on *Ixora* (FL, Mar).

Plate 59

Ascochyta Blight of Lilac (Plate 60)

Little known and seldom noticed, Ascochyta blight is occasionally destructive to common lilac. The pathogen, *Ascochyta syringae* (Deuteromycotina, Coelomycetes) was described in the late 1800s in Europe, where it is widespread. In the Western Hemisphere it has been recorded in Chile and in scattered locations from coast to coast in the northern United States and Canada. In the East it occurs from North Carolina northward.

Ascochyta blight has two phases, one on shoots in spring and the other on leaves later in the season. Shoot blight develops as elongate lesions girdle soft green twigs and flower stalks, causing the parts beyond to droop, shrivel, and turn brown. Lesions are usually restricted to tissues of the current season but may spread into apical parts of 1-year-old twigs, which then die back to the next branch. Lesions caused by *A. syringae* are prominent, even on brown dead shoots, because of their distinct lower margins, tan to grayish color, and shriveled appearance. During wet weather or when specimens are incubated in high humidity, white mycelium and dark gray pycnidia of the causal fungus break through the surface, giving the lesion a speckled gray appearance (B). This disease is readily distinguished from bacterial blight (Plate 75) or Phytophthora dieback in that the grayish lesions remain distinct on dead shoots.

The foliar phase develops in late summer and autumn as leaf spots or as spreading, coalescing lesions. Lesions are initially olive green and water-soaked, but they soon turn tan to light brown with indefinite margins. Lesions centered on major veins tend to become elongate, but most are more or less round.

The pathogen produces pycnidia prolifically on the brownish parts of foliar lesions, mainly on the upper surface. Young pycnidia lack pigment and can best be perceived by viewing lesions under magnification with a bright light behind the leaf. Thus illuminated, the pycnidia appear as minute light spots. They darken with age, however, becoming grayish brown and readily visible (E). The two-celled conidia are colorless with blunt ends and measure $8–10 \times 3.0–3.5$ μm.

The disease cycle is not known from research, but its main elements are apparent from observations of symptoms and signs. Pycnidia of *A. syringae* were observed on bud scales during winter in Italy, and shoot blight developed soon after buds opened. Thus the pathogen overwinters in buds, and shoot blight probably results either from latent infections in buds or, more likely, from spores dispersed from infected bud scales during wet weather in early spring. Rain splash is the principal means of local dispersal of conidia of *Ascochyta* and allied fungi. By this means conidia from lesions on shoots are dispersed to leaves, and foliar infection develops in late summer or early autumn. Whether buds or bud scales are infected simultaneously with leaves or by spores dispersed from leaves in autumn is unknown.

No sexual state of *A. syringae* has been described. It has been suggested, however, that the conidial fungus *Phyllosticta syringae*, characterized by colorless, unicellular spores, is a young form of *A. syringae*.

A. syringae can readily be isolated from diseased tissues on laboratory media. Its pathogenicity has been proved by experimental inoculation.

References: 250, 425, 436, 672

A–C. Shoot blight of common lilac caused by *Ascochyta syringae*. A. Lesions in soft parts of green twigs cause shoots to droop and shrivel. B, C. Closer views of lesions, showing abrupt lower margins, tan surfaces, and fruiting of *A. syringae* (NY, Jun).

D, E. Foliar phase of Ascochyta blight on common lilac. D. Upper and lower surfaces of leaves with spreading lesions. E. Part of a lesion, magnified to show gray-brown pycnidia of *A. syringae* (NH, Sep).

Plate 60

Sirococcus Canker of Butternut (Plate 61)

Butternut, throughout most of its range in eastern North America, is subject to cankers and dieback that shorten its life span, suppress natural reproduction, and greatly devalue the wood for lumber. The problem was recognized early in the 20th century, but the true cause was not learned until the late 1970s. Then a fungus new to science was described and shown to cause cankers and dieback. It was named *Sirococcus clavigignenti-juglandacearum* (Deuteromycotina, Coelomycetes). No sexual state of the pathogen has been reported. Sirococcus canker is now blamed for the rapid deterioration that makes butternut unsuited for landscape planting or timber production. This disease has virtually eliminated butternut in the Carolinas and threatens the species in many other areas.

Sirococcus cankers arise on all aboveground parts of the trees, even the buttress roots, and affected trees usually have multiple cankers. The fungus infects young twigs via leaf scars and lenticels. It enters older stems via adventitious buds, natural bark cracks, and wounds of all kinds. Infection in spring often leads to canker formation within 3 weeks, but often also the fungus remains quiescent at the point of infection for several weeks before beginning to colonize and kill bark rapidly. If infection occurs in autumn through leaf scars, symptoms do not usually appear until the next spring.

Cankers are elliptical; they expand more rapidly along the axis of a twig or branch than around it. Single cankers may girdle twigs or small branches, but large limbs and trunks succumb only as the result of the girdling effect of multiple coalescing cankers.

Newly infected bark, and soon the wood beneath, turn dark brown to nearly black. Hyphae of the parasite grow via the medullary rays from bark into sapwood and then up and down the stem in xylem vessels and parenchyma cells. Hyphae in sapwood grow beyond the limits of cankers in the bark, and they cause new cankers where they grow outward and kill the cambium. If bark is shaved from stems with multiple cankers, many will be found connected by dead dark streaks in the outermost sapwood. The pathogen can be isolated readily from diseased bark or wood.

During lesion formation, the inner, initially healthy bark is degraded to a dark brown mush. Liquid oozes through small openings to the bark surface and spreads out as shiny black spots. After drying, these spots remain conspicuously brownish black on smooth gray bark. The bark in cankers later splits and shreds as the unaffected part of the stem continues growth. Adventitious sprouts commonly grow from the edges of cankers on trunks or limbs, and these sprouts are soon killed as a result of new infections or by enlargement of nearby cankers.

Cankers on limbs and trunks are perennial, enlarging in a stop-and-go manner characterized by alternating lesion expansion and callus formation at the margins. After disappearance of bark, the concentric folds of callus give cankers a targetlike appearance. Eventually, coalescing cankers on the trunk or buttress roots kill the tree. Death typically comes many years after the first twig or branch infections.

Within a few weeks after lesion formation, *S. clavigignenti-juglandacearum* begins to produce a thin black stroma and irregularly shaped black fruit bodies (pycnidia) in the outer bark. Hyphae of the stroma differentiate into tiny pillars, sometimes called hyphal pegs, about 0.5 mm thick and 1.5–2.0 mm tall, that lift and rupture the papery outermost bark layer. In this function they are similar to the pillars illustrated for *Hypoxylon mammatum* (Plate 106). The pycnidia, usually about 100–375 μm in length, develop on the stroma or sometimes in the tips of the hyphal pillars. Conidia, individually colorless, two-celled, and spindle shaped, are extruded from the pycnidia in light tan tendrils (cirrhi) during wet weather from early spring until midautumn. They are dispersed by splashing rain. Sporulation on dead limbs or trunks continues for up to 20 months.

Conidia have been trapped at distances up to 45 meters from diseased trees, but most are deposited close to their sources. Thus Sirococcus canker intensifies most rapidly in low branches subject to rain splash from lesions above or in adjacent trees. Modes of long-distance dispersal of *S. clavigignenti-juglandacearum* are mostly unknown, but transmission by seed to young seedlings does occur.

Wet weather is required for conidial germination and for infection. Conidia remain germinable for only a few hours except under conditions of very high humidity without direct sunlight. Fungal growth and lesion development are most rapid at temperatures of 24–28°C.

Although Sirococcus canker has heretofore caused significant damage only to butternut, experiments have shown that most species and hybrids of walnut are susceptible to some degree. Therefore *S. clavigignenti-juglandacearum* is regarded as a potential threat to walnuts. Black walnut seedlings are highly susceptible if inoculated artificially, but only insignificant twig infections have been found in nature. Those were on saplings growing among severely affected butternuts. Heart nut is the least susceptible of species and varieties that have been tested.

S. clavigignenti-juglandacearum went undetected for many years because it is relatively inconspicuous and because a secondary fungus, *Melanconis juglandis* (Diaporthales, Melanconidaceae), was blamed for the dieback and cankers. For many years the disease now known as Sirococcus canker was called Melanconis dieback or Melanconis canker.

M. juglandis is quick to invade butternut twigs or the bark of branches and trunks weakened or newly killed by other agents. This fungus is a proven pathogen but only to predisposed tissues. It colonizes most or all species of walnut, and a variety of the fungus occurs also on mockernut and pignut hickories. In pathogenicity tests it has infected seedlings of black, Japanese, and Persian walnuts as well as butternut.

M. juglandis is usually encountered in its conidial state, *Melanconium oblongum*, which is characterized by subepidermal acervuli that produce prominent black tendrils or rounded black masses of spores on the bark surface. The conidia are dark brown, ovoid, and unicellular and are dispersed by water. Long-necked perithecia follow the acervuli but are less conspicuous. The bodies of the perithecia are embedded in the bark, and their necks protrude.

References: 107, 228, 676, 783, 1016, 1376, 2014–16, 2135

A. Twig and branch dieback of butternut (NY, Aug).

B, C. The main stem of a butternut sapling killed by Sirococcus canker. Cracked, shredding bark, dark brown stain on the bark, and sprout growth below cankers are typical of this disease. Tiny black lumps on the bark are spore masses of the secondary fungus, *Melanconis juglandis* (MN, Aug).

D. A typical Sirococcus canker with brownish black wood beneath, exposed by removal of bark. Coalescing cankers have girdled the stem above the light-colored sapwood (MN, Aug).

E. A butternut twig with its bark surface broken by hyphal pillars and pycnidia of *Sirococcus clavigignenti-juglandacearum.* More pycnidia would be found by peeling back the papery surface layer of bark (MN, Aug).

F. A twig of little walnut with black, rounded masses of conidia of *M. juglandis,* for comparison with (E) (IL, Jun).

Plate 61

133

Sirococcus Blight of Conifers (Plate 62)

Sirococcus conigenus (Deuteromycotina, Coelomycetes) causes shoot blight and seedling mortality of conifers in North America, Europe, and North Africa. The causal fungus, described in the mid-1800s, is also known as *S. strobilinus* and was formerly known in North America as *Ascochyta piniperda*. The disease occurs in nurseries, plantations, natural forests, and occasional landscape plantings coast to coast in Canada and the northern half of the United States. Plants most commonly damaged are western hemlock; Jeffrey, ponderosa, and red pines; and blue and Sitka spruces. Other hosts include Douglas-fir; white fir; eastern and European larches; Aleppo, Coulter, jack, Scots, sugar, and whitebark pines; and black, Engelmann, Norway, red, and white spruces.

Succulent shoots and often 1-year-old twigs are killed. Damage on large trees is confined mainly to low branches and has slight effect on general health. In exceptional cases, however, branch mortality in large red pines spreads upward gradually during a series of years with weather conducive to the disease.

Small susceptible trees in the understory beneath larger diseased trees are at high risk of damage. Small red pines in the Great Lakes region and the East, and western hemlock and Sitka spruce in the Pacific Northwest, have been grossly disfigured or killed by repeated infection of new shoots and in some cases woody twigs. In localities of greatest damage, the natural reproduction of these species has been temporarily curtailed. Fortunately, however, such damage is sporadic in time and location. In forests near the Pacific coast, *S. conigenus* has been found causing foliar blight and twig dieback of western hemlock, lodgepole and ponderosa pines, and, rarely, Douglas-fir. In the Midwest and East, red pine in natural forests and blue spruce in plantations have been damaged similarly. Blue spruce in Christmas tree plantations occasionally sustains such severe damage that trees become unsalable. On the basis of field observations, it has been suggested that strains of *S. conigenus* from pines and spruces are more or less host-specific, but evidence is contradictory.

Sirococcus blight in nurseries usually appears in patches that enlarge as the result of rain-splash dispersal of conidia from diseased seedlings. Oubreaks follow two distinct patterns. In pines, the disease may arise in seedbeds, transplants, or container trees adjacent to large diseased trees. In spruces, symptoms show up randomly in crops of seedlings to which *S. conigenus* was transmitted in infested seed. First-year seedlings are often killed, and larger plants may become so deformed that they are unfit for planting. Outbreaks of Sirococcus blight involving up to 30% of spruce seedlings in seedbeds have been traced to seed lots in which only 0.1–3% of seeds were infested. Seed infestation has in turn been traced to the colonization of spruce cones by *S. conigenus* in forests of the interior West.

Infection occurs on or adjacent to needle bases on new shoots. The first symptom on pine shoots may be a small purplish lesion with a drop of resin at the point of infection. When infection occurs in the region of shoot elongation, growth stoppage on the infected side causes the shoot tip to bend downward. Red pine needles infected before reaching maturity droop conspicuously. Shoot death and needle droop on established plants occur from June into August, depend-ing upon the plant species and region. On 1st-year seedlings, symptoms may appear in late summer and autumn. Within several days to a few weeks, the lesion girdles the shoot, causing the tip to die and turn brown. The damage may be confined to a portion of the new shoot but often extends into the 1-year-old twig. Death of 1-year-old twigs and needles is common in western hemlock, red pine, and blue spruce.

Pycnidia appear on the bases of killed needles and along the stem of the killed shoot in late summer to early autumn or the next spring, depending upon the climatic zone and the time when infection occurred. On red pine, pycnidia develop most abundantly beneath the basal sheaths of drooping needles. On spruces and hemlock, pycnidia are most numerous on the stem. They also occur on spruce cones and bud scales. They are brown initially and blacken with age. They are irregularly round, average about 100 μm in diameter (often much larger on the bases of red pine needles), and open by a wide irregular pore that can often be perceived with a hand lens. The conidia are colorless, two-celled, and irregularly spindle shaped. They measure about 12–15 × 3 μm. Some pycnidia may contain a small proportion of three- and four-celled conidia that are longer than normal. *S. conigenus* can readily be isolated from infected tissues and in culture forms pycnidia within 2 weeks.

The disease cycle is completed in 1 year, although spore dispersal from killed parts may continue for an additional 10 months. The fungus overwinters in killed shoots and cone scales. Conidia are dispersed by splashing water during spring and summer, with a peak of dispersal during the period of host shoot growth. Most new infections develop within a few meters of spore sources. Infections develop readily if conidia are deposited on succulent plant parts that remain wet for at least 24 hours at 10–25°C. Longer periods of wetness favor increasingly severe disease. Temperatures of 16–20°C are most favorable for disease development. Symptoms appear about 2 weeks after infection, and lesion expansion on a given shoot may be complete after 4–6 weeks. Under optimum circumstances, pycnidia may form within an additional 2 weeks. Often, however, pycnidial formation is delayed until the next spring.

Long-distance dispersal of *S. conigenus* is by means of infested seed and perhaps infected planting stock. Outbreaks of Sirococcus blight occur in localities with mild, moist weather, such as the fog belt along the coast of northern California, or elsewhere during unusually cool wet summers. Low light intensity, for example beneath the forest canopy, enhances susceptibility and thus promotes severity of Sirococcus blight.

Little attention has been given to detecting plants resistant or immune to *S. conigenus*. Austrian and white pines and balsam fir exposed to spores from diseased red pines have remained healthy. Populations of lodgepole pine seedlings naturally infected in a Canadian nursery showed differences in disease incidence related to seed source. Those from Washington and southern British Columbia were most commonly blighted.

References: 610, 926, 1418, 1499, 1751, 1849, 1935, 1936, 2092, 2187

A–C. Sirococcus shoot blight of red pine. A. Dead branch tips with brown needles on current-season and 1-year-old twigs. Needles infected before reaching maturity droop, die, and adhere for a year or longer. White pine, its branches intermingled with diseased red pine, remains unaffected (NY, Jul). B. Close view of a blighted branch tip in early spring. Needles 1 and 2 years old when photographed were killed the previous summer (NY, Jun). C. Pycnidia of *Sirococcus conigenus* on the base of a dead needle 1 year after infection.

D–G. Sirococcus shoot blight of blue spruce. D. A 2-year-old branch with most twig tips killed. Diseased twigs slowly die back to the parent branch. E. Some needles are discolored and others are missing at a lesion midway between the base and the terminal bud of a 1-year-old shoot. F. A twig tip killed during growth the previous year shows a characteristic crook. G. Pycnidia of *S. conigenus* on a twig tip that became infected 11 months earlier (NH, Jun).

Plate 62

135

Sphaeropsis Blight of Pine and Other Conifers (Plate 63)

Sphaeropsis blight, also called Diplodia blight, is a sometimes devastating disease of pines, especially two- and three-needle pines growing under stress. The causal fungus, *Sphaeropsis sapinea* (syn. *S. ellisii, Diplodia pinea*) (Deuteromycotina, Coelomycetes), is common in the northern and southern temperate regions around the world. In North America it occurs in California and western Canada and throughout the region delimited by Maine, Ontario, Montana, Oklahoma, and South Carolina. It causes tip blight, resinous cankers on mainstems and branches, misshapen tops, death of cones, blight of seedlings, dieback or basal cankers accompanied by gray to black stain of sapwood, and sometimes death of entire trees. In a given region, Sphaeropsis blight may cause severe damage to pine species introduced from elsewhere, but it seldom does much damage to trees planted within the natural range of their species. It rarely causes damage in natural forests.

Many pines and several other conifers are reported hosts of *S. sapinea*. Austrian, ponderosa, and Scots pines are most often damaged in North America. In the Southern Hemisphere, plantings of loblolly, Monterey, and slash pines are often infected. Other pine hosts include: Aleppo, Benguet, big-cone, Bishop, Canary Island, chir, cluster, Cuban, digger, loblolly, mugo, pinyon, red, Italian stone, Swiss stone, eastern white, Himalayan white, Mexican yellow, and tablemountain pines. Hosts other than pines include Portuguese cedar; Italian, Lawson, and Monterey cypresses; Douglas-fir; noble and silver firs; eastern larch; hoop pine (*Araucaria cunninghamii*); and blue, Norway, and white spruces. These lists serve as a guide to hosts already reported, but *S. sapinea* should be considered likely to colonize any pine and many other conifers that grow under unfavorable environmental conditions and among diseased, highly susceptible species such as Austrian pine. The susceptible species are sources of large numbers of infectious spores. Similarly, although *S. sapinea* is sometimes seedborne, it has caused much damage in nurseries only where old infected pines near seedbeds served as sources of spores.

The most common symptoms in North America are tip blight and death of low branches, the latter caused both by cankers and by cumulative destruction of buds and shoots. Annual destruction of many buds and shoots causes gradual decline of ornamental trees. Infections leading to tip blight begin on buds, on succulent stems of elongating shoots, and sometimes on immature needles during a 2- to 3-week period in spring when shoots start to grow. The fungus penetrates succulent stems through the intact epidermis and enters needles through stomata. The first symptom may be exudation of a drop of resin from a small lesion. Lesions enlarge quickly, and infected buds or shoots cease growth before or during needle elongation. Dying shoots turn shades of yellow green to straw color. Tissues in lesions are resin-soaked and discolored dark reddish brown, and they often exude resin. In time the resin crystallizes, making the dead shoot hard and brittle. When disease is severe, clusters of shoots are blighted and branches deformed. Stunted, straw-colored shoots with short needles glued in place by resin are diagnostic markers for Sphaeropsis blight. These can be distinguished from insect damage by the resin infiltration of the killed part, persistent needles, absence of tunnels, and, beginning in autumn, presence of black fruit bodies (pycnidia).

Pycnidia of *S. sapinea*, typically about 200 μm in diameter, break through the surfaces of killed needles, fascicle sheaths, cone scales, and the bark of twigs or branches beginning in late summer of the year of infection and continuing the next spring. Pycnidia on needle bases may be exposed by pulling off the fascicle sheath. Short, straw-colored to gray needles adjacent to those held in place by resin are most suitable for this test.

Conidia are dispersed from early spring until late autumn by splashing rain. The spores are dark brown and large enough to see with a hand lens as dark specks around the tips of pycnidia. They measure 30–45 × 10–16 μm. Most conidia are unicellular, but a few are two-celled.

Cones become infected while growing rapidly in the spring of their 2nd year of development. Cone infections are inconsequential for general health of the trees but are important for their contribution to epidemics because *S. sapinea* sporulates prolifically on cone scales. The pathogen often builds up on cones before extensive infection of new shoots. Severe tip blight on pines of less than cone-bearing age is uncommon.

In trees that are relatively free from stress, *S. sapinea* kills only current-season buds and shoots and 2nd-year cones. Older twigs and branches are damaged only if the trees are predisposed by such conditions as water shortage, compacted soil, root injury, excess shade, or heat reflected from nearby roofs. Invasion of older parts is facilitated by growth and persistence of the parasite in the pith, in which it often grows a short distance into 1-year-old twigs that appear healthy. From this advanced position it moves out into sapwood, bark, and needle bases if host defenses are impaired. Tip blight and low branch mortality tend to increase during successive years of drought and to subside thereafter.

Cankers, marked by exuding resin and persistent dead needles, arise either as nongirdling shoot infections or as the result of infection through fresh wounds. Dead tops, resinous cankers on mainstems, or scattered dead branches in otherwise green crowns indicate wound infection. Wounds made by insects, hail, or other agents are suitable. Austrian and Scots pines attacked by the pine spittlebug, and digger pine infested by pitch nodule moth (see Johnson and Lyon, 2nd ed., Plates 36 and 29), have sustained much damage attributed to infection through wounds.

Top killing and death of whole trees result from massive invasion of the wood and bark of trunks and, rarely, of the major roots of trees severely stressed by drought or other agents. *S. sapinea* causes a gray to black stain in the sapwood and produces pycnidia beneath the loose outer bark scales of trunks and branches killed in this manner. Pycnidia beneath bark scales, easily detected by knocking off the superficial scales, appear as small flattened black pads accompanied by grayish mycelium.

Given predisposed trees, epidemics are promoted by wet spring weather favorable for spore production, dispersal, and infection. Conidia are capable of germinating within a few hours at 12–36°C. A wet period of 12 hours suffices for germination and infection. In warm weather, symptoms appear 3–4 days later. The pathogen grows most rapidly at about 28°C.

Trees with Sphaeropsis blight may be pruned to improve their appearance, but this practice does not affect the likelihood of new infections because great numbers of conidia are released from diseased cones on green branches.

References: 78, 261, 357, 387, 708, 1218, 1444, 1491–93, 1614, 1824, 1941, 2101, 2110, 2218

A. An Austrian pine with tips blighted and low branches killed by *Sphaeropsis sapinea*. The disease spreads upward gradually in stressed trees (NY, Sep).

B. A shoot tip of mugo pine killed by *S. sapinea* the previous summer (NY, May).

C, E. Basal infection of partly grown needles of Austrian pine has halted needle elongation and caused changes in color through shades of yellowish green to straw (NY, Jun).

D. A red pine needle fascicle from a twig killed by *S. sapinea* the previous year; the basal sheath is cut away to reveal pycnidia breaking through the epidermis (NY, May).

F, G. Cones of Austrian pine, killed during their 2nd season of development and observed the next spring, display tightly closed scales with profuse pycnidia of *S. sapinea* (NY, May).

Plate 63

Phomopsis, Kabatina, and Sclerophoma Blights of Juniper and Other Gymnosperms (Plate 64)

Many gymnosperms, especially members of the cypress family, are hosts of *Phomopsis juniperovora* (Deuteromycotina, Coelomycetes), which causes shoot blight, twig cankers, and dieback collectively called Phomopsis blight. This is one of the most damaging disease of junipers in North America. The main aspects of damage are disfigurement and occasional failure of landscape plantings, death of seedlings or young grafted stock in nurseries, and impaired survival of nursery-grown seedlings after outplanting. Junipers and other hosts in natural stands sustain no significant damage.

Phomopsis blight occurs throughout the eastern half of the United States and Canada, in the Pacific Northwest, and also in Europe, Africa, Australia, and New Zealand. Major damage is limited to members of the cypress family. Species most commonly infected are eastern red cedar; Arizona cypress; and creeping, Rocky Mountain, and savin junipers. Other species susceptible to various degrees include: American and Oriental arborvitae; Japanese, southern red, western red, and southern white cedars; Arizona, smooth-barked Arizona, Bhutan, Gowen, Hinoki, Italian, Lawson, Modoc, Monterey, mourning, Portuguese, and Sawara cypresses; Douglas-fir and true firs; Ashe, cherrystone, Chinese, common, creeping, needle, shore, and Utah junipers; *Juniperus formosana* and *J. squamata*; European larch; jack pine; English yew; and Harrington plum yew. Even within highly susceptible species, cultivars vary widely in susceptibility. Several resistant cultivars and varieties are listed in the next column.

The usual symptom is tip blight. This begins with infection of immature scale leaves or needles. Mature leaves are resistant. Lesions first appear as tiny yellowish spots. Soon, as the fungus advances into the xylem, diseased shoots fade to light green, then turn reddish brown. A grayish band, which marks the site of infection, appears at the base of the killed part of the shoot. Killed shoots usually remain on the plant for several months, turning weathered gray during that time. When side shoots of highly susceptible plants become infected, the parasite grows into stems and causes cankers that may girdle stems smaller than about 1 cm in diameter. Seedlings of eastern red cedar in nurseries are often killed in this manner. Nongirdling cankers may extend to several centimeters in length.

Infection occurs any time during the growing season when succulent young foliage is available. During warm, wet weather, the first symptoms may appear 3–5 days after infection, but much longer incubation periods have also been reported.

Fruit bodies (pycnidia) develop in the gray band at the base of the killed part of the shoot or in the bark of cankers and killed stems. They appear on shoots beginning 3–4 weeks after infection. Pycnidia appear gray to black, depending upon location, and they range in size up to about 400 μm in diameter. They are embedded in host tissue initially, but they break through the surface while maturing. During wet weather, pale yellowish to cream-colored masses of conidia in a mucilaginous matrix extrude from pycnidia, either as tiny slimy blobs or as hairlike tendrils. Plate 67 shows tendrils of spores of another *Phomopsis* species. Conidia can tolerate temporary drying and still remain germinable. They are dispersed by splashing or wind-driven rain. *P. juniperovora* persists and sporulates as a saprobe in killed host material for as long as 2 years.

Infection in spring is caused by conidia from shoots or twigs killed the previous year. Late-season infection may be caused by conidia from shoots killed in spring, especially if plant growth is prolonged by shearing or by a high fertility level and plenty of water. Infection can begin within as few as 7 hours on continuously wet plant surfaces at 20–24°C. Longer wet periods promote increasingly severe disease. Temperatures as low as 8–12°C delay but do not prevent spore germination and infection. After infection, severe symptoms are promoted by relatively high temperatures—26–32°C.

P. juniperovora produces two types of single-celled, colorless spores: ellipsoid conidia measuring about 7.5–10 × 2.2–2.8 μm, with a prominent oil droplet in each end of the spore, and filamentous, slightly curved cells measuring 20–27 × 1 μm. Only the ellipsoid cells germinate. The oil droplets help diagnosticians distinguish *P. juniperovora* from other microfungi on junipers and related plants. Pycnidia in diseased plant material often contain only the ellipsoid spores, which complicates diagnosis, but both types of spores are usually abundant in pycnidia that form in laboratory cultures.

P. juniperovora grows readily on laboratory media, in which it usually produces a deep yellow pigment accompanied by orange-red crystals consisting primarily of a tetrahydroquinone. The pigment and crystals are useful diagnostic characters.

Few if any junipers are immune to *Phomopsis juniperovora*. Varieties and cultivars considered resistant on the basis of field performance include the following. Those marked by asterisk have also been reported resistant to cedar-apple and cedar-hawthorn rusts.

Juniperus chinensis	*J. horizontalis*
cv. 'Foemina'*	cv. 'Procumbens'
cv. 'Iowa'	*J. sabina**
cv. 'Keteleeri'	cv. 'Broadmoor'
cv. 'Pfitzeriana Aurea'	cv. 'Knap Hill'
cv. 'Robusta'	cv. 'Skandia'
var. *sargentii**	*J. scopulorum*
var. *sargentii* cv. 'Glauca'	cv. 'Silver King'
cv. 'Shoosmith'	*J. squamata*
J. communis	cv. 'Campbellii'
cv. 'Ashfordii'	var. *fargesii**
cv. 'Aureospica'*	cv. 'Prostrata'
var. *depressa**	cv. 'Pumila'
cv. 'Depressa Aurea'	*J. virginiana*
cv. 'Hulkjaerhus'	cv. 'Tripartita'*
cv. 'Prostrata Aurea'	
cv. 'Repanda'	
var. *saxatilis**	
cv. 'Suecica'*	

Two additional fungi, both coelomycetes, cause symptoms very similar to those of Phomopsis blight, but only on twigs a year or more old. *Kabatina juniperi* kills 1-year-old twigs in spring, and *Sclerophoma pithyophila* attacks 1-year-old and sometimes older parts, also in spring. These fungi, apparently incapable of penetrating intact healthy foliage, enter through wounds made by insects or injuries caused by severe winter weather. Both fungi are widely distributed in North America and in Europe, but they cause less damage than *P. juniperovora*.

In Kabatina blight and Sclerophoma blight, grayish lesions at the bases of blighted portions of shoots are the same as in Phomopsis blight. At low magnifications, the fruit bodies found there are indistinguishable from those of *P. juniperovora*. On microscopic examination, however, the three fungi differ conspicuously. The fruit bodies of *K. juniperi* are erumpent acervuli, numerous in spring and diminishing in number during the remainder of the season. Pycnidia of *S. pithyophila* (not shown) closely resemble those of *Phomopsis* but never contain two types of spores. The conidia of both *K. juniperi* and *S. pithyophila*, although similar to the colorless, single-celled ellipsoid conidia of *Phomopsis*, lack the distinctive oil droplets.

K. juniperi and its close relative, *K. thujae*, infect many of the same arborvitae, cypress, and juniper hosts attacked by *P. juniperovora* (*K. thujae* also causes tip blight in Alaska yellow cedar). *S. pithyophila*, on the other hand, occurs on more diverse conifers: arborvitae, Douglas-fir, true fir, hemlock, juniper, larch, pine, and spruce.

In pure cultures, *K. juniperi* and *S. pithyophila* form slimy colonies and produce conidia without fruit bodies. This cultural state in each case appears essentially identical to the cosmopolitan *Aureobasidium pullulans*, which is usually dismissed by diagnosticians as a saprobe.

S. pithyophila alone among the fungi considered with this plate has a sexual state, *Sydowia polyspora* (Dothideales, Dothioraceae). This is occasionally found on dead coniferous twigs.

References for Phomopsis blight: 387, 709, 711, 715, 1481, 1488, 1493, 1498, 1499, 1714
References for Kabatina and Sclerophoma blights: 248, 387, 611, 612, 855, 872, 1428, 1469, 1483, 1493, 1614, 1711, 1941

A–E. Phomopsis blight of juniper. A, D. A blighted twig, killed by a girdling lesion (grayish band) at the base of the browned part. The pathogen, *Phomopsis juniperovora*, fruits only on the lesion (NY, Jun). B, C. Blight in landscape plants. New infections, indicated by fading foliage in (C), are confined to current season's growth (TN, Apr; NY, Jun). E. Magnified view of a diseased twig showing pycnidia of *P. juniperovora* (NY, Aug).

F. Kabatina blight of juniper. Erumpent acervuli of *Kabatina juniperi* on a dead, 1-year-old twig of eastern red cedar (NE, Jun).

Plate 64

Phomopsis Canker of Russian Olive (Plate 65)

Russian olive, or oleaster, introduced to North America from Asia, has been planted widely because of its winter hardiness, tolerance to drought and salt spray, and attractive silvery foliage. Initially it seemed relatively free from pests and diseases, but three serious canker diseases were eventually noticed in the Midwest. Phomopsis canker is one of them. The others are caused by *Lasiodiplodia theobromae* (Plate 85) and *Tubercularia ulmea* (Plate 99).

Within 15 years after Phomopsis canker was first described in 1967, it was found in several midwestern and eastern states, Ontario, Quebec, and British Columbia. In the United States, infected nursery stock was planted in widely scattered landscapes where wilting, dieback, and cankers attracted attention. In Canada the disease was noted only in nurseries, usually on plants imported from Europe. In some instances, stock that appeared healthy on arrival developed cankers during the next year.

Phomopsis canker is caused by *Phomopsis arnoldiae* (syn. *P. elaeagni*) (Deuteromycotina, Coelomycetes). This fungus kills seedlings and saplings and causes dieback and cankers on larger plants. Shriveled, faded foliage clinging to killed branches or to entire small trees from midsummer through autumn calls attention to the disease. Seedlings, twigs on larger plants, and occasionally saplings up to 2 meters tall may wilt and die without the formation of distinct cankers. Usually, however, a definite lesion 10–30 cm long is visible at the base of the dead or dying part.

Young cankers in smooth-barked branches or stems are varying shades of reddish brown to nearly black, and the underlying sapwood is reddish brown. Gum often exudes, forming prominent amber-brown blobs. These spread out somewhat during wet weather and in time darken and dry to a nearly black incrustation. Cankers on trunks and scaffold limbs appear as dark, depressed areas where the bark begins to split as the result of drying in the lesion and growth of the stem beyond the edges of the lesion. Whether cankers enlarge during more than one season has not been reported.

Within 1–4 weeks after infection, pimplelike eruptions appear on the surface of a lesion. The bark soon breaks at these points, and prominent clusters of fruit bodies (pycnidial stromata) emerge. The stromata, initially grayish tan, darken with age and eventually become nearly black. Usually they are abundant by the time the disease is detected, and they remain prominent for at least a year. Pycnidia have even been found on roots of diseased seedlings.

Individual stromata are nearly 1 mm in diameter and contain one to several fertile cavities with two types of colorless, unicellular spores: spindle-shaped conidia and curved, filamentous, nongerminating cells. The production of pycnidia with two types of spores, one filamentous and the other variable in form, is characteristic of the genus *Phomopsis*. In laboratory cultures, *P. arnoldiae* commonly forms pycnidia within 3 weeks at 24–30°C. Lower temperatures allow slower development.

Transmission of *P. arnoldiae* in nature has not been studied. Fungi similar to this species are dispersed as spores in splashing or wind-driven rain droplets and may be transmitted by contact when plants are wet. Artificial inoculations have shown that various fresh wounds in Russian olive stems or roots are readily infected by *P. arnoldiae* at any time during the growing season. Lesions become apparent within 1–2 weeks, and small stems wilt within 3–4 weeks. Severe outbreaks of Phomopsis canker on saplings, as occasionally seen in nurseries, could result from transmission to pruning wounds. This is a matter for research.

P. arnoldiae has also been reported to cause dieback of black walnut seedlings in nurseries. Lesions form at terminal buds or leaf scars, kill only a portion of the shoot, and expand during only 1 season. Most seedlings survive but develop multiple stems unless pruned back to leave only one or two buds. Whether the fungus from walnut can infect Russian olive or vice versa is unknown.

P. arnoldiae may be the conidial state of *Diaporthe elaeagni* (Diaporthales, Valsaceae), which is widely distributed in Europe, but this connection has not been proved. In North America, *D. elaeagni* has been found on dead branches of cherry elaeagnus and silverberry but has not been associated with Phomopsis canker of Russian olive or black walnut.

References: 52, 53, 332, 678, 1348, 1941

A. A young Russian olive with its top killed by a girdling canker on the mainstem.

B, E. Cankers on smooth bark of young mainstems or branches show varying shades of reddish brown to nearly black. The smooth surface in the center of the canker is soon roughened by developing fruit bodies of the pathogen.

C. During the year after infection, bark in nongirdling cankers splits.

D. Gum, brown and translucent when fresh, exudes from the margins of some expanding cankers and, on drying, persists as a smooth, hard, black deposit.

F. Magnified view of young, grayish tan pycnidial stromata of *Phomopsis arnoldiae* that have recently broken through the bark surface.

G. Old pycnidial stromata appearing as clusters of tiny black cushions on the bark of a 1-year-old canker (all NY, Sep).

Plate 65

Diaporthe and Phomopsis Blights, Cankers, and Diebacks (Plates 66–67)

Many species of *Diaporthe* (Diaporthales, Valsaceae) cause or are found associated with blight, cankers, and dieback of woody plants. These fungi range from aggressive parasites to saprophytes that only colonize dead plant parts. Many are opportunistic invaders of bark and sapwood predisposed or injured by transplanting shock, drought, freezing, or other pests and pathogens.

Diaporthe species have black perithecia shaped more or less like flasks with spherical bodies and cylindrical necks. The bodies, often considerably flattened, are embedded in bark or the outermost wood, sometimes in a distinct stroma, and the necks protrude. Two-celled, colorless ascospores are produced in the perithecia. The conidial states of these fungi are in the form-genus *Phomopsis* (Deuteromycotina, Coelomycetes). Pycnidia of the *Phomopsis* state precede perithecia in killed tissues.

Diagnosticians commonly encounter only the *Phomopsis* state in plant specimens or in cultures from diseased material. In a typical *Phomopsis* species, each pycnidium produces two kinds of colorless, unicellular conidia—short ellipsoid to spindle-shaped cells and curved linear ones. The former germinate readily; the latter do not.

One basis for distinguishing between the aggressive parasites and the opportunistic colonists of debilitated tissue is the time of year when lesions develop. Aggressive parasites such as *P. arnoldiae* (Plate 65), *P. juniperovora* (Plate 64), and *P. macrospora* (Plate 66) may cause lesions that expand any time of year except when limited by low winter temperature. Most of the fungi mentioned with Plates 66–67, however, are opportunists that are likely to infect only while the host's defenses are minimal during dormancy or drought or after freeze damage or other stress. Even so, the opportunistic pathogens occasionally cause much damage.

These pathogens are dispersed as conidia and ascospores in splashing and wind-blown water droplets. When the bark or other substrate is moist, masses of conidia in a mucilaginous matrix extrude from pycnidia. Under wet conditions the spore mass appears as a blob that spreads out when diluted with water. If the substrate is somewhat drier, the spore mass emerges as a tendril (Plate 67G, H). Ascospores mixed with partially dissolved asci extrude similarly from perithecia. The sticky spore masses are also well suited for dispersal by vagrant insects, but the importance of this mode of dispersal is unknown for most species. Except for transport with diseased plants, modes of long-distance dispersal are also generally unknown.

Diaporthe and *Phomopsis* species survive the winter as mycelium and fruit bodies in killed bark and sapwood and sometimes leaves or fruit. Infection may be caused by either conidia or ascospores. Fresh wounds, including natural ones such as leaf scars, are common sites of infection. Some species, however, can penetrate intact young leaves, shoots, or fruits. *D. citri* on citrus and *D. alleghaniensis* on yellow birch are examples. Secondary cycles of disease are generally unimportant because most plants are susceptible during only one part of the year, either the period of dormancy or the first part of the growth period.

Diseases of rhododendron. Rhododendrons are occasionally affected in spring by foliar lesions in which a species of *Phomopsis* fruits (Plate 66A, D). The lesions develop on 1-year-old leaves as more or less circular brown areas with a narrow purple-brown line at the junction of healthy and diseased tissue. In time the center whitens. Pycnidia break through the surface in both the white and the brown parts of the lesion. Relationships of this disease to others described for rhododendron are unknown because *Diaporthe* and *Phomopsis* species pathogenic to rhododendrons have not been studied as a group. *P. rhododendri* was described from foliar lesions in catawba rhododendron in New Jersey, and *P. ericaceana* from leaves of West Coast rhododendron in California. *Phomopsis* species have been associated with twig blight and dieback of rhododendron in both East and West and of azalea in the South. The azalea fungus, characteristic of opportunistic pathogens, was inconsistently pathogenic when inoculated into azalea stems and was described as mildly pathogenic to stems of farkleberry, mountain laurel, leucothoe, and rhododendron. The ubiquitous *D. eres*, a minor pathogen of many plants, occurs on rhododendron branches and may be related to some of the conidial fungi just mentioned.

Melanose of citrus. Melanose (Plate 66B), caused by *Diaporthe citri* (conidial state *Phomopsis citri*), is one of the significant diseases of citrus. Its impact is related less to killing of plant parts than to disfigurement and consequent downgrading of fresh fruit. Most or all types of citrus are susceptible to some extent; grapefruit is the most susceptible. *D. citri* is part of a group species named *D. medusaea* (conidial state *P. cytosporella*). Plant pathologists, however, treat the citrus pathogen as a distinct species. *D. citri* is both a primary pathogen of immature leaves, stems, peduncles, and fruit and a secondary invader of wounds and tissues damaged by other pests or pathogens.

Melanose is characterized by numerous tiny brown eruptive lesions, sometimes called pustules, with yellow halos on leaves, green twigs, and fruit. Only immature tissues are susceptible. The lesions begin as tiny water-soaked areas in which cells degenerate and a dark gummy material accumulates. Extensive colonization of tissue by the pathogen does not occur at this stage. Rather, the host produces a layer of cork beneath the degenerated tissue. As the cork layer thickens, the entire lesion is forced outward until the cuticle breaks. On exposure to air, the gummy substance hardens in a mahogany brown bump. Numerous lesions greatly roughen the surface, may distort leaves, and cause premature yellowing. When they are numerous, lesions also tend to coalesce and produce continuous areas of superficial scar tissue. Such scars on fruit crack and create a roughened condition called sharkskin or mudcake melanose. Fruit infected while very young may be dwarfed and shed prematurely.

Lesions on green twigs apparently contribute to minor twig dieback, and *D. citri* produces pycnidia in the killed twigs. Conidia from this source start new infections. Perithecia form in dead twigs and branches on the ground. Ascospores from this source are infectious, but their importance as inoculum is considered to be slight. Melanose is usually diagnosed on the basis of symptoms, since isolation of the causal fungus from the gummy superficial lesions is difficult.

Phomopsis dieback of poplar. *Phomopsis macrospora* (Plate 66C, G) causes elliptical cankers and dieback on various poplars in nurseries and young plantations. Outbreaks in eastern cottonwood and other poplars have been reported from Mississippi to Minnesota and also in Japan. Loss of bare-root trees in winter storage has been reported also. Infection occurs through wounds and perhaps at leaf scars. Fresh wounds are suitable for infection at any time of year in warm areas. Trees are more susceptible during dormancy than at other times. Those stressed by water shortage are especially vulnerable. Cankers on thin stems cause dieback; those on larger stems usually do not girdle, but stem breakage may result at the point of infection. Stem growth around nongirdling cankers makes them appear greatly sunk-

---

A, D. Lesions apparently caused by *Phomopsis* sp. on 1-year-old leaves of rhododendron. A. Young lesions are brown; older ones develop grayish white centers. D. Close view of a lesion bearing numerous pycnidia (NY, May).

B. Melanose, caused by *Diaporthe citri,* on a grapefruit leaf. Minute brown lesions with yellowish halos are typical. The pathogen does not fruit in lesions in living leaves (TX, Apr).

C, G. Dieback of a cottonwood twig caused by *P. macrospora*. C. Pycnidia breaking through the surface of dead, tan bark. The advancing edge of the lesion displays gradual color change from normal green to tan. G. Magnified view of bark showing pimplelike swellings where black pycnidia have developed. The ostioles (mouths) of the pycnidia have broken through the surface, and off-white, translucent masses of conidia are extruding (MS, Apr).

E, H. Branch dieback of weeping fig caused by *D. cinerascens*. The bark surface, broken by the ostioles of numerous pycnidia of the *Phomopsis* conidial state of the pathogen, is shown in successively magnified views. Lens-shaped transverse protrusions are normal lenticels (indoor plant; NY, Oct).

F, I. Cankers on tulip tree saplings. F. A 1-year-old canker appearing sunken and with cracked bark at edges as the result of growing callus folds. I. Transverse view of a canker that formed between periods of host growth; sapwood beneath the canker is dead or dying. The *Phomopsis* species associated with this damage was an opportunistic invader of bark and wood predisposed by an unknown factor (NY, Sep).

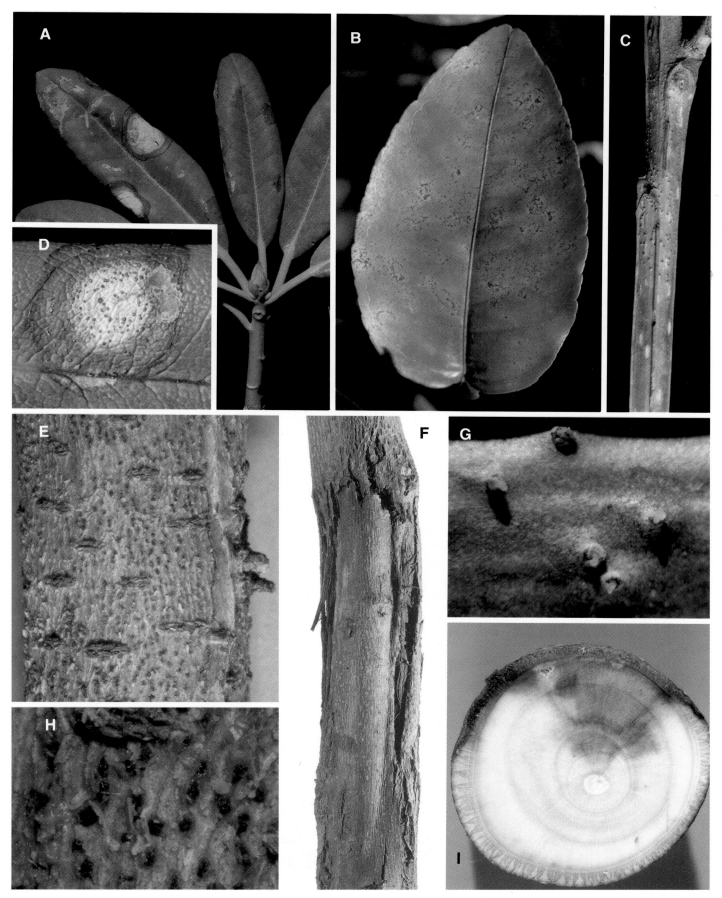

Plate 66

143

en within 1 year. Susceptible plants include eastern cottonwood, Carolina and 'Robusta' poplars, *Populus caudina*, *P. maximowiczii*, and several hybrid clones. No doubt many more species and cultivars will be found susceptible.

Phomopsis dieback of weeping fig. Weeping fig, which grows naturally in tropical areas, is widely used as a long-lived indoor ornamental plant because it can tolerate low light intensity. When stressed by inadequate light, water, or fertility, however, it becomes susceptible to *Diaporthe cinerascens*, which in its conidial state, *Phomopsis cinerascens* (Plate 66E, H), causes branch cankers and dieback. The disease is usually noticed when leaf drop and twig dieback begin on one or more branches. The cankers are diffuse and initially undetectable except by shaving the surface to reveal brown discoloration of bark and the underlying sapwood. In time the bark surface becomes rough as the ostioles (mouths) of black pycnidia break through the surface in profusion. The parasite probably becomes established in pruning wounds or other minor injuries while the plants are being grown for eventual sale, but it causes no damage as long as the plants are vigorous. Long known as a pathogen of fig trees in orchards, *D. cinerascens* causes trouble there after trees have been stressed by drought or winter injury.

A canker of tulip tree. Plate 66F and I illustrate a canker and a common fate of sapwood beneath cankers. This disease was encountered in a nursery. Saplings had elongate cankers in which a *Phomopsis* species was fruiting. When isolated from canker margins and inoculated into healthy trees, however, the fungus did not cause disease. Therefore, either the fungus had been a saprophytic invader of bark killed by another agent or it had killed tissues in which defensive responses were temporarily impaired by freezing injury or transplanting stress. Tulip tree is one of the hosts of *Diaporthe eres*.

D. eres, including its conidial state *P. oblonga*, occurs around the world in association with cankers and the dieback of hundreds of plant species. *D. eres* is really a complex of species that vary in host preference and in minor characters of the *Phomopsis* state but are indistinguishable in the perithecial state. Diseases attributed to *D. eres*, in addition to those mentioned elsewhere with these plates, include canker and dieback of elms, mock orange, paulownia, and rose, and dieback and sapwood stain of maples. This ubiquitous fungus has been recorded on plant species in more than 60 genera. Since most records do not distinguish between saprophytic and pathogenic occurrence, a complete list of host plants would be of little value. The symptoms and signs on fir, mock orange, and rose in Plate 67, although caused by forms of *Phomopsis* that differ in appearance, could all be attributed to *D. eres*.

Phomopsis cankers of conifers. Conifers in North America are subject to twig cankers and occasional top killing caused by *Diaporthe lokoyae* and *D. conorum* in their conidial states, *Phomopsis lokoyae* and *P. occulta*, respectively. *D. lokoyae* occurs on Douglas-fir and rarely on western hemlock in forests and nurseries near the Pacific coast. *D. conorum* is widespread on many conifers. A third fungus, *P. porteri*, causes twig and branch canker of Douglas-fir in western Canada. These fungi are opportunistic pathogens likely to infect wounds on trees predisposed by other agents. *D. conorum* is part of the *D. eres* complex but is commonly listed separately. In pathogenicity tests it has either failed to cause disease or has done so only if given a start in tissues killed by some other agent. Even so it attracts

diagnostic attention because of common association with damaging cankers such as the one illustrated (Plate 67A, B).

Two other genuine pathogens that cause twig and branch cankers in eastern and western North America must be mentioned here because they have been confused with *Phomopsis*. *Phacidium coniferarum* (syn. *Potebniamyces coniferarum*) attacks many conifers. *Phacidium balsamicola* (syn. *Potebniamyces balsamicola*) infects only true firs. These are apothecial fungi (Phacidiales, Phacidiaceae) unrelated to *Diaporthe*. Their pycnidial states, which never produce two types of conidia, were classified first in *Phomopsis*, then in *Phacidiopycnis* as *Phacidiopycnis pseudotsugae* and *Phacidiopycnis balsamicola*, respectively, and now in the form-genus *Apostrasseria*. The plethora of names is the consequence of an early, erroneous mycological judgment that the conidial states of these fungi were merely atypical representatives of *Phomopsis*.

More cankers and diebacks caused by *Diaporthe* and *Phomopsis*. In addition to the pathogens and diseases illustrated or discussed above, the following are noteworthy.

Diaporthe alleghaniensis (conidial state, *Phomopsis* sp.): leaf and shoot blight and canker of yellow birch; associated with birch dieback (Plate 218); occasionally destructive in seedling stands; widespread from the Great Lakes region eastward.

D. ambigua (*P. ambigua*): canker of pear, fruit rot of apple and quince, twig dieback of Persian walnut and red horse-chestnut; widespread in North America and Europe.

D. crustosa (*P. crustosa*): canker of English holly in the Northwest.

D. diospyri (*P. diospyri*): twig dieback of common persimmon in the Southeast.

D. dubia (*Phomopsis* sp.): canker and dieback of sugar maple in Canada; also occurs on other maples but not necessarily as a pathogen.

D. japonica (*P. japonica*): twig blight of Japanese rose in the eastern and southern United States.

D. kalmiae (*P. kalmiae*): leaf and twig blight of mountain laurel in the eastern United States.

D. oncostoma (*P. oncostoma*): canker and dieback of black locust in the eastern United States.

D. perniciosa (*P. mali*): cankers and dieback on apple, mountain ash, cherry, maple, peach, and pear; also infects apple leaves and fruit; widespread in Europe and North America.

D. pruni (*P. pruni*): twig blight and dieback of black cherry in the northeastern United States.

D. scabra (*P. scabra*): canker and dieback of sycamore in the South.

D. vaccinii (*P. vaccinii*): canker and dieback of highbush blueberry; occasionally epidemic in northern production areas.

Of the species listed above, *D. ambigua*, *D. crustosa*, *D. diospyri*, *D. japonica*, *D. perniciosa*, and *D. scabra* are members of the *D. eres* complex but are usually listed as separate species.

References for Diaporthe and Phomopsis: 51, 77, 210, 225, 256, 340, 387, 566, 608, 611, 674, 710, 711, 723, 1038, 1041, 1042, 1317, 1377, 1608, 2134, 2140, 2211, 2224, 2230

References for Phacidium and Potebniamyces: 387, 488, 609, 611, 710, 712, 1827, 1831, 2185

A, B. Top kill of balsam fir, the result of a girdling canker apparently caused by *Phomopsis* sp. A. A damaged tree. B. The lower edge of the canker (NY, Apr).

C, D, E, I. Cankers and dieback of mock orange caused by *Diaporthe eres*. C. Fading and brown foliage on affected branches. D. The lower edge of a canker; both bark and sapwood have been colonized by the fungus and are discolored brown. Black specks on light brown background at the edges of the shaved area are pycnidia and perithecia of the pathogen. E. The bark surface in a canker, showing black necks of perithecia protruding from a crack. I. Magnified view of killed bark, shaved to show embedded bodies of perithecia (NY, Aug).

F–H. Dieback of a rose cane apparently caused by *Phomopsis* sp. F. The surface of the lower part of a girdling canker shows gradation from brown and speckled with pycnidia through purplish dying tissue to normal green below. G, H. Successively magnified views of the canker surface with numerous masses of conidia extruded as tendrils from pycnidia in the bark (NY, June).

Plate 67

Phomopsis Galls (Plate 68)

In the course of early studies of crown gall (Plate 73) and other plant tumors during the 1920s and 1930s, many gall and tumor diseases were discovered that could not be attributed to bacteria. Most such diseases, economically insignificant, were never satisfactorily diagnosed (see Plate 241), but some galls, as shown on Plates 68–69, were found to be caused by fungi. In particular, one or more unnamed species of *Phomopsis* (Deuteromycotina, Coelomycetes) were found by means of isolation from diseased tissues to be associated with galls on highbush blueberry, cranberry bush, American elm, hickories, winter jasmine, maples, oaks, and common privet. Galls on hickories, maples, and oaks occur throughout the central and eastern United States, more commonly in the North than in the South. Black, bur, chestnut, post, red, and white oaks are susceptible. Galls on highbush blueberries occur in several northern states across the continent.

Phomopsis galls range from pea size up to 25 cm or more in diameter, depending on plant species, age, and location on the plant. Most are more or less spherical, but the largest ones appear as hemispheric protrusions from the lower parts of tree trunks. Affected plants typically occur singly or in small clusters and bear many galls while neighbor plants of the same species have none. Galls often develop for several years and then die. Multiple galls seem to cause general loss of vigor of the affected plant, and gall formation on twigs and small branches apparently causes dieback.

A Phomopsis gall may appear as a cluster of nodules pressed tightly together (C, F). Beneath the roughened bark, each nodule consists of hard wood that is somewhat disorganized in comparison with the anatomy of normal wood. Intercellular mycelium has been observed in young, enlarging galls. The galls on maple differ in form from those on other hosts. They begin as smooth swellings on which the bark eventually cracks and roughens, and they lack the nodular construction. Internally, however, they also are composed of hard wood.

Pycnidia of the gall-inducing *Phomopsis* species have been observed on galls on cranberry bush and privet but are uncommon or perhaps do not form on living galls on elm, hickory, maple, or oak. The fungus was originally detected when it grew from bits of internal gall tissue onto laboratory media, where it produced fruit bodies (pycnidia) with colorless, single-celled conidia typical of the genus *Phoma*. When cultures of the fungus were chilled in a refrigerator for several months, however, pycnidia with the two kinds of conidia typical of *Phomopsis* (Plates 66–67) formed. Similarly, fresh cultures from chilled galls formed pycnidia with two kinds of conidia.

The *Phomopsis* species isolated from galls on hickory and oak was reported to cause galls after inoculation into wounded twigs of cranberry bush, jasmine, and privet in addition to the original hosts. An isolate from oak also caused galls on highbush blueberry. *Phomopsis* isolates from galls on elm were reported to cause small, slow-growing galls on elm, jasmine, and privet, and an isolate from maple also caused galls on privet. Thus it seemed possible that one fungus lacking significant host specificity was responsible for galls on these and other hosts. These early reports were never corroborated, however. Other investigators of galls on maples failed to detect an associated *Phomopsis* species. This subject deserves fresh investigation.

The disease cycle has not been studied. From observations it seems that infection may start only on young twigs and that most galls found on major limbs and trunks were initiated on lateral twig bases. Modes of transmission of the pathogen(s) are unknown except as may be inferred from knowledge of other *Phomopsis* species (Plates 66–67). No sexual state of a gall-inducing *Phomopsis* species has been found. *References:* 266, 267, 448

A, D, G. Galls attributed to *Phomopsis* sp. on azalea. A. An affected shrub at the beginning of the growing season shows retarded bud development and loss of vigor as indicated by slow twig growth on a branch that bears several galls. D. Close view of galls on a 6-year-old branch segment; bark is roughened and dead twigs are associated with galls. G. Close view of small galls on a 3-year-old branch segment (NY, May).

B, C. Galls attributed to *Phomopsis* sp. on forsythia. The galls are nodular, each with multiple centers of enlargement, and are often associated with twig dieback (NY, Jun).

E, F. Phomopsis galls on bitternut hickory. E. Affected trees typically bear many galls that arise at twig bases. F. Close view of active (brown) and apparently dead (brownish black) galls (NY, May).

Plate 68

Phomopsis Canker of Gardenia and Nectriella Gall of Ornamental Plants (Plate 69)

The diseases illustrated here and on Plate 70 are characterized by both cankers and galls. The pathogens kill some tissues and induce overgrowth of others, usually in close proximity.

Phomopsis canker of gardenia. This major disease of gardenia causes damage in nurseries and landscapes in Florida and California and in greenhouses in the cooler parts of North America. It occurs also in Europe and South America. The pathogen, *Diaporthe gardeniae* (Diaporthales, Valsaceae), is usually found in its *Phomopsis* state, for which the disease is named. It causes cankers and cankerous galls on branches and mainstems, commonly near the soil line. The cankers are perennial, enlarge slowly, and eventually girdle stems. Diseased branches lose vigor, drop leaves prematurely, and die back (A). Slow growth and reversible wilting may call attention to cankers on mainstems.

Lesions first appear as small brown spots, usually centered on wounds, around which callus soon begins to form. As a lesion expands into callus tissue, stimulating more overdevelopment beyond, a dark brown, elliptical canker with greatly roughened surface and furrowed edges develops. Bark in the canker tends to break away and leave the wood exposed. If the surface of the last-formed callus roll at the edge of a canker is cut away, living inner bark there may be bright yellow or yellow-orange, a useful diagnostic character. Normal inner bark is greenish white.

Infection on branches (A, C, E) often results in distinctive cankerous galls several times the diameter of the branch. These galls have one flat or concave face with an irregular outline corresponding to the margin of a sunken lesion. Callus develops massively along the edge of the lesion and is the main basis of gall enlargement. Yellowish orange inner bark may extend a few centimeters beyond the gall but not without swelling and superficial roughening that indicates infection (F). Roughened scaly bark and enlarged nodal areas (F) occur occasionally in the absence of cankers. Leaves and roots are also susceptible to *D. gardeniae* if the parasite is introduced into wounds. Brown necrotic blotches form on leaves, and cankers develop on roots. Normally, however, only stem infections are significant.

Small black fruit bodies (pycnidia) develop just beneath the surface of dead bark in cankers. These pycnidia are 300–500 μm in diameter. During wet weather, conidia in mucilaginous masses or tendrils extrude from pycnidia, as illustrated in Plates 66 and 67. Conidia (of the two types discussed with Plate 66) are dispersed by running and splashing water and would be suited to dispersal by vagrant insects as well. Shipment of infected plants accounts for long-distance dispersal.

Perithecia of *D. gardeniae* have occasionally been found on dead galls and may develop in laboratory cultures. The two-celled ascospores are capable of causing infection, but most infections are caused by conidia.

D. gardeniae is a wound invader that causes problems primarily in gardenias propagated by cuttings. Swellings and cankers develop near the bases of cuttings, above or below the soil line, where leaves were removed before the cuttings were stuck in the rooting medium.

Cultivars of gardenia apparently vary in susceptibility to *D. gardeniae*, but this subject has not been studied enough to provide a basis for recommending resistant or tolerant cultivars. Sanitary propagation procedures reduce the chance of outbreaks.

Nectriella gall. *Nectriella pironii* (Hypocreales, Hypocreaceae) causes galls or cankers on many woody and herbaceous plants in Florida. Susceptible woody plants include aralia; bottlebrush; ceniza (Texas sage); chinaberry; wild coffee; croton; *Dombeya;* dracena; southern elderberry; American elm; Japanese fatsia; common and strangler figs; forsythia; glory-bower; false heather; West Indian holly; hydrangea; English ivy; jasmine; Carolina jessamine; *Lansinum domesticum;* cape leadwort; lebbek; Bahama lysiloma; mahonia; sweet olive; orchid tree; pittosporum; poinsettia; Chinese, glossy, and waxleaf privets; rubber plant; Jerusalem thorn; tree tomato; and willow.

The disease was first noted in 1955 on croton, on which the symptoms are irregular, corky, greatly roughened proliferations of callus up to 5 cm in diameter along the stems and on petioles and midveins of leaves. Galls are light tan while expanding but turn brown when mature. Numerous galls cause loss of vigor and then dieback of affected stems. Old galls are frequently covered with algae and support large populations of saprophagous mites. *N. pironii* causes similar galls on its other host plants apart from glory-bower, on the stems of which it causes deep, roughened cankers.

N. pironii is a wound parasite and is apparently unable to penetrate intact surfaces. Fresh leaf scars and wounds made by removing cuttings are common sites of infection.

The fungus produces two fruiting stages in corky superficial gall tissues, often on the same gall. White tufts (sporodochia) of the conidial state, *Kutilakesa pironii* (Hyphomycetes), arise first. Yellow to nearly white perithecia of the sexual state follow. Sporodochia, 250–750 μm in the longest dimension, are the more conspicuous. They produce on their surfaces unicellular, ellipsoidal, pale orange conidia and pale yellow hairs 50–150 μm long. The pear-shaped perithecia are immersed to various degrees in gall tissue. Some have only the tips exposed, but others are nearly superficial. They develop singly or in groups of two to six and measure 165–235 μm high and 115–185 μm wide. The ascospores are pale orange, spindle shaped, and equally two-celled. Superficial gall tissues containing sporodochia and perithecia tend to flake off easily, so these fruiting structures are sometimes difficult to find.

The means of transmission of *N. pironii* in nature are unknown. Fungi of this sort are adapted for transmission by rain splash or by vagrant insects or mites. Two species of saprophagous mites have been found contaminated with *N. pironii,* but it has not been claimed that they serve as vectors.

References: 18–21, 730, 731, 1267, 1305

A, C, E, F. Diaporthe canker and gall of gardenia. A. Leaf cast and dieback as secondary symptoms on small branches bearing galls. C. Close view of a cankerous gall on a small branch. Prolific callus formation at the edge of an initially small canker resulted in the formation of the gall. E. Enlarged view of the same gall shown in (C), the face cut away to reveal diagnostic orange-yellow color of infected inner bark. F. Cutaway view of one end of a roughened infected area, showing contrast between swollen, orange-yellow bark in the diseased region and normal greenish white inner bark beyond (FL, Mar).

B, D, G. Nectriella gall of croton. B. Galls and dieback on stems in a landscape planting. Green appearance of old galls is due to superficial growth of algae. D, G. Successively closer views of typical, rough-textured galls. New gall tissue is light in color. Dead parts turn dark brown and eventually weather to grayish or to grayish green if algae develop (FL, Mar).

Plate 69

Sphaeropsis Knot (Plate 70)

Sphaeropsis tumefaciens (Deuteromycotina, Coelomycetes) causes galls on various woody plants in Florida and the Caribbean region. Sphaeropsis knot, as the disease is called, is best known for its damage to citrus, especially in Jamaica. It was once common although not important in citrus groves in Florida, but now it is confined mainly to ornamental or wild plants in that state. It is the most important disease of oleander there. In commercial citrus it was easy to prevent by using proper hygiene in propagation activities and to control in groves by sanitary pruning and destruction of galls. Principal citrus hosts are calamondin, rough lemon, key lime, sweet orange, and ortanique (a natural hybrid between tangerine and orange). Ornamental hosts other than citrus include bottlebrush, American and dahoon hollies, oleander, Brazilian pepper tree, and Natal plum. The diversity of these hosts indicates that *S. tumefaciens* is relatively nonspecialized and is therefore likely to be found on additional plants.

Symptoms on all hosts include localized swellings (hypertrophy), broom formation by growth of multiple shoots from hypertrophic tissue, and dieback of twigs and branches. Specific symptoms vary with host.

On holly (A–D), symptoms range from swelling of elongate segments of young twigs (B, D) to irregular knobby swellings on mature branches. Numerous stunted, sometimes leafless, twigs arise close together, forming loose brooms (A, C, D). Foliage on diseased branches turns yellow and drops, and dieback ensues.

Typical symptoms on weeping bottlebrush are globose galls 1–9 cm in diameter on twigs and small branches (E), sometimes in a row along a branch. The galls become darker than normal bark and have knobby surfaces and deep indentations. Galls at nodes often give rise to multiple buds that produce brooms consisting of somewhat stunted shoots. Galls on internodes frequently remain devoid of shoots. Occasionally cankers develop instead of galls. In these cases, hypertrophy and roughening occur at canker margins. Bark within cankers separates from the wood at an early stage. In time, diseased branches die back to below the galls or cankers.

Symptoms on citrus, oleander, and Brazilian pepper tree are similar to those on bottlebrush—dieback and knots that often serve as loci of brooms. The knots are woody, nearly the color of normal bark while young, and become rough, fissured, and dark colored with age. Eventually the bark sloughs from dead galls. Natal plum forms either greatly roughened, often elongate galls (F) or sunken cankers in response to *S. tumefaciens*.

Sapwood within galls or beneath cankers often becomes darkly stained as a result of colonization by dark hyphae of the parasite. The staining sometimes extends several centimeters in each direction from the original locus of infection, and galls that arise in a series on a branch may be connected by hyphae in the wood. Internal spread and development of secondary knots up to 90 cm from the point of infection in citrus branches have been documented.

Black pycnidia of *S. tumefaciens*, 180–220 μm in diameter, arise singly or in small groups in dead outer bark of cankers, galls, and dead twigs. They produce large (20–34 × 6–10 μm), colorless to pale yellow, mostly unicellular conidia. Usually, however, pycnidia are hard to find. Firm diagnosis may then depend upon isolation and identification of the pathogen in pure culture. It grows readily from gall or canker tissue onto laboratory media, producing black mycelium, pycnidia, and conidia as above and also microconidia measuring 3–5 × 1.5 μm. Two-celled conidia also are somewhat common in the pycnidia that form in pure cultures. Isolates from one host plant readily infect wounds in other reported hosts.

The mode of transmission of *S. tumefaciens* in nature is unknown. Wounds are common infection courts, but unwounded succulent stem tissues or buds may also be susceptible. These matters need study. After infection, the parasite may persist several years, first in living stem tissue and then for a time as a saprophyte in dead tissue, where it fruits. Experimental inoculation of various hosts has resulted in galls or cankers that become noticeable after 6–15 weeks in most cases. Galls may enlarge for several months before attaining final size.

Plants affected by relatively few knots or cankers can be cured by removing the affected parts, provided that each cut is made at least 15 cm below a swelling or gall.

In the North, *Diplodia tumefaciens* (syn. *Macrophoma tumefaciens*) causes galls and rough bark on various species of aspen and poplar from Ontario westward through Canada and the northern United States to British Columbia. This fungus is similar to *S. tumefaciens* in appearance, cultural characteristics, and symptoms induced in host plants. It has not been reported to invade sapwood, however.

References: 20, 387, 577, 770, 1221, 1222, 1609, 1941, 2267, 2268

A–D. Brooms and dieback of dahoon holly, caused by *Sphaeropsis tumefaciens*. A. Dieback of twigs and branches on a large tree. B. A diseased branch tip with upturned dark brown leafless shoots arising from the infected portion and prematurely senescent leaves beyond the point of infection. C. A branch tip killed by *S. tumefaciens,* showing swelling of the once-infected portion of the branch axis and blackened upright twigs of a broom. D. A diseased branch tip infected at two points: on a lateral twig, causing dieback, and at the branch tip, causing broom formation (FL, Apr).
E. Sphaeropsis knot of bottlebrush. Multiple shoots growing from the knot are typical of the disease on this plant (FL, Apr).
F. Stem galls caused by *S. tumefaciens* on Natal plum. Numerous small galls cause enlarged and greatly roughened stems (FL, Apr).

Plate 70

Black Knot of *Prunus* (Plate 71)

Black knot affects at least two dozen species of cherries, plums, and other members of the genus *Prunus*. It occurs throughout Canada and the United States where hosts grow in regions with moist climate, and it has been reported from New Zealand. The pathogen is *Apiosporina morbosa* (syn. *Dibotryon morbosum*) (Pleosporales, Venturiaceae). It causes rough, more or less elongate galls (knots) on twigs and branches of all sizes. At maturity, the knots' surfaces are covered with the pathogen's black fruit bodies (pseudothecia). Infection of limbs or trunks leads to the formation of cankerous swellings and dieback. Outbreaks of black knot are common on wild chokecherry and plum trees and on domestic plums in orchards where no disease control measures are practiced. Severely diseased fruit trees become worthless because of dieback. Large galls or cankerous knots 60 or more cm long may form on trunks of black cherry trees, reducing their value for lumber.

Susceptible plants include flowering almond; apricot; blackthorn; cherries such as bird, bitter, black, mahaleb, Nanking, pin, sand, western sand, sour, and sweet; chokecherry; peach; and plums such as American, beach, Canada, common, damson, Japanese, myrobalan, and Sierra.

Knots first appear in autumn as swellings on twigs of the current season that became infected the preceding spring. Enlargement is interrupted by winter dormancy. In spring the knot resumes growth in girth and length. Often the diseased twig bends sharply at the knot because of one-sided overgrowth. Surface bark on the knot splits, revealing an olive green fungal stroma. The stroma, corky in texture when first exposed, darkens during the season and turns hard and black during the next winter. The outer portion is composed entirely of fungal tissue; the inner part is a mixture of host and fungal cells. Black pseudothecia mature on the surface of the stroma and liberate their ascospores the following spring. The black part of the knot then dies. Often the twig beyond the knot either fails to leaf out or wilts suddenly in early summer. If the twig remains alive, the margins of the knot, still covered with bark, continue growth and stromal production. The new portion of the knot will bear pseudothecia the next spring. Thus the knot becomes perennial and elongates in annual increments toward the base of the twig. On reaching the next branch, the parasite slowly spreads along and around it. The branch is likely to be girdled and die, but if it survives, a swollen, rough, often gum-encrusted canker with a sunken center eventually forms. The fungus continues its annual developmental cycle at the advancing margin. After the 1st year of knot growth, pseudothecia in various stages of formation and degeneration can be seen throughout the year.

The surface of the olive green stroma on a knot in early summer is covered with conidiophores and light brown, single-celled or occasionally two-celled conidia measuring 4–9 × 3.5–5.5 μm. Pseudothecia mature on the same stroma in early spring. They are more or less globose, 150–300 μm in diameter, and produce unequally two-celled, club-shaped, olivaceous ascospores 13–18 × 4.5–7.5 μm. Ascospores are ejected into the air throughout the spring in response to rain. When ascospores are present on new shoots, a wet period as brief as 6 hours at the optimum temperature of 21–24°C is sufficient for infection.

Some infections begin at wounds on stems, but the great majority occur on green shoots. Cankerous galls on limbs and trunks most often result from invasion via adjacent twigs. Infections on limbs or trunks of black cherry sometimes result in galls on which the bark remains intact and stromata of the pathogen emerge as small black cushions.

Most infections are considered to be initiated by ascospores. Conidia are also infectious, but their importance is unknown. Plants are less susceptible during the time of conidial abundance in summer than during the period of ascospore abundance in spring. *A. morbosa* apparently also grows saprophytically on twig and branch surfaces and produces dark chlamydospores there. Their role in the disease cycle is unknown.

After germination of ascospores on new shoots, the parasite penetrates the cuticle and grows intercellularly in the cortex, cambial region, and xylem, causing no symptoms during the first several months. It secretes indoleacetic acid and other growth-regulating chemicals that eventually stimulate cambial activity and swelling of the infected area. The amount of new xylem and phloem produced, the proportion of parenchyma in the xylem, and the size of parenchyma cells are all abnormally great.

The duration of the disease cycle is normally 2 years, but pseudothecial maturation 1 year after infection has been noted occasionally in southern Ontario and is presumably common in regions with a long growing season.

A. morbosa exists as several host-specialized strains. An early study revealed, for example, that spores from black knot on chokecherry would not infect wild plum, and vice versa. Susceptible domestic plums and cherries near thickets of diseased wild species have sometimes remained free from disease. Black cherry may be severely damaged by black knot in some areas but may remain free from it in other places where the disease is present on other hosts.

The black knot fungus is host to various mycoparasitic fungi and predaceous insects and mites that tend to suppress production of pseudothecia. Some insects also inhabit the knots for shelter. The most conspicuous fungus is *Trichothecium roseum*, which forms white to pink patches of mycelium and spores. Natural control of black knot by these organisms is insufficient for practical purposes, however. In plum orchards, black knot is suppressed by sanitation and fungicidal sprays. Effective sanitation is possible by removing knots, cutting at least 10 cm beyond the swelling before spore dispersal begins in early spring. Since the fungus continues to develop in pruned material, this must be destroyed or removed from the site.

References: 387, 679, 1044, 1045, 1377, 2088

A. Black knot on twigs and branches of a wild plum (NY, Feb).

B, C, E. Symptoms and signs on chokecherry (NY, Jun). B. A 2-year-old knot that caused a sharp bend in the twig because of one-sided overgrowth in the 1st year of development. The twig tip and host tissues in the black part of the knot have died. The living part of the knot, where the pathogen has not yet sporulated, represents 1 year's spread toward the twig base. C. Close view of the surface of a living portion of a knot. The outer bark has split, exposing olive green fungal tissue that darkens slowly as pseudothecia mature. E. Magnified view of the surface of a knot covered with mature pseudothecia of *Apiosporina morbosa*.

D, F, G. Black knot on black cherry (NY, Jun). D. A gall about 40 cm in diameter on a limb. F. Multiple knots developing into swollen cankers and causing dieback in a young tree. Large stems were probably invaded via small twigs. G. Close view of the surface of a large knot similar to that in (D), with scattered black stromata of *A. morbosa*, the larger one about 1 cm long.

Plate 71

Bacterial Galls of Olive, Oleander, and Douglas-fir (Plate 72)

Olive knot, oleander gall, and bacterial knot of ash. Galls on olive trees were recorded by Theophrastus more than 2000 years ago. Thus the disease that we now call olive knot ranks with a select few known since ancient times. Olive knot, oleander gall, and bacterial knot of ash and other plants in the olive family are caused by strains of the bacterium *Pseudomonas syringae* pathovar *savastanoi* (formerly *Pseudomonas savastanoi*), herein called pv. *savastanoi*. The olive pathogen was undoubtedly carried around the world with its host plants.

Aside from the obvious galls or knots, the disease on olive trees results in dieback of twigs and branches and imparts to the fruit a bitter, salty, sour, or rancid taste. Disfigurement and dieback are the main effects on other host plants. In North America, the disease is significant only in California on common olive and oleander. In Europe, bacterial knot of European ash is widespread, and primrose jasmine and wax-leaf privet are also affected. Other woody plants found susceptible in experiments include *Chionanthus virginicus* (fringe tree), *Forestiera* spp. (desert olive and swamp privet), two species of *Forsythia*, three species of *Fraxinus* (flowering, velvet, and white ash), California privet and one other species of *Ligustrum*, four little-known species of *Olea*, and *Osmanthus* spp. (American, fragrant, and holly olive).

Knots on olive are globose to irregular, roughened, eventually fissured tumors that may enlarge for several years and may attain sizes of several centimeters in the largest dimension. They arise anywhere on twigs and branches at wounds, freezing injuries, bark cracks where adventitious shoots emerge, and especially leaf scars. A beadlike arrangement of knots along a twig (E) indicates that infections began at successive leaf scars. Initially light tan in color, knots become dark gray to nearly black as the surface tissues die. Leaves, peduncles, and fruit are also susceptible, but infection on these is usually inconsequential.

On oleander, galls and lesions damage stems, flower parts, seed pods, peduncles, and young leaves. Mature galls on stems are 1–2 cm in diameter. Galls on succulent twigs may appear cankerous because the tissues first infected die, and the stem may split at that point (B). Leaves, flowers, and seed pods develop wartlike galls and become distorted. The seed pods may be twisted and stunted. The diseased parts turn brown within the first season, and severely infected parts die back. Lesions in succulent tissue sometimes spread upward and, less often, downward, causing a streak of swollen tissue or numerous smooth, secondary knots or tubercles along the path of infection. This spread occurs internally via laticifers (latex-containing ducts). Olive and ash, which lack laticifers, show little evidence of internal spread.

If infection occurs during a period of host growth, symptoms appear within 10–14 days. If infection occurs during winter, however, galls do not appear until spring. The first symptom is often a small, water-soaked lesion. A cavity 2–5 mm long develops and becomes filled with a mixture of bacteria and cellular debris. Abnormal plant cell enlargement and division begin adjacent to the lesion and spread outward, with the result that galls form. The hypertrophy is induced by indole-3-acetic acid and cytokinins produced by the bacteria. Gall tissue, initially spongy, becomes woody, with a disorganized arrangement of xylem, phloem, and parenchyma. While galls remain alive, bacteria can be found in creases formed by the infolding of proliferated tissues. Bacteria also multiply in tiny irregular cavities much like the original infection site, located near the surface of the gall or close to apparently healthy tissues. Fissures extend from cavities to the surface of the knot. Bacteria thus reach the surface, where they are available for dispersal by rain splash or by contaminated hands or implements. Rains followed by high relative humidity and temperature near 23–24°C are most favorable for dispersal and infection. In Italy the olive fly, *Daucus olea*, carries olive knot bacteria, but transmission by insects is not reported in California.

Only fresh wounds are suitable infection courts. The susceptibility of leaf scars, for example, diminishes rapidly during the first day after leaf fall, and by the ninth day the scars are not susceptible. Plants remain susceptible throughout the year, but in areas with distinct wet and dry seasons nearly all infection occurs during the wet season—approximately October to May in California.

Bacterial knot of ash, which to date occurs only in Europe, differs from the diseases of olive and oleander in that lesions with roughened, hypertrophied bark at their margins form on twigs, branches, and trunks. The lesions enlarge into the hypertrophied bark; thus the cankerous area becomes raised, greatly roughened, and fissured. These distinctive symptoms are attributable to the strain of the bacterium rather than to the host. When strains of pv. *savastanoi* from ash, oleander, and olive were inoculated into ash, the ash strains caused cankerous swellings in the bark, while the strains from olive and oleander caused knots similar to those on their original hosts. Some strains are capable of infecting an array of hosts in the olive family, while other strains are pathogenic to only one or a few species.

Bacterial knot on specimen plants, if not too severe, can be suppressed by pruning infected parts during dry weather.

Bacterial gall of Douglas-fir. Galls putatively caused by bacteria occur on Douglas-fir in scattered localities from the mountains of southern Arizona to British Columbia and Alberta, especially in crowded stands on moist or mountainous sites where the host grows poorly. Globose galls up to 30 cm (mostly 1–10 cm) in diameter develop on twigs, branches, and mainstems of young trees and apparently suppress growth and cause dieback. The galls have rough, fissured surfaces and in young stages a distinctive cross-shaped mark where dead surface tissues split as the gall expands. Galls originate at the vascular cambium and are distinctly woody.

This disease was described during the 1930s and was attributed to a new bacterial species then called *Bacterium pseudotsugae* and later *Agrobacterium pseudotsugae*. Bacteria isolated from galls caused hypertrophy after inoculation to puncture wounds into the wood of twigs, and an insect, *Adelges cooleyi* (see Johnson and Lyon, 2nd ed., Plate 49), was suggested to be a vector. Unfortunately, the early work was not confirmed by repetition. Bacterial strains labeled "*Agrobacterium pseudotsugae*" in contemporary culture collections are neither pathogenic nor representative of the genus *Agrobacterium*. Thus although the disease is real, the pathogen is unknown. This subject needs reinvestigation.

References for olive knot: 533, 796, 949, 1818, 1833, 1834, 1931, 1932, 1937, 2207, 2209–11, 2265
References for bacterial gall of Douglas-fir: 468, 732

A–D. Bacterial knot of oleander. A. Distortion and necrosis of flower parts (CA, Oct). B. Galls on stems, the small galls having originated at leaf scars. The elongate gall may have arisen from multiple infections along a wound or as the result of internal spread of bacteria from one point of infection (CA, Apr). C. Young galls on a leaf (CA, Apr). D. Galls on seed pods and stems (CA, Oct).

E–H. Olive knot. E. A series of knots that originated at leaf scars, resulting in dieback of the horizontal twig. F. Close view of a knot on a 1-year-old twig. Light color indicates recent development. G. Continuous knots extending several centimeters along an olive twig. H. Old, blackened, apparently dead knots along a branch about 6 cm in diameter. These knots apparently developed at wounds (CA, Jul).

I. Bacterial galls, 1–2 cm in diameter, on a Douglas-fir stem (OR, Jun).

Plate 72

Crown Gall (Plate 73)

Crown gall is the most famous plant tumor and is one of the most studied plant diseases. It occurs around the world, causing economically significant damage to fruit and nut trees and ornamental plants. The disease is caused by *Agrobacterium tumefaciens*, a soil-inhabiting bacterium that belongs to the family Rhizobiaceae. *A. tumefaciens* has the broadest host range of any bacterial plant pathogen. More than 600 plant species in over 90 families are susceptible, although relatively few species sustain significant damage.

Galls form on roots and stems, especially at the root collar, or root crown—the junction of roots and stem. Young plants with large or numerous galls tend to be stunted and predisposed to drought damage or winter injury. Floral display or fruit production may be suppressed. Damage is greatest when galls encircle the root crown, but few plants are killed by crown gall alone. Mature trees often seem able to support numerous large, often grotesque, galls without noticeable debilitation. Severely diseased plants, however, are subject to attack by secondary pathogens such as *Armillaria* (Plates 148–149) that enter through decaying galls. This problem was severe in walnut orchards of California, but it lessened when growers began using California black walnut and Hinds walnut as rootstocks for the highly susceptible Persian walnut. The greatest economic losses now occur in nurseries that produce stone and pome fruit trees such as almond, apple, cherry, peach, and plum. Galled plants, even though not obviously impaired, must be culled because regulations prohibit their shipment.

Galls enlarge during periods of host growth and vary in diameter from a few millimeters to 30 cm or more. Gall size depends upon plant species and size and growth rate of the infected plant part. Aerial galls are common on such highly susceptible plants as poplar, rose, willow, and wintercreeper. Galls are the color of the infected plant part—cream colored on young roots, greenish white to green on young stems, and the color of normal bark on older roots or stems. Galls on woody plants are initially spongy but become quite hard as disorganized clusters of xylem elements differentiate in them. Gall surfaces are typically rough and become fissured with age. Dead surface tissues, especially of galls in soil, decay and slough off.

The cycle of crown gall is simple yet elegant. Bacteria in soil or water or on implements infect fresh wounds of any sort and occasionally enter roots through lenticels. Suitable wounds arise during propagation, transplanting, cultivation, frost heaving of soil, or feeding by soil insects or nematodes, especially root knot nematodes (Plate 144). As surviving plant cells in a wound begin the reparative processes that would normally isolate the wound from healthy tissues, the cells become receptive to crown gall bacteria, which attach to the plant cell walls. Cells in wounds remain receptive for periods ranging from a few days during active plant growth to months while plants are dormant. Next, a large bacterial plasmid (circular extrachromosomal molecule of DNA) is released within the plant cell. A plasmid fragment with genes that code for tumor production is inserted among the nuclear genes of the plant cell. Once in place, the bacterial genes redirect the synthetic processes of the plant cell toward unregulated division and growth. Thus transformed, tumor cells produce abnormal concentrations of auxins and cytokinins (growth-regulating chemicals) and multiply autonomously.

Crown gall bacteria, although no longer necessary for gall development, remain active in young galls, multiplying and migrating intercellularly. This migration results in the induction of secondary growth centers that contribute to the irregular shape and rough, fissured surfaces of old galls. Secondary galls may encircle stems at the root crown and sometimes form on rose or bramble canes as the result of bacterial movement within the plant upward from the point of initial infection. Such galls are arrayed in a row or appear as a linear roughened swelling (G).

Tumors become visible 2–4 weeks after inoculation if plants are growing, but the onset of gall formation may be delayed months in dormant plants. Tumor development is most favored by temperature near 22°C. Temperature above 30°C prevents transformation of normal cells to tumor cells but does not prevent gall growth after transformation.

As surface tissues of galls weather or decay, crown gall bacteria are returned to soil, thus completing the disease cycle. Soils where plants were previously infected may support high populations of *A. tumefaciens* on root fragments left after removal of plants.

Crown gall bacteria are dispersed in soil or irrigation water, on horticultural implements, and on or within plants. Plants with latent infection are a big problem because dormant nursery stock that became infected at the time of harvest may not develop galls until after planting in a new location.

Populations of *A. tumefaciens* in soil diminish markedly during the first several months after removal of plants but then persist at low levels for extended periods. Long-term survival, however, depends upon association with plant roots. *A. tumefaciens* colonizes the root surfaces of many plants without regard to their susceptibility to infection and gall formation.

A. tumefaciens is part of a cluster of species of aerobic, Gram-negative, rod-shaped bacteria that are motile by one to five flagella and are readily cultivable on various bacteriological media. (The pathogen can be isolated from young galls easily but often cannot be isolated from old galls.) In nature, *A. tumefaciens* exists as a mixture of saprophytic and tumorigenic strains. The latter contain the tumor-inducing plasmid and have host ranges that are also determined by plasmid genes. The saprophytic strains have been considered by some authorities to be a distinct species, *A. radiobacter*, but since the differences between it and the pathogen depend upon a plasmid that may be lost or gained, most contemporary bacteriologists consider *A. radiobacter* and *A. tumefaciens* one species. The older name, *A. radiobacter*, has priority.

A. rhizogenes, another member of the cluster, causes hairy root disease, which is occasionally important on apple, pear, raspberry, rose, and walnut. About 35 plant species, mainly in the rose and aster families, are susceptible. Hairy root is characterized by dense clusters of small roots up to about 25 cm long protruding from roots, stems, or galls. A third member of the cluster, *A. rubi* (pathogenic to brambles), is regarded by some authorities as indistinguishable from *A. tumefaciens*.

References: 30, 466, 467, 478, 480, 641, 865, 981, 1011, 1012, 1037, 1165, 1344, 1369, 1617, 2265

A, B. Large old galls at the root crown and on the trunk of a mature butternut tree (NY, Jul).

C. Crown gall on the base of a dying, 1-year-old rhododendron cutting. Infection occurred at the time of propagation (NY, Jul).

D, F. Crown gall on 2-year-old mazzard cherry seedlings. Numerous galls on lateral roots may only retard growth, while one gall encircling the root collar may be lethal (NY, Jul).

E. Crown gall on wintercreeper. Galls formed during the current season are light brown and somewhat spongy; those 1 or more years old are dark brown and hard (NY, Oct).

G. Crown gall on a rose cane. The streaks of gall tissue could have arisen either from multiple infections along a scratch or scrape or from internal spread of bacteria from an initial point of infection (NY, Sep).

H, I. Successively closer cutaway views of a stem of common privet with crown gall developing at the site of a scrape wound. Gall tissue differs from normal bark and wood in texture and anatomy (NY, Sep).

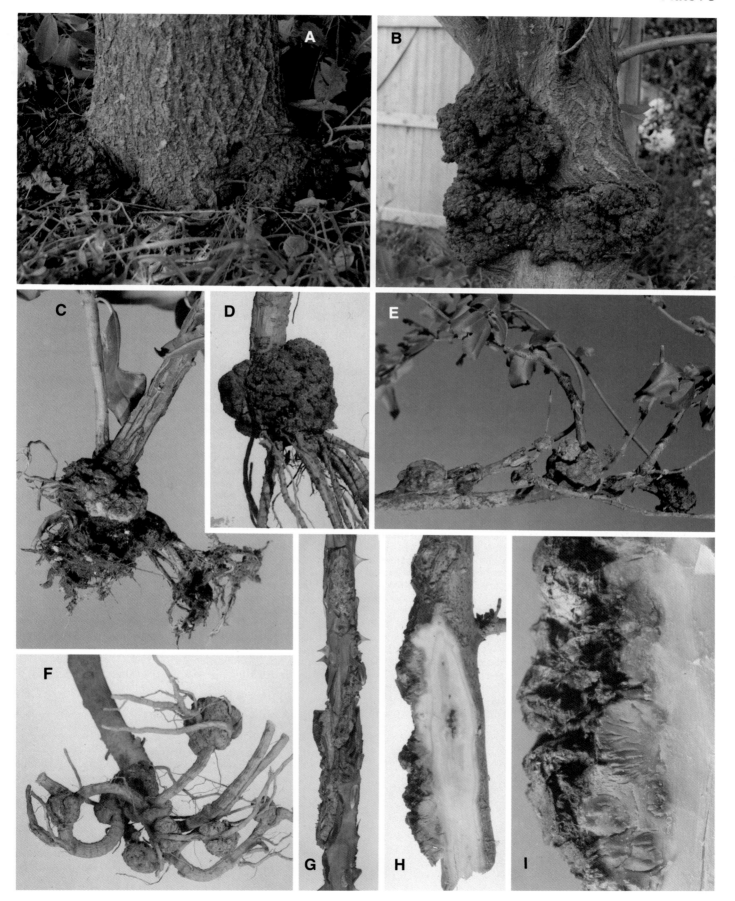

Plate 73

Bacterial Spots of English Ivy and California Laurel and Bacterial Rots of Cacti (Plate 74)

This plate presents two foliar diseases typical of those caused by bacteria and two bacterial rots. The latter are important in plants, such as cacti, that have a large amount of fleshy tissue.

Bacterial spot of English ivy. This disease, known around the world where the host plant is grown, is caused by *Xanthomonas campestris* pathovar *hederae* (syn. *X. hederae*). Severe infection results in leaf distortion, blight, and premature defoliation. Epidemics may develop after propagation of ivy by cuttings from infected stock plants. Damage is likely in nurseries or landscape beds where the pathogen is present and plants receive frequent rain or overhead irrigation. During early research, 12 varieties of English ivy were found to be equally susceptible.

Leaf blades, petioles, and the succulent parts of stems are affected. The bacteria usually enter stomata, but they may also colonize wounds and the tender tips of growing shoots. Lesions, visible first as greenish brown water-soaked spots, expand to 2–10 mm in diameter, and their centers become rust brown to brownish black. Lesions bounded by veins have straight sides and appear angular. A chlorotic halo 1–2 mm wide frequently surrounds a lesion. Under moist conditions, some lesions exude orange-red fluid containing bacteria. In final stages the water-soaked margins disappear, and cracks form in the dry tissue. Some varieties of ivy develop reddish color around old lesions in mature leaves.

Most lesions remain nearly round, but elongate ones form along veins and may extend into petioles, thus killing entire leaves. Similarly, stems may be invaded from petioles. A soft, dark brown decay develops rapidly in succulent stems, but in mature stems the bacteria are restricted to small, dark brown, nongirdling cankers.

Pathovar *hederae* is dispersed by water from lesions on leaves or stems. Symptoms develop 5–21 days after inoculation, depending upon the age of inoculated tissue, temperature, and humidity. The causal bacteria multiply most rapidly at temperatures of 22–26°C. Abnormally severe symptoms sometimes develop when pv. *hederae* is accompanied by certain nonpathogenic bacteria in lesions. Careful hygiene and maintenance of dry surfaces of stock plants for propagation are keys to prevention.

Leaf spot of California laurel. A bacterium once named *Pseudomonas lauraceum* causes leaf spot and, occasionally, severe leaf blight of California laurel. Small, black angular spots delimited by veinlets arise in mature leaves. Lesions may enlarge through coalescence, and yellow halos often develop. Young moist lesions often have a swollen central portion that when broken exudes bacteria-laden fluid. This mechanism permits dispersal by water. Minor wounds seem to be common sites of infection, since lesions are often found on leaves injured by wind. Severe outbreaks have occurred during exceptionally rainy winters in California. The location and mode of survival of the bacteria from year to year have not been ascertained.

The pathogenicity of *P. lauraceum* to California laurel and to avocado was proved by experimental inoculations, but unfortunately, reference cultures were not maintained. Thus under rules of bacterial taxonomy *P. lauraceum* is not a recognized species or pathovar. The pathogen awaits renaming on the basis of contemporary criteria.

Bacterial rots of cacti. At least three species of bacteria cause rot in cacti, macerating the extensive parenchyma tissue in stems of these plants. From an ecological standpoint, the most important disease of this type is bacterial necrosis of saguaro, which threatens the existence of the giant cactus in its natural range. The disease, known since before 1920, is caused by *Erwinia carnegiana*. This bacterium infects wounds made by various agents anywhere on the trunk or branches. When first noticed, lesions may be light-colored or water-soaked spots beneath which the tissues turn mushy and brown to nearly black. The internal extent of a lesion may be indicated by purple discoloration on the surface. If decay is rapid, a brown liquid may exude and run down the plant or drip onto the soil. Bacteria spread throughout large limbs or trunks, degrading them except for woody ribs or causing them to break off at the base. Entire large plants may be killed within 2–3 weeks and small ones in only a few days.

Saguaros often grow in groups in which the infection of one member leads to involvement of all. Those infected secondarily have lesions near the soil line. Contaminated soil is a likely source of inoculum, since *E. carnegiana,* in common with other soft rot bacteria, infests soil during plant decomposition and can survive several months in either dry or moist soil.

Not all infections are lethal. Rot is often arrested by a cork barrier that forms in healthy tissue surrounding a lesion. If the barrier breaks, however, bacteria move farther into susceptible tissue.

Boring insects such as the moth, *Cactobrosis fernaldialis,* are important in this disease. The adults are often contaminated with bacteria, and the larvae tunnel extensively in saguaro, providing infection courts and contributing to the internal spread of decay.

E. carnegiana, once thought to be restricted to saguaro, also infects cholla and prickly pear cacti. In tests it has caused soft rot in diverse cacti, succulents, and vegetable fruits.

E. carotovora subsp. *carotovora* infects cacti also. This bacterium causes soft rot in prickly pear and cholla as well as in dozens or perhaps hundreds of other plants worldwide. The lesions in prickly pear appear as enlarging water-soaked areas that eventually become blackish with purple margins. Occasionally the epidermis ruptures, and a dark orange to brown fluid leaks out. Often, however, the affected segments simply dry out. Some remain attached; others fall. The causal bacteria usually do not spread downward into additional segments. *E. carotovora* subsp. *carotovora* is likely to be present in any natural soil in which there is decaying plant material. Wounds are required for infection, and spreading lesions are most likely in a weakened plant. Healthy, growing plants usually exclude the bacteria by laying down protective corky tissue in response to wounding.

A soft rot of cacti other than prickly pear, caused by a nonfluorescent strain of *Pseudomonas fluorescens,* has been described from England and is suspected to occur naturally in the southwestern United States and Mexico, which export susceptible species. This disease, called orange soft rot, starts at the base of the stem and spreads upward, causing affected tissues to turn orange or orange-brown. Copious orange fluid may ooze from the decaying part. Susceptible cactus groups indigenous to North America include hatchet cactus, hook cactus, living-rock cactus, star cactus, pincushion, visagna, *Pediocactus,* and *Strombocactus.*

References for English ivy and California laurel: 285, 532, 533, 751, 1457, 2173

References for cacti: 7, 46, 236, 387, 533, 967, 1139, 1818

A, E. Bacterial spot of English ivy, caused by *Xanthomonas campestris* pv. *hederae.* A. Necrotic spots and blight in a landscape bed. E. Close view of typical lesions with rust brown centers and water-soaked margins (MS, May).

B. Bacterial spot of California laurel, caused by "*Pseudomonas lauraceum.*" Typical lesions are small, black, and angular, bounded by veinlets of the leaf (CA, Mar).

C, D. Bacterial rot of saguaro, caused by *Erwinia carnegiana.* The dark streak downward from a hole excavated by birds indicates extensive infection (AZ, Apr).

F, G. Bacterial soft rot of prickly pear. F. Grayish brown rotting segments in contrast to normal. G. Close view of collapsing segments. Bacteria-laden fluid from a diseased segment stains the ground below (AZ, Apr).

Plate 74

Bacterial Blights and Cankers Caused by *Pseudomonas syringae* pv. *syringae* and *Xanthomonas campestris* pv. *juglandis* (Plate 75)

P. syringae pv. *syringae* is an unspecialized pathogen that attacks many plants, while *X. campestris* pv. *juglandis* is a host-specialized bacterium that causes blight, or bacteriosis, of walnuts. The abbreviation *pv.* in these names indicates pathovar—a strain that differs from others in the species on the basis of distinctive pathogenicity to one or more plants. Most pathovars were formerly considered to be separate species.

Blights and cankers caused by *Pseudomonas syringae* pv. *syringae*. This bacterium, known since the turn of the 20th century, is distributed globally. First described in connection with blight of lilac, it causes leaf spots, blossom blight (usually called blast), shoot blight, gummosis (of citrus and stone fruit trees), and cankers and dieback of many plants. It is most important on sweet cherry and other stone fruit trees, citrus, peach, and pear.

Symptoms vary with the host. Bacterial blight of lilac (A, B) is characterized by death and shriveling of leaves, shoots, and sometimes flower clusters. Chinese, Japanese, Persian, and common lilacs, especially white-flowering types, are susceptible. The first symptoms are usually irregular to circular dark brown spots with yellowish halos on leaves. Spots coalesce, and blight lesions develop or entire leaves die. Petioles or succulent stems with girdling lesions droop, and the leaves or shoots distal to lesions wither. Flower buds are killed by early-season infections. Parts invaded by the bacteria turn black, and other parts distal to girdling lesions may wilt and turn brown. If a stem becomes woody before infection, it remains erect and may escape girdling. Lesions on woody green twigs often take the form of black streaks (B). Lesions in stems a year or more old are uncommon.

Bacterial blast of apple and pear occurs on many commercial varieties as brown to black lesions on flower parts, green stems, and the bases of fruit clusters. Blossoms are affected (blossom blast) most often on trees predisposed by frost or a cold period in spring. Infection is usually restricted to current growth but sometimes extends into fruit spurs and 1-year-old twigs, causing dieback or cankers. Brown streaks on the surface of the sapwood may extend several centimeters beyond the margin of necrosis in the bark of apple twigs. Blast cankers on pear twigs are usually tan to light brown. The periderm of diseased apple or pear shoots and spurs tends to separate from underlying tissues, so that a superficial layer appears papery, a diagnostic symptom. Infected pear fruits show discrete black, depressed lesions that occasionally involve the entire fruit. Lesions do not usually exude drops of fluid as is characteristic of fire blight (Plates 76–77).

On stone fruit trees such as almond, apricot, sweet cherry, peach, and plum, pv. *syringae* kills buds in winter and blossoms in spring, causes necrotic spots on leaves and fruit, and causes cankers that enlarge on branches and trunks during host dormancy. Blossom blight is most severe in years when frost occurs during bloom. Cankers usually enlarge from the bases of infected spurs. Narrow brown streaks extend into normal tissue at the upper and lower margins of cankers. Cankers tend to be roughly elliptical and to exude gum or sour-smelling liquid. The extent of gummosis varies with host species. It is most profuse on apricot.

Other diseases caused by pv. *syringae* include blossom blast of grapefruit, lime, tea rose, and Persian and Manchurian walnuts; leaf blight and brown rot of mandarin orange; blossom blast and black pit of the fruit of sweet orange; leaf spots, vein blackening, and tip dieback of Japanese, Norway, and red maples; leaf and shoot blight of mock orange; leaf spots of basswood, hibiscus, and saucer magnolia; fruit spot of avocado; stem cankers (frost cankers) on black poplar; and dieback of Monterey pine seedlings. In nurseries in northwestern states, pv. *syringae* also causes various of the above symptoms on trembling aspen, flowering dogwood, golden-chain tree, and callery pear. Other woody hosts include flowering and Himalayan ash, forsythia, primrose jasmine, coast live and interior live oaks, oleander, orach, Natal plum, and black and Carolina poplars. Many herbaceous plants are attacked also.

P. syringae is an aerobic, Gram-negative, rod-shaped bacterium with rounded ends. It is motile by one to several polar flagella. When growing on certain diagnostic media it produces a substance that fluoresces greenish blue when viewed under ultraviolet radiation. This species, including pv. *syringae*, commonly inhabits plant surfaces in the absence of disease. Strains of pv. *syringae* vary in pathogenicity; no single pathogenic isolate will infect all recorded hosts.

Two intrinsic features of pv. *syringae* greatly enhance its pathogenic ability. In culture and in infected plant tissue it produces syringomycin, a broad-spectrum toxin that destroys host cell membranes and thus contributes to symptom development. The toxin is not the sole determinant of pathogenicity, however, since the tissues of plants both susceptible and resistant to pv. *syringae* are sensitive to the chemical in the absence of bacteria. The second and more unusual feature is the capability of some strains to act as nuclei for formation of ice crystals. At temperatures a few degrees below 0°C, most plants that grow in temperate zones escape significant frost damage because water in their tissues remains in liquid form, supercooled. If pv. *syringae* or other ice-nucleating organisms or substances are present, however, ice crystals form and disrupt plant tissue, resulting in typical symptoms of frost or freeze damage. Pathovar *syringae* in combination with frost or freezing causes severe damage to plants that would not be harmed by either agent alone. In addition, the bacterial population rises rapidly in host tissues that have been chilled but not visibly damaged by exposure to temperature below 0°C. Freezing and untimely pruning (in autumn or winter) often predispose plants to bud blight or cankers caused by pv. *syringae*. Bacterial canker and freeze damage are major factors in the failure of young peach trees in the southeastern United States.

Bacterial blight of walnut. This blight, also known as walnut blight or walnut bacteriosis, is caused by *Xanthomonas campestris* pv. *juglandis* (syn. *X. juglandis*). It is one of the most serious diseases of walnut trees in commercial orchards. The disease in North America was first noted in California in about 1891. The causal agent, not then known, was thought to have been imported with nursery stock. Now it occurs around the world where Persian walnuts are grown. Although widespread in the eastern and southern United States, walnut blight is important mainly in the commercial orchards of the West and Pacific Northwest. Persian walnut is most susceptible; black, Hinds, Japanese and 'Paradox' hybrid walnuts are less so.

Economic loss is due to fruit infection, but leaves, buds, catkins, and succulent twigs of the current season are also affected. Twigs a year or more old are not susceptible. The first infections each year usually arise on catkins and young leaves. Catkins become infected as soon as they emerge from the bud, and they show symptoms after elongation. Diseased florets droop and appear water-soaked, then turn black. One floret or all on a portion of the catkin may be infected. Lesions on the rachis of the catkin cause deformity. Bacteria multiply in the diseased portions and contaminate the pollen produced by normal parts of the catkin.

Lesions develop on all parts of leaves. Those on leaf blades begin as water-soaked, pale yellowish green spots that enlarge up to 4 mm and at maturity are angular and brown with yellowish green margins. They are usually most numerous near leaflet edges and often cause deformity but seldom cause abscission. Lesions on veins, petioles, and leaf rachises are angular and dark brown to black. Infection of flower or vegetative buds leads either to tiny inconspicuous lesions or to death of the buds.

Fruits become infected by pv. *juglandis* at any time during growth. The first symptom is usually a black spot at the blossom end, which may enlarge to involve the entire young fruit. Infections on partly

---

A–E. Diseases caused by *Pseudomonas syringae* pv. *syringae*.

A, B. Blight of common lilac. A. Blighted shoots. B. An elongate, black canker on a current season's twig (NY, Jun).

C. Pear blast on 'Bartlett' pear. One-year-old twigs were killed just as new growth started (OR, Jul).

D, E. Canker and gummosis on 'Sam' sweet cherry. Extent of the canker is revealed by shaved bark at margins (OR, Jul).

F, G. Bacterial blight, caused by *Xanthomonas campestris* pv. *juglandis*, on Persian walnut fruit. F. A mature lesion on the side of a full-grown fruit. G. A young lesion at the blossom end (CA, Jul).

Plate 75

(Plate 75, continued)

grown fruit result in black depressed areas where the epidermis ruptures and drops of black liquid containing bacteria may exude. Fruits infected during or soon after pollination time usually drop; those attacked later remain attached, but the meats frequently fail to develop.

The pathogen survives winter on plant surfaces and in and on buds—both symptomatic and apparently noninfected buds. Survival in twig lesions and diseased fruit on the tree is relatively unimportant. Inoculum for both primary and secondary infections is dispersed by running and splashing water. In addition, contaminated pollen results in aerial transmission not only to the stigmas of flowers but to all susceptible parts. Diseased leaves are a source of secondary inoculum for late-season infection of nuts and infestation of developing buds and catkins.

X. campestris pv. *juglandis* is a rod-shaped, Gram-negative bacterium, motile by one polar flagellum, that forms glistening yellow colonies on laboratory media. It infects through stomata and wounds and is confined for the most part to parenchyma. The incubation period is 5–34 days (normally 10–15 days), depending upon environmental conditions and the age of host parts. The bacteria at first spread between host cells, then cause cellular collapse and the formation of cavities that fill with bacteria in a slimy matrix. This is the substance that exudes to the surface, where the bacteria are available for dispersal. Disease can appear at temperatures of 4–30°C and develops most rapidly at 20–30°C after the infection of succulent organs. Rainfall from the time of bud opening until nuts are about half grown favors epidemics. Wetness of plant surfaces for only several minutes is sufficient for infection. In the West, early-blooming walnut cultivars tend to be damaged most because rainfall and fog are abundant in the early part of the growing season.

References for Pseudomonas syringae: 272, 298, 299, 316, 342, 387, 426, 456, 533, 546, 577, 987, 1005, 1035, 1105, 1145–47, 1634, 1738, 1810, 2126, 2211

References for Xanthomonas campestris: 425, 533, 1316, 1367, 2211

Fire Blight (Plates 76–77)

Fire blight is the most important bacterial disease of rosaceous plants. Known in the northeastern United States since before 1800, it now occurs wherever susceptible plants grow in Canada and the United States. Only members of the rose family are affected. Apple, mountain ash, cotoneaster, firethorn, hawthorn, pear, and quince are most commonly damaged, but the host range includes over 130 species in nearly 40 genera. Many countries have enacted regulations to prevent importation of the causal bacterium, *Erwinia amylovora*, with plant materials, but somehow the pathogen has slipped through. It now occurs in New Zealand, Chile, Egypt, England, and parts of northern Europe. Fire blight was the first plant disease shown to be caused by bacteria, and *E. amylovora* was the first plant pathogen shown to be transmitted by insects.

The genera with plants susceptible to fire blight are listed below.

Amelanchier serviceberry	*Exochorda* pearlbush	*Potentilla* cinquefoil
Aronia chokeberry	*Fragaria* strawberry	*Prinsepia* —
Aruncus goatsbeard	*Geum* avens	*Prunus* apricot, cherry, plum
Chaenomeles flowering quince	*Heteromeles* toyon	*Pyracantha* firethorn
Cotoneaster cotoneaster	*Holodiscus* creambush	*Pyrus* pear
Cowania cliff rose	*Kageneckia* —	*Raphiolepis* Indian hawthorn
Crataegomespilus —	*Kerria* Japanese rose	*Rhodotypos* jetbead
Crataegus hawthorn	*Malus* apple, crabapple	*Rosa* rose
Cydonia quince	*Mespilus* medlar	*Rubus* brambles
Dichotomanthes —	*Osteomeles* —	*Sorbaria* false spirea
Docynia —	*Peraphyllum* —	*Sorbus* mountain ash
Dryas mountain avens	*Photinia* photinia	*Spiraea* spirea
Eriobotrya loquat	*Physocarpus* ninebark	*Stranvaesia* —

The principal symptoms are blight of blossoms, fruitlets, spurs, and leafy shoots and the formation of cankers that cause dieback of twigs and branches. The first symptom annually is usually blossom blight. Individual flowers or entire flower clusters appear water-soaked and then quickly droop, shrivel, and turn brown. Some fall, but many remain attached. Lesions progress from blossoms into peduncles and fruit spurs (Plate 77D) and thence often into twigs. Fruitlets that escape infection via blossoms may be invaded by way of peduncles on blighted spurs. Spur leaves become infected as bacteria pass from the spur into petioles.

In regions with cool weather during the period of bloom in spring, blossoms may escape infection because temperature is not high enough for much bacterial activity. The first symptom may then be shoot blight (Plate 76C). The succulent tips of blighted shoots often droop. Blighted parts are nearly black in pear, brown in most other common hosts. If many shoots are blighted simultaneously, the tree or shrub may appear scorched as if by fire; thus the name *fire blight*. Internally the woody parts of newly infected shoots or twigs and the surface of the sapwood beneath canker margins are discolored reddish brown (Plate 76D). This diagnostic feature is useful for differentiating fire blight in pear from blast disease caused by *Pseudomonas syringae* (Plate 75).

The progress of infection slows markedly in woody twigs and branches, where discrete cankers form. These are at first small, brown to black, slightly sunken areas. During the dormant season, the edges often crack, and where the margins remain inactive, cankers become distinctly sunken as adjacent healthy tissues grow. During the growing season, the active margins of cankers may appear slightly raised or blistered (Plate 76F).

The most damaging infections are those arising on the trunk or root collar as the result of wound infection by bacteria carried in water from diseased parts above. Cankers in these locations may kill entire trees.

During moist weather, droplets of cream- to honey-colored, turbid fluid called ooze emerge from lenticels and tiny wounds on the surfaces of recently infected parts. Ooze consists of a dense suspension of bacteria in a watery polysaccharide matrix. Hydrostatic pressure forces it out of infected tissues. The bacteria thus become available for dispersal. Under somewhat drier conditions, strands of bac-

Fire blight is most destructive in apple and pear orchards. It was responsible for major shifts in fruit-growing practices (away from highly susceptible apple and pear varieties) in both eastern and western North America. Economic loss is caused by death of blossoms, fruit spurs, branches, and major limbs. Entire trees are sometimes killed, but orchardists usually destroy unproductive trees before they succumb to the disease. Damage to ornamental plants is similar to that of fruit trees—disfigurement or death. In the northern Great Plains where severe winters limit the choice of flowering ornamentals mainly to rosaceous plants, fire blight is their most important disease.

A. Shoot blight and branch dieback in common pear, caused by the fire blight pathogen, *Erwinia amylovora* (NY, Sep).

B, F. Dieback in European mountain ash, a result of fire blight cankers on the trunk and a major limb. Note blistered appearance of bark at the edge of the canker (NY, Sep).

C, D. Shoot blight in *Malus baccata* 'Gracilis'. Note reddish brown discoloration of inner bark and the cambial region exposed by shaving bark (NY, Sep).

E. A fire blight canker on a young branch of 'Ida Red' apple. Bacteria entered the branch from a killed twig at upper right (NY, Aug).

Plate 76

teria in a semisolid matrix may be forced out. The strands either disperse in water or, if they are dry, may be airborne for short distances.

The polysaccharide, produced by the bacteria, apparently plugs xylem of shoots and twigs and thus plays a role in inducing the wilt portion of the fire blight syndrome. The biochemical mechanism of cellular death, however, is not known with certainty. A lipopolysaccharide from *E. amylovora* has been implicated as a host-selective toxin, but this finding awaits corroboration.

The annual cycle of fire blight is uncomplicated. Bacteria overwinter at the margins of cankers. With the onset of warm weather in spring, renewed bacterial activity results in formation of ooze droplets. Bacteria in ooze or strands are transmitted by insects, splashing or running water, birds, or humans. The most common vectors are pollinators (bees and flies), but flying and crawling insects in 77 genera have been implicated. In addition, wind-driven rain splash can disperse bacteria, a significant means of spread in nurseries and landscape plantings. Infection courts include stigmas and the nectaries of blossoms, fresh wounds on any plant part, and natural openings such as stomata and hydathodes in leaves and lenticels in succulent twigs. Bacteria in these places multiply rapidly and move intercellularly en masse in a fluid matrix of bacterially produced polysaccharide and degraded host cell wall material. Under favorable conditions the bacterial population can double every 2 hours. Within several days, the bacteria may spread 15–30 cm in shoots. All living tissues are invaded, and blight appears within 1–3 weeks, depending upon temperature and moisture.

Fire blight bacteria have occasionally been detected in symptomless apple and pear buds, but the epidemiological significance of this niche is unknown.

E. amylovora, a member of the Enterobacteriaceae, is a Gram-negative, rod-shaped bacterium that is motile by peritrichous flagella. Strains of *E. amylovora* vary in virulence, and this variation may apparently be conditioned by the level of host resistance. Strains that were repeatedly inoculated into and reisolated from highly resistant apple selections gradually increased in ability to cause blight in those selections.

Epidemics are favored by warm, humid (relative humidity above 60%) weather. The bacteria can multiply at temperatures of 15–32°C but do so most rapidly at 27–29°C. Symptoms usually appear 1–3 weeks after inoculation, depending on temperature and humidity. Devastating epidemics sometimes follow hail storms because bacteria enter wounds. Conditions of plant culture also influence the severity of blight. Practices that promote succulent growth—sprinkler irrigation in nurseries and high fertility or severe pruning—favor severe disease. For a given level of inherent susceptibility, plants growing in well-drained, moderately fertile soil with a balanced supply of nutrients are least damaged.

Plants that resist the fire blight bacterium are known within even the most susceptible genera and species. The 'Bradford' flowering pear, for example, is rarely affected, and then sustains only minor twig blight that does not noticeably disfigure the tree.

Crabapple cultivars that have been observed for many years and found to be relatively free from fire blight in addition to other diseases (see Plate 43) include 'Baskatong', 'Centurion', 'David', 'Dolgo', 'Henry Kohankie', 'Liset', 'Professor Sprenger', 'Sugartyme', and *Malus sargentii* 'Tina'. Among species of *Malus*, high levels of resistance have been found in selections of *M. fusca*, *M. sargentii*, *M. sieboldii*, and *M. yunnanensis*. Unfortunately, resistance to both fire blight and scab in crabapple cultivars is uncommon.

Resistant firethorns include *Pyracantha coccinea* cv. 'Sensation', *P. koidzumii* and its cultivar 'Santa Cruz Prostrata', and the hybrids 'San Jose' and 'Shawnee'. Again, resistance to both scab and fire blight is uncommon.

Among cotoneasters, the following are reported to be resistant: *Cotoneaster adpressus*, *C. apiculatus*, *C. dielsianus*, *C. foveolatus*, *C. franchetii*, *C. integerrimus*, *C. nitens*, and *C. zabelii*.

Resistant apple cultivars include 'Delicious', 'Red Delicious', 'Northwestern Greening', 'Liberty', 'Macfree', 'Nova Easygro', 'Prima', 'Priscilla', 'Stayman', and 'Winesap'. The pear cultivars 'Mac', 'Maxine', 'Moonglow', and 'Giant Seckel' are also somewhat resistant.

Destruction of ornamental plants by fire blight can be delayed or halted by carefully pruning out all blighted twigs or branches, making cuts at least 20 cm below cankers or margins of dieback while plant surfaces are dry, preferably during cold weather in late winter. Tools must be disinfested between cuts.

References: 10, 458, 619, 1403, 1633, 1796, 2004, 2284

A. Bacterial ooze issuing from lenticels in bark of a diseased limb of 'Ida Red' apple (NY, Jun).

B. Typical winter appearance of a fire blight canker on a young branch of 'Ida Red' apple. Bacteria are more likely to resume pathogenic activity in spring at the smooth part of the canker margin than where it is cracked (NY, Feb).

C. Fire blight on firethorn. Spurs and spur leaves have died as a result of bacterial movement downward from blighted blossoms. Firethorn leaves are seldom infected directly (NY, Aug).

D. Blossom blight on common pear. The black lesion on the spur indicates internal bacterial movement. Although not yet symptomatic, the twig is already infected (NY, May).

E, F. Blossom and shoot blight phases of fire blight on loquat (FL, Apr).

Plate 77

Bacterial Cankers (Plate 78)

Only a few canker diseases caused by bacteria in trees and shrubs of North America have been well characterized. Without exception these are diseases of fruit and nut trees. Fire blight (Plates 76–77) and bacterial blight and canker caused by *Pseudomonas syringae* (Plate 75) affect many woody plants in both orchards and landscapes, but aside from these, no bacterial canker disease important in forest, shade, or ornamental trees is known on this continent. This plate presents deep bark canker of walnut. Three additional canker diseases apparently caused by bacteria, but not well characterized, are shown for comparison.

Deep bark canker of walnut. Persian walnut trees in California are subject to deep bark canker, a bleeding lesion caused by *Erwinia rubrifaciens* in the inner bark of trunks and scaffold branches. This disease, unknown until 1962, seems to have been insignificant before the advent of mechanical harvesters in walnut orchards. Apparently infection begins in minor wounds such as those caused by the pads of machines that shake the trees to dislodge ripe fruit. The bacterium infects nonconducting secondary phloem primarily and causes elongate lesions that are not visible on the bark surface. Their presence is indicated by bleeding of dark reddish brown, slimy liquid from small cracks, primarily in summer. On drying, the fluid leaves a prominent brown streak. When severe, the disease destroys the productivity of the trees.

If one cuts into bark at a point of bleeding, dark brown to black streaks of varying width are found aligned with the axis of the trunk or branch. These result from the movement of *E. rubrifaciens* in the sieve tubes of 1-year-old or older phloem. This phloem is no longer functional in food conduction. Internal spread, detected up to 3 meters beyond visibly infected areas, explains the development of cankers on limbs distant from sites of primary infection. Dead streaks often merge into bands a meter or more in length. The innermost (functional) phloem and the cambium remain alive and free from bacteria except in spots 1–2 mm in diameter where infection has spread inward in the parenchyma cells of medullary rays. Localized killing of the cambium there results in dark brown pits on the surface of the sapwood. Occasionally, dead streaks involve inner bark, cambium, and superficial sapwood. Callus formation at the edges of dead streaks results in slight swellings and cracking of the bark surface. The dark fluid comes from these cracks.

Only Persian walnut is known to be affected. The cultivar 'Hartley' is most susceptible. 'Franquette' and 'Payne' are sometimes attacked. 'Gustine' and 'Howe' were susceptible when inoculated experimentally.

E. rubrifaciens, a member of the Enterobacteriaceae, is a Gram-negative, rod-shaped bacterium, motile by peritrichous flagella. It produces a water-soluble red pigment in culture media, which distinguishes it from the related but less important walnut pathogen, *E. nigrifluens.* The latter causes extensive superficial necrosis of Persian walnut bark (shallow bark canker). Both diseases may occur in a given tree, and both pathogens in a single lesion.

E. rubrifaciens may be dispersed as far as 6 meters by splashing and wind-blown water, but the more important means of dispersal is harvesting equipment that becomes contaminated with bacteria from the dark deposits on bark. Bacteria remain viable in dried deposits for up to 4 months. Sap-sucking birds are also suspected as vectors because cankers develop at their feeding wounds. Growth cracks and pruning wounds (if they are fresh) are also sites of infection.

Only mature parts of susceptible trees become infected. One-year-old seedlings or 1-year-old twigs of mature trees did not develop cankers when inoculated experimentally.

Most infections and canker elongation occur during summer. This seasonal trend is related to the temperature preference of *E. rubrifaciens,* which multiplies most rapidly at 26–32°C. The inner bark of walnut trunks in California reaches this temperature range during summer afternoons.

Other bacterial cankers. Bleeding cankers on many shade, ornamental, and forest trees, apparently caused by unidentified bacteria, are common but seldom seem to persist or cause much damage. Bacteria are blamed on the basis of isolation from dying or recently killed bark, but tests of pathogenicity have seldom been made. Some fungi such as *Cryptosporiopsis* sp. on red maple (Plate 96) and *Phytophthora cactorum* on maples and other trees (Plate 137) also cause bleeding cankers, so this symptom should be interpreted cautiously.

Bacterial infections that cause bleeding, usually localized around small wounds, often occur in the bark of trees affected by another disease, injury, or environmental stress. Galleries made by bark beetles or ambrosia beetles in previously stressed trees seem to be common sites of infection. Bacteria are associated with localized death of inner bark and often the cambium. Liquid flowing from the necrotic tissue forms a wet spot or streak on the surface, and this may be the only external symptom. Often, but by no means always, the liquid has a distinctly sour odor characteristic of the organic acids produced by many bacteria. Bleeding of sour liquid from small cankers should not be confused with that from wetwood (Plate 185).

A second type of bacterial canker, also seemingly associated with previous stress to the tree, results in production of white froth with an alcoholic odor at cracks on the bark surface. The microorganisms in these cankers apparently produce both gas and alcohol during fermentation of sap. This "alcoholic flux" is reportedly common on elm, sweet gum, and oak in the Midwest.

References: 330, 387, 616, 1107, 1421, 1690–92, 2211, 2212

A, B. Deep bark canker of Persian walnut, caused by *Erwinia rubrifaciens.* Brown streaks on bark are dried, bacteria-containing fluid that exudes from cracks overlying elongate cankers in inner bark (CA, Jul).

C, G, H. Unidentified bacterial canker of black oak. Bleeding spots (G) were associated with tiny wounds apparently made by insects. Lesions involved cambium and outer sapwood (C) as well as bark (NY, May).

D. Alcoholic flux at small cracks in bark of mimosa. Lesions in this case followed systemic infection of sapwood by *Fusarium oxysporum* f. sp. *pernicosum,* causing mimosa wilt (Plate 183). Whether the fungal pathogen or bacteria caused this flux is unknown (NY, Jun).

E, F. A bleeding canker apparently caused by bacteria on horse-chestnut. Extensive staining of bark (E) is caused by exudation from a canker about 10 cm long. Water-soaked sapwood beneath canker, exposed by removing bark, is dead and stained dark brown (NY, Jul).

Plate 78

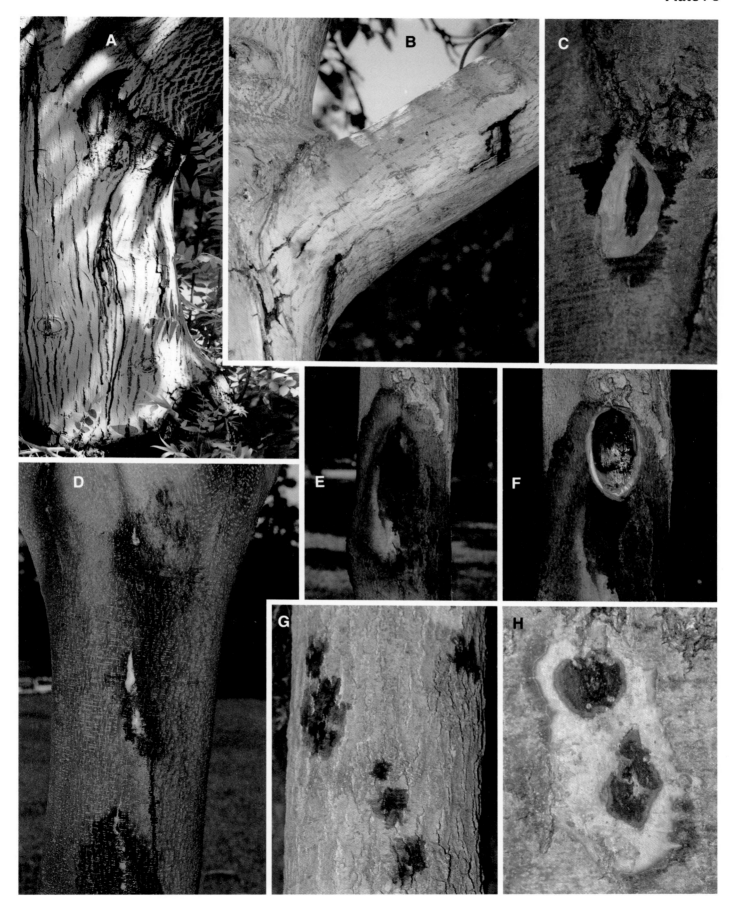

Smooth Patch and Bark Rot and Hendersonula Dieback (Plate 79)

This plate presents two unrelated topics, the decay of bark on living trees by saprophytic fungi and a dieback disease that affects trees and shrubs in regions with very hot summer weather.

Smooth patch and bark rot. Several fungi colonize and decompose the dead, corky, outer layers of bark of living tree trunks. They cause no known harm to the trees, but their sporocarps on the bark may be misinterpreted as indicating the presence of a canker or an extensive dead area with decaying wood beneath.

Smooth patch (A, B), also known as white patch and bark patch, is caused by species of *Aleurodiscus, Dendrothele,* and *Hyphoderma* (Aphyllophorales, Corticiaceae)—fungi that form smooth, more or less disklike white fruit bodies (basidiocarps) that may aggregate or coalesce into larger structures on the bark. Of several dozen species in North America, only two are plant pathogens. *A. amorphus* causes cankers on trunks and branches of unthrifty firs and other conifers, and *A. canadensis* causes twig blight of red spruce. The rest are saprophytic colonists of bark and wood, common on dead twigs. The best known member of the saprophytic group is *A. oakesii,* which causes smooth patch on American elm; American hornbeam; hop hornbeam; sugar maple; bur, Oregon, post, white, and Arizona white oaks; and occasionally yellow birch and willow.

Smooth patch results from the decomposition and sloughing of rough or fissured outer bark in a slowly expanding area on the trunk. The resulting slightly sunken, relatively smooth light gray area contrasts with normal bark texture and color. The largest patches apparently form by confluence of small ones. Basidiocarps of *A. oakesii* are visible all year, but in dry weather they may curl or shrivel and become inconspicuous. Individual basidiocarps are 1.5–5.5 mm in diameter but appear much larger where they are confluent or growing in clusters.

Other smooth-patch fungi include *A. diffisus* on madrone; *D. acerina* on ash, beech, sugar maple, and oaks (including live oak); *D. candida* on hickory, mango, red maple, Oregon oak, several eastern oaks, and pecan; *D. griseo-cana* on elm, grapevine, hop hornbeam, linden, and bur oak; *D. macrodens* on ash and oak; *D. nivosa* on arborvitae, Atlantic white cedar, and species of *Juniperus* (particularly eastern red cedar); *D. strumosa* on pignut hickory; and *H. baculorubrense* on live oak.

Perenniporia phloiophila is one of several pore fungi (Aphyllophorales, Polyporaceae) that decay the outer bark of living trees but do not cause smooth patches. It apparently occurs only on the trunks and large limbs of live oak and is found from South Carolina to Texas and Mexico. Its perennial basidiocarps develop more or less flat against the bark as units up to 6 cm wide. They may become confluent and up to 1 meter long or wide. The pore surface is cream colored to pale buff, and the circular to angular pore mouths number three to five per linear millimeter. This fungus is easily confused with *P. medulla-panis,* which decays the bark and wood of various deciduous trees as well as fence posts and structural timbers throughout eastern North America and in California. Substrates of the latter fungus include ash, basswood, beech, cherry, chestnut, elm, hop hornbeam, black locust, honey locust, maple, oak, poplar, sagebrush, sumac, and tulip tree. Microscopic examination of spores and anatomical characters is necessary to distinguish *P. medulla-panis* from related fungi.

Hendersonula dieback. This disease, also known as branch wilt, limb wilt, or sooty canker, occurs in deciduous trees with smooth or thin bark growing where summer heat becomes intense. The disease is characterized by cankers, wilting, and dieback of scattered limbs. It occurs in various parts of the world including western North America as far north as British Columbia. The pathogen, *Hendersonula toruloidea* (Deuteromycotina, Coelomycetes), became well known in Arizona and California because of damage in walnut and citrus groves. Suscepts include almond, apple, apricot, European chestnut, chinaberry, citrus (grapefruit, lemon, lime, and mandarin orange), euonymus, fig, grapevine, guava, madrone, mango, white mulberry, plum, poplars (black and white poplars and Fremont cottonwood), Para rubber tree, sagebrush, and Hinds and Persian walnuts. The majority of plants listed here have been found infected in the United States. Many other plants are also affected.

Symptoms appear during summer. Leaves on scattered branches wilt, turn brown, and cling to dead twigs or senesce and drop prematurely. Branches die back to below the locations of sooty-appearing cankers. The sootiness is due to dry black masses of fungal spores that develop just beneath the periderm and cause it to peel off as a thin layer. Bark beneath the spore mass is brown to nearly black, and the underlying sapwood is stained gray to black. Sapwood may be killed considerably beyond the margins of lesions in bark. Citrus and stone fruit trees often exude gum at the margins of lesions. Sunken cankers may develop on branches that survive infection.

Infection occurs through cracks or wounds and becomes serious only in tissues stressed by heat, freezing, or other factors. Most infections are thought to begin during mild wet weather in winter and to remain latent or restricted until host defenses are impaired in summer. Branches with sunburn (Plate 233) become highly susceptible even if the injury is mild. Waterlogging of soil, fluctuating water supplies, and predisposition by other pathogens have also been linked to susceptibility.

The sooty spore mass arises as a dense mat of branched, septate brown hyphae the cells of which become somewhat rounded and separate, producing a powdery layer of one- and two-celled spores (thallospores). These are dispersed by air or water, can survive hot dry weather for long periods, germinate readily in fresh wounds, and cause new infections.

Small pycnidial stromata may also form beneath the periderm in a minority of cases. The stromata, up to 0.5 mm in diameter, contain one or more cavities in which conidia form. Mature stromata break through the bark surface, and conidia in black blobs or cirrhi exude from stromata during wet weather. Conidia are three-celled, usually $10–14 \times 4–6$ μm, and narrowed at both ends. The central cell is darker than the others. Mycelium that breaks up into thallospores (form-genus *Scytalidium*) also precedes formation of pycnidia in culture.

Most strains of *H. toruloidea* are adapted for growth and reproduction at relatively high temperatures. The thallospores and conidia of isolates from regions with hot dry summers germinate most readily at 30°C or perhaps higher. Mycelial growth in culture is most rapid at 33–36°C and more so at 39° than at 18°C. Isolates from madrone in the Pacific Northwest, however, grew most rapidly and caused most rapid canker expansion at temperatures near 25°C.

Application of whitewash to bark has been shown to prevent infection by *H. toruloidea* in experimentally inoculated trees. Presumably whitewash reflects radiation that would otherwise cause heat stress or sunburn.

References for smooth patch and bark rot: 389, 425, 639, 640, 672, 727, 783, 1093, 1126, 1127, 1436, 1759, 1829, 1979

References for Hendersonula dieback: 295, 387, 461, 545, 1379, 1467, 1861, 2208, 2211

A, B. Smooth patch caused by *Aleurodiscus oakesii* on Arizona white oak. A. A large smooth patch on a mature tree. B. Close view of basidiocarps of a young colony of *A. oakesii,* approximately natural size (AZ, Jul).

C. Sporocarps of *Perenniporia phloiophila,* shown about one-half natural size, on the bark of live oak (TX, May).

D–F. Hendersonula dieback of white mulberry. D. Dieback. E, F. Diseased branches with periderm cracked and peeling, exposing sooty masses of spores (AZ, Apr).

Plate 79

Seiridium Canker of Cypress (Plate 80)

Seiridium canker of cypress, also known as Coryneum canker, is one of the famous examples of a disease innocuous to a tree species in the natural forest but devastating to the same species planted beyond its natural range. The tree, Monterey cypress, grows naturally in groves along the central Pacific coast of California. The pathogen is *Seiridium cardinale* (syn. *Coryneum cardinale*) (Deuteromycotina, Coelomycetes), a fungus that began to cause damage to planted Monterey and Italian cypresses in California in about 1915. Its geographic origin is unknown. During the 1920s to 1940s it destroyed the majority of Monterey cypress trees on inland sites in central and southern California. These sites were characterized by a prolonged dry period with high daytime temperatures each summer. Trees in the coastal groves with moderate temperature and frequent fog remained unaffected.

S. cardinale infects several other members of the cypress family as well, mainly in landscapes and plantations but sometimes in natural forests. In North America it is confined to western areas, occurring as far north as British Columbia, where it causes twig blight of young western red cedar in natural stands. It also occurs in Europe, the Middle East, South Africa, Argentina, Australia, and New Zealand. Italian, Leyland, and Monterey cypresses are most susceptible. Alaska cedar, eastern red cedar, and Gowen cypress, although not damaged naturally, were quite susceptible when tested. Plants only occasionally infected in nature or relatively resistant in tests include Oriental arborvitae; incense cedar; western red cedar; Arizona, Lawson, Mac-Nab, Mendocino, Modoc, Portuguese, and tecate cypresses; and Chinese and savin junipers. American arborvitae, Japanese cedar, Lawson cypress, and Sierra juniper are resistant. Resistant trees halt infection rapidly and then form a cork barrier that isolates the lesion.

Symptoms are described and illustrated for Monterey cypress, the species that sustains the greatest damage. The most conspicuous symptom, fading and death of foliage on twigs, branches, or tops of trees, occurs in all seasons but most commonly in spring during rapid tree growth. Yellowing and browning of old foliage usually precede fading and death of new foliage. Dead foliage is shed slowly during the ensuing year.

Branch or top death results from girdling by lesions that commonly start in wounds on twig or branch bases. Infection of apparently undamaged parts has also been noted but is unconfirmed. Lesions can first be detected when they are 10–15 mm long. The bark turns brown, and drops of resin exude. Resin also accumulates in irregular streaks in killed bark. Typical cankers are lens shaped and about 3.5 times as long as wide. As diseased bark dries and growth of surrounding tissues continues, cankers become distinctly sunken. Resin exudation continues at the margins of expanding cankers, but the volume of resin varies with age and with the vigor of the tree. Old or slowly growing trees may exude little resin.

Cankers on young, rapidly growing trees are most common on mainstems and may attain lengths of 60–90 cm before trees are girdled. Branch cankers predominate on old or crowded trees. Cankers increase in length 10–20 cm per year, depending on climate—faster in warm inland locations than near the ocean. Cankers on mainstems often extend several centimeters out on the bases of branches, where fruiting of the pathogen is subsequently prolific. Very small trees may be killed within 1 year. Large ones typically die in 5–12 years from the cumulative effects of many cankers. Death is hastened by water shortage and sometimes by insects, especially cypress bark beetles (see Johnson and Lyon, 2nd ed., Plate 24), that attack weakened trees.

If weather is mild and moist, *S. cardinale* produces acervuli within 4–8 weeks after death of bark. The acervuli appear as scattered black pustules 0.3–1.5 mm in diameter that break through the bark surface and open widely during wet weather. During hot dry weather, however, their development is arrested. Masses of conidia exude from acervuli while bark is moist. Conidia are six-celled, with the majority measuring 21–26 × 8–10.5 μm. The end cells are pointed and colorless; the four central cells are olive brown. Conidia are capable of surviving several weeks on exposed cypress foliage and for 2 years or more in protected dry situations.

The sexual state of *S. cardinale* was discovered and found to be widespread in California in cypress bark dead 1 year or more. It was tentatively classified in the genus *Leptosphaeria* (Pleosporales, Pleosporaceae) but was not named or fully described.

Most infections are presumed to be started by conidia during the rainy season (autumn through spring in California). Branches 1–4 cm in diameter are commonly the first infected and later serve as sources of conidia that cause a buildup of disease on other branches and on mainstems. The parasite colonizes both bark and outer sapwood. Parenchyma cells are killed, apparently by a toxin, as much as 1 mm in advance of hyphae.

S. cardinale is dispersed locally as conidia in splashing or running water. Airborne ascospores are presumed to function for long-distance dispersal. Transmission on pruning tools and long-distance movement on infected planting stock have also been noted. In addition, transmission by insects is possible. The cypress bark moth (see Johnson and Lyon, 2nd ed., Plate 23) lays eggs preferentially on bark that is diseased or otherwise damaged, and in so doing the insect becomes contaminated with conidia. *S. cardinale* has been isolated from various body parts of egg-laying adults, but whether they introduce spores to wounds or other potential sites of infection is unreported. Moth larvae frequently tunnel at the edges of cankers, and reddish brown resinous frass may extrude from their galleries and cling to the bark.

S. cardinale grows readily in pure culture. Isolates from California grew most rapidly at 26°C, slowly at 6° and 30°C, and not at all at 35°C.

S. cardinale is one of three fungi that cause similar diseases of trees and shrubs in the cypress family. The other two are *S. cupressi* and *Lepteutypa cupressi* (Sphaeriales, Amphisphaeriaceae; conidial state, *S. unicorne*). The latter fungus causes Monochaetia canker. It is widespread in the United States and in many other countries. Its conidial state was long known as *Monochaetia unicornis*. *S. cupressi* was differentiated from the other species in New Zealand, where all three occur. Its distribution is unknown.

References: 128, 212, 387, 588, 701, 1923, 1941, 1948, 2081

Seiridium canker of Monterey cypress.
A. A young tree with dead branches and a dying top (CA, Sep).
B. A tree nearing death due to multiple cankers (CA, Sep).
C. A canker with dripping resin on the trunk of young tree (CA, Sep).
D–F. Successively closer views of bark of a diseased branch with exuded resin and black acervuli of *Seiridium cardinale* (CA, Jul).

Plate 80

Botryosphaeria and Botryodiplodia Cankers and Diebacks (Plates 81—86)

This plate leads a group of six dealing with diseases caused by *Botryosphaeria dothidea*, *B. obtusa*, *B. quercuum*, *B. rhodina*, and *Botryodiplodia hypodermia*. Species of *Botryosphaeria* (Pleosporales, Botryosphaeriaceae) have conidial states in the form-genera *Botryodiplodia*, *Diplodia*, *Dothiorella*, *Lasiodiplodia*, and *Sphaeropsis*. Thus we present Botryodiplodia canker and dieback of elms (Plate 86) in this disease group even though no *Botryosphaeria* state of the elm pathogen is known. The conidia of these fungi have distinctive microscopic characteristics that allow diagnosis.

Cankers, dieback, and blight caused by *Botryosphaeria dothidea*. *B. dothidea* (syn. *B. berengiana*, *B. ribis*) is a nonspecialized pathogen, widespread in temperate and tropical regions around the world, with a reputation for attacking plants predisposed by other agents. It is commonly found associated with annual or perennial cankers and dieback in wounded plants or in those stressed by drought, freezing, or defoliation. It also colonizes dying or freshly killed twigs and branches that it has not previously infected, such as those pruned from orchard trees. In dead bark it grows and produces spores for 1 to several years. Because it attacks weakened plants and is commonly unable to infect healthy, vigorous tissues, *B. dothidea* has been called a minor pathogen, but this label is misapplied. Rather, the fungus is opportunistic.

B. dothidea attacks plants, mostly woody, belonging to more than 100 genera. In North America, it is perhaps most important in orchards, causing cankers in almond and apple trees, white rot of apple fruits, canker and gummosis of peach trees, dieback and nut rot of pecans, and stem blight of currants and blueberries. In landscapes and forests, it causes cankers and dieback of sweet gum, leaf blight and dieback of rhododendron, and cankers and dieback of dogwood, elm, linden, redbud, and sycamore. It also attacks conifers where these are grown far from their native habitat. Thus it has been found causing twig blight of Douglas-fir in the Great Plains, dieback and mortality of North American pines in dry regions of Hawaii and East Africa, and cankers in dawn redwood in the vicinity of Washington, D.C. A partial list of host groups follows. Whether *B. dothidea* is pathogenic to plants in all of these groups is unknown.

Partial host list for *Botryosphaeria dothidea*

alder; almond; apple; *Araucaria*; ash; mountain ash; avocado; *Baccharis*; barberry; basswood; bean tree; birch; bittersweet; blueberry; brambles; buckeye; buckthorn; butternut; buttonbush; camellia; camphor tree; carissa; carob; cassava; catalpa; Japanese cedar; chestnut; chinaberry; *Chrysophyllum*; citrus; cotoneaster; crabapple; currant; false cypress; dogwood; Douglas-fir; elder; elm; eucalyptus; eugenia; fig; filbert or hazel; firethorn; frangipani; fringe tree; fuchsia; grapevine; guava; sweet gum; hawthorn; Hercules club; hibiscus; hickory; holly; hop hornbeam; horse-chestnut; juniper; katsura tree; kudzu; mountain laurel; lead tree; linden; black locust; honey locust; loquat; macadamia; madrone; magnolia; mango; maple; mimosa; mulberry; oak; olive; Russian olive; coconut, date, fishtail, rock, and Washington palms; papaya; peach; pear; pecan; pepper tree; persimmon; phoenix tree; photinia; pieris; pine; screw pine; pistachio; poinsettia; plane tree; dwarf poinciana; pomegranate; poplar; privet; quince and flowering quince; redbud; redroot; dawn redwood; rhododendron; *Rhodoleia*; rose; salt bush; senna; giant sequoia; silk-oak; spice bush; sumac; sweetfern; sycamore; Chinese tallow tree; tree-of-heaven; toyon; tulip tree; tung-oil tree; viburnum; walnut; waxmyrtle; willow; wingnut; yellowwood

Symptoms caused by *B. dothidea* vary with the kind of plant, the size of the part attacked, and the extent of predisposition. Lesions vary from tiny, superficial spots in bark, soon isolated by the formation of protective cork, to sunken cankers delimited by vigorous callus ridges, to spreading lesions without marginal callus. Usually diseased twigs die, but trunks and large branches contain the pathogen within discrete cankers. Survival of small branches depends upon the number of lesions, their proximity to one another, and the degree of vigor before infection. Leaf blight, as shown on rhododendron, occurs also on macadamia. It is likely to be diagnosed only on plants with thick leaves, in which the pathogen fruits in a manner similar to its habit in bark.

On redbud, multiple cankers of varying size develop on branches and sometimes on the trunk, eventually encircling and killing the infected parts. This process may occur during or between growing seasons. Leaves on some branches wilt, and on other branches the buds simply fail to open in spring. Cankers are usually centered on twig or branch stubs and sometimes become elongate. The killed bark becomes depressed, roughened, and darkened where fungal stromata break the surface. An old canker may be surrounded by a conspicuous callus ridge. The wood beneath a canker is discolored brown, and this discoloration extends several centimeters above and below the canker.

On apple, in addition to fruit rot and death of shoots and spurs, water-soaked, blisterlike cankers up to 15 cm long may develop on scaffold limbs, most commonly on winter-damaged trees.

On peach trees the disease is often called gummosis because infected tissues produce copious gum. Lesions originating at wounds and lenticels become impregnated with gum, and the excess accumulates in blobs on the surface of the bark. Gum exudation is not diagnostic, however; other species of *Botryosphaeria* and a number of unrelated bark parasites also induce this symptom.

Sweet gum also responds with gum production. Thus *B. dothidea* is said to cause bleeding necrosis when it attacks the trunk. If trunk lesions are numerous, they may coalesce and kill the tree, but they usually do not. Rather, lesions eventually become partly or completely overgrown by callus. Lesions result in defective lumber if the trees are later harvested for wood products. Fresh lesions on sweet gum have a phenolic odor, and the killed tissue has a reddish brown color that extends into the wood and sometimes to the pith of small stems. Brown leaves adhere to twigs and branches that die during the growing season.

On almond trees, bandlike or irregular cankers form on the trunks or scaffold limbs, and distal parts die.

On pecan, lesions 1.5—3 cm long may girdle twigs and thus destroy their potential for nut production. In addition, *B. dothidea* causes nut rot—dark brown lesions that begin on the shuck and progress through the shell into the kernel.

On currant and blueberry plants, lesions develop in both new shoots and older wood, and branches often die during the growing season, as indicated by wilting and browning of leaves.

On pussy willow, elongate depressed lesions may enlarge on branches at the bases of killed twigs. Cankers are sharply demarcated, and the bark cracks after drying. Eventually the bark is shed, exposing the wood. Lesions on willow trunks often remain 6—12 mm in size and cause significant damage to the plant only when they coalesce. The inner bark becomes purplish or reddish for several centimeters beyond the margins of the lesions because of the accumulation of anthocyanin pigment. Pigment also accumulates adjacent to lesions in red-osier dogwood.

On some species of linden, in addition to cankers of the types shown here, perennial trunk cankers develop. These have rough, cracked bark and are flanked by callus ridges so large that the stem may appear swollen at the level of the canker.

Diagnosis of disease caused by *B. dothidea* requires recognition of its conidial (*Dothiorella*) state, which is characterized by black pycnidial stromata that differentiate beneath the surface of killed bark and break through at maturity. Stromata vary in shape and are 1—4 mm in the longest dimension. Each contains one to several pycnidial cavities 150—250 μm in diameter, with colorless contents that appear white when sliced (Plates 81E, 82H). When diseased bark is moist or wet, one-celled, colorless conidia 17—25 × 5—7 μm, somewhat narrowed at each end, extrude in a white mass from the pore at the top of each pycnidium. The pycnidial stromata develop throughout the year as temperature permits, usually beginning within days to weeks after

---

Cankers and dieback caused by *Botryosphaeria dothidea*.

A—E. Dieback and cankers on redbud. A. Branches girdled by cankers. B. An inconspicuous canker, its presence revealed by cracked bark at its edges. C, D. Cankers with prominent callus at margins. Killed bark remains in place. E. Magnified view of the surface of a canker on redbud; the bark has been shaved to reveal clusters of pycnidia with white contents (OH, Jun).

F—H. Twig cankers on blue ash. F. Magnified view of stromata, each containing several pycnidia, bursting through the bark surface in a canker. G. A canker shown at twice actual size. Pycnidial stromata appear as black dots. H. A twig with several cankers (OH, May).

Plate 81

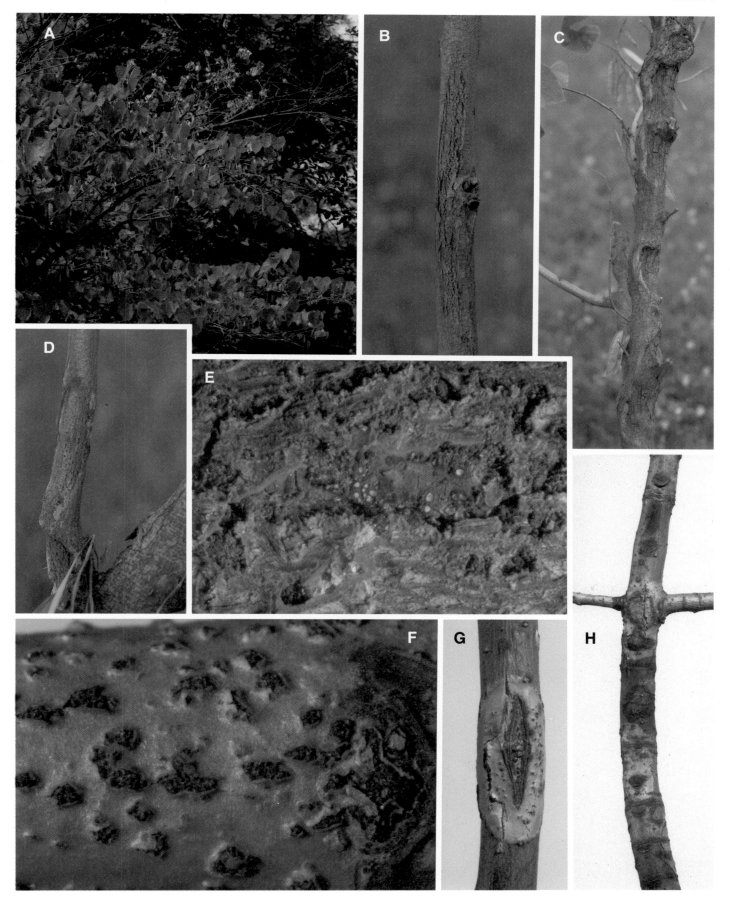

diseased tissue dies. Stromatal development depends upon sufficient moisture in the killed tissue and upon the kind of plant. Cankers on dawn redwood, for example, may lack fruit bodies, although isolates of B. dothidea from these cankers readily produce pycnidial stromata in culture or after inoculation into the bark of other hosts.

Pseudothecia of B. dothidea, preceded by spermagonia, may develop in the same stromata that previously produced conidia or in new stromata, usually on tissue dead for several months to a year or more. They may be overlooked among other saprophytic fungi in dead bark. They are not found on current season's twigs or on recently killed parts of cankers. Sometimes both the conidial and ascigerous stages can be found together, either in separate stromata or in the same one, and sometimes the conidial state is apparently not followed by pseudothecia.

Probably most infections are initiated by conidia. These are dispersed with dripping or splashing rainwater during much of the year but in greatest numbers in late spring and early summer. Where pseudothecia develop, ascospores are dispersed by air and water during much of the growing season and, like conidia, are most abundant during late spring to early summer. The fungus may also be transmitted by pruning tools.

B. dothidea commonly infects through wounds, but in addition, lenticels and growth cracks in bark are natural avenues of entry in many plants. Lesions in stems may develop within days to weeks after inoculation, but in some cases a long incubation period precedes symptoms. Elm twigs experimentally inoculated in June, for example, had no symptoms until October at the earliest, and some were asymptomatic until the following April. Incubation periods of 3–13 months have been recorded for peach twigs inoculated with conidia. The fungus has been reported to kill drought-stressed pines after entering roots, but how it reached the roots is unknown.

During parasitic attack as well as when acting as a saprophyte, the fungus colonizes both bark and sapwood. It produces macerating enzymes that are thought to contribute to cellular death. The sapwood at an infected wound or beneath a bark lesion that involves the cambium dies rapidly, leading to wilt and dieback. Many incipient infections in bark, on the other hand, are halted by host defenses before noticeable lesions form. Slight swelling around a lenticel may be the only evidence of infection.

Isolates of B. dothidea vary greatly in virulence but display little evidence of host specialization. An isolate from one host is normally capable of causing significant disease in an array of unrelated plants. Correlation between pathogenicity and pigment production in culture, widely reported in early literature on B. dothidea, proved unreliable as more information accumulated. The fungus grows rapidly at 25–35°C in laboratory cultures, which suggests that it is often favored by temperatures above the optimum range for plant growth.

Plants such as apple, blueberry, or peach that are intensively selected for horticultural characteristics may be susceptible to B. dothidea under a wide array of circumstances, but for most host plants, resistance is the rule. In birch, for example, nonstressed plants resist the pathogen by biochemical processes that inhibit its growth. Stressed plants lack this inhibitory ability.

Environmental stress insufficient to cause overt symptoms may predispose plants to infection by B. dothidea. For example, cankers and dieback of redbud and sweet gum are more common and severe on plants growing in shade than on those in the open. In experiments, drought stress insufficient to cause wilting permitted invasion of European mountain ash, European white birch, red-osier dogwood, and sweet gum by B. dothidea, while comparable nonstressed plants remained resistant. Resistance returned to stressed plants within days after they were resupplied with water. Similarly, rapid freezing such as may occur after mild weather in winter caused temporarily elevated susceptibility of the same test species in the absence of visible damage. Thus prevention of infection depends mainly upon avoiding or minimizing environmental stresses that induce susceptibility.

References for Botryosphaeria dothidea: 106, 255, 263, 264, 387, 425, 672, 849, 1180, 1270, 1302, 1383, 1717, 1718, 1835, 1943, 2089, 2122, 2127, 2154, 2232

Dieback, cankers, and leaf blight caused by Botryosphaeria dothidea.
A. Dieback of flowering dogwood (NY, Jun).
B. Leaf blight of 'Nova Zembla' rhododendron (NY, Aug).
C. A canker encircling a branch of red-osier dogwood (NY, Jun).
D, E. Wilting of 'Nova Zembla' rhododendron caused by a canker on a mainstem. In (E), the bark has been removed to show the lesion responsible for wilting and dieback (NY, Aug).
F, G. The killed portion of a rhododendron twig about natural size (F) and magnified (G) to show pycnidial stromata breaking through the surface (NY, Aug).
H. A magnified portion of a blighted rhododendron leaf showing inconspicuous pimplelike pycnidia, a cluster of which at center has been sliced at the level of the leaf surface to reveal the white contents (NY, Aug).

Plate 82

Diseases caused by *Botryosphaeria obtusa*. *B. obtusa* (syn. *Physalospora obtusa*), known primarily as a pathogen of fruit trees, is a nonspecialized opportunistic fungus that colonizes the bark of many kinds of plants saprophytically and attacks living parts if they are predisposed by environmental stress, wounds, or intrinsic high susceptibility. This fungus occurs in temperate zones around the world on woody and nonwoody host plants in at least 55 families. Some of the host groups are listed below. In North America, *B. obtusa* is best known for causing dieback of grapevine, canker and gummosis of peach trees, cankers and dieback of oaks, fruit rot of citrus, and three distinct disorders of apples and crabapples: frogeye leaf spot, black rot of fruit, and black rot canker.

Except for frogeye leaf spot on apple or crabapple, symptoms of diseases caused by *B. obtusa* are not diagnostic. Diebacks, cankers, and fruit rots caused by this fungus are indistinguishable from those caused by such other fungi as *B. dothidea* and *B. rhodina*. Diagnosis usually depends upon recognition of the conidial state of the fungus, a *Sphaeropsis* characterized by black pycnidia up to 0.5 mm in diameter. They break through the bark surface and produce single-celled, brown spores (usually 20–26 × 9–12 μm). In dead bark, pycnidia may be followed by black stromata up to 3 mm wide, in which pseudothecia form.

Symptoms are best known on apple or crabapple. In leaves, which the fungus penetrates via stomata, it causes circular spots with indefinite purple margins. These appear beginning shortly after leaves unfold, expand to 2–10 mm (average 4 mm) in diameter, and by early summer are brown with abrupt margins. Many spots develop no further, but in others, secondary enlargement begins at one or more points on the margin, resulting in a lobed brown lesion with a purple margin. As the brown lobes surround the original spot, which has weathered to grayish, a concentric pattern emerges—frogeye leaf spot. If the lobes fail to encircle the original spot, the result is an irregular blotch. Pycnidia sometimes form in the centers of the lesions and produce conidia that infect fruit or woody parts. Severe leaf infection causes premature yellowing and defoliation.

Fruit infection occurs any time during the season and on green fruit appears as tiny purple pimples or slowly enlarging black spots. Rapid rot begins as fruit ripens. Fruit infected when ripe may rot completely in 3–5 days in part because of macerating enzymes secreted by the fungus. Diseased fruit may change color and drop prematurely, may rot in place and then fall, or may shrivel and remain on the tree. The name *black rot* refers to the formation of dense masses of black pycnidia on the surface of the decayed fruit.

Black rot canker is more noticeable on limbs than on twigs or trunks. Lesions commonly develop at wounds, around the bases of dead twigs, and on upper or southwest-facing sides of limbs. These observations indicate the importance of wounds or lesions caused by other pathogens as sites of infection and of temperature extremes as predisposing stresses. Black rot canker has long been known to follow winter injury in apple orchards. Lesions in smooth bark are initially reddish brown to purple-brown and may later become tan near their margins. Some lesions remain confined to superficial bark and never involve the cambium. Others extend to the wood and either cease enlargement during the 1st year or develop perennially, attaining lengths to 30 cm or in extreme cases to 1 meter. They enlarge during the growing season, most rapidly during periods of water stress. Cracks in the bark at the margins of a canker indicate that it has at least temporarily ceased enlarging. Points of new or expanding infection at canker margins give cankers a lobed appearance. The fungus penetrates the wood only slightly except in twigs, where both bark and wood are extensively invaded. Dead bark remains attached for at least a year, then slowly cracks and falls away. Cankers that kill limbs during the growing season cause either wilting or premature yellowing and casting of leaves.

On peach trees *B. obtusa* causes cankers and gummosis, but gum production is a nonspecific response. *B. dothidea* and *B. rhodina* cause symptoms identical to those caused by *B. obtusa* and may be found together with *B. obtusa* in peach orchards in the Southeast. Spores of these fungi often lodge in peach buds, so that lethal bud infections occur as tree growth begins in spring.

B. obtusa overwinters as mature or partly developed pycnidia and pseudothecia in the bark of cankers and in the bark of dead twigs or branches that it has invaded either parasitically or as a saprophyte. It also survives in mummified fruit of some apple cultivars, such as 'Cortland'.

Pycnidia may develop during the 1st season of canker formation and nearly always do so during the 2nd season. When pycnidia break through the bark surface, they cause conspicuous roughening and darkening. Under moist conditions, tendrils of brown conidia ooze from the pycnidia. The spores are dispersed locally by splashing water and over distances by insects. The convergent ladybird beetle, *Hippodamia convergens,* was found carrying conidia of *B. obtusa* in New Hampshire. Pseudothecia develop in the 2nd and subsequent seasons after infection and produce ascospores that are borne by both water and air.

Both conidia and ascospores are dispersed throughout the season but most abundantly in spring. Spore release is favored by periods of wet weather with cool, stable temperatures for 12–24 hours or more. Nighttime rain stimulates spore dispersal more than does rain in daylight. Temperature of 20°C for 24 hours seems optimum for the initiation of infection, and leaf spots may be observed 2–3 weeks after the first large release of conidia in spring.

B. obtusa builds up in neglected or environmentally stressed plantings, colonizing and reproducing on such material as pruned twigs or those killed by other diseases, especially fire blight (Plates 76–77). Thus control depends upon avoiding or minimizing stress caused by water shortage or temperature extremes. Sanitary pruning is also practiced to eliminate twigs and branches upon which the fungus sporulates. Lists of disease-free or resistant crabapples are available, but direct tests of their susceptibility to *B. obtusa* have not been made.

A second black rot fungus, *B. stevensii,* occurs less commonly than *B. obtusa* on apple, grapevine, oak, and other plants in North America and Europe. It has the same general appearance and causes the same symptoms as *B. obtusa* but differs in producing mostly colorless, sometimes two-celled conidia. The name *Sphaeropsis malorum,* widely but incorrectly used for the pycnidial state of *B. obtusa,* was first applied to the pycnidial state of *B. stevensii* and is now properly used for neither fungus.

Some host plants of *Botryosphaeria obtusa*

alder, ampelopsis, apple, arborvitae, ash, mountain ash, avocado, azalea, *Baccharis,* beautyberry, birch, blueberry, brambles, broom, buckeye, buckthorn, buttonbush, carissa, cherry, chestnut, chinaberry, citrus, clematis, cotoneaster, crabapple, crape myrtle, cross vine, currant, cypress, false cypress, dogwood, elder, elm, fig, filbert, firethorn, golden-rain tree, grapevine, sweet gum, hackberry, hawthorn, hazel, hibiscus, hickory, holly, honeysuckle, hornbeam, hop hornbeam, false indigo, English ivy, jasmine, juniper, kudzu, larch, lead tree, leatherleaf, lilac, linden, lyonia, honey locust, loquat, magnolia, maple, mulberry, myrtle, ninebark, oak, oleander, Russian olive, osage orange, trifoliate orange, *Osteomeles,* Mazari palm, paulownia, peach, pear, prickly pear, pecan, persimmon, *Phyllanthus,* pine, plane tree, plum, poplar, privet, quince and flowering quince, redbud, redroot, rhododendron, rose, sagebrush, sassafrass, serviceberry, snowberry, sourwood, spice bush, spirea, sugarberry, summer sweet, sweet shrub, sycamore, teak, tree-of-heaven, tulip tree, viburnum, Virginia creeper, walnut, waxmyrtle, willow, wisteria, yellow-jessamine, yucca.

References for Botryosphaeria obtusa: 255, 263, 387, 425, 792, 869, 870, 1102, 1403, 1766, 1790, 1899, 1900, 1943

A, B, D, E, G. Dieback and black rot cankers caused by *Botryosphaeria obtusa* on crabapple cultivar 'Rosybloom'. Pruning wounds on scaffold limbs in (B) were sites of infection that killed one branch (arrow) and would soon girdle the limb at left. The branch at top center in (B) has a linear canker that developed after nearby branches were removed, exposing the bark to bright sun. D, E. Killed bark turns tan, then dark and rough where pycnidial stromata of *B. obtusa* break the surface. Cracks in bark (E) identify the margin of a canker that has temporarily ceased expansion. G. Magnified view of pycnidial stromata breaking the bark surface (ND, Jul).

C. Gummosis on a peach limb identifies the location of a canker caused by *B. obtusa* (GA, May).

F. Frogeye leaf spot, caused by *B. obtusa,* on 'Rosybloom' crabapple (ND, Jul).

Plate 83

Diseases of oak caused by *Botryosphaeria quercuum* and related fungi. *B. quercuum* (syn. *Physalospora glandicola*) is perhaps the most important of several related fungi that cause twig cankers and dieback of eastern oaks. The fungus is usually encountered in its pycnidial form, *Dothiorella quercina* (syn. *Sphaeropsis quercina*), in the bark of cankers on twigs and small branches. Lesions develop in summer and, by killing segments of twigs or small branches, cause wilting and browning of fully expanded leaves. This flagging is indistinguishable from symptoms caused by oak twig pruners or cicadas (see Johnson and Lyon, 2nd ed., Plates 124 and 236), but distinct lesions indicate a fungal disease.

Diseased bark is dark brown to almost black, becomes shriveled, and eventually cracks at the junction with living tissue. The outer sapwood beneath a bark lesion is discolored brown to black, and brown streaks in the sapwood extend up and down the twig or branch for 2–25 cm. In transverse view, individual streaks may be up to 2 mm across, and they commonly coalesce to form dark bands. The pathogen can be isolated from discolored bark or wood.

Most lesions remain localized in twigs and small branches, but occasionally the fungus advances downward perennially, causing progressive dieback or, on large stems, sunken strips of dead bark bordered by callus ridges. Sprouts sometimes develop along trunks or branches of trees that have sustained dieback. *B. quercuum* has been blamed for the death of entire trees, but this is uncommon.

Pycnidia develop in bark lesions on twigs beginning soon after leaves wilt. They are discernible first as minute pimplelike swellings, usually beneath lenticels. The bark surface at these points soon ruptures to reveal black stromata 0.6–1.3 mm across, protruding up to 0.5 mm above the contour of the twig, each containing one to several cavities in which conidia form. If the tops of mature stromata are shaved off, pycnidial cavities appear as white spots on an otherwise dark background, as illustrated for *B. dothidea* (Plate 81). The conidia are thick-walled, wide in relation to their length (18–25 × 12–17 μm), and are at first colorless and unicellular. After discharge, some conidia turn light brown and become two-celled. Pycnidia also occur on dead oak galls and acorns. Pseudothecia differentiate and produce ascospores in overwintered stromata in bark.

The relative roles of conidia and ascospores of *B. quercuum* in causing new infections are unknown, and spore dispersal, presumably similar to that of *B. dothidea* and *B. obtusa*, has not been studied. The fungus can infect wounds, but whether it also enters twigs by other routes such as flower buds or lenticels is unknown. The timing of infection is also unknown.

Several other fungi related to *B. quercuum* cause similar diseases or at least develop similarly on dead oak twigs. Unfortunately, the identities and host records of members of the group are jumbled. *B. quercuum* is reliably reported to attack black, blackjack, bur, yellow chestnut, northern pin, post, northern red, Texas red, shingle, white, and swamp white oaks. Additional reported hosts are green ash; American beech; American chestnut; grapevine; hickory; chestnut, southern live, scarlet, and water oaks; and *Prunus* sp. In the southern and southeastern United States, oak twig canker and dieback are often caused by *Diplodia longispora*, a pycnidial fungus quite distinct from *Dothiorella quercina* but sometimes erroneously listed as a conidial state of *B. quercuum*. *Botryodiplodia gallae* causes cankers and dieback of several oak species in the Great Lakes region and the Northeast. Other fungi in this complex on oaks are *Botryosphaeria melanops*, *Botryosphaeria obtusa*, and forms of *Botryodiplodia*, *Dothiorella*, and *Sphaeropsis* with unknown relationships to the fungi already mentioned.

A twig canker and dieback caused by a species of *Diplodia* occurs on drought-stressed California live oaks.

Twig cankers and flagging caused by *Botryosphaeria quercuum* tend to occur in conspicuous outbreaks that subside after 1 or more years. Knowledge of the related pathogens, *B. dothidea* and *B. obtusa*, leads us to suspect that outbreaks are triggered by environmental or biotic factors that promote oak susceptibility. *B. quercuum* and *Botryodiplodia gallae* have been associated with the golden oak scale (see Johnson and Lyon, 2nd ed., Plate 168) in dieback of chestnut and white oaks.

References for Botryosphaeria quercuum *and related pathogens of oak:* 60, 106, 228, 327, 389, 423, 766, 783, 928, 1706, 1790, 1898

Twig canker and dieback of northern red oak caused by *Botryosphaeria quercuum* (NY, Aug).

A, B. Wilting or browned foliage identifies diseased twigs.

C. Bark removed to show the location of a lesion.

D. Bark in a lesion on a 1-year-old twig is nearly black in contrast to normal dark green; necrotic streaks in xylem extend several centimeters beyond the edge of the canker.

E. Mature pycnidial stromata of *B. quercuum*.

F. A lesion that did not kill the twig is partly covered by callus folds 1 year later.

G. Smooth bark in a lesion at a fruit spur on a 1-year-old twig is roughened by pimplelike swellings where pycnidia are developing, especially beneath lenticels. The tattered membrane is the original epidermis.

Plate 84

Cankers and diebacks caused by *Botryosphaeria rhodina*. *B. rhodina* (syn. *Physalospora rhodina*) is a versatile, opportunistic, and cosmopolitan pathogen that attacks woody and herbaceous plants in at least 280 genera. It occurs around the world, mainly in the belt between 40° north and 40° south latitudes. By *opportunistic*, we mean likely to attack trees weakened by heat, drought, or other pathogens. *B. rhodina* is usually found in its pycnidial state, *Lasiodiplodia theobromae* (syn. *Botryodiplodia theobromae, Diplodia natalensis*), which develops on diseased or dead leaves, stems, or fruits. Depending on the host plant, the fungus causes cankers, twig blight, fruit rot, seed rot, leaf blight, gummosis, root rot, collar rot, or black stain of wood. It causes elongate cankers on tung-oil trees in southern states and widespread death of Russian olive in shelterbelts in the Great Plains. It contributes to decline of live oak and sycamore in the South. In live oak it often follows or accompanies the oak wilt pathogen (Plate 176). In sycamore it may follow the fastidious bacterium that causes bacterial scorch (Plate 187), and it often collaborates with *Phomopsis scabra* in causing dieback of sycamore. It is one of three species of *Botryosphaeria* that cause identical cankers, twig dieback, and gummosis on peach trees. Other woody hosts grown in the United States include apple, prickly ash, avocado, azalea, baccharis, bead tree, camphor tree, carissa, catalpa, chestnut, dogwood, eucalyptus, fig, grapevine, sweet gum, holly, linden, honey locust, macadamia, mallow, mango, mimosa, trifoliate orange, coconut palm, pear, prickly pear, pecan, persimmon, Norfolk Island pine, pittosporum, rose, sassafras, silk-oak, sumac, tree-of-heaven, viburnum, willow, and yellow-jessamine.

B. rhodina enters stems through wounds, especially pruning wounds. The degree of stress imposed on the plant by other factors then influences the extent of damage. Plants capable of producing gum or resin do so at sites of infection. The gum or resin permeates infected tissue and may exude at the margins of lesions. This is common on peach, Norfolk Island pine, and Russian olive.

Cankers caused by *B. rhodina* are elongate; the fungus moves along a stem much more rapidly than around it. Several years may pass before a diseased stem is girdled. Cankers on stems and branches of Russian olive consist of strips of dead bark up to a few meters long. Unless exuding gum marks the edges of such lesions, they may be indistinct because the dead bark persists and little callus forms. The only marker may be small dead branches along the killed strip. Pycnidia develop in the bark and break through its surface in the dead strips and on dead branches.

On sycamore, *B. rhodina* typically causes dieback. Infection begins on broken or wounded twigs and advances into progressively larger branches. The fungus grows in the sapwood, killing and staining it darkly in advance of progress in the bark. In experiments, cankers elongated four times as rapidly as they widened, and stained sapwood extended 30–40 cm beyond cankers.

On tung-oil trees, *B. rhodina* causes black, sunken cankers on trunks, limbs, shoots, and petioles. Affected parts wilt quickly when girdled. Pycnidia develop on killed parts in autumn, and perithecia form on ash gray bark the next spring.

The black pycnidia may arise singly but are usually found in groups 2–4 mm across. They extrude conidia in a mass that soon appears black. Perithecia, also black, are much less common, or are at least less commonly found, than pycnidia. They usually appear in crowded groups, sometimes in rows oriented with minute ridges and furrows in the bark. If perithecia or young pycnidia are cut through with a sharp blade and examined with a hand lens, they exhibit white contents and black walls. This feature, typical of *Botryosphaeria* species and related fungi, is illustrated in Plates 82 and 86.

Conidia start most new infections. These spores usually come from pycnidia on dead twigs and branches on the same tree or an adjacent one. The fungus can live saprophytically and may produce spores for a year or longer on woody plant parts that it has killed. Conidia are discharged during or after rainfall and are dispersed by running or splashing water and also by air. They may be found singly or in clumps. When released they are one- or two-celled and often colorless, but they soon turn dark brown. Brown conidia tolerate desiccation and are capable of surviving from one growing season to the next on the bark of citrus trees. The importance and role of ascospores are unknown. *B. rhodina* is seedborne in some crop plants and in loquat.

In tropical forests where *B. rhodina* causes black stain in the wood of certain trees, the stain is associated with wounds made by boring insects. Insect vectors are suspected of being important in temperate regions, but none has been associated with tree diseases caused by this fungus.

Strains of *B. rhodina* vary in virulence and show little host specificity. An isolate from one plant is usually able to cause disease in several unrelated plants. Some strains are able to cause cankers on apparently nonstressed trees, but damage by all strains is favored by stress—heat stress and water shortage. The fungus grows at temperatures up to nearly 40°C and grows most rapidly at 30–34°C, well above the optimum temperature for growth of most plants. Cankers develop most rapidly in hot summer weather (30–35°C) and faster in drought-stressed trees than in nonstressed trees. In experiments, the fungus caused little or no damage to inoculated plants kept at 20–25°C. A natural outbreak of dieback caused by *B. rhodina* in sycamore in the Mississippi Delta in the 1950s was apparently triggered by drought, and it subsided with a return to normal precipitation in that region. *References for* Botryosphaeria rhodina: 60, 263, 265, 387, 425, 567, 569, 1108, 1133, 1135, 1136, 1489, 1614, 1897, 1941, 2078

A. Dieback in sycamore (FL, May).

B. Dieback in Russian olive (NE, Jun).

C. An elongate perennial canker that originated at a wound on a trunk of sycamore; bark was shaved at the top of the canker to show its extent (MS, Apr). Either *Botryosphaeria rhodina* or *Ceratocystis fimbriata* f. sp. *platani* (Plate 174) can cause such cankers.

D. The mainstem of a young, top-killed sycamore with bark shaved to display the edge of the lesion. *Phomopsis scabra* and *B. rhodina* were both associated with dieback in this tree (MS, Apr).

E. Callus ridges at edges of a canker on sycamore indicate that *B. rhodina* was halted after moving into the mainstem from a branch (MS, Apr).

F. Close view of sycamore bark with black pseudothecia of *B. rhodina*. Microscopic examination is required to differentiate this fungus from others of similar appearance found on sycamore (MA, Apr).

Plate 85

Botryodiplodia canker and dieback of elm. In the Great Plains, Siberian elm is planted along streets and in windbreaks and shelterbelts because of its reputation for hardiness and drought tolerance. Actually the tree is often stressed by unfavorable weather, and when stressed it is subject to cankers and dieback caused by *Botryodiplodia hypodermia, Tubercularia ulmea* (Plate 99), and several other fungi. By killing branches and entire trees, these fungi reduce the protective influence of the trees on nearby fields or homes.

B. hypodermia (Deuteromycotina, Coelomycetes) was first described from Massachusetts in 1931, but it attracted little attention until the 1970s when plant pathologists began to investigate the deterioration of shelterbelts in the Great Plains. There it was found to be the most common cause of cankers and dieback in Siberian elm, and it has reportedly limited the usefulness of this species in windbreaks. It occurs now from Montana and Saskatchewan to Massachusetts. It may be more widespread, but records are confused with those for a separate fungus, *B. ulmicola* (syn. *Sphaeropsis ulmicola*), which also causes twig blight and dieback in elms. Apparently *B. hypodermia* causes significant damage only to drought-stressed elms in the Great Plains. American, Chinese, English, Scotch, Siberian, and smooth-leaf elms are known suscepts. Other elm species have not been tested for susceptibility.

The cankers develop on twigs, branches, or trunks and may girdle small limbs within 1 season. In Chinese and Siberian elms, newly infected bark is water-soaked, soft, and reddish brown to brownish black. The junction of diseased and normal tissue is sharply demarcated. The cambium and sapwood beneath diseased bark also turn reddish brown. During the year of infection, discolored wood does not extend much beyond the killed bark. In subsequent years, discolored wood extending from cankers may indicate colonization by secondary organisms. The foliage above girdling cankers wilts and dies, sometimes turning yellow first. Adventitious sprouts often develop below cankers and create a bushy appearance. On Chinese elm, the bark sloughs from old cankers, leaving conspicuous scars.

On American elm, cankers caused by *B. hypodermia* result in dieback of branches up to 10 cm in diameter. The leaves on affected limbs turn bright yellow in mid to late summer and are shed prematurely but do not wilt. The surface of the wood beneath new cankers is chocolate brown.

Pycnidia of *B. hypodermia* develop in killed bark in autumn, causing a prominent roughening of the bark surface (E). When pycnidia are cut through by shaving the surface of colonized bark, they appear black-walled with white contents (F, G). The conidia are initially single-celled and colorless, but they eventually become brown, and some become two-celled. Mature spores are present all year, and masses of them exude from pycnidia after wetting. They are dispersed by splashing or running water. Other possible modes of transmission are unexplored. No ascigerous state of the fungus is known.

B. hypodermia infects through wounds. Winter-injured twigs and branches of American elm are commonly infected. The fungus causes most damage if infection occurs during the growing season while the trees are stressed by insufficient water. In experiments, lesions up to 72 cm long developed during the 1st season in Siberian elms infected during July–September, and many of the lesions enlarged and caused dieback and death of trees during the 2nd season. By contrast, infection at other times led to small cankers that became inactive after 1 season and callused over. In stressed trees, lesions expand rapidly at temperatures of 15–30°C. *B. hypodermia* also colonizes bark at the ends of cut or broken branches and may thus be present in trees without cankers or dieback.

Some observers have speculated that herbicides, especially 2,4-D, may predispose Siberian elm to infection, but experimental evidence is lacking.

In American elm but not Siberian elm, Botryodiplodia dieback is of minor importance and can be managed by pruning.

References for Botryodiplodia hypodermia: 278, 1062, 1063, 1611, 1614, 1882, 1909

Dieback and cankers caused by *Botryodiplodia hypodermia*.
A. Dieback and mortality in a Siberian elm windbreak (ND, Jul).
B–D. Cankers and adventitious shoots, some shoots wilting as the result of canker extension, on Chinese elm (IL, Jun).
E. Pycnidia developing in Siberian elm bark, causing numerous minute, pimplelike swellings (ND, Oct).
F, G. Successively magnified views of pycnidia exposed by shaving the surface of dead bark. Pycnidial walls are dark, the contents white (ND, Oct).

Plate 86

Cryptodiaporthe Cankers (Plates 87–88)

Cryptodiaporthe (Dothichiza) canker of poplar. Cryptodiaporthe canker, also known as Dothichiza canker, is one of the major diseases of poplars. Nearly all species are susceptible to some extent. The pathogen is *Cryptodiaporthe populea* (Diaporthales, Valsaceae), which is usually encountered in its conidial state *Discosporium populea* (syn. *Chondroplea populea, Dothichiza populea*). The disease occurs in Europe, the Middle East, and the Americas. Known in North America since 1915, it is widespread from the maritime region of Canada to Florida and the Southwest, and it occurs also in the Pacific Northwest. It is most common and destructive in trees weakened by transplanting, drought, freezing, waterlogged or infertile soil, or wounding of stem or roots. Damage is most severe in landscape plantings of black poplar, especially the cultivar 'Lombardy', which is often killed at a young age. Poplars in forest plantations sustain relatively little damage from this disease, and those in natural forests are virtually unaffected. In Europe, on the other hand, damage from Cryptodiaporthe canker is widespread in nurseries and plantations of several species and cultivars. Most information about the disease comes from studies there.

Many poplar and aspen species may be infected, but the degree of susceptibility varies greatly. Species in *Populus* section Leuce (e.g., bigtooth, European, and trembling aspens) are generally resistant, as are various cultivars of *P. ×berolinensis, P. ×canadensis* (Carolina poplar), *P. canescens* (gray poplar), *P. fremontii* (Fremont cottonwood), *P. maximowiczii, P. trichocarpa* (black cottonwood), and assorted hybrids. A partial list of hosts in addition to those just mentioned includes eastern and Great Plains cottonwoods; balsam, black, Simon, and white poplars; also *P. ×acuminata, P. cathayana, P. laurifolia, P. ×petrowskiana, P. ×robusta,* and *P. tristis.*

The disease at first kills scattered twigs and is not conspicuous, but it increases rapidly after spores have dispersed from the first lesions. Cankers form in stems of all sizes and ages, most often in twigs and small branches, especially in branch axils. Cankers range from tiny brown spots in bark just beneath the periderm to spreading lesions that involve both bark and cambium. The bark surface is only slightly or not at all discolored, but the tissue beneath appears brown, grayish brown, or black, depending on tree species and time since death. The sapwood beneath lesions in bark is also invaded and discolored light brown. Many cankers girdle twigs and branches, causing dieback, especially on trees with numerous small upright branches. The fungus may then grow downward into larger stems and cause perennial cankers. Coalescing cankers result in dieback of limbs and trunks that before death may show premature yellowing and dropping of leaves. Adventitious sprouts commonly develop below large cankers and are soon also diseased.

Many infections are halted. Even highly susceptible poplars often resist infection and produce barriers of lignified parenchyma, then cork, that isolate lesions. A callus ridge develops at the margin of most cankers that involve the cambium. Many small arrested cankers are covered by callus during the 1st season. Callus formation and drying of killed bark lead to cracking and eventual sloughing of bark. The wood is thus exposed.

From early spring until late autumn, scattered black pycnidia of the *Discosporium* state of *C. populea* develop just beneath the periderm in lesions and cause a pimplelike appearance where they raise and break the bark surface (G). Pycnidia are 700–800 μm in diameter and open widely at maturity. While bark is moist, greenish to amber blobs or tendrils of conidia issue from pycnidia and dry to a brown color or are dispersed by water. Viewed microscopically, the single-celled conidia appear colorless and are 9–11 × 6–8 μm in size. These spores can survive for several months in nature.

Perithecia develop after pycnidia but are apparently uncommon. They are long-necked, 500–600 μm in diameter, and tend to form in groups. They push up the periderm in tiny mounds.

Dying and dead twigs and branches beyond cankers are quickly colonized by other fungi, especially *Valsa* species (Plate 95).

The pathogen survives winter as mycelium in cankers and as conidia in late-developing, unopened pycnidia. Infection occurs via buds and bud scale scars in spring, via leaf scars in autumn, and via lenticels and wounds at various times of year. The fungus may remain latent up to 2 years after infection but normally causes lesions during the dormant period. A toxin produced by *C. populea* is implicated in lesion formation. Lesions may elongate as much as 30 cm during the dormant season.

Temperature and moisture relationships of poplar bark are correlated with seasonal trends in susceptibility. Bark is less well hydrated and more susceptible during the dormant period than during the growing season. In certain poplars, resistant reactions including cork formation were found to be weak or absent and lesions developed at 12°C or less, but resistance was strong and lesions did not develop at 16°C or higher. Similarly, lesions expanded more rapidly in bark stressed by water shortage than in well-hydrated tissues.

Strains of *C. populea* differ in virulence and in behavior in culture. On average, cultures grow most rapidly at temperatures near 20°C, and spore germination is most rapid and complete at 16–20°C. These responses contrast with maximum canker expansion at 4–10°C in nondormant bark. Conidia germinate at temperatures of 5–28°C and have been reported capable of germinating at relative humidities as low as 90.5%, an unusual capability in comparison with the usual requirement for free water or 100% relative humidity for germination of fungal spores.

On most poplar hosts as mentioned, severe outbreaks of Cryptodiaporthe canker require some predisposing stress on the trees. One of the few reported outbreaks in North America other than on 'Lombardy' poplar involved hybrid poplars stressed by drought. The intrinsic susceptibility of 'Lombardy' poplar may be so great that little predisposition is needed for damage. An interesting interaction reported in Europe is the infection of leaf scars that result from defoliation by a Melampsora rust (Plate 123).

Cryptodiaporthe canker of willow. One of several diseases responsible for dieback of willow twigs and branches, Cryptodiaporthe canker is usually of minor importance, but occasionally it causes severe twig dieback, especially on pussy willow. The pathogen, *Cryptodiaporthe salicella* (Diaporthales, Valsaceae), is usually found in its conidial state, *Diplodina microsperma* (Coelomycetes). It occurs across North America and is common also in Europe. It is generally considered to be an opportunist that causes damage to plants stressed by environment, injuries, or other diseases. *C. salicella* infects many species including bebb, black, crack, goat, heart-leaved, Hooker, osier, pussy, sandbar, Scouler, and weeping willows and several varieties of white willow. It is common also as a saprobe on dead twigs, sometimes including those of alder, maple, and poplar.

Cankers arise during host dormancy at lenticels, nodes, and wounds made by insects or other agents. Expanding cankers girdle twigs and small branches, which die back to the next living side branch. Symptoms are noticed in spring as buds on diseased twigs fail to open and the twigs appear somewhat dried and shriveled. Cankers are yellow in contrast to the normal green or are brown in contrast to the normal yellow bark of willow twigs. Bark remains on killed parts but loosens and cracks at the junction with living tissues. In spring, black, disk-shaped pycnidia form beneath the periderm. Pycnidia measure 0.5–1 mm in diameter. Their tips break the surface and release masses of conidia in blobs during wet weather or in tendrils under somewhat drier conditions. Conidia are equally two-celled, colorless, and spindle shaped, 13–18 × 3.5–5 μm. They are dispersed by rain splash or by water running on plant surfaces. Perithecia form later, singly or in small groups in the oldest parts of cankers. They are long-necked, black, and nearly 0.5 mm in diameter and produce colorless, two-celled ascospores.

Cryptodiaporthe canker of 'Lombardy' poplar.

A. Damage to mature trees (NY, Oct).

B, C. Cankers on a trunk and a branch. Swelling and cracks in bark result from formation of callus at canker margins (NY, Mar).

D. Dead twigs and foliage, the early stages of damage to trees in a windbreak (NY, Aug).

E. A small branch with bark shaved to reveal a girdling canker at center and numerous small cankers below, the latter having been invisible while covered by periderm (NY, Aug).

F. Pycnidia of *Cryptodiaporthe populea* in dead bark near an old canker from which bark has sloughed (NY, Aug).

G. Magnified view of a bark surface raised or broken by pycnidia (NY, Jul).

Plate 87

(Cryptodiaporthe Cankers, continued, Plate 88)

Susceptibility of willows during dormancy is linked to impairment of defense reactions in bark. These are rapid and effective during the growing season but are not consistently so during dormancy. Research on canker development in Hooker and Scouler willows showed that dormant twig cuttings were resistant when well hydrated but became susceptible as their water content diminished to less than about 80% of capacity. Temporary partial drying of twig or branch bark is normal during dormancy but apparently impairs host defense mechanisms and thus permits canker development.

C. salicella is mentioned in some references as one of two similar fungi on willows; the second is *C. salicina*. Most contemporary mycological authorities treat these as one species, *C. salicella*. Synonyms for the conidial state include *Discella carbonacea*, *D. salicis*, and *Diplodina salicis*.

References for Cryptodiaporthe canker of poplar: 288, 387, 685, 900, 1521, 1614, 1941, 2114, 2152
References for Cryptodiaporthe canker of willow: 172, 425, 783, 1521, 1941

Cryphonectria Cankers (Plates 88–90)

Cryphonectria cankers of live oak and Chinese chestnut. *Cryphonectria parasitica* (Diaporthales, Valsaceae), the chestnut blight pathogen (Plate 89), infects several species of chinkapin and oak in addition to chestnuts. Allegheny and bush chinkapins are highly susceptible. Giant chinkapin proved susceptible in pathogenicity tests, although the disease has not been found in the natural habitat of this species in California. Oak hosts include live, post, scarlet, and white. On rare occasions *C. parasitica* infects shagbark hickory, red maple, and staghorn sumac. More commonly it colonizes these species and various oaks as a saprophyte.

Live oaks from Virginia to Florida and Mississippi are subject to cankers caused by *C. parasitica* on trunks and limbs. The cankers often presage decline and sometimes death. They enlarge very slowly, eventually developing or coalescing into broad longitudinal bands of dead bark. Cankers are initially inconspicuous because the thick, fissured outer bark remains in place for many years. Callus formation at canker margins is often so slight that the location and extent of the canker may not be apparent. The pathogen forms pycnidia and perithecia in cracks between bark plates, but these fruit bodies may be so inconspicuous as to escape detection. The crowns of severely affected trees decline slowly, displaying sparse, yellowish green foliage and dead and dying branches.

Post oak and scarlet oak, unlike live oak, produce considerable callus at the edges of perennial cankers caused on their trunks by *C. parasitica*. The fungus invades each annual callus ridge in turn, and the canker thus expands. Limbs or entire tops of post oaks sometimes die back, but discrete cankers on trunks are more common. Cankers are either superficial, with the parasite confined to outer bark, or deep, involving the vascular cambium. Dead bark eventually sloughs from deep cankers, exposing the wood. Such cankers have been reported from Connecticut to Alabama. In scarlet oak observed in North Carolina, cankers are most common on the butt and remain covered with bark. Swelling due to callus formation around cankers is prominent, giving rise to involuted bark and to general swelling of the butt. Butt swelling also occurs in infected white oak. Buff-orange mycelial fans, diagnostic for *C. parasitica*, develop in the diseased bark of oaks.

All chestnut species are more or less susceptible to *C. parasitica*, but Chinese and Japanese chestnuts and hybrids involving these species generally resist infection. The level of resistance varies among individual trees. In resistant plants the fungus behaves as an opportunist, causing most damage to those predisposed by frost or freezing, drought, or other factors. Chinese chestnuts commonly display twig and branch dieback or targetlike perennial cankers on limbs and trunks. Cankers become prominent as old bark sloughs off. Pycnidia arise singly or in orange stromata on killed bark of the most recent callus ridge at the edge of a canker. Adventitious sprouts do not generally form below cankers unless the trunk or limb is girdled.

A–D. Cryptodiaporthe canker of goat willow. A. A canker causing dieback of a small branch. B. The basal part of a twig canker with cracked, peeling periderm and black pycnidia of the *Diplodina* state of *Cryptodiaporthe salicella*. C, D. Magnified views of pycnidia, dry (C) and producing masses of conidia while bark is moist (D) (NY, May).

E–G. Cryphonectria canker of live oak. E. A large old canker on a trunk. F. A canker on a large limb, inconspicuous except where bark has begun to fall away. G. Callus delimits the margin of a canker (VA, May).

H. A perennial canker caused by *Cryphonectria parasitica* on Chinese chestnut. Infection began at a pruning wound (NY, Jul).

Plate 88

Chestnut blight. One of the most famous plant diseases in North America, chestnut blight was discovered in 1904 in New York City. Within 50 years it spread to the extremes of the natural range of the American chestnut, destroying the economic and aesthetic value of one of America's most versatile trees. American chestnut was reduced in most areas to a scattered population of sprout stems in the forest understory. Other tree species, notably oaks and hickories, have taken over most of the spaces formerly occupied by chestnuts.

The blight is caused by *Cryphonectria parasitica* (syn. *Endothia parasitica*) (Diaporthales, Valsaceae). The fungus was apparently introduced with chestnut seeds or young plants from the Orient, as searches in China and Japan revealed its presence in both countries. Native chestnut species in Asia are not severely damaged.

Blight also appeared in scattered chestnut plantings in California, Oregon, and British Columbia but was suppressed there by intensive eradication and sanitation efforts. An epidemic began in southern Europe in the 1930s but subsequently subsided as a result of natural biological control.

American chestnut is the most susceptible species, followed by European chestnut. Chinese and Japanese chestnuts and hybrids with a large proportion of genes from these species are generally resistant, although the level of resistance varies widely among individuals. Other plants susceptible to varying degrees are listed with Plate 88.

Chestnut blight is characterized by cankers that kill bark and usually cambium and sapwood of twigs, branches, and trunks but not roots. Leaves and shoots wilt and die after the sapwood at the site of a girdling lesion ceases to conduct water. From a distance, the prominent symptoms are yellow or brown leaves on one or more branches and eventually dead, leafless branches. Branches girdled shortly before spring growth may produce short-lived, dwarfed, chlorotic leaves. Dead leaves and burrs usually remain on killed branches through the first winter.

Typical cankers on young, smooth-barked stems appear yellowish brown to orange-brown or reddish brown in contrast to normal greenish brown bark. Within only a few weeks they encircle and kill the affected part. Cankers vary from slightly depressed to somewhat raised and often show both characters in different parts of the same lesion. Cankers on large stems with thick corky bark are inconspicuous unless the bark begins to swell or crack. Many cankers are characterized by localized swelling and appear as spindle-shaped bands on small branches or as raised elongate areas with longitudinal cracks in thick bark of limbs and trunks. Cankers caused by virulent strains of *C. parasitica* expand laterally at rates of 14–25 cm per year, depending on average temperature at the site. Dead bark eventually cracks and falls away from old cankers. Adventitious sprouts (water sprouts, suckers) commonly develop below cankers, often from roots of American chestnut, but do so infrequently on resistant Chinese and Japanese chestnuts. Sprouts have the same inherent susceptibility as their parent trees.

Lesions initiated on twigs or small branches often expand into larger branches or trunks. Thus many cankers are found centered at branch axils. When a limb or tree dies, the fungus spreads into bark beyond the original lesion and sporulates there as a saprobe.

Nuts often become infected, but the epidemiological significance of this is unclear. Apparently the lesion is confined to the shell, where the pathogen may sporulate. In one study of nut infection, diseased nuts germinated, but blight was not transmitted to the seedlings.

Within a few weeks after lesion formation, numerous orange to yellowish brown pycnidial stromata break through the surface of killed bark or develop in crevices of thick fissured bark. Stromata are up to 3 mm long or wide and about 0.5 mm deep, with multiple fertile chambers where conidia are produced. Solitary pycnidia, about 250 μm in diameter, are common on wood or on the inner surface of bark, especially on plants colonized saprophytically. The conidia are rod shaped, colorless, and unicellular and measure only 3–5 × 1–1.5 μm. They ooze in yellow to buff tendrils from pycnidia when bark is moist, are carried down the tree in dripping or running water, move laterally by rain splash, and are carried by many insects and birds. Pycnidia mature throughout the growing season, and conidia disperse and are present in large numbers on bark of diseased trees throughout the year.

Perithecia also form throughout the growing season in the same stromata that gave rise to pycnidia. Both types of fruit bodies may be present simultaneously in one stroma. Perithecia are flask shaped, with their bodies embedded in the stroma. They open by means of long necks that extend to the stromal surface. The interval between appearance of stromata and formation of perithecia in them varies from several weeks to more than a year. Perithecial stromata are usually reddish brown. Perithecia eject ascospores into the air from early spring until late autumn after being moistened by rain. The ascospores are colorless, 7–12 × 3.5–5 μm, equally two-celled with rounded ends, and slightly constricted in the middle. Airborne ascospores and conidia carried by birds and insects were primarily responsible for overland spread of blight as the edge of the North American epidemic advanced at an average 30 km per year.

Mild wet weather is conducive to spore dispersal and infection. The optimum temperature range for ascospore expulsion is 20–25°C, although some spores are expelled at temperatures as low as 10°C. Ascospore discharge is prevalent as bark begins to dry within the first 5 hours after cessation of rainfall.

Infection occurs when ascospores or conidia germinate in fresh wounds that penetrate to living bark. Insects are probably responsible for most of the wounds. Hyphae grow into and between cortical cells and also into the outer sapwood beneath bark lesions, but the fungus does not cause wood decay. In the bark, hyphae soon organize into flat, spreading, buff to orange-buff masses (mycelial fans) that disrupt living bark and cambium by combined chemical and mechanical action. Cellular death is attributed to acidification of tissue by oxalic or possibly gallic acid and to the action of macerating enzymes. The tannins in chestnut bark, which would inhibit growth of many microorganisms, support growth of *C. parasitica*.

Almost as soon as the North American epidemic began, Chinese and Japanese chestnut trees were observed to resist infection. Breeding programs were then begun to produce resistant trees with the fruiting and growth characteristics of American chestnut. The work went slowly because resistance in the Asiatic species was usually associated with poor form and slow growth. Eventually a number of promising hybrids were produced, but none has shown broad adaptability to various soils and climatic zones. As for blight-resistant American chestnuts, a few trees have been found that remain alive although disfigured by blight. These may indeed be somewhat resistant, but most instances of apparent resistance really represent either escape or infection by hypovirulent strains of *C. parasitica*.

Trees resist *C. parasitica* in various ways. The bark of Chinese chestnut contains factors that inhibit growth of the parasite. Both American and Chinese chestnut barks contain chemicals that inhibit fungal macerating enzymes. This inhibitory activity, however, is much stronger in the Chinese trees. Resistant and susceptible trees both respond to infection by forming a zone of lignified cells and subsequently a cork barrier around the lesion. These responses proceed faster and more effectively in resistant trees. Mycelial fans of *C. parasitica* break through incompletely formed cork barriers in susceptible trees. Cankers in resistant stems are halted, sometimes before reaching the cambium, or enlarge slowly and become deeper year by year because of callus formation around the edges. Typically the blight pathogen kills some portion of the new callus roll each year, so the canker remains open, as shown in Plate 87.

Blight of American chestnut.

A. A young sprout recently killed (NY, Jul).

B. A sprout clump that arose from roots where a previous tree was killed. Girdling cankers are apparent on two stems (NY, Jul).

C. Swollen, longitudinally cracked bark in a blight canker (PA, Aug).

D, E. Distant and close views of typical cankers with perithecial stromata on a smooth-barked sprout about 15 cm in diameter (NY, May).

F, G. Close views of the advancing margin of a canker with pycnidial stromata. A buff-colored mycelial fan of *Cryphonectria parasitica* (arrow in G) is visible where outer bark was shaved (NY, Jul).

Plate 89

189

In southern Europe, and also in parts of Michigan, susceptible chestnut populations survived the blight epidemic because naturally occurring viruslike agents (double-stranded RNA, or dsRNA) diminished the virulence of the blight fungus. The natural defense systems of the trees were then sufficient to halt disease. Infections were arrested, and lesions were either overgrown by callus or confined, harmless, in outer bark.

Hypovirulence was found to be caused in most instances by dsRNA that was carried in hyphal cytoplasm and could be transmitted by hyphal fusion. Virulent C. parasitica in cankers could be rendered hypovirulent by applying mycelium or conidia of a compatible hypovirulent isolate. These discoveries prompted extensive searches for hypovirulence in the American population of C. parasitica. Many types of infectious dsRNA that could convert virulent to hypovirulent strains were found. Unfortunately, hypovirulent strains in most cases either lacked sufficient vitality to increase and disperse naturally or lacked compatibility factors necessary for promiscuous transmission of the dsRNA to virulent strains. Even so, the natural suppression of chestnut blight by infectious hypovirulence in some areas indicates that an effective system of biological control can be devised for regions where virulent strains still predominate.

Cryphonectria canker of eucalyptus. Many species of Eucalyptus have been introduced as shade or ornamental trees or have been established in experimental forest plantations in southern Florida and the warmer parts of California. Plantations of E. grandis (rose gum) and E. camaldulensis (Murray red gum) in Florida are subject to cankers that develop on the trunks near ground level. The pathogen, Cryphonectria cubensis (syn. Diaporthe cubensis) (Diaporthales, Valsaceae), is widespread on various eucalypts in South America and the Caribbean region. It occurs also in Hawaii. In South America, this disease sometimes intensifies in eucalypt stands that regenerate as sprouts from stumps of the original plantation. The fungus spreads from the stump into the sprout.

Many eucalypt species are susceptible to some extent. Several have been ranked for resistance as follows: E. maculata and E. saligna, highly susceptible; E. grandis, E. propinqua, and E. tereticornis, moderately susceptible; E. microcorys, E. paniculata, and E. robusta, moderately resistant; E. citriodora, E. torelliana, and E. urophylla, highly resistant. E. grandis, the only species damaged appreciably in Florida, shows much variation in resistance of trees from various seed sources.

Symptoms on E. grandis in Florida are similar to those on various other species in South America and the Caribbean region. Lesions develop superficially in bark near the bases of trunks less than 2 years old and slowly deepen and expand around the stem. The only external symptom in early stages may be a slightly depressed area. Later, however, the bark surface cracks longitudinally and becomes roughened, and superficial strips begin to fall away. The canker margin is abrupt but barely apparent unless the bark surface is shaved to reveal brown dead tissue in the lesion. The canker is most extensive in the outer bark (old secondary phloem) and may involve much of this tissue without damaging the cambium. Diseased outer bark may be sloughed from some trees. On others, lesions involve the cambium and may either become delimited by a callus ridge or kill the tree by girdling. Girdling is rare in Florida, however. Where the cambium is killed, sectors of discolored wood, wedge shaped in cross-section, extend inward to the pith. Lesions 1 meter or more in length are common, and multiple lesions may occur on a given tree. Occasional cankers occur above the trunk base, often centered on branch stubs. Many cankers cease development by the time trees are 6–8 years old, but others continue much longer.

Pycnidia and perithecia, sometimes both at the same time, form in abundance in killed bark, especially in crevices and on inner surfaces of loosened bark strips overlying cankers. The pycnidia are superficial or nearly so, more or less pear shaped, and solitary or fused at their bases in groups of two to five. Reddish brown at first, they darken to nearly black except at the tip, which remains brown (F). They measure 0.4–1.2 mm tall and 0.2–0.8 mm in diameter. The conidia, extruded in yellowish tendrils under moist conditions, are individually colorless, ovoid, unicellular, 2.5–4 × 1.8–2.2 μm.

The perithecia are black, partially immersed in bark, solitary or in small groups, and sometimes aligned in a crevice. Their necks vary in length up to about 0.5 mm in humid locations near ground level but are much shorter where the bark remains relatively dry. Ascospores are colorless, equally two-celled, and elliptical in outline and measure 5.8–8.2 × 2.2–3.0 μm. The epidemiological roles of ascospores and conidia have not been reported.

As an indication of its preference for warm climates, C. cubensis grows rapidly in pure culture at 30°C. Observations in South America have indicated that Cryphonectria canker is most severe in regions with consistent warm weather and abundant rainfall. South Florida is considered to be somewhat unfavorable for the disease because winters are relatively cool (by tropical standards) and summers, although hot, are somewhat dry.

References for Cryphonectria cankers of oak and Chinese chestnut: 491, 760, 951, 970, 1188, 1517, 1625, 1765, 1842
References for chestnut blight: 43, 387, 491, 501, 605, 606, 688, 752, 765, 951, 1066, 1188, 1245, 1396, 1625, 1666, 1667, 1842
References for Cryphonectria canker of eucalyptus: 17, 209, 848, 850

Cryphonectria canker of Eucalyptus grandis (FL, Apr).

A. A 12-year-old plantation affected by Cryphonectria canker in South Florida.

B. The base of a healthy tree. The bark surface is mostly smooth, as old bark naturally falls away in strips.

C. A tree with a basal canker extending nearly 2 meters up the trunk. Roughened bark is typical.

D. A basal canker with longitudinally cracked outer bark. The junction of healthy and diseased bark is revealed where bark was shaved.

E, F. Successively closer views of pycnidia of Cryphonectria cubensis on the inner surface of a dead bark plate.

Plate 90

Endothia Canker (Plate 91)

Endothia gyrosa (Diaporthales, Gnomoniaceae) causes twig and branch dieback and perennial cankers on branches, trunks, and exposed roots of beech, sweet gum, oaks, and occasionally other trees. It is widespread in the eastern United States and is destructive in the Southeast from Virginia to Texas. It also occurs in California. Europe, Asia, and New Zealand are represented in records from abroad. The fungus infects broken branches or wounds such as pruning wounds or damage by lawn mowers at trunk bases. In most circumstances it causes localized, slowly expanding cankers, but it may girdle and kill branches and trunks of trees under stress. It is best known as the cause of a syndrome called pin oak blight in the Tidewater region of Virginia.

Hosts include American beech, European beech, American chestnut, American elm, fox grape, Formosan gum, karri gum, sweet gum, American holly, silver maple, and several oaks: black, cork, English, California live, southern live, pin, post, northern red, southern red, scarlet, water, and willow oak.

Cankers on oaks and sweet gum are usually depressed or sunken, but cankers on slowly growing limbs of European beech cause scant change in bark contours (B, C). On beech the diseased bark sometimes turns orange-pink in contrast to normal gray or gray-green. Killed bark dries and cracks (D) but remains in place for several years. Callus rolls at canker margins vary from absent to prominent. On pin oak predisposed to severe disease, girdling cankers on trunks and branches result in scorch and browning of leaves, premature defoliation, dieback, and general decline. On sweet gum and Formosan gum, purple staining and gum exudation at canker margins are common. In large branches and trunks the fungus causes a yellow-orange decay of wood beneath the canker.

E. gyrosa fruits on killed bark, commonly near the edges of wounds, producing conspicuous persistent stromata that are orange initially and darken to cinnamon brown. Their internal tissue remains bright orange. Stromata measure 1–2 mm in diameter and 1.5–2 mm high. Often they are arrayed in parallel rows (C, E) like hobnails. The name *orange hobnail canker* was coined because of this feature. Young stromata contain irregular chambers where conidia form. When stromata are moist, amber blobs of conidia ooze out (F). The conidia are rod shaped, $3–4 \times 0.75–1$ μm.

At a later stage in autumn or winter, long-necked perithecia form in the same stromata, the surfaces of which by then have dark, nipplelike projections where the mouths of pycnidia or perithecia open. The perithecia are globose and 250–300 μm in diameter, with necks to 500 μm long. The ascospores are unicellular, curved with tapering ends, and measure $6–10 \times 2–2.5$ μm. An unusual diagnostic feature of the stroma is the instant release of beet red pigment where it is touched by dilute potassium hydroxide solution.

Dispersal of *E. gyrosa* is presumed to be similar to that of its much-studied relative, *Cryphonectria parasitica* (Plate 89). The conidia may be dispersed by rain splash, insects, and possibly pruning tools. Ascospores are probably airborne.

Infection can occur at any time of year in a warm climate, and cankers develop most rapidly during summer. Small oak or beech trees are at risk of girdling by *E. gyrosa* during the period of stress after transplanting. Whether cankers remain localized or girdle and kill the affected part depends on the vitality of the host. Early reports characterized *E. gyrosa* as a weak parasite because it colonized artificially inoculated branches or roots slowly. Later research showed that the fungus causes rapidly expanding lesions in the bark of trees stressed by water shortage. Rates of canker elongation up to 9 cm per month in pin oak have occurred in experiments. The formation of cankers 1.0–1.6 meters long and 8–12 cm wide in a single year has been recorded for oaks in Texas, but this may represent simultaneous formation of coalescing cankers.

E. gyrosa grows readily in culture, producing orange pigment in most media. Such pigment production is characteristic of several species of *Endothia* and *Cryphonectria*.

References: 49, 107, 387, 922, 1625, 1626, 1765, 1854, 1910, 2059

A. Dieback caused by *Endothia gyrosa* on pin oak. This canker-dieback syndrome has been called pin oak blight (VA, May).
B, C. Cankers caused by *E. gyrosa* at pruning wounds on branches of European beech (NY, Aug).
D, E. Close views of cankers on European beech, with persistent dead, cracked bark and numerous orange to brown stromata (PA, Jun).
F. Magnified view of stromata on bark of European beech. Amber blobs of conidia are visible on stromata (PA, Jun).

Plate 91

Leucostoma and Valsa Cankers (Plates 92–96)

This group of plates illustrates cankers and dieback caused by species of *Leucostoma*, *Valsa*, and their conidial states, *Leucocytospora* and *Cytospora*, respectively. The pathogens are ascomycetes in the order Diaporthales, family Valsaceae. They are opportunistic invaders of bark on twigs, branches, or mainstems that have been weakened by drought, freezing, heat, poor nutrition, mechanical injuries, or other diseases. These fungi are also ubiquitous invaders of dying bark and are thus commonly found on dead twigs and branches. Sometimes they colonize cankers caused by other agents (e.g., see Plate 96). Parasitic species show limited host specialization, usually occurring on several related (and in some cases many unrelated) host plants. In one case recorded from British Columbia, *C. pulcherrima* was found on 14 species of shrubs and trees that had been damaged but not killed by a ground fire.

Species of *Leucostoma* or *Valsa* produce conidia and later perithecia and ascospores in stromata that develop just beneath the bark surface. Stromata are roughly conical and short in relation to their diameter. Their tips break through the bark surface and appear as black to gray or even white dots. Each conidial stroma contains a single large, irregularly lobed fertile chamber that opens through a pore at the top. Perithecia, numbering 5–30 in a stroma, are arranged in circular fashion, with their necks converging at the top of the stroma. With a hand lens and sharp knife, the diagnostician can see these features after shaving off a thin layer of superficial bark (D; see also Plates 95–96). All members of this group were once assigned to the genus *Valsa*. Now those species having perithecial stromata with a black basal layer adjacent to the underlying bark are segregated in the genus *Leucostoma*.

Conidia produced by these fungi are tiny (e.g., 1.5 × 4 μm), individually colorless, single-celled, and sausage shaped. When a mature conidial stroma is moistened, a yellow to orange mass of conidia in a gelatinous matrix extrudes as a curled tendril, often called a spore horn, from the opening in the top of the stroma. When wet, the conidial mass disperses over the bark surface as a film of spore-laden water. Conidia thus become available for dispersal by dripping and splashing rain.

Perithecia follow pycnidia, often in the same stromata. Ascospores are released in response to moisture. Studies of *V. ceratosperma* and *L. persoonii* (Plate 94) revealed two mechanisms of ascospore release. If a stroma remains moist for a long time, a mass of asci containing ascospores extrudes from the perithecium in a manner resembling the release of conidia. If dry bark with stromata is saturated, however, ascospores may be forcibly expelled into the air. Ascospores are colorless, unicellular, sausage shaped, and 5–30 μm long, depending upon species.

Valsa cankers of conifers. Many species of *Cytospora* and *Valsa* have been described on the basis of collections from coniferous trees or shrubs, but the species are poorly delimited, and only a few are proven pathogens. Identification is an uncertain business because each species varies considerably in characteristics of stromata, asci, and ascospores. Diagnosticians often find only the *Cytospora* state of a pathogen in this group and usually identify it only to the generic level. Species for which pathogenicity to one or more coniferous hosts has been proven are listed here with host plants.

V. abietis (conidial state *C. abietis*) is common in both western and eastern coniferous forests. It is a proven pathogen of red and white firs and Douglas-fir in the West and of balsam fir, eastern hemlock, eastern and European larches, and black, Norway, red, and white spruces in the East. This fungus also occurs on arborvitae; Alaska cedar; grand, noble, Pacific silver, and subalpine firs; common juniper; Japanese and western larches; certain pines; and Colorado blue spruce. Its perithecial state is reportedly rare in the West but is common on several hosts in Quebec. *V. abietis* causes branch mortality and dieback of large trees and sometimes death of seedlings and saplings of grand, noble, California red, and Pacific silver firs. Damage is common in forests on the east side of the Cascades and Sierra Nevada, where trees are often subject to drought stress. Symptoms, the progression of disease, and the influences of predisposing factors are similar to those described for Leucostoma canker of spruce (Plate 93) except that sunken cankers are relatively more common. One singular aspect, however, is the frequent occurrence of cankers on red fir and to a lesser extent on white fir in association with dwarf mistletoe infections. The pathogen grows in laboratory cultures at 5–30°C and grows most rapidly at 25–30°C.

V. friesii (conidial state *C. pinastri*) is a proven pathogen of balsam fir; eastern hemlock; creeping and ground junipers; eastern and European larches; jack, lodgepole, red, Scots, and eastern white pines; and black, Norway, red, and white spruces.

V. pini (conidial state *C. pini*) is a proven pathogen of the same species as *V. friesii* with the exception of larches and junipers. In the West this fungus has been reported to cause cankers on grand fir. In the East it occurs mainly on pines but also on common juniper and European larch.

L. kunzei, the most important pathogen of the group on conifers, is discussed with Plate 93.

A–D. Cytospora canker of eastern hemlock. These specimens came from a salt-stressed hedge. A. A canker with a dark, resin-coated surface at the base of a dead lateral branch on a young tree. B. A canker similar to that shown in (A), with superficial bark shaved to show that the lesion is restricted to superficial tissues. The dark brown line at the edge of the lesion is a zone where host defenses have halted the parasite (see Plate 242). C. Magnified view of the bark surface raised in minute mounds where pycnidia are developing just beneath the periderm. D. Pycnidial stromata of *Cytospora* sp., viewed after slicing off their tops. Fertile chambers with multiple lobes are revealed (NY, Sep).

E, F. Cytospora canker of balsam fir. E. A sapling girdled by a canker at a branch whorl. F. A closer view showing erumpent pycnidia near the upper margin of the original canker. Bark above is dead and discolored because of the girdling action of the canker (NY, Jun).

G, H. Perithecial stromata of *Valsa* sp. in bark of a red pine branch. Flask-shaped black perithecia embedded in stromata have long black necks that emerge in clusters at the tops of the stromata. This fungus was associated with dieback of drought-stressed trees (MN, Aug).

Plate 92

Leucostoma canker of spruce and other conifers. Leucostoma canker, known also as Cytospora canker, is the most common and damaging disease of spruces used for ornament or windbreak in the Midwest and East. The pathogen is *Leucostoma kunzei* (syn. *Valsa kunzei*) (conidial state *Leucocytospora kunzei*, syn. *Cytospora kunzei*). It attacks not only spruces (black, Colorado blue, Engelmann, Norway, Oriental, red, and white), but also western red cedar; Douglas-fir, balsam fir; eastern hemlock; eastern, European, and Japanese larches; and red, eastern white, and Himalayan white pines. Its pathogenicity to spruces and Douglas-fir has been proved. This is the best known, most common, and most destructive species of *Leucostoma* or *Valsa* on conifers. Three varieties have been recognized: *L. kunzei* var. *piceae* on spruces, var. *superficialis* on pines, and var. *kunzei* on other conifers. These designations, however, are not widely used.

Colorado blue spruce sustains the greatest damage, being attacked wherever it is grown east of its natural range in the Rocky Mountains. Within that range, however, Leucostoma canker is rare. The disease seldom kills trees outright but disfigures them by killing branches and causing profuse exudation of resin from cankers on branches or trunks. Infections on Oriental and Colorado blue spruces are usually confined to branches. Branches and trunks are damaged in black, Engelmann, Norway, red, and white spruces. The disease also occurs in forest plantations and sometimes in natural stands where trees are stressed by drought or poor site conditions.

Dying or dead branches call attention to Leucostoma canker. Old branches are more susceptible than young ones. In spring and early summer, the foliage on one or more branches fades and turns brown, an indication that the branch has been girdled or that a canker on the trunk has encircled the branch. Brown needles persist during much of the growing season and drop off during winter, leaving bare twigs and branches. This process recurs more or less annually, moving from low branches to higher ones and destroying the symmetry of ornamental trees. Unsightly dead twigs and branches persist for many years. The disease occasionally starts in a high branch and, exceptionally, may kill the top of an otherwise green tree, but attack of low branches is the rule. Damage usually does not begin until trees are at least 10–15 years old. In landscape nurseries, however, small branches of young Colorado blue spruce or occasionally white spruce may be killed.

Lesions often start at the bases of small twigs and expand along the stem or branch more rapidly than around it. This process results in elliptical or sometimes diamond-shaped cankers. Cankers that originate on branches within several centimeters of the mainstem may spread into it, but most infections on branches are confined there. Internally, bark killed by *L. kunzei* is brown to reddish brown and infiltrated with resin. The underlying sapwood, although killed and colonized by the pathogen, is scarcely discolored. Clear amber resin exudes copiously from the edges of cankers, runs down the bark, or drips onto lower branches or the ground and hardens in a conspicuous white crust. Except for the resin, canker locations on spruces remain obscure for several years because diseased bark becomes infiltrated and glued in place by resin, and callus formation at canker margins is slight or absent. Cankers on trunks eventually appear sunken because of the growth of surrounding tissues. Several years or sometimes decades may pass before a trunk or large limb is completely girdled. By that time the stem may be deformed because of one-sided growth.

Symptoms caused by *L. kunzei* on other conifers are similar to those on spruces except that resin exudation is usually less prominent. Cankers form on branches and trunks of balsam fir, Douglas-fir, eastern hemlock, and eastern larch. Cankers on hemlock are rare except in association with pruning wounds. Inconspicuous branch cankers are the rule on pines.

Stromata of the *Leucocytospora* state of *Leucostoma kunzei* form in killed bark, beginning during the 1st season of canker enlargement. They are found near the edges of old cankers. After a branch or stem has been girdled, the pathogen rapidly colonizes large areas of bark beyond the girdle, forming numerous pycnidial and later perithecial stromata.

Pycnidial stromata are 1–2 mm in diameter with fertile chambers radiating from the center and opening through a common pore at the top of the stroma. When they are moistened, pycnidial stromata produce yellow tendrils of conidia. An individual stroma does this only once, however. Conidia, $4–6 \times 0.5–1$ μm, are released during wet weather throughout spring, summer, and autumn. Conidial production is greatest during spring. Conidia can withstand freezing, and they germinate at temperatures of 20–33°C. The optimum temperature for both conidial germination and initial growth of the fungus is near 27°C.

Perithecial stromata, which mature in spring, are 1–2 mm in diameter and prominent in dead bark beyond cankers. The interior tissue is pale yellow to grayish brown with 5–30 black perithecia embedded in it. Perithecia are 200–600 μm in diameter, and their necks converge at the disclike top of the conical stroma. The disc is usually 200–1000 μm in diameter and gray to black on the surface. Ascospores, measuring $5–8 \times 1–2$ μm, are released in spring and early summer.

Dispersal of both conidia and ascospores by running and splashing water undoubtedly accounts for the intensification of Leucostoma canker within individual trees. Both spore types can also be trapped from the air in the vicinity of diseased trees. Conidia are more numerous than ascospores in both water and air. Just how conidia become airborne is uncertain, since they are not forcibly expelled from pycnidia. Perhaps the spores are carried on tiny rain-splash droplets. Airborne spores and insects account for spread from tree to tree.

Conidia, ascospores, or mycelium from cultures of *L. kunzei* are capable of causing cankers or dieback if they are introduced to wounds. Most infections are presumed to occur in early spring when conidia and ascospores are abundant, shortly before symptoms appear. Tiny breaks in the outer bark, caused by mechanical stresses such as the weight of ice or snow, are possible sites of infection. However, *L. kunzei* can be found in outer bark of apparently healthy branches. Latent infection may therefore occur long before lesions develop.

Of the factors that predispose conifers to infection by *L. kunzei*, drought is most commonly named. Damage to white spruce in Ontario and outbreaks of a canker called pitch girdle (= Leucostoma canker) of Douglas-fir in the Rocky Mountains were preceded by years of low rainfall. When drought-stressed and nonstressed blue spruce were inoculated with *L. kunzei* in experiments, the number and rate of development of cankers were greater in the stressed trees. On the other hand, freezing shock in early autumn did not predispose trees to unusually severe disease. Other factors reported to predispose to infection include hail damage and injury to roots by plant-parasitic nematodes.

Leucostoma canker of Colorado blue spruce.

A, B. Leafless twigs and faded or brown foliage on branches girdled by cankers (NY, Jun).

C. Twig dieback caused by *Leucostoma kunzei* on a young tree in a nursery (NY, Sep).

D. Resin exuded from diseased bark. Fresh resin is amber fluid. With time it crystallizes into hard white deposits. Arrow indicates a portion of the branch, beyond the canker, where fruit bodies of the pathogen may be found most readily (NY, Mar).

E. Magnified view of bark, similar to that identified by arrow in (D), beyond a girdling canker. After removal of loose, outermost bark scales, the tips of pycnidial or perithecial stromata are apparent (NY, May).

F. Magnified view of moist bark at the margin of a canker. The white deposit is crystallized resin. The amber blob is a mass of conidia that have oozed up from a pycnidium in the bark. The black spot at left is the tip of a pycnidium that has ceased producing spores (NY, Oct).

Plate 93

Leucostoma cankers of *Prunus*. *Leucostoma cincta* and *L. persoonii* cause twig dieback and perennial cankers on branches and trunks of many species of *Prunus* (stone fruit trees) in orchards and landscapes and sometimes in forests. The diseases are known as Cytospora canker, Leucostoma canker, peach canker, perennial canker, Valsa canker, and gummosis. When severe in orchard trees, they cause marked decline in productivity. Both fungi occur across North America and also in Japan and Europe. They are usually found in their conidial states, *Leucocytospora cincta* and *L. leucostoma*, respectively. They were long known as *Valsa cincta* and *V. leucostoma*, and their conidial states as *Cytospora cincta* and *C. leucostoma*, respectively.

These fungi are opportunistic, quick to colonize dying twigs or bark, and able to use these substrates as beachheads for invasion of healthy bark. They are aggressive in plants weakened by environmental insults. Strains found on stone fruit trees vary from strictly saprophytic to virulently pathogenic. A given isolate is usually able to infect several *Prunus* species.

The host ranges of *Leucostoma cincta* and *L. persoonii* are similar. Therefore host plants of the individual pathogens are not distinguished in the list below. Also, many records of Cytospora canker on stone fruit trees do not include specific diagnoses. Such records represent several pathogens in addition to the species named above. *V. ambiens* (Plate 96), for example, occurs on various species of *Prunus*. Host plants of *Leucostoma* and *Cytospora* fungi that infect *Prunus* species include apple; apricot; Sitka mountain ash; blackthorn; black, Japanese flowering, pin, sour, and sweet cherries; chokecherry; Russian olive; peach; pear; common, damson, and wild plums; prune; serviceberry; and golden willow.

Affected trees usually have multiple cankers. Common sites of infection are leaf scars, scars left after fruit is picked, wounds made during pruning and other cultural operations, blossom and fruit spurs, fruit racemes (of black and pin cherries), winter injuries, lesions caused by brown rot fungi (Plate 58), and insect injuries. Insects that provide sites of entry include the peach tree borer, the lesser peach tree borer, and the shothole borer (see Johnson and Lyon, 2nd ed., Plates 117 and 121), the peach twig borer, the Oriental fruit moth, and in black cherry a cambium miner, *Phytobia pruni*.

Symptoms include dead buds, twigs, and branches; cankers on stems of all sizes, gummosis at sites of current infection, and yellowing or wilting and browning of leaves on diseased branches. Dead twigs or spurs often extend from the centers of cankers, indicating a common route of invasion of branches. Newly infected inner bark turns reddish brown and contains darker lines that mark former temporary boundaries of the lesion. With time the inner bark collapses, causing the canker surface to appear depressed. The pathogens invade bark, wood, and pith of twigs and sapwood beneath and beyond cankers. In one study, *Leucostoma* species were isolated from wood 2 cm beyond canker margins on twigs.

Several hosts, notably black cherry and peach, produce gum at sites of infection. Gum accumulates in cavities in infected inner bark and oozes out through cracks or ruptured lenticels. During rainy weather, gum deposits swell, dissolve partially, and spread over nearby bark. Debris and dark-colored saprophytic fungi accumulate in the gum, which eventually appears as a blackened crust on and around the canker. Gum also infiltrates sapwood beneath and beyond cankers, causing the wood to cease sap conduction. Parts distal to a canker may lose vigor and show premature yellowing or interveinal chlorosis of leaves or wilting and defoliation. These symptoms frequently precede girdling. Concentrations of several nutrient elements, especially calcium, are subnormal in foliage beyond cankers.

Cankers develop most commonly and rapidly in autumn to early spring, but infection occurs at any time of year if temperature and impaired host resistance permit. Lesions develop least rapidly or not at all during periods of host growth when cork barriers and callus ridges form at canker margins. At other times of year, the parasites breach these barriers and kill bark on and beyond the callus. This sequence is repeated annually at canker margins. During the 1st year or so, the bark surface remains intact except where gum exudes or pycnidial stromata cause minute ruptures. Eventually, swelling due to callus formation results in torn or broken bark. Typical old cankers on limbs or trunks are elongate, rough, sunken, and black, with prominent concentric ridges of callus. Eventually they girdle limbs.

Pycnidial stromata form in cankers and on girdled twigs and branches distal to cankers beginning several weeks after death of bark. Stromata are superficially black, gray to grayish brown internally, and underlain by a black layer. Stromal development is first evident as pimplelike swellings at the tips of which the bark ruptures to expose the top, or disc, of the stroma. The disc of *L. persoonii* is white or frosted in appearance, while that of *L. cincta* is gray to brownish gray. When pycnidia are mature and bark is moist, a flesh-colored to orange tendril of conidia in a gelatinous matrix extrudes from each stroma. Conidia of both *L. cincta* and *L. persoonii* are individually colorless and measure $5-10 \times 1-2$ μm.

Perithecia form much later, sometimes 2–3 years after branch death, also within stromata. Ascospores are produced during spring and early summer. Asci and ascospores of *L. cincta* are larger than those of *L. persoonii*: asci 45–80 versus 35–55 μm long and spores $15-30 \times 4-8$ versus $10-18 \times 2-5$ μm.

Conidia cause the majority of infections. They are most abundant in spring but are present throughout the year. Late winter or early spring is the most common time of infection. Both conidia and ascospores are dispersed by dripping or wind-blown water, and ascospores of *L. persoonii* are dispersed also by air during or after rain. Possible air dispersal of *L. cincta* has not been studied. Infection resulting from contaminated pruning tools has also been reported.

Temperature preferences of *L. cincta* and *L. persoonii* are related to their seasonal behavior as pathogens. The temperature most favorable for growth of *L. cincta* is 18–24°C, depending upon strain of the fungus, while *L. persoonii* grows most rapidly at 27–33°C. *L. persoonii* may be actively parasitic whenever the air temperature is above freezing, whereas *L. cincta* is inactive during hot summer weather.

Host vigor influences damage from Leucostoma canker. Factors that impair vigor and predispose peach trees to infection under field conditions include water shortage, deficient supply of potassium, and freeze damage. The last is important throughout peach-growing areas in the East. Peach rootstocks that enhance the winter hardiness of stems also enhance resistance to Leucostoma canker.

A–F. Leucostoma canker of 'Kwanzan' flowering cherry. A. Stunted shoots and delayed opening of buds suggest presence of a canker (NY, May). B, C. Closer views of a canker site on the branch shown in (A). Intact bark hides the canker (B), but extensive necrosis is revealed by shaving off the periderm (C). D. Pycnidial stromata beginning to raise the periderm several weeks after canker formation (NY, May). E, F. Foliar yellowing and death on a girdled branch. The girdling canker, its margin located by an arrow, was initiated the previous year at the base of the dead twig at upper right (NY, Jun).

G–I. Leucostoma canker of peach. G. Gum exuding from a lesion at a twig base. H. Bark shaved to reveal a small brown lesion. Reddish discoloration of sapwood is a symptom of freeze damage that may have predisposed to infection (NY, May). I. Pycnidial stromata of *Leucostoma persoonii* on a dead peach twig (GA, May).

Plate 94

Cankers of poplar and willow. Poplars and willows are subject to cankers and dieback caused by *Cytospora chrysosperma* and *Leucocytospora nivea*, the conidial states of *Valsa sordida* and *Leucostoma niveum* (syn. *V. nivea*), respectively. These fungi are opportunistic pathogens, quick to attack plants that have been predisposed by heat, drought, winter damage, or infection by other pathogens. They also colonize dying or dead bark as saprophytes. The two fungi cause indistinguishable symptoms: rapidly spreading necrosis of weakened bark or, on stems of intermediate susceptibility, localized annual cankers or slowly expanding perennial cankers. Both fungi occur around the Northern Hemisphere, and *V. sordida* has accompanied its host plants to Australia. Both are inconsequential in natural forests, but *V. sordida*, the more common and aggressive species, may cause devastating losses in nursery seedbeds, storage, newly established forest plantations, and landscape or shelterbelt plantings. This fungus often inhabits apparently healthy bark and buds and is thus in a position to colonize weakened tissues quickly and massively. *Leucostoma niveum* usually causes slowly expanding cankers.

Experts have suggested that all willows and poplars are at least somewhat susceptible to *V. sordida*, although in reality only a few dozen host species have been recorded. *L. niveum* affects fewer hosts. Both fungi occur on bigtooth and trembling aspens, balsam and white poplars, and the following willows: basket, creeping, purple osier, white, *Salix aurita*, *S. daphnoides*, *S. glauca*, and *S. nigricans*. *V. sordida* also occurs on European aspen, eastern and Great Plains cottonwoods; black poplar; white poplar; *Populus ×acuminata*, *P. angustifolia*, *P. macdougalii*, *P. wilsonii*, and *P. wislizeni*; and bay-leaved, black, peach-leaved, shining, and weeping willows. Ash; mountain ash; paper birch; elder; Rocky Mountain, Norway, and sugar maples; and Russian olive are also colonized on occasion. In addition, there are numerous reports of unnamed *Valsa* or *Cytospora* species on various poplars and willows.

Stems of all sizes become infected. Definite cankers, usually irregular in outline and more or less elongate, form on trunks, limbs, and small branches, but twigs are killed without the formation of discrete lesions. Cankers on branches often arise at the bases of dead twigs. New infections in smooth bark appear as brownish sunken patches that may girdle the stem. The inner bark turns nearly black and gives off a foul salty odor. The sapwood appears reddish brown and water-soaked. Cankers on large stems with thick, rough bark may be imperceptible except for spore horns (described below) in bark fissures. Elongate cankers 1 meter or more in length may form in white poplar. Severely affected trees often die branch by branch.

Blackstem disease of cottonwood seedlings and cuttings, attributed primarily to *V. sordida*, causes severe losses in storage and sometimes in nursery beds. Blackstem symptoms arise in autumn as tiny lesions at the ends of cuttings or at leaf scars and lenticels, usually on stems but sometimes on roots. Lesions enlarge during winter, becoming dark brown to black and water-soaked with sharply defined margins.

Within a few weeks after death of bark in or distal to cankers, fruit bodies (conidial stromata) of *V. sordida* or *L. niveum* form in outer bark just beneath the periderm. The stromata are shaped like short cones the flattened tips of which break through the bark surface. The exposed tip, termed the disc, is typically gray-brown to black in *V. sordida* and white to gray-brown (usually white) in *L. niveum*. Each stroma at maturity is 1–2 mm in diameter and gray-brown internally. It contains a complex fertile chamber with several irregular lobes that radiate downward and outward from a single opening in the disc. Masses of conidia in a sticky matrix ooze from stromata during wet weather or periods of high atmospheric humidity and are available for dispersal by water, insects, or birds. If the bark is moist but not wet, the spore masses exude as ribbons or tendrils that are often several millimeters long (E, H). These are known as cirrhi or spore horns. They vary in color from reddish orange through yellow and eventually bleach nearly white if they are not first dispersed by water.

Perithecia of *V. sordida* and *L. niveum* are less common than pycnidia. Perithecia form in the same stromata with pycnidia or in new stromata beginning in autumn and winter after pycnidial formation. Perithecial bodies, approximately 0.5 mm in diameter, are black and spherical and are arranged several in a ring in the lower, outer part of the stroma. Their necks converge to a circle of openings on the disc. Sometimes a pycnidial cavity persists in the center of the stroma. Ascospores are liberated as described with Plate 92.

V. sordida and *L. niveum* are highly variable in morphological features and cultural characteristics. This presents a problem in identification of *V. sordida* because it resembles the ubiquitous *V. ambiens* (Plate 96), which occurs on several of the same hosts.

Most infections are assumed to be initiated by conidia because conidia are much more abundant than ascospores and are liberated throughout the year except perhaps during midwinter. Buds, nodes, lenticels, and wounds of all kinds are sites of infection. Cankers usually develop during the dormant season and enlarge at rates related to temperature in the range of 2–30°C. Rates of canker enlargement up to 40 mm per day have been recorded.

The speed of canker development and the annual or perennial nature of the canker depend in large measure on effectiveness of host defenses. The blackstem symptom in cottonwood indicates a virtual absence of defense reactions in bark. The cortex is rapidly colonized and cellular contents digested. In a slowly enlarging canker, in contrast, the pathogen is continually confronted and inhibited by tannins and other phenolic compounds secreted by the host in cells at the margin of the lesion. These reactions are somewhat weaker in sapwood than in bark; thus the fungus is able to advance in the cambium and sapwood beyond the externally visible edge of the lesion. If fungal advance is slowed sufficiently or halted, a periderm differentiates around the lesion in the bark in the same manner as around a wound (Plate 242), and the killed sapwood becomes compartmentalized (Plates 162–163). The lesion in bark and wood is thus isolated from normal tissues. The canker may resume enlargement if the periderm is breached by new wounds or by wedges of aggregated hyphae of *V. sordida* in the vicinity of phloem fiber bundles.

Predisposing environmental or cultural factors are paramount in the susceptibility of poplars to *V. sordida*. Bark susceptibility may be induced by heating to approximately 40°C, such as may occur during summer days. Rapid shifts between warm and subfreezing bark temperature in spring also predispose to infection. Blackstem symptoms invariably develop in cottonwood cuttings physiologically weakened by prolonged or improper storage. Damage to poplars by *V. sordida* has often been linked to unfavorable sites for growth and to water shortage, but no simple direct relationship exists between susceptibility and plant water content. In fact, epidemics of Valsa canker in the absence of drought have been recorded. Temperatures of 20–30°C favor rapid canker formation in inoculated plants that are in a susceptible state, whether they are well watered or not. Both *V. sordida* and *L. niveum* also grow most rapidly in culture at about 25°C.

V. sordida and *L. niveum* often colonize bark weakened or killed by other pathogens, such as *Septoria musiva* on cottonwood (Plate 32), and may obscure the primary pathogen.

Tree breeders have searched for poplar species and clones resistant to *V. sordida*, and some workers have reported heritable resistance. To date, however, heritable resistance has not been distinguished from heritable avoidance of predisposing stress at test sites. The best approach to control of this disease remains the planting of stock known to be well adapted to the planting site.

A, F, G. Cankers and dieback caused by *Leucostoma niveum* in a fastigiate cultivar of white poplar (NY, Jul). A. Dieback, the result of cankers that girdled twigs and branches. F. New (left) and year-old cankers on small branches. The canker at right is studded with raised white dots, which are the discs of stromata of *L. niveum* that have broken through the bark surface. G. Magnified view of bark in a year-old canker, showing white discs of perithecial stromata and, where bark was sliced, the circle of perithecia within one stroma.

B. An elongate Cytospora canker that arose at a pruning wound on a mature white poplar (NY, Jun).

C. Blackstem symptoms on cuttings of eastern cottonwood (ND, May).

D. Cankers caused by *Valsa sordida* on goat willow (NY, Jun).

E. Cirrhi (spore horns) of *Cytospora chrysosperma*, the conidial state of *V. sordida*, on trembling aspen (NY, May).

H. Magnified view of a cirrhus of conidia of *C. chrysosperma* emerging from smooth bark of willow. Dark color of bark around the base of the cirrhus indicates the location and size of the conidial stroma beneath the periderm. Younger stromata are developing nearby (NY, Jun).

Plate 95

Valsa cankers of maple. Maples in forests, landscapes, and nurseries in eastern North America commonly develop cankers and branch dieback with which *Valsa ambiens, V. ceratosperma,* and their conidial states (*Cytospora* species) are associated. Branch death may occur any time of year and is conspicuous during spring through early autumn when foliage on scattered branches wilts and turns brown. Leaves remain attached; thus the affected branch is sometimes called a flag. Occasional trees with flagged branches can be seen every year; outbreaks occur in occasional years. Distinct girdling cankers sometimes cause flagging and dieback, but often no clearly defined lesion is apparent. The branch simply dies back to its junction with a parent limb or to a healthy lateral branch. A dark olive green to greenish black line usually separates healthy sapwood from dead at that point (B).

Pycnidial stromata of *Cytospora,* usually *C. annulata* (D, E), develop in recently killed bark and may be found at any time of year. The perithecial state of this pathogen, *V. ambiens* subsp. *leucostomoides* (syn. *V. leucostomoides*), develops in spring to early summer. This fungus also invades and stains freshly wounded maple sapwood. For example, it is commonly associated with columns of dark olive to greenish black stain that start from holes drilled in sugar maples for sap extraction. Other hosts include ash, box elder, and Norway, red, and silver maples.

V. ambiens subsp. *ambiens* (syn. *V. myinda*), *V. ceratosperma,* and their conidial states, *C. leucosperma* and *C. sacculus,* respectively, occur as saprophytes on diverse woody plants, including red and silver maples, in North America and northern Europe. Both subspecies of *V. ambiens* are opportunistic pathogens and saprophytic colonists of maples weakened by other diseases or injuries. The association of *V. ambiens* subsp. *leucostomoides* with Cryptosporiopsis canker of red maples is an example. *V. ambiens* subsp. *ambiens* is also suspected to be pathogenic to such other plants as apple, rose, serviceberry, and willow.

The general morphology and biology of *Valsa* species on maple are as described for related fungi on other plants (Plates 92–95). Strains of both subspecies of *V. ambiens,* isolated from maples, caused small cankers in experimentally inoculated branches of Norway, red, and sugar maples, but neither subspecies was an aggressive colonist of healthy bark. Several strains of *V. ceratosperma* in the same tests were nonpathogenic.

References for Leucostoma *and* Valsa *in general:* 107, 465, 469, 1008, 1671, 1830, 1870
References for Valsa *cankers of conifers:* 571, 1585, 1695, 1830, 2246
References for Leucostoma *canker of spruce and other conifers:* 107, 687, 974, 988, 1008, 1114, 1719, 2113, 2247
References for Leucostoma *cankers of* Prunus: 162–64, 173, 178, 479, 693, 724, 774, 775, 818, 1008, 1656, 1727, 1980, 2151, 2211, 2220
References for Leucostoma *and* Valsa *cankers of poplar and willow:* 177, 199, 202, 359, 566, 830, 901, 1008, 1009, 1713, 1722, 1969, 2094
References for Valsa *canker of maple:* 1869, 1870

Cryptosporiopsis Canker of Maple (Plate 96)

This disease of red maples, recognized in 1977, is common in certain years in nurseries and landscapes in the northeastern United States. The pathogen, an unidentified species of *Cryptosporiopsis* (Deuteromycotina, Coelomycetes) infects oviposition wounds made by the narrow-winged tree cricket, *Oecanthus angustipennis,* in trunks and low branches. The cankers are annual, usually less than 20 cm long and often only several millimeters wide, tapering toward the ends. They reach their greatest length and width in the cambial region, and each is centered on a tiny hole from which dark brown fluid bleeds. Streaks of discolored sapwood extend as far as 50 cm above cankers. Small trees may be girdled by these cankers, and larger stems may be greatly disfigured by multiple cankers.

Infection occurs at the time of cricket oviposition in autumn, cankers develop mainly during the dormant season, and bleeding is conspicuous in spring. Vigorous trees produce prominent callus rolls at canker margins by late summer and may enclose the smallest cankers beneath callus within 1 year. Often, however, opportunistic secondary fungi, especially *Valsa ambiens,* colonize the outer bark of cankers and perhaps cause some canker expansion.

The relationship between insect and fungus in this case is not fully understood, but inoculation is believed to occur as the female tree cricket scrapes fragments of superficial bark, infested with the fungus, into the oviposition hole to form a plug. Oviposition wounds made by other insects such as buffalo treehoppers do not become infected with *Cryptosporiopsis* sp. Incidence of this disease varies greatly from year to year.

References: 1976, 1977

A–F. Cankers on Norway and sugar maples caused by *Valsa ambiens* subsp. *leucostomoides.* A. A diseased Norway maple with dead branches bearing faded and brown leaves (NY, Jun). B. The base of a diseased sugar maple branch with a characteristic dark line between healthy and dead tissues (NY, Sep). C. Norway maple bark with rows of white tips (discs) of pycnidial and perithecial stromata, approximately natural size (NY, Jul). D. Magnified view of sugar maple bark with rows of stromata of the conidial state, *Cytospora annulata,* breaking through the surface (NY, Sep). E. Magnified view of Norway maple bark with the surface shaved off to reveal dark gray-brown pycnidial stromata 2–3 mm in diameter (NY, Sep). F. Magnified view of perithecial stromata in bark of red maple; gelatinous contents of perithecia are visible where tops of fresh stromata were sliced (NY, Apr).

G–J. Cryptosporiopsis canker on red maple cultivar 'October Glory' (NY, Jun). G. A mature canker that developed during the dormant season. A wound at the center of the canker was the apparent site of infection. The bark surface in this canker is roughened by pycnidia of *V. ambiens* subsp. *leucostomoides,* a secondary invader. H, I. A stem with two small bleeding cankers before (G) and after (H) removing surface bark of one canker to reveal a dead yellowish brown zone in the cambial region. J. A 1-year-old inactive canker, sunken as the result of callus development at margins.

Plate 96

Thyronectria Canker of Honey Locust (Plate 97)

Honey locust was once reputed to be a drought-tolerant tree with no significant pest problems. This was perhaps true in its natural habitat, but important canker and root rot problems, aggravated by water and heat stress, have developed in landscape and street-side plantings. The first important disease to be publicized was Thyronectria canker, reported from Massachusetts in 1939. The causal fungus, *Thyronectria austro-americana* (Hypocreales, Hypocreaceae), is found in a region extending from Massachusetts to Colorado and the Gulf states. It is a common saprophyte, producing pycnidial stromata (*Gyrostroma* state) and perithecia on dead trunks and branches. In its parasitic phase the fungus is most aggressive in the Great Plains, where its damage to honey locust has been linked to drought stress. It is cited as a major cause of decline of thornless honey locust in urban plantings in Illinois, but it causes only branch cankers and minor dieback on honey locust in natural wooded areas in the same region. The fungus also occurs on Japanese honey locust and mimosa, causing wilt and death of the former species. On mimosa it is restricted to senescent branches or to injured trees or those affected by mimosa wilt. It has been reported to occur on acacia in South America.

Symptoms on honey locust include annual and perennial cankers on trunks and branches of all sizes. Trees or branches girdled by cankers develop yellow or wilted foliage and dieback. Trees with trunks girdled by cankers do not usually produce epicormic sprouts. The cankers are usually elongate and slightly depressed when young and become conspicuously sunken if callus ridges develop at the edges. Callus development varies from none to prominent, according to the vigor of the tree. The surface of killed bark on young trunks or limbs is often orange-brown initially. It bleaches somewhat with time, often to bright yellow-orange, and the periderm cracks and begins to peel after the emergence of fruit bodies. Thick bark on old stems obscures cankers. Reddish brown discoloration develops in sapwood beneath and near the cankers and may extend to the heartwood. In the South, a gummy substance sometimes exudes from cankers. Trunk cankers are often associated with pruning wounds or sunburn.

T. austro-americana has been characterized as causing two distinct diseases, canker of native honey locust and wilt of Japanese honey locust. The parasite invades the current season's sapwood of Japanese honey locust systemically, causing reddish orange streaks and wilt similar to the syndrome of mimosa wilt. The streaks correspond to clusters of xylem vessels in which the parasite moves. The disease develops rapidly and may kill an entire tree within 1 season. All foliage may wilt suddenly in spring, or if the disease develops later in the season, yellowing and defoliation may precede branch death. If the tree lives more than 1 season, excessive fruit production may signal impending death. In native honey locust, in contrast, the fungus is usually restricted to lesions in bark and to the sapwood beneath and immediately adjacent to diseased bark.

Numerous clusters of stromata of the pycnidial state of the pathogen, *Gyrostroma austro-americana* (syn. *Kaskaskia gleditsiae*), break through the bark surface, mainly through lenticels, in a canker or on a killed branch several weeks after the death of bark. Stromata are pinkish at first and gradually darken through shades of tan and brown to almost black. They contain irregular fertile chambers from which pinkish orange to cream-colored masses of conidia are released after wetting. Conidial masses sometimes take the form of cirrhi (spore horns). The conidia are individually colorless and one-celled and measure $1.8-3.6 \times 0.6-1.6$ μm. They are unusually resistant to desiccation and heat; in one test they survived 3 weeks at 32°C and 0% relative humidity.

Clusters of perithecia, 10–100 or more in a group, develop on the same stromata previously occupied by pycnidia in dead bark beginning in autumn and are mature the next spring. Perithecia, 200–450 μm in diameter, are distinctively two-toned with yellow-brown bodies and black tips, easily recognized with a hand lens. The yellowish ascospores, $8-16 \times 4.5-9$ μm and elliptical in outline, are divided irregularly into many cells. They are forcibly discharged up to 2 mm into the air.

Either conidia or ascospores may cause infection if introduced into wounds less than approximately 3 weeks old. Wounds become progressively less susceptible during this period. Pruning wounds and sunburned bark on the southwest side of the trunk are common sites of infection of landscape trees. Cankers may elongate as much as 30 cm per year. Cankers have in some cases been found associated with wounds made by flatheaded borers (see Johnson and Lyon, 2nd ed., Plate 127). A vector relationship of *T. austro-americana* with these insects has been suggested but not confirmed.

T. austro-americana is adapted to a variety of climatic regions, including dry areas with hot summers such as the western Great Plains. It can grow in pure culture at temperatures of 10–40°C, and it does so most rapidly at 28–32°C. Within the temperature range 16–28°C, it causes most rapid canker expansion at 28°C. Conidia in pycnidia can survive several months of desiccation and apparently do not require free water for germination if the relative humidity is greater than 75%. Conidia germinate at 15–40°C, and germination is maximal at 25–35°C. The fungus produces pink-orange mycelium and dark pycnidial stromata in pure culture.

References: 136, 221, 427, 908, 941, 1614, 1736, 1737

A–C. Dieback and canker caused by *Thyronectria austro-americana* on thornless honey locust (MN, Aug). A. Dieback. B. A canker with prominent conidial stromata on the trunk of a young street tree. C. Close view of conidial stromata near a margin of the canker.

D–F. Thyronectria canker and associated reddish stain of sapwood on a honey locust limb about 8 cm in diameter. D. A canker with typical yellowish bark roughened by conidial stromata. E, F. Successively closer views of conidial stromata (CO, Jul).

G. Magnified view of wet conidial stromata producing cream-colored masses of conidia (NY, Aug).

H. Magnified view of clusters of perithecia on dead bark (NY, Jun).

Plate 97

Volutella Blight of Pachysandra (Plate 98)

Japanese pachysandra, a low-growing semiwoody plant, is widely planted as an ornamental ground cover. Its most destructive disease is blight and stem canker caused by *Volutella pachysandricola,* the conidial state of *Pseudonectria pachysandricola* (Hypocreales, Hypocreaceae). Outbreaks, said to be favored by humid weather, disfigure or kill many plants at the same time. The disease has been reported from Kansas, several eastern states, and Ontario and also from England.

Wilting and dying plants call attention to Volutella blight. Lesions develop on leaves, stems, or stolons. Infection often begins in damaged or senescent parts such as sunburned leaf tips and spreads into previously normal tissues. Leaves may be killed by expansion of irregular tan to brown blotches, often with concentric lighter and darker zones, with dark brown margins. Most leaf death, however, results from stem girdling by cankers. Cankers, at first greenish brown and water-soaked, shrivel, turn brown, and usually encircle the stem, causing wilt and death of parts above. Lesions on stolons often originate at points of root emergence.

P. pachysandricola is a wound parasite, able in some cases to girdle stems within 2 weeks after infection. During late spring and summer under humid conditions, pink, cushionlike fruiting structures (sporodochia) of the *Volutella* state of the parasite develop on the surfaces of cankers or on the undersurfaces of recently killed leaves, frequently in concentric rings. The sporodochia are up to 400 μm in diameter and produce conidia that are pink in mass. Individual conidia are single-celled and colorless, measuring 14–24 × 2–4 μm, with somewhat pointed ends. Each spore contains one or two oil droplets. Red perithecia form on stems during summer and early autumn, singly or in clusters, either replacing sporodochia or developing on new reddish stromata that break through the epidermis. Perithecia are more or less globose, 200–250 μm in diameter. The single-celled ascospores are colorless and measure 10–15 × 3–4.5 μm.

Most reports of outbreaks of Volutella blight mention the association of previous winter damage, infestation by scale insects, recent transplanting or shearing, or exposure to bright sun. Thus *P. pachysandricola* is considered to be an opportunistic parasite that causes little or no damage to vigorous plants but aggressively colonizes those previously injured or stressed.

The fungus grows readily in pure culture, and its pathogenicity has been proved by several investigators through artificial inoculation.

A related, similar-appearing fungus, *P. rouselliana* (conidial state *V. buxi*), is widespread on dead leaves and stems of common boxwood, on which it is the apparent cause of cankers and dieback.

References: 505, 1528

Symptoms and signs of Volutella blight of pachysandra.

A, B. Blight lesions on leaves and shriveled girdling cankers on stems (NY, Aug).

C, E. Damage in a landscape planting. Most of the wilted or dead leaves were not infected directly but died when stems were girdled (NY, Jun).

D. Sporodochia of the *Volutella* stage (left) and young perithecia (right) of *Pseudonectria pachysandricola* (NY, Aug).

Plate 98

Tubercularia and Nectria Cankers and Diebacks (Plates 99–101)

Tubercularia canker and dieback. *Tubercularia ulmea* (Deuteromycotina, Hyphomycetes) causes annual cankers and dieback of alder buckthorn, Siberian elm, winged euonymus, and Russian olive in the Great Plains. This fungus is the second most common cause of cankers and dieback of Siberian elm in the prairie region; only *Botryodiplodia hypodermia* (Plate 86) is more common. *T. ulmea* is reported to occur all across Canada on Siberian elm, but these reports cannot reliably be separated from those for *T. vulgaris* (Plate 100). The few known hosts of *T. ulmea* are diverse; therefore additional hosts are probable.

Symptoms develop similarly on all known hosts. Cankers form in spring, enlarge during a period of several weeks and then cease enlargement permanently. Often they are not noticed until they girdle and kill twigs and branches. The resulting dieback becomes apparent when buds fail to open or when foliage wilts in late spring and early summer. Nongirdling cankers are oval to elongate, often taking the form of longitudinal strips. New lesions in smooth bark of young stems appear reddish brown and slightly shriveled, but lesions in older stems with a thick periderm overlying the inner bark are imperceptible until cracks form at margins. Cankers on Russian olive sometimes exude gum at the margins. Callus formation at the margins of nongirdling cankers causes the bark to crack there, and eventually the bark shreds and peels from the canker. Small cankers on vigorous stems may be enclosed by callus within 1 year, but usually several years are required. Vigorous sprouts may develop below girdling cankers. Cankers on Siberian elm are most common on small branches and twigs in the interior of the canopy. Alder buckthorn (cultivar 'Tallhedge') is subject to girdling cankers near stem bases.

Cushionlike fruiting structures (sporodochia) of *T. ulmea* begin to break through the surface of killed bark during spring and early summer. In experiments with Siberian elms artificially inoculated in April, cankers began to form within 10 days after inoculation, and sporodochia began to emerge within an additional 10 days. The sporodochia, initially cream colored, slowly darken through shades of rust-pink to brown and finally black. At maturity they are up to 1.5 mm in diameter and 0.9 mm high. Unicellular, ovoid, colorless conidia, mostly 4.6–6.2 × 1.5–2.5 μm, form on their surfaces. Conidia are dispersed locally by water.

Sporodochia of *T. ulmea* may be found in dead bark throughout the summer and autumn. Beginning in autumn and continuing the next spring, sporodochia are sometimes followed by pale pinkish stromata on which bright red perithecia form. This sexual state cannot be distinguished from *Nectria cinnabarina* (Plate 100). Therefore *T. ulmea* may be only a variant of *T. vulgaris*, the conidial state of *N. cinnabarina*. Research is needed to resolve this question.

Conidia of *T. ulmea* germinate within 24–40 hours on laboratory media and produce rapidly growing white colonies.

T. ulmea is an opportunistic fungus that exists mainly as a saprophyte in dead bark. Apparently it is pathogenic only to trees or branches weakened by injuries or adverse environment, especially freezing. A rapid drop in temperature to below −20°C, as may happen after relatively mild weather in late autumn or late winter, has been shown to result in the girdling cankers on 'Tallhedge' alder buckthorn stems noted above. The cankers were caused in part by the direct effects of freezing and in part by *T. ulmea*. Tubercularia dieback has also been observed on heavily fertilized 'Tallhedge' plants and on those growing near lights that artificially extended the photoperiod during winter. These conditions, by preventing normal cold hardiness, may predispose plants to infection.

A–E. Tubercularia canker and dieback of Russian olive. A. A tree with two mainstems girdled by cankers. Wilted foliage adheres to twigs on the stem that died early in the growing season. Epicormic sprouts are developing below the cankers (ND, Jul). B. Gum exuding from longitudinal cracks at the margin of a Tubercularia canker. Gum exudation may be the only external indicator of disease on a stem with thick bark (ND, Jul). C. A branch canker, with bark shaved off to show the location of the margin (CO, Jul). D. A branch canker with the bark surface about to be broken by developing sporodochia (CO, Jul). E. Young sporodochia of *Tubercularia ulmea*, rust-pink in color, breaking through the bark surface in a canker (ND, Jul).

F–H. Magnified views of *T. ulmea* and its perithecial state, *Nectria cinnabarina*, on the bark of Siberian elm (ND, May). F. Old sporodochia (black cushion-shaped structures) and young perithecial stromata (pink structures). G. Sporodochia of various ages, the oldest nearly black. H. Clusters of perithecia.

Plate 99

Coral spot Nectria canker. *Nectria cinnabarina* (Hypocreales, Hypocreaceae), the coral spot fungus, is a cosmopolitan saprophyte on many plant species, mostly angiosperms, worldwide. The fungus is also an opportunistic pathogen that colonizes the bark, cambium, and outer sapwood of stems weakened by freezing, water shortage, mechanical injuries, or other diseases. It may cause significant damage to recently transplanted trees or shrubs on landscape or forest sites. Cankers and dieback of elms, honey locust, maples, and Japanese zelkova in landscapes, apple trees in orchards, and black locust in forest plantations are examples. The pathogenicity of *N. cinnabarina* has been proven by experimental inoculations to several plant species, but most host records merely report the fungus fruiting on dead parts without reference to possible pathogenic behavior.

Angiosperm host groups of *N. cinnabarina* or its conidial state, *Tubercularia vulgaris*, include alder, ampelopsis, apple and crabapple, ash, mountain ash, barberry, basswood and other lindens, beautyberry, beech, birch, boxwood, brambles, buckthorn, butternut, catalpa, cherry and other stone fruit trees, chinaberry, cotoneaster, currant, dogwood, elder, elm, euonymus, common fig, hickory, honeysuckle, horse-chestnut, hydrangea, California laurel, black locust, honey locust, maple, mimosa, mulberry, paper mulberry, oak, Russian olive, Mexican orange, Japanese pagoda tree, pear, privet, flowering quince, rose, Japanese rose, serviceberry, spirea, tree-of-heaven, viburnum, willow, and Japanese zelkova. The fungus has also been found on eastern and European larches and jack pine in Quebec. No doubt this list is incomplete.

Symptoms of coral spot canker or dieback include failure of twigs or entire branches to produce leaves in spring or sudden wilting of shoots on twigs and branches or occasionally on entire plants soon after growth begins in spring. Branches and small trunks may be girdled. Cankers often appear as slightly sunken areas associated with wounds. The killed bark of young stems without a corky surface is discolored brown to dark brown. If the outer bark is corky, the dead area remains imperceptible until the fungus begins to produce pink sporodochia or the growth of the surrounding tissues leads to cracks at canker margins. Cankers usually expand during host dormancy in only 1 year and then become sunken as callus forms around the lesions and slowly encloses them. In severely stressed trees, however, cankers may develop during the growing season and may enlarge during more than 1 year.

Lesions caused by *N. cinnabarina* are seldom noticed before the fungus begins to form the coral spots for which it is named. These are prominent cushionlike sporodochia of the conidial state, *T. vulgaris*. They arise most commonly in spring and early summer as masses of mycelium (stromata) that break through the bark surface at points of natural weakness such as lenticels in young bark or cracks in corky bark. Development in cracks often causes sporodochia to be aligned in rows. Sporodochia range from less than 0.5 mm to about 1.5 mm in diameter and height. They vary in color with host and stage of maturity from creamy or coral pink to pink-orange or light purplish red when young and to tan, brown, or nearly black in aging specimens, especially after exposure to frost. Conidia, individually colorless and single-celled, measuring $5-7 \times 2-3$ μm, are produced in a gelatinous mass on the surface of the cushion. Sporodochia swell when wet, and conidia are dispersed by splashing and dripping water and perhaps on pruning tools. During dry weather, on the other hand, the surface of the sporodochium is crusty, and spores are probably not dispersed.

In summer and autumn, globose orange-red perithecia form singly and in clusters of up to 15 either around the sporodochia or replacing them on the same stromata. Perithecia have rough surfaces, average $375-450$ μm in diameter, and turn dark reddish brown with age. They may persist through winter on dead bark. The ascospores are colorless, two-celled with a slight constriction at the septum, and narrowed at the ends, mostly $12-20 \times 4-7$ μm. During wet weather they are extruded in a mass from a pore in the top of the perithecium and are dispersed mainly by water.

N. cinnabarina invades dead buds and injuries from frost, hail, insects, and mammals. It is associated with girdling cankers on branches or on stem bases of small plants weakened by freezing. It infects pruning wounds and cracks in bark caused by the weight of ice or snow.

Both conidia and ascospores are infectious, and infection may occur throughout the year when temperature and moisture permit. Dispersal of ascospores continues even in winter during wet weather at temperatures above 0°C. Ascospores germinate slowly at 0°C and most rapidly at 21–26°C. Infection is more likely at injuries sustained in autumn and winter than in those of late spring and summer when wound-isolation responses are strong. In currants and probably in other hosts, the advance of mycelium in sapwood precedes that in the cambium and phloem.

Virulence of isolates of *N. cinnabarina* varies widely, but host specialization has not been reported. The form occurring on currants may be a distinct species, *N. ribis*. The relationship between *N. cinnabarina* and *T. ulmea* (Plate 99) is unclear.

Predisposition, or the weakening of host defenses, determines the occurrence and severity of coral spot Nectria canker. Untimely pruning (e.g., in late summer or autumn), root pruning, transplanting, and, most commonly, freeze damage predispose to infection. *N. cinnabarina* can cause cankers or dieback of cold-conditioned stems that lack any evidence of direct injury from the previous low temperature. Similarly, if the fungus invades stems dying back from freeze damage, it will increase the amount of dieback. Rapid drops in air temperature during late autumn and late winter when woody plants are not maximally cold hardy apparently predispose trees and shrubs to the coral spot fungus under natural circumstances. Fortunately this predisposition is temporary. Normal resistance to the parasite returns after several days.

From information about the importance of predisposition, it follows that much damage by the coral spot fungus can be prevented by planting trees or shrubs known to be well adapted to the climate of the planting site, by minimizing water stress, and by avoiding untimely pruning or other wounding.

A–C, E. Coral spot Nectria canker of sugar maple. A, B. Two views of a stem transplanted the previous year, nearly girdled by cankers initiated at pruning wounds. The outer bark at canker margins is cracked due to pressure from callus forming around the lesion. The bark of newly formed callus, visible through cracks in outer bark, is orange-brown (NY, Sep). C. Close view of a canker with pink sporodochia and, in the oldest part of the canker, red perithecia of *Nectria cinnabarina* (NY, Jun). E. Part of a cross-section of the stem shown above. A very narrow growth ring (arrows) indicates severe stress during the year of transplanting when infection began. Cracks in the wood, unrelated to disease, formed as the specimen dried.

D. Coral spot dieback of rose. A rough wound near the top of the pruned cane was the site of infection (NY, Nov).

F. Magnified view of young sporodochia of *N. cinnabarina* on sugar maple bark (NY, Sep).

G. Magnified view of overwintered sporodochia on honey locust bark (NY, Jun).

H. Magnified view of sporodochia and young perithecia on wet bark of sugar maple (NY, Jun).

Plate 100

211

Perennial Nectria canker. The canker caused by *Nectria galligena* (Hypocreales, Hypocreaceae) is one of the most common and readily recognized diseases of deciduous trees in North American and European forests. It occurs in temperate regions worldwide except in Australia, and it affects more than 60 species of trees and shrubs in North America. In forests, Nectria canker is more common and damaging in the East than in the West. It is the most important stem disease of sweet birch, yellow birch, and black walnut, reducing or destroying the value of the butt logs of these species for lumber. The disease also affects apple and pear trees in orchards and on these hosts is often called European canker. *N. galligena* may also rot apple fruit in storage.

Host plants represent more than 30 genera: alder, apple, ash, mountain ash, aspen and poplar, basswood, birch, alder buckthorn, cherry, dogwood, elm, filbert, sweet gum, hawthorn, hickory and pecan, holly, hornbeam, hop hornbeam, horse-chestnut, California laurel, magnolia, mahogany, maple, mulberry, oak, pear, quince, redbud, sassafras, serviceberry, sourwood, sumac, tupelo, walnut and butternut, and willow.

Cankers in their first year of development are small, dark, depressed areas on young, smooth-barked stems, but they are seldom noticed except on apple trees, where they girdle twigs. Cankers more than 1 year old are shaped like targets, round to elongate, with concentric ridges of wood exposed where successive rolls of callus were killed and the bark decayed and weathered away. Most cankers are centered on a small branch stub or the remains of one, and a diseased tree usually has more than one canker. Cankers expand slowly, often less than 1 cm tangentially per year, during host dormancy. During the next growing season a callus ridge is produced at the canker margin. The callus is invaded and killed, wholly or in part, during the succeeding dormant period. This alternation may continue for decades. If callus is greatly stimulated, as in maples and black walnut, the stem becomes markedly swollen at the level of the canker. If callus formation is slight, as in basswood, cankers become craterlike. As long as the parasite is active all around the periphery, the canker remains symmetrical. Where cankers coalesce, elongate compound cankers develop. Stems larger than twigs are rarely girdled unless by multiple cankers. If the fungus dies or is effectively resisted at one or more points on the canker margin while expansion continues elsewhere, the canker becomes irregular in outline. Some cankers become inactive and are then slowly enclosed by callus.

In the West, Nectria cankers on bigleaf maple sometimes expand not only during host dormancy but also during the summer dry season. Elongate cankers on this species may girdle small trunks or branches, causing dieback or conspicuous flags as foliage wilts or turns prematurely yellow and dies. Canker development also occurs on apple twigs during the growing season in the West.

Bright red or reddish orange, lemon-shaped perithecia 250–425 μm in diameter develop in autumn through spring. They arise singly or in groups on the surfaces of young cankers still covered with bark or on callus rolls and in bark crevices at the margins of old cankers. New perithecia may form up to 30 months after death of the bark; they darken with age. The ascospores are colorless, equally two-celled, oval to somewhat spindle shaped, and slightly constricted at the septum. They measure 14–22 × 6–9 μm.

Inconspicuous creamy white, cushionlike sporodochia of the conidial state, *Cylindrocarpon mali*, precede perithecia during moist weather, often protruding from lenticels. Sporodochia produce colorless conidia of two types: nearly cylindrical macroconidia with rounded ends, 10–65 × 4–7 μm, with two to five or more cells; and ellipsoid microconidia. Length of macroconidia varies with the number of cells in the spores.

Ascospores may be expelled from perithecia and dispersed by wind or water any time of year during rain or moist periods when the temperature is above freezing. Most discharge occurs in spring and autumn. In free water, ascospores can germinate at temperatures of 0–30°C and do so most rapidly at 21–26.5°C. Conidia are dispersed by rain splash throughout the year in areas of warm climate and during the growing season in colder areas. Conidia are the most important type of inoculum in some western apple-growing areas.

Infection occurs through wounds of various kinds: leaf scars, points of fruit detachment, cracks in twig axils caused by weight of ice or snow, sunscald lesions, and senescing low branches. Infection of apple trees is most frequent in autumn, but symptoms may not appear until spring.

To become permanently established, the parasite must reach the cambium. Infections restricted to phloem are soon isolated by cork barriers. In an established canker, *N. galligena* is able to breach the cork barrier on the callus roll annually; thus canker expansion resumes each dormant season. The fungus kills bark, cambium, and the outermost sapwood in advance of its hyphae. Pectic enzymes are presumed responsible in part for the characteristic soft decay of diseased bark. The fungus also produces indoleacetic acid, which may stimulate the formation of the prominent marginal callus that typifies the cankers. *N. galligena* does not decay the wood. In fact, wood decay beneath Nectria cankers on hosts other than birch is uncommon. Nectria cankers on birches are sometimes sites of infection for the canker-rot fungus, *Inonotus obliquus* (Plate 170).

N. galligena displays limited host specificity. When a given isolate is tested for pathogenicity to several plant species already known to be hosts, the typical result is canker formation on the host of origin plus several others but not on all hosts tested. Isolates pathogenic to birch and aspen, for example, did not infect apple trees when leaf scars were inoculated. Isolates from apple trees were pathogenic to apple, beech, hawthorn, and poplar but not to ash or maple. In Ireland, two formae speciales, *N. galligena* f. sp. *mali* and f. sp. *fraxini*, have been proposed on the basis of differing patterns of host specificity, but this scheme is not recognized in North America.

Perennial Nectria cankers on members of the magnolia family such as Fraser magnolia and tulip tree are often caused by *N. magnoliae* rather than *N. galligena*. The two fungi differ microscopically but have the same general appearance and cause the same symptoms. Two other species in the same group, *N. mammoidea* and *N. coccinea*, also occur on bark and wood of various trees, have *Cylindrocarpon* states, and may be confused with *N. galligena*. Both are apparently pathogenic, but little information about them is available. Most early North American reports about *N. coccinea* and *N. ditissima*, the latter a European species, would now be referred to *N. galligena*.

Perennial Nectria canker is most common in regions with cool, humid climate. Where it occurs at relatively high elevations, it is most common on exposed slopes with shallow or infertile soils. At low elevations the disease is common in cold pockets and on poorly drained soils.

References for Tubercularia canker and dieback: 329, 1614, 1715, 1717, 1909

References for coral spot Nectria canker: 134–36, 215, 387, 969, 1636, 1717, 1718, 1720, 2071

References for perennial Nectria canker: 62, 215, 228, 239, 387, 522, 575, 675, 1158, 1168, 1169, 1212, 2146, 2211, 2270

A. Old confluent Nectria cankers on sweet birch. The large callus roll at left indicates an inactive portion of the canker margin (NY, May).

B. Typical craterlike Nectria cankers on basswood (NY, Apr).

C–E, H. Nectria canker of bigleaf maple. C. Dead and dying branches, girdled by cankers (WA, Aug). D, E. A girdling canker on the trunk of a young tree. Cracked bark marks the location of a callus ridge formed the previous year. Dark red perithecia of *Nectria galligena*, visible as tiny dots several centimeters above the canker margin, are shown close up in (H) (OR, Sep).

F. A typical "target" canker on a trunk of black walnut (NY, Apr).

G. Magnified view of young perithecia on bark of Norway maple (NY, Sep).

Plate 101

Twig Blight of Mulberry and Fusarium Cankers of Golden-chain and Other Trees (Plate 102)

Twig blight of mulberry. Red mulberry in the northeastern United States is commonly affected by a syndrome involving twig blight, canker, and dieback that has occasionally been confused with the widespread twig blight of white mulberry caused by *Fusarium lateritium* f. sp. *mori*. Symptoms on red mulberry include failure of buds to open in spring, wilting and death of shoots in late spring, and cankers centered at dead buds or shoots on twigs and small branches. The dieback and wilt result from girdling lesions at buds or shoot bases. The outermost sapwood beneath cankers is discolored buff brown, and brown streaks in outer sapwood extend several centimeters beyond cankers toward the twig base. The brown streaks are associated with gum-plugged vessels. Repeated twig dieback and growth from surviving buds results in clusters of dead and living twigs near branch ends.

We have not found *Fusarium* consistently associated with these symptoms. Instead, *Nectria cinnabarina* (Plate 100) is most common in cankers and *Cytospora* sp. is common in dead twigs. Often, no fungi or bacteria can be isolated from streaks in the xylem except very close to the canker or dead tip. This syndrome seems to be caused primarily by freeze damage and secondarily by opportunistic fungi, especially *N. cinnabarina*, that invade weakened bark.

Fusarium cankers. *Fusarium lateritium* and *F. solani* (Deuteromycotina, Hyphomycetes) are associated with twig dieback and cankers on many trees and shrubs worldwide. Often these fungi are saprophytes or secondary invaders of tissues weakened or dying from other causes, but some strains act as primary pathogens. *F. lateritium* often produces prominent peach-colored sporodochia at lenticels on cankers or on dead and dying twigs or branches. *F. solani*, on the other hand, seldom produces macroscopic signs and is usually detected by isolation from diseased tissues. The latter fungus also causes root rot and damping-off of hundreds of plant species. Both fungal species include many biologically distinct, host-specialized, look-alike strains called formae speciales. *F. lateritium* f. sp. *mori* (also designated as variety *mori*), for example, is pathogenic to black locust, mimosa, and mulberry but not to assorted other plants that have been tested. The sexual states of *F. lateritium* and *F. solani*, *Gibberella baccata* and *Nectria haematococca*, respectively (both Hypocreales, Hypocreaceae), are found on dead plant material (*G. baccata*) or develop in laboratory cultures (*N. haematococca*) but are not commonly associated with disease symptoms in plants.

Species of *Fusarium* have colorless, elongate, multicelled macroconidia. These are slightly curved along their length and more or less strongly curved and pointed at the ends, canoe shaped in outline. The species are distinguished from one another on the basis of characteristics in culture: pigment formation; macroconidial size, shape, and septation; formation of microconidia, and formation of chlamydospores. Sporodochia such as those formed by *F. lateritium* on many host plants are useful only for tentative field diagnosis.

In North America, *F. lateritium* is associated with twig dieback or cankers on apple, green and red ash, American mountain ash, trembling aspen, chinaberry, citrus, cotoneaster, black cottonwood, winged euonymus, fig, golden-chain tree, hibiscus (including rose-of-sharon), American hornbeam, black locust, mimosa, white mulberry and other mulberry species, peach, Japanese pagoda tree, tree-of-heaven, black and Persian walnuts, and willow. Pathogenicity of *F. lateritium* to several of these hosts has been proved but should not be assumed for all hosts because this fungus is an invader of twigs and bark damaged by other agents.

Trunk and branch cankers caused by *F. lateritium* on Japanese pagoda tree are well documented. These are initiated at wounds and are annual, which is to say that they expand during only 1 year, primarily during host dormancy. They resemble the canker caused by the same fungus on golden-chain tree (F) in being elliptical and tan with a purplish brown margin in contrast to the green color of healthy smooth bark of young stems. Fusarium cankers on Japanese pagoda tree or golden-chain tree often form on branches damaged by freezing. Sporodochia become prominent on recently colonized bark. *F. lateritium* also causes cankers on dormant nursery stock of black cottonwood in British Columbia.

In North America, *F. solani* causes cankers or twig dieback on ash; black cherry; eastern cottonwood; sour gum; sweet gum; American holly; Norway, red, and sugar maples; paper mulberry; Nuttall, red, water, and willow oaks; black poplar; and tulip tree. The fungus has been found associated with stain in wood of American elm and silver maple. In Japan it causes twig blight of black locust.

F. solani is a relatively important canker pathogen throughout the eastern half of the United States, causing elongate annual cankers on trunks of cottonwood, maples, various oaks, tulip tree, and swamp tupelo. The cankers are initiated and expand during 1 dormant season, then are slowly enclosed beneath callus rolls. Eventually they are "buried" beneath a layer of sound wood. Trees are seldom killed, but the butt logs are devalued for lumber because of the defects left in the wood. These defects in cross-section are T-shaped. The top of the T represents the location of the vascular cambium when a portion of it was killed. The leg of the T, pointing toward the surface of the stem, is the mark left where callus rolls met and enclosed the lesion.

F. lateritium and *F. solani* are symbionts of certain ambrosia beetles. The latter fungus has caused girdling cankers around holes bored by *Xyloborus sayi* and *Xylosandrus germanus* on young tulip trees. Young black walnut trees in midwestern plantations are occasionally damaged or killed by the combined attack of *F. lateritium* and *X. germanus*. In each reported case the attack was initiated first by the insects, presumably because the trees were stressed in some way. Ambrosia beetles normally bore in weakened or dying trees.

F. lateritium and *F. solani*, especially the latter, are common in soils and in or on the corky outer bark of healthy trees. They are dispersed mainly as conidia by water, air, and insects. Thus they are able to enter wounded or otherwise predisposed bark under a variety of circumstances. In Mississippi, for example, cankers caused by *F. solani* developed on the trunks of cottonwood, oaks, and swamp tupelo following prolonged immersion during springtime flooding of lowland forests. In Ohio, cankers were caused by *F. solani* in tulip trees following drought. *F. lateritium* in black cottonwood cuttings and *F. solani* in white mulberry seedlings become aggressive colonists when plants are stressed by water shortage.

References: 45, 50, 172, 216, 233, 387, 497, 499, 1014, 1232, 1501, 1817, 2025, 2026, 2137

A–E. Twig dieback and cankers of red mulberry apparently caused by freeze damage followed by opportunistic fungi. A–C. Wilted shoots, twigs with unopened buds, and cankers at bases of dead twigs or buds; dead and living twigs are clustered because of repeated twig dieback (NY, Jun). D. Twig cankers in July, no longer expanding. *Nectria cinnabarina*, *Fusarium* sp., and other fungi were associated with the twig cankers (NY, Jul). E. Sapwood near a girdling lesion on a small branch shows brown streaks extending beneath healthy bark toward the base of the branch; no organisms were isolated from such streaks (NY, Jun).

F, G. A canker caused by *F. lateritium* on a branch of golden-chain tree (NY, Jul). F. Sporodochia erumpent at lenticels in the canker. The mature lesion is tan with a purple border, in contrast to normal green bark on young branches. G. Magnified view of sporodochia.

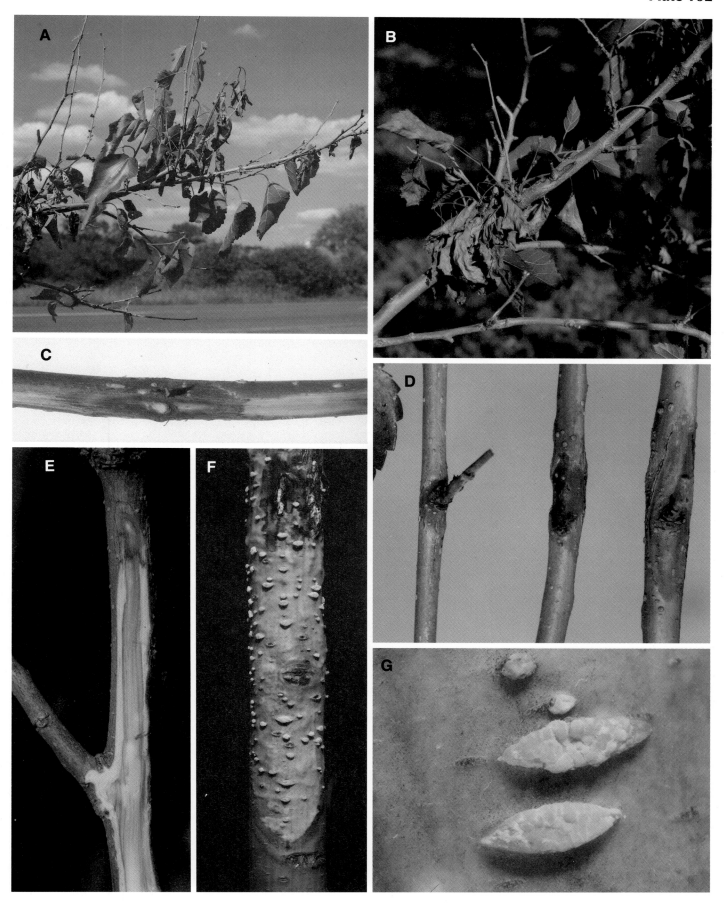

Plate 102

Pitch Canker of Pine (Plate 103)

Pitch canker, named for the copious exudation of resin from infected parts, is one of the major diseases of pines in the southeastern United States. It affects many pine species and occurs from northern Virginia to Florida and Texas. The disease deforms or kills trees in plantations and landscapes, suppresses the height and volume growth of survivors, and kills seedlings in nurseries. The greatest damage occurs among saplings and young pole-sized trees in seed orchards and plantations. Mortality rates of 25% or higher have occurred in some plantations and seed orchards of slash and shortleaf pines. Death is more often due to multiple branch infections than to girdling trunk lesions.

Pitch canker was discovered in the Appalachian region in the mid-1940s, and a fungus new to science, *Fusarium lateritium* f. sp. *pini*, was described as the cause. No great attention was paid to the disease for nearly 3 decades thereafter until a widespread epidemic began in the mid-1970s. This led to the discovery of new symptoms, new hosts, much epidemiological information, and even a new identity for the pathogen: *F. moniliforme* var. *subglutinans* (syn. *F. subglutinans*) (Deuteromycotina, Hyphomycetes).

Hosts in nature include loblolly, longleaf, pitch, sand, shortleaf, slash, South Florida slash, table-mountain, Virginia, and eastern white pines. Monterey and Scots pines were also susceptible when tested. *F. moniliforme* var. *subglutinans* is a widespread pathogen on many other plants, but except for isolates from gladiolus, strains from other plants do not infect pines. Isolates from gladiolus corms in Florida have been found pathogenic to pines, and vice versa.

Infection occurs in wounds caused by various agents on stems of all sizes at any time of year. Most infections probably occur in late summer and autumn. The primary symptoms are resin-soaked lesions that encircle twigs and small branches or develop as perennial cankers on large branches and trunks. A perennial canker may eventually girdle the affected part. Bark remains on the canker face. Diseased bark turns dark reddish brown and the underlying sapwood light yellowish brown or darker where infiltrated with resin. Resin-soaked wood beneath cankers may extend to the pith. Resin usually exudes from the lesion and may run down the bark or drip on foliage. The degree of resin soaking and resin exudation depends on the pine species, the part infected (trunk versus twigs), and the time of year.

Regardless of the time of infection, most shoot and branch dieback develops during autumn through early spring. Needles on girdled twigs and branches turn yellow, then brown. Dead needles, glued in place by crystallized resin, often hang from killed shoots or branches for more than a year, slowly weathering to dull grayish brown. Shoot elongation in spring may be arrested at any stage. Succulent shoots, deprived of normal water supply because of girdling lesions and resin-soaked wood in branches, wilt and the tips droop, assuming the shape of a shepherd's crook.

The most common symptom may be shoot dieback or limb and trunk cankers, depending in part on the host species. Cankers on limbs and trunks are common on Virginia and eastern white pines, while loblolly and slash pines usually have shoot dieback. In seed orchards, trees injured by harvesting equipment develop a high proportion of infections on limbs and trunks. The pitch canker fungus infects cones and seed, and it is probably transmitted into nurseries with seed. Pine seedlings in nurseries and new plantations may be killed as resinous lesions develop on the lower stem, root collar, or tap root. Diseased seedlings often occur in clusters where the pathogen has spread from an initially diseased individual.

F. moniliforme var. *subglutinans* produces macroconidia on tiny cushionlike, salmon orange sporodochia that are usually 1–2 mm broad and up to 1 mm high on killed shoots and twigs. Sporodochia are most common at scars left by needle fascicles that have fallen. In Florida, sporodochia may be visible at any time of year and are most abundant in autumn and winter on branches that still have reddish brown needles. Apparently they are rare, or at least are rarely noticed, on trunk and limb cankers. Conidia are colorless and one- to four-celled. Four-celled conidia are curved and elongate with pointed ends. Conidia measure 12–36 (average 19) × 6 μm and are dispersed by air, water, and insects throughout the year. Spores have been trapped from air in greatest numbers during turbulent, rainy weather.

Wounds suitable for infection include those made on current-year shoots by insects such as tip moths (*Rhyacionia* sp., see Johnson and Lyon, 2nd ed., Plate 17), a needle midge (*Contarinia* sp.), and the deodar weevil (see Johnson and Lyon, 2nd ed., Plate 20). The last named also acts as a vector. Injuries from hail or wind and galls caused by fusiform rust (Plate 132) also become infected. In seed orchards, cankers form at wounds made by twisting and tearing cones from loblolly pine and at injuries made on limbs and trunks of slash pine by tree shakers used to dislodge cones. Injuries remain susceptible at least 2 days, and lesion formation begins soon after inoculation. The optimum temperature for growth of the fungus is near 24°C. Water stress and fertilization at high rates may enhance pine susceptibility. Isolates of the pitch canker fungus vary in virulence, but host-specialized strains among those capable of infecting pines are unknown. Isolates from one pine species can readily infect others.

F. moniliforme var. *subglutinans* is the conidial state of *Gibberella fujikurai* var. *subglutinans* (Hypocreales, Hypocreaceae). Perithecia of this fungus, however, have not been found associated with pitch canker in pines. The association between conidial and perithecial states is known only from experiments in which fertile perithecia formed in cultures after isolates of the pitch canker fungus were paired with single-ascospore isolates of *G. fujikurai* var. *subglutinans*.

Pine species vary in susceptibility to pitch canker. Observations and experiments indicate that shortleaf and Virginia pines are highly susceptible, slash pine is intermediate, and loblolly and eastern white pines are somewhat resistant. More important, heritable variation in resistance within susceptible pine species has been demonstrated. In seed orchards, certain clones have been observed to be severely damaged by pitch canker while others remained healthy. Differences in susceptibility of pine families have been noted after seedlings, saplings, and mature trees were inoculated. Thus suppression of pitch canker through utilization of seed from resistant parents can be expected.

References: 114, 188, 529, 531, 573, 787, 1068, 1069, 1170, 1394

A–C. Young slash pines deformed by pitch canker. A. Damage to terminal shoots and upper lateral branches. The tree will survive but has lost several years' height growth and will have a crooked stem. B. As mainstems are killed, lateral branches turn upward and assume apical dominance. Branches girdled during the period of shoot elongation in spring display shepherd's crook as succulent shoots wilt and dry. C. The top of a diseased tree showing the sequence of leader death and assumption of dominance by a lateral branch. The dominant shoot at left replaces that at right, killed the previous year, which in turn replaced the original leader at center (FL, Apr).

D. Resin dripping from a trunk canker on eastern white pine (TN, Apr).

E. Close view of typical symptoms of pitch canker on a twig of slash pine: a dark, resin-infiltrated lesion, resin-soaked wood below the lesion, dripping and crystallized resin on the surface, and dead needles glued in place by crystallized resin (FL, Apr).

F. Magnified view of part of a killed slash pine twig showing pink sporodochia of *Fusarium moniliforme* var. *subglutinans* (arrows) on scars left by fallen needle fascicles (FL, Apr).

Plate 103

Cryptosphaeria Canker of Aspen (Plate 104)

Cryptosphaeria canker is a recently described, destructive disease of trembling aspen. It occurs in the West from Arizona to Alaska and in the East from Quebec and New York State to beyond the Great Lakes. The causal fungus is *Cryptosphaeria populina* (Sphaeriales, Diatrypaceae), which for many years was known only as a saprophyte on branches of aspen and black cottonwood. Its conidial state, *Libertella* sp. (Coelomycetes), had previously been found associated with brown stain and yellow decay of aspen wood in Ontario and Colorado and is now known to cause decay. Pathogenicity tests have shown that the fungus is able to kill trees and that it does so mainly by parasitism of sapwood rather than bark. Thus this disease is similar to the cankerrots caused by basidiomycetes (Plates 169–173).

In Colorado, Cryptosphaeria canker is the second most common canker disease of trembling aspen. It was found at over 80% of survey locations examined, affecting 1.1% of the trees but accounting for 26% of aspen mortality. Young pole-sized stems were most frequently infected. In New York State the disease was found causing up to 14% mortality in certain aspen families in an experimental plantation. Variation in canker incidence among families indicated genetic variation in susceptibility.

Hosts of *C. populina* include bigtooth and trembling aspens and aspen hybrids; black, eastern, and yellow cottonwoods; balsam and 'Lombardy' poplars; and purple willow. Pathogenicity has been proved only in trembling aspen.

The cankers, frequently associated with trunk wounds, are always elongate, following the grain of the wood. Often they are only 5–10 cm wide but up to 3 meters long. Many are centered on branch stubs, an indication that the fungus enters trunks via branches. Cankers on branches have been observed to girdle the branch and spread into the trunk. Cankers on trunks expand tangentially only several millimeters per year and may not girdle. Trees up to 15 cm in diameter may be killed within 3 years, however, not because of bark necrosis, but because the pathogen kills a large volume of sapwood.

Young cankers and the margins of older ones are orange-brown on the surface. They become light gray brown with time. Dead bark adheres tightly to the wood, the surface of which is discolored gray. Internal bark turns black and sooty except for prominent fibers and small lens-shaped light-colored marks 0.5–2 mm in size (at center in C). Successive callus ridges become visible at the edges of cankers after 2 or more years. Reddish brown fluid sometimes bleeds from canker margins.

Light orange acervuli of the *Libertella* state (not shown) develop near the edge of a canker and produce colorless, elongate conidia that are single-celled and strongly curved. Perithecia develop singly in a grayish pseudostroma (a tissue composed of compacted hyphae and the remains of dead bark cells) 1–2 cm wide and up to 30 cm long beneath the surface of bark dead at least 1 year. The pseudostroma and black perithecia impart a gray color to bark in the center of the canker (B, C). Pseudostromata vary in size in relation to the size of the canker zone that died in a given year. Perithecial bodies are nearly spherical, about 0.4 mm in diameter, and perithecial necks protrude above the bark surface. The ascospores are yellowish, curved, and single-celled and measure 8–12 × 2–3 μm. Ascospores are presumed to be dispersed by wind, and conidia mainly by water. The respective roles of these spore types in the disease cycle have not been described, but they are presumed to infect wounds.

After artificial inoculation to a wound, the pathogen colonizes sapwood rapidly. In aspens inoculated in Colorado, columns of discolored wood up to 4 meters long developed within 24 months, while associated cankers extended 33 cm or less. Apparently the cambium and bark die after much underlying sapwood has been killed.

Dead sapwood, discolored various shades of gray, brown, yellow, orange, or pink, extends a meter or more above and below the limits of a canker. This sapwood, especially near the interface with normal wood, fluoresces yellow when viewed under ultraviolet radiation. The fluorescence, however, is not a specific diagnostic character, since several other fungi that cause trunk decay in aspen also cause fluorescence under ultraviolet. Brown, mottled decay develops in the central part of the column of discolored wood. Decay columns apparently caused by *C. populina* averaged 4.6 meters in length in one study in Colorado. Most such columns were found in the middle and upper parts of the trunk. The *Libertella* state of *C. populina* can readily be isolated from cankers and associated discolored and decaying wood. Its ability to decay both wood and bark has been confirmed experimentally.

References: 611, 829, 830, 980

Cryptosphaeria canker and trunk rot of trembling aspen (CO, Jun).

A. An aspen with coalescing, elongate cankers. Note bleeding at canker margins.

B. Close view of coalescing cankers on a pole-sized tree. A large canker on the right side of the trunk has joined with the smaller, spindle-shaped canker centered on the branch stub.

C. A portion of a canker with periderm removed to show blackened cortex, tiny lens-shaped white spots, and the gray surface of dead sapwood. The gray color of the bark surface at bottom center indicates the presence of a pseudostroma where perithecia develop.

D, E. Close view and magnified view, respectively, of bark with perithecia of *Cryptosphaeria populina*. Black dots in (D), magnified in (E), are the protruding tips of perithecia. Round, black embedded perithecial bodies are revealed where the bark surface was removed in (E).

F. Cross-section of a diseased trunk, showing the relationship between cankers and internal stain and decay caused by *C. populina*.

Plate 104

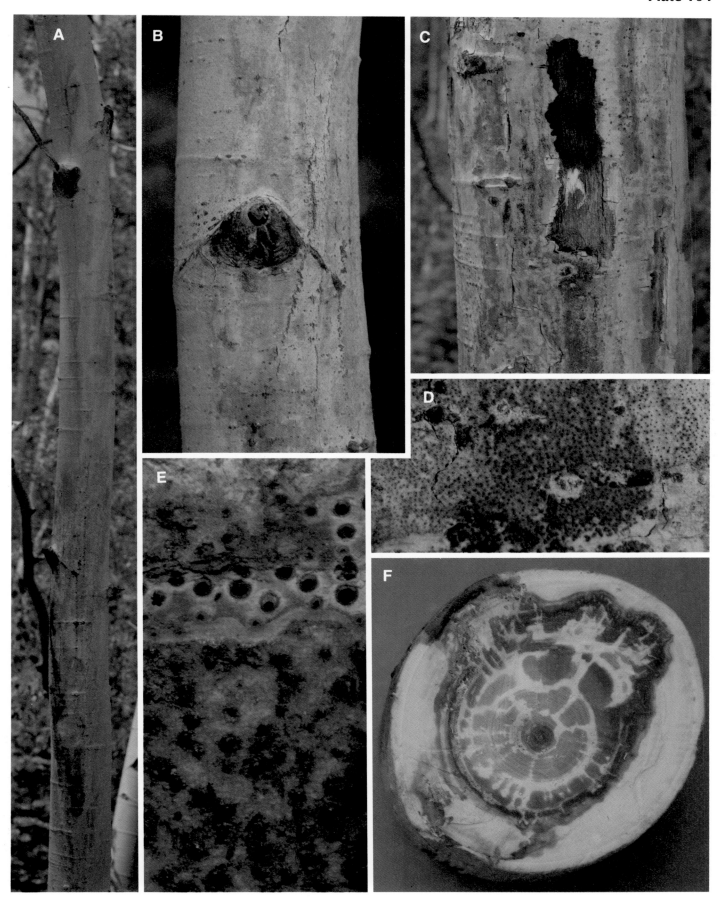

Eutypella Canker of Maple (Plate 105)

Eutypella parasitica (Sphaeriales, Diatrypaceae), the cause of Eutypella canker, infects several maple species in forests and landscapes. This disease, recognized since the mid-1930s, occurs from Minnesota eastward to Quebec, Maine, and Pennsylvania. Affected species include box elder and black, Norway, red, silver, sugar, and sycamore maples. Eutypella canker is most common on sugar maple in forests and on Norway maple in landscapes. Striped maple is also susceptible when experimentally inoculated. Only one report exists of *E. parasitica* on a nonmaple host: pin cherry in Quebec. We view this report with suspicion.

Incidence of Eutypella canker in sugar maple stands is typically 0–4%, but much higher levels are common. Many small trees less than about 10 cm in diameter are girdled and killed, and wood decay associated with the cankers leads to breakage of many trees. Since the canker affects the valuable butt log, it causes significant economic loss. By one assessment in Ontario, defective wood in the average tree with Eutypella canker amounted to 12% of total wood volume but 49% of potentially merchantable wood volume.

Cankers are initiated primarily on young trees, mostly within 3 meters of the ground, and are perennial, often remaining active for decades. Their appearance and shape vary with age and with the maple host. During the first several years of development, cankers appear as depressed or flattened areas tightly covered with bark (A), usually with the remains of a branch stub in the center. The margin of an expanding canker has white to buff mycelial fans in recently killed inner bark closely adjacent to healthy tissue (E). These fans are diagnostic. Old cankers (C, D, H) may have a rough exposed face of decaying wood and are often outlined partially or completely by a large callus ridge. The margins of large old cankers that are still active may be outlined by cracks in thick outer bark (D). Cankers on box elder, Norway maple, and sycamore maple tend to be oval to nearly circular, while those on red and sugar maples are elliptical to elongate, sometimes as long as 1.5 meters.

Cankers form by an alternation of growth of host and parasite. A cork barrier and callus form around the lesion during each growing season, and the parasite breaches or evades the barrier and expands the lesion during host dormancy. How it passes the cork barrier is not known. Callus development at active canker margins in any 1 year is relatively slight, but affected trunks often become swollen at the level of the canker because of the aggregate effect of callus production around a slowly expanding canker for many years. A prominent callus roll develops only where the parasite has died. Elongate cankers lengthen 1–2 cm annually and widen at a slower rate.

E. parasitica colonizes sapwood beneath cankers and causes a slow brown decay. This commonly leads to stem breakage, although the internal defect seldom extends more than 30 cm beyond the canker.

The bark at the center of cankers at least 6–8 years old is darkened by the black necks of numerous perithecia that form close to the bark surface, initially in groups a few millimeters in diameter (F) but later almost continuous over the center of the canker face. They seldom form on callus ridges dead for less than 4–5 years. Sometimes perithecia form in the wood of the central branch stub. Young, jet black perithecia are interspersed with old, dull black ones, and some with

viable ascospores can be found throughout the year. Their bodies are 0.6–1.0 mm in diameter, and their necks vary in length from less than 1 mm to as long as 5 mm, depending upon distance of the body from the bark surface. They eject ascospores in groups of eight. Ascospores are single-celled, 8–11 × 2–2.3 μm, slightly curved, and brown.

The fungus also produces an inconspicuous conidial state referable to either *Libertella* or *Cytosporina* (Coelomycetes) on stromata that break the bark surface in cankers (not shown). Conidia, produced in irregular fruit bodies intermediate between acervuli and pycnidia, are colorless and crescent shaped and measure 17–32 × 1.2–1.8 μm. These have not germinated when tested on a variety of media, including maple tissues. Therefore they are considered to be unimportant in the disease cycle.

Ascospores, on the other hand, are germinable and infectious. They are ejected from perithecia several millimeters into the air soon after the onset of rain at moderate temperatures (e.g., 12–25°C) and are dispersed by wind. Ascospore octads have been trapped 25 meters downwind from cankers. Ascospores are released from a given portion of a canker for several years regardless of whether the tree remains alive. Given moist conditions, these spores germinate within 10 hours at 28°C, a temperature that is near the optimum for mycelial growth.

Small branches, if bruised, broken, or cut experimentally, are susceptible to infection by germinating ascospores. Many infections centered at branch stubs are thought to begin on small lateral branches. Cankers on trunks are initiated at either branch stubs or wounds that penetrate to the xylem. *E. parasitica* grows intracellularly in both wood and bark and can apparently grow from wood into bark at the edges of wounds. The point of advance in bark is in inner phloem, where host cells are killed slightly ahead of the mycelial fans. A second layer of mycelial fans sometimes forms in old phloem closer to the bark surface and slightly behind the advancing fans in the inner phloem. Cambial necrosis also occurs slightly behind the limit of hyphal advance in phloem.

Pure cultures of *E. parasitica* can be isolated readily from diseased bark or wood. The fungus grows well at 20–35°C and produces pycnidiumlike structures with crescent-shaped conidia within 1 month.

Cankers on landscape trees can be arrested at least temporarily by carefully removing a narrow strip of bark all around the canker. This measure denies the fungus its customary route from dead tissues into living tissues. Long-term observations to learn whether such cankers might be reactivated by growth of the parasite through wood to adjacent bark have not been made.

References: 451, 598, 963, 964, 1015, 1036, 1084–86

Eutypella canker of maple.

A. A red maple, about 8 cm in diameter, swollen at the site of a canker about 9 years old that has girdled the stem. Indistinct callus ridges and a branch stub in the center of the canker are typical (NY, Apr).

B. Cankers on box elder. The blackened center of the lower canker indicates the presence of perithecia of *Eutypella parasitica* (NY, Apr).

C. Prominent callus ridges at the edges of a canker on sugar maple indicate portions of the canker margin where the parasite is no longer active (NY, Jul).

D. An old canker on Norway maple. Once partially enfolded by callus, the canker resumed expansion several years before the photograph was made. The approximate location of its current margin is indicated by cracks in the bark (NY, Jul).

E. Close view of white to buff mycelial fans, diagnostic for Eutypella canker, at the margin of a canker on red maple. Diagonal parallel lines in dead bark show the limits of successive annual advances of the parasite. Two age classes of mycelial fans, current and 1 year old, are visible (NY, Apr).

F. Magnified view of necks of perithecia of *E. parasitica* protruding from red maple bark (NY, Apr).

G. Magnified view of perithecial bodies embedded in bark, revealed where surface bark was shaved (NY, Apr).

H. An old canker on sycamore maple (NY, Jan).

Plate 105

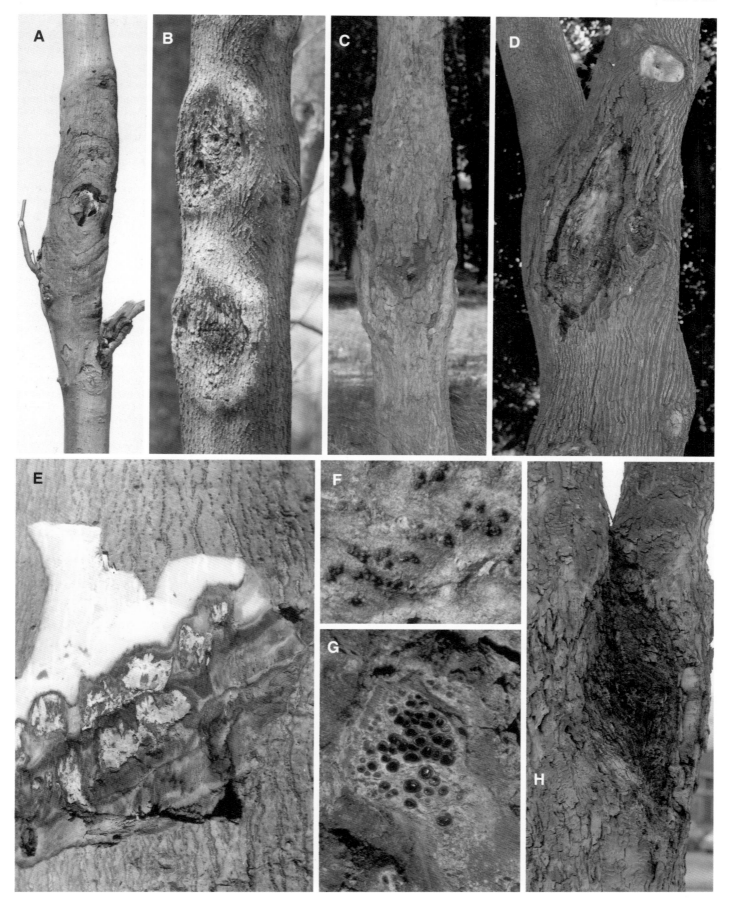

221

Hypoxylon Cankers and Diebacks (Plates 106–108)

Hypoxylon (Sphaeriales, Diatrypaceae) is a large genus of fungi that decay the bark and wood of various trees and shrubs. Many of these fungi are strictly saprophytic so far as is known, but *H. mammatum* (Plate 106) is a primary pathogen, and others, such as *H. atropunctatum* (Plate 107), *H. deustum* (Plate 146), and *H. tinctor* (Plate 108) are opportunists that attack trees weakened by other factors.

Hypoxylon canker of aspen. Hypoxylon canker is the most important disease of aspens, especially trembling aspen, in eastern North America. The disease is important mainly in forests, but aspens in landscapes and occasionally in nurseries may also be infected. The causal fungus, *Hypoxylon mammatum*, was known by mycologists for nearly 100 years before the first report, in 1924, of its association with cankers. The fungus occurs from British Columbia and Arizona in the West to Newfoundland and Pennsylvania in the East, but it is common and destructive only east of the Great Plains. It is also widespread in Europe.

Cankers develop on branches and trunks. Trees with trunk cankers die within 3–8 years as the result of girdling or trunk breakage due to wood decay (by secondary fungi as well as by *H. mammatum*) and to borer activity at cankers. Hypoxylon canker is the greatest cause of premature mortality of trembling aspen in Great Lakes region, where 12% of these trees are estimated to be infected at a given time. The annual loss is estimated to be 30% of the net growth of aspen in that region. The disease is nearly as severe in the East, where aspen stands have been found with up to 70% of the stems cankered.

Several species and hybrids of aspens and poplars are susceptible to *H. mammatum,* but trembling aspen is by far the most severely damaged. Cankers develop occasionally on bigtooth aspen and rarely on balsam poplar. European aspen is affected in Europe and also in North America. Other suscepts in nature are Chinese aspen, white poplar, and *Salix daphnoides. H. mammatum* also occurs as a saprophyte or inconsequential pathogen on several other kinds of trees: alder, apple, mountain ash, beech, birch, elm, hornbeam, maple, oak, pear, spruce, sycamore, and willow. Most strains that cause typical cankers on aspen are apparently host-specialized—capable of infecting aspens under a wide range of conditions but unable to infect other hosts except when inoculated into water-stressed tissues under artificial conditions. Bark of black and eastern cottonwoods, bigleaf maple, and willow (probably Scouler willow) has been parasitized under such conditions, although these species have not been found infected in nature. An isolate from a canker on *S. daphnoides,* on the other hand, caused typical cankers after inoculation into trembling aspen in the field. Isolates from aspen and alder, however, did not cause cankers on alder.

Flagged branches (those with dead adhering leaves) often call attention to diseased aspens. When trunks or branches are encircled by cankers during spring or summer, the leaves wilt, turn brown, and remain attached for nearly a year before dropping. Leaves on a diseased branch sometimes appear yellowish and undersized before death. Branches girdled during dormancy fail to produce leaves in spring. Typical cankers several years old are irregularly elongate and are usually centered on branch stubs. They display a rough central area with irregular yellowish gray lines and patches on a nearly black background. From a distance the diseased trunk appears mottled. The outer zone of the canker consists of blistered yellowish orange to orange-brown bark with a smooth, irregularly lobed margin. Some cankers have one or more callus ridges and cracked bark at the margin. The majority of trunk cankers develop within 3 meters of the ground. The wood beneath a canker several years old is always invaded by boring insects and decay fungi.

The developmental sequence of Hypoxylon canker requires 3 growing seasons. The canker first appears as a yellowish orange to orange-brown area in contrast to the normal green or grayish bark of a young stem. The lesion enlarges more rapidly along the axis of the stem than around it and becomes irregularly lobed. White mycelial fans (not shown) can be found in the cambial region approaching but not reaching the edges of the canker. In spring and summer of the 2nd year, the dead bark becomes blistered and superficially broken as the periderm (the thin outermost layer) raises slightly and begins to flake or peel off. This process exposes a gray mat of fungal tissue on which tiny pillars (sometimes called hyphal pegs) composed of parallel hyphae have formed (E). The pillars, 0.2–0.5 × 0.5–2.5 mm, raise the periderm and cause the blistered appearance. Tiny (5.5–8 × 1.5–4 μm) one-celled, nearly colorless conidia form in great numbers on the mat and pillars. Conidia are capable of germination but are apparently incapable of causing new infections.

The conidial mat and pillars are replaced within several months by perithecial stromata that develop on blackened, cracked bark from which the periderm has disappeared. Stromata are more or less cushionlike, whitish to gray when young and hard and black when mature. They are more or less round, 2–5 mm in diameter and 1–2 mm thick. A stroma contains up to 30 perithecia, each 0.7–1.0 mm in diameter and with a prominent nipplelike projection where spores emerge. Ascospores are unicellular, oblong-elliptic, and dark brown and measure 20–33 × 9–12 μm.

Ascospores are discharged to the air throughout the year after rain or when the stromata are moist, even when the air temperature is as low as −4°C. This low temperature capability is probably unimportant, however, for ascospores germinate best at 28–32°C but germinate slowly or not at all at temperatures below 16°C. Saturated air for 24–48 hours is necessary for ascospore germination.

Natural infections are probably initiated by ascospores that infect either wounds or dying 1-year-old twigs at their junctions with young mainstems or branches. For wounds to become infected, they must apparently penetrate to the wood; minor abrasions or natural wounds such as fresh leaf or bud scars are not suitable. In fact, living aspen bark is strongly inhibitory to growth of *H. mammatum*. Also, wounds more than several days old are unsuitable for infection.

At least four insects cause wounds that serve as infection courts. The periodical cicada (*Magicicada septendecim;* see Johnson and Lyon, 2nd ed., Plate 236) and three roundheaded borers (the poplar gall saperda, *Saperda inornata;* the poplar borer, *S. calcarata;* and *Obera schaumi*) are involved (see Johnson and Lyon, 2nd ed., Plates 130–131). Woodpeckers create additional suitable wounds by drilling into galls in search of larvae. Insect population fluctuations may be why many new infections occur in occasional wave years.

One or 2 years (average 22 months in one study) may elapse between wounding or twig death and the onset of canker development, apparently because incipient infections remain latent in stems that appear healthy. Once lesion development commences, the parasite grows intracellularly in the bark, cambial region, and outer sapwood. Its advance in sapwood precedes that in bark, and it causes the death of host cells in advance of its own progress. Toxins have been implicated in this process, but their precise role remains unclear. Cankers expand throughout the year, elongating 7–8 cm per month in summer and a few millimeters per month in winter. Mycelium of the fungus grows most rapidly at 25–30°C. Branch cankers initiated close to the trunk commonly spread into it, but those at a distance are inconsequential because girdled branches soon die back to the trunk and *H. mammatum* does not grow through the dead wood.

Strains of *H. mammatum* vary in virulence, and clones and families of aspen vary in susceptibility. Resistant trees slow the advance of the pathogen or halt it by rapid branch death. Callus forms at the edges of inactive lesions.

Hypoxylon canker of trembling aspen.

A. Dying and dead trees at a forest edge (NY, Jul).
B. A canker that originated at a branch axil (NY, Apr).
C. Close view of a portion of a canker with perithecial stromata and cracked, blackened inner bark visible where the papery periderm has peeled off (NY, Apr).
D. Close view of a canker margin showing the extension of fingerlike necrotic areas in the cambial zone beyond the margin of the canker in outer bark (NY, Apr).
E. Magnified view of the gray fungal mat and conidial pillars exposed by peeling back the periderm in the 1-year-old portion of a canker (NY, Jul).
F, G. Close (F) and magnified (G) views of perithecial stromata of *Hypoxylon mammatum*. Young stromata are grayish; mature ones are black. Perithecia, visible in (G), have a prominent black nipplelike tip or papilla with an opening through which ascospores are ejected (NY, Jul).

Plate 106

Environmental and cultural factors affecting the incidence and severity of Hypoxylon canker have been much studied, but few reliable statements can be made. Water stress enhances aspen susceptibility under experimental conditions, but neither wet nor dry sites seem unusually favorable for the disease in the field. Open stands are likely to be damaged more than fully stocked ones, but after infection, cankers enlarge more rapidly on shaded stems than on sunlit ones. Nitrogen fertilization or prior defoliation by insects may enhance susceptibility. Thinning leads to increased infection, and pruning diminishes the incidence of trunk cankers.

Hypoxylon dieback of oak. In the southern United States, oaks stressed by heat, drought, wounds, root injury, toxic chemicals, or other diseases often develop dieback, with which several species of Hypoxylon are associated. Three of the most common species are *H. atropunctatum* and *H. mediterraneum*, discussed here, and *H. punctulatum* (Plate 108). These are opportunistic fungi, unable to cause disease in trees of normal vitality but quick to colonize weakened or dying bark and wood.

H. atropunctatum, known only in North America, occurs across the United States, especially in the South. Oak species commonly colonized include black, blackjack, laurel, live, post, southern red, Texas red, white, and Oregon white. Species in the red oak subgenus (*Erythrobalanus*) are more frequently affected than those in the white oak subgenus (*Leucobalanus*). *H. atropunctatum* is also found as a saprophyte on basswood, beech, hickory, hornbeam, maple, and sycamore. This fungus was associated with widespread dieback and mortality of oaks in 1981, following the intense heat and drought that afflicted the South in 1980. *H. atropunctatum* never acts alone, however, so it is usually mentioned as contributing to dieback or death of predisposed trees rather than as a primary cause.

Yellowing or wilting of oak leaves may signal the onset of dieback, but these symptoms are merely general indicators of distress, not necessarily caused by fungi. *H. atropunctatum* invades weakened trunks and limbs, decays the inner bark and sapwood, and forms extensive stromata in the cambial zone. During the growing season, conidial stromata may appear within 2 months after a predisposing event, and perithecia develop within an additional 2–3 months. Pressure from a stroma and decay of the inner bark cause the corky outer bark to slough off, sometimes in small patches but characteristically in strips up to several meters long. The fungus initially induces a light brown discoloration of sapwood. This precedes yellow decay accompanied by black zone lines.

Stromata exposed by sloughing outer bark are thin (0.5–1 mm) and tan to silver gray. Powdery, spherical conidia 3–4 μm in diameter are produced on tan stromata and may be released as a cloud if a stroma is disturbed. As perithecia form, stromata become hard, and their color turns to silver gray on the surface and black within. On close inspection, the black tips of perithecia can be seen on the silvery surface. Perithecia liberate dark brown, elliptical to somewhat spindle-shaped single-celled ascospores that measure 24–33 × 11–16 μm. Old stromata eventually lose the silvery surface layer and appear black.

The quickness with which *H. atropunctatum* attacks stressed oaks is due to its presence as a latent colonist in normal tissues of healthy trees. In Arkansas, *H. atropunctatum* was isolated from 57% of the branches and 11% of the trunks of apparently healthy oaks. The fungus was also isolated from seemingly healthy stems of seedlings of several oak species and from leaves of white and black oak seedlings in the forest understory. Thus *H. atropunctatum* apparently enters an intimate association with southern oaks at an early stage of their growth and is perennially in position to colonize weakened bark or wood.

Latent colonization and rapid invasion of stressed tissues by *H. atropunctatum* are useful on occasion. In Arkansas and Texas this fungus rapidly colonizes trunks of oaks killed by oak wilt (Plate 176), depletes the supply of starch and sugars in the sapwood, and thus deprives the oak wilt pathogen of substrate needed for colonization and sporulation. This biological control is credited in part for the general lack of importance of oak wilt in Arkansas. In Texas, where local intensification of oak wilt depends primarily upon transmission of the pathogen through root grafts rather than as spores on insects, *H. atropunctatum* seems less important as an agent of control.

H. atropunctatum is favored by warm climate. The fungus grows at temperatures of 10–40°C and grows most rapidly at temperatures near 30°C.

H. mediterraneum is a globally distributed fungus best known for causing charcoal disease of cork oak in the Mediterranean region. Stressed or injured trees are most susceptible. The name *coal canker* has been applied to the association of this fungus with oaks in Texas, although pathogenicity there has not been proved. It is found as a saprophyte on various deciduous trees, most commonly oaks. Host plants in North America include red alder; American beech; flowering dogwood; American and rock elms; fox grape; vine maple; black, bur, chestnut, English, Gambel, California live, overcup, post, northern red, Texas red, scarlet, white, Oregon white, swamp white, and water oaks; and willow.

H. mediterraneum colonizes the inner bark and sapwood and forms stromata in the cambial zone. Its perithecial stromata are coal black, shiny, 2–5 cm long and 1–1.5 mm thick, with raised edges that cause the overlying bark to crack and fall away. The stromatal surface is studded with pimplelike tips of perithecia. The single-celled ascospores are dark brown, oblong to elliptic, 16–23 × 6–10 μm.

A. Dieback in a large water oak (MS, Apr).

B–D. Stromata of *Hypoxylon atropunctatum*. B. A tan conidial stroma on black oak (AR, winter). C. A silver gray perithecial stroma on white oak (AR, winter). D. Maturing perithecial stromata on water oak (MS, Apr).

E, F. *H. mediterraneum* on Texas red oak. Close (E) and magnified (F) views of a perithecial stroma. Black pimplelike projections are the tips of individual perithecia (TX, Mar).

G, H. Sapwood of Texas red oak colonized by *H. atropunctatum*. Black zone lines are typical (TX, Mar).

Plate 107

Hypoxylon dieback of plane tree and sycamore. *Hypoxylon tinctor* is considered to be an opportunistic parasite that colonizes stressed or declining London plane trees and sycamores, causing cankers and dieback. Proof of its pathogenicity has not been reported, however. It has been known for many years as a saprophyte on dead sycamore and occasionally on other trees, such as sour gum, hornbeam, and maple. It is widespread in forests and landscapes of central and southeastern United States and occurs on various other angiosperms in Central and South America. Cankers bearing stromata of *H. tinctor* were first reported from Georgia, where they developed on residual sycamore trees in a thinned forest. Heat stress caused by the loss of shade on tree trunks was presumed to have predisposed them to infection. Urban shade trees, shown here, may also be affected, but factors responsible for their predisposition have not been described.

The cankers are perennial and develop on trunks or large limbs as bands that follow the grain of the wood. If the grain of the wood turns in a helix, so does the canker. The margins are often imperceptible beneath scaly bark. Successive callus ridges sometimes mark annual increments of canker expansion, and the centers of cankers may appear sunken. Bark or exposed wood in cankers may have longitudinal cracks. Callus is barely perceptible at canker margins in declining trees such as those pictured here. Dull dark brown stromata 2–15 cm long and 1–1.5 mm thick appear on smooth bark or beneath the outer scales of rough bark near the canker margin. Stromata are studded with dark raised dots, which are the mouths of embedded perithecia. Perithecia produce ascospores that are light brown, single-celled, 13–20 × 5–8 μm, and tapered toward each end. Sometimes stromata of several age classes are visible. Beneath the youngest, which are darkest in color, the inner bark and cambial region are discolored orange-brown. As stromata age, the underlying tissue turns darker brown, as may be seen where portions of stromata were removed in (C). Wood in the vicinity of stromata is often stained bright yellow or orange and contains prominent black zone lines similar to those shown in Plate 107. Except for the dark stromata, these cankers are superficially similar to those of the canker-stain disease of sycamore (Plate 174).

Two prominent saprophytes. As mentioned with Plate 106, most species of *Hypoxylon* are found on dead, decaying bark or wood and are considered to be primarily or exclusively saprophytic. They enter trunks, limbs, and sometimes roots weakened or killed by drought, insects, lightning, and other diseases. Species that have been tested, including *H. atropunctatum* and *H. mediterraneum* (Plate 107), *H. mammatum* (Plate 106), and *H. punctulatum* (Plate 108), are able to decay bark and wood although at rates slower than basidiomycete decay fungi. Panels D and F–H display *H. punctulatum* and a colonist of beech, *H. cohaerens*.

H. punctulatum is an aggressive primary colonist of bark and sapwood of dying trees. The fungus is found only in the United States, usually on oaks. Other occasional hosts include beech, birch, chestnut, elm, hickory, and silverbell. Stromata of this fungus are dark brown, appearing black from a distance, less than 1 mm thick, and usually somewhat elongate. They measure 5–20 × 3–15 cm. They develop in bark just beneath the corky outer layers and, through the force applied by a ridge of gelatinous tissue around the edge of the stroma, cause the outer bark to break and drop away. The stromatal surface is dotted with the mouths of perithecia that produce brown, elliptic, unicellular ascospores measuring 7–9 × 3–4 μm. Stromata begin to appear as soon as 6–7 months after the death of a limb or tree.

A gray conidial stage precedes the perithecial stromata. The conidia form on mats of mycelium in tangential cavities between inner and outer bark. Tiny pillars 1–1.5 mm tall composed of parallel hyphae form on the mats, forcing the overlying bark outward and causing cracks in its surface. Conidia also form on the pillars. This mode of conidial production is similar to that of *H. mammatum* (Plate 106).

In the Appalachian region, *H. punctulatum* invades trunks and limbs of oaks dying from oak wilt (Plate 176) and suppresses saprophytic growth and sporulation of the oak wilt pathogen through competition for nitrogenous compounds and sugars.

H. cohaerens colonizes American and European beech and rarely maple in the East and occasionally hazelnut and oak in the West. This fungus also occurs in Europe. It is one of the common invaders of beech trunks dead or dying from beech bark disease (Plate 215). It is sometimes associated with discrete cankers, but no direct evidence for its pathogenicity has been reported. Perithecial stromata form on the bark surface as reddish brown to pale brown cushions 2–4 mm in diameter and 1–1.5 mm high, often in clusters over large areas. They darken with age, becoming purplish black. Each stroma contains 6–15 perithecia that produce dark brown, spindle-shaped, single-celled ascospores measuring 9–12 × 4–5 μm.

References for Hypoxylon canker of aspen: 31, 36, 42, 44, 80, 81, 152, 171, 172, 270, 387, 593, 597, 624, 830, 1209, 1281, 1306, 1429, 1702, 1895, 2056, 2238
References for Hypoxylon diebacks and cankers of oak: 118, 387, 973, 1281, 1306, 1963, 1965, 2001, 2059
References for Hypoxylon tinctor, H. punctulatum, and H. cohaerens: 103, 460, 569, 973, 1242, 1281, 1306, 1632

A–C, E. Association of *Hypoxylon tinctor* with dieback and canker of sycamore (IL, Jul). A. A street tree with some branches dead and others showing sparse, tufted leaves, indicating slow growth. B, C. Elongate cankers with dark brown stromata of *H. tinctor* on declining trees. Parallel arrangement of stromata in (C) indicates tangential enlargement of the canker from right to left. Orange-brown discoloration of the inner bark and cambial region beneath active stromata is typical of tissues colonized by *H. tinctor*. E. Magnified view of the surface of a perithecial stroma. Dark dots mark the locations of openings where ascospores are liberated from perithecia.

D, G. Perithecial stromata of *H. punctulatum* on a recently dead oak branch (NY, Oct). Dots on the surface in magnified view (G) are openings of perithecia.

F, H. Perithecial stromata of *H. cohaerens* on a canker on American beech (NY, May). Stromata of this species are cushionlike and reddish brown when young and become blackish with age. Each stroma contains several perithecia. The cause of this canker is unknown.

Plate 108

Sooty-bark Canker of Aspen (Plate 109)

In the Rocky Mountain region, sooty-bark canker, caused by *Encoelia pruinosa* (Helotiales, Leotiaceae), is the most damaging canker disease of trembling aspen. It affects mature trees primarily, on sites of both high and low quality for aspen growth, and is lethal within several years. Surveys in Colorado revealed the presence of sooty-bark canker on over 90% of sites examined, affecting 1–2% of living aspens. The pathogen, found on over half the dead trees, was the most common cause of aspen mortality. The disease occurs in the west from Arizona and New Mexico to Alaska, usually on trembling aspen. Black cottonwood and balsam poplar are occasionally affected. *E. pruinosa* has been found on trembling aspen in the Great Lakes Region and on bigtooth aspen in New Hampshire, but it apparently does not cause perennial cankers in central and eastern locations.

E. pruinosa was known as a saprophyte on aspen bark for nearly 70 years before it was associated with a destructive disease. Sooty-bark canker was described and the causal fungus renamed in 1956. From then until the late 1970s the pathogen was known as *Cenangium singulare*.

Sooty-bark cankers are unusual for their rapid development and lethal impact. In Colorado they elongate an average 45 cm and increase an average 16 cm in width per year. Canker expansion up to a meter in length and 35 cm in width in the 1st year has been recorded. Young cankers appear as slightly sunken elliptical areas with nearly normal surface color and blackened internal tissues. Marginal callus development is slight or absent in most cases, but variations in bark color reveal concentric zones of annual expansion. After bark has been dead 2–3 years, the periderm begins to slough off, revealing blackened inner bark. A band of light gray periderm usually remains attached at the margin of each annual zone of expansion. These bands provide a record of the age of the canker and the rate of its enlargement.

Blackened inner bark, crumbly (sooty) on the surface, usually remains tightly attached to the wood for several years, held in place by mycelial mats of *E. pruinosa*. Eventually, however, it begins to break away, revealing distinctive patterns of gray to black spots on the wood surface where fungal mats formerly adhered.

Apothecia develop in great numbers on blackened bark and are present throughout the year, either exposed or beneath loose periderm. They are leathery, silver gray, disc or cup shaped, and about 3 mm in diameter when moist. They curl into angular or elongate shapes when dry. The outer surface is encrusted with crystals. Ascospores are forcibly ejected and are dispersed by air. These spores are colorless, ellipsoid, and single-celled and measure $8–12 \times 2.5–3.5$ μm.

Infection occurs in wounds of various kinds at any height on the trunk. Superficial wounds as well as those reaching the xylem are suitable. The fungus advances in the inner bark and cambium and often girdles the tree within 4–5 years. Some cankers resulting from experimental inoculations have ceased expansion after 1 or more years, as evidenced by a callus ridge at the margin. The same has not been reported for naturally occurring cankers.

E. pruinosa grows readily in pure culture after isolation from cankers or ascospores. Ascospores germinate at temperatures of 5–30°C and do so most rapidly at 20–25°C. The fungus produces tiny spherical microconidia in culture. These do not germinate, and their function is unknown.

Disturbance such as selective logging and campsite construction and use in aspen stands in the Rocky Mountain region may cause increased incidence of sooty-bark canker and also Ceratocystis canker of aspen (Plate 175).

References: 450, 826, 830, 832, 979, 980, 2100

Sooty-bark canker of trembling aspen.

A. A mature tree, girdled and killed 8 years after infection. Light gray arcs mark the limits of annual increments of canker expansion.

B. Part of the canker shown in (A). Bark here has been dead 4–6 years. The light gray periderm is absent except in arcs between annual zones of expansion. Sooty-appearing inner bark is beginning to crack and fall away.

C. Close view of sooty-appearing dead bark killed about 3 years earlier. The large white patch is periderm. Tiny silver gray structures all over the black surface are apothecia of *Encoelia pruinosa*.

D. Characteristic dark spots on the surface of wood where mycelial mats of *E. pruinosa* held dead inner bark in place for several years (A–D: CO, Jun).

E. Magnified view of dry apothecia (CO, Oct).

Plate 109

Cenangium Dieback of Pine (Plate 110)

Pine branches with brown needles often signal Cenangium dieback. The disease is caused by either of two ubiquitous, opportunistic fungi, *Cenangium ferruginosum* (syn. *C. abietis*) and *C. atropurpureum* (Helotiales, Leotiaceae), which kill the bark and cambium of twigs and branches weakened by environment, other pests and pathogens, or natural senescence. These fungi are also competent saprobes, common on pine twigs killed by various agents. *C. ferruginosum* occurs all around the Northern Hemisphere. *C. atropurpureum* is widespread in North America, but its true distribution is unknown because records of its occurrence are often lumped with those for *C. ferruginosum*.

The disease is most common after severe winters, especially if winter is preceded by unusually mild autumn weather. When host defenses are impaired, as by sublethal winter injury, these fungi invade and kill branches that would otherwise remain alive. Foliage on scattered twigs and branches turns drab green, then yellowish and finally brown in late winter and spring. By early summer, fruit bodies (apothecia) of the fungus are conspicuous on killed twigs or branches up to about 5 cm in diameter.

Disorders attributed to *Cenangium* species include twig blight of lodgepole, ponderosa, and eastern white pines; girdling cankers on branches of Austrian, Japanese black, and Japanese red pines; branch death of ponderosa pine following infestation by the scale insect *Matsucoccus vexillorum;* and death of eastern white pine saplings in the Appalachian forest understory after severe winters. Although some species of spruce are also susceptible to *C. ferruginosum,* they are rarely much damaged by it.

Because *Cenangium* species fruit only on dead bark, records of strictly saprobic occurrence cannot be distinguished from cases where the fungi once parasitized the bark. Moreover, since *C. ferruginosum* and *C. atropurpureum* are widely considered to be one species, published host records are muddled. We therefore present them together. In pathogenicity tests, *Cenangium* (apparently *C. ferruginosum*) has caused cankers or twig dieback on jack, lodgepole, ponderosa, red, Japanese red, Scots, and eastern white pines and on black and white spruces. Observations indicate pathogenicity of *C. ferruginosum* or *C. atropurpureum* to the trees named above and also to Austrian, Japanese black, cluster, digger, Jeffrey, Swiss stone, sugar, and western white pines and to Norway spruce. Other hosts of one or both fungi, listed without regard to possible pathogenicity of the fungi, include balsam fir; Douglas-fir; Chinese, limber, loblolly, Monterey, mugo, pitch, shortleaf, slash, Mexican stone, table-mountain, and Virginia pines; and Engelmann spruce.

Outbreaks of Cenangium dieback, although conspicuous, are of little consequence in forests. Indeed, the killing of weak, low branches on ponderosa pines, formerly called pruning disease, is clearly beneficial for the production of knot-free logs. Landscape trees may become conspicuously flagged, but the disease usually occurs only once in several years. After the brown needles drop or the killed branches are pruned, disfigurement is slight. Chronically diseased trees on which branches die each year are exceptional and presumably indicate chronically stressful sites. The Japanese red pine shown in panel A lost branches every year after construction of the adjacent building until its ornamental value was destroyed. In Christmas tree plantations Cenangium dieback of Scots pine sometimes appears after severe winters and makes trees unsalable. Where the disease appears, random individuals are affected, and only a small minority are severely damaged.

Diagnostic features of Cenangium dieback, in addition to those already mentioned, are a sharply delimited boundary between brown, dead bark and normal tissues, browning of needles from bases toward tips, loosening and casting of needles during the summer after branch death, little or no resin production on surfaces of infected parts, and resin impregnation of tissues only at margins of killed bark. *Cenangium* species cause neither the dark staining characteristic of Atropellis cankers (Plate 112) nor the yellow-green discoloration that sometimes distinguishes Scleroderris cankers (Plate 111). Also, unlike *Atropellis* or *Ascocalyx, Cenangium* species produce apothecia on bark during the first summer after it dies.

The apothecia arise from black stromata that form during spring in newly killed bark just beneath the corky outer layer (H). Foliage distal to the canker may still be green when these form. Dark brown, globose tubercles of fungal tissue then break through the bark surface in clusters (D) and become larger and lighter in color until they mature during summer. When on twigs, the apothecia commonly emerge at needle bases or scars. The surface of bark that is normally smooth, as on white pines, becomes shriveled during the period of apothecial development. Mature apothecia of *C. ferruginosum* are yellowish brown, cup shaped when moist, and 2–5 mm in diameter, with a yellow to buff interior surface from which ascospores are discharged during rainy weather. The apothecia of *C. atropurpureum* are similar but have a distinctly purplish exterior. After discharging spores, apothecia slowly become black and brittle.

Cenangium species produce no infectious asexual spores. Thus new infections must be initiated by ascospores, presumably in summer or early autumn soon after they are liberated. The saprobic existence of *Cenangium* species ensures a supply of spores. Early stages of infection have not been studied, but it seems likely that incipient infections begin each year in summer or autumn and are held in check by host defenses unless these are defeated by environmental damage or by other pests. This pattern would explain the sporadic appearance of symptoms.

References: 228, 334, 389, 425, 561, 783, 1043, 1161, 1266, 1806, 1831, 2142

A, B, E. Branch mortality in Japanese red, Austrian, and Japanese black pines, respectively (NY, May–Jun).

C, F, H. A branch of Japanese red pine partly killed by *Cenangium ferruginosum.* At the branch whorl beyond which needles are brown, the surface of recently killed bark appears normal. Black stromata of the fungus can be seen at right in (H) where surface bark was shaved (NY, Jul).

D. Apothecia in an early stage of development in bark of Japanese red pine (NY, Jul).

G, I. Apothecia on Japanese black pine just before and just after opening. The fertile layer is buff color (NY, Jun).

Plate 110

Scleroderris (Ascocalyx) Canker of Conifers (Plate 111)

Scleroderris canker of conifers, although important in Europe for several decades, was unknown in North America until destructive outbreaks began in pine plantations and nurseries in the Great Lakes region in the 1950s and 1960s. By the mid-1980s, the disease had been found in several species of pine and spruce throughout the region extending from Nova Scotia and Maine to western Ontario and Minnesota, and outlying disease centers were known in eastern Newfoundland and in the mountains of British Columbia and Alberta. The pathogen is *Ascocalyx abietina* (Helotiales, Dermateaceae), formerly known as *Gremmeniella abietina* and before that as *Scleroderris lagerbergii*.

Scleroderris canker causes economic loss by killing seedlings and saplings, by causing growth loss and deformity of survivors as the result of trunk cankers and dead leaders, and by reducing log quality because of deformities and delayed self-pruning of lateral branches on surviving trees in understocked stands. Red pine is damaged most, and jack and Scots pines to a lesser degree. The disease is not severe in other species unless they are exposed to intense deposition of spores from diseased red pine. Other hosts of *A. abietina* in North America include Austrian, lodgepole, pitch, ponderosa, eastern white, and whitebark pines; black, Norway, and white spruces; and, rarely, eastern larch and balsam fir. Many other conifers are known to be susceptible on the basis of experimental inoculations in North America or field experience in Europe. In general, two- and three-needle pines are most susceptible, five-needle pines and spruces somewhat susceptible, and all other conifers highly resistant or immune to the North American and European races of the pathogen.

Asian, North American, and European races of *A. abietina* are known. The Asian race damages Todo fir in Japan but is unknown elsewhere. The North American race occurs in western Canada, the Great Lakes region, and northern Quebec. It is destructive to seedlings and young trees up to about 2 meters tall and causes inconsequential disease on low branches of taller trees. It attacks terminal and lateral shoots and slowly spreads in branches to the trunk, causing cankers that deform or girdle trees. The European race, the North American race, and strains intermediate between them (possibly hybrids) occur from eastern Ontario and northern New York eastward. The European race kills shoots and causes twig cankers and dieback on low branches and also in the crowns of tall trees, but it seldom causes cankers on mainstems. It devastated several plantations of pole-sized red and Scots pines in New York State in the 1970s, and this damage caused concern that widespread devastation would follow. Fortunately it did not.

The North American and European races of *A. abietina* cause essentially the same symptoms on small pines: needle browning and casting and death of buds, shoots, and branches. Infection occurs primarily in spring and summer, sometimes as late as October. Foliar browning and shoot death sometimes begin in late summer but are usually delayed until early the next spring. Bracts (scales) at the bases of short shoots (needle clusters) are a common site of infection on red pine. Spores germinate on the surface, and germ tubes enter stomata, then produce mycelium within the bract. The pathogen typically remains quiescent there until late winter. Then it grows into the base of the short shoot and the adjacent part of the long shoot and causes a resinous brown lesion. A different infection scheme has been reported for Austrian pine in Europe. Spores germinate and mycelium persists beneath bracts until autumn, when the hyphae penetrate epidermal cells of long or short shoots directly.

Lesions in cortical tissue spread around the long shoot and into needle bases. Orange-brown discoloration develops on needle bases and progresses toward the tips. This discoloration is visible in early spring. Needle clusters with this symptom can be pulled from the shoot readily. Buds on twigs with discolored needles fail to open, and the needles begin to drop during late spring and summer, leaving bare twigs. Resinous lesions can be found by slicing into the bark of twigs near the junction of green and brown foliage. The cambial region in the lesion is often discolored yellow-green by a pigment secreted by the pathogen. On small trees twig lesions caused by the North American race often eventually spread to the trunk and develop into elongate, resinous perennial cankers with yellow-green discoloration in the cambial region.

The pycnidial state of *A. abietina*, named *Brunchorstia pinea*, develops in the bark of killed twigs or sometimes in needle bases during the first summer. Pycnidia form singly or in clusters. They are irregular in shape with one or more fertile cavities, nearly black on the surface and white within, up to 1 mm wide. They may break through the bark surface anywhere but are most common at the former locations of needle clusters. Pycnidia that have not yet broken the bark surface may be found by shaving superficial bark. They produce masses of conidia in a mucilaginous matrix. Conidia are individually colorless, mostly four-celled (some with five to seven cells), about 3×30 μm, with pointed ends. Conidia are dispersed mainly during periods of rainfall, most abundantly in late spring and early summer, and may be carried many meters in mist droplets. Small numbers of viable conidia have also been trapped from the air during dry weather. They can survive up to 20 days on dry plant surfaces.

Brown, cuplike apothecia appear in spring on twigs dead 1–2 years. Apothecia are about 1 mm in diameter and slightly more than 1 mm tall, with short stalks. The fertile inner surface is cream colored. The ascospores are colorless, ellipsoidal, four-celled with rounded ends, and often slightly curved and measure $15–22 \times 3–5$ μm. Apothecia of the North American race are common; those of the European race in North America are rare. Thus in some areas ascospores are unimportant as inoculum.

Scleroderris canker is a disease of cold snowy places. The pathogen does much of its damage under snow cover during host dormancy while defensive reactions are minimal. Disease builds up on sites such as frost pockets and cold air drainages where plant surfaces stay wet for long periods and airborne spores become concentrated. Large-scale outbreaks may occur in years with long periods of cool moist weather in spring, but new infection is minimal in many years. Disease prevention in northern nurseries now precludes the distribution of *A. abietina* with seedlings. Damage to new plantations can be minimized by planting resistant species where the planting site is known to be conducive to infection.

References: 387, 513–15, 907, 1103, 1211, 1364, 1465, 1819, 1820, 1823, 2252, 2253

A. Death of low branches on young jack pines caused by the North American race of *Ascocalyx abietina* (Que, Aug).

B. A canker caused by *A. abietina* (North American race) on the mainstem of a young jack pine (Ont, Jul).

C. Pole-sized red pines devastated by the European race of *A. abietina* (NY, Jun).

D. Typical early foliar symptoms of Scleroderris canker: orange-brown discoloration beginning at needle bases and spreading outward, illustrated on Scots pine (NY, Jun).

E. A red pine branch on which, within 2 years, all growing points have been killed. Needles are shed from killed twigs within the 1st year after display of symptoms (NY, Jun).

F. A resin-infiltrated canker on a small branch of red pine. The lesion was barely perceptible before removal of outer bark. Yellow-green pigment in the cambial zone of the lesion (arrow) is diagnostic (NY, Jun).

G. Pycnidia of the *Brunchorstia* state of *A. abietina* on scars where short shoots were shed from a diseased twig of Scots pine (NY, Jun).

H. Apothecia of *A. abietina* on a red pine twig (NY, Jun).

Plate 111

Atropellis Cankers of Pine (Plate 112)

Native and introduced pines across North America develop perennial cankers on twigs and branches or trunks when infected by species of *Atropellis* (Helotiales, Dermateaceae). Four fungi, all apparently native to North America, are involved. *A. tingens* attacks and kills small branches and mainstems of at least 17 pine species throughout the eastern half of the continent, most commonly in Appalachian and Piedmont areas. It causes economic damage by disfiguring pines in Christmas tree plantations. *A. pinicola* acts similarly, killing branches and occasionally small mainstems of six or more pine species in the Northwest and California. *A. piniphila,* the most important fungus of the group, also infects branches of several pine species but is best known for causing elongate, resinous cankers on trunks of lodgepole and ponderosa pines in northern California, the Northwest, and the northern Rocky Mountain regions. It also attacks ponderosa pine in the mountains of Arizona and New Mexico. The cankers, seldom found on large trees, reduce the value of stems for pulpwood because bark in cankers sticks to the wood and the diseased wood requires extra cooking and bleaching. *A. piniphila* is inexplicably rare or absent from the central Rocky Mountains but is occasionally found in southeastern states. The fourth species, *A. apiculata,* is known only in North Carolina and Virginia, where it kills twigs and branches of Virginia pine.

Hosts of *A. tingens* include Austrian, cluster, jack, loblolly, lodgepole, pitch, pond, ponderosa, red, Japanese red, sand, Scots, shortleaf, slash, table-mountain, Virginia, and eastern white pines. *A. pinicola* infects Austrian, lodgepole, Scots, sugar, eastern white, and western white pines. *A. piniphila* attacks jack, loblolly, lodgepole, ponderosa, shortleaf, Virginia, western white, and whitebark pines.

Atropellis cankers, whether on branches or trunks, can be confused with rust cankers from a distance because both kill branches or small mainstems, and resin drips down the bark of stem cankers. But unlike rusts or other fungi that cause cankers on pines, all *Atropellis* species stain the wood beneath cankers darkly. *A. apiculata* causes a chocolate brown stain, and the other three species cause gray to bluish black discoloration that is easy to detect by cutting into the edge of a canker. These fungi also produce numerous small but conspicuous black apothecia in cankers on bark that has been dead 2–3 years.

Cankers caused by *A. tingens* are small and elliptical and originate at the bases of needle fascicles. They enlarge rapidly in the 1st year but slowly thereafter and become inactive after about 10 years. Typical mature cankers on a surviving branch may be only about 2 cm long. Dead needle clusters glued in place by resin remain on cankers until they slowly weather away. Multiple lesions girdle twigs and small branches, causing foliage to fade and turn brown in spring and early summer. Cankers several years old have sunken centers and raised margins due to callus formation. Callus rolls sometimes enclose old, inactive cankers on mainstems.

Cankers caused by *A. pinicola* are most common on trees less than about 13 meters tall. They originate at leaf scars or, in the case of trunk cankers, at branch bases. In the smooth bark of western white pine, they appear as elongate, flattened depressions.

The disease caused by *A. piniphila* is best known. Most cankers start in branch axils on stems 5–30 years old and thus have a branch stub in the center. They elongate about 5 cm per year on lodgepole pine in Alberta but much faster on ponderosa pine in the Southwest, where cankers sometimes attain lengths of 3 meters. Some cankers start on internodes, reflecting the ability of the fungus to infect wounds or

bruised bark as well as apparently undamaged bark of young stems. The first symptoms are tiny brown spots, each with a resin drop on its surface. New resin appears at the margin of colonized bark as long as the canker is active. Young, dense stands of ponderosa and lodgepole pines sustain the greatest damage, partly through tree death, but mostly through deformity of trunks with multiple cankers. These appear as elongate depressions covered with bark and, in lodgepole pine, with dripping resin. Dark brown mycelium permeates the bark and wood, creating a zone of bluish black stain that in cross-section is initially wedge shaped, pointing toward the center of the stem, and becomes irregular with age. The stain does not enter heartwood already formed, although it may be surrounded by heartwood in later years. Multiple cankers on a given stem are sometimes united internally by the discolored wood, and the fungus appears able to grow longitudinally through wood far beyond canker margins. Cankers and stain caused by *A. piniphila* near ground level can be confused with black stain root disease (Plate 178).

Atropellis species, except *A. apiculata,* produce two kinds of fruit bodies, pycnidial stromata and apothecia, on killed bark. Pycnidial stromata develop first. These are black, about 1 mm across, and contain multiple chambers where spores form. The spores are noninfectious and perhaps function as male gametes in the sexual cycles of these fungi. Apothecia form soon afterward, arising for the first time on cankers 2 or more years old and developing annually thereafter on bark killed 2–3 years previously. On lodgepole pines, the larger the cankered stem, the longer the fruiting of *A. piniphila* is delayed, from 4–5 years after infection on small stems to 25 years or more.

Apothecia of all *Atropellis* species are black, cuplike bodies with a brown interior, measuring 2–3 mm or up to 4 mm across when moist and expanded, depending on species and environment, or smaller and globose when dry. They arise singly or in groups and are present throughout the year. Ascospores are dispersed by wind, mainly in summer to early autumn, and infections apparently begin then.

Apothecia of *Atropellis* species can be distinguished from those of other dark-colored fungi on pine bark not only by the presence of stained wood beneath but also by a simple diagnostic test. If tissue from an apothecium of *Atropellis* is placed in a drop of 5% potassium hydroxide solution, it releases blue-green or, for *A. apiculata,* chocolate brown pigment. Species of *Atropellis* are distinguished from one another by microscopic characters of their ascospores.

Individual lodgepole pines that suppress both number and size of cankers have been found in severely damaged stands, but no information is available about possible resistance of other pines to *Atropellis* species.

Epidemics are documented for cankers caused by *A. tingens* and *A. piniphila.* The latter occur in dense, slowly growing stands of lodgepole and ponderosa pines on dry sites. Damage in lodgepole pine stands is related to previous forest fires. Dense young stands of saplings growing up after a fire are in some cases uniformly infected as the result of inoculation by spores from cankers on older trees that survived the fire. Cankers are relatively more damaging to trees growing slowly as the result of competition in dense stands than on those growing normally. Site factors are generally unrelated to outbreaks of *A. tingens.* Weather patterns that favor outbreaks of these diseases are unknown.

References: 228, 490, 783, 880, 1140, 1592

A–C. Atropellis canker of lodgepole pine caused by *Atropellis piniphila.* A. Resin dripping profusely from an old trunk canker (Alta, summer). B. The margin of a trunk canker with bark removed to show darkly stained wood (BC, Jul). C. Cross-section of a small stem with staining deep in the wood (Alta, summer).

D–H. *A. tingens* causing branch cankers on Austrian pine (VA, May). D. Disease in low branches has destroyed the value of this tree in a Christmas tree plantation. E. Cankers that arose at bases of needle fascicles have encircled this branch. F, G. Dark bluish gray to black stain in diseased bark and wood is diagnostic. G, H. Black apothecia 2–3 mm across, in clusters on bark of cankers 2 or more years old.

Plate 112

Strumella Canker of Oak and Lachnellula Cankers of Conifers (Plate 113)

Strumella canker. This perennial canker is common on various oaks and sometimes on other deciduous trees in eastern North America. The causal fungus is *Urnula craterium* (Pezizales, Sarcosomataceae). Its conidial state is *Conoplea globosa* (Hyphomycetes) (syn. *Strumella coryneoidea*).

The pathogen, native to North America, occurs in the region from northern New England to Minnesota, Missouri, and North Carolina and also in Oregon. It causes cankers most commonly on black, red, scarlet, and scrub oaks. Other oaks affected include blackjack, bur, chestnut, chinkapin, Oregon, overcup, pin, post, water, white, and swamp white. Cankers caused by *U. craterium* have also been found on basswood, American beech, American chestnut, sour gum, pignut and shagbark hickories, hop hornbeam, and red maple.

Strumella canker seldom affects more than 2–3% of oaks in a forest stand but has been reported to cause death or severe damage to up to 63% of trees in certain young oak plantations. Most infections on mainstems develop within 4 meters above ground on trees less than 25 years old. Trees less than about 10 cm in diameter may be girdled, and many pole-sized stems break because of decay at cankers. Surviving trees are devalued for lumber. The disease is uncommon on oaks in landscapes.

Infection begins at branch axils or possibly on branches. The first symptom on the trunk is yellowish brown discoloration of bark around a dead branch or branch stub. Cankers on young stems often enlarge rapidly and girdle them. Typical cankers, however, enlarge slowly (1–3 cm in width per year) during many years, becoming irregular, depressed areas with concentric ridges of callus. The parasite kills bark during the period of host dormancy, and the tree halts this advance and produces a callus ridge around the lesion each growing season. White mycelial strands or mats can be seen by cutting into dead bark near the canker margin. Adventitious sprouts often develop just below large cankers. Killed bark remains tight to the wood for several years but eventually decays and sloughs off, revealing a yellowish decay caused by the same fungus in underlying wood.

Diseased trunks are often flattened, widened, and concave at cankers because of decay in killed tissues and stimulated growth near lateral margins of the canker. Strumella cankers up to 1.5 meters long and 60 cm wide have been recorded.

Dark brown to black nodules of sterile mycelium a few millimeters in length and width form on the bark during the 1st year after its death and persist for several years. Sometimes dark brown to black sporodochia with great numbers of tiny brown spores form in cankers, but usually this sporulation is restricted to dead sprouts or branches or to the trunk after the tree dies. Sporodochia on dead trunks develop in and above the cankers. Sporodochia are rounded and 1–3 mm in diameter. The spores are spiny and irregularly globose to pear shaped and measure 6.7–8.1 × 4.7–5.8 μm. They are windborne but have not been observed to germinate. Their role is unknown.

Cuplike, stalked apothecia of *U. craterium* form in early spring on decaying wood of stumps and fallen trunks or limbs. Apothecia are 3–4 cm wide and 4–6 cm deep, grayish on the surface and black inside. The colorless ascospores, which are windborne and germinate readily, are smooth and one-celled and measure 12–14 × 24–35 μm.

Lachnellula cankers. These cankers are, with one exception, inconsequential lesions on twigs of conifers caused by opportunistic fungi that produce small, usually white apothecia on dead tissues. European larch canker, caused by *Lachnellula willkommii* (Helotiales, Hyaloscyphaceae), is the damaging exception. The first North American record of it came in 1927 from Massachusetts, where it was subsequently eradicated. Next, in the early 1980s, it was found to be widespread on eastern larch in Nova Scotia, New Brunswick, and Maine.

The disease in Maine and eastern Canada, as in Europe, is most common and severe in coastal areas with oceanic climate—abundant moisture and moderate winter temperatures. Single or multiple cankers occur on branches and mainstems, reducing wood quality and killing stems up to 10 cm in diameter. Incidence of infection averaged 59% of trees in a large sample of natural eastern larch stands in coastal New Brunswick.

All species of larch and golden larch are at least somewhat susceptible. Dunkeld, eastern, European, Siberian, and western larches are rated highly susceptible, while Dahurian and Japanese larches are less so. Japanese larch is damaged by *L. willkommii* in Japan, however.

The cankers are perennial and begin during host dormancy as circular to elliptic depressions in the bark of twigs and small mainstems or branches, nearly always surrounding a dead dwarf shoot or twig. They form by alternation of host and pathogen activity, as described for Strumella canker. Shoots on girdled stems and branches shrivel and die in spring, or the needles turn yellow prematurely in late summer. Resin exudes from recently infected bark. Old cankers are rough, dark, and resinous, with many concentric callus ridges and cracked bark at the margins. Dead bark eventually falls away from the center of a canker, but the wood remains more or less sound. A diseased stem or branch appears swollen at the site of the canker because of stimulated growth of tissues adjacent to the lesion. Many cankers cease enlargement after a few years, apparently because of competitive microorganisms that displace *L. willkommii*.

Apothecia of *L. willkommii* are similar in appearance to those of *L. agassizii* (H). They grow from killed bark in the centers of young cankers and near the edges of old cankers but not along dead twigs as do the apothecia of *L. occidentalis* (see below). Apothecia, visible throughout the year, are 1–6 mm broad, cup shaped, hairy, and white with an orange-yellow to pale buff fertile layer within. They discharge ascospores to the air throughout the year when moistened by rain. The ascospores are colorless, single-celled, and oblong with somewhat pointed ends. They measure 15–28 × 6–9 μm.

Ascospores cause new infections, probably in autumn. The precise sites of infection are not known but are suspected to be moribund dwarf shoots in which the periderm that would normally exclude the parasite is absent or only partially formed. Frost injury may aggravate damage by *L. willkommii* but is not required for it.

Many more *Lachnellula* species occur on conifers in North America. Those shown to be pathogenic, and hosts on which they apparently cause cankers or twig dieback in nature, are *L. agassizii* on balsam and Fraser firs and eastern white pine, *L. arida* and *L. gallica* on balsam fir, and *L. laricis* and *L. occidentalis* on eastern larch. Each of these fungi occurs as a saprophyte on several additional conifers.

References for Strumella canker: 92, 169, 228, 761, 888, 911, 1032, 2240

References for Lachnellula cankers: 92, 275, 276, 387, 481, 713, 714, 937, 1192, 1193, 1214, 1427, 1521, 1831, 1954, 2240, 2258–61

A–D. Strumella canker of red oak (NH, Mar). A. A canker that is expected to girdle a young pole-sized tree. Concentric callus ridges, adhering bark, and a branch stub in the center are typical. B. A canker that has become inactive around most of its margin. Dead bark has sloughed off, exposing decayed wood. C. Close view of sterile nodules of fungal tissue on the bark in a canker. D. Transverse section through a canker showing, from center, decayed wood and bark, successive dead callus ridges, and the current canker margin in bark at the upper right.

E. Apothecia of *Urnula craterium* on a fallen oak branch (NH, spring).

F. A canker caused by *Lachnellula willkommii* on a branch of eastern larch. The branch is swollen and exuding resin at the canker; white apothecia are visible (ME, Jul).

G, H. A canker caused by *L. agassizii* on a Fraser fir seedling (NY, Aug). G. The diseased part of the stem is slightly shrunken and bears white apothecia. H. Magnified view of apothecia of *L. agassizii*, showing their orange-yellow interior and hairy white surface.

Plate 113

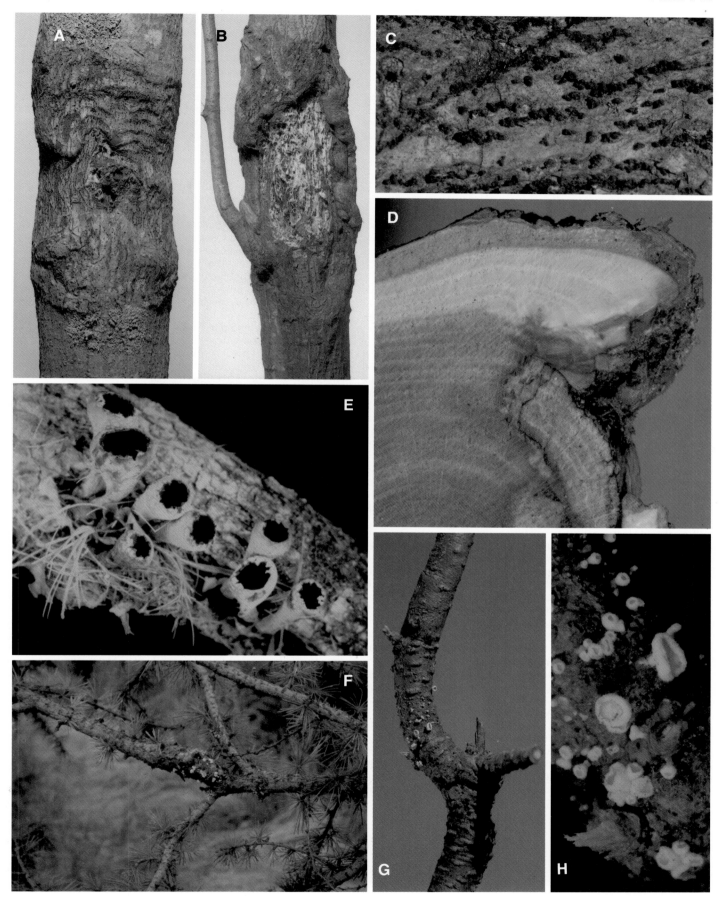

Rose Rust and Fig Rust (Plate 114)

This plate introduces a section on rusts—diseases caused by obligately parasitic, host-specialized fungi of the order Uredinales in the Basidiomycotina. Rust fungi may infect any plant part except roots. Their hyphae grow inter- and intracellularly, and they produce intracellular haustoria. Rust fungi have complex life cycles that may involve alternate host plants and may include up to four different spore-producing stages and five functionally different kinds of spores. A rust fungus that completes its life cycle on one host is termed autoecious. Fungi requiring two host species are termed heteroecious. Readers are referred to general references among those listed below for information on rust morphology, biology, and terminology. Names and functions of life stages of a generalized rust fungus are outlined here in order of their appearance in the life cycle.

Spermagonia (syn. pycnia) are structures that produce spermatia—noninfectious spores that function as gametes. Spermagonia develop after infection by basidiospores.

Aecia are cuplike, blisterlike, or cylindrical asexual fruiting structures that follow spermagonia on the same host and produce nonrepeating, dry, light yellow to orange wind-dispersed aeciospores. In a heteroecious rust, aeciospores usually do not infect the host on which they are produced.

Uredinia (syn. uredia) are asexual, blisterlike pustules that produce urediniospores (syn. uredospores). These spores are dry, orange to rust-colored (purplish in some species), wind-dispersed, and sometimes long-lived. They cause repeating (secondary) cycles on the host on which they form.

Telia are sexual structures of various forms that follow uredinia on the same host and produce teliospores that vary greatly in color and form. Teliospores of most rusts are not dispersed or are dispersed only short distances and in most species are not infectious. Many rusts survive winter as teliospores. Teliospores of most rusts germinate to produce basidiospores. Basidiospores are sexually produced, non-repeating, colorless, short-lived, and wind-dispersed spores that in a heteroecious rust do not infect the host on which the teliospores were formed.

Rusts that produce all the stages above are termed macrocyclic. Many rusts omit one or more stages, however, and some rusts produce fruiting structures and spores with the appearance of one stage and the function of another. Rust structures with massed spores are often called sori.

Rose rust. At least nine species of *Phragmidium* cause rust on roses in North America. These are *P. americanum, P. fusiforme, P. montivagum, P. mucronatum, P. rosae-californicae, P. rosae-pimpinellifoliae, P. rosicola, P. speciosum,* and *P. tuberculatum.* Each rust species is specialized for parasitism of certain rose species, hybrids, and cultivars. All these rusts are autoecious, and most are macrocyclic, producing yellow to bright orange powdery masses of thick-walled, spiny (or sometimes warty) aeciospores and urediniospores that break through the epidermis of various plant parts. Spores of both types are subglobose to somewhat ellipsoid with dimensions in the range 15–35 μm. Teliospores follow urediniospores, often in the same pustules. Teliospores of most species are various shades of brown to nearly black and more or less cylindric with thick walls and rough surfaces. They have 4–11 cells, with a long, colorless stalk cell and a thornlike pointed apex. The body of the teliospore of most species is in the range 30–40 × 60–120 μm, and the stalk cell is 70–180 μm long.

P. mucronatum and *P. tuberculatum* are the best known and most common species on cultivated roses. Both may produce spermagonia and aecia on stems and leaves and uredinia and telia on stems and the undersides of leaves of cultivated roses. *P. tuberculatum* is distinguished from *P. mucronatum* on the basis of its warty (versus spiny) aeciospores. The description of symptoms, damage, and disease cycle applies to *P. mucronatum.*

Rust appears in spring or may be found throughout the year in areas with mild winter weather. It intensifies on mature leaves throughout the growing season. Spermagonia on stems and upper surfaces of leaves are seldom noticed. Powdery orange masses of aeciospores develop on stems and other green parts except upper surfaces of leaves and are followed within about 2 weeks by orange urediniospores. Young stems and sepals may be distorted by rust. The upper surfaces of leaves show yellow to orange-brown or dark reddish brown spots above the rust pustules. When rust is severe, a deposit of orange urediniospores may discolor the plant surface and even the ground beneath. As rust intensifies, severely diseased leaves of highly susceptible cultivars may turn yellow or may become brown and shriveled and then drop. The growth of plants used as understocks may be severely retarded.

Black telial pustules supplant the uredinia outdoors in late summer and autumn in cold climates and during winter in mild climates. The teliospores, readily dislodged from leaves, alight on various surfaces and become glued in place by a gelatinous substance released from the stalk cell after it imbibes water. Teliospores may overwinter and in spring germinate to produce basidiospores. The rust fungus may also overwinter in stems as mycelium capable of producing aecia and (in warm areas such as coastal California, where roses retain leaves all year) on leaves as uredinia. Thus the first new infections in spring may be initiated by basidiospores liberated from overwintered teliospores, by aeciospores formed by overwintered mycelium in stems, or by urediniospores. The latter infect only the lower surfaces of leaves, which are penetrated through stomata.

Conditions most favorable for development of *P. mucronatum* are frequent wet periods (rain, dew, or fog) of at least 4 hours' duration (optimum 12 hours) and temperatures of 18–21°C. Thus rose rust flourishes in coastal California and in Europe but is not severe in areas with hot or dry summers.

Fig rust. This rust, caused by *Cerotelium fici,* occurs widely on common fig and strangler fig in the southeastern United States and is found occasionally in California. It occurs around the world in tropical to warm temperate regions on members of the mulberry family. Other reported hosts in the United States are red mulberry and osage orange. Only the uredinial (repeating) stage of *C. fici* is known in North America. The telial stage, apparently rare, has been reported from India. This rust is confined to leaves and, when severe, causes yellowing and defoliation in late summer or early autumn. Fruit drop sometimes follows defoliation. Defoliated trees may produce new leaves in autumn, and these are at risk of being killed by frost.

Uredinia, initially salmon colored and later golden yellow because of their masses of spores, are found scattered or in irregular groups on small yellowish green spots on the undersurfaces of leaves. The spots progress to angular, rusty brown lesions that extend to the upper surface, where they appear purplish to brown. The disease cycle is imperfectly known. In the southern United States, the fungus is believed to overwinter as urediniospores that are in some manner dispersed to developing leaves in spring. About 14 days elapse between infection and the production of new uredinia.

References for rusts in general: 54, 434, 623, 836, 1153, 1154, 1484, 1732, 2213, 2281
References for rose rust: 371, 387, 883, 898, 927, 1756
References for fig rust: 536, 1061, 1987

A–C. Rust caused by *Phragmidium mucronatum* on *Rosa floribunda* cv. 'Sparta'. A, B. Uredinia on the undersurface of a leaf, magnified in (B). C. Chlorosis and necrotic spots caused by the rust on the upper surface of the same leaf (NY, Sep).

D. Aecial stage of *P. mucronatum* on a rose cane. Particles of orange powder are aeciospores (NY, May).

E–G. Rust caused by *Cerotelium fici* on common fig. E, F. The lower surface of a diseased leaf, showing uredinia and associated necrotic lesions, magnified in (F). G. Lesions viewed on a portion of the upper surface of the same leaf (FL, Mar).

Plate 114

Gymnosporangium Rusts (Plates 115–119)

This plate and the four that follow present Gymnosporangium rusts. These rusts alternately infect trees or shrubs in the cypress family and, in most cases, the rose family. On evergreen hosts they cause galls, stem swellings, witches'-brooms, and dieback of twigs and branches. Rosaceous hosts develop colorful spots and localized swellings on leaves, fruits, and green twigs, followed often by casting or distortion and death of these parts. We illustrate these rusts first on juniper or cedar hosts and then on rosaceous plants. Because of this arrangement, descriptive information about a particular rust may not face the plate that displays it.

The genus Gymnosporangium includes some 57 species worldwide that produce telia on members of the cypress family, mainly junipers. Thirty-six species occur in North America. Nearly all species lack a repeating (uredinial) stage and require an alternate host to complete their life cycles and perpetuate themselves. Two exceptions that form uredinia are G. nootkatense on Alaska cedar in the Pacific Northwest and G. gaeumannii on common juniper in the Canadian Rockies. The spermagonia and aecia of Gymnosporangium species (Plates 117–119) form in most cases on members of the apple tribe in the rose family. The only known exceptions are G. speciosum (Plate 117), which produces its aecia on fendlera and mock orange; G. ellisii, which produces aecia on bayberry, sweet gale, wax myrtle, and sweetfern; and the autoecious species G. bermudianum, which produces aecia and telia on eastern and southern red cedars in southeastern states.

Some Gymnosporangium species complete their life cycles in 1 year, but many require 2 years. Spermagonial and aecial stages form in spring through late summer on colorful (yellow to reddish orange) spots on leaves or on chlorotic to variously colored spots on green stems and fruits of the rosaceous host. Spermagonia on leaves open to the upper surface, and aecia usually form on the undersurface. Spermagonia are flask shaped, immersed in host tissue, and visible only as tiny yellow to black dots on the lesions (Plate 118). Aecia are white or light-colored hornlike or cylindric (rarely cup-shaped) structures (Plates 117–119) that produce powdery yellow to brown aeciospores. Aeciospores are windborne to the evergreen host during summer and autumn. Rust in the aecial host is shed with leaves or fruit, or it dies out in stems during winter in most cases.

Gymnosporangium species overwinter in their evergreen hosts, where they produce telia in spring. In rusts with a 2-year cycle, 21–22 months elapse between infection and telial formation. Telia (Plates 115–116) appear as hornlike or cushionlike projections from galls, bark, or leaves. They consist primarily of large masses of long-stalked teliospores. During wet weather in spring, telia become swollen and gelatinous, and teliospores germinate in place, producing basidia and basidiospores. The basidiospores are dispersed by air and infect succulent parts of the rosaceous host. Many Gymnosporangium species become perennial in juniper or cedar branches and produce telia annually on galls, swellings, or witches'-brooms.

Cedar-apple rust. This is the best known and most economically important of the Gymnosporangium rusts. In apple-growing areas where eastern red cedar or Rocky Mountain juniper is abundant, this disease has the potential to cause severe crop reduction through fruit infection and premature defoliation of the trees. In commercial orchards such damage is prevented by fungicidal sprays. The pathogen, Gymnosporangium juniperi-virginianae, is one of the few to have triggered legislation for the purpose of disease suppression. Several eastern states enacted laws requiring the destruction of the principal telial host, eastern red cedar, near commercial apple orchards. The cedars in some areas were valued for ornament and for their fragrant lumber, however, and the eradication efforts often led to hard feelings among neighbors.

Cedar-apple rust is indigenous to North America. It is widespread on apples and crabapples, eastern red cedar, southern red cedar, and Rocky Mountain juniper throughout the United States east of the Rockies and in southern Ontario and Quebec. This rust also has been reported from California and Washington but is uncommon in the West. Red-berry and Utah junipers and a few cultivars of Chinese and prostrate junipers are also susceptible. Symptoms on junipers and red cedars are galls and associated twig dieback. Dieback is most prominent on Rocky Mountain juniper. Junipers become infected in late summer and autumn, and galls begin development in spring as minute green swellings on leaves, usually on the upper and inner surface. When at full size in autumn, the galls are greenish brown, globose to kidney shaped, and mostly 10–30 mm in diameter (extremes 2–50 mm). Galls consist mainly of parenchyma cells and abundant intercellular hyphae of the parasite. (It was from galls of eastern red cedar that G. juniperi-virginianae was first isolated in pure culture; the report, in 1951, was a scientific milestone because no rust fungus had previously been induced to grow apart from its host.) Circular depressions appear on the gall surface in autumn, and telia emerge from these the next spring. The telia are golden brown hornlike projections 10–20 mm long and 1–2 mm wide and are composed primarily of teliospores. During warm spring rains they swell to 2–3 times their former size, becoming gelatinous and bright orange-yellow (Plate 115A–C). Telia gelatinize and dry several times in response to intermittent rains. During each period of gelatinization some of the teliospores germinate and liberate basidiospores, primarily at night. An overnight period of high humidity after rain is sufficient for gelatinization of telia, germination of teliospores, formation and dispersal of basidiospores, and infection of the alternate host. Telia finally disintegrate into a slimy mass (Plate 115E) that may obscure the gall and may cover adjacent foliage and twigs. The slime eventually dries and falls away. The galls soon die but remain attached to twigs for a year or more.

Basidiospores infect primarily young leaves and sometimes young fruit and green stems of apples and crabapples. Infection of hawthorn has also been reported but is uncommon. Basidiospores may remain germinable while being carried several kilometers in air, but most infections occur within a few hundred meters of the source cedars. Given wet plant surfaces, infection can occur at temperatures of 2–24°C. A wet period of 4–6 hours at 10–24°C is sufficient for severe infection. Basidiospores require free water for germination. Those deposited on dry foliage remain germinable for 1 or more days if humidity remains high and the spores are not exposed to sunlight. Lesions on apple or crabapple leaves are at first greenish yellow, then orange-yellow, often bordered by a red band, and swollen. Lesion size varies with host susceptibility and inversely with the frequency of spots. Numerous spots tend to be individually small. Spermagonia appear on the upper center of each lesion as yellow dots that exude tiny droplets of yellow fluid containing spermatia. After several days, spermagonia darken and exudation ceases.

A–C, E. Gelatinized telia of Gymnosporangium juniperi-virginianae.
  A. On Rocky Mountain juniper (NE, Jun). B, C. On eastern red cedar, showing variation in gall size and number of telia per gall (NY, May–Jun). E. Collapsed telia after final gelatinization (NY, Jun).
D. Gelatinizing telia of G. globosum: short, blunt, and reddish brown in comparison with those of G. juniperi-virginianae (NY, May).
F. Young, growing galls of hawthorn rust, caused by G. globosum, on common juniper. A dead awn-shaped leaf remains attached to each gall (NY, Sep).
G. A living hawthorn rust gall, several years old, on eastern red cedar. Red-brown color is typical (NY, Mar).
H. Dead cedar-apple rust galls and associated twig dieback on Rocky Mountain juniper (IL, Jun).

Plate 115

Aecia of the cedar-apple rust fungus appear in midsummer, 2–4 weeks after spermagonia, and may be scattered or arranged in a circle on the lower surface of the lesion. If they are on fruit or a twig, aecia appear among the spermagonia. The aecia, similar to those shown in Plate 118, are off-white cylindrical structures 1–2 mm long, the walls of which split into strips with ends turned down. The strips flare out, exposing a reddish brown mass of aeciospores. Aeciospores are produced for several days and drop out of the aecia chiefly during morning hours in response to drying. Wetness or high humidity inhibits dispersal of aeciospores of the cedar-apple rust fungus because the wall of the aecium absorbs water and swells, restricting the opening. Aeciospores, windborne to cedar or juniper, cause new infections from midsummer into autumn, completing the 2-year disease cycle.

Races of *G. juniperi-virginianae* that vary in virulence to various apple cultivars are known, but no comparable information exists regarding reactions of juniper hosts. Several old apple cultivars, such as 'Baldwin', 'Delicious', 'Gravenstein', and 'Rhode Island Greening', are resistant to Gymnosporangium rusts. Resistance is expressed as limitation of infection to small flecks or mottled areas or to small spots where only spermagonia develop. Newer apple cultivars in general are somewhat susceptible. Crabapples observed extensively and found rust free or only slightly affected by cedar-apple rust are listed opposite Plate 43. Junipers resistant to cedar-apple and hawthorn rusts are listed opposite Plate 65.

Hawthorn rust. This disease (Plates 115, 118, 119), caused by *Gymnosporangium globosum,* is similar to cedar-apple rust in most respects. We emphasize the differences. The fungus is indigenous to North America. It is distributed with its juniper hosts from southeastern Quebec to Georgia, Texas, and Saskatchewan. It has also been reported from Alaska. The most common juniper hosts are eastern red cedar and Rocky Mountain juniper; others affected are southern red cedar and common and prostrate junipers. *G. globosum* has a wider rosaceous host range than does the cedar-apple rust fungus. *G. globosum* infects apple and crabapple, many hawthorns, and occasionally pear, quince, and serviceberry. Medlar and mountain ash were also susceptible when inoculated in experiments.

Hawthorn rust most often affects leaves, causing yellow spots (Plate 119B, D). When this rust is severe, all the foliage of a hawthorn tree may turn bright yellow and drop prematurely. Hawthorn rust occasionally occurs also on fruit or green stems, causing deformity. Black flattened lesions, usually lacking aecia, form on quince fruit. Spots on pear leaves (Plate 118) are dark brown to nearly black with a reddish border on the upper surface. Aecia form on pear petioles and on the lower surfaces of the largest leaf spots. The interval from infection to aecial maturation of hawthorn rust in northern areas is 80–95 days, about 10 days longer than for cedar-apple rust. Precipitation causes immediate release of aeciospores, but in the absence of rain these spores are released during morning hours as humidity diminishes.

Spermagonia and aecia of *G. globosum* are similar to those of the cedar-apple rust fungus, but the aecia of *G. globosum* may be distinguished by their greater length (often 3–4 mm), irregular mode of splitting, and relatively straight (rather than turned-down) strips of wall cells after splitting. Viewed microscopically, *G. globosum* has smaller aeciospores. Also, the aecial wall cells adhere to one another rather than separating in a water mount, and the tip cells of the walls are nearly straight rather than strongly curved.

G. globosum induces spheroid galls on juniper hosts (Plate 115). The galls originate on leaves, grow slowly, become firmly attached to the twigs, and often become flattened on the side next to the twig. They differ from galls of cedar-apple rust in being smaller (3–15 mm diameter), perennial, and reddish brown (rather than greenish) when young. Hawthorn rust galls in winter display domelike swellings where telia will emerge in spring. Old galls turn grayish brown and display scars where telia were produced previously. Telia of *G. globosum* are chestnut brown, conic with blunt tips, and 3–12 mm long and 1.3 mm wide before gelatinization. They double in size when gelatinized, and in moist air they produce and liberate basidiospores within a few hours. When humid weather enhances survival of basidiospores, these spores can infect alternate hosts at considerable distances from source cedars. In Ontario, severe infection of hawthorn has been recorded as far as 24 km from telia-bearing cedars. Sometimes telia of this fungus emerge from the upper surfaces of discolored, slightly swollen leaves. *G. globosum* causes little or no dieback of woody twigs.

Junipers resistant to *G. globosum* are listed opposite Plate 65. Hawthorns reported to be resistant include English hawthorn, Cockspur thorn, yellow-fruited thorn, *Crataegus intricata,* and *C. pruinosa.*

The hawthorn rust fungus can be confused with *G. connersii,* which induces similar symptoms and produces similar structures on junipers and hawthorns in the northern Great Plains and eastern Canada. Its telia, however, occur primarily on prostrate and Rocky Mountain junipers, and the aecia mature 2–4 weeks before those of *G. globosum.* These fungi also differ microscopically.

Juniper broom rust. The causal fungus is *Gymnosporangium nidus-avis.* This North American fungus occurs from coast to coast and from northern Alberta to Arizona and Florida. Its telial state develops on eastern red and southern red cedars and prostrate and Rocky Mountain junipers, causing witches'-brooms and stem swellings. The spermagonial and aecial states occur on leaves, fruits, and green stems of various species of serviceberry and on apple, quince, and flowering quince. Mountain ash and hawthorn are also susceptible by artificial inoculation. Severe infection of serviceberry may kill fruit and may cause conspicuous leaf and shoot blight as lesions encircle petioles and green twigs. Leaf infections result in red-bordered brown lesions that enlarge from the point of infection to the leaf edges.

Infection of cedar or juniper occurs on leaves or green twigs and becomes perennial if the host part remains alive. Growth of rust mycelium near host growing points leads often to formation of witches'-brooms, while infection of a stem at a distance from its apex may lead to swelling but no broom formation. Brooms are most conspicuous on eastern red cedar and Rocky Mountain juniper. Most brooms die while small, but some attain a diameter of 50–60 cm or more and live 15 years or longer. Branches within brooms tend to be somewhat enlarged, with roughened bark, and often they bear only awn-shaped (juvenile) leaves.

A. Juniper broom rust, caused by *Gymnosporangium nidus-avis,* on Rocky Mountain juniper (CO, Jun).

B. Stem swelling with roughened bark caused by the medlar rust fungus, *G. confusum,* on savin juniper (CA, Aug).

C–G. Stem swellings caused by the quince rust fungus, *G. clavipes,* on eastern red cedar. C. Orange-red, cushionlike telia protruding from roughened bark of a cedar twig in early spring (NY, Apr). D. Spindle-shaped swelling with roughened bark on a branch in summer; no signs of the pathogen are visible at this time of year (NY, Aug). E. Mature telia just before gelatinization (NY, May). F. Gelatinized telia (NY, Jun). G. Old telia, depleted of germinable teliospores, visible as a dry orange scum on the branch. Dead twigs apparently killed by rust extend from the infected region. The larger dead twig was once the main axis of the branch (NY, Jun).

Plate 116

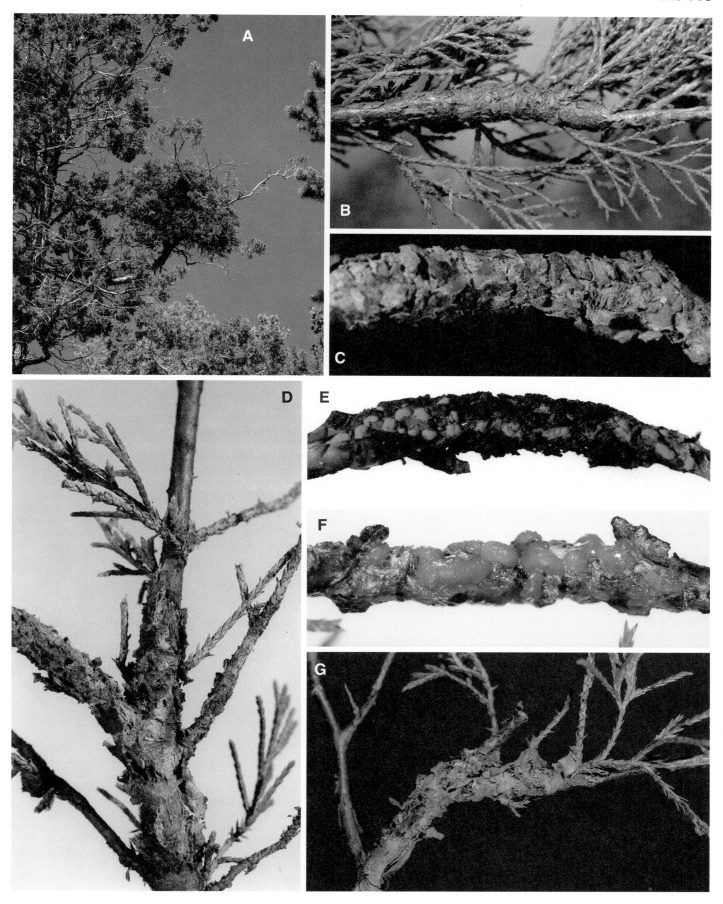

The rust fungus often grows many centimeters in living bark along a branch from a broom toward or into the mainstem. Rust growth at a rate of 5 cm per year in eastern red cedar has been measured. Swellings on nonbroomed branches may be spindle shaped, encircling small branches, or elliptic, appearing as bulges on large branches or trunks. Such infections may persist and enlarge for 3 decades or more.

The bark of junipers or cedars infected by *G. nidus-avis* usually becomes roughened and in time develops deep longitudinal fissures and corky parallel ridges where the fungus produces telia annually. The ridges are smooth in contrast with the fibrous texture of normal outer bark. Occasionally long strips of diseased bark die, and the fungus produces telia only beyond the margins of the dead strip. Whether *G. nidus-avis* or a secondary pathogen causes such cankers has not been ascertained.

Orange-brown telia somewhat similar to those of *G. clavipes* (Plate 116D–F) protrude in spring from abnormally roughened bark in brooms and on swollen parts of nonbroomed twigs, branches, and even trunks. Telia may also form in the axils of awn-shaped leaves or between scale leaves. Telia vary in shape and size from tiny cushions to elongate mounds 3–5 mm long × 3–6 mm wide × 2–9 mm high. When gelatinized they may rise as high as 16 mm. Teliospores germinate in place and liberate basidiospores that infect rosaceous hosts. Old telia become cinnamon brown and eventually drop from the host, leaving oval scars that are at first yellowish orange, then brownish.

Basidiospores of *G. nidus-avis* germinate within a few hours on wet surfaces of succulent parts of rosaceous hosts. The fungus then penetrates directly through the cuticle, forming intercellular hyphae that send haustoria into host cells. Spermagonia develop within 9–14 days after infection in swollen yellow spots on soft twigs, petioles, or leaf veins and on yellow to yellow-orange leaf spots with red borders. Rust spots on leaves are 1–5 mm in diameter, depending on host susceptibility. Vein infections cause leaf distortion. Spermagonia appear as yellow dots from which droplets containing yellowish spermatia exude for 5–10 days, after which the spermagonia and often the lesions turn black. Aecia (Plate 117F–H) 0.3–0.6 mm in diameter and up to 5 mm long form after an additional 3–4 weeks and open by lateral slits. Aecia tend to retain their cylindrical shape, and their wall cells remain straight when wet. Aecia on leaf veins sometimes develop on upper leaf surfaces as well as on lower. The aeciospores are golden brown. In the North, aecia maturing on serviceberry and associated with leaf and shoot blight in early to mid-July are likely to be those of *G. nidus-avis*.

Juniper broom rust on serviceberry is somewhat similar to European hawthorn rust caused by *G. clavariiforme*, aecia of which mature at about the same time. The latter rust, however, does not usually cause prominent foliar necrosis. Its aecia are short (to 1.5 mm high), and they split open to the base. The telia of *G. clavariiforme* occur on common juniper and are cylindric.

Medlar rust. Europeans gave this name to the disease caused by *Gymnosporangium confusum* (Plates 116, 118). This fungus is indigenous to Eurasia, where it shows broad host adaptation. Its telial stage develops on eight species of juniper in both sections of *Juniperus* (Oxycedrus and Sabina), and its spermagonia and aecia form on plants in seven rosaceous groups: mountain ash, cotoneaster, hawthorn, medlar, pear, quince, and flowering quince. In 1966 this rust was found on savin juniper and 'Paulii' English hawthorn in California—the first record in North America. It is still restricted to that state. The common name, *hawthorn-juniper rust,* has been used, but since this could also refer to diseases caused by *G. clavipes* and *G. globosum,* we prefer *medlar rust.* This disease causes little damage to junipers, but when it is severe it can destroy foliage of highly susceptible hawthorns (Plate 118A). Hawthorns with brown, shriveled foliage produce a second complement of shoots and leaves in late spring. The new foliage remains rust free because inoculum from junipers is no longer available.

G. confusum infects current-season shoots of savin juniper and persists for many years as perennial mycelium. No symptoms or signs are apparent until telia emerge from 1-year-old twigs in early April. In subsequent years the infected region develops into a spindle-shaped swelling with abnormally rough bark (Plate 116B), from which telia protrude annually in spring. Telia of this species are horn shaped, velvety, dark reddish brown, and up to 8 mm long before gelatinization. When wet they expand and gelatinize, their teliospores germinate, and basidiospores are released. After being depleted of germinable teliospores in mid-April, telia deliquesce to a slime that coats the bark and surrounding foliage.

Orange spermagonia of *G. confusum* develop in late April in brown thickened spots with yellow margins on both surfaces of hawthorn leaves as well as in swellings on succulent stems, fruit, fruit stalks, and calyxes. Aecia develop among the spermagonia except on leaves, where they are restricted to the lower surface. Aecia mature in late May. They are cylindrical, 2–4 mm long, and off white, becoming tan and tattered. Sometimes they cover the surfaces of infected fruit. The aeciospores, infectious to juniper, are reddish brown. Aeciospores on a wet surface may germinate within 4 hours at 24°C.

Broom rust of incense cedar. *Gymnosporangium libocedri* causes witches'-brooms on incense cedar in Nevada and the Pacific coast states. The spermagonial and aecial states of the fungus develop on leaves and fruit of various rosaceous plants—mostly hawthorn and serviceberry, but also apple and crabapple, mountain ash, pear, and quince. *G. libocedri* has occasionally caused severe infection of pear and quince fruits in orchards in the Northwest where incense cedar was growing nearby. Spermagonia occur on the upper sides of colorful swollen leaf spots, and white aecia develop mostly on the undersides. Most infections on incense cedar are confined to leaves and are scarcely noticeable. Telia develop only on the leaves. Before gelatinization they are cushionlike, reddish brown, and 0.8–1.5 mm in diameter. The long-stalked brown teliospores of this rust are distinctive for having up to five cells. Germinated teliospores produce basidiospores that can cause infection of rosaceous hosts at distances as great as 12–16 km from incense cedars.

A–C. Witches'-brooms on incense cedar caused by *Gymnosporangium libocedri*. A. A dying tree with several brooms clearly visible because of the absence of normal host foliage. B. A closer view of the topmost broom shown in (A). C. A rust-induced broom in which a large number of branches radiate from a gall (CA, Jul).

D. A spindle-shaped gall caused by *G. speciosum* on a branch of alligator juniper. The basal end of the branch is at right (AZ, Jul).

E–H. Leaf and shoot blight of serviceberry caused by *G. nidus-avis*. E. Dead leaves and twigs. F, G. Necrotic lesions and leaf distortion caused by foliar infection; also, aecia on swollen lesions on petioles, associated with death of entire leaves. H. Close view of aecia on swollen petiolar lesions (Ont, Jul).

Plate 117

Mycelium of *G. libocedri* sometimes becomes perennial in the wood of twigs and then causes spindle-shaped swellings or witches'-brooms. Telia develop on leaves of the brooms. As branches grow, and the rust with them, swellings or witches'-brooms come to be located on large branches and sometimes on the mainstem. Infected wood in swellings or brooms has numerous dark brown flecks arranged in radiate patterns as viewed in transverse section. By one estimate from the study of infected wood, living mycelium in incense cedar may persist more than 200 years. Trees with many brooms eventually decline and die.

Rust galls on southwestern junipers. *Gymnosporangium speciosum* causes witches'-brooms, spindle-shaped swellings (Plate 117), and sometimes large woody galls on branches of alligator, cherry-stone, Sierra, and Utah junipers in the southwestern United States. Galls on alligator juniper are usually oblong but sometimes nearly globose, 3–4 times the diameter of the branch, and may develop to 35 cm or more in diameter. They have rough corrugated surfaces. Witches'-brooms caused by *G. speciosum* are most common on Utah juniper, and they are associated with spindle-shaped stem swellings on alligator juniper. Telia form in spring in more or less distinct rows on the swollen parts. Occasional specimens of Utah juniper have been observed that seem to be completely infected, with telia appearing in small patches all over the trunk and branches. The telia are orange-yellow, elongate in basal outline and 3–4 mm high before gelatinization. The teliospores have either two or three cells.

This rust fungus is unusual among *Gymnosporangium* species in that it alternates not to rosaceous plants but rather to several species of fendlera and mock orange. Spermagonia of *G. speciosum* open to the upper surface and aecia to the lower surface of yellow leaf spots. The aecia are cylindrical, white, and 2–3 mm high. Their side walls split open at the top as masses of yellow aeciospores are produced. The aecial hosts do not sustain significant damage. In New Mexico, the rust fungus *Uredo apacheca* causes on alligator juniper rough galls similar to those caused by *G. speciosum*.

Quince rust. This disease (Plates 116, 119) is the most damaging of the Gymnosporangium rusts to rosaceous plants because it affects primarily fruit, green stems, and petioles. It stunts and kills fruit and causes swelling, distortion, and death of twigs and petioles. It sometimes causes cankers on small branches. Apples are somewhat resistant, but infection of apple fruit is occasionally important because infected parts of the fruit stop growing, with the result that distortion and loss of commercial value occur. Several old apple cultivars such as 'Baldwin', 'Ben Davis', 'Northern Spy', 'Twenty Ounce', and 'Wealthy' are resistant to the quince rust fungus. Most contemporary cultivars, however, are considered to be slightly to moderately susceptible.

On junipers and red cedar the causal fungus infects leaves and soft shoots, becomes perennial in the living bark, and causes dark brown, flaky-barked, spindle-shaped swellings that encircle twigs and small branches. Diseased twigs and branches of red cedar often die, and trees with numerous infections display slow loss of vigor and gradual thinning of foliage. Red cedar also develops circular to elliptic swollen patches with flaky bark on the trunk, presumably the result of invasion via twigs many years earlier.

The pathogen is *Gymnosporangium clavipes*. It occurs only in North America, ranging from Newfoundland to Alaska and south to Florida, Texas, and Mexico. This transcontinental distribution notwithstanding, the fungus is common only in eastern regions. *G. clavipes* has a much broader host range than most members of its group. Rosaceous hosts include more than 480 species in 11 generic groups: apple, mountain ash, chokeberry, cotoneaster, hawthorn, medlar, pear, photinia, quince, flowering quince, and serviceberry. Telia form on plants in both the Oxycedrus and Sabina sections of *Juniperus*: eastern red cedar and common, prostrate, Rocky Mountain, and savin junipers. A host-specialized strain, *G. clavipes* f. sp. *cupressi*, infects Portuguese cypress and hawthorn in Mexico.

On rosaceous hosts, chlorotic spots or swellings may be visible 7–10 days after infection, spermagonia after 10–15 days, and aecia after 30–60 days (45–60 days in the north). Leaves commonly become infected on veins, where spermagonia but no aecia develop except occasionally on the large veins. Infected veins of quince leaves swell to nearly twice their original size, and the leaves curl and eventually drop. Infected hawthorn leaves curl and die. Swollen spots with spermagonia and later aecia develop on quince fruit. Fruits of serviceberry and hawthorn often become uniformly covered with aecia (Plate 119) and then die and dry out. Rust development on apple or crabapple fruit varies with susceptibility of the species or cultivar. Some cultivars show no reaction or only flecks; others develop spots (either swollen from overgrowth or sunken because of suppressed growth) and spermagonia but no aecia. Still others support normal aecial development. In some instances where aecia fail to appear on apple fruit, they have formed but remain covered by the epidermis. Apple fruit is highly susceptible for only about 2 weeks when young.

Petioles and green twigs or thorns begin to swell soon after infection. The rust fungus commonly spreads up to 3 cm in each direction from the point of infection, causing a spindle-shaped swelling (Plate 119) or uncommonly a distinct gall on which spermagonia and aecia develop. Newly formed buds become infected via internal mycelium, then begin growth when they would normally remain closed. The resulting shoots swell and elongate only a few centimeters, producing stunted, distorted leaves. Spermagonia and later aecia may develop on the shoot axis. Stems or shoots encircled by rust die after aecial production, turn weathered gray, and persist for 1 or more years. In highly susceptible hosts, *G. clavipes* sometimes grows from a thorn or stunted shoot into the 1-year-old branch, causing a canker where aecia form until midautumn. The fungus survives winter at the edge of

A, B. Medlar rust, caused by *Gymnosporangium confusum*, on 'Paulii' English hawthorn. A. Foliage completely shriveled and brown as the result of severe infection. B. Aecia on fruit (CA, Aug).

C–F. Hawthorn rust, caused by *G. globosum*, on 'Bradford' callery pear. C. Magnified view of a petiole with spermagonia (dark dots) and aecia on the swollen infected portion. Pimplelike bulges near edges of the swelling are developing aecia. Aecia of this rust fungus tend to retain cylindrical form, at least near the base. D. Diseased leaves with brown spots typical of hawthorn rust on pear. E. Magnified view of the upper surface of a foliar lesion showing spermagonia (yellow to dark dots) on swollen tissue. Resistance of this pear cultivar to the rust is indicated by the small size of the lesion and by the apparent death of some of the infected tissue (dark brown area). F. Magnified view of two rust spots on the undersurface of a leaf. Much leaf tissue in one spot has died. Aecia in the normal spot display side walls torn from the top toward the base, characteristic of *G. globosum*. Orange-brown aeciospores color the leaf surface among the aecia (NY, Jun).

Plate 118

the colonized area, and a new crop of aecia forms there in spring. Most branches with cankers are girdled by rust during the 2nd season and then die back to a healthy bud or lateral twig.

Spermagonia of *G. clavipes* are flask shaped and immersed in host tissue except for an opening at the apex. Viewed from above they appear as reddish dots that blacken with time. Aecia form in cortical tissues, where they are embedded to a depth of about 1 mm. They are white, tubular, and up to 3 mm long × 0.3–0.5 mm wide and become torn at the apex or split along the sides as aeciospores mature. Aecia on twigs always precede aecia on fruit. The aeciospores of *G. clavipes* are bright orange or orange-yellow in mass. This character and the distinct preference of the fungus for stems and fruit distinguish quince rust from cedar-apple and hawthorn rusts.

Aeciospores are released not only in response to rain but also in dry air during the period of diminishing atmospheric humidity in the morning hours. They are dispersed by wind and infect juniper hosts during late summer and autumn. Given free water these spores germinate at temperatures of 3–30°C; the optimum temperature is near 16°C.

Leaves and green stems of eastern red cedar and junipers become infected, and the rust overwinters in symptomless tissue. Cushionlike telia arise on leaves or on stems in leaf axils the next spring or after a lapse of 1 year. Accounts of telial development differ on this point. Prior to the first gelatinization, telia are 1–5 mm broad and high, brick red when young, and orange-brown when mature. They turn bright orange and swell to 5–10 times their initial size when wet and gelatinized. Gelatinized telia sometimes coalesce and form an irregular ring around a diseased twig. The teliospores are two-celled, with long stalks that swell greatly when wet. This is the principal basis of telial swelling and gelatinization.

While telia are gelatinized, teliospores germinate in place within several hours, producing basidiospores that are released into the air and infect rosaceous hosts. The optimum temperature for teliospore germination is 22–24°C. Basidiospore dispersal is more or less synchronized with the onset of growth in rosaceous hosts, and the spores are capable of germinating within 2 hours on a wet surface. Thus an overnight period of wet weather when telia are mature in spring is sufficient for infection. Individual telia gelatinize 6–8 times before collapsing as an orange slime that soon weathers away.

Mycelium of *G. clavipes* in cedar or juniper is confined mainly to epidermal cells in leaves and to the outermost living bark in twigs. Infected leaves and many small twigs die during the season after first telial formation, but some twigs survive and the fungus remains alive in them, growing at most a few centimeters each year and causing slight swelling in an encircling band of bark. Telia develop annually on the infected area, and the bark there becomes roughened. Most diseased twigs die within 4–6 years, but a minority of infections on eastern red cedar persist many years, becoming swollen, dark, scaly-barked patches up to 15–20 cm in diameter on limbs and trunks. These are most noticeable when the bark is drying after rain because the infected area remains wet longer than normal bark. Bark scales on the patches are loose and easily rubbed off. Telial production may occur annually for 20 years or longer.

After telial production each spring, a cork layer forms beneath the telial stroma, isolating most of the infected tissue and causing both rust and host cells to die by attrition. Before this barrier forms, however, rust hyphae grow inward to the region of the cork cambium and persist there, where they form a new telial stroma the next spring. The tissues isolated by the cork barrier become the loose, superficial bark scales.

Unless telia are present, trunk and branch infections by *G. clavipes* are easily confused with those caused by several other *Gymnosporangium* species, especially *G. nidus-avis* and *G. effusum*.

Infection by the quince rust fungus apparently impairs the winter hardiness of eastern red cedar. Many severely infected trees in the northeastern United States died during the intensely cold winter of 1933–1934, while adjacent lightly infected or noninfected trees survived.

Additional Gymnosporangium rusts. Those significant or conspicuous in North America include the following:

G. bethelii, causing dieback associated with irregular galls along spindle-shaped swellings on branches or trunks of Rocky Mountain and Sierra junipers; aecia on fruit and leaves of hawthorn.

G. biseptatum, causing broadly spindle-shaped galls on southern white cedar; aecia on leaves of serviceberry.

G. clavariiforme, causing fusiform swellings on stems of common juniper across the continent; aecia on leaves, fruits, and green stems of mountain ash, chokeberry, pear, quince, and serviceberry.

G. cornutum, causing spindle-shaped swellings on branches of common juniper; aecia on mountain ash and, rarely, apple.

G. cupressi, causing spindle-shaped swellings and subglobose galls on branches and trunks of Arizona cypress in the Southwest; aecia on serviceberry.

G. effusum, causing long slender swellings on small branches and occasionally trunks of eastern red cedar; aecial stage unknown.

G. ellisii, causing witches'-brooms on southern white cedar; aecia on leaves of bayberry, sweet gale, and sweetfern.

G. fuscum, causing fusiform swellings on branches of savin juniper; aecia on pear; responsible for pear trellis rust in California and British Columbia; introduced from Europe.

G. kernianum, causing witches'-brooms on alligator, Sierra, and Utah junipers; aecia on leaves and fruit of hawthorn, pear, quince, and serviceberry.

G. tremelloides, causing small subglobose galls on branches or hemispherical galls on limbs of common juniper in the West; aecia on leaves of apple, mountain ash, and quince.

References for Gymnosporangium rusts as a group: 54, 387, 1007, 1377, 1452, 1454, 2281
References for cedar-apple and hawthorn rusts: 8, 9, 11, 12, 197, 387, 428, 825, 886, 1189, 1322, 1377, 1403, 1442, 1443, 1453, 1473, 1474, 1996, 2138, 2139
References for juniper broom rust and medlar rust: 97, 387, 502, 1377, 1454, 1510, 1555, 1556, 2213, 2281
References for broom rust of incense cedar and rust galls on southwestern junipers: 54, 228, 387, 939, 1007, 1510
References for quince rust: 8, 387, 429, 503, 1050, 1321, 1377, 1443, 1452, 1994, 1996, 2281
References for additional Gymnosporangium rusts: 54, 228, 387, 1007, 1377, 1452, 1454, 1510

A, D. Quince rust, caused by *Gymnosporangium clavipes*, on fruit, and hawthorn rust, caused by *G. globosum*, on leaves of hawthorn. A. White, cylindric aecia of the quince rust fungus have emerged from fruit. D. Yellow spots of hawthorn rust do not yet bear aecia (NY, Jul).

B. Spermagonial stage of the hawthorn rust fungus, appearing as yellow-orange dots on the upper surfaces of yellow spots on leaves of *Crataegus mollis* (IL, Jun).

C, E, F. Quince rust on stems of English hawthorn. C. Magnified view of a 2-year-old segment of stem at the margin of a canker caused by *G. clavipes*. The roughened, gray-brown, dead area is where aecia were produced the previous year. The fungus overwintered and produced a second crop of aecia. Orange color is the result of many aeciospores among tattered remains of aecia. E. Cankers caused by the quince rust fungus. The canker at left is active for a 2nd year, as indicated by orange aecia at the margins. The canker at right caused dieback during winter. F. Shoots infected the previous year (gray) became swollen and distorted and died back after aecial production. The rust fungus overwintered at the base of the lower shoot and continued to produce aecia and cause swelling and distortion (NY, Aug).

Plate 119

Puccinia Rusts of Currant and Buckthorn and Cumminsiella Rust of Oregon Grape (Plate 120)

Rust fungi of the genus *Puccinia* are known primarily for the damage caused by the uredinial (repeating) stages of several species to cereal grain crops. The aecial states of some species cause conspicuous but inconsequential infections on woody plants. *P. graminis* (which causes black stem rust of wheat and other cereals) on European barberry and Oregon grape, and *P. coronata* (which causes crown rust of oats and many other grasses) on various species of buckthorn, are examples. The latter rust is presented here. A few species of *Puccinia* cause important diseases of trees or shrubs. Cluster-cup rust of currant and gooseberry, caused by *P. caricina*, is one example. Ash rust (Plates 121–122) is another.

Cluster-cup rust of currant and gooseberry. This rust is caused by *Puccinia caricina* (syn. *P. caricis*, *P. karelica*, *P. pringsheimiana*), which is globally distributed in the temperate zones, occurring from coast to coast in Canada and the United States. Its spermagonia and aecia develop on various currant and gooseberry species, chickweed wintergreen and starflower (*Trientalis*), and nettles (*Urtica*). Its uredinia and telia are found on many sedges (*Carex* species). Currant and gooseberry hosts in North America include *Ribes alpinum*, *R. americanum*, *R. aureum*, *R. bracteosum*, *R. cereum*, *R. cognatum*, *R. cynosbati*, *R. divaricatum*, *R. glandulosum*, *R. ×gordinianum*, *R. grossularia*, *R. hirtellum*, *R. howellii*, *R. hudsonianum*, *R. inerme*, *R. lacustre*, *R. laxiflorum*, *R. nigrum*, *R. petiolare*, *R. sanguineum*, *R. sativum*, *R. setosum*, and *R. triste*. Botanical rather than common names are listed here because several species are known by the same common names, such as red or black currant.

On currants or gooseberries *P. caricina* causes yellow to orange spots on leaf blades, petioles, green stems, fruits, and fruit stalks. Severe infection kills affected parts. Fruit losses as great as 75% of the crop have been reported from scattered locations in Europe and eastern Canada where winter weather is not extremely severe. The rust fungus survives winter on sedge hosts as teliospores that are apparently intolerant of temperatures below −30°C. In spring the teliospores germinate and liberate basidiospores coincident with the unfolding of gooseberry and currant leaves. The basidiospores infect these plants, leading to formation of colorful spots with spermagonia on the upper surfaces and, several days later, aecia on the lower surfaces of leaves. The disease is named for the arrangement of the tiny cuplike aecia in clusters. Like other rusts, this species consists of strains with varying host specificity. Currants or gooseberries resistant to one strain may be highly susceptible to another. Several other rusts, including the white pine blister rust (Plate 130), infect currant and gooseberry leaves but do not form spermagonial and aecial stages on these plants, as does *P. caricina*.

Crown rust on buckthorn. Many species of buckthorn (*Rhamnus*) and also buffalo berry and silverberry serve as hosts for the spermagonial and aecial states of *Puccinia coronata*. This cosmopolitan rust causes yellowish green to yellow, swollen spots on leaves, petioles, and green stems. Severe rust causes distortion. The severity of rust on buckthorn depends upon the successful overwintering of the rust as teliospores on oat or rye stubble or on any of hundreds of species of susceptible grasses. Infection is caused by basidiospores liberated in spring from germinated teliospores. The germ tube penetrates the epidermis directly, allowing the establishment of a mostly intercellular mycelium that gives rise to spermagonia and, several days later, aecia. In leaves, spermagonia open to the upper surface and aecia to the lower surface. The aecia develop in clusters as short, open cylinders with white sidewalls enclosing a mass of powdery, orange aeciospores. The uredinial state of the rust on cereals and grasses causes yellowing and premature necrosis of leaves. *P. coronata* may overwinter as uredinia in some areas of mild climate.

P. coronata includes many varieties, or formae speciales, and hundreds of races distinguished from one another mainly by virulence on different arrays of host plants, both grasses and *Rhamnus*. On *Rhamnus* all formae speciales and races appear nearly identical. Moreover, their hybridization on *Rhamnus* results in reassortment of virulence factors. In North America the most common aecial hosts are alder, alder-leaved, California, Carolina, cascara, European, and lance-leaved buckthorns. Others affected include Avignon berry, rock buckthorn, green indigo, redberry, *R. rubra*, *R. smithii*, *R. tinctoria*, and *R. utilis*. These and other species are also affected in Eurasia.

Rusts on *Mahonia*. At least three rusts, all members of the Pucciniaceae, attack Oregon grape and other species of *Mahonia*. *Cumminsiella mirabilissima* (syn. *C. sanguinea*), presented here, produces spermagonia, aecia, uredinia, and telia on leaves. *Puccinia graminis* and *P. brachypodii* (syn. *P. koeleriae*) produce only spermagonia and aecia on Oregon grape; their alternate hosts are cereals and grasses. *C. mirabilissima* is indigenous and widespread in the West on Oregon grape; Cascades, cluster, and creeping mahonias; *M. dictyota*, *M. nervosa*, *M. piperana*, and *M. pumila*. *Berberis atrocarpa* is also reported to be a host. This rust also occurs on Oregon grape in scattered locations in the Great Plains and eastern Canada and is widespread on this plant in Europe, where the rust was apparently introduced early in the 20th century.

The spermagonial and aecial stages of *C. mirabilissima* are apparently uncommon in North America, having been clearly documented only in British Columbia on yellow swollen spots on leaves and fruits of Oregon grape. Aecial, uredinial, and later telial pustules develop on the undersurfaces of leaves. The uredinial state causes small, angular spots that become reddish brown and visible on both leaf surfaces. Uredinia are reddish brown and telia dark brown. Both uredinia and telia can be found on overwintered leaves. Urediniospores are capable of withstanding temperatures of −10° to −20°C for at least 1 month. Infection occurs on young leaves and may occur late in the season if mahonia is still growing then. The rust is capable of developing at temperatures only a few degrees above 0°C; its optimum temperature range is only 8–12°C, and it does not develop at temperatures above 20°C.

References for Puccinia caricina: 389, 623, 1688, 1759
References for Puccinia coronata: 23, 431, 489, 548, 1515, 1801, 2281
References for Cumminsiella mirabilis: 54, 387, 623, 722, 1034, 1377, 1566

A–C. Cluster-cup rust, caused by *Puccinia caricis,* on skunk currant. A. Yellow to orange rust spots on leaves. B. Close view of the upper surface of a rust spot with tips of spermagonia visible as black dots. C. The lower surface of a leaf with two aecial sori (NY, Jun).

D–F. Crown rust, caused by *P. coronata,* on common buckthorn. D. Aecial sori on leaves and a green stem. E. Spermagonia visible as orange dots in the raised center of the upper surface of a greenish yellow infected spot. F. An aecial sorus with clustered, cuplike aecia containing orange aeciospores. Aeciospores are also scattered as orange powder adjacent to the sorus (NY, May).

G–J. Rust caused by *Cumminsiella mirabilissima* on Oregon grape. G. Dark rust spots and rust-induced discoloration on 1-year-old leaves; current year's leaves are healthy. H. Lower (left) and upper surfaces of diseased leaves with rust-colored uredinial sori on the lower surface and reddish brown to dark brown spots on the upper surface. I. Magnified view of uredinia with powdery, rust-colored urediniospores. J. Magnified view of the undersurface of a leaf with dark brown to black telia (WA, Jun).

Plate 120

Ash Rust and Birch Rust (Plates 121–122)

Ash rust. *Puccinia sparganioides* (syn. *P. peridermiospora*) causes one of the most spectacular and destructive rust diseases of deciduous trees or shrubs. It occurs in the region from Nova Scotia to Saskatchewan, Montana, Texas, and Florida and has been reported from Brazil. Susceptible tree species include Berlandier, black, blue, Carolina, green, pumpkin, red, velvet, and white ash. In the South, swamp privet and Florida privet (*Forestiera* species) are affected also.

The spermagonial and aecial stages of this rust develop on ash and swamp privet in spring and early summer, and the uredinia and telia occur on various species of cordgrass (*Spartina*) during summer and autumn. Along the New England coast, this rust infects both cordgrass and another grass species that grows in marshes, *Distichlis spicata*. Epidemics on the grass hosts are scarcely noticed, but those on ash result in striking foliar distortion and browning, and death of green twigs. Severe rust in successive years weakens trees and makes them susceptible to winter damage, causes dieback, and has been reported to kill small white ash in Nova Scotia.

The disease is most severe in coastal regions where the rust fungus reproduces and overwinters on grasses in salt marshes. In spring, severe rust may develop on ash after a period of foggy weather with onshore air flow. Clouds of basidiospores rise from the marshes and move inland under these conditions. In epidemic years, severe ash rust has been noticed as far as 48 km inland.

Infection of ash occurs on leaves, petioles, and green twigs. Spermagonia appear from mid-April in the South to mid-June in the North on yellow to yellow-orange spots on upper surfaces of leaves and on chlorotic spots on petioles and stems. Clusters of aecia become abundant 10 days to 2 weeks later as bright orange spots on petioles, stems, and undersurfaces of leaves. Diseased tissues swell markedly, causing distortion of leaves, sharp bends in petioles, and elliptic to hemispheric wartlike galls on twigs. Spots on leaves may enlarge to several millimeters in diameter, and swellings with aecia on stems and petioles may grow to 2–3 cm long. Individual aecia are cuplike, with white walls around a bright orange mass of aeciospores (Plate 122).

Tissues bearing aecia, and often the surrounding zone as well, die soon after aecial maturation; thus cankers form on twigs, and petiole infections lead to withering and browning of leaves in early summer. Severely affected trees look scorched. Brown leaves drop off during mid to late summer. Rust spots on stems often remain superficial and, after aecial production, become separated from normal tissues by a layer of cork. This accounts for the wartlike appearance of the swellings in (D).

Aeciospores, windborne from ash in late spring to early summer, infect leaves of cordgrass, causing elliptic yellowish spots, or when severe, general yellowing similar to that shown in Plate 122. Severely infected leaves die prematurely. Uredinia appear in May–July, depending upon latitude. They form beneath the epidermis, oriented with the veins of the leaf as light-colored pustules 2–3 mm long that split open to reveal a mass of yellowish urediniospores. Urediniospores reinfect grass hosts, resulting in rust increase during mid to late summer. In some coastal marshes, this increase has been shown to be influenced by the degree of salinity of the water. Salt deposition on the grass apparently inhibits spore germination and multiplication of the rust. In autumn, brownish black telia replace the uredinia.

The rust fungus normally overwinters as teliospores on marsh grasses. During warm wet weather in spring, teliospores germinate and produce basidiospores. In the North, this requires only 3–4 hours at temperatures of 12–21°C. Infection of ash occurs within another 3–4 hours if moist air moves basidiospores from the marshes to wet ash surfaces. In Texas, the optimum temperature for germination of all spore types except spermatia was found to be near 25°C. Aeciospores germinated at 10–30°C, urediniospores at 15–30°C, and teliospores and basidiospores at 15–25°C.

In Connecticut, germinable urediniospores were found to survive the winter on smooth cordgrass. These spores seem potentially capable of starting new infections on the grass in spring, which would allow the rust fungus to persist in years unfavorable for infection of ash trees. Whether they actually do so, however, has not been established.

Rust caused by *Puccinia sparganioides* on white ash (CT, Jun).
A. A large tree with nearly all foliage killed by rust; green foliage of a different tree provides contrast.
B. Close view of foliage that is wilting or dead because of infection of petioles.
C. Twig cankers and petiole lesions caused by the rust fungus. The canker at right is 1 year old and the rust is no longer active in it.
D. Wartlike swellings on a green twig.
E. Multiple infections girdling petioles and a green twig.
F. Close view of aecia on swollen, bent midveins of leaves.
G. Spots caused by rust on the upper surface of an ash leaflet. An orange deposit of aeciospores from the sorus on the petiole is visible on the base of the leaf blade.

Plate 121

Birch rust. This disease, caused by *Melampsoridium betulinum*, is common and usually inconsequential in forests around the Northern Hemisphere. In nurseries, however, it is occasionally destructive, causing defoliation that predisposes young plants to winter damage or to attack of twigs by facultatively parasitic fungi. In North America this rust occurs from Newfoundland to Alaska and as far south as California, Indiana, and New Jersey. It also occurs in New Zealand and Australia where the pathogen was carried with its hosts.

M. betulinum is the best known species in a small genus of heteroecious rusts that produce spermagonia and aecia on larch needles and uredinia and telia on leaves of plants in the birch family. Other members of the group are *M. alni* and *M. hiratsukanum* with uredinia and telia on alders and *M. carpini* on hop hornbeam in the northeastern United States and on hornbeam in Europe and Japan.

The aecial stage of *M. betulinum*, apparently rare in North America, has been recorded on eastern larch in Connecticut and Wisconsin. European and Japanese larches are susceptible in Europe. Aecia arise in spring on the undersides of slightly chlorotic spots on larch needles as white blisters about 1.5 × 1 mm, 0.5 mm high. The aeciospores have nearly colorless walls but appear orange-yellow because of their contents. This aecial stage has not been reported from Canada.

M. betulinum can persist indefinitely in its uredinial stage on birches, which accounts for its occurrence in areas far removed from larch trees. Hosts of the uredinial stage of this rust in North America include dwarf, gray, paper, swamp, water, European white, and yellow birches, also *Betula nana* and *B. pubescens*. The relative susceptibilities of these species have not been recorded. Species judged somewhat resistant on the basis of European observations include Japanese white birch and sweet birch, the latter native to North America but not reported to be infected here. The Himalayan species *B. utilis* and the east Asian species *B. ermanii* are resistant.

Birch rust causes small, yellow, angular spots on leaves. Where spots are numerous and coalesce, brown lesions with yellow borders, several millimeters in diameter, may develop. Severe rust causes leaves to turn brown and fall prematurely. Yellow to orange uredinia appear in mid to late summer as tiny pustules 0.5 mm in diameter on the undersides (occasionally on the upper surface) of the leaf spots, on green stems, and sometimes on catkins. Uredinia open by a central pore from which urediniospores issue in chains. These spores are oblong-elliptic and distinctly spiny except for the apical end. Their walls are colorless, and their contents are orange-yellow. Because urediniospores reinfect birch foliage during cool weather, the rust typically increases rapidly during late summer and early autumn. Foliage wet for several hours by rain or dew is suitable for spore germination and infection.

Telia arise in autumn as waxy, yellow, flat, crustlike areas among the uredinia. Telia gradually turn dark brown and may eventually cover nearly the entire undersurface of the leaf. The teliospores are arranged in a single palisadelike layer beneath the epidermis, raising it slightly. The epidermis ruptures when teliospores in fallen leaves germinate in spring.

The annual cycle of birch rust is simple because the causal fungus can persist in its uredinial (repeating) stage. Urediniospores infect leaves, resulting in the production of new generations of these spores at intervals of 2 weeks or less during favorable (cool) weather. Some urediniospores lodge among bud scales and survive winter there, especially on low branches where axillary buds receive many spores produced on adjacent leaves or twigs. When infested buds open in spring, a few young leaves become infected, and the cycle is complete. This scheme was verified by observing urediniospores among scales of dormant buds and by observing that rust developed on leaves of birches that were collected from a nursery in winter and forced in a greenhouse.

The preference of the birch rust fungus for low temperatures may be why the rust does not intensify greatly until early autumn, even though urediniospores from spring infections are available as inoculum. Daytime temperatures in summer are likely to inhibit birch rust by preventing or greatly suppressing spore germination. In Oregon, urediniospores germinated at temperatures of 1–20°C and did so most rapidly (within 3 hours) at temperatures near 10°C. Uredinia developed at 10–20°C. When birch leaves were inoculated and held at 12°C, uredinia developed within 13–14 days. Urediniospores did not germinate at 25°C, and they were killed by exposure to 30°C for 6 hours. In Pennsylvania, although uredinia were found as early as June 20, the major period of spore release was from mid-August to mid-October.

Individual European white birch trees free from rust or only slightly affected have been observed among populations of severely rusted trees. This indicates the possibility of selecting rust-resistant birches from within susceptible species.

In Europe *M. betulinum* and *M. alni* are considered to be one species. This judgment is based upon experiments in which urediniospores from birch infected several species of alder, urediniospores from alder infected birches, and basidiospores derived from telia on either birch or alder infected larches. If this single-species concept is accepted, the rust on birches and alders should be called *M. alni*.

The possible coidentity of birch and alder rusts has not been studied in North America, but collection records here suggest that distinct fungi, or at least distinct host-specialized strains, infect the leaves of birches and alders. Rust on alder occurs in places as diverse as California, Texas, and Alberta but is unreported from eastern and western forests where birch rust is common and alders and birches grow close together. The alder rust in Alberta is *M. hiratsukanum*. That in Texas and California is reported as *M. alni*.

References for ash rust: 27, 54, 420, 1459, 2062, 2065, 2281
References for birch rust: 54, 512, 1469, 1631, 2281

A, B. Ash rust. A. Aecial sori on the undersurface of an ash leaf. B. Magnified view of aecia on a petiole (CT, Jun).

D, F. Rust caused by *Puccinia* sp., similar to the ash rust fungus, on smooth cordgrass. D. Yellow spots and general yellowing associated with uredinial sori. F. Magnified view of uredinia (CT, Jun).

C, E, G. Birch rust. C, E. Angular yellow to brown spots as viewed on upper and lower surfaces of leaves of water birch (CA, Sep). G. Slightly magnified view of uredinia (yellow) and telia (brown) on the undersurface of a leaf of European white birch (NY, Oct).

Plate 122

Melampsora Rusts (Plates 123–124)

Melampsora rusts affect plants as diverse as cottonwood, Douglas-fir, flax, and onion. Few general statements about this group can be made because they vary so much in life cycles and host preferences. Tree rusts in this group affect conifers, currants, poplars, willows, and various herbaceous plants. One species, *Melampsora farlowii* on eastern and Carolina hemlocks, is autoecious and produces only telia and basidiospores. The other well-known tree pathogens are heteroecious, although several can persist as uredinia or uredinial mycelium in the absence of the aecial host. The most important pathogens are those causing premature defoliation and growth suppression of poplars. Uredinia of these species develop on leaves and sometimes on green stems. Telia follow uredinia on poplar leaves. Spermagonia and aecia form on leaves, green shoots, and cones of conifers and, in Europe, other plants. The aecia of *M. allii-populina* occur, for example, on leaves of wild onion. Uredinial and telial states on poplar and willow are shown here, and aecial states on conifers are shown in Plate 124.

Melampsora medusae. The most widespread and important of the Melampsora rusts in North America is *M. medusae* (syn. *M. albertensis*), which occurs throughout nearly the entire range of poplars around the world. It causes the leaves of highly susceptible clones to shrivel and drop prematurely, reducing growth. Growth loss due to rust, often masked by the intrinsically rapid growth of poplars, can be detected by comparing growth of infected trees to that of noninfected ones protected by fungicides. In one test involving natural infection, the average annual growth loss (wood volume) of five clones was 31–42%, and the volume loss in highly susceptible clones ranged up to 57%.

The spermagonial and aecial states of *M. medusae* develop on Douglas-fir, eastern and western larches, and lodgepole and ponderosa pines. In addition, the following were found susceptible in tests: grand and white firs; mountain hemlock; European, Japanese, and Lyall larches; jack, Monterey, red, Scots, and sugar pines; and Sitka spruce. Douglas-fir and larches are highly susceptible; jack pine, sugar pine, and Sitka spruce are only slightly susceptible. The most common aspen and poplar hosts are trembling aspen in the West and eastern cottonwood in the East. Others affected in North America include bigtooth aspen; Great Plains, Rio Grande, and yellow cottonwoods; balsam and Lombardy poplars; and many hybrid species and clones.

The cycle and features of *M. medusae* typify host-alternating *Melampsora* species. In spring, basidiospores from telia in dead poplar leaves on the ground infect young current-year needles of coniferous hosts. Spermagonia and yellow aecia develop within approximately 2 weeks on the undersides of slightly chlorotic portions of needles. After aecial sporulation, needle discoloration intensifies, and diseased needles shrivel. Orange-yellow aeciospores, produced in chains, are dispersed by wind and infect aspen and poplar leaves. Golden yellow uredinia appear on both sides of yellow leaf spots within 2 more weeks, and their numbers increase throughout the summer during humid or wet weather as urediniospores reinfect poplar. When rust is severe, leaves become completely yellow with uredinia and spores. Rust covering half or more of the leaf surface causes leaves to shrivel and fall. In late summer, brown, crustlike telia form in place of uredinia, and the surrounding leaf tissue dies, so that irregular gray-brown lesions form. Telia arise beneath the epidermis and are composed of a single layer of brown, oblong teliospores oriented perpendicular to the plant surface and stuck together at their sides. Dormant telia, appearing black, overwinter in fallen poplar leaves and produce basidiospores in spring, completing the cycle. In areas of mild winter climate, *M. medusae* may overwinter as uredinial mycelium, allowing the perennial development of rust on poplar in the absence of a conifer host. In the North, however, host alternation seems necessary for year-to-year survival.

In the region from Texas to the Canadian border, the timing of rust appearance and buildup on poplar varies with latitude and the presence or absence of rust on conifer hosts. In the North, where larch grows, rust appears in May on larch and soon thereafter on poplar. In Texas, where coniferous hosts are absent, rust appears on poplar in April. Rust on poplar appears progressively later as one moves north from Texas or south from the range of larch to the latitude of southern Illinois. This indicates two patterns of dispersal of rust spores related to the mode of winter survival. Where only telia survive, no rust appears on poplars until after aeciospores are dispersed from conifers. Urediniospores from this region of coniferous forest are blown southward beyond the range of larches, starting successive new cycles on poplar. In warm areas where uredinia survive, new rust develops soon after leaves grow, and urediniospores spread northward, also causing successive new cycles on poplar.

Mild wet weather favors infection of both conifers and poplars. At 18°C, more than 24 hours with free moisture on needles are necessary for infection of larch needles by basidiospores, and more than 48 hours are necessary for maximum infection. Production of urediniospores on poplar is favored by humid weather and temperatures of 15–20°C.

Hot dry summer weather limits rust intensification. Perhaps for this reason, *M. medusae* in most areas tends to increase in late summer and early autumn after the host has completed most of its year's growth. Severe rust in late summer suppresses synthesis of carbohydrates needed for early growth the next year.

M. medusae enters poplar leaves through stomata. In susceptible clones it establishes a feeding relationship via haustoria. Then after several days it produces uredinia. In some resistant clones, however, infected leaf cells die soon after invasion by haustoria, and the rust fungus dies as well.

Resistance of poplars to *M. medusae* is under strong genetic control. Therefore, selection and breeding programs now emphasize rust resistance as well as rapid growth. In general, resistance is most strongly expressed in warm weather, for example at 25°C. Many races of *M. medusae* exist that vary in virulence patterns on poplar clones.

A–D, F. Rust caused by *Melampsora occidentalis* on black cottonwood. A. Young trees with foliage turned yellow by rust. Douglas-fir at the forest edge is an alternate host. B, C. Upper and lower surface views of angular yellow spots caused by rust. Areas of sporulation on the lower surface correspond to the locations of yellow spots on the upper surface. Brown spots on the lower surface are telia. D. Enlarged view of uredinia (yellow) and telia (brown) on the undersurface of a leaf (CA, Sep). F. Necrotic blotches developing where telia have formed (OR, Sep).

E, H. Rust caused by *M. epitea* on an unidentified willow. E. Magnified view of uredinia on the undersurface of a leaf. H. Yellow spots on apical leaves and browning and withering of severely diseased older leaves. The older leaves will soon drop (OR, Sep).

G. Uredinia and telia of *M. medusae* on the undersurface of an eastern cottonwood leaf. Leaf tissue where uredinia are most numerous is beginning to die (brown spots) (NY, Aug).

Plate 123

Melampsora occidentalis and *M. populnea*. In the West from Alaska to Saskatchewan, Wyoming, and California, *M. occidentalis* causes yellow leaf spot, premature browning, and death of leaves of several poplar species and premature yellowing and death of Douglas-fir needles. Black, Fremont, lanceleaf, and yellow cottonwoods and bálsam and Carolina poplars are reported hosts for the uredinial and telial states. For many years Douglas-fir was the only known coniferous host; thus this rust is often called Douglas-fir rust. When exposed to basidiospores in tests, however, seedlings of European, Japanese, Lyall, and western larches and lodgepole, Monterey, and ponderosa pines were also susceptible. White fir was somewhat susceptible, and sugar pine, western white pine, and Sitka spruce were slightly susceptible.

The annual cycle of *M. occidentalis* is as described for *M. medusae* except that winter survival of the former in its uredinial state has not been reported. As *M. occidentalis* occurs only in areas where its coniferous hosts grow, host alternation is probably necessary.

M. occidentalis produces aecia and uredinia that are macroscopically indistinguishable from those of *M. medusae*. The latter, however, has smaller aeciospores and urediniospores. The two species can be differentiated by measurements of these spores if multiple collections are available. In the West, rust of trembling aspen is usually caused by *M. medusae*, while rust on black cottonwood and other balsam poplars is usually caused by *M. occidentalis*.

Another poplar rust, *M. populnea* (syn. *M. aecidioides*), occurs on white poplar in coastal northeastern and northwestern United States, British Columbia, and Colorado. The pathogen is a European species that was probably introduced to North America with its host plant. Only uredinia and telia are known in North America.

Leaf rust of willows. Willows are hosts of several formae speciales of *Melampsora* that cannot yet be differentiated in their uredinial or telial states. All North American willow rusts are currently assigned to the group species *M. epitea*. This includes the former *M. abieti-caprearum* with aecia on fir, *M. arctica* with aecia on *Saxifraga*, *M. epitea* with aecia on hemlock, *M. paradoxa* with aecia on larch, and *M. ribesii-purpureae* with aecia on currant and gooseberry. While convenient for diagnosis, this grouping may not represent a natural species. Willow rusts have received little research attention beyond mycological description and demonstration of alternate host relationships. Most members of the group conform to the annual cycle described for *M. medusae* (Plate 123), and some apparently overwinter in the uredinial state in willow buds and twigs. Thus they have the potential to occur far from alternate hosts.

Melampsora rusts on eastern and Carolina hemlocks. *Melampsora abietis-canadensis* infects cones, needles, and green stems of eastern and Carolina hemlocks from Nova Scotia to North Carolina and Wisconsin. The uredinial and telial states of this fungus develop on leaves of aspens and poplars from Newfoundland to North Carolina, Montana, and Colorado—a region larger than that where the rust is known on hemlock. Aspen and poplar hosts include bigtooth and trembling aspens; eastern, Great Plains, and swamp cottonwoods; and balsam and white poplars.

Aecia, so numerous that they seem continuous, appear in late spring on hemlock. They develop on slightly swollen, curled, current-season shoots, mainly on the stem portion but also on undersurfaces of needles. Orange-yellow aeciospores form a powdery coating on the infected area. The shoot tip dies during summer. Cones are infected systemically in many cases, although part of the cone may remain green. Aecia develop all over the infected scales, which soon die. Infected cones become attractive to boring insects. This rust and that caused by *M. farlowii* occasionally become severe enough to cause noticeable damage to hedges and nursery plants.

The cycle of *M. abietis-canadensis* apparently conforms to the scheme described for *M. medusae* (Plate 123): overwintering of dormant telia, infection of hemlock by basidiospores from telia on fallen poplar leaves, infection of poplar by aeciospores from hemlock, and intensification of rust in the uredinial state on poplar leaves. Also, since this rust occurs far to the west of hemlocks, it perhaps survives winter as urediniospores or uredinial mycelium on or in poplar. The frequency of *M. abietis-canadensis* in comparison to other rusts on poplars is unknown because few observers bother to identify poplar leaf rusts.

M. farlowii occurs on shoots and cones of eastern and Carolina hemlocks in the same region where *M. abietis-canadensis* occurs. The two rusts cause similar symptoms but are readily differentiated because the former is autoecious and produces only telia. Its annual cycle is simple. Telia survive winter in twigs killed the previous year. Teliospores germinate in place in spring, producing basidiospores that infect hemlock needles, cones, and perhaps succulent stems. Soon a new generation of telia forms on shoots that twist, droop, defoliate, and die. These shoots remain on the plants, and the telia remain dormant until the next spring. Carolina hemlock is less susceptible than eastern hemlock to *M. farlowii*.

References for Melampsora medusae: 387, 1257, 1430, 1499, 1812, 1813, 1984, 2190, 2191, 2279, 2281

References for other Melampsora species: 54, 623, 788, 1171, 2143, 2213, 2279, 2281

A–D. Twig and cone rust of eastern hemlock caused by *Melampsora abietis-canadensis*. A, B. Views of a deformed shoot that will soon die. Numerous aecia and aeciospores give the stem a powdery yellow appearance. C. Cones killed by rust. Aeciospores in mass are responsible for the yellowish appearance. Frass on the cone at left indicates tunneling by a borer that was attracted to the diseased cone. D. Magnified view of part of a diseased shoot with aecia on the undersurface of one needle (upper left) and a mass of aeciospores on the stem (NY, Jun).

E, F. Close and magnified views of young aecia of *M. occidentalis* on current-season needles of Douglas-fir. These aecia will soon open and appear powdery with aeciospores. Distortion of needles was caused mainly by adelgids (see Johnson and Lyon, 2nd ed., Plate 49) rather than by rust (ID, Jul).

Plate 124

Pucciniastrum Rusts (Plates 125–126)

Pucciniastrum is a genus of heteroecious rust fungi that produce spermagonia and aecia on needles of fir, hemlock, or spruce and uredinia and telia on leaves or stems of various woody and nonwoody dicots. The aeciospores and urediniospores are both yellow. These fungi overwinter in needles of the aecial (coniferous) host or as telia in fallen leaves of the telial host or as perennial mycelium in stems or rhizomes of the telial host. Some species, such as *P. pustulatum* on fuchsia, may continue indefinitely in the uredinial stage without host alternation. Plate 125 presents the aecial states of two species: *P. hydrangeae* (syn. *Thekopsora hydrangeae*) on hemlock and *P. epilobii* on fir. Plate 126 shows the uredinial state of *P. pustulatum* on fuchsia and the telial state of *P. goeppertianum* (syn. *Calyptospora goeppertiana*) on huckleberry.

Hemlock-hydrangea rust. *Pucciniastrum hydrangeae* causes rust on eastern and Carolina hemlocks, wild hydrangea, and panicle hydrangea. Aside from descriptions of the fungus and records of occurrence, the pathogen and disease are little known. This rust occurs from New York to Illinois, Arkansas, and Georgia. Within that region it has been reported more widely on hydrangea than on hemlock.

Inconspicuous spermagonia develop in spring on current-season needles of hemlock, mostly on the undersides, and are followed there in late spring to early summer (late June in New York) by prominent cylindric, white aecia arranged along the entire length of the needle on the white rows of stomata. The aecia are 160–220 μm in diameter and 1.0–1.5 mm long. They split open from the tips and sometimes along the sides as well. The aeciospores are orange-yellow in mass and finely verrucose and have an elongate smooth spot that distinguishes them from aeciospores of *P. vaccinii* (below). Diseased needles turn chlorotic to distinctly yellow by the time aecia form. Needles die, shrivel, and drop soon after aeciospores are liberated. Thus except for missing needles, hemlock is asymptomatic most of the year.

The aeciospores infect hydrangea leaves, and uredinia develop within 2 weeks. Uredinia appear on the undersides of indefinite yellowish spots as tiny blisters barely protruding above the epidermis, opening by an apical pore, and releasing urediniospores that are orange-yellow in mass. Urediniospores cause repeating rust cycles on hydrangea leaves.

Flat, reddish brown telia develop in the epidermis of both leaf surfaces of hydrangea in late summer and autumn. The teliospores are brown, globoid, and several-celled with vertical walls and are arranged in a single layer. In spring, they germinate in place, liberating basidiospores that infect hemlock needles.

Hemlock-blueberry rust. *Pucciniastrum vaccinii* (syn. *Thekopsora vaccinii*) also causes yellowing and premature shedding of hemlock needles. This rust produces aecia on Carolina, eastern, mountain, and western hemlocks. Its uredinia and telia develop on various members of the Ericaceae (blueberries, azalea, etc.). On blueberries and related plants, *P. vaccinii* causes yellow leaf spot and leaf blight. This rust occurs across Canada and the northern United States, and in the East it is found on ericaceous hosts as far south as Florida and Arkansas. *P. vaccinii* is a group species (biologically distinct fungi with the same name); thus descriptions of its aecial state vary. The aecia of *P. vaccinii*, like those of *P. hydrangeae*, develop along lines of stomata on undersurfaces of needles. Aecia of the western form of the fungus resemble uredinia (Plate 126) in being tiny yellow conical structures that open by a central pore. On the other hand, an eastern form previously known in its aecial state as *Peridermium peckii* has cylindric aecia similar to those of *P. hydrangeae*.

Fir-fireweed rust. *Pucciniastrum epilobii* causes fir-fireweed rust. The fungus occurs around the Northern Hemisphere and in New Zealand. In North America it occurs across Canada and as far south as Tennessee and New Mexico. It produces spermagonia and aecia on the undersides of the current year's fir needles and causes bending or curling, yellowing and death of these needles on alpine, balsam, grand, noble, Pacific silver, and white firs. Occasionally aecia form also on cone scales of alpine fir. Severe foliar browning of balsam fir and alpine fir has been caused by this rust in localities where fireweed became abundant after fire or logging. Fortunately, the intensity of infection varies from year to year, and older needles are not susceptible. Thus fir trees escape lethal defoliation.

Aecia arise in early summer to midsummer along the lines of stomata on the undersides of the needles in indefinite yellowish spots or along the length of the needle. Aecia are cylindric or nearly so, white, up to 0.3 mm in diameter, and approximately 1 mm high. They produce yellow verrucose aeciospores in chains. The aeciospores have a smooth spot along one side. This species on firs is easily confused with *P. pustulatum*, which is considered by some experts to be synonymous with *P. epilobii*, and with *P. goeppertianum* (Plate 126). *P. epilobii* and *P. pustulatum* differ, however, in several morphologic characters (the latter has larger aecia) and in hosts of the uredinial state. *P. goeppertianum* has no uredinial state, and its aecia on current-year fir needles do not mature until autumn.

Aeciospores of *P. epilobii* infect several species of fireweed (*Epilobium*), notably *E. angustifolium*, causing yellow-brown leaf spot or leaf blight. Uredinia, similar to those shown in Plate 126, form on the undersides of fireweed leaves about 2 weeks after infection. Brown subepidermal, crustlike telia follow the uredinia, and the fungus overwinters as teliospores in dead leaves of fireweed. In spring, teliospores germinate and produce basidiospores that infect fir needles. *P. epilobii* is probably also capable of overwintering in the uredinial state on fireweed. Uredinia have been observed on fireweed shoots as early as mid-April, long before aeciospores mature on fir trees. Such shoots must have become infected while in the bud or underground the previous year.

A–D. Rust caused by *Pucciniastrum hydrangeae* on eastern hemlock. A, B. Views of upper (A) and lower (B) surfaces of a small branch with scattered diseased current-season needles. White, cylindrical aecia are visible on yellowish needles. C. Mature aecia extending from the lower surface of a hemlock needle. Numerous aeciospores are responsible for the yellowish color. D. A diseased needle dying back from the tip (NY, Jun).

E, F. Rust caused by *Pucciniastrum* sp., probably *P. epilobii*, on current-season needles of grand fir. E. Diseased needles are chlorotic or have indefinite chlorotic bands and are bent slightly. F. Close view of the undersurfaces of diseased needles with chlorotic bands and young aecia of the rust fungus (ID, Jul).

Plate 125

Fir-blueberry rust. *Pucciniastrum goeppertianum* (syn. *Calyptospora goeppertiana*) causes fir-blueberry rust, also known as blueberry witches'-broom rust. This fungus occurs around the Northern Hemisphere. It is found coast to coast in Canada and the northern United States, and in western mountains it extends south to Mexico. It causes yellowing and premature shedding of fir needles, and witches'-brooms on various species of bilberry, blueberry, cranberry, and huckleberry. Hosts for the aecial stage include alpine, balsam, grand, noble, California red, Pacific silver, and white firs. Damage to these plants is usually inconsequential, but severe rust on balsam fir grown for Christmas trees occasionally causes such needle loss that crowns appear thin and tree value is reduced. Fir-blueberry rust has suppressed growth of young balsam fir under experimental conditions, but this rust ordinarily affects such a small proportion of foliage that the impact on growth is negligible. When severe rust on fir foliage is encountered, the observer should be aware that any of several rusts may be involved.

Spermagonia of *P. goeppertianum* develop in summer on both surfaces of the current season's needles as tiny dome-shaped yellow pustules that turn brown with age. Sometimes, however, spermagonia do not form. Aecia mature on the undersurface in late summer or autumn. In the West, aecia also develop on 2nd-year needles in spring and early summer. Thus this rust often overwinters in diseased needles. The aecia are cylindric or columnar and produce yellow aeciospores in chains. Aeciospores infect blueberries and related plants beginning in late summer and continuing the next spring and early summer.

P. goeppertianum has no uredinial stage. Hosts for the telial stage include bog bilberry, lowbush and northern highbush blueberries, mountain cranberry, grouseberry, evergreen huckleberry, tall red huckleberry, thinleaf huckleberry, and blue whortleberry. Infection leads to extensive perennial growth of the fungus within phloem of stems, to swelling of current season's stems, and to formation of witches'-brooms. Leaves within brooms are stunted, and the branches are swollen, spongy, and distinctively yellowish to reddish brown in contrast to the greenish color of normal twigs and young branches. The brooms produce no fruit, which causes occasional economic damage in fields of cultivated blueberries in the East.

Telia mature in spring as a conspicuous reddish brown layer in the epidermis, encircling the swollen stems of shoots colonized the previous year. The teliospores are one- to four-celled with dark brown walls. They germinate in place under wet conditions during the period of shoot growth of fir trees, and they liberate basidiospores that infect current-season fir needles. Peaks of basidiospore discharge occur just after rain. Discharge occurs at 10–25°C and is most rapid near 18°C. At this temperature under moist conditions, 18 hours is sufficient for basidiospore production, germination, and infection of fir needles. Balsam fir shoots are most susceptible during the first 2 weeks of growth, and susceptibility declines rapidly after 3–5 weeks.

P. goeppertianum becomes perennial in rhizomes of lowbush blueberries. Although blueberry fields are burned at intervals of 3 years as a regular cultural measure, this does not kill the rust fungus in rhizomes.

In the absence of a repeating (uredinial) stage, host alternation of *P. goeppertianum* would seem necessary. Observers in both Europe and North America, however, have noted abundant broom rust on *Vaccinium* species far removed from fir trees and have suggested that basidiospores may be capable of infecting the host on which they originate. This question has not been resolved.

Fuchsia rust. This rust is caused by *Pucciniastrum pustulatum*, which occurs around the world on many species and cultivars of fuchsia and other members of the Onagraceae: *Epilobium* (fireweed), *Clarkia*, and *Godetia*. The fungus is classified by some experts as a host-specialized form of *P. epilobii* (Plate 125). It apparently infects the same array of fir species as does *P. epilobii*, but separate records for the two fungi are not available. Similarly, the geographic range of this rust on wild plants in North America is not distinguished from that of *P. epilobii*. *P. epilobii* and *P. pustulatum* produce aecia on current year's fir needles in late spring and early summer. Their aecia resemble those of *P. goeppertianum* but mature many weeks earlier each year. *P. pustulatum* differs from *P. epilobii* in being able to infect the fireweed species *E. adenocaulon* and *E. glandulosum* and unable to infect *E. angustifolium*. Also, it overwinters not only as teliospores in dead fireweed leaves but also in the uredinial state on fuchsia and in the rosettes of fireweed species. This rust can maintain itself indefinitely in the uredinial state, independent of alternation to fir trees. Fuchsia rust thus occurs in warm regions far from fir trees. It is common and locally severe on fuchsia in California every summer, and it breaks out sporadically in greenhouses around the world as the result of transportation of the pathogen in diseased cuttings.

Uredinia of *P. pustulatum* develop in small irregular groups on the undersurfaces (sometimes later on the upper surfaces as well) of pale yellow to reddish spots on fuchsia or fireweed leaves. These uredinia are minute yellow conic or hemispheric swellings, 0.1–0.2 mm in diameter, which at first open by an apical pore, then break open widely. They become distinctly yellow and powdery in appearance as urediniospores are released. On fuchsia the pale or yellow spots may progress rapidly to tan or brown necrotic spots or blotches. In such cases the tiny uredinia mature only at the edges of necrotic areas and may easily be overlooked (E). Severely diseased leaves dry and drop. If the combination of rust strain and fuchsia cultivar is somewhat more compatible, uredinia may appear all over the leaves, but such leaves also die and fall. Old leaves are more severely affected than young ones. Low light intensity favors this rust.

Telia of *P. pustulatum* are tiny flat, brown, subepidermal structures, scarcely wider than uredinia. They form on leaves of fireweed and have developed on rusted fuchsia leaves in a refrigerator but have not been reported to occur on fuchsia in nature.

References for hemlock-hydrangea, hemlock-blueberry, and fir-fireweed rusts: 2, 54, 315, 389, 425, 434, 834, 835, 1759, 2143, 2213, 2279, 2281

References for fir-blueberry rust: 54, 434, 834, 835, 839, 1059, 2068–70, 2072, 2073

References for fuchsia rust: 54, 434, 834, 835, 839, 1167, 1269, 1921, 2143, 2213, 2281

A–C. Fir-blueberry rust, caused by *Pucciniastrum goeppertianum*, on evergreen huckleberry. A. A typical witches'-broom with swollen, reddish brown twigs and stunted leaves. B. Normal branch for comparison. C. Close view of a diseased branch along which the rust has deformed all but one shoot (OR, Jul).

D–G. Fuchsia rust, caused by *P. pustulatum*. D. Spreading lesions caused by the rust fungus. E. Close view of the lower surface of a diseased leaf with tiny, inconspicuous uredinia (arrow) at one edge of a lesion. F, G. Magnified views of uredinia. Immature uredinia (G, left) are blisterlike (CA, Jul).

Plate 126

Broom Rust of Fir and Fir-Fern Rusts (Plate 127)

Broom rust. *Melampsorella caryophyllacearum* causes broom rust, or yellow witches'-broom, of firs. This disease occurs around the Northern Hemisphere. It is common and conspicuous where firs grow in cool, moist places across North America, and it extends beyond the range of firs on its alternate hosts (chickweeds). The pathogen has also been introduced to South America. Conifer hosts in North America are alpine, balsam, grand, noble, California red, Pacific silver, and white firs. The disease reaches greatest severity on alpine fir in the northern Rocky Mountains, where in some localities 70–95% of the trees may be diseased, and individual trees may bear several dozen brooms.

This rust fungus infects fir buds in spring, invades the young shoots, becomes perennial, and causes witches'-brooms. It grows slowly and causes only slight elongate swellings on diseased shoots during the 1st season. The next spring, however, buds on the diseased twig produce upright shoots that are thicker and shorter than normal. Needles on these shoots are stunted, thickened, pale green, and arranged in a helix. This aberrant growth during several years results in a witches'-broom. Systemic intercellular mycelium of the pathogen is abundant in needles but sparse in stems of brooms.

Brooms may arise on branches or mainstems. Old brooms such as that shown on California red fir (C) occasionally grow to lengths approaching 2 meters, but most remain less than half that size. Brooms bear only current year's needles, which are initially pale green but become distinctly yellow during summer. In autumn, needles die and drop, leaving the broom devoid of foliage during winter. A new crop of pale green needles forms in spring. The infected branch at the base of the broom becomes swollen into a spindle-shaped to globose gall on which cankers may develop. Trees with multiple brooms often decline and die, and the growth of surviving trees is suppressed. In the West, wind or snow often causes trunk breakage at rust cankers.

Spermagonia and aecia develop on all needles of a broom. Orange spermagonia arise between the cuticle and epidermis in spring. Light yellow, blisterlike aecia develop in summer on the undersurfaces along the lines of stomata and produce orange-yellow aeciospores in chains. The spore walls are verrucose with rodlike warts.

The aeciospores are windborne and infect leaves of several species of chickweed (*Cerastium* and *Stellaria*). Severe rust on these plants causes leaf or shoot blight. Uredinia may develop on both sides of the leaves as tiny orange-red pustules that open by a central pore, releasing orange-yellow urediniospores that have spiny walls. These spores are windborne and cause repeating cycles on chickweed. *M. caryophyllacearum* overwinters as systemic perennial mycelium in perennial species of chickweed. Thus the pathogen persists in the absence of fir trees in some areas. In spring it grows up in chickweed shoots and extends into leaves, where telia and later uredinia mature in spring. Telia cause the leaves to appear orange. Teliospores are arranged in a layer one cell thick within the lower epidermis. They germinate in place, producing basidiospores that infect fir. This completes the 2-year life cycle of the fungus. Urediniospores from the same leaves that liberated basidiospores cause more rust on chickweed.

Aeciospores and urediniospores of *M. caryophyllacearum* are capable of germinating at temperatures of 5–30°C. They germinate most rapidly at 20–25°C. Free moisture is required.

Fir-fern rusts. Rust fungi that infect fir needles include species of *Uredinopsis, Milesina, Hyalospora, Melampsora, Melampsorella,* and *Pucciniastrum.* For diagnostic purposes, these fall into two groups: those with white aeciospores (*Uredinopsis* and *Milesina,* with which we deal here) and those with yellow aeciospores (all others). The white-spored rusts cannot be differentiated in their aecial states on firs. All are characterized by white cylindric or tongue-shaped aecia that form on green to yellow needles. All produce their uredinia and telia on ferns, hence the common name, *fir-fern rust.* Those with acutely pointed, spindle-shaped urediniospores and subepidermal telia are classified in *Uredinopsis. Milesina* species produce telia within the epidermis, and their urediniospores are broadly rounded at the apex or, if pointed, are broadest above the middle.

At least 10 species of *Uredinopsis* and 4 of *Milesina* occur on fir in Canada and the United States. Each rust species is able to infect several species of fir. Needle browning and (or) defoliation by these rusts occasionally causes economic damage in Christmas tree plantations of grand, red, and white firs in the West and balsam fir in the East. All species infect current-year needles, and most species produce aecia on these needles in summer to early autumn. Diseased needles then die and drop. Three species—*U. hashiokai, U. pteridis,* and *M. pycnograndis*—become perennial in fir needles, producing aeciospores on the same needles each spring for several years until the needles die. The aeciospores infect ferns, resulting in pale green to yellow spots that later become necrotic and are typically bounded by veins. Severe rust may blight the fern.

Fir-fern rust fungi pass the winter by various combinations of four survival modes: as dormant telia in dead fir fronds, as aecial mycelium in living fir needles, as uredinial mycelium in living fir fronds, and on fern as resting spores called amphispores that reinfect fern in spring. Amphispores are essentially urediniospores specialized for winter survival. The first two modes, singly or together, lead to production of spores that infect the alternate host—basidiospores from telia on fern to fir and aeciospores from fir to fern. The second two modes allow these rust fungi to persist indefinitely on ferns. Telia are still produced, leading to infection of fir trees where there are present, but annual infection of fern in the absence of fir is assured.

Several of the white-spored fir rusts and their alternate hosts and regions of occurrence are listed below.

M. fructuosa: wood fern, east.
M. laeviuscula: polypody, west.
M. marginalis: wood fern, east.
M. pycnograndis: polypody, east; causes brooms on fir.
U. atkinsonii: wood fern, east.
U. ceratophora: bladder fern, east.
U. hashiokai: bracken fern, west.
U. longimucronata: lady fern, transcontinental.
U. mirabilis: sensitive fern, east.
U. osmundae: flowering fern, east.
U. phegopteridis: oak fern, transcontinental.
U. pteridis: bracken fern, mainly west.
U. struthiopteridis: ostrich fern, transcontinental.

References for broom rust: 623, 1440, 1507, 1508, 2164, 2281
References for fir-fern rusts: 555–57, 589, 1240, 2281

A–D, F–H. Broom rust, or yellow witches'-broom, of fir, caused by *Melampsorella caryophyllacearum.* A. A typical compact broom on balsam fir, with foliage only on the current growth of upright twigs. Foliage early in the season is pale green. B, D. Young aecia on the undersurfaces of stunted, abnormally thickened leaves of the broom shown in (A) (NY, Jun). C. A large broom, partly dead, in the crown of a California red fir (CA, Jul). F. A broom on balsam fir in late summer. Foliage has turned yellow and is beginning to die. G. Close view of a twig from the broom in (F), with stunted, thickened, yellow needles on only the youngest segments of upturned twigs. H. Scars where aecia were produced earlier in the season on the lower surface of a needle from the broom in (F) (F–H: Que, Aug).

E, I. Needle rust caused by *Uredinopsis* sp. on balsam fir. E. Yellow diseased needles contrast with normal green foliage. I. White cylindrical aecia along the lines of stomata on the undersurface of a needle (Ont, Aug).

Plate 127

Chrysomyxa Rusts of Spruce (Plate 128)

Rust fungi of the genus *Chrysomyxa* infect the current year's spruce needles or cones wherever spruces grow naturally in the Northern Hemisphere. Eleven species of *Chrysomyxa* occur on spruce in Canada and the United States, of which 10 are heteroecious, producing spermagonia and orange-spored aecia on spruce and uredinia and telia on various plants, mainly members of the heath family (Ericaceae). Each of these fungi seems capable of attacking several spruce species but only one or a few closely related dicot species. The 11th fungus, *C. weirii,* is autoecious on spruce. Information about *Chrysomyxa* species (including one member not known to be associated with spruce) is summarized here, and diseases caused by two species are illustrated.

Spermagonia and aecia of *Chrysomyxa* develop beneath the epidermis. Aecia are white and blister- or bubblelike or tongue shaped. They produce chains of orange-yellow aeciospores with verrucose walls. The aeciospores infect dicot hosts. Uredinial pustules form on the upper or lower surfaces of dicot leaves and rupture irregularly to reveal chains of orange-yellow urediniospores that resemble the aeciospores. Telia mature on overwintered persistent leaves and break through the epidermis as cushionlike or conic structures with teliospores also arranged in chains. Teliospores germinate in place, producing basidiospores that infect spruce. Host alternation is optional in several species that persist as mycelium in dicot hosts. Their urediniospores reinfect the same plants each summer.

The most widely noticed disease in this group is broom rust, or yellow witches'-broom, caused by *C. arctostaphyli*. This occurs across Canada and the northern United States on black, blue, Engelmann, Norway, red, Sitka, and white spruces. In the mountainous West it occurs from Alaska into Mexico. Brooms commonly grow to lengths of 2 meters. Spruces with numerous large brooms often have dead or broken tops, grow slowly, and die prematurely. Spermagonia and aecia arise in summer on all needles of the broom. Spermagonia appear as tiny reddish eruptions that emit a putrid odor discernible many meters from the source. Aecia develop in mid to late summer, and the needles are shed in autumn, leaving the broom devoid of foliage in winter. New pale green needles emerge in spring and turn yellow with the formation of spermagonia. The rust fungus overwinters not only in brooms but in leaves of bearberry where telia mature in spring on the undersurfaces of reddish brown spots. Since there is no uredinial state, host alternation is necessary.

C. chiogenis is an uncommon rust fungus that lacks a spermagonial stage. It produces aecia on current-year needles of black and white spruces in eastern Canada and the northeastern United States. The alternate host is creeping snowberry (*Gaultheria hispidula*), on which uredinia and telia occur in the East. *C. chiogenis* also occurs in its uredinial (but not telial) state on creeping snowberry in British Columbia. This rust is unknown on spruce in the West.

C. empetri causes premature needle loss from Engelmann, red, Sitka, and white spruces in eastern and western Canada and the northeastern United States. Uredinia and telia develop on upper surfaces of leaves of black crowberry, causing leaf blight.

C. ilicina produces uredinia and telia on the undersurfaces of leaves of American holly in Tennessee and West Virginia. This rust has not been associated with spruce.

C. ledi infects needles of black, blue, Engelmann, red, Sitka, and white spruces, causing yellowing and defoliation. Several varieties of *C. ledi* are recognized that differ in their alternate hosts. *C. ledi* var. *ledi* occurs on spruce across North America. Its uredinia and telia develop on undersurfaces of 2nd-year leaves of crystal tea (*Ledum palustre*), Labrador tea (*L. groenlandicum*), and *L. glandulosum*, causing brown leaf spots. *C. ledi* var. *cassandrae* causes brown leaf spot on cassandra. It occurs across the continent on this plant but on spruces is recorded only from Manitoba and Minnesota eastward to Nova Scotia and Connecticut. *C. ledi* var. *rhododendri* produces uredinia (but no telia) on Lapland rosebay and on various cultivated introduced rhododendrons across Canada and in the northwestern United States. This variety has not been detected on spruce in North America. *C. ledi* var. *vaccinii* produces uredinia and telia and causes purple-brown leaf spot of red huckleberry in British Columbia. Its spermagonial and aecial states are unknown but are suspected to occur on spruce because the rust occurs only where spruce is abundant.

C. ledicola causes one of the most prominent rust diseases in the North American boreal forest. This fungus is common on black, blue, Engelmann, red, Sitka, and white spruces across Canada and the northern United States. Its spermagonia and white aecia with orange spores develop on all faces of pale green to yellow current-year needles and rarely on cones. Diseased needles are shed in autumn. Epidemics are occasionally so severe that needle yellowing is conspicuous from aircraft. Diagnosis with a microscope is usually necessary, however, because *C. ledi* may also cause conspicuous yellowing of spruce foliage in southern Canada and the northern United States. Uredinia and telia of *C. ledicola* develop on upper surfaces of leaves of crystal tea and Labrador tea, causing brown leaf spot.

C. monesis becomes systemic in cones of Sitka spruce in British Columbia, destroying the seed and causing cones to open prematurely. This species and *C. pirolata* occasionally disrupt seed collection for forestry purposes. The uredinia and telia of *C. monesis* occur on leaves of single-delight (*Moneses uniflora*) from Washington to Alaska.

C. piperiana infects needles of Sitka spruce and leaves of California rosebay and cultivated rhododendrons in Oregon and California. On rhododendrons it occurs as far north as British Columbia. It causes premature defoliation of spruce and yellow to brown leaf spot of rhododendron.

C. pirolata infects cones of black, blue, Engelmann, Norway, red, Sitka, and white spruces across North America as far south as Colorado. This rust kills seeds and causes cones to turn brown and open prematurely. Seeds that do mature in infected cones tend to be lighter and to germinate less well than normal. In occasional years and localities, cone rust has destroyed the majority of the spruce seed crop. Uredinina and telia develop from systemic perennial mycelium in various species of shinleaf (*Pyrola*) and in single-delight.

C. roanensis occurs in North Carolina and Tennessee on catawba and Piedmont rhododendrons and on red spruce.

C. weirii is autoecious on black, blue, Engelmann, red, Sitka, and white spruces. Its orange telia form on 1st- and 2nd-year needles. Spermagonial, aecial, and uredinial states are unknown. This rust occurs across Canada and also in northwestern, northeastern, and Appalachian states.

C. woroninii infects opening buds, produces aeciospores on stunted needles, and then causes shoot blight of black spruce and white spruce in northwestern and northeastern forests of Canada. It produces telia but not uredinia from perennial mycelium in crystal tea.

References: 54, 389, 425, 1506, 1507, 1685, 1686, 1811, 1934, 2281

A, C–E. Needle rust caused by *Chrysomyxa ledicola*. A. A young black spruce with nearly all the current year's foliage diseased (Que, Aug). C. Close view of needles with aecia on red spruce (NY, Aug). D. Enlarged view of aecia on black spruce (Que, Aug). E. Dying foliage invaded by secondary molds after dispersal of aeciospores (Que, Aug).

B. A rust broom about 2 meters tall caused by *C. arctostaphyli* on blue spruce (CO, Jun).

Plate 128

Needle Rusts of Pine (Plate 129)

Needle rusts of pine in North America are caused by more than 20 species of *Coleosporium*. With but few exceptions these fungi produce spermagonia and aecia on pine needles and uredinia and telia on leaves of various monocotyledonous and dicotyledonous plants, especially members of the daisy, or aster, family (Compositae). Coleosporium rusts occasionally destroy enough foliage to disfigure or retard the growth of young pines, but usually they cause little damage.

The generalized cycle for pine needle rusts requires 1 year. Basidiospores from the telial host infect pine needles of various ages in late summer to early autumn, causing no symptoms that season. The pathogen overwinters in living needles. By the next spring, chlorotic to yellow spots have appeared, and spermagonia develop beneath the epidermis in the spots. White, bubblelike, columnar or tongue-shaped aecia follow the spermagonia and split open, exposing orange aeciospores. Aecia disappear by the end of summer, leaving tiny scars on yellow to brown spots or bands on green or partly yellowed needles. Aeciospores, dispersed by wind, infect leaves of the alternate host. Within several days, tiny orange uredinial pustules develop on the undersides of the leaves of this plant. Urediniospores cause repeating cycles and intensification of rust on the angiosperm host during summer. Telia develop subepidermally among the uredinina in late summer and autumn. Telia appear as dark crusts or cushions somewhat larger than uredinia and are composed of a single layer of obovoid teliospores. Teliospores germinate in place without a rest period, producing basidiospores that infect pine needles and complete the cycle. Some species of *Coleosporium* persist 2–3 years as mycelium in living pine needles, and some also survive winter in the uredinial state on perennial herbaceous hosts.

The best known rust in this group is *C. asterum* (*C. solidaginis*), which occurs coast to coast in Canada and the United States, southward into Mexico, and also in China and Japan. It occasionally causes severe browning and loss of needles on low branches of young trees. Many two- and three-needle pines—including Austrian, Japanese black, jack, loblolly, lodgepole, longleaf, mugo, pitch, ponderosa, red, Scots, shortleaf, table-mountain, and Virginia pines—are susceptible. Needles of all age classes become infected. Spermagonia appear as reddish orange dots on yellow spots or bands. The aecia develop about a month later as flat-sided white columns 0.5–1 mm high, which break open at the sides, releasing aeciospores. Many 1- and 2-year-old needles persist although infected, and perennial rust mycelium in them produces aecia during 1 or 2 more summers.

Uredinial pustules, initially bright orange but fading to yellow, develop 10–15 days after infection of wild or cultivated species of *Aster* (aster), *Solidago* (goldenrod), and several other members of the Compositae: *Callistephus* (China aster), *Erigeron* (daisy fleabane), *Grindelia* (gumweed), *Haplopappus*, *Heterotheca*, and *Machaeranthera*. Severe rust causes leaf blight of aster and goldenrod. Reddish brown crustlike telia, about 0.5 mm in diameter when dry, develop among the uredinia and produce basidiospores in early autumn. This rust overwinters not only as mycelium in pine needles but also, except in areas with harsh climate, as uredinial mycelium in the rosettes of perennial herbaceous hosts. All spore types germinate most rapidly at temperatures near 20°C. The three infectious spore types of some strains of the rust fungus are capable of starting new infections within 24 hours under wet conditions at this temperature.

C. asterum is apparently a group species or at least a complex of host-specialized races. An eastern and a western form are differentiated on the basis of microscopic characters of aecia and aeciospores. In the Great Lakes region, three races were differentiated on the basis of differing aster and goldenrod hosts. These races also differed in optimum time and temperature for urediniospore germination (45 hours at 24°C for one race versus 20 hours at 20°C for another).

Pine-tarweed rust, caused by *C. pacificum* (syn. *C. madiae*), is common in California and Oregon, where it attacks Coulter, Jeffrey, and Monterey pines. Its conspicuous spermagonia and tongue-shaped aecia occur on all needle surfaces in February–May, depending on locality and year. Aecia have been found as early as January. The aeciospores are uncommonly large (40–45 × 25–29 μm). Uredinia and telia develop on species of *Gaillardia*, *Hemizonia*, *Lagophylla*, *Madia* (tarweed), and *Tagetes* (marigold), all in the Compositae. This rust occurs on tarweed as far north as British Columbia.

Not all *Coleosporium* species that infect pines in the United States and Canada can be listed here. Well-known species and their hosts, geographic distributions, and special characteristics include:

C. apocynaceum on loblolly, longleaf, and slash pines. The alternate hosts are *Amsonia* species.

C. campanulae on jack, pitch, red, Scots, and Virginia pines in the eastern half of the United States. Alternate hosts are *Campanula* species (bluebell), *Lysimachia* species, and *Specularia* species. This rust occurs on bluebell in western Canada, but is unknown on pine there.

C. crowellii on limber and pinyon pines in the Southwest. This autoecious fungus produces only telia from which basidiospores are dispersed that reinfect pine needles.

C. delicatulum on several two- and three-needle pines from Maine to Florida, Texas, and Kansas. Alternate hosts are species of goldenrod.

C. helianthi on several two- and three-needle pines from New York and Georgia west to Wisconsin and Oklahoma. Alternate hosts are Jerusalem artichoke and several species of *Helianthus* (sunflower), *Parthenium*, and *Silphium*.

C. inconspicuum on longleaf, shortleaf, and Virginia pines in the region from Maryland and Ohio to Georgia. Aecia are longer and narrower (narrowly tongue shaped) than those of most other rust fungi in this group. Uredinia and telia form on species of *Coreopsis*.

C. ipomoeae on Chihuahua pine in the Southwest and on several two- and three-needle pines in the East. Alternate hosts are species of *Ipomoea* (morning-glory) and other members of the Convolvulaceae.

C. jonesii on pinyon pine and singleleaf pinyon. Uredinia and telia form on *Ribes* species, especially *R. inebrians*.

C. laciniariae on loblolly, longleaf, and pitch pines from New Jersey to Florida and Arkansas. Uredinia and telia develop on *Liatris* species (gay-feather).

C. minutum on loblolly and spruce pines in the Southeast. The alternate host is *Forestiera* (swamp privet).

C. pinicola on Virginia pine in the Southeast. This autoecious rust produces orange telia and then basidiospores on 1-year-old yellow needles in spring. The basidiospores infect new needles.

C. vernoniae (syn. *C. elephantopodis*) on many species of two- and three-needle pines from Massachusetts and Nebraska to Florida and Central America. Alternate hosts are species of *Elephantopus* (elephant's-foot) and *Vernonia* (ironweed).

C. viburni on jack pine from Quebec to Manitoba. The alternate hosts are several species of *Viburnum*.

References: 227, 228, 432, 1198, 1398, 1434, 2281

A, E. *Coleosporium asterum* on red pine. A. Aecia on yellow spots on 1-year-old needles. E. Enlarged view of aecia from which the white coverings have been mostly removed by wind and abrasion (NY, Jun).

B, D. Needle rust of Monterey pine caused by *C. pacificum*. B. Foliage with yellow bands. D. Close view of rust lesions with the remains of aecia being colonized by secondary molds (CA, Apr).

C, F. Uredinia and telia of *C. asterum* on undersurfaces of goldenrod leaves. C. Magnified view of yellow uredinia and rust brown telia. F. Rust-affected leaves (NY, Sep).

G. Tongue-shaped aecia of *Coleosporium* sp. on loblolly pine (MS, May).

Plate 129

Stem and Cone Rusts of Pine (Plates 130–136)

The stem and cone rusts of pines, caused by *Cronartium* and *Endocronartium* species, are among the most widely recognized and destructive tree diseases and certainly are the most important rusts of forest trees. Approximately 15 species of *Cronartium* are recognized worldwide, and most of them occur in North America. The spermagonial and aecial states occur on pines, associated with galls, cankers, and dieback of branches and trunks or deformity and death of cones. These diseases, also known as blister rusts, kill and deform so many trees that they act as strong agents of selection in forests. The uredinial and telial states of the causal fungi develop on various dicotyledonous plants in the Fagaceae (chiefly oaks), Myricaceae (sweet gale, sweetfern), Santalaceae (*Buckleya,* comandra, northern comandra), Saxifragaceae (currants and gooseberries), and Scrophulariaceae (lousewort, paintbrush, and relatives). Cronartium rusts on uredinial and telial hosts may cause yellow leaf spots or necrotic blotches and premature defoliation, but this is insignificant apart from the relationship to destructive diseases of pines. Representative uredinial and telial states of the group are shown here. Aecial states, all in the form-genus *Peridermium,* are presented in Plates 131–136. Diseases caused by autoecious rust fungi believed to have evolved from *Cronartium* are also presented. The following occur in North America.

C. appalachianum causes Appalachian blister rust on Virginia pine in the Appalachian Mountains. The telial host is the parasitic plant *Buckleya.*

C. arizonicum causes limb rust on ponderosa pine in the Southwest. Telial hosts are species of *Castilleja* (paintbrush).

C. coleosporioides causes stalactiform rust, also known as cow wheat rust, on various two- and three-needle pines across Canada and in the north-central and western United States. It also causes limb rust on Jeffrey and perhaps ponderosa pines in the West. Telial hosts are *Castilleja, Melampyrum* (cow wheat), *Orthocarpus* (yellow owl's clover), *Pedicularis* (lousewort), and *Rhinanthus* (yellow rattle), all of which are members of the Scrophulariaceae.

C. comandrae causes comandra rust on several species of two- and three-needle pines across Canada and in eastern, south-central, and western states. Telial hosts are *Comandra* (comandra, or bastard toadflax) and *Geocaulon* (northern comandra).

C. comptoniae causes sweetfern rust on several species of two- and three-needle pines across Canada and the northern United States. Telial hosts are *Comptonia* (sweetfern) and *Myrica* (sweet gale).

C. conigenum causes conigenum rust (southwestern cone rust) on two- and three-needle pines in the Southwest and southward into Central America. Telial hosts are oaks.

C. occidentale causes pinyon blister rust on pinyon pine and singleleaf pinyon in the West. Telial hosts are *Ribes* species (currants and gooseberries).

C. quercuum causes pine-oak gall rust on many two- and three-needle pines, mainly in eastern and central North America. Telial hosts are oaks.

C. quercuum f. sp. *fusiforme* causes fusiform rust on two- and three-needle pines in the Southeast. Telial hosts are oaks.

C. ribicola causes white pine blister rust. Telial hosts are currants and gooseberries and, for a strain that occurs in Japan and eastern Asia, lousewort and paintbrush. This rust occurs throughout most of the North American range of white (five-needle) pines.

C. strobilinum causes southern cone rust on longleaf and slash pines in the Southeast. Telial hosts are oaks.

Endocronartium harknessii causes pine-pine gall rust (western gall rust) on many two- and three-needle pines. The pathogen is autoecious. It occurs across Canada and the United States except in the Southeast. Some strains in central and western United States may have the capability of spreading both from pine to pine and from pine to *Castilleja.*

Peridermium filamentosum causes limb rust on Jeffrey and ponderosa pines in western United States and on additional species in Mexico. Two races of this apparently autoecious rust are known, one on Jeffrey pine in California and Nevada and the other on ponderosa pine in the Rocky Mountain region.

The characteristic structure of all *Cronartium* species is the columnar or hairlike telium, which emerges usually from the lower surface of a host leaf. The infected area may be chlorotic but is often essentially free from symptoms. Telia are composed of chains of one-celled, thin-walled, short-lived teliospores joined side to side. When wet, telia swell somewhat, and the teliospores germinate in place, each producing a basidium and four colorless, single-celled basidiospores. These short-lived spores, produced and released to the air at night, are susceptible to death by drying or from the sun's radiation. They infect pine needles, succulent shoots, or cones. The interval between infection and formation of spermagonia varies from several weeks in the case of southern cone rust to 2 or more years for some of the canker and gall rusts.

Most species of *Cronartium* are perennial in living pine tissues. In stems they grow as intercellular mycelium in the cortex and secondary phloem, sending simple haustoria into parenchyma cells. Most species also grow into the outer rings of sapwood, mainly in the medullary rays. Limb rust fungi are exceptional for invading deep-lying wood of the trunk. Spermagonia of the blister rust fungi develop beneath the periderm as flat, irregularly shaped fertile zones of interwoven rust hyphae up to a few millimeters in diameter. They produce spermatia in a fluid matrix that exudes in drops from minute ruptures in the bark or cone surface. In spring, several weeks to nearly a year after spermagonial activity, prominent light yellow or cream to white, blisterlike aecia break through the bark surface where spermagonia appeared previously. The wall (peridium) of the aecium is fragile and soon breaks, releasing powdery yellow to orange (rarely white) aeciospores with thick verrucose walls. Aeciospores resist drying, and those of some species can germinate after being windborne for long distances. Aecia gradually disappear during spring or summer. Aecia may be produced in the same bark area on stem galls in each of several years, but cones die after aecial fruiting. In canker rusts, bark broken by aecial fruiting soon dies, but the pathogen survives by hyphal colonization of bark beyond the zone of sporulation. Aeciospores infect leaves of the angiosperm host, and uredinia are produced within approximately 2 weeks. Uredinia are tiny yellow dome-shaped structures that open by a central pore to release yellow urediniospores with thick, spiny walls. Urediniospores reinfect the same host species on which they are produced. Thus the uredinial stage of the rust may intensify during spring or summer. Telia follow the uredinia, completing the cycle.

References for Cronartium rusts as a group: 228, 579, 1058, 1511, 1513, 2235, 2281

A, D, F. *Cronartium ribicola* on an unidentified currant. A. Yellow leaf spots and necrotic blotches caused by rust. D. Uredinia (powdery yellow pustules) and immature telia on the undersurface of a leaf. F. Mature, dry telia on a rust-induced lesion on the undersurface of a leaf (NY, Sep).

B, E, H. *C. quercuum* on Shumard oak. B. Small yellow spots with brown centers are the only symptoms on the upper surface of a leaf. E. Hairlike brown telia on the undersurface. H. Magnified view of telia (FL, Apr).

C. Magnified view of young uredinia of *C. comptoniae* on the undersurface of a sweetfern leaf. Each uredinium will open by a central pore, visible as a dark dot (NY, Jun).

G. *Castilleja* sp. (paintbrush), a host of *Cronartium coleosporioides* and *C. arizonicum* (AZ, Jul).

I, J. Brown, hairlike telia of *C. conigenum,* the southwestern cone rust fungus, on the undersurfaces of leaves of silverleaf oak. The telia are shown magnified in (J) (AZ, Jul).

Plate 130

White pine blister rust. This devastating disease is caused by a Eurasian fungus, *Cronartium ribicola,* that was introduced in diseased planting stock from Europe to both eastern and western North America near the beginning of the 20th century. It occurs around the Northern Hemisphere and has spread throughout nearly the entire range of susceptible (white, or five-needle) pines in North America.

North American species of white pines are, in general, highly susceptible. Those affected are bristlecone, limber, sugar, eastern white, southwestern white, western white, and whitebark pines. European and Asiatic white pines are less susceptible.

Many species of *Ribes* (currant and gooseberry, Plate 130) serve as alternate (telial) hosts of *C. ribicola. R. bracteosum, R. cynosbati, R. inerme, R. nevadense, R. nigrum, R. petiolare, R. prostratum, R. roezli,* and *R. rotundifolium* are highly susceptible. In Japan and eastern Asia a white pine blister rust fungus known as *C. kamtschaticum,* which some specialists regard as a strain of *C. ribicola,* infects both *Ribes* and members of the Scrophulariaceae—lousewort (*Pedicularis*) and paintbrush (*Castilleja*). Infection of paintbrush by *C. ribicola* has been reported from greenhouse experiments in Canada but has not been corroborated, and no evidence exists for natural infection of members of the Scrophulariaceae by this fungus in North America. Many races of *C. ribicola* exist that vary in virulence patterns on both pine and *Ribes.*

The symptoms of blister rust most noticeable from a distance are dead branches (flags) or a branch or tree with chlorotic foliage and drastically slowed growth. The limb or trunk in such cases has already been girdled. The girdling lesion is resinous internally and often exudes resin. A lesion on a limb or trunk is usually centered at a dead twig or branch that has somewhat swollen, roughened bark at its base. Nongirdling cankers are usually diamond shaped to elliptic, 2.5 to 3 times longer than broad, with a rough dead central area surrounded by a band of yellowish green infected bark several centimeters broad. This border blends indistinctly into normal greenish bark beyond. These color patterns are apparent only in smooth bark; canker margins in old rough bark are indistinct. Rust-infected bark is often gnawed off by rodents.

In spring when aecia are present (summer at high elevation or latitude), blister rust cankers are unmistakable (B). The aecia are prominent light yellow-orange blisters that arise deep in living bark, rupture the thin periderm, break open irregularly, and release powdery, bright yellow aeciospores. Remnants of the white walls (peridia) of aecia persist for several weeks. Bark in the zone of aecial production dies soon after spore release and remains swollen, roughened, and cracked. Spermagonia develop in summer as shallow, irregular to circular structures a few millimeters in diameter, just beneath the periderm surrounding the zone of recent aecial production. Yellow-green infected bark extends beyond the zone of spermagonial production. Spermagonia produce drops of sticky yellowish fluid for several days in summer and then dry out, leaving numerous small brown superficial scars on living bark (D).

Callus is usually absent from canker margins, but prominent callus may form where the rust fungus becomes inactive along a margin as the result of debilitation by mycoparasitic fungi or bark removal by rodents. The purple mold fungus, *Tuberculina maxima,* frequently parasitizes *C. ribicola* and other blister rust fungi in pines, suppressing aecial production and inhibiting canker enlargement.

The cycle of white pine blister rust typically lasts 3–6 years. Infection begins on pine needles from midsummer to early autumn when nighttime air currents carry basidiospores from *Ribes.* Usually the source plants are within a few hundred meters of pines that become infected, but patterns of air movement around lakes and in mountain valleys can carry infectious spores several kilometers. Basidiospores

germinate on wet needles, and the parasite enters through stomata. The only symptom initially is a yellow to reddish spot at the site of infection. Infected needles turn yellow and drop prematurely but often not before the fungus has grown down the needle and entered the twig or young mainstem. This occurs late in the year of infection or early the next season. The rust fungus grows intercellularly in bark and in the rays of the outermost sapwood, sending simple haustoria into parenchyma cells. It grows along the branch at rates up to approximately 8 cm per year. Rust mycelium can be found in bark 1–5 cm beyond surface discoloration. Infected, discolored bark swells because of the volume of rust mycelium and also because of mild hypertrophy and hyperplasia. Swelling is most apparent on twigs and small branches. Spermagonia are produced in summer of the 1st or 2nd year after infection of bark, and aecia develop in spring (or summer at high elevations) about 10 months after spermagonia. Aeciospores are then airborne locally and over long distances to infect *Ribes.* Within about 2 weeks, yellow uredinia form on the undersides of *Ribes* leaves, producing urediniospores that reinfect *Ribes* and may cause a buildup of rust during the summer. As the temperature and day length diminish from midsummer into early autumn, telia replace uredinia. Teliospores germinate in place during wet weather at night and produce basidiospores that infect pines, completing the cycle.

Not all infections are lethal. Several resistance mechanisms are known, and many trees within susceptible species are resistant. Infection in susceptible trees is often halted by death of diseased twigs or branches before the rust fungus reaches the trunk. Small trees are at greatest risk of lethal infection because in them the path from needles to mainstem is shortest. After the branches of trees in fully stocked stands begin to intermingle, low branches die naturally, and the hazard of lethal infection declines markedly.

The main problem after trees attain crown closure is "top rust." Basidiospores from sources outside the pine stand cause infections on branches above the canopy. Since these branches often remain alive except for the distal part, the fungus eventually reaches the trunk and kills the top of the tree.

Regions of high blister rust hazard are in northern and high-elevation areas where the average temperature in July is below 21°C. If relatively cool weather prevails, telia form beginning in midsummer, but if the weather is hot, their formation is delayed. After telial formation, at least 60 hours of wet weather with temperatures not exceeding 20°C are required for basidiospore formation, dispersal, and infection of pine.

The eradication of *Ribes,* long advocated for control of white pine blister rust, was gradually abandoned because it was ineffective in zones of high rust hazard and because it was unnecessary elsewhere. Pruning low branches of young trees protects against trunk infections. Similarly, pruning diseased limbs removes potentially lethal infections.

References for white pine blister rust: 180, 181, 228, 345, 384, 579, 838, 842, 854, 916, 917, 1030, 1058, 1224, 1261, 1291, 1462, 1513, 1863, 1892, 1925, 2057, 2058, 2063, 2145, 2186, 2189, 2235, 2281

A–D, G, H. Blister rust on eastern white pine. A. A mature tree with top and branches killed (Ont, Jul). B. A typical blister rust canker girdling a young tree near its base. Blisterlike aecia are conspicuous (NY, May). C. Close view of the margin of a blister rust canker in summer. Drops of spermagonial fluid are visible on the most recently diseased bark (left), and remnants of aecia remain on dying bark at right (NY, Jul). D. A typical canker in summer. Colorless, dripping resin and white, crystallized resin are prominent on dying and dead bark. Small brown scars (arrows) where spermagonia were recently active are visible on diseased bark near the canker margin (NY, Aug). G. Rust on a small branch. Infection began 2–3 years earlier in a needle near the branch junction (NY, Jun). H. Enlarged view of open aecia of *Cronartium ribicola* (NY, May).

E, F. Blister rust on whitebark pine. E. Bark is swollen and cracked at the location of a canker that girdled a branch. F. Massed aeciospores still impart prominent yellow color to a canker in late summer at high elevation (OR, Sep).

Plate 131

Fusiform rust. Caused by *Cronartium quercuum* f. sp. *fusiforme*, fusiform rust is one of the most economically important tree diseases in North America. It occurs in forests and landscapes from Maryland to Florida, Arkansas, and Texas. It is most severe on loblolly and slash pines, killing seedlings and young trees, causing cankers that lead to trunk breakage, or causing galls, cankers, and bushiness that destroy tree form and value. Austrian, Cuban, longleaf, pond, pitch, and South Florida slash pines are also affected in nature, and many other two- and three-needle pines are susceptible by inoculation. Shortleaf pine, although occasionally infected, is quite resistant and can therefore be used as a source of resistance in breeding programs.

The uredinial and telial stages of the parasite develop on oak leaves, usually causing inconspicuous symptoms (Plate 130). Oak species most important as sources of spores that infect pines are in the red/black oak group, especially cherrybark, blue jack, laurel, water, and willow oaks. Other hosts in nature include black, blackjack, southern live, post, southern red, scarlet, shingle, Shumard, turkey, and white oaks. Many other species of oak, chestnut, and chinkapin are susceptible by artificial inoculation.

Fusiform rust caused no great damage until after millions of hectares of pine plantations, mainly slash pine, had been established across the Southeast in a rust-conducive climate in the presence of susceptible oaks. Rust increased gradually and began to cause consistently great annual losses. In 1981 the annual loss due to devaluation of living trees, exclusive of mortality, was estimated at $75 million. Although rust-resistant planting stock is now produced through selection and breeding, severe losses are expected to continue for many years because of the time required for large-scale conversion of southern pine forests to resistant trees. Also, development of resistant trees is complicated by variation in virulence of the rust fungus. A pine selection resistant to one rust strain may be quite susceptible to another. Fertilization of pine to promote growth, a common practice in the Southeast, generally increases susceptibility.

The fusiform rust pathogen requires 2 years or longer to complete its life cycle. Needles and succulent stem tissue are infected in spring (March–May) by basidiospores released from telia on oaks. Within a few weeks a purplish spot usually appears at the site of infection. The fungus grows from needles (or cotyledons of 1st-year seedlings) into stems. Stem swelling, mainly the result of stimulated wood production, begins 4–6 months after infection and typically results in spindle-shaped (fusiform) galls. Young seedlings are often killed in the 1st year of infection, but many survive 3–4 years. Adventitious branches often arise at galls, giving the appearance of witches'-brooms. The rust fungus overwinters in the pine host. Spermagonia form in October–December, occasionally in the 1st year after infection but usually 1 year later. Blisterlike aecia break through the bark surface the next spring (March in most areas) and liberate powdery yellow aeciospores that infect expanding oak leaves. Orange uredinia develop on the undersurfaces of oak leaves within 7–10 days and produce urediniospores that are capable of reinfecting oaks. Repeating cycles of rust on oak are unimportant, however, because they contribute little to the increase of rust on pines. Telia develop 2–3 weeks after the first uredinia or may develop without prior uredinial formation. This development occurs in early spring while pine shoots are succulent. Teliospores germinate in place, producing basidiospores that complete the cycle.

Pine bark often dies after aecial production, resulting in dieback of small stems or cankers on larger ones. Aecial production recurs annually on living parts of rust galls and on swollen tissue at margins of cankers. Galls enlarge slowly. Those on branches within 40 cm of the trunk often progress into the trunk. Insects and wood-decay fungi invade cankers, and stem breakage often results.

Conditions that promote dispersal of basidiospores and infection of pines are relative humidity above 97%, temperatures of 15–27°C, and wet pine leaves and shoots. Eighteen hours is sufficient for severe infection. The necessary conditions occur so frequently in spring that for practical purposes the density of the local population of oak hosts is the most important determinant of rust intensity on pines. *C. quercuum* in oak leaves becomes inactive or dies during hot weather—for example, several days at temperatures above 29°C.

Pine-oak gall rust. This disease, also known as eastern gall rust, is caused by forms of *Cronartium quercuum* other than f. sp. *fusiforme*. These include f. sp. *banksianae*, primarily pathogenic on jack pine; f. sp. *echinatae*, primarily pathogenic on shortleaf pine; and f. sp. *virginianae*, primarily pathogenic on Virginia pine. The host and geographic ranges of the named formae speciales and possible others not yet described remain to be learned. Therefore we consider together all forms of *C. quercuum* that cause globose galls. The oak host range of these forms is similar to that of the fusiform rust pathogen. North of the range of fusiform rust, however, dwarf chinkapin and bur, chestnut, pin, northern pin, and northern red oaks are hosts. *C. quercuum* induces globose galls on Austrian, jack, loblolly, mugo, pitch, pond, ponderosa, red, Japanese red, sand, Scots, shortleaf, slash, spruce, and Virginia pines. Most records of this disease on pines in eastern Canada, northeastern United States, and the West are suspect because the galls are indistinguishable from those caused by the autoecious *Endocronartium harknessii* (Plate 136). From records of telia on oaks in northern and western areas and globose galls on pines in the South, *C. quercuum* occurs from Massachusetts to the limits of oaks in Ontario and South Dakota and south to Florida and Texas, also in California and Oregon (on California scrub oak and Oregon white oak).

Damage by pine-oak gall rust varies with pine species and age. Galls that form on the mainstems of seedlings are usually lethal within 4 years. Thus this rust has caused severe losses in nursery stock and young plantations of jack and Scots pines. Galls on jack, pitch, and Virginia pines commonly become cankerous, and those on mainstems stunt tree growth. Branch galls have little effect on overall growth. In Ontario, stunting due to trunk galls in fully stocked jack pine stands was found to have no effect on forest productivity because diseased trees merely lost out to others in competition for growing space.

The cycle of pine-oak gall rust is similar to that of fusiform rust. In the North, however, aecia and telia are delayed until late spring and summer, respectively. The galls are woody and vary from globose to fusiform, often with bark collars at the edges. Yellowish aecia form an anastomosing or cerebroid network on the gall surface in spring. Although galls caused by *C. quercuum* are indistinguishable from those caused by *E. harknessii*, diagnosis is possible if aeciospores are available (details with Plate 136).

References for fusiform rust: 181, 282, 437, 492, 527, 528, 864, 1318, 1319, 1511, 1518, 1551, 1552, 1707, 1880, 2148

References for pine-oak gall rust: 33, 38, 282, 694, 745, 962, 1411, 1511

A–E. Fusiform rust, caused by *Cronartium quercuum* f. sp. *fusiforme*, on slash pine. A. Damage in a young plantation (GA, Aug). B. A fusiform (spindle-shaped) swelling with a prominent yellow region where aecia have just matured (GA, Mar). C. Fusiform swellings and dieback (GA, Mar). D. Multiple shoots growing from a rust gall (MS, May). E. A young tree in which rust low on the mainstem has induced fusiform swelling and multiple leaders (GA, Aug).

F. A fusiform rust canker on a pole-sized loblolly pine. The large callus roll at the right margin indicates inactivation of the pathogen there (GA, Aug).

G. Pine-oak gall rust, caused by *C. quercuum* f. sp. *virginianae*, on a branch of Virginia pine. Blisterlike aecia are prominent (TN, Apr).

Plate 132

Sweetfern blister rust. Known only in North America, this rust came to prominence early in the 20th century when it began to cause damage in young plantations of two- and three-needle pines ("hard pines") in the Great Lakes region. Foresters had established experimental plantations of Austrian, lodgepole, and ponderosa pines on sand plains that were carpeted with sweetfern. The pines soon began to die, and some plantations were completely destroyed. The cause was readily apparent—a blister rust that caused stem swellings and perennial cankers that girdled branches and mainstems. Similar epidemic occurrences on other pine species have since been reported from the Maritime Provinces, northeastern United States, and British Columbia. Plantations tend to be more severely affected than natural stands. The disease has also occasionally caused severe losses in nurseries, and it is a continuing problem in jack pine and lodgepole pine forests.

The pathogen is *Cronartium comptoniae*. It produces spermagonia and aecia on pines, and uredinia (Plate 130) and telia on the undersides of leaves of sweetfern and sweet gale. Wax myrtle is also susceptible by artificial inoculation. Host alternation is obligatory. The fungus causes inconspicuous leaf spots on telial hosts. It occurs transcontinentally in Canada and the United States wherever sweetfern or sweet gale grow in association with two- and three-needle pines from Nova Scotia to northern Georgia and Missouri in the East and from Alaska to northern Oregon in the West. Hosts include Austrian, Bishop, cluster, Coulter, jack, Jeffrey, loblolly, lodgepole, Monterey, mugo, pitch, ponderosa, red, Japanese red, Scots, shortleaf, tablemountain, and Virginia pines. Telial hosts are absent from most of the natural range of ponderosa pine; thus this highly susceptible species escapes severe damage.

Nearly all infections of pines begin before trees reach age 10, usually before age 5. Seedlings and the trunks and branches of saplings develop spindle-shaped swellings or commonly galls. Aecia form on the swollen area in spring, and bark in the area of sporulation dies subsequently. Branches and even mainstems of small trees may be girdled, the foliage turning yellow and then brown. Sometimes the trunk of a small tree is girdled above a whorl of living branches, leading to survival of a shrubby, multistemmed individual. Seedlings that become infected in their 1st year usually die within 3–4 years. Trees infected at a somewhat older age may persist for decades with elongate trunk cankers. Canker margins often have small, irregular galls with dead areas between. Some cankers, hidden by scaly bark, escape detection until trees are more than 25 years old, but most become apparent sooner as sunken areas up to 2 meters long near or extending to ground level. The sunken appearance is accentuated by development of swollen ridges along the sides of the cankers. Insects, particularly larvae of the Zimmerman pine moth (see Johnson and Lyon, 2nd ed., Plate 18), tunnel in infected bark, and resin oozes from the insect wounds. Trunk deformity, resin-soaked wood beneath cankers, and wood decay associated with cankers devalue the butt log for pulpwood or lumber. Height and diameter growth of jack pine are reduced by sweetfern rust, and trees with cankers eventually die prematurely.

Infection occurs in late summer and autumn, and swellings develop beginning 12–15 months later at branch whorls or on segments between whorls. Intercellular hyphae become abundant in bark and extend in rays across the cambium into the outer sapwood, sending simple or twisted haustoria into parenchyma cells of cortex, phloem, and xylem. Mycelium in bark and wood extends about 1 cm and 2 cm, respectively, beyond externally visible symptoms. In lodgepole pine the infected area extends 5–15 cm per year. Swelling is due to hyperplasia and hypertrophy in the cambial region, resulting in aberrant wood production. In jack and lodgepole pines the infected tissue contains abnormally large and numerous rays.

Spermagonia similar to those of the white pine blister rust (Plate 131) develop on swollen bark in late summer or autumn, beginning the year after infection. They are noticeable, however, only on small, smooth-barked stems. There they appear as yellowish patches up to about 3 mm in diameter, from which orange droplets exude for a few days. Later they darken. Blisterlike aecia, cream colored to nearly white and often confluent, break through the bark in the same area the next spring and rupture irregularly. Aecia on stems with scaly bark are often hidden by loose bark scales. The orange aeciospores are dispersed by wind at about the time that leaves of sweetfern and sweet gale are expanding.

Aeciospores require only a few hours to infect the lower surfaces of sweetfern or sweet gale leaves under wet conditions at the most favorable temperatures of 12–16°C, but infection can occur (although more slowly) within the temperature range of 0–28°C. Uredinia develop about 2 weeks later and are present beginning in June in most areas. Given free water, urediniospores and teliospores germinate most rapidly at 16–20°C. Urediniospores cause repeating cycles on sweetfern or sweet gale during summer. Hairlike orange to cinnamon telia about 2 mm long, similar to those of other *Cronartium* species (Plate 130), develop 6–7 weeks after the first uredinia. Teliospores germinate in place, producing basidiospores that are dispersed short distances to pines.

In eastern and central regions, sweetfern rust, like sweetfern, is most common on sandy or droughty, relatively infertile soils. In the West, the disease is most common in moist habitats, reflecting the site preference of sweet gale.

Jack and lodgepole pines, the species in which sweetfern rust is most important, vary considerably in susceptibility. This variation offers the possibility of long-term disease management through selection and breeding.

Sweetfern rust and stalactiform rust (Plate 134) are similar in diagnostic characters, pine hosts, and geographic ranges. When found in branches or in young cankers, the causal fungi cannot be differentiated with certainty except by inoculating alternate hosts with aeciospores. For tentative diagnosis, however, stalactiform rust cankers originate at any height on the trunk, whereas sweetfern rust cankers are always near the ground. Swollen ridges and irregular bumps are common at the edges of sweetfern rust cankers but are not features of stalactiform rust. If abundant sweetfern or sweet gale is associated with a blister rust on hard pines, it is likely to be the sweetfern rust, because *C. comptoniae* usually spreads only a short distance (a few tens of meters) between pines and alternate hosts. The abundance of sweetfern is the strongest determinant of intensity of the rust on pines.

References for sweetfern rust: 32, 39, 128, 228, 387, 696–98, 840, 920, 1058, 1088, 1342, 1511, 1826, 1865, 2067, 2279, 2281

A–E, G. Sweetfern blister rust on pitch pine (NY, Jun). A. Dead and dying saplings. B. Close view of a portion of a rust canker on a small stem. Aecia at left were exposed by removing outer bark scales. Rodents gnawed bark from the portion at right. C, D. Elongate, sunken, bark-covered cankers outlined by ridges of swollen tissue, aligned with the grain of the underlying wood. The arrow in (D) locates a small gall caused by the rust fungus. E. Close view of the gall located in (D), breaking through the outer bark scales and bearing aecia. G. A cross-section of a young stem showing marked swelling and yellowish discoloration of wood caused by a rust infection that reached this part of the stem when it was in its 2nd year of growth.

F. Aecia of *Cronartium comptoniae* on a small branch of Scots pine (VT, May).

Plate 133

Stalactiform rust. Cronartium coleosporioides, the cause of stalactiform rust, and C. comandrae, the cause of comandra rust, are widespread on many two- and three-needle pines in Canada and the United States. These native fungi kill seedlings and cause branch dieback and resinous, perennial trunk cankers that girdle and kill trees of all ages and degrade the trunks for lumber or pulpwood. They cause greatest damage in western forests. Stalactiform rust, also known as cow wheat rust, is important primarily on lodgepole pine, while comandra rust is a major pathogen of lodgepole and ponderosa pines. Stalactiform rust cankers are long (up to several meters) and narrow, while comandra rust cankers are usually only two to three times as long as wide and seldom extend more than 1.2 meters. Squirrels commonly remove bark infected by either pathogen, and the trees respond to this injury by copious production of yellowish resin. In some cases this bark removal arrests canker enlargement. Host alternation is obligatory for both rust fungi; thus they occur only where pines grow in association with alternate hosts.

C. coleosporioides occurs across Canada and in north-central and western United States as far south as Colorado and southern California. Telial hosts are Castilleja species (paintbrush), Melampyrum lineare (cow wheat), Orthocarpus luteus (yellow owl's clover), Pedicularis bracteosa (lousewort), and Rhinanthus crista-galli (yellow rattle), all in the Scrophulariaceae. Cow wheat is the principal telial host in the East, paintbrush in the West. Spermagonia and aecia form on Coulter, jack, Jeffrey, knobcone, lodgepole, and ponderosa pines in nature and on other species after artificial inoculation. Outbreaks in nurseries and young plantations have caused losses of 50–80% of lodgepole pine seedlings in extreme cases. On older trees, girdling is usually confined to limbs. On Jeffrey and ponderosa pines, the pathogen may cause limb rust (see Plate 135). Stalactiform rust on pines can be confused with sweetfern rust and with the limb rust caused by Peridermium filamentosum; for differences see Plates 133 and 135. The stalactiform rust fungus on paintbrush is similar to the limb rust fungus C. arizonicum (Plate 135) and to heteroecious strains of the pine-pine gall rust fungus (Plate 136).

The disease cycle requires 2 years or longer. Basidiospores from telia on the angiosperm host infect pine needles in summer, and the rust fungus grows into branches or the mainstems of seedlings and young trees. Infected bark swells the next year, and spermagonia develop in late summer and autumn, producing orange droplets that contain spermatia. White aecia break through the bark surface the following spring and disperse orange aeciospores. (A white-spored form of this rust occurs in Alberta.) After aecial sporulation, the bark dies but remains swollen and roughened. Aecial sporulation is often suppressed or apparently absent at the edges of old trunk cankers. Aeciospores infect members of the Scrophulariaceae, resulting in uredinial formation after 2–3 weeks and telia after 5 weeks on both surfaces (mainly the undersurface) of leaves and on stems. All spore types require wet plant surfaces for germination and infection. The optimum temperature for aeciospore germination is near 15°C and that for teliospores and basidiospores is 15–21°C.

C. coleosporioides invades sapwood as well as phloem. In lodgepole pine its hyphae may be found in tracheids as far as 20 cm beyond the point of a canker and 2–5 cm in advance of hyphae in bark. Apparently it grows in the rays from sapwood into bark, which hastens the extension of cankers. Cankers elongate 16 cm or more per year and may reach lengths up to 9 meters before girdling occurs.

Comandra rust. This rust occurs from New Brunswick to British Columbia in the North and extends southward in eastern and western mountains to Alabama and southern California. Telial hosts are the herbaceous plants comandra or bastard toadflax (Comandra umbellata) and northern comandra (Geocaulon livida). On both plants the rust causes pale yellow leaf and stem spots and leaf abscission. Comandra rust rapidly causes death of young pines of several species and causes significant loss of growth and volume in western forests by killing the tops of lodgepole and ponderosa pines. Other hosts in nature include jack, Jeffrey, knobcone, loblolly, mugo, pitch, pond, Scots, shortleaf, slash, spruce, and Virginia pines and Pinus eldarica.

Many additional species have proved susceptible when inoculated experimentally.

Symptoms on pines begin with slight spindle-shaped swellings or, occasionally, distinct galls on branches and mainstems. Infected parts die after aecial production. Trunk cankers result from infection through needles while the stem is young or from growth of the fungus along branches. Differences between these cankers and those caused by the stalactiform rust fungus are discussed above.

The disease cycle in the North takes 2 or more years but in the South may require only 1 year. Pines become infected in summer by basidiospores liberated from telia on the angiosperm host. The spores germinate on wet needles, and the parasite enters through stomata. It then grows into the twig or stem, the bark of which swells to 2–4 times normal thickness. Intercellular mycelium becomes abundant in bark and the outermost sapwood, extending up to 2 cm beyond the swollen area. In lodgepole pine in the Rocky Mountain region, rust mycelium advances about 2 cm per year toward the branch or trunk base. Spermagonia, similar to those of C. ribicola (Plate 131), arise in the swollen bark in late summer or autumn of the year of infection or after a lapse of 1–2 years, and aecia develop during the next spring (and summer in the North). Bark becomes roughened as the result of aecial production. Aecial blisters rupture irregularly, revealing orange to red-orange aeciospores that slowly fade to yellow. The aeciospores are shaped like teardrops, a form that distinguishes C. comandrae from other pine stem rusts. Aeciospores, dispersed in air during daylight, can survive several days on dry comandra surfaces and then infect within as few as 6 hours if the surface becomes wet and temperature is in the favorable range of 5–22°C (optimum 15–20°C). Longer wet periods allow more severe infection. Uredinia develop within 1–2 weeks, and telia after 2–6 weeks, depending on temperature. Uredinia and telia are similar to those shown in Plate 130. Rust increases on comandra during summer as the result of reinfection by urediniospores. Teliospores germinate in place, liberating basidiospores that infect pines. Wet weather at 13–23°C for 24 hours or longer is sufficient for infection.

Basidiospores may infect pines at distances greater than 1.6 km from comandra, but aeciospores only infect comandra close to the source pines—less than 120 meters in one study. In the South, the humid climate is suitable for pine infection every year, but in the West, suitable conditions over large areas occur infrequently. As a result there is a high incidence of new infections on pines in only a few years each century. Because of perennial canker activity, however, residual damage continues for decades.

The purple mold fungus, Tuberculina maxima, parasitizes both C. coleosporioides and C. comandrae, suppressing or inactivating aeciospore production in some cankers, but it does not prevent significant losses in areas where it occurs.

References for stalactiform rust: 40, 228, 387, 1057, 1293, 1511, 2084, 2279, 2281

References for comandra rust: 153, 154, 228, 387, 507, 768, 840, 993, 1057, 1058, 1274, 1296, 1511, 1548–50, 1960, 2281

A–D. Stalactiform rust, caused by Cronartium coleosporioides, on lodgepole pine. A. A mature tree killed by the rust. The yellow appearance of the trunk indicates presence of a rust canker (OR, Sep). B. Resin exudation identifies the location of a canker (CO, Jun). C. Old cankers from which bark was removed by rodents (CA, Jul). D. Part of a stalactiform rust canker that has engulfed a circular canker that developed at a site of infection by the pine-pine gall rust fungus, Endocronartium harknessii. Crystallized resin colors the wood yellow where bark was removed by rodents. Linear ridges at left indicate annual expansion of the stalactiform rust canker (OR, Sep).

E, F. Comandra rust cankers on ponderosa pine (WA, Jul). E. A canker on the mainstem, visible as a yellowish area where bark was removed by rodents, has nearly girdled the stem, resulting in slow growth and twig dieback. The top of this tree will die. F. Bark of a young tree is roughened where broken by aecia of C. comandrae. Massed aeciospores impart yellow color. Branch bases at the infected whorl are swollen.

Plate 134

Limb rusts. These rusts of pines occur from the Black Hills of South Dakota to northern California and south into Mexico. They are the most destructive rust diseases of Jeffrey and ponderosa pines. Additional pine hosts in Mexico include Apache and rough-barked Mexican pines as well as *Pinus cooperi, P. durangensis,* and *P. michoacana.* The term *limb rust* refers to the progressive invasion and killing of branches by rust mycelium that is perennial and systemic in the trunk. From an initially infected branch, the parasite grows into the trunk and advances as much as 20–25 cm per year in each direction, growing out into each branch as encountered. It produces aecia only on twigs or small branches, which subsequently die. Progressive twig and branch mortality over many years suppresses growth and leads eventually to death of the tree. Secondary agents such as bark beetles are often the ultimate cause of death.

Limb rusts are caused by four distinct pathogen populations that appear similar on pines. The stalactiform rust fungus, *Cronartium coleosporioides* (Plate 134), is a member of this group. *C. arizonicum,* with aecia on ponderosa pine and uredinia and telia on paintbrush (*Castilleja* species), is another member. *C. arizonicum* is widespread in the Southwest. Its aecial state is one of three populations that are grouped under the name *Peridermium filamentosum.* Inoculation of aeciospores of *C. arizonicum* to leaves of *Castilleja miniata* resulted in development of uredinia after 18–20 days and telia after 39–50 days. Two other populations, or races, of *P. filamentosum* are known, which differ from each other in host specialization, seasonal development, and temperature optima for aeciospore germination. These races have no known alternate hosts and are thought to be autoecious. The germ tubes of their aeciospores are shorter and stouter than those of *C. arizonicum* and resemble germ tubes of the autoecious *Endocronartium harknessii* (Plate 136). One of these strains attacks ponderosa pine in Utah and adjacent areas. The other infects Jeffrey pine in California and Nevada.

Limb rust caused by *C. coleosporioides* can be distinguished from that caused by *P. filamentosum* on the basis of differing symptoms and differing macroscopic and microscopic characteristics of aecia. For field diagnosis, the bark is more conspicuously roughened where aecia of *C. coleosporioides* have been produced than where other limb rust fungi have fruited. Aecia of *P. stalactiforme,* the aecial state of *C. coleosporioides,* typically measure 1 × 2 mm but often 5 × 6 mm or more where they are confluent, and they have a low profile. Aecia of *P. filamentosum* are nonconfluent and tongue shaped or narrowly conical, often 8 mm or more tall and usually less than 2 × 4 mm at the base. Peridia (aecial walls) of some strains of *P. filamentosum* are tough, and remnants of peridia may persist several months or occasionally 2–3 years. Aeciospores of *C. coleosporioides* are infectious to paintbrush. Limb rust caused by this fungus is common in the Sierra Nevada region.

Limb rust infection of pine begins on needle-bearing stems. Trees of all ages and sizes are affected. The causal fungus grows intercellularly in rays into the xylem, sending haustoria into ray parenchyma. In the xylem it grows longitudinally within tracheids, entering and killing small branches and twigs. On reaching large branches or the trunk, it grows along rays into inner sapwood, where it spreads up and down the trunk. In transverse view, infected sapwood is discontinuous around the trunk. Limb rust mycelium has been detected in sapwood more than 60 years old. It does not approach the cambium or grow into bark except in small branches and twigs where spermagonia and aecia form. Aecia appear in spring to midsummer, depending on rust strain and locality, and disperse spores for several weeks.

Cone rusts. Southwestern cone rust, caused by *Cronartium conigenum,* occurs in southern Arizona and southward through Mexico into Central America. In Arizona the fungus produces spermagonia and aecia in July and August on swollen deformed cones of chihuahua pine and uredinia and telia throughout the summer on the undersides of leaves of Mexican blue, Dunn, Emory, gray, canyon live, netleaf, silverleaf, and Arizona white oaks. Telia are up to 6 mm long and may be so numerous that the lower leaf surface appears clothed with a loose brown wool (Plate 130). Cones become infected during their 1st year of development, swell into misshapen galls of various sizes, produce no seeds, and do not open. Aecia develop 2–3 years after infection. Galls die after aecial production but remain on the trees. Large galls usually terminate the branches that bear them. During periods of outbreak, cone rust may kill more than 50% of the cones on groups of trees. Other pine species affected farther south include Apache, Cuban, rough-barked Mexican, ponderosa, twisted-leaf, and Mexican yellow pines; also *Pinus cooperi, P. durangensis, P. hartwegii, P. lawsonii, P. lumholtzii, P. michoacana, P. oocarpa,* and *P. pseudostrobus.* The cone rust fungus causes stalked stem galls on several of these species.

Southern cone rust, caused by *C. strobilinum,* occurs from North Carolina to Florida and Louisiana. It kills 1st-year cones of longleaf, slash, and South Florida slash pines. In wet years conducive to infection it may destroy as much as 90% of the cone crop in some localities. Young cones become infected during the period of pollination in winter, swell during the spring to several times the volume of healthy cones of the same age, turn reddish during the period of spermagonial activity in March, then turn yellow as aecia mature in April and May. As aecia develop, the cone surface ruptures irregularly and falls away, exposing a layer of aeciospores that are dispersed by wind and infect various oaks. After spore dispersal, the cones die, dry, and fall. The cone moth *Dioryctria abietivorella* (see Johnson and Lyon, 2nd ed., Plate 17) is attracted to diseased cones at the time aecia develop. Its larvae feed on both fungus and host tissues, and the expanding insect population later attacks healthy cones.

Uredinia develop in autumn, and telia develop from December to February on evergreen oaks in the Gulf States. Many deciduous oak species also become infected, but because their leaves drop before pine cones become susceptible, they are not important in the disease cycle. Teliospores germinate in place during wet weather and liberate basidiospores that infect young pine cones. Telial hosts in nature include bluejack, bur, Chapman, chestnut, laurel, dwarf live, southern live, myrtle, post, running, turkey, water, white, and swamp white oaks. Uredinia and telia diagnosed as *C. strobilinum* have been found on oaks in areas remote from pines with infected cones. On this basis it has been suggested that *C. strobilinum* may in some cases overwinter in the uredinial state on green sprout stems of oaks. This point, however, has not been corroborated by research.

References for limb rusts: 228, 433, 1292, 1509, 1511, 1512, 1514, 2084, 2279

References for cone rusts: 228, 767, 769, 785, 1511

A–C. Limb rust on ponderosa pine. A. Dead and dying branches on a diseased tree. B, C. Successively closer views of aecia on a small branch. The aecia resemble those of *Peridermium stalactiforme,* the aecial state of *Cronartium coleosporioides* (CO, Jun).

D. Southwestern cone rust, caused by *C. conigenum,* on chihuahua pine. A healthy cone is at left and a diseased cone, swollen and deformed, at right (AZ, Jun).

E, F. Southern cone rust, caused by *C. strobilinum,* on South Florida slash pine (FL, Apr). E. A diseased cone (yellow object), swollen and prominent against green foliage, is distinct from normal dead cones of a previous year. F. A diseased cone (left), greatly enlarged in comparison to its healthy companion. The surface of such a cone will soon burst and liberate aeciospores. No seed is produced.

Plate 135

Pine-pine gall rust. This disease, also known as western gall rust, occurs transcontinentally on two- and three-needle pines in Canada and southward to Virginia, Nebraska, and northern Mexico. The pathogen, *Endocronartium harknessii,* is autoecious and is apparently restricted to North America. It induces the formation of globose or sometimes pear-shaped woody galls. Cankers develop at loci of gall rust infection in some hosts, but the cankers are apparently caused by secondary fungi. Galls usually form on branches but are common on mainstems of some species, especially jack and lodgepole pines. Galls enlarge from year to year until they or the supporting branches die. Trees with numerous galls grow slowly, often develop witches'-brooms (this varies with host), sustain dieback of twigs and branches, and in some cases are killed. Secondary organisms may be the direct cause of death of branches or trees debilitated by the rust.

Species that sustain the greatest damage in natural forests are Bishop, jack, lodgepole, Monterey, and ponderosa pines. Damage in plantations has been severe in Scots pine grown for Christmas trees. Other species affected in nature include Aleppo, Austrian, Canary Island, cluster, Coulter, digger, Jeffrey, knobcone, and mugo pines. Several other pines are susceptible by inoculation.

Infection occurs on succulent stems, which usually have no symptoms in the year of infection. Swelling begins the next year. In Bishop and lodgepole pines, the peduncles of young female cones also become infected, and the pathogen grows down the peduncle into the stem, where a gall forms. Galls typically enlarge to diameters of 1–10 cm before dying. Those on mainstems may be globose or hemispheric and sometimes attain diameters of 20–30 cm. Beginning in the 2nd or 3rd year, galls may develop bark collars at one or both ends where thick, normal periderm has been raised by gall expansion along the stem. The xylem in galls consists of hyperplastic and hypoplastic aggregates of unusually short tracheids and abundant rays. In Scots pine, the relatively thin sapwood region of a gall encloses a resinous, reddish brown core. Gall mortality is caused mainly by secondary fungi and insects that invade rust-infected tissues. Several hyperparasitic fungi, especially *Scytalidium uredinicola,* exert some natural control of sporulation of *E. harknessii* and are potentially useful for practical biological control, but secondary fungi that kill rust-infected host tissue probably suppress disease more than do hyperparasites.

Cankers, most common on lodgepole pine, may originate at obvious galls or in their absence at the bases of small branches that become infected close to the mainstem. In some trees, wood formation at lateral margins of a canker is stimulated, so that the stem expands and flattens. Cankers with bulging flanks are called hip cankers.

Western gall rust is often lethal to seedlings, killing the entire seedling or the part above the point of infection. Unfortunately, many diseased seedlings survive. Thus *E. harknessii* has been dispersed widely in seedlings shipped from nurseries, leading to buildup of gall rust in young plantations.

Aecia break through the surfaces of galls annually during the period of pine shoot elongation, beginning 2–4 years after infection. Aecia are low in profile, confluent, and often cerebroid in surface view (A) but are more or less continuous beneath the periderm (B). They rupture irregularly, and remnants of the peridia (aecial walls) soon disappear. The aeciospores, yellow-orange when fresh (except for an albino strain that is widespread in the Rocky Mountain region), are verrucose and variable in shape.

Aeciospores are liberated primarily during periods of decreasing atmospheric humidity each morning for 2–3 weeks in spring. Germination and infection occur on wet plant surfaces at 10–30°C. The range of optimum temperature varies with the strain of the fungus in the overall range of 15–28°C. Prolonged cool, wet weather during the period of pine shoot growth is conducive to severe infection. In the East, these conditions occur in most years. In central and western regions, however, highly conducive weather occurs at intervals of several years, resulting in "waves" of infection.

Within a highly susceptible species such as ponderosa or Scots pine, susceptibility varies among individuals and also with geographic origin of seed. Healthy trees within severely damaged populations are common. Artificial inoculations into Japanese black, loblolly, longleaf, Japanese red, sand, shortleaf, slash, and Virginia pines have resulted in no disease or in resistant reactions.

Galls caused by *E. harknessii* are indistinguishable from those of pine-oak gall rust (Plate 132). Where globose galls form on pines in the absence of oaks, *E. harknessii* is the cause. Where oaks are present, however, *Cronartium quercuum* may be the pathogen. These fungi can be differentiated on the basis of germination behavior of their aeciospores. Those of *C. quercuum* will infect oak leaves and, when allowed to germinate on water agar, produce long, slender germ tubes with a tendency toward apical branching. Aeciospores of *E. harknessii* will not infect oak but will infect succulent pine shoots. Their germ tubes are short (up to half as long as those of *C. quercuum*), stout, and septate, with a tendency to produce branches near the proximal end.

The name *Endocronartium* signifies that, while possessing only spermagonia (uncommonly seen) and aecia, the pathogen nevertheless completes its nuclear cycle, including nuclear fusion and meiosis. These events occur in germinating aeciospores and their germ tubes. The rust is therefore said to be endocylic. In *Cronartium*, by contrast, nuclear fusion and meiosis occur in teliospores and basidia. Because aecia and aeciospores of *Endocronartium* function like telia and teliospores in the life cycle of the rust, the terms *aecidioid telia* and *aecidioid teliospores* have been proposed.

We have presented pine-pine gall rust as autoecious and endocyclic, but there are discrepancies in this pattern. Independent researchers in the north-central and western United States have reported infection of paintbrush (*Castilleja* species) by aeciospores of this fungus, resulting in production of uredinia and (or) telia indistinguishable from those of *Cronartium coleosporioides*. In two cases, aeciospores from the same source or line were also capable of infecting pine. Therefore this rust was considered to be part of a *C. coleosporioides* complex and to include both autoecious and heteroecious strains. In the literature, pine-pine gall rust was distinguished from other diseases in the complex—stalactiform rust and limb rust—by using the names of the aecial forms of the pathogens, *Peridermium harknessii, P. stalactiforme,* and *P. filamentosum,* respectively. Strains of *P. harknessii* capable of alternating to other plants have not been detected in eastern United States or Canada.

References for pine-pine gall rust: 34, 35, 292, 293, 745, 836, 840, 1275, 1282, 1487, 1493, 1499, 1503, 1511, 1983, 1991, 2038, 2045, 2273, 2281

A, B. Sporulating galls on Scots pine branches (NY, May). A. Aecia of *Endocronartium harknessii* breaking through the bark surface. B. Galls in which a layer of yellow aeciospores has developed beneath the periderm, causing the outer bark scales to break away.

C. Infections close together on a Scots pine twig have resulted in nearly continuous twig swelling, induction of multiple terminal shoots, and dieback (NY, Apr).

D. A "hip canker," several decades old, on a large lodgepole pine (CA, Jul).

E. Branch mortality caused by western gall rust and secondary organisms on ponderosa pine (ID, Jul).

F. Swelling and upturned multiple shoots induced by *E. harknessii* at a branch tip of Aleppo pine (CA, Feb).

G. A mature Scots pine, killed by severe gall rust (NY, Sep).

H. Galls on the trunk of a young lodgepole pine (OR, Sep).

I. A loose witches'-broom resulting from multiple gall rust infections on a Scots pine limb (NY, Sep).

Plate 136

Diseases Caused by *Phytophthora* Species (Plates 137–140)

The name *Phytophthora* means plant destroyer. This genus of Oomycetes (Peronosporales, Pythiaceae) contains about 40 species, many of global importance. They attack all major groups of higher plants, causing blight, dieback, fruit rot, cankers, collar rot, root rot, and necrosis of feeder roots. Some species, such as *P. cactorum* and *P. cinnamomi*, attack hundreds of plant species and cause several different symptoms. Other species, such as *P. inflata*, have narrow host ranges and cause one characteristic syndrome. These plates present representative diseases caused by *Phytophthora* species.

Field diagnosis of tree diseases caused by *Phytophthora* species depends entirely upon interpretation of symptoms, for these fungi produce no fruit bodies, and their reproductive structures are microscopic. Their mycelium, and spores when present, can easily be seen by microscopic examination of infected tissues, but these fungi are usually detected by growth on agar media from dying or recently dead plant tissues. Specific diagnosis requires microscopic examination of the isolated fungus, for several *Phytophthora* species as well as unrelated organisms may be capable of causing similar symptoms on the same host plant.

Phytophthoras exist as quiescent spores in soil or killed plant material, but most species do not grow to a significant extent as free-living saprophytes in nature. Those that infect trees and shrubs are dispersed in soil, diseased plants, dead infested plant debris, and splashing and running water.

These fungi produce stout, aseptate, freely branching, colorless hyphae that grow inter- and intracellularly. Some species send haustoria into host cells. All species produce three kinds of infective structures: sporangia, zoospores, and oospores. Some species also produce chlamydospores. Spores of all types require free water for germination.

Sporangia are short-lived vegetative structures that under some conditions behave like simple asexual spores, germinating by a germ tube that gives rise to a new mycelium. More often, however, they "germinate" by converting their contents to wall-less, kidney-shaped, motile spores (zoospores) that escape through a pore in the tip of the sporangium. Several zoospores are produced by each sporangium; this mechanism amplifies the potential of the fungus to infect host tissue. Zoospores move in water by means of flagella. They swim more or less randomly for a few minutes to a few hours at most, changing directions frequently. Thus their net movement is seldom more than a few centimeters unless they are in flowing water. They are attracted by amino acids and other chemicals that exude from roots or other plant parts. Wounds and the surfaces of tender, succulent plant parts such as shoot or root tips are suitable for infection. On reaching such a surface, zoospores round up, lose their flagella, quickly secrete a cell wall, and produce a germ tube that penetrates the plant and gives rise to mycelium.

Oospores are products of the union of gametangia called oogonia and antheridia, and they serve as resting structures. Chlamydospores are vegetative cells, within hyphae or at their tips, that develop thick walls and also serve as resting structures. Oospores and chlamydo-spores germinate by germ tubes from which in some species either mycelium or sporangia and then zoospores may differentiate. In general, warm temperature (e.g., above 18°C) favors mycelium, and cool temperature favors production of zoospores.

Pit canker of elm. This disease, caused by *Phytophthora inflata*, occurs in the northeastern United States and Quebec on American and, less commonly, red elms and was significant on elm shade trees before the population of native elms was overtaken by Dutch elm disease. Although known by arborists since early in the 20th century, pit canker went undescribed until 1949. It was reputed to be most common on trees growing under adverse conditions, such as in compacted soil. Affected trees develop multiple perennial cankers, often sunken and partly hidden by thick outer bark, on the trunk and scaffold limbs. The phloem, cambium, and outermost sapwood at the canker margin are discolored red-brown. Reddish brown fluid may bleed from small fissures near canker margins. Narrow bands of discolored tissue often connect closely adjacent cankers. After the bark begins to fall away, concentric markings that indicate intermittent enlargement are visible on the wood. The parasite is active primarily while the host is dormant. Callus development at the canker margin, responsible for the concentric markings, varies from weak, as in the specimen shown here, to vigorous. Trunks with multiple cankers become distorted because of growth in asymmetric bands between cankers. As cankers coalesce and increase in size and number, trees lose vigor, have undersized chlorotic leaves that droop and drop prematurely, and begin to die back. *P. inflata* can readily be isolated from the margins of active cankers. In pure culture it produces distinctively inflated antheridia, for which it is named.

Dieback of rhododendron and related ericaceous plants. Rhododendron, azalea, and Japanese andromeda may be attacked by *Phytophthora cactorum*, *P. citricola*, *P. citrophthora*, *P. heveae*, or *P. parasitica*, which are often followed by opportunistic fungi such as *Botryosphaeria dothidea* (Plate 82). These *Phytophthora* species, most common in nurseries, kill succulent young leaves and twigs but do not usually advance far into woody stems. They cause necrotic blotches that are at first olive colored and later brown and sometimes have a dark red margin. Infection occurs when zoospores are splashed from soil to plant surfaces, and lesions develop within 2–3 days during warm moist weather. Sporangia and later oospores and sometimes chlamydospores form in the dead tissues. Diseased leaves abscise, returning the fungi to the soil, where they survive in plant debris. *P. citricola*, *P. citrophthora*, and *P. cryptogea*, in addition to *P. cinnamomi* (Plate 140), *P. gonapodyides*, *P. lateralis* (Plate 139), and *P. megasperma* (Plate 138) may also cause root and crown rot of rhododendron. This phase of disease is characterized by stunting, epinasty, chlorosis, dieback, and eventual plant death.

Fruit rot. *Phytophthora* species cause fruit rot of diverse plants. The usual symptom is a soft brown decay that spreads from one spot to involve the entire fruit. Dead fruits drop or shrivel. *P. cactorum* is a common cause of fruit rot of cotoneaster (shown here), firethorn, pear, and other plants.

A. Pit canker of American elm, caused by *Phytophthora inflata*. Coalescing perennial cankers have extended more than halfway around the trunk of this tree (NY, Jun).

B. Shoot blight and dieback of rhododendron caused by *Phytophthora* sp. Infection has spread from the stem apex into leaves, causing olivaceous lesions (NY, Jul).

C. Fruit rot of rock cotoneaster caused by *P. cactorum*. Dark lesions expand to involve entire fruits, which eventually shrivel (CA, Oct).

D–F. Collar rot (crown canker) of dogwood. D. Drooping leaves, chlorosis, and dieback of top branches associated with a large basal canker on Pacific dogwood (OR, Sep). E. Large basal cankers on Pacific dogwood, their extent partially revealed where bark has cracked and fallen away (WA, Jun). F. Crown canker on flowering dogwood. Old lesions, previously inactive for a time, as evidenced by large callus ridges, are expanding beneath the outer scaly bark (NY, May).

Plate 137

Phytophthora collar rots, foot rots, and bleeding cankers. The terms *collar rot, crown canker, basal canker,* and *foot rot* refer to large lesions that develop at the trunk base and on the adjacent parts of major roots of trees or shrubs. Fungi that commonly cause such symptoms are *Phytophthora cactorum, P. cambivora, P. cinnamomi, P. citricola, P. cryptogea* (including *P. drechsleri*), *P. megasperma,* and *P. syringae.* Specific pathogens or combinations of them vary with host and geographic area. Some of them also cause bleeding trunk cankers and other symptoms such as tip blight, fruit rot, or necrosis of feeder roots. The incidence and importance of *Phytophthora* species were not fully apparent until contemporary times because the causal fungi were difficult to isolate by common procedures. Fast-growing fungi often obscured the slower phytophthoras. Bait techniques and selective culture media now allow the routine isolation of these fungi from soil and infected plants.

Collar rots are among the most important lethal disorders of fruit and nut trees, especially on sites that are poorly drained or are irrigated by flooding. Shade and ornamental trees are less often damaged, although significant collar rot or bleeding canker diseases occur in many species.

Phytophthora species that cause collar rots and bleeding cankers are soilborne and are distributed widely with nursery stock. They enter host tissues near the root collar via wounds or the succulent parts of small roots. They cause brown or reddish brown water-soaked lesions with abrupt margins. If a lesion enlarges for several years, only the marginal area shows the typical color and texture of newly killed tissue. Reddish brown liquid sometimes exudes from this area. Gum may also exude from plants such as citrus or cherry. Cankers often go unnoticed until foliar symptoms develop, by which time a lesion may extend halfway around the butt. As the pathogen destroys roots or the physiological connection between roots and stem, symptoms of distress appear on branches above. Leaves become stunted, sparse, and chlorotic. Leaves of deciduous trees develop premature autumn color, and twigs and branches begin to die back.

The most common, widespread, and important pathogen causing collar rot and bleeding canker on shade and orchard trees is *P. cactorum.* This fungus attacks plants in over 80 genera, causing fruit rot, twig dieback, and root rot in addition to cankers. It occurs from coast to coast in Canada and throughout the United States. Plants such as cotoneaster, firethorn, and pear are subject to fruit rot. Fruit and nut trees such as apple, avocado, cherry, and walnut are subject to collar rot. *P. cactorum* also causes bleeding cankers on trunks of various trees. Shade trees affected by bleeding canker or basal canker caused by this fungus include American and European beech; sweet birch; flowering and Pacific dogwoods; sweet gum; horse-chestnut; madrone; black, Norway, red, silver, sugar, and sycamore maples; California live, southern live, pin, and red oaks; tulip tree; and weeping willow.

Crown canker of dogwood. This disease, caused by *P. cactorum,* is destructive to landscape specimens of flowering and Pacific dogwoods. In early stages the disease is not apparent except for occasional bleeding of reddish brown fluid from bark at the base of the trunk. Eventually, however, the dead bark dries, cracks, and begins to fall away. Portions of the canker often become temporarily or permanently inactive, and a prominent callus roll may form at the edge. Thus the dead area may appear sunken. Leaves on branches above the

canker become slightly chlorotic and smaller than normal, with a tendency to bend downward (epinasty), fold along the midrib, and, in flowering dogwood, turn red in late summer. Dieback ensues. This same fungus also causes root rot of dogwood growing in poorly drained soils in nurseries.

The death of tissues infected by *P. cactorum* has been attributed to toxins that it produces. In laboratory assays this fungus does produce substances capable of killing various plant organs, but the role of these toxins in nature has not been ascertained.

Collar rot of apple. Any of several *Phytophthora* species may kill bark and cambium at the root collar of apple trees. *P. cactorum* is the most common pathogen, but *P. megasperma* caused the canker shown here. Nearly all common apple rootstocks are susceptible to some extent. Trees are most susceptible in spring near the time of flowering. They become relatively resistant beginning during the period of shoot growth. Accordingly, collar rot lesions expand rapidly during spring and slowly during the remainder of the warm season. The period of spring susceptibility coincides with the period of greatest activity of the pathogen in soil. Wet soil and moderate temperatures at that time favor germination of oospores or chlamydospores and local dispersal of zoospores.

Foot rot, or collar rot, of citrus. This disease, also known as gummosis because of the exudation of gum from small fissures in dying or recently killed bark, is caused primarily by *P. parasitica* and occasionally by *P. citrophthora.* Lesions usually begin near the graft union and spread upward more than down. Oozing gum marks their location until the bark finally cracks and begins to fall away. The aboveground parts of a lesion may cease expansion and become delimited by callus, but continued expansion below the soil line often leads to girdling and death. Partial girdling causes one-sided stunting and dieback. *P. parasitica* is favored by high temperatures (30–32°C) and water-saturated soil. The fungus becomes inactive as soil dries within a range that is still favorable for root growth. Thus trees may grow satisfactorily in infested soil much of the time. In contemporary times foot rot in citrus has been suppressed by using resistant rootstocks, but highly susceptible scions become infected through wounds near the graft union.

Collar rots and root rots caused by *Phytophthora* species are favored by flooded or water-saturated soil. Such conditions not only promote reproduction and dispersal of the causal fungi but promote the susceptibility of plant roots. Roots stressed by reduced oxygen supply in waterlogged soil exude more amino acids and other substances attractive to zoospores of *Phytophthora,* and root resistance to pathogens is impaired. Saline soil or previous water shortage also lowers plant resistance.

Surgical tactics and heat treatment for preservation of orchard, shade, and ornamental trees with basal cankers are available but not often used. These techniques require excavation to expose the lesion below soil line. One technique reported to halt the advance of the citrus foot rot pathogen is to heat the cambial region at the canker margin to approximately 60°C by lightly charring the surface of diseased bark with a kerosene blowtorch. Such techniques have not become popular because of impracticality in orchards and failure to reverse the decline of branches in landscape specimens. Also, these techniques are laborious, destroy healthy tissues around cankers, and leave unsightly scars.

A–E. Foot rot of citrus caused by *Phytophthora parasitica.* A. Sparse, chlorotic foliage on a severely affected grapefruit tree. B. A large, bleeding lesion near the trunk base, partly exposed by removal of bark. C. Dieback in Mandarin orange associated with foot rot. D, E. The root collar of a diseased Mandarin orange tree on which a foot rot lesion was revealed by excavation. The sunken canker on the subterranean part of the butt extends to the soil line (D). The canker margin was exposed by shaving surface bark (E) (TX, May).

F. Collar rot of apple caused by *P. megasperma.* Excavation at the base of a young, declining tree reveals that all the major roots on one side of the tree are dead. The swelling just above ground level is the graft union (NY, Aug).

Plate 138

Phytophthora root rot of Port Orford cedar. Most or all of the soilborne phytophthoras kill roots of some of their host plants whether or not they also cause other types of damage. Root rot of Port Orford cedar caused by *Phytophthora lateralis* is a particularly destructive example. Port Orford cedar is native to a small region in southwestern Oregon and northern California but grows satisfactorily in a much larger area of the Pacific Northwest. Phytophthora root rot of this species was noticed first in nursery stock in Washington in 1923 and appeared to spread north and south from there, becoming epidemic in the main range of the host plant near the Oregon coast. The epidemic intensified in forests and landscapes, especially along watercourses and roads, wherever infected plants or soil with spores of the causal fungus might move. The disease spread slowly from the lowlands toward the limits of the range of Port Orford cedar in the mountains. It occurs now from northern California to southern British Columbia, and it has greatly diminished the population of Port Orford cedar. Only the part of the population growing in undisturbed wilderness has escaped damage.

P. lateralis was described in 1942. It is suspected to be of Eurasian origin because Asiatic species of *Chamaecyparis* resist it; resistance may have arisen through coevolution with the pathogen. *P. lateralis* attacks trees of any size or age, entering succulent (nonsuberized) roots, foliage, or wounds to inner bark and spreading in the inner bark and cambial region. It kills small seedlings within a few weeks, and large trees within 2–4 years. Infected inner bark quickly turns cinnamon brown in contrast to the normal cream color of healthy tissue. The usual progression is from tiny roots to increasingly large ones until the tree is girdled at its butt. All of the foliage then withers and discolors simultaneously, turning successively chlorotic, bronze, and brown. Foliar discoloration is often preceded by slight wilting during warm days in spring.

Infection may also begin on leaves or stems. Symptoms there are initially localized, usually on low branches where foliage may brush against the ground when weighted by rain. Foliar lesions proliferate as zoospores from the initial lesion are splashed about and initiate new infections. Lesions also expand as mycelium of the pathogen grows toward the trunk. More branches become involved gradually, and the tree is slowly killed.

Although infection can occur at temperatures of 3–25°C, the pathogen is most favored at 15–20°C. Both aerial and root infections occur beginning in late autumn and are most frequent in early spring. At temperatures of 10–20°C, sporangia form on foliage within 36–48 hours after infection. Sporangia release zoospores but apparently are not themselves dispersed. Foliage that is wet for 2 hours or less may become infected if zoospores are present.

After infection, mycelium of *P. lateralis* grows in bark at rates averaging 5 cm per month during late autumn through early spring, but chlamydospores form and growth ceases as temperature rises above 25°C. Mycelial growth through root grafts between trees is thought to contribute to upslope movement of the pathogen and local intensification of the disease.

Oospores and chlamydospores in killed tissue of roots and foliage act as survival structures during summer. They require a low-temperature stimulus for germination. During wet weather after a cold period in autumn, they germinate and give rise to sporangia and zoospores. Whether mycelium survives summer in aboveground parts of infected trees has not been reported. The fungus has not been isolated from dry, formerly diseased tissue during summer, but lesions that are halted in summer resume expansion in autumn.

Port Orford cedar normally grows in association with several other coniferous species that are not harmed by *P. lateralis*. Alaska cedar, closely related to Port Orford cedar, grows farther north along the Pacific coast and is not significantly damaged. It was somewhat resistant to *P. lateralis* when tested experimentally. *P. lateralis* has been found associated with root rot of rhododendron in Ohio and Pennsylvania, but pathogenicity of rhododendron strains to cedar, or vice versa, has not been reported. Port Orford cedar may also sustain root rot by *P. cinnamomi* in areas of mild climate as far north as British Columbia.

Diseases caused by *Phytophthora cinnamomi*. Originally isolated from cinnamon trees in Sumatra, *P. cinnamomi* is one of the best known and most important members of its genus. It is considered to have originated in southeastern Asia or northeastern Australia. Now it occurs in tropical and warm temperate areas around the globe, attacking more than 900 species and varieties of plants. In North America it is common along the Pacific coast as far north as British Columbia and in eastern areas as far north as Ohio and New York. It is primarily a pathogen of trees and shrubs. Major hosts include avocado, chestnut, eucalyptus, fir, pine, and ericaceous ornamentals, especially azalea and rhododendron. This fungus causes root rot and death of many woody species in forest and ornamental nurseries, decline (littleleaf disease) of pines in forests of the southeastern United States, and a devastating decline and death of eucalyptus (jarrah dieback) in Western Australia.

P. cinnamomi attacks primarily the small absorbing roots but in many hosts also invades larger roots or the mainstem, growing mainly in the inner bark and cambial region. It derives nourishment from recently killed as well as living tissues, and it reproduces in the dead material. It produces four types of spores: oospores, chlamydospores, sporangia, and zoospores. Spores of the first two types allow the fungus to survive environmental adversity or persist several years in soil in the absence of hosts. Under favorable conditions in the presence of host roots, oospores and chlamydospores give rise to mycelium or to sporangia and zoospores. When water content of soil is insufficient for zoospore activity, hyphae may infect roots.

P. cinnamomi requires warm, wet soil for significant pathogenic activity. The fungus is intolerant of freezing and does not persist where soils regularly freeze deeply in winter. Along the Pacific coast, *P. cinnamomi* is important mainly in irrigated soils of nurseries and orchards. It is unimportant in western forests, perhaps because when the soils are warm enough for growth and sporulation of the fungus, they become too dry.

A, B. Phytophthora root rot of Port Orford cedar, caused by *Phytophthora lateralis* (OR, Jul). A. Dead and dying trees in a pocket of severe disease. Dying trees have chlorotic foliage. B. A basal canker on a dying tree. The pathogen has spread up from roots and is girdling the butt. Bark has been shaved off to reveal orange-brown phloem and cambium in the canker. Brown tissue above the healthy white inner bark is normal dead, corky outer bark.

C–F. Root rot of Fraser fir seedlings caused by *P. cinnamomi* (NC, May). C. Mortality in a nursery bed of 1-year-old seedlings. D. Dead and dying 5-year-old transplants. E. Dead foliage associated with dark brown dead roots on one side of a 5-year-old plant. F. A reddish brown lesion extending up the mainstem from dead roots. The tree at left was already dead when collected; the pathogen apparently spread from it into the living tree via root contacts.

Plate 139

Temperature optima for such functions as sporulation, infection, and mycelial growth are usually similar within an isolate but may vary between isolates in the range 20–32°C. Temperature minima for these functions are 5–15°C, and maxima are 30–36°C. Plants that become infected at low temperatures may lack foliar symptoms, or symptoms may be delayed until after a period of higher temperature or an environmental change stressful to the plant.

Soil water content and nature of soil microbe populations are the most important factors influencing infection. Disease is most severe in imperfectly drained soils because water-saturated soil promotes formation and dispersal of sporangia and zoospores. The interval between infection and sporulation of *P. cinnamomi* is only a few days, so disease can increase explosively if infested soil is flooded or waterlogged. Microbial populations influence the conduciveness or suppressiveness of soil to disease caused by *P. cinnamomi*. Most nursery soils and those formerly used for agricultural crops are likely to be conducive.

Pathogenic specialization has been detected in strains of *P. cinnamomi* from various hosts, but in general an isolate from one kind of plant is capable of infecting many others.

Phytophthora root rot of Fraser fir. This disease (Plate 139) is typical of those caused by *P. cinnamomi* in forest nurseries and young plantations. Symptoms begin with reddish brown decay of rootlets and of the cortex and secondary phloem of woody roots. Root loss leads to cessation of growth and then chlorosis, drooping, and browning of foliage. Symptomatic plants occur singly and in enlarging groups.

Fraser fir, native at high elevations in the southern Appalachian mountains, is grown in nurseries and plantations at lower elevations. In the nurseries, seedlings are attacked by *P. cinnamomi* and other *Phytophthora* species that cause root rot and death. At a given time, many seedlings with diseased roots may lack foliar symptoms. The shipment of infected seedlings from nurseries results in establishment of *P. cinnamomi* in plantations where it may kill the seedling that carried it and may then attack adjacent healthy trees. Once established at a site conducive to infection, this pathogen may persist indefinitely by means of its resting spores and its ability to infect the roots of a broad array of plant species.

Fortunately, many forest soils are suppressive to *Phytophthora* species. In the Pacific Northwest, for example, *P. cinnamomi* and other phytophthoras that cause root rot in nurseries are regularly carried in infected seedlings to forest planting sites. Seedlings with root disease at the time of outplanting may die, but disease does not spread or intensify, and new roots generated by surviving trees remain healthy. Root rot of Port Orford cedar caused by *P. lateralis* (Plate 139) is an important exception.

Littleleaf disease of pines. In southeastern states, *P. cinnamomi* in combination with a stressful environment causes long-term damage to shortleaf and loblolly pines. The disease became of concern during the 1930s and 1940s, when pole-sized and mature pines began to decline. Sites conducive to littleleaf tend to be severely eroded and characterized by poor aeration (and associated periodic waterlogging), low fertility, and periodic water shortage. Low fertility and water shortage impair the recuperative ability of the trees. Symptoms above ground are slow growth; chlorotic, stunted, tufted foliage; and progressive dieback. Early in this sequence the tree may signal distress by producing an abnormally large seed crop. Diseased trees live an average of 6 years after the onset of symptoms.

In littleleaf disease, *P. cinnamomi* kills numerous feeding root tips but not structural roots. The root system then cannot adequately supply water and nutrients to the foliage and branches. Chlorosis is due mainly to impaired uptake of nutrients, especially nitrogen. Littleleaf symptoms in landscape trees can therefore often be arrested or reversed by application of nitrogen fertilizer. Forest stands with severe littleleaf disease must be converted to resistant species. Ectomycorrhizal fungi apparently protect some pine rootlets from *P. cinnamomi*, but this is not sufficient to suppress littleleaf on soils with conducive physical characteristics.

Root rot and wilt of ericaceous plants. Root rot and wilt (one disease) is important in Japanese andromeda, azaleas, and rhododendron in nurseries. The disease is caused primarily by *P. cinnamomi*, but other *Phytophthora* species are often also involved. Many plants become infected in propagation beds and remain symptomless until after transplanting, and spot outbreaks in ground beds or container-grown crops result. Symptoms appear on one plant and subsequently on those surrounding it.

In typical cases fine roots become infected and turn brown. If infection is confined to roots, plants become stunted and chlorotic with dieback of scattered branches. The pathogen usually spreads toward the root crown and there either girdles the plant or advances up the stem in the cambial region and phloem, eventually colonizing the xylem and pith and reaching green shoots. Affected plants or branches cease growth and the leaves droop, turn chlorotic to dull yellowish green, and wilt permanently. All colonized tissues turn brown, and chlamydospores form in dead cortex and phloem.

Rhododendrons and azaleas resistant to *P. cinnamomi* have been identified but have not replaced highly susceptible cultivars in commerce. Unfortunately, flooding, water shortage, or salinity stress can compromise the resistance of rhododendron cultivars. Root rot and wilt in nurseries can be suppressed by growing plants in root substrates that are suppressive to the pathogen (e.g., composted hardwood bark or a substrate at pH near 3.5) and by placing container-grown plants on beds of stones such that water drains away from containers and spores do not splash up into them.

References for pit canker and Phytophthora diebacks and fruit rots: 150, 323, 387, 547, 625, 1080, 1081, 1604
References for collar rots of orchard trees: 84, 185, 387, 887, 976, 1038, 1233, 1234, 1330, 1538, 1747, 1748, 2175, 2211
References for bleeding cankers and collar rots of shade trees and woody ornamentals: 185, 241, 320, 387, 418, 519, 896, 952, 1327, 1926
References for Phytophthora lateralis: 1499, 1653, 2035, 2047
References for Phytophthora cinnamomi: 71, 147, 149, 186, 187, 306, 387, 721, 728, 729, 858–61, 1002, 1003, 1187, 1229, 1499, 1599, 1721, 1770, 1771, 2172, 2277

A, B. Littleleaf disease of pines. A. Shortleaf pine: declining trees in contrast to one relatively healthy individual. B. Loblolly pine with chlorotic, stunted, tufted foliage and a distress crop of cones (GA, Aug).

C, D. Root rot of azalea caused by *Phytophthora cinnamomi*. C. The stem base of a potted plant, with bark shaved to reveal brown, necrotic xylem and inner bark. D. Aboveground symptoms in contrast to the appearance of one healthy plant of 'Hershey' azalea (NY, Aug).

E. Phytophthora wilt of rhododendron (NY, Aug).

F–H. Phytophthora root rot and wilt of Japanese andromeda. F. A stunted chlorotic plant in contrast to healthy plants in a nursery. G. Chlorosis and wilt. H. The stem base of a diseased plant, with bark shaved to show brown, necrotic bark and wood (NY, Jul).

Plate 140

Phymatotrichum and Charcoal Root Rots (Plate 141)

Phymatotrichum root rot. *Phymatotrichum omnivorum* (syn. *Phymatotrichopsis omnivora*) (Deuteromycotina, Hyphomycetes) causes Phymatotrichum root rot, also known as Texas root rot, in the region extending from western Louisiana and Arkansas to southern California and Nevada, also in Utah and Mexico. This pathogen attacks over 2000 plant species, mainly dicots. It causes significant economic damage to crop plants, especially cotton, and is lethal to many species of woody plants in orchards, landscapes, and shelterbelts. Highly susceptible woody plants include almond, apple, apricot, heavenly bamboo, western catalpa, chinaberry, cottonwood and poplar, elm, Japanese euonymus, fig, ginkgo, grape, lilac, black locust, honey locust, weeping myall, paper mulberry, white mulberry, cork oak, silk-oak, pear, pecan, pepper tree, Japanese pittosporum, privet, quince, rose, silverberry, tree-of-heaven, and willow.

The disease tends to occur in expanding patches that involve many plants. *P. omnivorum* causes multiple, coalescing necrotic lesions on roots of all sizes. It enters through lenticels, wounds, and other breaks in the periderm. Host cells are killed in advance of hyphae, apparently by a heat-labile toxin. The cortex and phloem become brown and mushy and are eventually decomposed by soil saprophytes. The point of lethal attack is the root collar or tap root within the upper 30 cm of soil. Trees girdled at the root collar may still have root ends that appear healthy. Infected roots still alive at the periphery of the patch of diseased plants bear small lesions characteristic of early stages of disease. Symptoms on aboveground parts often do not appear until the plant is dying. Leaves may then become slightly yellow or bronze before they wilt and dry. Small or highly susceptible plants may die within several weeks after first infection. Diseased trees, depending upon level of susceptibility, may live for several years and show stunted new growth, yellowing, and dieback for 1 or more years before death. Resistant plants, although infected, may persist indefinitely.

P. omnivorum spreads along root surfaces and through soil at rates up to 9 meters per year by means of brown mycelial strands, as illustrated on a cotton root. The strands, consisting of a large central hypha surrounded by many small interwoven hyphae, may form a network on the root surface. The tips of mycelial strands differentiate into individual hyphae that penetrate the root at numerous points.

Mycelial strands are a useful diagnostic feature, but microscopic examination is necessary to complete the diagnosis. The observer looks for cross-shaped brown hyphal structures (cruciform hyphae) arising at right angles from a mycelial strand. The tip and arms of the cruciform structure are pointed (acicular). Only *P. omnivorum* is known to have such structures. *P. omnivorum* also produces sclerotia (resting structures) that are resistant to environmental extremes and microbial attack. These form along mycelial strands in soil at depths to 2 meters or more. The sclerotia are 1–3 mm in diameter, initially light colored but eventually black, and can survive several years in the absence of hosts. In moist soil in the presence of host roots sclerotia produce new infectious mycelium.

Under warm moist conditions, for example after abundant summer rain, the pathogen may produce on the soil surface around the stem of a diseased plant a buff-colored mat ("spore mat") of mycelium bearing conidiophores and conidia. This is more common beneath the canopy of a field crop than around the exposed bases of woody plants. The function of the conidia in nature has not been learned.

P. omnivorum was once thought to be the asexual state of a basidiomycete, *Sistotrema brinkmanii,* but this was an error. No sexual state of the pathogen is known.

P. omnivorum is favored by warm, alkaline (pH 7.2–8.5), calcareous soils with high content of montmorillonite clay, conditions that in North America apparently exist only in the southwestern United States and Mexico. Acid soils or those with high sodium content inhibit the fungus. In soils conducive to its survival, *P. omnivorum* inhabits the roots of many kinds of plants that do not have symptoms above ground. The fungus grows and produces sclerotia at temperatures of 15–35°C and does so most rapidly at 28°C.

Where soil characteristics are conducive to Phymatotrichum root rot, resistant plants may be grown. Monocots and many dicots native to the Southwest are resistant. Resistant plants include sweet acacia, bamboo, century plant, blue elderberry, the firethorn *Pyracantha atalantioides,* desert gum, Murray red gum, Chinese and Rocky Mountain junipers, wax mallow, mesquite, oleander, Russian olive, palms, Siberian pea tree, western soapberry, Arizona sycamore, desert willow, and yucca.

Charcoal root rot. *Macrophomina phaseolina* (Deuteromycotina, Coelomycetes), the cause of charcoal root rot, affects more than 300 plant species, including many agricultural crops as well as forest seedlings, in warm temperate and tropical regions around the globe. It occurs in nurseries, agricultural fields, and uncultivated soils (on weed hosts) in the southern and southwestern United States. We illustrate the disease and pathogen on caper spurge, a plant cultivated experimentally as a possible source of hydrocarbons for fuel. Trees are damaged by *M. phaseolina* while in the seedling stage, primarily in nurseries. Highly susceptible species include Fraser, red, and white firs; Douglas-fir; many pines; giant sequoia; and Engelmann spruce. Jeffrey and ponderosa pines are somewhat resistant.

M. phaseolina was formerly known by the name of its sclerotium-forming stage, *Sclerotium bataticola.* In forest nurseries it is often accompanied by *Fusarium* species (Plate 143).

Symptoms on nursery seedlings are necrosis and blackening of roots, leading in severe cases to stunting, chlorosis, and death of the seedling. Infection progresses up the root system, resulting in distinctive roughening and blackening of patches of bark on the tap root and major laterals. The woody cylinder, although dead, remains white. Roughening results from production of abnormally abundant cork and phelloderm cells. The fungus produces numerous tiny black sclerotia in killed bark of roots and the lower stem. The sclerotia remain in soil after dead roots decay, germinate when new roots grow close to them, and are responsible for new infections.

Charcoal root rot intensifies during hot summer weather. The pathogen can cause severe damage at soil temperatures above 15°C, but it grows most rapidly at 28–32°C and causes most damage in soil above 30°C. The reason may be that heat stress lowers the resistance of plant roots. Some strains of *M. phaseolina* that attack agricultural crops produce pycnidia and conidia, but no conidial state has been associated with tree diseases.

References for Phymatotrichum root rot: 254, 1183, 1479, 1919, 1974, 2248
References for charcoal root rot: 482–84, 570, 845, 1654, 1749, 1847, 1850, 2109, 2264

A. Death of sour orange trees along a residential street typifies the patchy occurrence of damage caused by *Phymatotrichum omnivorum* (AZ, Jul).

B–D. Signs and symptoms of Phymatotrichum root rot on woody taproots of cotton. B. Brown strands of intertwined hyphae on the root surface. C. A diseased tap root, its tip dead and the periderm shaved off to reveal numerous small lesions that will coalesce. D. A tap root in which the pathogen has advanced nearly to the soil line; killed tissue is brown (AZ, Jul).

E–H. Macrophomina root rot of caper spurge. E. Wilting and brown foliage on a dying plant; the root system is already dead. F. Darkly discolored internal tissues of the stem base, exposed where the periderm has peeled away. G, H. Successively enlarged views of microsclerotia (black objects), exposed by removal of periderm, responsible for the discoloration visible in (F) (AZ, Jul).

Plate 141

Cylindrocladium Root Rot and Blight and Thielaviopsis Root Rot (Plate 142)

Cylindrocladium root rot and blight. Several species of Cylindrocladium and Cylindrocladiella (Deuteromycotina, Hyphomycetes) cause root rot and blight. These fungi affect crop plants and nursery stock around the world. The two species most widespread and important on woody plants in North America are Cylindrocladium floridanum and C. scoparium. Others are C. avesiculatum, C. crotalariae, C. ellipticum, C. theae, and Cylindrocladiella parva. All are capable of infecting diverse plants, and most also cause diverse symptoms such as damping-off, rot of cuttings, root rot, hypocotyl rot, leaf spots, foliar blight, lesions on green stems, twig dieback, and wilting and death. Some species, such as Cylindrocladium ellipticum on Mahonia bealei and C. avesiculatum on holly, are known mainly as foliar pathogens. In aggregate, these fungi attack plants in more than 65 genera.

Cylindrocladium diseases are indistinguishable from one another on the basis of symptoms. In most situations diagnosis can be made and the specific pathogen identified only through isolation and microscopic examination of the fungus.

Severe losses to Cylindrocladium root rot may occur in nurseries that produce either forest tree seedlings or ornamental and fruit trees. C. scoparium attacks coniferous species in forest nurseries from Quebec to Minnesota, and it attacks both coniferous and deciduous species in the southeastern United States. Species reported damaged are Douglas-fir, eucalyptus, balsam and Fraser firs, sweet gum, pines (Austrian, jack, mugo, red, Scots, and eastern white), spruces (black, blue, Norway, and white), tulip tree, and black walnut. Among coniferous species tested for susceptibility to C. scoparium, only arborvitae was resistant. In nurseries that produce fruit trees and ornamentals, apple, apricot, azalea, bottlebrush, silver buttonwood, cedar (Cedrus), cherry, fig, holly, leucothoe, lilac, mahonia (M. bealei), peach, plum, redbud, rhododendron, and rose may be damaged.

Significant damage by Cylindrocladium to woody plants outside nurseries and greenhouses is unusual, but C. scoparium has been found associated with root rot and decline of peach trees in orchards, of pole-sized tulip trees in a plantation, and of sweet gum in a natural forest stand.

Root infections are characterized by multiple dark brown to nearly black lesions, often with longitudinal cracks, on lateral and tap roots. Coalescing lesions may girdle the tap root or root collar. Severe root disease leads to stunting, yellowing, wilting, and death of seedlings. If damage is sublethal, new roots may proliferate above the killed part of the root system.

Often several types of symptoms caused by Cylindrocladium occur in the same nursery at one time. Dark lesions form on leaves of broad-leaved plants. Infected conifer needles turn yellow and then, in some pines, red-brown. Stem infections of both conifers and broad-leaved plants arise at leaf bases and may girdle small stems. Infection of aboveground parts is fostered by warm moist conditions such as are provided in summer by overhead irrigation and close spacing of plants. Cylindrocladium species grow most rapidly at 24–28°C.

Azalea affected by Cylindrocladium in nurseries or greenhouses displays two distinct syndromes: root rot and wilt, ending with plant death, and leaf blight. Leaf blight leads to defoliation, but stems remain alive and the plants produce new leaves.

Under hot humid conditions, wefts of brown mycelium and/or tiny white tufts of conidia may be produced on lesions on leaves or stems. The conidia are colorless, long-cylindrical, and two- to several-celled. They are dispersed by air or splashing water and infect roots or aerial parts of plants. Germ tubes from conidia penetrate conifer needles via stomata.

Cylindrocladium species produce brown, irregularly shaped microsclerotia, 40–200 μm in the longest dimension, in killed roots, leaves, flowers (of azalea), and the bark of stem lesions. As a rule, microsclerotia can be detected only by microscopic examination. Microsclerotia tolerate environmental extremes and are resistant to microorganisms. Survival of microsclerotia for 7 years in fallow soil has been reported. Microsclerotia are dispersed with soil or seedlings. In soil they germinate when stimulated by chemicals exuded from nearby roots, producing mycelium that causes new infections.

C. scoparium is capable of competitive saprophytic colonization of plant materials in soil, and it can persist in association with roots of plants that show no symptoms. A high population of C. scoparium was found in the soil and associated with clover roots in an abandoned Minnesota nursery 14 years after production of tree seedlings ceased.

Several species of Cylindrocladium are the conidial states of perithecial fungi in the genus Calonectria (Hypocreales, Hypocreaceae). Confirmed relationships include Cylindrocladium crotalariae–Calonectria crotalariae, Cylindrocladium floridanum–Calonectria kyotensis, and Cylindrocladium theae–Calonectria theae. The perithecial state of Cylindrocladium scoparium, noted occasionally in laboratory cultures, is unnamed. These fungi produce orange to yellow-orange perithecia 0.25–0.45 mm in diameter and height on killed leaves and bark under moist conditions. Blobs of colorless ascospores extrude from perithecia and are dispersed by splashing water. Ascospores of Calonectria theae have been shown to be infectious; however, ascospores are probably much less important than microsclerotia and conidia as inoculum.

Thielaviopsis root rot. This disease is caused by Chalara elegans (Deuteromycotina, Hyphomycetes), formerly known as Thielaviopsis basicola. This fungus is confined to roots, soil, and sometimes graft unions close to the soil. It occurs around the temperate zones of Earth, associated with plants in more than 130 genera, mainly herbaceous species. It causes black root rot of various plants, but the majority of its plant associations are apparently benign. Japanese holly in nurseries and landscapes may sustain growth loss, sparse foliage, poor foliar color, and dieback after the death of fibrous roots. Other woody hosts affected by root rot include western catalpa, citrus, American elm, garland flower, black locust, and poinsettia. In roots the fungus grows intracellularly and causes many small dark lesions that may coalesce and blacken much of the root system. Microscopic, colorless conidia and brown cylindrical, multicelled chlamydospores form on the lesions, and chlamydospores also form within lesions. Chlamydospores serve as survival structures in soil. When stimulated by chemicals that exude from plant roots, chlamydospores germinate and may cause new infections.

C. elegans has also been blamed for graft failure in common camellia, honeysuckle, and tree peony. It grows on the cut surfaces, producing a gray mat of mycelium and conidiophores. Apparently it kills enough tissue there to prevent union.

References for Cylindrocladium root rot: 100, 211, 277, 387, 405, 406, 413, 577, 594, 648, 1000, 1067, 1141–43, 1277, 1499, 1642, 1647, 1859, 1860, 1986, 2009
References for Thielaviopsis root rot: 387, 404, 1094, 1095, 1578, 2184, 2202, 2256

A, C. Wilting and dieback of azalea caused by Cylindrocladiella parva. Shoots on diseased branches wilt, shoot bases turn brown, and buds remain closed (NY, greenhouse).

B, F. Root rot of a yellow poplar seedling caused by Cylindrocladium scoparium. Dark brown lesions on otherwise light-colored roots are typical (MS, greenhouse).

D, E. Thielaviopsis root rot of Japanese holly. D. A diseased plant, its growth suppressed by root infection. E. Numerous brown lesions revealed in a magnified view of fibrous roots of the plant shown in (D) (VA, greenhouse).

Plate 142

Southern Blight and Fusarium Root Rot (Plate 143)

Southern blight. This disease is caused by *Sclerotium rolfsii* (Deuteromycotina, Hyphomycetes), a cosmopolitan fungus first described in Florida in 1892. It attacks plants of over 500 species in the tropical and temperate regions of Earth. In the United States southern blight occurs sporadically as far north as the states of Washington and New York but is important mainly in the Southeast. It is common also in California, but diagnostic signs at the bases of infected plants are often absent there because of dry weather.

Although capable of killing woody plants up to several years old, *S. rolfsii* is important mainly on herbaceous plants and on young woody plants not yet protected by corky bark in the root-collar region. Woody hosts include apple, Japanese aucuba, avocado, azalea, western catalpa, Japanese cedar, century plant, citrus, daphne, fig, grape, hydrangea, jasmine, loquat, white mulberry, olive, Russian olive, oil palm, papaya, pawpaw, peach, pittosporum, quince, rose, schefflera, tung-oil tree, and black walnut. With but few exceptions involving young transplanted trees, trees such as apple are subject to southern blight only while in the nursery.

The symptoms and signs of southern blight are similar for all hosts. The disease typically appears on scattered plants interspersed among healthy ones. It may be noticed only when a plant suddenly wilts and dies. Infection occurs near the soil line, and death results from girdling at the stem base or root collar. The pathogen rapidly kills cortex and secondary phloem a few centimeters above and below the soil line. If the stem base is not covered by corky bark, the lesion is visible as a darkly discolored area. As attack begins, a coarse web of white mycelium grows conspicuously on the stem base and on the surrounding soil. The fungus penetrates succulent or wounded tissues by growth of hyphae inward from appressoria that develop a short distance behind the advancing edge of the mycelium. The interval from infection until plant death is usually less than 1 month. At about the time of plant death, sclerotia form on the surface mycelium, which then disappears. The sclerotia, spherical and 1–2 mm in diameter, are white at first but soon become brown. The surface mycelium and sclerotia are diagnostic.

The sclerotia have a brown rind and white internal tissue that stores organic compounds to be used later for growth. Sclerotia persist on crop debris and weed hosts and are dispersed with soil or flowing water. Moist sclerotia germinate when stimulated by chemicals such as volatile aldehydes that emanate from plant materials. Sclerotia germinate either by the growth of individual hyphae through the rind or by eruptive germination. The latter process involves rupture of the rind by a white mass of mycelium that grows toward the chemical stimulus. Mycelium from an eruptively germinated sclerotium may infect plant tissue as far as 3.5 cm distant. Alternate wetting and drying may stimulate sclerotial germination in the absence of susceptible plant material, but this is to the detriment of the fungus because its mycelium in soil may be killed by parasitic or antagonistic microorganisms. The fungus is capable of saprophytic growth, however, and may spread in surface soil. In fact, a dead organic substrate as a food base for *S. rolfsii* often seems to be a prerequisite for infection.

Southern blight is favored by hot (30–35°C), wet weather and acidic (pH 3–6), well-aerated soils. As wet soil begins to dry, however, the fungus may continue growth and attack plants below the soil line.

S. rolfsii is the sclerotium-forming stage of a basidiomycete, *Athelia rolfsii* (syn. *Corticium rolfsii, Pellicularia rolfsii*) (Aphyllophorales, Corticiaceae). The basidial stage occasionally develops on killed leaves of herbaceous plants near the soil line. Basidiospores, although infectious, are considered to play only a minor role in epidemiology of southern blight.

Fusarium root rot. *Fusarium* species (Deuteromycotina, Hyphomycetes) are significant pathogens in nurseries, causing seed rot, damping-off, hypocotyl rot, cotyledon blight, stem rot, and root rot of coniferous seedlings. These several disorders tend to be caused by distinct fungal strains that display specialized behavior. We deal here with Fusarium root rot, which is caused most commonly by strains of *F. oxysporum* and *F. solani*. These fungi are invariably accompanied by other more or less pathogenic fusaria and often act in concert with other fungi or with nematodes. Therefore the symptoms displayed by seedlings in nurseries (stunting, chlorosis, root necrosis, death) represent responses to a combination of organisms as conditioned by local environment and cultural practices. Many species of conifers are affected. The disease is usually most significant in the 1st growing season, but root rot and mortality of seedlings up to 3 years old have occurred in some eastern nurseries. Fortunately, most forest planting sites are unfavorable for fusaria that cause damage in nurseries.

Fusarium root rot of Douglas-fir and other coniferous seedlings in British Columbia and the northwestern United States has been characterized rather fully and is representative of diseases in this group. The disease is important mainly during the 1st year of seedling growth, and *F. oxysporum* is the primary pathogen. The fungus is soilborne, persisting as microscopic chlamydospores in decaying plant debris. Chlamydospores germinate in the presence of seedling roots, and infection occurs during the first 6 weeks after seed is sown. The tap root is killed, and lateral roots also perish soon after they emerge. Adventitious roots may form close to the root collar. Severely affected seedlings either die during hot weather in late summer or survive stunted until winter and then perish as the result of frost heaving or infection by other opportunistic pathogens during mild weather in late winter. Chlamydospores form in dead root tissue, thereby providing inoculum for disease in the next seedling crop.

In the western version of Fusarium root rot, seedlings that survive to the 2nd growing season usually outgrow the effects of 1st-year infection, although their roots still harbor the pathogen. Many such seedlings do not attain sufficient size for outplanting with others of the same age, however, and are therefore culled.

Fusarium root rot is promoted by urea fertilizer or unrotted organic amendments. Therefore the disease is managed in part by regulating nitrogen source and applying only the minimum necessary for seedling growth during the 1st year.

References for southern blight: 74, 75, 387, 396, 577, 1144, 1563–65
References for Fusarium root rot: 200, 201, 203, 269, 670, 1287, 1394, 1499, 1805, 1936

A, C, F. Southern blight, caused by *Sclerotium rolfsii* (FL, summer). A. White mycelium, a diagnostic sign, on the stem base of a diseased schefflera and on surrounding soil. C. Small sclerotia, white at first, brown when mature, at the base of an apple seedling. F. Close view of mycelium and sclerotia at the base of a dying bugleweed.

B, D, E, G. Fusarium root rot of Douglas-fir seedlings caused by *Fusarium oxysporum*. B. Dead and distressed 1st-year seedlings; curling of top needles indicates distress but is not diagnostic (WA, Aug). D. First-year seedlings stunted by severe root rot, in comparison with a normal seedling of the same age (WA, Aug). E. Enlarged view of the severely damaged root system of a 1st-year seedling. Major lateral roots are absent, and short roots as well as the tap root are decayed. G. Diseased 2-year-old seedlings showing stunting, chlorosis, sparse lateral root development, and brown decayed roots (NY, Aug).

Plate 143

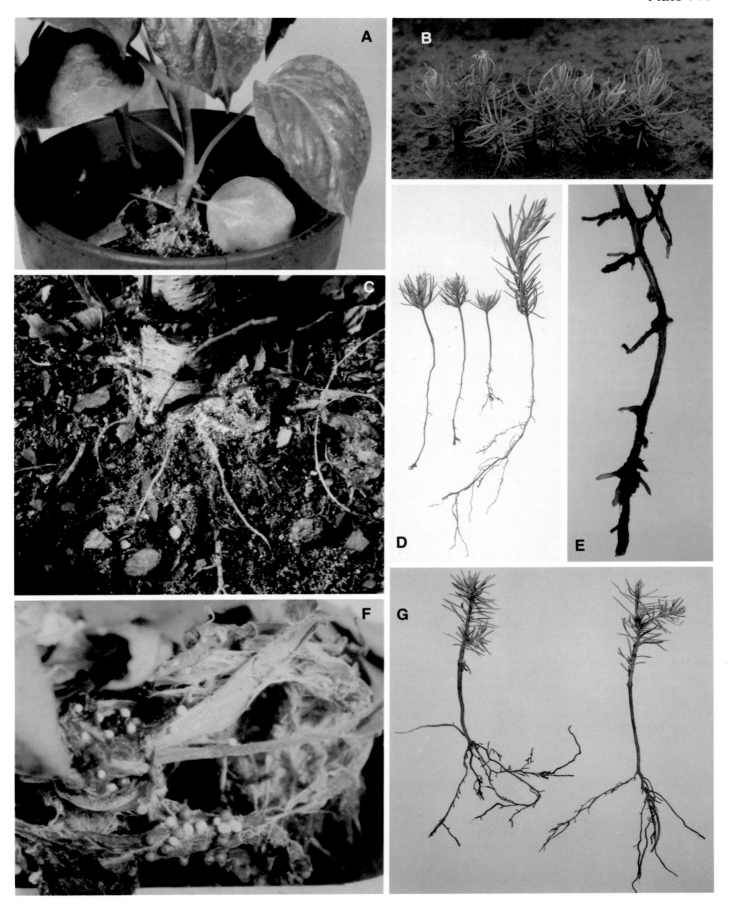

Nematode Diseases of Roots (Plates 144–145)

Here we provide a glimpse of diseases caused by root-feeding nematodes, a topic that could occupy an entire book. Nematodes (phylum Nematoda) are soft-bodied, tubular animals that inhabit many ecological niches: water, soil, plants, and other animals. Most species are free-living and consume various microorganisms as food, but many hundreds (perhaps thousands) of species parasitize plants or other animals. Nematodes reproduce by eggs and pass through four larval stages before adulthood. Each larval stage terminates when the nematode molts or sheds its outer covering (cuticle). Males of some species are unnecessary for reproduction, and some species apparently lack males. More than 70 genera of plant-parasitic nematodes are recognized on the basis of conservative criteria; we deal with representatives of only 5 genera.

Plant-parasitic nematodes are small (most species less than 2 mm long), usually colorless, and worm shaped in one or more stages of development. They move like eels. They require wet surfaces for self-propulsion and are dispersed with soil, plants, and flowing water. Some species attack succulent aboveground plant parts, and a few species such as the pine wood nematode (Plate 184) invade plants systemically. Most plant-parasitic nematodes attack only roots—small, succulent roots not protected by periderm.

Plant-parasitic nematodes puncture the walls of plant cells by means of a sclerotized spear, or stylet, and suck out cytoplasm. During or preparatory to feeding, they produce secretions that aid digestion or alter the plant in various ways favorable for the nematode: dissolve intercellular cement, degrade cell walls, suppress cell division, induce cell proliferation, or induce formation of giant cells that serve as long-term sites of feeding. Nematodes infest mycorrhizal as well as nonmycorrhizal roots. Some plant parasites feed exclusively from outside the root (ectoparasites); others enter and remain within roots for much of their development (endoparasites). Some move from one feeding site to another (migratory); others are sedentary. The females of some sedentary types become swollen and pear shaped to spheroid. The duration of the life cycle varies with species from a few days to a year, and the life span varies from several weeks to 5 years. More types of plant-pathogenic nematodes are important in tropical and warm temperate regions than in cool regions.

Symptoms on roots are sometimes diagnostic, as with root knot, but are usually not. Nematodes may induce various swellings, cause lesions, suppress root elongation or differentiation, or interact with other microorganisms to cause symptoms such as root rot or wilt where neither type of organism alone would do so.

Symptoms on aboveground parts of plants with infested roots are of little or no diagnostic value and are not conspicuous until much root damage has occurred. Infested plants may grow slowly and become chlorotic or have other symptoms of nutrient deficiency associated with loss of root function. Leaves may wilt temporarily during the heat of the day or may develop marginal browning, or the plant may be abnormally subject to winter damage or attack by opportunistic fungi or bacteria.

Specific diagnosis of most nematode diseases requires microscopic examination of adult nematodes. This entails extracting juvenile and adult nematodes from soil or plant material, identifying and counting the types present, and interpreting the data. For diseases such as root knot, in which the parasite causes diagnostic symptoms, the counting step may not be necessary. In either case, only a person trained in plant nematology can complete the diagnosis.

The interpretation of nematode numbers is essential for several reasons. (1) Plants vary in suitability as hosts for a particular nematode; plants capable of supporting only a small population may not be damaged. (2) Plants vary in ability to tolerate parasitism by a given nematode; a nematode population large enough to suppress growth of one plant may not affect a different plant. (3) Plant-parasitic nematodes vary in ability to damage a particular plant; a small population of one nematode might cause significant root damage, but a much larger population of a different nematode might cause little harm.

Control, other than with pesticides, is difficult because most root-infesting plant-parasitic nematodes can reproduce on a wide array of plants, including common weeds. Thus they persist at a particular site. Damage to container-grown plants can be prevented by utilizing sterilized or soilless root media and avoiding contamination from nearby soil. Selection or breeding of resistant plants has not yet been emphasized for woody ornamental or forest tree species.

Root knot nematodes. *Meloidogyne* species cause root knot and are globally important plant pathogens. More than 35 species are recognized. Several occur in North America, and four are widespread and important on woody plants. These are *M. hapla* (northern root knot nematode) and three species that inhabit warm regions: *M. arenaria*, *M. incognita*, and *M. javanica*. Woody plants subject to severe root knot include abelia, azalea, barberry, boxwood, deutzia, flowering dogwood, fig, forsythia, gardenia, Japanese holly, lilac, peach, rose, schefflera, weigela, and willow.

Root knot nematodes are sedentary endoparasites. Only males and second-stage larvae are mobile, and only the second-stage larvae and adult females feed. Eggs are produced at the surfaces of knots by stationary females embedded in the knot, and the first larval stage develops within the egg. Second-stage larvae hatch out, migrate a short distance in soil, enter succulent roots near the root tip, and migrate to the region of differentiating xylem. Larvae come to lie with their bodies in the cortex and their heads within the vascular cylinder, where they feed. This feeding stimulates knot formation; plant cells in the outermost stele and the cortex undergo abnormal division and growth. Also, several abnormally large, multinucleate cells (giant cells) rich in cytoplasm differentiate around the nematode's head. By moving its head only slightly, the animal can ingest food from each cell. Giant cells remain alive throughout the life of the nematode. The larva swells to sausage shape, then undergoes three molts within only a few days and becomes an adult male or female. If male, the animal is worm shaped. It leaves the root, mates, and remains free in soil. If female, the nematode remains in place, and her body swells, becoming nearly spherical except for the head region. The swollen white female, more than 0.1 mm in diameter (Figure 14C), can be exposed by careful dissection. Her posterior is near the root surface, where she deposits several hundred eggs into a gelatinous egg sac.

Under favorable conditions in warm soil (25–30°C), a new generation may be produced in 3–4 weeks. On suitable hosts the population increases until it is limited by food supply. Severe infestation may result in conversion of the feeding root system to a mass of short, thick, bumpy roots subject to decay by secondary pathogens.

Root knot nematodes interact with other pathogens to induce more damage than either would cause alone. Crown gall (Plate 73) and several fungal root rot and wilt diseases may be aggravated by root knot nematodes.

A–C, F. Root knot of gardenia (FL, Apr). A. Diseased container-grown plants in a nursery; the plant at left is chlorotic and severely stunted. B. Roots of a severely diseased plant are distorted by large numbers of small galls ("knots"). Normal rootlets are scarce. C. Close view of a severely diseased root. F. Magnified view of knots on a small portion of a root.

D, E. Root knot caused by the northern root knot nematode, *Meloidogyne hapla*, on forsythia (NY, Feb). D. Distorted, knotted roots. E. Close view of knots. Note browning of tissue and scarcity of tiny rootlets.

Plate 144

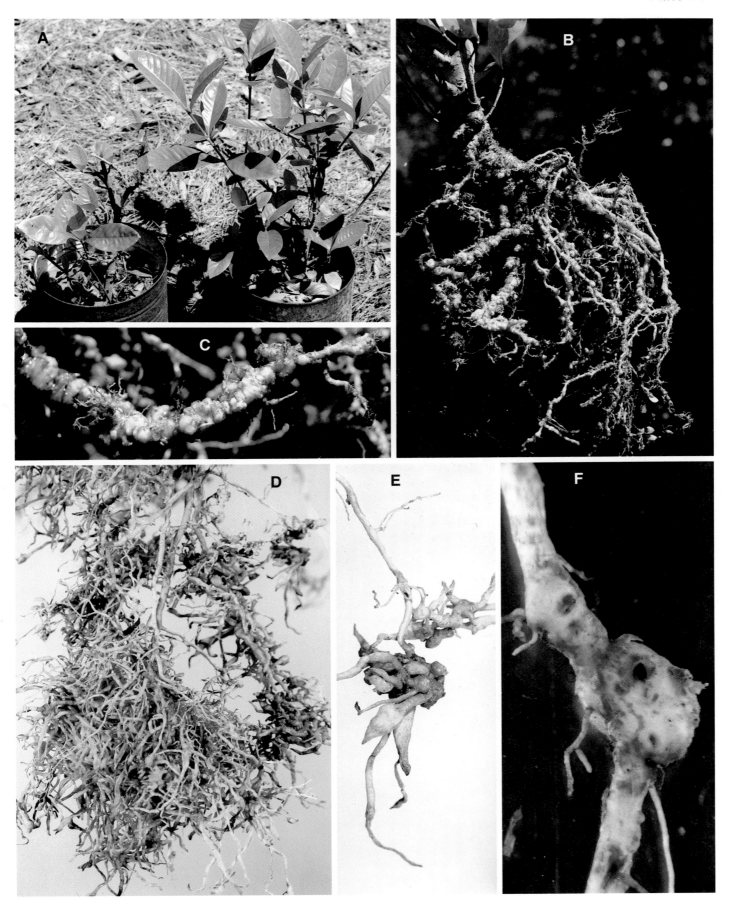

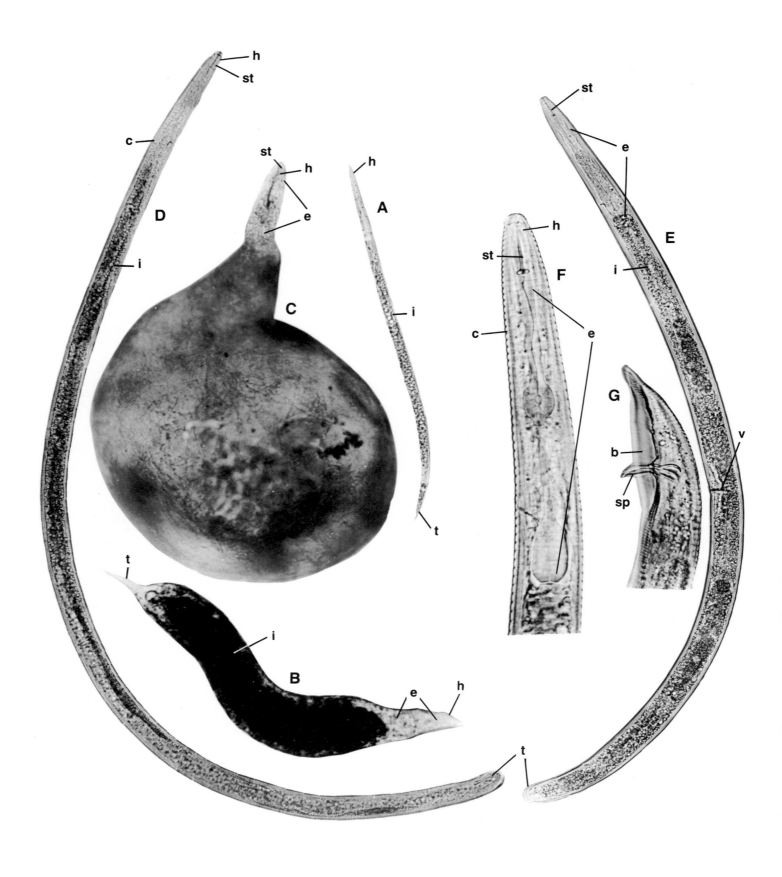

Figure 14A–D. Root knot nematode, *Meloidogyne* sp. A. A vermiform larva. B. A sausage-shaped larva extracted from a root. C. Mature female. D. Mature male.

Figure 14E–G. Stunt nematode, *Tylenchorhynchus* sp. E. Mature female. F. Detail of anterior portion. G. Detail of male tail and genital organs. b = bursa (clasping organ), c = cuticle, e = esophageal region, h = head region, i = intestine, sp = spicules, st = stylet, t = tail, v = vulva.

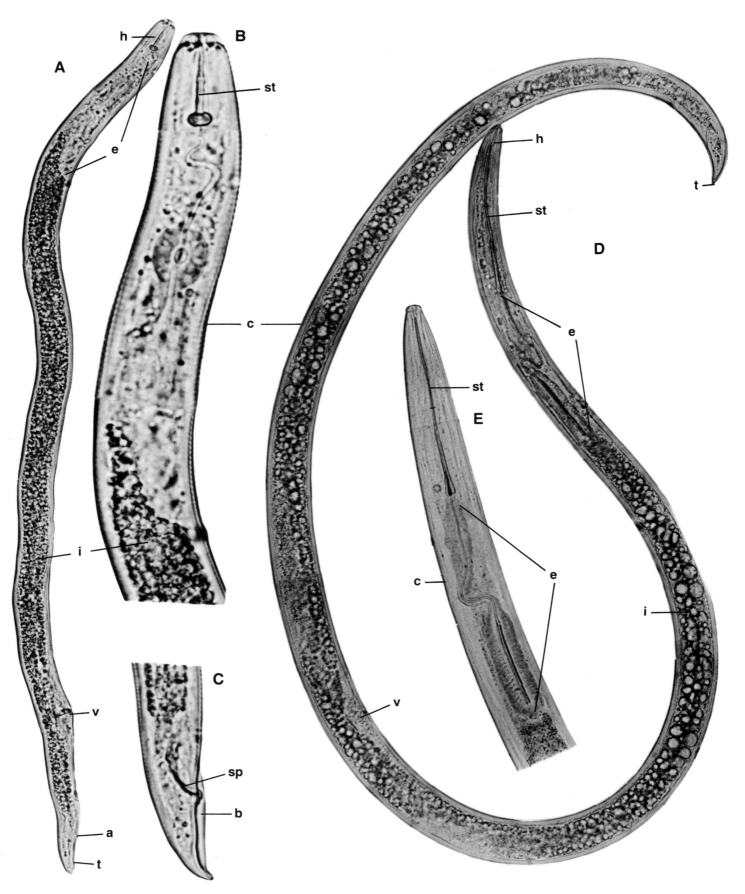

Figure 15A–C. Lesion nematode, *Pratylenchus penetrans*. A. Mature female. B. Detail of anterior portion. C. Detail of male tail and genital organs.

Figure 15D, E. Dagger nematode, *Xiphinema americanum*. D. Mature female. E. Anterior portion. a = anus, b = bursa (clasping organ), c = cuticle, e = esophageal region, h = head region, i = intestine, sp = spicules, st = stylet, t = tail, v = vulva.

Burrowing nematode. *Radopholus similis* is a migratory endoparasite of succulent root tips and other fleshy subterranean organs. It occurs around the world in tropical and subtropical areas, parasitizing plants of more than 250 species. It is best known for damage to banana, citrus, and black pepper trees. Two races occur, differentiated by the ability of one to infect citrus. The citrus race is known only in Florida. It also infects avocado, mango, trifoliate orange, peach, pecan, persimmon, and loblolly and slash pines in addition to many herbaceous and woody tropical plants. It kills citrus root tips, inducing the disease called spreading decline. Affected trees lose vigor; foliage becomes somewhat chlorotic, sparse, and undersized; shoots may wilt temporarily during the day; twigs die back, and productivity is lost. Symptoms slowly spread across the orchard. Grapefruit is more severely affected than orange.

Other host plants of the burrowing nematode in Florida do not show a decline syndrome similar to that in citrus. This difference is perhaps associated with plant rooting habits. Citrus is deep rooted, and the nematode prefers deep-lying roots. It causes most damage at soil depths greater than 50 cm and no damage in the topmost 25 cm. Plants with root systems principally in the top 50 cm of soil would escape severe damage.

The life cycle of the burrowing nematode is completed in about 3 weeks. All life stages can be found within roots, and all stages except eggs are mobile. The feeding and movement of larvae and females within roots cause lesions that coalesce and destroy the root tip. The nematodes then leave the root and migrate to another. Males, however, do not feed and, after leaving a root, do not enter another. The migration of nematodes, up to 15 meters per year, is the basis of spreading decline in orchards. Long-distance dispersal occurs in infested plant materials. Resistant rootstocks appear promising for control in citrus orchards.

Stunt nematodes. *Tylenchorhynchus* species (Figure 14E–G) are migratory ectoparasites that feed on epidermal cells and root hairs of various plants, suppressing root growth and causing sufficient root damage to stunt plant growth. Roots become discolored light to dark brown. Black locust, rose, and spruce are particularly favorable hosts for nematode reproduction. Japanese andromeda and azalea become stunted and develop chlorosis, tip burn, and (in azalea) leaf drop. Magnolia and pine may also be damaged.

Lesion nematodes. *Pratylenchus* species (Figure 15A–C) are migratory endoparasites with a broad plant host range. They occur in cool and warm temperate regions. Larvae and adults feed in cortical parenchyma of roots, and females lay eggs in small clusters within the cortex. Feeding causes cell death and development of numerous small lesions. The nematodes can feed only on living cells and so must move frequently within roots or exit and reenter at another point. In so doing, they leave in the cortex tracks of ruptured cells that are soon colonized by saprophytic and pathogenic bacteria and fungi. The life cycle is completed in 1–3 months, depending upon nematode species, host, and temperature. Root damage by lesion nematodes, usually in concert with fungi and bacteria, causes growth loss of forest tree seedlings, woody ornamentals in nurseries, fruit trees in orchards, and many agricultural crops. Woody hosts include, but are not restricted to, almond, apple, avocado, boxwood, eastern red cedar, Japanese cedar, cherry, cotoneaster, Douglas-fir, fig, forsythia, ginkgo, grape, sweet gum, juniper, magnolia, maple, olive, peach, pine, privet, rose, smoke bush, spruce, tulip tree, walnut, and willow. Diseases such as Fusarium root rot of coniferous seedlings (Plate 143) and Verticillium wilt (Plates 181–182) are aggravated if roots are infested by *Pratylenchus*.

Dagger nematodes. *Xiphinema* species (Figure 15D, E) are among the largest (to 5 mm) and include some of the slowest developing and longest lived plant-parasitic nematodes. *X. americanum* undergoes only one generation per year, and females of the European species *X. diversicaudatum* may live as long as 5 years. These times are the extremes, however. Dagger nematodes feed from outside the root through a hollow spear that may be nearly 200 μm long, enabling the nematode to reach deep-lying root cells. When feeding at a root tip these nematodes inhibit root elongation but stimulate cell division. The result is stunted, swollen root tips that are attractive feeding sites for other nematodes of the same species. Formation of short lateral roots may be suppressed also, resulting in an abnormally coarse root system. Secondary pathogens kill many rootlets damaged by the nematodes. *Xiphinema* species important in North America have many hosts. They are among the most common plant-parasitic nematodes found in eastern forests, and they have been associated with decline of shelterbelt trees (green ash, eastern red cedar, cottonwood, elms, hackberry, Russian olive, and American plum) in the Great Plains; white ash and sugar maple in northeastern states; and fruit trees in eastern orchards. Root damage to blue spruce in midwestern nurseries led to abnormal winter injury and an outbreak of Leucostoma canker (Plate 93). *X. bakeri* causes corky root disease of Douglas-fir and other coniferous seedlings in western Canada. *X. index* damages grapevine and fig in California and elsewhere around the world.

X. americanum, *X. californicum*, *X. rivesi*, and *X. index* are important in part because of their roles as virus vectors. *X. index*, the vector of grapevine fanleaf virus in California, was the first nematode vector of a plant virus to be identified. *X. americanum* (in its former broad sense—i.e., including *X. californicum* and *X. rivesi*) is a vector of tomato ringspot and tobacco ringspot viruses, both of which are associated with ash decline in northeastern states. Tomato ringspot virus also causes graft union necrosis and decline in apple and stem pitting and decline in peach and cherry.

References for nematodes in general: 386, 520, 576, 1096, 1194, 1406, 1413, 1660–63, 1665, 1684, 1862, 1922
References for root knot nematodes: 148, 386, 520, 1610, 1613, 1922
References for other nematodes discussed with Plates 144–145: 98, 386, 494, 576, 687, 944, 1197, 1199, 1365, 1413, 1416, 1547, 1767, 1933, 2251

A. At left, sparse foliage and dieback on a 'Valencia' orange tree on 'Rough' lemon rootstock damaged by the burrowing nematode, *Radopholus similis*. At right, 'Valencia' orange on 'Milam' rootstock, resistant to the nematode. Burrowing nematodes enter but do not reproduce in the resistant rootstock (FL, Apr).

B. Chlorosis, tip browning of leaves, and reduction of leaf size associated with damage to roots of Japanese andromeda by a stunt nematode, *Tylenchorhynchus* sp. (NY, Nov).

C, D. Damage to apple seedlings by the lesion nematode, *Pratylenchus penetrans*. C. A plant growing in nematode-infested soil (right) in comparison to a plant in noninfested soil. D. Root systems of the two plants shown in (C). The damaged root system is stunted and darkened by many small brown lesions (NY, greenhouse).

E, F. Apple roots damaged by the dagger nematode, *Xiphinema americanum*. E. Proliferation of short lateral roots above the point where two long roots were killed. Rootlets or root tips fed upon by the nematodes are stunted and swollen. F. Swollen rootlets damaged by the nematode turn dark brown when killed by secondary microbial pathogens (NY, Jun).

Plate 145

Root and Butt Rot and Basal Canker Caused by *Hypoxylon deustum* (Plate 146)

Hypoxylon deustum (Sphaeriales, Xylariaceae) causes a white rot of the major roots, butt, and trunk of many species of angiosperm trees and some monocots around the world. The fungus is also known as *Ustulina deusta*, *U. vulgaris*, or *Kretzschmaria deusta*. A closely related fungus, *K. clavus*, causes root rot and decline of macadamia in Hawaii. In Eurasia *H. deustum* causes charcoal base rot of oil palm and rubber trees, root rot of tea and teak, and butt and trunk rot of beech and linden. The dark brown to black stromata of *H. deustum* are found in eastern North America on the butts, dead roots, and stumps of white ash; basswood and other lindens; American and European beech; paper and yellow birches; box elder; American elm; American holly; horse-chestnut; red and sugar maples; black, red, scarlet, and white oaks; plane tree; sugarberry; and tulip tree. In the West, the fungus occurs on alder, box elder, and bigleaf maple. Usually it appears to be merely a saprophyte, but its association with root, butt, and trunk rot and basal cankers in living trees is well documented.

Infection and decay commonly begin at wounds caused by fire or mechanical equipment at the trunk base or on major roots. Observations and experiments have indicated that *H. deustum* may be pathogenic to sapwood. When ascospores were introduced to fresh wounds in sapwood of European beech and linden, the fungus survived and could be isolated 1 year later from stained wood near the wound. Mycelium of *K. clavus* introduced to wounds in living macadamia roots caused white rot of the wood. From observation, *H. deustum* infects sprout maples via the parent stump or the stumps of companion sprouts, and the decay column may extend more than 2 meters above ground in trunks only 20–25 cm in diameter. *H. deustum* occasionally sporulates on basal cankers on red and sugar maples, and observation of dissected trees suggests that it may have caused the cankers by spreading outward from a central column of decayed wood.

Wood decayed by *H. deustum* or *K. clavus* is typically light in color and brittle to crumbly and contains numerous irregular black sheets (pseudosclerotial plates) that appear as lines (zone lines) on broken or cut faces. Decaying stumps and logs eventually decompose except for a persistent mass of pseudosclerotial plates that appear as brittle black sheets. Pseudosclerotial plates are produced by many wood-decaying fungi and are not diagnostic. They are shown in Plate 107.

Root and butt rot and basal cankers caused by *H. deustum* are common on mature sugarberry trees in southern Louisiana, where infection is presumed to be fostered by a high water table. This may weaken or kill roots and dispose them to attack. Diseased trees, if not cut first, eventually topple because of decay of buttress roots on one side.

European beech is subject to root rot, basal cankers, and trunk rot caused by *H. deustum*. Prominent stromata of the fungus develop on exposed major roots and on the butt at the edges of basal wounds and cankers.

H. deustum is dispersed as airborne ascospores and conidia. Both types of spores are reported to be infectious to freshly wounded sapwood, but information on their relative importance as inoculum is lacking. It has been suggested that tree-to-tree transmission in tea plantations occurs via root contacts and by hyphal growth through soil, but these possibilities have not been studied in North America.

After an indefinite interval of wood decay activity, *H. deustum* forms first a conidial stroma and then a perithecial stroma on the surface of exposed wood or dead bark. In North America the conidial stroma forms in winter to spring (depending on latitude), and the perithecial stroma ripens in early summer to autumn. The conidial stroma is an effuse grayish white to grayish brown layer about 0.5 mm thick consisting of vertically oriented tufts of branched conidiophores on a mycelial base. Conidia are nearly colorless, single-celled, and wider at one end than the other, measuring 5–9 × 2–3.5 μm. The conidial state forms readily on agar media.

Perithecial stromata of *H. deustum*, massive in comparison with stromata of many related fungi (Plates 106–108), develop on the same area previously occupied by the conidial state. These stromata may take the form of discrete cushions or a sheet of undulating to lumpy tissue, its topography determined in part by that of the host surface. Stromata are grayish white and leathery at first, become black and brittle with age, and can readily be detached from the substrate. Stromata often become perennial and then develop concentric zones. For example, portions of old black stromata extend beyond the gray-brown young stroma shown in (B).

Scattered black, unusually large perithecia (1.5 × 1 mm), develop in the stroma, immersed in its whitish interior tissue, their nipplelike tips protruding slightly above the surface. The ascospores at maturity are single-celled, dark brown to black, and irregularly elliptic with pointed ends, measuring 26–40 × 6–13 μm. Ascospores germinate readily in water at temperatures of 15–30°C. Germination and mycelial growth are most rapid at 20–30°C, at which temperature most spores germinate within 24 hours.

References: 310, 387, 973, 1039, 1040, 1223, 1226, 1306, 1557, 2198

A, C. Butt rot and basal canker of sugarberry. A. The butt of a mature landscape tree is dead around nearly half its circumference. A large, rough black stroma of *Hypoxylon deustum* is visible at ground line. C. Closer view of the stroma (LA, Dec).

B, D, E. Butt rot and basal canker of European beech. B. Close view of the edge of a large basal canker. A maturing stroma of *H. deustum* (gray-brown to dark brown) lies at the junction of bark and exposed wood. Reddish brown stains on bark result from bleeding at points of attack by ambrosia beetles that tunnel into weakened trees. D. A large brownish black stroma of *H. deustum* on the soil surface (growing from roots) between brace roots of a large tree. E. Closer view of the lumpy surface of the stroma (NY, Sep).

Plate 146

Dematophora (Rosellinia) and Xylaria Root Rots (Plate 147)

Dematophora root rot. *Dematophora necatrix,* the conidial state of *Rosellinia necatrix* (Sphaeriales, Xylariaceae), causes Dematophora root rot, also known as white root rot, of apple and many other woody and herbaceous plants around the world. The disease is widespread in Europe but in North America is important only in California in apple orchards. Early reports of *D. necatrix* on fig in North Carolina and on grapevine in several eastern states have not been followed by contemporary records. Other woody plants found naturally affected in California include apricot, barberry, California buckthorn, big-pod ceanothus, *Cotoneaster salicifolius,* currant, *Echium* sp., holly osmanthus, black poplar, privet, and viburnum. In addition, the following plants became diseased when planted in infested experimental plots: almond, apricot, brambles, cherry, fig, grapevine, peach, pear, quince, and Hinds walnut. More than 170 species in 30 families are known hosts worldwide. Several other *Rosellinia* species also cause root rot of various plants or are found as saprophytes on stems or branches.

Aboveground symptoms of Dematophora root rot may include slow growth and sparse, chlorotic, undersized foliage for 1 or more years before death. Symptoms may involve the entire plant or initially just one side. Plants killed quickly may retain dry foliage for some time. Killing is possibly aided by toxin(s). Cytochalasin E, produced by *D. necatrix,* caused wilt in apple cuttings that were allowed to take up solutions containing $1-10\ \mu g \cdot ml^{-1}$.

Infection begins on small roots and progresses to larger ones, eventually killing the plant by girdling at the root-collar region. Hyphae invade both bark and wood, growing inter- and intracellularly. Fibrous roots rot off, often leaving a residue of white mycelium in soil. White mycelium also grows on or beneath the bark of larger roots, forming plaques and loosely aggregated white strands that grow into adjacent soil. A dark crust sometimes forms on the dead root or at the root collar. *D. necatrix* also produces dark microsclerotia about 0.1 mm in size. These and mycelium in root fragments allow the fungus to survive in soil. Mycelial strands growing from decayed roots and hyphae growing from microsclerotia penetrate small roots. Transmission also occurs where roots touch one another. Infected planting stock is responsible for long-distance dispersal of *D. necatrix.*

During wet weather, white mycelium of *D. necatrix* may appear at the soil surface above killed roots, or the fungus may form at the stem base a dark mat on which synnemata and conidia develop. Synnemata are upright aggregates of conidiophores. The conidia, however, have no known role in the disease cycle. White mycelium and synnemata also form on diseased root segments that are incubated under moist conditions for diagnosis. Viewed microscopically, each hyphal cell is distinctly swollen at its distal end. The perithecial state, *Rosellinia necatrix* (syn. *Hypoxylon necatrix*), has not been found in nature in North America.

D. necatrix is favored by relatively cool (<20°C), moist soils. Its mycelium grows most rapidly at 22–25°C and is intolerant of heat. Some apple growers delay the loss of diseased trees by removing soil from around the root-collar region, which allows the exposed bark to reach temperatures inhibitory to the pathogen.

Apple rootstocks vary in susceptibility to *D. necatrix,* but white root rot occurs in such a limited area in North America that little attention has been given to resistant rootstocks here. Susceptible almond, apple, and peach have been shown to contain lower concentrations of certain phenolic compounds to which *D. necatrix* is sensitive than do resistant plants such as pecan and persimmon.

Xylaria root rot. *Xylaria* (Sphaeriales, Xylariaceae) is a relatively large, globally distributed genus of wood-decaying Ascomycotina. Of the several species in North America, two are recognized pathogens. *X. mali* and *X. polymorpha* cause black root rot of apple and other trees. Probably other North American species are also pathogenic, but their habits have not been studied critically.

Black root rot is named for a black stromal sheath that develops on decaying roots. *X. mali* and *X. polymorpha* cause an off-white decay of wood beneath the black sheath, and they produce clusters of black, fingerlike to irregularly club-shaped fruit bodies (perithecial stromata) that originate on the butt or on major roots and extend a few centimeters above the soil line. These structures are the basis of the common name of *X. polymorpha,* "dead man's fingers." Diagnosis is usually based upon recognition of the stromata.

Because *X. mali* and *X. polymorpha* are of similar appearance and have several hosts in common, we consider them together. *X. polymorpha* occurs across North America from the Gulf states northward, but it is rare in the Rocky Mountain region and the far West. *X. mali,* the more aggressive parasite, is reported only from the East. *X. mali* is a proven pathogen of apple, mahaleb cherry, American elm, honey locust, Norway maple, and common and sand pears. *X. polymorpha* is a proven pathogen only of apple. In addition, these fungi fruit on stumps or on the butts of living beech, butternut, currant, hickory, black locust, sugar maple, oak, trifoliate orange, sassafras, spruce (on the basis of one Canadian record), black walnut, and yellowwood.

The "fingers" of mature perithecial stromata of *X. mali* or *X. polymorpha* consist of white to pale buff fleshy internal mycelium, often with a central cavity, covered with a thin black crust. Black perithecia 0.7–0.8 mm in diameter form just beneath the crust, and at maturity they discharge to the air dark brown, single-celled ascospores through minute pores in the crust. Young stromata, initially nearly white, become covered with a layer of brown conidiophores and unicellular conidia that cause the stroma to appear tan. The conidial state is present in spring, and stromata mature during summer.

Aboveground symptoms, similar to those of Dematophora root rot, do not appear until root damage is extensive. Apple trees stressed by root damage may set an abnormally large crop of small fruit.

Roots are killed and then decayed by the fungus. A mycelial encrustation, white at first but turning black, develops on the surface, and a network of tiny black rhizomorphs extends from the edge of the encrustation several centimeters along the root. Decay may extend into the butt, and basal cankers sometimes develop on the trunk above dead roots.

Cultural practices and fertility have little influence on Xylaria root rot except that, if susceptible trees are planted in soil where others have been killed by *X. mali,* severe disease in the new planting is likely to develop within 2–3 years.

Wood undergoing decay by *Xylaria,* especially *X. polymorpha,* contains an irregular lattice of black sheets (pseudostromatal plates) that appear as black lines (zone lines) on cut or broken faces. Many other wood-decaying fungi also produce these structures. For an illustration, see Plate 107. Pseudostromatal plates are composed of dark, thick-walled bladderlike cells of the fungus within wood cells plus a dark substance that impregnates the wood cell walls. The plates are refractory to microbes that ordinarily decompose wood.

References for Dematophora root rot: 291, 387, 982, 1225, 1689, 1957–59, 2211

References for Xylaria root rot: 303, 387, 517, 602, 816, 1225, 1226, 1629

A, C, F, G. Dematophora root rot of apple (CA, Jul). A. A diseased tree with sparse, chlorotic, undersized leaves. C. Close view of the root-collar region of the tree; superficial bark has been shaved to reveal the margin of necrosis near the soil line. Necrotic bark contains tiny pockets of white mycelium. F, G. A dead root extending from the butt shown in (C), excavated to reveal white superficial mycelium of *Dematophora necatrix.*

B, D. Xylaria root rot of yellowwood (NY, Jun). B. A dying tree with thin, chlorotic foliage and dieback. D. Young stromata of *Xylaria polymorpha* visible as a cluster of rounded tan objects at ground level adjacent to bare wood.

E, H, I. Mature perithecial stromata of *X. polymorpha* (NY, Aug). E. Stromata at the base of a pear stump. H. A stroma cut down the middle, exposing soft white interior tissue, a hollow center, and a black surface layer covering the black perithecia. I. Magnified view of perithecia.

Plate 147

Armillaria Root Rot (Plates 148–150)

Armillaria root rot, also called shoestring root rot or, in some areas, mushroom root rot, is one of the best known and most damaging diseases of forest, shade, and ornamental trees and shrubs around the world in regions ranging from subtropical to boreal. Hundreds of plant species, mainly but not exclusively woody, are attacked. Various plants differ in susceptibility, but this is conditioned so much by plant age and predisposing stress that lists of susceptible and resistant plants (see references) are useful only for general guidance. Also, some species of *Armillaria* are more virulent than others on a given plant species.

We use the term *Armillaria root rot* in a generic sense to designate a group of similar diseases caused by a complex of at least 11 fungal species that have been known by the names *A. mellea* and *A. tabescens* (Agaricales, Tricholomataceae). The name *A. mellea* is often still used to refer to the entire group (except *A. tabescens,* Plate 150) in North America because the species here have not been studied sufficiently for practical differentiation. Differences in their host ranges and virulence are imperfectly known. Two of the seven or more members of the group that occur in western forests have been designated *A. bulbosa* and *A. ostoyae.* The former is a weak pathogen of broad-leaved species, and the latter is an aggressive pathogen of conifers. *A. bulbosa* is also a member of the species complex in the East. The other species in North America have not been named. In Europe, five species are recognized: *A. borealis, A. bulbosa, A. cepistipes, A. mellea,* and *A. obscura.*

Armillaria species occur in all forested or once-forested areas except in arctic and hot lowland tropical regions. They are sometimes called honey mushrooms, honey agarics, or shoestring fungi. These names indicate the color of basidiocarps and the appearance of the rhizomorphs by which some strains spread in the vegetative state. Rhizomorphs are thick strands of hyphae that resemble shoestrings and grow in the manner of roots. We will refer to these fungi collectively as *Armillaria.*

Armillaria invades the bark and cambial region of roots and the root collar, killing roots and trees of all sizes. Some species or perhaps strains within species are virulent parasites, but others are opportunists that act selectively on small or weak individuals such as those shaded by taller plants, damaged by mechanical equipment, defoliated by insects, or weakened by freezing, drought, or polluted air. *Armillaria* also colonizes the declining root systems of trees felled or killed by other agents. Wood decay by *Armillaria* usually follows cambial attack, and the wood serves as a source of energy necessary for infection of new hosts. The fungus can persist for decades in decaying wood in soil.

Armillaria is the most common root-rotting fungus in the boreal and northern hardwood forests of North America. It is also common in landscapes, orchards, and gardens, more so in the West than the East. Damage is difficult to estimate, however, because *Armillaria* often follows other malefactors and is often associated with other root-infecting fungi and with secondary insects. In northern regions both east and west, it is apparently responsible for many deaths among coniferous saplings, killing these at rates averaging about 2% during the first several years of forest growth. Tree resistance to killing increases with age, so high rates of sapling mortality in dense young stands do not necessarily presage significant loss of timber volume. Root rot can be severe, however, in young plantations of coniferous species established where broad-leaved trees, especially oaks, formerly grew. Trees are at high risk for 10–15 years because felling and other disturbances that kill trees provide new food bases in the form of stumps and roots from which the fungus later initiates parasitic attack on nearby trees. Young trees have also been severely damaged after

harvest or thinning of coniferous forest crops in western North America. Severe mortality in disease centers up to 3–4 hectares in size has occurred in some inland forests in Oregon and Washington.

Diseased trees or shrubs occur singly or in groups. Symptoms above ground are growth reduction, which may be abrupt or gradual, yellowish or undersized foliage, premature leaf drop, branch dieback in the upper crown, or rapid browning and death of the entire plant during summer. Growth reduction becomes noticeable after more than half the root system is already dead. Dying conifers sometimes produce a distress crop of numerous small cones. Fanlike, veined white mycelial sheets invade and kill cambial tissue of major roots and the root collar. These sheets, called fans or plaques, may girdle roots or the butt or may cause discrete cankers. Decayed wood may be present at the root collar at the time of tree death, but decay is usually confined to roots or interior wood of the butt until after death.

Small trees or shrubs often die quickly, but large ones may be chronically diseased, sustaining growth loss and decay of roots and a portion of the butt. Such trees are subject to windthrow. Large trees often fight a seesaw battle with *Armillaria* and are able to confine the fungus within cankers on roots or the root collar for many years. When thus restricted, *Armillaria* may decay the wood beneath the lesion but is usually compartmentalized there.

In resinous conifers, infection stimulates heavy flow of resin, which may saturate the bark and wood near points of attack, accumulate in pockets between the bark and wood, and seep out onto the surface and into surrounding soil and litter, causing incrustation. Similarly, gum-producing broad-leaved trees such as cherries and their relatives may respond by producing gum in and on infected regions, but this is less dramatic than resin production in conifers.

In most geographic areas, *Armillaria* produces branched, brown to black rhizomorphs that grow along the surfaces of living or dead roots, outward into soil or humus, or up between the bark and wood of a tree trunk already dead. Rhizomorphs under the bark are flattened and anastomose frequently, but they are round where free in soil, litter, or decayed wood. They are 1–2 mm in diameter with a white interior and a rind that is reddish brown when young and black when mature. Although some strains or species of *Armillaria* seldom or never form rhizomorphs in the field, most strains will produce them in laboratory cultures. This aids diagnosis.

Wood decayed by *Armillaria* is at first slightly water-soaked and light brown. With time it becomes light yellow to nearly white, soft and spongy, often stringy in conifers, and marked on the surfaces by black lines (zone lines; see Plate 107 for an example). Decay in the butt and major roots of birches, firs, and perhaps other trees results in vertical cracks at the root collar. The cracks arise as trees weakened by internal decay are stressed by wind or the weight of snow or ice. Decay caused by *Armillaria* seldom extends as much as 2 meters above ground.

The mushroom stage of the fungus develops annually in autumn and, in regions with mild climate, the first part of winter. Mushrooms appear on or near wood that is undergoing decay, often arising from rhizomorphs connected to decaying wood.

A, E. Heather attacked by *Armillaria.* A white mycelial plaque between bark and wood of a recently killed root (E) is exposed by removing bark. Rhizomorphs (arrow) are difficult to distinguish from thin roots (WA, Sep).

B. A dying Norway maple, roots of which were infected by both *Armillaria* and *Phytophthora* sp. (NY, Sep).

C, D. Mushrooms of *Armillaria* on the base of a dead American elm and on a log of yellow birch, respectively (NY, Oct).

F. Rhizomorphs between bark and wood on the butt of a dead sugar maple (NY, May).

G. A mycelial plaque in the cambial region at the base of a weeping willow; bark has been cut away to show the limit of necrosis (arrow) (NY, Sep).

H. Fanlike development of a mycelial plaque in the cambial zone at the base of a chestnut oak that was defoliated by insects in 2 previous years (NY, Aug).

Plate 148

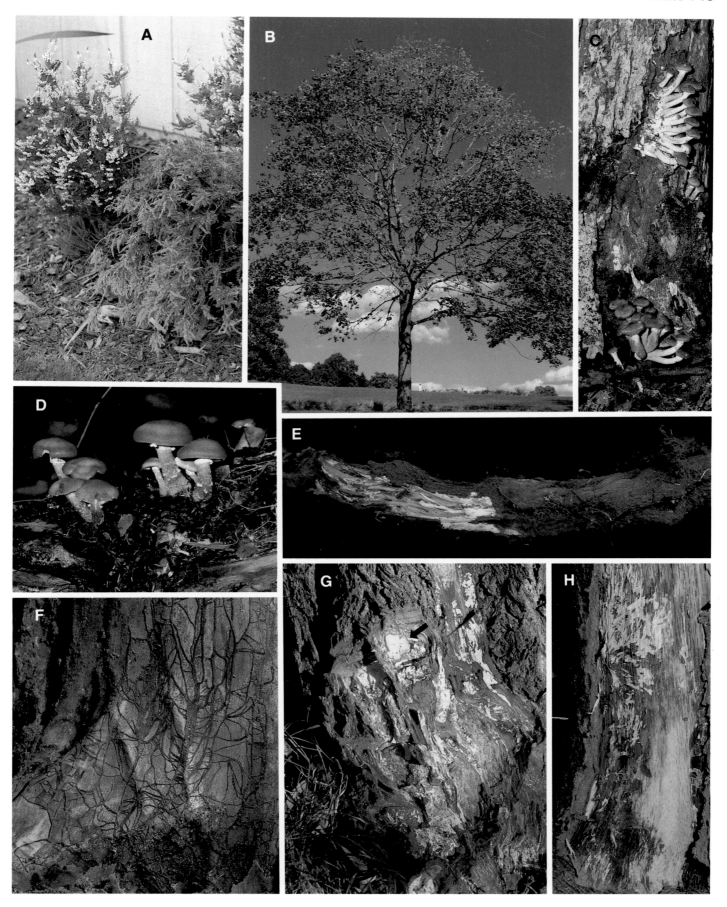

Mushrooms grow in groups of several to 100 or more. The cap, initially globose, expands to form a flat upper surface 4–15 cm across, often with a depressed center, varying from honey colored to tawny or brown, sometimes dotted with brown scales, and gelatinous when moist. The flesh is white and, in some strains, luminescent. The gills are white to flesh colored, and the spores in mass are light cream color. The stem, 1–2 × 5–15 cm long (exceptionally 25 cm), often has a persistent yellowish to whitish ring encircling its upper part. In areas of low rainfall, mushrooms may not develop consistently, but rhizomorphs (when present) and thick mycelial fans between the bark and the wood at the root collar or in roots are diagnostic.

Armillaria perpetuates itself mainly by mycelium and rhizomorphs. Rhizomorphs are responsible for local spread in most parts of North America. In inland areas of the Northwest, however, hyphal growth across root contacts is important for new infections. Rhizomorph tips, or hyphae at contacts between infested and healthy roots, penetrate intact bark. Rhizomorphs typically grow over the root surface, become attached at intervals, and produce short penetrating branches at points of attachment. Hyphae growing from rhizomorph tips proliferate into mycelial fans that kill cambial tissue in advance of their own progress.

Rhizomorphs of opportunistic, weakly virulent species or strains commonly colonize root surfaces of healthy trees but only penetrate roots after stress-induced chemical changes occur that cause a shift from resistance to susceptibility. Defoliation of broad-leaved trees encourages such attack by causing depletion of starch, increased levels of glucose and fructose, and changes in amino acids in roots. Elevated glucose levels not only stimulate growth of the pathogen but enable it to tolerate concentrations of host-produced phenolic compounds that would normally be inhibitory.

Armillaria grows much more rapidly toward the tip of a girdled root than toward the butt of a living tree, but when these differential rates are averaged, or when growth of rhizomorphs is considered, the fungus spreads 1–2 meters per year in many temperate areas. In warm regions, rates up to 5 meters per year have been recorded.

Even susceptible plants have an array of resistance mechanisms. These are most effective in well-nourished plants and at soil temperatures most favorable for root growth. Trees that successfully resist the pathogen produce callus at canker margins. Callus growth may eventually enclose small cankers together with associated pockets of decaying wood beneath a layer of sound wood in the butt or a major root. Infections are sometimes arrested and die out even when extensive cambial killing has occurred in the root-collar region. Old lesions persist as scars that after many years are difficult to distinguish from fire scars or frost cankers. Some old Armillaria cankers can be identified by the images of former mycelial fans etched darkly on the scar face.

Eventually, after predisposing events or because of old age and declining vigor of a tree or shrub, *Armillaria* has the chance to invade weakened tissue and kill at least part of the root system, which then serves as a new food base. Within 2 years, the fungus colonizes the sapwood, and most strains then begin to send out rhizomorphs. This activity reaches a peak after about 10–15 years (less time if the root system is small) and then declines markedly.

Although rhizomorphs require a woody food base for sustained growth, they derive substantial nourishment from other decomposing organic matter. Rhizomorphs are invariably more abundant and more highly branched in the humus layer than in mineral soil. They are sensitive to variation in oxygen and carbon dioxide concentrations and perhaps for this reason are usually not found more than about 30 cm deep in soil. Where soil conditions permit deep rooting, however, mycelium of *Armillaria* grows readily. Diseased roots have been found at soil depths greater than 2 meters in California.

Armillaria is generally inhibited by soil temperatures above 26°C, so plants that grow where soil temperature commonly exceeds this level may escape damage.

Armillaria disperses airborne basidiospores from mushrooms, but the epidemiological role of these spores is apparently slight. The spores of some strains can germinate and establish colonies in recently wounded sapwood, especially in stumps. Occasional infections of wounded roots presumably also occur. Spores of other strains are capable of germinating but when tested have been incapable of establishing persistent colonies in natural substrates. Once established in a locality, a given clone of the fungus persists for decades or perhaps centuries, occupying successive woody substrates as the result of parasitic activity. One clone may occupy a territory ranging from a few square meters to hundreds of hectares. A typical infested area may contain several clones of *Armillaria* in a sort of subterranean mosaic.

Evidence about host-specialized strains or species of *Armillaria* is conflicting. Some strains seem specialized for attack of conifers, others for broad-leaved trees, but in most tests strains from a given host have been found capable of infecting several different kinds of plants. Strains do vary considerably in virulence, however, as reflected by speed of attack and killing under standard conditions. Reports of host specialization are confounded by the fact that wood of broad-leaved trees is superior for vegetative growth and rhizomorph production by the fungus. Thus, if other conditions are equal, an infested oak or maple root may be more effective inoculum than a conifer root of the same size.

Armillaria often acts in conjunction with other secondary pests and pathogens. In western forests, ponderosa pines with root rot caused by *Armillaria* or other fungi are attacked by the western pine beetle and the mountain pine beetle, and white fir trees with root rot are attacked by the fir engraver. In the Northeast, oaks with root rot are attacked by the two-lined chestnut borer (see Johnson and Lyon, 2nd ed., Plates 28, 117, and 127). Whether insect or fungus comes first in these cases is unimportant because each agent is acting as an opportunist on trees weakened by other factors. In western forests, *Armillaria* is often associated with other root-infecting fungi, particularly *Heterobasidion annosum* (Plates 151–152), *Phellinus weirii* (Plates 156–157), and *Ceratocystis wageneri* (Plate 178). *Armillaria* seems to act as a primary pathogen mainly in the attack of small or young trees or shrubs close to relatively large food bases such as stumps.

Prevention of Armillaria root rot is difficult because the fungus is so common in woody debris in soils and has such a broad host range and a nonspecialized mode of parasitism. The most important strategies for protection of susceptible shade and ornamental trees and shrubs are to promote vigor, minimize stress, and deny the fungus large food bases by removing infested stumps and roots from soil or promoting development of antagonistic organisms in these. In California, various fruit and ornamental trees and shrubs that have been observed to resist *Armillaria* can be grown on infested sites, but reliable information about resistant plants is not available for other regions. In orchards in areas where summer is hot and dry, diseased trees have been saved in some instances by removing soil from around the root buttresses and root collar. The pathogen does not survive in tissues thus exposed.

A, C, D. Armillaria root rot of ponderosa pine. A, C. Young trees dying after infection by root contact with older, previously killed trees. D. Mycelial fans extend about 40 cm above soil line in the cambial zone of a recently killed tree (WA, Jul).

B, E, F. Armillaria root rot of Austrian pine in a Christmas tree plantation. B. A recently killed tree. E. A tree girdled by *Armillaria* at the root collar exuded resin that caused incrustation of soil. Where encrusted soil and bark have been removed, white mycelial fans of the fungus are visible in the cambial zone. Needles on the tree are still green. F. Another girdled tree, with a pocket of crystallized resin (arrow) between bark and wood above the mycelial fan (NY, May).

Plate 149

Mushroom root rot. In the southern United States from Oklahoma eastward, *Armillaria tabescens* (syn. *Clitocybe tabescens*) causes root rot of more than 200 plant species in about 60 families. The disease is called either mushroom root rot, because clusters of mushrooms often grow at the bases of diseased plants, or Clitocybe root rot. The pathogen is part of the *Armillaria* complex discussed with Plate 148, and the disease conforms to the general description of Armillaria root rot. We present this topic separately because previous authors have differentiated *A. tabescens* from *A. mellea*. *A. tabescens* has been collected as far north as Michigan and New York, but the northernmost records of its parasitic behavior are from southern Illinois and localities eastward to Maryland.

A. tabescens is an opportunistic fungus best able to attack plants already killed or weakened by other agents. In sand pine in Florida, for example, it is commonly found in roots also infected by *Inonotus circinatus* (Plate 153). Thus although it has been credited with killing up to 25% of the trees in some sand pine plantations, the amount of damage it does alone is open to question. In highly susceptible plants, however, it seems to be an aggressive killer.

Plants most commonly infected are macranthum azalea, common guava, horsetail tree and other so-called Australian pines (*Casuarina* species), rough lemon, peach, Japanese photinia, Rottnest Island pine, sand pine, India rubber tree, and tung-oil tree. Other plants frequently infected vary from one locality to another but include apple; American and Oriental arborvitae; common camellia; camphor tree; sweet cherry; flowering dogwood; elms; grapevine; Chinese hibiscus; laurustinus; loquat; southern magnolia; silver maple; Russian olive; sweet olive; laurel; post, red, turkey, water, and willow oaks; sand pear; Brazilian pepper tree; loblolly pine; Japanese pittosporum; 'Lombardy' poplar; and Turk's-cap waxmallow.

Trees and shrubs of all ages are attacked, singly or in groups, and trees with severely decayed roots are subject to windthrow. Symptoms before death include yellowing, sparse or undersized leaves, premature defoliation, branch dieback, decayed roots, and lesions that partly or completely girdle the plant at the root collar. Basal lesions often extend 10–30 cm, rarely 60 cm, above soil level and are contiguous with dead roots. In some species of Australian pine, basal girdling by *A. tabescens* leads to pronounced stem swelling above the girdle before the tree dies. Diseased sand pines often have sunken cankers at the root collar and produce resin profusely at sites of infection. The resin saturates bark and wood, accumulates in pockets in the cambial region, and exudes around diseased roots and the root collar, causing encrustations as it hardens. Symptoms on aboveground parts usually do not appear until the fungus has invaded the root collar; then the entire plant may wither and die quickly.

Excavation of the root collar and larger roots is necessary for diagnosis if the fungus has not progressed above soil level. Beneath the bark of infected roots or the root collar, mats of white, cream, or chamois-colored mycelium form that kill cambial tissue in advance of their own progress. Elliptic perforations 0.5–1.0 mm long in the mats are said to be diagnostic but may be lacking in very young and very old mats. The surfaces of diseased roots often display narrow black ridges consisting of fungal tissue that develops in bark fissures above mycelial mats.

In autumn, clusters of mushrooms grow at the bases of diseased plants or occasionally emerge from cracks in the bark on the butt of a tree. Their remains sometimes persist for several weeks, a useful diagnostic indicator. On dry sites and in dry years, however, mushrooms form infrequently. As a result the occurrence and significance of the disease are underestimated. In Georgia, for example, mushroom root rot was for many years thought to be restricted to a few localities and host plants, but widespread fruiting of the pathogen after a particularly wet summer in 1959 showed that the disease was common there on many kinds of plants. *A. tabescens,* like *Armillaria* species in the North, is a common colonist of oak roots and causes greatest damage to other plants on sites where oak trees grow or grew.

A. tabescens persists in dead, decaying roots for many years, and it attacks new hosts by growing as hyphae across points of contact with roots of living plants. It infects most readily and then grows rapidly in the new host if attack begins in root tissue already dying or recently dead. Symptoms in woody ornamentals frequently arise 3–4 years after transplanting, an indication that wounded roots may be important avenues of entry. Vigorous plants with undamaged roots resist attack. Experimental inoculations of peach roots, for example, resulted in only localized lesions after 4 years.

A. tabescens produces rhizomorphs in culture but does not commonly do so in nature. Rhizomorphs have been noted under the bark of dying water oak, and they have been infective in pathogenicity tests.

The role of basidiospores in allowing the fungus to colonize new substrates is unknown. Presumably airborne spores alighting on fresh sapwood do initiate some colonies as is the case for other *Armillaria* species.

The mushrooms of *A. tabescens* resemble those of other strains or species of *Armillaria* (Plates 148–149) in growth habit, color, texture, and general appearance but lack a ring around the stem. Typical North American strains of *A. tabescens* form mushrooms readily in pure cultures, possess binucleate cells and clamp connections in the hyphae of mushrooms, are not luminescent, grow at temperatures up to 32–34°C, and grow optimally at 25–30°C. The perforations in mycelial fans of *A. tabescens* are said to be diagnostic, but *Armillaria* that kills ponderosa pines in the Pacific Northwest sometimes also has perforated mycelial fans, so the usefulness of this character to differentiate species is in question. Species of *Armillaria* can be distinguished one from another by serological tests, but this knowledge does not help in field diagnosis.

Plants found to be infected in only a few roots or a small part of the root collar can be saved for a time by exposing the root collar to aeration and drying, which halts the pathogen.

References for Armillaria *species formerly called* A. mellea: 37, 198, 228, 387, 1118, 1353–55, 1370, 1461, 1569–73, 1621–23, 1630, 1761, 1789, 1995, 2104–6, 2121, 2211
References for Armillaria tabescens: 632, 672, 1175, 1602, 1603, 1621, 1643, 2121, 2125

Mushroom root rot (FL, Mar–Apr).

A. Dying and recently killed sand pine trees 9 years old. Shoots are not elongating and foliage is thin on the dying tree at right center.

B. Roots of dying sand pine. Deep roots are dead and decayed. Resin exudation at the root collar is a response to the pathogen.

C. Necrotic bark and cambium at the base of a dying hibiscus. Necrotic tissue extends about 5 cm above the limit of the mycelial plaque (arrow). Recently killed sapwood exposed by cutting through the plaque is only slightly darker than normal.

D–F. Mycelial plaques of *Armillaria tabescens* in the cambial region of a dead root of sand pine. Perforations, 0.5–1.0 mm long, are diagnostic.

Plate 150

313

Annosum Root and Butt Rot (Plates 151–152)

Annosum root rot is one of the most important and most studied forest diseases in the temperate zones of Earth. The pathogen is *Heterobasidion annosum* (syn. *Fomes annosus*) (Aphyllophorales, Polyporaceae). It occurs in most regions where coniferous forests grow, and it causes damage in both managed and unmanaged forests after thinning and harvesting. Root rot, blowdown due to rotted roots, butt rot, suppressed growth, and tree mortality are all significant loss factors. In addition, root and butt decay of trees in recreational areas creates hazard and sometimes causes damage to people and property because diseased trees fall unexpectedly. Firs, ponderosa pine, and giant sequoia in California have been notable in this respect.

H. annosum occurs throughout the United States, including Alaska, and in British Columbia and southern Ontario. It has not been reported from the boreal forests of central and eastern Canada. Damage to forest productivity in North America has been most severe in the Southeast, where intensive forestry has been practiced longest. Surveys there have indicated the presence of the pathogen in a majority of loblolly and slash pine plantations. Fortunately, only a minority of these plantations are on sites conducive to severe damage. On some conducive sites mortality rates of slash pine in excess of 40% due to root rot have followed thinning, and living infected trees have sustained growth losses of 20–32% in diameter and 40% in height during the first 6 years after thinning. In the West, *H. annosum* often acts in concert with other root pathogens that collectively cause tree mortality or growth loss. Other pathogens commonly associated with *H. annosum* are *Armillaria* (Plates 148–149), *Phellinus weirii* (Plates 156–157), and *Verticicladiella wageneri* (Plate 178). *H. annosum* is also common as a saprophyte on logs and stumps in areas such as the mountains of Arizona, where it causes little damage. Landscape and ornamental plantings are seldom affected, apparently because thinning and harvesting activities do not occur and environmental conditions are not conducive to infection and local increase of the pathogen.

H. annosum has been reported as colonizing and producing basidiocarps on approximately 200 diverse species of plants, including a few that are nonwoody. The fungus is an important pathogen only on gymnosperms, but angiosperms may be lethally attacked where they grow in soil that contains a large residue of gymnosperm roots colonized by the pathogen. The gymnosperm substrate apparently provides an energy supply sufficient for mycelium to overcome the normally effective defenses of angiosperm roots. The only report of continuing independent pathogenicity on an angiosperm species concerns madrone in California.

Gymnosperm host groups include arborvitae, cedar, incense cedar, Japanese cedar, cypress, false cypress, Douglas-fir, fir, hemlock, juniper, larch, monkey-puzzle, pine, redwood, giant sequoia, spruce, and yew. Angiosperm host groups include alder, apple, ash, mountain ash, beech, birch, blueberry, brambles, buckthorn, cherry, chestnut, crowberry, dogwood, elm, gorse, sweet gum, hawthorn, hazel, heather, honeysuckle, hornbeam, horse-chestnut, laurel, mountain laurel, lilac, madrone, manzanita, maple, ninebark, oak, pear, persimmon, poplar, privet, rhododendron, rose, serviceberry, spirea, sumac, tupelo, and willow. These lists represent European as well as North American records.

All hosts are subject to root killing and decay by *H. annosum*, but the nature of further damage varies with host group. Pines sustain cambial death and may be killed outright while they are young, but the heartwood in pine trunks usually resists decay by *H. annosum*. Young firs and incense cedar may also be killed within a few years after infection, and older firs (e.g., 50–150 years) are subject to decay of both heartwood and sapwood in the butt. Infection of sapwood causes decline and death. Larch, giant sequoia, and spruces resist early killing

attack but are subject to root decay leading to blowdown and to butt and heart rot that may extend several meters up the trunk. Incense cedar is affected similarly if it escapes early death. Western hemlock is subject to trunk rot beginning at mechanical injuries at any level on the trunk.

Symptoms of annosum root rot, except for the decay, are not diagnostic. Other root pathogens may cause similar symptoms: dead and declining trees alone or in groups and windthrown trees with decayed roots. In young trees, particularly pines, foliage may turn brown and drop rapidly, with no prior indication of distress. This leaf drop indicates rapid killing of bark and cambium at the root collar. Older trees may decline during 1 or more years before death. Such trees often have unusually short, sometimes chlorotic needles in tufts at twig tips. Pines and other resinous trees may exude resin at the butt, especially below the soil line. Similarly, resinous lesions develop on roots at points of attack and spread along roots as the pathogen grows toward the butt. Resin-soaked wood may extend for some distance beneath and beyond the point of advance of the fungus in sapwood of roots and the butt. Many trees with diseased roots do not have obvious symptoms above ground, although a careful comparison may indicate some suppression of growth. In general, pines can tolerate the loss of half their roots before growth suppression becomes noticeable. If *H. annosum* is present in a given stand, many more trees are infected than can be detected on the basis of symptoms and signs.

Recognition of the basidiocarps of *H. annosum* is useful for diagnosis. Basidiocarps develop on decaying parts of dead or sometimes living trees: on fallen logs or the bases of stumps or butts, on superficial roots, on the undersides of windthrown root systems, in hollow butts or stumps, under the bark of stumps, and on roots in animal burrows. In California and elsewhere in the West where the summer is dry, basidiocarps rarely form on exposed surfaces. Fresh basidiocarps are leathery and are white when young or when they grow in continual darkness. Those exposed to light turn buff to brown with a white margin as viewed from above. The lower surface, poroid where a layer of tiny tubes opens to the surface, is cream colored. Basidiocarps vary in size and shape, from tiny white plaques consisting of little more than a tube layer (Plate 151C, E) to bracketlike conks up to 25 cm across, as shown in Plate 152. Some are perennial, but most live less than 1 year. They may form on stumps as soon as 18 months after cutting and then develop more or less annually until the substrate is exhausted. Basidiocarps are preceded by white mycelial pustules or plaques that form between bark scales and on root surfaces (Plates 151C, E, 152A). Growing basidiocarps tend to envelop rather than push aside needles and other forest litter. Basidiospores, liberated from a fertile layer that lines the tubes, drop out and are carried away by air currents.

There are several other diagnostic characters. Delicate wefts of white mycelium (Plate 152B) form between bark and wood of parts colonized by the fungus except where the tissues are infiltrated with resin. As decay begins, elliptic to elongate white pockets, often containing single black specks, form by dissolution of host tissues in the cambial zone (Plate 151E). The wood gradually turns weak, soft, and stringy but is never brittle. It may also have tiny elongate pockets that sometimes contain black specks. The decay is called a white pocket rot (Plate 152D, G), a white stringy rot, or sometimes (in fir in the West) a laminated rot, according to the host affected.

Annosum root rot of red pine.

A. Part of a red pine plantation devastated by annosum root rot. Surviving trees at the edge of a disease gap display slow growth and tufted foliage as the consequence of root damage (RI, Aug).

B. Windthrow following decay of roots by *H. annosum* (NY, Sep).

C. Irregular sterile white masses of packed mycelium that will differentiate into basidiocarps at the base of a stump, about one-quarter natural size (NY, Jul).

D. Young basidiocarps of various sizes up to 9 cm across, revealed by removing needle litter at the base of a recently killed tree (NY, Sep).

E. The base of the stump shown in (C), cut at soil line, with a bark chip removed to reveal diagnostic white pockets that form in the cambial zone (about one-third natural size).

Plate 151

A distinctive conidial stage (*Spiniger meineckellus*, syn. *Oedocephalum meineckella*, Plate 152E), visible with a hand lens, develops under moist conditions on surfaces of colonized wood: on broken roots, under debris on stumps, in insect galleries, and in cavities formed by shrinkage of soft tissue at the bark-wood interface. Conidia are borne on the swollen heads of unbranched conidiophores. Each head bears many spores and appears as a white speck, giving the colony a granular appearance. The conidial state develops within 2 days on newly cut surfaces of wood previously colonized by *H. annosum* if samples are incubated at room temperature in a moist atmosphere. This is useful for diagnosis.

Infection may be initiated by basidiospores, conidia, or mycelium. The most common means of initial entry of *H. annosum* into a stand is via airborne basidiospores that germinate on freshly cut stumps or other wounds that expose sapwood. Freshly cut sapwood of conifers acts as a selective medium that allows growth of few fungi other than *H. annosum*. This selectivity lasts for only a few weeks or less in warm weather, after which various saprophytic organisms begin to grow and compete for the substrate. Given the temporary advantage, however, mycelium of *H. annosum* colonizes the stump and grows out into its roots. Transmission to adjacent trees occurs via root contacts or grafts; the fungus does not grow free for more than a few millimeters in soil. It grows in roots at average rates of 0.5 –2 meters per year, depending on strain of pathogen, host material, temperature, and soil conditions. Parasitic growth is slow because of host resistance, but growth along dying or recently dead roots is relatively rapid. Centrifugal growth results in the arrangement of diseased and dead trees in groups that enlarge for 10–30 years before stabilizing. Multiple infection centers and delayed stabilization result in severe damage.

The hazard of stump infection varies with season and climate, reflecting trends in basidiospore dispersal. Spore dispersal and stump infection occur all year in the Pacific Northwest, but mainly in spring and autumn in the Northeast and during autumn and winter in the South. Basidiospore dispersal is inhibited during winter in the North and during the heat of summer in the South. In parts of the South also, the temperature of stump surfaces in summer rises high enough to prevent colonization by *H. annosum*.

Basidiospores and conidia also percolate with rain water into coarse soil and there may initiate infection on wounded roots of vigorous trees or on unwounded roots of stumps or weak or suppressed trees. This mechanism allows *H. annosum* to enter stands where no cutting has been done. Conidia can survive up to 10 months in some soils, and relatively few conidia are required to infect nonwounded stump roots. Conidia are not airborne, however, and their role in the epidemiology of annosum root rot is unknown. Dispersal of conidia by water and insects has been suggested.

Colonization of a woody substrate usually depends initially upon parasitic activity in the course of which *H. annosum* must overcome host defenses. One mechanism of attack is the secretion in host tissues of a dihydrobenzofuran toxin called fomannoxin. Fomannoxin is produced in pure cultures, has been isolated from naturally infected Sitka spruce heartwood, and was lethal to Sitka spruce seedlings in toxicity assays.

Host defenses are based in part on secretion of oleoresin into infected tissues, which slows growth of the parasite, and in part on production of fungitoxic compounds. Pinosylvin and related compounds are significant inhibitors of *H. annosum* in pines, and these chemicals are responsible for the resistance of pine heartwood to decay. Lignans are important in active resistance of Norway spruce sapwood to the pathogen. Vigorous trees able to resist *H. annosum*

tend to compartmentalize infected tissues (see Plates 162–163) in both roots and stem.

Trees weakened by adverse environment or suppressed by dominant neighbors are most likely to be successfully attacked by *H. annosum*. Ponderosa pines weakened by air pollution, for example, have been shown to be abnormally susceptible. Weak trees, usually with depleted reserves of starch and other energy-storage compounds, are unable to synthesize sufficient fungitoxic compounds to slow or halt the parasite.

H. annosum growing as a saprobe is incapable of replacing other microorganisms within dead wood. Parasitic attack gives this fungus an important head start in colonization of a new substrate. Once established, the fungus is an able competitor and may persist for many years or until the wood is reduced to mush.

Longevity of *H. annosum* in stumps is greatest in regions of cool climate and in large stumps such as those of white fir and ponderosa pine in California. The presence of *H. annosum* in larch stumps up to 63 years old has been recorded in Great Britain. On the other hand, the wood in pine stumps in the southern United States may be completely decayed in less than 10 years, leaving bark-lined holes in the ground—a hazard for humans but not for the new crop.

Annosum root rot intensifies in a stand after the second and subsequent thinnings because of local production of spores. It causes greater damage in the second crop than the first on a given site because of spread from residual roots of the first crop into young trees of the second crop. As an exception, however, second-crop buildup has not occurred in the South because stumps decay rapidly there. Annosum root rot also predisposes trees to other pests, such as bark beetles, and to opportunistic pathogens such as *Armillaria*.

In eastern states, annosum root rot becomes most severe on deep, well-drained sandy or sandy loam containing little organic matter. Soils with relatively high content of clay or organic matter, poor internal drainage, or a high water table are not conducive to the disease. Infection may occur frequently on such sites, but damage is slight.

In general, a given isolate of *H. annosum* is pathogenic to a wide array of plants. Isolates do vary in virulence, however. Also, some strains show pathogenic specialization. Isolates from white fir and from ponderosa pine, for example, were more virulent on the host of origin than on the other host.

Annosum root rot provides one of the most often quoted examples of biological control of a plant disease. Pine stumps can be inoculated with spores of a competitive saprophyte, *Phlebiopsis gigantea* (syn. *Peniophora gigantea*), which prevents colonization by *H. annosum* and tends to replace it in roots.

References: 14, 601, 640, 726, 793, 820, 847, 851, 852, 918, 947, 1065, 1070, 1620, 1623, 1644, 1669, 1752, 1753, 1755, 1780, 1802, 2245

A. Sterile white pustules of packed mycelium of *Heterobasidion annosum* at the root collar of a recently killed eastern white pine sapling; scale is graduated in millimeters (NY, Sep).

B. Magnified view of the characteristic thin weft of white mycelium between bark and wood of a red pine root colonized by *H. annosum* (NY, Sep).

C. New (white) and dead 1-year-old basidiocarps of *H. annosum* at the duff line on a red pine stump (NY, Sep).

D. Characteristic stringy white pocket rot caused by *H. annosum* in red pine (NY, Sep).

E. Magnified view of the conidial state of *H. annosum* growing on the transverse surface of a wood section from a diseased pine root. Each conidiophore is surmounted by a white head on which conidia form (NY, Sep).

F. A small eastern white pine root killed by *H. annosum*. Red-brown, resin-infiltrated internal tissue and soil-encrusted resin deposits on the root surface indicate previous infection rather than colonization after death (NY, Sep).

G. Large, bracket-type basidiocarps on a decaying stump of eastern white pine (WV, Oct).

Plate 152

Red Root and Butt Rot of Conifers, Caused by *Inonotus circinatus* and *I. tomentosus* (Plates 153–154)

Inonotus circinatus and *I. tomentosus* (Aphyllophorales, Hymeno-chaetaceae), formerly assigned to the genus *Polyporus*, cause stand-opening disease of spruce and red root and butt rot of various conifers across Canada and the United States. *I. circinatus* was formerly considered to be a variety of *I. tomentosus*. These fungi also occur widely in Eurasia. The diseases are named for typical symptoms—slow development of openings caused by mortality in spruce stands and reddish brown stain of wood in the incipient stage of decay. The advanced decay is a white pocket rot.

Hosts of *I. circinatus* and *I. tomentosus* are seldom recorded separately except in incomplete modern lists. Therefore we present a combined list. Hosts in North America include western red cedar; Douglas-fir; alpine, balsam, grand, noble, and Pacific silver firs; eastern and western hemlocks; eastern and western larches; jack, loblolly, lodgepole, Monterey, pitch, ponderosa, red, shore, shortleaf, slash, eastern white, and western white pines; and black, blue, Engelmann, red, Sitka, and white spruces. Norway spruce, Austrian pine, and Scots pine are often attacked in Europe. Records of occurrence on angiosperms are rare. *I. tomentosus* has been found on alder and birch in British Columbia.

In Canada and the Great Lakes region of the United States, eastern larch, red pine, black spruce, and white spruce are often damaged. *I. tomentosus* is the second most common cause of root and butt rot in fir and spruce forests of central and eastern Canada; it is outranked only by *Armillaria* (Plates 148–149), which often occurs in the same stands. Severe butt rot or mortality caused by *I. tomentosus* has affected over 80% of the trees in the most heavily damaged spruce stands. Sand and slash pines are often attacked in the Southeast. In sand pine in Florida, *I. circinatus* may occur in association with *Phytophthora cinnamomi* (Plates 139–140) or less frequently with *A. tabescens* (Plate 150) or *Verticicladiella procera* (Plate 179).

I. circinatus and *I. tomentosus* cause four types of direct damage: growth suppression, cull of the butt log at harvest, blowdown, and tree mortality. In addition, susceptible seedlings planted on sites of previous mortality are presumed to be subject to lethal attack after their roots grow into contact with infested roots of the previous crop. This has been demonstrated experimentally for spruce replanted in areas of stand-opening disease in Canada.

Diseased trees occur singly or in groups. Infection occurs at wounds on roots or the root collar, and the fungus spreads both outward in roots and up into the butt. Diseased roots have dead, decayed distal portions and red-brown, resin-soaked wood extending into the living proximal portion and the butt. The reddish stain evolves to advanced decay and slowly expands to involve both deep-lying wood and sapwood near the cambium. Mycelium and dark zone lines become prominent in the cambial region of killed spruce roots. Resin exudes from the butt or root buttresses where the cambium and bark are attacked. Resin exuding from roots may mix with soil and accumulate as an incrustation. Aboveground symptoms develop over a period of many years in the North and over 2 years or more in the South. The growth rate of diseased trees diminishes, and lower branches begin to die prematurely. Foliage becomes thin, and that on upper branches becomes chlorotic. In time the tree dies, but it often breaks at the butt or blows over while alive because of decayed wood in the major roots or butt.

Although infection may occur in young trees, mortality in natural conifer stands in Canada is usually delayed until trees are more than 50 years old. Spruce affected by stand-opening disease occurs in patches that may slowly expand and coalesce to involve 0.5 hectare or more. The fungus grows linearly only 1–10 cm per year in spruce in Canada; thus 15–20 years usually elapse between infection and death. The disease progresses much more rapidly in the South; blowdown and mortality in the sand pine stand illustrated here began before age 20.

Trees of all vigor classes are at risk but not equally so. For example, in white spruce plantations affected by *I. tomentosus*, observed for 4 years in Wisconsin, mortality rates in dominant, intermediate, and suppressed trees were 2%, 15%, and 48%, respectively, versus 0%, 0%, and 29% in plots not affected by the fungus. The 29% mortality of suppressed trees in the absence of *I. tomentosus* presumably represented normal attrition of small, weak individuals.

The decay caused by *I. circinatus* or *I. tomentosus* is classified as a white pocket rot. The early stage is dark reddish brown. The late stage is also reddish brown but appears lighter because of numerous small elongate pockets with pointed ends, filled with white fibers (Plate 154G). Decaying wood slowly softens and eventually becomes mushy but never brittle. The decaying portion of major roots or the butt typically expands nearly to the vascular cambium before tree breakage or death.

Field diagnosis usually depends upon discovery of basidiocarps. These are annual, developing in summer in the North and in autumn in the South. They arise on stumps, trunks, or roots of living or dead coniferous trees or from soil containing diseased roots. They are nearly white when young and become yellowish, then turn tan to rusty brown. The margin remains light colored during growth. Basidiocarps are somewhat spongy to leathery when fresh but are rigid when dry. Their caps have a slightly velvety surface and are 3–18 cm broad and 0.3–4 cm thick, often with concentric zones of varying color (Plate 154C). The poroid undersurface of a basidiocarp is gray-brown, and the mouths of the tubes average two to four per linear millimeter. Basidiocarps arising from soil or roots have short central or lateral stalks, but those on the base of a trunk lack a stalk. The stalked basidiocarp is typical of *I. tomentosus*, and the laterally attached type on the trunk base is typical of *I. circinatus*. Exceptions are common, however. Old, darkened, inactive basidiocarps may persist through winter beneath snow in the North and decompose during the next summer. In the South, however, basidiocarps disappear in early spring. They may be scarce in dry seasons.

The microscopic characters of basidiocarps of *I. tomentosus* and *I. circinatus* include brown setae (pointed structures) that protrude from the hymenium (basidial layer) lining the tubes. We mention this because setae are the basis for distinguishing *I. tomentosus* from *I. circinatus*. The former has nearly straight setae, while those of the latter are strongly curved or hooked. Since this is the only consistent morphological difference, many mycologists consider *I. circinatus* to be a variety of *I. tomentosus*.

Root rot of sand pine caused by *Inonotus circinatus* (FL, Apr).

A. Dead and dying trees in a 20-year-old stand. One dead tree bears a distress crop of cones. A dying tree has thin, chlorotic foliage.

B. A disease gap at the edge of which infected trees exhibit sparse, chlorotic foliage.

C, D. The butt of a diseased tree before and after removal of bark from a resinous basal canker.

E. A preserved basidiocarp of *I. circinatus* has been posed as it would normally appear at the base of a diseased tree in winter. Resin exudation indicates possible infection.

F. Excavation and removal of bark show the association of diseased roots (wood stained reddish brown) with a resinous basal canker.

G. Close view of typical reddish brown stain (incipient decay) within a root.

Plate 153

319

I. tomentosus (including *I. circinatus*) is sometimes called the false velvet-top fungus, reflecting the occasional similarity of its basidiocarps to those of *Phaeolus schweinitzii* (Plate 155). The latter is known as the velvet-top fungus. Basidiocarps of *I. circinatus* on slash pine are sometimes so similar to those of *P. schweinitzii* that microscopic examination is necessary to differentiate between them.

Infection by *I. tomentosus* in spruce may be initiated by either mycelium or basidiospores. Mycelium is effective where healthy roots grow in contact with dead roots infested with the pathogen. Basidiospores are effective if introduced to wounds that penetrate into sapwood or heartwood of the roots or butt. Wounds to heartwood are most suitable except that freshly cut stump tops are not colonized. Certain root-tunneling weevils (*Hylobius* species) cause wounds that also serve as points of infection in both pine and spruce. Roots as small as 1 cm in diameter may be colonized by *I. tomentosus* following damage by these insects. Also, the declining roots of stumps or weakened trees seem to serve as sites of entry.

Infection in southern pines has not been studied experimentally, but circumstantial evidence indicates that, in slash pine, basal cankers caused by the fusiform rust fungus (Plate 132) may be sites of entry for *I. circinatus*. In one study, basal cankers could be found without associated root rot, but roots decayed by *I. circinatus* were found only beneath basal cankers. Sand pine, however, is not affected by fusiform rust, although it often has basal cankers associated with red root and butt rot. Early reports of *I. circinatus* on western white pine mentioned an association with fire scars.

I. tomentosus may persist in spruce roots for 15–20 years after death of a tree. Presumably its longevity in southern pine roots is less because of rapid decay in the warm climate.

I. tomentosus and *I. circinatus* are proven pathogens of sapwood, which partially explains why their attack is lethal. These fungi act relatively slowly, however, and must therefore possess some as yet undescribed mechanism for combating or avoiding host responses that would normally compartmentalize infection. When Canadian isolates of these fungi were inoculated to seedlings of 11 coniferous species, isolates of *I. tomentosus* were more virulent than isolates of *I. circinatus* on every species.

In Canada, red root and butt rot occurs on a variety of sites. It is most common and severe in spruce growing on soils that are acidic (pH 4–5), relatively infertile, and with low water-holding capacity or where shallow soil overlies rock or hardpan. In sand pine in Florida, the disease is most severe where the soil is poorly drained or the water table or an impermeable soil layer lies close to the surface. In mixed coniferous forests in Idaho, the disease is most common in stands at elevations above 1500 meters.

Widespread occurrence in northern forests notwithstanding, *I. tomentosus* seems well adapted to moderate rather than low temperatures. Its mycelium grows most rapidly at temperatures near 20°C. Its basidiospores are discharged most rapidly in moist air (85–100% relative humidity) at 16–22°C. They germinate well at temperatures of 15–27°C; germination is most rapid at 20–24°C.

I. tomentosus grows readily but slowly on agar media, the mycelium extending only about 2 mm per day. It forms thick-walled chlamydospores in culture, but whether similar spores occur in nature is unknown. The fungus does not produce clamp connections (typical of many basidiomycetes) in its mycelium in culture.

References: 101, 213, 230, 636, 640, 707, 844, 1374, 2180–83

A, B. Root and butt rot of white spruce caused by *Inonotus tomentosus*. A. Resin exudation is the only external symptom on the butt of a dying tree. Dead wood in a root buttress is revealed where a strip of bark has been removed. B. A basidiocarp of the previous year, on a decaying root of the tree shown in (A) (Ont, Jul).

C–E. Basidiocarps of *I. tomentosus* associated with roots of western conifers. C. Variation in basidiocarps associated with western white pine and grand fir, collected from soil within an area only 10 meters square. D. Small, stalked basidiocarps on the forest floor. E. The lower, poroid surface of a stalked basidiocarp (ID, Jul).

F. White pocket rot typical of that caused by *I. tomentosus* in western white pine (ID, Jul).

G. Close view of white pocket rot caused by *I. circinatus* in sand pine (FL, Apr).

Plate 154

Brown Root and Butt Rot of Conifers, Caused by *Phaeolus schweinitzii* (Plate 155)

Brown root and butt rot, also known as brown cubical butt rot, is caused by the velvet-top fungus, *Phaeolus schweinitzii* (Aphyllophorales, Polyporaceae). This fungus, long known as *Polyporus schweinitzii*, occurs throughout the forested regions of North America and is also widespread in Eurasia. It is one of the most common root and butt pathogens in both natural and planted coniferous forests, and it also causes damage to conifers in landscapes. It is considered to be the most serious butt rot pathogen of conifers in old-growth forests in the United States. It was associated with butt rot in 5–30% of old-growth Douglas-fir harvested in various parts of the Northwest, for example, and with 66% of all storm-broken butts of Douglas-fir in the mountains of Arizona.

P. schweinitzii commonly causes damage in Douglas-fir, true firs, larches, eastern and western white pines, and spruces. Probably all conifers are susceptible to some extent. In addition to the species above, those reported to be hosts in North America include American arborvitae; Atlantic white, incense, and western red cedars; Lawson cypress; alpine, balsam, Fraser, grand, noble, red, Pacific silver, and white firs; eastern, mountain, and western hemlocks; eastern, Japanese, and western larches; digger, jack, Jeffrey, knobcone, limber, loblolly, lodgepole, longleaf, Monterey, mugo, pinyon, singleleaf pinyon, pond, ponderosa, red, sand, Scots, shortleaf, slash, sugar, tablemountain, Virginia, and whitebark pines; giant sequoia; black, Engelmann, Norway, red, Sitka, and white spruces; and Pacific yew. Angiosperms are rarely infected; there are records of the occurrence of *P. schweinitzii* on cherry and silverbell in the East, on paper birch, blue gum, and Oregon white oak in the West, and on koa in Hawaii.

P. schweinitzii is often considered to cause significant butt rot mainly in mature trees. This, however, is a misconception based on delayed discovery. The fungus may infect roots of any age and may enter the stems of young trees through roots or basal wounds. The Japanese larch trunk shown in (E) was less than 15 years old at the time of wounding that led to decay by *P. schweinitzii*. Trunk rot without a connection to decaying roots is rare. Decay eventually involves most of the cross-section of the butt and usually extends less than 3 meters up the trunk before the tree breaks or is windthrown during a storm. Decay columns extending more than 10 meters into the trunks of ponderosa pines and as high as 30 meters in Douglas-fir have been reported, however.

Wood in the incipient stages of decay is stained pale yellow or yellow-brown to reddish brown. It losses strength rapidly, and it gradually becomes dry and crumbly, with a pronounced tendency toward transverse fracture. In late stages it turns darker brown and cracks into more or less cubical pieces, with white to yellowish mycelium in the cracks. Wood undergoing decay by *P. schweinitzii* often has a pungent odor that has been likened to turpentine and to aniseed.

Basidiocarps grow annually from decaying wood and live for 2–3 months in late summer and autumn. In California and other parts of the West where little or no rain falls during summer, basidiocarps form only in autumn. They develop on stumps of trees that were diseased before felling, on the bases of living diseased trees near ground level, on roots, or on soil, connected to a root by a stout strand of mycelium. Basidiocarps growing from roots often arise 1–2 meters or more from the trunk, grow to 30 cm across, and stand 15 cm tall on a thick velvety stalk. Specimens growing from trunk bases take the form of brackets lacking a stalk. The upper surface of a mature basidiocarp is dark rust brown and has a rough velvety texture when fresh. Young, expanding specimens have a thick yellow rim. The poroid underside is gray-green to gray-brown when fresh and turns brown where bruised. The angular mouths of the tubes number one to three per linear millimeter. Active basidiocarps are water-soaked to soft-corky and often exude droplets on the undersurface. In the North, the annual activity of basidiocarps is terminated by frost, after which they turn brownish black and corky and, after drying, become very light in weight. Old dead basidiocarps may persist a year or more.

Infection biology of *P. schweinitzii* has been studied more in Great Britain than in North America. Most infections are probably initiated by spores. Basidiospores that alight on the forest floor are capable of establishing persistent infestation of soil. Although the fungus does not grow freely as a soil saprophyte, it has been isolated from forest soils, including those where no root rot has occurred. Whether the basidiospores persist or produce persistent resting spores is unknown. *P. schweinitzii* does produce chlamydospores. These are common and diagnostic in agar cultures and have been found in soil.

There is but little evidence for root-to-root mycelial growth of *P. schweinitzii*, although its mycelium has been detected in soil. Affected trees are usually scattered rather than in groups, and adjacent decaying trees contain different clones of the fungus. Adjacent trees would be expected to contain the same clone if mycelial growth were a principal means of local spread. Furthermore, damage occurs not only in long-time coniferous forests but also in first-rotation conifers on sites where nonhost hardwoods formerly grew. There is, however, one report of butt rot in young slash pine (10–22 years old) associated with stumps of previous pine crops. The stumps were found to be infested with *P. schweinitzii*.

P. schweinitzii often enters damaged or perhaps dead superficial roots or wounds such as fire scars at the trunk base. Because severe butt rot often develops in trees that lack such external markers, however, wounds may not be required. *P. schweinitzii* has been reported to infect and kill root tips and induce swelling at the ends of the resulting root stubs of Douglas-fir. In experiments with pine seedlings it penetrated suberized or nonsuberized rootlets and subsequently killed the seedlings. It has caused severe damage to red pine and eastern white pine planted on shallow soils subject to periodic waterlogging and summer drying, which weaken or kill many roots. Roots previously colonized by *Armillaria* (Plates 148–149) have also been suggested to be avenues of infection because the two pathogens are often associated and *P. schweinitzii* is capable of growing through wood colonized by *Armillaria*. In Douglas-fir and grand fir in Idaho, however, *Armillaria* was considered to be secondary to *P. schweinitzii* because the former fungus was found at the root collars of trees with severe root rot caused by the latter.

Site factors that influence infection are poorly understood. Shallow, poorly drained soils apparently predispose to damage in red pine and eastern white pine. The most common sites of occurrence on various conifers in the Northwest are at elevations above 1500 meters.

References: 109–12, 128, 189, 228, 477, 640, 844, 1521, 1669, 2124, 2263

Brown butt rot of Japanese larch, caused by *Phaeolus schweinitzii* (NY, Jun–Oct).

A. A living tree with severe butt rot, broken by wind. Brown cubical rot is visible. Dark brown basidiocarps at the base of the trunk are diagnostic (Oct).

B. Decaying wood in the butt of a tree that broke while alive; transverse lines of fracture are evident.

C. Dead basidiocarps of the previous year on a wound scar at the base of a trunk (Jun).

D. Basidiocarps at the base of a diseased tree, appearing dark reddish brown after the onset of freezing weather in autumn (Oct).

E. Cross-section of a butt with decay caused by *P. schweinitzii*. The decay, associated with an old basal wound, has been compartmentalized within wood present at the time of wounding (Jun).

F. A cluster of mature basidiocarps of *P. schweinitzii*, collected from soil at the base of a tree. A stout stalk is apparent at one side of the cluster (Oct).

G. Close view of the lower, poroid surface of an active basidiocarp (Oct).

Plate 155

Laminated Root Rot of Conifers, Caused by *Phellinus weirii* (Plates 156–157)

Laminated root rot, also known as yellow ring rot, is caused by *Phellinus weirii* (syn. *Poria weirii*) (Aphyllophorales, Hymenochaetaceae). It affects conifers in Japan, Manchuria, and western North America from southern Oregon to southern British Columbia and eastward into northern Idaho. It occurs in many soil types from sea level to the upper limits of forests that are managed for timber production. Annual losses have been estimated at more than 4 million cubic meters of wood, and increased losses are expected where susceptible species are regenerated after diseased crops. Two strains of the fungus are recognized. One forms perennial basidiocarps, causes butt rot in western red cedar, and is common in the eastern (Rocky Mountain) portion of the range of the disease. The other strain forms annual basidiocarps principally on hosts other than western red cedar and is responsible for most of the damage in lowland and Cascade forests.

Mountain hemlock is perhaps the most susceptible species naturally affected in North America. Douglas-fir (in which economic loss is greatest), grand fir, red fir, Pacific silver fir, and white fir are also highly susceptible. Alpine fir, western larch, lodgepole and western white pines, and Engelmann and Sitka spruces are attacked when associated with the species named above. Western hemlock is highly susceptible when growing in mixture with Douglas-fir of the same age, but it is rarely infected in pure stands and seems little damaged when growing in the understory of dominant Douglas-fir. Western red cedar and ponderosa pine are resistant; incense cedar is highly resistant. Species such as cedars or pines, resistant to killing, may be subject to butt rot. Hardwoods such as red alder and bigleaf maple that commonly grow in association with susceptible conifers are highly resistant to *P. weirii*.

P. weirii attacks trees of all ages, but diseased trees may show no symptoms above ground for 5–15 years. Eventually, however, most begin to grow slowly and produce thin or asymmetric crowns, chlorotic foliage, and distress crops of numerous but abnormally small cones. These symptoms, although indicative, are not diagnostic. Trees with black stain root disease (Plate 178), for example, have similar symptoms. Declining trees are often attacked by bark beetles: fir by the fir engraver (*Scolytus ventralis*) and Douglas-fir by the Douglas-fir bark beetle (*Dendroctonus pseudotsugae*) (Johnson and Lyon, 2nd ed., Plate 28).

Laminated root rot is seldom noticed until a forest stand is at least 10 years old. It is most destructive in stands 25–125 years old. This disease is more important as a cause of death than of growth suppression or cull because young diseased trees often die within a few years after their growth rate diminishes, before significant extension of decay into the butt log. In large old trees, however, cull due to decay may extend as much as 4 meters up the trunk.

Growth losses have been measured in diseased trees that persisted several years before death. In one locality in Oregon, average volume losses in diseased trees due to retarded growth during the 10 years before death averaged 32%. Trees that remained alive 11–15 years after the onset of growth suppression accumulated 18% less volume on the average than comparable healthy trees during that period. Observations in British Columbia indicated annual height growth losses of 0.9–1.7% in diseased trees.

The two major aspects of disease in a given tree, cambial necrosis and decay, lead to different modes of killing. A minority of trees are killed by progressive advance of the parasite in the cambial zone until the tree is girdled at the butt. The majority fall or blow down while still alive, sometimes before crown symptoms develop, because of decay in the major roots and butt. These trees resist the pathogen well enough to halt cambial killing permanently or intermittently, but decay continues in root and butt wood present at the time of cambial killing. Decayed roots typically break at the butt and do not lift from the soil as the tree falls.

The early stages of decay caused by *P. weirii* are reddish brown to brown streaks or bands as viewed longitudinally or crescent-shaped to circular areas in cross-sections. In late stages, the wood tends to separate into sheets at the junctions of annual rings, and it contains numerous tiny cavities about 0.5 × 1.0 mm. In the final stages, the wood breaks down into a loose stringy mass.

Thin layers or tufts or brown mycelium are usually present in cracks in decayed wood, and thin brown fungal crusts often form on root surfaces and on the butt in root crotches. On microscopic examination, the mycelium lacks clamp connections. Laminated root rot can be diagnosed in the field by examining decaying wood with a hand lens. Stout, reddish brown hairs (setal hyphae) (Plate 157C) are visible, extending from the mycelium in cracks and cavities. Setal hyphae are easy to find just beneath the bark of a decaying butt.

Basidiocarps appear on decaying wood as gray-brown crusts with a poroid surface and white to cream-colored margins. They form in root crotches or on the lower sides of fallen trunks during late summer and early autumn. Crusts on fallen trunks are typically 30–60 cm long, but larger specimens are common. The tubes are tiny (five to seven pores per linear millimeter) and oriented vertically. By the next spring most basidiocarps are dead and moldy (notice mold on the specimen in Plate 157G). Although basidiocarps are relatively common in some years and localities, they are never numerous or conspicuous enough to be useful for detection or assessment of disease.

Infection in a young forest stand often begins in 15–20-year-old trees as their roots grow into contact with perennial mycelium that has persisted in and on the remains of roots of the previous forest. Once established in a living tree, the pathogen spreads along roots and infects other trees at points of contact. Therefore diseased or dead trees usually occur in circular to irregular patches. These range in area from several square meters to half a hectare or more. Continuous enlargement of disease patches for up to 185 years has been documented.

A, B. Disease gaps caused by root-infecting fungi in a northwestern mixed conifer forest. *Phellinus weirii* is the most common and important pathogen involved. Douglas-fir has died in the gap in (B); western larch and western white pine, somewhat less susceptible, survive in a state of low vigor (ID, Jul).

C. Young Douglas-fir, normal and dying of laminated root rot. Slow growth, thinning of foliage, chlorosis, and abnormal cone production are typical symptoms (OR, Jul).

D, F. Resinosis at the butt of a Douglas-fir infected by *P. weirii*. D. Resin exudation at bark surface. F. Pockets of resin and necrotic tissue in bark, exposed by cutting away superficial bark just below soil line (ID, Jul).

E. Portion of a Douglas-fir root killed by root-infecting fungi. Dark, resin-soaked sapwood indicates previous host response to infection and is typical of incipient decay by *P. weirii*. A white mottled rot is developing below the resin-soaked sapwood. *P. weirii* is often preceded or accompanied by other fungi in such roots in Idaho, but causes typical decay (Plate 157) above ground level (ID, Jul).

Plate 156

Mycelium of *P. weirii* grows across root contacts between standing or recently cut trees of the same or different species. Root grafts are not necessary. Roots of Douglas-fir remain susceptible at least 12 months after trees are cut. Transmission from a root colonized only internally by *P. weirii* is much less likely than from a root with superficial mycelium. Superficial mycelium grows profusely on roots of Douglas-fir or western hemlock, is limited on diseased roots of lodgepole, ponderosa, or western white pine, and is nearly absent on diseased roots of western red cedar.

After initial contact, the mycelium grows along the surface and penetrates at numerous points through intact or injured bark. Multiple infections where roots of a single tree contact one another facilitate the destruction of the root system. In highly susceptible species, superficial mycelium often advances considerably beyond the limits of internal decay.

As mentioned, even highly susceptible species resist the advance of the parasite and may halt it at least temporarily. Adventitious roots sometimes grow from callus that forms where the pathogen has been halted. Inner bark and outer wood responding to attack become infused with resin and dark in color. Deep-lying wood colonized after death of the cambium and outer wood remains light in color. Resinous wood decays slowly in comparison with wood colonized while dying or dead. Therefore resin is presumed to enhance long-term survival of the pathogen in dead roots.

P. weirii can survive more than 50 years in stumps, and its viable mycelium has been found in roots as small as 2 cm in diameter 11 years after cutting a diseased tree. The fungus is capable of growth as far as 5 cm through soil from infested wood, and it grew superficially across "bridges" of experimentally buried Douglas-fir wood from previously colonized to healthy roots. These attributes illustrate the versatility of the perennial mycelium associated with woody substrates in soil. Disease patches expand radially at rates of 20–40 cm per year (average 32–35), which corresponds to measured rates of growth of *P. weirii* along roots.

Typical disease patches contain the fallen or erect remains of trees dead for various lengths of time. For every tree killed by *P. weirii* at the edge of a disease patch, many more are likely to be infected. When one severely diseased stand in Oregon was harvested, for instance, *P. weirii* was visible in decay in 62% of the stumps within 9 meters of trees previously killed and in 20% of stumps 9–15 meters distant from dead trees. In British Columbia, dead and declining trees in disease patches were found to account for only about one-half the area of infection and one-half the number of trees actually diseased.

Old disease patches often appear like rings with green centers where young trees have developed from seed. The coniferous trees that develop in old patches tend to be more diverse than those in the surrounding forest, and a higher proportion are of species more or less resistant to *P. weirii*.

Mycelium of *P. weirii* in roots and stumps exists as long-lived clones that perpetuate themselves by colonizing new roots. A given clone tends to be incompatible with most others. Thus the mycelium is initially uniform throughout a disease patch. If disease patches caused by different clones merge, the clones do not overgrow one another, but each in its own sector contributes to the expansion of the composite patch. Nearly all disease patches in contemporary forests are considered to be caused by mycelium that has survived in buried wood of trees that were infected in previous forests. One study of disease progression indicated infection of naturally occurring young trees in old disease patches 88–165 years after the old trees died.

Mycelium of *P. weirii* is considered to be active in roots all year in the moderate climate of the Pacific Northwest. The fungus grows at temperatures from below 5° to 30°C; growth is most rapid near 25°C.

No role for basidiospores in the disease cycle has been confirmed. Stump infection by spores apparently does not occur, inasmuch as experimental inoculations of fresh-cut stumps failed. Decay attributed to *P. weirii* has occasionally been found in the lower trunks of trees without decayed roots; this indicates the possibility of occasional wound infection by basidiospores. Those of *P. weirii* were found capable of germinating and establishing colonies on surface-sterilized wood of several species, but possible colonization of wounds in nature has not been studied. Direct infection of roots by spores in soil is another unstudied possibility.

Damage by laminated root rot becomes more severe, and disease patches expand more rapidly in pure stands of susceptible species than in stands of mixed species. The presence of red alder (highly resistant to *P. weirii*) seems particularly beneficial, perhaps because the microbial and chemical conditions in its root zone are relatively unfavorable for long-term survival of *P. weirii* in buried coniferous wood.

P. weirii is often accompanied in disease patches by other root pathogens such as *Armillaria* (Plates 148–149) or *Verticladiella wageneri* (Plate 178). One survey in Oregon and Washington revealed two or more root pathogens in 16% of the patches surveyed. In northern Idaho the situation is even more complex; multiple root pathogens are present in most patches.

Local spread of *P. weirii* can be halted by trenching, which is potentially useful to prevent expansion of a disease patch into landscape trees at the forest edge, but it is impractical for forestry. Planting of resistant tree species or species mixtures offers the possibility of continued forest productivity on sites where laminated root rot has been severe. Such schemes require long-term tests, which have not been completed.

References: 204, 206, 274, 354, 387, 572, 656, 844, 1246, 1393, 1425, 1985, 2095–97

A. Excavation at the base of a young Douglas-fir, dying from attack by *Phellinus weirii*, reveals resin exudation from the butt, also whitish mycelium plus brown incrustation on surfaces of decaying roots (OR, Jul).

B. Close view of a brown crust of mycelium, diagnostic for *P. weirii*, on one of the roots excavated in (A) (OR, Jul).

C. Magnified view of reddish brown setal hyphae (sterile hairs) on a mat of mycelium between bark and wood of a tree recently killed by *P. weirii*. Setal hyphae are diagnostic (WA, Jun).

D–F. Typical decay caused by *P. weirii* in Douglas-fir. Decayed wood is soft and yellowish, with numerous tiny cavities and a tendency for the annual layers to separate (ID & WA, Jun–Jul).

G. Basidiocarps of *P. weirii*, formed the previous autumn, on the butt of a small, fallen Douglas-fir (OR, Jun).

Plate 157

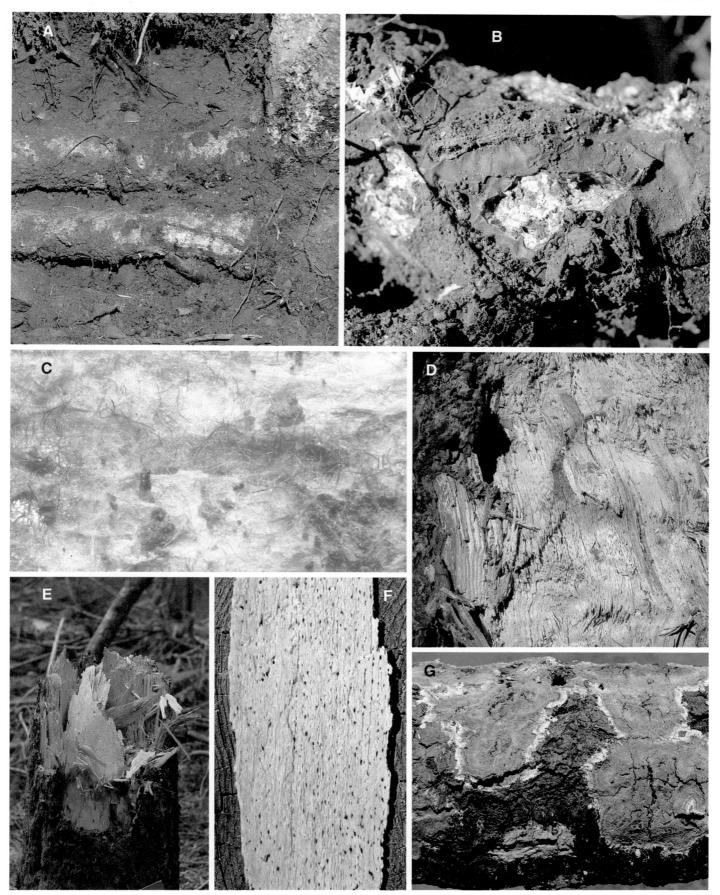

Root and Butt Rots Caused by *Inonotus dryadeus* and *Oxyporus latemarginatus* (Plate 158)

Root rot caused by *Inonotus dryadeus*. This fungus, also known as the weeping conk, is a member of the Aphyllophorales, Hymenochaetaceae. It is among the minority of wood-decaying fungi capable of parasitic colonization of both angiosperm and gymnosperm trees. It is widespread in Eurasia and in North America from British Columbia to California and New Mexico in the West to New York and Florida in the East. It is often found as a saprophyte on stumps or snags but also causes root rot of various trees, predominantly oaks in the East and South and conifers in the West. Coniferous hosts in North America include grand and white firs, mountain and western hemlocks, and Engelmann and Sitka spruces. Angiosperm hosts include American elm, bigleaf and Rocky Mountain maples, and many oaks (black, chestnut, pin, northern pin, post, northern red, southern red, Texas red, scarlet, water, white, and Oregon white). The foregoing lists include records for both living and dead hosts.

Trees affected by *I. dryadeus* often show no aboveground symptoms for many years or until they are blown down because of decayed roots. Roots of trees that topple because of root rot usually break within 30–60 cm of the trunk, and the remainder of the decayed roots stay in the soil. Necrosis of bark and cambium precedes decay of sapwood at a given location on the root system. Irregular white patches of mycelium may form on the outside of decaying roots. Trees that remain standing eventually show symptoms of decline: sparse foliage, poor color, and dieback. Oaks with many dead roots sometimes die suddenly during periods of summer heat and dryness. Diseased trees ordinarily occur singly rather than in clusters.

I. dryadeus causes a white mottled rot that begins in roots, first in the sapwood and later in deep-lying wood, and spreads toward the butt. In oaks the rot does not spread appreciably above ground level. Superficial roots may be affected only on the lower side or may escape decay altogether. In western conifers such as grand fir, the decay may spread into the base of the trunk.

Irregular, often massive basidiocarps arise annually from exposed roots or the trunk at or near the ground or from the ground, attached to roots in the soil. The basidiocarps are light yellow to yellow-brown initially, often with numerous drops of liquid on the surface. They turn brown to chestnut brown at maturity and after several months become blackish with rough, cracked surfaces that eventually weather to gray-brown. The interior is reddish brown to chestnut brown and fibrous-corky in texture. Basidiocarps vary in size and may be up to 40 cm long, 35 cm wide, and 10 cm thick, with rounded margins. The lower surface is poroid, but the tubes in old, inactive specimens are usually obscured by mycelial growth. Mycelium of *I. dryadeus* may permeate sandy soil for several centimeters beneath a fruit body and may hold the soil in a hard mass. Multiple basidiocarps are a sign of massive decay that could cause a tree to topple during a wind storm.

Basidiospores are assumed to play a role in dispersal and establishment of *I. dryadeus* in new substrates, but the site of infection (wounds versus intact roots) and rate of invasion are unknown. No evidence for tree-to-tree spread has been reported. In Mississippi, basidiospores are released from November to February when average daily temperatures are in the range 10–16°C. Peak release occurs during daylight and coincides with daily maximum temperature of 22–24°C. This temperature is close to the reported optimum (25°C) for mycelial growth of *I. dryadeus* on agar media.

Root rot caused by *I. dryadeus* in oaks is most common in old or suppressed trees and in those growing on sites unfavorable for vigorous growth.

Rot caused by *Oxyporus latemarginatus*. This fungus (syn. *Poria latemarginata, P. ambigua*) (Aphyllophorales, Polyporaceae) causes a white rot in the sapwood of roots and stems of diverse angiosperm trees around the world. It is a common invader of wood behind fire scars on living trees. It is also a common saprophyte on logs, snags, and logging slash. In North America it occurs from Nova Scotia to British Columbia, Florida, Texas, and California.

O. latemarginatus is a proven root pathogen. Root and root-collar rots caused by it have been noted on eastern cottonwood, peach, and sycamore in the South and on almond, apple, mahaleb cherry, citrus, peach, pear, and Persian walnut in California. The fungus occasionally causes significant damage in orchards, but its pathogenic importance in forests and landscapes is unknown. Other hosts, mainly those on which *O. latemarginatus* occurs as a saprophyte or is associated with trunk decay, include acacia (Cootamundra wattle and Sydney golden wattle), alder, apple, black ash, bigtooth aspen, sweet bay, American beech, yellow birch, chinaberry, black cottonwood, blue elder, American elm, sour gum, sweet gum, tupelo gum, hackberry, netleaf hackberry, Hercules' club, shagbark hickory, California laurel, common lilac, black locust, honey locust, red and sugar maples, mimosa, oaks (swamp chestnut, Nuttall, shingle, water, white, Oregon white, and swamp white), balsam poplar, privet, serviceberry (*Amelanchier canadensis*), black walnut, black willow, and yellowwood.

Slow growth and mortality of young trees are the aboveground symptoms of root rot but are not diagnostic. Signs of the pathogen are usually present, however. *O. latemarginatus* produces either a sterile white mat of mycelium or a white to dirty white or pale yellow basidiocarp at the soil line of a dying or recently dead plant. The basidiocarp, active for only 1 season, may surround litter and blades of grass as it grows. The fungus also infests the soil around diseased roots and the root collar and may form a white crust or interlaced white mycelial strands on dead parts of roots or on the soil at the stem base.

O. latemarginatus appears to kill small roots first and then spread toward the root collar. It colonizes the outer bark and then invades the inner bark, cambial region, and sapwood, causing enlarging cankers and a soft white rot of the wood. Seedlings may be killed within 9 months. Mature (16-year-old) cherry trees have succumbed within 2 years in California. Parasitic attack proceeds most rapidly during warm summer weather.

White mycelium of this fungus grows rapidly on agar media. Growth is maximal at temperatures near 30°C and in one study was 88–90% of maximum at temperatures of 25–35°C.

Although root rot by *O. latemarginatus* has frequently been linked to waterlogged soil or other conditions that might damage roots, cherry in California was severely diseased in orchards with well-drained soils.

References for Inonotus dryadeus: 314, 449, 563, 636, 640, 1160, 1162, 1255, 1669
References for Oxyporus latemarginatus: 126, 476, 568, 640, 1159, 1745

A–C. Root rot of willow oak caused by *Inonotus dryadeus* (MS, Apr). A. Irregular dark brown to black basidiocarps in the root crotches of a mature tree, an indication of severe root rot. Interior tissue of the basidiocarp is reddish brown. B. Basidiocarps extending from the base of a diseased tree. The gray-brown basidiocarp against the bark is old and dead. The blackish brown one is alive. C. A small, clinkerlike basidiocarp of *I. dryadeus*, growing from a decaying root beneath the soil surface.

D, E. Root rot of peach and cottonwood seedlings, respectively, caused by *Oxyporus latemarginatus*. The fungus produces white mycelial sheets and poroid basidiocarps near the bases of recently killed plants (MS, Apr).

F, G. Year-old bracket-shaped basidiocarps of *I. dryadeus* on the butt of a dead grand fir (OR, Jul).

Plate 158

Ganoderma Root and Butt Rots and Trunk Decay (Plates 159–161)

Ganoderma (Aphyllophorales, Ganodermataceae) is a genus of wood-decaying fungi that occur around the world in tropical to cool temperate habitats on monocots, dicots, and gymnosperms. These fungi cause decay of the white rot type (see Plate 165). Some species are only saprophytic, but several are pathogens that cause decay in roots, butts, and trunks of living trees. *G. lucidum, G. zonatum,* and other species that cause root rot sometimes kill their hosts, but often a diseased tree is windthrown while still alive. Decay of the butt and the base of the trunk of deciduous trees by *G. applanatum* often results in breakage. Usually the basidiocarps of a *Ganoderma* are the first external indicators of disease. The basidiocarps of many species have a distinctive varnishlike crust on the upper surface. All species produce thick-walled brown truncate basidiospores that often accumulate in conspicuous deposits on and around basidiocarps.

Root and butt rot of palms. At least three species of *Ganoderma* attack palms *G. zonatum* (syn. *G. sulcatum*) is blamed for damage in Florida and Georgia and throughout tropical Africa. *G. boninense,* which causes significant economic loss of oil palms in Malaysia, is distributed from Sri Lanka eastward through the Indian and south Pacific regions. *G. tornatum* occurs throughout the tropical belt of the world. This species attacks not only palms but various dicotyledonous plants (*Casuarina* in Florida). Records of all these fungi are confused with those of *Ganoderma* species of the temperate zone. *G. boninense* was often misidentified as *G. lucidum* (Plate 160), while *G. zonatum* and *G. tornatum* have at times been considered forms of *G. applanatum* (Plate 161).

Species of *Ganoderma* are differentiated on the basis of microscopic characteristics of the basidiospores and of the cutis, or crust, on the upper surface of the basidiocarp, but mycologists have not agreed on narrow as opposed to broad species concepts. For practical purposes, several *Ganoderma* species cause similar diseases that can be considered together.

Root and butt rot of palms in the United States, attributed usually to *G. zonatum,* affects at least the following palm types: Arikury, cabbage, Canary Island, coconut, Senegal date, jelly (pindo), oil, queen, thread, yellow (Madagascar), and saw palmetto.

Loss of vigor, undersized fruit and fronds, and premature yellowing and death of the oldest fronds first call attention to damaged roots. As the disease progresses, basidiocarps that often become fan or kidney shaped or semicircular as viewed from above grow out more or less horizontally from the base of the palm trunk or from its adventitious roots. Overhead, fronds die, and the crown eventually breaks off. If a declining palm that did not previously bear basidiocarps is felled, these fruit bodies often arise on the stump or the fallen trunk.

The basidiocarps of *G. zonatum* are up to 40 cm wide and 9 cm thick at the base, lack a stalk, and have a hard, thin, varnishlike reddish brown to gray-brown layer on the mature parts of the upper surface. The upper surface often has irregular lumps and usually has concentric ridges and furrows that indicate previous alternating periods of growth and quiescence. The margin is white and somewhat swollen while growing, and a dark interior tissue shows where the surface is bruised. Most of the lower surface is covered with nearly circular pores, 3–4 per mm if a straightedge is placed across them. These are the mouths of tubes that extend vertically one-half to two-thirds of the distance through the basidiocarp. When mature, each tube is lined with a spore-producing layer, the hymenium, from which yellow to brown basidiospores are released, fall out of the basidiocarp, and are dispersed by air currents. These spores often collect beneath basidiocarps, or on their upper surfaces if dropped from others above, and impart a dusty brown appearance.

Basidiocarps indicate advanced decay of adjacent parts of the roots and butt, which renders the surface roots dry and brittle and the underlying tissues weak and crumbly. Palm butts long diseased commonly have masses of white mycelium in the decayed tissues and sometimes have hollow centers. Tissues in early stages of decay are dark brown and sometimes water-soaked.

Diseases caused by *G. boninense* have received the most study. Infection presumably occurs either by root contact or by deposition of basidiospores in wounds above ground. In coconut palm and oil palm plantations in Sri Lanka and Malaysia, for example, stumps apparently become infected by spores after old palms are cut. Young palms planted among the stumps then become infected and begin dying at an early age. Excavation and dissection have indicated that decay spreads up into the butts from roots; thus root contact with old stumps is implicated as a likely means of inoculation. Moreover, identical isolates of the oil palm pathogen may be obtained from each plant in a cluster of diseased palms. This indicates that *G. boninense* establishes clones that each occupy discrete subterranean territories in colonized roots and butts. If each plant had been independently infected by spores, each plant would have yielded a distinct isolate. Although basidiospores undoubtedly play a role in establishing *Ganoderma* in some new substrates, this is not well documented. Some writers have suggested that the spores may infect not only stumps but frond bases and assorted wounds at the base of the palm. This suggestion deserves study.

G. boninense is a proven pathogen, but *G. zonatum* and *G. tornatum* have not been tested. Their pathogenicity must be assumed from the circumstances of their occurrence and from knowledge of *G. boninense.* For infection to occur by root contact, it seems that one of these pathogens must be growing from a relatively large food base such as a stump, as indicated by experiments in which blocks of various sizes from palm butts killed by *G. boninense* were used as inoculum for infection of seedlings. Only seedlings whose roots contacted the largest blocks became infected; *G. boninense* was replaced by other saprophytic fungi in the smaller blocks. Such information indicates that it is advisable to remove stumps when replanting palms, but small root fragments, although perhaps infested with *Ganoderma,* may not be a significant threat to new palms.

Butt rot of palms caused by *Ganoderma zonatum* (FL, Mar–Apr).

A, D. Yellowing and premature death of fronds, and young basidiocarps on the butt of a dying oil palm. Growing basidiocarps have thick, light-colored margins.

B. Large, old, inactive basidiocarps on the stump of a queen palm.

C. Small inactive basidiocarps on the butt of a living queen palm.

E. The upper surfaces of young basidiocarps, about one-third natural size. X-shaped scratch on the white, growing margin of one fruit body exposes dark brown interior tissue.

F. Magnified view of the lower surface of a basidiocarp, enlarged about 10 times, showing the pores or mouths of tubes from which spores are released to the air.

Plate 159

Root and butt rots of deciduous trees. The pathogens in the temperate zone of North America are *Ganoderma applanatum* (Plate 161), *G. brownii*, and *G. lucidum*. *G. brownii* occurs on California laurel in California and has been reported to occur on a *Prunus* species in British Columbia but is otherwise obscure. *G. lucidum* is the principal root pathogen. Synonyms of *G. lucidum* include *G. curtisii*, *G. resinaceum*, *G. sessile*, and *Polyporus lucidus*. The distinctive brown to reddish brown or dark red basidiocarps of this fungus, coated on the upper surface with a varnishlike crust, permit Ganoderma rot to be diagnosed readily. *G. lucidum* occurs across North America and in Europe, Venezuela, the West Indies, Africa, and Asia.

Host plants of *G. lucidum* include sweet acacia, apple, ash (green and white), European beech, yellow birch, boxwood, black cherry, golden chinkapin, citrus, Kentucky coffee tree, elms (American, water, and winged), sour gum, sweet gum, hickories, European hornbeam, black and honey locusts, magnolias, maples (Norway, red, silver, and sugar), mesquite, mimosa, oaks, Everglades palm, peach, pepper tree, persimmon, bullace plum, common privet, redbud, sassafras, sugarberry, and willows.

The basidiocarps of *G. lucidum* are annual, up to 35 cm across, leathery to corky when fresh, and irregularly humped, with or without lateral stalks. They develop singly or overlap one another in clusters. The upper surface and the stalk when present have a smooth, varnishlike coating that varies in color from mahogany to blood red or blackish red. The margin and the poroid (lower) surface are white during growth and tawny with age. Where basidiocarps are clustered, their upper surfaces and often the surrounding grass or bark may be discolored brown by a deposit of basidiospores.

G. lucidum is a proven pathogen of bark and sapwood of several broad-leaved tree species, and it can kill even large trees. Its basidiocarps are commonly noticed at or near the bases of trees along streets and in landscapes, and it is apparently one of the common causes of premature death of these trees in the eastern United States. It was found, for example, in about 20% of a sample of over 300 declining red and Norway maple trees in the New York metropolitan area. In forests, on the other hand, the fungus is most often found associated with heart rot in trunks of various trees.

Trees affected by root rot eventually lack vigor and bear undersized leaves, dead branches, and sometimes yellow or wilting leaves. Often, however, basidiocarps arise from the roots or butts of trees with healthy crowns. This occurs because many trees can tolerate the loss of up to half of their root systems before beginning to decline. Basidiocarps that develop at the soil surface away from the butt are connected to decaying roots.

Rapid decline and death of honey locust and sassafras infected by *G. lucidum* have been reported, but the fungus usually works slowly. In pathogenicity tests, it has taken 3–5 years to kill sweet gum and mimosa seedlings. Discrete lesions formed on roots of the inoculated seedlings before general root necrosis and decay killed them. Diseased red oaks sometimes exhibit no change in general health of the crown during 15 years or more after basidiocarps first appear at the ground line.

Lethal infection is apparently characterized by parasitism of the sapwood of all major roots and the butt. These parts of honey locust, Norway maple, and sassafras in diverse places have been found extensively decayed at the time of tree death. The advanced stage of the decay is light colored and spongy. Living trees with decayed wood in the roots and butt are subject to windthrow.

The means of natural transmission of *G. lucidum* are unknown. The association of basidiocarps with old wounds on roots and trunk bases of trees along streets and in landscapes indicates that wounds are infection courts. Basidiospores, dispersed throughout the summer, are released in greatest numbers during evening hours when the air is humid. Infection by root contact with previously colonized wood is also possible, as judged from the results of experiments, but patterns of disease incidence do not suggest tree-to-tree spread.

Observations suggest that environmental stress in addition to wounding may predispose trees to major damage by *G. lucidum*. In fern nurseries in Florida, for example, oak trees used as a canopy have shown unusual incidence of root rot after the use of herbicides and soil fumigants. *G. lucidum* is among the fungi commonly associated with decline and mortality of oaks during periods of drought in the southeastern United States.

Ganoderma species on conifers. *G. oregonense* (syn. *Polyporus oregonense*) and *G. tsugae* (syn. *P. tsugae*) occur on Douglas-fir, firs, hemlocks, pines, and Sitka spruce, causing a white spongy rot of roots and butts of dead trees. *G. tsugae* is apparently not pathogenic, but *G. oregonense* occasionally causes heart rot in living western hemlock trees. *G. tsugae* is widespread in the East and Southwest and *G. oregonense* in the West. They differ in that the heavily varnished basidiocarps of *G. oregonense* attain larger size, commonly 40 cm or more across, and produce larger spores.

A. Dieback associated with root rot by *Ganoderma lucidum* in honey locust (MS, Apr).

B, E. Basidiocarps of *G. lucidum* associated with root and butt rot of northern red oak. The cluster of basidiocarps in (E) was about 30 cm in diameter (NY, Aug–Sep).

C. Young basidiocarps of *G. lucidum* developing as irregular white masses at the base of a diseased honey locust. Several initially discrete masses may coalesce to form large compound basidiocarps as in (B) (MS, Apr).

D. Mature basidiocarps of *G. tsugae* on a decaying stump of eastern hemlock (NY, Sep).

F. A young basidiocarp of *G. oregonense*, about 12 cm across, growing from a decaying root of western hemlock (ID, Jul).

G. Typical stringy white rot caused by *G. oregonense* in western hemlock (ID, Jul).

Plate 160

Ganoderma applanatum. *G. applanatum* (syn. *Fomes applanatus*) is one of the fungi most often recognized in deciduous forests of the north temperate zone. Its basidiocarps are popular with artists for etching because the white lower surface of a fresh specimen turns dark brown where marked by a stylus. When dry, this surface is suitable for painting. The fungus is an important decomposer of logs and stumps. It also colonizes wounds, kills sapwood of some species, and causes decay of sapwood and heartwood in roots, butts, and trunks of a wide array of trees in forests and landscapes.

Whether *G. applanatum* is a primary colonist or ordinarily succeeds other organisms in wounds is unknown, but the relationship of wounds to subsequent decay is clear. So also is the relationship between a basidiocarp on a living tree, extensive decay within, and likelihood of windthrow or trunk breakage. Of windthrown trembling aspen trees in a storm-damaged stand in Colorado, for example, 86% had basidiocarps that indicated root and butt rot by *G. applanatum*. In fact, this fungus is the most important cause of root and butt rot of aspen in the central Rocky Mountain region, is a significant cause of trunk rot in cottonwoods across North America and northern hardwoods in the East, and was one of the three most important causes of heart rot in old-growth western hemlock.

G. applanatum is recorded as fruiting on living trees in the following groups in North America: apple, aspen, basswood, beech, birch, cherry, citrus, cottonwood, elm, sweet gum, hemlock, hornbeam, horse-chestnut, black and honey locusts, maple, mulberry, oak, poplar, spruce, sycamore, tulip tree, and willow. Other hosts, whether living or dead, include acacia, alder, arborvitae, ash, butternut, chestnut, Douglas-fir, eucalyptus, fir, hickory, honeysuckle, magnolia, pear, pepper tree, pine, Japanese raisin tree, and tree-of-heaven. Records of *G. applanatum* from subtropical regions and along the Pacific coast of North America are confounded with records for *G. zonatum* or *G. tornatum* (Plate 159). Thus reports of *G. applanatum* on citrus, *Coccoloba*, eucalyptus, California laurel, cabbage palmetto, and pepper tree are suspect.

Because diseased trees often break or fall before death, the only overt indicator of disease in most standing trees is the basidiocarp. This is usually found near ground level or on the lower part of the trunk, often but not always at an old wound. In old citrus trees as an exception to the rule, *G. applanatum* is reported to cause a sapwood rot that is contiguous with perennial, concentrically ridged cankers, on which the fungus produces basidiocarps. The cankers are usually centered on branch stubs or old wounds, and the adhering bark is broken by cracks that follow the ridges. Progressive destruction of the sapwood leads to decline or breakage of the tree.

Basidiocarps of *G. applanatum* become hard and woody at an early age and can attain impressive strength and dimensions (Figure 16). They are roughly semicircular or fan shaped as viewed from above and have a white margin and poroid surface and a flat to convex upper surface that is initially brown and weathers to gray, often becoming cracked and concentrically ridged. The interior tissues, including one or more layers of tubes 5–10 mm long, are dark brown. Individual basidiocarps sometimes live 8–10 years; 5–6 years is normal in southern Ontario. From early spring until frost in autumn, they produce enormous numbers of basidiospores that tend to collect as brown deposits on nearby bark or litter or on the upper surfaces of other basidiocarps. By one estimate, a basidiocarp with a poroid surface of only 100 cm², comparable to those illustrated, may liberate 4.65 billion basidiospores in a 24-hour period! The tubes from which basidiospores fall must be vertical for effective release of the spores. If the orientation of a basidiocarp shifts when a tree falls or a log rolls, spore production ceases until the fungus organizes a new layer of tubes.

G. applanatum colonizes new substrates by means of airborne basidiospores. These germinate readily only in the presence of other microorganisms, presumably because the latter provide necessary nutrients or stimulants. Wood undergoing decay by *G. applanatum*

Figure 16. *Ganoderma applanatum* and a man on a living American beech.

remains light colored and exhibits a mottled appearance. In advanced stages of decay, the wood remains firm for a time but readily fractures across the grain, and thin black lines (zone lines or pseudosclerotial plates, Plate 107) often develop in it. Eventually it becomes soft and spongy. A dark band several millimeters wide separates healthy sapwood from that undergoing decay. Columns of decaying wood usually extend 120–200 cm above and below basidiocarps of *G. applanatum* in trunks of beech, birch, or maple and an average of 3 meters above basidiocarps in old-growth western hemlock. In roots, the fungus is usually restricted within 1 meter of the soil line.

References for Ganoderma root rot of palms: 387, 408, 577, 1380, 1414, 1480, 1904, 1905, 1907, 2049

References for Ganoderma brownii, G. lucidum, and species on conifers: 194, 387, 389, 577, 640, 704, 783, 1255, 1438, 1529, 1759, 1906, 2027

References for Ganoderma applanatum: 228, 268, 352, 387, 408, 544, 640, 1280, 1438, 1645, 1669, 1907, 2165

Trunk decay by *Ganoderma applanatum*.

A. A large eastern cottonwood broken by wind after decay of one side of the trunk (NY, Sep).

B, C. A broken trunk of a sugar maple bearing typical basidiocarps. Brown color of bark in the vicinity of basidiocarps indicates a deposit of brown basidiospores. Decayed wood shows transverse fracture (B), in contrast to splintered fracture of nondecayed wood (NY, Sep).

D. The trunk of a dead American beech, broken after basal decay. The wood shows mottled white rot typical of that caused by *G. applanatum* (NY, May).

E. Basidiocarps about 2 years old on a stump of sugar maple. During growth, these fruit bodies have a prominent white margin (MI, Aug).

F. Basidiocarps on the base of the cottonwood shown in (A); these gave at least 2 years' warning of the impending crash (NY, Aug).

Plate 161

Compartmentalization of Wound-associated Discolored and Decayed Wood (Plates 162–163)

For this topic we depart from the usual format of the book. The text contains a summary of the main concepts, but the details are presented in expanded captions of the photographs.

The central concepts are straightforward. In presenting them, we disregard exceptions, fine points, and special cases. For these, see the references.

1. The sapwood of a tree is a living tissue capable of isolating damaged portions of itself and capable of dynamic interaction with invading microorganisms.

2. Normal wood is divided anatomically into numerous elongate compartments that are delimited by concentrations of meristematic or parenchyma cells. The meristematic cells are those of the vascular cambium, the outer tangential boundary of a compartment. Parenchyma cells are concentrated mainly in medullary rays, the radial boundaries of a compartment. Smaller numbers of parenchyma cells are associated with vessels of angiosperm wood and with resin canals in many gymnosperm woods. These parenchyma cells tend to be concentrated near the junctions of annual layers of wood and constitute an inner tangential boundary of the compartment. This is poorly defined in some gymnosperm woods.

3. When sapwood is wounded, the tree initiates processes, collectively called compartmentalization, that isolate the wound from normal tissues. Parenchyma cells near the wound produce and secrete substances that make the wood toxic or inhibitory to microorganisms and more or less impervious to water. These substances include gums, phenolic compounds, and suberin in angiosperms; terpenes, resins, and polyphenolics in gymnosperms. Vessels or tracheids that would normally conduct sap become plugged with these substances above and below the wound and near the wound edges. Pits that connect tracheids in gymnosperm wood close. Parenchyma cells may also produce tyloses that plug vessels and tracheids. (A tylosis is a balloonlike extension of the elastic wall of a parenchyma cell into a vessel or tracheid via a natural opening, or pit, in the thick wall of the latter cell.) The parenchyma cells that produce antimicrobial substances become infused with the inhibitory compounds, die, and become part of a zone of altered wood. Oxidation of phenolic compounds results in darkening of this wood.

The region of sapwood response around the wound is called the reaction zone. It extends farther up and down the stem than inward or to either side because of the way the wood elements are arranged. Elongate conduits (vessels or tracheids) extend up and down but not in other directions. Parenchyma cells are less abundant above and below the wound than beside it or toward the center of the stem.

Wounding also stimulates the vascular cambium nearby to produce abnormal xylem with reduced numbers and sizes of tracheids or fibers and conducting elements and a high proportion of parenchyma cells. This parenchyma produces chemicals like those mentioned above, resulting in a thin layer of new xylem more impervious than preexisting wood altered merely by infusion. The newly formed abnormal xylem, called a barrier zone, separates sapwood present before wounding from that produced subsequently. If wounding occurs during annual radial growth, the barrier zone forms within a growth ring. If wounding occurs between growth periods, the barrier zone forms at the beginning of the next annual ring.

A. A mature slash pine with trunk decay that originated at a fire scar many years ago. The tree survived repeated fires, one of which burned elliptic craters into the decaying wood. Nearly the entire cross-section of the trunk is contained within the compartment defined by wound responses of the tree after a portion of the cambium was killed by heat. Decay will proceed within wood formed before the fire, that is, within the cylinder defined by the barrier zone, as long as the wound remains open. Wood formed after the fire, outside the barrier zone, will remain sound (FL, Mar).

B. A small American beech with trunk decay that originated at a wound where a falling neighbor tree smashed off a strip of bark many years earlier. Wood that was present at the time of wounding is now degraded and is disappearing as the result of decay followed by insect, bird, and rodent activity. A cavity will result. Wood produced since the time of wounding remains sound. This wound is slowly being enclosed by healthy wood and bark produced by the vascular cambium in callus rolls that originated in undamaged tissues at the edges of the wound. The advancing callus rolls will eventually fuse, restoring the vascular cambium around the entire trunk. Thus the damaged area will regain cylindrical form. Decay within the compartment will essentially cease when wound closure is complete. Note that where decayed wood has been removed, the callus rolls wrap around into the cavity, producing a small amount of bark-covered wood inside. Similar wraparound callus growth at the edge of a trunk cavity is shown more fully in (C). Wood and bark on the callus rolls within the cavity will slowly die of attrition after the callus rolls fuse at the surface. The outlines of the original wound and a line where callus rolls meet will remain on the bark surface. After restoration of a continuous vascular cambium across the former wound site, trunk growth will result in an increasingly thick layer of normal sapwood between the superficial scar and the compartmentalized cavity in the center of the stem (NY, May).

C–E. Compartmentalization of wound-associated discoloration and decay in a red maple trunk. The tree, growing slowly in the forest understory, was about 70 years old when cut. C. Longitudinal section through a portion of the trunk at the base of the wound where decay began, about one-half natural size. A wound caused by a falling tree occurred on the right side of the trunk 31 years before the tree was cut. At the level of the wound, discoloration and decay expanded to involve all wood present at the time of wounding, as delimited by the barrier zone (bz). A cavity developed at the wound site, and a column of defective wood extended above and below the wound site. The cavity (cav) is lined with decayed wood. Discolored wood (di), the precursor of decay, lies adjacent to the barrier zone. Callus (c) that began growth at the edge of the original wound has wrapped around the edge of the cavity. A small wound (w) occurred 9 years before the tree was cut and was completely covered by callus after 7 years. During that time little discoloration and no decay originated at the small wound, and further development of discoloration or decay was precluded by wound closure. The point of fusion of callus rolls covering the small wound is marked by (f). D. An enlarged view of the region where callus rolls fused to close the small wound in (C). The year-to-year advance of the callus rolls, resulting eventually in wound closure, is marked by faint lines that delimit annual layers of wood. Discoloration extended only three annual layers into the wood at the time of wound closure. E. The column of defective wood at the top of the cavity, approximately 90 cm above the location shown in (C) and about one-half natural size. The tree was not wounded at this level, but a barrier zone (bz) corresponding to the location of the vascular cambium at the time of wounding restricts the defect to wood present at the time of wounding. Decaying wood (de) is surrounded by discolored wood (di). The darkly stained wood at the bulge in the stem is a compartmentalized defect near a branch stub that was overgrown by the expanding trunk (NY, Apr).

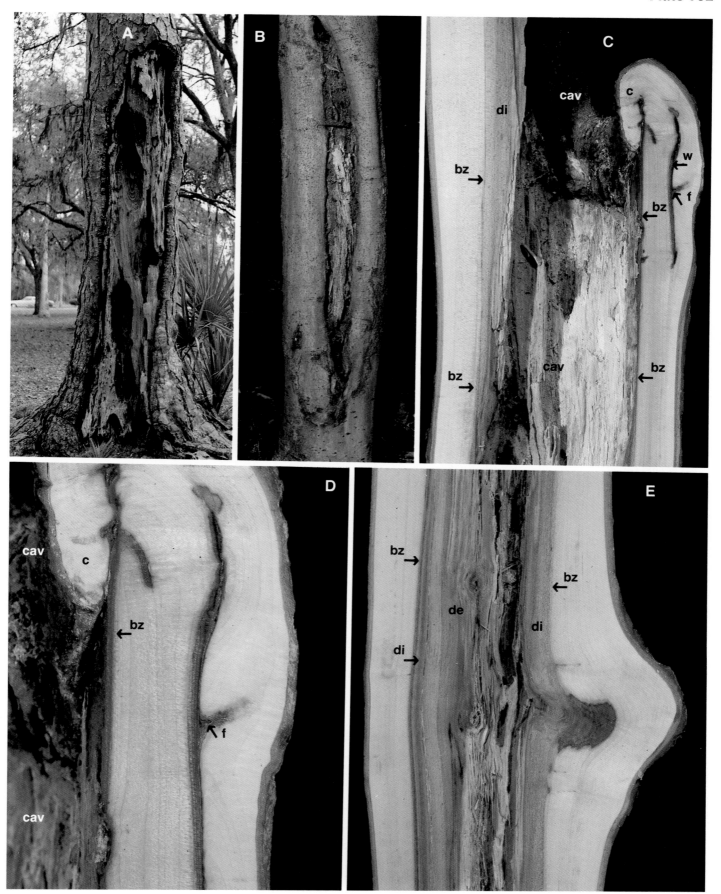

Plate 162

337

During compartmentalization the tree sacrifices energy and some nondamaged tissue in the formation of a perimeter, analogous to compartment walls, that isolates the wound. The isolation response within preexisting wood is strongest in young sapwood and diminishes with increasing depth into older wood. Similarly, the degree of physical and chemical aberration of xylem in the barrier zone is greatest near the wound and diminishes with circumferential and vertical distance from the wound. Sapwood beneath the fully formed barrier zone slowly dies, apparently because the barrier zone interrupts the normal inward flow of nutrients in rays to xylem parenchyma. Perhaps the barrier zone also inhibits gas exchange.

4. As soon as the wound occurs, different microorganisms in succession begin to invade it and to spread into surrounding wood. Microorganisms cannot be kept out of wounds; hence reaction of the tree to the wound versus reaction to invading organisms cannot be separated. Bacteria and molds that grow rapidly by utilization of simple carbon compounds in the wound are the first or primary colonists. They include or are followed by organisms, chiefly ascomycetous fungi, that are able to grow into the reaction zone and detoxify inhibitory chemicals or use them as nutrients. These organisms are resisted but not halted. Their advance causes continual expansion of the reaction zone up, down, and inward from the wound. These organisms cause dark discoloration of the wood by converting chemicals in the reaction zone to colored products, but they do not cause decay. After staining fungi detoxify the wood, the decay fungi move in, months to years after wounding. Most decay fungi are relatively intolerant of fungitoxic phenolics or terpenes produced by sapwood parenchyma and do not rapidly detoxify these chemicals. But decay fungi are able to degrade complex polymers—lignin, cellulose, and hemicelluloses—and use them as carbon sources. This is the main event in wood decay. Microorganisms in succession, continually confronted by the reaction of sapwood parenchyma, slowly advance and cause discoloration and decay within wood that was present before wounding. Microorganisms do not breach the barrier zone; therefore sapwood formed after wounding remains healthy.

If a wound penetrates heartwood (normal, usually colored nonliving wood in the center of the tree), invading organisms are also resisted by means of chemicals in the wood, but this inhibition is attributable mainly to chemicals previously secreted by parenchyma cells as the final process in their genetic program during formation of heartwood.

Some pathogenic wood-decay fungi, chiefly basidiomycetes but also some ascomycetes, are specialized for invasion through fresh wounds as primary colonists of sapwood. These fungi, tolerant of chemicals in the reaction zone, play all the roles that primary colonists, detoxifiers, and decay fungi play in the successions mentioned above. *Cryptosphaeria populina* (Plate 104), *Heterobasidion annosum* (Plates 151–152), *Hypoxylon mammatum* (Plate 106), *Trametes*

versicolor (Plate 165), and canker-rot fungi (Plates 169–173) are examples. Many wood-decay fungi that parasitize sapwood are poor competitive saprophytes and are unable to replace staining fungi or other decay fungi in established successions. *Heterobasidion annosum* and *Hypoxylon mammatum* are in this group.

5. Wound closure proceeds as the result of callus formation by the vascular cambium. Cambial activity near a wound is typically more intense than normal, leading to more wood formation than occurs elsewhere at the same level on a stem. Callus (bark plus sapwood produced at the edge of a wound) expands faster tangentially than radially; thus the wound closes while the stem expands in girth. When a wound closes by fusion of callus rolls, the advance of discoloration and decay from that wound ceases.

6. A tree may sustain wounds at various times. Therefore it may contain multiple, overlapping regions of discoloration and decay.

7. When a barrier zone is breached by a new wound or by a crack developing from within the column of defective wood, sapwood outside the barrier zone is subject to discoloration and decay caused by organisms that spread outward from the previously compartmentalized column.

8. Trees vary in ability to compartmentalize wound-associated discoloration and decay. This ability is under genetic control.
References: 5, 26, 195, 208, 613, 621, 814, 930, 1109, 1123, 1279, 1280, 1363, 1471; 1680, 1776, 1781–87, 1792, 2012, 2013, 2136

A, E. A cross-section of the red maple trunk shown in Plate 162, about 1.5 meters below the wound. The column of discolored and decaying wood at this level occupies nearly the entire cross-section of the stem as it existed 31 years earlier at the time of wounding. The barrier zone associated with the old wound above is indicated by the dashed line (bz). This section shows how discoloration and decay may spread outward when a barrier zone is breached by a new wound or crack. Discoloration beyond the barrier zone is extensive at three cracks and is just beginning at the other two marked (di). Discoloration has spread far outward along the three large cracks, but tangential spread from the cracks was resisted by ray parenchyma. Therefore, as viewed in cross-section, the defect associated with a crack extends spokelike from the central core and is compartmentalized near the plane of the crack. The crack at right never extended to the vascular cambium. The crack at left reached the vascular cambium 8 years before the tree was cut but did not break the bark. This injury to the cambium temporarily stimulated sapwood formation that resulted in a bulge on the side of the tree. At the time the tree was cut, this crack extended to within four annual rings of the cambium. The largest crack, shown approximately 1.5 times actual size in (E), extended intermittently, as indicated by dark color patterns in the wood. The crack first broke through the cambium and bark 4 years before the tree was cut, and it opened and closed annually thereafter. The opening induced callus growth and the development of a prominent rib on the trunk (NY, Apr).

B. A red maple similar in external appearance to the tree from which the section in (A) was cut. Callus growth at the edge of the crack resulted in formation of a prominent rib that follows the grain of the wood (MI, Aug).

C, D. Part of a cross-section of a sugar maple trunk repeatedly tapped (bored) to extract sap. The "shadows" of multiple tap holes slightly above and below the plane of the section are visible. Wound-associated discoloration and decay are extensive. White mycelium of a wood-decay fungus grew from colonized wood onto the surface of the section during several days of incubation. Columns of discoloration and decay, initially compartmentalized behind closed wounds, expanded and coalesced because repeated tapping broke compartment boundaries. Part of a closed tap wound is shown close up in (D). A T-shaped mark, with the leg of the T pointing out, remains where callus rolls closed the wound and then fused to produce an uninterrupted covering of sapwood (NY, Sep).

F, G. Examples of basal wounds that lead to extensive decay in the butts of trees. F. A honey locust repeatedly assaulted by automobile bumpers. G. A black oak assaulted by a beaver (GA, May).

Plate 163

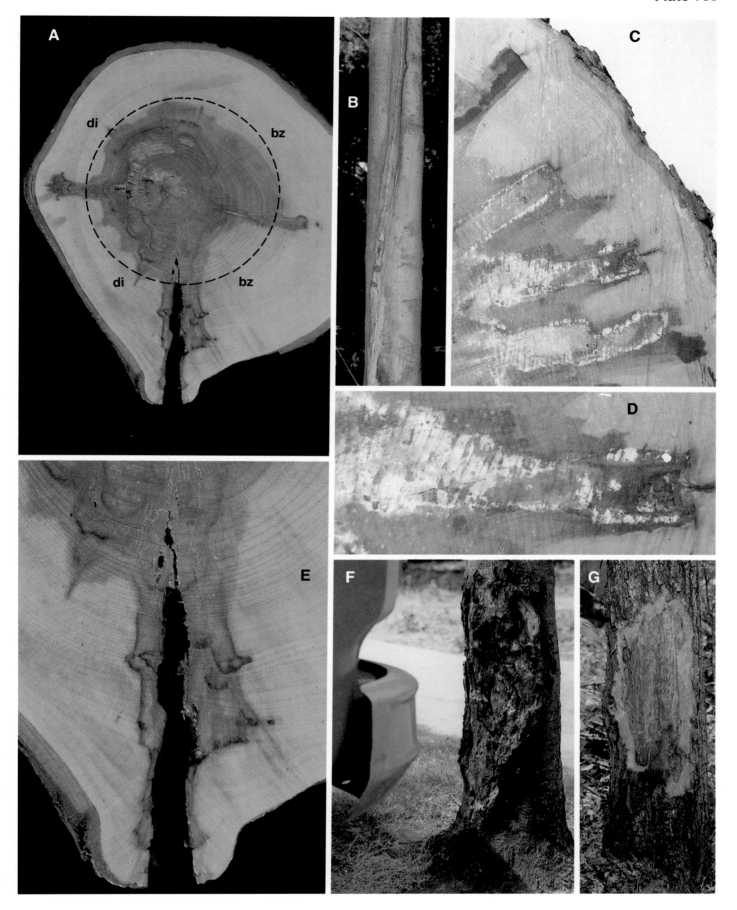

Drought Cracks and Frost Cracks in Relation to Internal Defect (Plate 164)

This plate completes a series of three about relationships between wounds and the development of discoloration and decay within tree trunks or roots. As with Plates 162–163, which present compartmentalization, we depart from the normal format of the book by summarizing concepts in text and giving details in expanded captions that interpret photographs.

First, some terms. Cracks in wood are often called checks or shakes. The former term is usually applied to small radial separations such as those in the ends of drying logs and pieces of lumber. Radial cracks that originate within standing trunks are called ray, or radial, shakes. Tangential cracks that separate annual layers of wood or that develop at sites of former cambial damage by heat or freezing, or at barrier zones associated with wounds, are called ring shakes. Mechanical stresses caused by frost or drying may cause or contribute to the enlargement of cracks, hence *frost crack* and *drought crack*.

The two main points here are simple and complementary. Radial and tangential cracks in wood usually develop at sites of former injury, thus usually in wood undergoing discoloration and decay. Discoloration and decay begin at or spread along cracks that open in previously sound wood.

Cracks in wound-associated discolored or decaying wood are far more common than in otherwise sound wood. Many people have heard sharp reports as cracks in hardwood trees pop open on frosty winter mornings. Cracks usually begin in dead wood at wounds or the stubs of branches and basal sprouts. After a radial crack breaks a barrier zone (Plate 163), the wood to either side of the crack is subject to discoloration and decay as microorganisms spread both along the crack and tangentially from it. Callus formation along cracks that break the vascular cambium may result in formation of prominent vertical ribs.

Tangential and radial cracks often develop near the bases of tree trunks that have sustained cambial damage by freezing. The sequence begins when a localized area of cambium or outermost sapwood is killed but the phloem survives. During the next growing season, surviving cambial cells or new vascular cambium differentiated from phloem parenchyma overlying the damaged zone begin to produce xylem parenchyma and callus, so that swelling results. Pressure from callus splits the bark and any new xylem parenchyma to the depth of the original injury and may cause the split to flare open widely. The cambium at each side of the split then produces callus that contributes to rapid swelling of the damaged part of the stem. After wound closure, the wood may separate tangentially along the plane of weakness where the cambium or sapwood was previously killed, or a radial crack may open at the line of fusion of callus rolls. The woody cylinder present at the time of damage is subject to discoloration and decay. Several of the points above are illustrated and explained in more detail with Plates 163, 235, and 236.

Frost cracks can be induced in sound wood by experimental freezing treatments. Therefore it seems likely that such cracks occasionally begin in sound wood in nature. Field evidence for this point is lacking, however.

Radial cracks apparently caused by drought in otherwise normal wood of conifers have been reliably reported from Alaska and Great Britain. These cracks originate in recently formed layers of sapwood and often extend only across the width of one annual layer of wood, appearing in transverse section as diamond-shaped clefts amid partially crushed tracheids. Usually they are entirely internal and are discovered only when trees are felled, but some cracks extend across several annual rings or through the bark, appearing on the surface as long lines that ascend the trunk in a high helix corresponding to the grain of the wood.

References: 290, 464, 1064, 1182, 1516

A–D. Relationships among drought stress, internal defect, and stem cracks in London plane trees (NY, Aug). A. Trees in this young landscape planting, subject to severe water stress during summer, display dieback, death of occasional individuals, and basal sprouting. Basal sprouts developed most often on trees with weakly growing tops and cracked, defective trunks. B. The trunk of the tree with basal sprouts in (A) has 2-year-old drought cracks and associated cankers, accentuated by callus formation. C. The trunk of a similar tree with discolored bark that indicates cambial dieback along drought cracks. Had this tree not been cut for examination, it would have survived with an elongate canker associated with the trunk cracks. D. Internal defects in the same stem segment shown in (C). The history of the tree was interpreted from examination of the annual rings and internal defects. The tree at the height of the cut had completed 5 seasons' growth. It was pruned and transplanted before the 3rd season. Wounds sustained at the time of pruning were poorly compartmentalized, leading in time to death and discoloration of all wood present at the time of pruning. One pruning wound at left center is still open. Three radial cracks formed in discolored wood that developed as the result of pruning wounds. During dry weather after the current season's growth was complete, one radial crack, pointing upward in the photo, opened to the stem surface, exposing sapwood that quickly became discolored. A second crack, at right, broke the barrier zone surrounding the wood of the first 2 years but did not break the vascular cambium or bark. Discoloration spread from the defective core outward along the crack. A third radial crack, pointing down in the photo, remained confined to the compartmentalized wood of the first 2 years. Closure of the two pruning wounds intersected by the diagonal cut became complete during the current season, but the wound on the near side of the stem reopened along the line of callus fusion. The pruning wound that remained closed, at top, corresponds to the short vertical seam on the bark surface in (C).

E. Vertical cracks and swelling of the butt of a young, recently transplanted red oak that sustained freeze damage to the cambium during the previous winter. The short cracks shown here closed completely within 1 year after the photograph was made, but one crack reopened after 6 years and then was 1.3 meters long. Cracks that begin in young trees such as this one persist internally and may reopen in old trees such as the black oak shown in (G). The volume of wood potentially subject to discoloration and decay expands in each year that a crack reopens (NY, Aug).

F. An Australian pine with a lens-shaped wound resulting from freeze damage to the vascular cambium several years before the photo was made. The wound began as a bark crack soon after freeze damage. Growth, as explained in the text, caused the crack to widen and the bark on the right side to flare out from the wood. The exposed woody cylinder constitutes a large site for invasion by microorganisms that cause discoloration and decay (FL, Mar).

G. A frost rib on a mature black oak. The "rib" is a callus ridge that developed where a crack opened through the cambium and bark. The tree had sustained a wound at the root collar many years before (scar visible as irregular bark surface below the rib), resulting in internal defect within which a radial crack formed and eventually extended to the surface (NY, Jul).

340

Plate 164

Trunk and Limb Rots of Hardwoods (Plates 165–167)

This plate leads a series of nine that survey heartwood and sapwood rots and canker-rots caused by basidiomycetes in the order Aphyllophorales, mainly pore fungi. The series begins with white rots caused by widespread decay pathogens of angiosperms: *Phellinus tremulae, P. igniarius,* and *Trametes versicolor.*

Before considering decay fungi, we review briefly the nature of sapwood, heartwood, and wood decay and introduce some terms used to describe and classify decaying wood. Normal sapwood is alive and involved in conduction of sap, storage of compounds that serve as energy reserves, compartmentalization of wounds, and defense against invading microorganisms (Plates 162–163). It is light in color and while alive is highly resistant to most wood-decay fungi. Dead sapwood has little resistance to wood-decay fungi.

Heartwood is dead, brightly to darkly colored wood that forms normally in the centers of trunks and large branches of many species of trees as parenchyma cells in old sapwood die. Heartwood provides mechanical support but is not involved in conduction or storage. It contains various chemicals (extractives) produced by parenchyma cells before their death that make it resistant to many wood-decay fungi. Heartwood differs in origin and chemical properties from wound-associated discolored wood. Some important groups of trees such as birches, maples, and poplars do not produce heartwood.

Nearly all wood decay is caused by fungi and is the result of digestion by extracellular enzymes and (in brown rots) hydrogen peroxide from the mycelium. The descriptive terms for decaying wood connote important chemical and physical characteristics.

White rots are characterized by degradation of all organic constituents of wood, often with more rapid degradation of lignin than of cellulose, so that the decaying wood becomes relatively poorer in lignin and richer in cellulose as it loses weight. Most white rot fungi leave no residue of colored breakdown products. When decay reaches an advanced stage, the predominantly cellulosic residue is light in color and may be spongy, stringy, laminated, or brittle and crumbly.

White stringy rots are characterized by relatively slow reduction in the length of cellulose polymers so that wood in early stages of decay does not rapidly lose strength. In advanced stages the wood may be soft and weak, but it does not fracture across the grain.

White rots with brittle or crumbly wood are characterized by random cleavage of cellulose molecules, so that much wood strength is lost in early stages of decay. Decaying wood easily fractures across the grain and becomes crumbly.

In brown rots, cellulose molecules are cleaved randomly, with loss of wood strength resulting in brittleness, as above, but lignin is used slowly and incompletely. Colored breakdown products remain and impart a dark color. Decaying wood shrinks and cracks into more or less cubical chunks.

Representative trunk and limb decays. *Phellinus tremulae* (Hymenochaetaceae) is restricted to aspens but attacks them wherever they grow in the Northern Hemisphere. It is their most important decay pathogen and is one of a group of closely related species that were all at one time considered to be forms or varieties of *P. igniarius* (syn. *Fomes igniarius*).

The yellowish white spongy decay caused by *P. tremulae* and other fungi in the *P. igniarius* complex is usually confined to a central core, contains black pseudosclerotial plates that appear as lines (zone lines) on cut or broken surfaces, and supports the development of basidiocarps at branch stubs, wound scars, and branch scars. Decay columns may extend 2–3 meters above and below basidiocarps. *P. tremulae* enters trunks via dead branch stubs and perhaps via fresh wounds. The fungus has been reported to be a primary parasite of aspen sapwood. It causes a sweet-smelling decay and after many years produces perennial basidiocarps or sometimes hard, blackish sterile masses of mycelium (sterile conks or punk knots) at branch scars. The basidiocarps (conks) produce a new layer of tubes on the lower surface during each of several years and attain dimensions up to 20 cm wide and 15 cm thick. Viewed from the side they appear triangular, with the upper and lower surfaces at angles nearly 45° from the horizontal. They are usually attached to the host by a granular core of tissue that continues into the decayed branch within the trunk. The upper surface is pale brown near the margin; older parts become blackened, crustlike, and cracked. The lower, poroid surface is purplish brown. Basidiospores are released throughout the growing season.

Other fungi that were once considered to be forms of *P. igniarius* include *P. arctostaphyli, P. laevigatus,* and *P. nigricans.* Because North American records seldom distinguish among these, we consider them together as a cause of white trunk rot in hardwoods. Species in this complex are the most important trunk decay pathogens in northern hardwood forests, causing extensive damage in beech, birch, and maple. Other angiosperm hosts include alder, apple, ash, mountain ash, birch, buckthorn, butternut, cherry, chinkapin, dogwood, elder, elm, hazelnut, hickory, hornbeam, hop hornbeam, honey locust, linden, madrone, manzanita, oak, pear, sassafras, walnut, and willow.

P. igniarius (narrow sense) succeeds pioneer organisms in wounds to sapwood but is not itself an aggressive colonist of fresh wounds. *P. laevigatus,* however, causes a canker-rot in yellow birch. *P. igniarius* produces basidiocarps similar to those of *P. tremulae* except for a tendency to be more hoof shaped with a thicker margin, a more nearly horizontal lower surface, and usually no granular core.

Trametes versicolor (syn. *Coriolus versicolor, Polyporus versicolor*) (Polyporaceae) occurs throughout the temperate zones of the world as a saprobe on sapwood of many angiosperms and occasionally conifers. This fungus is an opportunistic pathogen, able to kill and colonize sapwood of trees and shrubs stressed by water shortage, freeze damage, or wounding. It causes a spongy white rot and also kills areas of cambium, causing cankers and dieback. Living trees or shrubs on which it has been reported to occur in North America include apple, western catalpa, black cherry, Chinese chestnut, California laurel, common lilac, European linden, sugar maple, London plane tree, and black willow. Cankers on apple are distinctive for the papery texture of the surface of killed bark. Common sites of infection include pruning wounds, broken limbs, frost cracks, and freezing or sunscald injuries. Thin, tough, bracket-shaped basidiocarps of *T. versicolor,* usually 2–5 × 2–7 × 0.2–0.5 cm, develop annually in summer and early autumn, always one above another in clusters. The upper surface is slightly

---

A, E. Heartwood rot caused by *Phellinus tremulae* in trembling aspen. A. Perennial basidiocarps on the trunk of a pole-sized tree (NY, Aug). E. Heartwood decay with black zone lines (CO, Jun).

B, C. Heartwood rot caused by *P. igniarius.* B. A typical perennial basidiocarp at a canker on a pole-sized American beech (NY, May). C. A young basidiocarp developing at a wound scar on butternut (NY, Apr).

D, F, G. Sapwood rot caused by *Trametes versicolor.* D. Decay indicated by basidiocarps on a purple osier stump. Wounds created by repeated harvesting of shoots weakened the plant and served as sites of infection (NY, Dec). F. Upper surfaces of basidiocarps, about one-half natural size, on a fallen red oak branch (NY, Sep). G. Poroid white undersurfaces of basidiocarps and typical white rot caused by *T. versicolor* in red oak sapwood (NY, Sep).

Plate 165

velvety, with concentric zones of various colors. The lower, poroid surface is white to cream color.

Perenniporia fraxinophila (syn. *Fomes fraxinophilus*) (Polyporaceae), a North American pathogen, causes white mottled heart rot of black, blue, green, Oregon, velvet, and white ash and rarely other trees, including eastern red cedar, box elder, American elm, hawthorn, alligator juniper, sycamore, and willow. This fungus occurs from Quebec and Washington in the North to Tennessee and Arizona in the South. It is a significant factor in the deterioration of green ash in shelterbelts on the Great Plains. Decay develops in trunks and major limbs. The advanced stage is straw yellow to yellowish white, soft, and crumbly. Branch stubs are the usual sites of infection and fruiting. The perennial basidiocarps are bracket shaped when fully developed and are dirty white, darkening and becoming cracked on the upper surface with age. They range in size up to 30 × 20 × 10 cm or larger and sometimes arise in groups on dead areas of bark that remains attached to the wood.

Stereum gausapatum (Stereaceae) causes a white mottled rot of sapwood and/or heartwood of many species of oak. It is the most common heartwood pathogen of oaks, and it occasionally colonizes other plants, such as alder, birch, chestnut, and hornbeam. It occurs from coast to coast in Canada and the United States. The fungus enters branch stubs and wounds such as fire scars. It also colonizes oak stumps and spreads into sprouts after these grow large enough to make a heartwood connection with the stump. Decay columns elongate approximately 10 cm per year in the South. Basidiocarps arise annually in dense clusters as thin (0.5–1 mm), brownish shelflike structures up to 6 cm or more wide on scars, stumps, and logging slash. Basidiocarps bleed red fluid where injured.

Oxyporus populinus (syn. *Fomes connatus*) (Polyporaceae) is a major heart rot fungus of maples, especially red and sugar maples, and occurs also on ash, aspen, basswood, beech, birch, buckeye, crabapple, dogwood, elder, box elder, elm, ginkgo, horse-chestnut, sour gum, sweet gum, hickory, hop hornbeam, honey locust, oak, peach, poplar, silverbell, sycamore, tulip tree, and willow. It is common east of the Great Plains from Georgia and Arkansas northward, occurs also in western Canada and the northwestern United States, and occurs uncommonly in the Southwest. It causes a spongy straw-colored white rot in heartwood and sapwood. Decay columns are usually localized within 1 meter above and below the perennial white basidiocarps. Basidiocarps develop singly or in clusters along seams, in open cracks, in basal wounds or Eutypella or Nectria cankers, or in knot holes. Green moss usually grows on the upper surface. Well-developed specimens are shelflike and up to 15 cm wide and 10 cm thick, but an irregular mass such as that shown in (C) is also typical. They are soft-corky when dry, spongy when wet. *O. populinus* follows bacteria and nondecay fungi that invade wounds.

Climacodon septentrionalis (syn. *Steccherinum septentrionale*, *Hydnum septentrionale*) (Hydnaceae) causes a spongy white rot of heartwood in various angiosperms in North America and in Eurasia. There is a report of its occurrence on aspen in Alberta, but the fungus is common only east of the Great Plains. It is most destructive in maples, which it invades via frost cracks and wound scars, less commonly via branches and branch stubs. Decay columns in sugar maple extend to 4 meters above and 1.5 meters below basidiocarps. The advanced decay usually contains narrow black zone lines and is surrounded by a zone of brownish wood. Other hosts in North America include apple, basswood, beech, birch, hickory, and oak. Elaborate cream-colored basidiocarps of *C. septentrionalis* arise annually in summer. They consist of fleshy to fibrous shelflike projections, one above another, joined along the tree trunk, each with the lower surface covered with slender teeth about 1 cm long. The compound structure is commonly 20–30 cm wide and 50–80 cm long.

Hericium erinaceus (syn. *Hydnum erinaceus*) (Hericiaceae), the hedgehog fungus, is a common cause of butt rot in southern hardwoods and occurs northward into Canada. It is most often found on oaks but attacks many other angiosperms and, rarely, gymnosperms as well. American beech; yellow birch; American chestnut; sour gum; hickories; black, silver, and sugar maples; common persimmon; sycamore; and tulip tree are known hosts. In the Northwest it occurs on Douglas-fir and Oregon oak. The decay is a white pocket rot that becomes soft and spongy before its disintegration results in a cavity. Fire scars, stem cracks, and branch stubs are common sites of infection. The decay advances approximately 5 cm per year in water hickory. *H. erinaceus* is named for the long (2–10 cm), slender teeth that extend from the lower surface of its soft, annual basidiocarps. Each tooth is covered with fertile tissue where basidiospores are produced. The basidiocarps develop in autumn and winter in the South as white globoid masses, commonly 10–12 cm but sometimes up to 30 cm across. Stumps, logging slash, and branch scars on living trunks are common sites of fruiting. Basidiocarps turn pale yellow or pale brown with age.

Phellinus robiniae (syn. *Fomes rimosus*) (Hymenochaetaceae) is restricted to black locust and New Mexico locust but occurs wherever these trees grow in North America from the Atlantic coast to Idaho and the desert Southwest. This is one of the few fungi able to decay the heartwood of black locust. It infects trunks through branch stubs and wounds, especially those made by the locust borer (see Johnson and Lyon, 2nd ed., Plate 240), and causes a spongy yellow heart rot. It produces perennial shelflike basidiocarps up to 30 cm wide, with concentric ridges, at old borer wounds or where cracks in decayed wood extend to the surface. The upper surface of a basidiocarp is at first yellowish brown, but it blackens and cracks with age. The poroid surface is yellowish to reddish brown.

Polyporus squamosus (Polyporaceae) causes a spongy or stringy white rot leading to cavity formation in heartwood of living angiosperm trees, including trembling aspen, black cottonwood, box elder, American elm, Rocky Mountain and sugar maples, and black willow in North America. Widespread also as a decomposer of logs and stumps, it occurs across Canada and the northern half of the United States, in the Southwest, and also in Eurasia and the Southern Hemisphere. Additional substrates in North America include alder, basswood, beech, birch, yellow buckeye, hackberry, horse-chestnut, California laurel, bigleaf maple, and tulip tree. *P. squamosus* produces large, fan-shaped annual fleshy-fibrous basidiocarps at pruning wounds, branch stubs, and cankers. The basidiocarps grow rapidly in spring and early summer from a short, stout, dark-colored lateral stalk and attain widths of 20–30 cm, occasionally 45 cm. The upper surface is yellow-tan with prominent dark brown scales. The lower surface is

---

A. Basidiocarps of *Perenniporia fraxinophila* at old wounds and branch scars on a mature green ash (MI, Aug).

B. Basidiocarps of *Stereum gausapatum* on a basal scar on water oak (MS, May).

C. A basidiocarp of *Oxyporus populinus* in a Nectria canker on sugar maple. Moss on the upper surface of the basidiocarp is typical (NH, Sep).

D. A typical basidiocarp of *Climacodon septentrionalis* on a mature sugar maple (NY, Aug).

E. The hedgehog fungus, *Hericium erinaceus*, fruiting at a branch scar on Nuttall oak (MS, May).

F. Basidiocarps of *Phellinus robiniae* on a trunk of black locust (MS, Apr).

G. Basidiocarps of *Polyporus squamosus* (with scaly surface) and *Bjerkandera adusta* (clusters at left, see also Plate 167) on an American elm stump (NY, Jul).

Plate 166

cream colored and perforated by large tubes with angular mouths 1.0–3.0 × 1.4–6.0 mm.

Globifomes graveolens (syn. *Polyporus graveolens*) (Polyporaceae) causes a straw-colored to yellowish brown heart rot in oaks (e.g., pin, northern red, southern red, and water oaks) in eastern United States from New York and Wisconsin southward. Its perennial basidiocarps have also been collected from beech, sweet gum, hickory, and maple. This fungus is unimportant in forests but causes damage to street-side and landscape oaks. Their decayed trunks either break during windstorms or are cut because they are judged to be unsafe. Decay begins at wounds, usually low on the trunk. *G. graveolens* is apparently able to grow outward from heartwood, perhaps along small cracks, and kill areas of sapwood and vascular cambium. This results in the linear, depressed cankers where it fruits. Basidiocarps also grow on dead trees. They are globular to ovate, woody in the center, and covered with many downturned, scalelike outgrowths each of which has a layer of tubes on the lower surface. Fresh basidiocarps display varied reddish to brown and gray colors and emit a distinctive fruity odor. Old basidiocarps are gray-brown and lack the odor.

Bjerkandera adusta (syn. *Polyporus adustus*) (Polyporaeae) is a common saprobe that causes a white rot in many angiosperms and occasionally gymnosperms throughout the United States and across Canada as well as on all other continents. Its host plants in North America represent at least 40 genera. This fungus also colonizes the sapwood of wounded or diseased living trees. The fungus annually produces thin, shelflike basidiocarps 1–6 × 3–10 × 0.1–0.8 cm, usually overlapping one another in dense clusters. The basidiocarps are usually slightly velvety, and the upper surface varies in color from nearly white through shades of tan or gray, sometimes with reddish areas. The lower surface is gray.

Piptoporus betulinus (syn. *Polyporus betulinus*) (Polyporaceae) occurs only on birches (black, gray, paper, European white, and yellow), usually on dead trees and rarely on living ones. Its distinctive pale brown to nearly white annual basidiocarps with smooth, rounded upper surfaces up to 25 cm wide are unlike those of any other fungus on birch. It occurs throughout the range of birches in North America and Eurasia, causing a red-brown cubical rot.

Phellinus everhartii (syn. *Fomes everhartii*) (Hymenochaetaceae) is a common heart rot pathogen of oaks. It has been reported to attack various other hardwoods as well, but those records are questionable because this fungus has been confused with other species such as *P. badius* on mesquite, *P. igniarius* on various hardwoods, and *P. weirianus* on walnut. *P. everhartii* occurs in California and in western mountains from Mexico to Idaho and Montana and in the East from Arkansas and Georgia to Ontario and Prince Edward Island. The decay is of the white rot type, appearing golden brown or lighter and typically associated with trunk cracks where basidiocarps form. The column of decaying wood expands within the trunk as the pathogen grows outward along radial cracks. Basidiocarps, similar to those of several other species of *Phellinus*, are perennial and hoof shaped, up to 6 × 13 × 8 cm. The upper surface is yellowish brown when young but becomes nearly black and extensively cracked with age. The lower, poroid surface is yellowish brown to reddish brown. Occasionally basidiocarps develop at cankers or cankerous swellings. Sometimes sterile blackish cracked masses of fungal tissue (sterile conks) develop instead of fertile basidiocarps.

Phellinus weirianus, a white rot fungus restricted to mountainous areas in the Southwest (except for rare occurrences in Texas and Oklahoma), mimics the appearance of *P. everhartii*. The two fungi are separable, however, on the basis of microscopic characteristics and host plants. *P. weirianus* is restricted to Arizona black walnut, and *P. everhartii* in the Southwest is restricted to oaks.

Fomes fomentarius (Polyporaceae) causes white mottled trunk rot of many angiosperm trees from coast to coast in Canada and the northern half of the United States. It is widespread also in Eurasia. Its woody, perennial, hoof-shaped basidiocarps are up to 20 cm broad and 15 cm thick. They are found on living or dead alder, apple, aspen, beech, birch, cherry, hickory, California laurel, maple, poplar, and willow and rarely on conifers such as Douglas-fir and hemlock. The fungus is most abundant in the East on birches and beech. It is among those apparently unable to colonize freshly wounded sapwood; rather, it succeeds bacteria and nonbasidiomycetes. It discharges basidio-spores primarily in spring, although some specimens may continue throughout the growing season. By various estimates a typical basidiocarp may release on the order of 10^{11} basidiospores per season—more than its own weight in spores.

Laetiporus sulfureus (syn. *Polyporus sulfureus*) (Polyporaceae) is one of the most important brown rot fungi that attack both angiosperms and gymnosperms. It occurs mainly on angiosperms in the East and Southeast and on gymnosperms in the West. It is found worldwide in temperate regions and is known as the sulfur fungus because of its brightly colored basidiocarps. Host plant groups include ash, beech, butternut, cherry, chestnut, Douglas-fir, elm, eucalyptus, fir, hackberry, hemlock, larch, black locust, honey locust, maple, oak, pepper tree, pine, spruce, tamarisk, tulip tree, and walnut.

The basidiocarps arise annually in summer and autumn on stumps, logs, living trees, or the ground at the base of a trunk. They grow in clusters 20–60 cm broad, varying from sulfur yellow to salmon or bright orange on the upper surface and sulfur yellow on the poroid lower surface. They bleach nearly white with age. Individual basidiocarps measure 5–25 × 4–30 × 0.5–2.5 cm. Edible when young, they are prized by mushroom fanciers. The fungus causes a brown cubical rot of roots, butt, and trunk. Sheets of white to pale yellow mycelium grow in shrinkage cracks in decaying wood. Basidiocarps often issue from elongate cankers that appear as depressed strips of seemingly normal bark, an indication that the fungus has grown outward along cracks from decaying heartwood and has killed a zone of sapwood and vascular cambium. Basidiocarps die and drop after the onset of freezing weather in the North but persist and disperse spores during winter in the South. Basidiocarps on a living trunk indicate massive internal decay and the likelihood of breakage or uprooting by wind. Because fruiting does not occur until many years after the onset of decay, however, many infected trees bear no diagnostic signs. The fungus enters tree trunks through branch stubs and wounds or via roots. The site of root infection has not been reported.

References for trunk decay in general: 26, 228, 814, 1281, 1785, 2085
References for Phellinus igniarius *and* P. tremulae: 116, 312, 387, 637, 640, 658, 830, 1148, 1213, 1408, 1618, 1669, 1774, 1779, 2041, 2091, 2194, 2195
References for Trametes versicolor: 155, 311, 459, 640, 1022, 1023, 1669, 1725
References for Perenniporia, Phellinus robiniae, *and* Polyporus: 279, 283, 304, 407, 637, 640, 669, 784, 1255, 1438, 1616, 1668, 1669, 1723, 1724, 2093
References for Stereum *and* Hericium: 160, 207, 446, 447, 449, 783, 784, 1129, 1253, 1472, 1649, 2024
References for Climacodon *and* Oxyporus: 92, 125, 127, 640, 1310, 1412, 1438, 1669, 1779, 1786
References for Globifomes *and* Piptoporus: 3, 228, 407, 449, 638, 640, 675, 1186, 1438, 1628, 1669
References for Laetiporus, Phellinus everhartii, *and* P. weirianus: 160, 228, 387, 407, 449, 583, 637, 638, 640, 784, 1255, 1412, 1438
References for Bjerkandera *and* Fomes: 228, 508, 817, 1280, 1412, 1438, 1536, 1628, 1669, 1786, 2046

A, B. Trunk decay of oaks caused by *Globifomes graveolens*. A. Old basidiocarps on an elongate canker on water oak. B. A young basidiocarp on a southern red oak (GA, May).
C. Basidiocarps of *Bjerkandera adusta* on a canker on goat willow (NY, May).
D. A typical basidiocarp of *Piptoporus betulinus* on a dead trunk of gray birch (NY, Aug).
E. Basidiocarps of *Phellinus everhartii* on Arizona white oak (AZ, Jul).
F. *Fomes fomentarius* fruiting on a dead trunk of yellow birch (NY, Aug).
G. *Laetiporus sulfureus* fruiting on black cherry. This trunk, still alive around half its circumference, was subject to breakage because of extensive heart rot, as indicated by basidiocarps and revealed by woodpecker holes (NY, Jun).
H. Basidiocarps of *Phellinus weirianus* on Arizona black walnut (AZ, Jul).

Plate 167

Trunk Rots of Conifers (Plates 168–169)

Echinodontium tinctorium (Aphyllophorales, Echinodontiaceae), the Indian paint fungus, is an important heartwood decay pathogen of conifers in the northwestern United States and western Canada. It occurs from California and Colorado northward to Alaska except near the Pacific coast. The common name of the fungus refers to the traditional use of the bright red-orange internal tissue of basidiocarps by Indians as a source of pigment. *E. tinctorium* causes a yellow to golden brown or rust red rot that even in early stages weakens the wood in a manner that causes separation along the annual rings (laminated rot). Wood in the early stage of decay is water-soaked and discolored brown. With advanced decay, it becomes stringy and often contains small white pockets with black centers. Narrow brown to red zone lines are often present. Typically, the entire center of the trunk decays to a height of several meters and may eventually become hollow. Decay extends 3–5 meters beyond basidiocarps and may enter main roots and large limbs. Its color notwithstanding, the decay is of the delignifying type and is classified as a white rot. It is most common in grand and white firs and western hemlock and is somewhat less so in alpine, red, and Pacific silver firs and mountain hemlock. It is uncommon in Douglas-fir, Engelmann spruce, and white spruce.

E. tinctorium produces woody, perennial hoof-shaped basidiocarps 4–20 cm in diameter on diseased trunks, usually at junctions with the undersides of branch stubs. Basidiocarps sometimes form at wounds or on limbs and in the latter case are bell shaped. The upper surface is rough and cracked, nearly black. The lower surface is gray-brown to nearly black, composed of thick blunt spines. The interior tissue is brick red to rust red. This pigment extends into the adjacent decayed wood.

Basidiospores are dispersed during cool wet weather in autumn but germinate poorly until after a period of freezing temperature. They remain viable during winter. Infection is considered to occur in spring.

E. tinctorium does not require large wounds to become established and eventually cause extensive heart rot. In fact, large wounds are unsuitable. In western hemlock, after trees attain age 40 or more, the fungus enters the stubs of tiny shade-killed twigs only about 1 mm in diameter along the mainstem. After 2–3 years these stubs and the pathogen are enclosed by the growing trunk. *E. tinctorium* then becomes quiescent and may remain so for up to 50 years or sometimes longer until some unknown event, possibly a trunk wound, triggers renewed growth of the fungus and the onset of heart rot. Infection and decay in Pacific silver fir and white fir are similar to those processes in western hemlock.

Fomitopsis pinicola (syn. *Fomes pinicola*) (Polyporaceae), called the red belt fungus, is one of the minority of significant heart rot pathogens that commonly colonize both angiosperms and gymnosperms. It occurs from coast to coast in Canada and the northern half of the United States and extends farther south in eastern and western mountains. It usually fruits on dead trees and logging slash but causes decay in many living mature trees that lack basidiocarps. The decay is a brown cubical rot with shrinkage cracks in which prominent white sheets of mycelium develop. In North America, gymnosperm host groups include arborvitae and western red cedar, incense cedar, Douglas-fir, fir, hemlock, larch, pine, sequoia, spruce, and, uncommonly, false cypress. The most common angiosperm hosts are birches and maples. Others include alder, apple, aspen, beech, cherry, chestnut, hickory, magnolia, oak, plum, poplar, and willow. Isolates from angiosperms are able to cause decay in living gymnosperms and vice versa.

F. pinicola is one of the most important brown rot pathogens of old-growth western conifers, but it acts slowly and perhaps for this reason is not among the major decay pathogens of second-growth forests. It is, however, an important member of the coniferous forest ecosystem because it decays dead trees and logging slash and leaves a lignin-rich residue that is very stable and is a major component of the organic matter on the forest floor and in the upper layers of soil. This residue enhances water-holding and cation exchange capacities of soil and is a favorable habitat for the development of ectomycorrhizae (Plates 244–245) and for nitrogen-fixing bacteria. Brown rot caused by *F. pinicola* is difficult to distinguish from that caused by *Laetiporus sulfureus* (Plate 167), which may occur in trees of the same species.

The perennial basidiocarps of *F. pinicola* are hard-corky to woody, shelflike to rounded on the upper surface, or sometimes somewhat hoof shaped and encrusted, with a red-brown band near the white to cream-colored edge. The oldest parts of the upper surface turn gray to dark brown or black. During growth, the white edge may be thick and prominent. The poroid undersurface is white to cream-yellow. Basidiocarps vary in size up to 30 × 40 × 20 cm.

F. pinicola fruits at wounds, but whether these are required for entry is unclear. Decay by this fungus in true firs, western hemlock, and Sitka spruce is most commonly associated with scars such as those made by falling trees. Branch stubs are also common sites of decay. In red fir, white fir, and western hemlock, decay by *F. pinicola* is often associated with broken tops and with cankers that develop at sites of dwarf mistletoe infection. Moreover, *F. pinicola* and *E. tinctorium* have both been isolated from pith of suppressed grand fir trees. Knowledge of the infection biology of *E. tinctorium*, outlined above, raises the possibility that *F. pinicola* also enters young trees through dead twigs or other small, natural avenues and remains quiescent until a later wound triggers growth and wood decay. This fungus is probably also introduced to dying or dead trees by insects. It has been isolated from Douglas-fir bark beetles (see Johnson and Lyon, 2nd ed., Plate 28) captured both in flight and from galleries where eggs are laid.

A, B, E. The Indian paint fungus, *Echinodontium tinctorium*. A. An intact, hoof-shaped basidiocarp several years old on western hemlock. B. The same basidiocarp viewed in section, showing red-orange interior tissue and toothlike projections from the lower surface. E. Close view of the lower surface (ID, Jul).

C, F, H. Heart rot caused by *E. tinctorium* in western hemlock. C. Stringy, golden brown decayed wood occupies nearly the entire cross-section of the trunk of a pole-sized tree. Wood at the periphery in the early stage of decay is water-soaked and discolored brown. F. Part of the diseased trunk in longitudinal section, showing at top the tendency of decayed wood to separate along the annual rings. H. Close view of wood in an advanced stage of decay near a basidiocarp, showing typical rust red color and white pockets with black centers (ID, Jul).

D. Basidiocarps of *Fomitopsis pinicola* on a wind-broken red spruce trunk. While it was alive the tree sustained extensive trunk decay that led to breakage. The transverse fracture without splintering is typical of wood with a brown cubical rot (NY, Oct).

G. Basidiocarps of *F. pinicola* at the base of a decaying mountain hemlock snag (OR, Sep).

Plate 168

Phellinus pini (Aphyllophorales, Hymenochaetaceae) is perhaps the most damaging trunk decay fungus of conifers throughout North America and elsewhere around the Northern Hemisphere. It causes heart rot and, in some hosts, cankers. The disease is known as red heart, red rot, ring scale, red ring rot, and white pocket rot. These names indicate decay features that vary with host and site of infection. Decay columns commonly extend 10 meters or more, rendering entire trunks useless for lumber. Damage is much greater in old trees in virgin and unmanaged forests than it is in managed forests where trees are harvested at relatively young ages.

Most conifers that grow in temperate regions are susceptible. In North America, Douglas-fir, larches, pines, and spruces sustain greatest damage. Firs, hemlocks, and western red cedar are commonly affected. Occasional infections develop in arborvitae, incense cedar, false cypresses, junipers, and yews. The few reports of this fungus on angiosperms are suspect.

Decay by *P. pini* is usually confined to the heartwood of mature trees, either in one central column or in several discrete columns that extend from branch stubs. Decay often begins near the junction of heartwood and sapwood and may extend into sapwood adjacent to wounds, near canker margins, and along the trunks of severely diseased pines nearing death. Decay typically occurs well up in the trunk, but butt rot is common, and the rot sometimes extends into major roots. Young trees become infected in special circumstances. For example, *P. pini* infects young jack pine at cankers caused by sweetfern rust (Plate 133) and then slowly spreads upward and inward in the trunk.

The decay is a white pocket rot characterized by selective removal of lignin. Wood in early stages of decay appears pinkish to reddish in pines and purplish in Douglas-fir and spruces. In pines it is sometimes bounded by a zone of resin-soaked wood. Small elongate pockets lined with white fibers appear later and slowly expand and merge. In advanced stages the wood is soft, usually light colored, and fibrous, often with irregular narrow black lines as seen on cut or broken surfaces. Sometimes the decay spreads tangentially, causing a crescent or ring of defective wood as seen in cross-section. The specimen of decayed wood shown in Plate 169G, while exhibiting pockets and the ultimate fibrous texture of decayed wood, lacks the typical white lining of pockets that could have been seen at an earlier stage of decay. The latter feature, as caused by *Inonotus circinatus*, is shown on Plate 154. Decay caused by *I. circinatus* in sand pine is nearly indistinguishable from that caused by *P. pini*.

External indicators of decay by *P. pini* include swollen knots where branches were shed many years earlier, irregular bulges with exuding resin, resin flow from knots, and brown basidiocarps at branch bases, branch stubs, knots, wounds, and cracks and sometimes on seemingly normal bark. Punk knots are common in Douglas-fir, western larch, pines, and some spruces. A punk knot is a mass of tightly packed sterile brown hyphae that extend from a decayed branch stub within the trunk to a local swelling on the surface, there appearing like a blackened knot. Swelling results from slight overgrowth of wood around the punk knot. Typical dark yellowish brown to reddish brown fungal tissue is exposed where a punk knot is cut.

On western firs affected by *P. pini* var. *cancriformans,* the trunk may swell at the edges of large, flat, bark-covered cankers that bear small basidiocarps. Such cankers, often attaining dimensions of 40 × 60 cm, occur on grand, noble, Shasta, Pacific silver, and white firs. The associated decay is a white rot that lacks distinct white pockets. Cankers form as the result of successive advances of the fungus in inner bark, leading to cambial death. This process repeatedly opens new wood near canker margins to invasion and decay. The avenue of

initial advance into bark is unknown. Small cankers, to 9 × 24 cm, caused by typical *P. pini* and associated with white pocket rot, have occasionally been noted at branch stubs on balsam fir in the Great Lakes region. *P. pini* does not usually act as a canker-rot pathogen in other conifers. In some dying trees, however, it does apparently kill sapwood and cambium and begins to fruit on extensive areas of the bark.

Basidiocarps of *P. pini* vary from nearly flat annual incrustations 2–5 cm in diameter to large perennial bracket- or hoof-shaped forms and sometimes conchlike specimens with wavy edges, to 30 cm or more in diameter. The upper surface of perennial specimens is gray-brown to brownish black, roughened by concentric ridges and sometimes also by radial cracks. The margin and nearby upper surface are slightly to distinctly velvety. The margin, poroid lower surface, and interior tissue are yellowish brown to reddish brown. The pores are round to radially elongate and average two to three per linear millimeter. Many diseased trees lack basidiocarps, but these develop rapidly and often in profusion after death of the trunk. Peak dispersal of basidiospores occurs in spring and autumn in eastern Canada and the northeastern United States. Basidiocarps of *P. pini* var. *cancriformans,* known only on cankers on fir in the West, are small (to 5.5 × 4.5 × 1.7 cm), more or less shelflike, perennial, and usually found in clusters.

P. pini apparently colonizes wounds such as broken limbs or tops, felling scars, or rust cankers in some hosts, but decay by this fungus is associated with dead branches and stubs in other hosts. Jack pine and black spruce are in the former category; western conifers and eastern white pine are in the latter group. The majority of infections by *P. pini* in eastern white pine occur through small branch or leader stubs. For example, decay often develops at points in heartwood where leaders were killed many years before by the white pine weevil (see Johnson and Lyon, 2nd ed., Plate 20) or where small branches near the trunk base died from various causes. Whether the decay fungus colonizes killed leaders soon after their death or only after several years (as is presumed to occur in small lateral branch stubs) is unknown. Once the site of entry is overtaken by heartwood, however, decay begins.

Decay by *P. pini* advances slowly, only 5–10 cm per year up or down in white pines. In white fir, bacteria including nitrogen-fixing types are associated with all stages of wood colonization and decay by *P. pini,* but the ecological role of these microbes is unknown.

Decay by *P. pini* in old-growth conifers formerly provided dependable habitat for cavity-nesting animals. In the Southeast, for example, the red cockaded woodpecker learned to create nest sites by excavating decayed wood from mature pines. This bird is now considered to be an endangered species because of the loss of nesting habitat as old-growth forests have been cut. Artificial inoculation of heartwood of southern pines with *P. pini* has been suggested as a means of providing more potential nest trees.

References for Echinodontium tinctorium: 4, 550, 551, 692, 1028, 1029, 1204, 1993
References for Fomitopsis pinicola: 273, 550, 583, 638, 640, 742, 843, 977, 1027, 1029, 1109, 1358, 1669
References for Phellinus pini: 5, 117, 190–92, 228, 470, 583, 640, 705, 706, 820, 1109, 1478, 1669, 2002

A, B. Basidiocarps of *Phellinus pini* on trunks of western white pine. A. Young basidiocarps issuing from scars of old branch stubs, indicating extensive heart rot. B. Perennial basidiocarps issuing from a flattened, dead area on the butt of another tree (ID, Jul).

C. *P. pini* var. *cancriformans* fruiting on a large canker on grand fir. Callus growth at canker margins has resulted in trunk swelling (OR, Jul).

D. A perennial basidiocarp of *P. pini* on sand pine (FL, Apr).

E, F. Close views of upper and lower surfaces of a typical basidiocarp of *P. pini* from western white pine (ID, Jul).

G. Pocket rot caused by *P. pini* in sand pine (FL, Apr).

H. Heart rot caused by *P. pini* in old-growth western white pine. The trunk of this tree was left to rot because of extensive decay found at the time of felling. Blue-stain fungi invaded the sapwood after the tree was cut (ID, Jul).

Plate 169

Canker-rots of Hardwoods (Plates 170–173)

Canker-rots are caused by wood-decaying fungi that also kill phloem and the vascular cambium, resulting in perennial cankers. Some canker-rot fungi, such as *Trametes versicolor* (Plate 165) and *Schizophyllum commune* (Plate 173), enter wounds, parasitize sapwood, and subsequently kill overlying areas of vascular cambium and bark. Other canker-rot fungi, exemplified by *Inonotus obliquus* here and by other pathogens in Plates 169–173, enter wounds, colonize wood within wound-associated compartments (Plates 162–163), and eventually produce at the edge of the original wound site a mass of mycelium that kills adjacent bark and cambium, resulting in a canker. This lesion serves as a new site of entry into sapwood. Cankers and trunk decay expand by repetition of this process.

Canker-rot of birch. Canker-rot associated with prominent sterile conks on birches is caused by *Inonotus obliquus* (syn. *Poria obliqua*) (Aphyllophorales, Hymenochaetaceae). This fungus occurs around the Northern Hemisphere. In North America it most commonly infects paper and yellow birches, and it occurs also on gray and sweet birches. Occasional records exist of it also on red alder, American beech, black cottonwood, and hop hornbeam. One report of occurrence on beech is of interest because it involved a basidiocarp on a living tree. On birches the fungus produces basidiocarps only after the trees are dead.

The disease is easy to diagnose on the basis of the sterile conks. These are hard, black perennial masses of fungal tissue, rough and cracked on the surface, that slowly erupt from large internal columns of decayed wood at old Nectria cankers, branch stubs, cracks, and wound scars. These locations, considered to be sites of infection, are listed in descending order of frequency. The trunk is often swollen adjacent to a sterile conk because of increased wood production and an increase in bark thickness apparently induced by the pathogen. Bark sometimes thickens to as much as three times normal. The internal tissue of a sterile conk is yellow-brown to rust brown and punky and contains small dark brown fragments of partly decayed bark that was killed and displaced by the growing mass. Diseased trees usually display one to six sterile conks, the presence of which indicate a decay column so large that at least half the trunk—often the entire trunk—will be useless for lumber. Trees often break at the locations of sterile conks. The decay is of the white rot type, appearing light reddish brown and mottled, with veins of white mycelium near the sterile conks. In advanced stages it becomes spongy.

A canker forms as massed hyphae at the edge of a sterile conk intermittently grow tangentially into the inner bark, killing it and the vascular cambium. The canker is thus always occupied by the sterile conk. After each episode of canker expansion, the tree reacts with compartmentalizing responses, including the formation of a barrier zone at the sapwood surface. Thus the defective portion of the trunk eventually consists of many overlapping columns of decayed wood delimited by discolored wood containing old barrier zones. Decay columns have been reported to elongate as much as 37 cm per year up or down the trunk. If this figure is corroborated, it would represent very fast development in comparison with decay caused by most other heart rot fungi. Rapid growth in wood might be expected, however, since *I. obliquus* acts as a primary colonist of sapwood, requiring no prior succession of microorganisms.

Over a period of many years the canker and the overlapping columns of decaying wood enlarge so much that little normal sapwood remains to sustain the tree. Death ensues, but whether it is caused directly by *I. obliquus* or by opportunistic organisms is unreported.

Basidiocarps of *I. obliquus* arise as sheets of brown tissue in the outermost sapwood (1–25 mm deep) 3–4 years after a tree with canker-rot dies, usually while the trunk is still erect. The interval between tree death and sporulation may be longer for trees that break or are felled while alive; it was 7–12 years in one study. The onset of fruiting is signaled by formation of an undulating brown sheet of mycelium at the boundary of the decay column. Elongate ridges of sterile fungal tissue develop parallel to the trunk axis at the edges of the mycelial sheet, exerting radial and tangential force that causes the covering wood and bark to crack along the axis of the trunk and then lift away from the fungal tissue. This tissue soon develops a gray poroid surface from which basidiospores are dispersed. It then turns dark brown, dries, and begins to disintegrate. The entire basidiocarp may be 30–150 cm long and 15–30 cm wide. Basidiocarps may be found in July–November. They are most abundant during summer but are short-lived. Insects quickly destroy them. A given tree usually produces basidiocarps in only 1 year. The felling of diseased trees, the cutting of trunks into bolts, deep girdling, or poisoning does not prevent eventual formation of basidiocarps.

In the Appalachian region, yellow birch is subject to canker-rot caused by *Phellinus laevigatus* (syn. *Poria laevigata, Fomes igniarius* var. *laevigatus*) as well as to that caused by *I. obliquus*. The diseases and pathogens differ in several respects. *P. laevigatus* induces rough, irregular, sunken, bark-covered perennial cankers, often with a mat of dark brown to black sterile fungal tissue on the surface. A branch stub or the remains of one is usually visible in the canker. Projecting sterile conks are not present. The internal appearance of the decay column differs from that of decay caused by *I. obliquus,* and the two fungi differ in cultural characteristics.

A, B. Sterile conks of *Inonotus obliquus* on yellow birch. A. At a canker caused initially by *Nectria galligena*. B. Emerging from old wound scars. Slight swelling of the trunk at the level of the sterile conks is typical. The edge of each canker, except for portions of the old Nectria canker, is within a few millimeters of the edge of the sterile conk (NY, Sep).

C, G. A basidiocarp of *I. obliquus* on a dead tree. C. A large gray fertile surface following the contour of the trunk is exposed where the bark and outermost sapwood have lifted away from underlying wood adjacent to an old canker. Dark brown internal tissue of the basidiocarp is revealed where the surface was bruised in four spots near the lower end of the canker. Remnants of sterile conks remain in the canker. G. Close view of the edge of the basidiocarp, showing vertically oriented tubes (NY, Sep).

D–F. Exterior and internal views of canker rot. D. Sterile conks at each end of a crack where callus rolls closed over an old trunk wound. E. The same trunk, split through the crack, revealing the rust brown interior of one sterile conk and a large column of mottled rot extending up and down the trunk. Dark brown wood beneath the old wound site is occupied by organisms other than *I. obliquus*. F. Close view of the edge of the canker showing normal bark and sapwood at left center, the margin of necrosis in bark (arrow), a zone of discolored sapwood being invaded from right to left by *I. obliquus*, decayed wood laced with white mycelium, and rust brown tissue of the sterile conk. Bark no longer adjoins the surface of the discolored sapwood except at the canker margin. Decayed bark is recognizable only as small dark brown fragments within the fungal mass. The cut surface of healthy bark at left, normally light in color, quickly became rust colored where it was exposed to air after being cut (NY, Sep).

Plate 170

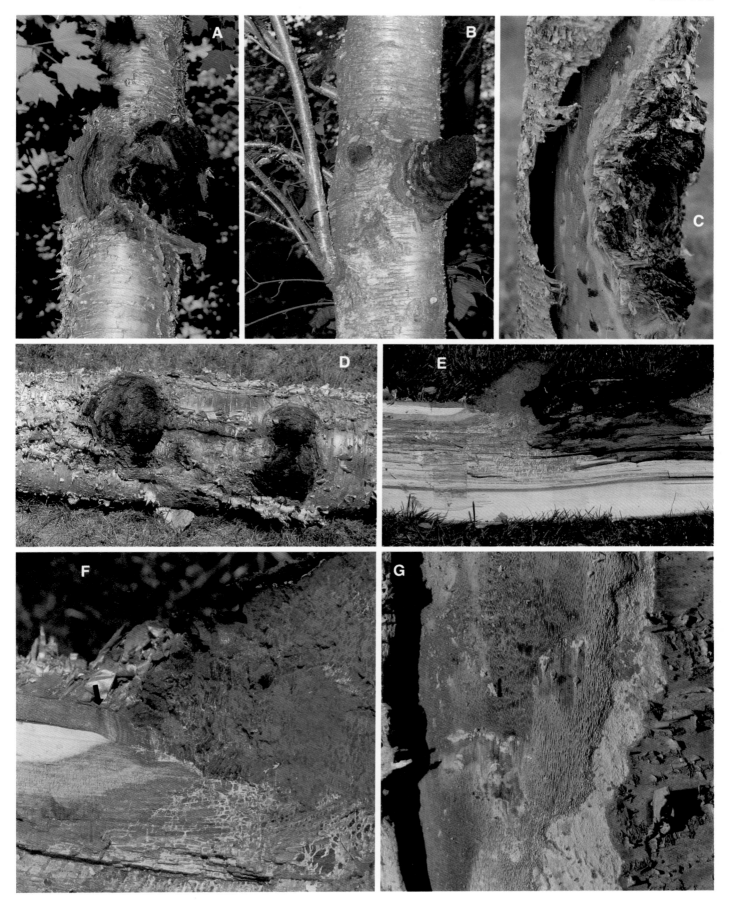

Spiculosa canker. *Phellinus spiculosus* (syn. *Poria spiculosa*) (Aphyllophorales, Hymenochaeteceae) causes white trunk rot and perennial cankers on hickories, honey locust, and oaks. The disease occurs from Pennsylvania and Delaware to the Gulf states. Hosts include mockernut, pignut, and shagbark hickories; honey locust; and blackjack, Nuttall, northern red, southern red, water, and willow oaks.

Infection begins at branch stubs, and cankers eventually form there. The cankers on hickories appear as rough, circular swellings on the trunk where callus rolls seem to have enclosed or nearly enclosed a wound. A diseased tree usually has only one canker, but occasional trees bear several cankers. Sometimes a canker bulges out as a burl, with one or more seams where callus rolls meet but apparently do not fuse. Open cankers, usually with depressed centers, develop occasionally. Cankers in early stages of development appear like branch stubs that have not yet been enclosed by growth of the trunk. Sterile brown fungal tissue, often called a punk knot, occupies the position of the former stub. This tissue often protrudes into cracks between callus folds at the surface of the swelling. Mycelium extends from the punk knot into a large column of decayed wood in the center of the trunk. An axe cut into the swollen area exposes the yellowish brown punk knot and the brown mycelium between callus folds.

Symptoms on oaks tend to be less prominent than on hickories. Often the only indicator of infection is an inconspicuous punk knot, black or the color of bark, appearing to be a branch stub that would soon be enclosed by the growing trunk. Such punk knots can be diagnosed by cutting into them, exposing brown fungal tissue. The cankers are circular, 5–15 cm in diameter, usually within 4 meters of the ground. Open cankers with depressed centers and moderate callus development around the margins occasionally form, especially on willow oak. Rough, bulging cankers with irregular callus folds on the surface, similar to those found on hickory, also form occasionally on oaks. Multiple cankers or punk knots on a tree do not necessarily indicate separate infections; all may be associated with one column of decay.

The advanced decay in all hosts is a soft crumbly white or slightly yellowish rot, sharply delimited from sound wood, typically confined to heartwood, and extending from ground level to a height of 3–4 meters or more. Decay columns caused by *P. spiculosus* in oaks in the South have been estimated to elongate at an average rate of 20 cm per year. These decay columns closely resemble those caused by *Inonotus hispidus* (Plate 172).

The function of the punk knot, it has been suggested, is maintenance of a connection for gas exchange between the central decay column and the exterior, thus allowing decay to progress. Most trunk decay fungi cease activity when trunk wounds close.

P. spiculosus occasionally fruits on the dead parts of living trees, but usually the perennial basidiocarps develop on dead standing trees and on logs on the ground. They arise in patches or as a continuous sheet up to 1.5 meters long between the bark and wood of decayed stems, pushing off the bark as they mature. The pore surface and interior are brown initially; the surface becomes whitish or grayish, cracked, and dry with age.

A, E. Spiculosa canker and associated trunk rot in water oak. A. A perennial targetlike canker, prominently raised because of callus growth around the margin, has developed at the site of a former branch stub. E. A section through the trunk several centimeters below the canker reveals advanced white crumbly rot throughout the heart of the tree, plus spokelike extensions of decay into outer sapwood. These extensions are the tips of secondary columns of decay that began as the canker enlarged. The column of advanced decay in this small tree was 3 meters long (GA, May).

B, D. Spiculosa cankers on Nuttall oak and willow oak, respectively. The cankers could be mistaken for raised, roughened branch scars. Sterile brown fungal tissue occupying the locations of former branch stubs is revealed by cutting into the raised knots (MS, Apr).

C, F, G. Dissection of the Nuttall oak shown in (B). C. The brown sterile conk is connected to a central column of advanced white rot bounded by a thin purple-brown zone of discolored wood. F. The central column of decay expands not only as the result of canker enlargement but also where wood-boring insects breach the barrier zone between decayed and sound sapwood. G. The butt log, approximately 4 meters long, with the canker at the midpoint and advanced decay throughout the heart of the log.

Plate 171

Hispidus canker. *Inonotus hispidus* (syn. *Polyporus hispidus*) (Aphyllophorales, Hymenochaetaceae) affects many angiosperms around the Northern Hemisphere. It occurs throughout most of the United States where hardwoods grow, and it causes large, elongate, bark-covered lesions on which it fruits conspicuously. Oaks are most commonly affected. Those in the red/black group (e.g., black, cherrybark, laurel, Nuttall, northern red, southern red, scarlet, turkey, water, and willow) are often infected in the East and South. White and chestnut oaks in the East and Oregon oak in the West are occasionally affected. Other susceptible trees and tree groups in North America include black ash, beech, yellow birch, sweet gum, hickory, honey locust, mulberry, pepper tree, walnut, and willow. This fungus has rarely been found on pine (unspecified) and grand fir, but canker formation on conifers is unreported.

Branch stubs less than 2.5 cm in diameter, usually within 5 meters of the ground, are the most common sites of infection. The pathogen first rots heartwood and then begins to kill sapwood and the vascular cambium, resulting in perennial cankers that usually increase to 1–1.5 meters in length, with bark remaining firmly attached. Cankers up to 4 meters long and 30 years old have been reported. Large callus ridges develop along the sides of a canker but are invaded and killed. More callus forms at the new canker edge, resulting eventually in a spindle-shaped swelling of the trunk. Cankers lengthen approximately 15 cm per year in the South, and heart rot extends, on average, 35 cm above and below the canker. Diseased trees in residential areas present hazards because their trunks are likely to break.

The decay is a white rot in which lignin is removed selectively. The wood becomes soft and spongy, straw-yellow to pale yellow-brown. Often a narrow, conspicuous black or dark purple-brown zone marks the abrupt transition between decayed and sound wood. Decay columns resemble those caused by *Phellinus spiculosus* (Plate 171).

Basidiocarps form annually in summer to early autumn, usually singly but occasionally in groups of two or three. They grow to full size within 1–2 weeks and are more or less shelflike and irregularly semicircular in outline as viewed from below. They are spongy, somewhat hairy, and yellowish brown to rust red on the upper surface and fawn (at first) to rust colored on the lower poroid surface. They measure 5–30 × 8–25 × 2–10 cm. They disperse spores for about 3 weeks, then slowly shrink and dry to a dark brown or black mass and eventually fall from the tree. Dead, shriveled basidiocarps on cankers or on the ground beneath cankers are diagnostic aids. The fungus may continue to fruit for up to 5 years after diseased trees are felled, but spore dispersal from basidiocarps on fallen logs is less efficient (fewer spores traverse a given distance) than dispersal from basidiocarps on standing trees.

Heart rot and canker-rot caused by *Inonotus andersonii*. This fungus (syn. *Poria andersonii*) is one of the major decay pathogens of oaks, occurring from coast to coast in the United States from Pennsylvania and Oregon southward. This fungus is an important cause of mortality of oaks in the Southwest. Hosts include blue, Mexican blue, Emory, Gambel, California live, Oregon, silverleaf, tanbark, valley, and Arizona white oaks in the West, and black, blackjack, northern red, scarlet, and water oaks in the East and South. Other hosts include hickory, poplar, and black willow. Branch stubs are the most common site of infection, but entry also occurs through fire scars and other injuries. When acting as a canker-rot pathogen, *I. andersonii* first causes a white rot of the heartwood and then moves outward, killing the sapwood and cambium. Trees often break at these cankers. The fungus produces annual sheetlike yellowish brown to dark dull brown basidiocarps that may exceed 50 cm in length, often with very rough surfaces with peglike outgrowths, beneath the bark or outermost wood on killed parts. Basidiocarps cause the overlying host tissues to fall away.

Canker-rot caused by *Inonotus glomeratus*. *I. glomeratus* (syn. *Polyporus glomeratus*) occurs from coast to coast in Canada and the northern United States and extends southward to Florida in the East. It is most abundant in the Great Lakes region and eastward, where it causes white to light brown spongy heart rot of beech and red and sugar maples. In the West it occurs on paper birch, black cottonwood, and Oregon oak. Laurel oak is a host in the South. This fungus is also reported to occur on box elder, poplar (unspecified), viburnum, and, rarely, hemlock. *I. glomeratus* is the most important trunk and butt decay pathogen of sugar maple in Ontario, accounting for 36% of the infections and 40% of decay volume in one study there. Most infections begin at dead branches and stubs, although other wounds are also suitable. In beech, the initial site of infection is often a branch stub on the upper trunk. After decay is advanced, this fungus forms sterile punky masses (sterile conks) that replace branch traces, extending from a central column of white rot to the trunk surface or, in beech, as much as 7–8 cm beyond the trunk. The surface of the sterile conk turns black, crusty, and cracked. Sterile conks are diagnostic. They are prominent on beech but are usually inconspicuous in irregular cankers with raised margins on maple trunks. Cankers are usually localized but occasionally become elongate on maples, perhaps reflecting the shape of the wound where infection began. Cankers on beech tend to be inconspicuous around sterile conks but sometimes also develop raised callus rolls at the margins. *I. glomeratus* causes cankers in the manner described and illustrated for canker-rot of birch (Plate 170).

I. glomeratus was formerly confused with *I. andersonii* because both may form yellow-brown sheetlike basidiocarps on dead trees or tree parts. Records of *I. glomeratus* on willow are thus suspect. Basidiocarps of *I. glomeratus* but not *I. andersonii* often grow shelflike, at least in part, and form on bark rather than beneath it. They appear 3–4 years or more after tree death, usually on fallen trunks. The two fungi are also distinct in cultural characteristics and temperature optima for growth, *I. andersonii* growing most rapidly near 35°C, while *I. glomeratus* grows best near 25°C.

A–D. Hispidus canker on oaks. A. A canker flanked by large callus rolls on willow oak. A rust brown basidiocarp remains from the previous autumn (MS, Apr). B. Large, active basidiocarps of *Inonotus hispidus* on a young canker not yet delimited by callus ridges (MS, Aug). C. A canker nearly enclosed by callus, appearing as a fissure. The canker has expanded several centimeters onto the original callus roll at the lower right side of the fissure, as evidenced by a sunken area bounded by secondary callus (MS, Apr). D. A blackened dead basidiocarp at the base of a diseased laurel oak (FL, Apr).

E, G. Canker-rot of blue oak, caused by *I. andersonii*. E. A partly killed tree with the dead top removed, showing white heart rot and a large brown basidiocarp that has caused the bark to fall away from the trunk. G. Close view of the basidiocarp (AZ, Apr).

F. An American beech with heart rot caused by *I. glomeratus*. The black sterile conk protruding from the location of a former branch stub is diagnostic. The old falling-tree scar, now evident only as a seam beside a large callus roll on the trunk, was possibly the site of infection (NY, Aug).

Plate 172

Sapwood rots and cankers caused by *Cerrena unicolor*. *C. unicolor* (syn. *Daedalea unicolor*) (Aphyllophorales, Polyporaceae) is a common and well-known saprophytic decayer of sapwood. It is also an opportunistic colonist of living sapwood, causing canker-rot in trees weakened by environmental stress. This fungus occurs throughout Canada and the United States except in southernmost areas.

C. unicolor has a broad host range consisting mainly of angiosperms; it occurs rarely on gymnosperms. The fungus fruits on dead branches or trunks, on discrete cankers, and on large areas of dead bark near trunk bases. Host groups include alder, apple, ash, beech, birch, buckthorn, cherry, chestnut, dogwood, elm, fir, sour gum, sweet gum, hackberry, hawthorn, hemlock, hickory, holly, hornbeam, hop hornbeam, horse-chestnut, juniper, linden, black locust, magnolia, maple, oak, pear, poplar, serviceberry, tree-of-heaven, tulip tree, and willow.

This fungus enters sapwood at wounds. If the tree is stressed in some way, for example by fire damage, sunscald, dense shade, or top breakage by ice, *C. unicolor* overcomes defensive responses and kills a large mass of sapwood in which it causes a crumbly white rot. By invasion of bark near the original wound site, it evades wound-associated compartmentalization and gains access to new layers of sapwood in the manner described for *Inonotus obliquus* (Plate 170). In maples this mode of attack results in distinctive color patterns in decaying wood—yellowish white decayed wood alternating with dark reddish brown, less decayed bands where living tissues once resisted invasion. When trees are greatly weakened and their compartmentalizing processes function poorly, the fungus may kill a large area of cambium in a single year, girdling the stem or causing a large, diffuse canker. It usually fruits on killed bark within the 1st year after cambial death.

C. unicolor is a common secondary factor in the decline of sugar maple (Plate 216), often causing relatively sudden death of major limbs or the tree top. The primary or predisposing influence is often unknown. Occasionally the fungus causes discrete cankers on sugar maples that appear vigorous. Advanced decay may be present 1–1.5 meters above the canker. Red maples overtopped by neighbor trees in the forest are also commonly attacked; sprout clumps on a common root system seem particularly vulnerable. The parasite spreads from one sprout stem to another at their common base, eventually killing all. On fruit trees *C. unicolor* often produces basidiocarps on perennial cankers centered on wounds, but orchardists, seeing the basidiocarps, tend to dismiss the fungus as "secondary" in the sense that it has invaded wood killed by some other agent. This interpretation is often erroneous; the fungus may have killed the sapwood beneath the site of fruiting.

The annual basidiocarps, 0.5–6 × 2–8 × 0.15–0.5 cm, develop in summer, usually in clusters. The upper surfaces are typically hairy, nearly white to greenish or greenish gray, often with concentric ridges. The pores on the lower surface are irregular in outline, and in old specimens the walls of the tubes break up into teeth. From a distance the weathered, nearly white basidiocarps of *C. unicolor* can be confused with those of *Irpex lacteus* (syn. *Polyporus tulipiferae*), which has similar hosts and opportunistic habits.

C. unicolor disperses basidiospores in the manner of many other wood-decaying basidiomycetes and is also carried by the pigeon tremex, *Tremex columba* (see Johnson and Lyon, 2nd ed., Plate 238). This insect possibly introduces the fungus into weakened trees not otherwise wounded.

Sapwood rot caused by *Schizophyllum commune*. *S. commune* (Aphyllophorales, Schizophyllaceae) is a ubiquitous saprophyte and opportunistic pathogen of a great many woody and herbaceous plants—monocots, angiosperms, and gymnosperms. It occurs around the globe, entering woody plants through various wounds and injuries, especially sunscald lesions and fire scars. As a pathogen it aggressively colonizes trees stressed by heat, drought, or major wounds, causing a white rot of sapwood. In temperate regions it is most common on apple, decaying both bark and sapwood, apparently causing cankers, and occasionally causing fruit rot. It is a proven pathogen of apple, peach, and white poplar. It causes sapwood rot in living sugar maple, entering trunks through dead branches and stubs or through tops broken by weight of ice. It attacks western catalpa and 'Greenspire' European linden via bark cracks or lesions caused by freezing. Apricot, cherry, and plum may be invaded through sunscald lesions, and hickory through fire scars. This fungus also infects acorns and stored Douglas-fir cones if they are not properly dried.

A partial list of other host groups includes acacia, arborvitae, ash, mountain ash, avocado, bamboo, beech, birch, blueberry, buckthorn, bunya-bunya, camphor tree, cherry and plum, citrus, bald cypress, false cypress, elder, elm, eucalyptus, fig, filbert, fir, grapevine, sweet gum, hemlock, hickory, holly, hornbeam, horse-chestnut, jacaranda, juniper, larch, California laurel, linden, black locust, honey locust, magnolia, mango, maple, mimosa, mulberry, oak, silk oak, oleander, olive, osage orange, palm, palmetto, paulownia, pear, pepper tree, persimmon, pine, pistachio, plane tree, poinciana, prickly pear, redbud, sequoia, serviceberry, spruce, sumac, tamarisk, tree-of-heaven, tulip tree, tung-oil tree, tupelo, walnut, willow, and yucca.

S. commune often begins to form basidiocarps within only several weeks after colonizing a substrate, but the presence of fruit bodies on a lesion is not a reliable indication of pathogenic attack because the fungus may colonize plant tissues killed by other organisms. Moreover, it has been shown to parasitize other fungi, which may explain its quick appearance on some cankers. The hairy white annual basidiocarps, to 5 cm wide, usually develop in clusters, are distinctly convex, broadly attached or sometimes with a short lateral stalk, and have bifurcate gills on the lower surface. White to pale brown when young, they turn grayish or darker with age.

S. commune grows at temperatures from near freezing to above 40°C and grows most rapidly at 30–35°C, which helps explain its aggressiveness in heat-stressed trees. The optimum temperature for spore discharge, however, is lower than that for growth. This fungus has been much studied in the laboratory because it fruits readily in culture and is amenable to genetic analysis. Considerable understanding of the genetics of higher basidiomycetes is based upon knowledge of *S. commune*.

References for canker-rot of birch: 193, 307, 640, 1669, 1777, 2041, 2266
References for spiculosa canker: 313, 640, 1254, 1408, 2022
References for hispidus canker: 387, 449, 577, 636, 640, 1254, 1258, 1259, 1438, 1669, 1825, 2023
References for Inonotus andersonii: 160, 308, 636, 640, 1438
References for Inonotus glomeratus: 308, 309, 636, 640, 657, 1412, 1422, 1438, 1777
References for Cerrena unicolor: 305, 640, 1438, 1669, 1864, 1908
References for Schizophyllum commune: 155, 394, 459, 549, 1377, 1412, 1568, 1864

A–D, F–H. Canker-rot caused by *Cerrena unicolor*. A. Basidiocarps of *C. unicolor* on a canker that has nearly girdled a scaffold limb of 'Montmorency' sour cherry. Only a narrow callus ridge remains alive (OR, Jul). B. *C. unicolor* fruiting on a large canker that originated at a branch stub on sugar maple (MI, Aug). C. Basidiocarps on a large basal canker on a living red maple. The parasite apparently spread from the dead trunk at right into its companion stem in the sprout clump (Ont, Jul). D, H. Dissection of a sugar maple with canker-rot. In longitudinal view the decaying wood is bounded by a dark reaction zone. At the canker margin (arrow), the fungus has entered and killed bark overlying sound wood, which will in turn be invaded and decayed. In cross-section a recently killed stem shows concentric reaction zones that formed as the pathogen advanced into younger sapwood. Part of a canker is visible at left (NY, Sep). F, G. Basidiocarps collected from the red maple in (C). The hairy, greenish, concentrically ridged upper surface is typical of basidiocarps grown in moist, shaded locations.

E, I. *Schizophyllum commune* on apple bark. E. Clusters of hairy white basidiocarps near the edge of a canker. I. Viewed from below, the basidiocarps have bifurcate gills (ND, Jul).

Plate 173

Ceratocystis Cankers and Sapstreak Disease (Plates 174–175)

Canker-stain of plane tree and sycamore. Canker-stain is a lethal disease caused by the wood-staining pathogen *Ceratocystis fimbriata* f. sp. *platani* (Microascales, Ophiostomataceae). The disease is characterized by spreading lesions that involve phloem, cambium, and extensive regions of sapwood. Infected wood, although darkly stained, is not decayed by the pathogen. It may, however, be invaded quickly by secondary wood-decay fungi. Canker-stain, destructive in eastern United States and in southern Europe, is unusual because the most severe outbreaks result from pruning and other human activities that wound trees. The causal fungus can infect even the tiniest fresh wounds (old wounds are not suitable) to bark or sapwood. It is transmitted in sawdust, in tree wound dressings that are applied by brushing, and on pruning saws, ropes, and other equipment used for tree care. Spores on equipment may remain infectious for a month or more. Natural transmission is by insects, primarily sap- and fungus-feeding beetles of the family Nitidulidae that visit wounds made by other agents. The beetles *Cryptarcha ampla* (Nitidulidae) and *Laemophloeus biguttatus* (Cucujidae) have transmitted the pathogen from diseased to healthy trees in experiments; however, these are probably not the only vectors.

Canker-stain became recognized as the most damaging disease of London plane trees when devastating outbreaks occurred in urban plantings in Pennsylvania and New Jersey during the 1930s and 1940s. Some plantings were destroyed in less than 20 years. Similar outbreaks began in Italy after World War II and later in France, possibly as a result of transporting the pathogen to Europe on green sycamore wood used for crating. When the principal means of local transmission was learned, epidemics in the eastern United States were halted by sanitary and hygienic procedures. Destruction of diseased trees, disinfestation of equipment after work on any sycamore or plane tree within the area of outbreak, and the addition of a fungicide to tree wound dressings became standard practice. Canker stain is now endemic or occurs in scattered outbreaks in both landscapes and forest stands. Native sycamore has been damaged less than London plane, although incidence rates up to 30% in some southern forest stands of sycamore have been reported.

The canker-stain pathogen is a host-specialized strain of *C. fimbriata* that is considered to be indigenous to the United States, where it occurs from New Jersey to Georgia, Louisiana, and Missouri. By most accounts it is capable of infecting only species of *Platanus*. London plane, Oriental plane, and sycamore are affected in nature. In one test, however, an isolate from diseased sycamore caused cankers in trembling aspen.

Cankers develop on trunks or branches, usually with only slight callus formation at the edges, and follow the grain of underlying wood, sometimes in a helix. Cankers in young, smooth bark appear as elongate dark areas. Those beneath scaly outer bark remain externally invisible for a time. The bark remains in place for a few years, so the canker is apparent only as a flattened or slightly sunken lens-shaped area or long strip. Diseased inner bark and cambial tissues turn nearly black, and reddish brown to blue-black discoloration develops in sapwood. Dead, stained sectors, lens shaped in cross-section, extend toward the center of the stem. By expansion and coalescence, these lesions in wood become wedge shaped as viewed in cross-section. In some cases stained sectors extend through the pith and expand toward the other side of the stem. The dark stain fades as secondary organisms, including wood-decay fungi, replace the pathogen in dead wood. In the South, dead or dying bark is often colonized by *Lasiodiplodia*

theobromae, the conidial state of *Botryosphaeria rhodina* (Plate 85), an opportunistic pathogen of stressed trees. The bark sloughs from old cankers, exposing decaying wood. Old cankers on plane trees appear more or less blackened and cracked. Diseased trees typically bear multiple coalescing cankers and decline during a period of several years. Leaves above the cankers may suddenly wither, but slow twig growth; sparse, stunted, and prematurely yellow foliage; and dieback usually develop first. Abundant epicormic sprouts often form below cankers, but these are killed as infection spreads down the stem.

Any fresh wound to living bark or sapwood is suitable for infection. In the North such wounds are suitable throughout the year except during December to mid-February. Symptoms may appear within 7 weeks after infection but are at first inconspicuous. The pathogen initially colonizes and stains phloem, cambium, or sapwood exposed by the wound and then spreads in the sapwood. Death of sapwood precedes death of overlying bark; thus stained streaks can be found in sapwood many centimeters above and below, and several millimeters to each side, of a canker. Radial spread occurs at rates up to 7.5 cm per year. In the south, cankers expand in width at an average rate of 9 cm per year (range 2.5–30 cm) and in length at rates up to 3 meters per year. Cankers as long as 20 meters have been noted. Pole-sized sycamores may die within 2 years after infection, but large mature sycamores or plane trees may persist several years before death. Disease progress is somewhat slower in the North, presumably because of lower average temperature. The pathogen grows relatively rapidly at temperatures of 15–30°C and most rapidly at temperatures near 25°C.

C. fimbriata f. sp. *platani* begins to sporulate in wounds within a few days after infection, before extensive colonization of bark or wood. Later it also produces spores within diseased tissues and on cut or broken surfaces of diseased wood or bark after further wounding. Asexual spores (endoconidia) are extruded from within conidiophores borne on the mycelium. Endoconidia are of two types, one thin-walled, colorless, and cylindric and the other thick-walled, light brown, and barrel shaped. The fungus also produces pale brown chlamydospores. The asexual state (in the form-genus *Chalara*) is followed or often accompanied by long-necked black perithecia. Perithecia develop within soft tissues or on a wound surface. When they are superficial they are readily visible with a hand lens. Their bodies are 120–255 μm in diameter, and their necks are up to approximately 1 mm long. The ascospores, shaped like hats with a gelatinous brim, extrude in a tiny blob of mucilaginous substance from the tip of the perithecial neck. Both ascospores and conidia are sticky and are readily transmitted by insects.

Mycelium and spore masses of this fungus and most others in the genus *Ceratocystis* emit fruity odors attractive to nitidulid beetles and other insects that act as vectors. Local transmission of spores by splashing or wind-driven water also occurs but is considered to be unimportant in comparison with transmission by humans and insects.

Canker-stain in sycamore (MS, Apr).

A. Sparse foliage and dieback in a diseased tree in contrast to healthy trees.

B. Flat to sunken cankers, some delimited by callus at margins. Growth of epicormic sprouts below cankers is typical.

C. Painted margins illustrate the vertical extent of a canker resulting from infection at ground level.

D, E. Portions of elongate cankers on pole-sized trees. Bark in the canker at left has been colonized by dark secondary fungi. Margins of the inconspicuous canker at right were exposed by removing superficial bark scales.

F. Cross-section of the trunk of a young tree with a typical darkly stained sector of wood extending from a canker face to the pith.

G. Part of the trunk of a mature tree half girdled by canker-stain. Wood long dead is decaying, and a transverse crack is evident in it. *Ceratocystis fimbriata* is active in dark brown and black discolored streaks at the canker margin.

H. Part of a transverse section above the canker shown in (G). Most of the sapwood is stained and dead as the result of vertical spread of the pathogen and secondary organisms. The canker will soon extend to the area where wood just beneath the bark is stained.

Plate 174

Ceratocystis canker of aspen. *Ceratocystis fimbriata,* the cause of Ceratocystis canker, occurs worldwide in tropical and temperate regions and infects diverse plant species. It causes cankers, usually distinctive for the dark stain in infected tissues, in aspen, cacao, coffee, *Gmelina arborea,* pimento, poplar, plane tree (including sycamore), rubber tree, and stone fruit trees (almond, apricot, peach, and prune). It is a significant pathogen of stone fruit trees in California. General characteristics of the fungus, its various spore stages, and modes of transmission are mentioned with Plate 174.

Ceratocystis canker of aspen occurs throughout the range of trembling aspen in the mountainous West and extends eastward to Quebec and Pennsylvania. The cankers originate at trunk wounds and at junctions of the trunk with small branches. They are similar to cankers caused on the same host by *Nectria galligena* (Plate 101). Indeed, the two diseases cannot be distinguished visually where both occur from Minnesota eastward. The cankers are perennial, sunken, and two to three times as long as they are wide. They are circular to oval when young but later tend to become diamond shaped to irregular in outline because of flaring dead bark at the sides. Old cankers contain concentrically arrayed ridges of dead callus. Killed bark eventually sloughs from the center of the canker. Viewed from the side, old cankers are concave, and the affected stem often has a crook at the level of the canker. Wood beneath the canker face and for a few centimeters above and below is darkly stained. Callus develops at the canker margin during the growing season, and *C. fimbriata* kills part or all of the new callus during the next dormant season. Amber to reddish brown fluid exudes from the canker margin and may cover the canker surface. This substance plus perithecia of various species of *Ceratocystis* and hyphae and spores of dark saprophytic fungi cause bark and exposed wood in the canker to become dark brown to black.

In the West, Ceratocystis cankers on aspen commonly enlarge for 30–40 years; specimens up to 78 years old have been found. On average, cankers on aspen in Colorado expand 1.3 cm in width and 2.8 cm in length per year. Individual cankers rarely girdle trees, but multiple coalescing cankers may do so, or diseased trees may break because of weakness at cankers. Cankers initiated at edges of large wounds such as those caused by logging equipment or falling trees tend to follow the outline of the wound.

Black perithecia of *C. fimbriata,* visible with a hand lens, can often be found in spring near the perimeter of the canker on bark killed 1 or more years earlier. Conidia and perithecia also form within 7–10 days on newly infected trunk wounds. Since several species of *Ceratocystis* occur in aspen cankers, however, specific identification requires expert examination.

C. fimbriata is capable of infecting and killing artificially inoculated aspen leaves and shoots but has not been found in association with leaf or shoot blight or twig dieback in nature.

Insects, especially sap beetles (Nitidulidae) in the genus *Epuraea,* and perhaps also a root-feeding beetle, *Colopterus truncatus,* and a rove beetle, *Nudobius corticalis,* are considered to be responsible for most natural transmission of *C. fimbriata* to aspen in Colorado. These insects become contaminated during visits to wounds and cankers where the pathogen is sporulating, and the fungus overwinters on or in adult insects in the ground as well as in cankers. Nitidulids also transmit *C. fimbriata* to stone fruit trees.

Sapstreak disease. Sapstreak is a systemic disease of sapwood of sugar maple and tulip tree caused by *Ceratocystis coerulescens.* It occurs in maple in forests and sugar bushes in scattered localities from Wisconsin to Vermont and in North Carolina and Tennessee. Tulip tree is also affected in the latter region. Diseased maples cease sap production and die, and logs and lumber from them are greatly devalued because of the stained wood.

C. coerulescens is widespread as a wood-staining organism in logs, stumps, and lumber of many kinds of trees, both angiosperms and gymnosperms, around the Northern Hemisphere. Strains that colonize conifers, however, differ physiologically from those colonizing hardwoods. Whether sapstreak of maple and tulip tree is caused by a strain with special pathogenic capability is unknown.

The first external symptoms of sapstreak in sugar maple are slow shoot growth and dwarfing and chlorosis of leaves on one or more major branches or throughout the crown. These symptoms intensify the next year, and dieback often begins. The entire tree dies within 2–4 years. The apparent cause of death is a massive water-soaked lesion that involves nearly the entire cross section of the lower trunk and major roots and may extend far up into branches at the time of death. This begins as dark green to black or sometimes reddish streaks that extend most rapidly along the grain and next most rapidly along the rays, producing a radiate pattern as viewed in cross-section. As streaks coalesce, sapwood beneath them toward the center of the tree turns yellow-brown and appears water-soaked. Dark streaks enveloped by the composite lesion lose color intensity, turning shades of brown or reddish brown against a yellow-green to yellow-brown background. Cambial death and the formation of elongate cankers occur as lesions approach the cambium. Stain patterns may also develop in yellow birch, but the disease in this species is little known. The fungus also colonizes and sporulates on the ends of freshly cut logs of beech, birch, maple, oak, and tulip tree.

If diseased wood is wounded or the tree cut, *C. coerulescens* may sporulate profusely, producing within several days a dark gray mat of mycelium, endoconidiophores, and black, long-necked perithecia on the moist wood surface. Two types of endoconidia (cylindric and barrel shaped) and the perithecia and ascospores are similar to those of *C. fimbriata* (Plate 174). Sap beetles are suspected to be vectors because they are involved with similar diseases: canker-stain of plane tree and oak wilt.

Infection is believed to occur primarily through wounds at the base of the trunk and on exposed roots. Most diseased trees occur adjacent to access roads where the probability of wounding is high, and nearly all such trees have wound scars on the lower trunk or exposed roots. Maples are most susceptible in spring; artificially inoculated saplings have died within 2 months after infection. Larger trees show symptoms 1–6 years after infection.

References for canker-stain: 387, 424, 569, 1256, 1445, 1595, 2054, 2098, 2099
References for other Ceratocystis cankers: 387, 686, 827, 828, 830–32, 980, 1212, 1559, 2054, 2100, 2132, 2211, 2237, 2269
References for sapstreak: 781, 1013, 1648, 2054

A, B, D, G. Sapstreak of sugar maple, caused by *Ceratocystis coerulescens.* A. Premature foliar color, defoliation, and dieback in the crowns of diseased trees. B. The scar of a wound that was possibly a site of infection on a surface root. D, G. Internal symptoms revealed by chopping into an exposed root at the base of the trunk. Water-soaked, brown-stained sapwood is dead. Dark greenish brown streaks adjacent to normal white sapwood are sites of current infection. The radiate pattern of discoloration results from growth of the parasite in rays (NY, Jul).

C, E, F. Ceratocystis canker of trembling aspen, caused by *C. fimbriata.* C. A canker approximately 3 years old on a young stem bleeds rust brown fluid. Blackened callus rolls nearly enclose the original site of infection and cause the stem to appear swollen. The lesion has expanded upward during the previous and current growing seasons as indicated by orange discoloration of the bark surface. E. A canker somewhat older than that in (C). The lesion is expanding upward (canker margin at top of photo), but large lateral callus rolls indicate inactivity at the sides. Infection apparently began at the branch stub. F. Part of an old targetlike canker with flaring bark at its sides (CO, Jun).

Plate 175

Oak Wilt (Plate 176)

Oak wilt is a systemic, lethal disease caused by *Ceratocystis faga-cearum* (Microascales, Ophiostomataceae). It occurs only in the United States in a region delimited by Minnesota, Texas, Pennsylvania, and South Carolina. About 20 species of oak and also Chinese chestnut have been found diseased. The oaks include black, blackjack, bur, chestnut, Durand, Lacey, southern live, West Texas live, overcup, pin, northern pin, post, northern red, southern red, Texas red, scarlet, shingle, turkey, water, and white. All oak species that have been tested are susceptible to some extent. Species in the red oak group are so susceptible that the oak wilt fungus has been proposed and demonstrated effective for use as a selective herbicide where these oaks occupy land desired for growth of more valuable tree species. Members of the white oak group (e.g., bur, overcup, post, and white oaks) are more or less resistant. Infections in these trees progress slowly and often are localized, resulting in remission of symptoms. Apple, chestnut, chinkapin, and tanoak are also susceptible by artificial inoculation.

Symptoms vary with oak species and region. In the North, trees of the red oak group (e.g., black, red, pin, and scarlet oaks) typically die within 1 year, often less than a month after symptoms appear. Wilt is first evident at or near the top of the tree. Leaves turn dull green, bronze, or tan, beginning along the tips and edges, often with abrupt transition from dying to green tissue. Or leaves may droop, curl lengthwise, and wilt. Yellow to brown color sometimes develops along the veins. Leaves at branch ends begin to fall soon after symptoms appear, and some leaves drop while they are green. Discoloration and defoliation progress through the crown within several days to several weeks, and twigs and branches die. Usually a few tan leaves cling throughout winter to killed trees of species that normally drop all their leaves. Brown streaks develop in outer sapwood throughout the tree coincident with foliar symptoms, but this symptom is sometimes hard to see when present only in the outermost sheath of sapwood. The streaks mark clusters of water-conducting vessels colonized by the pathogen. Fungal colonies develop within a vessel after spores, moving passively with sap, lodge in pits and at end walls of vessel elements. Such vessels soon become plugged with tyloses and dark gummy substance, and adjacent parenchyma cells die and discolor.

Symptoms in trees of the white oak group are similar to those in red oaks but are less dramatic, progress relatively slowly, and are sometimes confined to one small branch. They may resemble normal autumn color changes.

On the southern fringes of the range of oak wilt, symptoms and the progress of disease differ from the description above. Leaves of live oak in Texas become chlorotic or bronze, often with yellow to brown color along the leaf veins and necrosis of the leaf tips. Diseased trees may defoliate and die quickly but commonly survive several years, displaying progressive dieback of twigs and branches and producing adventitious sprouts with small leaves on trunks and large limbs. Some trees enter remission. Similar slow, progressive dieback or remission occurs in turkey oak in South Carolina.

At the time of wilting (late spring to late summer), *C. fagacearum* is present throughout the symptomatic part of the tree and in other parts as well. After leaves die, the pathogen also often dies in twigs and small branches, apparently because of heating (the fungus is intolerant of temperatures above 32°C), drying, and competition from secondary fungi. But *C. fagacearum* may remain alive until the next year in the trunk and up to 4 years in roots, initially growing both toward inner sapwood and outward to the cambial region and inner bark.

Beginning 2–3 months after defoliation (in northern areas but not commonly in the South), *C. fagacearum* may form opposing mats of gray mycelium with raised black centers on the surface of the wood and the inner surface of the bark of the dead trunk. Mats are elliptical, 2.5–20 × 1–10 cm. Cylindrical endoconidia of the asexual stage (form-genus *Chalara*) are produced on the mats. If a tree wilts in late summer, sporulation is delayed until the next spring. If two compatible strains of the fungus are present, long-necked black perithecia also form and produce colorless, elongate-ellipsoid ascospores. The thick ridges of fungal tissue in the centers of the mats, called pressure pads, cause the overlying bark to crack open, allowing access to the mat by insects. Mat formation is suppressed in dry years. Mats are uncommon on species in the white oak group.

Oak bark beetles (Scolytidae) and sap beetles (Nitidulidae) transmit the oak wilt fungus. The oak bark beetles, *Pseudopityophthorus minutissimus* and *P. pruinosus,* breed in recently killed oaks throughout the range of oak wilt. After egg laying, adults that emerge from trees killed by oak wilt are often contaminated with spores of *C. fagacearum*. These insects fly to healthy oaks and feed in cavities that they excavate in twigs, thereby transmitting the pathogen. The next generation of adults, after emerging from the brood tree late in the same season or the next spring, may transmit the pathogen in like manner.

Transmission by sap beetles is significant in the northern part of the range of oak wilt. This occurs primarily in May–June. Species of *Carpophilus, Colopterus, Cryptarcha, Epuraea,* and *Glischrochilus* visit mats of *C. fagacearum* on trunks and limbs of killed trees, attracted there by a fruity odor that the fungus emits, and become contaminated with conidia and ascospores. Then they may transmit the fungus to healthy oaks during visits to fresh wounds. Pruning wounds are important sites of infection in residential trees where mats of *C. fagacearum* occur nearby. Wounds more than a few days old or covered with paint are not suitable, however. The hazard of oak wilt infection at pruning wounds may be minimized by pruning during the dormant season or during summer. Oak wilt hazard may also be reduced by girdling or felling and cutting newly diseased trees into firewood. This hastens drying and suppresses or prevents mat formation.

After the first tree in a new location becomes diseased, *C. fagacearum* may spread to adjacent oaks of the same subgenus (red/black oaks, *Erythrobalanus,* or white oaks, *Leucobalanus*) via naturally grafted roots. Adjacent trees wilt 1–6 years after infection of the original tree. Such tree-to-tree spread may progress 6–8 meters per year, although usually much less, and may result in very large foci of disease—tens of hectares in the upper Midwest and in Texas. Smaller foci prevail in the East. Trenching or chemically killing roots between diseased and healthy trees can prevent transmission through root grafts.

Oaks are most susceptible in spring as new wood is forming. The mechanism of killing is not fully understood, although it is clear that water conduction ceases in diseased xylem. Toxins and growth-regulating chemicals are apparently involved in abscission and death of leaves and in formation of abnormal xylem outside the infected portion of a xylem ring.

Isolates of *C. fagacearum* vary in virulence, but no evidence of host specificity has been reported. Geographic variation in frequency of mat formation is due to environment.

References: 47, 48, 368, 460, 595, 596, 627, 779, 893, 942, 978, 1133, 1134, 1273, 1333, 1596, 1924, 1962–64, 2039, 2054

A–C, F, G. Oak wilt in black oak. A. Recently killed trees (MN, Aug). B. Drooping, curled, drying leaves shortly before defoliation (MO, Jul). C. Brown streaks following the grain in sapwood of a branch (MN, Aug). F. A diseased leaf at the time of abscission (MN, Aug). G. Brown streaks in outer sapwood of a small trunk (MN, Aug).

D. Fading and browning of distal parts of diseased red oak leaves at the time of leaf casting (PA, Aug).

E. Fading and veinal browning in leaves of live oak (TX, Apr).

H–J. Mats and pressure pads produced by *Ceratocystis fagacearum* beneath the bark of a recently killed black oak. H. Cracks in bark indicate presence of pressure pads. I. Opposing pressure pads on bark and wood. J. Gray-black pressure pads on a gray mycelial mat where conidia are abundant (MN, Jul).

Plate 176

Dutch Elm Disease (Plate 177)

This famous disease is caused by *Ceratocystis ulmi,* which entered North America in elm logs from Europe before 1930. Carried by elm bark beetles (see Johnson and Lyon, 2nd ed., Plate 116) and in elm wood by people, the fungus spread within 5 decades from multiple points of introduction in eastern states and provinces to most places where elms grow in North America.

Infection usually begins when leaves reach full size, and symptoms develop rapidly during a period of 4–6 weeks. The primary symptoms are internal: death of xylem parenchyma cells, loss of water-conducting ability of xylem vessels, and browning of infected sapwood in narrow streaks that follow the grain. The streaks are closely associated with vessels in the innermost part of the youngest annual sheath of wood. Wood produced after vessel formation, closer to the bark, remains normally colored for a time. Infected wood in the current season's shoots may show diffuse brown color associated with the scattered arrangement of xylem vessels. Secondary symptoms are the familiar wilting and browning early in the season, yellowing and defoliation during summer, and branch death either during the growing season or during dormancy.

Symptoms usually arise on one branch and gradually involve the entire crown. All branches may die within a few weeks or during several years. In a tree that is somewhat resistant, branch death may be localized and intermittent as the tree sustains and resists repeated infections. In a branch or trunk that survives infection, normal sapwood forms outside the sheath containing the brown streaks. Trees that sustain repeated infections have brown streaks in more than one annual sheath of wood. The pathogen remains alive in these streaks for years. Symptoms are caused in part by toxins (three identified heretofore) secreted by the fungus. In western areas where summers are dry, symptoms of Dutch elm disease are often masked by those of heat stress and water shortage. Yellowing, premature defoliation, and branch dieback are common there, whether caused by *C. ulmi* or by other factors.

Structures of the pathogen are inconspicuous and vary with its life stage and environment. In xylem vessels in early stages of disease, *C. ulmi* exists mainly as spores that reproduce by budding. During saprophytic growth as mycelium, it produces conidia on simple conidiophores or on black-stalked synnemata (H) in galleries made by elm bark beetles. Synnemata consist of multiple conidiophores that produce spores in slimy drops up to 1 mm across. Sometimes perithecia develop among or following synnemata, but these are seldom noticed. Perithecia produce ascospores that, like the conidia, are in a mucilaginous drop and are thus well suited to adhere to bark beetles during dispersal. Diagnosis usually depends on isolation and identification of the fungus on a laboratory medium.

C. ulmi is distributed long distances in elm logs and firewood. Elm bark beetles transmit it locally and for distances up to several kilometers. The native elm bark beetle, *Hylurgopinus rufipes,* and the smaller European elm bark beetle, *Scolytus multistriatus,* are the vectors in North America. *S. multistriatus,* inadvertently introduced in elm logs from Europe before 1900, predominates in most areas. *H. rufipes* is more important in Canada and in the northern parts of border states. The insects are attracted to healthy elms by volatile chemicals produced by the trees. Beetles bore into the inner bark, cambial region, and outermost sapwood and, while feeding, deposit spores of *C. ulmi* that cause infection. *S. multistriatus* feeds in twig crotches; thus most infections begin in twigs. *H. rufipes* bores in the bark of branches and small trunks, initiating infections that involve major branches from the outset.

From the point of inoculation, the fungus moves upward and apparently also downward by two alternating modes: passive transport of spores in liquid within xylem vessels and growth as hyphae between vessels after spores germinate at points of vessel contact. *C. ulmi* regularly reaches the roots of large trees within the 1st season of infection. The fungus proliferates in the roots and then ascends the trunk in a wave of systemic infection that kills the tree or a major part of it. Where elms are spaced 10 meters or less from one another and their roots are grafted together, *C. ulmi* may move from one tree into the next through roots.

As parts of an elm die, whether from Dutch elm disease or other causes, elm bark beetles aggregate and use stems or branches 3 cm or more in diameter as breeding sites. Usually some of the arriving beetles are contaminated with spores of *C. ulmi.* These germinate in egg galleries that the beetles excavate in the bark, and the fungus grows saprophytically. If the breeding site is a branch or trunk killed by Dutch elm disease, *C. ulmi* is already on hand in the outer sapwood and cambial region when the insects arrive. The fungus produces spores in the beetles' galleries, and as the insects molt to adulthood, they become contaminated. On emergence from the site of breeding, they carry the fungus either to sites of feeding, where new infections begin, or to breeding sites where saprophytic growth of the fungus resumes.

The severity and pace of a Dutch elm disease epidemic depend upon the strain of *C. ulmi,* the reproductive success of the vectors, and the level of elm susceptibility. *C. ulmi* varies in aggressiveness, or the ability to colonize and damage elm trees. Aggressive strains, which have replaced nonaggressive ones in most North American and European localities, kill elms rapidly and thus favor vector populations and fast epidemic development. The origins of aggressive versus nonaggressive strains of *C. ulmi* are unclear.

Vector populations rise to high levels in early years of an epidemic and remain high until nearly all large elms in a locality are gone. Natural enemies of the vectors are not significantly effective while breeding sites are abundant. Reproduction of the smaller European elm bark beetle is limited by severe winters in parts of the northern United States and in Canada.

Elm species vary from extremely susceptible to nearly immune. Highly susceptible species include American (or white), Belgian, English, red, rock, September, European white, and winged elms. Those of intermediate susceptibility are cedar, European field (or smoothleaf), and wych (or Scots) elms. Chinese, Japanese, and Siberian elms as well as many lesser-known Asiatic species are resistant. Even within susceptible species, however, resistant individuals occur. In addition, tree breeders have developed several resistant clones suitable for temperate regions of North America. These include, but are not limited to, the cultivars 'Dynasty', 'Groeneveld', 'Homestead', 'Jacan', 'Pioneer', 'Regal', 'Sapporo Autumn Gold', 'Thompson', and 'Urban', all derived from Asiatic and European elm species. A group of American elms collectively called 'American Liberty', introduced in 1985, have less resistance than the cultivars listed above but may be useful in areas where elm yellows (Plate 188) does not occur. Elms resistant to *C. ulmi* are able to restrict the pathogen to small brown streaks near the point of introduction.

Control of Dutch elm disease in susceptible elms depends mainly upon denying elm bark beetles places to breed. This means a continuing program of detection, removal, and burial or destruction of all dying or recently dead elm wood in and near elms that are to be preserved. Once this basic requirement is met, many additional techniques can be applied for higher levels of protection. Elms resistant to *C. ulmi* do not require protective measures.

References: 144, 145, 240, 280, 302, 317, 626, 628, 725, 1048, 1395, 1435, 1606, 1788, 1804, 1909, 1966–68, 2054, 2129

Dutch elm disease in American elm.

A, B. Wilting, yellowing, browning, and dieback.

C. Wilting, shriveling, and browning, most common in late spring and early summer.

D. A diseased branch at midsummer, with droopy leaves that may either wilt and die or turn yellow and drop.

E. Streaks in outer sapwood of a small trunk, exposed by removing bark (A–E: NY, Jun–Jul).

F. A slant cut through a twig that became infected in its 4th year of growth. Discolored streaks appear as brown dots in the most recently formed wood.

G. A slant cut through a diseased shoot in its first season of growth, revealing diffuse brown color in xylem.

H. Magnified view of *Ceratocystis ulmi* on elm wood. Synnemata are up to 1.5 mm tall. Hyphae have a granular appearance because of the presence of conidiophores with clusters of colorless conidia (F–H: NY, Oct).

Plate 177

Black Stain Root Disease of Conifers (Plate 178)

Black stain root disease is a debilitating, usually lethal disorder characterized by extensive chocolate to black staining and death of sapwood in roots and the lower stem of various conifers. The causal fungus is *Verticicladiella wageneri* (Deuteromycotina, Hyphomycetes). The disease occurs in a region extending from New Mexico and southern California to Colorado, Montana, and British Columbia. It is found in diverse habitats from sea level to nearly 3000 meters elevation and from the semiarid Southwest to the moist coastal forests of the Northwest. In areas where summers are hot and dry, however, it is most common on cool, wet sites.

V. wageneri consists of at least three host-specialized strains: one that inhabits Jeffrey, lodgepole, and ponderosa pines; another in pinyons; and the third in Douglas-fir. Other hosts occasionally to rarely infected, regardless of the strain of the pathogen, include grand and white firs, mountain hemlock, and knobcone, sugar, eastern white, and western white pines.

Black stain root disease is now considered a major threat to Douglas-fir and ponderosa pine under management in western forests. Where the disease is severe in these species, forest managers must either abandon the site or devise a scheme for conversion to resistant or nonhost species. The disease typically occurs in enlarging foci, beginning in single trees and spreading at rates up to 7 meters (average 1 meter) per year until patches several hectares in extent are involved. Coalescing patches may devastate entire forest stands. Mortality is common in Douglas-fir up to 80 years old and in pine to 100 years old. Indirect losses may result from the buildup of bark beetles in diseased trees. The insects then attack and kill other trees that are stressed by various agents. Seedlings of Douglas-fir planted among the stumps of former diseased trees soon become infected and may begin to die within 1 year after planting.

External symptoms that may be apparent several years before death are reduced height growth, subnormal needle size, and premature abscission of old needles, so that the crown appears thin. The tree may produce an abnormally large crop of small cones and may exude resin along the lower trunk. Resin exudation varies with tree species and vigor, bark thickness, and insect attack. Foliage turns yellowish and, after one or more summers, brown. These symptoms are similar to those of yellow laminated root rot (Plates 156–157). Internal water stress and reduced transpiration and photosynthesis precede changes in foliar color. Bark beetles attacking the stressed tree are most frequently the direct cause of death.

The causal fungus infects roots and spreads up the stem to a height of 2 meters or more in pines and as high as 10–15 meters in Douglas-fir before tree death. While the tree is alive, the fungus is restricted to xylem tracheids, both axial and ray tracheids, growing from one to another through bordered pits. It follows the grain of the wood in both directions from the point of infection and also spreads tangentially but grows little radially. The result is characteristic bands of stained wood that are sometimes widest at the root collar and taper into the trunk and roots. Discolored bands in cross-section appear as arcs in one or more annual rings. In pines the stained wood may be impregnated with resin, and many tracheids near those colonized become occluded with tyloses. Cankers may form near the base of the tree where the parasite approaches the vascular cambium. Although *V. wageneri* apparently causes some of these cankers, others may be caused by *Armillaria*, which is often found in them. After invasion of roots by insects, usually after root death, *V. wageneri* sporulates in insect galleries.

The dark stain is due to a combination of large, amber to brown much-branched hyphae of the parasite, a thick amber sheath that surrounds each hypha and sometimes fills the lumen of the tracheid,

and amber to brown discoloration of tracheid walls. Cells at a distance from hyphae may also be discolored. Stain patterns caused by *V. wageneri* are unusual because the fungus colonizes only tracheids in bands within annual rings. Most other fungi that cause stain in conifer sapwood colonize ray parenchyma and grow radially across several rings, staining the wood in sectors that appear wedge shaped in cross-section (see Plate 184D).

As judged from the presence of stain in deep-lying rings of wood overlain by normal sapwood, some ponderosa pines may survive one attack only to be killed by another advance of the fungus from the roots. In Douglas-fir, however, stain may spread from a root into sapwood of various ages, so stain in deep-lying sapwood does not necessarily mean that a tree has survived previous infections.

Local spread of *V. wageneri* occurs via root grafts and contacts, by the transport of conidia by insects and perhaps water, and by hyphae growing as far as a few centimeters through soil from diseased roots to small healthy roots of adjacent trees. The fungus produces minute sticky conidia in droplets on brushlike branched conidiophores in insect galleries and perhaps elsewhere on diseased or dead roots or in soil adjacent to roots. Conidia are possibly transported by water as indicated by the spread of disease along watercourses. Insects that may act as vectors of *V. wageneri* in Douglas-fir forests include the weevils *Steremnius carinatus* and *Pissodes fasciatus* and the root-feeding bark beetle *Hylastes nigrinus*. *H. macer* is a probable vector in stands of ponderosa pine. These animals colonize declining roots of diseased trees beginning long before tree death, become contaminated with *V. wageneri*, and are subsequently attracted to other trees that have been injured or predisposed in various ways. New foci of disease are more common along forest roadsides and in recently logged areas than on undisturbed sites.

The host-specialized strains of *V. wageneri* differ in appearance in culture, temperature optima for growth, and tolerance of high temperatures. All strains grow most rapidly and are most infectious at temperatures of 15–21°C and are inhibited or killed above 21–24°C.

A perithecial state of *V. wageneri*, named *Ceratocystis wageneri* (Microascales, Ophiostomataceae), has been found in insect galleries in diseased ponderosa pine but not in other hosts. Relationships of the pinyon and Douglas-fir strains to *C. wageneri* are not yet established.

Several additional species of *Verticicladiella* and the related genus *Leptographium* inhabit and apparently parasitize roots of conifers in western North America. These include *V. abietina*, *V. penicillata*, *V. procera* (Plate 179), *V. serpens*, and *L. terebrantis*. All cause resinous lesions or dark staining in wounded roots under some circumstances, but none causes symptoms similar to or as severe as those caused by *V. wageneri*.

V. wageneri often precedes or occurs with other root rot pathogens, especially *Armillaria* (Plates 148–149). Trees killed by *V. wageneri* alone or in collaboration with other pathogens are soon colonized by blue-stain and sap rot fungi that obscure black stain symptoms.

References: 161, 369, 655, 656, 738, 739, 741, 773, 794, 919, 1001, 1297, 1848, 1851, 2086, 2200, 2221

A. Singleleaf pinyons killed by *Verticicladiella wageneri*, the cause of black stain root disease. Trees with tan foliage wilted in early summer; those devoid of foliage died in previous years (CA, Jul).

B, D, E, F. Black stain root disease in Douglas-fir. B, D. The trunk of a young tree showing resin exudation and, where bark was cut away, streaks of black stain ascending in sapwood from roots; the trunk is about 15 cm in diameter. Where the trunk is notched (D), fresh black stain can be seen in sapwood at lower center, and faded stain appears gray in older, dead wood (OR, Jul). E, F. Successively closer cross-sectional views of the trunk of a dying tree. Stain in crescent patterns following the annual rings is typical. The three youngest growth rings are very narrow. Stain shows in rings up to 7 years old. Infection near or at the cambium has caused a canker at top left, and growth abnormalities are evident as wavy lines in outermost wood (CA, Jul).

C. Dark stain, from which *V. penicillata* was isolated, in a root of lodgepole pine. This root was about 6 cm in diameter (ID, Jul).

Plate 178

Procera Root Disease of Conifers (Plate 179)

Verticicladiella procera (Deuteromycotina, Hyphomycetes) infects the inner bark and sapwood of roots and the stem base of various conifers, causing basal cankers and decline or wilt. The species most commonly affected is eastern white pine, in which the disease was called white pine root decline. Beginning in the 1950s, young eastern white pines in plantations in the Appalachian region and the Ohio River basin were observed to decline and die. Girdling cankers were present at the root collars of such trees, and a fungus identified as *Leptographium* sp. was shown to be responsible. At nearly the same time, resinous lesions on roots of red pine were under study in failing plantations on poorly drained sites in New York State, and the same species of *Leptographium* was found associated with the lesions. This fungus was described and named *V. procera* in 1962. It was reported then to occur in Canada and Europe as well as in the United States. Later it was found associated with root diseases and basal cankers of conifers in plantations and natural stands from New York to Florida and Minnesota and rarely in northwestern states.

Hosts in addition to eastern white pine include Austrian, jack, loblolly, sand, Scots, shortleaf, slash, and Virginia pines and Douglas-fir, at least one species of fir, and blue spruce. In most species and situations, *V. procera* does not seem to be an aggressive pathogen. Pathogenicity tests on various hosts in New York, Minnesota, and the West showed the fungus to be a weak parasite able to cause only small annual cankers or localized resin-soaking of sapwood in roots.

In the Ohio River basin and the Appalachian region from New York State southward, however, *V. procera* is a proven killer of eastern white pines and occasionally other pine species. Trees up to 20 years old are most commonly affected, and 20–50% incidence of mortality in young plantations has been reported. Landscape trees are also at risk. The first symptom visible from a distance is reduced height growth. This is apparent 1–3 years before other symptoms. In the year of death, buds open late, shoot growth is arrested, and the mature foliage fades, droops, and turns brown, usually in May–July. Some trees discolor before the time of bud break. Dead foliage clings to the twigs for a year or longer. A tree with fading foliage usually has a resinous, girdling canker extending from below soil level to a height of 10–45 cm on the butt. Bark in the canker and wood beneath the canker face are soaked with resin, and the bark is chocolate brown to black. The girdling canker is often subtended by smaller resinous cankers and dieback on main roots. In some cases, especially in the Appalachian region, blackish streaks are present in the wood of roots and the butt. *V. procera* can easily be isolated from diseased bark and from black-streaked wood but less readily and often not at all from resin-soaked wood.

Infection often begins at the root collar or near it on roots. Cankers enlarge until a major proportion of the root system is cut off or the tree is girdled at the butt. Resin exudation from the roots, root collar, and the base of the trunk is often visible for a few years before the onset of foliar symptoms.

Austrian and Scots pines may be infected and killed much as is eastern white pine, but the stiff needles of these species do not droop before death. Scots pine may be diseased for several years without showing external symptoms. In red pine, which is not killed by *V. procera*, sapwood beneath cankers on roots may be resin-soaked to the center of the root but is not darkly stained.

V. procera produces minute, colorless conidia in cream-colored drops of mucilaginous fluid on brushlike multiple branches atop stout, black-stalked conidiophores. Conidiophores develop in wounds and tunnels made by insects in cankers and on dead roots and also form quickly when the pathogen is isolated on laboratory media. The sticky spores are adapted for transmission by insects. Propagules presumed to be conidia are numerous in soil adjacent to diseased roots and are most abundant in late summer. The fungus overwinters, most likely as mycelium, in diseased trees and in stumps.

Various insects that cause damage to the roots and root collar of pines possibly act as vectors of *V. procera*. The fungus has been isolated from the pine root-collar weevil, the pales weevil, and the pitch-eating weevil (*Pachylobius picivorus*) as well as from several bark beetles (see Johnson and Lyon, 2nd ed., Plates 21 and 27). Insects that attack living trees and breed in roots or root collars of moribund trees are the best candidates for vectors.

V. procera seems to be an opportunist that causes most damage to plants stressed by adverse sites, root wounds, insects, polluted air, or other root pathogens. Several examples may be cited. Damage to red pine is limited to resin-soaked cankers on roots in poorly drained soils. Cankers may girdle small roots, but most root death and the gradual decline in health of diseased trees is considered to be caused primarily by poor site conditions. *V. procera* was isolated from cankers at the bases of eastern white pine saplings damaged by mound-building ants, but it was the least common and least virulent of three fungi shown to cause the cankers. The fungus has been found associated with wounds made during root pruning and cultivation of several species of pines in southeastern seed orchards but has not been credited with significant damage there. It was commonly isolated from eastern white pines showing symptoms of air pollutant injury in Virginia but was not detected in pollution-tolerant trees to which the symptomatic group was compared. *V. procera* has also been found associated with root-rotting fungi such as *Heterobasidion annosum* in plantations and natural stands of loblolly and sand pines. Whether it precedes the root rotters is unknown.

References: 161, 738, 889, 1001, 1089–91, 1806, 1947, 2214

Symptoms caused by *Verticicladiella procera* in eastern white pine.

A, B. Fading and browning foliage in spring. A. In a natural old-field stand (NY, Jun). B. In a Christmas tree plantation (VA, May).

C. Exuded resin and constriction indicating a girdling basal canker at the base of a diseased tree (NY, Jun).

D. Darkly stained sapwood in the stump of a tree killed by *V. procera* (VA, May).

E. An upper branch of a diseased tree, showing reduced height growth before other external symptoms. The current-year and 1-year-old branch segments together are not as long as the 2-year-old segment (NY, Sep).

F. Part of a tree that died the previous year, amid others apparently unaffected in a hedge. Dwarfed shoots on the dead tree indicate arrested growth early in the season of death (NY, Apr).

G, H. Resin-infiltrated lesions and dieback on roots of an affected tree (NY, Sep).

Plate 179

Persimmon Wilt (Plate 180)

Persimmon wilt is a lethal, intermittently devastating, systemic disease of common persimmon, caused by *Acremonium diospyri* (syn. *Cephalosporium diospyri*) (Deuteromycotina, Hyphomycetes). The disease was first noted when an epidemic broke out in the mid-1930s in central Tennessee and in Florida. The majority of large persimmon trees in the Tennessee area soon died. Eventually the disease was found from North Carolina to Florida, Texas, and Oklahoma. Now it is endemic throughout that region.

Symptoms may appear at any time during the growing season but usually develop beginning in late spring. At that time, leaves at branch tips wilt and shrivel or turn yellow and drop. These symptoms soon involve the entire plant. After the onset of hot weather the plant may stand with branches nearly bare except for partly grown fruit. Entire trees sometimes die within 2 months. Trees wilting at the end of 1 growing season fail to grow the next season or may produce a sparse complement of undersized, chlorotic leaves. Such trees either die during the summer or persist in a declining state for 1–2 more years. Dying or recently dead trees are soon colonized by wood-decay fungi, notably *Schizophyllum commune* (Plate 173).

Diagnostic internal symptoms are narrow dark brown to black streaks that follow the grain of the wood in roots, trunks, and branches. Sapwood of the youngest age class in a wilting plant is always streaked. Numerous streaks may nearly coalesce into bands that appear as arcs following the annual rings in cross-sections. Usually streaks also are apparent in several older annual rings. With time the streaks fade to gray.

In autumn of the year that a stem dies, the bark surface cracks and loosens, exposing irregularly shaped patches covered with pinkish orange masses of conidia. These patches arise at the position of either the cork cambium or the vascular cambium and are often many square centimeters in extent. Patches form beneath dark reddish blisters on smooth-barked stems but cannot be detected beneath old rough bark until the cork layers crack and fall away. Spore-bearing patches may form on branches but are most common near the base of the trunk. Clouds of spores rise when dry spore masses are disturbed.

The first infection in a persimmon grove occurs in wounds made by insects, animals such as cattle, or storm breakage. Thereafter, the pathogen spreads among adjacent trees through root grafts as well as by airborne spores. Insects that create suitable wounds include the twig girdler, *Oncideres cingulata* (see Johnson and Lyon, 2nd ed., Plate 124), and a powder post beetle, *Xylobiops basilaris*. Stubs left by the twig girdler have been found infected before the fungus could be isolated from other parts of the trees.

Trees are susceptible throughout the year, but the interval from infection to wilting varies. Trees artificially inoculated from late winter to early summer wilted and produced spore masses in the same year. Trees inoculated in summer and autumn wilted and produced spores a year later. Most natural infections by spores presumably occur in late summer and autumn; the most common timing of infection through root grafts is unknown.

Once in the xylem, the parasite produces spores that spread in water-conducting vessels throughout the tree. Parenchyma cells that surround colonized vessels are killed and colonized and become discolored brown. Vessels become plugged with tyloses, and both vessels and nearby intercellular spaces also become filled with dark, gumlike deposits that are responsible for the appearance of streaks in the wood. Mechanisms of symptom induction are unknown. Death of a branch or the entire plant occurs when the pathogen approaches the vascular cambium.

A. diospyri can readily be isolated in pure culture from discolored wood. It produces pinkish spore masses after 7–10 days or common media. It grows most rapidly at temperatures near 30°C but grows only slowly at 20°C.

Factors that triggered the persimmon wilt epidemics of the 1930s and 1940s remain unknown. The importance of wounds was made clear, however, when an epidemic occurred in Mississippi in 1970 among trees damaged the previous summer by a hurricane. The disease had not previously been prominent where the outbreak occurred.

Among persimmons and other species of *Diospyros* tested for reaction to *A. diospyri,* common persimmon was most susceptible and Texas or black persimmon somewhat less so. Japanese persimmon, the species usually cultivated for fruit, is highly resistant unless it is grafted on common persimmon rootstock. Ebony (*D. ebenaster*), naturalized in Mexico and Central America, is highly susceptible. The Oriental species, *D. lotus,* is resistant. Inoculations to *D. discolor, D. montana,* and *D. rosei,* native to the Philippines, India, and Mexico, respectively, resulted in no symptoms.

Common persimmon has both friends and foes, and persimmon wilt has been important to both. The tree is useful for ornament. Its fruit is eaten or made into preserves, and the wood is used for golf club heads and specialty furniture items. On the other hand, persimmon is an aggressive colonist of open land. Persimmon groves encroaching on pastureland in Oklahoma and Arkansas prompted farmers to attempt to spread the wilt by introducing wood from diseased groves to healthy ones. Plant pathologists showed the farmers how to inoculate unwanted persimmons with spore suspensions of the pathogen. The procedure was simple: squirt the spore suspension into axe frills. If half the trees in a grove were thus treated in spring at about the time of bud burst, all or nearly all the trees would eventually die because the pathogen could spread from tree to tree via root grafts. Entire groves could be killed within 3 years in this way.

A. diospyri under the name *Cephalosporium diospyri* was reported to cause wilt diseases of live oak and elms as well as persimmon, but this information has been discredited. The decline and dieback of live oak once attributed to *C. diospyri* in Texas is caused primarily by the oak wilt fungus, *Ceratocystis fagacearum* (Plate 176). The so-called *Cephalosporium* isolated from live oaks was proven to be nonpathogenic and is now classified as a strain of *Phialophora parasitica.* The elm fungus, originally designated *Cephalosporium* and now known as *Dothiorella ulmi,* does cause a wilt and dieback disease in elms (symptoms similar to Dutch elm disease and Verticillium wilt) but is not identical to *A. diospyri. D. ulmi* produces pycnidia on killed branches and in cultures on laboratory media, but the persimmon pathogen does not.

References: 416, 417, 419, 614, 662, 753, 1134, 2076, 2204, 2205

Persimmon wilt in common persimmon.

A. A grove of diseased trees (OK, Jul).

B. Wilting and yellowing leaves on a small, systemically infected tree (MS, Jul).

C. Suppressed twig growth and stunted leaves with uprolled edges cause foliage on a systemically infected tree to appear sparse. Yellowing and dieback are beginning.

D, E. Gray to dark brown or black streaks in outer sapwood are diagnostic. The orange line between bark and wood in (D) is the cambial zone, not a symptom of disease.

F. Shriveled brown leaves that wilted in late spring cling to dead twigs. Leaves becoming symptomatic at midsummer turn yellow and drop, leaving bare twigs.

G. End view of the stem shown in (D); discoloration caused by *Acremonium diospyri* is visible in several annual rings of wood (C–G: OK, Jul).

Plate 180

Verticillium Wilt (Plates 181–182)

The soilborne fungi *Verticillium dahliae* and *V. albo-atrum* (Deuteromycotina, Hyphomycetes) invade the xylem and cause death or disfigurement of plants belonging to hundreds of species, both woody and herbaceous, around the world. These fungi are significant pathogens of crop and landscape plants, but only two reports exist of Verticillium wilt in North American forests: *V. albo-atrum* was isolated from wilting seedlings of tulip tree in Delaware and from wilting ceanothus in California. *V. dahliae* is the species that most commonly attacks trees and shrubs throughout the United States and southern Canada. It is distributed widely with plants and plant parts that, although infected, may not show symptoms.

Woody plants or plant groups with species susceptible to *Verticillium* include ash, avocado, azalea, barberry, boxwood, brambles, buckeye, camphor tree, carob, carrotwood, catalpa, ceanothus, cherry and other stone fruit trees, Kentucky coffee tree, cork tree, creosote bush, currant and gooseberry, daphne, elder, elm, erigonum, weeping fig, flannelbush, golden-rain tree, grapevine, guayule, heath, hebe, hibiscus, honeysuckle, hop seed bush, horse-chestnut, India-hawthorn, jasmine, lilac, black locust, magnolia, maple, nandina, olive, Russian olive, osage orange, osmanthus, Japanese pagoda tree, peony, pepper tree, persimmon, photinia, pistache, privet, rabbitbrush, redbud, rock-rose, rose, sage brush, salt bush, sassafras, serviceberry, smoke tree, spirea, sumac, tree-of-heaven, tulip tree, tupelo, viburnum, wiegela, winter fat, and yellowwood.

Woody plants and plant groups resistant or immune to *Verticillium* as indicated by testing or practical experience include all gymnosperms (cypress, fir, gingko, juniper, larch, pine, spruce, and yew), all monocots (bamboo, palm, and palmetto), cacti, and the following dicots: apple and crabapple, mountain ash, beech, birch, boxwood, butternut, ceanothus, chestnut, citrus, dogwood, eucalyptus, firethorn, sweet gum, hackberry, hawthorn, certain species of hebe (*Hebe anonda, H. ×franciscana, H. menziesii,* and *H. salicifolia*), hickory, holly, katsura tree, California laurel, linden, honey locust, manzanita, mulberry, oak, oleander, pawpaw, pear, pecan, plane tree and sycamore, poplar, quince and flowering quince, rhododendron, certain species of rock-rose (*Cistus ×hybridus, C. incanus* subsp. *tauricus,* and *C. salvifolius*), sugarberry, walnut, willow, and Japanese zelkova. Some plant groups above have been listed with both susceptible and resistant plants because some genera include both susceptible and resistant species and because resistance may vary from one region to another and according to virulence of the strains of *Verticillum* where the plants grow. Rare reports of Verticillium wilt in some plants listed as resistant—pin oak, for example—do not offset their general usefulness as resistant plants.

Verticillium wilt has acute and chronic phases. Acute symptoms include curling, drying, or abnormal red or yellow color of leaves or areas between leaf veins; defoliation; wilting; dieback; and death. These symptoms, which indicate infection of the current season's sapwood, may be restricted to one branch or may involve an entire plant. Symptoms appear from early summer into autumn, depending upon the type of plant and the geographic region. Plants may die suddenly, but they more often show progressive or intermittent symptoms. Yellowing and defoliation often progress upward. Plants may have acute symptoms in consecutive growing seasons or may skip 1 to several years. In the interim, depending on the severity of infection, the plant may appear normal, or one or more branches may have chronic symptoms: slow growth, sparse foliage, stunted leaves and twigs, leaf scorch, abnormally heavy seed crops, and dieback. Chronic symptoms indicate stress caused by the previous death of sapwood; infection is confined to wood 1 year or more old. Acute and chronic symptoms often occur simultaneously on one plant.

Infected sapwood may have dark streaks or bands that follow the grain, but this discoloration may be absent during the earliest stages of infection, and it does not develop in some susceptible species such as olive. Color of the streaks varies with the type of plant from light tan in ash trees to dark reddish brown in black locust, yellowish brown in cherry and smoke tree, and greenish to nearly black in maples. Hyphae are microscopically visible in vessels of the discolored wood.

Infection begins in roots. When roots in infested soil grow close to or into contact with survival structures of *Verticillium* (microsclerotia or resting hyphae, discussed later), these structures germinate and hyphae penetrate intact feeder roots. The fungus also enters wounds.

Hyphae grow in the root cortex and, in susceptible species, enter the stele. Hyphal growth may continue there, or more important, the fungus may produce spores that move upward passively in the sap stream. Mycelium of *Verticillium* grows only a few millimeters per day, but spores in the sap stream quickly traverse the length of a xylem vessel, a few centimeters to over 5 meters, depending upon the kind of plant and the location within it. Where the spores lodge and germinate, new hyphae grow and intensify the infection. By repetition of this process, the fungus can spread within large plants. Necrosis develops at loci of spore germination and hyphal growth. Expansion and coalescence of necrotic tissue give rise to the streaks and bands of discolored sapwood. Dead regions first appear slightly discolored and watersoaked, then assume a darker color characteristic of the plant species. Seen in cross or slant cuts, the necrotic tissue appears as arcs that follow growth rings. Stems infected more than 1 year may show these symptoms in more than one growth ring. In maples, *Verticillium* can be isolated from dark streaks or arcs in wood up to several years old. Wood beneath the dark streaks toward the center of the stem or root also dies but apparently not from direct attack by the pathogen. This centrally located dead wood is lighter colored than that recently invaded by *Verticillium,* and the pathogen cannot be isolated from it. (In species such as cherry or smoke tree that produce colored heartwood, that tissue is distinct from discolorations caused by the pathogen.)

The upward spread of *Verticillium* in xylem as described above has been studied in various herbaceous and woody plants, and the persistence of the pathogen has been studied in several woody species. But other aspects of colonization in trees are known mainly from research on maples. In these trees, *Verticillium* progresses around a growth ring by a combination of upward spread and tangential growth of hyphae; thus the cross-sectional extent of necrotic xylem increases with distance above its origin. Downward movement, however, is negligible. The fungus often grows outward to the vicinity of the vascular cambium and upward to bud traces. If the stem remains alive, the growth of *Verticillium* close to these meristematic regions allows continuous infection from the wood of 1 growing season to that of the next. In maples and tulip tree, elongate strips of the vascular cambium sometimes die, resulting in long cankers. *Nectria cinnabarina* (Plate 100) and *Cytospora* sp. (Plate 96) are common in the bark of these cankers in maples. If the pathogen fails to cross from 1 season's wood to the next, the result is remission of acute symptoms and compartmentalization of the diseased wood. The severity of chronic symptoms then depends upon the extent of damage in the old wood. Acute symptoms that recur after 1 or more years of remission indicate a new upward thrust of infection from the roots.

Verticillium wilt of maple caused by *Verticillium dahliae* (NY, Sep–Oct).

A. Diseased Norway maples. The tree at left shows acute symptoms: withering and death of foliage on a major branch. The tree at right shows chronic symptoms: slow growth, sparse foliage, and leaf scorch.

B. Symptoms on red maple leaves: dwarfing, chlorosis, yellowing, and scorch.

C. Dying branches in a young sugar maple, typical of symptoms in nurseries.

D. Symptoms on sugar maple leaves: dwarfing, mottling, scorch at tips and margins, water-soaking and then scorching of interveinal areas, and premature red coloration.

E. Withering of scattered leaves at the top of a young sugar maple. These branches remained alive but stunted.

F. Stunting, curling, and withering of leaves on a diseased red maple sprout.

G. Segments of diseased stems of sugar (top and center) and red maple saplings. Segments at top and bottom were from trees with acute symptoms. Segments at center were from a tree with chronic symptoms (stunting, scorch, etc.). All but the outermost sapwood was dead. Although the infection had been compartmentalized, the volume of normal sapwood was insufficient to sustain the tree.

Plate 181

Infection of the current season's wood can occur without causing foliar symptoms. A hidden infection of this sort may, however, result in branch dieback during the dormant season or failure of buds to open in the spring. Foliar symptoms usually precede dieback and bud failure, however.

Wood killed by *Verticillium* does not conduct water or nutrients to leaves. Thus noninfected twigs and leaves above necrotic xylem are subject to water stress, nutrient shortage, abrupt cessation of growth, and perhaps death. Wilting and death of all leaves on a branch indicate infection of the current season's wood near the branch base or in the trunk below. The infection may extend into twigs, buds, and leaf bases in a small plant or into sprouts low on the trunk and main branches of a large plant. Usually, however, the extremities of trees are not invaded.

Verticillium produces or causes the production of enzymes and toxic chemicals that kill plant cells at some distance from those directly invaded. Thus the fungus often cannot be isolated from the apex of a streak of discolored wood, and the symptoms on leaves suggest toxicity in addition to simple water and nutrient stress.

Isolates of *Verticillium* vary in pathogenic capability. Some cause severe disease, others mild. Some cause wilt in one plant species but not in another. In general, any isolate from a damaged plant is capable of causing some wilt in each of several other plant species and is likely to be most aggressive in the original host. Isolates of *V. albo-atrum* tend to cause more severe disease than do isolates of *V. dahliae* when these fungi are tested in plants of the same type.

The survival ability of *V. dahliae* is a key to its success. The fungus persists within plants and is frequently transmitted by using scions, buds, or rootstocks from diseased plants in grafting or budding. After diseased plant parts die, the fungus survives in the form of microsclerotia (Figure 17). These are black cellular aggregates 15–100 μm in diameter, formed by hyphae within diseased tissue. In some plants the fungus may move into leaves, produce microsclerotia, and be dispersed when leaves drop. When diseased roots or other plant parts die and decompose in soil, the microsclerotia remain. Their numbers rise to high levels in soils planted annually to susceptible crops. Many weeds are susceptible and thus are important for increase and dispersal of the pathogen. Microsclerotia have even been found in mature weed fruits, available there for dispersal by the same mechanisms that disperse seeds. Microsclerotia also form on and within the fine roots of many kinds of resistant plants that have no symptoms above ground. Thus a population of the fungus can persist indefinitely in the absence of susceptible plants. Accordingly, crop rotation is not effective for control. Because of the frequent occurrence of *Verticillium* in plants without external symptoms, nursery crops on land formerly used for susceptible vegetable or fruit crops are at risk, and plants from such nurseries may develop wilt after they are transplanted to landscapes.

Microsclerotia move with soil particles in water or wind, on mechanical equipment, or with animals, including humans. Earthworms, for example, move large quantities of soil locally. Microsclerotia pass through these animals without loss of viability. Microsclerotia are resistant to most environmental stresses, toxic chemicals, and microbial attack. They can survive in soil for years in the absence of plants. Only in warm, waterlogged soils do microsclerotia, like plant roots, die rapidly.

When stimulated by nutrients that leach from decaying organic matter or roots, microsclerotia germinate and produce mycelium and perhaps spores (conidia) capable of causing new infections. Each microsclerotium is capable of germinating several times. If stimulated to germinate when conditions favorable for growth are not sustained, the microsclerotium retains the ability to germinate upon the return of favorable conditions.

V. albo-atrum depends upon specialized dark-colored resting hyphae for survival. Apparently these hyphae have attributes and func-

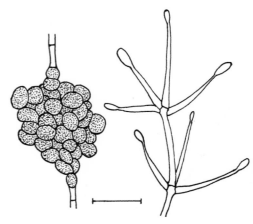

Figure 17. Microscopic structures of *Verticillium dahliae*: microsclerotium at left and conidiophore at right. The whorled arrangement of conidiophore branches typifies the genus *Verticillium*. Swollen branch tips are developing conidia. Scale bar = 20 μm.

tions similar to those of the microsclerotia of *V. dahliae*, although much less is known about them. Conidia of either species have only slight survival value. They cannot tolerate sunlight or drying, and they are rapidly destroyed by other microorganisms in soil.

The severity of Verticillium wilt in a given plant reflects not only the virulence of the strain of pathogen and the level of susceptibility of the host but also the influence of environmental factors. Some of these factors can be manipulated for disease management. For example, if other conditions are equal, generously watered plants are invaded less extensively than those under moderate to severe water stress. Stress induced by salt or transplanting has a similar adverse effect. Trees that become infected while in nurseries often develop symptoms during the first 1–2 years after transplanting. Nutrient balance is important also. Within limits suitable for good plant growth, the most resistant plant is that grown in moderately fertile soil in which the balance of major nutrients is tipped slightly toward high potassium and low nitrogen.

Verticillium species do not tolerate high temperatures. *V. dahliae* in diseased branches and trunks of orchard trees in California dies during the hot dry season. Recurrence of disease in these trees depends upon reinfection of roots during the cool winter and spring. In hot, sunny regions, *V. dahliae* can be partially controlled by covering the soil with a film of transparent plastic, but in most regions where Verticillium wilt is a problem, solar heat is insufficient for disease suppression.

Nematodes that attack plant roots may aggravate Verticillium wilt. Lesion nematodes and root knot nematodes (Plates 144–145) are noteworthy in this respect. The corollary is that, in soils where *Verticillium* species occur, control of nematode pests helps to suppress Verticillium wilt.

The most promising long-term approach to control of Verticillium wilt in landscapes and orchards is the use of resistant or tolerant plant types. Tolerant individuals or strains occur even within plant groups such as maples that are known for their susceptibility. Resistant or tolerant plants can be selected or bred to produce new cultivars or new rootstocks for susceptible cultivars.

References: 64, 66, 93, 220, 321, 322, 553, 740, 823, 932, 1203, 1244, 1349, 1350, 1451, 1475, 1658, 1746, 1807, 1843, 1844, 2196

Verticillium wilt of fragrant sumac, pistache, smoke tree, and cherry.

A. Wilt of fragrant sumac (NY, Jul).

B. Wilted sprouts on a scaffold limb of pistache (CA, Aug).

C, F, G. Internal symptoms and wilt in smoke tree. Brown streaks caused by *Verticillium dahliae* lie in outer sapwood between white inner bark and white inner sapwood. The yellowish to brown color of heartwood is normal (NY, Aug).

D, E. Brown streaks and central discoloration in young branches of sour cherry infected by *V. dahliae*. The center of the branch in (D) died after being surrounded by a sheath of infected outer sapwood. Although heartwood of cherry is colored somewhat like the central wood in (D), there is no basis for confusion, because heartwood does not form in such young branches as these. Brown pith, appearing as a broad line in the center of the split stem, is normal (NY, Sep).

Plate 182

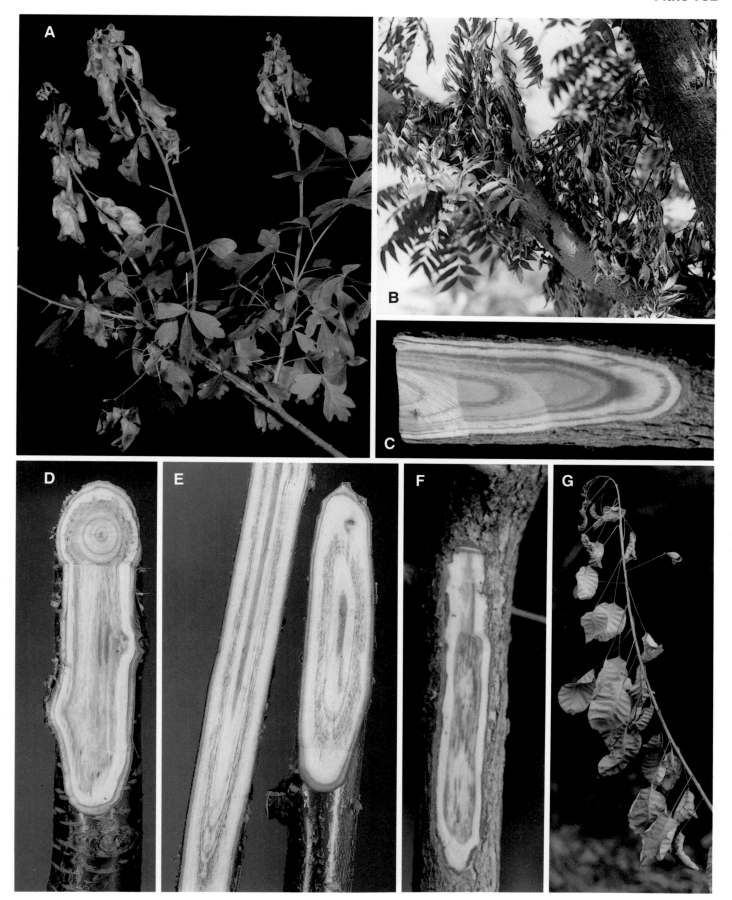

Mimosa Wilt (Plate 183)

Wilt caused by *Fusarium oxysporum* f. sp. *pernicosum* (Deuteromycotina, Hyphomycetes) is the most serious disease of mimosa, limiting the use of this ornamental tree in areas from New York to Florida, Louisiana, and Arkansas. The disease was first noticed around 1930 when an epidemic began in North Carolina. Its origin is unknown, but similar diseases had been reported in the Soviet Union and Japan.

The first external symptoms are chlorosis and drooping of leaves on one or more branches, usually in early summer to midsummer. Thereafter, many leaves may turn yellow and drop without wilting, or they may shrivel and fall. Symptoms spread throughout the tree. Entire trees sometimes die within a month after first wilting, but most die branch by branch over several months. Gum may ooze from tiny cracks that open in the bark of trees with advanced symptoms. Adventitious sprouts often develop along the trunk but are killed later by frost. Trees infected before winter sometimes produce dwarfed chlorotic leaves before dying the next summer. Most diseased trees die, at least to ground level, within a year after first wilting. Parts of the root system may survive longer, however, and sprouts may grow from the root collar.

White froth (Plate 78), often with the consistency of a gel and with a sweet fermentative odor, occasionally exudes from cracks in bark on the trunk and major limbs. Whether the wilt pathogen, which we will refer to as *F. o. pernicosum,* or some secondary organism is responsible for this "alcoholic flux" is unknown.

Brown streaks develop and begin to coalesce in outermost sapwood beginning before leaves wilt. Streaks are most prominent, often coalescing to a brown arc or ring beneath the bark, in the roots and lower trunk on the side where branches first wilt. The streaks eventually extend into small branches but may not be present there when wilting begins.

Streaks in sapwood are evidence of infection that began in roots. The pathogen resides in soil as chlamydospores that germinate in the presence of host rootlets. It penetrates intact or wounded rootlets, and its hyphae grow into the xylem. There it produces spores (microconidia that measure 6–12 × 2–3.5 µm) that are carried upward in the sap stream. Spores lodge in pits and at vessel end walls, where new colonies grow and the fungus penetrates adjacent vessels and parenchyma cells. Systemic movement is accomplished by alternating active growth where spores lodge, and passive distribution of spores in the sap stream of successive vessels. Parenchyma cells around colonized xylem vessels and in nearby rays are soon killed and turn brown. A brown gummy substance, apparently secreted by parenchyma cells before their death and perhaps derived in part from degraded pectic substances in cell walls, appears in pit apertures, where spores may have germinated. The substance completely coats or occludes vessel lumens at the time of wilting. The brown color is macroscopically evident as streaks. Mycelium develops in vessels that have ceased to conduct sap, and hyphae then grow from cell to cell via pits, colonizing xylem parenchyma. After the diseased stem has defoliated, the fungus grows from wood into bark and eventually produces orange-pink sporodochia at lenticels on the surface. Masses of colorless macroconidia form on the sporodochia and are dispersed by water. Macroconidia typical of *F. o. pernicosum* are colorless, multicellular, and canoe shaped in outline, measuring 23–60 × 3–4.5 µm. Other fusaria, capable of causing diagnostic confusion but incapable of causing wilt, also sporulate on the bark of recently killed mimosa.

Mimosa wilt seemed to spread rapidly throughout the Southeast during the first several years after its discovery. Therefore humans were suspected to be vectors. Now the modes of rapid spread are known. The pathogen produces great numbers of macroconidia at lenticels on bark of killed trees beginning soon after tree death and continuing up to 2 years. Macroconidia are washed to the soil, where one or more cells per spore develop into chlamydospores with thick, brownish walls, adapted for survival. Soil containing chlamydospores may be transported by surface water, earth-moving equipment, or other agents. Long-distance transportation in seeds is also possible because diseased trees often produce a seed crop before death, and the pathogen does enter seeds. Seedlings grown from infested seed soon become diseased.

In 1945 a Fusarium wilt of *Albizzia procera* was reported from Puerto Rico, and the pathogen was considered to be *F. o. pernicosum.* Isolates from *A. procera,* however, would not infect common mimosa (*A. julibrissin*), and vice versa. This was the basis for distinguishing two races of *F. o. pernicosum:* Race 1 is pathogenic on *A. julibrissin* and Race 2 on *A. procera.*

The severity of mimosa wilt may be influenced by root knot nematodes, *Meloidogyne incognita* and *M. javanica* (Plate 144). In experiments, mimosa seedlings exposed to either of these nematodes and to *F. o. pernicosum* developed more severe wilt than did seedlings exposed only to the fungus. At first this difference seems irrelevant, since mimosa wilt is eventually lethal in either case. The practical implication, however, is that root knot nematodes plus *F. o. pernicosum* might cause damage to a mimosa cultivar that would resist the fungus alone. Fungus-nematode interactions that cause breakdown of resistance are known to occur in Fusarium wilt of tomato. The mimosa pathogen is able to infect roots of resistant trees that do not wilt; thus it is in position for systemic spread if some third agent compromises host resistance.

A search for wilt-resistant mimosa began soon after the disease was discovered. Seedlings from various sources were grown in soil infested with *F. o. pernicosum,* and the survivors were repeatedly challenged with the pathogen. Resistance of the survivors was corroborated, and two resistant clonal cultivars, 'Charlotte' and 'Tryon', were released. Unfortunately, their resistance broke down after several years. Another resistant cultivar, 'Union', was released in 1979.

F. o. pernicosum is but one of several formae speciales of *F. oxysporum* that cause wilt diseases. The pathogens, although morphologically indistinguishable, seem to be reproductively isolated and are thus the biological equivalents of species. They have no known sexual state, and to the extent that they have been tested, the formae speciales are vegetatively incompatible with one another. Thus transfer of genetic information among them must be limited or absent. Other formae speciales of *F. oxysporum* that cause wilt diseases of woody plants in North America include *albedinis* on Canary Island date palm, *callistephi* Race 3 (syn. f. sp. *rhois*) on sumac, *citri* on Mexican lime and some other citrus rootstocks, *hebe* on hebe, and *pyracanthae* on firethorn. Several more diseases in this group occur in woody hosts on other continents. Wilt-inducing strains of *F. oxysporum* are generally found in warm temperate and tropical regions.

References: 644, 645, 647, 689, 780, 1272, 1394, 1524, 1562, 1574, 2008, 2020, 2021

A, B. External symptoms of mimosa wilt: one-sided wilting, defoliation, and dieback (NY, Jun).

C, D. Internal symptoms: brown streaks in the outermost sapwood. The entire current-season ring of sapwood in the stem shown in (D) is discolored (MD, Sep).

C, E–G. Signs of *Fusarium oxysporum* f. sp. *pernicosum:* sporodochia and masses of conidia at lenticels and on adjacent bark. These are visible macroscopically, as shown in (C) and (F), but magnification is needed for interpretation. Sporodochia erupting from lenticels are orange pink, and masses of colorless conidia washed from the sporodochia appear as white deposits. Brown eruptions at left in (G) are normal lenticels.

Plate 183

Wilt of Conifers Caused by the Pine Wood Nematode (Plate 184)

The pine wood nematode, *Bursaphelenchus xylophilus* (syn. *B. lignicolus*), kills conifers, especially pines, of many species. The pest is unusual among plant pathogenic nematodes in that it is transmitted by insects, moves throughout the infected plant, often kills its plant host rapidly, and reproduces using either plant cells or fungal mycelium as a food source. It is endemic throughout the eastern half of the United States and occurs in scattered localities in western states, Manitoba, and Ontario. Scots pine is the host most commonly reported. Locally destructive outbreaks occur in forest and landscape plantings of introduced pine species in the midwestern United States. In Japan, to which the nematode may have been introduced from the United States, it has devastated forests of Japanese red and Japanese black pines.

In the United States and Canada, *B. xylophilus* has been found in many pines: Aleppo, Austrian, Japanese black, jack, loblolly, lodgepole, longleaf, Monterey, mugo, pitch, ponderosa, red, Japanese red, sand, Scots, shortleaf, slash, Swiss stone, Virginia, and eastern white. It has also been found in diseased Atlas cedar, deodar, balsam fir, eastern larch, European larch, blue spruce, and white spruce. Additional species have proved susceptible when artificially inoculated. Species that seem resistant are Douglas fir; Jeffrey, pitch, Virginia, and eastern white pines; and Colorado blue spruce. Populations of *B. xylophilus* from balsam fir and from pine have shown host specificity.

Diseased pines most commonly die in early summer to early autumn, singly or in small groups and without relationship to previous vigor. Early symptoms are often inconspicuous or are missed because death comes quickly. The first visible symptom may be arrested growth (C), fading green color, or slight yellowing. This may be evident on one or a few branches or may develop simultaneously on all. Trees often die so rapidly that brown needles cling to the twigs. In large trees, branch dying sometimes begins in the autumn of 1 year and resumes the next spring. Yellowing or wilting and browning begin soon after color begins to fade. When the disease develops slowly, the oldest needles turn yellow and fall before the remaining foliage turns brown. Resin-soaked wood in the trunk is an intermittent feature of pine wilt. In the true cedars, some branches die while others remain green. Larch foliage turns yellow and falls prematurely.

Before visible symptoms arise in resinous hosts, parenchyma cells lining resin ducts die, oleoresin pressure diminishes so that resin no longer flows at wounds, and transpiration slows or ceases. Symptoms in pine are caused in part by toxins produced during the interaction of the nematodes, associated bacteria, and sapwood. Bacteria alone, however, do not cause pine wilt. In experiments, small seedlings have been killed after inoculation with nematodes alone. Even large trees have died within 2 months after inoculation. Under cool, moist conditions the interval between inoculation and wilt is extended or wilt may not occur.

B. xylophilus is transported long distances in logs or wood chips and is transmitted by certain wood-boring beetles of the family Cerambycidae, the longhorned borers. In the United States, *Monochamus carolinensis* is an important vector, and *M. titillator* (the southern pine sawyer) has transmitted the pine wood nematode in experiments. Other cerambycid beetles found to carry *B. xylophilus* and in several cases to transmit nematodes into dead pine stems during oviposition include *Arthropalus rusticus obsoletus, Monochamus mutator, M. notatus, M. obtusus,* and *M. scutellatus* (white spotted sawyer).

Insects that could fulfill the vector role are those that breed in logs or killed trees and, as adults, feed upon healthy pines. *M. carolinensis,* for example, develops in pine stems dead less than 1 year, and the adult beetles become infested with *B. xylophilus* just before emerging from these breeding sites. Individual beetles may carry thousands of nematodes on their bodies and also within their bodies in trachea, especially beneath the elytra (wing covers). The beetles feed for a time on bark of healthy pine twigs before seeking a dying or dead pine in which to breed. They inadvertently deposit nematodes while feeding and also during oviposition. The nematodes, by killing trees, provide a place suitable for reproduction of the beetles.

The pine wood nematode has two modes of development, termed *propagative* and *dispersive*. During rapid reproduction (propagative mode), four larval stages functionally similar to one another precede the adult stage. When reproduction diminishes after the host plant dies, development switches to the dispersive mode, in which third- and fourth-stage larvae are distinctive. These are adapted to withstand starvation and, in the fourth, or dauerlarval, stage, drying. Third-stage larvae in the dispersive mode migrate to the pupal chambers of insect vectors, molt to dauerlarvae, and enter the bodies of new adult beetles just before the beetles emerge from the bark. The beetles introduce dauerlarvae into healthy pines via feeding injuries. There the larvae molt to adulthood and reproduce in the propagative mode. They move vertically and horizontally within the tree's bark and resin ducts, kill cells that line the ducts, and then may increase in numbers rapidly. Populations of 10^3 to 10^4 nematodes per gram of wood (dry weight basis) are the rule when foliar symptoms appear, and the nematodes are found throughout the trunk, branches, and large roots of pines. In larches, which have resin canals only in the bark, the nematodes are confined to this tissue.

B. xylophilus has enormous reproductive potential. It can complete its life cycle in 12 days at 15°C and in only 4–5 days at 25°C. The females, after insemination, each lay several dozen eggs during a period of 2–4 weeks before they die.

Trees dying as the result of nematode infection are attacked not only by longhorned beetles but also by bark beetles and other secondary invaders. These bring or permit the entry of wood-staining fungi (D; also Plate 169) and nematodes of both fungus-feeding and free-living types. The additional nematodes complicate diagnosis. In the early stages of symptom development, however, *B. xylophilus* is the only nematode consistently present. Another diagnostic complication is the frequent transmission (during beetle oviposition) of *B. xylophilus* to pines and other conifers already damaged by agents such as root-rotting or needle-blighting fungi, rusts, root-collar weevils, or shoot moths. Discovery of the pine wood nematode in dead trees or branches is no assurance that it has killed them.

Little is known of factors that favor or prevent epidemics of pine wilt. Merely putting *B. xylophilus* into susceptible plants does not assure lethal disease. Nematode populations within trees sometimes remain stable or decrease under cool, moist conditions, and the trees do not wilt. High temperatures, water stress, and attack by other pests and pathogens seem to favor increase of the nematode population and development of wilt.

References: 521, 530, 1047, 1200, 1205–8, 1406, 1423, 1946, 2215–17

A, B. Japanese black pines recently killed by the pine wood nematode, showing uniform (A) and variable development of symptoms among branches (NY, Aug).

C. A diseased branch of Scots pine showing arrested growth. Some branches on this tree were dead when the photo was made (MO, Jul).

D. A cross-section of the trunk of a recently killed Japanese black pine, showing uniform annual growth until the year of death, stain caused by secondary fungi, and sectors of resin-soaked wood (NY, Aug).

E. *Monochamus carolinensis,* a vector of the pine wood nematode, on a branch of Scots pine (MO, Apr).

Plate 184

Bacterial Wetwood and Slime Flux (Plate 185)

Wetwood, a water-soaked condition of wood, occurs in trunks, roots, and branches of many kinds of trees. In some trees, notably elms and poplars, bacteria are consistently associated with the condition and apparently cause it. Water-soaked wood with large numbers of bacteria (bacterial wetwood) is dead, usually discolored, and contains fatty acids that give it a sour or rancid odor.

Foliage in the tops of trees severely affected by bacterial wetwood sometimes wilts, and branches may die back. Usually, however, wetwood in landscape trees is unimportant except for the disfiguring appearance of vertical light or dark streaks where liquid seeps out of cracks or wounds and runs down the bark. In the wood, the sour liquid is colorless to brown and may contain up to 10^9 bacteria per milliliter. It darkens on exposure to air. On the plant surface, the liquid supports growth of many other kinds of bacteria, yeasts, and filamentous fungi that give it a slimy texture and an often fetid color and odor. This foul brew, which may bubble at a wound, is called slime flux or wetwood slime. Wetwood slime is distinct from the white or "alcoholic" flux that marks sites where microorganisms infect shallow wounds in the bark and cambial region (see Plate 78). Alcoholic flux is nearly colorless and acidic, often gives off a pleasant fermentative odor, and persists only a short time in summer.

Wetwood is most important in forest trees that are cut for lumber. Abnormal color and moisture cause lumber to be devalued. In hemlock and oak, affected wood tends to crack along or perpendicular to the growth rings. Poplar wetwood is inferior to normal wood in crushing stength and toughness. Such weakness is attributed to the enzymatic degradation by wetwood bacteria of part of the binding substance between cells. In the standing tree this is thought to promote cracks caused by bending action, differential expansion along temperature gradients, and freezing and expansion of water within the degraded part of the cell wall. The wet zones are more rigid than normal wood when frozen. Often a log with wetwood has no evident defect until lumber sawn from it cracks during kiln drying. Even when no cracks develop, wetwood dries only about one-third to one-half as fast as normal wood and requires twice as much energy.

In normal sapwood and heartwood of most trees, bacteria are scarce and fungi essentially absent while the wood is young. These organisms eventually enter through assorted wounds above and below soil line, even in young seedlings or cuttings. Wetwood, however, does not immediately or automatically follow bacterial entry. Typically it develops in wood several years old where organisms may have been present in small numbers, inactive during the years of greatest physiological activity of living wood cells.

Not all wetwood is caused by bacteria. Wet zones in pine wood are nearly devoid of organisms. In grand fir, wetwood may form before bacteria invade.

Bacterial wetwood in comparison with normal sapwood or heartwood has higher mineral content and specific gravity, less oxygen and more carbon dioxide, often contains methane, and has elevated gas pressure. It is usually malodorous and discolored and is often more alkaline by as much as 1 pH unit. Water content varies from near normal (e.g., in white fir) to twice the norm (in elm and cottonwood). In one analysis, liquid from elm wetwood contained 3, 8, and 11 times as much calcium, magnesium, and potassium, respectively, as sap of healthy sapwood. These differences together with elevated carbonate content are sufficient to account for elevated pH and for movement of water along an osmotic gradient from normal wood into the wetwood zone. Alternate (but less satisfactory) explanations for wetwood involve inward movement of water from external sources such as branch stubs wet by rain. Gas pressure up to 0.7 kg $\cdot$ cm^{-2} is common in wetwood. Extremes to 4.2 kg $\cdot$ cm^{-2} have been detected in elms. Internal pressure causes liquid to seep from wounds and cracks. The gas in bacterial wetwood typically consists of 45–60% methane, 0–7% oxygen (often none), 6–16% carbon dioxide, 23–34% nitrogen, and about 1% hydrogen. Air in normal wood contains about 19% oxygen, 75% nitrogen, and up to 5% carbon dioxide. Methane and hydrogen are produced by certain bacteria under anaerobic conditions and are never found in normal wood. The nearly anaerobic nature of wetwood prevents decay by fungi.

Wetwood liquid under pressure sometimes spreads to outer sapwood, where in elms and poplars it may cause gray-brown streaks and bands extending into small branches and twigs. These streaks are associated with scorch, wilting, yellowing, defoliation, and dieback that resemble the symptoms of Dutch elm disease in elms. Small diseased elms often wilt, and large ones may develop general dieback. Wetwood wilt and dieback of elms are most common in the Great Plains. In poplars, wetwood may develop in trees as young as 2 years. It is associated with branch death in large cottonwoods and with dieback and premature death of Lombardy poplars. Small trees may wilt and die suddenly, but most die back over a period of years.

The diverse bacteria in wetwood are able to tolerate low oxygen concentrations, and some are obligately anaerobic. Several species often inhabit a single tree. *Methanobacter arbophilicum*, which produces methane in poplars, and *Clostridium* species in oaks are anaerobes. Bacteria of the sort found in wetwood are common in soils and on plant surfaces. Soil is therefore thought to be a source of populations that build up within trees. Examples are *Corynebacterium humiferum* in poplars, a similar but unnamed organism in white fir, and the following in elms: *Bacillus megaterium*, *Enterobacter agglomerans* (syn. *Erwinia herbicola*), *E. cloacae* (syn. *Erwinia nimipressuralis*), *Pseudomonas fluorescens*, and *Klebsiella oxytoca*. Many wetwood bacteria remain unidentified.

Virtually all large elms and poplars have bacterial wetwood. Wetwood is also common in aspen, fir, hemlock, maples (including box elder), mulberry, oak, and white pine. Less frequently affected are apple, mountain ash, paper birch, butternut, western red cedar, dogwood, sweet gum, magnolia, mesquite, Russian olive, pines native to western North America, redbud, sycamore, tulip tree, and walnut. Many other kinds of trees are affected occasionally.

No controls for wetwood are known. Slime flux can be alleviated for cosmetic purposes by installing drain tubes (plastic is satisfactory) that allow wetwood liquid to drop to the ground rather than run down the bark.

References: 328, 382, 415, 747, 930, 1371, 1372, 1439, 1646, 1670, 1741, 1909, 2007, 2064, 2102, 2243, 2244, 2276

A, B. Discolored streaks caused by wetwood slime on bark of sugar maple and horsetail casuarina, respectively. Brown liquid seeps from a crack in the maple trunk and from an old pruning wound on the horsetail casuarina (NY, May; FL, Mar).

C. Slime flux at a pruning wound on Siberian elm. Cambial dieback, as at left and lower right sides of the wound, is sometimes caused by toxic chemicals in the fluxing liquid (IL, Jun).

D. Cross-section of an American elm trunk about 20 years old. A brown core of wetwood occupied the center of the trunk. Discolored spokelike marks show the pattern of holes bored 2–3 years earlier near the level of the slice to allow injections for disease control. Wherever the drill bit entered wetwood, the discolored core began to expand and involve all wood present at the time of wounding. In this case, the injection wounds closed and became covered with normal sapwood. Often, however, injection wounds that penetrate wetwood become sites of flux and do not close normally (NY, Aug).

E. Cross-section of a 7-year-old eastern cottonwood trunk. A core of wetwood occupies portions of five annual layers of wood (MS, Apr).

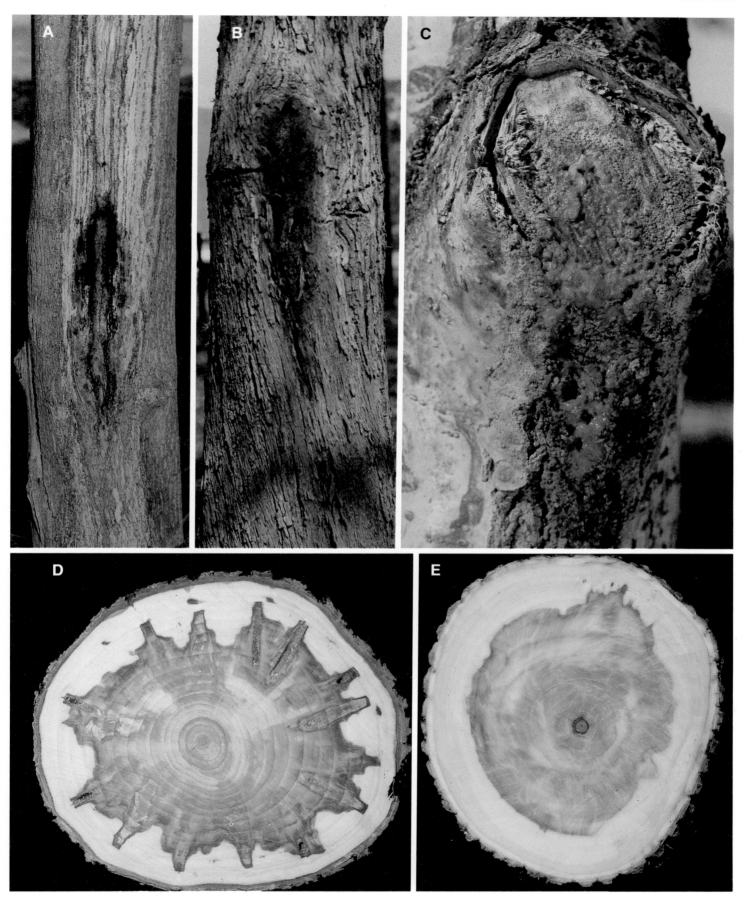

Plate 185

Bacterial Scorch Diseases (Plates 186–187)

Bacterial scorch refers to yellowing and browning of leaf tips and margins, caused by fastidious, xylem-inhabiting bacteria (FXIBs). Fastidious bacteria in the present context are those that in nature are found only within plants, where they act as obligate parasites, or within insect vectors and can be cultivated only on special media, if at all. Recognition of FXIBs as plant pathogens first occurred in the early 1970s. The diseases that they cause were previously believed to be caused by viruses. FXIBs infect plants systemically, moving and multiplying only in xylem. There they can be examined in detail by electron microscopy, and they can be extracted in xylem sap. They may occur in leaves, stems, and roots, not only within vessels, but also (less frequently) in intercellular spaces around vessels. They measure 0.25–0.4 × 1–4 μm (smaller than other plant pathogenic bacteria), and their walls tend to be rippled in outline. Strains known to cause disease in trees and shrubs are all Gram-negative rods and are transmitted by xylem-feeding leafhoppers (family Cicadellidae) and spittlebugs (Cercopidae). In 1986 a proposal was made that strains of FXIBs that cause several diseases of trees in the United States and elsewhere be united under the name *Xylemella fastidiosum*. We have adopted this name although, at the time of our writing, it had not yet gained official status in systematic bacteriology. *X. fastidiosum* causes almond leaf scorch, Pierce's disease of grapevine, phony disease of peach, plum leaf scald, and bacterial scorch of elm, red maple, mulberry, oak, and sycamore (Plate 187). Probably more diseases will be added to this list. Bacterial scorch diseases are known mainly in landscape and orchard trees. Their incidence and importance in forests remain to be learned.

Diseases caused by FXIBs are characterized by disruption of water transport, decline of vigor, marginal reddening or yellowing followed by browning of leaves, decreased fruit production, dieback, and usually eventual death. Leaves begin development more or less normally each year, and symptoms develop in late spring and summer. Disease occurs only after FXIBs are placed within xylem, for example by vectors or by grafting techniques that transfer xylem along with other tissues. Strains of FXIBs that infect trees have several vecto s and infect various nonwoody plant species, often without causing external symptoms. Bacterial scorch diseases are limited to regions of moderate winter temperature, but whether this reflects vector ranges or other factors is unknown.

Pierce's disease of grapevine and almond leaf scorch. The best understood disease caused by *Xylemella fastidiosum* is Pierce's disease of grapevine. Pierce's disease, widespread in the Southeast and known in California since the late 1800s, is one of the most important diseases of grapevine. Knowledge of it is a major basis for understanding other bacterial scorch diseases. The symptoms are as described for bacterial scorch diseases in general. In addition, leaves often turn distinctly red before or in association with marginal scorch. Lower leaves on shoots may have chlorotic mottling, and part or all of the vine may wilt and the leaves wither.

The interval between infection and appearance of foliar symptoms varies with plant size and time of year from several weeks to several months. Most infections are believed to occur in spring. Xylem vessels in diseased plants become plugged with tyloses and with gummy material containing bacteria. Plugging impedes water flow and causes water stress in leaves. Although tyloses form normally as old xylem vessels become nonfunctional, this process is accelerated in plants with Pierce's disease. The pathogen also produces one or more phytotoxic substances that may contribute to yellowing and scorch of leaves. Muscadine grapes are generally tolerant of the Pierce's disease

bacterium. Most grape cultivars used within the range of severe Pierce's disease in the South are derived from wild muscadine types.

Strains of *X. fastidiosum* that infect grapevine have a wide plant host range including species in 28 families, both monocots and dicots, most of which remain symptomless. Symptomless plants, including many grasses, are important as reservoirs of the pathogen near almond, grapevine, and other plants that are damaged by infection. Alfalfa, in which the Pierce's disease bacterium causes dwarf disease, is also a reservoir. In Florida *X. fastidiosum* occurs in beautyberry, blackberry, citrus, American elder, pepper vine, and woodbine. It is associated with the disease called blight, or young tree decline, of citrus, but it is not known to cause that disease.

X. fastidiosum has many vectors, both leafhoppers and spittle bugs. The most abundant vectors vary with locality and season. The pathogen does not circulate within vector bodies and probably does not multiply in them.

Almond leaf scorch, which appeared in epidemic proportions in central and southern California in the 1970s, is also caused by *X. fastidiosum*. This disease causes significant economic damage in orchards and aesthetic damage in landscapes where almond orchards have been supplanted by residential development. Several cultivars are susceptible. Marginal yellowing and necrosis of leaves appear annually after mid-June. Dead tissue is tan, often with zonate markings. A yellow band often separates green from brown areas, and tissue along the secondary veins may be yellowish. Symptoms appear initially on single branches, spreading and intensifying for 3–8 years until the entire canopy is affected and dieback begins. Scorch caused by buildup of salt in irrigated soil can mimic the symptoms of bacterial leaf scorch, but salt-injured leaves usually lack a yellow band between green and brown areas. Also, all leaves on a salt-injured shoot usually show scorch, whereas some leaves on a shoot with bacterial leaf scorch usually remain green. Scorched leaves remain on the tree until the normal time of defoliation in autumn.

Foliar symptoms are often sufficient for diagnosis from midsummer into autumn. At other times, a chemical test is useful. Diseased sapwood develops diffuse dark pink to reddish purple streaks within 10–30 minutes after immersion in acidified alcohol. The almond leaf scorch bacterium has been found in xylem of leaves and stems but not in roots.

Bacterial scorch of California buckeye. This disease occurs in central and northern California but has not been described in detail, and the identity of the causal agent is unreported. The disease is common in wild trees and those planted in landscapes, but it was long overlooked because during summer the symptoms caused by bacteria often become masked by other injuries (e.g., foliar injury by leafhoppers) and by the normal response of California buckeye to hot dry summer weather. These trees naturally drop their leaves and enter dormancy before autumn. Symptoms of bacterial leaf scorch become apparent soon after leaves unfold. Leaves on affected branches or entire trees are somewhat dwarfed and may appear crinkled. Tips and margins of leaflets develop purplish pigment that progresses through shades of reddish brown to typical brown scorch. Symptoms recur yearly, but trees remain alive.

A, B. Leaf scorch of California buckeye. A. Scorched, partially defoliated buckeye trees among normal oaks in summer. Trees with bacterial leaf scorch cannot be distinguished from those with scorch caused by environmental factors. B. Leaves with scorch caused by fastidious bacteria. The brown areas began as reddish discoloration (CA, Jul).

C–E. Almond leaf scorch. C, D. Symptoms in the cultivar 'Jordanolo'. All foliage of a tree with advanced disease (C) is yellow or scorched, and dieback has begun. A tree in an early stage of disease (D) shows yellow and scorched foliage mainly on one major branch. E. Close view of affected leaves on almond cultivar 'Long IXL'. Brown areas have a diffuse yellow border (CA, Jul).

F. Pierce's disease of grapevine. Leaves on a diseased shoot vary from nearly normal to severely scorched. Brown areas often have a diffuse yellow border (CA, Jul).

Plate 186

Leaf scorch of elm, oak, and sycamore. Elms, oaks, sycamore, and most recently red maple and red mulberry in eastern and southern states have been found with leaf scorch and dieback caused by strains of *Xylemella fastidiosum*. These diseases, except for that of elm, were unreported until 1980 or later. Little is known about them other than symptoms and the nature of the causal agents. Serological and physiological tests indicate that the causal FXIBs are closely related to one another, but electron microscopic observations of bacteria in diseased plants indicate that they may be nonidentical. Bacterial cells in each host were 0.3–0.4 μm wide with rippled walls. Lengths were 0.9–2.4 μm in elm, 1.0–2.0 μm in oak, and 1.0–1.8 μm in sycamore. Reciprocal pathogenicity tests have not been reported.

Knowledge of bacterial scorch diseases in shade trees began in the 1950s with research on a graft-transmissible disorder of elms. Symptoms could be transmitted if buds or twigs were used as inoculum. No transmission resulted when bark patches from diseased trees were grafted to healthy ones. This difference indicated the presence of a causal agent in xylem but not phloem. The bacterial nature of the pathogen was reported in 1980.

Bacterial scorch diseases of elm and sycamore occur from Maryland to Louisiana, and the sycamore disease extends into Texas. The elm disease is most common and prominent in American elm but has also been found in Scots and Siberian elms. The oak disease has been observed from southern New York to Georgia, with up to 50% incidence in some landscape plantings. Pin, red, and scarlet oaks are susceptible. Bacterial scorch of red mulberry, affecting trees of all ages from seedlings to mature specimens, occurs from near New York City to Louisiana.

Several symptoms and the general progression of symptoms are common to all of these diseases. Trees begin to show scorch in early summer to midsummer, and it increases during late summer, apparently enhanced by hot, dry weather. Scorch first appears on one branch or on branches in one part of the crown, and additional branches show symptoms in subsequent years. Scorch develops first on the oldest (basal) leaves on a shoot and progresses toward the tip, where some leaves may remain green. Diseased branches may produce leaves more or less normally the next spring, but scorch appears again in late June or July. Diseased trees lose vigor, and branches or entire trees with severe leaf scorch eventually die.

Leaf scorch on oak appears first on outer and upper branches and often involves all leaves on a shoot. Chlorosis and then faded green color develop at leaf tips and margins, and the discolored area then dries, fades to drab green, and turns brown. A reddish brown band may separate scorched tissue from green. Leaves remain attached until autumn. Foliar symptoms of bacterial scorch of oak are similar to those of oak wilt (Plate 176). Scorch, however, can be distinguished on the basis of symptoms that develop over several years and the retention of symptomatic leaves until autumn. Trees with oak wilt usually drop most of their leaves and die soon after the first symptoms appear. Outer sapwood in oaks with bacterial scorch never has dark streaks like those in oaks with oak wilt. Also, the two diseases tend to occur in different geographic areas: bacterial scorch is most common in coastal regions and oak wilt in the interior.

On sycamore, leaf margins and interveinal areas dry out and turn drab olive green, then tan, usually with a brown band between tan and green tissue. Severely scorched leaves curl upward from the edges but remain attached to the twigs until autumn. Diseased trees produce leaves later than normal in spring, and some produce abnormally small leaves. Progression of dieback is more rapid in the South than in the northern range of the disease. The reason may in part be the colonization of declining trees by opportunistic pathogens such as *Botryosphaeria rhodina* (Plate 85) in the South.

Leaves on diseased elms show undulating marginal zones of discolored tissue. Affected tissues first fade and dry, then turn brown. A yellow band usually separates brown from green tissue. Scorched leaves often curl up (sometimes down) at the edges and are likely to drop prematurely. Stem water potentials (measured as the pressure needed to equal the tension in water columns of the sap stream) are abnormally negative throughout the growing season, an indication that leaves are not receiving water as rapidly as it is lost through transpiration. Occasionally leaves on affected branches wilt and die. Starch storage in stem wood is suppressed. FXIBs have been extracted from stems of elms down to the soil line but not from elm roots.

FXIBs causing leaf scorch of elm, oak, and sycamore are assumed to be transmitted by leafhoppers and perhaps spittlebugs, which are vectors of similar diseases of fruit trees and grapevine. No facts are available, however. The time of year when infection occurs and the incubation period (interval from infection to display of symptoms) in nature are unknown. Sycamore seedlings developed leaf scorch 4 months after they were allowed to take up a suspension of FXIBs through severed roots. American elm seedlings grafted in spring with scions from diseased trees had no symptoms during the ensuing growing season but developed scorch 1 year later. On the other hand, after healthy scions were grafted on diseased branches, symptoms developed in midsummer as in natural shoots. Apparently extensive colonization of stem xylem precedes scorch symptoms in leaves.

The physiological cause of scorch symptoms is unknown. Water stress is implicated by abnormally negative xylem pressure in diseased elms and also because an amorphous matrix containing bacteria often lines tracheary elements or fills pits and the ends of tracheary elements in scorched leaves of elm, oak, and sycamore. Toxins of bacterial origin perhaps also cause symptoms, as reported for Pierce's disease of grapevine (Plate 186).

Bacterial leaf scorch predisposes elms to attack by elm bark beetles, which carry the Dutch elm disease fungus (Plate 177). Dutch elm disease in an elm population observed for several years in the District of Columbia was about 12 times as common in scorch-affected elms as in others. Over 40% of all cases of Dutch elm disease occurred in trees already affected by bacterial leaf scorch.

References for Pierce's disease of grapevine and almond leaf scorch: 457, 592, 615, 875–79, 1125, 1329, 1340, 1567, 1580, 1581, 2219
References for scorch of elm, mulberry, oak, and sycamore: 764, 1049, 1768, 1909, 2155, 2156

A, D, E. Bacterial scorch of red oak. A. Scorched leaves throughout the canopy of a diseased tree remain attached to twigs. D, E. Close views of symptomatic leaves. Leaf tips and margins dry and fade to drab light green, then turn brown (DC, Aug).

B, C. Bacterial scorch of sycamore. Diseased leaves fade to light green and then tan at the edges and between veins, often with a narrow brown band between tan and green tissue. Scorched areas show zonate markings, indicating intermittent enlargement. Edges of scorched leaves roll upward (DC, Aug).

F, G. Bacterial scorch of American elm. Tips and margins of leaves dry, fade to drab light green, then turn brown, often with darkest brown near the inner edge of the scorched area, and show a bright yellow band between brown and green tissues (DC, Aug).

Plate 187

Elm Yellows (Plate 188)

This plate introduces a series about systemic infections of phloem caused by prokaryotic microorganisms called mollicutes. Two broad groups of mollicutes that infect plants are known: those with helical, motile cells, called spiroplasmas, and those with nonhelical cells of variable shape, called mycoplasmalike organisms (MLOs). Some spiroplasmas can be grown in laboratory media, but the requirements of MLOs for growth in pure cultures are not yet known. Mycoplasmas are unicellular, pleomorphic microorganisms that have the general characteristics of bacteria but lack cell walls. In shape they range from spheres to branched filaments, and they appear to reproduce by fission. Details of their size, form, and structure can be seen only with the aid of an electron microscope (Figure 18). In strict usage, the term *mycoplasma* is reserved for certain mollicutes that can be cultivated and characterized apart from plant or animal hosts. Unicellular plant-infecting organisms that look like mycoplasmas but are not yet well characterized are called MLOs. Plant-pathogenic MLOs are obligate parasites that invade only phloem sieve cells. Sieve cells are organized into tubes that serve as conduits for translocation of photosynthetic products and growth regulators. MLOs multiply in sieve tubes and are translocated with photosynthetic products throughout the plant.

Elm yellows is a systemic, lethal disease formerly called phloem necrosis. It occurs in the eastern half of the United States and in southern Ontario, erupting in localized epidemics. It is caused by an unnamed MLO that is transmitted by leafhoppers. Highly susceptible plants die rapidly. Tolerant ones become stunted and may develop chlorosis and witches'-brooms. Diseased trees never recover. Natural infections are known only in five North American elm species: American, or white, elm (*Ulmus americana*); cedar elm (*U. crassifolia*); red, or slippery, elm (*U. rubra*); September elm (*U. serotina*); and winged elm (*U. alata*). Elm species of European or Asiatic origin and hybrids between these and native species are tolerant or immune. Such trees have survived in localities where epidemics killed all large native elms.

Symptoms usually develop in mid to late summer. External symptoms include yellowing, epinasty (drooping or downward bending of the petioles of turgid leaves), and premature casting of leaves, then death of branches. The sequence takes only a few weeks. All branches usually show symptoms at once, but occasionally in American elm and commonly in southern species, yellowing develops first in one branch system of a tree and spreads to other parts during 2 or more seasons. Bright yellow leaves may be interspersed with green ones on a single branch, but more often all leaves become yellowish green, then yellow. When these symptoms arise in late summer or early autumn, only the timing distinguishes them from normal leaf senescence. Diseased trees that survive a period of dormancy may open buds at the normal time but may grow only enough to produce chlorotic, dwarfed shoots and leaves that soon wilt or turn yellow and drop. Trees sometimes wilt and die quickly without prior external symptoms. Shriveled brown leaves may adhere to such trees for several weeks.

By the time foliar symptoms appear, root mortality and the degeneration of phloem in the roots and base of the tree are extensive. Fine roots die first, then successively larger ones succumb. Within roots and the basal parts of trunks and low branches, the innermost bark and the cambial zone change color from nearly white to yellow, then butterscotch or tan, sometimes with darker flecks, and finally dark brown. Then other tissues die. The surface of the wood may also be discolored light butterscotch where pigment diffuses from degenerating phloem. On exposure to air, the inner phloem and cambial region turn brown much more rapidly in diseased elms than in healthy ones.

Diseased, living phloem of American, cedar, September, and winged elms produces methyl salicylate (oil of wintergreen). This chemical occurs in elms only as a symptom of yellows. It can be detected by sniffing at the surface of freshly exposed inner bark or at the mouth of a vial enclosing a sample of inner bark. Diseased red elms produce a characteristic aroma somewhat like that of maple syrup. The odor emanates from newly killed bark and leaves.

Red elms usually show symptoms in 2 seasons before death. Witches'-brooms form during the final season. The brooms, which begin as stunted, branching shoots, are ordinarily a few centimeters long but sometimes grow much larger (Plate 192). Red × Siberian elm hybrids and Chinese elm also produce brooms but remain alive. The response of Chinese elm is known from experiments in which the

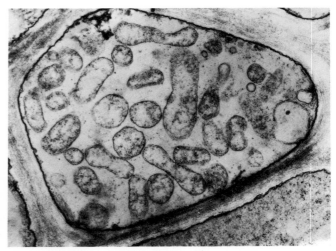

Figure 18. Cells of the mycoplasmalike organism (MLO) that causes elm yellows, viewed in a transverse section of an elm sieve cell. The widest MLO profiles here measure about 0.5 μm.

yellows agent was transmitted by grafting bark patches from diseased red elm.

Populations of the elm yellows agent become most dense in the petioles and stems of brooms. The MLOs are easier to find in tolerant species than in those that are rapidly killed. They occur both in living, discolored phloem and in apparently normal phloem of diseased trees.

Water movement in xylem and translocation in phloem diminish before any visible symptoms of yellows develop. Stomata remain partially or completely closed, beginning some weeks before discolored phloem can be found. How the MLOs cause these malfunctions and kill elm tissues is unknown.

One vector of the elm yellows agent is the white-banded elm leafhopper, *Scaphoideus luteolus*. Adults of this species are widespread in some areas from early summer until frost in autumn. Additional vectors are probable, moreover, because epidemics of elm yellows sometimes develop and persist where *S. luteolus* is scarce. Two possible vectors are the meadow spittlebug (see Johnson and Lyon, 2nd ed., Plate 202) and a leafhopper, *Allygus atomarius*, each of which after capture in the vicinity of naturally diseased trees has transmitted yellows to test seedlings.

The yellows agent overwinters in its plant hosts: in roots of American elm and perhaps also in brooms on hosts that produce them. Vectors can acquire the pathogen from elm phloem beginning after the first flush of shoot growth. Few, if any, trees develop symptoms in the year of inoculation. The incubation period is at least 3 months in very small trees and 9–10 months or more in large ones. Elm yellows spreads among closely spaced trees of the same species via root grafts.

Epidemics of elm yellows, although locally spectacular, do not spread rapidly. The edges of an area of outbreak in New York State, observed for 15 years, advanced at rates up to 6 km per year. The disease can be endemic for many years between flareups in a given locale. Spot outbreaks and single-tree occurrences beyond the main range of the disease presumably develop after long-distance transport of vectors by wind.

References: 91, 242, 243, 331, 633, 634, 1235, 1237, 1909, 1952

Symptoms of elm yellows in American elm (NY and PA, Aug).

A, B. Dying trees.

C. Chlorosis and epinasty on the branch at right, contrasted with normal foliage.

D. Brown discoloration of the innermost phloem and cambial region exposed by a slanting cut into a small stem.

E. Foliage on a low branch of the tree shown in (A).

F, H. Normal (above) and discolored inner phloem exposed by peeling bark from small stems. Wintergreen odor comes from discolored phloem.

G. Dead rootlets on a pencil-sized root. Rootlet mortality precedes foliar symptoms.

Plate 188

Lethal Yellowing of Palm (Plate 189)

Lethal yellowing of palms is one of the most explosively destructive tree diseases. It occurs in Florida, Texas, the islands of the northern Caribbean region, and west Africa. Until the mid-1950s it was apparently restricted to the Caribbean and a few west African locations. In 1955, coconut palms began dying at Key West, and in 1961 an epidemic broke out in eastern Jamaica, killing a million or more coconut palms there and spreading to other islands and the mainland within the next decade. Within 5 years after it appeared in mainland Florida in 1971, the malady had destroyed an estimated 300,000 of the 1.5 million coconut palms in southern Florida and threatened to make short work of the entire population. Streets, parks, beaches, and yards once shaded by these palms were left with stark, lifeless trunks or only stumps. By 1980 the disease was recognized in 25 additional species of palms in Florida and had been discovered in date palms in south Texas.

Most of the palms killed in Jamaica and also in Florida during the early phases of the epidemics were coconut palms of one cultivar, 'Jamaica Tall', which had been widely planted because of easy maintenance, rapid growth, and desirable horticultural characteristics. Christmas, date, and *Pritchardia* palms were also killed in large numbers. What triggered the spread and intensification of the disease is unknown. Affected palm species have origins around the globe, so lethal yellowing has become a worldwide threat in the tropical and subtropical belt.

Lethal yellowing is caused by an unnamed mollicute that has not been cultivated apart from plant hosts or insect vectors. The cause was unknown until 1972, when multiple reports established the constant association of mycoplasmalike organisms (MLOs) with typical symptoms. Progress toward understanding the disease was slow because MLOs were not recognized as plant pathogens until 1967, and palms could not successfully be grafted in attempts to transmit an infectious agent. No insect vector was confirmed until the late 1970s.

Susceptible palms in the United States include Arikury, cabada, Christmas (or Manila), coconut, date, Canary Island date, Senegal date, wild (or Sylvester) date, Englers, Chinese fan, Fiji fan, Burmese fishtail, cluster fishtail, Puerto Rican gaussia, Gebang (or Buri), Hildebrand's, Kona, Latan, Mazari, Montgomery, Palmyra, princess, seashore, Belmore sentry, spindle, Thurston, and windmill palms; also *Aiphanes lindeniana* and *Neodypsis decaryi*. All of these are introduced types; palms native to Florida have remained unaffected. A lethal decline similar to that in palms and apparently caused by MLOs occurs, however, in common screw pine (*Pandanus utilis*) in the area where lethal yellowing is severe.

The name *lethal yellowing* or *lethal decline* indicates progressive yellowing and browning of foliage leading to death of the palm. On average, 4–6 months elapse from the onset of overt symptoms until death, but sequences both faster and slower are common. Palms of all ages and sizes are affected. A very few cases of natural remission of symptoms have also been recorded.

Diseased palms show four general symptoms in variable order and combinations before death: nutfall, yellowing and/or browning of mature fronds, blackening of flower parts, and necrosis of the spear leaf. Premature nutfall is usually noticed first. Nuts in all stages of development fall within 1–4 weeks.

Foliar discoloration usually begins on mature fronds, progressing from the tip toward the base and often from the oldest toward the youngest fronds. The discoloration begins with loss of normal greenness and then, according to palm species and locality, either progresses through shades of yellow, orange yellow, or bronze to brown or is followed by withering and browning. Prominent yellowing sometimes begins on only one frond high in the green canopy. This is the flag-leaf symptom. Christmas palms and date palms are among those species in which fronds turn brown and wither. Brown fronds either fall or droop and hang from the dying crown.

Blackened flower parts, which soon shrivel, are usually noticeable in recently opened inflorescences. In advanced stages of disease, flower parts begin to degenerate while still enclosed in the spathe, and the spathe itself may discolor and split abnormally.

Spear-leaf necrosis is the death of a young upright leaf. In normal palms, young leaves elongate in an upright position above the canopy before unfolding. Necrosis that begins in the spear leaf spreads to nearby parts in the center of the crown and eventually involves the growing point. Moist dead tissues there are often invaded by secondary microorganisms that cause a slimy, putrid decay. Browning of mature fronds is usually complete by that time, and as their sheathing bases no longer strengthen the trunk, it collapses near the growing point. The remainder of the crown then topples, leaving a denuded pole. Roots begin to die as discoloration develops in the crown, but some roots remain alive until late stages of canopy symptoms.

Disturbed water relations are an early, consistent feature of lethal yellowing. Stomata close and transpiration slows greatly while leaves are still green.

The interval from infection until onset of overt symptoms is 3–15 months, depending upon the age and size of palm. During this period, MLOs become distributed systemically in the phloem, aggregating at sites of growth such as young leaves and inflorescences, root tips, and the stem apex just below the growing point. They do not enter developing nuts, however. Thus nuts from diseased plants, if capable of germinating, produce healthy seedlings.

A planthopper, *Myndus crudus* (Homoptera, Cixiidae), previously known as *Haplaxius crudus,* is apparently the principal vector in Florida, Jamaica, and south Texas. This insect feeds extensively on palms, and it breeds among the roots of grasses. In South Florida the breeding sites are in St. Augustine, Bahia, and Bermuda grasses, which are widely used for turf. Adults often migrate to palms at night and return to the grass by day. Whether these grasses serve as reservoirs of the palm pathogen is unknown.

Outbreaks of lethal yellowing typify the jump-spread pattern of disease occurrence and intensification associated with airborne vectors. One tree may become diseased at a distance of several kilometers from known sources of infection. Then new cases develop randomly near the first one.

Many types of palms resist the lethal yellowing agent or tolerate infection and can be planted where susceptible trees have died. Among coconut palms, 'Malayan Dwarf', 'Fiji Dwarf', 'Panama Tall', and various hybrids such as 'Maypan' are useful. In addition, areca, cabbage, pindo, pygmy date, MacArthur, paurotis, queen, royal, solitaire, and Washington palms are resistant or tolerant. In outbreak areas, injections of antibiotics have been used to preserve parts of the susceptible population while young resistant palms grow large enough to take their places.

References: 510, 511, 897, 1215, 1250–52, 1593, 1992

A. Symptoms of lethal yellowing in coconut palm, progressing from old to young fronds (FL, Apr).

B, C. Canary Island date palms, exhibiting withering and browning of foliage but little yellowing (TX, Apr).

D. A Christmas palm affected by lethal yellowing has a dying spear leaf (FL, Apr).

E. Necrosis of flower parts, a diagnostic symptom, in coconut palms (FL, Apr).

F. A normal inflorescence (left) produced by a diseased palm after injection of an antibiotic, in contrast with a dead inflorescence (right), affected before treatment (FL, Apr).

Plate 189

X-disease of Cherry and Peach (Plate 190)

X-disease is a lethal, economically important mycoplasmal disease of cherries and peach. The pathogen has not been cultivated apart from plants or insect vectors. Therefore it has not been characterized or named, and its relationship to other mycoplasmalike organisms (MLOs) that cause plant diseases is unknown. It occurs from coast to coast in the northern half of the United States and in Canada. Its economic importance is associated with loss of productivity of diseased fruit trees.

X-disease was unknown before the early 1930s, when it was detected in California and Connecticut. Soon thereafter it spread or was found throughout a large region of eastern and western United States and Canada. It became a major limiting factor in peach and cherry production in both East and West. Major outbreaks in orchards have occurred at intervals of several years, presumably reflecting trends in vector activity.

Known hosts in nature, in addition to chokecherry and peach, include mahaleb, sour, and sweet cherries; nectarine; and Japanese plum. In addition, the X-disease agent has been transmitted experimentally by grafts, dodder, or leafhoppers to many other species of *Prunus* and to diverse herbaceous plants in which it causes stunting, chlorosis, and various deformities.

Nearly all information about X-disease has come from research on the disease and the vectors in orchards. X-disease in chokecherry, however, is also common and prominent in many areas, including those lacking stone fruit orchards. The disease in chokecherry is important mainly because this plant acts as a reservoir from which leafhoppers carry the pathogen to orchard trees. In the Great Plains, however, chokecherry is planted in windbreaks, for wildlife habitat, and occasionally for ornament. X-disease limits its longevity and usefulness. We feature the disease in chokecherry.

Eastern and western manifestations of X-disease differ somewhat and may be separate diseases. In the West, moreover, three or more variations of X-disease occur, and each is characterized by different symptoms in various hosts. The variants are called peach yellow leafroll, Napa Valley X-disease, and Green Valley X-disease. The degree of relatedness of the pathogens is unknown.

The first symptom in chokecherry is abnormal leaf color. Beginning in early summer, leaves turn light green and then chlorotic, often with green tissue remaining along the veins. Many chlorotic leaves drop, but enough remain to call attention to the greenish yellow plant in contrast to surrounding green vegetation. Leaves nearest the shoot tip may be thicker than normal, somewhat leathery in texture, and crinkled. Chlorotic spots often develop, turn brown, and drop out of fully expanded leaves, leaving shot holes. Shot holes are not diagnostic, however, because chokecherry, like many other *Prunus* species, develops them in response to various stresses. Plants with leaves fully exposed to sun often turn reddish purple to bright red or yellow-red beginning in midsummer. Stomata begin to close, and transpiration is retarded beginning when the color of leaves changes. Transpiration through red leaves is negligible. Blossom clusters on diseased chokecherry are undersized, and many blossoms abort. Fruit are undersized, less colorful, and more pointed than normal, and their seeds do not form normally. In one study, less than 10% of seeds on diseased plants formed normally, versus 99% on healthy plants.

Many diseased chokecherry plants, instead of setting winter buds in the normal fashion, produce a tuft or rosette of stunted, abnormally upright leaves very close together at the tip of the shoot. These often remain green longer than leaves of normal form below. Buds in leaf axils of the rosette sometimes open and produce very short secondary shoots. Stems with rosetted shoot tips and red foliage often die during

winter. Entire diseased plants sometimes die after the 2nd season but more commonly die after 3 or more seasons with symptoms. Chokecherry propagates itself by root sprouts as well as seeds. Therefore all stems in a clump may be on a common root system, leading to disease in all of them once the first becomes infected. The roots of diseased plants often survive for several years, however, repeatedly sending up new sprouts.

Symptoms of X-disease on peach include formation of shotholes in leaves (often so severe that leaves become tattered), chlorosis, premature defoliation, reduced fruit yield, rosetting at twig tips, dieback, and eventual death. The yellow leafroll type of X-disease is characterized by chlorosis and upturned leaf edges. Symptoms of all types of X-disease usually appear first on one scaffold limb and spread throughout the tree. Affected cherry trees on mazzard rootstock grow poorly, bloom late, produce small deformed fruit with poor flavor, display dull or sometimes bronzed foliage, and may begin to die back after several years. Cherry on mahaleb rootstock may either decline slowly or wilt and die within 1 season.

Natural infection occurs during summer when the population of MLOs in aboveground parts of host plants is highest and vectors are abundant. Symptoms appear the year after infection.

MLOs occur only within phloem sieve tubes of diseased plants, and when examined by electron microscopy, they have variable form, from spherical to ovate to tubular, depending on the plane of the section. MLOs observed in peach with western X-disease were 0.12–0.36 μm in diameter and up to 5.4 μm long.

Several species of leafhoppers act as vectors of the X-disease agent. Some vectors such as *Paraphlepsius irroratus* occur in both East and West and have vector roles in more than one plant disease. *P. irroratus* is a probable vector of the ash yellows agent (Plate 191) in the East. Vector insects require feeding periods of several hours to days to acquire MLOs from infected plants and do not become capable of transmission until 4 or more weeks after acquisition. This delay reflects the time required for multiplication and circulation of MLOs into an insect's salivary gland.

X-disease spreads rapidly among chokecherry plants. In an experimental planting in Nebraska, disease incidence increased from 5% to over 80% within 5 years, and more than half the plants were dead within 8 years. In eastern orchards, most cases of X-disease in fruit trees seem to result from transmission of MLOs from chokecherry. In the West, however, MLOs are carried from peach to peach and from cherry to cherry. The peach yellow leafroll type of X-disease is often most severe adjacent to pear orchards where pear decline (Plate 193) occurs.

In California, the citrus stubborn agent, *Spiroplasma citri*, occasionally infects X-diseased cherry and was for a time thought to be involved in X-disease, but it apparently causes little or no damage in cherry.

Several additional mycoplasmal diseases of *Prunus* have been described: for example, peach rosette, peach yellows, albino disease of cherry, and cherry buckskin. Cherry buckskin is considered to be caused by the X-disease agent. Relationships among the other diseases are unclear.

References: 819, 958, 1184, 1215, 1237, 1485, 1496, 1525, 1638, 1639, 1815

A–D, G. Eastern X-disease in chokecherry. A. Plants showing loss of apical dominance and yellow-red leaves with edges rolled upward; the plant at right has small rosettes of stunted light green leaves at its shoot tips (PA, Aug). B. Close view of a rosette (PA, Aug). C. Close view of red leaves and rosettes on a stem that has lost apical dominance. Dense foliage in rosettes results partly from growth of secondary shoots from buds in leaf axils (NY, Aug). D. Shot hole formation in light green leaves of a plant in early stages of disease (NY, Jul). G. A diseased shoot tip in winter, with dwarfed twigs that grew in leaf axils the previous summer (NY, Mar).

E. Shot holes forming in leaves from a peach tree in early stages of X-disease (NY, Jul).

F. Yellow leafroll symptoms on peach, considered to be caused by a strain of the western X-disease pathogen (CA, Jul).

Plate 190

Ash Yellows (Plate 191)

Ash yellows, caused by mycoplasmalike organisms (MLOs) that infect trees systemically, is responsible for decline and premature death of ash in parts of the north-central and northeastern United States and adjacent Canada. The causal organism(s) awaits full description and naming because it has not yet been isolated from infected plants or from insect vectors. White ash sustains the greatest damage. Red ash and its variety green ash are also affected, but trees of this species often tolerate infection without progressive deterioration. Infected black, blue, European (cv. 'Hessei' and 'Nana'), flowering, Oregon, and pumpkin ash, as well as *Fraxinus angustifolia, F. bungeana,* and *F. potamophila,* have also been found, but experience with these species is too limited to permit statements about their relative susceptibilities.

The geographic range of ash yellows is only partially known. It has been found from eastern Iowa and southern Wisconsin to southern Ontario, southwestern Quebec (to 45°30'N latitude) and New England, and as far south as approximately 39°N in Indiana. Witches'-brooms similar to those caused by the ash yellows agent have been noted on green ash in Georgia, and a lethal disease similar to ash yellows was found in Berlandier ash in Louisiana in the 1940s. The relationship of the symptoms noticed in the South to ash yellows as it occurs in the North is unknown, however.

An epidemic of ash yellows has persisted since the 1950s or earlier in New York and adjacent areas, but the nature of the disease was not understood until the early 1980s (see Plate 217 and related text). Understanding came late because MLOs were not recognized as pathogens of any plants until 1967 and because, before the 1980s, witches'-brooms were the only symptoms known to be caused by MLOs in ash. Only a minority of ash trees affected by yellows have brooms at a given time, so the impact of MLOs on ash populations was underestimated.

Trees of all sizes and ages are susceptible, but symptoms vary, reflecting differing stages of the disease and differing ability of the trees to tolerate infection. The first symptom, apparent 1 or more years after infection, is subnormal growth. Twig elongation and trunk growth become permanently suppressed, usually to less than half their former rates. Highly susceptible trees die to ground level within 1–3 years. Trees slightly less susceptible may remain alive for several years, during which many or most of their twigs may elongate less than 2 cm annually and their trunks may increase in diameter less than 1 mm annually. Twigs that grow very slowly for many years may twist upward or to one side. Because of slow twig growth, foliage may be arranged in tufts or, in extreme cases, rosettes at branch tips. Leaves often become chlorotic, but chlorosis is an inconsistent symptom. Leaves often fail to attain normal size, and the leaflets may be partially folded lengthwise so that they are V-shaped in cross-section and appear abnormally narrow as viewed from the ground. Slow growth and small, folded leaves give an overall appearance of sparse foliage.

Dieback develops on scattered branches either high or low in the crown, progresses intermittently, and gradually involves the entire crown. Nearly all dieback occurs during dormant seasons, but occasional branches die during the growing season, and brown leaves cling to them. Also, occasional trees in late stages of decline produce a set of feeble leaves that all drop during the growing season. A pinkish gray color of smooth bark (in contrast to normal drab gray or gray-green) often precedes or accompanies dieback in white ash and red ash that are exposed to sun and wind, but pinkish bark is an inconsistent symptom. Some diseased trees, apparently tolerant of MLOs, continue slow to moderate growth for many years while showing only sparse foliage and perhaps dieback of scattered twigs.

Trees with advanced dieback often produce epicormic sprouts at the root collar and along the trunk and major limbs. Many of these sprouts die back during winter (many at the root collar are also browsed by animals). Surviving sprouts at ground level or on the trunk may develop deliquescent branching, and some of those within approximately 3 meters of ground level may eventually become converted into witches'-brooms. A deliquescent branch is one on which the terminal shoot has lost dominance, so that the branch axis becomes lost among several twigs of more or less equal length. Brooms form from deliquescent branches as buds in leaf axils open prematurely and multiple upright shoots grow from the axes of parent shoots that were produced earlier in the same year. Repetition of this process yields dense clusters of spindly twigs with dwarfed leaves. A broom may also develop within a single season as a cluster of spindly shoots growing from a common locus, usually low on the trunk or at the root collar. Leaves on brooms tend to be chlorotic, dwarfed, narrower than normal in relation to their length, often simple rather than compound, and retained longer than is normal in autumn. On a chlorotic leaflet, yellow color gradually intensifies from the veins into interveinal areas and toward the margins. The stomata of leaves on brooms or deliquescent branches remain partially to completely closed.

Brooms form most often at the root collar, frequently on the trunk, and rarely on other branches or at branch ends. Those at branch ends appear as small rosettes with chlorotic leaves. Brooms on branches and rosettes at branch ends occur only on small trees. Brooms at the root collar or on the lower part of the trunk frequently do not develop until 1–3 years after the crown has become leafless. Sometimes single or multiple scattered tiny sprouts (3–10 cm tall) develop at the root collar, but these often die before the end of the growing season, and they seldom survive winter.

White ash trees affected by yellows begin growth 1–2 weeks earlier than normal in spring. This behavior is most noticeable on witches'-brooms. Diseased trees also tend to display autumn color prematurely, and they do not become normally cold hardy. Shoots in brooms, for example, usually sustain dieback every winter. Freezing damage to the vascular cambium at the base of the tree is common also and may lead to split bark and sometimes to basal cankers extending from the soil line upward 50 cm or more (Plate 234). Damage by freezing is most common in saplings and small pole-sized trees.

Diagnosis of ash yellows is often difficult, especially in isolated landscape specimens, because many trees affected by this disease decline without producing either of the symptoms most useful for diagnosis—witches'-brooms and deliquescent branching. If a declining ash lacks these symptoms, but nearby trees of the same species (especially seedlings and saplings in a forest understory) have symptoms of yellows, a presumptive diagnosis of yellows in the declining tree is reasonable. For cases in which field diagnosis is only presumptive, laboratory diagnosis is possible by electron microscopy or by light microscopic examination of stained sections of phloem.

Two probable vectors of the ash yellows pathogen are the leafhopper *Paraphlepsius irroratus* and the meadow spittlebug (see Johnson and Lyon, 2nd ed., Plate 202). These insects, collected in the vicinity of diseased white ash saplings, transmitted a yellows-inducing agent to previously healthy white ash seedlings. Nothing is known, however, about their efficiency as vectors or about possible additional vectors. Symptoms of yellows appear 1 year after MLOs are transmitted to small trees by insects or by grafting. An incubation period longer than 1 year is likely in large trees.

No plant-parasitic mollicute is known to be transmitted through seed. Moreover, although diseased white ash often produce heavy crops of fruit, the seeds have failed to germinate when tested. Therefore, seed transmission is probably not involved in the spread of ash yellows. The quality of seed produced by tolerant trees should be studied, however, for some of their progeny would possibly be useful in horticulture and forestry in areas where ash yellows is severe.

References: 809, 1235, 1237, 1238

Yellows in white ash.
A. Stunted leaves and negligible twig growth contribute to open appearance of the crown of a dying tree (NY, Jun).
B. A broom about 5 years old at the base of a tolerant tree (NY, Mar).
C. Deliquescent branching and brooming in a small tree. Symptoms as distinct as these are uncommon (NY, Apr).
D. A normal (left) and a slowly growing, diseased twig (right) of white ash. The portion above the tape marker represents 2 years' growth of the normal twig and 15 years' growth of the diseased one (NY, Aug).
E. Magnified view of part of a cross-section of the trunk of a tree affected by yellows for about 12 years. The annual layer of wood at the bottom of the picture is of normal thickness (NY, Oct).
F. Brooms and deliquescently branched sprouts on the mainstem of a young tree (NY, Aug).
G. The base of a diseased tree with several brooms and split bark due to freezing injury (NY, May).
H. Deliquescent branches (NY, Apr).

Plate 191

Witches'-brooms Caused by Mycoplasmalike Organisms (Plate 192)

This plate presents six of the many witches'-broom diseases believed to be caused by mollicutes called mycoplasmalike organisms (MLOs). Cellular characteristics of MLOs are described with Plate 188.

The first report of mollicutes associated with plant diseases, published in Japan in 1967, dealt with four diseases in which witches'-brooms were prominent symptoms. Two tree diseases, mulberry dwarf and paulownia witches'-broom, were included. Within the ensuing decade, hundreds of plant diseases of the "yellows" type, formerly thought to be caused by viruses, were reinvestigated, and MLOs were found associated with them. The pathogens are transmitted by leafhoppers and other suctorial insects. Witches'-brooms are a feature of most such diseases either in the natural hosts or in indicator plants, such as periwinkle, to which MLOs may be transmitted by insects or dodder. Mycoplasmal diseases of monocots are exceptions, however, because the morphology of monocots precludes broom formation, and experimental transmission of MLOs from them to dicotyledonous plants is difficult.

Pathogenicity of plant-infecting MLOs remains unproved because none of the organisms has been isolated in pure culture on an artificial medium. When this can be done, it will be possible to test pathogenicity by inoculating MLOs into healthy plants, either directly or by means of insect vectors. Healthy insects in such cases would be injected with a preparation of MLOs and then allowed to feed on healthy plants. If the plants became diseased and the same type of MLOs could be reisolated from them, pathogenicity would be proved. In the interim, strong evidence for a causal role of MLOs is available on the basis of (1) constant presence of MLOs in phloem sieve tubes of diseased but not healthy plants, (2) transmissibility of symptoms and of MLOs by grafting, insects, or dodder, (3) temporary remission of symptoms and temporary reduction of MLO populations in plants treated with antibiotics to which bacteria and mycoplasmas are sensitive, (4) destruction of MLOs and permanent remission of symptoms in plants given suitable heat therapy, and (5) lack of any other known type of infectious agent associated with symptoms.

The interrelatedness of MLOs that cause various plant diseases is unknown. These organisms cannot adequately be characterized and their relatedness tested until it is possible to grow them in pure culture. Since MLOs from different hosts often cause somewhat different symptoms in indicator plants, however, more than one type of MLO is probably involved. The assignment of Latin names to MLOs will be deferred until they can be grown in pure culture and the relatedness of MLOs from various plants is known.

Witches'-brooms caused by MLOs begin with precocious growth from buds in leaf axils, resulting in a cluster of shoots all produced in the same year as the central twig. Apical growth of the central twig is usually suppressed, so the secondary shoots all arise close together and the branch axis is lost among these shoots. Additional generations of shoots may form similarly, contributing to the denseness of the broom. Twigs and often leaves in brooms tend to be abnormally upright, although brooms that weigh heavily on branch ends may become prostrate or pendant, with only the twig tips turned up. Twigs in brooms tend to continue growth or retain leaves beyond the normal time, and they lack normal cold hardiness. Brooms or branches that support them often die back in winter. Surviving parts resume growth before the normal time in spring, producing dwarfed leaves on spindly, usually greatly stunted shoots. In general, no flowers develop on brooms.

Witches'-brooms are a common feature of elm yellows (Plate 188) in red elm. They form in the 2nd year that a tree shows foliar symptoms, which is usually its final year of life. Brooms caused by the elm yellows agent are usually small, similar to the cluster of spindly twigs with brown curled leaves on the branch at lower right in (A). Large, dense brooms are uncommon.

Bunch disease of walnut and butternut occurs throughout the range of these species in eastern United States. Its effects vary from merely growth suppressive to lethal, depending on host species. Little walnut, shown in (E), and Japanese walnut are highly susceptible, while eastern black walnut is tolerant. Other hosts include heart nut and Japanese, Manchurian, and Persian walnuts. Virtually all twigs on small, highly susceptible trees may be converted to brooms, while large trees tend to retain normal overall form but produce brooms that begin as sprouts along the trunk and limbs. Leaflets in brooms tend to be abnormally narrow, curled or cupped, and often chlorotic. Nuts fall from diseased trees prematurely or fail to fill normally, the kernel turning black and shriveled. Diseased black walnut trees often show no symptoms other than slow growth until the tree is cut back or felled. Then fast-growing brooms unusual for their long internodes may grow from the stub or stump. Some diseased trees may live 40 years or longer. Recovery of black walnut from bunch symptoms has been reported, but this probably represents only a temporary increase in growth in response to favorable environment or fertilization, not a spontaneous cure. Vectors of bunch disease are unknown.

Witches'-broom of lilac is widespread in the midwestern and eastern United States. Highly susceptible cultivars produce dense brooms of short, thin twigs. Leaves on witches'-brooms are variously distorted, usually stunted, and often chlorotic. Leaf scorch and dieback sometimes develop on witches'-brooms during the growing season, but dieback usually occurs during winter. Entire plants die within a few years after the onset of brooming. Vectors are unknown. Severe disease has been noted in *Syringa josikaea* (Hungarian lilac), *S. reticulata* (Japanese tree lilac), *S. sweginzowii,* and hybrids with these species and *S. reflexa* or *S. villosa* as parents. Tolerant lilacs may retain nearly normal twig form except for retarded growth and short internodes near shoot tips, resulting in bunched foliage. Some display only premature swelling of buds and elongation of shoots. Common lilac is tolerant; proliferation occurs but is not marked.

MLOs have been detected in witches'-brooms in panicled, red-osier, and silky dogwoods, and a similar disease occurs in flowering dogwood. The witches'-broom of willow, noted in black and heart-leaved willows in scattered eastern localities, is unusual because of the form of the twigs in brooms—long and spindly with stunted leaves. Witches'-brooms apparently caused by MLOs also occur in North America on apple, ash, blueberry, Chinese elm, black locust, honey locust, papaya, peach, and sassafras. Many other witches'-brooms are undiagnosed.

References: 251, 253, 506, 802, 866, 1185, 1215, 1579, 1742, 2163

A, B. Witches'-brooms on red elm affected by elm yellows. A. An unusually large broom that formed in the final year of life of the tree (NY, Jul). B. Details of growth preceding broom formation in an experimentally infected seedling. A winter bud opened prematurely, producing a short side shoot during the 1st growth period of the mainstem. Buds on the side shoot then opened and produced stunted shoots with dwarfed leaves. Buds in leaf axils would soon have opened, resulting in a dense cluster of shoots (NY, greenhouse).

C. Bunch disease in butternut (NY, May).

D. Witches'-broom of Hungarian lilac. This portion of the plant died during the preceding winter (NY, Jul).

E. Bunch disease in little walnut. Dense brooms, lacking normal cold tolerance, died the previous winter. One branch not yet severely diseased bears normal leaves (IL, Jun).

F. Spindly witches'-brooms with greatly dwarfed leaves (Salix yellows) in an unidentified willow (NY, Sep).

G. Detail of the onset of broom formation in flowering dogwood. Spindly shoots with dwarfed, distorted leaves have grown from buds in the current season's leaf axils. Such buds would normally remain dormant until the next spring (PA, Jun).

Plate 192

Pecan Bunch, Pear Decline, and Citrus Stubborn (Plate 193)

This plate presents aspects of three unrelated diseases caused by mollicutes that reduce or destroy commercial productivity of their host plants. Pecan bunch and pear decline are caused by unnamed, non-cultivable mycoplasmalike organisms (MLOs), while stubborn disease of citrus is caused by *Spiroplasma citri,* a motile organism of helical form. *S. citri* was the first plant pathogenic mollicute to be isolated, characterized, and named.

Pecan bunch. Bunch disease of pecan occurs from North Carolina to Arkansas and Texas. It affects trees in orchards, landscapes, and the wild. Nut production ceases on broomed parts, and nut yield, size, and quality on nonbroomed branches of diseased trees diminish. Diseased trees typically have one to several brooms on scattered branches or on sprouts on the trunk. Trees of any size may become systemically infected, however, and may produce brooms on all branches. Catkins sometimes become broomed also. See Plate 192 for description and illustrations of broom development.

Foliage on diseased parts of pecan trees varies from dark green to chlorotic, and leaflets are sometimes distorted or shorter and broader than normal. Small trees may produce brooms with a profusion of slender sprouts with stunted, light green leaves. Sprouts and twigs within brooms stand more erect than normal, produce leaves about 2 weeks earlier than normal in spring, and tend to continue growth beyond the normal time. Leaflets produced in spring, however, tend to drop prematurely. Broomed twigs or branches often die back during winter. Systemically diseased trees of highly susceptible cultivars eventually decline and die. Pecan cultivars 'Stuart' and 'Success' are resistant.

Bunch disease of pecan is somewhat unusual among mycoplasmal diseases of trees in that infection may temporarily be localized in particular branches. Eventually, however, MLOs spread throughout the tree. They are most abundant in phloem sieve tubes in brooms.

Observers have long suspected that bunch diseases of pecan, other hickories, and perhaps walnut and butternut (Plate 192) are caused by the same agent. Bitternut, mockernut, shagbark, and water hickories are affected in the same region where bunch disease is common on pecan. Bunch disease in pecan orchards has developed near diseased hickories in several areas.

MLOs that cause bunch disease can be transmitted experimentally by grafting or by dodder. The only known host other than hickory and pecan is periwinkle, to which MLOs may be transmitted experimentally by dodder. Suctorial insects such as leafhoppers are assumed to transmit the pathogen in nature, but specific vectors have not been reported.

Pear decline. First noticed in the 1940s, pear decline is now a major disease in pear-growing areas of western North America, the northeastern United States, and Europe. The pathogen is transmitted by the pear psylla (see Johnson and Lyon, 2nd ed., Plate 137), which became established in western pear-growing regions several years before the disease did. Symptoms of pear decline appear in three general patterns, depending upon the scion-rootstock combination: collapse, decline, or leaf curl. Collapse, or rapid decline, occurs in cultivars of common pear on rootstocks of the Asian species *Pyrus pyrifolia* or *P. ussuriensis,* in which phloem just below the graft union dies. Although abundant replacement phloem may form, this too becomes diseased. Affected trees wilt and die during summer or decline and die within 1–2 years. Decline is characterized by slow growth, stunted leaves with upturned margins, red leaf color in summer, premature leaf drop, and progressive dieback. Trees with declining tops also have more dead rootlets than is normal. Trees on tolerant rootstocks such as *P. betulifolia,* Callery pear, or seedlings of common pear may show only slow growth or downward curling, purplish color, and early shedding of leaves.

MLOs associated with pear decline spread throughout plants in phloem sieve tubes. The MLOs are presumed to produce or induce the host to produce a toxic substance that causes collapse of infected sieve tubes and, in highly susceptible rootstocks, death of all phloem cells at the graft union.

Natural remission of symptoms of pear decline sometimes occurs, apparently as the result of death of MLOs during winter. Remission might be expected in aboveground parts because most phloem sieve tubes in pear stems degenerate during autumn and winter. Phloem of roots remains intact, however, and MLOs usually survive winter there. They colonize new phloem above ground during the growing season.

Pear decline in California has been linked to X-disease of *Prunus* (Plate 190) on the basis of a high incidence of the yellow leafroll type of X-disease on peach adjacent to pear orchards where decline occurs. The degree of relatedness of the pathogens is unknown.

Stubborn disease of citrus. Stubborn disease, caused by *Spiroplasma citri,* affects several types of citrus and is a major disorder of grapefruit, sweet orange, and tangelo in California and Arizona. The disease occurs also in Mediterranean and Middle Eastern countries. Stubborn has been known since early in the 20th century, but its true cause remained unknown until 1970, when a mollicute of helical form was isolated and its pathogenicity proved. Symptoms include suppressed growth, shortened internodes, multiple axillary buds that produce abnormally numerous shoots, and a compact, bushlike appearance. Shoots tend to grow abnormally upright, and the affected tree makes most growth during the autumn flush rather than in spring as normal trees do. Leaves are undersized and show various chlorotic patterns. Fruits and seeds are abnormally small and variously deformed; fruit is abnormally bitter or sour, and yield is suppressed. Symptoms begin on one branch and spread to all within 1 to several years. Infection usually occurs within the first 5 years after propagating; thus diseased trees are markedly stunted. Diseased trees seem abnormally sensitive to heat and frost, and dieback occurs in some of them. Helical mollicutes (spiroplasmas) can be observed by electron microscopy in phloem sieve tubes of diseased trees, and spiroplasmas can be isolated on special media.

Several insects have been found to harbor *S. citri,* and at least three leafhoppers act as natural vectors. The beet leafhopper (*Circulifer tenellus*) and two species of *Scaphytopius* transmit the pathogen from diseased to healthy citrus and also carry it to and among an array of herbaceous plants in which it causes wilting and death. *S. citri* occurs in herbaceous plants in central and eastern states also, but infection of trees there is unreported.

S. citri has been isolated also from pear trees affected by pear decline, but it does not cause pear decline. The symptoms it may cause in pear, if any, are unknown.

References: 222, 381, 443, 1155, 1215, 1401, 1694, 1739, 1743, 2159, 2163

A, B, D. Bunch disease of pecan. A. Bunch symptoms in spring on one branch system in a tree that otherwise appears normal. The diseased portion at left began growth earlier than normal; thus its foliage, although bunched at branch ends, appears fuller and darker green than the young foliage on normal branches. Brownish catkins are visible on normal branches but absent from those with bunch symptoms. B, D. A systemically diseased, stunted tree about 4 meters tall. Bunched foliage and multiple spindly stems in place of single branches are visible throughout the tree. The most severely affected portion has stunted, chlorotic leaves (MS, Apr).

C. Stubborn disease of citrus. The lemon trees shown are all the same age. Diseased plants in the left foreground are stunted in comparison with normal plants at right and in the background (CA, Oct).

E. An orchard of 'Bartlett' pear trees devastated by pear decline (CA, Nov).

F. 'Bartlett' pear trees, all infected with the pear decline agent but tolerant of it, on tolerant rootstocks. The trees show no symptoms other than slow growth. Rootstocks of the two end trees are more tolerant, hence support more rapid growth, than rootstocks of the four center trees (CA, Jul).

Plate 193

Viral Diseases of Aspen and Poplar (Plates 194–195)

This plate introduces a series of 10 that illustrate diseases caused by viruses. Viruses are infectious particles that consist of a nucleic acid core with a protein coat. They replicate in nature only within living cells, controlling synthetic processes of the host so that it produces more virus. Various physiological disturbances and morphologic symptoms accompany viral infections of plants, but the mechanisms of symptom induction are generally unknown. Hundreds of plant viruses are known, and these infect thousands of species. Many viruses that cause loss in agricultural crops occur in weeds and in trees and shrubs. Viral diseases of gymnosperms seem to be rare, however. None has been confirmed to be present in North America, and European reports await corroboration. Viruses in trees and shrubs often cause no visible symptoms, or they may cause symptoms ranging from slight foliar markings and mild growth suppression to dramatic colorful patterns on leaves, stem distortions and cankers, twig and branch dieback, graft union necrosis, and slow or rapid decline ending in death. Two or more viruses or viruses plus other pathogens such as mycoplasmalike organisms may infect a plant simultaneously, which complicates both symptoms and diagnosis.

Viruses are classified according to physical and chemical characteristics of the particles: shape, size, nucleic acid type and content, unified or divided genome (i.e., one particle or more than one required to contain the entire genetic code of the virus), presence or absence of a membranous envelope around the particle, and other characters. Some viruses are very stable, remaining infectious in soil or plant debris in the absence of hosts, while others lose infectivity rapidly. Plant viruses are classified in more than 20 groups, but no system of Latin nomenclature is in general use. Plant virologists have adopted colloquial names (e.g., poplar mosaic virus) and have coined group names that in most cases reflect the name of a well-known typical member of the group. For example, members of the carlavirus group, including poplar mosaic virus, are related to carnation latent virus.

All plant viruses are so small that they are visible only with an electron microscope, although virus-induced structures (called inclusions) in infected cells may be seen with a light microscope. The nucleic acid of most plant viruses is single-stranded RNA, but double-stranded RNA and both single- and double-stranded DNA viruses occur. Plant virus particles range in shape from polyhedra to rigid or flexuous rods; those with envelopes appear bacillar or spherical. (There is another group of noncellular submicroscopic plant pathogens called viroids. A viroid consists of a small loop of infectious RNA without any associated protein. Avocado sun blotch and citrus exocortis are diseases caused by viroids, but these are the only known viroid diseases of trees or shrubs in North America.)

Plant viruses that cause significant disease become distributed systemically in their hosts, but this distribution may be erratic and may vary seasonally. Most plant viruses invade phloem and parenchyma tissues, but some are limited to phloem. Viruses are transported widely within infected, often symptomless plants or seeds and are transmitted in nature by various insects (mainly those that suck plant juices), certain mites, certain nematodes and fungi, pollen, grafts, and contact. The most common modes vary with the virus. A virus that is transmitted by one group of organisms (e.g., aphids) is not transmitted by other kinds of organisms.

Most viral diseases of trees or shrubs cannot be diagnosed on the basis of symptoms alone. Symptoms caused by different viruses may appear similar, and many nonviral agents or conditions cause symptoms that mimic those caused by viruses. Virus and viruslike symptoms are relatively well known and understood in species cultivated for orchard crops, but most viral disorders in forest, shade, and ornamental trees and shrubs have not been studied. Therefore the identities of the pathogens are often unknown. These plates illustrate both known and unknown viral and viruslike disorders. Some aspects of diagnosis and identification of plant viruses are described with Plate 197.

Poplar mosaic. Poplar mosaic virus (PMV) was originally discovered in Europe and for a time was thought to be confined to Eurasia. It was discovered later, however, in Japan, Canada, and the United States. Probably it had been transported in poplar and aspen propagation materials. Poplar mosaic affects nearly all of the commercially important cultivars of European-American hybrid poplars as well as black cottonwood, eastern cottonwood, balsam poplar, and black poplar. European and trembling aspens have been found infected in Europe. PMV infects only poplar and aspen in nature as far as is known, but it can be transmitted in sap to various herbaceous plants that are used as diagnostic indicators. PMV causes damage through growth suppression. Height growth may be suppressed 0–50%, depending on tolerance of the cultivar, and shoots of highly susceptible plants may die back. These severe symptoms, noted only in young trees, subside within a few years after infection. Other clones tolerate infection with no measurable growth loss. The severity of symptoms also varies with strain of PMV.

Foliar symptoms of poplar mosaic include diffuse chlorotic spots, mosaics of chlorotic and green tissue, starlike chlorotic spots with points extending along fine veins (asteroid spots), vein-associated chlorosis, veinal reddening and necrosis, and necrotic lesions on petioles and main veins. Leaves that develop lesions on petioles and midribs tend to be deformed and brittle and to abscise prematurely from nodes near the stem apex. Symptoms usually appear in late spring on the first fully expanded leaves, and foliar markings may disappear during summer. High temperature tends to mask foliar symptoms. Photosynthesis is retarded and respiration increased in diseased trees as compared with healthy trees. Specific gravity and strength of wood are only slightly affected.

PMV invades both parenchyma and phloem. It is perpetuated by vegetative propagation and is transmitted by grafting. Indicator plants (e.g., cowpea and tobacco) may be inoculated by rubbing leaves with sap from poplar. Vectors or other modes of transmission in nature are unknown, although natural spread in plantations has been observed. PMV does not seem to be borne in seed or pollen.

PMV is a member of the carlavirus group, whose particles are flexuous rods. Particles of PMV are slightly flexuous and measure 15 × 675–685 nm. The nucleic acid core consists of single-stranded RNA. The virus is detectable by serological tests at concentrations below those at which it can be transmitted in poplar sap to indicator plants. It attains highest concentrations in symptomatic leaves but can be detected in symptomless leaves from the same plants. In winter it is easiest to detect in buds.

Rhabdovirus in poplar. A rhabdovirus that has not been fully characterized infects balsam poplar in Ontario. It is associated with yellowing and necrosis along leaf veins. Its geographic distribution, host range, and impact on host growth are unknown. Rhabdoviruses, transmitted in nature by suctorial insects (aphids, leafhoppers, and piesmids) or mites, are unusual among viruses in that each rod-shaped particle is enveloped by a membrane. The nucleic acid is a single-stranded RNA. In other parts of the world, rhabdovirus particles have also been found associated with fasciation in Japanese spindletree and veinal chlorosis or vein clearing in elder, golden-chain, Chinese hibiscus, and pittosporum.

A–D, F. Poplar mosaic in eastern cottonwood. A. Dwarfed and distorted terminal leaves and missing leaves at subterminal nodes on diseased saplings. B–D. Chlorotic spots and abnormal reddening of veins. F. Reddening and necrosis on major veins and the petiole, also abnormal bending of the midrib, causing leaf distortion (Ont, Jul).

E. Vein-associated chlorosis caused by poplar mosaic virus in the hybrid poplar cultivar 'Dorskamp' (Ont, Jul).

G, H. Foliar symptoms caused by a rhabdovirus in balsam poplar. G. Chlorosis along veins. H. Close view of chlorosis and necrotic flecks along veins (Ont, Jul).

Plate 194

Other aspen and poplar viruses. Several viruses in addition to poplar mosaic virus (Plate 194) infect aspens and poplars, but their identities, geographic distributions, and effects on the hosts are not well known. Known viruses include tobacco necrosis virus, found in hybrid poplar clones in the Great Lakes region and associated with decline of aspen in the central Rocky Mountains, a potyvirus associated with decline of aspens and hybrid poplars in the Great Lakes region, arabis mosaic virus, and raspberry ringspot virus. The last two are known only in Europe. Several additional viral disorders, including some shown in Plate 195, have not been characterized.

A potyvirus informally dubbed poplar potyvirus (PPV) was detected in declining trembling and bigtooth aspens and hybrid poplars in Wisconsin and is apparently widespread in the Great Lakes region. Declining trembling aspen develops chlorotic to necrotic leaf spots early in the growing season, and some trees show bronzing of leaves on scattered branches in summer. Chlorotic to necrotic leaf spots (similar to those shown in Plate 195D), slow growth, and dieback of twigs and branches have been noted in European-American poplar hybrids infected by PPV. Declining bigtooth aspen shows bronzed leaves in the lower crown in summer. These leaves die, and the branches that bear them die during winter. Bronzing develops in the upper crown in subsequent years, and affected trees eventually die. This bronzing is associated with PPV, but proof that the virus causes it is lacking. Severe bronzing somewhat resembles the disease caused by *Apioplagiostoma populi* (Plate 56).

PPV has been transmitted in sap from aspen and poplar to cowpea, bean, and other indicator plants and from cowpea to trembling aspen, which developed chlorotic to necrotic leaf spots similar to those originally observed. Bronzing and dieback did not occur in inoculated seedlings.

PPV, in common with other potyviruses, is a single-stranded RNA virus whose particles are flexuous rods. They measure 13 × 800–810 nm. In indicator plants PPV induces intracellular structures (inclusions) of the pinwheel type as do some other potyviruses. The mode of transmission in nature is unknown, but aphids transmit many other potyviruses, and some are dispersed with seed. Natural hosts other than aspen and poplar are also unknown. Many members of the potyvirus group have restricted host ranges limited to a few plant species. Other viruses in or related to the potyvirus group that infect trees and shrubs include bean yellow mosaic virus, causing veinal chlorosis, mottle, and deformity of yellowwood leaves; Daphne virus Y, causing mosaic in *Daphne* species; Maclura mosaic virus, causing mosaic in osage orange; palm mosaic virus, causing mosaic in palms; papaya ringspot virus, causing ringspot in papaya; and plum pox virus, causing plum pox disease.

Tobacco necrosis virus (TNV) has been found in aspen clones affected by an unexplained disorder termed *aspen deterioration* in the Rocky Mountain region, but whether it plays a role in the deterioration is unknown. Aspen reproduces naturally by root suckers as well as seed. Therefore it occurs in genetically uniform (clonal) clusters, with the oldest trees in the center. Aspen deterioration is characterized by low vigor; crooked, branchy stems; abnormally high mortality rate of stems; and suppressed production of root suckers (sprouts from roots) within a clone. Deterioration primarily affects clones in which the original trees have reached ages of 80–100 years. Stands damaged by grazing animals are often affected. Viruslike symptoms have been found on leaves in both deteriorating and normal clones. Symptoms begin in spring as faint chlorotic spots, mottling, and chlorosis along the fine veins, then progress to light green or chlorotic patterns that are often reddish brown by August. When leaves turn yellow in autumn, the symptom patterns may remain orange with brown flecks. Faint chlorotic spots but no vein mottling developed after TNV from aspen leaves was transmitted in sap to leaves of an indicator plant and was then inoculated in sap of that plant into healthy aspen leaves.

TNV is a single-stranded RNA virus whose particles are polyhedra about 30 nm in diameter. It occurs around the world. The virus is soilborne, transmitted by zoospores of the root-infecting fungus *Olpidium brassicae*, and has been detected in roots of many plants. Apparently it does not infect most hosts systemically, and in most plants it does not cause severe damage; indeed it often causes no noticeable symptoms. TNV has been detected in symptomless blossoms, leaves, and fruits of apple and pear and in symptomless grape leaves. Many strains have been detected that vary in host range and in the symptoms they induce in artificially inoculated plants. TNV can be transmitted in sap to various herbaceous plants in the leaves of which it usually causes local lesions. Beans, chenopodium, and tobacco are used as indicators.

Vein mottling of aspen (Plate 195G, H) is a viruslike disorder characterized by fine chlorotic patterns with diffuse margins along major veins early in the season, followed by light green mottling that may extend over a large part of the leaf blade. Light green areas sometimes turn yellow. These symptoms have been noted in Ontario and in the northern Rocky Mountains. The cause is unknown.

Trembling aspen and hybrids between this species and European aspen are also subject to a necrotic leaf spot of undetermined cause. Symptoms consist of chlorotic spots 2–3 mm in diameter that develop necrotic centers when conditions favor severe symptoms. Marked hypertrophy occurs in certain cells within the chlorotic zone, and unusual viruslike particles have been observed in diseased tissue. The symptoms may be transmitted through grafts and seed, but no pathogen has been characterized. The disease has been noted in Ontario and Quebec.

References for plant viruses in general and for poplar mosaic: 166–68, 341, 388, 398, 400, 1079, 1381, 1382
References for other aspen and poplar viruses: 235, 398, 801, 1079, 1227, 1381, 1537, 1558, 1701

A–C. Symptoms caused by unidentified virus(es) in the hybrid poplar cultivar 'Robusta'. A. Growth suppression in a diseased plant at right, compared with a healthy plant at left. B. Stunting and distortion of subterminal leaves. C. Vein-associated chlorotic blotches in subterminal leaves (Ont, Jul).

D. Small chlorotic spots, similar to those caused by poplar potyvirus, on a leaf of eastern cottonwood (Ont, Jul).

E. Vein-associated yellowing (vein banding) in the hybrid poplar cultivar 'Northwest', caused by an unidentified virus (Ont, Jul).

F. Hybrid aspen leaves with bronzing similar to that associated with poplar potyvirus (MI, Aug).

G, H. Aspen vein mottling, an undiagnosed viruslike disorder (Ont, Jul).

Plate 195

Cherry Leafroll Virus in Walnut and Dogwood (Plate 196)

Cherry leafroll virus (CLRV), first described in Great Britain in 1961, is widespread in Europe, New Zealand, and North America in various woody hosts. In most of these it causes either no obvious symptoms or no great damage. Curled leaves or leaf markings such as ringspot, chlorotic spots, or yellow veins (yellow net) are often the only symptoms. In sweet cherry, however, leaves roll upward from margins, trees become stunted, and they eventually die. Blackberry develops line patterns on leaves, becomes stunted, and dies. European reports are the main sources of information about CLRV in hosts other than walnut.

The most important disease known to be caused by CLRV in North America is blackline of Persian walnut, a lethal disorder of trees grafted on rootstocks of certain other walnut species. First noted in Oregon in the 1920s, blackline became widespread there and in California and began to limit the profitability of walnut orchards. The disease appeared after growers began to shift away from Persian walnut rootstocks because of their high susceptibility to *Armillaria* (Plate 148). Blackline, which occurs also in Europe, is characterized by decline and death of trees on rootstocks of northern California black walnut (Hinds walnut), 'Paradox' hybrid walnut, wingnut, or hybrids between Persian walnut and other walnut species. The first indication of trouble is gradual loss of vigor and early shedding of leaves in autumn. Symptoms typically appear on one part of the tree and gradually involve all limbs. Slow twig growth and consequent short internodes lead to bunching of leaves at branch tips and a general openness of the canopy. Twigs and branches die back, and the entire top of the tree may be dead 2–5 years after symptoms first appear. Profuse sprouts grow from the understock as the top begins to decline. Decline coincides with development of a dark brown or black corky band of dead cambial and phloem tissue at the graft union. This begins to form on one side of the trunk and gradually girdles it. Decline of the top reflects the progress of girdling. The dead tissue appears as a transverse line 3–6 mm wide in trees on Hinds walnut rootstock but may extend several centimeters below the graft union on 'Paradox' rootstock. Tissue of the black line extends slightly into sapwood because cambial activity and differentiation of wood continue for a time on both sides of the girdle.

The cause of blackline disease was unknown for many years, although the nature of the symptoms was clearly evident. The rootstock and scion apparently became incompatible, resulting in death of tissue at the graft union. The mystery was why this incompatibility should be delayed until 10–20 years after grafting. The eventual explanation was that Hinds, 'Paradox', and other walnut rootstocks subject to blackline are hypersensitive to CLRV, resulting in death of cells infected by the virus. Persian walnut scions are tolerant of CLRV, and the virus slowly becomes systemic in them. On reaching the graft union, often several years after first infection, the virus causes localized death of rootstock cells, halting further internal spread but girdling the tree. The rootstock remains uninfected. The only symptoms caused by CLRV in Persian walnut on its own roots or on rootstocks of the same species are chlorotic spots, chlorotic to necrotic rings, yellow blotches, and vein necrosis on undersides of leaflets of occasional trees. The majority of infected trees at a given time are symptomless.

Blackline disease seldom appears until trees are at least 10 years old and have begun to produce blossoms. The reason is that the causal virus is carried in and on pollen. Blackline disease in walnut serves as a possible model for understanding delayed expression of graft incompatibility in other orchard trees and in shade trees and woody ornamentals.

Flowering dogwood infected by cherry leafroll virus may grow slowly and develop twisted bracts; transient chlorotic spots, ringspots, and arcs along the major veins of leaves; or abnormally narrow leaves with uprolled wavy margins, light green foliar color, and general thinness of the canopy. Chlorotic spots apparent in spring may disappear during summer. Not all these symptoms develop on a given tree, and their diagnostic usefulness for CLRV is not established. Other viruses that infect flowering dogwood (arabis mosaic, broad bean wilt, cucumber mosaic, tobacco ringspot, and tomato ringspot viruses known in North America) may contribute to these symptoms. Dogwood infected with CLRV or other viruses often has no symptoms.

Other symptoms caused by CLRV in woody plants include yellow net of European elder, chlorotic blotches and necrotic flecks or rings in leaves of American and European red elders, elm mosaic (Plate 197), and chlorotic spots and mottle evolving to yellow line patterns, rings, and patches in various birch species. The diseases of elder and elm occur in North America. CLRV also occurs in forsythia, lilac, and privet, but the symptoms it may cause in these plants have not been separated from symptoms caused by other viruses. CLRV is readily transmissible in sap to a large number of herbaceous plant species, among which bean, chenopodium, cucumber, and tobacco are used as diagnostic indicators.

CLRV is a member of the nepovirus group—single-stranded RNA viruses with a divided genome. The complete virus consists of two similar polyhedral particles, each containing a portion of the RNA necessary for viral replication and induction of symptoms. Particles of CLRV are 28–30 nm in diameter. Several nepoviruses, including CLRV, induce within host cells inclusion bodies consisting of membrane-bound tubules or vesicles containing rows of viruslike particles.

Nepoviruses are transmitted in nature by certain plant-parasitic nematodes and also through seed and pollen. The nematode *Xiphinema diversicaudatum* is a reported vector of cherry strains of CLRV in Europe, but attempts to demonstrate nematode transmission of other strains have been largely unsuccessful. Most natural transmission of CLRV occurs through seeds and pollen. Transmission rates through seed range from less than 1% to 100%, depending on host and virus strain.

Strains of CLRV vary in virulence, some strains causing severe symptoms in a given plant, others mild. Strains from different hosts also differ in serological properties.

References: 398, 399, 401, 402, 1079, 1231, 1331, 1332, 1587, 1655, 1733, 2117, 2211, 2283

A–C. Blackline disease of Persian walnut, caused by cherry leafroll virus. A. A diseased tree showing dieback and light green foliage bunched at the ends of slowly growing twigs. B. The butt of a diseased tree on California black walnut understock. Profuse sprouting of the understock is a symptom of the disorder. Bark at the graft union has been shaved off to reveal a transverse black line, the principal diagnostic symptom. C. Close view of the black line, a narrow band of necrotic cambium and phloem at the graft union (CA, Jul).

D–I. Symptoms associated with cherry leafroll virus in flowering dogwood. D, E. Slow growth and undersized, light green leaves give a severely diseased landscape tree an unusual light color and sparse-appearing canopy. F. Leaves of another infected tree are nearly normal in length and color but are somewhat narrow, with edges upturned and fluted; compare with normal leaves in (I) (SC, May). G. Twisted bracts such as those shown here have been noted on trees infected with CLRV; compare with normal bracts in (H) (NC, May).

Plate 196

Viruslike Symptoms Associated with Decline of Redbud, and Elm Mosaic (Plate 197)

Viruslike disorders of redbud. Panels A–D of this plate show two undiagnosed disorders of redbud that include viruslike symptoms on leaves. We display these diseases in the absence of diagnostic information because they cause progressive dieback of redbud in landscape plantings; therefore they are potentially important. Also, their inclusion here provides an opportunity to describe practical aspects of the diagnosis of viral diseases.

The disorder shown in panels A and B is characterized by chlorotic ringspots and short bands of chlorotic tissue that arise along primary or secondary veins in scattered leaves. These symptoms are visible in spring soon after leaf expansion. Brown flecks of necrotic tissue develop and eventually coalesce in a line near the margin of each chlorotic band or spot, producing brown ringspots or short parallel brown bands along veins. Tissue enclosed by necrotic lines dies and splits, leaving ragged holes in the leaves. Some spotted leaves and others on the same branches are deformed by wrinkles and bulges. These symptoms are fully developed by early May in Georgia. Branches with foliar lesions lose vigor, produce undersized leaves, and die back during winter. Small leaf size plus slow twig growth results in abnormal openness of the canopy on symptomatic branches. A similar disorder, dubbed *yellow ringspot of redbud,* was noted in Arkansas, and particles 120–150 nm in diameter, resembling those of an enveloped virus but larger than those of any known virus, were detected in cytoplasm of cells of diseased leaves. No further report has appeared.

The disorder shown in panels C and D is similar in many respects to that described above. The main difference is that foliar markings develop on a grander scale. Symptoms appear in early summer as slightly chlorotic rings or irregular figures on scattered leaves. Reddish brown necrotic flecks develop within the chlorotic tissue, often in two or more concentric lines. Colors within the lesions then intensify, and a greenish black border may develop, so that striking two- or three-tone line patterns appear by late summer. Symptoms are confined to certain branches, and these branches lose vigor and die back during winters. This sequence results in branch-by-branch death of the plant.

To judge from the symptoms alone, these diseases may be caused by one or more viruses. Diagnosis would require the following.

1. Attempts to detect cultivable pathogens such as fungi or bacteria consistently associated with symptoms. A negative result would be expected.

2. Demonstration of transmissibility of symptoms from diseased to healthy plants. A preferred scheme involves rubbing the leaves of herbaceous indicator plants with juice from pulped, diseased leaves. If symptoms develop on indicators, this provides evidence of transmissibility and furnishes preliminary information on host range of the pathogen and diagnostic symptoms in indicators. If the original disease can be reproduced by inoculating redbud with juice from indicator plants, this would confirm transmissibility. Many viruses, however, are not transmissible in sap. In such cases, grafts or insects may be used for attempted transmission. The incubation period (i.e., time from inoculation to appearance of symptoms) is months to years in some tree-virus interactions.

3. Detection and purification of virus. Virus particles can often be detected by electron-microscopic examination of diseased tissue or sap squeezed from it. Serological tests are also sometimes used for detection. Purification involves physical and chemical procedures to separate virus particles from plant materials and to separate different viruses from one another and concentrate them. Concentrated preparations can be subjected to physical and chemical tests and to examination with an electron microscope to "see" virus particles.

4. Inoculation of healthy plants with a purified or partially purified preparation of viruslike particles. Reproduction of the original symptoms in an experiment with appropriate controls would be strong evidence for a viral cause of the disease.

5. Virus identification. This would be based on physical, chemical, and biological properties such as particle size and form, nucleic acid type, proportion of particle weight in nucleic acid, retention of infectivity in heated or aged crude plant juice, host range and symptoms induced in diagnostic plant species, and reactions with antisera to known viruses.

The complexity of these procedures and the special equipment and training required for them are reasons why many viruslike disorders of woody plants remain undiagnosed. Unless a disease causes considerable economic or aesthetic damage or for some other reason attracts attention from a plant virologist, complete diagnosis is not made.

Elm mosaic. This disease is widespread in North America, probably beyond the range indicated by records from Nova Scotia, Wisconsin, Oklahoma, and Virginia. It is reported to occur only in American elm and is caused by a strain of cherry leafroll virus (CLRV, Plate 196). Leaves of diseased trees are usually abnormally small and stiff, and some are distorted by curling in various directions. Small leaves often have yellow mottling or mosaic, ringspots, and abnormal ridges and wrinkles (rugosity). Enations (ragged elongate projections similar to those shown on cherry in Plate 198) have been observed along the midvein and between lateral veins on leaves of the cultivar 'Moline'. Twigs may appear spindly and may be produced in abnormal abundance, giving the tree a bushy appearance. Diseased trees lose vigor, and scattered twigs die, so that the canopy appears ragged and somewhat open. The tree may persist for decades in this condition.

The causal virus is seed- and pollenborne in nature. In one test, 1–3.5% of seedlings from diseased American elm seedlots developed mosaic. CLRV can be transmitted among elms by grafting, including bark patch grafts. Various herbaceous plants are susceptible when their leaves are rubbed with sap from diseased elm. French bean, cucumber, and species of chenopodium and tobacco are used as diagnostic indicators.

The elm strain of CLRV has the same physical properties as strains from other hosts, but it differs from them serologically. Strains of CLRV from several hosts have been found to differ from one another in serological tests.

Elm mosaic is at times associated with cankers, called zonate canker, on American elm. The agent that causes zonate canker is unknown but is assumed to be viral because it may occur in and be transmitted from symptomless trees.

Elm mosaic was once thought to be caused by tomato ringspot virus (TmRSV), but serological tests and mode of transmission indicate that the two viruses are unrelated. TmRSV is transmitted by the nematode *Xiphinema americanum,* whereas CLRV is not.

Elm mottle virus, known in Europe, causes yellow and green mottling and chlorotic spots, lines, and rings that sometimes contain necrotic lesions, on leaves of Scots and European field elm. This virus also naturally infects lilac, causing white mosaic symptoms, and it causes white mosaic when inoculated into *Forsythia intermedia.*
References: 388, 398, 580, 607, 1025, 1079, 1704, 1953

A, B. An undiagnosed viruslike disorder of redbud. A. Undersized, deformed leaves on a declining branch of a tree with advanced dieback. B. Necrotic ringspots and vein banding on leaves from the plant shown in (A) (GA, May).

C, D. Another undiagnosed viruslike disorder of redbud. C. Large necrotic ringspots and line patterns on a leaf viewed in July. D. Line patterns on a leaf viewed in October. Had this leaf been photographed in July, it would have appeared similar to that in (C) (NY).

E, F. Mottling, leaf distortion, and chlorotic spots on leaves of American elm affected by elm mosaic (NY, Jun).

Plate 197

Some Viral Diseases of Cherry (Plate 198)

Many viral diseases affect cherries and other stone fruit trees in orchards and home plantings. Some of the causal viruses are also important in ornamental plants. This plate provides glimpses of yellows and green ring mottle of sour cherry; rusty mottle, rasp leaf, and necrotic ringspot of sweet cherry; and an undiagnosed viruslike disorder of chokecherry.

Sour cherry yellows. Prune dwarf virus (PDV) causes this disease, which occurs worldwide in stone fruit trees and limits sour cherry production in some regions. PDV also causes dwarfing and stunting of peach, plum, and prune, mainly in the West. The wide distribution of PDV and other fruit tree viruses is no doubt a result of virus transportation in infected seeds and plants. Mahaleb and mazzard cherries, used as rootstocks, and Damson plum may be symptomless carriers of PDV. The virus has been transmitted experimentally to over 100 species of *Prunus* and to many herbaceous plants, but only a small fraction of the woody plants and none of the others have been found naturally diseased. Other natural woody hosts of PDV include paper birch, sweet birch, and European white birch. Effects of PDV on birch have not been described.

Sour cherry trees infected by PDV grow slower than normal and produce more than the normal number of flower buds at the expense of vegetative buds on terminal twigs. The system of fruit spurs, and fruit yield, are thus reduced, although fruit quality remains high. Leaves at basal and intermediate positions along new shoots show bright yellow and green mottle or various yellow patterns in late spring and early summer. Symptomatic leaves soon drop, and the twig remains bare except for green leaves near the tip. Diseased trees may lose 30–50% of their leaves by midsummer. Leaves on newly infected trees may show acute symptoms consisting of chlorotic mottle or chlorotic rings and flecks, and later holes where tissue within chlorotic rings dies and drops out. These symptoms give way in subsequent years to chronic leaf yellowing and casting. Yellowing and leaf cast are promoted by cool weather or at least cool nights after bloom. Yellows is most severe in trees simultaneously infected with PDV and Prunus necrotic ringspot virus (PNRSV).

PDV causes ringspot symptoms in some sweet cherry cultivars, but other cultivars remain symptomless although infected. Diseased plum and prune may be dwarfed with narrow, rugose, leathery leaves, often only on certain branches. PDV is seedborne and is also carried in pollen from tree to tree in orchards.

PDV and PNRSV are members of the ilarvirus group—single-stranded RNA viruses with genome divided among three types of particles. Particles of PDV vary from isometric, about 19–20 nm in diameter, to short rods up to 73 nm long. Those of PNRSV are asymmetric polyhedra, commonly about 23 nm in diameter.

Rusty mottle. This name refers to a group of diseases of sweet cherry, widespread in western North America, presumed to be caused by related viruses. Necrotic rusty mottle, shown here, is attributed to an uncharacterized agent called necrotic rusty mottle virus. Angular necrotic spots form in leaves of susceptible cultivars 3–6 weeks after full bloom and cause distortion of leaves not yet fully expanded. Lesions may remain discrete or may coalesce to involve large areas of the leaf, often inducing leaf drop. Severe symptoms are promoted by prolonged cold weather in spring. Yellow and green mottling with green islands or bands develops in summer. During a period of years, buds and twigs die, leaving branches with terminal tufts of foliage. Cankers and gum blisters form on branches. The causal agent is readily transmitted by grafting, and most cases of the disease in orchards are considered to result from propagating diseased plants in nurseries. The natural mode of transmission is unknown.

Rasp leaf. Ragged or rasplike projections of green tissue (enations) may form on the undersides of cherry leaves. Cherry rasp leaf virus (CRLV), although not the only cause of enations in cherry, is the most common cause in North America. It is widespread in the West. Prominent dark green rounded to pointed enations grow from the undersides of twisted, abnormally narrow leaves, especially along the midvein and in rows between major lateral veins. Upper surfaces of leaves are rough and have depressions that correspond to the enations below. Many spurs and branches low in the tree die, giving the canopy an open appearance and reducing fruit yield. CRLV is one of the nepoviruses: nematode-transmitted polyhedral viruses with divided genome of single-stranded RNA. The dagger nematode, *Xiphinema americanum,* is a vector. Weeds including dandelion and plantain have been found infected and may serve as natural reservoirs of the virus.

Green ring mottle. This disease occurs in apricot; Japanese flowering, sour, and sweet cherries; nectarine; and peach in eastern and western North America, Europe, and New Zealand. The first two hosts are damaged economically; the others are symptomless carriers. In 'Montmorency' sour cherry, a mottle consisting of persistent green islands or rings on a chlorotic to yellow background appears on scattered leaves 4–6 weeks after flower petals fall. Affected leaves soon drop, and few foliar symptoms are visible during the remainder of the season. Fruit on diseased trees may show mottled and pitted surfaces, and the flesh is bitter. Diseased 'Kwanzan' and 'Shirofugen' flowering cherries display epinasty (downward bending of petioles), and parts of the midrib and lateral veins of leaves may die, resulting in twisting and curling. Twig growth is suppressed, and strips of maturing bark on twigs die and crack. Dieback ensues. The causal agent, presumed to be viral and called green ring mottle virus, is transmitted by grafting and has been dispersed widely in propagating materials. Natural spread occurs in orchards, but no vector has been identified.

Prunus necrotic ringspot virus. This virus occurs worldwide in rosaceous fruit trees and ornamental plants as well as other woody hosts. Besides causing necrotic ringspot in many species of *Prunus,* PNRSV causes almond calico, rugose mosaic of sweet cherry, line patterns in plum, ring and band mosaic of hop, and part of the rose mosaic syndrome (Plate 199). The virus occurs in pin cherry in the East and in desert almond in the West. PNSRV has also been detected in sweet birch and wax-leaf privet, but symptoms in these plants are unreported. First-year, or shock, symptoms in *Prunus* occur on single or scattered branches and often consist of chlorotic to necrotic rings from which the centers drop, leaving shot holes. In some cases, buds or twigs and branches may die back; this symptom varies with host and strain of the virus. In subsequent years, foliar symptoms caused by most strains are mild or obscure, but tree growth and fruit yield are suppressed. PNRSV spreads from tree to tree on pollen that is superficially contaminated although not infected. The virus is also seedborne.

References: 300, 377, 388, 401, 1079, 1525, 1949

A, B. Sour cherry yellows in 'Montmorency' cherry. A. Sparse foliage due to premature abscission, and sparse fruit set due to subnormal number of fruiting spurs. B. The tip of a diseased branch bears only one fruit spur on the previous year's wood, and nearly half the leaves on the current season's shoot have dropped by harvest time (OR, Jul).

C. Necrotic spots and distortion caused by necrotic rusty mottle virus in 'Sam' sweet cherry (OR, Jul).

D. Viruslike symptoms (left) in an ornamental chokecherry. In comparison to normal (right), the affected plant has no lateral twigs or fruit spurs, shoot growth is suppressed, and leaves are undersized and chlorotic, with edges uprolled (ND, Jul).

E. Leaf distortion and enations (green projections) caused by cherry rasp leaf virus in 'Lambert' sweet cherry (OR, Jul).

F. Green ring mottle on 'Montmorency' sour cherry (NY, Sep).

G. Chlorotic to necrotic ring spots in a mazzard cherry leaf infected with Prunus necrotic ringspot virus (OR, Jul).

Plate 198

Rose Mosaic Complex and Viral Diseases of Hibiscus (Plate 199)

Rose mosaic. Once thought to be caused by one virus, rose mosaic is a complex of viral diseases that occur worldwide on wild and cultivated roses. The principal pathogens in North America are Prunus necrotic ringspot virus (PNRSV) and apple mosaic virus (ApMV). Tobacco streak virus has been isolated from prairie rose in the Northwest and from other roses in Europe. In California and Oregon, rose ring pattern caused by an unknown viruslike agent is also part of the mosaic complex. Several additional viruses, among which arabis mosaic virus is most common, have been isolated from rose with mosaic symptoms in Europe. Arabis mosaic virus is uncommon in North America and is not known to contribute to rose mosaic here.

Mosaic disease diminishes the vigor of rose bushes and renders them abnormally sensitive to winter damage. The quality and numbers of blossoms are also diminished in some cultivars infected with ApMV. PNRSV interferes with establishment of bud grafts of some cultivars. In 'Fragrant Cloud', PNRSV delays the onset of flowering, reduces size and number of blossoms, and increases the proportion of deformed blossoms.

Mosaic symptoms vary with strains or combinations of viruses, rose species or cultivar, and environmental conditions. The cultivars 'Madame Butterfly', 'Ophelia', and 'Rapture' are among the most highly susceptible on the basis of dramatic foliar symptoms. Foliar markings range from chlorotic mottles and ringspots through light green or chlorotic line patterns to vein clearing (yellow net), vein banding (at temperatures above 27°C), and mosaics of green and yellow or white. Leaflets may be puckered. Infected plants, even those of highly susceptible cultivars, however, often lack obvious symptoms. In California, mosaic symptoms develop during cool (e.g., 15°C) weather in spring, but no new symptoms appear during hot summer weather.

In general, infection caused by a particular virus cannot be reliably diagnosed on the basis of field symptoms. When isolates of PNRSV and ApMV have been transmitted to certain rose cultivars by grafting, however, each virus has caused distinct symptoms. ApMV caused chlorotic mottle and a mosaic of white or yellow and green patches or fused chlorotic rings, while PNRSV induced light green to chlorotic line patterns that varied from 1–2 mm up to several millimeters in width. Symptoms caused by ApMV were most distinct during summer when temperatures ranged above 21°C. Those caused by PNRSV were most distinct during relatively cool weather in early and late summer. PNRSV has also been isolated from rose plants showing yellow net symptoms similar to those illustrated in panel A.

ApMV and PNRSV are ilarviruses. ApMV causes mosaic and line patterns in apple and diverse other plants (Plate 200). It can be mechanically transmitted with difficulty in sap from rose or apple leaves, but inadvertent mechanical transmission in the field is unknown; transmission during propagation by budding is the rule. The nucleic acid of ApMV is single-stranded RNA, distributed in two or more asymmetrically polyhedral particles 25–29 nm in diameter. Characteristics of PNRSV are similar, as reported with Plate 198. PNRSV isolates from cherry, plum, rose, and ornamental *Prunus* are indistinguishable from one another but are serologically and symptomologically distinct from ApMV.

Rose ring pattern is caused by an unknown graft-transmissible agent thought to be viral but distinct from the other viruses of the rose mosaic complex. Symptoms in hybrid tea roses include irregular rings, fine line patterns, and chlorotic flecks on leaves, but these symptoms are often indistinct. Color-break rings develop in the petals of some cultivars. 'Queen Anne' plants infected with the ring pattern agent show yellow blotches on leaflets. The agent is transmitted by grafting, but no mode of natural transmission has been identified. Multiflora cultivar 'Burr' is a reliable indicator; leaflets on diseased plants become stunted, deformed, rugose, and mottled.

Diseased rootstocks are a common source of rose mosaic viruses. PNRSV, readily transmitted by grafting, has been detected in multiflora rose rootstocks. This virus is also carried on pollen of roses as well as susceptible stone fruit trees. Animal vectors other than pollinating insects are unknown.

Rose mosaic viruses common in North America can be inactivated in young rooted cuttings by holding them at temperatures near 38°C for 4 weeks.

Roses in North America are subject to several additional viral diseases: rose streak, rose rosette, rose wilt, rose spring dwarf, and rose leaf curl. Tobacco streak virus causes irregular chlorotic areas, vein chlorosis, and twisted leaves in prairie rose.

Viral diseases of hibiscus. The viruslike symptoms in Chinese hibiscus shown in panels C and E are undiagnosed. The symptoms appear in spring on the oldest leaves, which develop a mosaic of green rings and blotches on a yellow background. Yellow leaves soon drop, giving the shrub an open appearance. Leaves formed later may show irregular chlorotic spots, puckering, and cupping.

Several viral diseases of Chinese hibiscus are known, but none is reported to cause the dramatic symptoms shown here. The best known disease in North America is chlorotic ringspot caused by Hibiscus chlorotic ringspot virus. This virus occurs worldwide in Chinese hibiscus, causing a generalized mottle or numerous tiny chlorotic spots, rings, or vein banding on most leaves of diseased plants. Plants continue to flower, however, and do not become stunted unless infected when very young. The virus, related to the tombusvirus group, is spread in commercial cultivars by grafting and in sap on pruning shears. The particles are isometric, about 28 nm in diameter, and contain single-stranded RNA.

References: 388, 398, 883, 925, 1079, 1735, 1949, 1988–90, 2118

A, B, D, F. Rose mosaic complex. A. Yellow net symptoms, also called vein clearing, possibly caused by Prunus necrotic ringspot virus (PNRSV). B. Yellow and green line patterns and puckering of leaflets, also possibly caused by PNRSV (A, B: NY, Jun). D. Mosaic possibly caused by apple mosaic virus (CA, spring). F. Chlorotic ringspots and mottle, similar to symptoms of rose ring pattern (NY, Jul).

C, E. Viruslike symptoms on Chinese hibiscus. Green blotches and ringspots are prominent on the oldest leaves on the shoot, but leaves formed later show only slight curling and rugosity (FL, Mar).

Plate 199

Diseases Caused by Apple Mosaic Virus, and Apple Flatlimb Disease (Plate 200)

Diseases caused by apple mosaic virus. This virus occurs around the world in the temperate zones in many rosaceous and some non-rosaceous woody plants. In addition to its causal role in apple mosaic disease, it causes line pattern in plum and flowering cherry and is one of the complex of viruses that cause rose mosaic (Plate 199). It also occurs naturally in almond, flowering almond, paper and yellow birches, European filbert, horse-chestnut and red horse-chestnut, and hops. A virus that causes symptoms like those of apple mosaic in indicator plants has been isolated from symptomless *Pyracantha rogersiana*. Also, an agent associated with mosaic in chokecherry in Maine caused mosaic when graft-inoculated to apple. ApMV in some of these hosts has been recorded only on other continents but is mentioned here because the same plants grow where the virus occurs in North America. ApMV can be transmitted experimentally to plants in many families. Plants used as diagnostic indicators include blue-wings, cowpea, cucumber, periwinkle, and the apple cultivars 'Golden Delicious', 'Jonathan', and 'Lord Lambourne'. ApMV is readily transmitted by grafting and thus spreads mainly in infected nursery stock. Transmission through natural root grafts also occurs in apple. Symptoms develop after a few weeks to a year, depending upon plant size and season of inoculation.

Symptoms caused by ApMV vary with host and strain of virus. In apple, the most common symptom is angular cream to yellow spots on a green background, often coalescing to produce large areas devoid of green. Chlorotic bands may develop along major veins. In other instances, leaves may show light and dark green mosaic or diffuse yellowish white patches. Severe strains of ApMV may induce chlorotic to white areas, whereas mild strains cause only chlorotic spots. Although all leaves on a branch may show symptoms, it is more common to find mosaic on only scattered leaves. Mild strains of the virus in apple protect against infection by severe strains. Apple mosaic has been described in detail only for commercial apple cultivars. Many other species of apple and crabapple are susceptible, however. Prairie crabapple and showy crabapple are highly susceptible. Mosaic symptoms also develop in mountain ash, hawthorn, pear, and plum after transmission by grafting. Horse-chestnut and red horse-chestnut found naturally diseased in England develop bright yellow and green mosaic. In birch, flowering cherry, filbert, plum, flowering quince, and rose, ApMV induces primarily line patterns rather than mosaic. The symptoms in rose cannot dependably be differentiated from those caused by Prunus necrotic ringspot virus.

Paper and yellow birches infected with ApMV may have foliar markings similar to those on filbert in panel D: chlorotic to white line patterns, often of the oak-leaf type, and chlorotic flecks. Birch may also develop chlorotic vein banding, concentric ring patterns, and sometimes mild mosaic. Symptoms usually appear on only a few leaves and not in every year. Young plants that developed line patterns within 2 years after inoculation by grafting were symptomless by the 4th year. The discovery of ApMV in birch was initially of interest because viruslike symptoms had been associated with the destructive birch dieback disease (Plate 218) that swept the Maritime Provinces and northeastern United States from the 1930s to the 1950s. Symptoms of the sort caused by ApMV occurred from Nova Scotia to Wisconsin, however, well beyond the region affected by birch dieback. Moreover, ApMV caused no dieback in experimentally inoculated birch seedlings. The possibility remains that ApMV may predispose birch to damage by other factors, but evidence is lacking.

European filbert is subject to infection not only by the common strain of ApMV, which causes the line patterns and flecks shown here, but in Europe also by Tulare apple mosaic virus. The latter, an ilarvirus unrelated to ApMV, has otherwise been detected in nature only in a single apple tree in Tulare County, California.

ApMV, typical of ilarviruses, is a single-stranded RNA virus with its genome divided among two or three asymmetrically polyhedral particles 25–29 nm in diameter. It is closely related to Prunus necrotic ringspot virus and occurs in several of the same hosts, notably rose. Although it becomes systemic in its woody hosts, invasion of apple or plum trees is often slow and erratic; symptoms and virus may not occur in all branches. Natural spread in orchards is slow and in many instances seems not to occur. ApMV can be inactivated in young apple plants by holding them near 36°C for 3–10 weeks.

The viruslike symptoms shown on flowering quince in photos A–C were not diagnosed but are shown here because ApMV would be one of the possible causes considered during diagnosis. Flowering quince is susceptible to ApMV, and in one test it developed line patterns after artificial inoculation with the virus. The development of mosaic on oldest quince leaves while young terminal leaves remain unaffected is similar to the development of mosaic symptoms in apple.

Apple flatlimb. Flatlimb disease is a graft-transmissible, presumably viral disease that affects many commercial apple cultivars. 'Ida Red' and 'Gravenstein' are particularly susceptible. The disease occurs around the world where 'Gravenstein' is grown. Flatlimb symptoms have also been observed in naturally diseased mountain ash, pear, and quince in Europe. Slight linear depressions or slight flattening develops in strips on shoots and branches where cambial activity slows or ceases. The depressions later become deep furrows or eventually cankers as surrounding tissues grow more or less normally, and branches may become flattened and twisted, often brittle. The causal agent may occur in symptomless rootstocks. Similarly, it may occur in symptomless trees, leading to flatlimb in scions after such trees have been "top worked" by grafting susceptible cultivars into them. Symptoms develop beginning 1 or more years after grafting.

The flatlimb agent is unusual in that it seems to move only upward in diseased trees. This has repeatedly been indicated by graft-transmission experiments in which buds or other grafting material from below sites of previous transmission did not carry the agent. The flatlimb agent has been transmitted experimentally to many woody rosaceous species and to walnut. It has not yet been characterized, however, and its relationships to other viruses are unknown.
References: 388, 398, 663, 1031, 1079, 1545, 1949, 1950

A, B. An undiagnosed mosaic in flowering quince. Leaves on spur shoots and the oldest leaves on long shoots develop yellow and green mottle or mosaic and drop, while leaves on terminals remain green (NY, Aug).

C. Undiagnosed line patterns in flowering quince. Necrotic flecks have formed within some lines (NY, Jul).

D. Chlorotic to white line patterns in European filbert caused by apple mosaic virus (OR, Jul).

E. Flatlimb disease in 'Gravenstein' apple. Branches have flattened areas, groovelike depressions, and cankers (WA, Jun).

Plate 200

Tobacco Mosaic, Tobacco Ringspot, and Tomato Ringspot Viruses in Ash and Other Woody Plants (Plate 201)

The three viruses named above are widespread in many woody and herbaceous plants in North America. They were detected in white ash during searches for pathogens associated with decline of this species (Plate 217) in the northeastern United States. To date, however, although at least two of these viruses are common in declining trees, none has been implicated as a primary cause of decline. Also, the range of foliar symptoms that each may cause in ash is unknown.

Tobacco ringspot virus (TbRSV) and tobacco mosaic virus (TMV) occur singly or together in white ash that shows assorted foliar markings: faint chlorotic spots and rings, irregular chlorotic sectors and line patterns along veins, mosaic, mottling, red spots and rings in late summer, green spots and rings in chlorotic or red leaves in autumn, and premature autumn color. Symptoms are variable and may be absent from most leaves on a diseased tree. Symptoms are presumed to vary with environment, strains or combinations of viruses, and strain of host.

Neither TbRSV nor TMV is consistently associated with slow growth and dieback of ash in the field. TbRSV suppressed the growth of artificially inoculated white ash seedlings, however, and caused chlorotic spots and blotches, mosaic, necrotic or light green line patterns along secondary veins, crinkling, and irregular margins on leaflets. Possibly TbRSV, alone or with other viruses, predisposes ash to damage by other pathogens. TbRSV and TMV have also been detected in green ash and 'Moraine' ash, associated in the latter with chlorosis, mosaic, and puckering of leaflets. In a test with green ash, mosaic developed, and newly produced leaflets were abnormally narrow after seedlings were inoculated with TbRSV. A strain of TMV isolated from white ash with chlorotic spots, rings, and line patterns caused chlorotic spots in artificially inoculated white and green ash seedlings and concentric chlorotic rings in European ash. TbRSV and TMV have been detected in seeds of white ash, but possible seed transmission of these viruses in ash has not been investigated.

Tomato ringspot virus (TmRSV), common in many woody and herbaceous plants in the region where ash decline occurs, has been found but has not yet been sought extensively in ash. The symptoms it causes in ash are unknown. Since both TbRSV and TmRSV are associated with debilitating diseases in other woody plants, they remain of interest as possible causes of or contributors to decline in ash.

In England, arabis mosaic virus (AMV) was isolated from white ash with puckered and twisted leaves and chlorotic blotches and ring patterns. By artificial inoculation it was found to cause such symptoms in white ash and European ash and to suppress growth of white ash on European ash rootstocks. AMV has been detected in North America in dogwood but not in ash. Many additional viruslike symptoms on various ash species have been reported in Europe, but the causes were undetermined in most cases.

TbRSV and TmRSV belong to the nepovirus group. Nepoviruses are nematode-transmitted, polyhedral, single-stranded RNA viruses whose genome is divided between two particles near 28 nm in diameter. TbRSV and TmRSV exist as many strains that differ in serological properties and symptoms induced in plants.

TMV is the type member of the tobamovirus group. It is a single-stranded RNA virus with its genome contained in one rigid rod-shaped particle about 18 × 300 nm. Many variants exist.

TbRSV, TmRSV, and TMV all infect many herbaceous and woody plants and have been carried in plants to many parts of the world. TMV is transmitted mainly by contact between plants, through seeds (e.g., in apple, grape, and pear), and by hands or implements during plant care. TbRSV and TmRSV spread in nature mainly in those parts of the United States and Canada where their dagger nematode vectors, *Xiphinema americanum, X. californicum, X. rivesi,* and related species (Plate 145), are common. These nematodes collectively have a broad host range, which accounts for the presence of the viruses in weeds as well as in many ornamental and crop plants. Virus is acquired and transmitted as nematodes feed on succulent root tips. Nematode adults and larvae both transmit virus, and the long-lived adults, once infective, remain so for several months. Seed transmission of TbRSV and TmRSV occurs in herbaceous plants but is unreported for woody plants except for TmRSV in brambles. TmRSV also spreads with pollen in some herbaceous plants.

TbRSV is associated with or may cause the following: necrosis at graft unions (brown line disease) in apple; necrotic ringspot of blueberry; mottling, mosaic, line patterns, distortion, and stunting of leaves of brambles; unspecified symptoms in daphne; chlorotic rings, mottling, leaf distortion, and general stunting of grapevine; ringspot, line pattern, and dark green mottle of elderberry; chlorosis, crinkled leaves, and stunting of French hydrangea; yellow net of forsythia (Plate 202); chlorotic to yellow leaf spots and irregular leaf margins in Japanese holly; mosaic in rose; and ring patterns in skimmia. In addition, TbRSV has caused various symptoms in artificially inoculated woody plants: mosaic and stunting of sweet cherry and American elm, mosaic and abnormally narrow leaves in box elder, ring spots and line patterns on American elder, and mild mosaic in Amur maple and rose-of-sharon. TbRSV has also been detected in naturally infected trembling aspen; cinquefoil; flowering, gray, and red-osier dogwoods; American elm; hawthorn; bush honeysuckle; jasmine (*Jasminum nudiflorum*); and black willow. Except for elm, however, symptoms caused by the virus in these hosts are undescribed.

TmRSV is associated with the following in woody hosts: tree decline because of necrosis at graft unions (brown line disease) in apple, cherry, peach, and prune; stem pitting and decline of cherry and peach; yellow bud mosaic of peach; ringspot and decline of brambles; leaf mottling of *Daphne mezureum*; ringspot, line pattern, and dark green mottle of elderberry; chlorotic rings, concentric ring spots, and oak-leaf patterns on *Euonymus fortunei* and *E. kiautschovica*; chlorosis with green blotches and stunting of French hydrangea; decline of grapevine; mosaic of rose.

Symptoms associated with TMV include chlorotic spots and lines and deformed leaves in American elder, chlorotic spots and mottling on sugar maple leaves; yellowing, rosetting, and necrotic spotting of tanoak; chlorotic spots and mosaic of wisteria. TMV is also widespread in many woody plants that show no viruslike symptoms—for example, apple, avocado, cherry, currant, grapevine, several oaks, pear, and plum.

References: 28, 252, 338, 388, 398, 581, 799, 800, 803, 804, 807, 899, 1079, 1098–1100, 1328, 1360, 1410, 1426, 1561, 1637, 1773, 2006, 2050, 2120

Virus and viruslike symptoms in leaves of white ash.

A. Wavy margins and abnormally narrow leaflets. Tobacco mosaic virus was detected in this plant (NY, May).

B. Mosaic and distortion. Virus content of these leaves was unknown; however, tobacco ringspot virus has caused such symptoms in artificially inoculated seedlings (NY, Jun).

C, E. Red ring spots (which appeared chlorotic earlier in the season) and chlorotic spots. Tobacco ringspot virus is associated with these symptoms (NY, Sep).

D. Light and dark green mottling and mild veinal chlorosis. Tobacco mosaic and tobacco ringspot viruses together have been detected in leaves with such symptoms (NY, Jul).

F. Chlorotic vein banding. Tobacco mosaic virus has been associated with this symptom and with line patterns of oak-leaf type that result when the edges of the chlorotic zone are more distant from the midvein (NY, Jun).

Plate 201

Viral Diseases of Azalea, Rhododendron, and Forsythia (Plate 202)

Viral diseases of azalea, rhododendron, and related plants. Although viral diseases of these plants occur widely, little is known about them. Mosaic-type diseases of rhododendron were noted in Germany and in New Jersey in the early 1930s but were not fully described or shown to be transmissible. *Rhododendron ponticum* and catawba hybrids were affected in New Jersey. This plate presents two syndromes of unknown cause. In that shown in (A) and (B), symptoms on expanded leaves include mosaic, chlorotic mottling, and rugosity. The leaves are undersized and often somewhat twisted, with edges rolled downward. Typically the entire plant is affected, loses vigor, and may be overgrown by healthy neighbors. The syndrome shown in (C) and (D) is characterized by chlorotic flecks on all parts of the leaf. In addition, leaves have chlorotic lines or zones where tissue differentiation ceased prematurely, resulting in wrinkling and deformity. Affected leaves tend to be undersized, and many are abnormally narrow in relation to their length. Severity of symptoms may vary from one branch to another.

One viral disease of rhododendron, not shown here, has been relatively well characterized, although the pathogen was not fully identified. Prominent concentric necrotic ring patterns develop in leaves. The disease occurs in many cultivated rhododendron cultivars and in mountain laurel in the northwestern United States and British Columbia and has been noted in mountain laurel in Great Britain. The causal virus, possibly a member of the potex group and transmissible by grafting, can be detected by electron microscopy in leaf cells as flexuous rods most commonly 13 × 500–510 nm.

Yellow net symptoms in forsythia. These symptoms, widespread in the northeastern United States and in Europe, may be caused by either tobacco ringspot virus (TbRSV) or arabis mosaic virus (AMV). TbRSV (see also Plate 201) was identified as the pathogen in New Jersey, and AMV in Europe. The disease in the United States is characterized by mild to dramatic yellowing along all veins and veinlets. Symptoms diminish in intensity from older to younger leaves on a shoot, and leaves near the tip may appear normal. Yellow net is associated with low vigor, dieback of canes during winter, and occasional death of canes during the growing season. Canes that die during the growing season first produce stunted, chlorotic shoots that in time wilt and turn brown. Generally all stems of a plant have symptoms, although their nature and intensity vary.

A strain of TbRSV from forsythia with yellow net in New Jersey was identified on the basis of characteristic symptoms and cross-protection in herbaceous indicator plants to which virus was transmitted mechanically in sap. *Cross-protection* refers to the failure of a virus to induce symptoms in a susceptible plant when the plant is already infected with a strain of the same virus that causes only mild symptoms or no symptoms. The forsythia strain caused yellow net when transmitted in sap from artificially infected cowpea to healthy forsythia. TbRSV has also been detected in symptomless *Forsythia ovata*.

Yellow net in Europe was first described as affecting *F. intermedia*. Tissue along the veins in leaves turned yellow in contrast to green areas between veins. Published photographs indicated more dramatic color changes along the major veins than along veinlets. Symptoms could be transmitted to healthy forsythia by grafting and to herbaceous indicator plants by sap or dodder. A strain of arabis mosaic virus was isolated and was shown to cause yellow net in artificially inoculated forsythia. AMV was subsequently found in *F. europaea* with bright chlorotic leaves. So far as is known, AMV occurs only rarely in North America. It was found in flowering dogwood in South Carolina, associated with mild yellow mosaic and chlorosis but has not been reported to occur in forsythia or other woody plants. Other viruses reported to occur in forsythia, all in Europe, include alfalfa mosaic, cherry leafroll, cucumber mosaic, raspberry ringspot, tobacco rattle, and tomato black ring viruses. Characteristic symptoms caused by these viruses separately or in combination are unreported.

References: 388, 398, 414, 1703, 1773, 2006, 2119

A–D. Viruslike symptoms of unknown cause in cultivated rhododendrons. A, B. Disease of the mosaic type causes a plant to appear stunted in contrast to a normal neighbor. Because of severe mosaic, leaves appear light green from a distance (NY, Jul). C, D. Another viruslike disease is indicated by chlorotic flecking and yellow-white lines and zones where arrested growth of leaf tissue resulted in distortion (WA, Jun).

E–F. Yellow net disease of forsythia. E. An affected plant with stunted new shoots amid dead and dying canes of the previous year. Because of the yellow net symptom, the entire plant has a yellowish hue as viewed from a distance. F. A diseased shoot (NY, Jul).

Plate 202

Viral Diseases of Cactus, Camellia, Magnolia, and Oak (Plate 203)

Viral diseases of cactus. Several viruses infect cacti, a fact not generally known until the 1960s because diseased cacti often have no overt symptoms. Cactus viruses include Sammon's Opuntia virus (SOV), cactus virus X (CVX), saguaro virus (SV), cactus virus 2 (CV2), Zygocactus virus (ZV), and Zygocactus virus X. The first three occur in North America. Although cactus viruses can be transmitted artificially to certain herbaceous indicator plants (useful for diagnosis), these viruses seem to be restricted to cacti in nature. Furthermore, viruses that in nature infect other kinds of plants have not been found in cacti. Cactus viruses are transmitted by grafting, but natural modes of transmission are unknown. A survey in southern Arizona revealed high incidence of viral diseases of cacti only in areas disturbed by humans.

SOV occurs widely in the western United States, where it is associated with prominent concentric and interlocking chlorotic rings on the pads of various species of prickly pear such as *Opuntia basilaris, O. chlorotica, O. ficus-indica, O. phaeacantha,* and *O. violacea.* Often some pads of a diseased plant lack the chlorotic rings, but all pads contain virus. SOV is believed to cause the chlorotic rings, although it has not been proved to do so. SOV closely resembles members of the tobamovirus group. Its particles are rigid rods 18 × 321 nm, similar to but slightly longer than those of tobacco mosaic virus. Tobamovirus particles contain single-stranded RNA, and the entire genome is contained in one particle.

Cactus virus X, which consists of flexuous, rod-shaped particles, is widespread in various species of cacti in Europe and North America. This virus is associated with a deformative disorder of California barrel cactus in southern California. Diseased barrel cacti have twisted and malformed spines, areoles arranged in disorganized manner along the affected stem, depressed necrotic spots, and systemic mottle. The basal part of the plant is usually free from deformity, presumably because it matured before infection, CVX was isolated from diseased plants and caused systemic mosaic when inoculated to healthy barrel cactus, but the entire syndrome was not reproduced after artificial inoculation. CVX has also been associated with stunting, malformation, and systemic mottling of night-blooming cereus and of other cacti grafted to it. External symptoms that CVX may cause in prickly pear cacti are unknown. When sap from a cactus with microscopic spindle-shaped intracellular bodies (indicative of viral infection in cacti) was injected into rooted segments of an *Opuntia* species (*O. brasiliensis* or *O. bahiensis*), however, sunken chlorotic flecks and mosaic developed after 18–22 months on segments that grew from those inoculated.

Intracellular spindle-shaped bodies containing virus result from infection by either SOV or CVX and perhaps other cactus viruses. Therefore, although the spindle bodies indicate viral infection, they are not useful for specific diagnosis.

Particles of CVX are about 13 nm wide and have a normal length of between 511 and 519 nm. CVX belongs to the potex virus group of single-stranded RNA viruses. Zygocactus virus and zygocactus virus X are similar to CVX, but no externally visible symptoms have been associated with them.

Saguaro virus causes no apparent symptoms and induces no spindle bodies in its only known natural host, the saguaro. The virus consists of isometric particles about 32 nm in diameter.

Cactus virus 2 is a carlavirus that occurs in symptomless cacti as flexuous rods 11–13 nm wide with normal length of 650 nm.

Variegation of camellia. Common and sasanqua camellias are subject to a disease of presumed viral cause that was once confused with genetically controlled variegation. The disease has been called leaf and flower variegation or camellia yellow spot. Symptoms on leaves include yellow mottling or circular to irregular or elongate yellow or white zones, sometimes in a mosaic but often concentrated along veins or margins; chlorosis or albinism of entire leaves; roughening or corkiness of the epidermis; and sometimes necrosis of white or chlorotic tissues. Ring spots have also been reported but may represent a different disease. Symptoms are most severe on leaves formed early in the season and may be absent from those formed last. Flowers show aberrant color patterns that range from scattered white flecks to large round or irregular white spots or elongate, radiating white zones following veins. Growth of diseased plants is not greatly affected, but leaves with prominent white or yellow zones are abnormally susceptible to sunburn and frost injury. Symptoms can be transmitted by grafting, but the causal agent has not been characterized. Several strains of the causal agent have been designated on the basis of different symptoms caused in indicator varieties of camellia to which diseased scions were grafted. Infected rootstocks and attractively variegated plants, although diseased, have been widely dispersed in horticultural trade. Thus it is not surprising that an infectious variegation of camellia has also been reported from Germany. Modes of natural transmission are unknown.

Viruses of magnolia. Except for one European report, no information is available about viral diseases of magnolia. The cause of the striking green and white mosaic and chlorotic rings shown here on great-leaved magnolia is unknown. In Europe, saucer and yulan magnolias are subject to a viral disease characterized by chlorotic lines, arcs, rings, and oak-leaf patterns on leaves. A strain of cucumber mosaic virus was isolated from the diseased plants, but its ability to cause the original symptoms was not tested.

Oak viruses. Several viral or viruslike diseases of oak are known, but none in North America is known to cause significant deformity or growth loss. In Arkansas, chlorotic ring spots, chlorotic zones along veins (vein banding), and line patterns of the oak-leaf type on black and blackjack oaks have been described and shown to be associated with flexuous rodlike particles resembling viruses. One instance of graft transmission to oak was reported, but attempts to transmit symptoms to herbaceous plants failed. A graft-transmissible yellow or chlorotic mottle of red oak was found in Denmark, but the nature of the causal agent is unknown. Also in Europe, three strains of tobacco mosaic virus (TMV) were isolated from English and sessile oaks with chlorotic flecks, mottle, mosaic, and ring spots. In addition, particles resembling those of TMV were associated with chlorotic lesions in narrow, deformed leaves of turkey oak (*Quercus cerris*) and Spanish chestnut. None of the symptoms mentioned here, however, has been shown to be caused by TMV. This virus is apparently widespread in symptomless oaks in California, where it was isolated from blue, Engelmann (mesa), English, holly, Kellogg, California live, Mexican live, white, and willow oaks and *Q. aegilops.* It has been suggested that TMV spreads with conidia of the powdery mildew fungus *Sphaerotheca lanestris* among oaks in California. The chlorotic ring spots shown here on myrtle oak indicate possible viral infection, but nothing more is known about this disorder or its possible relationship to ring spots or other viruslike symptoms on other oaks.

References: 25, 72, 73, 237, 336, 348–50, 398, 603, 653, 884, 1024, 1079, 1299, 1300, 1409, 1410, 1534, 1535, 1705

A, B. Chlorotic ring mottle on prickly pear (*Opuntia ficus-indica*) infected with Sammons' Opuntia virus (AZ, Apr).

C. Leaves of common camellia showing mosaic and chlorosis representing the foliar phase of leaf and flower variegation, a graft-transmitted disorder believed to be caused by a virus (MS, May).

D, F. Chlorotic ringspots indicating possible viral infection of myrtle oak (FL, Mar).

E. Mosaic and chlorotic rings indicating possible viral infection of a leaf of great-leaved magnolia (NY, Jun).

Plate 203

American Mistletoes (Plates 204–206)

Mistletoes and dwarf mistletoes are shrubby or dwarfed, photosynthetic, plant-parasitic seed plants that belong to the families Eremolepidaceae, Loranthaceae, and Viscaceae. Mistletoes in three genera of Viscaceae but none of the other families occur in North America. *Arceuthobium* (dwarf mistletoes), *Phoradendron* (American mistletoes), and *Viscum album* (a European species introduced to California) are represented. All are parasitic on the stems of woody plants, from which they derive water, mineral nutrients, and presumably small amounts of the organic compounds carried in xylem sap. The so-called true mistletoes (*Phoradendron* and *Viscum*) derive most of their organic nutrients from the products of their own photosynthesis and obtain little carbon from the host. The dwarf mistletoes (Plates 207–211), although photosynthetic, extract most of their carbon from the host. This plate and the following two introduce the American mistletoes.

General characteristics and habits. *Phoradendron* is a large genus (perhaps 170 species) of primarily tropical and subtropical evergreen plants restricted to the Americas. Twelve species occur in the United States and 11 in the temperate zone, their northern limit describing a skewed arc from Oregon (45°N) through southern Kansas (37°N) to New Jersey (40°N). They parasitize only angiosperms in the East but attack both gymnosperms and angiosperms in the West. The northern limits of mistletoes seem to be determined mainly by winter temperature, since host plants of a particular mistletoe species always occur farther north than does the mistletoe itself. Economic damage by *Phoradendron* is considered to be slight, although these parasites cause decline of many trees, especially across the Southwest. Fruit-bearing branches of the eastern mistletoe, *P. serotinum,* and a southwestern species, *P. tomentosum,* are collected and marketed for Christmas ornament. The fruit, although eaten by birds, is reported to be toxic to humans and livestock.

Phoradendron species have simple leaves, oppositely arranged, either expanded or scalelike, depending upon species. The stem is woody, at least near the base, and the nodes are closely spaced and somewhat swollen, so that the plants sometimes appear jointed. The stem contains chlorophyll and is the principal site of photosynthesis in species with scalelike leaves. Stems are much branched and densely clustered in comparison with those of host plants; thus the mistletoe often appears as a spherical bunch of dense vegetation, prominent on deciduous hosts during winter. Some mistletoes are monoecious, but all species that occur in the United States are dioecious (male and female flowers on separate plants). Flowers are whitish and generally inconspicuous. Fruits are nearly spherical white or straw-colored to reddish berries 3–6 mm in diameter. Each contains a single seed surrounded by viscid pulp. Mistletoe plants develop well in full sunlight and reach most extensive development high in the crowns of large trees.

Initial infection usually occurs on a small branch and is followed by multiple infections on the same tree after the initial plant produces fruit. The dominant symptom caused by mistletoe is atrophy and dieback of branch ends beyond the point of attachment of the parasite.

Galls, elliptic swellings, or clusters of host twigs sometimes develop at loci of infection. Trees growing in the open or in disturbed forests with open canopies are infected more frequently than those in undisturbed forests with closed canopies. This reflects the roosting preference of birds that eat mistletoe berries and disperse the seeds. Mistletoe may increase dramatically within a single tree where birds feed on berries, roost, and deposit seeds on twigs and branches. Multiple infections result in loss of vigor, dieback, and often death.

Birds that disperse mistletoe seeds include robins, thrushes, bluebirds, phainopeplas, and cedar waxwings. The birds ingest the fruit and digest the pulp, but the seed passes quickly through the intestinal tract, retaining a sticky covering of hairlike threads that serve to glue it to the surface on which it falls. Seeds can germinate anywhere if temperature and moisture are suitable, but only seeds that lodge on thin bark of twigs and small branches of a suitable host will cause infection.

Upon germination, the radicle flattens itself against the bark, forming an attachment disc, or holdfast. A multicellular projection called the primary haustorium grows from the undersurface of the holdfast and penetrates the bark, often through lenticels or axillary buds. Once beneath the periderm in living cortical tissue or secondary phloem, the primary haustorium produces a radiating system of branches termed cortical strands or cortical haustoria. Wedge-shaped projections called sinkers grow from the cortical strands and pass through the cambium to the outer surface of the lignified xylem. Certain cells within the sinker differentiate into water-conducting tracheids and vessels. Some of these come into intimate contact with vessels or tracheids of the host such that open pits and perforations connect the water-conducting systems of the two plants. This assures transport of water and minerals to the parasite.

Sinkers elongate perennially by growth from a basal (intercalary) meristem that differentiates where the sinker passes through the vascular cambium of the host. Activity of this meristem is synchronized with that of the host so that the sinker elongates as the host stem increases in radius. Sinkers thus become deeply embedded in wood. The oldest and most deeply embedded sinkers occur beneath the original site of infection. Invasion of the host stem does not trigger compartmentalizing responses; therefore the parasite may persist more or less in harmony with its host for many years.

Aerial shoots begin to grow after the system of cortical strands and sinkers is initiated. The first shoots arise from buds on the holdfast, and they grow only a few millimeters during the 1st year. Additional shoots may grow later from buds that develop on the outer surfaces of cortical strands. The rate of shoot growth increases to several centimeters per year after the system of cortical strands and sinkers (endophytic system) is well established.

The longevity of the endophytic system seems limited only by that of the host and may extend to hundreds of years. Aerial shoots and stems, however, usually survive only a few years before being broken off or killed by freezing. When a mistletoe stem dies, multiple new shoots often develop from adventitious buds on the cortical strands.

A. Eastern mistletoe (*Phoradendron serotinum*) on sour gum trees. Branches of adjacent ash trees are unaffected (WV, Dec).

B. *P. tomentosum* on live oak (TX, Apr).

C–G. Eastern mistletoe on oak. C. A vigorous mistletoe growing from a small oak branch. D. An unusually large mistletoe on the trunk of a young tree. E, F. Close views of the attachment of the plant shown in (C). Bark tissues of host and parasite meet in a convoluted line at the swollen union. Dissection shows continuity of xylem of host and parasite. Greenish mistletoe xylem extends nearly to the center of the oak branch, where infection apparently occurred at a twig axil when the branch was 1 year old. G. A transverse section through a mistletoe attachment 5 years after infection. The mistletoe, at top, has light-colored wood. Infection occurred on the base of a 1-year-old sprout on a stem that was 7 years old. The parent stem (lower portion) continued normal growth. Abnormally rapid wood production at the site of infection caused the sprout base to swell to nearly the same size as the parent stem. Mistletoe sinkers extend as light-colored wedges to the oldest annual ring of wood in the sprout (FL, Mar).

Plate 204

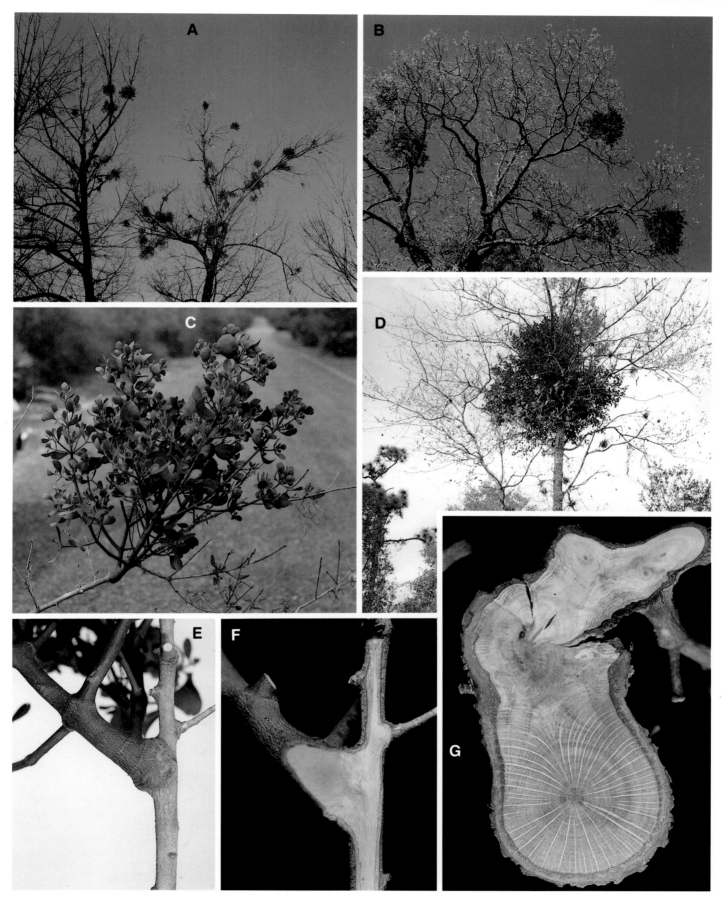

Mistletoe tissues are capable of maintaining greater osmotic potential than tissues of the host; thus the parasite preferentially receives water during times of water shortage. Moreover, mistletoes waste water by continuing to transpire even when under water stress. This causes abnormally severe water stress in hosts and is considered to contribute to loss of vigor and dieback. Organic nutrients are translocated from mistletoe leaves and stems to the cortical strands and sinkers but do not pass to the host. When a leafless host branch terminates in a mistletoe plant, the branch receives water and minerals as the result of the mistletoe's transpiration. Organic nutrients necessary to sustain the branch are translocated from other parts of the host.

Mistletoes on angiosperms. Six American mistletoes and the European species *Viscum album* parasitize angiosperms in the United States. In general, these mistletoes may be separated on the basis of geographic and host ranges.

Eastern mistletoe, *Phoradendron serotinum* (syn. *P. flavescens*) (Plate 204), occurs throughout eastern and southern states south of a line from southern New Jersey to southeastern Kansas. Local northward or southward deflections of the line occur in river valleys or along mountain chains, respectively. Over the greater part of its length the line corresponds to the isotherm for mean minimum January temperature of −4.5°C.

Eastern mistletoe, superficially similar to *V. album*, is the species used most often by Americans for Christmas ornament. It parasitizes about 110 host species in 50 genera. Host groups include ash, basswood, beech, birch, buckeye, camphor tree, cherry, chestnut, chinaberry, dogwood, elm, sour gum, sweet gum, hackberry/sugarberry, hickory, honey locust, maple, oak, osage orange, pecan, persimmon, sassafras, sycamore, tulip tree, walnut, and willow. In a given locality, *P. serotinum* often displays distinct host preference that cannot be explained simply on the basis of host abundance. The host most commonly parasitized in southern Mississippi, for example, is water oak. Sour gum is preferred on the Allegany Plateau in southern Ohio, while silver maple and American elm are preferred in several Ohio River valley drainages, and shagbark hickory or elms are most commonly attacked in parts of North Carolina, Tennessee, and Georgia. Across much of the South, this mistletoe is most abundant in moist lowland areas.

P. serotinum produces flowers from October to late November in the northern part of its range and into January in Florida. The male flowers are jointed spikes with a few tiny white flowers per segment. Insects act as pollinators. White berries 4 mm in diameter mature in autumn or winter 1 year after pollination.

In most areas of occurrence, *P. serotinum* does not cause great damage, perhaps because acute water stress is seldom prolonged in the humid Southeast. Sugarberry and water oak in eastern Texas and water oak in Mississippi, however, have reportedly been killed within several years as branches died back to sites of mistletoe infection successively closer to the trunk.

Seeds of *P. serotinum* germinate readily in bright light and are inhibited by darkness. Only young stems are susceptible. Trunks and major limbs do not become infected directly but may be invaded by cortical haustoria of mistletoe established on sprouts close to the trunk or limb.

The only other mistletoe in the East is the West Indian species, *P. rubrum,* which parasitizes mahogany in southern Florida.

P. tomentosum (Plate 204) occurs from central Oklahoma south and southwestward into Mexico on elm, mesquite, oak, osage orange, and sugarberry. It has been reported as having killed sugarberry in Oklahoma and Texas. This mistletoe, with moderately to densely pubescent segments in its floral spikes, grows to a diameter of 1 meter or more. It blooms in winter, the male inflorescence being larger and with more flowers per segment than that of *P. serotinum,* which occurs on some of the same hosts in Oklahoma. The berries are nearly white, 4–6 mm in diameter.

P. macrophyllum ranges from northern California into Mexico and across the Southwest to western Texas. It parasitizes at least 60 species in about 30 genera, including ash, cottonwood, black locust, sycamore, walnut, and willow but not oaks. This mistletoe causes galls on branches of black locust. On Fremont cottonwood it causes galls on branches and large burls on the trunk. It reportedly kills cottonwoods in southern New Mexico.

P. villosum, the so-called hairy mistletoe, occurs from northwestern Oregon through California into Mexico. Its leaves and young stems are covered with whitish or yellowish hairs. Flowers are produced in July to September. The berries are white to pink, about 3 mm in diameter. This mistletoe prefers oaks and usually grows as spherical clumps on high branches, but it may also cause elliptic swellings on limbs or trunks as shown here on Oregon white oak. *P. villosum* also parasitizes California buckeye, manzanita, walnut, and other plants. This species also has unexplained host preferences that vary from one locality to another. California live oak, for example, is commonly parasitized in southern California but not in the San Francisco Bay region, while blue and valley oaks are parasitized in both areas.

P. coryae occurs on various oaks and occasionally other plants from Arizona to western Texas and in adjacent Mexico. This species has dense clusters of star-shaped hairs on its leaves. It may cause large galls and is the apparent cause of dieback of oak limbs parasitized for many years, as shown here on Emory oak.

P. californicum, sometimes called desert mistletoe, is a nearly leafless species common on leguminous trees and shrubs, especially mesquite and paloverde, throughout the deserts of the Southwest and northwestern Mexico. This mistletoe produces long pendulous stems and, in time, dense bushes that attain lengths greater than 1 meter. Its leaves are minute and scalelike. The stems vary from green to reddish. Small white flowers are produced in December–March. The berries are white to red, about 3 mm in diameter.

V. album, the Christmas mistletoe of Europe, was apparently introduced to California in 1900 by the famous horticulturist Luther Burbank. It occurs now on more than 20 native and introduced trees and shrubs within a few kilometers of Burbank's experimental farm in Sebastopol, California.

Damage by mistletoe to valuable trees can be retarded or prevented by removing or killing the parasite. Details are presented with Plate 206.

A, B. *Phoradendron coryae* on Mexican blue oak and Emory oak, respectively. Galls 6–25 cm or more in diameter, caused by the mistletoe, are visible on the dead limb of Emory oak. The mistletoe is presumed to have contributed to death of the branch (AZ, Jul).

C. Shoots of hairy mistletoe, *P. villosum,* growing from the edges of a large swelling on a trunk of Oregon white oak. The original site of infection may have been a small sprout from which cortical strands of the mistletoe invaded the trunk (OR, Jul).

D–F. Desert mistletoe, *P. californicum,* on mesquite. D. Dense growth of mistletoe on deformed host plants. E. A typical mistletoe plant with long, slender, pendant, leafless branches. F. Close view of mistletoe twigs and branches (AZ, Jul).

Plate 205

Mistletoes on gymnosperms. Five species of *Phoradendron* occur on gymnosperms, all in western states. These parasites display greater host specificity than do the common mistletoes on angiosperms in the United States. One infects white fir, and the others attack members of the cypress family. They cause only slight economic damage; therefore researchers and teachers have paid less attention to them than to the dwarf mistletoes (Plates 207–211). The general characteristics, life history, and mode of parasitism of mistletoes on gymnosperms are as described for American mistletoes as a group (Plate 204).

P. bolleanum subsp. *pauciflorum* grows as prominent dense, dark green, often globose bushes 60–100 cm in diameter in the tops of white fir trees from central California to Mexico and in one locality in Arizona. Its stems are smooth to slightly hairy, with internodes 6–21 mm long, and they become woody with age. Leaves are 15–30 mm long and up to 8 mm wide. The male inflorescence has one or two segments, each with several tiny flowers. The female inflorescence has one or two segments, each with two flowers. The fruit is white and smooth, about 4 mm in diameter. Fir trees with large or multiple mistletoes often die back from the top, and they are abnormally susceptible to attack by the fir engraver beetle, *Scolytus ventralis*. Cone production is suppressed also. In the United States, white fir is the only member of the pine family to be parasitized by a true mistletoe and the only conifer subject to infection by both true and dwarf mistletoes. Cold winters in the northern part of the range of white fir prevent the mistletoe from succeeding there.

P. densum occurs on cypresses in northern California and central Arizona and on junipers from southern Oregon southward to Baja California, Mexico. It sometimes causes atrophy and death of juniper branches beyond the point of attachment and thus comes to reside on the clubbed ends of living branches that lack juniper foliage. Its leaves are smooth, 10–20 × 3–5 mm. Observations of *P. densum* on Sierra juniper in northern California after the unusually cold winter of 1932 furnished compelling evidence of the limitation of leafy mistletoes by low temperature. In an area where this mistletoe was abundant, 90% of the plants, including both aerial shoots and endophytic systems, were killed.

P. juniperinum, the most widespread western *Phoradendron,* is a nearly leafless species that parasitizes junipers and incense cedar from Oregon and Colorado southward into Mexico. This species grows into bushes 40–80 cm in diameter. The mature stems are woody and their surfaces smooth. The leaves are scalelike, only about 1 mm long. Flowers are produced in July–September. The male inflorescence consists of one or two segments, each with five to nine flowers. Female inflorescences have one segment with two flowers. Fruits are pinkish white and smooth, about 4 mm in diameter.

P. juniperinum includes two subspecies that are differentiated on the basis of host preference and morphology. Subspecies *juniperinum* is the most common mistletoe on juniper and occurs rarely on Arizona cypress. Its internodes are usually less than 1 cm long, and the plants usually remain erect. It occurs from central Oregon and Colorado to southern California, western Texas, and Mexico. In Arizona it has been associated with, and presumed responsible for, an abnormally high mortality rate of Utah juniper. Other hosts include alligator, one-seed, Rocky Mountain, and Sierra junipers and, rarely, fernbush.

P. juniperinum subsp. *libocedri,* restricted to incense cedar, occurs intermittently from Oregon through California into Mexico. Its internodes are usually more than 1 cm long, and the plants tend to become pendulous with age. Normally this mistletoe is found as bunches of vegetation on branches throughout the crowns of incense cedars, associated with thin-appearing host foliage and occasionally with host dieback or death. Trunk infections also occur, presumably when the parasite grows as cortical haustoria from a small side branch into the trunk. This leads to spindle- or barrel-shaped swellings up to 2 meters long, usually on upper parts of the mainstem, from which mistletoe shoots grow. With time, the mistletoe may cease producing shoots, perhaps because it is unable to force its way through the thick covering of corky bark on old stems. Its endophytic system remains alive, however, nourished by the tree. This mistletoe, persisting as an endophytic system in a swelling on the trunk of an old incense cedar, holds the North American record for longevity—over 400 years as estimated by counting annual rings of host wood that developed around sinkers.

Junipers in the Southwest are also hosts for two little-known mistletoes, *P. capitellatum* in southern Arizona and southern New Mexico and *P. hawksworthii* in southern New Mexico and western Texas. The ranges of both species extend into Mexico. *P. capitellatum* has hairy leaves measuring 8–14 × 1–2 mm. *P. hawksworthii* has smooth leaves 6–20 × 1.4–3 mm.

Mistletoe on either gymnosperms or angiosperms can be controlled by removing infected branches, cutting at least 30 cm proximal of mistletoe shoots. Mistletoe shoots on limbs or trunks may be broken off or cut off. This reduces the parasite's demand for water but must be done repeatedly because new sprouts will grow from the cortical haustoria. Treatment of the mistletoe with certain herbicides while the host is dormant has also proven effective for control.

References for mistletoes in general: 296, 297, 650, 863, 913, 1071–74, 1130, 1373, 1674, 1696, 2192

References for mistletoes on angiosperms: 297, 539, 863, 1071, 1072, 1674, 1696, 1697, 1875, 2005, 2192

References for mistletoes on gymnosperms: 560, 755, 2080, 2083, 2192

A, B. Fir mistletoe, *Phoradendron bolleanum* subsp. *pauciflorum,* in the tops of white fir trees. A. Foliage of the parasitized tree is thin, and twig dieback is evident. B. Mistletoe has colonized both tops of a twin-stemmed tree, and the limber tops have bent under the weight of the mistletoe, augmented by snow in winter (CA, Jul).

C, E, F. *P. juniperinum* subsp. *juniperinum* on junipers. C. Mistletoe (dense olive green masses) associated with sparse foliage and dieback on Rocky Mountain juniper. E. Close view of a small mistletoe plant on Sierra juniper. The branch tip beyond the parasite has died (CA, Jul). F. Close view of mistletoe twigs and fruit. Scalelike leaves are visible as pairs of tiny opposing projections at intervals along the twigs (AZ, Apr).

D, G. *P. juniperinum* subsp. *libocedri* on incense cedar. D. A large cluster of shoots on the mainstem in the top of a declining tree. Host foliage is sparse, and dieback has begun. G. Pendant mistletoe plants, typical of this subspecies, are visible on low branches (CA, Jul).

Plate 206

Dwarf Mistletoes (Plates 207–211)

Dwarf mistletoes (genus *Arceuthobium* of the Viscaceae) are small, leafless, chlorophyllous seed plants that parasitize stems of gymnosperms, extracting water, minerals, and carbon compounds, especially sugars. These are considered to be the most evolutionarily advanced of the mistletoes. We first consider the group as a whole and then present information about the species that are illustrated. Because of this arrangement, information about a specific dwarf mistletoe may not face the plate that displays it. Discussion of *A. americanum* begins opposite Plate 208.

Dwarf mistletoes occur around the Northern Hemisphere. At least 34 species plus 6 infraspecific taxa are recognized. There are 5 species in Canada, and these plus 11 more occur in the United States. Dwarf mistletoes range from Newfoundland to southern Alaska in the boreal forests of the North, throughout the coniferous forest zones of western North America, and into Central America and the Caribbean region, but they do not occur in the pine forests of the southeastern United States.

A dwarf mistletoe plant consists of nonwoody shoots anchored by an endophytic system composed of a rootlike array of strands in the inner bark with extensions (sinkers) into the wood. These plants are unusual in having stems that lack a central vascular cylinder and sieve tubes, fruits that explode when ripe, and seeds that contain chlorophyll in the endosperm and produce radicles with stomata. The major symptoms caused by dwarf mistletoes are witches'-brooms, loss of vigor, dieback, and death.

Although the dwarf mistletoes of North America infect only members of the pine family, they cause more damage than any other group of pathogens in forests of western North America. The annual loss of wood due to mortality and growth suppression by dwarf mistletoes in the western United States was estimated in 1984 to be 1.7×10^7 cubic meters. Height growth is suppressed relatively more than growth in girth. Direct damage also includes reduction of tree seed quality and quantity and reduction of wood quality. Infected wood has subnormal strength and poor pulping quality due to abnormally short, distorted tracheids and a high proportion of ray tissue. Diseased trees may also have distorted trunks and abnormally large knots, which are undesirable in lumber. In addition, dwarf mistletoes predispose trees to attack by opportunistic insects and fungi (which are usually the direct causes of death), and they influence forest succession through their effects on host growth and flammability. Dead branches and trees in stands damaged by dwarf mistletoes are an abnormally great fire hazard. Dwarf mistletoes also weaken, disfigure, and kill landscape trees in residential or resort areas in coniferous forest zones.

Most dwarf mistletoes are host-specialized. Each species parasitizes one or a few principal and secondary hosts, but each also has limited capability to infect diverse conifers (occasional and rare hosts) where these grow among severely infected trees of a principal host species. Dwarf mistletoes have common names that reflect their principal hosts or region of occurrence. Dwarf mistletoes in Canada and the United States and their principal and secondary hosts, but not occasional or rare hosts, are as follows:

A. abietinum, fir dwarf mistletoe, from Washington to southern California and also in scattered localities in Nevada, Utah, and Arizona; includes host-specialized forms: f. sp. *concoloris*, primarily on grand and white firs and secondarily on Pacific silver fir and Brewer spruce, and f. sp. *magnificae*, on noble and California red firs.

A. americanum, lodgepole pine dwarf mistletoe (Plate 207), on jack and lodgepole pines, secondarily on ponderosa pine, common from British Columbia to Manitoba and from California to Colorado.

A. apachecum (Apache dwarf mistletoe, Plate 210) and *A. blumeri*, on southwestern white pine in Arizona, New Mexico, and Mexico.

A. californicum, sugar pine dwarf mistletoe, on sugar pine and western white pine in California and Oregon.

A. campylopodum (western dwarf mistletoe, Plate 208) on Jeffrey, knobcone, and ponderosa pines, secondarily on Coulter pine from Washington to Baja California and inland to Idaho and Nevada.

A. cyanocarpum, limber pine dwarf mistletoe, on bristlecone, limber, and whitebark pines from California to Montana and Colorado.

A. divaricatum, pinyon dwarf mistletoe, on pinyon, Mexican pinyon, and singleleaf pinyon from California and Baja California to Colorado and western Texas.

A. douglasii, Douglas-fir dwarf mistletoe, on Douglas-fir (Plate 209) from southern British Columbia to California and central Mexico.

A. gillii (Plate 210) on Chihuahua pine in Arizona and New Mexico and on Chihuahua and other pines in Mexico.

A. laricis, larch dwarf mistletoe (Plate 210), on western larch and secondarily on alpine fir, mountain hemlock, and lodgepole pine in British Columbia and northwestern states.

A. microcarpum, western spruce dwarf mistletoe, on bristlecone pine and blue and Engelmann spruces in Arizona and New Mexico.

A. occidentale, digger pine dwarf mistletoe (Plate 208), on Bishop, digger, and Monterey pines, and secondarily on Coulter and knobcone pines, in California.

A. pusillum, eastern dwarf mistletoe (Plate 211), on black, red, and white spruces from Newfoundland to Pennsylvania, Minnesota, and Saskatchewan.

A. tsugense, hemlock dwarf mistletoe (Plate 209), on mountain and western hemlocks, secondarily on alpine, noble, and Pacific silver firs and shore, western white, and whitebark pines from southeastern Alaska to California.

A. vaginatum subsp. *cryptopodum*, southwestern dwarf mistletoe (Plate 210), on the Arizona variety of ponderosa pine from Utah and Colorado southward into Mexico, where it occurs also on Apache pine. Two other subspecies occur on pines in Mexico.

Most dwarf mistletoes remain localized near the point of infection and cause spindle-shaped swelling of the host stem. Some species, however, notably *A. americanum*, *A. douglasii*, and *A. pusillum*, grow systemically in the host branch and induce little or no swelling. Most species also induce witches'-brooms. Brooms are of two types, called systemic if the endophytic system keeps pace with apical growth of the branch, resulting in a systemically infected broom, or nonsystemic if the endophytic system remains concentrated near the original site of infection. Aerial shoots develop along the branches of systemically infected brooms but only near the original site of infection on nonsystemic brooms. Cones usually do not form on brooms. If the tree supports multiple infections, the noninfected branches gradually weaken and produce less foliage than normal so that the contrast between mistletoe brooms and the remainder of the crown becomes dramatic. Dieback ensues.

Lodgepole pine dwarf mistletoe, *Arceuthobium americanum* (description begins on page 428).

A. Witches'-brooms caused by *A. americanum* on a declining lodgepole pine. The top of this tree, including brooms on upper branches, is dying. Brooms on lower branches appear dense and dark green in comparison with parts above (OR, Sep).

B. Witches'-brooms caused by *A. americanum* on ponderosa pine, an occasional host in areas where the mistletoe is common on lodgepole pine (OR, Sep).

C. Twig mortality, the result of rodents gnawing infected bark, in brooms on lodgepole pine (CO, Jun).

D, E. Female and male plants, respectively, of *A. americanum* on lodgepole pine. The branch in (E) has a spindle-shaped swelling at the site of infection (OR, Sep).

Plate 207

The aerial shoots of dwarf mistletoes are perennial and branched at least once. They appear jointed because each segment originates in a more or less cuplike depression at the node. Shoots of various species range in height from a few millimeters to over 70 cm and in habit from dense clusters to scattered individuals. Each shoot grows from a basal cup that remains visible on host bark after the shoot dies and/or breaks off. Flowers on both male and female plants are minute. The fruits, 3–5 mm long in most species, contain viscid pulp and a single seed.

Dwarf mistletoe seeds are dispersed from midsummer to late autumn, depending on the species. The seeds are shot as projectiles from ripe fruit that burst with explosive force. Seeds of species that have been studied are discharged at velocities of 22–26 meters per second! Most seeds fall within 5 meters of the source tree, but some travel up to 15 meters or more. Seeds have a viscid coating that causes them to stick to twigs, foliage, or other objects. Seeds intercepted by conifer foliage remain there until lubricated by rain. Then they either fall to the ground or slide down the needle and stick to the twig. Dwarf mistletoe populations increase, and damage develops most rapidly in relatively open, unevenaged forests or on steep slopes where seeds are scattered from diseased trees in the overstory upon young trees below.

Squirrels and birds including chickadees, jays, nuthatches, warblers, and woodpeckers also disperse dwarf mistletoe seeds that stick temporarily to their fur or feathers. This occurs when animals disturb ripe fruits and cause seed expulsion. Later the seed may be wiped off on a healthy shoot or twig, especially where birds preen. This accounts for dwarf mistletoe infection in the tops of tall trees and for new foci of infection far from sources of seeds.

The time of seed germination varies with species from autumn to spring. Germinating seeds carry on photosynthesis, which presumably helps the young plant sustain itself until it parasitizes the host. The short radicle produces a holdfast where it contacts a needle base or other obstruction. A penetrating wedge then grows from the holdfast into the twig. Most infections occur on stems less than 5 years old, although some dwarf mistletoes have the ability to penetrate older parts.

From the penetrating wedge, a system of cortical strands grows in all directions in the living bark, producing sinkers at irregular intervals. Sinkers are wedge-shaped protrusions that grow along or within host rays toward the center of the stem until their tips pass the vascular cambium and encounter lignified xylem. Sinkers consist of parenchyma or parenchyma plus tracheary elements. Some of the latter differentiate in contact with host tracheids, creating a route for the flow of water and minerals from host to parasite. Sinkers elongate and increase in width perennially by growth from a meristem located where the sinker passes through the host cambium. Growth of a sinker is synchronized with that of the host, and the tip of the sinker becomes deeply embedded in wood as the radius of the parasitized stem increases.

The endophytic system of a dwarf mistletoe usually develops 2–5 years before aerial shoots begin to form, and it may persist indefinitely without shoots if the host lacks vigor. Aerial shoots begin to produce flowers, male and female on separate plants, after 1–2 years and fruit 5–19 (usually 12) months later. Insects and wind pollinate the flowers.

Dwarf mistletoes maintain greater osmotic potential than do their hosts. This and the intrinsically higher rate of transpiration from parasite than from host permit the parasite to draw water from the host and continue to transpire even when the host is under such water stress that its stomata close. The transpiration rate of a dwarf mistletoe may be many times greater (surface area basis) than that of its host.

Dwarf mistletoes assimilate some carbon by photosynthesis but at low rates compared with their hosts, and they respire at higher rates than their hosts. Therefore they must extract from the host most of the carbon necessary for growth and respiration. In those species studied, virtually all of the carbon in the cortical strands and sinkers comes from the host. The pathway of carbon from host to dwarf mistletoe is unreported.

Various species of dwarf mistletoe exhibit an unexplained mutual exclusion. In a region where several species of dwarf mistletoe occur, if the principal parasite of a given host is present in a locality, other dwarf mistletoes tend to be excluded from that host. Conversely, other dwarf mistletoes are likely to parasitize that host species if the principal parasite is absent. In stands containing both limber and lodgepole pines, for example, if *A. americanum* is present on its principal host, lodgepole pine, *A. cyanocarpum* (principal host limber pine) does not parasitize lodgepole pine, although it commonly does so in the absence of *A. americanum*. The reciprocal relationship is also known. Infection of a given tree by two species of dwarf mistletoe is rare.

Resistance to a particular dwarf mistletoe occurs not only in its occasional or rare hosts but also in individuals of the principal host species. The bases of resistance are poorly known, however. Rare hosts may be quite susceptible, as evidenced by multiple infections on individual trees.

Several fungal parasites and insects may destroy shoots or fruits of dwarf mistletoes, but no dwarf mistletoe seems significantly limited by biotic agents. Practical control of dwarf mistletoes in forests depends upon cutting practices that remove diseased trees and favor evenaged stands. Clearcutting is often appropriate to sanitize an area where an evenaged stand will be planted or established from seed. In residential areas, damage to individual trees of high value can be suppressed or delayed by pruning infected branches.

Lodgepole pine dwarf mistletoe. *Arceuthobium americanum,* the dwarf mistletoe of jack pine and lodgepole pine, is the most widespread of the western species, occurring from northern Alberta to southern California and central Colorado and from central British Columbia to Manitoba (and formerly at one location in western Ontario) at elevations of 200–3350 meters above sea level. In some large

---

A–F. Western dwarf mistletoe, *Arceuthobium campylopodum,* on ponderosa pine (description begins on p. 430). A, B. Witches'-brooms in mature trees. The tree in (A) shows dense foliage on brooms in the lower crown and thinning and retarded growth of parts above (WA, Jul). B. The degree of proliferation of twigs in a broom is revealed when needles drop after death (OR, Sep). C. Aerial shoots growing from an extensive endophytic system on the trunk of a small tree. Infection began many years earlier near the center of the area of darkened bark. Shoots are sparse near the original site of infection but abundant near the edges of the infected region (OR, Sep). D, E. Female dwarf mistletoe shoots bearing fruit, growing from spindle-shaped swellings on branches. Basal cups where previous shoots died and/or broke off are apparent on the branch in (E) (OR, Sep). F. A relatively young male plant growing as a cluster of shoots from a common origin at the original site of infection (OR, Sep).

G. Digger pine dwarf mistletoe, *A. occidentale,* on digger pine (description on p. 432). Two plants of different sexes (the male darker in color) have developed close together, causing an elongate spindle-shaped swelling on the host branch. Brown basal cups show the locations of former shoots (CA, Sep).

Plate 208

forest areas it causes annual loss equivalent to 25% of annual growth. Its principal host from Saskatchewan eastward is jack pine. *A. americanum* frequently parasitizes ponderosa pine growing in association with lodgepole pine. Occasional or rare hosts of this parasite include Douglas-fir; bristlecone, limber, knobcone, and whitebark pines; and blue, Engelmann, and white spruces. Scots pine in experimental plantings has also been found naturally infected. Shore pine, although susceptible, is seldom infected, because it occurs primarily along the Pacific coast where *A. americanum* does not occur.

Aerial shoots of *A. americanum* are yellowish to olive green with branches in whorls. They average 6 cm high, with extremes to 20 cm. The life cycle requires at least 5 years. In the shortest possible cycle, seeds germinate in spring, and infection begins during summer of the 1st year. Aerial shoots appear 2 years later, and flowers open in March–June, depending on locality, of the 4th year. The blue-green fruit matures, and seeds are scattered in late summer of the 5th year. Often the cycle is extended as several years elapse between infection and growth of aerial shoots.

Infections by *A. americanum* may be localized, inducing stem swellings from which aerial shoots emerge, or may be systemic in a given branch, resulting in witches'-brooms. Aerial shoots on brooms appear first at branch whorls. This species is unusual in its ability to penetrate stem segments of its principal host, lodgepole pine, that are 60 or more years old. Most trunk infections, however, result from growth of the endophytic system from a branch. Cambial activity slows or ceases at the centers of old trunk infections, resulting in trunk deformity. Severely infected branches often die and appear as conspicuous reddish brown "flags." Rodents gnawing infected bark are often the direct cause of branch death.

When a young lodgepole pine stand is subject to infections by seeds from scattered trees of a previous generation, the rate of dwarf mistletoe increase may be rapid. In areas of Alberta where this was studied in trees aged 16–23, infections increased exponentially, with an average doubling time of 1.25 years. The average rate of spread from tree to tree within a locus of infection, however, is only about 0.5 meter per year.

Western dwarf mistletoe. *Arceuthobium campylopodum* parasitizes principally the western variety of ponderosa pine, *Pinus ponderosa* var. *ponderosa*, from Washington and Idaho to southern California. Although the host variety extends into Canada and east of the continental divide in Idaho and Montana, *A. campylopodum* is unknown in those areas. This mistletoe, however, occurs beyond the range of the principal host on Rocky Mountain ponderosa pine (*P. ponderosa* var. *scopulorum*) in southern Nevada and on Jeffrey pine in Baja California. It is found at elevations from 30 meters above sea level in the Columbia River valley to about 2380 meters in the mountains of southern Nevada. Jeffrey pine is damaged as much as ponderosa pine in California and is classed as a principal host there. Coulter and knobcone pines are also commonly infected where they are accompanied by one of the principal hosts. Lodgepole pine is an occasional host, and sugar pine a rare one. Aleppo, cluster, and Scots pines in arboreta or experimental plantings have also been found infected.

Grand fir, white fir, and western larch are susceptible to *A. campylopodum* on the basis of artificial inoculation with seeds, but these species have not been found naturally infected.

Damage by *A. campylopodum* is more severe in southern California than in northern areas and more severe on the eastern than the western slopes of the Cascade-Sierra Nevada cordillera. These trends are associated with moisture and temperature differences: damage is most severe where hosts are most stressed by heat and dryness.

Aerial shoots of *A. campylopodum* are olive green to yellow or yellowish brown, up to 13 cm tall (average 8 cm), with branches in a fanlike arrangement (flabellate branching). Within the color range mentioned, male plants tend toward yellow-brown and female toward green, especially in northern areas. Fruits are bluish to olive green, about 5 mm long.

A. campylopodum has at minimum a 5-year life cycle, as indicated by the fact that swellings with aerial shoots may be found on host branch segments as young as 4 years. Latent infection leading to a much longer life cycle is common, however. In research plots where all aerial shoots and stem swellings were removed by pruning, new swellings or aerial shoots representing over 350 previously latent infections per hectare appeared during the next 9 years.

Flowers of *A. campylopodum* open from August to October, most commonly in mid-September, and seeds mature the next year in late August to late November, most commonly in mid-September. The seeds require an afterripening period before germination, and they germinate the next spring. After penetration, the endophytic system at first radiates from a compact perennial mass of cells in the outer bark, causing the typical spindle-shaped swelling. If the branch remains alive, a witches'-broom may develop at the point of initial infection. Brooms often attain large size (more than 1 meter) on old branches. Infections on mainstems may become more extensive than those on branches. A parasite located on the mainstem may develop an elliptic endophytic system many centimeters long, with aerial shoots arising first at the center but eventually only near the edges, as shown in Plate 208C. This process takes many years, however, because the endophytic system extends less than 2 cm per year on average in each direction.

Pathogenicity of *A. campylopodum* was demonstrated by grafting a population of ponderosa pine seedlings with scions either healthy or infected with dwarf mistletoe. The trees with infected scions subsequently grew only half as much as those with healthy scions, and one-third of the infected trees, versus none of those initially healthy, died within 12 years.

The susceptibility of ponderosa pine to *A. campylopodum* is greatest while trees are young. When scions from trees of various ages were grafted on seedling rootstocks and were inoculated uniformly with dwarf mistletoe seeds, the greatest number of infections developed on scions from the youngest trees, and fewest infections occurred on scions from trees more than 50 years old. Two types of susceptibility associated with damage in old trees have been recognized: susceptibility to infection, allowing the accumulation of many mistletoe plants, and susceptibility to extensive invasion after infection, permit-

A, B, D, G, H. Hemlock dwarf mistletoe, *Arceuthobium tsugense* (description on p. 432). A. Large witches'-brooms caused by dwarf mistletoe on a dead mountain hemlock; dwarf mistletoe was the apparent cause of death (OR, Sep). B. Typical witches'-brooms in the lower crown of western hemlock (WA, Jun). D. A male plant in blossom on a spindle-shaped swelling on a branch of western hemlock (WA, Aug). G. A female plant with ripening fruit on western hemlock (WA, Aug). H. Part of a male plant as it appears in spring or early summer (WA, Jun).

C, E, F, I. Douglas-fir dwarf mistletoe, *A. douglasii* (description on p. 432). C. A large old witches'-broom caused by dwarf mistletoe on a mature Douglas-fir (OR, Sep). E. A small witches'-broom (AZ, Jul). F. Female shoots of *A. douglasii* on the mainstem of the broom shown in (E). I. Shoots of a male plant on a small branch of a similar broom (AZ, Jul).

Plate 209

ting the development of large endophytic systems in trees that may support relatively few mistletoe plants.

Strains of ponderosa pine with drooping needles are less frequently infected by *A. campylopodum* than are trees of typical form because most seeds of the parasite, although intercepted by foliage, slide off when needles are wet. Seeds intercepted by twigs or the basal parts of needles above twigs remain on the tree, capable of normal penetration and infection.

Digger pine dwarf mistletoe. *Arceuthobium occidentale* (Plate 208) is limited to California, where it parasitizes Bishop, digger, and Monterey pines as principal hosts and Coulter and knobcone pines secondarily. Aleppo, Italian stone, and unidentified black pines (Austrian or Japanese) have also been found infected adjacent to infected Bishop or digger pines. This parasite occurs from sea level to above 1400 meters elevation. Aerial shoots vary in color, according to the host: straw colored to light brown on digger pine, olive green on Monterey pine, and dark brown on Bishop pine. Shoots range up to 17 cm (average 8 cm) in height, and branches are arranged in fanlike manner (flabellate branching). Flowers open in autumn, and seeds mature about 13 months later. *A. occidentale* is the most conspicuous of the dwarf mistletoes in the United States because the plants are relatively large and produce dense clusters of shoots on the widely spaced branches of the principal host, digger pine. This dwarf mistletoe has slight economic importance, however, because its hosts have limited commercial value.

The endophytic system of *A. occidentale* in digger pine grows in each direction from the point of infection at an average rate of 2.0–2.3 cm per year, the rate varying with host vigor. The extent of the endophytic system is nearly the same as the zone of stem swelling. *A. occidentale* does not usually induce witches'-brooms, although branches may proliferate somewhat, especially on Bishop and Monterey pines. The presence or absence of a stimulus to broom formation is suspected to be related to the composition of cytokininlike substances (cell-division factors) produced by the parasite. *A. occidentale* and the broom-inducing species, *A. vaginatum,* differ in these substances.

A. occidentale increases exponentially after a lag period of several years. On digger pines artificially inoculated with seeds, for example, an average of four plants with aerial shoots developed per tree within 5–6 years, but the average number rose to 250 per tree during the next 10 years.

Hemlock dwarf mistletoe. *Arceuthobium tsugense* occurs from southeastern Alaska to central California at elevations from sea level to 2460 meters. This species has a broader host range than most dwarf mistletoes. The principal hosts are mountain and western hemlocks. Secondary hosts include alpine, noble, and Pacific silver firs and whitebark pine. Grand fir and Brewer spruce are occasionally infected, and Douglas-fir, western white pine, and Engelmann and Sitka spruces rarely so. In addition, a race that parasitizes shore pine, and western white pine secondarily, is recognized in the vicinity of Vancouver Island. Many additional conifers have been proved susceptible by artificial inoculation.

Aerial shoots of *A. tsugense* range to 13 cm tall (average 5 cm) with fanlike branching, and vary in color from green through straw-colored to reddish, darker in winter than in summer. The life cycle requires 4–6 years. Flowers open in late summer to late autumn, depending on locality, and seeds are dispersed 13–14 months later. Seeds germinate in February to May, depending on local climate, and stem swelling

appears after 1–2 years. Aerial shoots appear after another 1–2 years, and fruit a year later.

A. tsugense induces prominent witches'-brooms. In addition, the trunk may swell and cankers may be caused by secondary fungi where the endophytic system grows from a branch into the trunk. Opportunistic fungi also cause branch dieback of fir infected by hemlock dwarf mistletoe. This parasite develops best on trees or branches in the open, on which the population may double every 2 years or less. A single tree may bear over 4000 infections. Seeds are expelled mainly during daylight and may travel as far as 15 meters, although most fall within 5 meters. New infections occur progressively higher on diseased hemlocks, ascending at 30–65 cm per year. Since vigorous trees often grow faster than this, however, they may outgrow the parasite unless they are overtopped by old, diseased trees. Mistletoe-induced brooms that become shaded on low branches, as shown in Plate 209D, eventually perish without doing much harm.

Douglas-fir dwarf mistletoe. *Arceuthobium douglasii* occurs from southern British Columbia into Mexico at elevations from about 275 meters in the North to over 3000 meters in the South. It does not occur in the coastal forests of the Pacific Northwest, where Douglas-fir grows most rapidly and attains greatest size. *A. douglasii* occurs secondarily on cork fir in Arizona, occasionally on grand fir, and rarely on alpine and white firs, limber pine, and blue and Engelmann spruces.

A new infection by *A. douglasii* may remain latent for some years, or the branch may begin to swell and a broom develop at the point of infection. Stem swelling, indicating growth of the endophytic system, extends slightly more than 1 cm per year. This parasite becomes systemic within brooms, reaching even into buds and the primary tissues of young twigs. Aerial shoots arise along host branches at least 3 years old within brooms. Brooms may become massive, sometimes involving the entire living part of the crown, and they are associated with severe growth reduction, dieback, and death.

Aerial shoots of *A. douglasii* are typically olive green but sometimes reddish, up to 7 cm tall but averaging only 2 cm. Flowers open in spring, and seeds are dispersed 17–18 months later, typically in September. They germinate the next spring. Aerial shoots arise 2 years later and produce flowers after another year; thus the life cycle is at least 6 years long.

Southwestern dwarf mistletoe. *Arceuthobium vaginatum* subsp. *cryptopodum* parasitizes and causes witches'-brooms on the Arizona and Rocky Mountain varieties of ponderosa pine from Colorado and Utah southward into Mexico (where Apache pine is also a principal host) at elevations of 1680–3000 meters. It also occurs occasionally on bristlecone and lodgepole pines and rarely on limber and southwestern white pines. This parasite is the most important pathogen of ponderosa pine in the Southwest, reducing radial growth rates by 35–52% in 55- to 140-year-old stands in one area studied. Dwarf mistletoe also predisposes ponderosa pine to attack by the mountain pine beetle (see Johnson and Lyon, 2nd ed., Plate 28). The witches'-brooms grow for decades and often reach diameters greater than 2 meters. Infection is localized in the branch bases of most brooms but becomes systemic in a small minority.

Aerial shoots of *A. vaginatum* subsp. *cryptopodum* are large—up to 10 mm thick at the base (average 4 mm) and up to 27 cm tall (average 10 cm). Usually they are orange to reddish brown, with branches in a fanlike arrangement. Stem colors sometimes vary to greenish, red, yellow, or (rarely) dark purplish. Flowers are pollinated in late spring,

---

A. Witches'-brooms caused by southwestern dwarf mistletoe, *Arceuthobium vaginatum* subsp. *cryptopodum,* on ponderosa pine (CO, Jun).

B. Female plant of Chihuahua pine dwarf mistletoe, *A. gillii* subsp. *gillii,* on the trunk of Chihuahua pine (AZ, Jul).

C. Witches'-brooms caused by larch dwarf mistletoe, *A. laricis,* on western larch (ID, Jul).

D–G. Apache dwarf mistletoe, *A. apachecum,* on southwestern white pine. D, E. Female plant bearing fruit. F, G. Male plants on spindle-shaped swellings on host branches. Numerous pine shoots, signaling the onset of broom formation, are beginning to grow from the swelling in (F) (AZ, Jul).

Plate 210

and fruits about 5 mm long are dispersed in summer of the next year. The average horizontal distance of unimpeded seed flight is 5 meters, but since many seeds are intercepted close to the source, and 6 years elapse between inoculation and production of new seeds, the average rate of horizontal spread of this dwarf mistletoe in forest stands is only 27–52 cm per year. New centers of infection are started by birds and squirrels that inadvertently carry seeds. Seeds germinate in late summer and may infect stems up to 9 years old. Two other subspecies of this dwarf mistletoe occur in Mexico.

The induction of witches'-brooms by *A. vaginatum* is suspected to be related to the composition of cytokininlike substances (cell-division factors) produced by the parasite. Cytokininlike substances have been detected in *A. vaginatum* subsp. *cryptopodum* and in ponderosa pine tissues infected by this parasite but not in comparable tissues of healthy ponderosa pine. Moreover, these substances differ from those of *A. occidentale,* which seldom induces brooms in its primary host.

Chihuahua pine dwarf mistletoe. *Arceuthobium gillii* subsp. *gillii* occurs from southern Arizona and New Mexico into Mexico at elevations of 1700–2650 meters. The only known host in the United States is Chihuahua pine, but the parasite also occurs on *Pinus cooperi, P. lumholtzii,* and the Arizona variety of ponderosa pine in Mexico. *A. gillii* subsp. *nigrum,* also on pines, is restricted to Mexico. The Chihuahua pine parasite produces greenish brown shoots up to 25 cm tall (average 11 cm), as much as 8 mm thick at the base. Flowers are pollinated in spring, and seeds mature on average 19 months later in autumn.

Larch dwarf mistletoe. *Arceuthobium laricis* occurs in a high proportion of stands of western larch (including Lyall larch) in northwestern states and southern British Columbia at elevations of 700–1980 meters. Alpine fir, mountain hemlock, and lodgepole pine are secondary hosts. Ponderosa and whitebark pines and Engelmann spruce are infected occasionally, and grand fir and western white pine rarely. Introduced tree species found infected alongside diseased larch include jack, red, and Scots pines and Norway spruce. This dwarf mistletoe causes witches'-brooms on branches, often infects the trunk, and causes slow growth and premature death. Its shoots are dark purple and up to 6 cm tall (average 4 cm), with a fanlike arrangement of branches. Flowers are pollinated in late July to early September, and seeds mature 13–14 months later. Like other dwarf mistletoes, this one increases exponentially after a lag period. When larches were artificially inoculated with seeds, for example, little intensification occurred during the first 6 years because the life cycle of the parasite approaches that interval, but a 5- to 6-fold increase in population occurred during the next 3 years. This species spreads between trees at average rates of 1.2–1.6 meters per year, depending on average spacing of trees; the rate is more rapid as tree spacing increases within the range of seed flight.

Apache dwarf mistletoe. *Arceuthobium apachecum* is restricted to southwestern white pine in the mountains of Arizona, New Mexico, and northern Mexico. Its shoots are yellow-green or sometimes reddish and up to 7 cm tall (average 3–4 cm), with fanlike branching. They develop on spindle-shaped swellings where host buds proliferate and give rise to witches'-brooms. Fruits are blue-green, about 4 mm long, and mature in September, 12 months after pollination. This parasite, like its host, is of little economic importance.

Eastern dwarf mistletoe. *Arceuthobium pusillum* parasitizes primarily black, red, and white spruces in a region from Newfoundland to northern Pennsylvania, Minnesota, and eastern Saskatchewan. It occurs from sea level to about 900 meters elevation. Eastern larch is occasionally infected. Balsam fir; jack, red, and white pines; and blue spruce are rare hosts. This parasite causes severe growth loss and mortality in stands of black spruce in many areas and is locally severe on white spruce in Manitoba and along the coasts of Maine and Nova

Scotia. It is common also in the mountains of northeastern states on mature red spruce, on which it causes trunk swellings and witches'-brooms. Although eastern larch is commonly infected where it grows with diseased black spruce, it is classed as an occasional host because the parasite does not seem to perpetuate itself on larch in the absence of spruce. Aerial shoots are often sparse or lacking on infected larch branches. Larch and jack pine often seem to resist the development of an endophytic system of *A. pusillum,* producing a cork barrier that isolates the would-be invader.

A. pusillum, although not the smallest mistletoe (*A. minutissimum* of the Himalayas is smaller), is diminutive by contrast with most species of western North America. Its aerial shoots occasionally attain a height of 3 cm but average only 1 cm and are often shorter than the needles of black spruce. The shoots are greenish to brown and simple or with short primary branches. The life cycle normally requires at least 4 years. Flowers are pollinated in spring by beetles, flies, wasps, and other insects. Fruits mature and disperse seeds in September or October of the same year. Seeds germinate and infection occurs the next spring. Aerial shoots appear beginning in autumn of the 2nd or 3rd year but may be delayed several years. Shoots develop vegetatively in the 3rd or 4th year and produce flowers and fruit in the 4th or 5th year.

A. pusillum induces loose to compact witches'-brooms that may become massive in relation to the size of the host tree. In early stages of infection, host growth is locally stimulated. The branch swells, and the bark often appears reddish. Twigs grow more rapidly than normal, and the foliage is dark green. After several years, however, growth slows and the broom loses its verdant appearance. Infection of the mainstem often causes it to be converted to a large erect broom that may persist for 40–50 years. Trees with large brooms usually decline and die prematurely, however. The endophytic system of the mistletoe is systemic within the broom, reaching even into the buds.

Black spruce twigs infected by *A. pusillum* have higher than normal levels of cytokinins and indoleacetic acid and less than normal concentration of abscisic acid. General knowledge of the effects of these growth regulators suggests that the imbalance is probably responsible for stem swelling, loss of apical dominance on the affected branch, and diversion of organic metabolites from host to parasite.

Seeds of *A. pusillum* are dispersed both as projectiles from ripe fruit and by squirrels and birds struck by the sticky seeds when they disturb ripe fruit. Autonomous dispersal occurs primarily during morning hours as temperature rises. Seeds are propelled less than 2 meters on average, but some travel more than 10 m. Animal vectors are responsible for new infections beyond the range of autonomously dispersed seeds. Gray jays are perhaps the most common bird vectors.

References for dwarf mistletoes in general: 13, 296, 650, 756, 758, 759, 913, 1075, 1130, 1373, 1698, 1881

References for Arceuthobium americanum: 94, 96, 756, 759, 1362, 1846

References for Arceuthobium apachecum, A. douglasii, A. gillii *subsp.* gillii, A. laricis, *and* A. occidentale: 756, 759, 1693, 1846, 2010, 2188

References for Arceuthobium campylopodum: 355, 373, 624, 756, 759, 1651, 1652, 1763

References for Arceuthobium pusillum: 41, 85–87, 363, 756, 757, 759, 906, 1431, 1432, 1961

References for Arceuthobium tsugense: 16, 205, 325, 571, 756, 759, 1309, 1607, 1764, 1845, 1846, 2003

References for Arceuthobium vaginatum *subsp.* cryptopodum: 754, 756, 759, 909, 1693

A, C–G. Eastern dwarf mistletoe on black spruce. A, C. Large, old witches'-brooms and associated mortality (NY, Jul). D. Two witches'-brooms, illustrating variation in branch proliferation and growth rate (NY, Jun). E, F. Male shoots, approximately actual size and enlarged, respectively, on 3-year-old twig segments from a broom (PA, Feb). G. Enlarged view of female shoots (NY, Jun).

B. A witches'-broom caused by *A. pusillum* on eastern larch. Stem swelling and abnormally reddish bark color are prominent on the main axis of the branch just below the region of proliferation (NY, Jun).

Plate 211

Cassytha and Dodder (Plate 212)

This plate shows representatives of two groups of leafless, vinelike epiparasitic seed plants, cassytha and dodder. Cassytha and dodder are strikingly similar in general form and habits, but the two groups are unrelated. Their similarities are considered to result from parallel evolution.

Cassytha. The genus *Cassytha* includes perhaps 20 species of perennial, plant-parasitic vines that are widespread in tropical and subtropical regions on diverse hosts. *Cassytha* species are classified in the Lauraceae (laurel family), where they are the only vinelike group and the only parasitic one. *C. filiformis*, a cosmopolitan species, is the only representative in the United States. It occurs in central and southern Florida, where it is known as dodder laurel, woe vine, or simply cassytha.

In Florida, *C. filiformis* occurs on several types of citrus: grapefruit, Mexican lime, sour orange, sweet orange, and shaddock. It is an occasional pest in commercial citrus groves. It also occurs in Florida on live oak and assorted other plants but attracts little attention because of its preference for infertile or barren sites where plant damage is unimportant or at least unnoticed. Elsewhere cassytha parasitizes a great array of hosts without apparent preference. Acacia, sea grape, juniper, mangrove, and pine are examples of host groups. On one island in the Bahamas, the parasite was found on 81 plant species representing 45 families.

C. filiformis has been aptly described as a sprawling vine. It possesses chlorophyll and may appear distinctly green, but this color is often masked by an orange pigment that makes the vine appear orange-brown. Each plant consists of a tangle of long, runnerlike stems (stolons) 2–3 mm thick, with widely separated tiny scalelike leaves. It grows initially from a seed that germinates in soil, but the root withers and disappears after several weeks, by which time the parasite has often made suitable connections to a host plant. This occurs as stolons entwine host stems and multicellular haustoria grow from the coils into the stems. Stolons always coil counterclockwise around the host part. Haustorial connections are made usually to petioles and leaf rachises and to young twigs that lack a periderm, but leaf blades may also be parasitized. Connections to twigs seem to be maintained for a time after periderm formation. The stimulus to entwine and penetrate is nonspecific; cassytha strands often entwine one another and produce self-parasitic haustoria. Tiny spikelike flower clusters are produced from haustorial shoots. Insects, perhaps thrips (which have been found in flowers), probably act as pollinators. The fruit is a small drupe, white at maturity, and contains a single seed that may remain viable for up to 2 years. Birds visit the fruit and are suspected of dispersing the seeds.

Cassytha seedlings may grow as much as 10–12 cm per week and reach heights of 25–30 cm while using their own roots, but subsequent growth slows markedly. Mature stolons have been observed to elongate at average rates of 2.3–6.3 mm per day under dry conditions. Stolons spread easily from one host plant to another, allowing a single cassytha plant to colonize a large area. Herbaceous plants may serve as initial hosts, but cassytha flourishes only on woody plants. Host parts intensively colonized seem to lose vigor and may die back, but firm evidence for pathogenicity is lacking. Twigs bearing coiled stolons (haustorial coils) may appear swollen between turns of the coil and sunken beneath the coil.

Cassytha prefers dry, infertile sites and is most common in coastal habitats. It develops profusely on low trees and shrubs in the open, seems intolerant of shade, and is considered to be incapable of maintaining itself in closed forest canopies.

Dodder. Dodders compose a large genus (*Cuscuta*) of annual plants in the Convolvulaceae (morning-glory family). More than 150 species are known, of which more than 20 occur in the United States and southern Canada. These parasites are most abundant and troublesome on herbaceous plants in tropical and warm temperate areas, but they often occur in cool temperate regions and may attack the green or succulent parts of low-growing woody plants.

Dodder plants consist of soft, branched, yellowish or reddish strands that grow in a tangle over the leaves and succulent stems of their hosts. The strands bear scattered minute scales that are vestigial leaves. When a strand encounters a host plant, it begins to curl around it and by subsequent growth forms a helical coil from which numerous haustoria penetrate the host. Dodder strands coil only counterclockwise and only upward. Because of their strong phototropism, they do not entwine horizontal objects. Dodder flowers are small, five-parted, and white in most species. They are usually produced in summer and autumn. The fruits are four-seeded capsules a few millimeters in diameter. No special mode of seed dispersal is known. Dodder seeds have been transported globally in seedlots of economic plants.

The seeds germinate in soil, and the young plants are nourished by their own roots for up to several weeks or until a haustorial connection to the host is established. Seedlings may attain lengths of more than 30 cm on their own roots, and they are attracted to hosts along gradients of water vapor and probably other volatile chemicals. The root then withers and disappears. If no host has been penetrated, the seedling dies.

Dodder may kill herbaceous plants and may greatly devitalize woody ones. Sometimes noticeable swelling of a host stem develops beneath dodder strands in response to haustorial penetration. When dodder strands break as the result of host growth or movement, the separate parts continue as individual plants. Instances of perennial development of dodder by regrowth of shoots from the endophytic system have been reported, but these seem rare.

Dodder extracts water and nutrients from its hosts by means of an intricate haustorial (endophytic) system that establishes intimate connections with parenchyma cells and phloem sieve tubes of the host. Initial penetration is accomplished by numerous intrusive organs that grow from the dodder coil where it touches the host. Once the outer layers of the host are breached, linear structures called search hyphae (not to be confused with hyphae of fungi) differentiate near the tip of the penetrating organ and grow both inter- and intracellularly. Some hyphae penetrate parenchyma cells, while others extend to the phloem and produce fingerlike branches that seem to clasp phloem sieve tube members. These "contact hyphae" differentiate at maturity into conductive cells that transport solutes from the host to the main body of the parasite. The dodder causes an unloading of sugars, amino acids, and certain ions from host phloem at the site of contact.

Dodder can simultaneously parasitize two or more plants. For this reason and because of its ability to extract substances from phloem, dodder is often used for experimental transmission of viruses and mycoplasmalike organisms from diseased to healthy plants. In nature, however, dodder has no significant role as a vector.

References: 1038, 1072, 1073, 1075, 1373, 2234

A, C, D. *Cassytha filiformis* on sand live oak. A. Cassytha appearing as a wiry tangle of light green strands on its shrubby host. A dead portion of the same vine at upper left appears brown. C. Close view of the parasite on an oak branch tip. Green immature fruit is present. D. A cassytha strand entwining an oak twig (FL, Apr).

B, E–H. Dodder. B. Unidentified dodder on senna (*Cassia armata*). The host is a nearly leafless leguminous shrub on which the parasite grows as a dense web (CA, Jul). E. Unidentified dodder on oleander (CA, Jul). F–H. *Cuscuta ceanothi* on periwinkle. A dense tangle of strands with maturing fruit entwine the top of the plant. Helical lines of scars representing former haustorial connections are visible on the stem in (H) where dodder strands were displaced (greenhouse plants).

Plate 212

Vines That Damage Trees (Plate 213)

Many species of climbing vines, both native and introduced, are capable of injuring or suppressing the growth of trees and shrubs. Relatively few types of vines cause significant damage over large regions, however. Of these, grapevines, Japanese honeysuckle, and kudzu are most destructive in the United States and Canada. The latter two were introduced from Asia. Other vines causing occasional or local damage include ampelopsis, Virginia creeper, English ivy, and wisteria. Poison ivy, although abundant throughout eastern North America, has not been blamed for damage to trees, although it obviously suppresses their reproduction where growing as a ground cover.

Relatively little reliable information has been published about damage to trees by vines; most reports are observational, and some claims of damage are only speculative. This page collates examples of damage. The plate shows the vine called kudzu and the constriction injury caused by climbing bittersweet.

Climbing vines are most destructive in the Appalachian and southeastern regions. They cause damage in several ways. Vines compete with trees for water and nutrients in soil. The weight of vines, causing twigs and branches to bend, twist, or break, deforms trees. By adding to the surface area and wind resistance of the tree, vines increase the likelihood of breakage by wind or the weight of snow and ice. By growing on low vegetation and as a ground cover, they suppress reproduction of other plants, thus maintaining forest openings and thwarting normal plant succession. By growing over branches atop the canopy, they cause dense shade, which suppresses photosynthesis and may eventually kill trees. Stems or tendrils of some vines grow in tight coils around twigs and young trunks, causing constriction and deformity as the supporting stems increase in girth. In theory, some vines may also suppress development of neighboring plants through release of toxic chemicals (allelopathy), but evidence has not been presented, and allelopathy probably plays at most a minor role.

Most vines grow well in full sun and are relatively intolerant of shade. Thus they are most common on south and southwest slopes and are quick to colonize old fields and forest openings and edges but are less successful or unsuccessful beneath the canopy of a high forest. They do not directly suppress tree growth unless they cover tree tops. Growth suppression by vines on trunks and low branches results from competition for water and nutrients in soil.

The leguminous kudzu vine, *Pueraria lobata,* was introduced from Southeast Asia to many places in the Southeast for purposes of erosion control and as forage for livestock. The vine literally became wildly successful and now occurs in many localities from Texas and Florida northward to Illinois and the Potomac region. Scattered infestations farther north and west have also been reported. In most localities where it has overgrown other vegetation, kudzu has come to be detested, although it still has many advocates because of its usefulness for the purposes noted above. Fanciers of herbs and natural foods also are aware of its medicinal and food value for humans. Its stems are hairy, become woody with age, and attain lengths of 20 meters or more. They entwine and climb on any object. Its fragrant reddish purple flowers are abundant in late spring, but it produces pods sparsely. In a given locality, kudzu spreads mainly by growth of vines that root at nodes when in contact with soil. The roots are swollen, woody, and perennial. The leaves are compound with three leaflets. Leaves and young stems are sensitive to frost, and the vine defoliates after the first severe frost in autumn. Woody stems, however, including those high in trees, survive winter in the South. Kudzu dies back to ground level during winter in the coldest parts of its range.

Climbing bittersweet, native and common throughout the Appalachian region and northward into eastern Canada, climbs stems by twining in a clockwise helix around them. Coils of the helix lie at angles of 40–70° from the vertical. Bittersweet plants are shown in Plate 55. The ripe fruit is bright orange, and stems stripped of leaves so as to display the fruit are often used for indoor ornament. Damage by bittersweet occurs because the vine coils allow no room for increase in girth of the host stem. As growth occurs, constriction begins and interrupts the downward translocation of photosynthate in the host stem. The stem enlarges markedly above the constriction and very little immediately below it. The host remains alive, however, because conductive elements of xylem and phloem produced by the cambium above the constriction become aligned with the coils of the vine's helix. Thus routes of transport between leaves and roots are maintained, although they are lengthened.

Grapevines of at least 15 species grow wild in the United States and Canada. They climb trees and shrubs by means of tendrils that encircle twigs or small branches, providing temporary suspension for the vine. The tendrils eventually die and break, and the mainstem of the vine hangs free from the canopy of the tree or shrub. Occasionally tendrils cause constriction deformities of young stems, but this injury is insignificant in comparison to breakage and growth suppression caused by the vines. Grapevines are a significant obstacle to production of high-quality hardwood timber on many Appalachian sites where they grow quickly on young trees. In experiments wherein crop trees were designated at a young age and surrounding trees were removed to reduce competition, grapevines nullified the benefit of release from other competition. In times of severe drought, trees with heavy grapevine growth may perish where trees without vines survive. This tendency was noted in Oklahoma during the intense drought of 1934–1936. At a site in Connecticut, breakage by a heavy ice-glaze was more severe among trees supporting grapevines than among noninfested trees.

Japanese honeysuckle, an evergreen vine, has become naturalized throughout much of the eastern United States from the latitude of Pennsylvania southward. It is one of the major weeds that interfere with reestablishment of forests after harvest. In experiments, vine removal from sweet gum trees resulted in no growth increase unless the vines, mainly Japanese honeysuckle, were removed from the ground around the trunk. Removal of vines only from the trees, even if the vines extended to three-fourths the height of the tree, had negligible effect on tree growth. This research showed the importance of competition between trees and vines for water and nutrients.

English ivy has escaped from cultivation in scattered locations from Missouri and the Potomac region southward. It may grow as a ground cover occupying many square meters, but it also climbs by means of specialized roots that form along the stem and adhere to rough surfaces. The aerial roots do not penetrate bark other than into natural crevices or cracks, and the vine does little harm unless it grows atop and shades the foliage of the supporting plant. In that event, however, it may kill trees and shrubs.

References: 462, 525, 559, 1097, 1181, 1307, 1793, 1794, 1997, 2034

A–D. Kudzu. A. Kudzu covering oak trees at the edge of a forest (MS, Jul). B. Kudzu causing damage to tulip tree. Limbs are broken or bent downward. Height growth of the tree has been halted by the weight of the vine and the dense shade it casts as it grows above most of the tree's foliage. Many of the bare kudzu stems died during winter, but some survived even at the top of the tree, where new kudzu leaves are visible (SC, May). C, D. Close views of kudzu leaves and young stems, showing the tendency of stems to twine even around themselves (SC, May).

E. The stem of a box elder sapling deformed by climbing bittersweet. The tree stem is swollen above the constriction caused by the vine. The bittersweet stem is visible in the highest turn of the helix (NY, Sep).

Plate 213

Beech Bark Disease (Plates 214–215)

Beech bark disease is a devastating disease of American and European beech trees. It results from sequential attack by scale insects, especially the beech scale, *Cryptococcus fagisuga,* and certain fungi in the genus *Nectria* (Hypocreales, Hypocreaceae). The fungus usually involved in North America is *N. coccinea* var. *faginata. N. galligena* is sometimes found with or in place of *N. coccinea* var. *faginata* in bark predisposed by scale infestation. *N. coccinea* prevails in Europe.

Known in Europe since the mid-1800s, beech bark disease appeared in Nova Scotia in about 1920, some 30 years after the beech scale was detected near Halifax. Apparently the beech scale was imported with planting stock of European beech. Since that time the beech scale, followed after an interval of several years by *N. coccinea* var. *faginata,* has spread west and south at an average rate of about 15 km per year. By the mid-1980s it had reached Virginia and Ohio, and an outlying scale infestation followed by *N. galligena* rather than *N. coccinea* var. *faginata* had been found in the Monongahela National Forest in West Virginia. Landscape specimens of American and European beech, although generally susceptible, have often escaped damage, either because isolated, exposed trees do not offer favorable habitat for the scale or because the wind-dispersed, first-stage larvae perish in transit outside the protection of the forest canopy.

Three geographic zones are recognized with respect to beech bark disease: the advance zone, where beech scale infests many trees but *Nectria* is generally absent; the killing front, where *N. coccinea* var. *faginata* lethally attacks large trees predisposed by the scale; and the aftermath zone, in which many beech trees persist as deformed survivors and sprout clumps growing from the roots of killed trees. Where the killing front has passed, relatively few large beech remain that have not been deformed by multiple cankers. The beech scale and the fungal pathogens are endemic in the aftermath zone and are expected to cause a new wave of mortality as susceptible beech sprouts grow large enough to provide favorable habitat for the scale. In the northwestern part of the range of beech bark disease along the St. Lawrence valley in Quebec, wavelike mortality has not occurred. Rather, although scale infestation is general there, the disease is discontinuous and has not intensified as it did farther south.

For many years beech bark disease caused no great alarm among North American forest managers because beech was not a commercially valuable species. It commonly grows in association with yellow birch and sugar maple but was left behind when these species were harvested. Now, owing to changed wood utilization practices, sound beech has considerable economic value, but little of it remains where beech bark disease occurs.

Colonies of the beech scale become prominent on beech trunks before any symptoms develop. At first the colonies appear as scattered white woolly tufts in tiny bark crevices. The woolly appearance is due to a protective covering of wax filaments that the insects secrete over their bodies. Each tuft covers a small colony. The scale population increases on a susceptible tree during a period of years, becoming dense first in protected areas beneath branches, in rough bark of scars, and under lichens or mosses, especially on the north side of the trunk. The infestation commonly appears as vertical lines of colonies in tiny crevices and later as strips virtually covered by a white wax felt.

The beech scale is soft bodied, yellow, elliptical in outline, and 0.5–1.0 mm long when mature. It has reddish brown eyes, rudimentary antennae and legs, and a stylet (piercing-sucking mouthpart) that can be extended as much as 2 mm. Adults and stationary nymphs secrete wax filaments from numerous small glands. The insect reproduces parthenogenetically; only females, eggs, and nymphs are known. Eggs are laid during mid to late summer, and the adults then die. First-stage nymphs (crawlers), about 0.3 mm long, the only mobile stage, remain under the wax or emerge and crawl about near the site of egg hatch. Many settle and feed nearby; others drop from the tree and are carried away by wind. Most are dispersed only several meters beneath the forest canopy, but a small minority (less than 1% in a British study) are lofted above the canopy and may then be blown several kilometers. The buildup of beech scale on islands along the Maine coast attests to effective dispersal over land or water. After settling and beginning to feed, nymphs secrete the woolly wax cover and pass the winter. In spring they molt to adults. There is one generation per year. Experiments in England showed variation among scale populations in their ability to attack individual trees.

A second scale insect, *Xylococcus betulae,* is also important in the beech bark disease complex in North America. This animal infests paper and yellow birches as well as beech. Adult females are soft bodied, reddish orange, and up to 4 mm long. They are found embedded, except for the posterior, in cystlike cells beneath or along the sides of fissures in rough bark. A wax tube 13–50 mm long for the conduction of honeydew extends from the anus, often to beyond the bark surface. A droplet of honeydew may hang from the tip of the wax tube. *X. betulae* can be detected by searching for wax tubes associated with bark fissures. The tubes occur singly or in clusters. A female feeds in living bark around the cell by means of flexible stylets at least twice as long as its body. This feeding induces swelling of bark, leading to embedment of the cell, and causes new or enlarged fissures, creating more habitat for offspring. Males have wings and do not feed. Presumably their only function is to fertilize the embedded females. Eggs are laid in the cell. After hatching and dispersal, crawlers settle in crevices in rough bark and begin to feed. Each forms a protective cell by secretion of wax around itself, and the cell becomes embedded as the surrounding bark swells and cracks. Bark fissures induced by *X. betulae* are also colonized by *C. fagisuga.*

Some bark necrosis is caused directly by the scales, but most is due to *Nectria* species that invade and kill bark stressed by scale infestation. The first evidence of bark colonization by one of the *Nectria* species may be discoloration or disappearance of scale colonies because scales cannot live on dead bark. Slimy red-brown fluid often bleeds from tiny breaks in the bark, calling attention to underlying areas of cambial necrosis. Cankers may be diffuse with imperceptible margins or may appear as clearly delimited patches or strips that

Beech bark disease: the scale insects.

A. Severe mortality of dominant beech trees, caused by beech bark disease, in a forest of mixed deciduous species (NY, Aug).

B–E. Beech scale, *Cryptococcus fagisuga,* on American beech. B, C. Heavy and moderate infestations on mature trees. Vertical white lines indicate the preference of the insects for tiny fissures in bark. D. Magnified view of mature insects partly exposed by removing the woolly cover of white wax filaments. E. Close view of colonies in minute bark crevices (NY, Sep–Oct).

F–J. *Xylococcus betulae.* F, G. Magnified views of bark crevices where clusters of curled wax tubes indicate presence of the scale in underlying bark. H. Typical habitat: crevices in roughened bark. The beech scale is also present. I. Bark in a crevice dissected to reveal a mature female, anterior still embedded, with wax tubes extending from the posterior. J. A female, about 4 mm long, pulled from a feeding site. Stylets with which the scale probes bark around the feeding site are visible as a tiny thread (arrow) extending into the hole (NY, Oct).

Plate 214

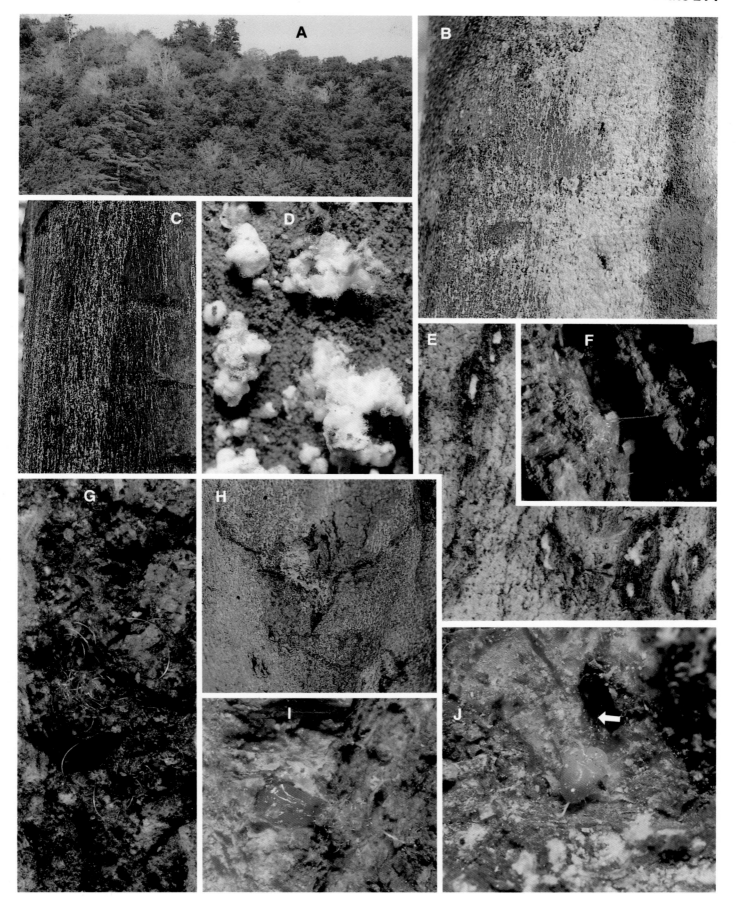

become raised and cracked at the edges due to the formation of cork barriers and callus. As large areas of bark die, often in vertical strips, branches above fail to produce leaves, or they produce sparse, chlorotic foliage that may wilt and turn brown during the growing season. Some trees have no foliar symptoms before sudden wilting and browning. Others remain partly alive for several years. Dead bark cracks and begins to fall away after 2–3 years.

Beech wood has negligible resistance to decay; thus many wood decaying fungi as well as ambrosia beetles and other secondary insects invade the dead parts of trunks. Living trees with defective trunks may break and fall without warning, which poses a hazard in recreational areas such as campgrounds. Beech trees dead for a year or more may be so decayed that they are unfit even for firewood.

Signs of the causal fungi become visible soon after bark dies. All Nectria species involved in beech bark disease produce sporodochia of their conidial stage, Cylindrocarpon (Hyphomycetes), in summer and autumn as tiny white tufts that on casual inspection may be confused with the remnants of scale colonies. The sporodochia of N. coccinea var. faginata are covered with colorless conidia that range from oval and single-celled, 11–14 × 3 μm (microconidia), to cylindrical or slightly curved macroconidia that are two- to eight-celled with rounded ends. Most macroconidia measure 40–110 × 5–7 μm. Red, lemon-shaped perithecia of N. coccinea var. faginata, 200–300 μm in diameter and 250–400 μm high, arise in groups and mature in autumn. During wet weather they expel masses of colorless ascospores that when dry appear as white spots on the tips of the perithecia. The ascospores are two-celled, 10.5–12.7 × 4.8–6.2 μm with broadly pointed ends. Ascospores are dispersed by wind or water, and this continues throughout winter and spring when the air temperature is above approximately 0°C.

Both types of spores are infectious if introduced to wounds in bark predisposed by scale infestation or drought. Wounds need only break the periderm; penetration to the cambium is not necessary. Most infections and canker expansion occur in autumn, although severely stressed bark can be colonized at any time if temperature permits. The precise mode of penetration of the fungus into predisposed bark is unknown. Stylet wounds and tiny cracks resulting from drying of severely infested bark have been suggested. There is some evidence that N. coccinea on European beech may colonize microsites in outer bark saprophytically and may thus occupy a niche from which expansion into stressed living tissues is facilitated. After infection the fungus colonizes and kills bark, cambium, and outermost sapwood. Neither N. coccinea nor its variety faginata is an aggressive parasite of nonstressed beech bark.

The nature of predisposition and the precise mode of parasitism of beech by Nectria species are unclear. The beech scale secretes pectinases that may play a role by degrading parenchyma cell walls and middle lamellae. Its saliva is suspected also of containing auxinlike growth-regulating chemicals that cause slight swelling and cracking of bark. N. coccinea secretes pectic and cellulolytic enzymes capable of degrading cell walls and, theoretically, of killing cells.

N. coccinea and its variety faginata, N. ditissima, and N. galligena are all capable of causing cankers in beech bark. These fungi can be distinguished from one another only on the basis of microscopic characters of perithecia, ascospores, and conidia. In Europe N. ditissima causes a separate canker disease that is unrelated to scale infestation. This fungus also causes perennial cankers on red alder in the Pacific Northwest. N. coccinea is apparently a minor canker pathogen of yellow birch and sugar maple in eastern North America. It and N. galligena have larger ascospores and smaller conidia than does N. coccinea var. faginata. The latter is known only on American beech. N. galligena causes perennial cankers on many deciduous trees and shrubs (Plate 101).

The susceptibility of European beech trees to bark killing by N. coccinea varies with the duration and severity of scale infestation. Similar relationships probably exist for American beech, since the beech scale precedes Nectria by several years. The susceptibility of European beech may also be increased by water shortage and nutritional deficiency.

In both Europe and North America, many beech trees within forests infested by beech scale are resistant to the insect and thus remain noninfested or only lightly infested. Therefore they do not become predisposed to infection by Nectria. In European beech, resistance to beech scale is associated with a high proportion of lignified cells (stone cells) in outer bark. American beech resistant to beech scale does not seem resistant to X. betulae.

Natural controls on the beech scale and associated Nectria species seem ineffectual. Larvae of the twice-stabbed ladybeetle, Chilocorus stigma, prey on the scale without seeming to influence its population density. The scale cannot withstand air temperatures lower than about −37°C, but this temperature is also near the limit for survival of American beech. A mycoparasite, Nematogonum ferrugineum (syn. Gonatorrhodiella highlei), often parasitizes Nectria coccinea and Nectria galligena and suppresses their fruiting on beech in North America. The buff-brown mycelium and spores of the mycoparasite may be diagnostically useful where red perithecia cannot be seen (Plate 215H), but the mycoparasite does not seem to influence the beech bark disease epidemic.

References: 215, 244, 387, 537, 611, 890, 894, 895, 929, 931, 1158, 1163, 1298, 1304, 1482, 1775, 1778, 2087

Beech bark disease: symptoms and fungal pathogen.

A. A dying American beech with dead twigs and sparse chlorotic foliage (NY, Jun).

B. Cankers marked by irregular rings of red perithecia of Nectria coccinea var. faginata on American beech (NY, Oct).

C. Superficial cankers on a somewhat resistant American beech. Infections, confined to outer bark, are delimited and isolated from healthy tissues by a layer of cork. Cracks in the dead bark of cankers result from mechanical stress caused by growth of tissues beneath. Cracks are preferred habitat for Xylococculus betulae, which by its feeding in underlying living bark may stimulate further roughening (NY, Jun).

D. Magnified view of white sporodochia of Cylindrocarpon faginatum, the conidial stage of N. coccinea var. faginata (NY, Oct).

E–G. Perithecia of N. coccinea var. faginata. E. Young perithecia on a small, circular canker. F. Young perithecia (bright red) adjacent to older, darker ones. G. Magnified view of young perithecia (NY, Oct).

H. Patches of buff-brown mycelium and spores of the mycoparasite Nematogonum ferrugineum on bark killed by Nectria coccinea var. faginata. Nematogonum ferrugineum parasitizes the beech pathogen (NY, Jun).

Plate 215

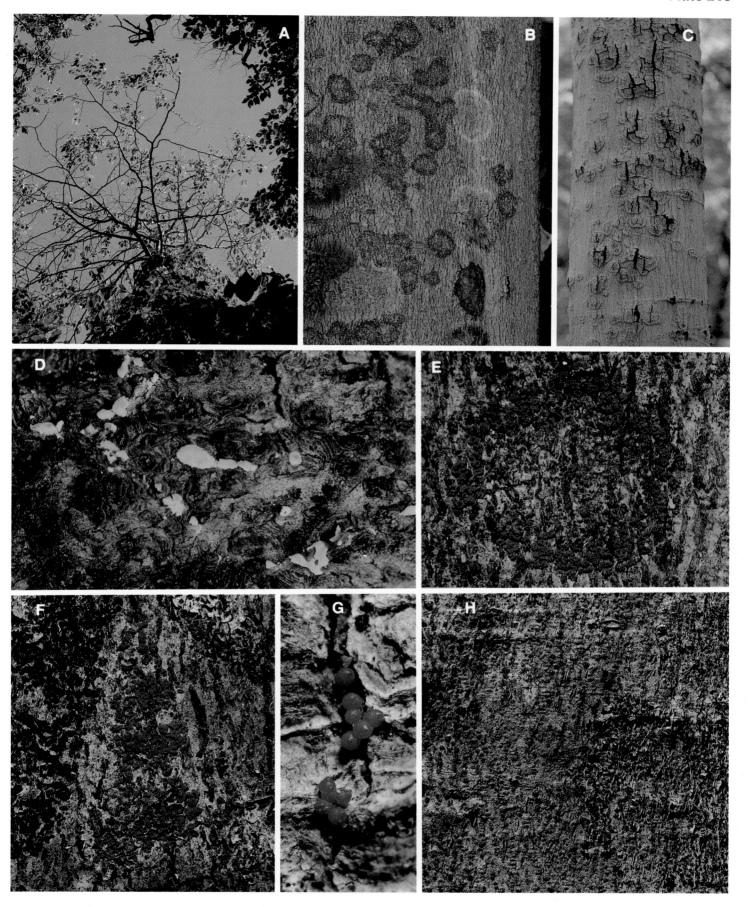

Decline Caused by Multiple or Unknown Factors (Plates 216–219)

Decline refers to progressive loss of vigor and health, not to any specific disease or disorder. Trees decline for many reasons, sometimes as the result of a single disease or damaging environmental factor but often as the result of several environmental and biotic factors acting in concert or in sequence. The key concept in either case is that decline results from the action of stressing factors over periods of years.

Symptoms of decline vary with cause and with tree species. They include slow growth; sparse or undersized or distorted, often chlorotic nutrient-deficient leaves; browning of leaf margins, premature autumn color, premature leaf drop, abnormally large crops ("distress crops") of fruit, subnormal storage of food reserves (especially starch), and progressive or intermittent dieback of twigs and branches and eventually the entire tree. Adventitious sprouts often develop for a time along the trunks of trees that have sustained branch dieback. Two general sequences of symptoms are recognized. If decline is initiated by a damaging event such as root cutting or severe defoliation, death of buds and twigs may occur as a shock response, and other symptoms follow. If decline results from chronic stress, for example by salt or water shortage associated with poor root development in compacted soil, foliar symptoms and slow growth are likely to precede dieback.

Three theoretical explanations of decline that seem applicable in different situations have been advanced. One scheme emphasizes interchangeable predisposing, inciting, and contributing factors. During predisposition, biotic and abiotic stressing factors cause partial loss of the tree's ability to tolerate environmental adversity, resist pests and pathogens, and respond to favorable factors. If stressing factors are removed or counterbalanced by favorable factors, the tree recovers full ability to respond to its environment. If the tree loses its ability to respond to favorable factors, decline begins. Decline may be incited either by the addition of a new factor or event or by the cumulative influence of the predisposing factors. Decline is accelerated and symptoms are intensified by the action of contributing factors such as opportunistic pathogens and secondary insects.

The second scheme is similar to the first but emphasizes an initial event that alters the tree and triggers processes that make it abnormally susceptible to further damage by the initial factor and by secondary factors. Severe defoliation by insects, for example, can lead to decline, as discussed below and with Plate 219.

The third scheme, like the first, involves a three-phase process but explains the predisposition phase as an irreversible group behavior of trees in forests. Individuals of the same species and position in the canopy first lose vigor simultaneously (cohort senescence) because of aging and gradually increasing environmental stress. This predisposes them to damage by an event or factor that causes dieback. The damage may be aggravated by contributing factors or retarded by mitigating factors.

The theories mentioned above are based on the following propositions.

- Tree growth and behavior may vary from site to site or may fluctuate widely on a given site and still be judged normal for the conditions present.
- Trees in forests alter their environment, and they tend to display group behavior.
- Trees growing in man-made landscapes and disturbed forest sites are subject to stressing factors not encountered by trees in undisturbed forests; individual trees vary in exposure to these factors and in response to them.
- A change in intensity or supply of an environmental factor may trigger decline.
- Introduction of a new biotic or abiotic factor into a tree's environment may trigger decline.
- A factor not damaging within a short time may predispose trees to damage by other factors or may cause damage if applied chronically. Also, repetitive minor injuries exert a cumulative effect.
- A factor that triggers decline may cease causing stress, but secondary or contributing factors may perpetuate stress and cause progressive decline.
- Old trees are less resilient than young ones.

Decline of maple. Sugar maple in forests and sugar bushes (maple stands managed for production of syrup and sugar) may decline after severe defoliation by insects, harvesting of timber, root damage by grazing livestock, or overzealous tapping and sap extraction, alone or in combination. Opportunistic fungal pathogens then move in. *Armillaria* (Plate 148) attacks roots, and fungi such as *Cerrena unicolor* (Plates 173 and 216), *Stegonsporium* species (Deuteromycotina, Coelomycetes) (Plate 216), and *Valsa* species (Plate 96) attack stems of weakened trees and may be direct causes of dieback.

Severe defoliation by insects after shoot growth is complete in early summer may cause winter buds to open and produce a new flush of shoots. This process nearly depletes stored food reserves, lowers resistance to fungal pathogens, and delays cold hardiness, thus predisposing the tree to frost damage in autumn. If the tree is further stressed the next season by drought, root damage, repeated defoliation, or other factors, it may begin to decline.

Timber harvesting promotes decline in the residual stand because tree trunks and soil covering feeder roots, previously shaded, are subject to abnormal heating and drying by direct sunlight. Trunks and roots of residual trees are also often wounded by logging equipment. Grazing cattle cause or contribute to decline because their hoofs break or cut small roots. Harvesting sap may contribute to decline if too many tap wounds are made or if chemicals are used that thwart compartmentalization of tap holes (causing death of a large volume of sapwood) or too much sap is extracted (depleting sugar that would be used as the energy source for growth).

Factors that often trigger decline of maples in landscapes include girdling roots (Plate 238), restricted rooting space with associated water shortage or water logging, cankers and collar rots caused by fungi (especially *Phytophthora* species), soil compaction leading to water shortage and rootlet mortality, deicing salt (Plate 221), chronic effects of *Verticillium dahliae* (Plate 181), severe trunk wounds, and severance of roots during excavation for utility channels, walks, and roadside drainage. The same opportunistic fungal pathogens found in forests contribute to decline of maples in urban plantings and are joined there by such fungi as *Ganoderma lucidum*, causing root rot (Plate 160), and *Botryosphaeria obtusa* (Plate 83) and *Nectria cinnabarina* (Plate 100), causing cankers and dieback.

Observations suggest that decline in urban maples can be avoided or delayed or its incidence suppressed by careful selection of planting sites for avoidance of environmental stress.

Decline of sugar maple.

A. Premature autumn color on the side of a tree overhanging a rural road (NY, Sep).

B, D. Contrast between twig growth and form on a healthy, vigorous tree and on a declining tree. Twigs of the declining tree grow slowly, and many have died, as evidenced by lack of swollen buds. Apical buds on living branches of the declining tree have lost dominance, so that all twigs are about the same length (NY, May).

C. Contrast in condition of roadside trees as related to position above or below highway elevation; trees below the highway receive more snow and meltwater laden with deicing salt than do trees above the highway (NY, Sep).

E. Dieback in a tree that sustained root death due to flooding (NY, Aug).

F. A tree declining because roots died after burial under fill in a landscape project. Bark has fallen from a dead area that developed on the trunk above dead major roots (NY, Aug).

G, J. Basidiocarps of the canker-rot fungus, *Cerrena unicolor*, an opportunistic pathogen, on the trunk of a dying tree (NY, Aug).

H. A forest tree, defoliated by insects 6 and 5 years earlier, that never regained vigor (note sparse foliage) and suddenly died during a period of summer water shortage (Ont, Jul).

I. Black spore masses of *Stegonsporium pyriforme* on a recently killed branch of a declining tree. This fungus is believed to be an opportunistic pathogen that kills weak branches (NY, Aug).

Plate 216

Decline of ash. White ash in the north-central and northeastern United States and parts of southeastern Canada is subject to a devastating disorder called ash dieback or ash decline, characterized by progressive loss of vigor during periods of 2–10 or more years before trees die. Green ash and red ash are also affected but less severely. Several biotic agents as well as water shortage and damage by freezing have been suggested as primary or secondary causes, and a close association of mycoplasmal infection with decline has been reported, but full understanding of ash dieback had not been reached at the time of this writing. Therefore, although we have emphasized the ability of mycoplasmalike organisms to cause decline as part of the ash yellows syndrome (Plate 191), we consider additional possible causes here. Other topics related to ash decline include viral diseases of ash (Plate 201), water stress (Plate 232), and freeze damage (Plate 236). General concepts of tree decline are presented with Plate 216.

Ash decline has been observed in eastern Canada and northeastern United States since the 1920s, but the same causal factors were not necessarily involved at each time and locality. Episodes of dieback in Quebec, for example, were attributed to freezing injury to roots during winters with subnormal snow cover. These episodes were perhaps unrelated to an outbreak of ash dieback that began in the northeastern United States in the 1930s and continues to the present.

The decline syndrome that has been called ash dieback includes reduced radial and apical growth, leading to short internodes and consequent tufting of foliage at branch ends. Leaves become undersized and pale green to chlorotic, and the canopy appears sparse because of the small size and abnormal arrangement of leaves. Premature autumn color and casting of leaves are common. Viruslike symptoms (chlorotic to reddish spots and rings, light green to pale cream line patterns, and mottling of leaves; Plate 201) are common. The smooth bark of affected trees often turns pinkish in contrast to normal gray, especially on trees exposed to sun and wind. Annual cankers form on branches and trunks. Cankers on stems with smooth bark are reddish brown to orange-brown; those in old bark with cork ridges are initially invisible. The bark in cankers begins to split, shred, and drop off 1–2 years after canker formation. Twigs and branches die back, usually during dormancy but also during the growing season in drought years. Epicormic sprouts form on the trunks of many trees in advanced stages of decline. These sprouts often eventually become deliquescently branched or develop into witches'-brooms, indicating mycoplasmal infection. Death occurs one to many years after the onset of slow growth. Roots appear normal until the last stages of decline. Recovery is uncommon, although decline may be temporarily arrested at various stages. Ash in early stages of decline, before major limbs begin to die, is capable of responding to favorable environmental conditions, as indicated by fluctuations in annual growth rate.

For many years the primary cause of ash dieback was considered to be environmental because no biotic agents other than opportunistic canker-causing fungi were found consistently associated with declining trees. Water shortage in particular was blamed because incidence and severity of dieback seemed greater during periods of drought in the 1930s, 1950s, and early 1960s than in preceding or intervening periods. Also, white ash was shown to be unusually sensitive to water shortage. This species naturally becomes partly dehydrated during dormancy and does not rehydrate until just before growth in spring. Therefore ash growth rate is highly sensitive to the soil moisture supply in spring and early summer. Ash growth during a 12-year period (1950–1962) that included episodes of severe ash dieback was highly correlated with rainfall during May–July. Ash stomata were found to be unusually sensitive to water stress, closing when xylem water

potential diminished to approximately −1.8 megapascals. This information supported an interpretation that ash dieback was caused primarily by drought and that biotic agents play secondary roles. The main problem with this hypothesis is that many ash undergo a sudden permanent reduction in growth rate (the first symptom of decline) during years of normal to abundant precipitation.

Other environmental factors that could possibly trigger or contribute to decline are polluted air and low temperatures. White ash is sensitive to ozone, but foliar symptoms typical of ozone injury are not common on declining trees. Freeze damage to roots during cold winters with subnormal snowfall could cause sporadic dieback, as mentioned, but it would not occur annually. Cambial damage by freezing is associated with and presumably contributes to decline in some areas, but this damage is also associated with mycoplasmal infection that apparently prevents ash from becoming fully cold hardy. On balance, the case for an environmental primary cause of ash dieback is weak.

On the other hand, a biotic primary cause of ash dieback is indicated by several observations. Affected trees rarely recover from even the earliest stages of decline, regardless of normal rainfall and temperature. New cases of decline and dieback occur annually in young, previously vigorous ash trees without relation to water shortage. Moreover, the onset of decline in young trees is random within a stand, and the most common years of onset vary from one stand to another even in the same locality. This variation is characteristic of infectious disease but uncharacteristic of wide-scale, environmentally induced disease.

Biotic agents associated with ash dieback that could induce or contribute substantially to it include mycoplasmalike organisms (MLOs) causing ash yellows, viruses (tobacco mosaic, tobacco ringspot, and tomato ringspot viruses detected heretofore), and two fungal pathogens that cause cankers and twig dieback. MLOs or viruses singly or in combination could induce decline and may in fact do so, since evidence exists for their pathogenicity to ash and other trees. Both types of agents have been detected in nondeclining trees, however, and they have not yet been shown by experiments to cause decline and dieback in large ash trees. The canker pathogens, often directly responsible for much dieback of twigs and branches, are *Cytophoma pruinosa* and an unidentified species of *Fusicoccum* (both in Deuteromycotina, Coelomycetes). *C. pruinosa* is the more common and was the more virulent in pathogenicity tests. These fungi cause secondary damage to trees already stressed by other agents. Both fungi, when inoculated into healthy white ash saplings, caused more or larger cankers on saplings stressed by water shortage than on saplings allowed to receive normal rainfall. This was true whether experiments involved deflecting rainfall by plastic barriers or simply inoculating trees in dry months or wet ones. *C. pruinosa* produces black pycnidia, 450–800 μm in diameter with a simple central fertile cavity, in recently killed bark. Its tiny single-celled conidia (3–9 × 1–2.5 μm) are colorless and curved.

A hypothesis that accommodates available information is that ash dieback is caused primarily by MLOs (perhaps interacting with viruses) and that symptoms are intensified by drought stress, freeze damage, and parasitism by opportunistic fungi.

A, B. Gross symptoms of ash decline: foliage tufted at twig tips, dwarfed leaves often light green in color, pinkish bark, and dieback of twigs and branches. Small clusters of brown leaves in the top of the tree in (B) indicate that dieback occurs during the growing season as well as during the dormant period (NY, Aug).

C–G. Cankers caused by *Cytophoma pruinosa*. C. Small, inactive, superficial cankers 1 year after formation. These cankers are raised with cracked margins due to periderm formation beneath them. D. A canker about 8 cm long, appearing sunken because all bark tissues and the cambium were killed and then shrank somewhat. Infection began at a wound. Black specks are pycnidia. E, F. Old, inactive cankers on smooth and rough bark, respectively. G. Magnified view of black pycnidia of *C. pruinosa*, with bark surface shaved at left center to reveal a simple, central fertile cavity in each pycnidium (NY, Oct).

Plate 217

Pole blight of western white pine. Pole blight was among the several decline-type forest diseases that came to prominence across North America and then subsided during a span of 3 decades beginning near 1930. Pole blight was eventually explained as primarily a response to a temporary climatic trend of diminishing rainfall and increasing temperature. First noticed in 1929, the disease occurred and may still be observed in the Rocky Mountain and Intermountain regions of the northwestern United States and adjacent British Columbia. It caused great damage in certain localities, but most stands of commercially valuable western white pine were spared. The disease afflicts primarily dominant and codominant trees that have reached pole size (stem diameter of 15–30 cm at breast height).

The first symptom of pole blight is death of fine rootlets. Height and diameter growth then diminish, larger roots die, and foliar symptoms appear, presumably as the consequence of impaired root function. Foliage becomes chlorotic, dwarfed, and tufted on branch ends. The tufted appearance results from production of a nearly normal number of needles on abnormally short twigs. Foliar symptoms progress downward, followed by dieback. Trees in distress often produce a large crop of cones before dying back. Long lesions involving phloem, cambium, and sapwood form on the trunk following the grain of the wood, primarily on the lower half of the tree. Resin exudes at the bark surface during lesion formation, and the underlying sapwood becomes resin-soaked. Blue stain is also common in sapwood beneath cankers. Affected trees die within 5–10 years.

Pole blight is most severe on shallow soils (30–45 cm rooting depth) with limited water-holding capacity (5–10 cm water in the top meter of soil). The role of water shortage in pole blight was indicated not only by this site relationship but also by the coincidence of the pole blight outbreak with a period of 3 decades of subnormal rainfall and above-normal summer temperatures and by the partial recovery of affected trees after normal rainfall resumed. Many affected trees then began to produce callus at margins of trunk lesions, indicating static or improving condition.

The decline of western white pine where other tree species remain healthy has been explained on the basis of less ability to compete for scarce moisture and less ability to adapt to dry conditions. This species was found to have fewer root tips, less length of fine roots per unit of soil volume, and a smaller root/shoot ratio than did other conifers under conditions of moderate water shortage. The white pine did not curtail top growth under conditions of moderate water shortage that suppressed growth of other conifers. Therefore, a root/shoot imbalance may have resulted such that during periods of severe water shortage, the roots could not meet the water requirements of the tops. Vigorous young dominant or codominant pole-sized trees were considered to be those most likely to develop such an imbalance. The physiological aberration resulting from acute water shortage was hypothesized to result first in rootlet mortality and then in decline. Symptoms of pole blight have not been induced experimentally.

No organism has been found consistently associated with pole blight. The fungus *Europhium trinacriforme* (Microascales, Ophiostomataceae) in its conidial state, *Leptographium* (Hyphomycetes), occurs commonly in trunk lesions. It caused elongate resinous lesions and resin-soaking of sapwood when inoculated to healthy trees, but it was considered to be only a secondary factor in pole blight because lesions always formed after marked growth reduction of the trees and on stem parts where prior growth reduction was most striking. The root-rotting fungus, *Armillaria* (Plates 148–149), is also a common invader of trees with pole blight.

Birch dieback. A striking decline and dieback of birch was first noticed in 1932 in New Brunswick, whence it appeared to spread in all directions. It caused spectacular devastation of paper and yellow birches throughout Nova Scotia, New Brunswick, northern New England, and much of eastern Quebec. In the areas first affected, over 80% of the birch otherwise suitable for harvest died within 12–13 years, and reproduction was scarce. Damage was less severe in areas affected later.

For many years before the outbreak, birch and to a lesser extent other northern hardwoods had been observed to decline and die back on sites of previous logging. This condition was called postlogging decadence; it was apparently caused by the sudden exposure of residual trees to full sun, higher air and soil temperatures, lower relative humidity, and higher wind velocity than occurred in uncut stands. Opportunistic organisms such as *Armillaria* and the bronze birch borer (see Johnson and Lyon, 2nd ed., Plate 128) attacked the disturbed trees and contributed to the damage. The widespread outbreak of dieback, however, was unrelated to logging or to any other known human disturbance.

Death of feeder roots and a more or less abrupt reduction in growth rate signaled the onset of birch dieback. The majority of fine roots were dead by the time a tree had severe symptoms above ground. Symptoms included undersized chlorotic leaves near branch tips; shriveled, rolled, or cupped leaves, sometimes with ring spots, vein clearing, or mild mosaic; swollen yellowish buds (versus normal brown), distorted shoots that withered early in the season, proliferation of buds and stunted shoots, sterile seeds, failure of buds to open in spring, dieback of twigs and branches, and in severe cases, death of the tree after 3–6 years. As dieback abated, the growth of surviving trees began to increase.

The search for causal factors focused initially on the bronze birch borer because this insect quickly attacks, girdles, and kills weak trees. The borer was not present in trees with early symptoms, however; therefore it was not the primary cause of birch dieback. No insects or fungi capable of causing dieback were found. Viruslike symptoms were common, however, and some symptoms were transmitted to healthy yellow birch by grafting. A strain of apple mosaic virus was eventually identified (symptoms described with Plate 200), but its role in birch dieback remained unclear.

A climatic warming trend was considered as a possible causal factor. Average summer temperatures in the Maritime Provinces rose 1.0–1.4°C during 1920–1950. Dieback was most severe where birch rooting depth was least, as would be expected if warming and drying of soil were important. When roots of yellow birch seedlings were subjected to soil temperature 2°C above normal for 1 growing season, populations of root-inhabiting fungi were altered, and rootlet mortality rose from a normal 6% to 60%. This shift supported a theory that climatic change was responsible for birch dieback. The growth reduction that signaled the onset of dieback in mature trees, however, was neither related to measured environmental factors nor correlated with growth perturbation in adjacent healthy trees, as would be expected if an environmental factor was the primary cause; but subsidence of the outbreak did coincide with a downturn in the long-term temperature trend.

A–E. Pole blight of western white pine. A. Healthy trees at left, in contrast to declining trees with chlorotic, tufted foliage at right center. A dead tree at left center bears a distress crop of cones. B, C. Young pole-sized trees, normal (B) and diseased (C). The diseased tree shows no new shoots; foliage is sparse, stunted, chlorotic, and tufted at branch tips. D, E. A large canker with resin exuding at its edge, on the base of a pole-sized tree. Bark was removed (E) to show extent of the lesion and associated resin-soaked sapwood (ID, July).

F, G. Postlogging decadence of paper birch. F. Dead trees and tree tops. G. Sparse, chlorotic foliage on a top branch of the tree at right indicates continuing decline (Ont, Jul).

Plate 218

Decline of oak. The term *oak decline* is not reserved for any single disorder or syndrome. Oaks decline for many reasons, and the most common causes of decline vary with region and oak species. General aspects of tree decline are discussed with Plate 216. The visible symptoms on oaks are similar regardless of cause: slow growth; sparse, undersized, sometimes distorted, often chlorotic leaves; death of scattered twigs either during the growing season, leading to prominent "flags" of brown foliage, or during dormancy; dieback of branches and major limbs; and often growth of adventitious sprouts along the trunk and large limbs following dieback (Plate 241). Leaf scorch and premature autumn color precede or accompany dieback in some circumstances. We consider first decline of individual oak trees or small groups in urban and landscape plantings, and then decline of oak populations.

Common factors that stress and shorten the lives of oaks as street or landscape trees are root wounds caused by excavation equipment; root death caused by oxygen deficiency in compacted or waterlogged soil; root death caused by soil changes around new structures (change of grade, change in height of water table, or heating and drying of surface soil after other vegetation is removed); defoliation by insects, discussed below; drought, discussed below and with Plate 232; massive or multiple trunk wounds leading to death of sapwood and eventual trunk decay; nutrient deficiency, especially iron deficiency induced by alkaline soil (Plate 231); salinity stress where salt from irrigation water accumulates in the root zone, as discussed with Plate 221 (oaks in the North are relatively tolerant of deicing salt); and for species in the white oak group (subgenus *Leucobalanus:* e.g., chestnut, white, and Oregon white oaks), chronic infestation by pit-making scales (see Johnson and Lyon, 2nd ed., Plate 168). Trees initially stressed by the factors noted above become abnormally susceptible to further damage or killing by opportunistic insects and fungi. The two-lined chestnut borer (see Johnson and Lyon, 2nd ed., Plate 127), for example, may girdle trunks or limbs of weakened trees. Bark-infecting fungi such as species of *Hypoxylon* (Plate 108), *Cryphonectria* and *Endothia* (Plates 88 and 91), and *Botryodiplodia, Diplodia,* and *Sphaeropsis* (Plates 84–85) may cause trunk and branch cankers and twig dieback. Root-infecting fungi such as *Armillaria* (Plate 148) and *Ganoderma* (Plate 160) may kill major roots and cause decay of root wood, and, in the West, *Phytophthora* species may cause cankers on the trunk and the root collar (Plate 138).

In large areas of the East, decline of oaks in both wild and urban forests is often triggered by defoliating insects, particularly oak leaf rollers and the gypsy moth (see Johnson and Lyon, 2nd ed., Plates 99 and 61–62). Severe defoliation—removal of three-fourths or more of the foliar surface—can incite decline directly or may predispose trees to damage by other agents. A single severe to complete defoliation will cause subsequent growth loss but will not in itself trigger decline unless it comes after full leaf expansion and results in a second flush of growth during the same season. This growth depletes the stored carbohydrate reserves of the tree and leaves it abnormally susceptible to attack by secondary insects and opportunistic fungi. A single complete defoliation in spring by insects or frost is of less consequence because regrowth comes early in the season, does not use up all stored energy reserves, and leads to photosynthetic accumulation of new reserves (although not to the normal level) during the late part of the season. Severe defoliation in 2 or more successive years (or twice in 1 year as by frost and then insects) will usually trigger decline and mortality; trees commonly wilt and die suddenly in August. The death rate begins to rise in the 2nd consecutive year and commonly exceeds 50% in the 3rd year. Defoliation followed or accompanied by severe drought will trigger decline, but drought alone does not do so except in the South where drought and heat stress occur together (Plate 232). Species in the red oak group seem more susceptible than members of the white oak group to decline and mortality associated with defoliation and drought.

Examples that involve factors listed above include (1) periodic decline and mortality of species in the red oak group (subgenus *Erythrobalanus:* e.g., black, live, red, scarlet, and water oaks) in the Midwest, Northeast, and Appalachian regions, caused by defoliation, drought, borers, and root-infecting fungi; (2) decline of southern live oak along the Gulf Coast, caused by construction damage, soil compaction, trunk wounds, hurricanes, *Endothia* sp., and *Botryosphaeria*

rhodina (Plate 85); (3) pin oak blight in eastern Virginia, caused by unsuitable soil, drought, and *Endothia gyrosa* (Plate 91); (4) decline and mortality of oaks across the South in 1978–1981, caused by drought, heat, borers, and *Hypoxylon atropunctatum* (Plate 107); (5) dieback of coast live oak in California in 1978–1981, caused by drought and *Diplodia quercina;* (6) decline of oaks and other trees in irrigated landscapes in Texas, caused by accumulation of soluble salts; (7) decline of oaks and other trees along city streets because of excess water in planting holes; and (8) decline of residual oaks and other trees wherever residential developments are implanted in forested landscapes, caused by root disturbances and, in the Midwest, alkalinization of a mantle of topsoil. Alkalinization occurs in two ways. Topsoil may be mixed with alkaline subsoil when disturbed by construction activity, and topsoil near concrete structures may receive alkaline substances that leach from the concrete.

Citrus blight. Also known as young tree decline and sandhill decline, citrus blight is a major disease of unknown cause, occurring in Florida but not in other North American citrus-producing regions. It affects many citrus types on various rootstocks. Annual losses in Florida exceed half a million trees per year, and annual incidence of new disease in a grove is commonly 10–15%. Contrary to the usual connotation of *blight,* the disease is characterized by slow decline beginning any time after trees reach the age of 6 years.

Symptoms of blight include slow growth and loss of productivity, chlorotic patterns typical of zinc or manganese deficiency on leaves, wilting despite adequate soil moisture, and branch-by-branch dieback. Trees with dieback usually produce adventitious sprouts on the trunk and scaffold limbs, but the sprouts also eventually die. Aerial symptoms are followed by root decay. The xylem vessels of major roots, the trunk, and major limbs but not small branches become plugged with an amorphous substance that greatly inhibits water movement. Zinc and water-soluble phenolic substances accumulate to levels at least three times normal in outer sapwood. Blight cannot be diagnosed from visible symptoms, but slow water uptake during diagnostic injection into the trunk (less than 10% of the normal rate), and elevated zinc and phenolic concentrations are diagnostic criteria. A biotic cause of blight is suspected because it occurs randomly within a grove and symptoms have been transmitted experimentally by grafting. The bacterium that causes Pierce's disease of grapevine (Plate 186) has been detected in citrus with blight but has not been shown to cause it. Affected trees never recover, and no control is known.

References for maple decline: 88, 475, 494, 519, 635, 690, 718, 719, 797, 798, 891, 892, 1082, 1191, 1210, 1361, 1377, 1460, 1605, 1657, 1803, 2074, 2105–7, 2157

References for ash decline: 338, 339, 469, 803–5, 809, 891, 892, 1100, 1238, 1426, 1541, 1542, 1640, 1641, 2017

References for pole blight of western white pine: 782, 1119–22, 1361, 1447

References for birch dieback: 151, 361, 362, 663, 681, 719, 782, 1361, 1543, 1590, 1866

References for oak decline: 49, 118, 157, 496, 766, 891, 892, 1076, 1131, 1132, 1134, 1187, 1188, 1327, 1402, 1884, 1894, 1965, 2060, 2103, 2104

References for citrus blight: 257–59, 375, 615, 671, 876, 1038, 1124, 2048, 2249

A, B. Black oaks in decline following repeated severe defoliation by insects. A. A dead tree and one with dead branches show adventitious sprouts that grew on major limbs after peripheral branches died; sprouts on the living tree now bear dense foliage. B. Dwarfed, distorted, chlorotic leaves, stunted twigs, and dieback (IL, Jun).

C, E. Decline in southern live oak. C. A large old tree during spring growth, with sparse-appearing, undersized foliage and early-stage dieback attributed to prior heat and drought stress; low branches bear denser foliage more nearly normal for the season (MS, Apr). E. Dieback and adventitious sprouts; the specific cause of this decline was unknown (TX, Nov).

D, F. Late-stage symptoms of citrus blight in grapefruit: sparse, dwarfed leaves, dieback, and (F) sprouts on the base of the trunk (FL, Apr).

Plate 219

Damage Caused by Salt Spray and by Salt and Other Inorganic Poisons in Soil (Plates 220–221)

Saline water retards growth and causes foliar browning, death of vegetative and flower buds, twig dieback, or outright death of plants when applied in sufficient quantity or frequency to aerial parts as a spray or to roots as soil solution. Damage from spray and saline meltwater occurs along northern highways where salt is used to melt snow and ice. Salt spray damage and occasional saltwater flooding during storms occur along the ocean coasts. Saline soils cause plant growth problems in many western and southwestern localities.

Salt spray. Prominent damage by wind-driven salt spray occurs occasionally to plants along ocean shores and occurs every winter to plants along highways in Canada and the northern United States. Chronic minor damage also occurs along the ocean shore as spray from bursting bubbles on the water is blown to the shore. Salt spray causes most damage within 50 meters of highways and 300 meters of the ocean, although damage from severe coastal storms may extend many kilometers inland. Airborne salt may also cause local damage around evaporative cooling towers and potash mines.

Leaves and twigs, especially those well separated or extending beyond the general contour of dense vegetation, are efficient collectors of spray droplets. Whenever the plant is wet at temperatures above freezing, sodium and chloride ions (and assorted other ions from seawater) enter and accumulate. External symptoms develop quickly during mild weather but are delayed in cold weather. Symptoms on evergreens along highways appear in late winter and intensify in early spring. Conifer needles turn brown beginning at the tips and progressing toward the base. Damaged needles fall prematurely. With injury in successive years, branches slowly become barren and die. Symptoms appear on deciduous species when growth begins in spring. Buds, especially flower buds, fail to open, and twig dieback becomes apparent. New growth occurs from adventitious or previously dormant buds on 2-year-old branch segments, producing a tufted growth pattern. Foliar browning develops within only a few days after a coastal storm during the growing season, and severely damaged leaves drop prematurely. Barren twigs may be apparent the next spring. The side of a plant away from the highway or shore receives less salt than the side facing the salt source and sustains less damage. As a result growth and flowering are one-sided. Repeated killing of exposed twigs or shoots of salt-sensitive woody species growing near the shore often trains them to shrubby or prostrate form.

The amount of salt absorbed and the intensity of damage increase with the amount of salt deposited on the plant. The type of salt, mainly sodium chloride along highways versus a complex mixture from seawater, seems unimportant. Damage at the cellular or tissue level is caused mainly by toxicity of the chloride ion. Chloride is translocated toward twig tips and toward the tips and margins of leaves, where it accumulates to lethal concentration. Sodium and other ions also enter and accumulate, but more slowly, and their concentrations in salt-damaged plants are not as closely correlated with the degree of damage as is the concentration of chloride.

Salt tolerance of many plants seems to be based more on ability to retard the internal accumulation of chloride than on intrinsic ability to tolerate it. For many woody species, foliar and twig damage appear when internal chloride concentrations rise above approximately 0.3% (dry weight basis). Plants with severe damage may have chloride contents of 1.25–2% or more in their twigs, while tolerant species, asymptomatic under the same conditions, have 0.5–0.9% chloride content. Normal values are under 0.1%.

Plants with thick wax layers on leaves and twigs, with leaves having a low surface-to-volume ratio (e.g., pines with thick needles), with large resinous buds (e.g., eastern cottonwood and horse-chestnut), or with buds submerged in the twig (e.g., black locust and honey locust) tend to accumulate salt relatively slowly. The relationship is not simple, however. Consider Austrian pine, for example. This species, tolerant of salt spray and bearing more wax than the salt-sensitive eastern white pine, accumulates chloride readily but remains apparently undamaged by internal concentrations that kill tissues of the latter species. Salt accumulation in plant tissues is facilitated by minute injuries where leaves and twigs abrade one another in the wind. Such injuries are presumably exaggerated during violent wind storms.

Salt accumulation causes hypertrophy of parenchyma in leaves of coastal plants such as live oak and yaupon, leading to greater apparent succulence of shore plants than of the same species inland. Species that tolerate the highest internal concentrations of chloride show the greatest hypertrophy, but hypertrophied leaves do not tolerate additional chloride as well as normal leaves on the same plant. Salt spray also selects for tolerant species at the shore, limiting the diversity of plant communities and resulting often in nearly pure stands of such tolerant trees as bear oak and southern live oak.

Salt in soil. Salt in the soil solution may cause damage in several ways. Chloride and other ions absorbed by roots and transported to leaves and twigs eventually accumulate to toxic levels, as do ions taken in from spray. As water evaporates from saline soil, the osmotic concentration of the remaining soil solution rises; this diminishes availability of water to roots and retards metabolic functions, photosynthesis, and growth. The availability of mineral nutrients is also diminished in saline soil because sodium ions may replace calcium and certain other nutrient elements on soil particles and because excess sodium in soil raises pH. Elevated pH may interfere with solubility of mineral nutrients such as iron. High sodium level also prevents normal aggregation of soil particles and thus contributes to soil compaction.

Salt pollution of soil along highways is greatest at the end of winter and subsides until salt applications begin the next winter. Chloride dissipates more rapidly than sodium. The degree of pollution also diminishes with distance from the pavement. Over several years, however, both sodium and chloride accumulate in the soil and within plants, and the chloride causes leaf scorch, premature defoliation, dieback, and decline of salt-sensitive trees such as sugar maple. Salt pollution of coastal soils as the result of storm-caused saltwater flooding is usually temporary because rains subsequently leach the soil.

In many western and southwestern locations, naturally saline soil inhibits or prevents growth of various plants that would otherwise be adapted to the region. Native plants in those places have undergone natural selection for salt tolerance. Salt accumulation is also a problem in irrigated horticultural landscapes in those regions. Mineral salts naturally present at low concentration in irrigation water slowly build up in surface soil unless excess water is applied to leach the soil. Leaching is usually impractical or impossible because of the amount

---

A, B. Foliar damage by salt spray blown inland by a hurricane. A. Mixed oak species in the background are completely brown, and a small red maple at left foreground is defoliated, in contrast to salt-tolerant Austrian and Japanese black pines. This site was approximately 10 km inland. B. A young sweet gum tree in a nursery 3 weeks after the hurricane shows brown original leaves and green new growth triggered by death of most of the original foliage (NY, Aug).

C, F, G. Damage to eastern white pine by deicing salt spray. C. Needle browning and suppressed growth, becoming less severe with distance along a row of trees planted perpendicular to the highway. F, G. Typical injury to needles, showing abrupt transition between dead and green parts (NY, Apr).

D. Damage by deicing salt spray to black cherry. Repeated killing of buds and twigs has led to branch dieback and to formation of clusters of adventitious shoots resembling witches'-brooms (NY, Apr).

E. Damage to arborvitae by dog urine (NY, May).

Plate 220

of water needed or because of impeded soil drainage. Slow growth, chlorosis, dwarfed or scorched leaves, and dieback then slowly develop. Also, if chloride occurs at concentrations greater than about 40 μg $\cdot$ g^{-1} in irrigation water, it may accumulate to injurious levels in plants without reaching high levels in soil.

Salt injury has deleterious secondary effects, including impairment of cold hardiness and of resistance to some fungal pathogens. Experiments with ash, lilac, and crabapple suggest that much twig dieback attributed to salt may be caused by freeze damage following salt-induced impairment of cold hardiness. Similarly, salted soil plus *Verticillium dahliae* killed experimentally treated young sugar maples when neither agent alone did so.

Salt-laden soil can be made more satisfactory for plant growth by working gypsum (calcium sulfate) into the surface layer. Appropriate rates vary with local conditions. Fertilizer also suppresses salt damage, provided that the nutrient ions, potassium and nitrate, are available to plants at the same osmotic concentrations as the sodium and chloride.

In the following lists, plants are arranged according to relative salt tolerance. Data for plant response to salt spray and to salt in soil have been merged. Species sensitive in one circumstance but tolerant in another are ranked according to tolerance in the most common circumstance of salt exposure. If this circumstance was unknown, we assigned the rank of less tolerance.

Relative tolerance of plants to salt

High tolerance (symptoms mild or uncommon): agave; apricot; Arizona ash; black ash; avocado; baccharis; red bay; bayberry; beautyberry; bottlebrush; bridal-wreath; buffaloberry; Japanese cedar; Sargent cherry; chokecherry; shrubby cinquefoil; citrus; eastern and Great Plains cottonwoods; European cranberry bush; alpine currant; devilwood; hybrid elms; Siberian elm; evergreen euonymus; golden-rain tree; groundsel tree; Chinese holly; European fly and Zabel's honeysuckles; horse-chestnut; eastern, European, and Japanese larches; cherry laurel; black locust; honey locust; Norway and sycamore maples; mesquite; bear, bur, English, northern red, southern live, white, and willow oaks; oleander; Russian olive; mock orange; Japanese pagoda tree; palmetto; palms; Siberian pea tree; callery pear; evergreen pear; Austrian, Japanese black, jack, longleaf, mugo, slash, and Italian stone pines; umbrella pine; beach plum; balsam, black, gray, hybrid, and white poplars; Turkestan rose; saltbush; blue spruce; staghorn sumac; Chinese tallow tree; tamarisk; tree-of-heaven; waxmyrtle; basket and purple willows; yaupon; southern yew; yucca.

Moderate tolerance: European, Italian, and speckled alders; eastern arborvitae; European, green, and white ash; European mountain ash; bigtooth and trembling aspens; Japanese barberry; gray, paper, sweet, and yellow birches; alder and European buckthorns; western catalpa; eastern red cedar; 'Kwanzan' cherry; bald cypress; American and wych elms; winged euonymus; firethorn; cockspur and other hawthorns; shagbark hickory; common juniper; Japanese lilac; large-leaved European linden; Amur, hedge, and silver maples; blackjack, post, and scarlet oaks; osage orange; loblolly, pitch, ponderosa, and Scots pines; Norfolk Island pine; pittosporum; London plane tree; Lombardy poplar; Amur privet; common quince; snowberry; sourwood; Norway spruce; smooth sumac; black and Persian walnuts; crack, goat, laurel, and white willows.

Low tolerance (foliar browning, bud death, and twig dieback, common and severe): apple; azalea; basswood; beautybush; American and European

beeches; European and river birches; boxwood; buttonbush; Atlantic white cedar; Atlas cedar; black cherry; cornelian cherry; cotoneaster; crabapple; American cranberry bush; Sawara cypress; blood-twig, flowering, gray, redosier, silky, and Tatarian dogwoods; Douglas-fir; box elder; elderberry; American and beaked filberts; balsam and white firs; forsythia; ginkgo; sour gum; sweet gum; hackberry; eastern hemlock; hickory; red horse-chestnut; American holly; Tatarian honeysuckle; American and European hornbeams; American hop hornbeam; common lilac; small-leaved European linden; magnolia; red, sugar, and Tatarian maples; mimosa; monkey puzzle; white mulberry; crape myrtle; ninebark; pin, southern red, water, and swamp white oaks; orchid tree; pecan; red, Virginia, and eastern white pines; cherry plum; flowering plums; flowering quince; dawn redwood; rhododendron; cultivated roses; sassafras; Allegheny serviceberry; Bumalda spirea; white spruce; sycamore; tulip tree; wayfaring tree; yew.

Other inorganic poisons in soil. Inorganic chemicals other than common salt that occasionally cause damage to woody plants include various salts and ions of arsenic, boron, copper, nickel, and zinc among others. Except for arsenic, trace amounts of these elements are essential for plant growth and are supplied by natural soils. Soil pollution associated with mining, smelter waste, municipal waste or pesticidal uses of these elements is the usual reason for their occurrence at phytotoxic levels. Symptoms of toxicity (chlorosis and necrosis of leaf tips and margins) are similar for several different minerals and are not diagnostic. Diagnosis depends upon chemical analysis of plant tissues and the rooting medium. We illustrate injuries attributed to arsenic and boron.

Salts of arsenic, especially lead arsenate and sodium arsenite, were formerly used as insecticides and herbicides, and arsenic accumulated to phytotoxic levels in some agricultural and nursery soils until the early 1970s. Arsenic residues in soil dissipate slowly. Residues of 40–80 μg $\cdot$ g^{-1} or more in soil, versus normal values less than 10 μg $\cdot$ g^{-1}, have been blamed for rootlet mortality and impairment of mycorrhizal function, leading to stunting of apple trees, to rootlet mortality and tip burn of red pine needles in a forest nursery, and to rootlet mortality, needle browning, and defoliation of hemlocks in a landscape nursery. Arsenic does not become highly concentrated in leaves; foliar concentrations less than 20 μg $\cdot$ g^{-1} may be associated with severe foliar browning.

Excessive boron in soil comes from irrigation water and from application or disposal of boron compounds used as herbicides, fire retardants, fertilizers, and laundry products. Boron taken up by roots is transported into leaves, where it normally occurs at concentrations of 15–100 μg $\cdot$ g^{-1} (dry weight basis). Other factors equal, the more boron available to roots, the greater the accumulation in foliage. Foliar chlorosis and necrosis in a wide array of plants have been associated with foliar concentrations above 75 μg $\cdot$ g^{-1}. The pine shown in panel D had a foliar boron content of 355 μg $\cdot$ g^{-1}.

References for salt damage: 88, 113, 137, 156, 231, 324, 493, 578, 586, 717, 718, 856, 867, 903, 1082, 1152, 1172, 1173, 1357, 1681, 1928–30, 2029, 2060, 2061, 2147, 2158, 2250
References for inorganic poisons in soil: 65, 587, 651, 652, 763, 1174, 1808, 1809, 1856, 1914, 1916, 2031, 2239

A, B. Deicing salt spray damage to Austrian pine. Individual sensitive trees of this normally tolerant species may be severely damaged by the combination of salt, wind, and thaw-freeze cycles in late winter. A. The tree in the foreground has many dead branches near the base, indicating recurrent damage. B. Green needle bases of 1- and 2-year-old needles are nearly hidden by dead needle tips; new green shoots provide contrast (Ont, Jul).

C, E. Needle browning and defoliation of Sargent's weeping hemlock, associated with high concentrations of arsenic in soil and roots (NY, May).

D. Needle browning of eastern white pine caused by borax from laundry waste water that drained from the pipe at left (arrow) into the root zone (NY, May).

F. Tip necrosis on leaves of 'Mission' almond, a salt-sensitive cultivar, caused by accumulation of salts from irrigation water (CA, Jul).

G. Marginal scorch, attributed to salt uptake from soil, on leaves of Norway maple (NY, Aug).

Plate 221

Damage Caused by Misapplied Pesticides (Plates 222–224)

Those who apply pesticides to plants or to soil in which plants grow rely on the principle of selective toxicity. The pesticide is intended to suppress or kill specific plants or plant pests while causing little or no injury to nontarget organisms. The validity of this principle is borne out by the generally successful use of pesticides without significant undesirable side effects. But environmental contamination and injury to nontarget organisms occur occasionally when all normal precautions are taken. Such contamination and injury are common when pesticides are mishandled or are applied under improper conditions. This plate and the two that follow show symptoms caused by pesticides, mainly herbicides, on nontarget trees and shrubs. The text deals first with herbicides, then with plant injuries caused by insecticides and fungicides. Photos of damage by a particular pesticide may not face the page with information about it.

Herbicides that have often caused injury to nontarget plants fall into two general groups as regards their intended purpose. One group is intended to kill established broad-leaved weeds selectively in turf or agricultural or forest crops. Well known herbicides in this group include the phenoxyacetic acids (2,4-D, 2,4-DP, MCPA, MCPP, and related compounds) and the benzoic acid derivative dicamba. These act as plant hormones that disrupt normal growth processes. In sublethal doses they cause growing plant organs to assume aberrant forms. Chemicals in this group may reach and enter nontarget plants by as many as three modes. The most common is absorption of herbicide from droplets of spray that drift away from the site of application. A second mode is absorption of herbicide in the gaseous phase from drifting vapor after the evaporation of spray droplets or of liquid on sprayed surfaces. This occurs with volatile herbicides in warm weather. The third route is through soil, where the herbicide is absorbed by roots. The phenoxy herbicide 2,4-D is the best known example of a chemical that may reach nontarget plants by any or all routes. Dicamba is more likely than other hormone-type herbicides to cause injury as the result of uptake by roots.

The other group includes chemicals intended to prevent emergence of weed seedlings (preemergence herbicides) or to kill all vegetation to which they are applied. Some of these herbicides persist in soil for a year or more. Simazine is a much used preemergence herbicide that sometimes causes unwanted residual effects after application to nursery soils. General vegetation killers that sometimes cause trouble for nontarget trees or shrubs include amitrole, dichlobenil, paraquat, picloram, and substituted urea compounds such as diuron. These chemicals are either sprayed on target plants or applied to soil and taken up by roots. They are used to control vegetation along highways, railroads, fences, power lines, and similar places. Those applied to soil are intended to have long-lasting residual activity; therefore misapplication also leads to persistent effects. We have seen century-old sugar maples and Norway spruces, for example, killed by picloram that was applied for control of woody weeds around gravestones in a cemetery. The inexperienced applicator did not realize that the roots of mature trees extend far from the trunks, and he may not have fully appreciated the potency of the herbicide. Amateur applicators of herbicides often cause plant problems by using more than recommended amounts.

Glyphosate is the best known contemporary example of a general-purpose herbicide that is sprayed on target plants without significantly polluting the soil or air. Glyphosate deposited on soil becomes tightly adsorbed to surface soil particles and is soon degraded by microorganisms. This chemical is also relatively nonvolatile; thus it is essentially unavailable to plants except by spray deposition.

Symptoms of injury by herbicides vary with the type of chemical but are not diagnostic except for the aberrant growth caused by hormone-type herbicides. Symptoms similar to those of herbicidal injury can be caused by some diseases and insects, insufficient or excess water or heat, deficiencies of certain nutrients, nonherbicidal poisons, and other kinds of misapplied pesticides.

Herbicides of the hormone type are translocated to growing points and cause multiple abberations in new leaves and shoots. Nontarget plants usually receive sublethal doses and outgrow the symptoms within 1–2 years. Symptoms develop several days to several weeks after exposure or may appear in spring following an autumn exposure. Symptoms include cupped leaves (convex surface either up or down); abnormal bending or drooping of green twigs, petioles, rachises, or main leaf veins; abnormally prominent veins; parallel leaf venation on plants that normally have net-veined leaves; underdevelopment of interveinal leaf tissue (resulting in abnormally narrow leaves); wavy, frilled, or curled leaf margins; and in extreme cases bud death and dieback. Additional symptoms include abnormally tough or leathery leaves, partial failure of chlorophyll development, swollen buds, delayed bud break in spring, loss of apical dominance, enlarged lenticels, abnormal purple coloration of normally green stems, lesions in stem bark, and suppressed root growth. Purple stem color induced by phenoxy herbicides has been observed on ash, sugar maple, pin oak, and walnut. Conifers are relatively tolerant of phenoxy herbicides, but their shoots may curl and needles drop.

Some woody plants are so sensitive to phenoxy herbicides that they usually show at least mild foliar symptoms of injury if these chemicals are used anywhere in the vicinity. Box elder, grapevine, and redbud are among the most sensitive species and are useful indicators if damage to other plants by phenoxy herbicides is suspected.

Herbicide injuries to coniferous trees.

A. Bleached needles on Scots pine, attributed to injury by amitrole (NY, Aug).

B. Injury to blue spruce by an unidentified herbicide. Shoots of the current season have ceased growth; some are curled; shoot tips have died; new needles are dying. Several buds, including apical buds on two twigs shown here, have differentiated abnormally into cones. All buds on this branch would normally have been vegetative (NY, Jun).

C. Injury to Atlas cedar by an unidentified herbicide. The growing points of the twig and of several short shoots have died, and needles are dying. Stem swelling near the twig apex is also abnormal (NY, Oct).

D. Injury to blue spruce by dicamba, which was applied to adjacent turf for control of broad-leaved weeds. The branch is dying; needles of the current year were the first to discolor and begin to drop (NY, Dec).

E. Injury to Japanese black pine by a mixture of two phenoxy herbicides, 2,4-D and MCPP, applied to adjacent turf at the beginning of the growing season for control of broad-leaved weeds. These branches are dying. Shoot growth and needle elongation were arrested early in the growing season; growing points are dead, and old foliage has become chlorotic (NY, Sep).

F, H. Mild (F) and severe (H) symptoms caused by simazine to young eastern white pines in a nursery. The trees were planted in soil to which the herbicide had been applied in a previous year to prevent weed emergence. Needles elongated normally but began to droop and turn yellow at the bases as the herbicide accumulated. Discolored needles will drop; the plant with severe symptoms will slowly die (NY, Aug).

G. White fir damaged by dichlobenil, which was applied to the ground in a nursery to suppress weeds. Firs are highly sensitive to this herbicide. Contrary to the logical expectation that similar symptoms would appear on most trees in the affected nursery block, some individual white firs typically remain apparently normal, while others die (NY, Oct).

Plate 222

Hormone-type herbicides (phenoxy and benzoic acid derivatives) applied to turf for control of broad-leaved weeds often injure trees and shrubs. These chemicals, especially dicamba, are taken up by tree or shrub roots growing in the root zone of the grass. Dicamba causes drastic growth suppression, leaf cupping, bending and sometimes coiling ("fiddlenecking") of shoot tips, yellowing of new growth, bud failure, browning or blackening of foliage (depending on plant type), defoliation, and sometimes death.

Chemicals in the second general group, preemergence herbicides and nonselective vegetation killers, tend to halt growth and cause chlorosis of new and old leaves. If the dose is sufficient they cause foliar browning, leaf cast, and dieback of twigs and branches. Amitrole causes bleaching of leaves or leaf parts to white or pinkish white. Simazine and atrazine cause marginal and interveinal chlorosis of broad-leaved plants, or yellowing, starting at needle bases, of conifers. Dichlobenil causes marginal chlorosis or, if injury is severe, general chlorosis and browning. Substituted urea compounds such as diuron cause veinal chlorosis or, if the dose is greater, general chlorosis and death. Picloram in low doses may cause abnormal bending or curling of young shoots, and at higher doses causes foliar browning, leaf cast, and dieback. Trees or shrubs injured by these herbicides are less likely to recover than are plants injured by hormone-type herbicides.

Contact herbicides that kill leaves and succulent shoots but not woody parts are a special diagnostic case. Paraquat is such an herbicide. Spray droplets containing paraquat cause brown lesions, often similar to leaf spots caused by fungi, or superficial lesions on green stems. Stems of green-barked trees in nurseries, for example golden-chain, linden, honey locust, oaks, or Japanese pagoda tree, may develop brown streaks of superficial cork or sometimes brown cracked bark after spray treatment with a contact herbicide to kill sprouts or control weeds at the base of the trunk.

Herbicides generally have greatest effects on plants treated or inadvertently exposed during growth in warm weather. Water shortage or heat stress may enhance toxic effects. Dormant plants or those in a resting phase are less sensitive. Herbicides may also trigger damage by secondary factors. For example, some herbicides reduce cold hardiness; dieback attributed to these chemicals may be caused in part by freeze damage and by opportunistic fungal pathogens that attack tissues weakened by cold injury. The complex association of herbicide injury with decline of broad-leaved trees in the Great Plains illustrates these propositions and also exemplifies the difficulty that diagnosticians may face in fixing blame for tree damage.

In the Great Plains of Canada and United States, injury to trees by spray and vapor drift of hormone-type herbicides occurs up to several kilometers from agricultural fields where the chemicals are applied. Foliar symptoms typical of those caused by 2,4-D are common every year on several species, notably box elder, Siberian elm, and hackberry. Branch dieback and premature death of various broad-leaved trees in hedgerows, shelterbelts, and farm woodlots are also common and widespread. Siberian elm and hackberry are most severely affected. The scene shown in panel A of this plate illustrates the problem in elm. Such decline has been reported to be more severe in areas where field crops are grown (where more herbicide is used) than in rangeland areas. Writers have attributed decline of Siberian elm and hackberry to herbicides, but direct evidence for a primary causal role of herbicides is lacking, and alternative explanations are available. Herbicides clearly cause foliar symptoms, but the doses of airborne herbicide that trees receive are small in comparison with doses that cause dieback or death in trees intentionally treated. Herbicides apparently do interact with other stressing factors, especially low temperature, to cause bud failure and dieback, but the climate of the plains is harsh for trees and may itself be sufficient to preclude long life of species planted in shelterbelts. Summer heat and dryness, winter cold and wind, and rapid temperature changes in early and late winter stress or injure trees and also predispose them to attack by opportunistic fungal pathogens. Much dieback in elms stressed by drought or freezing is caused directly by *Botryodiplodia hypodermia* (Plate 86) and *Tubercularia ulmea* (Plate 99). Herbicides possibly also play a role in susceptibility to pathogens, as has been demonstrated in experiments with field crops, tree seedlings, and small pines, but evidence for this role in trees of the Great Plains has not been presented.

Phenoxy herbicides and dicamba are the weed killers that most commonly injure nontarget trees and shrubs in horticultural landscapes. The following ratings for tolerance to these herbicides were collated from listed references and authors' and colleagues' observations.

Herbicide injury to deciduous trees.
A. Dieback and premature death of Siberian elm in a woodlot in a wheat-producing area of the Great Plains. Such damage has been linked to chronic exposure to phenoxy herbicides that drift from adjacent fields (CO, Jul).
B. Typical symptoms of injury to box elder by 2,4-D. This branch began to grow normally, then received 2,4-D as aerosol or vapor from nearby turf to which the herbicide was applied. Leaves that were fully expanded at the time of exposure remained normal. Leaves just completing growth at the time of exposure have leaflets with midveins bent and edges turned downward. Leaves that were in an early stage of growth became stunted and distorted and did not attain normal green color (NY, Jun).
C. Symptoms of phenoxy herbicide injury in Siberian elm. The sequence of normal growth, herbicide exposure, and symptom development was similar to that described for box elder (ND, Jul).
D. Injury to red oak by 2,4-D. The tree acquired the herbicide via roots and had begun to outgrow its effects when the photo was made. Deformed leaves have petioles and veins of normal length, but interveinal tissues did not develop, and the veins grew curved (NY, Jul).
E. Severe stunting and distortion of new growth on a bigleaf maple sapling treated the previous year with glyphosate. This plant received spray intended for roadside weeds (OR, Jul).
F. Irregular contours on the trunk of a honey locust tree damaged by a mixture of phenoxy herbicides applied to surrounding turf. The tree showed severe foliar symptoms. Trunk abnormalities resulted from localized growth adjacent to areas of inhibited cambial activity (NY, Aug).

Plate 223

Relative tolerance of plants to phenoxy herbicides and dicamba

Plants tolerant of phenoxy herbicides: eastern arborvitae; western azalea; basswood; American beech; catalpa; eastern red cedar; bald cypress; Douglas-fir; fir; sour gum; hawthorn; juniper; bigleaf, red, and silver maples; southern live oak; common pear; pine; rosebay.

Plants of intermediate sensitivity: red alder; blue, green, and white ash; trembling aspen; barberry; buckeye; cherry and chokecherry; Kentucky coffee tree; eastern cottonwood; American and winged elms; sweet gum; hemlock; hickory; common lilac; black locust; sugar maple; mulberry; Kellogg, pin, red, water, and white oaks; Russian olive; peach; balsam poplar; privet; rhododendron; blue spruce; yew.

Plants sensitive to phenoxy herbicides: apple, Oregon ash, hoary azalea, birch, Amur cork tree, Great Plains cottonwood, dogwood, box elder, elderberry, Siberian elm, forsythia, grapevine, hackberry, horse-chestnut, European linden, honey locust, madrone, Amur and Norway maples, mimosa, black oak, plane tree, redbud, rose, serviceberry, sumac, sycamore, tree-of-heaven, tulip tree, walnut, willow, wisteria, yellowwood.

Plants tolerant of dicamba: hoary azalea, eastern red cedar, eastern cottonwood, black and southern live oaks.

Plants of intermediate sensitivity: green and white ash; American beech; buckeye; American and winged elms; hackberry; hickory; Norway, red, and silver maples; red oak; black walnut.

Plants sensitive to dicamba: red alder, apple, eastern arborvitae, trembling aspen, barberry, basswood, birch, catalpa, Great Plains cottonwood, Douglas-fir, box elder, sour gum, sweet gum, hawthorn, eastern hemlock, juniper, lilac, black locust, honey locust, sugar maple, pin and white oaks, shortleaf pine, redbud, rosebay, serviceberry, blue and white spruces, sycamore, tree-of-heaven, tulip tree, walnut, Japanese and related hybrid yews.

Plant injury by insecticides and fungicides is infrequent compared to that by herbicides, partly because the former are selectively toxic to animals or fungi rather than to plants and partly because they are used less in horticultural landscapes. When visible symptoms do occur, a diluent or solvent of the toxicant is often responsible. The active ingredients in modern organic insecticides and fungicides seldom cause visible symptoms unless deliberately applied at rates higher than specified or under improper environmental conditions or to plants not specified on product labels. Other factors equal, the likelihood of pesticide injury is greatest for plants stressed by heat (temperatures above 30–32°C) or water shortage, or for plants sprayed under poor drying conditions such that spray liquid remains on plant surfaces. Plants sensitive to particular pesticides are usually identified on the labels. An extensive list of plant sensitivities to insecticides and acaracides is compiled in one of the references.

Symptoms of plant injury by insecticides or fungicides are not usually diagnostic and therefore cannot reliably be interpreted unless the history and treatment of the plant are known. Many pesticides and antitranspirants temporarily suppress photosynthesis and growth of treated plants without causing visible symptoms, but such suppression of woody ornamental plants is usually unimportant. Visible symptoms include yellow to brown leaf spots; chlorosis of leaf tips, margins, or interveinal areas; general chlorosis; browning of leaf margins or interveinal areas; stunted shoots; and abnormal crinkling or curling of leaves. Foliar yellowing or browning is often followed by premature leaf drop.

Sprays that contain oil, although among the safest from the standpoint of toxicity to mammals, occasionally cause injury to plants because the oil film interferes with gas and heat exchange at plant surfaces. Symptoms on leaves may range from small dark spots to general chlorosis, browning, and leaf cast. Twigs may also die back.

In times past, inorganic pesticides such as lead arsenate, lime-sulfur, and copper fungicides, including Bordeaux mixture, caused injury to many kinds of plants. The copper compounds are still used and may cause injury if they are applied to sensitive plants such as holly or Norway maple. Red spots with associated hypertrophy may develop on the lower surfaces of holly leaves. Norway maple leaves sprayed with Bordeaux mixture or fixed copper may develop numerous minute black flecks on the lower surface. If injury is severe, flecks coalesce into necrotic blotches.

Wood preservatives may also cause plant injury, although in landscapes this is usually restricted to foliage and shoots touching or closely adjacent to treated wood. Wood preservatives are most hazardous to plants in confined spaces or growing in preservative-treated containers. Creosote is a wood preservative commonly associated with plant injury. Creosote releases toxic volatile chemicals, especially during warm, sunny weather. These cause uprolled leaf edges, chlorosis, and premature leaf drop. Treated wood below the soil level does no known harm to nearby plants, however.

References for herbicide damage: 140, 141, 223, 326, 498, 810, 1156, 1387, 1433, 1523, 1576, 1614, 1769, 1838
References for injury by other pesticides: 76, 452, 582, 791, 811, 934, 971, 1419, 1971

Herbicide and insecticide injuries.

A, D. A London plane tree severely injured by dicamba, which was applied to surrounding turf for control of broad-leaved weeds. Shoot elongation and leaf expansion were arrested; many buds did not open; leaves are dwarfed, distorted, and chlorotic (NY, May).

B. Silver maple and pistachio trees killed or severely damaged by atrazine, which was applied to soil in their root zones (CA, Nov).

C, F. Paper birch injured by a spray of dimethoate insecticide, which was applied at the beginning of the growing season to prevent damage by birch leaf miner. Shoot growth and leaf expansion on injured stems were suppressed, and expanded leaves developed marginal and interveinal browning. The stem at left, part of the same birch clump, remained uninjured (Ont, Jul).

E. Japanese tree lilac injured by the herbicide amitrole, which was applied for control of weeds in the root zone of the tree. Bleached leaves or areas on leaf blades, preceding leaf browning, are typical of injury by this herbicide (NY, Jun).

G. Interveinal yellowing and marginal browning on a leaf of European white birch injured by the systemic insecticide disulfoton. The insecticide was applied to the soil for uptake by roots (NY, Aug).

H. Veinal chlorosis and general bleaching of apple leaves, caused by the herbicides diuron and terbacil, which were absorbed by roots of nontarget trees in an orchard after application of an inappropriately heavy dose to soil to kill roots of trees that had been removed. Some branches and entire trees failed to grow the next year, while those with lesser injury appeared like this shoot. Veinal chlorosis may be caused by either diuron or terbacil (NY, Jun).

Plate 224

Diseases and Injuries Caused by Air Pollutants (Plates 225—229)

Many books and many thousands of research articles deal with the effects of polluted air on plants. This plate and the four that follow present glimpses of the impact of air pollutants on forests and on woody plants in landscapes. We discuss the subject in several segments: the problem of polluted air for plant health, major phytotoxic air pollutants, effects of oxidants on conifers, effects of oxidants on broad-leaved plants, effects of sulfur dioxide on coniferous and broad-leaved plants, and plant damage by fluoride and minor air pollutants. Lists of plants sensitive or tolerant to major air pollutants are also provided. Because of this arrangement, descriptive information about a particular disorder may not face the plate that displays it.

The first examples of plant damage by man-made air pollutants were localized: decline and death of vegetation around smelters, brick works, and other industries and failure of some trees to thrive in urban and industrial areas. Sulfur dioxide and fluorides were the main pollutants. Later, with burgeoning needs for power and industrial products, came the continual massive release of oxides of sulfur and nitrogen as well as other pollutants from the stacks of factories, refineries, and coal- or oil-fired power-generating stations. Beginning after World War II, air pollution by ozone and other oxidants became a problem as the exhaust gases of tens of millions of automobiles polluted large areas. In the 1950s to 1970s, many plant injuries from short-term, acute exposure to air pollutants were described, and some diseases caused by chronic exposure were recognized as well. Forests in southern California began to decline as the result of smog damage, and depressed tree growth in the East was linked to polluted air. Beginning in the 1950s, rain and snow around the Northern Hemisphere became increasingly acidic, mainly because of the atmospheric burden of oxides of nitrogen and sulfur. By the mid-1980s these atmospheric changes were associated with and were suspected of causing widescale forest decline in Europe and in some mountainous areas in eastern United States, but proof of a causal role of air pollutants had not been presented.

The major classes of phytotoxic air pollutants, in descending order of direct damage caused, are oxidants (ozone, oxides of nitrogen, and peroxyacyl nitrates), sulfur dioxide, and fluorides (hydrogen fluoride and silicon tetrafluoride). Ammonia, chlorine, and hydrogen chloride cause occasional local damage. Metallic elements from polluted air also accumulate in plants, but their effects at the concentrations encountered in most localities are minor or unknown.

Ozone is the most damaging oxidant. Ozone and nitrogen oxides are constituents of photochemical smog (the sort that became infamous in southern California). Nitrogen oxides are primary pollutants generated during combustion of many substances in air. They are also released naturally from soils. The main anthropogenic sources are factories, power-generating stations, and internal combustion engines. Ozone is a secondary pollutant generated spontaneously in polluted air by the action of ultraviolet radiation on molecular oxygen and oxides of nitrogen. Ozone generation is enhanced by the presence of organic pollutants from automobile exhaust. Ozone commonly builds up to phytotoxic levels in the atmospheres of urban areas during warm, sunny weather. It may cause damage to plants far from the source of its precursors as masses of polluted air move overland. Ozone from natural sources (carried from the stratosphere or generated during electrical storms) is alone inconsequential for plant health but may be added to that generated by humans.

Oxides of nitrogen are less toxic than ozone and cause much less direct damage but are important as precursors of ozone and of acids found in acid rain. Peroxyacetyl nitrate (PAN) and related compounds also form as secondary pollutants by the action of ultraviolet radiation on precursors in photochemical smog. These compounds are potentially highly toxic to plants but do not regularly accumulate to toxic concentrations. Natural damage to trees by PAN compounds has not been documented, and we do not consider them further.

Sulfur dioxide (SO_2) is formed primarily during combustion of coal and oil. The main sources are power-generating stations, heavy industries, and smelters that reduce metallic ores. Until the 1960s, when air pollution regulations began to take force, damage by sulfur dioxide was primarily localized near such sources because the gas was released near ground level. Now, however, most SO_2 is released from tall stacks and becomes greatly diluted before reaching ground level. Therefore, acute injury to plants by this gas is less common. SO_2 remains important, however, for its role in acid deposition.

Fluorides are constituents of many substances mined from earth and are released to air during aluminum reduction, petroleum refining, and the manufacturing of steel, phosphate fertilizers, bricks, glass, and ceramics. Plant damage by fluorides occurs only near such sources. Damage by chlorine compounds, ammonia, and other minor air pollutants is localized mainly around chemical spills.

Air pollutant impact on plants is seldom apparent to the casual observer except near sources of severe pollution. Most plant species in a given region and most individuals of a given species are either tolerant of pollutants at the concentrations prevailing in ambient air or escape conspicuous injury because of their location or growth stage at the time of a pollution episode. Atmospheric turbulence and winds cause rapid dilution and movement of pollutants away from sources. This dispersion may prevent conspicuous visible injury to plants near the sources, although the cumulative burden of pollutants from many sources apparently suppresses plant growth over large regions. This suppression has been demonstrated for various agricultural plants and some tree species by comparing growth in air filtered to remove pollutants with growth in nonfiltered air. Growth suppression may occur in the absence of diagnostic symptoms of pollutant injury. Tree species as diverse in type and location as white fir; chestnut and white oaks; ponderosa, shortleaf, and eastern white pines; and red spruce have displayed growth reductions unrelated to climate but correlated with increasing air pollution since approximately 1950. In most instances, however, only circumstantial evidence exists that air pollutants have caused the growth reductions.

Noticeable injury to plants occurs during warm weather when pollutants accumulate in stagnant air. Accumulation often occurs during atmospheric inversions in valleys and basins bounded by mountains but may also occur in large regions surrounding metropolitan areas when normal west-to-east air movement is interrupted by a stationary high-pressure system. When a mass of polluted air moves from the source locality, pollutant concentrations rise temporarily downwind. Thus plants in areas with a normally low pollutant burden may develop symptoms of injury.

Oxidant damage to ponderosa pine and winter fleck of Austrian pine.
A. Photochemical smog in the foothills of the San Bernardino Mountains of southern California (CA, Jul).
B–E. Damage by ozone to ponderosa pine in the San Barnardino Mountains. B. A declining ozone-sensitive tree with stunted shoots and yellow foliage, flanked by a mildly affected ponderosa pine (left) and an apparently normal sugar pine (right). C. A ponderosa pine forest in decline. Normal appearance of the species is shown by the dense crown of one ozone-tolerant tree at right center; all others have open crowns resulting from premature loss of needles 2 or more seasons old and tufting of undersized needles on stunted twigs. D. A branch tip from a diseased tree. Current season's foliage is not yet symptomatic; 1-year-old needles are chlorotic, and needles 2 and 3 years old have fallen prematurely. E. Typical chlorotic mottling caused by ozone on 1-year-old needles (CA, Jul).
F. Austrian pine needles, magnified to show chlorotic lesions of undetermined cause, called winter fleck. This disorder and others like it on ponderosa and other pines are apparently caused by environmental factors but not by air pollutants (NY, Feb).

Plate 225

Acid-forming pollutants, mainly NO_2 and SO_2, carried from urban and industrial areas have been blamed for decline of red spruce on mountains of eastern North America and for decline of Norway spruce and other forest species in Germany. Red spruce displays thin crowns, needle browning, and twig dieback that progresses downward and inward from the top and periphery of the crown until limbs and entire trees die. Although these declines have occurred coincidentally with an increase in acidic deposition, the scientific evidence for a causal role of "acid rain" is equivocal, and other explanations for decline are available (see Plate 216).

Plant disorders caused by gaseous air pollutants vary with concentration of pollutant, duration of exposure, intrinsic sensitivity of the plants, and environmental conditions affecting the plants before and during exposure. All major air pollutant gases enter plants via stomata and are capable of suppressing the rate of photosynthesis, stimulating respiration, and ultimately suppressing growth when administered at doses insufficient to cause visible symptoms on leaves. This information, from controlled experiments, supports evidence from field observations and experiments that air pollutants have suppressed tree growth in forests in recent decades. Fortunately, tree-to-tree variation in tolerance of air pollutants occurs in every species observed, and this tolerance is heritable in species that have been studied.

Visible injuries by air pollutants are classed as acute or chronic. Acute injury, associated with brief episodes of greater than normal air pollution, results from rapid absorption of enough toxicant to kill parts of leaves, causing characteristic markings. Chronic injury results from repeated absorption of a pollutant in amounts that do not kill tissues but are sufficient to cause cumulative physiological disturbance. Symptoms include chlorosis and premature foliar senescence or death, diminished growth, and in some cases progressive decline in plant health. Sensitive trees typically sustain both types of injury during a growing season, and the same trees tend to have symptoms year after year. The minimum concentrations of ozone and sulfur dioxide that cause measurable effects in sensitive plants are on the order of 0.05 and 0.10 ppm (volume basis), respectively, but concentrations up to 10 times greater are required to cause visible injury or growth suppression in many woody plants tested. Injury by fluorides usually occurs as the result of prolonged foliar absorption and accumulation from air containing low concentrations of toxicant—as low as a few parts per billion.

Mixtures of air pollutants often cause different or more severe symptoms than would be caused if the same plants were exposed to the individual pollutants at the same dose levels. Moreover, interacting air pollutants sometimes exert greater than additive effect on symptom severity. Such interactions complicate diagnosis. A tip necrosis of current-season needles of Austrian pine, common in New Jersey, for example, was reproduced experimentally by fumigation with a mixture of 0.2 ppm ozone and 0.1 ppm SO_2 for 6 hours. Grafted plants originating from sensitive and tolerant trees in the field had corresponding sensitivity or tolerance to experimental fumigation. In the field, the injury is probably caused by lower concentrations of pollutants absorbed during longer periods than were tested to reproduce the symptoms.

Air pollutants also modify the interactions of plants with biotic pathogens. In general, plants injured by air pollutants become abnormally susceptible to opportunistic, nonspecialized pathogens and to secondary insects. Highly specialized pathogens such as rust fungi are themselves adversely affected by air pollutants and cause less disease in polluted air than in clean air.

Damage by ozone. Ozone causes symptoms on sensitive trees in areas where daytime average atmospheric concentrations in summer often exceed 0.05 ppm (98 $\mu g \cdot m^{-3}$) or where the maximum hourly average often exceeds 0.1 ppm. Ozone at concentrations near 0.05 ppm may cause acute symptoms in highly sensitive plants, but similar concentrations have also stimulated growth of seedlings of white ash and tulip tree. Stimulation by ozone is not understood. At concentrations of 0.1 ppm or higher, however, all measured effects of ozone are deleterious. This highly reactive gas breaks down during reactions with plant constituents; it leaves no residue or diagnostically useful reaction products.

Visible foliar injury by ozone results from more or less rapid cellular death, sometimes preceded or accompanied by pigment accumulation. Ozone causes membranes (plasmalemma, tonoplast, and chloroplast envelopes) to lose integrity; thus sensitive cells collapse. Damaged cells typically collapse in clusters that are visible externally as white to tan flecks or dark stipples. In conifer needles, mesophyll cells adjacent to stomata are affected first, but two rather different patterns of symptoms occur. In one, typified by ozone needle mottle of ponderosa pine, mature needle tissue is the most sensitive. Chlorotic spots with diffuse margins develop irregularly along nearly the entire length of needles 1 or more years old, giving them a mottled appearance. The second pattern is typified by pollutant injuries that cause tip necrosis of eastern white pine needles. The most sensitive tissue in this case is that in an intermediate stage of maturation (semi-mature tissue) near the base of current-season needles. Mature tissues toward the needle tip and the youngest tissues just emerged from the sheath of basal scales are tolerant. Initial symptoms, caused by collapse of mesophyll cells, are tiny bleached flecks and pinkish yellow spots along the lines of stomata. If injury is severe, many cells in the sensitive zone collapse, and a band of pinkish yellow, eventually brownish tissue separates the green needle base from the green tip. The portion beyond the dead band then dies, necrosis progressing from the injured zone toward the tip.

Minor injuries caused by pollutants to conifer needles during the growing season often intensify and lead to needle browning during winter. This browning is presumed to result from drying and freeze-thaw stresses on previously injured tissue rather than from more pollutant injury, as dormant plants are tolerant of polluted air. Needle markings caused by air pollutants can be distinguished from spots,

---

Oxidant damage and similar injuries of eastern white pine.

A, C, D, G. Injury of eastern white pine, typical of that caused by either ozone or sulfur dioxide. A. Foliage of a pole-sized tree appearing faded as the result of injuries like those shown in (C) and (D) (NY, Sep). C. A twig from a highly sensitive tree. By the end of the growing season, chlorotic spots and bands occur all along the needles except on the basal few millimeters. Lesions that involve the entire cross-section of a needle cause tip browning (NY, Jan). D. Enlarged view of needles with pinkish spots with diffuse margins along the lines of stomata (NY, Jan). G. Twigs from (left to right) a tree tolerant of air pollutants, a tree of moderate sensitivity, and a highly sensitive declining tree (NY, Sep).

B, E, F. "Semimature tissue needle blight" of eastern white pine. B. Scattered trees in a Christmas tree plantation display uniformly browned needle tips, while the majority of trees remain unaffected. E. Close view of affected shoots. F. Needles from three fascicles, showing that necrosis begins as a pinkish tan band 1–2 cm from the needle base and spreads to the tip. Mild injury results in slight discoloration at the same level on the needle (IL, Jun).

H. Enlarged view of chlorotic spots that often arise during winter (winter fleck) along needle edges of eastern white pine, unrelated to stomata and apparently unrelated to injury by air pollutants (NY, Jan).

Plate 226

flecks, and tip necrosis caused by stresses of winter weather on the bases of external appearance and microscopic characteristics of injured tissue. In conifers such as eastern white pine that have one needle face lacking stomata, the contrast between flecked or spotted stomatal surfaces and the uninjured nonstomatal surface of pollutant-injured needles is diagnostically useful.

The best known single example of forest damage by oxidant air pollutants is the decline of Jeffrey and ponderosa pines in the San Gabriel and San Bernardino mountains of southern California. Photochemical smog from the Los Angeles metropolitan region drifts eastward into the mountains and causes chlorotic mottling and premature abscission of needles, slowed growth, decline, and premature mortality. Ozone-damaged trees become abnormally susceptible to attack by the root-rotting fungus, *Heterobasidion annosum* (Plates 151–152), and by bark beetles. These biotic agents are often the direct causes of death.

In the East, eastern white pine develops foliar markings like those in Plate 226. In some areas it also shows general yellowing and decline caused by mixed pollutants, chiefly ozone and sulfur dioxide. This damage is documented best along the Blue Ridge in Virginia. Fortunately, a high proportion of trees of this species are tolerant enough to persist with damage no more severe than chlorotic spotting and premature casting of needles. Trees tolerant enough to retain needles for the normal 27 months, however, are now uncommon in many areas.

Eastern white pine is subject also to a "semimature tissue needle blight" (SNB), attributed by various authors to ozone, to SO_2, and to heat and water stress on succulent needle tissue during the time of needle elongation in late spring. Sensitive trees develop blight in occasional years and may not have the chlorotic flecks and pinkish yellow spots typical of injury caused by low doses of ozone or sulfur dioxide. Necrosis spreads upward from pinkish yellow bands that develop in semimature tissue near needle bases. Critical experimental tests of the ability of SO_2 or ozone, alone or combined at concentrations likely to occur in the field, to induce SNB on grafted plants derived from those naturally susceptible to SNB have not been reported.

In typical broad-leaved plants, the cells most sensitive to ozone are in the palisade parenchyma of leaves that have just completed expansion. Therefore symptoms appear first on subterminal leaves of growing shoots. Affected cells collapse in small groups bounded by the smallest veinlets. Clusters of collapsed cells appear as light flecks or dark stipples (depending on species) on the upper surface 24–48 hours after a pollution episode, but no symptoms are visible from below. Flecks or stipples usually occur in interveinal areas, while the tissue along major veins remains green. Severe injury leads to either a bleached or a darkly pigmented appearance of the upper surface. Flecks or stipples sometimes fade and may be replaced by chlorosis. Injured leaves often drop prematurely.

In experiments involving constant or daily fumigation of growing seedlings of deciduous trees with ozone, growth of green ash, eastern cottonwood, silver maple, loblolly pine, hybrid poplars, sycamore, and tulip tree has been suppressed in the absence of visible symptoms.

Virtually all plants are tolerant of air pollutants at various times of year or stages of their life cycles. Evergreens as well as deciduous plants are tolerant during dormancy. Also, little injury can occur at night or during times of water stress when in most species stomata are closed and gas exchange inhibited. Typical minimum ozone doses (concentration × time) capable of causing visible injury to intrinsically sensitive plants in a sensitive condition range from 0.03–0.1 ppm for 8 hours to 0.1–0.25 ppm for 1 hour. The following lists classify woody plant species according to the most sensitive condition of the average plant. Readers are reminded that individual plants of a species vary in tolerance.

Relative tolerance of plants to ozone

Tolerant: European alder; Japanese andromeda; eastern arborvitae; blue, European, flowering, 'Summit' green, pumpkin, and 'Autumn Purple' white ash; European mountain ash; avocado; European beech; river and European white birches; incense cedar; Virginia creeper; cucumber tree; bald cypress; flowering and gray dogwoods; Douglas-fir; hybrid elms; balsam fir; 'Lalandei' firethorn; fuchsia; ginkgo; sour gum; hawthorn; eastern hemlock; American and Japanese hollies; European hornbeam; Sierra juniper; mountain laurel; lemon; littleleaf and silver lindens; black locust; honey locust cultivars 'Emerald Lace', 'Majestic', 'Moraine', 'Rubylace', and 'Skyline'; saucer and other magnolias; Norway, red, sugar, and sycamore maples; mimosa; bur, 'Sovereign' pin, northern red, shingle, and Shumard oaks; pachysandra; 'Regent' Japanese pagoda tree; paulownia; peach; callery pear; 'Bartlett' common pear; Japanese black, bristlecone, digger, red, sugar, and southwestern white pines; singleleaf pinyon; Amur privet; redwood; dawn redwood; Carolina rhododendron; giant sequoia; spirea; blue, Norway, and white spruces; viburnum; Korean spice viburnum; eastern black walnut; yew.

Intermediate: apple and crabapple (some cultivars sensitive); green ash; basswood (tolerance varies with clone); catalpa; 'Lambert' sweet cherry; box elder; European elder; Chinese elm; white fir; 'Lynwood' forsythia; goldenrain tree; sweet gum; Zabel's honeysuckle; Japanese larch; 'Shademaster' honey locust; silver maple; black oak; English oak (tolerance varies with cultivar); pin and scarlet oaks; white oak (some cultivars sensitive); sweet mock orange; Austrian, knobcone, lodgepole, pitch, Scots, shortleaf, and eastern white pines; Torrey pine; London plane tree (some cultivars sensitive); poinsettia; common privet; redbud; catawba hybrid rhododendrons; silverberry; Persian walnut.

Sensitive: white ash; trembling aspen; 'Campfire', Hinodegiri, Korean, and snow azaleas; bridal-wreath; Ohio buckeye; black cherry; 'Bing' sweet cherry; citrus; Kentucky coffee tree; rock and spreading cotoneasters; Siberian crab; bigcone Douglas-fir; grapevine (some cultivars tolerant); European larch; common and Japanese lilacs; Crimean linden; 'Fastigiata' and 'Orebro' cultivars of large-leaved European linden; 'Imperial' honey locust; white mulberry; 'Fastigiata' English oak; Gambel oak; Coulter, jack, Jeffrey, loblolly, Monterey, ponderosa, Virginia, and Japanese white pines; 'Bloodgood' London plane tree; hybrid poplars; Chinese redbud; western serviceberry; snowberry; sumac; eastern and California sycamores; tree-of-heaven; tulip tree; weeping willow; Japanese zelkova.

Oxidant damage to broad-leaved trees.

A–C. Naturally occurring oxidant injury on white ash. A. Purple stippling on the upper leaf surface (NY, Sep). B. White flecks on the upper leaf surface, coalescing where numerous, and then causing a bleached appearance (NY, Aug). C. Magnified view of a portion of a severely injured leaflet, showing purple-brown discoloration in islets delimited by the smallest veins (NY, Sep).

D–H. Symptoms on apple seedlings artificially fumigated with ozone, illustrating general diagnostic features of ozone injury on broad-leaved plants. D. Symptoms on a fully expanded subterminal leaf and on the apical (oldest) part of a slightly younger leaf show that recently expanded leaves are most sensitive. E. Tiny reddish brown spots or, more commonly, dark stippling or white flecks develop on the upper leaf surface. Tissue adjacent to large veins typically remains green. F, G. Magnified views with light behind injured leaves to show (F) flecks consisting of clusters of bleached cells, and, where injury was more severe (G), tiny angular lesions bounded by veinlets. H. Flecks as seen in magnified view with reflected light (greenhouse plants).

466

Plate 227

Damage by sulfur dioxide. Damage to plants by sulfur dioxide (SO_2) was formerly concentrated around industrial sources and power-generating stations. As one approached a major pollution source from downwind, tree mortality increased and the total number of plant species diminished. All plants immediately adjacent to major industrial sources were killed, and the surface soil eroded away. Famous ore smelters at Anaconda, Montana; Copper Hill, Tennessee; Sudbury, Ontario; and Trail, British Columbia; and the iron sintering plant at Wawa, Ontario, are well-documented examples. Where such emitters of pollutants have either closed or erected tall stacks, intense local damage has ceased, and revegetation has begun. SO_2 from tall stacks becomes greatly diluted before reaching ground level. Therefore acute injury by SO_2 has become uncommon. The effects of chronic exposure to low concentrations of the gas are apparently limited to growth suppression of highly sensitive plants.

Acute foliar injury by SO_2 is indicated by bleached or pigmented (tan to reddish brown or dark brown, depending on species) necrotic interveinal areas on broad-leaved plants and chlorotic spots and bands or brown tips on needles of conifers. Young leaves that have just attained full size are generally most sensitive. Typical minimum doses (concentration × time) of SO_2 that may cause acute symptoms on sensitive plants range from 0.05–0.10 ppm for 8 hours to 0.5 ppm or higher concentrations for 1 hour. Chronic injury causes chlorosis, premature senescence of leaves, and depressed growth.

For many woody species that have been tested, chronic exposure to SO_2 at low concentrations—for example, 0.1 ppm (262 $\mu g \cdot m^{-3}$) or less—causes no measurable effect. Such exposures stimulate the growth of some plants, at least for the 1st year.

SO_2 causes injury when it and its sulfite derivative (SO_3^{2-}) are absorbed (through stomata) faster than they are detoxified. The gas is oxidized successively to sulfite and sulfate, which interfere with photosynthesis and energy metabolism. For equivalent amounts, sulfite is about 30 times more toxic than sulfate. The latter in modest doses is used as a sulfur source for normal growth and development. Plants injured by SO_2 have an abnormally high foliar sulfur content for a time, which is diagnostically useful if appropriate standards are available for comparison.

Plants vary widely in tolerance to SO_2 for several reasons. First, plants vary in the efficiency with which they absorb the gas; damage is least likely if plants absorb SO_2 slowly. Second, plants vary in ability to detoxify SO_2 and to dispose of excess sulfur. Excess sulfur is both incorporated into organic compounds and reduced to hydrogen sulfide gas, which escapes from leaves. Much sulfur absorbed by leaves is also translocated to other plant parts, thus diluting it.

Tolerance of trees to SO_2 is roughly correlated with shade tolerance. Maples, for example, are generally tolerant, while birches are sensitive.

The following classification of plants according to relative tolerance to SO_2 was constructed from reports of naturally and artificially induced injury. The lists do not necessarily indicate which species are most likely to sustain growth suppression by chronic exposure to low doses of SO_2, as this has been found to be not well correlated with sensitivity to acute injury. Also, readers are reminded that sensitive individuals may occur within intermediate or tolerant species.

Relative tolerance of plants to sulfur dioxide

Tolerant: arborvitae; common boxwood; yellow buckeye; buffalo berry; Japanese, eastern red, western red, and Alaska yellow cedars; black cherry; chokecherry; citrus; bald and sawara cypresses; flowering dogwood; European and European red elders; Fraser and Pacific silver firs; gingko; Oregon grape; black hawthorn; shellbark hickory; English holly; Tatarian honeysuckle; English ivy; Chinese, common, Rocky Mountain, savin, Sierra, and Utah junipers; kinnikinnik; Amur, hedge, Norway, silver, sugar, and sycamore maples; black, chestnut, Durmast, Gambel, live, northern red, scarlet, and white oaks; Japanese pagoda tree; Austrian, lodgepole, mugo, and pinyon pines; London plane tree; myrobalan plum; balsam and Carolina poplars; California and common privets; dawn redwood; catawba rhododendron; silverberry; snow brush; spirea; blue spruce; English and Pacific yews.

Intermediate: black alder; apple; apricot; European, green, red, and white ash; European mountain ash; Japanese barberry; basswood; American and European beeches; bridal-wreath; alder and common buckthorns; catalpa; redstem ceanothus; cornelian, Japanese flowering, sour, and sweet cherries; spreading cotoneaster; eastern cottonwood; European cranberry bush; Virginia creeper; alpine currant; Lawson cypress; blood-twig and red-osier dogwoods; blue elder; American, Chinese, smooth-leaf, and wych elms; European filbert; firethorn; alpine, balsam, grand, and white firs; forsythia; golden-chain; sour gum; sweet gum; Columbia, English, and Washington hawthorns; witch hazel; eastern and western hemlocks; European hornbeam, horse-chestnut; panicled hydrangea; katsura tree; European and Japanese larches; common and Japanese lilacs; littleleaf and large-leaved European lindens; black locust; honey locust; red maple; white mulberry; pin oak; Russian olive; mock orange; Siberian pea tree; peach; common pear; limber, Macedonian, ponderosa, Swiss stone, eastern white, and western white pines; plum; Lombardy poplar; common quince; redbud; smoke tree; snowberry; blue, Engelmann, and white spruces; smooth sumac; tree-of-heaven; tulip tree; eastern black and Persian walnuts; wayfaring tree; weigela; goat willow.

Sensitive: mountain alder, Greene and Sitka mountain ash; big-tooth and trembling aspens; gray, paper, river, sweet, water, European white, and yellow birches; lowbush blueberry; bitter cherry; black and narrowleaf cottonwoods; red currant; Douglas-fir; box elder; beaked filbert; gooseberry; grapevine; Morrow honeysuckle; eastern and western larches; Japanese and Rocky Mountain maples; Texas mulberry; mallow and Pacific ninebarks; bur oak; oceanspray; Lewis mock orange; Cuban, jack, Jeffrey, pitch, red, Scots, and Virginia pines; flowering quince; sassafras; serviceberry (several species); mountain snowberry; Norway spruce; staghorn sumac; sycamore; bay-leaved, black, crack, purple, and weeping willows.

Sulfur dioxide damage.

A. Devegetation around a smelter, caused mainly by sulfur dioxide. New damage close to the smelter has ceased or greatly diminished because phytotoxic gases are now discharged from the tall stack and become diluted before returning to ground level. Railroad cars in the foreground provide a size standard for the stack (ID, Jul).

B, G. Acute injury by sulfur dioxide to leaves of paper birch growing near an iron sintering plant. B. A young plant with more than half its foliar area killed. G. Close view of interveinal necrosis of original leaves and green new growth produced after an episode of severe air pollution (Ont, Jul).

C, D. Injury by sulfur dioxide to eastern white pine. C. Sensitive and tolerant trees, the former showing sparse, faded foliage and slow growth. The principal source of pollutant was a steam plant 1 km upwind. D. Foliage from the sensitive tree, with chlorotic spots and bands and tip necrosis. Symptoms of SO_2 injury on this species are essentially indistinguishable from those caused by ozone (NY, Apr).

E, F. Injury by sulfur dioxide to foliage of slash pine. E. The crown of a sensitive tree is tinged with yellow. F. Viewed closely, injured foliage has numerous yellow bands. The tree was growing near an oil-fired power-generating station (FL, Mar).

Plate 228

Damage by fluorides and minor pollutant gases. Fluoride air pollutants affect the health of plants and animals near sources of emission. Plants absorb fluoride from both air and soil but generally accumulate toxic quantities only from air. Excess fluoride acquired from plants may cause disease in grazing animals whether or not the plants show symptoms of injury. Fluorides occur as natural air contaminants in windborne soil particles and in ash and gas released from volcanos and fumaroles, but these natural sources account for only a small fraction of airborne fluoride. Most is released during the manufacture of various products for which raw materials are mined from earth. Aluminum reduction plants, producers of phosphate fertilizer, steel mills, brick works, glass and ceramics factories, and installations where large amounts of coal are burned are major sources of atmospheric hydrogen fluoride, and, to a lesser extent, silicon tetrafluoride.

Plant injury by fluorides is most common and severe near sites where phosphate fertilizer and aluminum are produced because these processes employ fluoride-rich materials. Phosphate rock contains fluoride (3–4%), which is released during heating, grinding, or acid treatment. Hydrogen fluoride is released from molten cryolite (sodium aluminum fluoride) and fluorite (calcium fluoride) during electrolytic reduction of alumina.

Fluoride enters leaves primarily through stomata and accumulates near needle tips of conifers, near leaf tips of plants with parallel leaf veins, and in interveinal areas as well as at tips and margins on broad-leaved plants. It causes breakdown of chlorophyll and inhibits various metabolic processes. Exposure of sensitive plants to fluoride at concentrations of 0.05–0.10 ppm for several hours may cause acute symptoms—interveinal necrosis on broad leaves or tip burn of conifer needles. Tissues that accumulate an injurious amount first turn yellowish, then some shade of tan to brown or reddish brown, often with a narrow darker band at the boundary of living tissue. As dead tissue no longer accumulates toxicant, the site of accumulation shifts to living tissue adjacent to that previously killed. In time this too dies. Thus symptoms of fluoride toxicity may develop incrementally from leaf tips and margins toward main veins and leaf bases. Slow accumulation over days or weeks leads to chronic symptoms—chlorosis at leaf tips and margins. This is the more common situation because pollution control devices and tall smoke stacks have greatly diminished the atmospheric fluoride concentration at ground level near many pollution sources. Given sufficient time—10 days to 1 or more growing seasons—sensitive plants can accumulate an injurious amount of fluoride from an atmosphere containing less than one part per billion (0.8 μg · m^{-3}). In conifers, the youngest needles are most sensitive, and new symptoms arise only on current-season needles. Symptoms may slowly intensify on older needles, however, as these continue to accumulate fluoride for 2 or more seasons. Such needles become senescent and drop prematurely. The growth of Douglas-fir and presumably other species may be depressed by fluoride at concentrations in tissue that do not cause visible symptoms.

Plant species and individual plants within species vary widely in tolerance to fluoride. Normal or expected concentrations of fluoride in plant foliage are on the order of 5–20 μg · g^{-1} for many species. Elevated fluoride content (e.g., 50 to several hundred micrograms per gram) of leaves with chlorotic or necrotic patterns in species known to

be sensitive is presumptive evidence of fluoride injury. Since plants vary in tolerance and fluoride is leached from leaves by rain, however, no clear-cut quantitative relationship between fluoride content and severity of symptoms has been found. Plants tolerant of fluoride may accumulate it to concentrations of several hundred micrograms per gram in leaves that appear healthy. Also, some plants such as camellia may apparently accumulate a large amount of fluoride from soil.

Woody plants that have been ranked for sensitivity or tolerance to atmospheric fluoride are listed here.

Relative tolerance of plants to fluorides

Tolerant: black alder, arborvitae, Modesto ash, basswood, bridal-wreath, camellia, cornelian cherry, flowering cherry, Virginia creeper, currant, dogwood, European and European red elders, American and Chinese elms, firethorn, sweet gum, juniper, mountain laurel, oak, Russian olive, mock orange, common pear, London plane tree, flowering plum, balsam poplar, privet, sycamore, tree-of-heaven, willow.

Intermediate: apple; European and green ash; European mountain ash; trembling aspen; azalea; European beech; European white birch; sweet cherry; chokecherry; European filbert; grand fir; grapefruit; English holly; European hornbeam; lemon; lilac; littleleaf linden; black locust; hedge, Norway, and silver maples; red mulberry; English oak; orange; peach; lodgepole and mugo pines; Carolina and Lombardy poplars; rhododendron; rose; white spruce; smooth and staghorn sumacs; tangerine; eastern black and Persian walnuts; Japanese yew.

Sensitive: apricot (some cultivars intermediate); blueberry; Douglas-fir; box elder; Oregon grape; grapevine (some cultivars intermediate); western larch; paulownia; loblolly, ponderosa, Scots, and eastern white pines; redbud; serviceberry; blue spruce.

Minor air pollutants that may cause injury to plants include chlorine, hydrogen chloride, ammonia, ethylene, hydrogen sulfide, and an array of metallic elements (arsenic, copper, mercury, nickel, lead, zinc, and others). Metallic pollutants are discharged in smoke from smelters and industrial sources (lead from automobile exhaust), and slowly accumulate in plants downwind, but their impact on plant health in the field has not been separated from that of major air pollutants. Gaseous minor pollutants are most likely to cause plant injury near sites of accidental leaks and spills from tanks or pipes. In these cases, prominent foliar injury occurs to many kinds of plants. Those nearest the leak or spill may have all foliage killed. The ammonia leak that caused the injury shown here occurred at a distribution facility for anhydrous ammonia, which is used as a fertilizer.

References for air pollutants in general: 79, 219, 319, 599, 943, 1083, 1113, 1164, 1248, 1268, 1313, 1359, 1375, 1560, 1731, 1852, 1896

References for oxidants: 146, 156, 249, 319, 370, 409–11, 454, 524, 590, 806, 943, 946, 947, 959, 960, 990–92, 1060, 1083, 1149, 1150, 1164, 1195, 1313–15, 1359, 1375, 1681, 1841, 1852, 1896, 2197, 2199, 2254

References for sulfur dioxide: 29, 156, 176, 249, 319, 351, 409, 410, 455, 590, 622, 660, 720, 943, 960, 961, 990–92, 1083, 1149, 1164, 1195, 1201, 1202, 1268, 1313, 1359, 1375, 1681, 1699, 1852, 1896, 2052, 2254

References for fluorides and minor air pollutants: 318, 319, 498, 620, 943, 1083, 1164, 1201, 1313, 1359, 1375, 1852, 1896, 2032, 2033

Damage by fluoride and by ammonia.

A, B. Foliar browning and death of slash pine, caused by fluoride accumulated from severely polluted air near a phosphate fertilizer plant. Tree-to-tree differences in pollutant tolerance are evident (FL, May).

C, D. Needle tip necrosis on mugo pine, believed caused by fluoride accumulated from air near a brick works. Only old symptoms on 2- and 3-year-old needles are visible (NE, Jun).

E. Necrosis caused by fluoride at margins of redbud leaves. The tree was growing near an aluminum reduction plant (SC, Apr).

F, G. Damage to eastern cottonwood, caused by an ammonia spill. The incident occurred 3 weeks before photographs were made. Most foliage present at the time of the spill was killed; leaves produced subsequently are green (MN, Jul).

Plate 229

Nutrient Deficiencies (Plates 230–231)

At least 20 elements are essential nutrients for plants, although not all of these have been proved essential for trees. All except carbon, hydrogen, and oxygen are normally supplied by soil. Elements required in relatively large amounts (macronutrients) include nitrogen, phosphorus, potassium, calcium, magnesium, and sulfur. The remaining essential elements (micronutrients) are needed in small to trace amounts. This plate and the next one show diseases caused by insufficiency of one or more nutrient elements normally supplied by soil. A pictorial survey of nutrient deficiencies of trees is beyond the intended scope of this book. The listed references contain more information and illustrations.

It should be understood at the outset that the growth of healthy plants is normally limited by the supply or availability of one or more nutrient elements, as may be inferred from the increase in growth that usually follows fertilization. A growth-limiting nutrient supply is considered to be deficient if the limitation causes abnormally slow growth, deformity, depressed yield, abnormal color, or necrosis.

Nutritional deficiencies cause varied symptoms and in most cases cannot be diagnosed reliably on the basis of symptoms alone. Different deficiencies may cause similar symptoms, and symptoms are often complicated by simultaneous deficiency of two or more elements. Also, many symptoms of nutrient deficiency are similar to those caused by other factors such as heat, drought, waterlogged soil, chemical injuries, or damage by insects or pathogens. Chemical analyses of plant tissues and/or the rooting medium are needed for accurate diagnosis in most situations, and a positive plant response to application of the limiting nutrient(s) is often necessary to complete the diagnosis. Moreover, the results of tissue analysis are reliable for diagnosis only if normal values and deficiency levels of nutrients in tissues have been reported for the affected plant species. These levels vary considerably from one species to another. Diagnosticians who identify nutrient deficiencies accurately by observation alone are relying upon personal experience with the species or cultivar in the locality where the disorder occurs. Nutrient elements, some of their functions, and some typical symptoms caused on aerial parts by deficiency are summarized below.

Nitrogen is a component of amino acids, structural proteins, enzymes, nucleic acids, purines and pyrimidines, chlorophyll, growth regulators, and alkaloids. It is required for diverse functions. A typical concentration in tree foliage is 1.5% of dry weight. Nitrogen supply limits growth of woody plants on most sites. Deficiency causes slow growth, diminished leaf size or number of leaflets on compound leaves, and subnormal synthesis of chlorophyll, resulting in general pale green color or chlorosis. This is usually more severe on old leaves than on young ones.

Phosphorus is a constituent of phospholipids and nucleoproteins, nucleic acids, energy transfer compounds, and buffer systems. It is mobile and occurs in both organic and inorganic forms within plants. A typical concentration in leaves is 0.12–0.15% of dry weight. Deficiency may cause general chlorosis or abnormal reddish purple pigment on petioles or in interveinal areas of broad-leaved species. These symptoms sometimes develop without noticeable change in foliar density or leaf size on established plants. Seedlings deficient in phosphorus become stunted and often chlorotic and may also develop abnormal reddish purple color. Purple pigmentation presages browning in leaves of coniferous seedlings. The threshold foliar concentration for deficiency is near 0.09–0.1%. Young plants lacking mycorrhizae (Plate 244) are most likely to display symptoms of phosphorus deficiency.

Potassium mediates activity of some enzymes and is involved in nitrogen metabolism, carbohydrate translocation, and regulation of transpiration through its function as an osmotic agent in the opening of stomata. This element remains in inorganic form and is highly mobile in plants; thus deficiency symptoms appear first in old tissues. Typical concentration of potassium in conifer leaves is 0.4–1.0% of dry weight, while broad-leaved woody plants typically contain 1.0–1.5% or more. Foliar concentrations below 0.2–0.4% in conifers or 0.6–1.0% in broad-leaved deciduous trees are associated with foliar symptoms. Deficiency causes slow growth, chlorosis and browning of tips, margins and interveinal areas of leaves, sparse-appearing foliage of the plant as a whole, and shoot dieback. Red or purple pigment may appear before browning of leaf margins. Potassium-deficient conifer needles display general chlorosis, subnormal length, tip browning in severe cases, and premature senescence and abscission. Potassium deficiency also increases sensitivity to freezing and promotes susceptibility to various pathogens, such as *Verticillium* species (Plates 181–182).

Calcium occurs in pectate polymers in primary cell walls and intercellular binding substance. It is essential for integrity of cell membranes, acts as a cation in buffer systems, and is an activator of enzyme systems, notably amylase. Typical concentration in tree leaves is 1.5% of dry weight. Calcium is immobile after incorporation into structural polymers; therefore, symptoms of deficiency appear first in young shoots. Deficiency is indicated by chlorosis and subnormal size of youngest leaves and marginal and sometimes veinal necrosis that progresses inward and toward the bases of chlorotic leaves. Severe deficiency results in death of shoot tips.

Magnesium is a constituent of the chlorophyll molecule and an activator of enzyme systems involved in phosphate transfer. It is involved in the maintenance of ribosome integrity and as a cation in buffer systems. It is relatively mobile in plants; therefore, deficiency symptoms appear first in old tissues. Typical concentration in tree leaves is 0.1–0.2% of dry weight. Deficiency symptoms in angiosperms and palms include chlorosis that develops first at margins and on apical parts of the oldest leaves, then extends between the veins and to younger leaves. The chlorotic region enlarges toward the midvein and leaf base, eventually leaving a green area (triangular in broad-leaved species) at the leaf base. More severe deficiency results in general chlorosis, irregular marginal necrosis, and premature leaf drop. Some plants produce purplish red pigment in interveinal areas of magnesium-deficient leaves. Magnesium deficiency in conifers causes premature yellowing and browning of 2nd-year or older needles, bright yellow discoloration of current-year needle tips in pines, and general yellowing of spruce needles. Needle tips turn brown in cases of severe deficiency. Foliar concentrations of this nutrient in foliage showing symptoms of deficiency are usually less than 0.08%.

Nutrient deficiencies in monocots and gymnosperms.

A. "Frizzle top" of royal palm, a distortion of leaflets attributed to deficiency of manganese (FL, Apr).

B. Lime-induced chlorosis of slash pine, attributed to deficiency of iron, in contrast to normal foliage in right foreground. Trees here became diseased after installation of turf over their roots. Fertilizer, lime, and irrigation water containing carbonate were applied to the turf. This induced lush grass growth but raised the pH of the soil, and the pines eventually showed deficiency symptoms (FL, Apr).

C. Chlorosis attributed to deficiency of iron in royal palm (TX, Apr).

D. Chlorosis induced primarily by nitrogen deficiency in young loblolly pines. These trees grew from seed on a site where topsoil had been removed. Normal trees of the same species in the background provide contrast (GA, May).

E. Chlorosis caused by iron deficiency in bamboo. Yellow color appeared first between leaf veins (TX, Apr).

F. Chlorosis of needle tips, induced primarily by deficiency of magnesium in a young Scots pine (NY, Sep).

G. Premature senescence of 2nd-year needles, induced by magnesium deficiency, in balsam fir (NY, Sep).

H. Premature senescence of 1-year-old needles in a young eastern white pine affected by deficiencies of magnesium and potassium (NY, Sep).

Plate 230

Sulfur is a constituent of certain amino acids and proteins, vitamins, and coenzyme A. Typical foliar concentration in trees is near 0.2% of dry weight, and deficiency symptoms are associated with concentrations below about 0.15%. Sulfur is relatively immobile within plants. Therefore, deficiency symptoms are likely to appear first in young shoots. Deficiency causes diminished protein synthesis, accumulation of amino acids, general chlorosis (less marked along than between leaf veins in some species), and subnormal leaf size.

Boron is required for normal differentiation of cells and tissues in meristematic regions of the plant, and it is involved in sugar transport. Typical concentrations in leaves are 15–100 $\mu g \cdot g^{-1}$ dry weight. Deficiency symptoms are associated with concentrations below 15 $\mu g \cdot g^{-1}$. Boron-deficient plants display chlorosis, subnormal leaf size, aberrant cellular differentiation and death of apical meristems, fusion of leaves or leaflets that would normally be separate, shoot dieback, proliferation of buds where shoots have died back, and blunt, much-branched roots. Boron deficiency in apple trees causes, in addition to some of the symptoms above, brittle deformed leaves and lesions in the cambial region. Boron deficiency can be alleviated by application of various boron-containing compounds to soil, but this must be done with care to avoid toxic overdose.

Copper is an activator of several oxidative enzymes and a constituent of enzymes such as tyrosinase and ascorbic acid oxidase. It is required for normal lignification, protein utilization, and chlorophyll formation. Typical concentrations in leaves are 5–15 $\mu g \cdot g^{-1}$ dry weight. Adequate copper is usually available from soil at pH 7 or less, although leaching from poorly buffered sandy soils may result in deficiency. Copper may become unavailable to plants in organic soils or soilless rooting media because of binding in insoluble organic complexes. Copper is relatively immobile in plants; thus young tissues show symptoms first. Copper content of <4 $\mu g \cdot g^{-1}$ in foliage is usually associated with symptoms. Deficiency may cause marginal chlorosis; general light green, gray-green, or darker color, sometimes with contrasting chlorotic veins; irregular marginal or interveinal necrosis; cupping of leaves; stunting of leaves and shoots; premature abscission of leaves; failure of new growth to develop; and twig dieback. Conifers may show prostrate growth or drooping shoots and brown tips on needles. Small shoots may proliferate below dead tips, and the new shoots may die in turn. Nitrogen application to copper-deficient plants may accentuate symptoms. A large proportion of copper absorbed by roots remains there, even in plants with symptoms of copper deficiency in foliage. Copper deficiency may be corrected by acidifying neutral or alkaline soil, fertilizing soil with copper sulfate, or spraying plants with copper-containing fungicides.

Iron functions as a prosthetic group or coenzyme in certain enzyme systems. It is involved in synthesis of chloroplast proteins, and it occurs in oxidases and respiratory enzymes. Typical concentrations in tree leaves are 40–100 $\mu g \cdot g^{-1}$ dry weight, and values of 25–40 $\mu g \cdot g^{-1}$ are associated with deficiency symptoms. Iron is most readily available from soils at pH 6 or less, and it is bound in insoluble forms in alkaline or calcareous soils. It is relatively immobile within plants; thus symptoms of deficiency tend to appear first on young shoots that cannot obtain sufficient iron from older tissues. Deficiency causes interveinal chlorosis and slow growth that is at first irregular from branch to branch within a tree and is variable among trees of the same species in a given locality. Affected leaves retain chlorophyll along the veins, tend to be undersized, and in severe cases develop irregular browning along the margins. Branches with severe foliar symptoms eventually die back. Deficiencies of manganese and zinc may cause similar symptoms, and subnormal levels of two or all of these elements may occur in chlorotic foliage. Application of a chelated iron compound or acidifying substance (e.g., elemental sulfur) to soil may be beneficial. Injection or implantation of iron salts or chelated iron into holes drilled in sapwood of the affected trees also often helps.

Pin oak and northern pin oak are notoriously prone to "iron chlorosis" resulting from iron deficiency. This proneness varies within species, however. In one study, seedling populations of pin oak derived from parents from upland sites in the north-central and northwestern parts of the geographic range of that species were less prone to iron chlorosis than were populations of other origins. There is thus the possibility of selection for ability to extract adequate iron from soils where iron chlorosis commonly develops.

Manganese is an activator of enzyme systems and is essential for synthesis of chlorophyll. Reported concentrations in normal tree leaves range from 50 to several hundred micrograms per gram, and deficient concentrations are in the range of 5–40 (usually less than 20) $\mu g \cdot g^{-1}$. Manganese is most readily available from soils at pH 6 or less, and deficiency is most likely in plants growing on alkaline or calcareous soils. Symptoms include interveinal chlorosis, undersized leaves, and, in some plants, wavy, crinkled, or curled leaf margins. Deficiency of iron causes similar chlorosis. Manganese deficiency can be corrected by application of manganese sulfate to infertile acidic soil, but implantation of this substance into sapwood of the trunk may be more effective where deficiency has developed in trees on alkaline or calcareous soil. Manganese in plant tissues may be toxic at relatively low concentrations; therefore, suspected manganese deficiency should be confirmed by tissue analysis before additional manganese is added by injection or implantation.

Molybdenum, associated with nitrate reductase, is essential for conversion of nitrate, the principal form of nitrogen absorbed by plants, to reduced forms that can be incorporated into organic compounds. Typical concentrations in leaves are 0.05–0.15 $\mu g \cdot g^{-1}$ dry weight. Natural deficiency has been described for only a few trees and shrubs. Molybdenum-deficient citrus develops large yellow spots between veins on leaves. Symptoms in Chinese hibiscus include foliar stunting and deformity resembling that caused by phenoxy herbicides.

Zinc is associated with various enzyme systems. It is involved in synthesis of the amino acid tryptophan, which is a precursor of auxin. Zinc is readily available from soils at pH 6 or less. Typical concentrations in leaves are 12–80 $\mu g \cdot g^{-1}$ dry weight, and levels of 5–20 $\mu g \cdot g^{-1}$ (40 $\mu g \cdot g^{-1}$ in pecan) are associated with deficiency symptoms. Zinc deficiency causes interveinal chlorosis or mottling, dwarfed leaves, crinkled leaf margins, and stunted or rosetted shoots. Symptoms tend to be variable within the tree and between adjacent trees. Zinc deficiency can be corrected by injection of zinc sulfate solution or, in some cases, by simply driving a few zinc-coated nails into the stem.

Deficiencies (and in some cases functions) of other confirmed or probable micronutrients—chlorine, cobalt, silicon, sodium, and vanadium—in trees or shrubs are unknown.

References: 65, 104, 158, 179, 196, 486, 577, 703, 771, 824, 945, 1116, 1288, 1477, 1840, 1912, 1913, 2077, 2090, 2162, 2241

Nutrient deficiencies in angiosperms.

A. Chlorosis induced by iron deficiency in pin oak (NY, Jun).

B. Chlorosis and twig dieback in a Chinese tallow tree deficient in iron (TX, Apr).

C. Interveinal chlorosis, indicating iron deficiency in rhododendron (TN, Apr).

D. Chlorosis induced by iron deficiency in red oak. Tissue along the veins typically retains green color longest (NY, Jun).

E, F. Marginal yellow and red pigmentation in black cherry, and chlorosis in paper birch, induced by deficiency of magnesium and potassium (NY, Sep).

G. Pale green color, stunted leaves and shoots, and twig dieback in a variegated privet (*Ligustrum sinense*), induced by copper deficiency, compared with normal at right (FL, Apr).

H. Interveinal chlorosis attributed to deficiency of manganese in red maple (MI, Aug).

Plate 231

Damage Caused by Drought, Heat, and Freezing (Plates 232–236)

This plate leads a series of five about injuries caused by water shortage and extremes of temperature. We present these topics together because in nature they are inextricably related to each other. Captions are expanded beyond the normal length in many cases in order to convey diagnostic features or the circumstances of injury. The order of text presentation is water shortage, heat stress, damage by frost during the growing season, frost rings in wood, damage by freezing in winter, and predisposition to attack by fungal pathogens. Because of this arrangement, text information relevant to a particular illustration may not face it. For cracks caused in wood of tree trunks by freezing or drying, see Plate 164.

Water shortage. At a given latitude and elevation, the amount and seasonal distribution of annual precipitation determine which woody species grow and how fast they grow. Native plants in a given area are adapted to seasonal and annual variations in water supply characteristic of the local climate. Accordingly, only unusually severe drought is likely to cause noticeable injury to plants that have grown naturally on a given site. Planted trees and shrubs, on the other hand, often show symptoms of severe water stress. Perhaps more important, water deficit predisposes trees to infection by opportunistic pathogens, to attack by secondary insects, and to injury by severe winter weather.

Water deficit develops as a normal phenomenon in plants when loss by transpiration exceeds the rate of absorption from soil, as it does almost daily during the growing season. The deficit is made up at night and during periods of rain or dew formation when transpiration slows or ceases. As soil dries, however, roots fail to extract as much water as has been lost, and physiological stress develops. If this intensifies sufficiently, or if in technical parlance the water potential of plant tissues decreases sufficiently, tissues and organs lose turgor, degenerate, and die. Water deficits also develop in dormant plants, especially evergreens, during warm weather in winter or early spring when water evaporates from leaves and stems while the soil is cold or frozen. Roots extract insufficient water from cold soil and none from frozen soil.

Plants vary in ability to tolerate low water potential. The minimum tolerated by most woody species that grow in mesic environments is −1.6 to −2.4 megapascals, but water potentials of −6 megapascals have been measured in desert plants such as mesquite, and −9 megapascals in Engelmann spruce during late winter.

Seedlings and recently transplanted plants are at greatest risk of direct damage or death caused by drying. Roots of seedlings occupy only the uppermost layer of soil, where drying is most rapid, and seedlings lack a significant internal reservoir such as that represented by sapwood in larger plants. Trees or shrubs recently transplanted have lost many absorbing roots, so the creation of an abnormal water deficit is unavoidable. Also, if the root ball contains a highly porous rooting medium instead of soil, it is likely to dry out, so that a water shortage occurs even though the surroundng soil contains water sufficient for plant growth. This problem continues until roots grow beyond the ball.

Established trees or shrubs respond to and cope with water deficits in various ways. Stomata in many species close after development of significant internal stress. Chlorophyll formation, photosynthesis, and growth slow or cease. Chlorosis may develop. In trees that produce latex or resin, the exudation pressure of the substance diminishes. Green leaves, stems, roots, and fruits shrink. Shrinkage of stems may continue for several weeks during drought, and it may lead to formation of radial cracks in the sapwood (Plate 164). The cambium in the lower part of the trunk of a severely stressed tree may fail to produce any new wood in a year of severe drought, or patches of cambium may die, resulting in cankers. Roots in drying soil become less permeable to water, and root tips may be damaged by drying.

Leaves of drought-stressed plants lose turgor and may droop, wilt, turn yellow, turn brown at the tips and margins, curl, or show all of these symptoms. When green leaves wilt and turn brown, the oldest leaves usually succumb first. Chestnut, flowering dogwood, and sourwood in the Appalachian region are among the species most likely to wilt and to develop foliar browning. Leaves of some deciduous trees such as ash, basswood, hickory, and black locust become prematurely senescent (usually yellow) and drop. Severely stressed deciduous species may drop all their leaves. This commonly occurs where trees have colonized pockets of soil on ledges or in any soil after severe mechanical damage to roots. Similarly, the oldest age class of needles on coniferous trees may turn yellow and drop.

Severe water deficit in pines during summer causes needles to lose turgor and to bend or droop at a point near the needle base where lignification is incomplete. Needles then either fade and turn brown or remain green and permanently bent.

A. Drought damage to oak and Ashe juniper on an Oklahoma hillside as observed 1 year after the intense heat wave and drought of 1980. Many plants are dead; others show extensive dieback. The initial display of brown foliage has given way to bare twigs and the gray-brown weathered appearance of adhering dead foliage. Junipers at the base of the slope where drying was least severe remained green (Jul).

B. Dieback of Portugal laurel exposed during winter to warm air discharged intermittently through a ground-level ventilator from the basement of an adjacent building. Defoliation and dieback resulted from freezing and desiccation. The plant escaped such damage in previous years because severe freezing did not occur and branches had not grown into the air stream from the ventilator (WA, Jun).

C. Interveinal chlorosis and necrotic spots on leaves of tulip tree. This unexplained disorder, apparently caused by environmental stress, commonly develops in landscape specimens during hot dry weather after midsummer. Affected leaves drop prematurely (NY, Jul).

D. Rhododendron leaves injured by winter desiccation. Such damage is often compounded by opportunistic fungi such as *Botryosphaeria dothidea* (Plate 82) or *Botrytis cinerea* (Plate 25) (NY, Jan).

E. Scorch of Norway maple leaves. This branch experienced severe water shortage because of heat reflected from pavement below and because the pavement had been installed over roots (NY, Aug).

F. Freeze damage to leaves of Oregon grape. Injured leaves died back from tips and margins. Tissues with sublethal injury produced purple pigment; normal leaf color is dark green (NY, Apr).

G. Winter injury to a recently transplanted London plane tree. Dieback of twigs and branches was followed by sprout growth at branch bases and along the mainstem. Symptoms above ground in this case were secondary to freezing of roots. The tree had been transplanted in autumn into a sandy nursery soil where severe freezing occurred in absence of snow cover during winter (NY, Aug).

H. Twig dieback and premature leaf color in spreading cotoneaster injured by freezing the previous winter. Many buds failed to open, and the twigs died. Other buds produced normal-appearing shoots that came under water stress and changed color during summer when injured sapwood failed to supply sufficient water (NY, Aug).

I. Winter injury to leaves of common boxwood. Tips and edges of injured leaves became chlorotic and produced reddish pigment. Injured leaves will drop prematurely (NY, May).

Plate 232

Similar symptoms may be caused by frost, as shown in Plate 233A. Drooping and bending of needles under severe water stress have been noted in ponderosa, red, eastern white, and western white pines. Dieback of twigs and then limbs may follow foliar symptoms in both conifers and broad-leaved trees. The more severe the stress, the larger the diameter of limbs that die.

After the addition of water to soil, absorption begins, and shrunken organs expand quickly, but there is a lag time of up to several days before roots become as highly permeable to water as they were before the onset of the dry period. Also, stomata that close during drought may recover normal function slowly or not at all, and the rate of photosynthesis may remain below the prestress level for a time. Behavior varies with plant species and the severity of water stress.

Many instances of severe drought damage to native trees have been reported. Dieback and death of Appalachian and southern oak species (black, laurel, northern red, southern red, scarlet, water, and willow oaks) have occurred during or after severe droughts throughout the 20th century, most recently after a series of droughts beginning in 1978. The relative roles of water stress and opportunistic organisms are unclear, however, because the fungus Hypoxylon atropunctatum (Plate 107) and the two-lined chestnut borer (see Johnson and Lyon, 2nd ed., Plate 127) quickly colonize weakened and dying oaks. Western conifers, especially incense cedar, Douglas-fir, grand fir, western larch, lodgepole pine, and western white pine, develop brown foliage and dieback in years of severe heat and dryness. Pole blight of western white pine (Plate 218) is considered to be primarily a syndrome of water stress. Drought-stressed white spruce may become chlorotic.

Drought damage usually appears on trees in groups because of common soil conditions and root grafts among trees of the same species. Root grafts are frequent in various oaks and pines, for example. These factors promote a similar level of water stress in most members of the group, but relatively small individuals in the group usually sustain greatest injury.

Intense, short-term water stress may develop in immature leaves exposed to drying during bright, hot windy weather in late spring or early summer, even when soil moisture is adequate to replace normal transpirational loss. Young leaves of sugar maple, for example, may suddenly die or may develop interveinal patches of dead light brown tissue (leaf scorch). Similarly, dramatic marginal browning of leaves of California black oak and to a lesser extent canyon live oak and interior live oak have been noted in early summer in northern California. Drooping and death of young Douglas-fir and white fir shoots have been observed during hot weather in late May and early June. Because injury from brief, intense water stress can occur to plants in moist soil, such injury was at one time attributed solely to heat, but this judgment preceded modern knowledge of plant-water relations. The relative roles of heat and drying in these disorders are not known.

Winter desiccation leads to chlorosis and foliar browning (in evergreens), to dieback of twigs, and in some cases to radial cracks in sapwood of the trunk. Browning of conifers, caused by winter desiccation, is common in red spruce and eastern hemlock in the East and in Douglas-fir, lodgepole and ponderosa pines, and blue and white spruces in the West. Symptoms, similar to those in yew in Plate 234, appear most commonly after a period of warm, dry, windy weather while the soil is still cold or frozen. The terms parch blight and red belt injury (referring to reddish brown foliage, often on trees in a narrow altitudinal stratum on mountainsides) have been applied to this desiccation in the West.

Heat stress. Few claims of heat injury to trees or shrubs, other than by fire, are well supported by evidence from experiments or temperature measurements. Heat injury to immature leaf tissues not yet active in transpiration has been suggested to be the cause of sudden foliar collapse during exceptionally hot weather in late spring or early summer (as in oaks in California, mentioned above), but no proof has been presented that heat rather than water shortage causes such symptoms. The stems of many woody plants can withstand temperatures somewhat above 50°C for brief periods.

Injury to tree trunks by solar heat (sunburn) probably occurs, but its importance is hard to assess because, in nature, trunks subject to intense heat in summer are also subject to freeze damage (winter sunscald) in winter. Injury similar to that shown on eastern white pine is common on many species of trees that have smooth, thin bark. It occurs on the south to southwest sides of stems that were formerly shaded by their own branches or those of adjacent trees. Bark becomes pinkish or reddish, and the surface slowly roughens and darkens as patches of outer bark die, dry, and crack. Some of this roughening is due to killing of the outermost tissue by heat or freezing, and much roughening is caused by formation of patches of periderm beneath stressed outer bark (details with Plate 242), which isolates the superficial tissues and causes their death. Direct heat injury is probably limited to outermost bark, the surface of which in brilliant sun may occasionally rise to 50°C or higher. Whether such tissue is killed

---

A. Injury to red pine by summer frosts that occurred in two successive years. Immature or recently matured tissue near needle bases was most sensitive. Injured tissues collapsed, and needle droop and death resulted. The current season's shoot is stunted, and 2nd-year needles are missing as the result of injury the previous year (NY, Oct).

B. Freeze injury to mature 1st-year needles of Japanese black pine. Injured tissues near needle bases became bleached and turned brown, but because the injury was superficial or limited to one side of each needle, needle tips remained green. Needles injured in this fashion usually become senescent and drop 1 or more years prematurely (NY, Jun).

C, F. Sunscald on the southwest side of the trunk and foliar symptoms of decline due to sudden exposure of a mature eastern white pine by removal of adjacent trees. Mild injury causes reddish discoloration of smooth bark that was formerly gray or gray-green. Severe injury leads to roughening of the bark surface (F) as damaged outer layers of secondary phloem die, shrink, split, and peel back in loose, ragged scales. Damaged tissues become isolated from living inner bark by formation of a new periderm. Slow growth and sparse, off-color foliage are presumed to result partly from injury along the trunk and partly from changes in the root environment. A normal eastern white pine in the background provides contrast (NY, Jun).

D. Winter freeze injury to needles of eastern white pine. Needles nearest the apical buds become chlorotic or mottled yellow to yellow-green in late winter or early spring, and they drop at the beginning of the growing season. Buds are usually not injured, and new growth develops normally. Injury of this type commonly follows a period of warm weather in late winter or early spring during which the youngest needles lose cold hardiness. A sudden return to freezing temperatures normal for the season causes the injury (VA, May).

E. Injury to grand fir by a late spring frost. New shoots were several centimeters long at the time of freezing. Year-old foliage and the basal parts of current shoots survived, but growing points and the youngest leaves were killed (WA, Jul).

Plate 233

directly or is just stressed sufficiently to trigger periderm formation beneath is unclear. Also unclear is whether injury to superficial bark may be caused by freezing of deacclimated tissue after warming by afternoon sun in winter. This is the general mechanism of cambial damage, but superficial bark injury without coincident cambial damage is unreported.

Solar heat at the soil line may cause the collapse of soft tissues of young coniferous seedlings in nurseries. Injured seedlings lacking lignified tissue bend at the point of the lesion, and the tops soon die. Older seedlings remain erect but girdled. The hypocotyl swells above the girdle, roots cease growth, and the top of the seedling turns chlorotic before death in late season, in winter, or rarely in the next season. Temperatures sufficient to kill succulent plant tissue (54°C or higher) do occur at the surface of dry soil, and this disorder has been reproduced by experimental heat treatments. Caution is necessary in diagnosis of girdling disorders of young seedlings, however, because certain pesticides (e.g., benzene hexachloride) can cause localized swelling similar to that which follows heat girdling.

Two aspects of damage to tree trunks by ground fire should be mentioned. First, the inner bark and sapwood of living trees are virtually incombustible because of their water content. Ground fires may char surface bark and kill underlying tissues, but the first such fire does not burn into the trunk. However, the heat may kill the bark and cambium of a tree with thin bark, even when the bark surface is not charred. Damage becomes apparent a few years later as dead bark cracks and begins to expose the wood. Trunk expansion and callus growth at the edges of the damaged area eventually cause the dead bark to loosen and fall. Second, ground fires occurring after the formation of a zone of dead, dry bark and wood may burn into the trunk, especially in resinous wood of conifers, as shown in Plate 162.

Injury by frost during the growing season. Solutes in water within plant tissues depress the freezing point somewhat below that of pure water in nearly all plants. Even during shoot elongation, most trees and shrubs can withstand air temperatures of −1° to −2°C without injury. At lower temperatures, water either becomes supercooled or freezes. Pure water can be supercooled to −38°C, but water in plants freezes at much higher temperatures because plants contain substances that act as nuclei for formation of ice crystals. When crystals begin to form, ice spreads rapidly. Intercellular ice (between cells) causes relatively little harm, but ice within cells causes lethal disruption of membranes and organelles.

Woody plants in temperate zones undergo seasonal changes in ability to tolerate low temperature. In autumn, perennial plants become acclimated to withstand low temperature. The degree of cold acclimation varies during winter, more or less in relation to ambient temperature. The hardiest trees or shrubs, fully acclimated, can withstand temperatures much lower than ever occur where they grow. In these plants (e.g., trembling aspen, birches, red-osier dogwood, willows), all freezable water is extracted from the cells and freezes intercellularly. Only chemically bound (unfreezable) water remains within living cells. The majority of woody plants of the temperate zone become acclimated to withstand minimum temperatures between −20° and −40°C. They avoid lethal freezing by combined mechanisms of withdrawal of water into intercellular spaces, where it freezes without consequence, and supercooling of the remaining intracellular water. Tissues in trees as diverse as apple and hickory may supercool

as low as −40°C without harm. Apparently the process of cold acclimation involves reduction or elimination of ice nucleating centers within cells and/or development of barriers to ice crystal formation.

As temperatures rise in late winter and early spring, plants deacclimate until, by the time growth begins, they can no longer tolerate more than a few degrees of frost. Most damage by freezing during winter follows untimely deacclimation during temporary warm weather. Frost injury during the growing season is possible because plants lack any cold acclimation at that time.

Leaves and stems killed by spring frost are usually small and succulent at the time of injury. After thawing, they at first appear water-soaked and soon become shriveled and reddish brown to dark brown or nearly black, depending on the species. Dead leaves or shoots break off or abscise during the ensuing several weeks. New shoots begin to grow from dormant or adventitious buds almost immediately and soon mask the early-season damage. By midsummer, only a trained observer is likely to notice the stubs of frost-killed spring shoots.

The likelihood that a particular tree species or an individual within a species will be damaged by spring frost is related to the date when it begins growth. Conifers that break bud earliest are most likely to sustain damage, which may involve browning of 1-year-old needles or collapse of new growth. Douglas-fir and European and Japanese larches planted in the Appalachian region, for example, begin growth earlier than other conifers planted there and are more frequently

---

A, B. Winter freeze injury to needles and twigs of blue spruce. This injury develops in late winter or early spring when freezing follows a period of warm weather. A. Foliar symptoms. Mild injury leads to symptoms resembling those of a needle cast disease. Scattered brown lesions appear on green needles, or entire needles turn brown while others remain green. Xylem at the base of the apical bud may turn light brown in contrast to the nearly white pith and green cortex, but the bud usually remains alive. Foliar symptoms later become masked by growth of new shoots and dropping of injured needles (NY, Mar). B. If injury is more severe, most needles on the twig turn brown, and xylem along the shaft of the twig, including xylem traces leading into needles, becomes distinctly discolored (NY, May).

C. Winter freeze damage to Japanese black pine following untimely pruning. The tree from which this twig came was pruned in August, which promoted growth of new shoots that did not acclimate to cold weather before winter. Foliage and stem tissues produced during the previous spring acclimated normally and remained unharmed (NY, Jun).

D. First-year seedlings of eastern hemlock, normal (left) and putatively injured by heat at the soil line. The injured seedling has been girdled by a band of dead tissue at the former location of the soil line. Swelling above the girdle resulted from growth stimulation associated with blocked downward translocation of organic compounds. Chlorosis reflects inadequate supply of minerals from starving roots. Symptoms as shown become apparent several weeks after the injury. The affected seedling will soon die (NY, Aug).

E–G. Damage to Douglas-fir by freezing in early spring. E. Top killing. F. Deformity of the trunk of a small tree, caused by freeze injury in previous years. The scar of an irregular canker, now covered by callus, is visible at center, and a lens-shaped split in the bark is present near the base of the stem portion shown. G. Magnified cross-section of a young stem at the base of a portion killed by freezing. The injury occurred after 2 years of normal growth. In the upper part of the photo, phloem, cambium, and sapwood were all killed. A prominent frost ring extends from the necrotic zone around the stem beneath living bark and cambium. Sapwood produced after the injury appears as a narrow, light-colored band between the bark and the frost ring (WA, Jun).

H. Foliar browning and twig mortality of Japanese yew caused by freezing in late winter. Foliage, warmed by afternoon sun, lost cold hardiness and then froze when the temperature dropped to a nighttime low that was normal for the season. Symptoms appeared during the next warm period (NY, May).

Plate 234

damaged by spring frost. On the other hand, deciduous trees that begin growth earliest are least frequently damaged. More than simple escape is involved; species such as black cherry or sugar maple that begin growth early may have a large amount of relatively succulent tissue exposed but may escape damage by a frost that kills all new growth of black locust or white ash in the same locality. Of 29 deciduous species studied in West Virginia, only American beech and Fraser magnolia ranked high in susceptibility to damage by spring frost while still being among the earliest species to begin spring growth. The order of frost susceptibility in terms of most frequent to least frequent damage in nature was: sassafras, sycamore, black locust, American beech, cucumber tree, tulip tree, hickories, black walnut, butternut, Fraser magnolia, white ash, white oak, chestnut oak, northern red oak, yellow birch, black birch, smooth alder, serviceberry, witch hazel, striped maple, pin cherry, red maple, red elm, willows, flowering dogwood, basswood, sugar maple, black cherry.

Damage by summer frost is less common and more serious than that by spring frosts. It occurs chiefly to pines in plantations on plains and in topographic depressions (frost pockets) in the boreal forest regions of Canada and northern United States. Naturally established trees usually escape damage because they do not successfully colonize sites prone to summer frost. Injury occurs on clear nights when, due to rapid radiation of heat from the soil, a layer of cold air forms at ground level and becomes 1 meter to a few meters thick, killing needles and succulent stem tissue. Needle tissue near the basal sheath collapses, and the needles bend down in a characteristic droop, soon drying and turning brown. If the frost is severe enough to kill entire shoots, drooping needles may remain in place for a year or more, slowly weathering to gray color. The plant resumes growth the next year from adventitious buds and shows bushy form. If the stem portion of a shoot survives, injured needles abscise, leaving a bare twig. Shoots of the next season may grow from surviving terminal buds or from adventitious buds near the stem apex, but this growth is likely to be slight, producing tufted foliage. In unusual cases, light frost during shoot elongation may damage stems enough to cause bending or drooping but not death, and the stem survives, permanently deformed. On most frost-prone sites, injury does not occur every year, and trees that grow above the level where cold air collects continue normal growth.

Frost injury to many plants, both woody and herbaceous, is aggravated by the presence of certain bacteria that act as nuclei for the formation of ice crystals. In the absence of these bacteria, nonacclimated plants can supercool several degrees and may thus escape damage from light frosts. The plant pathogen *Pseudomonas syringae* and some strains of the saprophytes *P. fluorescens* and *Erwinia herbicola* are active in ice nucleation. Ice nucleation is the basis of the reported interaction of *P. syringae* with freeze damage in formation of bacterial canker of stone fruit trees (Plate 75). Large populations of this bacterium may also be present on plant surfaces in the absence of disease. *P. syringae* has been found associated with frost-induced shoot dieback in many species of woody plants in nurseries. If ice-nucleating bacteria are suppressed by bactericides or competition from other bacteria lacking ice nucleation properties, frost hardiness is enhanced. This enhancement has been demonstrated with both agricultural plants and stone fruit trees.

Frost rings in wood. Injury to the xylem or cambium by freezing leads to anatomical aberrations in wood that is forming at the time and in the first several tiers of wood cells formed after injury. Injury between growing seasons leads to aberrant wood formation in the early part of the next season. The aberrant tissue extends as a sheet partly or completely around the stem, appearing as a ring (frost ring) in cross-section. Frost rings have different structures, depending on species, severity of injury, and degree of cambial activity at the time of injury. Injury during the growing season kills differentiating xylem elements, leaving a sheet of dead undifferentiated cells overlying crumpled, partly lignified xylem elements. The vascular cambium usually survives and produces an abnormal amount of xylem parenchyma and abnormally shaped tracheary elements. Gum or resin and dark pigment may be deposited. Injury during a period of cambial inactivity causes frost rings that lack remnants of undifferentiated or partly differentiated xylem. Slight freezing injury to the cambium and/or differentiating xylem commonly causes frost rings in the absence of external symptoms. More severe injury, usually accompanied by foliar symptoms or dieback, leads to formation of small pockets of xylem parenchyma (which may eventually turn brown) along the frost ring or to blisterlike tangential separation of wood rings. Frost rings may also form in phloem, but they disappear from that tissue because of natural rearrangement of cell positions with age and stem growth.

A, D. Winter freeze damage to English holly. A. All leaves and many buds present during winter were killed, and most leaves had dropped when the photograph was made. By mid-June, scattered new shoots, mostly stunted, had grown from surviving buds. The tree died during summer, of water stress resulting from damage to sapwood. D. Cross-section of an injured branch about 1 cm in diameter, with xylem discolored pinkish while the cambial region and bark appear normal. The discolored xylem is dead or dying, its water-conducting capacity already impaired (WA, Jun).

B, F, G. Freeze damage to 'Kwanzan' flowering cherry. B. Stunted, wilting, and dead foliage on a tree that sustained trunk damage of the sort shown in panels F and G. Because of water stress caused by death of sapwood, some leaves failed to attain full size, and others wilted after growing normally (NY, Jun). F, G. The trunks of young trees damaged by freezing in late winter. Both stems show lens-shaped cracks in the bark near ground line where callus formation at the edges of killed areas caused the rigid outer bark to split. Above the level of the cracks, elongate splits developed in bark, and the edges flared back. Flaring bark results from mechanical stresses imposed by drying of killed tissues, coupled with growth of tissues in surviving parts of the phloem. The tree in (F) shows abundant sprout growth from the uninjured rootstock (WA, Jun). The dying stem in (G) has discolored, dead sapwood, and the flaring bark shows a reddish brown inner surface where a periderm formed after the bark separated from the wood (NY, Jul).

C, E. Freeze damage to peach. C. An injured twig has dead buds, stunted growth from one bud that survived, and freeze-induced reddish discoloration in sapwood (NY, Jun). E. Magnified view of a cross-section of a 2-year-old stem that sustained freeze damage after its 1st year of growth. Wood that was present at the time of freezing is dead, discolored light brown, and enclosed within a dark brown sheet (frost ring, also a barrier zone, as explained with Plates 162–163) that was produced by the cambium after it sustained sublethal injury (NY, May).

H, I. Internal symptoms of previous freeze injury to winter creeper (evergreen euonymus). H. The dark brown line marks the boundary between living 2-year-old wood and older wood killed by freezing. Apparently the injured xylem temporarily conducted enough water to allow leaf functions, which in turn allowed formation of new sapwood outside the damaged tissues. I. Tissues have been separated at the brown line to show in tangential view that it is a sheet of dead tissue. The sheet is composed of parenchyma that differentiated in place of normal xylem and phloem during the beginning of the growing season after the freeze (NH, Aug).

Plate 235

Injury by freezing during dormancy. As mentioned previously, the cold acclimation of woody plants fluctuates during dormancy. Acclimation occurs gradually in response to diminishing temperature in autumn, and deacclimation occurs with rising temperature in spring. Unusually warm weather in autumn will retard acclimation, and warm weather in winter or early spring will induce partial deacclimation. Sapwood, cambium, and phloem acclimate to different degrees. During midwinter, the sapwood is least tolerant of cold. That is, intracellular freezing of sapwood parenchyma occurs at a higher temperature than does freezing of cambial or phloem cells. During autumn and early spring, although all tissues are less tolerant than in winter, sapwood is relatively more tolerant than cambium and phloem. If after a period of unusual warmth the temperature drops rapidly to a level normal or subnormal for the season, severe freeze damage may occur to plants that are nominally hardy in a given region. Mechanisms of injury by freezing are discussed above with injury by frost during the growing season.

The most common external symptoms caused by winter freezing are dieback, foliar browning, sunscald, and bark splitting near the base of the trunk. Dieback of twigs and branches, and foliar browning in evergreens, commonly follow freeze injury when winter temperatures arrive suddenly after warm autumn weather. Plants that survive produce new branch systems from dormant or adventitious buds. Severe twig dieback (and foliar browning in conifers) caused by freezing may follow late summer pruning, which retards cold acclimation. Freeze damage to roots is sometimes also associated with dieback and foliar symptoms of distress, but practical information about this, except for plants in containers, is scant. Roots become less acclimated to cold than do parts above ground, and roots may sustain damage when severe freezing weather occurs in the absence of snow cover.

Sunscald is a local injury that develops on the south and southwest sides of the trunk or on the upper surfaces of limbs exposed to sun. The cambial temperature of the sun-warmed sides of limbs or trunks may exceed 20°C in late winter when the air temperature and the temperature of shaded bark barely exceed 0°C. This heating causes deacclimation, which is followed by lethal freezing when the temperature drops at night. Damaged bark and cambium dry out, crack, separate from the wood, and eventually fall away, exposing dead sapwood. Trees with thin, smooth bark are most susceptible to this type of injury. Sunscald has been noted on trunks of American beech, yellow birch, flowering cherries, maples, callery pear, white pines, weeping willow, and many kinds of fruit trees and on limbs of European beech, catalpa, and red oak. As a variation on this type of injury, the bark may remain alive after the death of cambium. A pocket consequently forms (and can be detected by tapping) between living bark and wood as the bark is lifted by growth of callus at the edges of the injured area. Usually the covering of this pocket splits as shown here and in Plate 235 (further illustration and discussion with Plate 164), but sometimes, as in "blister shake" of tulip tree, it remains intact, creating a cavity between two age classes of wood.

Major injury that kills the sapwood or cambium leads to cankers, dieback, and often wilting and death during the next growing season as water lost by transpiration cannot be sufficiently replaced by conduction through damaged wood. Externally visible injury caused by freezing is invariably accompanied by a frost ring in the wood of less severely damaged parts of the tree or shrub. Freeze-killed sapwood in living stems becomes compartmentalized (Plates 162–163) and usually turns dark in color. It is then sometimes called black heart. Sapwood exposed by bark splitting or branch dieback is soon invaded by decay fungi. In trees prone to bacterial wetwood—elms and sycamore, for example—this condition becomes more widespread within trees after damage by freezing.

Predisposition to attack by opportunistic pathogens. Drought or freezing insufficient to cause evident stress in plants may interfere with defense against opportunistic fungal pathogens that cause cankers, dieback, sapwood decay, or root rot. For example, when young European mountain ash, European white birch, and red-osier dogwood were inoculated with *Botryosphaeria dothidea* (Plates 81–82),

then allowed to dry to internal water potentials below approximately −1.2 to −1.3 megapascals (insufficient stress to cause wilting), and were maintained in this condition, cankers developed after several days. This susceptibility disappeared and canker expansion ceased a few days after plants were rewatered, and susceptibility never occurred in normally watered plants. Similarly, when partially cold-acclimated plants of the same species plus sweet gum were inoculated, cooled to between −20° and −30°C (insufficient freezing to cause measurable tissue injury), and were then slowly thawed and observed for 2 weeks, cankers developed. Tallhedge buckthorn cooled and inoculated with *Tubercularia ulmea* (Plate 99) and winged euonymus cooled and inoculated with *Nectria cinnabarina* (Plate 100) behaved similarly. Susceptibility did not develop in plants cooled to temperatures above −20°C, and plants made susceptible by cooling recovered normal resistance within 2 weeks after thawing. These examples provide experimental evidence to support many field observations of environmental predisposition to fungal attack. Other diseases for which predisposition is important include Cytospora and Valsa cankers (Plates 92–96), dieback of sycamore and oaks caused by *Botryosphaeria rhodina* (Plate 85), Hendersonula dieback of walnut and fruit trees (Plate 79), Seiridium canker of Monterey cypress (Plate 80), Sphaeropsis blight of pines (Plate 63), Thyronectria canker of honey locust (Plate 97), and canker rots of various trees caused by *Cerrena unicolor, Trametes versicolor,* and *Schizophyllum commune* (Plates 165, 173).

Secondary insects often join opportunistic fungi in the attack on stressed trees. Drought-stressed conifers, for example, may be attacked by bark beetles and blue-stain fungi. The two-lined chestnut borer and *Hypoxylon* species (Plates 107–108) collaborate in causing death of drought-stressed oaks in the South.

References for water shortage: 139, 599, 924, 1051–53, 1117, 1295, 1368, 1463, 1635, 1965, 2040, 2193
References for heat stress: 435, 746, 902, 934, 1087, 1295
References for injury by frost and freezing: 63, 139, 284, 415, 435, 459, 463, 599, 654, 743, 786, 902, 1021, 1145–47, 1238, 1337, 1450, 1469, 1476, 1544, 1600, 1635, 1672, 1890, 1911, 2042, 2043, 2082, 2130, 2149, 2271, 2272, 2275
References for predisposition to attack by opportunistic organisms: 415, 459, 1716–20

A, B, H, I. Winter freeze damage to common lilac. A. Wilting, foliar browning, and dieback in young landscape plants. These secondary symptoms, indicating water stress, developed in early summer because damaged sapwood was unable to conduct enough water to new shoots. B. Curled chlorotic leaves on a plant that grew normally in late spring but came under water stress as the result of previous freeze damage to sapwood at the stem base. H, I. Symptoms at the stem base. A lens-shaped split in bark is evident externally. Removal of bark and outermost sapwood reveals a frost canker and numerous small lesions resulting from localized killing of the cambium. Sapwood of the previous year, beneath these marks, is dead (NY, Jul).

C. A 'Crimson King' Norway maple leaf appears tattered because interveinal tissues failed to develop as the leaf expanded. This type of injury occurs while the leaf is still enclosed in the bud and is commonly attributed to freeze damage, although experimental proof of cause is lacking (NY, Jun).

D. Late frost injury to tulip tree. A light frost when leaves were nearly full grown killed most leaves but spared others on the same limb. Petioles were not killed except adjacent to the leaf blades; thus dead leaves droop at the tips of green petioles (NC, May).

E–G. Cankers caused by freeze damage at the bases of pole-sized white ash trees. Trees lacked normal cold hardiness because of infection with the ash yellows agent (Plate 191), which induced the witches'-brooms at the base of the tree shown in (E) and (F). E. Bark has split as the result of pressure exerted by callus rolls forming at edges of an area where the cambium was killed. F. Removal of loose bark reveals an extensive area of dead sapwood and a callus roll at its edge. G. An old frost canker from which the bark has long since fallen (NY, May).

Plate 236

Damage Caused by Excess Water (Plate 237)

Roots in flooded or waterlogged soils often die of anoxia (oxygen deficiency). Damage occurs not only to plants on obviously wet sites but also to those in planting holes along city streets and in landscapes where soil drainage is impeded by high clay content.

Most trees and shrubs cannot grow for long in waterlogged soil, and some perish if flooded for only a few days during the growing season. Plant roots and soil organisms quickly use up gaseous oxygen in waterlogged soil. Carbon dioxide increases; redox potential diminishes; iron and manganese are reduced to soluble, toxic forms; and sulphides, ethylene, fatty acids, phenolic compounds, cyanogenic compounds (from roots of various plants, such as cherries and their relatives), and other toxic organic products accumulate.

Oxygen deficiency in roots causes a switch from aerobic to anaerobic respiration, which is much less efficient in utilization of organic compounds as energy sources and results in accumulation of toxic end products such as ethanol. Roots soon lose some of their permeability to water, retarding uptake of water and minerals. After a few days, internal water shortage, stomatal closure, and depressed photosynthesis and translocation of organic compounds occur. Stomata either remain closed or reopen only after root absorption resumes. If the plant remains alive, diminished root function soon causes foliar nitrogen deficiency. Synthesis and translocation of growth regulators (giberellins and cytokinins) in roots slows, and concentrations of auxins and ethylene in stems increase. Mycorrhizal fungi, which associate with plant roots symbiotically (Plates 244–245), are also adversely affected, further suppressing plant uptake of mineral nutrients, especially phosphorus. Internal water deficit in some plants increases until they die, but many kinds of plants regain the normal degree of hydration while their stomata remain closed during flooding. Stomata of some tolerant plants reopen as the plant adapts to flooding.

External symptoms of injury include various of the following: epinasty (downward bending of leaf petioles), stem swelling (particularly in small plants), chlorosis, edema (discussed below with adaptations), red or purple pigmentation in leaves (of pear and some other plants), browning of leaf margins, reduction or cessation of growth (more pronounced in roots than stems), twig dieback, absence or diminished yield of fruit, death of roots, wilting, leaf drop, and death of the entire plant. Seedlings develop symptoms more quickly than do large plants. Plants with roots injured by waterlogged soil may subsequently suffer drought stress or death when, after the soil drains, the root system is unable to meet transpirational demands of the top. This occurs in trees planted in holes in concrete landscapes and has been noted in plantations of loblolly and red pines on poorly drained sites.

Plants stressed or injured by waterlogging also become abnormally susceptible to certain fungal pathogens. *Phytophthora* species (Plates 139–140), for example, cause root rot most often in soils that are periodically waterlogged. Excess water not only promotes susceptibility of roots but acts as a dispersal medium for zoospores of *Phytophthora*. Root rot of conifers by *Phaeolus schweinitzii* (Plate 155) is also favored by prior root damage in periodically waterlogged soils.

Dormant woody plants can tolerate flooding or waterlogging for several weeks without much harm. This tolerance is usually associated with low oxygen demands of roots and organisms in cold soil, but dormant roots also display tolerance at moderate temperatures (e.g., 20°C).

Growing plants that tolerate flooding or waterlogged soil do so by means of adaptive changes in form or function. Lenticels on submerged roots and stems become enlarged; parenchymatous masses (called edema) that resemble lenticels may erupt on the lower sides of leaves; adventitious roots may grow from the lower stem or the upper main roots at the level of the water table; and large intercellular spaces may form in the cortex of roots. These features promote gas exchange and the diffusion of oxygen into roots. Edema develops in such plants as camellia, hibiscus, privet, schefflera, and yew when soil is waterlogged and transpiration is impaired. Root tissue with large intercellular air spaces, called aerenchyma, is normally present in some plants, such as sour gum and willows, that inhabit wet sites, and it is formed by other plants, such as lodgepole pine and poplar, in response to flooding. When induced by flooding, aerenchyma is created by the dissolution or collapse of certain cells and the swelling of others, resulting in large spaces. Functional adaptations include increased anaerobic metabolism during inhibition of the normal aerobic pathway, avoidance of ethanol accumulation, and tolerance of high carbon dioxide concentration.

Most angiosperms tolerate flooding better than most gymnosperms. The following classification of trees and shrubs according to tolerance of flooded or waterlogged soil has been compiled from several published sources. Where available, the ratings are for established trees rather than for seedlings.

Relative tolerance of plants to flooded or waterlogged soil

Tolerant: black, green, and pumpkin ash; buttonbush; bald cypress; red-osier dogwood; tupelo gum; water hickory; deciduous holly; eastern larch; water locust; red maple; overcup oak; swamp privet; black and sandbar willows.

Intermediate: speckled alder; eastern arborvitae; arrowwood; white ash; trembling aspen; Japanese barberry; red bay; sweet bay; river birch; box elder; Atlantic white cedar; cornelian cherry; eastern cottonwood; 'Dolgo' crabapple; American cranberry bush; panicled dogwood; American and cedar elms; balsam fir; sweet gum; hackberry; Pfitzer juniper; honey locust; silver maple; nannyberry; bur, swamp chestnut, Nuttall, pin, water, and willow oaks; osage orange; callery, common, and Oriental pears; persimmon; loblolly, lodgepole, pond, and slash pines; balsam poplar; Regel's privet; black spruce; sycamore; pussy willow; white willow.

Intolerant: European mountain ash; basswood; American beech; gray, paper, and European white birches; Oriental bittersweet; yellow buckeye; eastern red cedar; black and Higan cherries; crabapples; flowering dogwood; red, Siberian, and winged elms; forsythia; black gum; Lavall's and Washington hawthorns; eastern hemlock; mockernut, pignut, and shagbark hickories; American holly; Morrow and tatarian honeysuckles; American hornbeam; hop hornbeam; black locust; saucer and southern magnolias; Norway and sugar maples; red mulberry; ninebark; blackjack, cherrybark, chinkapin, laurel, southern live, post, Northern red, southern red, shumard, and white oaks; mock orange; pawpaw; pecan; peach; jack, red, shortleaf, Virginia, and eastern white pines; Amur and common privets; redbud; sassafras; sourwood; blue, Norway, Sitka, and white spruces; sugarberry; tulip tree; black walnut; wintercreeper; yellowwood; Japanese and hybrid yews.

References: 55, 128, 157, 260, 403, 412, 518, 577, 642, 873, 994, 995, 1053–55, 1187, 1520, 1721, 1917, 2036, 2167

A, B. Damage caused to a hedge of Japanese yew by waterlogged soil. A. Chlorosis, dieback, and stunted shoots resulting from death of roots. B. A declining plant, pulled from the soil, has dead, decaying roots. Living fibrous rootlets were found only near the soil surface (NY, May).

C. Sugar maples and adjacent turf killed by flooding. Water covered the root zones of the trees for several days in summer. Vegetation in the background at slightly higher elevation escaped damage (NY, Jul).

D, E. Edema (eruptions of corky tissue on the lower surfaces of leaves) on sasanqua camellia (FL, Apr).

F. Root necrosis and butt swelling of Shasta red fir, caused by waterlogged soil. The tree had been planted on a wet site. When foliage began to fade, the tree was uprooted. All roots and internal tissue below the center of the basal swelling were found to be dead (WA, Jun).

G. Edema on lower surfaces of leaves of Japanese yew. The upper surfaces of leaves showing edema are normal in form but are often chlorotic (NY, Sep).

Plate 237

Damage Caused by Girdling Roots, Hail, and Sheet Ice (Plate 238)

Girdling roots. Roots that constrict some part of the butt of a tree are called girdling roots. The constriction is usually self-inflicted but also occurs occasionally in woodlands where one tree grows between brace roots at the butt of another tree. Trees that grow from seed in nature are rarely self-girdled. This problem is confined to planted trees, mainly those raised originally in nurseries. Nursery practices such as cultivation close to the stem, and especially the growing of trees in containers, cause small roots to grow tangentially to the butt. Years later, when the butt has enlarged enough to contact a tangential root, constriction begins. Trunk diameter is often 16–20 cm at that time. One or more branches then lose vigor and begin to die back. Symptoms usually result from the action of multiple girdling roots rather than just one.

Aerial symptoms of damage by girdling roots are most common in maples, especially Norway and red maples. Root-girdled trees of other species seem to tolerate the constriction better than do maples but may be mechanically unstable. We have observed trees as diverse as American elm and Austrian pine blown over by wind, exposing a mass of roots coiled around a greatly constricted butt, the mass resembling the shape of a container in which the tree once grew. Similar damage has been observed in pines that as seedlings were carelessly planted or had their roots deliberately wound in a coil.

Aerial symptoms caused by girdling roots include slow growth, subnormal leaf size, and premature autumn color before branches die back. Because the constriction is usually applied only to part of the butt, symptoms do not usually progress branch by branch. Some branches may die, while others appear normal for many years. Aerial symptoms of the sort caused by girdling roots can be induced by diverse injuries and some diseases; therefore reliable diagnosis depends on detecting the offending root(s), usually by excavation.

Girdling roots below the soil line commonly occur at depths less than 8 cm but sometimes much deeper. Their presence is often indicated by absence of normal flaring of the butt at the junction of stem and roots on one side of the tree; instead, the trunk ascends more or less vertically from the soil. A girdling root at the soil line usually causes some swelling where the stem begins to overgrow the root. This swelling is not to be confused with normal flaring of the butt.

Most details of injury by girdling roots await study; we therefore present an interim hypothesis. Pressure from the girdling root is believed to disrupt and eventually halt cambial activity in a band partway around the butt. The resulting aberrations in phloem and outer sapwood are such that translocation of carbohydrates and growth regulators in phloem, and conduction of sap and nutrients in xylem, are inhibited or blocked. Roots below the girdle decline in growth and absorptive function, and the branches most directly served by the isolated roots decline in vigor. Multiple girdling roots may encircle the butt like a noose.

Damage to xylem beneath a girdling root has been documented for Norway maple. In one tree examined in detail, sapwood beneath a girdling root contained abnormally small xylem vessels whose lumens occupied only 10% as much cross-sectional area as those of vessels in normal wood. The abnormal vessels were aligned more with the constricting root than with the trunk axis. This was interpreted as evidence for disrupted sap flow that could contribute to symptoms of distress in leaves and branches. Phloem formation and function in relation to girdling roots have not been reported.

Symptoms in branches do not appear until girdling roots have done irreversible harm to part of the tree. The surgical removal of girdling roots may prevent further damage, but proof that it does so is lacking. Only one adequately designed study of therapeutic treatments for trees with girdling roots has been reported. In that instance neither fertilization nor removal of girdling roots, alone or in combination, caused measurable benefit to Norway maples within 2 years after treatment.

Preventive treatments of young planting stock have been suggested but have not been put to long-term test. These include cutting off spiral roots that have grown around the sides of containers, making four equally spaced vertical slits to a depth of 2–3 cm in the root ball, and growing young trees in containers with grooved sides or corners that train roots up or down rather than around the plant's axis.

Damage by hail and sheet ice. Injuries by hail, similar to those shown here, occur commonly but in small geographic areas. Hail-stones lacerate leaves, defoliate branches, remove twigs, bruise or break the bark of twigs and small branches, and kill small trees. By coalescence, severe hail wounds may kill all the bark on one side of a stem. In extreme cases the bark may be pounded off the windward sides of stems. Bruises and wounds tend toward elliptical shape and vary in length from a few millimeters to 10 cm or more. All occur on the upper sides of branches and on the side of the tree facing the storm. Bruised bark may crack after the storm as the result of drying and mechanical stress from callus growth at edges of the injured area. All tree species in the locality display similar symptoms, although the severity varies among species. Bruises or wounds result in dieback of twigs and branches if tissues around the injuries dry out. Dieback is more likely if injury occurs during dormancy rather than during the growing season. Similarly, secondary damage by canker fungi is more likely if the injury occurs during dormancy.

Injury by sheet ice, as shown in panels F and G, where stems worked against a stationary ice sheet, is uncommon. Probably it occurs only to young trees with thin bark. In such cases, some degree of injury can be found at the same level on all trees where water was impounded. Ice crusts on snow do not cause this symptom because snow usually melts away from stems before the crust forms.

Ice is most destructive by far when a heavy glaze forms on trees during freezing rain. The weight of a glaze more than about 1 cm thick breaks twigs, branches, and trunks or uproots trees. Supple young trunks may become permanently bent. Glaze reaching 2.5 cm thickness occurs occasionally and simply destroys trees. Damage increases if the wind rises before the ice melts. Severe breakage leads in turn to severe decay in surviving trees. Many hardwood stands in the eastern United States and southern Canada contain trees with grotesquely deformed tops and much internal defect resulting from breakage by ice. Other things being equal, the largest trees suffer the most severe damage. Conifers sometimes escape with slight damage, while hardwoods break. Experience with Douglas-fir suggests that glaze damage in coniferous stands with closed canopies may be less severe than in stands opened by thinning or partial harvest. Only a slight relationship has been detected between wood strength of a tree species and its susceptibility to breakage by ice glaze, but wood flexibility is an obvious asset.

Rankings of susceptibility to breakage by ice glaze vary with region and circumstance but have been compiled here as a general guide. Species judged by one observer to be highly susceptible and by another to be resistant are ranked intermediate here.

A, B. Injury to red maple by girdling roots. A. Dieback of twigs and branches. B. A girdling root at the soil line. A tree such as this one often also has several additional girdling roots below the soil line (NY, Apr).

C, D. Hail injury to twigs of sugar maple and crabapple, respectively. The storm that caused these wounds occurred early in the growing season. The initial injury, consisting mainly of bruises, was not readily apparent. During the growing season, however, bruised tissue died, dried, and split because of the mechanical stress imposed by callus growth at the edges of killed tissue. Small bruises caused only minor splits in the outer bark, and new bark soon covered these injuries. The largest bruises killed the cambium, resulting in exposure of sapwood. By autumn, when the photos were made, callus had overgrown many small wounds, but the largest remained open. Had such severe injury occurred at the end of the growing season, the injuries would have resulted in twig death from desiccation during winter (NY, Oct).

E. Twig dieback caused by hail injury to alpine fir. Split bark is evident at the junction of living and dead parts. Branch tips died during the winter after injury (CO, Jun).

F, G. Girdling of young Norway maples by bruises made by sheet ice. These trees grew in a nursery in a depression where a pool of water occasionally formed and froze during winter. Wind action caused bruises where the stems worked against the ice. The trees began growth normally the next spring but wilted after leaves expanded (NY, Aug).

Plate 238

Relative susceptibility of plants to breakage by ice glaze

Most damaged: black ash, bigtooth and trembling aspens, butternut, black cherry, eastern cottonwood, American and red elms, hackberry, silver maple, chestnut and northern red oaks, sassafras, tamarack, tulip tree, willow.

Intermediate: white ash; basswood; American beech; gray, European white, and yellow birches; box elder; eastern red cedar; cucumber tree; black gum; black locust; red maple; black and scarlet oaks; red, Scots, and eastern white pines; sycamore.

Least damaged: apple, eastern arborvitae, black walnut, catalpa, flowering dogwood, Douglas-fir, eastern hemlock, hickories, honey locust, hop horn-beam, Norway and sugar maples, white oak, common pear, persimmon, Austrian pine, Norway and red spruces, tree-of-heaven, umbrella-tree.

References for girdling roots: 664, 868, 905, 1972, 1973, 2011
References for damage by hail and ice glaze: 271, 311, 333, 430, 516, 1128, 1260, 1265, 1586, 1619, 1794, 1864, 2179

Lightning Injury (Plate 239)

Lightning is most important as a cause of forest fires, but it also causes significant damage to forest and landscape trees in the absence of fire. Few data are available, but in one study during the 1940s, lightning was found to be responsible for one-third of all mortality in stands of ponderosa pine in northern Arizona. Trees struck but not killed are likely to be disfigured by death of limbs or the top. In addition, wounds and killed parts provide entry for borers and for fungi that decay wood. Conifers weakened by lightning strikes may be attacked and killed by bark beetles. Lightning scars on surviving trees in forests cause logs to be downgraded at harvest time.

Lone trees, tall trees on the windward edges of groups or stands, and dominant trees in woodlands are struck most often. The charge follows the most conductive path between top and roots, sometimes along the surface but often in the outer sapwood. Apparently steam or possibly gas from electrolytic dissociation of water is generated instantly in tissue that conducts the charge, for a strip of bark and sapwood are often blown off the trunk, leaving a continuous or intermittent rough groove that follows the grain of wood. Sometimes the trunk simply shatters; fragments may be propelled several meters. Sometimes the bark and covering soil are blown off a root before the charge dissipates in soil.

A high proportion of trees struck by lightning eventually die because of the injury or because of attack by opportunistic organisms. Reliable data on the mortality rate of lightning-struck trees are not available.

Trees that survive a lightning strike may develop various abnormalities. Injury during the annual period of cambial activity may cause the formation of a sheath of abnormal wood (a ring as viewed in cross-section). Since such rings occur in nonwounded branches as well as grooved trunks, they probably reflect a cambial response to the electrical charge or associated heat. Lightning rings in resinous conifers contain an abnormally large number of resin ducts, and injured conifers may exude resin. Galls or short ridges may develop along the

trunks of lightning-struck trees, sometimes attaining diameters of several centimeters. If root necrosis has occurred, adventitious roots sometimes grow from the trunk above, usually beneath the protective cover of dead bark.

Lightning occasionally causes group death. Trees in patches up to 30 meters in diameter suddenly turn brown. In some of these cases, no external marks implicate lightning as the cause, but in other instances, typical grooves may be found on the trunk of one or more central trees. The unmarked trees are presumed to die because of root killing where the electrical charge dissipates in soil. Similar damage has been noted in nurseries, where all plants in the center of a patch die, and those around the edges remain stunted for a time. The deepest roots of surviving seedlings may be found dead. Few eyewitness accounts exist of lightning strikes causing patch dying in forests or nurseries, but the assignment of lightning as the cause of damage found later is probably correct, because similar damage has occurred in agricultural crops where lightning strikes were observed.

Trees of some species are struck more frequently than others, but the reasons are unknown. No investigation has been reported in which frequency of lightning strike could be related to tree species apart from tree location, tree size, and frequency of each species in the local tree population. In eastern United States, the frequency of lightning strikes has been reported to diminish among species in the following order: oak, elm, pine, tulip tree, poplar, ash, maple, sycamore, hemlock, and spruce. Black walnut, although not listed among species above, is often struck. Beech, birch, and horse-chestnut are not often struck.

Lightning rods are sometimes installed on large valuable trees, with copper cables leading down the trunk and through the soil to grounding rods driven into soil beyond the branch spread of the tree. The degree of protection thus given has not been assessed objectively.
References: 315, 421, 422, 591, 940, 1468, 1601, 2079

A. A mature red oak killed by lightning (MI, Aug).
B. Lightning injury to the trunk of an American elm. Strips of bark and sapwood were blown off the trunk as the charge traveled through the cambial region and outer sapwood. The tree died (NY, Jul).
C. An eastern white pine, its top killed by lightning (NY, Apr).
D, F. Branch mortality of red pine caused by lightning. A strip of exposed sapwood where bark was blown off the trunk (D) implicates lightning as the cause of branch death (NY, Aug).
E. Growth abnormalities on a mature Norway maple resulting from lightning injury many years earlier. The electrical charge killed broad strips of cambium as well as major roots on one side of the tree, but the top remained alive. Dead bark remained in place temporarily, and numerous adventitious roots grew from above the killed area downward beneath the killed bark. Dead bark eventually broke away, exposing the roots. An adventitious root about 8 cm in diameter is visible at right (NY, Aug).
G. The base of a longleaf pine struck by lightning. The electrical charge grounded through one major root, simultaneously blowing away its bark and the covering soil (MS, May).

Plate 239

Noninfectious and Unexplained Growth Abnormalities (Plates 240–241)

Fasciation. This term connotes abnormal flattening of plant organs, usually stems, resulting from a change in the form of the apical growing point from a minute dome to a jagged row of generative cells perpendicular to the stem axis. The width of the stem generated along the row usually varies, so that the stem has a ribbed appearance, as if several stems had fused laterally. Foliage is more dense than normal because each rib bears leaves in the manner of a separate stem. Leaves are usually normal in shape but may be undersized. Unequal elongation of young stem tissues below the row meristem may lead to grotesque bending and coiling of the stem tip. Often the row meristem cleaves into several parts, so that the ribs separate into discrete stems, some fasciated, others normal. Dwarfed, malformed shoots that abort at an early stage may develop along the row, and fasciated branch tips often die during winter.

Fasciation occurs in many kinds of plants, woody and nonwoody. In some instances it apparently results from a mutation in reproductive cells and is transmitted as a heritable character to progeny. In other cases a change occurs only in vegetative cells and can be perpetuated by vegetative propagation but not through seed. In still other instances a single plant will alternately produce normal and fasciated twigs, one growing from the other. The imbalances in growth regulators that cause this condition have not been described.

Graft union abnormalities. When freshly cut tissues of compatible plants are grafted together by appropriate methods, a strong union forms. Initially, thin layers of necrotic cells form on both cut surfaces, but callus from both stock and scion readily breaks through these and grows as a new common tissue along the graft interface. This process is followed by the formation of a cork cambium and periderm at the outer surface of the union and by differentiation of the vascular cambium, leading to the formation of phloem and xylem across the interface. A strong union is assured by interlocking growth of xylem cells derived from stock and scion. The entire process requires only several weeks.

Many variations of abnormal growth or growth failure at graft unions are lumped under the general term *graft incompatibility*. In the simplest case, no tissues of stock and scion unite, and no composite plant is produced. We are concerned, however, with delayed or partial incompatibility—cases in which stock and scion apparently unite, but subsequent growth is aberrant, leading to mechanical failure or symptoms of physiological distress. In one type of incompatible union, stock and scion become united by callus but not by conducting tissues. Growth is possible initially, but the union remains weak and usually fails within 1–2 seasons. In one common type of delayed incompatibility, a xylem union forms between stock and scion, but phloem connections are not established. The result is swelling of the scion above the union and slow starvation of the stock. In another type, no interlocking growth of xylem cells occurs; instead xylem elements of stock and scion grow parallel to one another at the union. Enough water moves across the union to satisfy mineral nutrient needs and transpirational demand of the scion, but the union remains physically weak and eventually breaks. There is one well-documented report, for example, of a white fir tree that broke cleanly off its rootstock at ground level after 40 years of apparently normal growth. Similar incompatibilities have been reported in various fruit and nut trees.

Graft incompatibility may be indicated by one or more of the following symptoms: markedly greater growth rate of the stock than the scion, or vice versa; difference between stock and scion in the onset or cessation of growth; local overgrowth above, below, or at the graft union; breakage of the grafted plant at the graft union; declining vigor, stunting, chlorosis, premature leaf shedding, or death of the scion or the entire plant.

Graft incompatibility may be due to intrinsic factors, but often it is caused by viral infection. One member of the grafted pair becomes infected but tolerates the virus, whereas the other member responds with cellular death, resulting in necrosis at the graft union. Blackline disease of walnut (Plate 196), exemplifies delayed incompatibility caused by a virus.

If a scion simply grows more rapidly than its understock, and no symptoms of distress occur in either grafted member, the two are not necessarily incompatible. Continuation of this inequality will result in a strangely shaped trunk, however.

Chimeras. A chimera is a plant or plant organ consisting of tissues of more than one genetic composition and origin. Natural chimeras apparently arise from mutations in certain cells at growing points or in the vascular cambium such that their vegetative progeny develop true to the new type, while the remainder of the plant, derived from unaltered cells, appears normal. Freezing can induce chimeras that originate in the cambium, but the inducing factors for most natural chimeras are unknown. Chimeras can be induced artificially by freezing, by certain chemicals such as colchicine, by radiation treatment, or by grafting into meristems. In the walnut leaf shown in Plate 240H, a change apparently occurred early in the formation of the leaf primordium such that cells differentiating from one side lost nearly all ability to synthesize chlorophyll (scattered angular green spots may be seen on the albino leaves). This chimera, derived only from a leaf primordium rather than from a permanent growing point, survived only 1 season. Variegated plants represent chimeras resulting from mutations (or in some cases grafting) in apical meristems. Unfortunately, variegated trees or shrubs tend to be poorly suited for horticultural use because the achlorophyllous tissues often die prematurely. Variegated Norway maples, for example, have a pronounced tendency to develop leaf scorch.

A, C. Fasciation in 'Kwanzan' flowering cherry. A. An affected branch from an otherwise normal tree. Leaves appear normal, while the stem is greatly distorted. A normal twig is growing from the coiled apex of the branch toward the upper left. C. The same branch with leaves and normal twigs removed, revealing the flattened stem, repeated division of the apical growing point, bending and coiling resulting from unequal growth below the various growing points, and necrosis of growing points at center and left center (NY, Aug).

B. Fasciation of Greek myrtle, in contrast to one normal twig at upper right. Foliage is abnormally dense, and leaves on the fasciated parts are undersized (CA, Oct).

D. Fasciation of European white birch. Except for young leaves, the parts shown are 1 year old. Two normal twigs developed from the fasciated stem. These twigs, but not the fasciated branch tip, survived winter. Buds on the normal twigs are late opening, however, perhaps because of cold injury (NY, May).

E. Fasciation of daphne. Transition from normal to fasciated stem form and clusters of dwarfed distorted shoots along the apical row of growing points are apparent. A fasciated branch has grown from the right end and a normal branch from the left end of the row (CA, Aug).

F. Gross swelling, indicating incompatibility, at the graft union of a hairy wattle tree (*Acacia pubescens*) on an unidentified acacia (PA, Jun).

G. A graft union between two species of *Prunus*. Radial growth rate of the stock and scion are dissimilar, so that swelling appears to have taken place above the union. Whether this is a case of delayed incompatibility or of intrinsically different radial growth rates in compatibly united stock and scion cannot be learned from external inspection at this stage (NY, Aug).

H. A chimera characterized by one-sided dwarfing and albinism in a black walnut leaf (NY, Jun).

492

Plate 240

Adventitious shoots and roots. Adventitious shoots or roots may arise as secondary responses of a tree or shrub to injury or disease. Adventitious shoots, also called water sprouts, grow from buds that are released or formed along the trunk and limbs when apical buds die or lose their normal dominance. Severe breakage of branches, dieback resulting from disease or defoliation, and loss of normal hormonal balance as the result of systemic disease are among the conditions that trigger formation of adventitious shoots. Numerous adventitious shoots, even if more or less vigorous, do not necessarily indicate intrinsic vigor of the tree as a whole. Sprout formation at the base of a tree often reflects chronic distress (for examples, see Plates 196, 219). Similarly, sprout formation along the trunk is often associated with dieback and often presages death. Such is the case with the chestnut oaks in Plate 241A. They were defoliated by cankerworms in 2 successive years, resulting in dieback followed by sprouting. While still in weakened condition during the year after the photograph was made, they were killed by opportunistic insects and fungi.

Norway maples occasionally develop more or less profuse adventitious sprouts from trunk swellings, as shown in Plate 241B. Adventitious sprouting in box elder and sycamore maple in Hungary was associated with infection by a virus that was tentatively named maple mosaic virus, but no viral association with these symptoms has been reported in North America.

Adventitious roots may form above wounded or killed roots or at the upper edges of stem wounds near ground level, provided that the root environment is moist. In nature, such roots often form on the upper main roots or lower stems of flooded plants or those in waterlogged soil, on branches continually pressed against moist ground, beneath the loosened bark of trees damaged by lightning (Plate 239) or freezing, and on the trunks of trees whose original roots were buried beneath additional soil.

The propensity to form adventitious roots varies greatly from one tree species to another. Willows are well known for their rooting ability, although they usually produce roots of more or less normal arrangement, unlike the dense mat shown in Plate 241C and F. A dense mat of adventitious roots is likely to be found where the roots grow in a confined space such as beneath bark.

Most adventitious roots form in response to stimuli by noninfectious agents. There is, however, one infectious disorder called hairy root that affects various plants and could be confused with wound-induced adventitious roots. Hairy root is caused by the bacterium *Agrobacterium rhizogenes*. Further information about it is presented with Plate 73.

Galls and burls. *Gall, tumor,* and *burl* are all terms used to denote swellings, usually but not necessarily on stems. *Burl* usually means a large woody swelling that is more or less hemispheric. *Gall* and *tumor* are terms applied to swellings of various sizes and shapes, woody or not. Galls and various other stem swellings may be induced by bacteria, fungi, insects, mistletoes, and environmental insults. Many stem swellings, including all those shown here, are unexplained.

The best known unexplained stem galls are those found on various conifers: white spruce in interior Alaska and along the coasts of Maine and the Maritime Provinces, Sitka spruce in western Canada, and lodgepole pine and other conifers in the northern Rocky Mountains. These tumors vary from tiny protuberances to globose masses nearly a meter in diameter, many times the diameter of the stem. They apparently originate in single cells near the pith and enlarge for many years. Because of their high incidence in some localities, they have been called epidemic, but no evidence for an infectious cause has been presented.

Burls often bear many buds or in some cases sprouts. When such burls are sawn and the sawn face smoothed, the wood grain is seen to swirl around each bud trace. This highly figured wood (burlwood) is prized by woodworking artisans. Burls on black cherry, sugar maple, redwood, and black walnut are highly valued.

The term *burl* has also been used to mean to any stem swelling characterized by swirling wood grain, regardless of the presence of buds. Freeze-induced burlwood, for example, may arise on small stems at the edges of cankers caused by freezing.

Some woody species produce burls with many buds (bud burls) at the root collar, partly or wholly subterranean, as an adaptation for regrowth after injury to the stem. If produced as a normal part of plant development, such burls are called lignotubers. They are common on birches, eucalypts, mountain laurel, manzanita, and rhododendron. Bud burls may be seen occasionally also on littleleaf linden and Norway maple.

Witches'-brooms. Brooms may be caused by fungi, mycoplasmalike organisms, mites, mistletoes, and environmental insults that kill growing points, resulting in the proliferation of new shoots. Many witches'-brooms on conifers and some on angiosperms apparently result from mutations in vegetative cells rather than irritation by any pathogen. Some brooms have been propagated by horticulturists for development of dwarf cultivars. The brooms shown here on pines illustrate the compact growth habit and dense foliage that propagators foresee in their new cultivars.

Pines with large brooms near their tops often decline and die prematurely even though no infectious agent is known to be present. The Scots pine shown in Plate 241D died a few years after the photograph was made. From a comparison of the vigorous broom with the sparse foliage on other branches, it seems likely that such a broom acts as a metabolic sink that retains or diverts to itself photosynthetic products that would normally be used for growth of other parts of the tree.

References for fasciation, graft abnormalities, and chimeras: 534, 748, 790, 1249, 1346, 1392, 2262, 2271, 2272
References for adventitious shoots and roots, galls and burls, and brooms: 228, 448, 923, 948, 1392, 1505, 1915, 1956, 2044, 2123, 2168, 2169, 2271, 2272

A. Dieback and adventitious sprouts on chestnut oaks. Defoliation by insects triggered the dieback, which in turn triggered the production of sprouts along the trunks (NY, Sep).
B. Burls and adventitious sprouts on a mature Norway maple. The cause of these abnormalities is unknown (NY, Mar).
C, F. Adventitious roots on a weeping willow. The roots began to develop between bark and sapwood after an unknown agent (probably either lightning or freezing) killed the vascular cambium around most of the butt of the tree. After the bark split open or was removed, the root tips dried and ceased elongation (NY, Aug).
D. An unusually large witches'-broom, not known to be caused by an infectious agent, on a mature Scots pine. The broom is apparently being nourished at the expense of the rest of the tree, which has begun to decline (NY, Apr).
E. Multiple burls, cause unknown, on the trunk of a mature hop hornbeam (NY, Aug).
G. A witches'-broom, viewed from below to reveal its form, on eastern white pine. Slight swelling of the supporting branch near the base of the broom indicates the original location of a presumed mutation in a meristem. Compared with normal branches, the broom has short stout twigs with a compact branching habit. One-year-old needles in the interior of the broom had turned brown (natural for the season) and were being shed at the time of photography (NY, Oct).
H. Large burls of unknown cause on the butt and trunk of a California pepper tree (CA, May).
I. A burl developing on the trunk of a young sugar maple (NY, May).

Plate 241

Bark Formation and Restoration (Plate 242)

Bark includes all tissues outside the vascular cambium. Inner bark (phloem) close to the cambium is the site of conduction of photosynthetic products and growth regulators, and corky outer bark serves as an inert protective layer. Between these zones is a layer consisting primarily of phloem parenchyma and the cork cambium. The cork cambium builds up the thickness of corky bark and restores the continuity of the cork layer after it is breached by wounds. The processes of formation of corky outer bark and restoration of a cork barrier between normal tissues and those wounded or diseased are the same and are homologous with compartmentalization of wounds or infections in xylem (Plates 162–163). Often after wounding or infection, corky outer bark is not renewed fast enough to prevent further damage by drying, or the process is disrupted by an insect or pathogen. Occasionally patches of normal corky bark on previously smooth green stems are mistaken for evidence of disease. This plate presents an overview of bark formation and responses of living bark to wounding or infection.

We begin with green bark like that pictured in panels A and B. Viewed in cross-section (panel C), the successive major tissues outside the vascular cambium (vc) are conductive secondary phloem (csp) consisting of sieve tubes, companion cells, phloem parenchyma, and phloem fibers (absent from white pine); nonconductive secondary phloem (nsp) formed in previous years, consisting of parenchyma cells, phloem fibers, and crushed remains of sieve tubes and companion cells; cortex (co), a parenchymatous tissue that was formed during the 1st year of stem growth and served for support and protection before wood and bark formed; the cork cambium (cc), known botanically as phellogen; and a layer of cork (ck), or phellem, derived from the cork cambium. The cork cambium arises in the cortex where a sheet of mature parenchyma cells returns to meristematic condition and the cells begin to divide. The cork cambium and cork collectively constitute a periderm that is covered by imperceptible remains of the original epidermis. The periderm, a barrier more or less impervious to water and microorganisms, protects tissues beneath.

Now consider the brown patches and the transition to mature bark (C and D). If we move outward from the vascular cambium as before, we encounter a new cork cambium that formed from parenchyma cells in the nonconductive secondary phloem beneath the original cork cambium. The edges of this new sheet of cork cambium meet the original cork cambium at the edges of the brown patch. All cells beyond the new cork cambium now constitute a plate or scale of dead outer bark. The plate consists of a new cork layer, to the outside of which are a zone of dead secondary phloem and cortex and the original periderm. As successive bark plates form, their edges overlapping, the stem becomes encircled by corky bark. Thereafter, new periderms form more or less regularly in the nonconductive secondary phloem, preventing indefinite thickening of that tissue and causing indefinite thickening of the outer corky bark. Bark plates are relatively inelastic and crack or flake off as the stem expands, producing the characteristic texture of mature bark.

Whenever a portion of the cork cambium becomes nonfunctional through wounding (as in panel E), infection, or natural senescence, this triggers the formation of a new periderm that restores the continuity of the protective layer. When the cork cambium senesces or is breached, the underlying parenchyma cells begin to senesce and die, producing fungitoxic metabolites as their final function. Parenchyma cells somewhat more distant secrete lignin and suberin in their walls before they too senesce and die. These changes occur within a few days after wounding (longer in cold weather) and create a temporary protective tissue that impedes water loss and is refractory to organisms (impervious tissue, labeled *it* in panel E). The new cork cambium differentiates after 2 to many weeks, depending on season, at the inner edge of the impervious tissue.

If a wound or infection reaches the vicinity of the vascular cambium, this meristem participates in restoration of the phloem and cork cambium. If the vascular cambium is impaired or breached, impervious tissue and periderm formation extend to the sapwood, and impervious tissue extends into the wood. There, through the activities of xylem parenchyma cells, wood elements become plugged with gums and toxic metabolites, and the walls of wood parenchyma cells become impregnated with suberin. Impervious tissue in xylem constitutes the boundary of a compartment within which nonfunctional xylem is isolated from normal tissues.

Leaf scars are natural wounds that are sealed off by the same processes discussed above. Xylem tracheary elements in the leaf trace become plugged, and parenchyma cells beneath the separation zone (where the leaf breaks off) produce first an impervious tissue and then a periderm that separates the surface of the scar from underlying normal tissues.

References: 173–75, 178, 904, 1151, 1366, 1727

A. Stages in formation of mature bark on a London plane tree. The smooth green bark has only a thin periderm overlying living cortex. Brown patches in otherwise smooth bark are zones where a periderm has formed within the cortex, separating senescing superficial tissue from inner bark. Cortical cells to the outside of the cork sheet have died and turned brown. This is the beginning of a bark scale. Mature bark scales at right have been pushed outward during trunk growth (NY, Dec).

B. Brown patches on the otherwise green bark of eastern white pine represent early stages of formation of mature bark (NY, Dec).

C. Magnified view of a cross-section through the pine bark shown in (B). Successive tissues from the bark surface inward in the green zone are: fp (first periderm, covered by remnants of the original epidermis), co (cortex composed of parenchyma), nsp (nonconductive secondary phloem), csp (conductive secondary phloem), vc (vascular cambium), and x (xylem). Beneath each brown zone, a new periderm separates living from dead secondary phloem. Cork of this periderm was generated by a cork cambium·(cc), which differentiated from phloem parenchyma cells. Resin pockets (rp) are visible in both living and brown tissues.

D. Magnified view of a cross-section of mature bark of eastern white pine. A thick plate of dead bark consisting of alternating zones of cork and secondary phloem has built up from successive events of periderm formation that isolated zones of nonconductive secondary phloem. The photograph was made at the edge of a crevice in the outer bark. The inner bark has remained at more or less constant thickness because of annual production of a layer of conductive phloem derived from the vascular cambium (NY, Oct).

E. Microscopic view of a cross-section of bark of a young eastern white pine stem, showing the isolation of a wound by formation of a periderm in the cortex (co) and secondary phloem (sp). The stem was wounded in early August and was sectioned and examined 3 weeks later. At the time of sampling, tissues near the wound were dead, as revealed by red staining in contrast to the blue of normal tissues. A cork layer (ck), visible as a thin layer of cells in regular tiers, was forming at the margin of the dead area, isolating it from normal tissues. Although the cells actually damaged during wounding were restricted to the wound edges, tissue in a much larger zone responded to the wound. Impervious tissue (it) developed around the wound as a temporary protective layer, and cells in that tissue died as the new periderm differentiated. The vascular cambium (vc) was not affected by the wound (magnification 75×; NY, Aug).

Plate 242

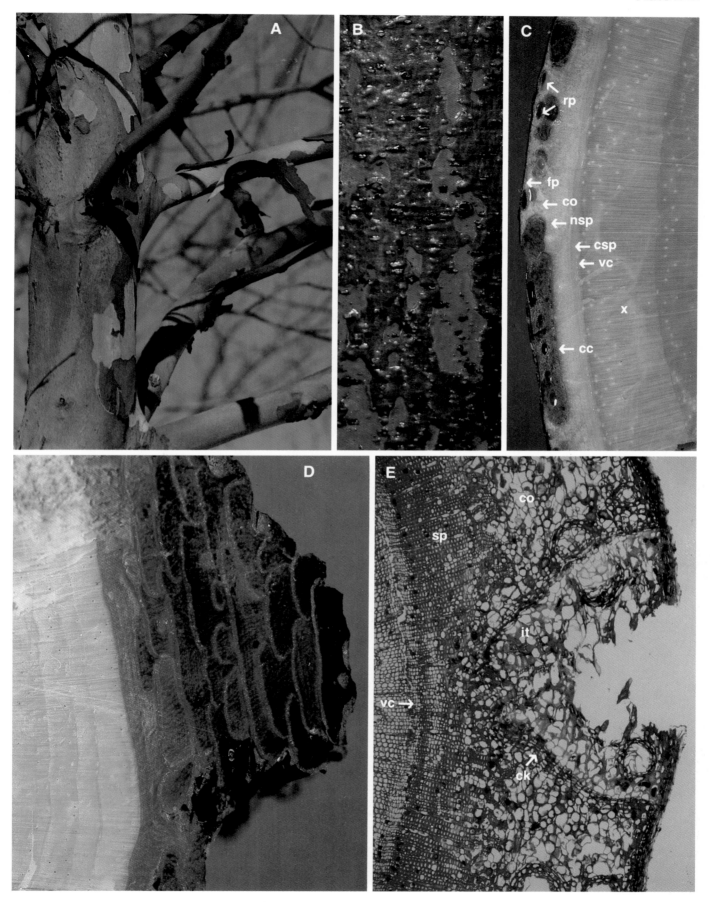

Color Changes Associated with Foliar Shedding (Plate 243)

All trees and shrubs renew their foliage annually, producing new leaves and then shedding old ones. The leaves of deciduous plants live only a few months, but those of evergreens live from one to several years, depending on the species. After producing new leaves, evergreens shed the oldest ones. Usually the old leaves turn yellow and then straw colored to brown before dropping. This display of normal senescence is often mistaken for malfunction.

Conifers normally shed their oldest foliage in autumn. Nearly all pines bear their needles in fascicles of two to five, and the needles remain together when they drop. If the preceding growing season or winter was particularly stressful, some needles may be shed in spring, but most drop later. Arborvitae and related species bear scalelike leaves covering tiny branchlets, and these trees shed the oldest branchlets in autumn. In conifers such as spruces and firs that bear several age classes of needles, shedding is not rigidly restricted to the oldest age class, although it is concentrated there. The foliage on a given branch segment may thin progressively over 2–3 years.

Broad-leaved evergreens vary in habits of foliar shedding. Some drop their old leaves as soon as new ones grow, while other species wait until autumn. The timing is often influenced by environmental stress. Early shedding may occur after a severe winter.

A, B. Normal autumn browning of red pine. Third-year (2-year-old) needles have turned brown and are beginning to drop (NY, Oct).

C. Yellowing and dropping of 2nd-year (1-year-old) leaves of southern magnolia. This species normally drops its 2nd-year leaves throughout spring and into summer. The yellow leaves in this case had been stressed by unusual winter cold and were dropping somewhat prematurely (GA, May).

D. Normal autumn browning of eastern arborvitae. Instead of shedding the oldest individual scale-leaves, this species sheds the oldest branchlets (NY, Oct).

Plate 243

Symbiotic Relationships of Roots (Plates 244–245)

Mycorrhizae. The term *mycorrhiza* means *fungal root*. A mycorrhiza is an absorbing organ composed of both root and fungal cells. It is the site of a symbiotic relationship between a plant and an ecologically specialized root-infecting fungus. The mycorrhizal relationship has been described as "balanced reciprocal parasitism." The fungal partner (mycobiont), unable to fix carbon as do photosynthetic organisms, obtains carbohydrates, vitamins, and other organic compounds from the plant. The plant receives from the fungus various mineral nutrients, especially phosphorus, that would be less available to it from soil via its own absorptive capability. The mycorrhizal fungus also enhances the plant's tolerance of environmental extremes and its resistance to or tolerance of pathogens. Most plants in the wild, including all trees and shrubs, are mycorrhizal, and many species are obligately so; that is, they cannot grow for long or complete their life cycles in a nonmycorrhizal state. Similarly, most mycorrhizal fungi are unable to persist in nature as saprobes. Some are obligately symbiotic and cannot be cultivated apart from living rootlets. Several thousand fungal species enter mycorrhizal associations.

All mycorrhizae consist of a root or rootlike structure plus external (extramatrical) mycelium that extends into the soil or other root medium. Most extramatrical mycelium is lost when mycorrhizae are removed from soil. Only primary plant tissues, usually the cortex of young rootlets, are colonized by mycorrhizal fungi. The stele and likewise the apical meristem of the root remain uninfected. In woody plants, usually only short roots, as opposed to those that grow extensively through the root medium, become mycorrhizal. The mycobiont may produce hormones that induce changes in root form and extend the life of cortical cells. Mycorrhizal roots of some types are not perceptibly different in form from uninfected rootlets, while other types are more or less swollen and highly branched.

Mycorrhizae are classified according to their anatomy, which in some types is linked to a particular group of plants or fungi. Endomycorrhizae (not shown) are those in which the fungal partner colonizes the root intracellularly, forming hyphal coils or arbuscules (highly branched absorptive structures) and vesicles within living cortical cells. Intracellular hyphal coils and arbuscules are eventually digested by the plant. Usually the endomycorrhizal root differs little from uninfected roots in form or color, and the endomycorrhizal condition can be detected only by microscopic examination. Major subgroups of endomycorrhizae include the vesicular-arbuscular (V-A) type, possessed by most plants, and ericoid mycorrhizae, formed by plants in the Ericales. Ectomycorrhizae are characterized by intercellular colonization and often envelopment of short roots by fungal hyphae. Living cortical cells become separated (except possibly for connections via plasmodesmata) by a network of hyphae called the Hartig net. This is continuous with a mantle of hyphae over the root surface. Ectomycorrhizae often become swollen in comparison with nonmycorrhizal short roots. Some develop elaborate branching, and those with thick mantles are often distinctively colored. Ectendomycorrhizae and arbutoid (for *Arbutus* and related plants) mycorrhizae have characteristics of both ecto- and endomycorrhizae.

Most plants are endomycorrhizal (some trees such as poplar and willow may form either ecto- or endomycorrhizae), and the great majority have V-A mycorrhizae. The V-A mycobionts are aseptate and reproduce asexually by chlamydospores or azygospores. They resemble members of the Zygomycete genus *Endogone* and therefore are usually referred to the family Endogonaceae. Many species produce globose chlamydospores that are gigantic in comparison with spores of other fungi, attaining diameters of $100–300$ μm or more. V-A mycorrhizal fungi are obligately symbiotic in nature and at the time of this printing had not been induced to grow on a laboratory substrate. Their spores germinate readily on various media, but no colony develops. V-A mycorrhizal fungi are dispersed with infected plants and as spores or hyphae with soil. In general they are unspecialized with respect to the identity of the plant partner. Any V-A mycorrhizal fungus can form mycorrhizae with many different plants, both herbaceous and woody, and any V-A mycorrhizal plant can form mycorrhizae with various fungi.

Ectomycorrhizae, featured on this plate, are formed only with certain groups of trees, notably conifers and members of the beech, birch, linden, and willow families, that grow in regions of temperate climate. The mycobionts are mostly mushroom-forming basidiomycetes (agarics and boletes, Plate 245) that are dispersed in nature as airborne basidiospores. They are also inadvertently dispersed in commerce as mycelium in mycorrhizal plants. Some ectomycorrhizal fungi are suitable partners for only a few species of trees, while others enter symbiosis with many diverse trees. Similarly, some trees are capable of forming ectomycorrhizae with hundreds of fungi. Mature trees are usually mycorrhizal with several fungi simultaneously. The diversity of mycobionts in the roots increases with a tree's age. Many or perhaps most ectomycorrhizal fungi are capable of growing in pure culture, but such growth in most cases is slow compared to that of saprophytic fungi.

The mycobionts of ericoid and arbutoid mycorrhizae are ascomycetes and basidiomycetes, respectively, and are cultivable apart from their plant partners.

Production of fruit bodies by a mycorrhizal fungus depends upon first developing an extensive mycelium and then extracting enough carbon and nitrogen from the plant to support reproductive growth. Fungi that are ectomycorrhizal with seedlings begin to fruit as early as the 2nd year of seedling growth and may fruit annually thereafter, the abundance of sporocarps varying with weather.

A–C. Magnified views of rootlets from 2-year-old seedlings of red pine. A. Ectomycorrhizal short roots, exhibiting the distinctive dichotomous branching typical of mycorrhizae on pines. These ectomycorrhizae, formed with an unknown fungal symbiont, extend about 2 mm from the mother root. B. Typical unbranched nonmycorrhizal short roots. C. Nonmycorrhizal short roots stimulated to unnatural extension by high fertility (NY, Aug).

D. Magnified view of mycorrhizal short roots on a 1st-year Douglas-fir seedling. A portion of the long root has been colonized by the ectomycorrhizal fungus *Laccaria bicolor,* which stimulated the formation of short lateral roots. All of these are ectomycorrhizal, as indicated by their relatively large diameters and rounded tips. The tip of the long root is nonmycorrhizal (WA, Aug).

E, G, H. Ectomycorrhizae formed by mugo pine roots in association with the fungus *Pisolithus tinctorius.* E. Young ectomycorrhizae appear white because of the mantle of colorless mycelium of the fungal symbiont. Mounted in water for photography, the mycelium lost the yellow to buff-brown pigment typical of *P. tinctorius.* These mycorrhizae, if allowed to continue development, would soon have become dichotomously branched. G, H. Successively magnified views of ectomycorrhizae at the surface of the root ball of a seedling grown in a container. Particles of the root medium (peat moss, vermiculite, and quartz sand) serve as size standards. Buff-brown strands of mycelium extend from the mycorrhizae into the medium (greenhouse plants).

F. Magnified view of typical black mantles formed by the fungus *Cenococcum geophilum* in ectomycorrhizal association with mugo pine. Stout black hyphae extend from the mycorrhizae (greenhouse plants).

I. Six-month-old Douglas-fir seedlings, mycorrhizal with the fungus *L. bicolor* (left) and nonmycorrhizal, showing the growth suppression that typically occurs in nonmycorrhizal conifers (greenhouse plants).

Plate 244

Mycorrhizal fungi absorb essential nutrient elements from soil and translocate them to roots. The accumulation of phosphorus by mycorrhizae is striking. Although total phosphorus in most soils is adequate to support plant growth, much of the supply is bound in insoluble forms that are scarcely available to nonmycorrhizal roots. Mycelium extending from a mycorrhiza into soil has a much greater absorbing surface and exploits a greater volume of soil and thus obtains phosphorus more readily than would the nonmycorrhizal root. Given adequate soil fertility, mineral nutrients other than phosphorus can usually be absorbed by nonmycorrhizal roots in amounts sufficient for normal growth, but mycorrhizae enhance this uptake. Mycorrhizal fungi also mingle with litter-decomposing organisms in the humus layer and there obtain nutrients from organic sources that would be unavailable to the nonmycorrhizal plant.

In many horticultural circumstances, fertilizer can be substituted for the nutritional benefits of mycorrhizae. The key nutrient is phosphorus in a form that can be dissolved by the weak acids associated with roots. High fertility levels such as those commonly used to promote plant growth in nurseries often mask any mycorrhizal contribution to plant nutrition and in some cases inhibit mycorrhizal formation or select for mycorrhizal fungi that tolerate high fertility. This varies with strain of mycorrhizal fungus.

Mycorrhizae enhance the tolerance of plants for adverse conditions such as drought, high temperature, salinity, acidity, or toxic elements in soil. The degree of enhancement varies with the mycobiont. It has been demonstrated most extensively in pines ectomycorrhizal with a puffball fungus, *Pisolithus tinctorius*. Enhanced environmental tolerance is important for successful revegetation of disturbed sites such as mine spoils and landfills and also for many planting sites in horticultural landscapes. Regarding drought tolerance, the mycobiont obviously cannot extract much additional water from a nearly dry soil. Rather, the mycobiont mediates resistance to water flow from soil into plant. The mycorrhizal plant maintains less resistance than does the nonmycorrhizal plant and is therefore more effectively able to utilize soil water.

Many reports exist of plant tolerance or resistance to pathogens as mediated or modified by mycorrhizal fungi. Most reported mycorrhizal effects are beneficial to the plant, but cases of increased susceptibility or damage have also been noted. Enhanced tolerance to pathogens presumably follows from the superior nutrition and water economy of the mycorrhizal plant. In addition, some ectomycorrhizal fungi produce antibiotics toxic to fungal pathogens, and various mycorrhizal fungi induce some degree of plant resistance to pathogens. In practical agriculture or forestry, however, there are no reports of disease control as the result of manipulating mycorrhizal fungi.

Practical applications of mycorrhizal technology are still limited in spite of burgeoning research in recent decades. Many writers have emphasized the necessity of mycorrhizae and the stimulation of plant growth by mycorrhizal fungi, but these concepts are misleading with respect to growth of trees and shrubs because the mycorrhizal condition is normal. A more useful concept is that mycorrhizal fungi vary in ability to colonize plants, to influence plant growth, and to condition tolerance or resistance of plants to various stressing factors. This concept is the basis of efforts to enhance plant performance by artificial inoculation with mycorrhizal fungi. The movement toward artificial inoculation is slow, however, because it is difficult to show cost effectiveness. Indigenous fungi enter mycorrhizal association with woody plants at most planting sites, so a mycorrhizal condition is nearly assured. Mycorrhizal deficiency is unlikely except in soils that have recently been fumigated or where trees have never grown or where topsoil has been removed. An introduced mycobiont must induce survival, growth, or yield that is clearly superior to that ex-

pected with native fungi, and the difference must be great enough to result in economic value that will offset the cost of developmental research and marketing of mycorrhizal inoculum. To date these conditions have not been met.

Actinorhizae. Nitrogen-fixing nodules called actinorhizae develop on the roots of diverse plants as the result of infection by actinomycetes in the genus *Frankia*. These nodules are perennial, coralloid structures as much as 3–4 cm in diameter, consisting of many lobes. Actinorhizae have been studied mainly on alder but also occur on antelope bush, bayberry, buffalo berry, casuarina, ceanothus, mountain mahogany, Russian olive, sweetfern, and other plants. Nodules on alder may live 3–8 years as the symbiotic microorganism, or endophyte, perennates within the lobes. Actinorhizae fix much more nitrogen than is used in their own growth and maintenance; the remainder is translocated to other parts of the plant. This mechanism allows nitrogen-fixing trees or shrubs to colonize nitrogen-poor sites. Nitrogen is then recycled at the site, becoming available to other plants there.

Infection leading to formation of actinorhizae occurs via root hairs, within which thin actinomycete hyphae develop. Cortical cells near the site of infection are stimulated to begin division, and these cells are soon entered by the endophyte. The root swells at the infection site, and the swelling (called a primary nodule) appears reddish if viewed in transmitted light. A lateral root is induced at the swelling, and it is immediately transformed into a true nodule, its cortex infected by the endophyte. The apical meristem of the nodule remains uninfected, and branching from this point results in the complex actinorhizal form. The endophyte forms intracellular hyphae, vesicles, and, in some strains, spores. Vesicles are the actual sites of nitrogen fixation. Spores often persist in old parts of the nodule where hyphae and vesicles have been digested by host cells. As lobes of a nodule decay, spores are released to soil, where they are capable of initiating new infections. Strains of *Frankia* show considerable plant specificity, and they vary in infectivity and efficiency of nitrogen fixation.

Nitrogen-fixing nodules on legumes. Symbiotic soilborne bacteria in the genus *Rhizobium* induce nodules on legumes. The bacteria enter a young root through root hairs and produce a strand that extends into cortical cells, which are stimulated to meristematic activity that results in nodule formation. Bacteria released within cortical cells of the developing nodule differentiate into swollen bacteroids, which are the sites of nitrogen fixation. Nodules on woody legumes apparently live more than 1 season. Strains of *Rhizobium*, like *Frankia*, are specialized for particular plants and vary in infectivity and efficiency of nitrogen fixation. A strain symbiotic with an agricultural plant would not be suitable for a leguminous tree. Trees with bacterial nodules are capable of fixing a significant quantity of nitrogen. In one study, 4-year-old stands of black locust were estimated to fix 30 kg per hectare per year. Virtually all of this fixation occurred during the growing season. The fixation process is inhibited at low temperatures.
References for mycorrhizae: 495, 509, 736, 1217, 1228, 1230, 1664, 1676, 1700, 1922, 1955, 2030
References for actinorhizae and rhizobial nodules: 218, 262, 509, 641, 661, 2028

A–F. Basidiocarps of ectomycorrhizal fungi. A. *Laccaria* sp. in association with Douglas-fir and ponderosa pine. One mushroom is upended to reveal the lower surface of its cap (AZ, Jul). B. *Suillus subaureus* fruiting near eastern white pine, eastern hemlock, and white oak (NY, Sep). C. *Amanita muscaria* in association with Douglas-fir and ponderosa pine. This fungus, known also as the fly agaric, is deadly poisonous if eaten by humans (AZ, Jul). D, F. *Suillus* sp. in association with ponderosa pine. The poroid lower surface of the basidiocarp is revealed in (F) (AZ, Jul). E. Basidiocarps of the panther fungus, *A. pantherina,* in association with Douglas-fir and ponderosa pine. This fungus is also poisonous to humans (AZ, Jul).
G. Actinorhizae on an alder root. Feeder roots of the alder, at lower left, provide a size standard (NY, Jul).
H. Nitrogen-fixing nodules formed by *Rhizobium* sp. on rootlets of black locust. Typical nodules are 3 mm long (NY, Oct).

Plate 245

Epiphytes and Lichens (Plates 246—247)

Epiphytes. Epiphytes are photosynthetic plants that grow upon other plants. They usually attach to the bark of trees by means of roots or rhizomes, gaining support from the tree but not parasitizing it. The water needs of epiphytes are met by rain or by water vapor in saturated air. Essential nutrients reach the epiphyte in water that has leached leaves and branches above, from air contaminants, and through microbial activity in organic debris that accumulates at the site of attachment.

Except for the mosses that are abundant on trees in the North, epiphytes occur mainly in warm to tropical regions, occupying various ecological niches that range in light exposure from full sunlight to deep shade. Epiphytic species have evolved in several plant families, notably the Bromeliaceae (pineapple family) and Orchidaceae.

Epiphytes that can utilize full sunlight sometimes develop luxuriantly on dead or dying trees and for this reason have often been assumed to be parasitic. In the southern United States from Texas eastward, Spanish moss, or graybeard, *Tillandsia usneoides* (Bromeliaceae), is a prominent example. In form, habit, and color it is unlike most seed plants, and to inexperienced observers it looks "as if it ought to be parasitic." Its stems are slender, branched, and pendant, attaining lengths of 6 meters or more. Its leaves are scattered, linear, up to 5 cm long. The flowers are borne inconspicuously in leaf axils. Some observers have noted diminished growth or subnormal leaf size on tree branches draped with Spanish moss and on this basis have suggested that the epiphyte is injurious, if not parasitic, but proof is lacking.

Ball moss, or bunch moss, *T. recurvata,* is another epiphytic bromeliad common from Texas eastward. This plant is found on bark, twigs, wires, poles, and even chain link fences. Ball moss appears as gray-green tufts that develop into dense clusters composed of numerous individual plants. Its stems are branched but short, and the ash-colored to reddish scaly leaves grow tc lengths of about 5 cm. When it grows on rough bark, ball moss produces rootlike holdfasts for attachment. Its small blue flowers and later the seeds are produced on stalks 8—10 cm long. The seeds are windborne. Plants found on wires have likely been dislodged from some other surface, for wires would seem to be inhospitable sites for seed germination. Ball moss is so unwanted in some areas that copper fungicides have been used to suppress it.

A, B. Spanish moss on slash pines. A. Abundant growth of the epiphyte on a dying tree in contrast to slight colonization of a healthier tree. B. Close view of the epiphyte on a small pine branch (FL, Mar).
C. *Tillandsia recurvata,* an epiphyte known as ball moss, on a twig of Australian pine (FL, Mar).
D. Lichens on branches of a dead white spruce give a gray color to the entire tree (Ont, Jul).
E. Two unidentified lichens on bark of a living pin cherry (Ont, Jul).

Plate 246

Lichens. A lichen is a perennial compound organism consisting of a fungus (usually an ascomycete) and a green or blue-green alga in symbiotic union. The terms *mycobiont* and *phycobiont* are often used for the fungal and the algal partner, respectively. The entire structure, called a thallus, is long-lived and morphologically unlike either algae or nonlichenized fungi. Fungal and algal components in the vegetative part of a lichen thallus can be distinguished from one another only by microscopic examination.

Lichens colonize various surfaces, often rocks and bark. The alga, through photosynthesis, supplies the fungus with carbohydrates and vitamins, and the fungus obtains water and minerals essential for both symbionts from the air and from the substrate. Some blue-green algae in the thallus or associated with it fix nitrogen that is utilized by both symbionts. In other lichens, nitrogen is obtained from leachate from plants, from bird excrement, or from organic debris.

Lichen thalli take various forms. Crustose types are appressed to the substrate. Foliose lichens have leaflike lobes borne above the substrate. Fruiticose lichens have linear, branched thalli that may be fingerlike, strap shaped, or hairlike. Examples of all three types are shown here. The mycobiont dominates the thallus, at least in terms of relative mass and abundance. The surface layer of the thallus, called the cortex, consists of gelatinous fungal hyphae that in some species are covered with an amorphous sheet of polysaccharide, perforated to permit gas exchange. Beneath the cortex is the medulla, where hyphae are relatively loosely arranged and less gelatinized. The medulla is a storage site for carbohydrate, lichen substances (secondary metabolites peculiar to lichens), and water. Cells of the phycobiont are typically arranged in a layer between the cortex and the medulla. Many lichens also have surface protrusions consisting of both algal and fungal cells.

Lichens reproduce vegetatively. Various substructures or fragments of thalli containing both symbionts are dispersed by wind or water and have the ability to colonize new substrates. Some lichens also reproduce by reassociation of the separate partners after spore dispersal, but this is a little-known phenomenon.

Lichens are classified primarily on the basis of morphological characteristics of the mycobionts and secondarily on the basis of thallus form and chemical components. Lichens bear the Latin names of their mycobionts; the identity of the phycobiont is not considered. Indeed, one species of alga can be the phycobiont in taxonomically unrelated lichens, and some lichens contain two different phycobionts.

Inexperienced observers often wonder whether lichens growing profusely on dead trees are pathogenic. One researcher studied a species of *Usnea* on bark of living trees and observed fungal hyphae, which he presumed to be those of the mycobiont, beneath the cork cambium. On this basis primarily, he concluded that the lichen was parasitic. Plant parasitism, however, is not among the lichen attributes discussed by contemporary authorities. Except for lichenized forms (*Strigula* species) of plant parasitic algae (*Cephaleuros* species, Plate 9), no claim of plant pathogenicity has been well supported. Many lichens grow rapidly when fully exposed to light, which seems to explain their profusion on dead trees.

References for epiphytes: 577, 1137, 2055
References for lichens: 716, 1115, 1522

A. Lichens on dead branches of an unidentified plant. The large foliose lichen along the left branch is *Parmotrema perforatum*. The lichens with many fine branches are primarily *Ramalina montagnei* and *R. willeyi* (MS, Aug).
B. Epiphytes (*Tillandsia* sp.) and lichens on a young oak. The red crustose lichen, *Chiodecton sanguineum*, is common on bark of many trees in swamp and floodplain habitats in the South and has been collected along the Atlantic coast as far north as southern Delaware (FL, Mar).
C. A lichen, *Hypogymnia physodes*, on a small branch of red pine (NY, May).
D. Lichens (*Usnea* sp.) hanging in festoons on a dead Douglas-fir (ID, Jul).

Plate 247

abiotic—lacking life or never having been alive.

abscission—of plants: the shedding of leaves or other parts as the result of physical weakness in a specialized layer of cells (abscission layer) that develops at the base.

acervulus (pl. -li)—a fruiting structure of certain microfungi, consisting of a layer of hyphae that bear conidiophores, lacking specialized wall structure, arising subcuticularly, subepidermally, or deeper in plant tissue, colorless to dark, visible with a hand lens and sometimes with the unaided eye, and often appearing like a tiny blister that opens wide at the plant surface (illustrated in Figures 10–12).

aeciospore—of rust fungi: a nonrepeating, asexual spore, often yellow to orange in mass, borne in chains in an aecium and incapable of infecting the host on which it is produced.

aecium (pl. -cia)—a fruiting structure that produces aeciospores of a rust fungus; forms after the spermagonium and before the uredinium in the life cycle, often cup shaped or in some cases blisterlike or tubular, often with colorless (white) walls.

agent of disease—an organism or abiotic factor that causes disease; a pathogen.

aggressiveness—of a plant pathogen: relative ability to colonize and cause damage to plants. See also *virulence*.

anaerobic—living, acting, or occurring in the absence of oxygen.

annual canker—see *canker*.

apothecium (pl. -cia)—the fruit body of certain Ascomycotina that are commonly called Discomycetes or cup fungi, in which the fertile layer of asci is exposed by a wide opening. Apothecia vary from linear and embedded in the substrate to sessile or stalked structures that may be shaped like a cup, saucer, or wine glass. In size they vary from nearly microscopic to several centimeters in diameter and height (see Figures 4–6).

appressorium—of fungi and some parasitic seed plants: a swollen or flattened portion of a germ tube (of fungus) or radicle (of seed plant), serving for attachment to the host, from which often the parasite penetrates the host.

arbuscule—of vesicular-arbuscular mycorrhizae: a much-branched, microscopic haustorial structure of the fungal symbiont that forms within living cortical cells of the root. The interface of arbuscule with plant protoplast is a site of exchange of nutrients and growth-regulating chemicals.

areole—of cacti: a small pit or cavity on the surface, from which a spine may arise.

arthrospore—of fungi: a spore resulting from the division of a hypha into separate cells.

ascigerous—of fungi: having asci.

ascocarp—a fruit body containing asci; the sexual fruit body of a fungus belonging to the Ascomycotina.

Ascomycotina—fungi that produce asci and ascospores; one of the major divisions of fungi; sac fungi.

ascospore—a spore produced within an ascus.

ascostroma (pl. -ata)—of certain Ascomycotina: a mass of hyphae within or on which asci form; often microscopic, dark in color, and partially to wholly embedded in dead plant tissue.

ascus (pl. -ci)—a saclike cell of the sexual state of a fungus belonging to Ascomycotina, in which ascospores form.

autoecious—of a rust fungus: completing its life cycle on one host (see also *heteroecious*).

auxin—any of a group of plant hormones related to indoleacetic acid that promote cell enlargement.

axil—of plants: the cavity or angle formed at the upper side of the junction of a petiole or stem with its parent stem.

axillary—pertaining to an axil, in an axil.

azygospore—a parthenogenetic zygospore; formed by some vesicular-arbuscular mycorrhizal fungi (family Endogonaceae).

bacillar—shaped like a short rod.

basidiocarp—a fruit body of a fungus belonging to Basidiomycotina, bearing or containing basidia.

bracket—of fungi: a shelflike basidiocarp.

broom—in plant pathology: short for *witches'-broom*.

bud trace—the microscopic vascular connection between stem and bud.

callus—of plants: undifferentiated tissue that proliferates at cut surfaces or wound edges; also, organized tissues including xylem, phloem, and periderms, usually growing more rapidly than normal, that proliferate at the edges of wounds and gradually cover wounds.

cambium, cork—a lateral meristem that produces cork.

cambium, vascular—a lateral meristem by growth from which the stem and root increase in girth.

canker—a necrotic lesion in bark of the stem or root, often extending to the xylem; also, the scar left after shedding of bark tissues killed by localized disease or environmental injury.

canker, annual—a canker that enlarges only once and does so within an interval briefer than the growth cycle of the plant, usually less than 1 year.

canker, diffuse—a canker that enlarges without characteristic shape or noticeable callus formation at margins.

canker, perennial—a canker that enlarges during more than 1 year.

canker, target—a canker that includes concentric ridges of callus.

carbonaceous—hard, black, and brittle, resembling charcoal.

cellulolytic—of certain enzymes: able to digest cellulose.

cerebroid—having a convoluted surface similar to that of the cerebrum of the brain.

chlamydospore—a thick-walled, nondeciduous spore that arises by the rounding up of one or more hyphal cells.

chlorotic—abnormally yellowish.

cirrus (or cirrhus, pl. -ri or -hi)—a mass of spores in the form of a ribbon or tendril, forced from the fruit body of a microfungus.

cladode—a stem with leaflike form; on certain cacti, a joint or stem segment.

clamp connection—of many fungi in Basidiomycotina: a clamplike hyphal outgrowth that at cell division connects the resulting two cells by fusing with the lower one.

cleistothecium (pl. -ia)—an ascocarp that lacks an opening, typical of powdery mildew fungi.

cm—centimeter.

compartmentalization—in trees: the processes that result in isolation of wounded or diseased xylem from normal xylem by the formation of chemically and anatomically specialized tissue around the damaged zone.

conidiophore—of fungi: a specialized hypha that bears or produces conidia.

conidium (pl. -ia)—an asexual spore that, when mature, separates from its conidiophore.

conk—the basidiocarp of a wood-decaying fungus, usually a polypore.

cortex—in plants: the outer primary tissue of a stem or root between the primary phloem or endodermis and the epidermis, composed mainly of parenchyma.

cortical haustoria—of mistletoes: cortical strands.

cortical strands—of mistletoes: radiating strands of mistletoe tissue that grow through cortex and secondary phloem of the parasitized tree stem.

cultivar—cultivated variety.

cuticle—of plants: the layer of waxy substance on the surface of a leaf, fruit, or young stem.

cutinolytic—of certain enzymes: able to digest cutin.

cutis—of basidiocarps of certain wood-decaying fungi: the outer layer consisting of compressed hyphae parallel to the surface, sometimes with varnishlike incrustation.

cv.—cultivar.

cytokinin—any of a group of plant hormones related to zeatin that promote cell division.

damping-off—of germinating seedlings: disease characterized by a lesion at or near the soil line, which prevents a seedling from growing above soil or causes a newly emerged seedling to fall over.

delignification—chemical, usually enzymatic, removal of lignin from xylem, leaving a cellulosic residue.

Deuteromycotina—asexual fungi; one of the major divisions of fungi.

dieback—dead apical parts, usually twigs or limbs; also the process of dying back.

disc—of *Valsa* and related fungi: a more or less flat apical part of a stroma that protrudes above the bark surface; also, of Discomycetes: the exposed fertile portion of an apothecium.

disease cycle—of a disease caused by a biotic agent: the cyclical sequence of host and parasite development and interaction that result in disease, in reproduction or replication of the pathogen, and in the readying of a new generation of the parasite for infection.

distal—of a stem or root: away from the base or origin.

DNA—deoxyribonucleic acid.

dwarf shoot—see *shoot, dwarf.*

effuse—spreading out loosely or flat.

endemic—of a disease: common in a place during an extended time and consistently causing little damage.

endoconidiophore—a conidiophore that produces conidia within itself.

endoconidium—a conidium produced within its conidiophore. Endoconidia are extruded from the tips of their conidiophores.

endophytic system—of mistletoes: the part of the parasite that grows within the host; cortical strands and sinkers.

epicormic sprouts—sprouts arising abnormally along a trunk or limb as the result of release of dormant buds or the differentiation of buds from callus; water sprouts.

epidermis—of plants: the outermost cell layer.

epinasty—downward bending of a turgid plant part, usually the petiole or midvein of a leaf.

epiparasite—an organism parasitic on another that parasitizes a third.

epiphyte—a plant that grows nonparasitically upon another plant.

epiphytic—upon a plant.

ericaceous—of the plant family Ericaceae.

erumpent—breaking through the surface; bursting forth.

extractive—any chemical that may be extracted by means of solvents.

fastigiate—of plants: having leaves or branches more or less parallel and pointed upward.

feeder root—a rootlet that absorbs water and minerals.

filamentous fungus—a fungus that produces hyphae that do not organize into distinct fruiting or resting structures.

first-year needles—of conifers: the age class of needles formed most recently; needles of the current year.

forma specialis—in certain fungi: an infraspecific population of a plant pathogenic species, distinguished by host preference (usually at the genus level) but scarcely or not at all by morphological criteria.

form-genus—in Deuteromycotina (asexual fungi): a genus defined arbitrarily on the basis of similar form of its member species regardless of their natural relatedness.

frass—particles remaining where insects, especially those boring in wood or bark, have chewed plant tissues.

frond—the leaf of a palm or fern.

fruit body—of fungi: a multicellular structure that produces spores.

f. sp.—forma specialis.

gall—a localized overgrowth.

gamete—sex cell; a reproductive cell capable of uniting with another, resulting in a mingling of their cytoplasm and nuclei.

germ tube—a hypha produced by a germinating spore or microsclerotium.

glaucous—having a bluish gray waxy surface.

group species—a cluster of populations that appear similar but are biologically distinct, all known by the same name, that have not been differentiated by traditional taxonomic criteria.

gummosis—exudation of gum.

haematochrome—a red-orange pigment produced by certain algae.

hard pine—a pine with two or three needles per cluster.

haustorium (pl. -ia)—of plant-parasitic fungi: a specialized hyphal branch within a host cell for absorption of nutrients. Of mistletoes and other parasitic plants: a multicellular, usually highly branched structure that differentiates in the cortex and/or secondary phloem of the host for anchorage and nutrient absorption.

heteroecious—of rust fungi: requiring two host species for completion of the life cycle.

hydathode—a tiny gland at the edge of a leaf, from which droplets of water may issue.

hymenium—of fungi: a layer of spore-producing cells, usually in or on a fruit body.

hypha (pl. -ae)—usually, one of the filaments of a fungus; also a linear extension of the haustorium of dodder that makes intimate contact with host phloem sieve cells.

hyphopodium (pl. -ia)—a short branch of one or two cells of the epiphytic mycelium of a black mildew fungus.

hysterothecium (pl. -ia)—an ascocarp that opens by a slit.

incubation period—in plant pathology: the interval between inoculation and display of symptoms.

infect—of a parasite: to begin or continue an interactive, usually pathogenic, relationship with the host.

infection—the interaction of parasite with host; or according to some authors, the beginning of that interaction.

infection court—in plant pathology: the site of the onset of infection; the site of inoculation.

infest—of a pest or pathogen: to populate or inhabit a thing or place.

infestation—the occurrence or development of an infesting population.

inflorescence—a flower cluster.

inoculation—in plant pathology: the placement of inoculum at a site (the infection court) where infection is possible.

inoculum—of a plant parasite: a unit or units capable of infecting.

isolate—in plant pathology: a culture or subpopulation of a microorganism separated from its parent population and maintained in some sort of controlled circumstance; also, to effect such separation and control, for example to isolate a pathogen from diseased plant tissue.

kg—kilogram.

knot—a localized abnormal swelling; a gall.

latent infection—infection unaccompanied by visible symptoms.

leader (of terminal shoots)—the topmost or dominant shoot of a tree or branch.

lenticel—a passage, loosely filled with parenchyma cells, between cortex of a stem or root and the outside environment; visible on the surface as a small, lens-shaped eruption.

lesion—a localized zone of dead or moribund tissue.

lignification—in plant tissue, mainly xylem: deposition of lignin within cell walls.

lignin—a complex structural polymer that imparts rigidity to certain plant cell walls, especially walls of wood cells.

Mastigomycotina—fungi with spores that are motile by means of flagella; one of the major divisions of fungi.

megapascal—10^6 pascals; approximately 7.5×10^3 mm mercury.

meq—milliequivalent weight.

meristem—a layer or zone of undifferentiated cells capable of division to produce more generative cells as well as cells that enlarge and differentiate, resulting in growth.

meristem, intercalary—a meristem located between differentiated tissues; for example, a mistletoe sinker elongates by growth from a meristem near the base of the sinker.

microsclerotium—a tiny mass of thick-walled, dark-colored fungal cells, specialized for survival, capable of germination to produce a mycelium (see Figure 17).

microzoospore—of certain algae: a microscopic, motile spore.

MLO—mycoplasmalike organism.

mollicute—a wall-less prokaryotic, pleomorphic microorganism.

morphologic—pertaining to form.

mycelium—an aggregation of hyphae, usually of a fungus.

mycobiont—the fungal symbiont in a lichen or mycorrhiza.

mycoplasma—a wall-less prokaryotic microorganism of the class Mollicutes.

mycoplasmalike organism—a prokaryotic microorganism resembling a mycoplasma but not yet isolable in pure culture or characterized taxonomically.

nanometer—1×10^{-9} meters.

necrosis—death.

nm—nanometer.

normal length—of virus particles: the most common length, believed to be the intrinsic natural length of a virus that consists of rod-shaped particles.

oleoresin—resin, pitch; a viscous, aromatic mixture of terpenes, resin acids, and fatty acids produced by various conifers.

olivaceous—olive colored.

oospore—a resting spore produced as the result of fertilization or a similar parthenogenetic process in a fungus belonging to Oomycetes.

ostiole—a more or less circular, differentiated pore in a fruit body of a microfungus, from which spores issue.

pathogen—an agent capable of causing disease.

pathogenicity—ability to cause disease.

pathovar—in plant-pathogenic bacteria: an infraspecific population distinguishable mainly or only on the basis of pathogenicity to certain plants.

pectolytic—of certain enzymes: capable of digesting pectic compounds.

peduncle—the stalk of a flower or fruit.

perennial canker—see *canker.*

periderm—a plant tissue external to the cortex and/or phloem of stems and roots, composed of cork and the cork cambium, more or less impervious to water, solutes, and organisms and serving to protect underlying tissues.

perithecium (pl. -ia)—a more or less globose to flasklike ascocarp with a pore at the top and a wall of its own.

peritrichous—of flagella of a microorganism: occurring all over the cell surface.

petiole—leaf stalk.

phellem—cork.

phellogen—cork cambium.

phloem—a plant tissue specialized for conduction of food and growth regulators, lying to the outside of the vascular cambium in stems and roots with secondary thickening.

phloem, secondary—phloem formed by the vascular cambium.

pinna (pl. -ae)—of palm fronds: the individual leaflet.

pl.—plural.

plasmid—a circular piece of cytoplasmic DNA.

polypore—one of a large group of wood-decaying fungi that produce basidiocarps with poroid lower surfaces.

primary cycle—of plant disease: the first cycle to begin in a given year.

primary infection—the first infection in a sequence or in any year.

prokaryote—a unicellular microorganism lacking an organized nucleus and organelles.

prokaryotic—having the characteristics of a prokaryote.

propagule—a structure by which an organism is propagated or multiplies.

pseudosclerotial plate—of certain fungi that colonize wood: a hard, dark plate formed within decaying wood, more or less impervious to water, solutes, and organisms, composed of large, thick-walled or encrusted fungal cells, affording protection to mycelium behind it; appearing as a black line (zone line) in transverse view.

pseudosclerotium—of fungi: a sclerotium that includes recognizable elements of the substrate (usually plant tissue).

pseudostroma—of fungi: a stroma that includes host tissue.

pseudothecium (pl. -ia)—of certain microfungi: a pseudoperithecium; an ascostroma that superficially resembles a perithecium.

punk knot—of a wood-decaying fungus in a trunk or limb: a dense sterile mass of mycelium issuing from the interior of the stem to the surface along a channel once occupied by a branch.

pv.—pathovar.

pycnidium (pl. -ia)—of fungi: a subglobose to cup- or flask-shaped fruit body containing conidiophores, producing conidia.

race—of a pathogen: an infraspecific population distinguishable only or mainly on the basis of its host range or (uncommonly) its aggressiveness.

rachis—the stemlike axis of a compound leaf or inflorescence.

ray, medullary ray—in woody stems and roots: a radial sheet of cells, mainly parenchyma but sometimes also tracheary cells, with their long axes (if any) perpendicular to the stem axis; serving for food storage and conduction from phloem to inner sapwood.

resinosis—exudation of oleoresin.

resistance—of a plant to a pathogen: the ability to retard, suppress, or prevent infection or colonization; also the functioning of attributes or processes that do so.

rhizomorph—a thick strand of fungal hyphae organized and capable of growth as a rootlike unit.

root hair—a hairlike projection from an epidermal cell of a root.

rot, brown—of wood: decay characterized by selective degradation of cellulose and hemicelluloses, leaving a crumbly brown residue rich in undigested lignin.

rot, white—of wood: decay characterized by degradation of lignin at a rate equal to or usually greater than the rate of degradation of cellulose and hemicelluloses, leaving a light-colored residue that usually contains relatively more cellulose and hemicelluloses than did the original wood.

rugose—of a surface: rough or wrinkled.

samara—a one-seeded, winged fruit, as of ash, elm, or maple.

saprobe—a microorganism that lives on and degrades dead organic matter; a saprophyte.

saprobic—pertaining to the microbial degradation of dead organic matter.

saprophyte—a microorganism that lives on and degrades dead organic matter; a saprobe.

secondary cycle—of plant disease: any cycle initiated by inoculum generated during the same season.

second-year, third-year (etc.) needles—of conifers: needles in their 2nd, 3rd (etc.) season of life. A 2nd-year needle is between 1 and 2 years old.

septum (pl. -ta)—in fungi: the wall or wall-like structure between two fungal cells.

seta (pl. -ae)—a stiff hair or bristle, macro- or microscopic.

shoot, dwarf—of gymnosperms such as pine or larch: a lateral shoot consisting primarily of a cluster of needles at the apex of a very short stem.

shoot, short—dwarf shoot.

sign—of a plant pathogen: any visible structure of the pathogen.

sinker—of a mistletoe: a wedgelike structure embedded in the wood of the host.

sorus (pl. -ri)—of rusts, smuts, and false smut: a mass of spores and the fruit body that produced them.

sp.—unspecified or unidentified species (singular).

spathe—of palms and other plants: a large bract or pair of bracts that sheath a flower cluster.

spermagonium (pl. -ia)—a fungal fruit body that produces spermatia.

spermatium (pl. -ia)—of fungi: a nonmotile male sex cell produced and dispersed in the manner of a spore; capable of emptying its contents into a receptive female structure.

sporangiophore—a sporangium-bearing hypha.

sporangium (pl. -ia)—of fungi: a microscopic saclike structure the contents of which become converted into a mass of spores.

spore—a minute propagule that functions in the manner of a seed but lacks an embryo. Most fungal spores are microscopic.

sporocarp—a fruit body that produces spores.

sporodochium (pl. -ia)—a tiny cushionlike fungal structure that protrudes above the surface of the substrate, similar in origin to an acervulus but consisting of a more or less thick pad of hyphae that bear conidiophores; white to pink, reddish, orange, or dark in color.

sporophore—a structure that bears spores.

sporophyte—of plants such as ferns that reproduce by alternation of generations: the stage that produces spores; also, the diploid part of the plant's life cycle.

sporulate—to produce spores.

spp.—species (plural)

sterile conk—a hard or tough, dense mass of sterile mycelium of a wood-decaying fungus on a trunk or limb.

stipule—an appendage at the base of the petiole.

stoma (pl. -ata)—a microscopic pore in a plant surface, for gas exchange.

strain—of a pathogen: an infraspecific population defined on any arbitrary basis, such as geographic or host origin.

stroma (pl. -ata)—of fungi: a mass of hyphae, with or without plant tissue or other substrate, in or on which spores or fruit bodies are usually produced; microscopic to macroscopic and colorless to bright or dark colored, forming within or on a substrate.

subcuticular—beneath the cuticle of a plant, on the epidermis.

subsp.—subspecies.

subspecies—an infraspecific population defined on the basis of one or more characters (morphologic for most organisms) that distinguish its members from typical representatives of the species.

suctorial—of certain insects: adapted for sucking host fluids by means of specialized mouthparts that pierce cells.

sugar bush—a grove of sugar maple trees from which sap is harvested.

suscept—in plant pathology: a plant susceptible to a given pathogen.

susceptibility—of a plant to a pathogen: a condition in which the plant is capable of interacting with the pathogen such that disease results.

symptom—abnormal appearance or function.

syn.—synonym.

synnema (pl. -ata)—a group of conidiophores cemented together to form an elongate spore-bearing structure, often large enough to be seen with the unaided eye.

teliospore—a thick-walled resting spore characteristic of rust and smut fungi; giving rise to a basidium and basidiospores upon germination.

telium (pl. -ia)—of rust fungi: a fruiting structure consisting of teliospores; produced after the uredinial stage in the life cycle of a rust.

thallospore—an asexual spore that has no conidiophore or is not separate from the hypha or conidiophore that produced it.

thallus—of a fungus: the vegetative body (somatic phase).

toruloid (also torulose)—cylindrical but with swellings at intervals.

transpiration—in plants: movement of water from the root medium through a plant into the atmosphere as the consequence of evaporation.

trichome—a tiny projection from a plant epidermis; may be hairlike, spiny, or glandular.

urediniospore—of rust fungi: an asexual, thick-walled spore capable of infecting the host on which it is produced.

uredinium (pl. -ia)—of rust fungi: a fruit body that produces urediniospores, formed after the aecium and before the telium in the life cycle of the fungus.

var.—variety.

vector—of a disease or pathogen: an agent, usually an animal, that transmits a pathogen.

vesicle—a microscopic, bladderlike sac.

virulence—the relative competence of a pathogen to cause disease in a given plant; see also aggressiveness.

witches'-broom—an abnormally dense cluster of twigs growing from a common locus or resulting from the proliferation of buds.

xylem—water-conducting tissue; wood.

yeast—a unicellular fungus that reproduces by budding, usually one of a group of Ascomycotina that form asci directly from zygotes or single cells; also, one of a group of Basidiomycotina related to smut fungi.

zone line—a black line visible in decaying wood where a cut or fracture crosses a pseudosclerotial plate.

zoospore—a spore that is motile by means of flagella.

zygospore—a thick-walled resting spore formed by the fusion of two gametes borne on somatic hyphae, characteristic of Zygomycetes.

511

References

1. Aa, H. A. van der. 1973. Studies in *Phyllosticta* I. Stud. Mycol. No. 5. 110 pp.
2. Adams, J. F. 1920. The alternate stage of *Pucciniastrum hydrangeae*. Mycologia 12:33–35.
3. Adams, T. J. H., Todd, N. K., and Rayner, A. D. M. 1981. Antagonism between dikaryons of *Piptoporus betulinus*. Trans. Br. Mycol. Soc. 76:510–513.
4. Aho, P. E., and Filip, G. M. 1982. Incidence of wounding and *Echinodontium tinctorium* infection in advanced white fir regeneration. Can. J. For. Res. 12:705–708.
5. Aho, P. E., Seidler, R. J., Evans, H. J., and Raju, P. N. 1974. Distribution, enumeration, and identification of nitrogen-fixing bacteria associated with decay in living white fir trees. Phytopathology 64:1413–1420.
6. Ainsworth, G. C., Sparrow, F. K., and Sussman, A. S. 1973. The fungi. An advanced treatise. Vols. IVA and IVB. Academic Press, New York. 621 and 504 pp.
7. Alcorn, S. M. 1961. Some hosts of *Erwinia carnegieana*. Plant Dis. Rep. 45:587–590.
8. Aldwinckle, H. S. 1974. Field susceptibility of 41 apple cultivars to cedar apple rust and quince rust. Plant Dis. Rep. 58:696–699.
9. Aldwinckle, H. S. 1975. Pathogenic races of *Gymnosporangium juniperi-virginianae* on apple. Phytopathology 65:958–961.
10. Aldwinckle, H. S., and Beer, S. V. 1979. Fire blight and its control. Hortic. Rev. 1:423–474.
11. Aldwinckle, H. S., Lamb, C., and Gustafson, H. L. 1977. Nature and inheritance of resistance to *Gymnosporangium juniperi-virginianae* in apple cultivars. Phytopathology 67:259–266.
12. Aldwinckle, H. S., Pearson, R. C., and Seem, R. C. 1980. Infection periods of *Gymnosporangium juniperi-virginianae* on apple. Phytopathology 70:1070–1073.
13. Alexander, M. E., and Hawksworth, F. G. 1975. Wildland fires and dwarf mistletoes: a literature review of ecology and prescribed burning. U.S. For. Serv. Gen. Tech. Rep. RM-14. 12 pp.
14. Alexander, S. A., Skelly, J. M., and Morris, C. L. 1975. Edaphic factors associated with the incidence and severity of disease caused by *Fomes annosus* in loblolly pine plantations in Virginia. Phytopathology 65:585–591.
15. Alexander, W. B. 1925. Natural enemies of prickly pear and their introduction into Australia. Aust. Inst. Sci. Indus. Bull. 29. 80 pp.
16. Alfaro, R. I., Bloomberg, W. J., Smith, R. B., and Thomson, A. J. 1985. Epidemiology of dwarf mistletoe in western hemlock stands in south coastal British Columbia. Can. J. For. Res. 15:909–913.
17. Alfenas, A. C., Jeng, R., and Hubbes, M. 1983. Virulence of *Cryphonectria cubensis* on *Eucalyptus* species differing in resistance. Eur. J. For. Pathol. 13:197–205.
18. Alfieri, S. A. Jr. 1979. *Kutilakesa pironii* sp. nov., a stem gall- and canker-inciting fungus new to the United States. Mycotaxon 10:217–218.
19. Alfieri, S. A. Jr., Knauss, J. F., and Wehlburg, C. 1979. A stem gall- and canker-inciting fungus new to the United States. Plant Dis. Rep. 63:1016–1020.
20. Alfieri, S. A. Jr., Langdon, K. R., Wehlburg, C., and Kimbrough, J. W. 1984. Index of plant diseases in Florida. Fla. Dep. Agric. & Consumer Serv. Div. Plant Indus. Bull. 11. 389 pp.
21. Alfieri, S. A. Jr., and Samuels, G. J. 1979. *Nectriella pironii* and its *Kutilakesa*-like anamorph, a parasite of ornamental shrubs. Mycologia 71:1178–1185.
22. Alfieri, S. A. Jr., Seymour, C. P., and Denmark, J. C. 1972. Aerial blight of *Carissa grandiflora* caused by *Rhizoctonia solani*. Plant Dis. Rep. 56:511–514.
23. Allen, R. F. 1932. A cytological study of heterothallism in *Puccinia coronata*. J. Agric. Res. 45:513–541.
24. Ambuel, B., Kuntz, J. E., Sarkis, E. H., and Worf, G. L. 1978. The effects of temperature and moisture on white oak anthracnose. (Abstr.) Proc. Am. Phytopathol. Soc. 4:85.
25. Amelunxen, F. 1958. Die Virus-Eiweißspindeln der Kakteen. Darstellung, elektronmikroskopische und chemische Analyse des Virus. Protoplasma 49:140–178.
26. American Phytopathological Society. 1979. Symposium on wood decay in living trees. Mechanisms of tree defense and wood decay. Phytopathology 69:1135–1160.
27. Amerson, H. V. 1976. A structural characterization of the uredial stage of *Puccinia sparganioides* and its associated interactions with *Spartina alterniflora*. Ph.D. thesis. N.C. State Univ., Raleigh. 116 pp.
28. Amico, L. A., O'Shea, M. T., and Castello, J. D. 1985. Transmission of tobacco mosaic and tobacco ringspot viruses from Moraine ash in New York. Plant Dis. 69:542.
29. Amiro, B. D., and Courtin, G. M. 1981. Patterns of vegetation in the vicinity of an industrially disturbed ecosystem, Sudbury, Ontario. Can. J. Bot. 59:1623–1629.
30. Anderson, A. R., and Moore, L. W. 1979. Host specificity in the genus *Agrobacterium*. Phytopathology 69:320–323.
31. Anderson, D. L., and French, D. W. 1972. Isolation of *Hypoxylon mammatum* from aspen stem sections. Can. J. Bot. 50:1971–1972.
32. Anderson, G. W. 1963. Sweetfern rust on hard pines. U.S. For. Serv. For. Pest Leafl. 79. 7 pp.
33. Anderson, G. W. 1965. The distribution of eastern and western gall rusts in the Lake States. Plant Dis. Rep. 49:527–528.
34. Anderson, G. W., and French, D. W. 1965. Differentiation of *Cronartium quercuum* and *Cronartium coleosporioides* on the basis of aeciospore germ tubes. Phytopathology 55:171–173.
35. Anderson, G. W., and French, D. W. 1965. Western gall rust in the Lake States. For. Sci. 11:139–141.
36. Anderson, G. W., and Martin, M. P. 1981. Factors related to incidence of Hypoxylon cankers in aspen and survival of cankered trees. For. Sci. 27:461–476.
37. Anderson, J. B., Korhonen, K., and Ullrich, R. C. 1980. Relationships between European and North American biological species of *Armillaria mellea*. Exp. Mycol. 4:87–95.
38. Anderson, N. A. 1970. Eastern gall rust. U.S. For. Serv. For. Pest. Leafl. 80. 4 pp.
39. Anderson, N. A., and French, D. W. 1964. Sweetfern rust on jack pine. J. For. 62:467–471.
40. Anderson, N. A., French, D. W., and Anderson, R. L. 1967. The stalactiform rust of jack pine. J. For. 65:398–402.
41. Anderson, N. A., and Kaufert, F. H. 1959. Brooming response of black spruce to dwarfmistletoe infection. For. Sci. 5:356–364.
42. Anderson, N. A., Ostry, M. E., and Anderson, G. W. 1979. Insect wounds as infection sites for *Hypoxylon mammatum* on trembling aspen. Phytopathology 69:476–479.
43. Anderson, P. J., and Rankin, W. H. 1914. Endothia canker of chestnut. Cornell Univ. Agric. Exp. Stn. Bull. 347:529–618.
44. Anderson, R. L., Anderson, G. W., and Schipper, A. L. 1979. Hypoxylon canker of aspen. U.S. For. Serv. For. Insect Dis. Leafl. 6. 7 pp.
45. Anderson, R. L., and Hoffard, W. H. 1978. Fusarium canker-ambrosia beetle complex on tulip poplar in Ohio. Plant Dis. Rep. 62:751.
46. Anson, A. E. 1982. A pseudomonad producing orange soft rot disease in cacti. Phytopathol. Z. 103:163–172.
47. Appel, D. N., Drees, C. F., and Johnson, J. 1985. An extended range for oak wilt and *Ceratocystis fagacearum* compatibility types in the United States. Can. J. Bot. 63:1325–1328.
48. Appel, D. N., and Maggio, R. C. 1984. Aerial survey for oak wilt incidence at three locations in central Texas. Plant Dis. 68:661–664.
49. Appel, D. N., and Stipes, R. J. 1984. Canker expansion on water-stressed pin oaks colonized by *Endothia gyrosa*. Plant Dis. 68:851–853.
50. Arnett, J. D., and Witcher, W. 1974. Histochemical studies of yellow poplar infected with *Fusarium solani*. Phytopathology 64:414–418.
51. Arnold, R. H. 1967. A canker and foliage disease of yellow birch. I. Description of the causal fungus, *Diaporthe alleghaniensis* sp. nov., and the symptoms on the host. Can. J. Bot. 45:783–801.
52. Arnold, R. H., and Carter, J. C. 1974. *Fusicoccum elaeagni*, the cause of canker and dieback of Russian olive, redescribed and redisposed to the genus *Phomopsis*. Mycologia 66:191–197.

53. Arnold, R. H., and Straby, A. E. 1973. *Phomopsis elaeagni* on Russian olive (*Elaeagnus angustifolia*) in Canada. Can. Plant Dis. Surv. 53:183–186.

54. Arthur, J. C. 1934. Manual of the rusts in the United States and Canada. Purdue Research Foundation, Lafayette, IN. 438 pp.

55. Arthur, J. J., Leone, I. A., and Flower, F. B. 1981. Flooding and landfill gas effects on red and sugar maples. J. Environ. Qual. 10:431–433.

56. Arx, J. A. von. 1949. Beiträge zur Kenntnis der Gattung *Mycosphaerella*. Sydowia 3:28–100.

57. Arx, J. A. von. 1957. Die Arten der Gattung *Colletotrichum*. Phytopathol. Z. 29:413–468.

58. Arx, J. A. von. 1957. Schurft op *Pyracantha*. Tijdschr. Plantenz. 63:198–199.

59. Arx, J. A. von. 1970. A revision of the fungi classified as *Gloeosporium*. 2nd ed. Bibl. Mycol. Vol. 24. 203 pp.

60. Arx, J. A. von, and Müller, E. 1954. Die Gattungen der amerosporen Pyrenomyceten. Beitr. Kryptogamenfl. Schweiz. Vol. 11. 434 pp.

61. Arx, J. A. von, Walt, J. P. van der, and Liebenberg, N. V. D. M. 1982. The classification of *Taphrina* and other fungi with yeast-like cultural states. Mycologia 74:285–296.

62. Ashcroft, J. M. 1934. European canker of black walnut and other trees. West Virginia Agric. Exp. Stn. Bull. 261. 52 pp.

63. Ashworth, E. N., Rowse, D. J., and Billmyer, L. A. 1983. The freezing of water in woody tissues of apricot and peach and the relationship to freezing injury. J. Am. Soc. Hortic. Sci. 108:299–303.

64. Ashworth, L. J. Jr., and Gaona, S. A. 1982. Evaluation of clear polyethylene mulch for controlling Verticillium wilt in established pistachio nut groves. Phytopathology 72:243–246.

65. Ashworth, L. J. Jr., Gaona, S. A., and Surber, E. 1985. Nutritional diseases of pistachio trees: potassium and phosphorus deficiencies and chloride and boron toxicities. Phytopathology 75:1084–1091.

66. Ashworth, L. J. Jr., Gaona, S. A., and Surber, E. 1985. Verticillium wilt of pistachio: the influence of potassium nutrition on susceptibility to infection by *Verticillium dahliae*. Phytopathology 75:1091–1093.

67. Atifano, R. A. 1981. Alternaria leaf spot of schefflera. Fla. Coop. Exten. Serv. Plant Pathol. Fact Sheet PP-17. 2 pp.

68. Atilano, R. A. 1983. Alternaria leaf spot of *Schefflera arboricola*. Plant Dis. 67:64–66.

69. Atilano, R. A. 1983. A foliar blight of Ming aralia caused by *Alternaria panax*. Plant Dis. 67:224–226.

70. Atkinson, G. F. 1894. Leaf curl and plum pockets. Cornell Univ. Agric. Exp. Stn. Bull. 73:319–355.

71. Atkinson, R. G. 1965. *Phytophthora* species inciting root rot of *Chamaecyparis lawsoniana* and other ornamentals in coastal British Columbia. Can. J. Bot. 43:1471–1475.

72. Attathom, S., Weathers, L. G., and Gumpf, D. J. 1978. Identification and characterization of a potexvirus from California barrel cactus. Phytopathology 68:1401–1406.

73. Attathom, S., Weathers, L. G., and Gumpf, D. J. 1978. Occurrence and distribution of a virus-induced disease of barrel cactus in California. Plant Dis. Rep. 62:228–231.

74. Aycock, R. 1966. Stem rot and other diseases caused by *Sclerotium rolfsii*. N.C. Agric. Exp. Stn. Tech. Bull. 174. 202 pp.

75. Aycock, R., West, E., Watkins, G. M., Cooper, W. E., Wilson, C., Boyle, L. W., Garren, K. H., and Harrison, A. L. 1961. Symposium on *Sclerotium rolfsii*. Phytopathology 51:107–128.

76. Ayers, J. C. Jr., and Barden, V. A. 1975. Net photosynthesis and dark respiration of apple leaves as affected by pesticides. J. Am. Soc. Hortic. Sci. 100:24–28.

77. Bach, W. J., and Wolf, F. A. 1928. The isolation of the fungus that causes citrus melanose and the pathological anatomy of the host. J. Agric. Res. 37:243–252.

78. Bachi, P. R., and Peterson, J. L. 1985. Enhancement of *Sphaeropsis sapinea* stem invasion of pines by water deficits. Plant Dis. 69:798–799.

79. Baes, C. F. III, and McLaughlin, S. B. 1984. Trace elements in tree rings: evidence of recent and historical air pollution. Science 224:494–496.

80. Bagga, D. K., and Smalley, E. B. 1974. The development of Hypoxylon canker of *Populus tremuloides*: role of ascospores, conidia, and toxins. Phytopathology 64:654–658.

81. Bagga, D. K., and Smalley, E. B. 1974. The development of Hypoxylon canker of *Populus tremuloides*: role of interacting environmental factors. Phytopathology 64:658–662.

82. Bagga, H. S., and Boone, D. M. 1968. Genes in *Venturia inaequalis* controlling pathogenicity to crabapples. Phytopathology 58:1176–1182.

83. Bagga, H. S., and Boone, D. M. 1968. Inheritance of resistance to *Venturia inaequalis* in crabapples. Phytopathology 58:1183–1187.

84. Baines, R. C. 1939. Phytophthora trunk canker or collar rot of apple trees. J. Agric. Res. 59:159–184.

85. Baker, F. A., and French, D. W. 1980. Spread of *Arceuthobium pusillum* and rates of infection and mortality in black spruce stands. Plant Dis. 64:1074–1076.

86. Baker, F. A., French, D. W., and Hudler, G. W. 1981. Development of *Arceuthobium pusillum* on inoculated black spruce. For. Sci. 27:203–205.

87. Baker, F. A., French, D. W., Kulman, H. M., Davis, O., and Bright, R. C. 1985. Pollination of the eastern dwarf mistletoe. Can. J. For. Res. 15:708–714.

88. Baker, J. H. 1965. Relationship between salt concentrations in leaves and sap and the decline of sugar maples along roadsides. Mass. Agric. Exp. Stn. Bull. 553. 16 pp.

89. Baker, K. F. 1953. Recent epidemics of downy mildew of rose. Plant Dis. Rep. 37:331–337.

90. Baker, K. F., and Dimock, A. W. 1969. Black spot. Pages 172–184 in: Roses. J. W. Mastalerz and R. W. Langhans, eds. Pennsylvania Flower Growers, New York State Flower Growers Assoc., Inc., and Roses, Inc. 329 pp.

91. Baker, W. L. 1949. Notes on the transmission of the virus causing phloem necrosis of American elm, with notes on the biology of its insect vector. J. Econ. Entomol. 42:729–732.

92. Baldwin, H. I., Boyce, J. S., Brown, R. C., Cline, A. C., Filley, W. O., Reynolds, H. A., and Turner, G. W. C., eds. 1940. Important tree pests of the Northeast. Massachusetts Forest and Park Association, Boston. 194 pp.

93. Banfield, W. M. 1941. Distribution by the sap stream of spores of three fungi that induce vascular wilt diseases of elm. J. Agric. Res. 62:637–681.

94. Baranyay, J. A. 1970. Lodgepole pine dwarf mistletoe in Alberta. Can. For. Serv. Publ. 1286. 22 pp.

95. Baranyay, J. A., and Hiratsuka, Y. 1967. Identification and distribution of *Ciborinia whetzelii* (Seaver) Seaver in western Canada. Can. J. Bot. 45:189–191.

96. Baranyay, J. A., and Safranyik, L. 1970. Effect of dwarf mistletoe on growth and mortality of lodgepole pine in Alberta. Can. For. Serv. Publ. 1285. 19 pp.

97. Barbe, G. D., McCartney, W. O., and Rosenberg, D. Y. 1966. *Gymnosporangium confusum* found in California. Plant Dis. Rep. 50:768–769.

98. Barker, K. R., Worf, G. L., and Epstein, A. H. 1965. Nematodes associated with the decline of azaleas in Wisconsin. Plant Dis. Rep. 49:47–49.

99. Barmore, C. R., Brown, G. E., and Youtsey, C. O. 1984. Fungicide control of albinism in citrus seedlings caused by *Alternaria tenuis*. Plant Dis. 68:43–44.

100. Barnard, E. L. 1984. Occurrence, impact, and fungicidal control of girdling stem cankers caused by *Cylindrocladium scoparium* on eucalyptus seedlings in a south Florida nursery. Plant Dis. 68:471–473.

101. Barnard, E. L., and Dixon, W. N. 1983. Insects and diseases: important problems of Florida's forest and shade tree resources. Fla. Dep. Agric. & Consumer Serv. Bull. 196-A. 120 pp.

102. Barnard, E. L., and Schroeder, R. A. 1984. Anthracnose of acacia in Florida: occurrence and fungicidal control. Proc. Fla. State Hortic. Soc. 97:244–247.

103. Barnett, H. L. 1957. *Hypoxylon punctulatum* and its conidial stage on dead oak trees and in culture. Mycologia 49:588–595.

104. Barney, D. L., Walser, R. H., Davis, T. D., and Williams, C. F. 1985. Trunk injection of iron compounds as a treatment for overcoming iron chlorosis in apple trees. HortScience 20:236–238.

105. Barr, M. E. 1968. The Venturiaceae in North America. Can. J. Bot. 46:799–864.

106. Barr, M. E. 1972. Preliminary studies on the Dothideales in temperate North America. Contrib. Univ. Mich. Herb. 9:523–638.

107. Barr, M. E. 1978. The Diaporthales in North America with emphasis on *Gnomonia* and its segregates. Mycol. Mem. 7. 232 pp.

108. Barr, M. E. 1984. *Herpotrichia* and its segregates. Mycotaxon 20:1–38.

109. Barrett, D. K. 1970. *Armillaria mellea* as a possible factor predisposing roots to infection by *Polyporus schweinitzii*. Trans. Br. Mycol. Soc. 55:459–462.

110. Barrett, D. K. 1985. Basidiospores of *Phaeolus schweinitzii*: a source of soil infestation. Eur. J. For. Pathol. 15:417–425.

111. Barrett, D. K., and Greig, B. J. W. 1985. The occurrence of *Phaeolus schweinitzii* in the soils of Sitka spruce plantations with broadleaved and non-woodland histories. Eur. J. For. Pathol. 15:412–417.

112. Barrett, D. K., and Uscuplic, M. 1971. The field distribution of interacting strains of *Polyporus schweinitzii* and their origin. New Phytol. 70:581–598.

113. Barrick, W. E., Flore, J. A., and Davidson, H. 1979. Deicing salt spray injury in selected *Pinus* spp. J. Am. Soc. Hortic. Sci. 104:617–622.

114. Barrows-Broaddus, J., and Dwinell, L. D. 1985. Branch dieback and cone and seed infection caused by *Fusarium moniliforme* var. *subglutinans* in a loblolly pine seed orchard in South Carolina. Phytopathology 75:1104–1108.

115. Barrus, M. F., and Horsfall, J. G. 1928. Preliminary note on snowberry anthracnose. Phytopathology 18:797–801.

116. Basham, J. T. 1958. Decay of trembling aspen. Can. J. Bot. 36:491–505.

117. Basham, J. T. 1975. Heart rot of jack pine in Ontario. IV. Heartwood-inhabiting fungi: their entry and interactions within living trees. Can. J. For. Res. 5:706–721.

513

118. Bassett, E. N., and Fenn, P. 1984. Latent colonization and pathogenicity of *Hypoxylon atropunctatum* on oaks. Plant Dis. 68:317–319.

119. Batista, A. C., and Ciferri, R. 1963. Capnodiales. Saccardoa 2:1–296.

120. Batko, S. 1956. *Meria laricis* on Japanese and hybrid larch in Britain. Trans. Br. Mycol. Soc. 39:13–16.

121. Batra, L. R. 1960. The species of *Ciborinia* pathogenic to *Salix, Magnolia,* and *Quercus*. Am. J. Bot. 47:819–827.

122. Batra, L. R. 1983. *Monilinia vaccinii-corymbosi* (Sclerotiniaceae): its biology on blueberry and comparison with related species. Mycologia 75:131–152.

123. Baudoin, A. B. A. M. 1986. Infection of photinia leaves by *Entomosporium mespili*. Plant Dis. 70:191–194.

124. Baudoin, A. B. A. M. 1986. Environmental conditions required for infection of photinia leaves by *Entomosporium mespili*. Plant Dis. 70:519–521.

125. Baxter, D. V. 1925. The biology and pathology of some of the hardwood heart-rotting fungi. Am. J. Bot. 12:522–576.

126. Baxter, D. V. 1936. Some resupinate polypores from the region of the Great Lakes. VII. Pap. Mich. Acad. Sci. 21:243–267.

127. Baxter, D. V. 1948. Some resupinate polypores from the region of the Great Lakes. XVIII. Pap. Mich. Acad. Sci. 32(1946):189–211.

128. Baxter, D. V. 1967. Disease in forest plantations: thief of time. Cranbrook Inst. Sci. Bull. 51. 251 pp.

129. Baxter, L. W. Jr., and Fagan, S. G. 1974. A comparison of the relative susceptibility of seedlings of *Camellia japonica* and *C. sasanqua* to dieback and canker caused by a strain of *Glomerella cingulata* pathogenic to camellias. Plant Dis. Rep. 58:139–141.

130. Baxter, L. W. Jr., Fagan, S. G., and Owen, M. G. 1982. Response of selected *Camellia reticulata* cultivars to *Glomerella cingulata*, cause of contagious camellia dieback and canker. Plant Dis. 66:1023–1024.

131. Baxter, L. W., and Plakidas, A. G. 1954. Dieback and canker of camellias caused by *Glomerella cingulata*. Phytopathology 44:129–133.

132. Bazzigher, G. 1976. Der schwarze Schneeschimmel der Koniferen [*Herpotrichia juniperi* (Duby) Petrak und *Herpotrichia coulteri* (Peck) Bose]. Eur. J. For. Pathol. 6:109–122.

133. Beaumont, A. 1950. Diseases of shrub Veronicas. Gard. Chron., Ser. 3, 128:128.

134. Bedker, P. J., and Blanchette, R. A. 1983. Development of cankers caused by *Nectria cinnabarina* on honey locusts after root pruning. Plant Dis. 67:1010–1013.

135. Bedker, P. J., Blanchette, R. A., and French, D. W. 1982. *Nectria cinnabarina*: the cause of a canker disease of honey locust in Minnesota. Plant Dis. 66:1067–1070.

136. Bedker, P. J., and Wingfield, M. J. 1983. Taxonomy of three canker-causing fungi of honey locust in the United States. Trans. Br. Mycol. Soc. 81:179–183.

137. Bedunah, D., and Trlica, J. M. 1981. Carbon dioxide exchange of ponderosa pine as affected by sodium chloride and polyethylene glycol. For. Sci. 27:139–146.

138. Bega, R. V. 1978. *Phomopsis lokoyae* outbreak in a California forest nursery. Plant Dis. Rep. 62:567–569.

139. Bega, R. V., tech. coord. 1978. Diseases of Pacific coast conifers. U.S. Dep. Agric. Agric. Handb. 521. 206 pp.

140. Beigl, H. J. 1977. Veränderung der Krankheitsdisposition von Forstpflanzen durch Herbizide. I. Freilandversuche mit *Lophodermium pinastri*. Eur. J. For. Pathol. 7:200–219.

141. Beigl, H. J. 1978. Veränderung der Krankheitsdisposition von Forstpflanzen durch Herbizide. III. Versuche mit Erregern der Umfallkrankheit. Eur. J. For. Pathol. 8:240–258.

142. Bender, C. L., and Coyier, D. L. 1983. Isolation and identification of races of *Sphaerotheca pannosa* var. *rosae*. Phytopathology 74:100–103.

143. Bender, C. L., and Coyier, D. L. 1985. Heterothallism in *Sphaerotheca pannosa* var. *rosae*. Trans. Br. Mycol. Soc. 84:647–652.

144. Benhamou, N., Lafontaine, J. G., Joly, J. R., and Ouellette, G. B. 1985. Ultrastructural localization in host tissues of a toxic glycopeptide produced by *Ophiostoma ulmi*, using monoclonal antibodies. Can. J. Bot. 63:1185–1195.

145. Benhamou, N., Ouellette, G. B., Lafontaine, J. G., and Joly, J. R. 1985. Use of monoclonal antibodies to detect a phytotoxic glycopeptide produced by *Ophiostoma ulmi*, the Dutch elm disease pathogen. Can. J. Bot. 63:1177–1184.

146. Benoit, L. F., Skelly, J. M., Moore, L. D., and Dochinger, L. S. 1982. Radial growth reductions of *Pinus strobus* L. correlated with foliar ozone sensitivity as an indicator of ozone-induced losses in eastern forests. Can. J. For. Res. 12:673–679.

147. Benson, D. M. 1982. Cold inactivation of *Phytophthora cinnamomi*. Phytopathology 72:560–563.

148. Benson, D. M., and Barker, K. R. 1982. Susceptibility of Japanese boxwood, dwarf gardenia, Compacta (Japanese) holly, Spiny Greek and Blue Rug junipers, and nandina to four nematode species. Plant Dis. 66:1176–1179.

149. Benson, D. M., and Cochran, F. D. 1980. Resistance of evergreen hybrid azaleas to root rot caused by *Phytophthora cinnamomi*. Plant Dis. 64:214–215.

150. Benson, D. M., and Jones, R. K. 1980. Etiology of rhododendron dieback caused by four species of *Phytophthora*. Plant Dis. 64:687–691.

151. Berbee, J. G. 1957. Virus symptoms associated with birch dieback. Can. Dep. Agric., Sci. Serv., For. Biol. Div. Bi-mon. Prog. Rep. 13(1):1.

152. Berbee, J. G., and Rogers, J. D. 1964. Life cycle and host range of *Hypoxylon pruinatum* and its pathogenesis on poplars. Phytopathology 54:257–261.

153. Bergdahl, D. R., and French, D. W. 1976. Relative susceptibility of five pine species, 2 to 36 months of age, to infection by *Cronartium comandrae*. Can. J. For. Res. 6:319–325.

154. Bergdahl, D. R., and French, D. W. 1976. Epidemiology of comandra rust on jack pine and comandra in Minnesota. Can. J. For. Res. 6:326–334.

155. Bergdahl, D. R., and French, D. W. 1985. Association of wood decay fungi with decline and mortality of apple trees in Minnesota. Plant Dis. 69:887–890.

156. Berrang, P., and Karnosky, D. F. 1983. Street trees for metropolitan New York. N.Y. Bot. Gard., Inst. Urban Hortic. Publ. No. 1. 177 pp.

157. Berrang, P., Karnosky, D. F., and Stanton, B. J. 1985. Environmental factors affecting tree health in New York City. J. Arboric. 11:185–189.

158. Berrang, P., and Steiner, K. C. 1980. Resistance of pin oak progenies to iron chlorosis. J. Am. Soc. Hortic. Sci. 105:519–522.

159. Berry, F. H. 1960. Etiology and control of walnut anthracnose. Md. Agric. Exp. Stn. Bull. A-113. 22 pp.

160. Berry, F. H., and Lombard, F. F. 1978. Basidiomycetes associated with decay of living oak trees. U.S. For. Serv. Res. Pap. NE-413. 8 pp.

161. Bertagnole, C. L., Woo, J. Y., and Partridge, A. D. 1983. Pathogenicity of five *Verticicladiella* species to lodgepole pine. Can. J. Bot. 61:1861–1867.

162. Bertrand, P. F., and English, H. 1976. Release and dispersal of conidia and ascospores of *Valsa leucostoma*. Phytopathology 66:987–991.

163. Bertrand, P. F., and English, H. 1976. Virulence and seasonal activity of *Cytospora leucostoma* and *C. cincta* in French prune trees in California. Plant Dis. Rep. 60:106–110.

164. Bertrand, P. F., English, H., and Carlson, R. M. 1976. Relation of soil physical and fertility properties to the occurrence of Cytospora canker in French prune orchards. Phytopathology 66:1321–1324.

165. Bertrand, P. F., English, H., Uriu, K., and Schick, F. J. 1976. Late season water deficits and development of Cytospora canker in French prune. Phytopathology 66:1318–1320.

166. Biddle, P. G., and Tinsley, T. W. 1968. Virus diseases of conifers in Great Britain. Nature 219:1387–1388.

167. Biddle, P. G., and Tinsley, T. W. 1971. Poplar mosaic in Great Britain. New Phytol. 70:61–66.

168. Biddle, P. G., and Tinsley, T. W. 1971. Some effects of poplar mosaic virus on the growth of poplar trees. New Phytol. 70:67–75.

169. Bidwell, C. B., and Bramble, W. C. 1934. The Strumella disease in southern Connecticut. J. For. 32:15–23.

170. Bier, J. E. 1939. Septoria canker of introduced and native hybrid poplars. Can. J. Res. C17:195–204.

171. Bier, J. E. 1940. Studies in forest pathology. III. Hypoxylon canker of poplar. Can. Dep. Agric. Publ. 691. 40 pp.

172. Bier, J. E. 1959–1961. The relation of bark moisture content to the development of canker diseases caused by native, facultative parasites. I. Cryptodiaporthe canker on willow. II. Fusarium canker of black cottonwood. VI. Pathogenicity studies of *Hypoxylon pruinatum* (Klotzsch) Cke. and *Septoria musiva* Pk. on species of *Acer, Populus,* and *Salix*. Can. J. Bot. 37:229–238; 37:781–788; 39:1555–1561.

173. Biggs, A. R. 1984. Boundary-zone formation in peach bark in response to wounds and *Cytospora leucostoma* infection. Can. J. Bot. 62:2814–2821.

174. Biggs, A. R. 1984. Intracellular suberin: occurrence and detection in tree bark. I.A.W.A. Bull., n.s. 5:243–248.

175. Biggs, A. R. 1985. Suberized boundary zones and the chronology of wound response in bark. Phytopathology 75:1191–1195.

176. Biggs, A. R., and Davis, D. D. 1981. Effect of SO_2 on growth and sulfur content of hybrid poplar. Can. J. For. Res. 11:830–833.

177. Biggs, A. R., Davis, D. D., and Merrill, W. 1983. Histopathology of cankers on *Populus* caused by *Cytospora chrysosperma*. Can. J. Bot. 61:563–574.

178. Biggs, A. R., and Miles, N. W. 1985. Suberin deposition as a measure of wound response in peach bark. HortScience 20:903–905.

179. Binns, W. O., Mayhead, G. J., and MacKenzie, J. M. 1980. Nutrient deficiencies of conifers in British forests. G.B. For. Comm. Leafl. 76. 23 pp.

180. Bingham, R. T. 1983. Blister rust resistant western white pine for the Inland Empire: the story of the first 25 years of the research and development program. U.S. For. Serv. Gen. Tech. Rep. INT-146. 45 pp.

181. Bingham, R. T., Hoff, R. J., and McDonald, G. I., sci. dir. and prog. coord. 1972. Biology of rust resistance in forest trees: Proceedings of a NATO-

IUFRO Advanced Study Institute August 17–24, 1969. U.S. Dep. Agric. Misc. Publ. 1221. 681 pp.

182. Björkman, E. 1948. Studier över snöskyttesvampens (*Phacidium infestans* Karst.) biologi samt metoder för snöskyttets bekämpande. Meddel. fran Statens Skogsforskningsinst. 37(2). 136 pp.

183. Black, W. M., and Neely, D. 1978. Effects of temperature, free moisture, and relative humidity on the occurrence of walnut anthracnose. Phytopathology 68:1054–1056.

184. Black, W. M., and Neely, D. 1978. Relative resistance of *Juglans* species and hybrids to walnut anthracnose. Plant Dis. Rep. 62:497–499.

185. Blackwell, E. 1943. The life history of *Phytophthora cactorum* (Leb. & Cohn) Schroet. Trans. Br. Mycol. Soc. 26:71–89.

186. Blaker, N. S., and MacDonald, J. D. 1981. Predisposing effects of soil moisture extremes on the susceptibility of rhododendron to Phytophthora root and crown rot. Phytopathology 71:831–834.

187. Blaker, N. S., and MacDonald, J. D. 1983. Influence of container medium pH on sporangium formation, zoospore release, and infection of rhododendron by *Phytophthora cinnamomi*. Plant Dis. 67:259–263.

188. Blakeslee, G. M., Foltz, J. L., and Oak, S. W. 1981. The deodar weevil, a vector and wounding agent associated with pitch canker of slash pine. (Abstr.) Phytopathology 71:861.

189. Blakeslee, G. M., and Oak, S. W. 1980. Residual naval stores stumps as reservoirs of inoculum for infection of slash pines by *Phaeolus schweinitzii*. Plant Dis. 64:167.

190. Blanchette, R. A. 1980. Wood decomposition by *Phellinus (Fomes) pini*: a scanning electron microscopy study. Can. J. Bot. 58:1496–1503.

191. Blanchette, R. A. 1982. *Phellinus (Fomes) pini* decay associated with sweetfern rust in sapwood of jack pine. Can. J. For. Res. 12:304–310.

192. Blanchette, R. A. 1982. Decay and canker formation by *Phellinus pini* in white and balsam fir. Can. J. For. Res. 12:538–544.

193. Blanchette, R. A. 1982. Progressive stages of discoloration and decay associated with the canker-rot fungus, *Inonotus obliquus,* in birch. Phytopathology 72:1272–1277.

194. Blanchette, R. A. 1984. Selective delignification of eastern hemlock by *Ganoderma tsugae*. Phytopathology 74:153–160.

195. Blanchette, R. A., Sutherland, J. B., and Crawford, D. L. 1981. Actinomycetes in discolored wood of living silver maple. Can. J. Bot. 59:1–7.

196. Blaser, H. W., Marr, C., and Takahashi, D. 1967. Anatomy of boron-deficient *Thuja plicata*. Am. J. Bot. 54:1107–1113.

197. Bliss, D. E. 1933. The pathogenicity and seasonal development of *Gymnosporangium* in Iowa. Iowa Agric. Exp. Stn. Res. Bull. 166:339–392.

198. Bliss, D. E. 1946. The relation of soil temperature to the development of Armillaria root rot. Phytopathology 36:302–318.

199. Bloomberg, W. J. 1962. Cytospora canker of poplars: factors influencing the development of the disease. Can. J. Bot. 40:1272–1280.

200. Bloomberg, W. J. 1971. Diseases of Douglas-fir seedlings caused by *Fusarium oxysporum*. Phytopathology 61:467–470.

201. Bloomberg, W. J. 1973. Fusarium root rot of Douglas-fir seedlings. Phytopathology 63:337–341.

202. Bloomberg, W. J., and Farris, S. H. 1963. Cytospora canker of poplars: bark wounding in relation to canker development. Can. J. Bot. 41:303–310.

203. Bloomberg, W. J., and Lock, W. 1972. Strain differences in *Fusarium oxysporum* causing diseases of Douglas-fir seedlings. Phytopathology 62:481–485.

204. Bloomberg, W. J., and Reynolds, G. 1982. Factors affecting transfer and spread of *Phellinus weirii* mycelium in roots of second-growth Douglas-fir. Can. J. For. Res. 12:424–427.

205. Bloomberg, W. J., Smith, R. B., and van der Wereld, A. 1980. A model of spread and intensification of dwarf mistletoe infection in young western hemlock stands. Can. J. For. Res. 10:42–52.

206. Bloomberg, W. J., and Wallis, G. W. 1979. Comparison of indicator variables for estimating growth reduction associated with *Phellinus weirii* root rot in Douglas-fir plantations. Can. J. For. Res. 9:76–81.

207. Boddy, L., and Rayner, A. D. M. 1982. Population structure, intermycelial interactions, and infection biology of *Stereum gausapatum*. Trans. Br. Mycol. Soc. 78:337–351.

208. Boddy, L., and Rayner, A. D. M. 1983. Origins of decay in living deciduous trees: the role of moisture content and a re-appraisal of the expanded concept of tree decay. New Phytol. 94:623–641.

209. Boerboom, J. H. A., and Maas, P. W. T. 1970. Canker of *Eucalyptus grandis* and *E. saligna* in Surinam caused by *Endothia havanensis*. Turrialba 20:94–99.

210. Boerema, G. H., and Verhoeven, A. A. 1972, 1973. Check-list for scientific names of common parasitic fungi. Series 1a, 1b: Fungi on trees and shrubs. Neth. J. Plant Pathol. 78 (suppl.):1–63; 79:165–179.

211. Boesewinkel, H. J. 1982. *Cylindrocladiella*, a new genus to accommodate *Cylindrocladium parvum* and other small-spored species of *Cylindrocladium*. Can. J. Bot. 60:2288–2294.

212. Boesewinkel, H. J. 1983. New records of the three fungi causing cypress canker in New Zealand: *Seiridium cupressi* (Guba) comb. nov. and *S. car-dinale* on *Cupressocyparis* and *S. unicorne* on *Cryptomeria* and *Cupressus*. Trans. Br. Mycol. Soc. 80:544–547.

213. Bohaychuk, W. P., and Whitney, R. D. 1973. Environmental factors influencing basidiospore discharge in *Polyporus tomentosus*. Can. J. Bot. 51:801–815.

214. Bolkan, H. A., Ogawa, J. M., and Teranishi, H. R. 1984. Shoot blight of pistachio caused by *Botrytis cinerea*. Plant Dis. 68:163–165.

215. Booth, C. 1959. Studies of Pyrenomycetes IV. *Nectria* (Part 1). Commonw. Mycol. Inst. Mycol. Pap. 73. 115 pp.

216. Booth, C. 1971. The genus *Fusarium*. Commonwealth Mycological Institute, Kew, Surrey, England. 237 pp.

217. Boothroyd, C. W. 1951. A new leaf spot of *Celastrus scandens* L., the climbing bittersweet. Mycologia 43:373–375.

218. Boring, L. R., and Swank, W. T. 1984. Symbiotic nitrogen fixation in regenerating black locust (*Robinia pseudoacacia* L.) stands. For. Sci. 30:528–537.

219. Bormann, F. H. 1982. The effects of air pollution on the New England landscape. Ambio 11:338–346.

220. Born, G. L. 1974. Root infection of woody hosts with *Verticillium alboatrum*. Ill. Nat. Hist. Surv. Bull. 31:205–249.

221. Born, G. L., and Crane, J. L. 1972. *Kaskaskia gleditsiae* gen. et sp. nov. parasitic on thornless honey locust in Illinois. Phytopathology 62:926–930.

222. Bové, J. M. 1984. Wall-less prokaryotes of plants. Annu. Rev. Phytopathol. 22:361–396.

223. Bovey, R. W. 1977. Response of selected woody plants in the United States to herbicides. U.S. Dep. Agric. Agric. Handb. 493. 101 pp.

224. Bowen, P. R. 1930. A maple leaf disease caused by *Cristulariella depraedens*. Conn. Agric. Exp. Stn. Bull. 316:625–647.

225. Boyce, J. S. 1933. A canker of Douglas fir associated with *Phomopsis lokoyae*. J. For. 31:664–672.

226. Boyce, J. S. 1940. A needle cast of Douglas-fir associated with *Adelopus gäumannii*. Phytopathology 30:649–659.

227. Boyce, J. S. 1943. Host relationships and distribution of conifer rusts in the United States and Canada. Trans. Conn. Acad. Arts Sci. 35:329–482.

228. Boyce, J. S. 1961. Forest pathology. 3rd ed. McGraw-Hill, New York. 572 pp.

229. Boyce, J. S. Jr. 1954. Hypoderma needle blight of southern pines. J. For. 52:496–498.

230. Boyce, J. S. Jr. 1967. Red root and butt rot in planted slash pines. J. For. 65:493–494.

231. Boyce, S. G. 1954. The salt spray community. Ecol. Monogr. 24:29–67.

232. Boyd, E. S. 1934. A developmental study of a new species of *Ophiodothella*. Mycologia 26:456–468.

233. Boyer, M. G. 1961. A Fusarium canker disease of *Populus deltoides* Marsh. Can. J. Bot. 39:1195–1204.

234. Boyer, M. G. 1961. Variability and hyphal anastomoses in host-specific forms of *Marssonina populi* (Lib.) Magn. Can. J. Bot. 39:1409–1427.

235. Boyer, M. G., and Navratil, S. 1970. Some aspects of transmission and electron microscopy of necrotic leaf spot of aspen. Can. J. Bot. 48:1141–1145.

236. Boyle, A. M. 1949. Further studies of the bacterial necrosis of the giant cactus. Phytopathology 39:1029–1052.

237. Brandes, J., and Bercks, R. 1962. Untersuchungen zur Identifizierung und Klassifizierung des Kakteen-X-Virus (cactus virus X). Phytopathol. Z. 46:291–300.

238. Brandt, R. W. 1960. The Rhabdocline needle cast of Douglas-fir. State Univ. Coll. For. (Syracuse, NY) Tech. Publ. 84. 66 pp.

239. Brandt, R. W. 1964. Nectria canker of hardwoods. U.S. For. Serv. For. Pest Leafl. 84. 7 pp.

240. Brasier, C. M. 1984. Inter-mycelial recognition systems in *Ceratocystis ulmi*: their physiological properties and ecological importance. Pages 451–497 in: The ecology and physiology of the fungal mycelium. D. J. Jennings and A. D. M. Rayner, eds. Cambridge University Press, Cambridge, England. 564 pp.

241. Brasier, C. M., and Strouts, R. G. 1976. New records of *Phytophthora* on trees in Britain. I. Phytophthora root rot and bleeding canker of horse chestnut (*Aesculus hippocastanum* L.). Eur. J. For. Pathol. 6:129–136.

242. Braun, E. J., and Sinclair, W. A. 1976. Histopathology of phloem necrosis in *Ulmus americana*. Phytopathology 66:598–607.

243. Braun, E. J., and Sinclair, W. A. 1979. Phloem necrosis of elms: symptoms and histopathological observations in tolerant hosts. Phytopathology 69:354–358.

244. Braun, H. J. 1976. Beech (*Fagus sylvatica* L.) bark disease, caused by *Cryptococcus fagi* Baer. I. Anatomy of bark of *Fagus sylvatica* L. as a basic factor. Eur. J. For. Pathol. 6:136–146.

245. Braun, U. 1981. Taxonomic studies in the genus *Erysiphe*. I. Generic delimitation and position in the system of the Erysiphaceae. Nova Hedgwigia 34:679–719.

246. Braun, U. 1982–83. Descriptions of new species and combinations in

Microsphaera and *Erysiphe*. I, II, III. Mycotaxon 14:369–374; 15:121–137; 16:417–424.

247. Braun, U. 1984. A short survey of the genus *Microsphaera* in North America. Nova Hedwigia 39:211–243.

248. Brener, W. D., Setliff, E. C., and Norgren, R. L. 1974. *Sclerophoma pithyophila* associated with a tip dieback of juniper in Wisconsin. Plant Dis. Rep. 58:653–657.

249. Brennan, E., Leone, I., Harkov, R., and Rhoads, A. 1981. Austrian pine injury traced to ozone and sulfur dioxide pollution. Plant Dis. 65:363–364.

250. Bresadola, J. 1894. Fungi aliquot saxonici novi vel critici a cl. W. Krieger lecti. Hedwigia 33:206–210.

251. Brierley, P. 1951. A witches'-broom of lilac. Plant Dis. Rep. 35:556.

252. Brierley, P. 1954. Symptoms in the florist's hydrangea caused by tomato ringspot virus and an unidentified sap-transmissible virus. Phytopathology 44:696–699.

253. Brierley, P. 1955. Dodder transmission of lilac witches'-broom virus. Plant Dis. Rep. 39:719–721.

254. Brinkerhoff, L. A., and Streets, R. B. 1946. Pathogenicity and pathological histology of *Phymatotrichum omnivorum* (the fungus causing cotton or Texas root rot) in a woody perennial—the pecan. Ariz. Agric. Exp. Stn. Bull. 111:103–126.

255. Britton, K. O., and Hendrix, F. F. 1982. Three species of *Botryosphaeria* cause peach tree gummosis in Georgia. Plant Dis. 66:1120–1121.

256. Britton-Jones, H. R. 1925. On the diseases known as "bark canker" and "die-back" in fruit trees. J. Pomol. Hortic. Sci. 4:162–183.

257. Brlansky, R. H., Lee, R. F., and Collins, M. H. 1985. Structural comparison of xylem occlusions in the trunks of citrus trees with blight and other decline diseases. Phytopathology 75:145–150.

258. Brlansky, R. H., Timmer, L. W., and Lee, R. F. 1982. Detection and transmission of a gram-negative, xylem-limited bacterium in sharpshooters from a citrus grove in Florida. Plant Dis. 66:590–592.

259. Brlansky, R. H., Timmer, L. W., Lee, R. F., and Graham, J. H. 1984. Relationship of xylem plugging to reduced water uptake and symptom development in citrus trees with blight and blightlike declines. Phytopathology 74:1325–1328.

260. Broadfoot, W. M., and Williston, H. L. 1973. Flooding effects on southern forests. J. For. 71:584–587.

261. Brookhouser, L. W., and Peterson, G. W. 1971. Infection of Austrian, Scots, and ponderosa pines by *Diplodia pinea*. Phytopathology 61:409–414.

262. Broughton, W. J., ed. 1981, 1982. Nitrogen fixation. Vol. 1: Ecology. Vol. 2: *Rhizobium*. Clarendon Press, Oxford. 306 and 353 pp.

263. Brown, E. A. II, and Britton, K. O. 1986. *Botryosphaeria* diseases of apple and peach in the southeastern United States. Plant Dis. 70:480–484.

264. Brown, E. A., and Hendrix, F. F. 1981. Pathogenicity and histopathology of *Botryosphaeria dothidea* on apple stems. Phytopathology 71:375–379.

265. Brown, G. E. 1971. Pycnidial release and survival of *Diplodia natalensis* spores. Phytopathology 61:559–561.

266. Brown, N. A. 1938. The tumor disease of oak and hickory trees. Phytopathology 28:401–411.

267. Brown, N. A. 1941. Tumors on elm and maple trees. Phytopathology 31:541–548.

268. Brown, T. S. Jr., and Merrill, W. 1973. Germination of basidiospores of *Fomes applanatus*. Phytopathology 63:547–550.

269. Brownell, K. H., and Schneider, R. W. 1983. Fusarium hypocotyl rot of sugar pine in California forest nurseries. Plant Dis. 67:105–107.

270. Bruck, R. I., and Manion, P. D. 1980. Interacting environmental factors associated with the incidence of Hypoxylon canker on trembling aspen. Can. J. For. Res. 10:17–24.

271. Bruederle, L. P., and Stearns, F. W. 1985. Ice storm damage to a southern Wisconsin mesic forest. Bull. Torrey Bot. Club 112:167–175.

272. Bryan, M. K. 1928. Lilac blight in the United States. J. Agric. Res. 36:225–235.

273. Buckland, D. C., Foster, R. E., and Nordin, V. J. 1949. Studies in forest pathology. VII. Decay in western hemlock and fir in the Franklin River area, British Columbia. Can. J. Res. C27:312–331.

274. Buckland, D. C., Molnar, A. C., and Wallis, G. W. 1954. Yellow laminated root rot of Douglas-fir. Can. J. Bot. 32:69–81.

275. Buczaki, S. T. 1973. A microecological approach to larch canker biology. Trans. Br. Mycol. Soc. 61:315–329.

276. Buczaki, S. T. 1973. Some factors governing mycelial establishment and lesion extension in the larch canker disease. Eur. J. For. Pathol. 3:39–49.

277. Bugbee, W. M., and Anderson, N. A. 1963. Infection of spruce seedlings by *Cylindrocladium scoparium*. Phytopathology 53:1267–1271.

278. Buisman, C. 1931. Three species of *Botryodiplodia* (Sacc.) on elm trees in the United States. J. Arnold Arbor. 12:289–296.

279. Buller, A. H. R. 1906. The biology of *Polyporus squamosus*, Huds., a timber destroying fungus. J. Econ. Biol. 1:101–138.

280. Burdekin, D. A., ed. 1983. Research on Dutch elm disease in Europe. G.B. For. Comm. Bull. 60. 113 pp.

281. Burdon, J. J., Seviour, R. J., and Fripp, Y. J. 1983. Electrophoretic patterns of soluble proteins of *Hendersonia* spp. Trans. Br. Mycol. Soc. 78:551–553.

282. Burdsall, H. H. Jr., and Snow, G. A. 1977. Taxonomy of *Cronartium quercuum* and *C. fusiforme*. Mycologia 69:503–508.

283. Burk, W. R., and Rex, R. E. 1974. *Polyporus squamosus* in Utah. Mycotaxon 1:135–136.

284. Burke, M. J., Gusta, L. V., Quamme, H. A., Weiser, C. J., and Li, P. H. 1976. Freezing and injury in plants. Annu. Rev. Plant Physiol. 27:507–528.

285. Burkholder, W. H., and Guterman, C. E. F. 1932. Synergism in a bacterial disease of *Hedera helix*. Phytopathology 22:781–784.

286. Burr, T. J., Hunter, J. E., Ogawa, J. M., and Abawi, G. S. 1978. A root rot of apple caused by *Rhizoctonia solani* in New York nurseries. Plant Dis. Rep. 62:476–478.

287. Burt, E. A. 1915. The Thelephoraceae of North America. IV. *Exobasidium*. Ann. Mo. Bot. Gard. 2:627–658.

288. Butin, H. 1958. Untersuchungen über ein Toxin in Kulturfiltraten von *Dothichiza populea* Sacc. et. Br. Phytopathol. Z. 33:135–146.

289. Butin, H. 1973. Morphologische und taxonomische Untersuchungen an *Naemacyclus niveus* (Pers. ex Fr.) Fuck. ex Sacc. und verwandten Arten. Eur. J. For. Pathol. 3:146–163.

290. Butin, H., and Shigo, A. L. 1981. Radial shakes and "frost cracks" in living oak trees. U.S. For. Serv. Res. Pap. NE-478. 21 pp.

291. Butler, E. J., and Jones, S. G. 1955. Plant pathology. Macmillan, London. 979 pp.

292. Byler, J. W., Cobb, F. W. Jr., and Parmeter, J. R. Jr. 1972. Effects of secondary fungi on the epidemiology of western gall rust. Can. J. Bot. 50:1061–1066.

293. Byler, J. W., Cobb, F. W. Jr., and Parmeter, J. R. Jr. 1972. Occurrence and significance of fungi inhabiting galls caused by *Peridermium harknessii*. Can. J. Bot. 50:1275–1282.

294. Byrde, R. J. W., and Willets, H. A. 1977. The brown rot fungi of fruit: their biology and control. Pergamon Press, New York. 171 pp.

295. Calavan, E. C., and Wallace, J. M. 1954. *Hendersonula toruloidea* Nattrass on citrus in California. Phytopathology 44:635–639.

296. Calder, D. M., and Bernhardt, P., eds. 1983. The biology of mistletoes. Academic Press, New York. 348 pp.

297. Calvin, C. L. 1967. Anatomy of the endophytic system of the mistletoe *Phoradendron flavescens*. Bot. Gaz. 128:117–137.

298. Cameron, H. R. 1962. Diseases of deciduous fruit trees incited by *Pseudomonas syringae* van Hall. Oreg. Agric. Exp. Stn. Tech. Bull. 66. 64 pp.

299. Cameron, H. R. 1970. *Pseudomonas* content of cherry trees. Phytopathology 60:1343–1346.

300. Cameron, H. R., Milbrath, J. A., and Tate, L. A. 1973. Pollen transmission of Prunus ringspot virus in prune and sour cherry orchards. Plant Dis. Rep. 57:241–243.

301. Camp, R. R., and Whittingham, W. F. 1974. Ultrastructural alterations in oak leaves parasitized by *Taphrina caerulescens*. Am. J. Bot. 61:964–972.

302. Campana, R. J. 1978. Comparative aspects of Dutch elm disease in eastern North America and California. Cal. Plant Pathol. No. 41:1–4.

303. Campbell, A. H. 1933. Zone lines in plant tissues. I. The black lines formed by *Xylaria polymorpha* (Pers.) Grev. in hardwoods. Ann. Appl. Biol. 20:123–145.

304. Campbell, A. H., and Munson, R. G. 1936. Zone lines in plant tissues. III. The black lines formed by *Polyporus squamosus* (Huds.) Fr. Ann. Appl. Biol. 23:453–464.

305. Campbell, W. A. 1939. *Daedalea unicolor* decay and associated cankers of maples and other hardwoods. J. For. 37:974–977.

306. Campbell, W. A., and Copeland, O. L. Jr. 1954. Littleleaf disease of shortleaf and loblolly pines. U.S. Dep. Agric. Circ. 940. 41 pp.

307. Campbell, W. A., and Davidson, R. W. 1938. A *Poria* as the fruiting stage of the fungus causing sterile conks on birch. Mycologia 30:553–560.

308. Campbell, W. A., and Davidson, R. W. 1939. *Poria andersonii* and *Polyporus glomeratus*, two distinct heart-rotting fungi. Mycologia 31:161–168.

309. Campbell, W. A., and Davidson, R. W. 1939. Sterile conks of *Polyporus glomeratus* and associated cankers on beech and red maple. Mycologia 31:606–611.

310. Campbell, W. A., and Davidson, R. W. 1940. *Ustulina vulgaris* decay in sugar maple and other hardwoods. J. For. 38:474–477.

311. Campbell, W. A., and Davidson, R. W. 1940. Top rot in glaze-damaged black cherry and sugar maple on the Allegany Plateau. J. For. 38:963–965.

312. Campbell, W. A., and Davidson, R. W. 1941. Cankers and decay of yellow birch associated with *Fomes igniarius* var. *laevigatus*. J. For. 39:559–560.

313. Campbell, W. A., and Davidson, R. W. 1942. A species of *Poria* causing rot and cankers of hickory and oak. Mycologia 34:17–26.

314. Campbell, W. A., and Miller, J. H. 1952. Windthrow of root-rotted oak shade trees. Plant Dis. Rep. 36:490.

315. Canadian Forestry Service, Forest Insect and Disease Survey. Forest insect

and disease conditions in Canada. (Issued annually beginning 1980; formerly titled Annual Report.)

316. Canfield, M. L., Baca, S., and Moore, L. W. 1986. Isolation of *Pseudomonas syringae* from 40 cultivars of diseased woody plants with tip dieback in Pacific Northwest nurseries. Plant Dis. 70:647–650.

317. Cannon, W. N. Jr., and Worley, D. P. 1980. Dutch elm disease control: performance and costs. U.S. For. Serv. Res. Pap. NE-457. 8 pp.

318. Carlson, C. E. 1978. Fluoride induced impact on a coniferous forest near the Anaconda aluminum plant in northwestern Montana. Ph.D. thesis, Univ. Montana. 176 pp.

319. Carlson, C. E., and Gilligan, C. J. 1983. Histological differentiation among abiotic causes of conifer needle necrosis. U.S. For. Serv. Res. Pap. INT-298. 16 pp.

320. Caroselli, N. E. 1953. Bleeding canker disease of hardwoods. Bartlett Tree Res. Lab. Sci. Tree Topics 2(1):1–6.

321. Caroselli, N. E. 1957. Verticillium wilt of maples. R.I. Agric. Exp. Stn. Bull. 335. 84 pp.

322. Caroselli, N. E. 1959. The relation of sapwood moisture content to the incidence of maple wilt caused by *Verticillium albo-atrum*. Phytopathology 49:496–498.

323. Caroselli, N. E., and Tucker, C. M. 1949. Pit canker of elm. Phytopathology 39:481–488.

324. Carpenter, E. D. 1970. Salt tolerance of ornamental plants. Am. Nurseryman 131(2):12, 54 . . . . 71.

325. Carpenter, L. R., Nelson, E. E., and Stewart, J. L. 1979. Development of dwarf mistletoe infections on western hemlock in coastal Oregon. For. Sci. 25:237–243.

326. Carpenter, P. L. 1972. Dicamba injury to *Taxus*. HortScience 7:573.

327. Carter, J. C. 1941. Preliminary investigation of oak diseases in Illinois. Ill. Nat. Hist. Surv. Bull. 21:195–230.

328. Carter, J. C. 1945. Wetwood of elms. Ill. Nat. Hist. Surv. Bull. 23:401–448.

329. Carter, J. C. 1947. Tubercularia canker and dieback of Siberian elm (*Ulmus pumila* L.). Phytopathology 37:243–246.

330. Carter, J. C. 1975. Diseases of Midwest trees. Univ. Ill. Coll. Agric. Spec. Publ. 35. 168 pp.

331. Carter, J. C., and Carter, L. R. 1974. An urban epiphytotic of phloem necrosis and Dutch elm disease, 1944–1972. Ill. Nat. Hist. Surv. Bull. 31:113–143.

332. Carter, J. C., and Sacamano, C. M. 1967. Fusicoccum canker, a new disease of Russian olive. Mycologia 59:535–537.

333. Carvell, K. L., Tryon, E. H., and True, R. P. 1957. Effects of glaze on the development of Appalachian hardwoods. J. For. 55:130–132.

334. Cash, E. K., and Davidson, R. W. 1940. Some new species of ascomycetes on coniferous hosts. Mycologia 32:728–735.

335. Cash, E. K., and Waterman, A. M. 1957. A new species of *Plagiostoma* associated with a leaf disease of hybrid aspens. Mycologia 49:756–760.

336. Casper, R., and Brandes, J. 1969. A new cactus virus. J. Gen. Virol. 5:155–156.

337. Casteldine, P., Grout, B. W. W., and Roberts, A. V. 1981. Cuticular resistance to *Diplocarpon rosae*. Trans. Br. Mycol. Soc. 77:665–666.

338. Castello, J. D., Amico, L., and O'Shea, M. T. 1984. Detection of tobacco mosaic and tobacco ringspot viruses in white ash trees by enzyme-linked immunosorbent assay. Plant Dis. 68:787–790.

339. Castello, J. D., Silverborg, S. B., and Manion, P. D. 1985. Intensification of ash decline in New York State from 1962 through 1980. Plant Dis. 69:243–246.

340. Cayley, D. M. 1923. Fungi associated with "die back" in stone fruit trees. I. Ann. Appl. Biol. 10:253–275.

341. Čech, M., Králík, O., and Blattný, C. 1961. Rod-shaped particles associated with virosis of spruce. Phytopathology 51:183–185.

342. Chandler, W. A., and Daniell, J. W. 1976. Relation of pruning time and inoculation with *Pseudomonas syringae* van Hall to short life of peach trees growing on old peach land. HortScience 11:103–104.

343. Chapman, R. L. 1976. Ultrastructure of *Cephaleuros virescens* (Chroolepidaceae; Chlorophyta). I. Scanning electron microscopy of zoosporangia. Am. J. Bot. 63:1060–1070.

344. Chapman, R. L. 1981. Ultrastructure of *Cephaleuros virescens* (Chroolepidaceae; Chlorophyta). III. Zoospores. Am. J. Bot. 68:554–556.

345. Charlton, J. W. 1963. Relating climate to eastern white pine blister rust infection hazard. U.S. For. Serv., E. Region, Upper Darby, PA. 38 pp.

346. Chastagner, G. A., and Byther, R. S. 1983. Infection period of *Phaeocryptopus gaeumannii* on Douglas-fir needles in western Washington. Plant Dis. 67:811–813.

347. Chastagner, G. A., Byther, R. S., MacDonald, J. D., and Michaels, E. 1984. Impact of Swiss needle cast on postharvest hydration and needle retention of Douglas-fir Christmas trees. Plant Dis. 68:192–195.

348. Chessin, M. 1965. Wild plant hosts of a cactus virus. Phytopathology 55:933.

349. Chessin, M., and Lesemann, D. 1972. Distribution of cactus viruses in wild plants. Phytopathology 62:97–99.

350. Chessin, M., Solberg, R. A., and Fischer, P. C. 1963. External symptoms and Geimsa-stainable cell inclusions associated with virus infection in cacti. Phytopathology 53:988–989.

351. Chiba, O., and Tanaka, K. 1968. The effect of sulfur dioxide on the development of pine needle blight caused by *Rhizosphaera kalkhoffii* Bubak (I). J. Jap. For. Soc. 50:135–139.

352. Childs, J. F. L. 1953. Concentric canker and wood rot of citrus associated with *Fomes applanatus* in Florida. Phytopathology 43:99–100.

353. Childs, T. W. 1968. Elytroderma disease of ponderosa pine in the Pacific Northwest. U.S. For. Serv. Res. Pap. PNW-69. 45 pp.

354. Childs, T. W. 1970. Laminated root rot of Douglas-fir in western Oregon and Washington. U.S. For. Serv. Res. Pap. PNW-102. 27 pp.

355. Childs, T. W., and Edgren, J. W. 1967. Dwarfmistletoe effects on ponderosa pine growth and trunk form. For. Sci. 13:167–174.

356. Childs, T. W., Shea, K. R., and Stewart, J. L. 1971. Elytroderma disease of ponderosa pine. U.S. For. Serv. For. Pest Leafl. 42. 6 pp.

357. Chou, C. K. S. 1978. Penetration of young stems of *Pinus radiata* by *Diplodia pinea*. Physiol. Plant Pathol. 13:189–192.

358. Christ, B. J., and Merrill, W. 1978. Effects of time and temperature on spore germination of *Rhizosphaera kalkhoffii*. (Abstr.) Proc. Am. Phytopathol. Soc. 4:185.

359. Christensen, C. M. 1940. Studies on the biology of *Valsa sordida* and *Cytospora chrysosperma*. Phytopathology 30:459–475.

360. Chupp, C. 1954. A monograph of the fungus genus *Cercospora*. Published by the author, Ithaca, NY. 667 pp.

361. Clark, J. 1961. Birch dieback. Pages 1551–1555 in: Recent advances in botany. University of Toronto Press, Toronto, Ont. 1766 pp.

362. Clark, J., and Barter, G. W. 1958. Growth and climate in relation to dieback of yellow birch. For. Sci. 4:343–364.

363. Clark, J., and Bonga, J. M. 1970. Photosynthesis and respiration in black spruce (*Picea mariana*) parasitized by eastern dwarf mistletoe (*Arceuthobium pusillum*). Can. J. Bot. 48:2029–2031.

364. Cline, M. N., Crane, J. L., and Cline, S. D. 1983. The teleomorph of *Cristulariella moricola*. Mycologia 75:988–994.

365. Cline, M. N., and Neely, D. 1979. *Cristulariella pyramidalis* and its pathogenesis on black walnut. Plant Dis. Rep. 63:1028–1032.

366. Cline, S., and Neely, D. 1983. Penetration and infection of leaves of black walnut by *Marssonina juglandis* and resulting lesion development. Phytopathology 73:494–497.

367. Clinton, G. P., and McCormick, F. A. 1929. The willow scab fungus. Conn. Agric. Exp. Stn. Bull. 302:443–469.

368. Cobb, F. W. Jr., and Fergus, C. L. 1964. Pathogenicity, host specificity, and mat production of seven isolates of the oak wilt fungus. Phytopathology 54:865–866.

369. Cobb, F. W. Jr., Slaughter, G. W., Rowney, D. L., and DeMars, C. J. 1982. Rate of spread of *Ceratocystis wageneri* in ponderosa pine stands in the central Sierra Nevada. Phytopathology 72:1359–1362.

370. Cobb, F. W. Jr., Wood, D. L., Stark, R. W., and Parmeter, J. R. Jr. 1968. Photochemical oxidant injury and bark beetle (Coleoptera: Scolytidae) infestation of ponderosa pine. IV. Theory on the relationships between oxidant injury and bark beetle infestation. Hilgardia 39:141–152.

371. Cochrane, V. W. 1945. The common leaf rust of cultivated roses, caused by *Phragmidium mucronatum* (Fr.) Schlecht. Cornell Univ. Agric. Exp. Stn. Mem. 268. 39 pp.

372. Coe, D. M., and Wagener, W. W. 1949. Ash anthracnose appears in California. Plant Dis. Rep. 33:232.

373. Cohen, L. I. 1954. The anatomy of the endophytic system of the dwarf mistletoe, *Arceuthobium campylopodum*. Am. J. Bot. 41:840–847.

374. Cohen, L. I. 1967. The pathology of *Hypodermella laricis* on larch, *Larix occidentalis*. Am. J. Bot. 54:118–124.

375. Cohen, M., Pelosi, R. R., and Brlansky, R. H. 1983. Nature and location of xylem blockage structures in trees with citrus blight. Phytopathology 73:1125–1130.

376. Colbaugh, P. F. 1983. Rose petal blight caused by *Alternaria alternata* (Fries) Keissler. (Abstr.) Phytopathology 73:499–500.

377. Cole, A., Mink, G. I., and Regev, S. 1982. Location of Prunus necrotic ringspot virus on pollen grains from infected almond and cherry trees. Phytopathology 72:1542–1545.

378. Cole, G. T. 1983. *Graphiola phoenicis*: a taxonomic enigma. Mycologia 75:93–116.

379. Cole, J. R. 1933. Liver-spot disease of pecan foliage caused by *Gnomonia caryae pecanae*, nov. var. J. Agric. Res. 47:869–881.

380. Cole, J. R. 1935. *Gnomonia nerviseda*, the perfect stage of the fungus that causes vein spot disease of pecan foliage. J. Agric. Res. 50:91–96.

381. Cole, J. R. 1937. Bunch disease of pecans. Phytopathology 27:604–612.

382. Coleman, J. S., Murdoch, C. W., Campana, R. J., and Smith, W. H. 1985. Decay resistance of elm wetwood. Can. J. Plant Pathol. 7:151–154.

383. Coley-Smith, J. R., Verhoeff, K., and Jarvis, W. R., eds. 1980. The biology of *Botrytis*. Academic Press, New York. 318 pp.

384. Colley, R. H. 1918. Parasitism, morphology, and cytology of *Cronartium ribicola*. J. Agric. Res. 15:619–659.

385. Collis, D. G. 1972. Pine needle casts in British Columbia. Can. For. Serv., Pac. For. Res. Cent., For. Insect Dis. Surv. Pest Leafl. 43. 9 pp.

386. Commonwealth Institute of Helminthology. 1972–present. C.I.H. descriptions of plant parasitic nematodes. Set 1– . Commonwealth Agricultural Bureaux, Kew, Surrey, England.

387. Commonwealth Mycological Institute. 1964–present. C.M.I. descriptions of pathogenic fungi and bacteria. No. 1– . Commonwealth Agricultural Bureaux, Kew, Surrey, England.

388. Commonwealth Mycological Institute and Association of Applied Biologists. 1970–present. Descriptions of plant viruses. No. 1– . Commonwealth Agricultural Bureaux, Kew, Surrey, England.

389. Conners, I. L. 1967. An annotated index of plant diseases in Canada. Can. Dep. Agric. Publ. 1251. 381 pp.

390. Conners, I. L., and Savile, D. B. O. 1953. Thirty-second annual report of the Canadian Plant Disease Survey, 1952. 123 pp.

391. Conti, G. G., Bassi, M., Maffi, D., and Bonecchi, R. 1985. Host-parasite relationship in a susceptible and a resistant rose cultivar inoculated with *Sphaerotheca pannosa*. I. Fungal growth, mechanical barriers and hypersensitive reaction. Phytopathol. Z. 113:71–80.

392. Converse, R. H. 1953. *Articularia* and *Microstroma* on pecan in Oklahoma. Plant Dis. Rep. 37:511–512.

393. Cook, R. T. A. 1981. Overwintering of *Diplocarpon rosae* at Wisley. Trans. Br. Mycol. Soc. 77:549–556.

394. Cooke, W. B. 1961. The genus *Schizophyllum*. Mycologia 53:575–599.

395. Cooke, W. B. 1962. A taxonomic study in the "black yeasts." Mycopathol. Mycol. Appl. 17:1–43.

396. Cooley, J. S. 1936. *Sclerotium rolfsii* as a disease of nursery apple trees. Phytopathology 26:1081–1083.

397. Cooley, S. J. 1984. *Meria laricis* on nursery seedlings of western larch in Washington. Plant Dis. 84:826.

398. Cooper, J. I. 1979. Virus diseases of trees and shrubs. Institute of Terrestrial Ecology, Oxford, England. 74 pp.

399. Cooper, J. I. 1980. The prevalence of cherry leaf roll virus in *Juglans regia* in the United Kingdom. Acta Phytopathol. Acad. Sci. Hung. 15:139–145.

400. Cooper, J. I., and Edwards, M. L. 1981. The distribution of poplar mosaic virus in hybrid poplars and virus detection by ELISA. Ann. Appl. Biol. 99:53–61.

401. Cooper, J. I., and Massalski, P. R. 1984. Viruses and virus-like diseases affecting *Betula* spp. Proc. R. Soc. Edinb. Sect. B, 85:183.

402. Cooper, J. I., Massalski, P. R., and Edwards, M. L. 1984. Cherry leaf roll virus in the female gametophyte and seed of birch and its relevance to vertical virus transmission. Ann. Appl. Biol. 105:55–64.

403. Copeland, O. L. Jr., and McAlpine, R. G. 1962. Soil characteristics associated with spot die-out in loblolly pine plantations. For. Sci. 8:12–15.

404. Corbaz, R. 1985. Pathotypes et variations du pouvoir pathogène chez *Chalara elegans* Nag Raj et Kendrick (*Thielaviopsis basicola*). Phytopathol. Z. 113:289–299.

405. Cordell, C. E., and Matuszewski, M. 1974. *Cylindrocladium scoparium*—damaging black walnut seedlings in Kentucky nurseries. Plant Dis. Rep. 58:188–189.

406. Cordell, C. E., and Rowan, S. J. 1975. *Cylindrocladium scoparium* infection in a natural sweetgum stand. Plant Dis. Rep. 59:775–776.

407. Corner, E. J. H. 1953. The construction of polypores—1. Introduction: *Polyporus sulfureus, P. squamosus, P. betulinus,* and *Polystictus microcyclus*. Phytomorphology 3:157–167.

408. Corner, E. J. H. 1983. Ad polyporaceas. I. *Amauroderma* and *Ganoderma*. Nova Hedgwigia 75:1–182.

409. Costonis, A. C. 1970. Acute foliar injury of eastern white pine induced by sulfur dioxide and ozone. Phytopathology 60:994–999.

410. Costonis, A. C. 1971. Effects of ambient sulfur dioxide and ozone on eastern white pine in a rural environment. Phytopathology 61:717–720.

411. Costonis, A. C., and Sinclair, W. A. 1969. Relationships of atmospheric ozone to needle blight of eastern white pine. Phytopathology 59:1566–1574.

412. Coutts, M. P., and Philipson, J. J. 1978. Tolerance of tree roots to waterlogging. II. Adaptation of Sitka spruce and lodgepole pine to waterlogged soil. New Phytol. 80:71–77.

413. Cox, R. S. 1954. *Cylindrocladium scoparium* on conifer seedlings. Del. Agric. Exp. Stn. Bull. 301 (Tech.). 40 pp.

414. Coyier, D. L., Stace-Smith, R., Allen T. C., and Leung, E. 1977. Viruslike particles associated with a rhododendron necrotic ringspot disease. Phytopathology 67:1090–1095.

415. Crandall, B. S. 1943. Bacterial infection and decay of the inner wood of winter-injured young London plane trees. Phytopathology 33:963–964.

416. Crandall, B. S. 1945. A new species of *Cephalosporium* causing persimmon wilt. Mycologia 37:495–498.

417. Crandall, B. S., and Baker, W. L. 1950. The wilt disease of American persimmon, caused by *Cephalosporium diospyri*. Phytopathology 40:307–325.

418. Creager, D. B. 1937. Phytophthora crown rot of dogwood. J. Arnold Arbor. 18:344–348.

419. Creager, D. B. 1937. The Cephalosporium disease of elms. Contrib. Arnold Arbor. 10. 91 pp.

420. Creelman, D. W. 1956. The occurrence of ash rust in western Nova Scotia. Plant Dis. Rep. 40:580.

421. Cripe, R. E. 1979. Lightning protection for trees and related property. J. Arboric. 5:145–149.

422. Cripe, R. E. 1985. Lightning protection for trees. Arbor Age 5(4):13, 14, 16, 18, 20.

423. Croghan, C. F., and Robbins, K. 1986. Cankers caused by *Botryodiplodia gallae* associated with oak sprout mortality in Michigan. Plant Dis. 70:76–77.

424. Crone, L. J., and Bachelder, S. 1961. Insect transmission of canker stain fungus, *Ceratocystis fimbriata* f. *platani*. (Abstr.) Phytopathology 51:576.

425. Crops Research Division, U.S. Agricultural Research Service. 1960. Index of plant diseases in the United States. U.S. Dep. Agric. Agric. Handbk. 165. 531 pp.

426. Crosse, J. E. 1966. Epidemiological relations of the pseudomonad pathogens of deciduous fruit trees. Annu. Rev. Phytopathol. 4:291–310.

427. Crowe, F., Starkey, D., and Lengleek, V. 1982. Honeylocust canker in Kansas caused by *Thyronectria austro-americana*. Plant Dis. 66:155–158.

428. Crowell, I. H. 1934. The hosts, life history, and control of the cedar-apple rust fungus *Gymnosporangium juniperi-virginianae* Schw. J. Arnold Arbor. 15:163–232.

429. Crowell, I. H. 1935. The hosts, life history, and control of *Gymnosporangium clavipes* C. & P. J. Arnold Arbor. 16:367–410.

430. Croxton, W. C. 1939. A study of the tolerance of trees to breakage by ice accumulation. Ecology 20:71–73.

431. Cummins, G. B. 1971. Rust fungi of cereals, grasses, and bamboos. Springer-Verlag, New York. 570 pp.

432. Cummins, G. B. 1978. Rust fungi on legumes and composites in North America. University of Arizona Press, Tucson. 424 pp.

433. Cummins, G. B. 1984. Two new rust fungi (Uredinales). Mycologia 20:617–618.

434. Cummins, G. B., and Hiratsuka, Y. 1983. Illustrated genera of rust fungi. Revised ed. American Phytopathological Society, St. Paul, MN. 152 pp.

435. Curry, J. R., and Church, T. W. Jr. 1952. Observations on winter drying of conifers in the Adirondacks. J. For. 50:114–116.

436. Curzi, M. 1927. Di uno speciale parassitismo dell' *Ascochyta syringae*. Riv. Patol. Veg. 17:22–23.

437. Czabator, F. J. 1971. Fusiform rust of southern pines—a critical review. U.S. For. Serv. Res. Pap. SO-65. 39 pp.

438. Czabator, F. J. 1976. A new species of *Ploioderma* associated with pine needle blight. Mem. N.Y. Bot. Gard. 28:41–44.

439. Czabator, F. J., Staley, J. M., and Snow, G. A. 1971. Extensive southern pine needle blight during 1970–71 and associated fungi. Plant Dis. Rep. 55:764–766.

440. Dance, B. W. 1957. A fungus associated with blight and dieback of hybrid aspen. Can. Dep. Agric., Sci. Serv., For. Biol. Div. Bi-mon. Prog. Rep. 13(6):1–2.

441. Dance, B. W. 1961. Leaf and shoot blight of poplars (Section Tacamahaca Spach) caused by *Venturia populina* (Vuill.) Fabric. Can. J. Bot. 39:875–890.

442. Dance, B. W. 1961. Spore dispersal in *Pollaccia radiosa* (Lib.) Bald. and Cif. Can. J. Bot. 39:1429–1435.

443. Daniels, M. J. 1983. Mechanisms of spiroplasma pathogenicity. Annu. Rev. Phytopathol. 21:29–43.

444. Darker, G. C. 1932. The Hypodermataceae of conifers. Contrib. Arnold Arbor. 1:1–131.

445. Darker, G. D. 1967. A revision of the genera of the Hypodermataceae. Can. J. Bot. 45:1399–1444.

446. Davidson, R. W. 1934. *Stereum gausapatum*, cause of heart rot of oaks. Phytopathology 24:831–832.

447. Davidson, R. W. 1935. Decay in living sprout oak trees. Plant Dis. Rep. 19:94–95.

448. Davidson, R. W., and Campbell, W. A. 1944. Observations on a gall of sugar maple. Phytopathology 34:132–135.

449. Davidson, R. W., Campbell, W. A., and Vaughan, D. B. 1942. Fungi causing decay of living oaks in the eastern United States and their cultural identification. U.S. Dep. Agric. Tech. Bull. 765. 65 pp.

450. Davidson, R. W., and Cash, E. K. 1956. A *Cenangium* associated with sooty-bark canker of aspen. Phytopathology 46:34–36.

451. Davidson, R. W., and Lorenz, R. C. 1939. Species of *Eutypella* and *Schizoxylon* associated with cankers of maple. Phytopathology 28:733–745.

452. Davies, W. J., and Kozlowski, T. T. 1974. Short- and long-term effects of

antitranspirants on water relations and photosynthesis of woody plants. J. Am. Soc. Hortic. Sci. 99:297–304.

453. Davis, B. H. 1938. The Cercospora leafspot of rose caused by *Mycosphaerella rosicola*. Mycologia 30:282–298.

454. Davis, D. D., Umbach, D. M., and Coppolino, J. B. 1981. Susceptibility of tree and shrub species and response of black cherry foliage to ozone. Plant Dis. 65:904–907.

455. Davis, D. D., and Wilhour, R. G. 1976. Susceptibility of woody plants to sulfur dioxide and photochemical oxidants. EPA Ecol. Res. Ser. EPA-600/3-76-102. 71 pp.

456. Davis, J. R., and English, H. 1969. Factors related to the development of bacterial canker in peach. Phytopathology 59:588–595.

457. Davis, M. J., Thomson, S. V., and Purcell, A. H. 1980. Etiological role of the xylem-limited bacterium causing Pierce's disease in almond leaf scorch. Phytopathology 70:472–475.

458. Davis, S. H., and Peterson, J. L. 1976. Susceptibility of cotoneasters to fire blight. J. Arboric. 2:90–91.

459. Davis, S. H. Jr., and Peterson, J. L. 1980. Trunk decay on Greenspire linden. J. Arboric. 6:258–260.

460. Davis, T. C. 1966. Appraisal of *Hypoxylon punctulatum* as a biological control agent of *Ceratocystis fagacearum* in oak-wilt trees. Phytopathology 56:772–775.

461. Davison, A. D. 1972. Factors affecting development of madrone canker. Plant Dis. Rep. 56:50–52.

462. Davison, S. E., and Forman, R. T. T. 1982. Herb and shrub dynamics in a mature oak forest: a thirty-year study. Bull. Torrey Bot. Club 109:64–73.

463. Day, W. R. 1928. Damage by late frost on Douglas fir, Sitka spruce, and other conifers. Forestry 2:19–30.

464. Day, W. R. 1954. Drought crack of conifers. G.B. For. Comm. For. Rec. No. 26. 40 pp.

465. Dearness, J., and Hansbrough, J. R. 1934. *Cytospora* infection following fire injury in western British Columbia. Can. J. Res. 10:125–128.

466. De Cleene, M., and De Ley, J. 1976. The host range of crown gall. Bot. Rev. 42:389–466.

467. De Cleene, M., and De Ley, J. 1981. The host range of infectious hairy-root. Bot. Rev. 47:147–194.

468. De Cleene, M., and De Ley, J. 1981. The reevaluation of pathogenicity of "*Agrobacterium pseudotsugae*." Phytopathol. Z. 101:185–188.

469. Défago, G. 1942. Seconde contribution à la connaissance des Valsées v. Höhnel. Phytopathol. Z. 14:103–147.

470. De Groot, R. C. 1966. Phenolic extractives in lateral branches and injured leaders of *Pinus strobus* L. Can. J. Bot. 44:57–61.

471. Deighton, F. C. 1973. Five North American *Cercospora*-like fungi. Trans. Br. Mycol. Soc. 61:107–120.

472. Deighton, F. C. 1976. Studies on *Cercospora* and allied genera. VI. *Pseudocercospora* Speg., *Pantospora* Cif., and *Cercoseptoria* Petr. Commonw. Mycol. Inst. Mycol. Pap. No. 140. 168 pp.

473. Demaree, J. B., and Cole, J. R. 1936. A disporous *Gnomonia* on pecan. Phytopathology 26:1025–1029.

474. Denham, T. G., and Waller, J. M. 1981. Some epidemiological aspects of post-bloom fruit drop disease (*Colletotrichum gloeosporioides*) in citrus. Ann. Appl. Biol. 98:65–77.

475. Dessureault, M., Lachance, D., Roy, G., Robitaille, L., and Gagnon, G. 1985. Symposium: maple decline in Quebec. Phytoprotection 66:69–99.

476. DeVay, J. E., Sinden, S. L., Lukezic, F. L., Werenfels, L. F., and Backman, P. A. 1968. Poria root and crown rot of cherry trees. Phytopathology 58:1239–1241.

477. Dewey, F. M., Barrett, D. K., Vose, I. R., and Lamb, C. J. 1984. Immunofluorescence microscopy for the detection and identification of propagules of *Phaeolus schweinitzii* in infested soil. Phytopathology 74:291–296.

478. Dhanvantari, B. N. 1978. Characterization of *Agrobacterium* isolates from stone fruits in Ontario. Can. J. Bot. 56:2309–2311.

479. Dhanvantari, B. N. 1978. Cold predisposition of dormant peach twigs to nodal cankers caused by *Leucostoma* spp. Phytopathology 68:1779–1783.

480. Dhanvantari, B. N., Johnson, P. W., and Dirks, V. A. 1975. The role of nematodes in crown gall infection of peach in southwestern Ontario. Plant Dis. Rep. 59:109–112.

481. Dharne, C. G. 1965. Taxonomic investigations on the discomycetous genus *Lachnellula* Karst. Phytopathol. Z. 53:101–144.

482. Dhingra, O. D., and Chagas, D. 1981. Effect of soil temperature, moisture, and nitrogen on competitive saprophytic ability of *Macrophomina phaseolina*. Trans. Br. Mycol. Soc. 77:15–20.

483. Dhingra, O. D., and Sinclair, J. B. 1977. An annotated bibliography of *Macrophomina phaseolina*, 1905–1975. Universidade Federal de Viçosa, Viçosa, Brazil, and University of Illinois, Urbana. 244 pp.

484. Dhingra, O. D., and Sinclair, J. B. 1978. Biology and pathology of *Macrophomina phaseolina*. Impresna Universitaria Universidade Federal de Viçosa, Viçosa, Brazil. 166 pp.

485. Diamandis, S. 1978. "Top-dying" of Norway spruce, *Picea abies* (L.) Karst., with special reference to *Rhizosphaera kalkhoffii* Bubak. II. Status of

R. *kalkhoffii* in "top-dying" of Norway spruce. Eur. J. For. Pathol. 8:345–356.

486. Dickey, R. D. 1977. Nutritional deficiencies of woody ornamental plants used in Florida landscapes. Fla. Agric. Exp. Stn. Bull. 791. 63 pp.

487. DiCosmo, F., Peredo, H., and Minter, D. W. 1983. *Cyclaneusma* gen. nov., *Naemacyclus*, and *Lasiostictis*, a nomenclatural problem resolved. Eur. J. For. Pathol. 13:206–212.

488. DiCosmo, F., Raj, T. R. N., and Kendrick, W. B. 1984. A revision of the Phacidiaceae and related anamorphs. Mycotaxon 21:1–234.

489. Dietz, S. M. 1926. The alternate hosts of crown rust, *Puccinia coronata* Corda. J. Agric. Res. 33:953–970.

490. Diller, J. D. 1943. A canker of eastern pines associated with *Atropellis tingens*. J. For. 41:41–52.

491. Diller, J. D., and Clapper, R. B. 1969. Asiatic and hybrid chestnut trees in the eastern United States. J. For. 67:328–331.

492. Dinus, R. J., and Schmidt, R. S., eds. 1977. Management of fusiform rust in southern pines. Symp. Proc., Univ. Fla., Gainesville. 163 pp.

493. Dirr, M. A., and Biedermann, J. 1980. Amelioration of salt damage to cotoneaster by gypsum. J. Arboric. 6:108–110.

494. Di Sanzo, C. P., and Rohde, R. A. 1969. *Xiphinema americanum* associated with maple decline in Massachusetts. Phytopathology 59:279–284.

495. Dixon, R. K., Pallardy, S. G., Garrett, H. E., and Cox, G. S. 1983. Comparative water relations of container-grown and bare-root ectomycorrhizal and nonmycorrhizal *Quercus velutina* seedlings. Can. J. Bot. 61:1559–1565.

496. Doane, C. C., and McManus, M. L., eds. 1981. The gypsy moth: research toward integrated pest management. U.S. Dep. Agric. Tech. Bull. 1584. 757 pp.

497. Dochinger, L. S. 1967. Occurrence of poplar cankers caused by *Fusarium solani* in Iowa. Plant Dis. Rep. 51:900–903.

498. Dochinger, L. S. 1973. Trees for polluted air. U.S. Dep. Agric. Misc. Publ. 1230. 12 pp.

499. Dochinger, L. S., and Seliskar, C. E. 1962. Fusarium canker found on yellow-poplar. J. For. 60:331–333.

500. Dodd, A. P. 1940. The biological campaign against prickly pear. Commonwealth Prickly Pear Board, Brisbane, Australia. 177 pp.

501. Dodds, J. A. 1980. Revised estimates of the molecular weights of dsRNA segments in hypovirulent strains of *Endothia parasitica*. Phytopathology 70:1217–1220.

502. Dodge, B. O. 1931. A destructive red-cedar rust disease. J. N.Y. Bot. Gard. 32:101–108.

503. Dodge, B. O. 1933. The orange-rust of hawthorn and quince invades the trunk of red cedar. J. N.Y. Bot. Gard. 34:233–237.

504. Dodge, B. O. 1938. A further study of the dry-rot disease of *Opuntia*. Mycologia 30:82–96.

505. Dodge, B. O. 1944. A new *Pseudonectria* on *Pachysandra*. Mycologia 36:532–537.

506. Doi, Y., Teranaka, M., Yora, K., and Asuyama, H. 1967. Mycoplasma- or PLT group-like microorganisms found in phloem elements of plants infected with mulberry dwarf, potato witches' broom, aster yellows, or paulownia witches' broom. Ann. Phytopathol. Soc. Jap. 33:259–266.

507. Dolezal, W. E., and Tainter, F. H. 1979. Phenology of comandra rust in Arkansas. Phytopathology 69:41–44.

508. Domanski, S. 1982. *Bjerkandera adusta* on young *Quercus rubra* and *Quercus robur* injured by late spring frosts in the Upper Silesia industrial district of Poland. Eur. J. For. Pathol. 12:406–413.

509. Dommergues, Y. R., and Krupa, S. V., eds. 1978. Interactions between non-pathogenic soil microorganisms and plants. Elsevier, Amsterdam. 475 pp.

510. Donselman, H. M. 1978. Palms resistant to lethal yellowing for Florida. Fla. State Hortic. Soc. Proc. 91:99–101.

511. Donselman, H. 1981. Lethal yellowing of palm trees in Florida. Univ. Fla., Fla. Hortic. OH-47. 4 pp.

512. Dooley, H. L. 1984. Temperature effects on germination of uredospores of *Melampsoridium betulinum* and on rust development. Plant Dis. 68:686–688.

513. Dorworth, C. E. 1971. Diseases of conifers incited by *Scleroderris lagerbergii* Gremmen: a review and analysis. Can. For. Serv. Publ. 1289. 42 pp.

514. Dorworth, C. E. 1972. Epidemiology of *Scleroderris lagerbergii* in central Ontario. Can. J. Bot. 50:751–765.

515. Dorworth, C. E. 1981. Status of pathogenic and physiologic races of *Gremmeniella abietina*. Plant Dis. 65:927–931.

516. Downs, A. A. 1938. Glaze damage in the birch-beech-maple-hemlock type of Pennsylvania and New York. J. For. 36:63–70.

517. Dozier, W. A. Jr., Latham, A. J., Kouskolekas, C. A., and Mayton, E. L. 1974. Susceptibility of apple rootstocks to black root rot and wooly apple aphids. HortScience 9:35–36.

518. Drew, M. C., and Lynch, J. M. 1980. Soil anaerobiosis, microorganisms, and root function. Annu. Rev. Phytopathol. 18:37–66.

519. Drilias, M. J., Kuntz, J. E., and Worf, G. L. 1982. Collar rot and basal canker of sugar maple. J. Arboric. 8:29–33.

520. Dropkin, V. 1980. Introduction to plant nematology. Wiley, New York. 293 pp.

521. Dropkin, V. H., Foudin, A., Kondo, E., Linit, M., and Smith, M. 1981. Pinewood nematode: a threat to U.S. forests? Plant Dis. 65:1022–1027.

522. Dubin, H. J., and English, H. 1975. Epidemiology of European apple canker in California. Phytopathology 65:542–550.

523. Dubreuil, S. H. 1981. Occurrence, symptoms, and interactions of *Phaeolus schweinitzii* and associated fungi causing decay and mortality of conifers. Ph.D. thesis, Univ. Idaho. 171 pp.

524. Duchelle, S. F., Skelly, J. M., and Chevone, B. I. 1982. Oxidant effects on forest tree seedling growth in the Appalachian Mountains. Water Air Soil Pollut. 18:363–373.

525. Duncan, W. H. 1975. Woody vines of the southeastern United States. University of Georgia Press, Athens. 76 pp.

526. Durrieu, G. 1957. Influence du climat sur la biologie de *Phaeocryptopus gäumannii* (Rohde) Petrak, parasite du *Pseudotsuga*. C. R. Hebd. Séances Acad. Sci. (Paris) D 244:2183–2185.

527. Dwinell, L. D. 1971. Interaction of *Cronartium fusiforme* and *Cronartium quercuum* with *Quercus velutina*. Phytopathology 61:1055–1058.

528. Dwinell, L. D. 1974. Susceptibility of southern oaks to *Cronartium fusiforme* and *Cronartium quercuum*. Phytopathology 64:400–403.

529. Dwinell, L. D. 1978. Susceptibility of southern pines to infection by *Fusarium moniliforme* var. *subglutinans*. Plant Dis. Rep. 62:108–111.

530. Dwinell, L. D. 1985. Relative susceptibilities of five pine species to three populations of the pinewood nematode. Plant Dis. 69:440–442.

531. Dwinell, L. D., Barrows-Broaddus, J. B., and Kuhlman, E. G. 1985. Pitch canker: a disease complex of southern pines. Plant Dis. 69:270–276.

532. Dye, D. W. 1967. Bacterial spot of ivy caused by *Xanthomonas hederae* (Arnaud, 1920) Dowson 1939, in New Zealand. N.Z. J. Sci. 10:481–485.

533. Dye, D. W., Bradbury, J. F., Goto, M., Hayward, A. C., Lelliott, R. A., and Schroth, M. N. 1980. International standards for naming pathovars of phytopathogenic bacteria and a list of pathovar names and pathotype strains. Rev. Plant Pathol. 59:153–168.

534. Eames, A. J., and Cox, L. G. 1945. A remarkable tree fall and an unusual type of graft-union failure. Am. J. Bot. 32:331–335.

535. Edgerton, C. W. 1908. The physiology and development of some anthracnoses. Bot. Gaz. 45:367–408.

536. Edgerton, C. W. 1911. Diseases of the fig tree and fruit. La. Agric. Exp. Stn. Bull. 126. 20 pp.

537. Ehrlich, J. 1934. The beech bark disease: a *Nectria* disease of *Fagus* following *Cryptococcus fagi* (Baer). Can. J. Res. 10:593–692.

538. Eisensmith, S. P., Sjulin, T. M., Jones, A. L., and Cress, C. E. 1982. Effects of leaf age and inoculum concentration on infection of sour cherry by *Coccomyces hiemalis*. Phytopathology 72:574–577.

539. Eleuterius, L. N. 1976. Observations on the mistletoe (*Phoradendron flavescens*) in South Mississippi, with special reference to the mortality of *Quercus nigra*. Castanea 41:265–268.

540. Ellis, J. B., and Kellerman, W. A. 1887. New Kansas fungi. J. Mycol. 3:102–105.

541. Ellis, M. A., Ferree, D. C., and Spring, D. E. 1981. Photosynthesis, transpiration, and carbohydrate content of apple leaves infected by *Podosphaera leucotricha*. Phytopathology 71:392–395.

542. Ellis, M. B. 1959. *Clasterosporium* and some allied Dematiaceae—Phragmosporae. II. Commonw. Mycol. Inst. Mycol. Pap. 72. 75 pp.

543. Ellis, M. B. 1971. Dematiaceous hyphomycetes. Commonwealth Mycological Institute, Kew, Surrey, England. 608 pp.

544. Englerth, G. H. 1942. Decay of western hemlock in western Oregon and Washington. Yale Univ. Sch. For. Bull. 50. 53 pp.

545. English, H., Davis, J. R., and DeVay, J. E. 1975. Relationship of *Botryosphaeria dothidea* and *Hendersonula toruloidea* to a canker disease of almond. Phytopathology 65:114–122.

546. English, H., Lownsbery, B. F., Schick, F. J., and Burlando, T. 1982. Effect of ring and pin nematodes on the development of bacterial canker and Cytospora canker in young French prune trees. Plant Dis. 66:114–116.

547. Erwin, D. C., Bartnicki-Garcia, S., and Tsao, P. H., eds. 1983. *Phytophthora*: its biology, taxonomy, ecology, and pathology. American Phytopathological Society, St. Paul, MN. 392 pp.

548. Eshed, N., and Dinoor, A. 1980. Genetics of pathogenicity in *Puccinia coronata*: pathogenic specialization at the host genus level. Phytopathology 70:1042–1046.

549. Essig, F. M. 1922. The morphology, development, and economic aspects of *Schizophyllum commune* Fries. Univ. Cal. (Berkeley) Publ. Bot. 7:447–498.

550. Etheridge, D. E. 1973. Wound parasites causing tree decay in British Columbia. Can. For. Serv., Pac. For. Res. Cent., For. Pest Leafl. 62. 15 pp.

551. Etheridge, D. E., and Craig, H. M. 1976. Factors influencing infection and initiation of decay by the Indian paint fungus (*Echinodontium tinctorium*) in western hemlock. Can. J. For. Res. 6:299–318.

552. Evans, H. C. 1984. The genus *Mycosphaerella* and its anamorphs *Cercoseptoria*, *Dothistroma*, and *Lecanosticta* on pines. Commonw. Mycol. Inst. Mycol. Pap. No. 153. 102 pp.

553. Farley, J. D., Wilhelm, S., and Snyder, W. C. 1971. Repeated germination and sporulation of microsclerotia of *Verticillium albo-atrum* in soil. Phytopathology 61:260–264.

554. Faull, J. H. 1930. The spread and control of Phacidium blight in spruce plantations. J. Arnold Arbor. 11:136–147.

555. Faull, J. H. 1934. The biology of Milesian rusts. J. Arnold Arbor. 15:50–85.

556. Faull, J. H. 1938. Taxonomy and geographical distribution of the genus *Uredinopsis*. Contrib. Arnold Arbor. 11:1–120.

557. Faull, J. H. 1938. The biology of rusts of the genus *Uredinopsis*. J. Arnold Arbor. 19:402–436.

558. Fayret, J. 1967. Action de la température et de la lumière sur la multiplication asexuée et de la reproduction sexuelle de *Gnomonia leptostyla* (Fr.) Cesat. et de Not., en culture pure. C. R. Hebd. Séances Acad. Sci. (Paris) D 265:1897–1900.

559. Featherly, H. L. 1941. The effect of grapevines on trees. Proc. Ok. Acad. Sci. 21:61–62.

560. Felix, L. S., Uhrenholdt, B., and Parmeter, J. R. Jr. 1971. Association of *Scolytus ventralis* (Coleoptera: Scolytidae) and *Phoradendron bolleanum* subspecies *pauciflorum* on *Abies concolor*. Can. Ent. 103:1697–1703.

561. Ferchau, H. A., and Johnson, T. W. Jr. 1956. Taxonomy of the Cenangium dieback fungus. For. Sci. 2:281–285.

562. Fergus, C. L. 1954. An epiphytotic of Phyllosticta leaf spot of maple. Plant Dis. Rep. 38:678–679.

563. Fergus, C. L. 1956. Some observations about *Polyporus dryadeus* on oak. Plant Dis. Rep. 40:827–829.

564. Ferrin, D. M., and Ramsdell, D. C. 1977. Ascospore dispersal and infection of grapes by *Guignardia bidwellii*, the causal agent of grape black rot disease. Phytopathology 67:1501–1505.

565. Ferrin, D. M., and Ramsdell, D. C. 1978. Influence of conidia dispersal and environment on infection of grape by *Guignardia bidwellii*. Phytopathology 68:892–895.

566. Filer, T. H. Jr. 1967. Pathogenicity of *Cytospora*, *Phomopsis*, and *Hypomyces* on *Populus deltoides*. Phytopathology 57:978–980.

567. Filer, T. H. Jr. 1969. Sycamore canker caused by *Botryodiplodia theobromae*. Phytopathology 59:76–78.

568. Filer, T. H. Jr., Davis, R. G., and Hegwood, C. P. 1979. Hardwood hosts for *Poria latemarginata*. (Abstr.) Phytopathology 69:527.

569. Filer, T. H., Solomon, J. D., McCracken, F. I., Oliveria, F. L., Lewis, R. Jr., Weiss, M. J., and Rogers, T. J. 1977. Sycamore pests: a guide to major insects, diseases, and air pollution. U.S. For. Serv. Southeast. Area, State & Priv. For., S. For. Exp. Stn. 36 pp.

570. Filho, E. S., and Dhingra, O. D. 1980. Survival of *Macrophomina phaseolina* sclerotia in nitrogen amended soils. Phytopathol. Z. 97:136–143.

571. Filip, G. M., Hadfield, J. S., and Schmitt, C. L. 1979. Branch mortality of true firs in west-central Oregon associated with dwarf mistletoe and canker fungi. Plant Dis. Rep. 63:189–193.

572. Filip, G. M., and Schmitt, C. L. 1979. Susceptibility of native conifers to laminated root rot east of the Cascade range in Oregon and Washington. For. Sci. 25:261–265.

573. Fisher, R. F., Garbett, W. S., and Underhill, E. M. 1981. Effects of fertilization on healthy and pitch canker-infected pines. S. J. Appl. For. 5:77–79.

574. Fitzpatrick, R. E. 1934. The life history and parasitism of *Taphrina deformans*. Sci. Agric. 14:305–326.

575. Flack, N. J., and Swinburne, T. R. 1977. Host range of *Nectria galligena* Bres. and the pathogenicity of some Northern Ireland isolates. Trans. Br. Mycol. Soc. 68:185–192.

576. Florida Department of Agriculture & Consumer Services, Division of Plant Industry. 1962–present. Nematology Circulars 1– .

577. Florida Department of Agriculture & Consumer Services, Division of Plant Industry. 1962–present. Plant Pathology Circulars 1– .

578. Flückiger, W., and Braun, S. 1981. Perspectives of reducing the deleterious effect of de-icing salt upon vegetation. Plant Soil 63:527–529.

579. Ford, D. H., and Rawlins, T. E. 1956. Improved cytochemical methods for differentiating *Cronartium ribicola* from *Cronartium occidentale* on *Ribes*. Phytopathology 46:667–668.

580. Ford, R. E., Moline, H. E., McDaniel, G. L., Mayhew, D. E., and Epstein, A. H. 1972. Discovery and characterization of elm mosaic virus in Iowa. Phytopathology 62:987–992.

581. Forer, L. B., Powell, C. A., and Stouffer, R. F. 1984. Transmission of tomato ringspot virus to apple rootstock cuttings and to cherry and peach seedlings by *Xiphinema rivesi*. Plant Dis. 68:1052–1054.

582. Forsyth, J., and Maynard, J. 1969. The sensitivity of ornamental plants to insecticides and acaricides. Commonw. Bur. Hortic. & Plantation Crops, Hortic. Rev. No. 1. 66 pp.

583. Foster, R. E., and Foster, A. T. 1951. Studies in forest pathology. VIII.

Decay of western hemlock on the Queen Charlotte Islands, British Columbia. Can. J. Bot. 29:479–521.

584. Fowler, M. E. 1947. Glomerella leaf spot of Magnolia. Plant Dis. Rep. 31:298.

585. Fowler, M. E., and Berry, F. H. 1958. Blossom-end rot of Chinese chestnuts. Plant Dis. Rep. 42:91–96.

586. Francois, L. E. 1982. Salt tolerance of eight ornamental tree species. J. Am. Soc. Hortic. Sci. 107:66–68.

587. Francois, L. E., and Clark, R. A. 1979. Boron tolerance of twenty-five ornamental shrub species. J. Am. Soc. Hortic. Sci. 104:319–322.

588. Frankie, G. W., and Parmeter, J. R. Jr. 1972. A preliminary study of the relationship between *Coryneum cardinale* (Fungi Imperfecti) and *Laspeyresia cupressana* (Lepidoptera: Tortricidae). Plant Dis. Rep. 56:992–994.

589. Fraser, W. P. 1914. Notes on *Uredinopsis mirabilis* and other rusts. Mycologia 6:25–28.

590. Freer-Smith, P. H. 1984. The responses of six broadleaved trees during long-term exposure to SO_2 and NO_2. New Phytol. 97:49–61.

591. Freier, G. D. 1977. Lightning and trees. J. Arboric. 3:131–137.

592. Freitag, J. H. 1951. Host range of the Pierce's disease virus of grapes as determined by insect transmission. Phytopathology 41:920–934.

593. French, D. W., Hodges, C. S. Jr., and Froyd, J. D. 1969. Pathogenicity and taxonomy of *Hypoxylon mammatum*. Can. J. Bot. 47:223–226.

594. French, D. W., and Menge, J. A. 1978. Survival of *Cylindrocladium floridanum* in naturally and artificially infested forest tree nurseries. Plant Dis. Rep. 62:806–810.

595. French, D. W., and Schroeder, D. B. 1969. Oak wilt fungus, *Ceratocystis fagacearum*, as a selective silvicide. For. Sci. 15:198–203.

596. French, D. W., and Stienstra, W. C. 1975. Oak wilt disease. Univ. Minn. Ext. Folder 310. 6 pp.

597. French, J. R., and Hart, J. H. 1978. Variation in resistance of trembling aspen to *Hypoxylon mammatum* identified by inoculating naturally occurring clones. Phytopathology 68:485–490.

598. French, W. J. 1969. Eutypella canker on *Acer* in New York. N.Y. State Coll. For. Tech. Publ. 94. 56 pp.

599. Friedland, A. J., Gregory, R. A., Kärenlampi, L., and Johnson, A. H. 1984. Winter damage to foliage as a factor in red spruce decline. Can. J. For. Res. 14:963–965.

600. Friend, R. J. 1965. What is *Fumago vagans*? Trans. Br. Mycol. Soc. 48:371–375.

601. Froelich, R. C., Cowling, E. B., Collicott, L. V., and Dell, R. R. 1977. *Fomes annosus* reduces height and diameter growth of planted slash pine. For. Sci. 23:299–306.

602. Fromme, F. D. 1928. The black rootrot disease of apple. Va. Agric. Exp. Stn. Tech. Bull. 34. 52 pp.

603. Fudl-Allah, A. E.-S. A., Weathers, L. G., and Greer, F. C. Jr. 1983. Characterization of a potexvirus isolated from night-blooming cactus. Plant Dis. 67:438–440.

604. Fukushi, T. 1921. A willow-canker disease caused by *Physalospora miyabeana* and its conidial form *Gloeosporium*. Ann. Phytopathol. Soc. Jap. 1(4):1–12.

605. Fulbright, D. W. 1984. Effect of eliminating dsRNA in hypovirulent *Endothia parasitica*. Phytopathology 74:722–724.

606. Fulbright, D. W., Weidlich, W. H., Haufler, K. Z., Thomas, C. S., and Paul, C. P. 1983. Chestnut blight and recovering American chestnut trees in Michigan. Can. J. Bot. 61:3164–3171.

607. Fulton, J. P., and Fulton, R. W. 1970. A comparison of some properties of elm mosaic and tomato ringspot viruses. Phytopathology 60:114–115.

608. Funk, A. 1968. *Diaporthe lokoyae* n. sp., the perfect state of *Phomopsis lokoyae*. Can. J. Bot. 46:601–603.

609. Funk, A. 1969. *Potebniamyces* (*Phacidiella*) disease of true firs in British Columbia. Can. J. Bot. 47:751–753.

610. Funk, A. 1972. Sirococcus shoot-blight of western hemlock in British Columbia and Alaska. Plant Dis. Rep. 56:645–647.

611. Funk, A. 1981. Parasitic microfungi of western trees. Can. For. Serv., Pac. For. Res. Cent. BC-X-222. 190 pp.

612. Funk, A. 1985. Foliar fungi of western trees. Can. For. Serv. BC-X-265. 159 pp.

613. Gallagher, P. W., and Snydor, T. D. 1983. Variation in wound response among cultivars of red maple. J. Am. Soc. Hortic. Sci. 108:744–746.

614. Gams, W. 1971. *Cephalosporium*-artige Schimmelpilze (Hyphomycetes). Gustav Fischer Verlag, Stuttgart. 262 pp.

615. Gardner, J. M., Feldman, A. W., and Stamper, D. H. 1983. Role and fate of bacteria in vascular occlusions of citrus. Physiol. Plant Pathol. 23:295–309.

616. Gardner, J. M., and Kado, C. I. 1973. Evidence for systemic movement of *Erwinia rubrifaciens* in Persian walnuts by the use of double-antibiotic markers. Phytopathology 63:1085–1086.

617. Gardner, M. W., and Yarwood, C. E. 1978. Host list of powdery mildews of California. Cal. Plant Pathol. No. 42. 9 pp.

618. Gardner, M. W., Yarwood, C. E., and Kuafala, T. 1972. Oak mildews. Plant Dis. Rep. 56:313–317.

619. Gardner, R. G., Cummins, J. N., and Aldwinckle, H. S. 1980. Fire blight resistance in the Geneva apple rootstock breeding program. J. Am. Hortic. Sci. 105:907–912.

620. Garrec, J. P., and Plébin, R. 1981. Etude de la relation entre la pluviosité et l'accumulation du fluor dans les forêts résineuses soumises à une pollution fluorée. Eur. J. For. Pathol. 11:129–136.

621. Garrett, P. W., Randall, W. K., Shigo, A. L., and Shortle, W. C. 1979. Inheritance of compartmentalization of wounds in sweetgum (*Liquidambar stryaciflua* L.) and eastern cottonwood (*Populus deltoides* Bartr.). U.S. For. Serv. Res. Pap. NE-433. 4 pp.

622. Garsed, S. G., and Rutter, A. J. 1982. Relative performance of conifer populations in various tests for sensitivity to SO_2, and the implications for selecting trees for planting in polluted areas. New Phytol. 92:349–367.

623. Gäumann, E. 1959. Die Rostpilze Mitteleuropas. Beitr. Kryptogamenflora Schweiz. Vol. 12. 1407 pp.

624. Gerhold, H. D., Schreiner, E. J., McDermott, R. E., and Winieski, J. A., eds. 1966. Breeding pest-resistant trees. Pergamon Press, New York. 505 pp.

625. Gerlach, W. W. P., Hoitink, H. A. J., and Ellett, C. W. 1974. Shoot blight and stem dieback of *Pieris japonica* caused by *Phytophthora citricola*, *P. citrophthora*, and *Botryosphaeria dothidea*. Phytopathology 64:1368–1370.

626. Gibbs, J. N. 1978. Intercontinental epidemiology of Dutch elm disease. Annu. Rev. Phytopathol. 16:287–307.

627. Gibbs, J. N., and French, D. W. 1980. The transmission of oak wilt. U.S. For. Serv. Res. Pap. NC-185. 17 pp.

628. Gibbs, J. N., Houston, D. R., and Smalley, E. B. 1979. Aggressive and non-aggressive strains of *Ceratocystis ulmi* in North America. Phytopathology 69:1215–1219.

629. Gibbs, J. N., and Reffold, T. C. 1982. *Gnomonia platani* and bark killing of London plane. Eur. J. For. Pathol. 12:395–398.

630. Gibson, I. A. S. 1972. Dothistroma blight of *Pinus radiata*. Annu. Rev. Phytopathol. 10:51–72.

631. Gibson, I. A. S. 1975. Impact and control of Dothistroma blight of pines. Eur. J. For. Pathol. 4:89–100.

632. Gibson, I. A. S., and Corbett, D. C. M. 1964. Variation in isolates from Armillaria root disease in Nyasaland. Phytopathology 54:122–123.

633. Gibson, L. P. 1973. An annotated list of the Cicadellidae and Fulgoridae of elm. U.S. For. Serv. Res. Pap. NE-278. 5 pp.

634. Gibson, L. P. 1977. Distribution of elm phloem necrosis in the United States. Plant Dis. Rep. 61:402–403.

635. Giese, R. L., Houston, D. R., Benjamin, D. M., Kuntz, J. E., and Skilling, D. D. 1964. Studies of maple blight. Wis. Agric. Exp. Stn. Res. Bull. 250. 129 pp.

636. Gilbertson, R. L. 1976. The genus *Inonotus* (Aphylloporales: Hymenochaetaceae) in Arizona. Mem. N.Y. Bot. Gard. 28:67–85.

637. Gilbertson, R. L. 1979. The genus *Phellinus* (Aphyllophorales, Hymenochaetaceae) in western North-America. Mycotaxon 9:51–89.

638. Gilbertson, R. L. 1981. North American wood-rotting fungi that cause brown rots. Mycotaxon 12:372–416.

639. Gilbertson, R. L., and Blackwell, M. 1984. Two new basidiomycetes on living live oak in the Southeast and Gulf Coast region. Mycotaxon 20:85–93.

640. Gilbertson, R. L., and Ryvarden, L. 1986, 1987. North American polypores. Vol. I, *Albatrellus–Lindtneria*. Vol. II, *Megasporaporia–Wrightoporia*. Fungiflora, Oslo.

641. Giles, K. L., and Atherly, A. G., eds. 1981. Biology of the Rhizobiaceae. Int. Rev. Cytol. Suppl. 13. 368 pp.

642. Gill, C. J. 1970. The flooding tolerance of woody species—a review. For. Abstr. 31:671–688.

643. Gill, D. L. 1953. Petal blight of azalea. U.S. Dep. Agric. Yrbk. 1953:578–582.

644. Gill, D. L. 1958. Effect of root-knot nematodes on Fusarium wilt of mimosa. Plant Dis. Rep. 42:587–590.

645. Gill, D. L. 1967. Fusarium wilt infection of apparently healthy mimosa trees. Plant Dis. 51:148–150.

646. Gill, D. L. 1977. Downy mildew of roses in Georgia. Plant Dis. Rep. 61:230–231.

647. Gill, D. L. 1979. 'Union' mimosa. HortScience 14:644.

648. Gill, D. L., Alfieri, S. A. Jr., and Sobers, E. K. 1971. A new leaf disease of *Ilex* spp. caused by *Cylindrocladium avesiculatum* sp. nov. Phytopathology 61:58–60.

649. Gill, D. L., and Sobers, E. K. 1974. Control of *Cercospora* sp. leafspot of *Ligustrum japonicum*. Plant Dis. Rep. 58:1015–1017.

650. Gill, L. S., and Hawksworth, F. G. 1961. The mistletoes: a literature review. U.S. Dep. Agric. Tech. Bull. 1242. 87 pp.

651. Gilliam, C. H., and Smith, E. M. 1980. Sources and symptoms of boron toxicity in container grown woody ornamentals. J. Arboric. 6:209–212.

652. Gilliam, C. H., Smith, E. M., Still, S. M., and Sheppard, W. J. 1981. Treating boron toxicity in *Rhododendron catawbiense*. HortScience 16:764–765.

653. Giri, L., and Chessin, M. 1975. Zygocactus virus X. Phytopathol. Z. 83:40–48.

654. Glerum, C., and Farrar, J. L. 1966. Frost ring formation in the stems of some coniferous species. Can. J. Bot. 44:879–886.

655. Goheen, D. J., Cobb, F. W. Jr., and McKibbin, G. N. 1978. Influence of soil moisture on infection of ponderosa pine by Verticicladiella wageneri. Phytopathology 68:913–916.

656. Goheen, D. J., and Filip. G. M. 1980. Root pathogen complexes in Pacific Northwest forests. Plant Dis. 64:793–794.

657. Good, H. M., and Nelson, C. D. 1951. A histological study of sugar maple decayed by Polyporus glomeratus Peck. Can. J. Bot. 29:215–223.

658. Good, H. M., and Nelson, J. I. 1962. Fungi associated with Fomes igniarius var. populinus in living poplar trees and their probable significance in decay. Can. J. Bot. 40:615–624.

659. Goos, R. D. 1974. A scanning electron microscope and in vitro study of Meliola palmicola. Proc. Iowa Acad. Sci. 81:23–27.

660. Gordon, A. G., and Gorham, E. 1963. Ecological aspects of air pollution from an iron-sintering plant at Wawa, Ontario. Can. J. Bot. 41:1063–1078.

661. Gordon, J. C., and Wheeler, C. T., eds. 1983. Biological nitrogen fixation in forest ecosystems: foundations and applications. M. Nijhoff/W. Junk, The Hague. 342 pp.

662. Goss, R. W., and Frink, P. R. 1934. Cephalosporium wilt and die-back of the white elm. Nebr. Agric. Exp. Stn. Res. Bull. 70. 24 pp.

663. Gotlieb, A. R., and Berbee, J. G. 1973. Line pattern of birch caused by apple mosaic virus. Phytopathology 63:1470–1477.

664. Gouin, F. R. 1983. Girdling roots: fact or fiction? Comb. Proc. Int. Plant Prop. Soc. 33:428–432.

665. Gould, C. T. 1945. The parasitism of Glomerularia lonicerae (Pk.) D. and H. in Lonicera species. Iowa State Coll. J. Sci. 19:301–331.

666. Gourbière, F., and Morelet, M. 1979. Le genre Rhizosphaera Mangin et Hariot 1. R. oudmansii et R. macrospora nov. sp. Rev. Mycol. 43:81–95.

667. Gourbière, F., and Morelet, M. 1980. Le genre Rhizosphaera Mangin et Hariot 2. R. pini, R. kobayashii, et R. kalkhoffii. Cryptog. Mycol. 1:69–81.

668. Graafland, W. 1960. The parasitism of Exobasidium japonicum Shir. on azalea. Acta Bot. Neerl. 9:347–379.

669. Graff, P. W. 1936. North American polypores—I. Polyporus squamosus and its varieties. Mycologia 28:154–170.

670. Graham, J. H., and Linderman, R. G. 1983. Pathogenic seedborne Fusarium oxysporum from Douglas-fir. Plant Dis. 67:323–325.

671. Graham, J. H., Timmer, L. W., and Young, R. H. 1983. Necrosis of major roots in relation to citrus blight. Plant Dis. 67:1273–1276.

672. Grand, L. F., ed. 1985. North Carolina plant disease index. N.C. State Univ. Agric. Res. Serv. Tech. Bull. 240. 157 pp.

673. Grand, L. F., and Menge, J. A. 1974. Sclerotia of Cristulariella pyramidalis in nature. Mycologia 66:712–715.

674. Grasso, S., and La Rosa, R. 1982. Cancri da Phomopsis incarcerata su rosa. Riv. Patol. Veg. 18:143–148.

675. Graves, A. H. 1919. Some diseases of trees in greater New York. Mycologia 11:111–124.

676. Graves, A. H. 1923. The Melanconis disease of the butternut (Juglans cinerea L.). Phytopathology 13:411–435.

677. Great Britain Forestry Commission. 1963. Keithia disease of Thuja plicata. G.B. For. Comm. Leafl. 43 (rev.) 6 pp.

678. Green, R. J. Jr. 1977. Dieback of black walnut seedlings caused by Phomopsis elaeagni. Plant Dis. Rep. 61:582–584.

679. Greene, G. L. 1962. Physiological basis of gall formation in black-knot disease of Prunus. Phytopathology 52:880–884.

680. Greenhalgh, G. N., and Bevan, R. J. 1978. Response of Rhytisma acerinum to air pollution. Trans. Br. Mycol. Soc. 71:491–494.

681. Greenidge, K. N. H. 1953. Further studies of birch dieback in Nova Scotia. Can. J. Bot. 31:548–559.

682. Gregory, C. T. 1915. Studies on Plasmopara viticola. Int. Cong. Viticult. Rep. 1915:126–150.

683. Gremmen, J. 1965. De Marssonina-ziekte van de populier 3. Het voorkomen van Marssonina brunnea (E. & E.) Magn. in Nederland. Ned. Bosbouw. Tijdschr. 37:196–198.

684. Gremmen, J. 1965. Three poplar-inhabiting Drepanopeziza species and their life-history. Nova Hedgwigia 9:170–176.

685. Gremmen, J. 1978. Research on Dothichiza-bark necrosis (Cryptodiaporthe populea) in poplar. Eur. J. For. Pathol. 8:362–368.

686. Gremmen, J., and de Kam, M. 1977. Ceratocystis fimbriata, a fungus associated with poplar canker in Poland. Eur. J. For. Pathol. 7:44–47.

687. Griffin, G. D., and Epstein, A. H. 1964. Association of dagger nematode, Xiphinema americanum, with stunting and winterkill of ornamental spruce. Phytopathology 54:177–180.

688. Griffin, G. J., Hebard, F. V., Wendt, R. W., and Elkins, J. R. 1983. Survival of American chestnut trees: evaluation of blight resistance and virulence in Endothia parasitica. Phytopathology 73:1084–1092.

689. Griffin, G. J., and Stipes, R. J. 1975. High populations of Fusarium oxysporum f. sp. pernicosum in mimosa bark lenticellar sporodochia and in soil. Plant Dis. Rep. 59:787–790.

690. Griffin, H. D. 1965. Maple dieback in Ontario. For. Chron. 41:295–300.

691. Grosclaude, G., Germain, E., Simone, J., and Leglise, P. 1976. La nécrose des inflorescences mâles du noisetier provoquée par Gloeosporium coryli (Desm.) Sacc. Ann. Phytopathol. 8:87–89.

692. Gross, H. L. 1964. The Echinodontiaceae. Mycopathol. Mycol. Appl. 24:1–26.

693. Gross, H. L. 1967. Cytospora canker of black cherry. Plant Dis. Rep. 51:941–944.

694. Gross, H. L. 1983. Negligible cull and growth loss of jack pine associated with globose gall rust. For. Chron. 59:308–311.

695. Gross, H. L., and Basham, J. T. 1981. Diseases of aspen suckers in northern Ontario. Can. For. Serv., Great Lakes For. Res. Cent., Inf. Rep. O-X-329. 16 pp.

696. Gross, H. L., Ek, A. R., and Patton, R. F. 1983. Site character and infection hazard for the sweetfern rust disease in northern Ontario. For. Sci. 29:771–778.

697. Gross, H. L., Patton, R. F., and Ek, J. R. 1978. Reduced growth, cull, and mortality of jack pine associated with sweetfern rust cankers. Can. J. For. Res. 8:47–53.

698. Gross, H. L., Patton, R. F., and Ek, A. R. 1980. Spatial aspects of sweetfern rust disease in northern Ontario jack pine–sweetfern stands. Can. J. For. Res. 10:199–208.

699. Grove, W. B. 1935, 1937. British stem- and leaf-fungi (Coelomycetes). Vols. 1 and 2. Cambridge University Press, Cambridge, England. 488 and 407 pp.

700. Groves, J. W., and Bowerman, C. A. 1955. The species of Ciborinia on Populus. Can. J. Bot. 33:577–590.

701. Guba, E. F. 1961. Monograph of Monochaetia and Pestalotia. Harvard University Press, Cambridge, MA. 342 pp.

702. Haasis, F. A. 1953. Flower blight of camellias. Am. Camellia Yearbk. 1953:111–119.

703. Hacskaylo, J., Finn, R. F., and Vimmerstedt, J. P. 1969. Deficiency symptoms of some forest trees. Ohio Agric. Res. Devel. Cent. Res. Bull. 1015. 68 pp.

704. Haddow, W. R. 1931. Studies in Ganoderma. J. Arnold Arbor. 12:25–46.

705. Haddow, W. R. 1938. On the classification, nomenclature, hosts, and geographical range of Trametes pini (Thore) Fries. Trans. Br. Mycol. Soc. 22:182–193.

706. Haddow, W. R. 1938. The disease caused by Trametes pini (Thore) Fries in white pine (Pinus strobus L.). Trans. R. Can. Inst. 29:21–80.

707. Haddow, W. R. 1941. On the history and diagnosis of Polyporus tomentosus Fries, Polyporus circinatus Fries, and Polyporus dualis Peck. Trans. Br. Mycol. Soc. 25:179–190.

708. Haddow, W. R., and Newman, F. S. 1942. A disease of the Scots pine (Pinus sylvestris L.) caused by the fungus Diplodia pinea Kickx associated with the pine spittle-bug (Aphrophora parallela Say.). I. Symptoms and etiology. Trans. R. Can. Inst. 24:1–17.

709. Hahn, G. G. 1920. Phomopsis juniperovora, a new species causing blight of nursery cedars. Phytopathology 10:249–253.

710. Hahn, G. G. 1930. Life-history studies of the species of Phomopsis occurring on conifers. Part I. Trans. Br. Mycol. Soc. 15:32–93.

711. Hahn, G. G. 1943. Taxonomy, distribution, and pathology of Phomopsis occulta and P. juniperovora. Mycologia 35:112–129.

712. Hahn, G. G. 1957. Phacidiopycnis (Phomopsis) canker and dieback of conifers. Plant Dis. Rep. 41:623–633.

713. Hahn, G. G., and Ayers, T. T. 1934. Dasyscyphae on conifers in North America. I. The large-spored, white-excipled species. Mycologia 26:73–101.

714. Hahn, G. G., and Ayers, T. T. 1943. Role of Dasyscypha willkommii and related fungi in the production of canker and die-back of larches. J. For. 41:483–495.

715. Hahn, G. G., Hartley, C., and Pierce, R. G. 1917. A nursery blight of cedars. J. Agric. Res. 10:533–539.

716. Hale, M. E. 1983. The biology of lichens. 3rd ed. E. Arnold, Baltimore, MD. 190 pp.

717. Hall, R., Hofstra, G., and Lumis, G. P. 1972. Effects of deicing salt on eastern white pine: foliar injury, growth suppression, and seasonal changes in foliar concentrations of sodium and chloride. Can. J. For. Res. 2:244–249.

718. Hall, R., Hofstra, G., and Lumis, G. P. 1973. Leaf necrosis of roadside sugar maple in Ontario in relation to elemental composition of soil and leaves. Phytopathology 63:1426–1427.

719. Hall, R. C. 1933. Post-logging decadence in northern hardwoods. Univ. Mich. Sch. For. & Conserv. Bull. 3. 66 pp.

720. Hällgren, J. E., and Fredriksson, S.-A. 1982. Emission of hydrogen sulfide from sulfur dioxide-fumigated pine trees. Plant Physiol. 70:456–459.

721. Hamm, P. B., and Hansen, E. M. 1982. Pathogenicity of Phytophthora species to Pacific Northwest conifers. Eur. J. For. Pathol. 12:167–174.

722. Hammarlund, C. 1930. Rostsvampar pa Mahonia (Puccinia mirabilissima Peck och P. graminis Pers.). Bot. Notis. 1930:380–407.

723. Hampson, M. C. 1981. Phomopsis canker on weeping fig in Newfoundland. Can. Plant Dis. Surv. 61:1–4.

724. Hampson, M. C., and Sinclair, W. A. 1973. Xylem dysfunction in peach caused by *Cytospora leucostoma*. Phytopathology 63:676–681.

725. Hanisch, M. A., Brown, H. D., and Brown, E. A., eds. 1983. Dutch elm disease management guide. U.S. Dep. Agric., For. Serv. & Exten. Serv., Bull. 1. 23 pp.

726. Hanlin, R. T. 1982. Conidiogenesis in *Spiniger meineckellus*. Mycologia 74:236–241.

727. Hansbrough, J. R. 1934. Occurrence and parasitism of *Aleurodiscus amorphus* in North America. J. For. 32:452–458.

728. Hansen, E. M., Hamm, P. B., Julis, A. J., and Roth, L. F. 1979. Isolation, incidence, and management of *Phytophthora* in forest tree nurseries in the Pacific Northwest. Plant Dis. Rep. 63:607–611.

729. Hansen, E. M., Roth, L. F., Hamm, P. B., and Julis, A. J. 1980. Survival, spread, and pathogenicity of *Phytophthora* spp. on Douglas-fir seedlings planted on forest sites. Phytopathology 70:422–425.

730. Hansen, H. N., and Barrett, J. T. 1938. Gardenia canker. Mycologia 30:15–19.

731. Hansen, H. N., and Scott, C. E. 1934. A canker and gall disease of *Gardenia*. Science 79:18.

732. Hansen, H. N., and Smith, R. E. 1937. A bacterial gall disease of Douglas fir, *Pseudotsuga taxifolia*. Hilgardia 10:569–577.

733. Hansen, H. N., and Thomas, H. E. 1940. Flower blight of camellias. Phytopathology 30:166–170.

734. Hansford, C. G. 1961. The Meliolineae. A monograph. Sydowia Beih. 2:1–806.

735. Hansford, C. G. 1965. Iconographia meliolinearum. Sydowia Beih. 5. Unpaged.

736. Harley, J. L., and Smith, S. E. 1983. Mycorrhizal symbiosis. Academic Press, New York. 483 pp.

737. Harniss, R. O., and Nelson, D. L. 1984. A severe epidemic of Marssonina leaf blight on quaking aspen in northern Utah. U.S. For. Serv. Res. Note INT-339. 6 pp.

738. Harrington, T. C., and Cobb, F. W. Jr. 1983. Pathogenicity of *Leptographium* and *Verticicladiella* spp. isolated from roots of western North American conifers. Phytopathology 73:596–599.

739. Harrington, T. C., and Cobb, F. W. Jr. 1984. Host specialization of three morphological variants of *Verticicladiella wageneri*. Phytopathology 74:286–290.

740. Harrington, T. C., and Cobb, F. W. Jr. 1984. *Verticillium albo-atrum* on *Ceanothus* in a California forest. Plant Dis. 68:1012.

741. Harrington, T. C., Cobb, F. W. Jr., and Lownsbery, J. W. 1985. Activity of *Hylastes nigrinus*, a vector of *Verticicladiella wageneri*, in thinned stands of Douglas-fir. Can. J. For. Res. 15:519–523.

742. Harrington, T. C., Furniss, M. M., and Shaw, C. G. 1981. Dissemination of hymenomycetes by *Dendroctonus pseudotsugae* (Coleoptera: Scolytidae). Phytopathology 71:551–554.

743. Harris, H. A. 1934. Frost ring formation in some winter-injured deciduous trees and shrubs. Am. J. Bot. 21:485–498.

744. Harrison, K. A. 1965. Willow blight and the survival of some *Salix* species in Nova Scotia. Can. Plant Dis. Surv. 45:94–95.

745. Hart, J. H., Mosher, D. G., and Ajewole, R. 1978. Occurrence of *Endocronartium harknessii* and *Cronartium quercuum* on Scots and jack pine in Michigan's Lower Peninsula. Plant Dis. Rep. 62:779–782.

746. Hartley, C. 1918. Stem lesions caused by excessive heat. J. Agric. Res. 14:595–604.

747. Hartley, C., Davidson, R. W., and Crandall, B. S. 1961. Wetwood, bacteria, and increased pH in trees. U.S. For. Prod. Lab. Rep. 2215. 34 pp.

748. Hartman, H. T., and Kester, D. E. 1975. Plant propagation: principles and practices. 3rd ed. Prentice-Hall, Englewood Cliffs, NJ. 662 pp.

749. Hartung, J. S., Burton, C. L., and Ramsdell, D. C. 1981. Epidemiological studies of blueberry anthracnose disease caused by *Colletotrichum gloeosporioides*. Phytopathology 71:449–453.

750. Harvey, G. M. 1976. Epiphytology of a needle cast fungus, *Lophodermella morbida*, in ponderosa pine plantations in western Oregon. For. Sci. 22:223–230.

751. Harvey, J. M. 1952. Bacterial leaf spot of *Umbellularia californica*. Madroño 11:195–198.

752. Havir, E. A., and Anagnostakis, S. L. 1983. Oxalate production by virulent but not by hypovirulent strains of *Endothia parasitica*. Physiol. Pl. Pathol. 23:369–376.

753. Hawksworth, D. L., Gibson, I. A. S., and Gams, W. 1976. *Phialophora parasitica* associated with disease conditions in various trees. Trans. Br. Mycol. Soc. 66:427–431.

754. Hawksworth, F. G. 1961. Dwarfmistletoe of ponderosa pine in the Southwest. U.S. Dep. Agric. Tech. Bull. 1246. 112 pp.

755. Hawksworth, F. G., and Scharpf, R. F. 1981. Phoradendron on conifers. U.S. For. Serv. For. Insect Dis. Leafl. 164. 7 pp.

756. Hawksworth, F. G., and Scharpf, R. F., tech. coords. 1984. Biology of dwarf mistletoes: proceedings of the symposium. U.S. For. Serv. Gen. Tech. Rep. RM-111. 131 pp.

757. Hawksworth, F. G., and Shigo, A. L. 1980. Dwarf mistletoe on red spruce in the White Mountains of New Hampshire. Plant Dis. 64:880–882.

758. Hawksworth, F. G., Wicker, E. F., and Scharpf, R. F. 1977. Fungal parasites of dwarf mistletoes. U.S. For. Serv. Gen. Tech. Rep. RM-36. 14 pp.

759. Hawksworth, F. G., and Wiens, D. 1972. Biology and classification of dwarf mistletoes (*Arceuthobium*). U.S. Dep. Agric. Agric. Handbk. 401. 234 pp.

760. Headland, J. K., Griffin, G. J., Stipes, R. J., and Elkins, J. R. 1976. Severity of natural *Endothia parasitica* infection of Chinese chestnut. Plant Dis. Rep. 60:426–429.

761. Heald, F. D., and Studhalter, R. A. 1914. The Strumella disease of oak and chestnut trees. Penn. Dep. For. Bull. 10. 15 pp.

762. Heald, F. D., and Wolf, F. A. 1911. New species of Texas fungi. Mycologia 3:5–22.

763. Heale, E. L., and Ormrod, D. P. 1982. Effects of nickel and copper on *Acer rubrum*, *Cornus stolonifera*, *Lonicera tatarica*, and *Pinus resinosa*. Can. J. Bot. 60:2674–2681.

764. Hearon, S. S., Sherald, J. L., and Kostka, S. J. 1980. Association of xylem-limited bacteria with elm, sycamore, and oak leaf scorch. Can. J. Bot. 58:1986–1993.

765. Hebard, F. V., Griffin, G. J., and Elkins, J. R. 1984. Developmental histopathology of cankers incited by hypovirulent and virulent isolates of *Endothia parasitica* on susceptible and resistant chestnut trees. Phytopathology 74:140–149.

766. Hecht-Poinar, E. I., Britton, J. C., and Parmeter, J. R. Jr. 1981. Dieback of oaks in California. Plant Dis. 65:281.

767. Hedgcock, G. G., and Hahn, G. G. 1922. Two important pine cone rusts and their new cronartial stages. Phytopathology 12:109–122.

768. Hedgcock, G. G., and Long, W. H. 1915. A disease of pines caused by *Cronartium pyriforme*. U.S. Dep. Agric. Bull. 247. 20 pp.

769. Hedgcock, G. G., and Siggers, P. V. 1949. A comparison of the pine-oak rusts. U.S. Dep. Agric. Bull. 978. 30 pp.

770. Hedges, F., and Tenny, L. S. 1912. A knot of citrus trees caused by *Sphaeropsis tumefaciens*. U.S. Dep. Agric Bull. 247. 74 pp.

771. Heiberg, S. O., and White, D. P. 1951. Potassium deficiency of reforested pine and spruce stands in northern New York. Soil Sci. Soc. Am. Proc. 15:369–376.

772. Heichel, G. H., Turner, N. C., and Walton, G. S. 1972. Anthracnose causes dieback of regrowth on defoliated oak and maple. Plant Dis. Rep. 56:1046–1047.

773. Helms, J. A., Cobb, F. W. Jr., and Whitney, H. S. 1971. Effect of infection by *Verticicladiella wageneri* on the physiology of *Pinus ponderosa*. Phytopathology 61:920–925.

774. Helton, A. W. 1961. First year effects of 10 selected *Cytospora* isolates on 20 fruit and forest tree species and varieties. Plant Dis. Rep. 45:500–504.

775. Helton, A. W., and Braun, J. W. 1970. Relationship of number of *Cytospora* infections on *Prunus domestica* to rate of expansion of individual cankers. Phytopathology 60:1700–1701.

776. Hemmi, T., and Kurata, S. 1931. Studies on septorioses of plants. II. *Septoria azaleae* Voglino causing the brown-spot disease of the cultivated azaleas in Japan. Mem. Coll. Agric. Kyoto Imp. Univ. No. 13, Art. 1. 22 pp.

777. Henderson, D. M. 1961. *Glomospora* and *Glomopsis*. R. Bot. Gard., Edinb. Notes 23:497–502.

778. Henry, B. W. 1951. Oak leaf blister severe in south Mississippi. Plant Dis. Rep. 35:384.

779. Henry, B. W., Moses, C. S., Richards, C. A., and Riker, A. J. 1944. Oak wilt: its significance, symptoms, and cause. Phytopathology 34:636–647.

780. Hepting, G. H. 1939. A vascular wilt of the mimosa tree (*Albizzia julibrissin*). U.S. Dep. Agric. Circ. 535. 10 pp.

781. Hepting, G. H. 1944. Sapstreak, a new killing disease of sugar maple. Phytopathology 34:1069–1076.

782. Hepting, G. H. 1963. Climate and forest diseases. Annu. Rev. Phytopathol. 1:31–50.

783. Hepting, G. H. 1971. Diseases of forest and shade trees of the United States. U.S. Dep. Agric. Agric. Handbk. 386. 658 pp.

784. Hepting, G. H., and Hedgcock, G. G. 1937. Decay in merchantable oak, yellow poplar, and basswood in the Appalachian region. U.S. Dep. Agric. Tech. Bull. 570. 29 pp.

785. Hepting, G. H., and Matthews, F. R. 1970. Southern cone rust. U.S. For. Serv. For. Pest Leafl. 27. 4 pp.

786. Hepting, G. H., Miller, J. H., and Campbell, W. A. 1951. Winter of 1950–51 damaging to southeastern woody vegetation. Plant Dis. Rep. 35:502–503.

787. Hepting, G. H., and Roth, E. R. 1946. Pitch canker, a new disease of some southern pines. J. For. 44:742–744.

788. Hepting, G. H., and Toole, E. R. 1939. The hemlock rust caused by *Melampsora farlowii*. Phytopathology 29:463–473.

789. Hermanides-Nijhof, E. J. 1977. *Aureobasidium* and allied genera. Stud. Mycol. No. 15:141–177.

523

790. Herrero, J. 1951. Studies of compatible and incompatible graft combinations with special reference to hardy fruit trees. J. Hortic. Sci. 26:186–237.

791. Herridge, E. A., and Lambe, R. C. 1960. A holly leafspot associated with the use of copper fungicides. (Abstr.) Phytopathology 50:84.

792. Hesler, L. R. 1916. Black rot, leaf spot, and canker of pomaceous fruits. Cornell Univ. Agric. Exp. Stn. Bull. 379:49–148.

793. Heslin, M. C., Stuart, M. R., Murchú, P. O., and Donnelly, D. M. X. 1983. Fomannoxin, a phytotoxic metabolite of Fomes annosus: in vitro production, host toxicity, and isolation from naturally infected Sitka spruce heartwood. Eur. J. For. Pathol. 13:11–23.

794. Hessburg, P. F., and Hansen, E. M. 1986. Soil temperature and rate of colonization of Ceratocystis wageneri in Douglas-fir. Phytopathology 76:627–631.

795. Hewitt, E. J., and Smith, T. A. 1974. Plant mineral nutrition. English Universities Press, London. 298 pp.

796. Hewitt, W. B. 1939. Leaf-scar infection in relation to the olive-knot disease. Hilgardia 12:41–66.

797. Hibben, C. R. 1959. Relations of Stegonosporium ovatum (Pers. ex Mérat) Hughes with dieback of sugar maple (Acer saccharum Marsh.). M.S. thesis, Cornell Univ., Ithaca, NY. 63 pp.

798. Hibben, C. R. 1964. Identity and significance of certain organisms associated with sugar maple decline in New York woodlands. Phytopathology 54:1389–1392.

799. Hibben, C. R. 1966. Transmission of a ringspot-like virus from leaves of white ash. Phytopathology 56:323–325.

800. Hibben, C. R., and Bozarth, R. F. 1972. Identification of an ash strain of tobacco ringspot virus. Phytopathology 62:1023–1029.

801. Hibben, C. R., Bozarth, R. F., and Reese, J. 1979. Identification of tobacco necrosis virus in deteriorating clones of aspen. For. Sci. 25:557–567.

802. Hibben, C. R., Lewis, C. A., and Castello, J. D. 1986. Mycoplasmalike organisms, cause of lilac witches'-broom. Plant Dis. 70:342–345.

803. Hibben, C. R., and Hagar, S. S. 1975. Pathogenicity of an ash isolate of tobacco ringspot virus. Plant Dis. Rep. 59:57–60.

804. Hibben, C. R., and Reese, J. A. 1983. Identification of tomato ringspot virus and mycoplasma-like organisms in stump sprouts of ash. (Abstr.) Phytopathology 73:367.

805. Hibben, C. R., and Silverborg, S. B. 1978. Severity and causes of ash dieback. J. Arboric. 4:274–279.

806. Hibben, C. R., and Walker, J. T. 1966. A leaf roll–necrosis complex of lilacs in an urban environment. Proc. Am. Soc. Hortic. Sci. 89:636–642.

807. Hibben, C. R., and Walker, J. T. 1971. Nematode transmission of the ash strain of tobacco ringspot virus. Plant Dis. Rep. 55:475–478.

808. Hibben, C. R., Walker, J. T., and Allison, J. R. 1977. Powdery mildew ratings of lilac species and cultivars. Plant Dis. Rep. 61:192–196.

809. Hibben, C. R., and Wolanski, B. 1971. Dodder transmission of a mycoplasma from ash witches'-broom. Phytopathology 61:151–156.

810. Hibbs, R. H. 1976. Decline of hackberry attributed to ambient herbicide drift. Proc. Iowa Acad. Sci. 72:187–190.

811. Hicock, H. W., and Olson, A. R. 1954. The toxicity to plants of wood preservatives and their solvents. Conn. Agric. Exp. Stn. Circ. 189. 4 pp.

812. Higgins, B. B. 1914. Contribution to the life history and physiology of Cylindrosporium on stone fruits. Am. J. Bot. 1:145–173.

813. Higgins, B. B. 1917. A disease of pecan catkins. Phytopathology 7:42–45.

814. Highley, T. L., Bar-Lev, S. S., Kirk, T. K., and Larsen, M. J. 1983. Influence of O_2 and CO_2 on wood decay by heartrot and saprot fungi. Phytopathology 73:630–633.

815. Higuchi, T., ed. 1985. Biosynthesis and biodegradation of wood components. Academic Press, New York. 679 pp.

816. Hilborn, M. T. 1937. The anatomy of a black zone caused by Xylaria polymorpha. Phytopathology 27:1177–1179.

817. Hilborn, M. T. 1942. The biology of Fomes fomentarius. Maine Agric. Exp. Stn. Bull. 409:161–214.

818. Hildebrand, E. M. 1947. Perennial peach canker and the canker complex in New York, with methods of control. Cornell Univ. Agric. Exp. Stn. Mem. 276. 61 pp.

819. Hildebrand, E. M. 1953. Yellow-red or X-disease of peach. Cornell Univ. Agric. Exp. Stn. Mem. 323. 54 pp.

820. Hiley, W. E. 1919. The fungal diseases of the common larch. Oxford University Press, Oxford, England. 204 pp.

821. Hill, S. A. 1975. The importance of wood scab caused by Venturia inaequalis (Cke.) Wint. as a source of infection for apple leaves in the spring. Phytopathol. Z. 82:216–223.

822. Himelick, E. B. 1961. Sycamore anthacnose. Proc. Natl. Shade Tree Conf. 37:136–143.

823. Himelick, E. B. 1969. Tree and shrub hosts of Verticillium albo-atrum. Ill. Nat. Hist. Surv. Biol. Notes No. 66. 8 pp.

824. Himelick, E. B., and Himelick, K. J. 1980. Systemic treatment for chlorotic trees. J. Arboric. 6:192–196.

825. Himelick, E. B., and Neely, D. 1960. Juniper hosts of cedar-apple rust and cedar-hawthorn rust. Plant Dis. Rep. 44:109–112.

826. Hinds, T. E. 1962. Inoculations with the sooty-bark canker fungus on aspen. Plant Dis. Rep. 46:57–58.

827. Hinds, T. E. 1972. Ceratocystis canker of aspen. Phytopathology 62:213–220.

828. Hinds, T. E. 1972. Insect transmission of Ceratocystis species associated with aspen cankers. Phytopathology 62:221–225.

829. Hinds, T. E. 1981. Cryptosphaeria canker and Libertella decay of aspen. Phytopathology 71:1137–1145.

830. Hinds, T. E. 1985. Diseases. Pages 87–106 in: Aspen: ecology and management in the western United States. U.S. For. Serv. Gen. Tech. Rep. RM-119. 283 pp.

831. Hinds, T. E., and Krebill, R. G. 1975. Wounds and canker diseases of western aspen (Populus tremuloides). U.S. For. Serv. For. Pest. Leafl. 152. 9 pp.

832. Hinds, T. E., and Ryan, M. G. 1985. Expansion of sooty-bark and Ceratocystis cankers on aspen. Plant Dis. 69:842–844.

833. Hirata, K. 1968. Notes on host range and geographic distribution of the powdery mildew fungi. Trans. Mycol. Soc. Jap. 9:73–88.

834. Hiratsuka, N. 1936. A monograph of the Pucciniastreae. Mem. Tottori Agric. Coll., Tottori, Japan. Vol. 4. 374 pp.

835. Hiratsuka, N. 1958. Revision of taxonomy of the Pucciniastreae with special reference to species of the Japanese archipelago. Faculty Agric., Tokyo Univ. of Educ., Tokyo. Contrib. No. 31. 167 pp.

836. Hiratsuka, Y. 1973. The nuclear cycle and the terminology of spore states in the Uredinales. Mycologia 65:432–443.

837. Hiratsuka, Y. 1984. New leaf spot fungus, Marssonina balsamiferae, on Populus balsamifera in Manitoba and Ontario. Mycotaxon 19:133–136.

838. Hiratsuka, Y., and Maruyama, P. J. 1976. Castilleja miniata, a new alternate host of Cronartium ribicola. Plant Dis. Rep. 60:241.

839. Hiratsuka, Y., McArthur, L. E., and Emond, F. J. 1967. A distinction between Pucciniastrum goeppertianum and P. epilobii, with clarification of status of Peridermium holwayi and P. ornamentale. Can. J. Bot. 45:1913–1915.

840. Hiratsuka, Y., and Powell, J. M. 1976. Pine stem rusts of Canada. Can. For. Serv., North. For. Res. Cent., Edmonton, Tech. Rep. No. 4. 103 pp.

841. Hirst, J. M., and Stedman, O. J. 1961–1962. The epidemiology of apple scab (Venturia inaequalis [Cke.] Wint.) I. Frequency of airborne spores in orchards. II. Observations on the liberation of ascospores. III. The supply of ascospores. Ann. Appl. Biol. 49:290–305; 50:525–550, 551–567.

842. Hirt, R. R. 1964. Cronartium ribicola: its growth and reproduction in the tissues of eastern white pine. State Univ. Coll. For. (Syracuse, NY) Tech. Publ. 86. 30 pp.

843. Hirt, R. R., and Eliason, E. J. 1938. The development of decay in living trees inoculated with Fomes pinicola. J. For. 36:705–709.

844. Hobbs, S. D., and Partridge, A. D. 1979. Wood decays, root rots, and stand composition along an elevation gradient. For. Sci. 25:31–42.

845. Hodges, C. S. 1962. Black root rot of pine seedlings. Phytopathology 52:210–219.

846. Hodges, C. S. 1962. Comparison of four similar fungi from Juniperus and related conifers. Mycologia 54:52–69.

847. Hodges, C. S. 1969. Modes of infection and spread of Fomes annosus. Annu. Rev. Phytopathol. 7:247–266.

848. Hodges, C. S. 1980. The taxonomy of Diaporthe cubensis. Mycologia 72:542–548.

849. Hodges, C. S. 1983. Pine mortality in Hawaii associated with Botryosphaeria dothidea. Plant Dis. 67:555–556.

850. Hodges, C. S., Geary, T. F., and Cordell, C. E. 1979. The occurrence of Diaporthe cubensis in eucalypts in Florida, Hawaii, and Puerto Rico. Plant Dis. Rep. 63:216–220.

851. Hodges, C. S. Jr., Koenigs, J. W., Kuhlman, E. G., and Ross, E. W. 1971. Fomes annosus: a bibliography with subject index—1960–1970. U.S. For. Serv. Res. Pap. SE-84. 75 pp.

852. Hodges, C. S., Rishbeth, J., and Yde-Andersen, A., eds. 1970. Fomes annosus. Proc. 3rd Int. Conf. on Fomes annosus, Aarhus, Denmark, 1968. U.S. For. Serv. Southeast. For. Exp. Stn. 208 pp.

853. Hoff, R. J. 1985. Susceptibility of lodgepole pine to the needle cast fungus Lophodermella concolor. U.S. For. Serv. Res. Note INT-349. 6 pp.

854. Hoff, R. J., Bingham, R. T., and McDonald, G. I. 1980. Relative blister rust resistance of white pines. Eur. J. For. Pathol. 10:307–316.

855. Hoffman, G. M., and Fliege, F. 1967. Kabatina juniperi als Ursache eines Zweigsterbens an verschiedenen Juniperusarten. Z. Pflanzenkr. Pflanzenschutz 74:587–593.

856. Hofstra, G., Hall, R., and Lumis, G. P. 1979. Studies of salt-induced damage to roadside plants in Ontario. J. Arboric. 5:25–31.

857. Höhnel, F. von. 1916. Über Cheilaria aceris. Sitzungsb. K. Akad. Wiss. Math-nat. Kl. I, 125:81–84.

858. Hoitink, H. A. J., and Schmitthenner, A. F. 1969. Rhododendron wilt caused by Phytophthora citricola. Phytopathology 59:708–709.

859. Hoitink, H. A. J., and Schmitthenner, A. F. 1974. Relative prevalence and

virulence of *Phytophthora* species involved in rhododendron root rot. Phytopathology 64:1371–1374.

860. Hoitink, H. A. J., and Schmitthenner, A. F. 1974. Resistance of *Rhododendron* species and hybrids to Phytophthora root rot. Plant Dis. Rep. 58:650–653.

861. Hoitink, H. A. J., VanDoren, D. M. Jr., and Schmitthenner, A. F. 1977. Suppression of *Phytophthora cinnamomi* in a composted hardwood bark potting medium. Phytopathology 67:561–565.

862. Holcomb, G. E. 1986. Hosts of the parasitic alga *Cephaleuros virescens* in Louisiana and new host records for the continental United States. Plant Dis. 70:1180–1183.

863. Hollinger, D. Y. 1983. Photosynthesis and water relations of the mistletoe *Phoradendron villosum* and its host, the California valley oak, *Quercus lobata*. Oecologia 60:396–400.

864. Hollis, C. A., and Schmidt, R. A. 1977. Site factors related to fusiform rust incidence in North Florida slash pine plantations. For. Sci. 23:69–77.

865. Holmes, B., and Roberts, P. 1981. The classification, identification, and nomenclature of agrobacteria, incorporating revised descriptions for each of *Agrobacterium tumefaciens* (Smith & Townsend) Conn 1942, *Agrobacterium rhizogenes* (Riker et al.) Conn 1942, and *Agrobacterium rubi* (Hildebrand) Starr & Weiss 1943. J. Appl. Bacteriol. 50:443–467.

866. Holmes, F. O., Hirumi, H., and Maramorosch, K. 1972. Witches'-broom of willow: Salix yellows. Phytopathology 62:826–828.

867. Holmes, F. W. 1976. Verticillium wilt of salt-injured sugar maple—preliminary study. (Abstr.) Proc. Am. Phytopathol. Soc. 3:305–306.

868. Holmes, F. W. 1984. Effects on maples of prolonged exposure by artificial girdling roots. J. Arboric. 10:40–44.

869. Holmes, J., and Rich, A. E. 1970. The influence of the retention of immature apple mummies by certain cultivars on the overwintering of *Physalospora obtusa*. Phytopathology 60:452–453.

870. Holmes, J., and Rich, A. E. 1970. Factors affecting release and dissemination of *Physalospora obtusa* spores in a New Hampshire apple orchard. Phytopathology 60:1052–1054.

871. Honey, E. E. 1940. *Monilinia* causing a brown rot and blight of the common azalea. Phytopathology 30:537–538.

872. Hoog, G. S. de, and Hermanides-Nijhof, E. J. 1977. The black yeasts and allied Hyphomycetes. Centralb. v. Schimmelcult., Baarn. Stud. Mycol. No. 15. 222 pp.

873. Hook, D. D. 1984. Waterlogging tolerance of lowland tree species of the South. S. J. Appl. For. 8:136–149.

874. Hooker, J., and Jackson, B. D. 1895. Index kewensis plantarum phanerogamarum. Vol. I. 1268 pp. Vol. II. 1299 pp. Supplementum I (1906)–XVI (1981). Clarendon Press, Oxford.

875. Hopkins, D. L. 1977. Diseases caused by leafhopper-borne, rickettsia-like bacteria. Annu. Rev. Phytopathol. 15:277–294.

876. Hopkins, D. L. 1982. Relation of Pierce's disease bacterium to a wilt-type disease in citrus in the greenhouse. Phytopathology 72:1090–1092.

877. Hopkins, D. L. 1983. Gram-negative, xylem-limited bacteria in plant disease. Phytopathology 73:347–350.

878. Hopkins, D. L. 1984. Variability of virulence in grapevine among isolates of the Pierce's disease bacterium. Phytopathology 74:1395–1398.

879. Hopkins, D. L. 1985. Physiological and pathological characteristics of virulent and avirulent strains of the bacterium that causes Pierce's disease of grapevine. Phytopathology 75:713–717.

880. Hopkins, J. C. 1963. Atropellis canker of lodgepole pine: etiology, symptoms, and canker development rates. Can. J. Bot. 41:1535–1545.

881. Horie, H., and Kobayashi, T. 1979–1980. Entomosporium leaf spot of Pomoidae (Rosaceae) in Japan. I. Distribution of the disease, morphology and physiology of the fungus. II. Parasitism and over-wintering of the fungus. III. Additional basis for identification of the fungus and distribution of the disease. Eur. J. For. Pathol. 9:366–379; 10:117–124, 225–235.

882. Horst, R. K. 1979. Westcott's plant disease handbook. 4th ed. Van Nostrand Reinhold, New York. 803 pp.

883. Horst, R. K. 1983. Compendium of rose diseases. American Phytopathological Society, St. Paul, MN. 50 pp.

884. Horvath, J., Eke, I., Gal, T., and Dezéry, M. 1975. Demonstration of virus-like particles in sweet chestnut and oak with leaf deformations in Hungary. Z. Pflanzenkr. Pflanzenschutz 82:498–502.

885. Hotson, J. W. 1927. A new species of *Exobasidium*. Phytopathology 17:207–215.

886. Hotson, H. H., and Cutter, V. M. 1951. The isolation and culture of *Gymnosporangium juniperi-virginianae* Schw. upon artificial media. Proc. Natl. Acad. Sci. 37:400–403.

887. Hough, A., Mulder, N. J., and LaGrange, J. M. 1979. Heat treatment for the control of Phytophthora gummosis in citrus. Plant Dis. Rep. 63:40–43.

888. Houston, D. R. 1963. Inoculation of oaks with *Urnula craterium* (Schw.) Fr. produces cankers identical to Strumella cankers. Plant Dis. Rep. 47:867–869.

889. Houston, D. R. 1969. Basal canker of white pine. For. Sci. 15:66–83.

890. Houston, D. R. 1975. Beech bark disease—the aftermath forests are structured for a new outbreak. J. For. 73:660–663.

891. Houston, D. R. 1981. Stress triggered tree diseases. The diebacks and declines. U.S. For. Serv. NS-INF-41-81. 36 pp.

892. Houston, D. R. 1985. Diebacks and declines of urban trees. Pages 120–137 in: Improving the quality of urban life with plants. D. F. and S. L. Karnosky, eds. N.Y. Bot. Gard., Inst. Urban Hortic. Publ. No. 2. 200 pp.

893. Houston, D. R., Drake, C. R., and Kuntz, J. E. 1965. Effects of environment on oak wilt development. Phytopathology 55:1114–1121.

894. Houston, D. R., Parker, E. J., and Lonsdale, D. 1979. Beech bark disease: patterns of spread and development of the inciting agent *Cryptococcus fasgisuga*. Can. J. For. Res. 9:336–344.

895. Houston, D. R., Parker, E. J., Perrin, R., and Lang, K. J. 1979. Beech bark disease: a comparison of the disease in North America, Great Britain, France, and Germany. Eur. J. For. Pathol. 9:199–211.

896. Howard, F. L. 1941. The bleeding canker disease of hardwoods and possibilities of control. Proc. W. Shade Tree Conf. 8:46–55.

897. Howard, F. W., and Thomas, D. L. 1980. Transmission of palm lethal decline to *Veitchia merrilli* by a planthopper *Myndus crudus*. J. Econ. Entomol. 73:715–717.

898. Howden, J. C. W., and Jacobs, L. 1973. Report on the rust work at Bath. Rose Annu. 1973:113–119.

899. Hoy, J. W., Mircetich, S. M., and Lownsbery, B. F. 1984. Differential transmission of Prunus tomato ringspot virus strains by *Xiphinema californicum*. Phytopathology 74:332–335.

900. Hubbes, M. 1959. Untersuchungen über *Dothichiza populea* Sacc. et Briard, den Erreger des Rindenbrandes der Pappel. Phytopathol. Z. 35:58–96.

901. Hubbes, M. 1960. Systematische und physiologische Untersuchungen an Valsaceen auf Weiden. Phytopathol. Z. 39:65–93.

902. Huberman, M. A. 1943. Sunscald of eastern white pine, *Pinus strobus* L. Ecology 24:456–471.

903. Hudler, G. W. 1980. Salt injury to roadside plants. Cornell Univ., N.Y. State Coll. Agric. & Life Sci. Inf. Bull. 169. 4 pp.

904. Hudler, G. W. 1984. Wound healing in bark of woody plants. J. Arboric. 10:241–245.

905. Hudler, G. W., and Beale, M. A. 1981. Anatomical features of girdling root injury. J. Arboric. 7:29–32.

906. Hudler, G., and French, D. W. 1976. Dispersal and survival of seed of eastern dwarf mistletoe. Can. J. For. Res. 6:335–340.

907. Hudler, G. W., Knudsen, G. R., and Beale, M. A. 1983. Dose-response relationships of five conifers to infection by conidia of *Gremmeniella abietina*. Plant Dis. 67:192–194.

908. Hudler, G. W., and Oshima, N. 1976. The occurrence and distribution of *Thyronectria austro-americana* on honeylocust in Colorado. Plant Dis. Rep. 60:920–922.

909. Hudler, G. W., Oshima, N., and Hawksworth, F. G. 1979. Bird dissemination of dwarf mistletoe on ponderosa pine in Colorado. Am. Midl. Nat. 102:273–280.

910. Hughes, S. J. 1952. Studies on micro-fungi. XIV. *Stigmella, Stigmina, Camptomeris, Polythrincium,* and *Fusicladiella*. Commonw. Mycol. Inst. Mycol. Pap. 49. 25 pp.

911. Hughes, S. J. 1960. Microfungi V. *Conoplea* Pers. and *Exosporium* Link. Can. J. Bot. 38:659–696.

912. Hughes, S. J. 1976. Sooty moulds. Mycologia 68:693–820.

913. Hull, R. J., and Leonard, O. A. 1964. Physiological aspects of parasitism in mistletoes (*Arceuthobium* and *Phoradendron*) I. The carbohydrate nutrition of mistletoe. Plant Physiol. 39:996–1007.

914. Humphreys-Jones, D. R. 1977. Leaf and shoot death (*Coniothyrium fuckelii* Sacc.) of *Juniperus communis* L. var. *compressa* Carr. Plant Pathol. 26:47–48.

915. Humphreys-Jones, D. R. 1977. Leaf blotch (*Coniothyrium viburni* Died.) of *Viburnum burkwoodii* L. Plant Pathol. 26:101.

916. Hunt, R. S. 1984. Inoculations of Scrophulariaceae with *Cronartium ribicola*. Can. J. Bot. 62:2523–2524.

917. Hunt, R. S. 1985. Experimental evidence of heterothallism in *Cronartium ribicola*. Can. J. Bot. 63:1086–1088.

918. Hunt, R. S., Cobb, F. W. Jr., and Parmeter, J. R. 1976. *Fomes annosus* stump colonization and fungus development in the California mixed-conifer type. Can. J. For. Res. 6:159–165.

919. Hunt, R. S., and Morrison, D. J. 1979. Black stain root disease in British Columbia. Can. For. Serv. FPL 67. 4 pp.

920. Hunt, R. S., and Van Sickle, G. A. 1984. Variation in susceptibility to sweet fern rust among *Pinus contorta* and *P. banksiana*. Can. J. For. Res. 14:672–675.

921. Hunt, R. S., and Ziller, W. G. 1978. Host-genus keys to the Hypodermataceae of conifer leaves. Mycotaxon 6:481–496.

922. Hunter, P. P., and Stipes, R. J. 1978. The effect of month of inoculation with *Endothia gyrosa* on development of pruned branch canker of pin oak (*Quercus palustris*). Plant Dis. Rep. 62:940–944.

923. Huppuch, C. D. 1960. Observations on white oak stem swellings. Plant Dis. Rep. 44:238–239.

924. Hursh, C. R., and Haasis, F. W. 1931. Effects of 1925 summer drought on southern Appalachian hardwoods. Ecology 12:380–386.

925. Ikin, R., and Frost, R. R. 1974. Virus diseases of roses I. Their occurrence in the United Kingdom. Phytopathol. Z. 79:160–168.

926. Illingworth, K. 1973. Variation in the susceptibility of lodgepole pine provenances to Sirococcus blight. Can. J. For. Res. 3:585–589.

927. Ingold, C. T., Davey, R. A., and Wakley, G. 1981. The teliospore pedicel of Phragmidium mucronatum. Trans. Br. Mycol. Soc. 77:439–442.

928. Ingram, D. E. 1914. A twig blight of Quercus prinus and related species. J. Agric. Res. 1:339–346.

929. Institut National de la Recherche Agronomique (Paris). 1980. Colloque sur la maladie de l'écorce du hêtre. Ann. Sci. For. 37:269–392.

930. International Association of Wood Anatomists. 1984. Special issue on discolored wood. IAWA Bull., n.s. 5:91–154.

931. International Union of Forest Research Organizations, Working Party on Beech Bark Disease. 1983. Proceedings of Beech Bark Disease Working Party Conference. U.S. For. Serv. Gen. Tech. Rep. WO-37. 140 pp.

932. Ioannou, N., Schneider, R. W., and Grogan, R. G. 1977. Effect of flooding on the soil gas composition and the production of microsclerotia by Verticillium dahliae in the field. Phytopathology 67:651–656.

933. Ito, K., Chiba, O., Ono, K., and Hosaka, Y. 1954. Pestalotia disease of Camellia japonica L. Bull. Gov. For. Exp. Stn., Tokyo 70:103–124.

934. Ito, K., Kobayashi, T., and Hayashi, H. 1972. Stem gall or stem swelling of larch seedlings caused by a phytotoxicity of benzene hexachloride (BHC). Bull. Gov. For. Exp. Stn., Tokyo No. 245. 19 pp.

935. Ito, K., and Shibukawa, K. 1956. Studies on some anthracnoses of woody plants—III. A new anthracnose of acacia with special reference to the life history of the causal fungus. Bull. Gov. For. Exp. Stn., Tokyo 92:51–64.

936. Ito, K., Shibukawa, K., and Kobayashi, T. 1974. Etiological and pathological studies on the needle blight of Cryptomeria japonica—IV. Blight and canker of the tree caused by Cercospora sequoiae Ellis et Everhart (C. cryptomeriae Shirai). Bull. Gov. For. Exp. Stn., Tokyo 268:81–134.

937. Ito, K., Zinno, Y., and Kobayashi, T. 1963. Larch canker in Japan. Bull. Gov. For. Exp. Stn., Tokyo. 155:25–47.

938. Ivory, M. H. 1972. Resistance to Dothistroma needle blight induced in Pinus radiata by maturity and shade. Trans. Br. Mycol. Soc. 59:205–212.

939. Jackson, H. S. 1914. A new pomaceous rust of economic importance, Gymnosporangium blasdaleanum. Phytopathology 4:261–268.

940. Jackson, L. W. R. 1940. Lightning injury of black locust seedlings. Phytopathology 30:183–184.

941. Jacobi, W. R. 1984. Optimal conditions for in vitro growth, asexual spore release, and germination of Thyronectria austro-americana. Phytopathology 74:566–569.

942. Jacobi, W. R., and MacDonald, W. L. 1980. Colonization of resistant and susceptible oaks by Ceratocystis fagacearum. Phytopathology 70:618–623.

943. Jacobson, J. S., and Hill, A. C., eds. 1970. Recognition of air pollution injury to vegetation: a pictorial atlas. Air Pollution Control Association, Pittsburgh, PA. 109 pp.

944. Jaffee, B. A., Abawi, G. S., and Mai, W. F. 1982. Role of soil microflora and Pratylenchus penetrans in an apple replant disease. Phytopathology 72:247–251.

945. James, D. W., et al., eds. 1984. Proceedings, Second International Symposium on Iron Nutrition and Interactions in Plants. J. Plant Nutr. 7:1–864.

946. James, R. L., Cobb, F. W. Jr., Miller, P. R., and Parmeter, J. R. Jr. 1980. Effects of oxidant air pollution on susceptibility of pine roots to Fomes annosus. Phytopathology 70:560–563.

947. James, R. L., Cobb, F. W. Jr., Wilcox, W. W., and Rowney, D. L. 1980. Effects of photochemical oxidant injury of ponderosa and Jeffrey pines on susceptibility of sapwood and freshly cut stumps to Fomes annosus. Phytopathology 70:704–708.

948. James, S. 1984. Lignotubers and burls—their structure, function, and significance in Mediterranean ecosystems. Bot. Rev. 50:225–266.

949. Janse, J. D. 1981–1982. The bacterial disease of ash (Fraxinus excelsior) caused by Pseudomonas syringae pv. fraxini. I. History, occurrence and symptoms. II. Etiology and taxonomic considerations. III. Pathogenesis. Eur. J. For. Pathol. 11:306–315, 425–438; 12:218–231.

950. Jarvis, W. R. 1977. Botryotinia and Botrytis species: taxonomy, physiology, and pathogenicity. A guide to the literature. Can. Dep. Agric., Res. Branch, Monogr. 15. 195 pp.

951. Jaynes, R. A., and Graves, A. H. 1963. Connecticut hybrid chestnuts and their culture. Conn. Agric. Exp. Stn. Bull. 657. 29 pp.

952. Jeffers, S. N. 1985. Phytophthora crown rot of apple trees: detection and occurrence of the pathogens and seasonal variation in colonization of two apple rootstocks. Ph.D. thesis, Cornell Univ., Ithaca, NY. 165 pp.

953. Jeger, M. J., Swait, A. A. J., and Butt, D. J. 1982. Overwintering of Venturia inaequalis, the causal agent of apple scab, on different cultivars. Ann. Appl. Biol. 100:91–98.

954. Jenkins, A. E. 1930. Sphaceloma symphoricarpi. Mycologia 22:106–110.

955. Jenkins, A. E., and Bitancourt, A. A. 1957. Studies in the Myriangiales. VII. Elsinoaceae on evergreen euonymus, rose, and English ivy. Mycologia 49:95–101.

956. Jenkins, A. E., and Jehle, R. A. 1951. An anthracnose of bittersweet in Maryland. Plant Dis. Rep. 35:413–414.

957. Jenkins, A. E., Miller, J. H., and Hepting, G. H. 1953. Spot anthracnose and other leaf and petal spots of flowering dogwood. Natl. Hortic. Mag. 32:57–69.

958. Jensen, D. D. 1971. Herbaceous host plants of western X-disease agent. Phytopathology 61:1465–1470.

959. Jensen, K. F. 1982. An analysis of the growth of silver maple and eastern cottonwood seedlings exposed to ozone. Can. J. For. Res. 12:420–424.

960. Jensen, K. F. 1983. Growth relationships in silver maple seedlings fumigated with O_3 and SO_2. Can. J. For. Res. 13:298–302.

961. Jensen, K. F., and Kozlowski, T. T. 1975. Absorption and transportation of sulfur dioxide by seedlings of four forest tree species. J. Environ. Qual. 4:379–382.

962. Jewell, F. F., and Walker, N. M. 1967. Histology of Cronartium quercuum galls on shortleaf pine. Phytopathology 57:545–550.

963. Johnson, D. W., and Kuntz, J. E. 1979. Eutypella canker of maple: ascospore discharge and dissemination. Phytopathology 69:130–135.

964. Johnson, D. W., and Kuntz, J. E. 1979. Eutypella parasitica: ascospore germination and mycelial growth. Can. J. Bot. 57:624–628.

965. Johnson, W. T., and Lyon, H. H. 1988. Insects that feed on trees and shrubs. 2nd ed. Cornell University Press, Ithaca, NY.

966. Johnston, J. C., and Trione, E. J. 1974. Cytokinin production by the fungi Taphrina cerasi and T. deformans. Can. J. Bot. 52:1583–1589.

967. Johnston, T. H., and Hitchcock, L. 1923. A bacteriosis of prickly pear plants (Opuntia spp.). Trans. R. Soc. S. Aust. 47:162–164.

968. Jokela, J. J., Paxton, J. D., and Zegar, E. J. 1976. Marssonina leaf spot and rust on eastern cottonwood. Plant Dis. Rep. 60:1020–1024.

969. Jones, A. L. 1963. Occurrence of Nectria twig blight of apple in western New York State in 1962. Plant Dis. Rep. 47:538–540.

970. Jones, C., Griffin, G. J., and Elkins, J. R. 1980. Association of climatic stress with blight on Chinese chestnut in the eastern United States. Plant Dis. 64:1001–1004.

971. Jones, L. H. 1943. Creosote injurious to plants. Horticulture, n.s. 21:462–463.

972. Jones, S. G. 1925. Life history and cytology of Rhytisma acerinum (Pers.) Fries. Ann. Bot. 39:41–75.

973. Jong, S. C., and Rogers, J. D. 1972. Illustrations and descriptions of conidial states of some Hypoxylon species. Wash. Agric. Exp. Stn. Tech. Bull. 71. 51 pp.

974. Jorgensen, E., and Cafley, J. D. 1961. Branch and stem cankers of white and Norway spruces in Ontario. For. Chron. 37:394–404.

975. Joubert, J. J., and Rijkenberg, F. H. J. 1971. Parasitic green algae. Annu. Rev. Phytopathol. 9:45–64.

976. Julis, A. J., Clayton, C. N., and Sutton, T. B. 1978. Detection and distribution of Phytophthora cactorum and P. cambivora on apple rootstocks. Plant Dis. Rep. 62:516–520.

977. Jurgensen, M. G., Larsen, J. J., Spano, S. D., Harvey, A. E., and Gale, M. R. 1984. Nitrogen fixation associated with increased wood decay in Douglas-fir residue. For. Sci. 30:1038–1044.

978. Juzwik, J., and French, D. W. 1983. Ceratocystis fagacearum and C. piceae on the surfaces of free-flying and fungus-mat-inhabiting nitidulids. Phytopathology 73:1164–1168.

979. Juzwik, J., and Hinds, T. E. 1984. Ascospore germination, mycelial growth, and microconidial anamorphs of Encoelia pruinosa in culture. Can. J. Bot. 62:1916–1919.

980. Juzwik, J., Nishijima, W. T., and Hinds, T. E. 1978. Survey of aspen cankers in Colorado. Plant Dis. Rep. 62:906–910.

981. Kado, C. I. 1976. The tumor-inducing substance of Agrobacterium tumefaciens. Annu. Rev. Phytopathol. 14:265–308.

982. Kahn, A. H. 1955. Dematophora root rot. Cal. Dep. Agric. Bull. 44:167–170.

983. Kais, A. G. 1971. Dispersal of Scirrhia acicola spores in southern Mississippi. Plant Dis. Rep. 55:309–311.

984. Kais, A. G. 1975. Environmental factors affecting brown spot infection on longleaf pine. Phytopathology 65:1389–1392.

985. Kais, A. G. 1977. Influence of needle age and inoculum spore density on susceptibility of longleaf pine to Scirrhia acicola. Phytopathology 67:686–688.

986. Kam, M. de. 1975. Ascospore discharge in Drepanopeziza punctiformis in relation to infection of some poplar clones. Eur. J. For. Pathol. 5:304–309.

987. Kam, M. de. 1982. Damage to poplar caused by Pseudomonas syringae in combination with frost and fluctuating temperatures. Eur. J. For. Pathol. 12:203–209.

988. Kamiri, L. K., and Laemmlen, F. F. 1981. Epidemiology of Cytospora canker caused in Colorado blue spruce by Valsa kunzei. Phytopathology 71:941–947.

989. Kapur, S., Kapur, S. P., and Rehlia, A. S. 1978. Studies on some of the biochemical changes occurring in date palm leaves infected by smut. Indian Phytopathol. 31:394–395.

990. Karnosky, D. F. 1977. Evidence for genetic control of response to sulfur dioxide and ozone in *Populus tremuloides*. Can. J. For. Res. 7:437–440.

991. Karnosky, D. F. 1981. Chamber and field evaluations of air pollution tolerances of urban trees. J. Arboric. 7:99–105.

992. Karnosky, D. F., and Steiner, K. C. 1981. Provenance and family variation in response of *Fraxinus americana* and *F. pennsylvanica* to ozone and sulfur dioxide. Phytopathology 71:804–807.

993. Kauffman, B. W., Applegate, H. W., Cordell, C. E., and Thor, E. 1980. Susceptibility of eight pine species to comandra blister rust in Tennessee. Plant Dis. 64:375–377.

994. Kawase, M. 1981. Anatomical and morphological adaptation of plants to waterlogging. HortScience 16:30–34.

995. Kawase, M. 1981. Effect of ethylene on aerenchyma development. Am. J. Bot. 68:651–658.

996. Keifer, H. H., Baker, E. W., Kono, T., Delfinado, M., and Styer, W. E. 1982. An illustrated guide to plant abnormalities caused by eriophyid mites in North America. U.S. Dep. Agric. Agric. Handb. 573. 178 pp.

997. Keitt, G. W., Blodgett, E. C., Wilson, E. E., and Magie, R. O. 1937. The epidemiology and control of cherry leaf spot. Wis. Agric. Exp. Stn. Res. Bull. 132. 117 pp.

998. Keitt, G. W., and Jones, L. K. 1926. Studies of the epidemiology and control of apple scab. Wis. Agric. Exp. Stn. Res. Bull. 73. 104 pp.

999. Kellerman, W. A., and Swingle, W. T. 1889. Branch knot of the hackberry. Kans. Agric. Exp. Stn. Rep. 1888:302–315.

1000. Kelman, A., and Gooding, G. V. Jr. 1965. A root and stem rot of yellow-poplar caused by *Cylindrocladium scoparium*. Plant Dis. Rep. 49:797–801.

1001. Kendrick, W. B. 1962. The *Leptographium* complex. *Verticicladiella* Hughes. Can. J. Bot. 40:771–797.

1002. Kenerley, C. M., and Bruck, R. I. 1981. Phytophthora root rot of balsam fir and Norway spruce in North Carolina. Plant Dis. 65:614–615.

1003. Kenerley, C. M., and Bruck, R. I. 1983. Overwintering and survival of *Phytophthora cinnamomi* in Fraser fir and cover cropped nursery beds in North Carolina. Phytopathology 73:1643–1647.

1004. Kenerley, C. M., Papke, K., and Bruck, R. I. 1984. Effect of flooding on development of Phytophthora root rot in Fraser fir seedlings. Phytopathology 74:401–404.

1005. Kennedy, B. W., Froyd, J., and Bowden, R. 1984. Bacterial blight of mock orange (*Philadelphus* spp.) caused by *Pseudomonas syringae*. Plant Dis. 68:916–917.

1006. Kennedy, L. L., and Stewart, A. W. 1967. Development and taxonomy of *Apiosporina collinsii*. Can. J. Bot. 45:1597–1604.

1007. Kern, F. D. 1973. A revised taxonomic account of *Gymnosporangium*. Pennsylvania State University Press, University Park, PA. 134 pp.

1008. Kern, H. 1955. Taxonomic studies in the genus *Leucostoma*. Mich. Acad. Sci. Pap. 40:9–22.

1009. Kern, H. 1957. Untersuchungen über die Umgrenzung der Arten in der Ascomycetengattung *Leucostoma*. Phytopathol. Z. 30:149–180.

1010. Kern, H., and Naef-Roth, S. 1975. Zur Bildung von Auxinen und Cytokininen durch *Taphrina*-Arten. Phytopathol. Z. 83:193–222.

1011. Kerr, A. 1980. Biological control of crown gall through production of Agrocin 84. Plant Dis. 64:24–30.

1012. Kerr, A., and Panagopoulos, C. G. 1977. Biotypes of *Agrobacterium radiobacter* var. *tumefaciens* and their biological control. Phytopathol. Z. 90:172–179.

1013. Kessler, K. J. Jr. 1972. Sapstreak disease of sugar maple. U.S. For. Serv. For. Pest Leafl. 128. 4 pp.

1014. Kessler, K. J. Jr. 1974. An apparent symbiosis between *Fusarium* fungi and ambrosia beetles causes canker on black walnut trees. Plant Dis. Rep. 58:1044–1047.

1015. Kessler, K. J. Jr., and Hadfield, J. S. 1972. Eutypella canker of maple. U.S. For. Serv. For. Pest Leafl. 136. 6 pp.

1016. Kessler, K. J. Jr., and Weber, B. C., coords. 1979. Walnut diseases and insects. U.S. For Serv. Gen. Tech. Rep. NC-52. 100 pp.

1017. Khairi, S. M., and Preece, T. F. 1978. Hawthorn powdery mildew: overwintering mycelium in buds and the effect of clipping hedges on disease epidemiology. Trans. Br. Mycol. Soc. 71:399–404.

1018. Khan, S. R., Kimbrough, J. W., and Mims, C. W. 1981. Septal ultrastructure and the taxonomy of *Exobasidium*. Can. J. Bot. 59:2450–2457.

1019. Kienholz, J. R., and Childs, L. 1937. Twig lesions as a source of early spring infection by the pear scab organism. J. Agric. Res. 55:667–681.

1020. Kienholz, J. R., and Childs, L. 1951. Pear scab in Oregon. Oreg. Agric. Exp. Stn. Tech. Bull. 21. 31 pp.

1021. Kienholz, R. 1933. Frost damage to red pine. J. For. 31:392–399.

1022. Kile, G. A., 1976. The effect of seasonal pruning and time since pruning upon changes in apple sapwood and its susceptibility to invasion by *Trametes versicolor*. Phytopathol. Z. 87:231–240.

1023. Kile, G. A., and Wade, G. C. 1974. *Trametes versicolor* on apple. I. Host-pathogen relationship. Phytopathol. Z. 81:328–338.

1024. Kim, K. S., and Fulton, J. P. 1973. Association of viruslike particles with a ringspot disease of oak. Plant Dis. Rep. 57:1029–1031.

1025. Kim, K. S., and Martin, E. M. 1978. Viruslike particles associated with yellow ringspot of redbud. (Abstr.) Phytopathol. News 12:199.

1026. Kimbrough, J. W. 1963. The development of *Pleochaeta polychaeta* (Erysiphaceae). Mycologia 55:608–618.

1027. Kimmey, J. W. 1964. Heart rots of western hemlock. U.S. For. Serv. For. Pest Leafl. 90. 7 pp.

1028. Kimmey, J. W. 1965. Rust-red stringy rot. U.S. For. Serv. For. Pest Leafl. 93. 8 pp.

1029. Kimmey, J. W., and Bynum, H. H. 1961. Heart rots of red and white firs. U.S. For. Serv. For. Pest Leafl. 52. 4 pp.

1030. Kimmey, J. W., and Wagener, W. W. 1961. Spread of white pine blister rust from ribes to sugar pine in California and Oregon. U.S. Dep. Agric. Tech. Bull. 1251. 71 pp.

1031. Kishi, K., Abiko, K., Takanashi, K., and Yano, R. 1973. Studies on the virus diseases of stone fruit. VII. Line pattern of plum and flowering cherry. Ann. Phytopathol. Soc. Jap. 39:288–296.

1032. Kistler, B. R., and Merrill, W. 1968. Effects of *Strumella coryneoidea* on oak sapwood. Phytopathology 58:1429–1430.

1033. Kistler, B. R., and Merrill, W. 1978. Etiology, symptomatology, epidemiology, and control of Naemacyclus needlecast of Scotch pine. Phytopathology 68:267–271.

1034. Klebahn, H. 1935. Einige Beobachtungen und Versuche über den Mahonia-Rost. Z. Pflanzenkr. Pflanzenschutz. 45:529–537.

1035. Klement, Z., Rozsnyay, D. S., Báló, E., Pánczél, M., and Prileszky, G. 1984. The effect of cold on development of bacterial canker in apricot trees infected with *Pseudomonas syringae* pv. *syringae*. Physiol. Plant Pathol. 24:237–246.

1036. Kliejunas, J. T., and Kuntz, J. E. 1972. Development of stromata and the imperfect state of *Eutypella parasitica* in maple. Can. J. Bot. 50:1453–1456.

1037. Knauf, V. C., Panagopoulos, C. G., and Nester, E. W. 1982. Genetic factors controlling the host range of *Agrobacterium tumefaciens*. Phytopathology 72:1545–1549.

1038. Knorr, L. C. 1973. Citrus diseases and disorders. University Presses of Florida, Gainesville. 163 pp.

1039. Ko, W. H., Ho, W. C., and Kunimoto, R. K. 1982. Relation of *Kretzschmaria clavus* to hypoxyloid stromata on diseased macadamia tissues. Phytopathology 72:1357–1358.

1040. Ko, W. H., Kunimoto, R. K., and Maedo, I. 1977. Root decay caused by *Kretzschmaria clavus*: its relation to macadamia decline. Phytopathology 67:18–21.

1041. Kobayashi, T. 1958. Phomopsis disease on Japanese cedar, *Cryptomeria japonica* D. Don, with special reference to the life-history and taxonomy of the causal fungus. Bull. Gov. For. Exp. Stn., Tokyo. No. 107:3–25.

1042. Kobayashi, T., and Chiba, O. 1961. Fungi inhabiting poplars in Japan I. Bull. Gov. For. Exp. Stn., Tokyo. No. 130:1–43.

1043. Kobayashi, T., and Mamiya, Y. 1963. A *Cenangium* causing dieback of Japanese pines. Bull. Gov. For. Exp. Stn., Tokyo 161:123–150.

1044. Koch, L. W. 1933–1935. Investigations on black knot of plums and cherries. I. Development and discharge of spores and experiments in control. II. The occurrence and significance of certain fungi found in association with *Dibotryon morbosum* (Sch.) T. & S. III. Symptomatology, life history, and cultural studies of *Dibotryon morbosum* (Sch.) T. & S. IV. Studies in pathogenicity and pathological histology. Sci. Agric. 13:576–590; 15:80–95, 411–423, 729–744.

1045. Koch, L. W. 1934. Studies on the overwintering of certain fungi parasitic and saprophytic on fruit trees. Can. J. Res. 11:190–206.

1046. Kohn, L. M. 1979. A monographic revision of the genus *Sclerotinia*. Mycotaxon 9:365–444.

1047. Kondo, E., Foudin, A., Linit, M., Smith, M., Bolla, R., Winter, R., and Dropkin, V. 1982. Pine wilt disease—nematological, entomological, and biochemical investigations. Univ. Mo. Agric. Exp. Stn. SR282. 56 pp.

1048. Kondo, E. S., Hiratsuka, Y., and Denyer, W. B. G., eds. 1982. Proceedings of the Dutch Elm Disease Symposium and Workshop. Environment Canada and Manitoba Department of Natural Resources. 517 pp.

1049. Kostka, S. J., Tattar, T. A., Sherald, J. L., and Hurtt, S. S. 1986. Mulberry leaf scorch, new disease caused by a fastidious, xylem-inhabiting bacterium. Plant Dis. 70:690–693.

1050. Kozar, F., and Netolitzky, H. J. 1975. Ultrastructure and cytology of pycnia, aecia, and aeciospores of *Gymnosporangium clavipes*. Can. J. Bot. 53:972–977.

1051. Kozlowski, T. T., ed. 1978. Water deficits and plant growth. Vol. V. Water and plant disease. Academic Press, New York. 323 pp.

1052. Kozlowski, T. T., ed. 1981. Water deficits and plant growth. Vol. VI. Woody plant communities. Academic Press, New York. 582 pp.

1053. Kozlowski, T. T. 1982. Water supply and tree growth. For. Abstr. 43:57–95, 145–161.

1054. Kozlowski, T. T., ed. 1984. Flooding and plant growth. Academic Press, New York. 356 pp.

1055. Kozlowski, T. T. 1984. Plant responses to flooding of soil. BioScience 34:162–167.

1056. Kraus, J. F., and Hunt, D. L. 1971. Inherent variation of resistance to *Hypoderma lethale* in slash and loblolly pines. For. Sci. 17:143–144.

1057. Krebill, R. G. 1968. *Cronartium comandrae* in the Rocky Mountain states. U.S. For. Serv. Res. Pap. INT-50. 28 pp.

1058. Krebill, R. G. 1968. Histology of canker rusts in pines. Phytopathology 58:155–164.

1059. Krebill, R. G. 1972. *Pucciniastrum goeppertianum* in *Abies lasiocarpa* in the Rocky Mountain states. Am. Midl. Nat. 87:570–576.

1060. Kress, L. W., and Skelly, J. M. 1982. Response of several eastern forest tree species to chronic doses of ozone and nitrogen dioxide. Plant Dis. 66:1149–1152.

1061. Krezdorn, A. H., and Adriance, G. W. 1961. Fig growing in the South. U.S. Dep. Agric. Agric. Handb. 196. 26 pp.

1062. Krupinsky, J. M. 1981. *Botryodiplodia hypodermia* and *Tubercularia ulmea* in cankers on Siberian elm in northern Great Plains windbreaks. Plant Dis. 65:677–678.

1063. Krupinsky, J. M. 1983. Variation in virulence of *Botryodiplodia hypodermia* to *Ulmus pumila*. Phytopathology 73:108–110.

1064. Kubler, H. 1983. Mechanism of frost crack formation in trees—a review and synthesis. For. Sci. 29:559–568.

1065. Kuhlman, E. G., ed. 1974. *Fomes annosus*. Proc. Fourth Int. Conf. on *Fomes annosus*, Athens, GA, 1973. U.S. For. Serv. Southeast. For. Exp. Stn. 289 pp.

1066. Kuhlman, E. G., Bhattacharyya, H., Nash, B. L., Double, M. L., and MacDonald, W. L. 1984. Identifying hypovirulent isolates of *Cryphonectria parasitica* with broad conversion capacity. Phytopathology 74:676–682.

1067. Kuhlman, E. G., Cordell, C. E., and Filer, T. H. Jr. 1980. Cylindrocladium root rots of sweetgum seedlings in southern forest tree nurseries. Plant Dis. 64:1079–1080.

1068. Kuhlman, E. G., Dianis, S. D., and Smith, T. K. 1982. Epidemiology of pitch canker disease in a loblolly pine seed orchard in North Carolina. Phytopathology 72:1212–1216.

1069. Kuhlman, E. G., Dwinell, L. D., Nelson, P. E., and Booth, C. 1978. Characterization of the *Fusarium* causing pitch canker of southern pines. Mycologia 70:1131–1143.

1070. Kuhlman, E. G., Hodges, C. S. Jr., and Froelich, R. C. 1976. Minimizing losses to *Fomes annosus* in the southern United States. U.S. For. Serv. Res. Pap. SE-151. 16 pp.

1071. Kuijt, J. 1964. Critical observations on the parasitism of New World mistletoes. Can. J. Bot. 42:1243–1278.

1072. Kuijt, J. 1969. The biology of parasitic flowering plants. University of California Press, Berkeley. 246 pp.

1073. Kuijt, J. 1977. Haustoria of phanerogamic parasites. Annu. Rev. Phytopathol. 15:91–118.

1074. Kuijt, J. 1982. The Viscaceae in the southeastern United States. J. Arnold Arbor. 63:401–410.

1075. Kuijt, J., and Toth, R. 1976. Ultrastructure of angiosperm haustoria—a review. Ann. Bot. 40:1121–1130.

1076. Kulman, H. M. 1971. Effects of insect defoliation on growth and mortality of trees. Annu. Rev. Entomol. 16:289–324.

1077. Kumi, J., and Lang, K. J. 1979. Susceptibility of various spruce species to *Rhizosphaera kalkhoffii* and some cultural characteristics of the fungus *in vitro*. Eur. J. For. Pathol. 9:35–46.

1078. Kurian, P., and Stelzig, D. A. 1979. The synergistic role of oxalic acid and endopolygalacturonase in bean leaves infected by *Cristulariella pyramidalis*. Phytopathology 69:1301–1304.

1079. Kurstak, E., ed. 1981. Handbook of plant virus infections. Comparative diagnosis. Elsevier/North Holland Biomedical Press, Amsterdam. 943 pp.

1080. Kuske, C. R., and Benson, D. M. 1983. Survival and splash dispersal of *Phytophthora parasitica*, causing dieback of rhododendron. Phytopathology 73:1188–1191.

1081. Kuske, C. R., and Benson, D. M. 1983. Overwintering and survival of *Phytophthora parasitica*, causing dieback of rhododendron. Phytopathology 73:1192–1196.

1082. Lacasse, N. L., and Rich, A. E. 1964. Maple decline in New Hampshire. Phytopathology 54:1071–1075.

1083. Lacasse, N. L., and Treshow, M., eds. 1977. Diagnosing vegetation injury caused by pollution. U.S. Environ. Protec. Agency, Air Pollut. Training Inst. Chapters paged separately.

1084. Lachance, D. 1971. Discharge and germination of *Eutypella parasitica* ascospores. Can. J. Bot. 49:1111–1118.

1085. Lachance, D. 1971. Inoculation and development of Eutypella canker of maple. Can. J. For. Res. 1:228–234.

1086. Lachance, D., and Kuntz, J. E. 1970. Ascocarp development of *Eutypella parasitica*. Can. J. Bot. 48:1977–1979.

1087. Lachmund, H. G. 1921. Some phases in the formation of fire scars. J. For. 19:638–640.

1088. Lachmund, H. G. 1929. *Cronartium comptoniae* Arth. in western North America. Phytopathology 19:453–466.

1089. Lackner, A. L., and Alexander, S. A. 1982. Occurrence and pathogenicity of *Verticicladiella procera* in Christmas tree plantations in Virginia. Plant Dis. 66:211–212.

1090. Lackner, A. L., and Alexander, S. A. 1983. Root disease and insect infestations on air-pollution-sensitive *Pinus strobus* and studies of pathogenicity of *Verticicladiella procera*. Plant Dis. 67:679–681.

1091. Lackner, A. L., and Alexander, S. A. 1984. Incidence and development of *Verticicladiella procera* in Virginia Christmas tree plantations. Plant Dis. 68:210–212.

1092. Lagerberg, T. 1949. Some notes on the Phacidiaceae and a new member of this family, *Lophophacidium hyperboreum* nov. gen. et sp. Svensk Bot. Tidskr. 43:420–437.

1093. Lair, E. D. 1946. Smooth patch, a bark disease of oak. J. Elisha Mitchell Sci. Soc. 62:212–220.

1094. Lambe, R. C., and Wills, W. H. 1978. Pathogenicity of *Thielaviopsis basicola* to Japanese holly (*Ilex crenata*). Plant Dis. Rep. 62:859–863.

1095. Lambe, R. C., and Wills, W. H. 1980. Distribution of dieback associated with Thielaviopsis black root rot of Japanese holly. Plant Dis. 64:956.

1096. Lamberti, F., and Taylor, C. E., eds. 1979. Root knot nematodes (*Meloidogyne* species). Systematics, biology, and control. Academic Press, New York. 477 pp.

1097. Lamson, N. I., and Smith, H. C. 1978. Response to crop tree release: sugar maple, red oak, black cherry, and yellow poplar saplings in a 9–year-old stand. U.S. For. Serv. Res. Pap. NE-394. 8 pp.

1098. Lana, A. F., Peterson, J. F., Rouselle, G. L., and Vrain, T. C. 1983. Association of tobacco ringspot virus with a union incompatibility of apple. Phytopathol. Z. 106:141–148.

1099. Lana, A. F., Thomas, O. T., and Peterson, J. F. 1980. A virus isolated from sugar maple. Phytopathol. Z. 97:214–218.

1100. Lana, A. O., and Agrios, G. N. 1974. Transmission of a mosaic disease of white ash to woody and herbaceous hosts. Plant Dis. Rep. 58:536–540.

1101. Landgraf, F. A., and Zehr, E. I. 1982. Inoculum sources for *Monilinia fructicola* in South Carolina peach orchards. Phytopathology 72:185–190.

1102. Landis, W. R., and Hart, J. H. 1967. Cankers of ornamental crabapples. Plant Dis. Rep. 51:230–234.

1103. Lang, K. J., and Schütt, P. 1974. Anatomische Untersuchungen zur Infektionsbiologie von *Scleroderris lagerbergii* Gr. [*Brunchorstia pinea* (Karst.) von Höhn]. Eur. J. For. Pathol. 4:166–174.

1104. Langford, M. H., and Keitt, G. W. 1942. Heterothallism and variability in *Venturia pirina*. Phytopathology 32:357–369.

1105. Langridge, Y. N., and Dye, D. W. 1982. A bacterial disease of *Pinus radiata* seedlings caused by *Pseudomonas syringae*. N.Z. J. Agric. Res. 25:273–276.

1106. Lanier, L., and Sylvestre, G. 1971. Epidémiologie du *Lophodermium pinastri* (Schrad.) Chev. Résultats de quatre années d'études à l'aide d'un piège à spores de Hirst. Eur. J. For. Pathol. 1:50–63.

1107. Large, J. R. 1944. Alcoholic flux or white slime flux of tung trees. Plant Dis. Rep. 28:35–36.

1108. Large, J. R. 1948. Canker of tung trees caused by *Physalospora rhodina*. Phytopathology 38:359–363.

1109. Larsen, M. J., Jurgensen, M. F., and Harvey, A. E. 1978. N$_2$ fixation associated with wood decayed by some common fungi in western Montana. Can. J. For. Res. 8:341–345.

1110. Larsen, M. J., Lombard, F. F., and Aho, P. E. 1979. A new variety of *Phellinus pini* associated with cankers and decay in white firs in southwestern Oregon and northern California. Can. J. For. Res. 9:31–38.

1111. Latham, A. J. 1969. Zonate leafspot of pecan caused by *Cristulariella pyramidalis*. Phytopathology 59:103–107.

1112. Latham, A. J. 1974. Effect of temperature and moisture on *Cristulariella pyramidalis*. Phytopathology 64:635–639.

1113. Laurence, J. A. 1981. Effects of air pollutants on plant-pathogen interactions. Z. Pflanzenkrankh. Pflanzenschutz 88:156–172.

1114. Lavallée, A. 1964. A larch canker caused by *Leucostoma kunzei* (Fr.) Munk ex Kern. Can. J. Bot. 42:1495–1502.

1115. Lawrey, J. D. 1984. Biology of lichenized fungi. Praeger, New York. 408 pp.

1116. Leaf, A. L. 1968. K, Mg, and S deficiencies in forest trees. Pages 88–122 in: Forest fertilization: theory and practice. Tennessee Valley Authority, Muscle Shoals, AL. 306 pp.

1117. Leaphart, C. D. 1959. Drought damage to western white pine and associated tree species. Plant Dis. Rep. 43:809–813.

1118. Leaphart, C. D. 1963. Armillaria root rot. U.S. For. Serv. For. Pest Leafl. 78. 8 pp.

1119. Leaphart, C. D., Copeland, O. L., and Graham, D. P. 1957. Pole blight of western white pine. U.S. For. Serv. For. Pest Leafl. 16. 4 pp.

1120. Leaphart, C. D., and Gill, L. S. 1959. Effects of inoculations with *Leptographium* spp. on western white pine. Phytopathology 49:350–353.

1121. Leaphart, C. D., and Stage, A. R. 1971. Climate: a factor in the origin of the pole blight disease of *Pinus monticola* Dougl. Ecology 52:229–239.

1122. Leaphart, C. D., and Wicker, E. F. 1966. Explanation of pole blight from responses of seedlings grown in modified environments. Can. J. Bot. 44: 121–137.

1123. Leben, C. 1985. Wound occlusion and discoloration columns in red maple. New Phytol. 99:485–490.

1124. Lee, R. F., Marais, L. J., Timmer, L. W., and Graham, J. H. 1984. Syringe injection of water into the trunk: a rapid diagnostic test for citrus blight. Plant Dis. 68:511–513.

1125. Lee, R. F., Raju, B. C., Nyland, G., and Goheen, A. C. 1982. Phytotoxin(s) produced in culture by the Pierce's disease bacterium. Phytopathology 72:886–888.

1126. Lemke, P. A. 1964. The genus *Aleurodiscus* (*sensu stricto*) in North America. Can. J. Bot. 42:213–282.

1127. Lemke, P. A. 1965. *Dendrothele* (1907) vs. *Aleurocorticium* (1963). Persoonia 3:365–367.

1128. Lemon, P. C. 1961. Forest ecology of ice storms. Bull. Torrey Bot. Club 88:21–29.

1129. Lentz, P. L. 1955. *Stereum* and allied genera of fungi in the upper Mississippi Valley. U.S. Dep. Agric. Monogr. 24. 74 pp.

1130. Leonard, O. A., and Hull, R. J. 1965. Translocation relationships in and between mistletoes and their hosts. Hilgardia 37:115–151.

1131. Lewis, R. Jr. 1981. Decline symptoms in Mississippi Gulf coastal live oaks. (Abstr.) J. Miss. Acad. Sci. 27 (Suppl.):6.

1132. Lewis, R. Jr. 1981. *Hypoxylon* spp., *Ganoderma lucidum*, and *Agrilus bilineatus* in association with drought related oak mortality in the south. (Abstr.) Phytopathology 71:890.

1133. Lewis, R. Jr. 1985. Temperature tolerance and survival of *Ceratocystis fagacearum* in Texas. Plant Dis. 69:443–444.

1134. Lewis, R. Jr., and Oliveria, F. L. 1979. Live oak decline in Texas. J. Arboric. 5:241–244.

1135. Lewis, R. Jr., and Van Arsdel, E. P. 1978. Vulnerability of water-stressed sycamores to strains of *Botryodiplodia theobromae*. Plant Dis. Rep. 62:62–63.

1136. Lewis, R. Jr., and Van Arsdel, E. P. 1978. Development of Botryodiplodia cankers in sycamore at controlled temperatures. Plant Dis. Rep. 62:125–126.

1137. Liberty Hyde Bailey Hortorium. 1976. Hortus third. Macmillan, New York. 1290 pp.

1138. Lightle, P. C., and Hawksworth, F. G. 1965. New hosts for broom-causing fungi in the Southwest. Plant Dis. Rep. 49:417–418.

1139. Lightle, P. C., Standring, E. T., and Brown, J. G. 1942. A bacterial necrosis of the giant cactus. Phytopathology 32:303–313.

1140. Lightle, P. C., and Thompson, J. H. 1973. Atropellis canker of pines. U.S. For. Serv. For. Pest Leafl. 138. 6 pp.

1141. Linderman, R. G. 1973. Formation of microsclerotia of *Cylindrocladium* spp. in infected azalea leaves, flowers, and roots. Phytopathology 63:187–191.

1142. Linderman, R. G. 1974. The role of abscised *Cylindrocladium*-infected azalea leaves in the epidemiology of Cylindrocladium wilt of azalea. Phytopathology 64:481–485.

1143. Linderman, R. G. 1974. Ascospore discharge from perithecia of *Calonectria theae*, *C. crotalariae*, and *C. kyotensis*. Phytopathology 64:567–569.

1144. Linderman, R. G., and Gilbert, R. G. 1969. Stimulation of *Sclerotium rolfsii* in soil by volatile components of alfalfa hay. Phytopathology 59:1366–1372.

1145. Lindow, S. E. 1983. Methods of preventing frost injury caused by epiphytic ice-nucleation-active bacteria. Plant Dis. 67:327–333.

1146. Lindow, S. E. 1983. The role of bacterial ice nucleation in frost injury to plants. Annu. Rev. Phytopathol. 21:363–384.

1147. Lindow, S. E., and Connell, J. H. 1984. Reduction of frost injury to almond by control of ice nucleation active bacteria. J. Am. Soc. Hortic. Sci. 109:48–53.

1148. Lindsey, J. P., and Gilbertson, R. L. 1978. Basidiomycetes that decay aspen in North America. J. Cramer, Vaduz, Austria. 406 pp.

1149. Linzon, S. N. 1966. Damage to eastern white pine by sulfur dioxide, semimature-tissue needle blight, and ozone. J. Air Pollut. Control Assn. 16:140–144.

1150. Linzon, S. N. 1967. Histological studies of symptoms in semimature-tissue needle blight of eastern white pine. Can. J. Bot. 45:133–143.

1151. Lipetz, J. 1970. Wound-healing in higher plants. Int. Rev. Cytol. 27:1–28.

1152. Little, S., Mohr, J. J., and Spicer, L. L. 1958. Salt-water storm damage to loblolly pine forests. J. For. 56:27–28.

1153. Littlefield, L. J. 1981. Biology of the plant rusts: an introduction. Iowa State University Press, Ames, IA. 103 pp.

1154. Littlefield, L. J., and Heath, M. C. 1979. Ultrastructure of rust fungi. Academic Press, New York. 277 pp.

1155. Liu, H.-Y., Gumpf, D. J., Oldfield, G. N., and Calavan, E. C. 1983. Transmission of *Spiroplasma citri* by *Circulifer tenellus*. Phytopathology 73:582–585.

1156. Lockerman, R. H., Putnam, A. R., Rice, R. P. Jr., and Meggitt, W. F. 1975. Diagnosis and prevention of herbicide injury. Mich. State Univ., Coop. Ext. Serv. Bull. E-809. 19 pp.

1157. Lockhart, C. L. 1958. Studies on red leaf disease of lowbush blueberries. Plant Dis. Rep. 42:764–767.

1158. Lohman, M. L., and Watson, A. J. 1943. Identity and host relations of *Nectria* species associated with diseases of hardwoods in the eastern states. Lloydia 6:77–108.

1159. Lombard, F. F., Davidson, R. W., and Lowe, J. L. 1960. Cultural characteristics of *Fomes ulmarius* and *Poria ambigua*. Mycologia 52:280–294.

1160. Long, W. H. 1913. *Polyporus dryadeus*, a root parasite on the oak. J. Agric. Res. 1:239–250.

1161. Long, W. H. 1924. The self pruning of western yellow pine. Phytopathology 14:336–337.

1162. Long, W. H. 1930. *Polyporus dryadeus*, a root parasite on white fir. Phytopathology 20:758–759.

1163. Lonsdale, D. 1980. Nectria infection of beech bark in relation to infestation by *Cryptococcus fagisuga* Lindiger. Eur. J. For. Pathol. 10:161–168.

1164. Loomis, R. C., and Padgett, W. H. 1974. Air pollution and trees in the East. U.S. For. Serv., State and Private For., Northeast. and Southeast. Areas. 28 pp.

1165. Loper, J. E., and Kado, C. I. 1979. Host range conferred by the virulence-specifying plasmid of *Agrobacterium tumefaciens*. J. Bacteriol. 139:591–596.

1166. Lorenz, D. H. 1976. Beiträge zur weiteren Kenntnis des Lebenszyklus von *Taphrina deformans* (Berk.) Tul. unter besonderer Berücksichtigung der Saprophase. Phytopathol. Z. 86:1–15.

1167. Loring, L. B., and Roth, L. F. 1964. *Pucciniastrum epilobii* on *Fuchsia* in Oregon. Plant Dis. Rep. 48:99.

1168. Lortie, M. 1964. Pathogenesis in cankers caused by *Nectria galligena*. Phytopathology 54:261–263.

1169. Lortie, M., and Kuntz, J. E. 1963. Ascospore discharge and conidium release by *Nectria galligena* Bres. under field and laboratory conditions. Can. J. Bot. 41:1203–1210.

1170. Lowerts, G. A., Zoerb, M. H. Jr., and Pait, J. 1985. Resistance to the development of pitch canker in open-pollinated slash pine families. Proc. S. For. Tree Impr. Conf. 18:334–340.

1171. Ludwig, C. A. 1915. Notes on some North American rusts with Caeoma-like sori. Phytopathology 5:273–281.

1172. Lumis, G. P., Hofstra, G., and Hall, R. 1973. Sensitivity of roadside trees and shrubs to aerial drift of deicing salt. HortScience 8:475–477.

1173. Lumis, G. P., Hofstra, G., and Hall, R. 1976. Roadside woody plant susceptibility to sodium and chloride accumulation during winter and spring. Can. J. Plant Sci. 56:853–859.

1174. Lumis, G. P., and Johnson, A. G. 1982. Boron toxicity and growth suppression of *Forsythia* and *Thuja* grown in mixes amended with municipal waste compost. HortScience 27:821–822.

1175. Lung-Escarmant, B., and Dunez, J. 1980. Les propriétés immunologiques, un critère possible de classification de l'Armillaire. Ann. Phytopathol. 12: 57–70.

1176. Luttrell, E. S. 1940. *Morenoella quercina*, cause of leaf spot of oaks. Mycologia 32:652–666.

1177. Luttrell, E. S. 1948. Physiologic specialization in *Guignardia bidwellii*, cause of black rot of *Vitis* and *Parthenocissus* species. Phytopathology 38:716–723.

1178. Luttrell, E. S. 1949. Horse chestnut anthracnose in Missouri. Plant Dis. Rep. 33:324–327.

1179. Luttrell, E. S. 1949. *Scirria acicola*, *Phaeocryptopus pinastri*, and *Lophodermium pinastri* associated with the decline of ponderosa pine in Missouri. Plant Dis. Rep. 33:397–401.

1180. Luttrell, E. S. 1950. *Botryosphaeria* stem canker of elm. Plant Dis. Rep. 34:138–139.

1181. Lutz, H. J. 1943. Injuries to trees caused by *Celastrus* and *Vitis*. Bull. Torrey Bot. Club 70:436–439.

1182. Lutz, H. J. 1952. Occurrence of clefts in the wood of living white spruce in Alaska. J. For. 50:99–102.

1183. Lyda, S. D. 1978. Ecology of *Phymatotrichum omnivorum*. Annu. Rev. Phytopathol. 16:193–209.

1184. MacBeath, J. H., Nyland, G., and Spurr, A. R. 1972. Morphology of mycoplasmalike bodies associated with peach X-disease in *Prunus persica*. Phytopathology 62:935–937.

1185. MacDaniels, L. H., Johnson, W. T., and Braun, E. J. 1976. The black walnut bunch disease syndrome. Annu. Rep. N. Nut Growers Assn. 66:71–87.

1186. MacDonald, J. A. 1937. A study of *Polyporus betulinus* (Bull.) Fries. Ann. Appl. Biol. 24:289–310.

1187. MacDonald, J. D. 1982. Role of environmental stress in the development of Phytophthora root rots. J. Arboric. 8:217–223.

1188. MacDonald, W. L., Cech, F. C., Luchok, J., and Smith, C. 1979. Proceedings of the American Chestnut Symposium. West Virginia University Books, Morgantown, WV. 122 pp.

1189. MacLachlan, J. D. 1935. The hosts of *Gymnosporangium globosum* Farl. and their relative susceptibility. J. Arnold Arbor. 16:98–142.

1190. MacLachlan, J. D. 1935. The dispersal of viable basidiospores of the *Gymnosporangium* rusts. J. Arnold Arbor. 16:411–422.

1191. Mader, D. L., and Thompson, B. W. 1969. Foliar and soil nutrients in relation to maple decline. Soil Sci. Soc. Am. Proc. 33:794–800.

1192. Magasi, L. P., ed. 1983. Proceedings of European Larch Canker Workshop. Can. For. Serv., Maritimes For. Res. Cent. MFRC Workshop Proc. No. 3. 46 pp.

1193. Magasi, L. P., and Pond, S. E. 1982. European larch canker: a new disease in Canada and a new North American host record. Plant Dis. 66:339.

1194. Maggenti, A. R. 1981. Nematodes: development as plant parasites. Annu. Rev. Microbiol. 35:135–154.

1195. Mahoney, M. J., Skelly, J. M., Chevone, B. I., and Moore, L. D. 1984. Response of yellow poplar (*Liriodendron tulipifera* L.) seedling shoot growth to low concentrations of O_3, SO_2, and NO_2. Can. J. For. Res. 14:150–153.

1196. Mahoney, M. J., and Tattar, T. A. 1980. Identification, etiology, and control of *Euonymus fortunei* anthracnose caused by *Colletotrichum gloeosporioides*. Plant Dis. 64:854–856.

1197. Mai, W. F., Bloom, J. R., and Chen, T. A., eds. 1977. Biology and ecology of the plant-parasitic nematode *Pratylenchus penetrans*. Penn. Agric. Exp. Stn. Bull. 815. 64 pp.

1198. Mains, E. B. 1938. Host specialization in *Coleosporium solidaginis* and *C. campanulae*. Pap. Mich. Acad. Sci. Arts Lett. 23:171–175.

1199. Malek, R. B. 1968. The dagger nematode, *Xiphinema americanum*, associated with decline of shelterbelt trees in South Dakota. Plant Dis. Rep. 52:795–798.

1200. Malek, R. B., and Appleby, J. E. 1984. Epidemiology of pine wilt in Illinois. Plant Dis. 68:180–186.

1201. Malhotra, S. S., and Blauel, R. A. 1980. Diagnosis of air pollutant and natural stress symptoms on forest vegetation in western Canada. Environ. Can. Inf. Rep. NOR-X-228. 84 pp.

1202. Malhotra, S. S., and Hocking, D. 1976. Biochemical and cytological effects of sulfur dioxide on plant metabolism. New Phytol. 76:227–237.

1203. Malia, M. E., and Tattar, T. A. 1978. Electrical resistance, physical characteristics, and cation concentrations in xylem of sugar maple infected with *Verticillium dahliae*. Can. J. For. Res. 8:322–327.

1204. Maloy, O. C. 1967. A review of *Echinodontium tinctorium* Ell. & Ev., the Indian paint fungus. Wash. Agric. Exp. Stn. Bull. 686. 21 pp.

1205. Mamiya, Y. 1983. Pathology of the pine wilt disease caused by *Bursaphelenchus xylophilus*. Annu. Rev. Phytopathol. 21:201–220.

1206. Mamiya, Y., and Enda, N. 1972. Transmission of *Bursaphelenchus lignicolus* (Nematoda: Aphelenchoididae). Nematologica 18:159–162.

1207. Mamiya, Y., and Furukawa, M. 1977. Fecundity and reproductive rate of *Bursaphelenchus lignicolus*. Jap. J. Nematol. 7:6–9.

1208. Mamiya, Y., and Tamura, H. 1977. Transpiration reduction of pine seedlings inoculated with the pine wood nematode, *Bursaphelenchus lignicolus*. J. Jap. For. Soc. 59:59–63.

1209. Manion, P. D. 1975. Two infection sites of *Hypoxylon mammatum* in trembling aspen (*Populus tremuloides*). Can. J. Bot. 53:2621–2624.

1210. Manion, P. D. 1981. Norway maple decline. J. Arboric. 7:38–42.

1211. Manion, P. D., ed. 1984. Scleroderris canker of conifers. Martinus Nijhoff/W. Junk, The Hague. 273 pp.

1212. Manion, P. D., and French, D. W. 1967. *Nectria galligena* and *Ceratocystis fimbriata* cankers of aspen in Minnesota. For. Sci. 13:23–28.

1213. Manion, P. D., and French, D. W. 1969. The role of glucose in stimulating germination of *Fomes igniarius* var. *populinus* basidiospores. Phytopathology 59:293–296.

1214. Manners, J. G. 1957. Studies on larch canker. II. The incidence and anatomy of cankers produced experimentally either by inoculation or by freezing. Trans. Br. Mycol. Soc. 40:500–508.

1215. Maramorosch, K., and Raychaudhuri, S. P., eds. 1981. Mycoplasma diseases of trees and shrubs. Academic Press, New York. 362 pp.

1216. Marks, G. C., Berbee, J. G., and Riker, A. J. 1965. Colletotrichum shoot blight of poplars. For. Sci. 11:204–215.

1217. Marks, G. C., and Kozlowski, T. T., eds. 1973. Ectomycorrhizae. Their ecology and physiology. Academic Press, New York. 444 pp.

1218. Marks, G. C., and Minko, G. 1969. The pathogenicity of *Diplodia pinea* to *Pinus radiata* D. Don. Aust. J. Bot. 17:1–12.

1219. Marlatt, R. B., and Alfieri, S. A. Jr. 1981. Hosts of a parasitic alga, *Cephaleuros* Kunze, in Florida. Plant Dis. 65:520–522.

1220. Marlatt, R. B., and Ridings, W. H. 1974. Sphaeropsis gall of bottlebrush tree, *Callistemon vimminalis*, a new host. Phytopathology 64:1001–1003.

1221. Marlatt, R. B., and Ridings, W. H. 1976. Sphaeropsis disease of *Carissa grandiflora*. Plant Dis. Rep. 60:842–843.

1222. Marlatt, R. B., and Ridings, W. H. 1979. Sphaeropsis gall of *Schinus terebinthifolius*, a new host. Plant Dis. Rep. 63:786–787.

1223. Marshall, R. P., and Waterman, A. M. 1948. Common diseases of important shade trees. U.S. Dep. Agric. Farmers' Bull. 1987. 53 pp.

1224. Martin, J. F., and Gravatt, G. F. 1954. Saving white pines by removing blister rust cankers. U.S. Dep. Agric. Circ. 948. 22 pp.

1225. Martin, P. 1967. Studies in the Xylariaceae. II. *Rosellinia* and the Primo Cinera section of *Hypoxylon*. J. South Afr. Bot. 33:315–328.

1226. Martin, P. 1970. Studies in the Xylariaceae. VIII. *Xylaria* and its allies. J. South Afr. Bot. 36:73–138.

1227. Martin, R. R., Berbee, J. G., and Omuemu, J. O. 1982. Isolation of a potyvirus from declining clones of *Populus*. Phytopathology 72:1158–1162.

1228. Marx, D. H., Cordell, C. E., Kenney, D. S., Mexal, J. G., Artman, J. D., Riffle, J. W., and Molina, R. J. 1984. Commercial vegetative inoculum of *Pisolithus tinctorius* and inoculation techniques for development of ectomycorrhizae on bare-root tree seedlings. For. Sci. Monogr. 25. 101 pp.

1229. Marx, D. H., and Davey, C. B. 1969. The influence of ectotrophic mycorrhizal fungi on the resistance of pine roots to pathogenic infections. IV. Resistance of naturally occurring mycorrhizae to infections by *Phytophthora cinnamomi*. Phytopathology 59:559–565.

1230. Marx, D. H., and Schenck, N. C. 1983. Potential of mycorrhizal symbiosis in agricultural and forest productivity. Pages 334–347 in: Challenging problems in plant health. T. Kommedahl and P. H. Williams, eds. American Phytopathological Society, St. Paul, MN. 538 pp.

1231. Massalski, P. R., and Cooper, J. I. 1984. The location of virus-like particles in the male gametophyte of birch, walnut, and cherry naturally infected with cherry leaf roll virus and its relevance to vertical transmission of the virus. Plant Pathol. 33:255–262.

1232. Massie, L. B., and Peterson, J. L. 1968. Factors affecting the initiation and development of Fusarium canker on *Sophora japonica* in relation to growth and sporulation of *Fusarium lateritium*. Phytopathology 58:1620–1623.

1233. Matheron, M. E., and Mircetich, S. M. 1985. Influence of flooding duration on development of Phytophthora root and crown rot of *Juglans hindsii* and Paradox walnut rootstocks. Phytopathology 75:973–976.

1234. Matheron, M. E., and Mircetich, S. M. 1985. Pathogenicity and relative virulence of *Phytophthora* spp. from walnut and other plants to rootstocks of English walnut trees. Phytopathology 75:977–981.

1235. Matteoni, J. A. 1983. Yellows diseases of elm and ash: water relations, vectors, and role of yellows in ash decline. Ph.D. thesis, Cornell Univ., Ithaca, NY. 152 pp.

1236. Matteoni, J. A., and Neely, D. 1979. *Gnomonia leptostyla*: growth, sporulation, and heterothallism. Mycologia 71:1034–1042.

1237. Matteoni, J. A., and Sinclair, W. A. 1983. Stomatal closure in plants infected with mycoplasmalike organisms. Phytopathology 73:398–402.

1238. Matteoni, J. A., and Sinclair, W. A. 1985. Role of the mycoplasmal disease, ash yellows, in decline of white ash in New York State. Phytopathology 75:355–360.

1239. May, C. 1961. Diseases of shade and ornamental maples. U.S. Dep. Agric. Agric. Handb. 211. 22 pp.

1240. Mazzola, M., and Bergdahl, D. R. 1983. Periods of spore dissemination for *Uredinopsis mirabilis* on *Abies balsamea* and *Onoclea sensibilis*. (Abstr.) Phytopathology 73:370.

1241. Mazzone, H. M., and Peacock, J. W. 1985. Prospects for control of Dutch elm disease—biological considerations. J. Arboric. 11:285–292.

1242. McAlpine, R. G. 1961. *Hypoxylon tinctor* associated with a canker on American sycamore trees in Georgia. Plant Dis. Rep. 45:196–198.

1243. McArthur, G. W. F. M. 1959. *Cercospora* leaf spot on rhododendron (*C. handelii* Bubak). N.Z. J. Agric. Res. 2:86–89.

1244. McCain, A. H., Raabe, R. D., and Wilhelm, S. 1979. Plants resistant or susceptible to Verticillium wilt. Univ. Cal., Div. Agric. Sci., Leafl. 2703. 10 pp.

1245. McCarroll, D. R., and Thor, E. 1985. Do "toxins" affect pathogenesis by *Endothia parasitica*? Physiol. Plant Pathol. 26:357–366.

1246. McCauley, K. J., and Cook, S. A. 1980. *Phellinus weirii* infestation of two mountain hemlock forests in the Oregon Cascades. For. Sci. 26:23–29.

1247. McClain, R. L. 1925. Scab of Christmas berry, *Photinia arbutifolia* Lindl., due to *Fusicladium photinicola* n. sp. Phytopathology 15:178–182.

1248. McClenahen, J. R., and Dochinger, L. S. 1985. Tree ring response of white oak to climate and air pollution near the Ohio River Valley. J. Environ. Qual. 14:274–280.

1249. McClintock, J. A. 1948. A study of uncongeniality between peaches as scions and the Marianna plum as a stock. J. Agric. Res. 77:253–260.

1250. McCoy, R. E. 1982. Use of tetracycline antibiotics to control yellows diseases. Plant Dis. 66:539–542.

1251. McCoy, R. E., ed. 1983. Lethal yellowing of palms. Florida Agric. Exp. Stn. Bull. 834. 100 pp.

1252. McCoy, R. E., Miller, M. E., Thomas, D. L., and Amador, J. 1980. Lethal decline of Phoenix palms in Texas associated with mycoplasmalike organisms. Plant Dis. 64:1038–1040.

1253. McCracken, F. I. 1970. Spore production of *Hericium erinaceus*. Phytopathology 60:1639–1641.

1254. McCracken, F. I. 1978. Canker-rots in southern hardwoods. U.S. For. Serv. For. Insect Dis. Leafl. 33. 4 pp.

1255. McCracken, F. I. 1978. Spore release of some decay fungi of southern hardwoods. Can. J. Bot. 56:426–431.

1256. McCracken, F. I., and Burkhardt, E. C. 1977. Destruction of sycamore by canker stain in the Midsouth. Plant Dis. Rep. 61:984–986.

1257. McCracken, F. I., Schipper, A. L., and Widin, K. D. 1984. Observations on occurrence of cottonwood leaf rust in central United States. Eur. J. For. Pathol. 14:226–233.

1258. McCracken, F. I., and Toole, E. R. 1969. Sporophore development and sporulation of *Polyporus hispidus*. Phytopathology 59:884–885.

1259. McCracken, F. I., and Toole, E. R. 1974. Felling infected oaks in natural stands reduces dissemination of *Polyporus hispidus* spores. Phytopathology 64:265–266.

1260. McCulloch, W. F. 1943. Ice breakage in partially cut and uncut second growth Douglas-fir stands. J. For. 41:275–278.

1261. McDonald, G. I., Hansen, E. M., Osterhaus, C. A., and Samman, S. 1984. Initial characterization of a new strain of *Cronartium ribicola* from the Cascade Mountains of Oregon. Plant Dis. 68:800–804.

1262. McDowell, J., and Merrill, W. 1985. *Rhabdocline* taxa in Pennsylvania. Plant Dis. 69:714–715.

1263. McGranahan, G. H., and Smalley, E. B. 1984. Influence of moisture, temperature, leaf maturity, and host genotype on infection of elms by *Stegophora ulmea*. Phytopathology 74:1296–1300.

1264. McGranahan, G. H., and Smalley, E. B. 1984. Conidial morphology, axenic growth, and sporulation of *Stegophora ulmea*. Phytopathology 74:1300–1303.

1265. McKellar, A. D. 1942. Ice damage to slash pine, longleaf pine, and loblolly pine plantations in the Piedmont section of Georgia. J. For. 40:794–797.

1266. McKenzie, H. L., Gill, L. S., and Ellis, D. E. 1948. The Prescott scale (*Matsucoccus vexillorum*) and associated organisms that cause flagging injury to ponderosa pine in the Southwest. J. Agric. Res. 76:33–51.

1267. McKenzie, M. A., Jones, L. H., and Gilgut, C. J. 1940. *Phomopsis gardeniae* in relation to gardenia culture. Plant Dis. Rep. 24:58–62.

1268. McLaughlin, S. B., McConathy, R. K., Duvick, D., and Mann, L. K. 1982. Effects of chronic air pollution stress on photosynthesis, carbon allocation, and growth of white pine trees. For. Sci. 28:60–70.

1269. McNabb, R. F. R., and Laurenson, J. B. 1965. A rust of cultivated fuchsias. N.Z. J. Agric. Res. 8:336–339.

1270. McPartland, J. M., and Schoeneweiss, D. F. 1984. Hyphal morphology of *Botryosphaeria dothidea* in vessels of unstressed and drought-stressed stems of *Betula alba*. Phytopathology 74:358–362.

1271. McRae, C. H., Rockwood, D. L., and Blakeslee, G. M. 1985. Evaluation of slash pine for resistance to pitch canker. Proc. S. For. Tree Impr. Conf. 18:351–357.

1272. McRitchie, J. J. 1973. Pathogenicity and control of *Fusarium oxysporum* wilt of variegated pyracantha. Plant Dis. Rep. 57:389–391.

1273. McWain, P., and Gregory, G. F. 1972. A neutral mannan from *Ceratocystis fagacearum* culture filtrate. Phytochemistry 11:2609–2612.

1274. Mead, M. A., Dolezal, W. E., and Tainter, F. H. 1978. Eighteen newly discovered pine hosts of comandra blister rust fungus. Plant Dis. Rep. 62:885–887.

1275. Meinecke, E. P. 1929. Experiments with repeating pine rusts. Phytopathology 19:327–342.

1276. Mence, M. J., and Hildebrandt, A. C. 1966. Resistance to powdery mildew in rose. Ann. Appl. Biol. 58:309–320.

1277. Menge, J. A., and French, D. W. 1976. Effect of plant residue amendments and chemical treatments upon the inoculum potential of *Cylindrocladium floridanum* in soil. Phytopathology 66:1085–1089.

1278. Menon, R. 1956. Studies on Venturiaceae on rosaceous plants. Phytopathol. Z. 27:117–146.

1279. Mercer, P. C., and Kirk, S. A. 1984. Biological treatments for the control of decay in tree wounds. II. Field tests. Ann. Appl. Biol. 104:221–229.

1280. Merrill, W., and Cowling, E. B. 1966. Role of nitrogen in wood deterioration: amount and distribution of nitrogen in fungi. Phytopathology 56:1083–1090.

1281. Merrill, W., French, D. W., and Wood, F. A. 1964. Decay of wood by species of the Xylariaceae. Phytopathology 54:56–58.

1282. Merrill, W., and Kistler, B. R. 1976. Phenology and control of *Endocronartium harknessii* in Pennsylvania. Phytopathology 66:1246–1248.

1283. Merrill, W., and Kistler, B. R. 1976. Seasonal development and control of *Lophodermium pinastri* in Pennsylvania. Plant Dis. Rep. 60:652–655.

1284. Merrill, W., and Kistler, B. R. 1978. Accelerated development of Rhizosphaera needlecast of blue spruce in Pennsylvania. Plant Dis. Rep. 62:34–35.

1285. Merrill, W., and Kistler, B. R. 1978. Needlecast and weeds interact to cause branch mortality in Scots pine. Plant Dis. Rep. 62:200–202.

1286. Merrill, W., Kistler, B. R., Zang, L., and Bowen, K. 1980. Infection periods in Naemacyclus needlecast of Scots pine. Plant Dis. 64:759–761.

1287. Merrill, W., McCall, K., and Zang, L. 1981. Fusarium root rot of Douglas-fir and Fraser fir seedlings in Pennsylvania. Plant Dis. 65:913–914.

1288. Messenger, A. S. 1984. Seasonal variations of foliar nutrients in green and chlorotic red maples. J. Environ. Hortic. 2:117–119.

1289. Michaels, E., and Chastagner, G. A. 1984. Distribution, severity, and impact of Swiss needle cast in Douglas-fir Christmas trees in western Washington and Oregon. Plant Dis. 68:939–942.

1290. Michaels, E., and Chastagner, G. A. 1984. Seasonal availability of *Phaeocryptopus gaeumannii* ascospores and conditions that influence their release. Plant Dis. 68:942–944.

1291. Mielke, J. L. 1943. White pine blister rust in western North America. Yale Univ. Sch. For. Bull. 52. 155 pp.

1292. Mielke, J. L. 1952. The rust fungus *Cronartium filamentosum* in Rocky Mountain ponderosa pine. J. For. 50:365–373.

1293. Mielke, J. L. 1956. The rust fungus (*Cronartium stalactiforme*) in lodgepole pine. J. For. 54:518–521.

1294. Mielke, J. L. 1957. Aspen leaf blight in the intermountain region. U.S. For. Serv. Intermount. For. Range Exp. Stn. Res. Note 42. 5 pp.

1295. Mielke, J. L., and Kimmey, J. W. 1942. Heat injury to the leaves of California black oak and some other broadleaves. Plant Dis. Rep. 26:116–119.

1296. Mielke, J. L., Krebill, R. G., and Powers, H. R. Jr. 1968. Comandra blister rust of hard pines. U.S. For. Serv. For. Pest Leafl. 62. 8 pp.

1297. Mielke, M. E. 1981. Pathogenicity of *Verticicladiella penicillata* (Grosm.) Kendrick to northern Idaho conifers. For. Sci. 27:103–110.

1298. Mielke, M. E., Haynes, C., and MacDonald, W. L. 1982. Beech scale and *Nectria galligena* on beech in the Monongahela National Forest, West Virginia. Plant Dis. 66:851–852.

1299. Milbrath, G. M., and Nelson, M. R. 1972. Isolation and characterization of a virus from saguaro cactus. Phytopathology 62:739–742.

1300. Milbrath, G. M., Nelson, M. R., and Wheeler, R. E. 1973. The distribution and electron microscopy of viruses of cacti in southern Arizona. Phytopathology 63:1133–1139.

1301. Miles, L. E. 1921. Leaf spots of the elm. Bot. Gaz. 71:161–196.

1302. Milholland, R. D. 1972. Histopathology and pathogenicity of *Botryosphaeria dothidea* on blueberry stems. Phytopathology 62:654–660.

1303. Millar, C. S. 1981. Infection processes on conifer needles. Pages 185–207 in: Microbial ecology of the phylloplane. J. P. Blakeman, ed. Academic Press, London. 502 pp.

1304. Miller, H. C., Allen, D., Silverborg, S. B., and Haines, J. H. 1981. Beech bark disease. N.Y. State Tree Pest Leafl. F32. Sci. Serv., N.Y. State Museum, Albany. 8 pp.

1305. Miller, H. N. 1961. Cause and control of diseases of potted plants. Pages 157–158 in: Fla. Agric. Exp. Stn. Annu. Rep. for Year Ending June 30, 1961. 399 pp.

1306. Miller, J. H. 1961. A monograph of the world species of *Hypoxylon*. University of Georgia Press, Athens. 158 pp.

1307. Miller, J. H., and Edwards, B. 1983. Kudzu: where did it come from? And how can we stop it? So. J. Appl. For. 7:165–169.

1308. Miller, J. H., and Wolf, F. A. 1936. A leaf-spot disease of honey locust caused by a new species of *Linospora*. Mycologia 28:171–180.

1309. Miller, J. R., and Tocher, R. D. 1975. Photosynthesis and respiration of *Arceuthobium tsugense* (Loranthaceae). Am. J. Bot. 62:765–769.

1310. Miller, L. W., and Boyle, J. S. 1943. The Hydnaceae of Iowa. Univ. Iowa. Stud. Nat. Hist. 18(2):1–92.

1311. Miller, P. M., and Rich, S. 1968. Reducing spring discharge of *Venturia inaequalis* ascospores by composting overwintering leaves. Plant Dis. Rep. 52:728–730.

1312. Miller, P. M., and Waggoner, P. E. 1958. Dissemination of *Venturia inaequalis* ascospores. Phytopathology 48:416–419.

1313. Miller, P. R., tech. coord. 1980. Proceedings of Symposium on Effects of Air Pollutants on Mediterranean and Temperate Forest Ecosystems. U.S. For. Serv. Gen. Tech. Rep. PSW-43. 256 pp.

1314. Miller, P. R., and Evans, L. S. 1974. Histopathology of oxidant injury and winter fleck injury on needles of western pines. Phytopathology 64:801–806.

1315. Miller, P. R., Parmeter, J. R. Jr., Taylor, O. C., and Cardiff, E. A. 1963. Ozone injury to the foliage of *Pinus ponderosa*. Phytopathology 53:1072–1076.

1316. Miller, P. W., and Bollen, W. B. 1946. Walnut bacteriosis and its control. Oreg. Agric. Exp. Stn. Tech. Bull. 9. 107 pp.

1317. Miller, S. B., and Baxter, L. W. Jr. 1970. Dieback in azaleas caused by *Phomopsis* species. Phytopathology 60:387–388.

1318. Miller, T., Cowling, E. B., Powers, H. R. Jr., and Blalock, T. E. 1976. Types of resistance and compatibility in slash pine seedlings infected by Cronartium fusiforme. Phytopathology 66:1229–1235.

1319. Miller, T., Patton, R. F., and Powers, H. R. Jr. 1980. Mode of infection and early colonization of slash pine seedlings by Cronartium quercuum f. sp. fusiforme. Phytopathology 70:1206–1208.

1320. Miller, V. M., and Bonar, L. 1941. A study of the Perisporiaceae, Capnodiaceae, and some other sooty molds from California. Univ. Calif. Publ. Bot. 19:405–427.

1321. Mims, C. W. 1981. Ultrastructure of teliospore germination and basidiospore formation in the rust fungus Gymnosporangium clavipes. Can. J. Bot. 59:1041–1049.

1322. Mims, C. W., and Glidewell, D. C. 1978. Some ultrastructural observations on the host-pathogen relationship within the telial gall of the rust fungus Gymnosporangium juniperi-virginianae. Bot. Gaz. 139:11–17.

1323. Minter, D. W. 1981. Lophodermium on pines. Commonw. Mycol. Inst. Mycol. Pap. 147. 54 pp.

1324. Minter, D. W., and Cannon, P. F. 1984. Ascospore discharge in some members of the Rhytismataceae. Trans. Br. Mycol. Soc. 83:65–92.

1325. Minter, D. W., and Millar, C. S. 1980. Ecology and biology of three Lophodermium species on secondary needles of Pinus sylvestris. Eur. J. For. Pathol. 10:169–181.

1326. Minter, D. W., Staley, J. M., and Millar, C. S. 1978. Four species of Lophodermium on Pinus sylvestris. Trans. Br. Mycol. Soc. 71:295–301.

1327. Mircetich, S. M., Campbell, R. N., and Matheron, M. E. 1977. Phytophthora trunk canker of coast live oak and cork oak trees in California. Plant Dis. Rep. 61:66–70.

1328. Mircetich, S. M., and Hoy, J. W. 1981. Brownline of prune trees, a disease associated with tomato ringspot virus infection of myrobalan and peach rootstocks. Phytopathology 71:30–35.

1329. Mircetich, S. M., Lowe, S. K., Moller, W. J., and Nyland, G. 1976. Etiology of almond leaf scorch disease and transmission of the causal agent. Phytopathology 66:17–24.

1330. Mircetich, S. M., and Matheron, M. E. 1976. Phytophthora root and crown rot of cherry trees. Phytopathology 66:549–558.

1331. Mircetich, S. M., and Rowhani, A. 1984. The relationship of cherry leafroll virus and blackline disease of English walnut trees. Phytopathology 74:423–428.

1332. Mircetich, S. M., Sanborn, R. R., and Ramos, D. E. 1980. Natural spread, graft transmission, and possible etiology of walnut blackline disease. Phytopathology 70:962–968.

1333. Mistretta, P. A., Anderson, R. L., MacDonald, W. L., and Lewis, R. Jr. 1984. Annotated bibliography of oak wilt, 1943–80. U.S. For. Serv. Gen. Tech. Rep. WO-45. 132 pp.

1334. Mitchell, C. P., Millar, C. S., and Minter, D. W. 1978. Studies on decomposition of Scots pine needles. Trans. Br. Mycol. Soc. 71:343–348.

1335. Mitchell, C. P., Millar, C. S., and Williamson, B. 1978. The biology of Lophodermella conjuncta Darker on Corsican pine needles. Eur. J. For. Pathol. 8:108–118.

1336. Mitchell, C. P., Williamson, B., and Millar, C. S. 1976. Hendersonia acicola on pine needles infected by Lophodermella sulcigena. Eur. J. For. Pathol. 6:92–102.

1337. Mix, A. J. 1916. Sun-scald of fruit trees. A type of winter injury. Cornell Univ. Agric. Exp. Stn. Bull. 382:235–284.

1338. Mix, A. J. 1925. Anthracnose of European privet. Phytopathology 15:261–272.

1339. Mix, A. J. 1949. A monograph of the genus Taphrina. Univ. Kansas Sci. Bull. 23, Pt. I, No. 1. 167 pp.

1340. Mollenhauer, H. H., and Hopkins, D. L. 1976. Xylem morphology of Pierce's disease-infected grapevines with different levels of tolerance. Physiol. Plant Pathol. 9:95–100.

1341. Moller, W. J. 1980. Effect of apple cultivar on Venturia inaequalis ascospore emission in California. Plant Dis. 64:930–931.

1342. Molnar, A. C. 1961. An outbreak of Cronartium comptoniae on Monterey and Bishop pines on Vancouver Island, British Columbia. Plant Dis. Rep. 45:854–855.

1343. Monod, M. 1983. Monographie taxonomique des Gnomoniaceae. Sydowia Ann. Mycol., Ser. II, Beiheft IX. 315 pp.

1344. Moore, L. W. 1976. Latent infections and seasonal variability of crown gall development in seedlings of three Prunus species. Phytopathology 66:1097–1101.

1345. Moore, M. H. 1963. A Gloeosporium bud-rot and twig-canker disease of cultivated hazel. J. Hortic. Sci. 38:109–118.

1346. Moore, R., and Walker, D. B. 1981. Studies of vegetative compatibility-incompatibility in higher plants. I. A structural study of a compatible autograft in Sedum telephoides (Crassulaceae). II. A structural study of an incompatible heterograft between Sedum telephoides (Crassulaceae) and Solanum pennellii (Solanaceae). Am. J. Bot. 68:820–830, 831–842.

1347. Moreau, M., Moreau, C., and Péresse, M. 1971. Une maladie chancreuse des genévriers d'ornement. Ann. Phytopathol. 3:233–241.

1348. Morehart, A. L., Carroll, R. B., and Stuart, M. 1980. Phomopsis canker and dieback of Elaeagnus angustifolia. Plant Dis. 64:66–69.

1349. Morehart, A. L., Donohue, F. M. III, and Melchior, G. L. 1980. Verticillium wilt of yellow poplar. Phytopathology 70:756–760.

1350. Morehart, A. L., and Melchior, G. L. 1982. Influence of water stress on Verticillium wilt of yellow-poplar. Can. J. Bot. 60:201–209.

1351. Morelet, M. 1985. Les Venturia des peupliers de la section Leuce I.—Taxinomie. Cryptogamie Mycol. 6:101–117.

1352. Morris, C. L. 1953. Chemical control of Hypoderma lethale on pitch pine. Plant Dis. Rep. 37:368–370.

1353. Morrison, D. J. 1981. Armillaria root disease. A guide to disease diagnosis, development, and management in British Columbia. Can. For. Serv. BC-X-203. 15 pp.

1354. Morrison, D. J. 1982. Effects of soil organic matter on rhizomorph growth by Armillaria mellea. Trans. Br. Mycol. Soc. 78:201–207.

1355. Morrison, D. J., Chu, D., and Johnson, A. L. S. 1985. Species of Armillaria in British Columbia. Can. j. Plant Pathol. 7:242–246.

1356. Morton, H. L., and Patton, R. F. 1970. Swiss needlecast of Douglas-fir in the Lake States. Plant Dis. Rep. 54:612–616.

1357. Moss, A. E. 1940. Effect on trees of wind-driven salt water. J. For. 38:421–425.

1358. Mounce, I. 1929. Studies in forest pathology. II. The biology of Fomes pinicola (Sw.) Cooke. Can. Dep. Agric. Bull. n.s. 111. 75 pp.

1359. Mudd, J. B., and Kozlowski, T. T., eds. 1975. Responses of plants to air pollution. Academic Press, New York. 383 pp.

1360. Mueller, W. C. 1967. Tobacco mosaic virus obtained from diseased wisteria and elder. Plant Dis. Rep. 51:1053–1054.

1361. Mueller-Dombois, D., Canfield, J. E., Holt, R. A., and Buelow, G. P. 1983. Tree group death in North American and Hawaiian forests: a pathological problem or a new problem for vegetation ecology? Phytocoenologia 11:117–137.

1362. Muir, J. A. 1972. Increase of dwarf mistletoe infections on young lodgepole pine. Can. J. For. Res. 2:413–416.

1363. Mulhern, J., Shortle, W., and Shigo, A. 1979. Barrier zones in red maple: an optical and scanning microscope examination. For. Sci. 25:311–316.

1364. Müller, E., and Dorworth, C. E. 1983. On the discomycetous genera Ascocalyx Naumov and Gremmeniella Morelet. Sydowia 36:193–203.

1365. Müller, J. 1977. Wechselwirkungen zwischen fünf Pratylenchus-Arten und Verticillium albo-atrum. Z. Pflanzenkr. Pflanzenschutz. 84:215–220.

1366. Mullick, D. B. 1977. The non-specific nature of defense in bark and wood during wounding, insect and pathogen attack. Recent Adv. Phytochem. 11:395–441.

1367. Mulrean, E. N., and Schroth, M. N. 1982. Ecology of Xanthomonas campestris pv. juglandis on Persian (English) walnuts. Phytopathology 72:434–438.

1368. Munger, T. T. 1916. Parch blight on Douglas fir in the Pacific Northwest. Plant World 19:46–47.

1369. Munnecke, D. E., Chandler, P. A., and Starr, M. P. 1963. Hairy root (Agrobacterium rhizogenes) of field roses. Phytopathology 53:788–799.

1370. Munnecke, D. E., Kolbezen, J. J., Wilbur, W. D., and Ohr, H. D. 1981. Interactions involved in controlling Armillaria mellea. Plant Dis. 65:384–389.

1371. Murdoch, C. W., Biermann, C. J., and Campana, R. J. 1983. Pressure and composition of intrastem gases produced in wetwood of American elm. Plant Dis. 67:74–76.

1372. Murdoch, C. W., and Campana, R. J. 1983. Bacterial species associated with wetwood of elm. Phytopathology 73:1270–1273.

1373. Musselman, L. J., Worsham, A. D., and Eplee, R. E. 1979. Proceedings. The Second International Symposium on Parasitic Weeds. North Carolina State University, Raleigh, NC. 296 pp.

1374. Myren, D. T., and Patton, R. F. 1971. Establishment and spread of Polyporus tomentosus in pine and spruce plantations in Wisconsin. Can. J. Bot. 49:1033–1040.

1375. Naegele, J. A., ed. 1973. Air pollution damage to vegetation. American Chemical Society, Washington, D.C. 137 pp.

1376. Nair, V. M. G., Kostichka, C. J., and Kuntz, J. E. 1979. Sirococcus clavigignenti-juglandacearum: an undescribed species causing canker on butternut. Mycologia 71:641–646.

1377. National Mycological Herbarium (Canada). 1973–present. Fungi Canadenses No. 1– . Agriculture Canada, Research Branch, Biosystematics Research Institute, Ottawa.

1378. Nattrass, R. M. 1928. The Physalospora disease of the basket willow. Trans. Br. Mycol. Soc. 13:286–304.

1379. Nattrass, R. M. 1933. A new species of Hendersonula (H. toruloidea) on deciduous trees in Egypt. Trans. Br. Mycol. Soc. 18:189–198.

1380. Navaratam, S. J., and Leong, C. K. 1965. Root inoculation of oil palm seedlings with Ganoderma sp. Plant Dis. Rep. 49:1011–1012.

1381. Navratil, S. 1979. Virus and virus-like diseases of poplar: are they threatening diseases? Rep. 19 in: Poplar research, management, and utilization in Canada. D. C. F. Fayle, A. Zsuffa, and H. W. Anderson, eds. Ont. Minist. Nat. Resour. For. Res. Inf. Pap. 102. 17 pp.

1382. Navratil, S. 1981. A rhabdovirus associated with vein yellowing and vein necrosis of balsam poplar. (Abstr.) Phytopathology 71:245.

1383. Neely, D. 1968. Bleeding necrosis of sweet gum in Illinois and Indiana. Plant Dis. Rep. 52:223–225.

1384. Neely, D. 1968. The somatic pressure cushion of Gnomonia platani. Mycologia 60:84–89.

1385. Neely, D. 1971. Additional Aesculus species and subspecies susceptible to leaf blotch. Plant Dis. Rep. 55:37–38.

1386. Neely, D. 1981. Application of nitrogen fertilizer to control anthracnose of black walnut. Plant Dis. 65:580–581.

1387. Neely, D., and Crowley, W. R. Jr. 1974. Toxicity of soil-applied herbicides to shade trees. HortScience 9:147–149.

1388. Neely, D., and Himelick, E. B. 1963. Aesculus species susceptible to leaf blotch. Plant Dis. Rep. 47:170.

1389. Neely, D., and Himelick, E. B. 1963. Temperature and sycamore anthracnose severity. Plant Dis. Rep. 47:171–175.

1390. Neely, D., and Himelick, E. B. 1965. Nomenclature of the sycamore anthracnose fungus. Mycologia 57:834–837.

1391. Neely, D., and Himelick, E. B. 1967. Characteristics and nomenclature of the oak anthracnose fungus. Phytopathology 57:1230–1236.

1392. Neilson-Jones, W. 1969. Plant chimeras. 2nd ed. Methuen, London. 123 pp.

1393. Nelson, E. E., Martin, N. E., and Williams, R. E. 1981. Laminated root rot of western conifers. U.S. For. Serv. For. Insect & Dis. Leafl. 159. 6 pp.

1394. Nelson, P. E., Toussoun, T. A., and Cook, R. J., eds. 1981. Fusarium: diseases, biology, and taxonomy. Pennsylvania State University Press, University Park, PA. 457 pp.

1395. Newbanks, D., Bosch, A., and Zimmermann, M. H. 1983. Evidence for xylem dysfunction by embolization in Dutch elm disease. Phytopathology 73:1060–1063.

1396. Newhouse, J. R., Hoch, H. C., and MacDonald, W. L. 1983. The ultrastructure of Endothia parasitica. Comparison of a virulent with a hypovirulent isolate. Can. J. Bot. 61:389–399.

1397. Ngo, H. C., Baxter, L. W. Jr., and Fagan, S. G. 1978. The status of our knowledge in 1978 of twig blight, canker, and dieback of camellias caused by a strain of Glomerella cingulata. Pages 75–91 in: American camellia yearbook 1978. American Camellia Society, Fort Valley, GA. 256 pp.

1398. Nicholls, T. H., Patton, R. F., and Van Arsdel, E. P. 1968. Life cycle and seasonal development of Coleosporium pine needle rust in Wisconsin. Phytopathology 58:822–829.

1399. Nicholls, T. H., Prey, A. J., and Skilling, D. D. 1974. Rhizosphaera kalkhoffii damages blue spruce Christmas tree plantations. Plant Dis. Rep. 58:1094–1096.

1400. Nicholls, T. H., and Skilling, D. D. 1972. Lophodermium needlecast disease of Scotch pine Christmas trees. Am. Christmas Tree J. 16(2):11–13.

1401. Nichols, C. W., Schneider, H., O'Reilly, H. J., Shalla, T. A., and Griggs, W. H. 1960. Pear decline in California. Cal. Dep. Agric. Bull. 49:186–192.

1402. Nichols, J. O. 1968. Oak mortality in Pennsylvania: a ten-year study. J. For. 66:681–684.

1403. Nichols, L. P. 1985. Disease resistant crabapples (results of 1984 survey). Penn. State Univ. Plant Pathol. Contrib. 1502. 7 pp.

1404. Nichols, L. P., and Peterson, D. H. 1971. Overwintering of Fusicladium dendriticum, the imperfect stage of the apple scab fungus, on twigs of flowering crab apples. Plant Dis. Rep. 55:509.

1405. Nicholson, R. L., Scoyoc, S. V., Williams, E. B., and Kuc, J. 1977. Host-pathogen interactions preceding the hypersensitive reaction of Malus sp. to Venturia inaequalis. Phytopathology 67:108–114.

1406. Nickle, W. R., ed. 1984. Plant and insect nematodes. Dekker, New York. 925 pp.

1407. Niedbalski, M., Crane, J. L., and Neely, D. 1979. Illinois fungi. 10. Development, morphology, and taxonomy of Cristulariella pyramidalis. Mycologia 71:722–730.

1408. Niemelä, T. 1972–1975. On Fennoscandian polypores. II. Phellinus laevigatus (Fr.) Bourd. & Galz. and P. lundellii Niemelä n. sp. III. Phellinus tremulae (Bond.) Bond. & Borisov. IV. Phellinus igniarius, P. nigricans, and P. populicola n. sp. Ann. Bot. Fenn. 9:41–59; 11:202–215; 12:93–122.

1409. Nienhaus, F. 1975. Viren und virusverdächtige Erkrankungen in Eichen (Quercus robur und Quercus sessiliflora). Z. Pflanzenkr. Pflanzenschutz 82:739–749.

1410. Nienhaus, F., and Yarwood, C. E. 1972. Transmission of virus from oak leaves fractionated with Sephadex. Phytopathology 62:313–315.

1411. Nighswander, J. E., and Patton, R. F. 1965. The epidemiology of the jack pine–oak gall rust (Cronartium quercuum) in Wisconsin. Can. J. Bot. 43:1561–1581.

1412. Nordin, V. J. 1954. Studies in forest pathology XIII. Decay in sugar maple in the Ottawa-Huron and Algoma Extension forest region of Ontario. Can. J. Bot. 32:221–258.

1413. Norton, D. C. 1978. Ecology of plant-parasitic nematodes. Wiley, New York. 268 pp.

1414. Norton, D. C., and Behrens, R. 1956. Ganoderma zonatum associated with dying mesquite. Plant Dis. Rep. 40:253–254.

1415. Nüesch, J. 1960. Beitrag zur Kenntnis der weidenbewohnenden Venturiaceae. Phytopathol. Z. 39:329–360.

1416. O'Bannon, J. H. 1977. Worldwide dissemination of Radopholus similis and its importance in crop production. J. Nematol. 9:16–25.

1417. Oberwinkler, F., and Bandoni, R. 1984. Herpobasidium and allied genera. Trans. Br. Mycol. Soc. 83:639–658.

1418. O'Brien, J. T. 1973. Sirococcus shoot blight of red pine. Plant Dis. Rep. 57:246–247.

1419. Oetting, R. D., Morishita, F. S., Helmkamp, A. L., and Bowen, W. R. 1980. Phytotoxicity of eight insecticides to some nursery-grown ornamentals. J. Econ. Entomol. 73:29–31.

1420. Ogawa, J. M., Bose, E., Manji, B. T., and Peterson, L. J. 1977. Life cycle and chemical control of Modesto tree anthracnose. Plant Dis. Rep. 61:792–796.

1421. Ogilvie, L. 1924. Observations on the "slime-fluxes" of trees. Trans. Br. Mycol. Soc. 9:167–182.

1422. Ohman, J. H. 1968. Decay and discoloration of sugar maple. U.S. For. Serv. For. Pest Leafl. 110. 7 pp.

1423. Oku, H., Shiraishi, T., Ouchi, S., Kurozumi, S., and Ohta, H. 1980. Pine wilt toxin, the metabolite of a bacterium associated with a nematode. Naturwissenschaften 67:198–199.

1424. Ondrej, M. 1972. Ein Beitrag zur Kenntnis der parasitischen imperfecten Pilze der Gattung Pollaccia Bald. et Cif. an Pappeln (Populus spp.). Eur. J. For. Pathol. 2:140–146.

1425. Oren, R., Thies, W. G., and Waring, R. H. 1985. Tree vigor and stand growth of Douglas-fir as influenced by laminated root rot. Can. J. For. Res. 15:985–988.

1426. O'Shea, M. T. 1981. Detection of viruses in Fraxinus using immunological methods. M.S. thesis, State Univ. Coll. Environ. Sci. & For., Syracuse, NY. 84 pp.

1427. Ostaff, D. P. 1985. Age distribution of European larch canker in New Brunswick. Plant Dis. 69:796–798.

1428. Ostrofsky, A., and Peterson, G. W. 1981. Etiologic and cultural studies of Kabatina juniperi. Plant Dis. 65:908–910.

1429. Ostry, M. E., and Anderson, N. A. 1983. Infection of trembling aspen by Hypoxylon mammatum through cicada oviposition wounds. Phytopathology 73:1092–1096.

1430. Ostry, M. E., and McNabb, H. S. Jr. 1985. Susceptibility of Populus species and hybrids to disease in the north central United States. Plant Dis. 69:755–757.

1431. Ostry, M. E., and Nicholls, T. H. 1979. Eastern dwarf mistletoe on black spruce. U.S. For. Serv. For. Insect Dis. Leafl. 158. 7 pp.

1432. Ostry, M. E., Nicholls, T. H., and French, D. W. 1983. Animal vectors of eastern dwarf mistletoe of black spruce. U.S. For. Serv. Res. Pap. NC-232. 16 pp.

1433. Otta, J. D. 1974. Effects of 2,4-D herbicide on Siberian elm. For. Sci. 20:287–290.

1434. Ouellette, G. B. 1966. Coleosporium viburni on jack pine and its relationship with C. asterum. Can. J. Bot. 44:1117–1120.

1435. Ouellette, G. B. 1978. Light and electron microscope studies on cell wall breakdown in American elm xylem tissues infected with Dutch elm disease. Can. J. Bot. 56:2666–2693.

1436. Overholts, L. O. 1923. Diagnoses of American Porias—II. Bull. Torrey Bot. Club. 50:245–253.

1437. Overholts, L. O. 1926. Mycological notes for 1924. Mycologia 18:31–38.

1438. Overholts, L. O. 1953. The Polyporaceae of the United States, Alaska, and Canada. Univ. Michigan Press, Ann Arbor. 466 pp.

1439. Owen, D. R., and Wilcox, W. W. 1982. The association between ring shake, wetwood, and fir engraver beetle attack in white fir. Wood Fiber 14:267–280.

1440. Pady, S. M. 1946. The development and germination of the intraepidermal teliospores of Melampsorella cerastii. Mycologia 38:477–499.

1441. Pady, S. M. 1972. Spore release in powdery mildews. Phytopathology 62:1099–1100.

1442. Pady, S. M., and Kramer, C. L. 1971. Basidiospore discharge in Gymnosporangium. Phytopathology 61:951–953.

1443. Pady, S. M., Kramer, C. L., and Clary, R. 1969. Aeciospore release in Gymnosporangium. Can. J. Bot. 47:1027–1032.

1444. Palmer, M. A., and Nicholls, T. H. 1985. Shoot blight and collar rot of Pinus resinosa caused by Sphaeropsis sapinea in forest tree nurseries. Plant Dis. 69:739–740.

1445. Panconesi, A. 1981. Ceratocystis fimbriata of plane trees in Italy: biological aspects and control possibility. Eur. J. For. Pathol. 11:385–395.

1446. Pantidou, M. E., and Korf, R. P. 1954. A revision of the genus *Keithia*. Mycologia 46:386–388.

1447. Parker, A. K. 1957. The nature of the association of *Europhium trinacriforme* with pole blight lesions. Can. J. Bot. 35:845–856.

1448. Parker, A. K. 1970. Effect of relative humidity and temperature on needle cast disease of Douglas fir. Phytopathology 60:1270–1273.

1449. Parker, A. K., and Reid, J. 1969. The genus *Rhabdocline* Syd. Can. J. Bot. 47:1533–1545.

1450. Parker, J. 1963. Cold resistance in woody plants. Bot. Rev. 29:124–201.

1451. Parker, K. G. 1959. Verticillium hadromycosis of deciduous tree fruits. Plant Dis. Rep. Suppl. 255:38–61.

1452. Parmelee, J. A. 1965. The genus *Gymnosporangium* in eastern Canada. Can. J. Bot. 43:239–267.

1453. Parmelee, J. A. 1968. Effective range of basidiospores of *Gymnosporangium*. Can. Plant Dis. Surv. 48:150–151.

1454. Parmelee, J. A. 1971. The genus *Gymnosporangium* in western Canada. Can. J. Bot. 49:903–926.

1455. Parmelee, J. A. 1977. The fungi of Ontario. II. Erysiphaceae (mildews). Can. J. Bot. 55:1940–1983.

1456. Parmeter, J. R. Jr., ed. 1970. *Rhizoctonia solani*: biology and pathology. University of California Press, Berkeley. 255 pp.

1457. Parmeter, J. R. Jr., Bega, R. V., and Hood, J. R. 1960. Epidemic leaf-blighting of California-laurel. Plant Dis. Rep. 44:669–671.

1458. Parris, G. K., and Byrd, J. 1962. Oak anthracnose in Mississippi. Plant Dis. Rep. 46:677–681.

1459. Partridge, A. D., and Rich, A. E. 1957. A study of the ash leaf rust syndrome in New Hampshire: suscepts, incitant, epidemiology, and control. (Abstr.) Phytopathology 47:246.

1460. Patterson, J. C. 1977. Soil compaction—effects on urban vegetation. J. Arboric. 3:161–167.

1461. Patton, R. F. 1978. Penetration and colonization of oak roots by *Armillaria mellea* in Wisconsin. Eur. J. For. Pathol. 8:259–267.

1462. Patton, R. F., and Johnson, D. W. 1970. Mode of penetration of needles of eastern white pine by *Cronartium ribicola*. Phytopathology 60:977–982.

1463. Patton, R. F., and Riker, A. J. 1954. Needle droop and needle blight of red pine. J. For. 52:412–418.

1464. Patton, R. F., and Spear, R. N. 1983. Needle cast of European larch caused by *Mycosphaerella laricina* in Wisconsin and Iowa. Plant Dis. 67:1149–1153.

1465. Patton, R. F., Spear, R. N., and Blenis, P. V. 1984. The mode of infection and early stages of colonization of pines by *Gremmeniella abietina*. Eur. J. For. Pathol. 14:193–202.

1466. Pawsey, R. G. 1960. An investigation into Keithia disease of *Thuja plicata*. Forestry 33:174–186.

1467. Paxton, J. D., and Wilson, E. E. 1965. Anatomical and physiological aspects of branch wilt disease of Persian walnut. Phytopathology 55:21–26.

1468. Peace, T. R. 1940. An interesting case of lightning damage to a group of trees. Q. J. For. 34:61–63.

1469. Peace, T. R. 1962. Pathology of trees and shrubs. Clarendon Press, Oxford. 722 pp.

1470. Peace, T. R., and Holmes, C. H. 1933. *Meria laricis*. The leaf cast disease of larch. Oxford For. Mem. 15. 29 pp.

1471. Pearce, R. B., and Holloway, P. J. 1984. Suberin in the sapwood of oak (*Quercus robur* L.): its composition from a compartmentalization barrier and its occurrence in tyloses in undecayed wood. Physiol. Plant Pathol. 24:71–81.

1472. Pearce, R. B., and Rutherford, J. 1981. A wound-associated suberized barrier to the spread of decay in the sapwood of oak (*Quercus robur* L.). Physiol. Plant Pathol. 19:359–369.

1473. Pearson. R. C., Aldwinckle, H. S., and Seem, R. C. 1977. Teliospore germination and basidiospore formation in *Gymnosporangium juniperi-virginianae*: a regression model of temperature and time effects. Can. J. Bot. 55:2832–2837.

1474. Pearson, R. C., Seem, R. C., and Meyer, F. W. 1980. Environmental factors influencing the discharge of basidiospores of *Gymnosporangium juniperi-virginianae*. Phytopathology 70:262–266.

1475. Pegg, G. F. 1974 Verticillium diseases. Rev. Plant Pathol. 53:157–182.

1476. Pellett, H., Gearhart, M., and Dirr, M. 1981. Cold hardiness capability of woody ornamental plant taxa. J. Am. Soc. Hortic. Sci. 106:239–243.

1477. Perala, D. A., and Sucoff, E. 1965. Diagnosing potassium deficiency in American elm, silver maple, Russian olive, hackberry, and box elder. For. Sci. 11:347–352.

1478. Percival, W. C. 1933. A contribution to the biology of *Fomes pini* (Thore) Lloyd (*Trametes pini* [Thore] Fries). N.Y. State Coll. For. Tech. Publ. 40. 72 pp.

1479. Percy, R. G. 1983. Potential range of *Phymatotrichum omnivorum* as determined by edaphic factors. Plant Dis. 67:981–983.

1480. Peries, O. S. 1974. Ganoderma basal stem rot of coconut: a new record of the disease in Sri Lanka. Plant Dis. Rep. 58:293–295.

1481. Pero, R. W., and Howard, F. L. 1970. Activity of juniper diffusates on spores of *Phomopsis juniperovora*. Phytopathology 60:491–495.

1482. Perrin, R., and van Gerwen, C. P. 1984. La variabilité du pouvoir pathogène de *Nectria ditissima* agent du chancre du hêtre. Eur. J. For. Pathol. 14:170–176.

1483. Perry, R. G., and Peterson, J. L. 1982. Susceptibility and response of juniper species to *Kabatina juniperi* infection in New Jersey. Plant Dis. 66:1189–1191.

1484. Petersen, R. H. 1974. The rust fungus life cycle. Bot. Rev. 40:453–513.

1485. Peterson, G. W. 1966. Western X-disease virus of chokecherry: transmission and seed effects. Plant Dis. Rep. 50:659–660.

1486. Peterson, G. W. 1967. Dothistroma needle blight of Austrian and ponderosa pines: epidemiology and control. Phytopathology 57:437–441.

1487. Peterson, G. W. 1973. Dispersal of aeciospores of *Periderium harknessii* in central Nebraska. Phytopathology 63:170–172.

1488. Peterson, G. W. 1973. Infection of *Juniperus virginiana* and *J. scopulorum* by *Phomopsis juniperovora*. Phytopathology 63:246–251.

1489. Peterson, G. W. 1976. Disease of Russian-olive caused by *Botryodiplodia theobromae*. Plant Dis. Rep. 60:490–494.

1490. Peterson, G. W. 1977. Epidemiology and control of a blight of *Juniperus virginiana* caused by *Cercospora sequoiae* var. *juniperi*. Phytopathology 67:234–238.

1491. Peterson, G. W. 1977. Infection, epidemiology, and control of Diplodia blight of Austrian, ponderosa, and Scots pines. Phytopathology 67:511–514.

1492. Peterson, G. W. 1981. Diplodia blight of pines. U.S. For. Serv. For. Insect Dis. Leafl. 161. 7 pp.

1493. Peterson, G. W. 1981. Pine and juniper diseases in the Great Plains. U.S. For. Serv. Gen. Tech. Rep. RM-86. 47 pp.

1494. Peterson, G. W. 1982. Dothistroma needle blight of pines. U.S. For. Serv. For. Insect Dis. Leafl. 143. 6 pp.

1495. Peterson, G. W. 1984. Resistance to *Dothistroma pini* within geographic seed sources of *Pinus ponderosa*. Phytopathology 74:956–960.

1496. Peterson, G. W. 1984. Spread and damage of western X-disease of chokecherry in eastern Nebraska plantings. Plant Dis. 68:103–104.

1497. Peterson, G. W., and Harvey, G. M. 1976. Dispersal of *Scirrhia* (*Dothistroma*) *pini* conidia and disease development in a shore pine plantation in western Oregon. Plant Dis. Rep. 60:761–764.

1498. Peterson, G. W., and Hodges, C. S. Jr. 1975. Phomopsis blight of junipers. U.S. For. Serv. For. Pest Leafl. 154. 5 pp.

1499. Peterson, G. W., and Smith, R. S. Jr., tech. coords. 1975. Forest nursery diseases in the United States. U.S. Dep. Agric. Agric. Handb. 470. 125 pp.

1500. Peterson, G. W., and Walla, J. A. 1978. Development of *Dothistroma pini* upon and within needles of Austrian and ponderosa pines in eastern Nebraska. Phytopathology 68:1422–1430.

1501. Peterson, J. L., and Davis, S. H. Jr. 1965. A Fusarium canker of *Sophora japonica*. Plant Dis. Rep. 49:835–836.

1502. Peterson, J. L., Davis, S. H. Jr., and Judd, R. W. Jr. 1976. Effect of fungicide and application timing on Cercospora leafspot of mountain laurel. Plant Dis. Rep. 60:138–140.

1503. Peterson, R. S. 1960. Development of western gall rust in lodgepole pine. Phytopathology 50:876–881.

1504. Peterson, R. S. 1960. Western gall rust on hard pines. U.S. For. Serv. For. Pest Leafl. 50. 8 pp.

1505. Peterson, R. S. 1961. Conifer tumors in the central Rocky Mountains. Plant Dis. Rep. 45:472–474.

1506. Peterson, R. S. 1961. Host alternation of spruce broom rust. Science 134:468–469.

1507. Peterson, R. S. 1963. Effects of broom rusts on spruce and fir. U.S. For. Serv. Res. Pap. INT-7. 10 pp.

1508. Peterson, R. S. 1964. Fir broom rust. U.S. For. Serv. For. Pest Leafl. 87. 7 pp.

1509. Peterson, R. S. 1966. Limb rust damage to pine. U.S. For. Serv. Res. Pap. INT-31. 10 pp.

1510. Peterson, R. S. 1967. Studies of juniper rusts in the West. Madroño 19:79–91.

1511. Peterson, R. S. 1967. The *Peridermium* species on pine stems. Bull. Torrey Bot. Club 94:511–542.

1512. Peterson, R. S. 1968. Limb rusts of pine: the causal fungi. Phytopathology 58:309–315.

1513. Peterson, R. S., and Jewell, F. F. 1968. Status of American stem rusts of pine. Annu. Rev. Phytopathol. 6:23–40.

1514. Peterson, R. S., and Shurtleff, R. G. Jr. 1965. Mycelium of limb rust fungi. Am. J. Bot. 52:519–525.

1515. Peturson, B. 1954. The relative prevalence of specialized forms of *Puccinia coronata* that occur on *Rhamnus cathartica* in Canada. Can. J. Bot. 32:40–47.

1516. Phelps, J. E., McGinnes, E. A. Jr., and Lieu, P. J.-Y. 1975. Anatomy of xylem tissue formation associated with radial seams and cracks in black oak. Wood Sci. 8:397–405.

1517. Phelps, W. R. 1974. Live oak in North Carolina and Florida infected by *Endothia parasitica*. Plant Dis. Rep. 58:596–598.

1518. Phelps, W. R., and Czabator, F. L. 1978. Fusiform rust of southern pines. U.S. For. Serv. For. Insect Dis. Leafl. 26 (rev). 7 pp.

1519. Phelps, W. R., Kais, A. G., and Nicholls, T. H. 1978. Brown-spot needle blight of pines. U.S. For. Serv. For. Insect Dis. Leafl. 44. 8 pp.

1520. Philipson, J. J., and Coutts, M. P. 1980. The tolerance of tree roots to waterlogging. IV. Oxygen transport in woody roots of Sitka spruce and lodgepole pine. New Phytol. 65:489–494.

1521. Phillips, D. H., and Burdekin, D. A. 1982. Diseases of forest and ornamental trees. Macmillan, London. 435 pp.

1522. Phillips, J. F. V. 1929. The influence of *Usnea* sp. (near *barbata* Fr.) upon the supporting trees. Trans. R. Soc. South Afr. 17:101–107.

1523. Phipps, H. M. 1963. The role of 2,4–D in the appearance of a leaf blight of some plains tree species. For. Sci. 9:283–288.

1524. Phipps, P. M., and Stipes, R. J. 1976. Histopathology of mimosa infected with *Fusarium oxysporum* f. sp. *pernicosum*. Phytopathology 66:839–843.

1525. Pine, T. S., Gilmer, R. M., Moore, J. D., Nyland, G., and Welsh, M. F., eds. 1976. Virus diseases and noninfectious disorders of stone fruits in North America. U.S. Dep. Agric. Agric. Handb. 437. 433 pp.

1526. Pinon, J., and Morelet, M. 1975. Le *Linospora ceuthocarpa* (Fr.) Munk ex Morelet, parasite foliaire des peupliers. Eur. J. For. Pathol. 5:367–376.

1527. Pinon, J., and Poissonnier, M. 1975. Etude épidemiologique du *Marssonina brunnea* (El. et Ev.) P. Magn. Eur. J. For. Pathol. 5:97–111.

1528. Pirone, P. P. 1942. Stem and stolon canker of pachysandra. N.J. Agric. Exp. Stn. Nursery Dis. Notes 14:40–43.

1529. Pirone, P. P. 1957. *Ganoderma lucidum,* a parasite of shade trees. Bull. Torrey Bot. Club. 84:424–428.

1530. Pirozynski, K. A., and Shoemaker, R. A. 1970. Some Asterinaceae and Meliolaceae on conifers in Canada. Can. J. Bot. 48:1321–1328.

1531. Plakidas, A. G. 1940. Angular leaf spot of *Pittosporum.* Mycologia 32:601–608.

1532. Plakidas, A. G. 1942. *Venturia acerina,* the perfect stage of *Cladosporium humile.* Mycologia 34:27–37.

1533. Plakidas, A. G. 1945. Blight of Oriental arborvitae. Phytopathology 35:181–190.

1534. Plakidas, A. G. 1954. Transmission of leaf and flower variegation in camellias by grafting. Phytopathology 44:14–18.

1535. Plakidas, A. G. 1962. Strains of the color-breaking virus of camellia. Phytopathology 52:77–79.

1536. Plank, S., and Wolkinger, F. 1976. Etudes du cours des hyphes de *Fomes fometarius* dans le bois d'*Aesculus hippocastanum* au microscope électronique à balayage. Can. J. Bot. 54:2231–2238.

1537. Plese, N., and Milicic, D. 1973. Two viruses isolated from *Maclura pomifera.* Phytopathol. Z. 77:178–183.

1538. Plich, M., and Rudnicki, R. M. 1979. Studies of the toxins of *Phytophthora cactorum* pathogenic to apple trees. 1. Isolation, some of the properties and activities of a toxin produced by the fungus cultured *in vitro.* Phytopathol. Z. 94:270–278.

1539. Pomerleau, R. 1937–1938. Recherches sur le *Gnomonia ulmea* (Schw.) Thüm. Nat. Can. 64:261–289, 297–318; 65:23–41, 253–279.

1540. Pomerleau, R. 1940. Studies on the ink-spot disease of poplar. Can. J. Res. C18:199–214.

1541. Pomerleau, R. 1944. Observations sur quelques maladies non parasitaires des arbres dans le Québec. Can. J. Res. C22:171–189.

1542. Pomerleau, R. 1953. History of hardwood species dying in Quebec. Can. Dep. Agric., Ottawa. Rep. of Sympos. on Birch Dieback, Part 1:10–11.

1543. Pomerleau, R., and Lortie, M. 1962. Relationships of dieback to the rooting depth of white birch. For. Sci. 8:219–224.

1544. Pomerleau, R., and Ray, R. G. 1957. Occurrence and effects of summer frost in a conifer plantation. Can. For. Serv. Tech. Note 51. 15 pp.

1545. Posnette, A. F., ed. 1963. Virus diseases of apples and pears. Commonw. Bur. Hortic. Plantation Crops Tech. Communic. 30. 141 pp.

1546. Potebnia, A. von. 1910. Beiträge zur Micromycetenflora Mittel-Russlands. Ann. Mycol. 8:42–93.

1547. Poucher, C., Ford, H. W., Suit, R. F., and DuCharme, E. P. 1967. Burrowing nematode in citrus. Fla. Dep. Agric. Div. Plant Indus. Bull. 7. 63 pp.

1548. Powell, J. M. 1971. Incidence and effect of *Tuberculina maxima* on cankers of the pine stem rust, *Cronartium comandrae.* Phytoprotection 52:104–111.

1549. Powell, J. M. 1972. Seasonal and diurnal periodicity in the release of *Cronartium comandrae* aeciospores from stem cankers of lodgepole pine. Can. J. For. Res. 2:78–88.

1550. Powers, H. R. Jr. 1972. Comandra rust on southern pines. J. For. 70:18–20.

1551. Powers, H. R. Jr., Matthews, F. R., and Dwinell, L. D. 1977. Evaluation of pathogenic variability of *Cronartium fusiforme* on loblolly pine in the southern USA. Phytopathology 67:1403–1407.

1552. Powers, H. R. Jr., Schmidt, R. A., and Snow, G. A. 1981. Current status and management of fusiform rust on southern pines. Annu. Rev. Phytopathol. 19:353–371.

1553. Pratt, R. M. 1958. Florida guide to citrus insects, diseases, and nutritional disorders in color. Fla. Agric. Exp. Stn., Gainesville. 191 pp.

1554. Price, T. V. 1970. Epidemiology and control of powdery mildew (*Sphaerotheca pannosa*) on roses. Ann. Appl. Biol. 65:231–248.

1555. Prince, A. E. 1943. Basidium formation and spore discharge in *Gymnosporangium nidus-avis.* Farlowia 1:79–93.

1556. Prince, A. E. 1946. The biology of *Gymnosporangium nidus-avis* Thaxter. Farlowia 2:475–525.

1557. Prljincević, M. B. 1982. Economic significance of the infection of beech forests by *Hypoxylon deustum* (Hoffm. et Fr.) Grev. at Šara mountain. Eur. J. For. Pathol. 12:7–10.

1558. Provvidenti, R., and Hunter, J. E. 1975. Bean yellow mosaic virus infection in *Cladrastis lutea,* an ornamental leguminous tree. Plant Dis. Rep. 59:86–87.

1559. Przybyl, K. 1984. Pathological changes and defense responses in poplar tissues caused by *Ceratocystis fimbriata.* Eur. J. For. Pathol. 14:183–191.

1560. Puckett, L. J. 1982. Acid rain, air pollution, and tree growth in southeastern New York. J. Environ. Qual. 11:376–381.

1561. Puffinberger, C. W., and Corbett, M. K. 1985. Euonymus chlorotic ringspot disease caused by tomato ringspot virus. Phytopathology 75:423–428.

1562. Puhalla, J. E. 1985. Classification of strains of *Fusarium oxysporum* on the basis of vegetative compatibility. Can. J. Bot. 63:179–183.

1563. Punja, Z. K., and Grogan, R. G. 1981. Eruptive germination of sclerotia of *Sclerotium rolfsii.* Phytopathology 71:1092–1099.

1564. Punja, Z. K., and Grogan, R. G. 1981. Mycelial growth and infection without a food base by eruptively germinating sclerotia of *Sclerotium rolfsii.* Phytopathology 71:1099–1103.

1565. Punja, Z. K., and Grogan, R. G. 1983. Germination and infection by basidiospores of *Athelia (Sclerotium) rolfsii.* Plant Dis. 67:875–878.

1566. Punja, Z. K., and Ormrod, D. J. 1979. New or noteworthy plant diseases in coastal British Columbia, 1975 to 1977. Can. Plant Dis. Surv. 59:22–24.

1567. Purcell, A. H. 1980. Almond leaf scorch: leafhopper and spittlebug vectors. J. Econ. Entomol. 73:834–838.

1568. Putterill, V. A. 1922. The biology of *Schizophyllum commune* Fries with special reference to its parasitism. South Afr. Dep. Agric. Sci. Bull. 25. 35 pp.

1569. Raabe, R. D. 1962. Host list of the root rot fungus, *Armillaria mellea.* Hilgardia 33:25–88.

1570. Raabe, R. D. 1965. Some previously unreported hosts of *Armillaria mellea* in California. Plant Dis. Rep. 49:812.

1571. Raabe, R. D. 1972. Variation in pathogenicity and virulence in single-spore isolates of *Armillaria mellea.* Mycologia 64:1154–1159.

1572. Raabe, R. D. 1979. Resistance or susceptibility of certain plants to Armillaria root rot. Univ. Cal. Div. Agric. Sci. Leafl. 2591. 11 pp.

1573. Raabe, R. D. 1979. Some previously unreported hosts of *Armillaria mellea* in California, III. Plant Dis. Rep. 63:494–495.

1574. Raabe, R. D. 1985. Fusarium wilt of *Hebe* species. Plant Dis. 69:450–451.

1575. Raabe, R. D., and Gardner, M. W. 1972. Scab of pyracantha, loquat, toyon, and kageneckia. Phytopathology 62:914–916.

1576. Radosevich, S. R., Roncoroni, E. J., Conrad, S. G., and McHenry, W. B. 1980. Seasonal tolerance of six coniferous species to eight foliage-active herbicides. For. Sci. 26:3–9.

1577. Raj, T. R. N. 1985. Redisposals and redescriptions in the *Monochaetia-Seiridium, Pestalotia-Pestalotiopsis* complexes. I. The correct name for the type species of *Pestalotiopsis.* Mycotaxon 22:43–51.

1578. Raj, T. R. N., and Kendrick, W. B. 1975. A monograph of *Chalara* and allied genera. Wilfred Laurier University Press, Waterloo, Canada. 200 pp.

1579. Raju, B. C., Chen, T. A., and Varney, E. H. 1976. Mycoplasmalike organisms associated with a witches'-broom disease of *Cornus amomum.* Plant Dis. Rep. 60:462–464.

1580. Raju, B. C., Goheen, A. C., and Frazier, N. W. 1983. Occurrence of Pierce's disease bacteria in plants and vectors in California. Phytopathology 73:1309–1313.

1581. Raju, B. C., and Wells, J. M. 1986. Diseases caused by fastidious xylem-limited bacteria and strategies for management. Plant Dis. 70:182–186.

1582. Ramsdell, D. C., Nelson, J. W., and Myers, R. L. 1975. Mummy berry disease of highbush blueberry: epidemiology and control. Phytopathology 65:229–232.

1583. Rathbun-Gravatt, A. 1927. A witches' broom of introduced Japanese cherry trees. Phytopathology 17:19–24.

1584. Rautenberg, E. 1973. Untersuchungen an *Exobasidium*-Gallen von *Rhododendron simsii* Planch. I. Der Erreger *Exobasidium japonicum* Shir. und Begleitpilze. II. Bedingungen der Gallentwicklung und die Wechselbeziehung zwischen Wirtspflanze und Cecidie. III. Fluoreszierende Phenolderivate in Gallen und Blättern. Phytopathol. Z. 78:2–13, 121–133, 214–226.

1585. Raymond, F. L., and Reid, J. 1961. Dieback of balsam fir in Ontario. Can. J. Bot. 39:233–251.

535

1586. Read, R. A., and Sprackling, J. A. 1981. Hail damage variation by seed source in a ponderosa pine plantation. U.S. For. Serv. Res. Note RM-410. 6 pp.

1587. Reddick, B. B., Barnett, O. W., and Baxter, L. W. Jr. 1979. Isolation of cherry leafroll, tobacco ringspot, and tomato ringspot viruses from dogwood trees in South Carolina. Plant Dis. Rep. 63:529–532.

1588. Reddick, D. 1911. The black rot disease of grapes. Cornell Univ. Agric. Exp. Stn. Bull. 293:287–364.

1589. Redhead, S. A. 1975. The genus Cristulariella. Can. J. Bot. 53:700–707.

1590. Redmond, D. R. 1955. Studies in forest pathology. XV. Rootlets, mycorrhiza, and soil temperature in relation to birch dieback. Can. J. Bot. 33:595–627.

1591. Reid, J., and Cain, R. F. 1962. Studies on the organisms associated with "snow-blight" of conifers in North America. II. Some species of the genera Phacidium, Lophophacidium, Sarcotrochila, and Hemiphacidium. Mycologia 54:481–497.

1592. Reid, J., and Funk, A. 1966. The genus Atropellis and a new genus of the Helotiales associated with branch cankers of western hemlock. Mycologia 58:417–439.

1593. Reinert, J. A. 1977. Field biology and control of Haplaxius crudus on St. Augustinegrass and Christmas palm. J. Econ. Entomol. 70:54–56.

1594. Reinert, R. A. 1984. Plant response to air pollutant mixtures. Annu. Rev. Phytopathol. 22:421–442.

1595. Rex, E. G., and Walter, J. M. 1946. The canker stain disease of planetrees, with recommendations for controlling it in New Jersey. N.J. Dep. Agric. Circ. 360. 23 pp.

1596. Rexrode, C. O. 1976. Insect transmission of oak wilt. J. Arboric. 2:61–66.

1597. Reynolds, D. R. 1978. Foliicolous Ascomycetes 1: the capnodiaceous genus Scorias, reproduction. Nat. Hist. Mus. L.A. County, Contr. Sci. 288:1–16.

1598. Reynolds, D. R. 1979. Foliicolous Ascomycetes. 3. The stalked capnodiaceous species. Mycotaxon 8:417–445.

1599. Reynolds, K. M., Benson, D. M., and Bruck, R. I. 1985. Epidemiology of Phytophthora root rot of Fraser fir: root colonization and inoculum production. Phytopathology 75:1004–1009.

1600. Rhoads, A. S. 1923. The formation and pathological anatomy of frost rings in conifers injured by late frosts. U.S. Dep. Agric. Tech. Bull. 1131. 15 pp.

1601. Rhoads, A. S. 1943. Lightning injury to pine and oak trees in Florida. Plant Dis. Rep. 27:556–557.

1602. Rhoads, A. S. 1950. Clitocybe root rot of woody plants in the southeastern United States. U.S. Dep. Agric. Circ. 853. 25 pp.

1603. Rhoads, A. S. 1956. The occurrence and destructiveness of Clitocybe root rot of woody plants in Florida. Lloydia 19:193–240.

1604. Ribeiro, O. K. 1978. A sourcebook of the genus Phytophthora. J. Cramer, Lehre, West Germany. 417 pp.

1605. Rich, S., and Walton, G. S. 1979. Decline of curbside sugar maples in Connecticut. J. Arboric. 5:265–268.

1606. Richards, W. C., and Takai, S. 1984. Characterization of the toxicity of cerato-ulmin, the toxin of Dutch elm disease. Can. J. Plant Pathol. 6:291–298.

1607. Richardson, K. S., and Van der Kamp, B. J. 1972. The rate of upward advance and intensification of dwarf mistletoe on immature western hemlock. Can. J. For. Res. 2:313–316.

1608. Richmond, B. G. 1932. A Diaporthe canker of American elm. Science 75:110–111.

1609. Ridings, W. H., and Marlatt, R. B. 1976. Sphaeropsis witches' broom of Nerium oleander. Proc. Fla. State Hortic. Soc. 89:302–303.

1610. Riffle, J. W. 1973. Histopathology of Pinus ponderosa ectomycorrhizae infected with a Meloidogyne species. Phytopathology 63:1034–1040.

1611. Riffle, J. W. 1978. Development of cankers on Ulmus pumila related to month of inoculation with Botryodiplodia hypodermia. Phytopathology 68:1115–1119.

1612. Riffle, J. W. 1981. Recovery of Herpobasidium deformans basidiospores from Lonicera tatarica leaves overwintered in nursery beds. (Abstr.) Phytopathology 71:251–252.

1613. Riffle, J. W., and Kuntz, J. W. 1967. Pathogenicity and host range of Meloidogyne ovalis. Phytopathology 57:104–107.

1614. Riffle, J. W., and Peterson, G. W., tech. coords. 1986. Diseases of trees in the Great Plains. U.S. For. Serv. Gen. Tech. Rep. RM-129. 149 pp.

1615. Riffle, J. W., and Peterson, G. W. 1986. Thyronectria canker of honeylocust: influence of temperature and wound age on disease development. Phytopathology 76:313–316.

1616. Riffle, J. W., Sharon, E. M., and Harrell, M. O. 1984. Incidence of Fomes fraxinophilus on green ash in Nebraska woodlands. Plant Dis. 68:322–324.

1617. Riker, A. J., Keitt, G. W., Hildebrand, E. M., and Banfield, W. M. 1934. Hairy root, crown gall, and other malformations at the unions of piece-root-grafted apple trees and their control. J. Agric. Res. 48:913–939.

1618. Riley, C. G. 1952. Studies in forest pathology. IX. Fomes igniarius decay of poplar. Can. J. Bot. 30:710–734.

1619. Riley, C. G. 1953. Hail damage in forest stands. For. Chron. 29:139–143.

1620. Rishbeth, J. 1951. Observations on the biology of Fomes annosus, with particular reference to East Anglian pine plantations. II. Spore production, stump infection, and saprophytic activity in stumps. Ann. Bot., n.s. 15:1–21.

1621. Rishbeth, J. 1978. Effects of soil temperature and atmosphere on growth of Armillaria rhizomorphs. Trans. Br. Mycol. Soc. 70:213–220.

1622. Rishbeth, J. 1978. Infection foci of Armillaria mellea in first-rotation hardwoods. Ann. Bot. 42:1131–1139.

1623. Rishbeth, J. 1979. Modern aspects of biological control of Fomes and Armillaria. Eur. J. For. Pathol. 9:331–340.

1624. Ritchie, D. F., and Werner, D. J. 1981. Susceptibility and inheritance of susceptibility to peach leaf curl in peach and nectarine cultivars. Plant Dis. 65:731–734.

1625. Roane, M. K., Griffin, G. J., and Elkins, J. R. 1986. Chestnut blight, other Endothia diseases, and the genus Endothia. APS Press, St. Paul, MN. 53 pp.

1626. Roane, M. K., Stipes, R. J., Phipps, P. M., and Miller, O. K. Jr. 1974. Endothia gyrosa, causal pathogen of pin oak blight. Mycologia 66:1042–1047.

1627. Roberts, J. W., and Dunegan, J. C. 1932. Peach brown rot. U.S. Dep. Agric. Tech. Bull. 328. 59 pp.

1628. Rockett, T. R., and Kramer, C. L. 1974. Periodicity and total spore germination by lignicolous basidiomycetes. Mycologia 66:817–829.

1629. Rogers, J. D., and Callan, B. E. 1986. Xylaria polymorpha and its allies in continental United States. Mycologia 78:391–400.

1630. Roll-Hansen, F. 1985. The Armillaria species in Europe. A literature review. Eur. J. For. Pathol. 15:22–31.

1631. Roll-Hansen, F., and Roll-Hansen, H. 1981. Melampsoridium on Alnus in Europe. M. alni conspecific with M. betulinum. Eur. J. For. Pathol. 11:77–87.

1632. Roncadori, R. S. 1962. The nutritional competition between Hypoxylon punctulatum and Ceratocystis fagacearum. Phytopathology 52:498–502.

1633. Roosje, G. S., ed. 1984. Third International Workshop on Fire Blight. Acta Hortic. No. 151. 348 pp.

1634. Rosen, H. R. 1935. Rose blast induced by Phytomonas syringae. J. Agric. Res. 51:235–243.

1635. Rosen, P. M., Good, G. L., and Steponkus, P. L. 1983. Desiccation injury and direct freezing injury to evergreen azaleas: a comparison of cultivars. J. Am. Soc. Hortic. Sci. 108:28–31.

1636. Rosenberger, D. A., Burr, T. J., and Gilpatrick, J. D. 1983. Failure of canker removal and postharvest fungicide sprays to control Nectria blight on apples. Plant Dis. 67:15–17.

1637. Rosenberger, D. A., Harrison, M. B., and Gonsalves, D. 1983. Incidence of apple union necrosis and decline, tomato ringspot virus, and Xiphinema vector species in Hudson Valley orchards. Plant Dis. 67:356–360.

1638. Rosenberger, D. A., and Jones, A. L. 1977. Seasonal variation in infectivity of inoculum from X-diseased peach and chokecherry plants. Plant Dis. Rep. 61:1022–1024.

1639. Rosenberger, D. A., and Jones, A. L. 1978. Leafhopper vectors of the peach X-disease pathogen and its seasonal transmission from chokecherry. Phytopathology 68:782–790.

1640. Ross, E. W. 1964. Cankers associated with ash die-back. Phytopathology 54:272–275.

1641. Ross, E. W. 1966. Ash dieback: etiological and developmental studies. State Univ. Coll. For. (Syracuse, NY) Tech. Publ. 88. 80 pp.

1642. Ross, E. W. 1967. Association of Cylindrocladium scoparium with mortality in a 27-year-old yellow-poplar plantation. Plant Dis. Rep. 51:38–39.

1643. Ross, E. W. 1970. Sand pine root rot—pathogen: Clitocybe tabescens. J. For. 68:156–158.

1644. Ross, E. W. 1973. Fomes annosus in the southeastern United States. U.S. Dep. Agric. Tech. Bull. 1459. 26 pp.

1645. Ross, W. D. 1976. Fungi associated with root diseases of aspen in Wyoming. Can. J. Bot. 54:734–744.

1646. Rossell, S. E., Abbott, E. G. M., and Levy, J. F. 1973. Bacteria and wood (a review of the literature relating to the presence, action, and interaction of bacteria in wood). J. Inst. Wood Sci. 6(2):28–35.

1647. Rossman, A. Y. 1979. A preliminary account of the taxa described in Calonectria. Mycotaxon 8:485–558.

1648. Roth, E. R., Hepting, G. H., and Toole, E. R. 1959. Sapstreak disease of sugar maple and yellow-poplar in North Carolina. U.S. For. Serv. Southeast. For. Exp. Stn. Res. Note 134. 2 pp.

1649. Roth, E. R., and Sleeth, B. 1939. Butt rot in unburned sprout oak stands. U.S. Dep. Agric. Tech. Bull. 684. 42 pp.

1650. Roth, L. F. 1959. Perennial infection of ponderosa pine by Elytroderma deformans. For. Sci. 5:182–191.

1651. Roth, L. F. 1971. Dwarf mistletoe damage to small ponderosa pines. For. Sci. 17:373–380.

1652. Roth, L. F. 1974. Juvenile susceptibility of ponderosa pine to dwarf mistletoe. Phytopathology 64:689–692.

1653. Roth, L. F., Bynum, H. H., and Nelson, E. E. 1972. Phytophthora root rot of Port-Orford-cedar. U.S. For. Serv. For. Pest Leafl. 131. 7 pp.

1654. Rowan, S. J. 1960. The susceptibility of twenty-three tree species to black root rot. Plant Dis. Rep. 44:646–647.

1655. Rowhani, A., Mircetich, S. M., Shepherd, R. J., and Cucuzza, J. D. 1985. Serological detection of cherry leafroll virus in English walnut trees. Phytopathology 75:48–52.

1656. Royce, D. J., and Ries, S. M. 1978. The influence of fungi isolated from peach twigs on the pathogenicity of Cytospora cincta. Phytopathology 68:603–607.

1657. Ruark, G. A., Mader, D. L., Veneman, P. L. M., and Tattar, T. A. 1983. Soil factors related to urban sugar maple decline. J. Arboric. 9:1–6.

1658. Rudolph, B. A. 1931. Verticillium hadromycosis. Hilgardia 5:197–361.

1659. Ruehle, G. D. 1941. Poinsettia scab caused by Sphaceloma. Phytopathology 31:947–948.

1660. Ruehle, J. L. 1962. Histopathological studies of pine roots infected with lance and pine cystoid nematodes. Phytopathology 52:68–71.

1661. Ruehle, J. L. 1964. Plant-parasitic nematodes associated with pine species in southern forests. Plant Dis. Rep. 48:60–61.

1662. Ruehle, J. L. 1967. Distribution of plant-parasitic nematodes associated with forest trees of the world. U.S. For. Serv., Southeast. For. Exp. Stn. 156 pp.

1663. Ruehle, J. L. 1968. Plant-parasitic nematodes associated with southern hardwood and coniferous forest trees. Plant Dis. Rep. 52:837–839.

1664. Ruehle, J. L., and Marx, D. H. 1979. Fiber, food, fuel, and fungal symbionts. Science 206:419–422.

1665. Ruehle, J. L., and Sasser, J. N. 1962. The role of plant-parasitic nematodes in stunting of pines in southern plantations. Phytopathology 52:56–58.

1666. Russin, J. S., and Shain, L. 1984. Initiation and development of cankers caused by virulent and cytoplasmic hypovirulent isolates of the chestnut blight fungus. Can. J. Bot. 62:2660–2664.

1667. Russin, J. S., and Shain, L. 1985. Disseminative fitness of Endothia parasitica containing different agents for cytoplasmic hypovirulence. Can. J. Bot. 63:54–57.

1668. Ryvarden, L. 1972. Studies on the Aphyllophorales of the Canary Islands with a note on the genus Perenniporia Murr. Norw. J. Bot. 19:139–144.

1669. Ryvarden, L. 1976, 1978. The Polyporaceae of North Europe. Vol. 1. Albatrellus–Incrustoporia. Vol. 2. Inonotus–Tyromyces. Fungiflora, Oslo. 577 pp.

1670. Sachs, I. B., Ward, J. C., and Kinney, R. E. 1975. Scanning electron microscopy of bacterial wetwood and normal heartwood in poplar trees. Pages 453–459 in: Proc. Workshop Scanning Electron Microscopy Plant Sci., I.I.T. Res. Inst., Chicago.

1671. Saito, I., Tamura, O., and Takakuwa, M. 1972. Ascospore dispersal in Valsa ceratosperma, the causal fungus of Japanese apple canker. Ann. Phytopathol. Soc. Jap. 38:367–374.

1672. Sakai, A. 1982. Freezing resistance of ornamental trees and shrubs. J. Am. Soc. Hortic. Sci 107:572–581.

1673. Saksena, H. K., and Vaartaja, O. 1961. Taxonomy, morphology, and pathogenicity of Rhizoctonia species from forest nurseries. Can. J. Bot. 39:627–647.

1674. Sallé, G. 1979. Le système endophytique du Viscum album: anatomie et fonctionnement des suçoirs secondaires. Can. J. Bot. 57:435–449.

1675. Salmon, E. S., and Ware, W. M. 1927. Leaf scorch of azalea. Gard. Chron. 81:286–288.

1676. Sanders, F. E., Mosse, B., and Tinker, P. B., eds. 1975. Endomycorrhizas. Academic Press, New York. 626 pp.

1677. Santamour, F. S. Jr. 1976. Resistance to sycamore anthracnose disease in hybrid Platanus. Plant Dis. Rep. 60:161–162.

1678. Santamour, F. S. Jr. 1977. The selection and breeding of pest-resistant landscape trees. J. Arboric. 3:146–152.

1679. Santamour, F. S. Jr. 1984. 'Columbia' and 'Liberty' planetrees. HortScience 19:901–902.

1680. Santamour, F. S. Jr. 1984. Wound compartmentalization in cultivars of Acer, Gleditsia, and other genera. J. Environ. Hortic. 2:123–125.

1681. Santamour, F. S. Jr., Gerhold, H. D., and Little, S., eds. 1976. Better trees for metropolitan landscapes. U.S. For. Serv. Gen. Tech. Rep. NE-22. 256 pp.

1682. Santesson, R. 1952. Foliicolous lichens. I. A Revision of the taxonomy of the obligately foliicolous lichenized fungi. Symb. Bot. Upsal. 12:1–590.

1683. Sartoris, G. B., and Kaufman, C. H. 1925. The development and taxonomic position of Apiosporina collinsii. Pap. Mich. Acad. Sci. 5:149–162.

1684. Sasser, J. N., Haasis, F. A., and Cannon, T. F. 1966. Pathogenicity of Meloidogyne species on Ilex. Plant Dis. Rep. 50:664–668.

1685. Savile, D. B. O. 1950. North American species of Chrysomyxa. Can. J. Res. C28:318–330.

1686. Savile, D. B. O. 1955. Chrysomyxa in North America—additions and corrections. Can. J. Bot. 33:487–496.

1687. Savile, D. B. O. 1959. Notes on Exobasidium. Can. J. Bot. 37:641–656.

1688. Savile, D. B. O. 1965. Puccinia karelica and species delimitation in the Uredinales. Can. J. Bot. 43:231–238.

1689. Sawai, K., Okuno, T., and Ito, T. 1982. The toxicity of cytochalasin E on plants. Ann. Phytopathol. Soc. Jap. 48:529–531.

1690. Schaad, N. W., Heskett, M. G., Gardner, J. M., and Kado, C. I. 1973. Influence of inoculum dosage, time after wounding, and season on infection of Persian walnut trees by Erwinia rubrifaciens. Phytopathology 63:327–329.

1691. Schaad, N. W., and Wilson, E. E. 1970. Pathological anatomy of the bacterial phloem canker disease of Juglans regia. Can. J. Bot. 48:1055–1060.

1692. Schaad, N. W., and Wilson, E. E. 1971. The ecology of Erwinia rubrifaciens and the development of phloem canker of Persian walnut. Ann. Appl. Biol. 69:125–136.

1693. Schaffer, B., Hawksworth, F. G., Wullschleger, S. D., and Reid, C. P. P. 1983. Cytokinin-like activity related to host reactions to dwarf mistletoes (Arceuthobium spp.). For. Sci. 29:66–70.

1694. Schaper, U., and Seemüller, E. 1982. Condition of the phloem and the persistence of mycoplasmalike organisms associated with apple proliferation and pear decline. Phytopathology 72:736–742.

1695. Scharpf, R. F., and Bynum, H. H. 1975. Cytospora canker of true firs. U.S. For. Serv. For. Pest Leafl. 146. 5 pp.

1696. Scharpf, R. F., and Hawksworth, F. G. 1974. Mistletoes on hardwoods in the United States. U.S. For. Serv. For. Pest Leafl. 147. 7 pp.

1697. Scharpf, R. F., and Hawksworth, F. G. 1976. Luther Burbank introduced European mistletoe into California. Plant Dis. Rep. 60:740–742.

1698. Scharpf, R. F., and Parmeter, J. R. Jr., tech. coords. 1978. Proceedings of the Symposium on Dwarf Mistletoe Control through Forest Management. U.S. For. Serv. Gen. Tech. Rep. PSW-31. 190 pp.

1699. Scheffer, T. C., and Hedgcock, C. G. 1955. Injury to northwestern forest trees by sulfur dioxide from smelters. U.S. Dep. Agric. Tech. Bull. 1117. 49 pp.

1700. Schenck, N. C. 1981. Can mycorrhizae control root disease? Plant Dis. 65:230–234.

1701. Schier, G. A. 1975. Deterioration of aspen clones in the middle Rocky Mountains. U.S. For. Serv. Res. Paper INT-170. 14 pp.

1702. Schipper, A. L. Jr. 1978. A Hypoxylon mammatum pathotoxin responsible for canker formation in quaking aspen. Phytopathology 68:866–872.

1703. Schmelzer, K. 1962. Untersuchungen an Viren der Zier- und Wildgehölze. 2. Mitteilung: Virosen an Forsythia, Lonicera, Ligustrum, und Laburnum. Phytopathol. Z. 46:105–138.

1704. Schmelzer, K. 1969. Das Ulmenscheckungs-Virus. Phytopathol. Z. 64:39–67.

1705. Schmelzer, K., and Schmelzer, A. 1968. Virusbefall an Magnolien (Magnolia spp.). Acta Phytopathol. Acad. Sci. Hung. 3:411–413.

1706. Schmidt, R. A., and Fergus, C. L. 1965. Branch canker and dieback of Quercus prinus L. caused by a species of Botryodiplodia. Can. J. Bot. 43:731–737.

1707. Schmidt, R. A., Powers, H. R. Jr., and Snow, G. A. 1981. Application of genetic disease resistance for the control of fusiform rust in intensively managed southern pine. Phytopathology 71:993–997.

1708. Schneider, A., and Dargent, R. 1977. Localisation et comportement du mycélium de Taphrina deformans dans le mésophyll et sous la cuticle des feuilles de pêcher (Prunus persica). Can. J. Bot. 55:2485–2495.

1709. Schneider, A., and Sutra, G. 1969. Les modalités de l'infection de Populus nigra L. par Taphrina populina Fr. C. R. Hebd. Séances Acad. Sci. (Paris) D 269:1056–1059.

1710. Schneider, R. 1961 Untersuchungen über das Auftreten der Guignardiablattbräune der Rosskastanie (Aesculus hippocastanum) in Westdeutschland und ihren Erreger. Phytopathol. Z. 42:272–278.

1711. Schneider, R., and Arx, J. A. von. 1966. Zwei neue, als Erreger von Zweigsterben nachgewiesene Pilze: Kabatina thujae n.g., n.sp., und K. juniperi n. sp. Phytopathol. Z. 57:176–182.

1712. Schneider, R., and Sauthoff, W. 1972. Absterbeerscheinungen an Carpinus betulus L. (Erreger: Monostichella robergei [Desm.] v. Höhn.). Nachrichtenbl. Deutsch. Pflanzenschutzd. 24:117–119.

1713. Schoeneweiss, D. F. 1967. Susceptibility of weakened cottonwood stems to fungi associated with blackstem. Plant Dis. Rep. 51:933–935.

1714. Schoeneweiss, D. F. 1969. Susceptibility of evergreen hosts to the juniper blight fungus, Phomopsis juniperovora, under epidemic conditions. J. Am. Soc. Hortic. Sci. 94:609–611.

1715. Schoeneweiss, D. F. 1974. Tubercularia ulmea canker of tallhedge: influence of freezing stress on disease susceptibility. Plant Dis. Rep. 58:937–941.

1716. Schoeneweiss, D. F. 1975. Predisposition, stress, and plant disease. Annu. Rev. Phytopathol. 13:193–211.

1717. Schoeneweiss, D. F. 1981. Infectious diseases of trees associated with water and freezing stress. J. Arboric. 7:13–18.

1718. Schoeneweiss, D. F. 1981. The role of environmental stress in diseases of woody plants. Plant Dis. 65:308–314.

1719. Schoeneweiss, D. F. 1983. Drought predisposition to Cytospora canker in blue spruce. Plant Dis. 67:383–385.

1720. Schoeneweiss, D. F., and Wene, E. G. 1977. Freezing stress predisposes *Euonymus alatus* stems to attack by *Nectria cinnabarina*. Plant Dis. Rep. 61:921–925.

1721. Schreiber, L. R., and Green, R. J. Jr. 1959. Die-back and root rot disease of *Taxus* spp. in Indiana. Plant Dis. Rep. 43:814–817.

1722. Schreiner, E. J. 1931. Two species of *Valsa* causing disease in *Populus*. Am. J. Bot. 18:1–29.

1723. Schrenk, H. von. 1901. A disease of the black locust (*Robinia pseudo-acacia* L.). Mo. Bot. Gard. Annu. Rep. 12:21–31.

1724. Schrenk, H. von. 1903. A disease of white ash caused by *Polyporus fraxinophilus*. U.S. Dep. Agric. Bur. Plant Indus. Bull. 32. 20 pp.

1725. Schrenk, H. von. 1914. A trunk disease of the lilac. Ann. Mo. Bot. Gard. 1:253–262.

1726. Schuldt, P. H. 1955. Comparison of anthracnose fungi on oak, sycamore, and other trees. Contrib. Boyce Thompson Inst. 18:85–107.

1727. Schulz, U. 1981. Histologische Untersuchungen der Eintrittspforten von *Cytospora*-Arten. Angew. Bot. 55:441–455.

1728. Schulz, U. 1981. Untersuchungen zur biologischen Bekämpfung von *Cytospora*-Arten. Z. Pflanzenkr. Pflanzenschutz. 88:132–141.

1729. Schütt, P. 1972. Untersuchungen über den Einfluss von Cuticular-wachsen auf die Infektionsfähigkeit pathogenen Pilze. 2. *Rhytisma acerinum*, *Microsphaera alphitoides*, und *Fusarium oxysporum*. Eur. J. For. Pathol. 2:43–59.

1730. Schütt, P. 1973. Die Wirkung gasförmiger Blättausscheidungen auf Sporenkeimung und Myzelentwicklung von *Botrytis cinerea*. Eur. J. For. Pathol. 3:187–192.

1731. Schütt, P., and Cowling, E. B. 1985. Waldsterben, a general decline of forests in central Europe: symptoms, development, and possible causes. Plant Dis. 69:548–558.

1732. Scott, K. J., and Chakravorty, A. K., eds. 1982. The rust fungi. Academic Press, New York. 288 pp.

1733. Scott, S. W., and Barnett, O. W. 1984. Some properties of an isolate of broad bean wilt virus from dogwood (*Cornus florida*). Plant Dis. 68:983–985.

1734. Seaver, F. J. 1922. Phyllostictales. N. Am. Flora 6 (Part 1):1–84.

1735. Secor, G. A., and Nyland, G. 1978. Rose ring pattern: a component of the rose-mosaic complex. Phytopathology 68:1005–1010.

1736. Seeler, E. V. Jr. 1940. Two diseases of *Gleditsia* caused by a species of *Thyronectria*. J. Arnold Arbor. 21:405–427.

1737. Seeler, E. V. Jr. 1940. A monographic study of the genus *Thyronectria*. J. Arnold Arbor. 21:429–460.

1738. Seemüller, E. 1980. Untersuchungen über die Populationsdynamik von *Pseudomonas syringae* an Sauerkirsche und die Toxizität von Syringomycin für Blatter isolierte Blattzellen. Z. Pflanzenkr. Pflanzenschutz 87:51–61.

1739. Seemüller, E., Schaper, U., and Zimbelmann, F. 1984. Seasonal variation in the colonization patterns of mycoplasmalike organisms associated with apple proliferation and pear decline. Z. Pflanzenkr. Pflanzenschutz 91:371–382.

1740. Seifers, D., and Ammon, V. 1980. Mode of penetration of sycamore leaves by *Gloeosporium platani*. Phytopathology 70:1050–1055.

1741. Seliskar, C. E. 1950. Some Investigations on the wetwood diseases of American elm and Lombardy poplar. Ph.D. thesis, Cornell Univ., Ithaca, NY. 148 pp.

1742. Seliskar, C. E. 1976. Mycoplasmalike organism found in the phloem of bunch-diseased walnuts. For. Sci. 22:144.

1743. Seliskar, C. E., KenKnight, G. E., and Bourne, C. E. 1974. Mycoplasma-like organism associated with pecan bunch disease. Phytopathology 64:1269–1272.

1744. Semeniuk, P. 1979. Spotless Gold, Spotless Yellow, and Spotless Pink Rose: blackspot resistant breeding lines. HortScience 14:764–765.

1745. Setliff, E. C., Hoch, H. C., and Patton, R. F. 1974. Studies on nuclear division in basidia of *Poria latemarginata*. Can. J. Bot. 52:2323–2333.

1746. Sewell, G. W. F., and Wilson, J. F. 1964. Occurrence and dispersal of *Verticillium* conidia in xylem sap of the hop (*Humulus lupus* L.). Nature 204:901.

1747. Sewell, G. W. F., and Wilson, J. F. 1973. Phytophthora collar rot of apple: seasonal effects on infection and disease development. Ann. Appl. Biol. 74:149–158.

1748. Sewell, G. W. F., Wilson, J. F., and Dakwa, J. T. 1974. Seasonal variations in the activity in soil of *Phytophthora cactorum*, *P. syringae*, and *P. citricola* in relation to collar rot disease of apple. Ann. Appl. Biol. 76:179–186.

1749. Seymour, C. P. 1969. Charcoal rot of nursery-grown pines in Florida. Phytopathology 59:89–92.

1750. Shabi, E., Rotem, J., and Lobenstein, G. 1973. Physiological races of *Venturia pirina* on pear. Phytopathology 63:41–43.

1751. Shahin, E. A., and Claflin, L. E. 1978. The occurrence and distribution of Sirococcus shoot blight of spruce in Kansas. Plant Dis. Rep. 62:648–650.

1752. Shain, L. 1967. Resistance of sapwood in stems of loblolly pine to infection by *Fomes annosus*. Phytopathology 57:1034–1045.

1753. Shain, L. 1971. The response of sapwood of Norway spruce to infection by *Fomes annosus*. Phytopathology 61:301–307.

1754. Shain, L., and Franich, R. A. 1981. Induction of Dothistroma blight symptoms with dothistromin. Physiol. Plant Pathol. 19:49–55.

1755. Shain, L., and Hillis, W. E. 1971. Phenolic extractives in Norway spruce and their effects on *Fomes annosus*. Phytopathology 61:841–845.

1756. Shattock, R. C., and Bhatti, M. H. R. 1983. The effect of *Phragmidium mucronatum* on rose understocks and maiden rose bushes. Plant Pathol. 32:61–66.

1757. Shaw, C. G. 1949. Nomenclatorial problems in the Peronosporaceae. Mycologia 41:323–338.

1758. Shaw, C. G. 1951. New species of the Peronosporaceae. Mycologia 43:445–455.

1759. Shaw, C. G. 1973. Host fungus index for the Pacific Northwest—I. Hosts. II. Fungi. Wash. Agric. Exp. Stn. Bull. 765 and 766. 121 and 162 pp.

1760. Shaw, C. G., and Leaphart, C. D. 1960. Two serious foliage diseases of western white pine in the Inland Empire. Plant Dis. Rep. 44:655–659.

1761. Shaw, C. G. III, and Roth, L. F. 1976. Persistence and distribution of a clone of *Armillaria mellea* in a ponderosa pine forest. Phytopathology 66:1210–1213.

1762. Shay, J. R., and Williams, E. B. 1956. Identification of three physiologic races of *Venturia inaequalis*. Phytopathology 46:190–193.

1763. Shea, K. R., and Lewis, D. K. 1971. Occurrence of dwarf mistletoe in sanitized ponderosa pine in south-central Oregon. Northwest Sci. 45:94–99.

1764. Shea, K. R., and Stewart, J. L. 1972. Hemlock dwarf mistletoe. U.S. For. Serv. For. Pest Leafl. 135. 6 pp.

1765. Shear, C. L., Stevens, N. E., and Tiller, R. J. 1917. *Endothia parasitica* and related species. U.S. Dep. Agric. Bull. 380. 77 pp.

1766. Shear, C. L., Stevens, N. E., and Wilcox, M. S. 1925. *Botryosphaeria* and *Physalospora* in the eastern United States. Mycologia 17:98–107.

1767. Sher, S. A. 1958. The effect of nematodes on azaleas. Plant Dis. Rep. 42:84–85.

1768. Sherald, J. L., Hearon, S. S., Kostka, S. J., and Morgan, D. L. 1983. Sycamore leaf scorch: culture and pathogenicity of fastidious xylem-limited bacteria from scorch-affected trees. Plant Dis. 67:848–852.

1769. Sherwood, C. H., Weigle, J. L., and Denisen, E. L. 1970. 2,4–D as an air pollutant: effects on growth of representative horticultural plants. HortScience 5:211–213.

1770. Shew, H. D., and Benson, D. M. 1981. Fraser fir root rot induced by *Phytophthora citricola*. Plant Dis. 65:688–689.

1771. Shew, H. D., and Benson, D. M. 1983. Influence of soil temperature and inoculum density of *Phytophthora cinnamomi* on root rot of Fraser fir. Plant Dis. 67:522–524.

1772. Shiam, R., and Sharma, O. P. 1975. Studies on *Graphiola phoenicis* (Moug.) Poit. I. Histopathological effects. Curr. Sci. 44:163–165.

1773. Shiel, P. J., and Castello, J. D. 1985. Detection of tobacco mosaic and tobacco ringspot viruses in herbaceous and woody plants near virus-infected white ash trees in central New York. Plant Dis. 69:791–795.

1774. Shigo, A. L. 1963. Fungi associated with the discolorations around rot columns caused by *Fomes igniarius*. Plant Dis. Rep. 47:820–823.

1775. Shigo, A. L. 1964. Organism interactions in the beech bark disease. Phytopathology 54:263–269.

1776. Shigo, A. L. 1967. Successions of organisms in discoloration and decay of wood. Int. Rev. For. Res. 2:237–299.

1777. Shigo, A. L. 1969. How *Poria obliqua* and *Polyporus glomeratus* incite cankers. Phytopathology 59:1164–1165.

1778. Shigo, A. L. 1970. Beech bark disease. U.S. For. Serv. For. Pest Leafl. 75 (rev). 8 pp.

1779. Shigo, A. L. 1974. Relative abilities of *Phialophora melinii*, *Fomes connatus*, and *F. igniarius* to invade freshly wounded tissues of *Acer rubrum*. Phytopathology 64:708–710.

1780. Shigo, A. L. 1979. Compartmentalization of decay associated with *Heterobasidion annosum* in roots of *Pinus resinosa*. Eur. J. For. Pathol. 9:341–347.

1781. Shigo, A. L. 1983. Tree defects: a photo guide. U.S. For. Serv. Gen. Tech. Rep. NE-82. 167 pp.

1782. Shigo, A. L. 1984. Compartmentalization: a conceptual framework for understanding how trees grow and defend themselves. Annu. Rev. Phytopathol. 22:189–214.

1783. Shigo, A. L., and Hillis, W. E. 1973. Heartwood, discolored wood, and microorganisms in living trees. Annu. Rev. Phytopathol. 11:197–222.

1784. Shigo, A. L., and Marx, H. G. 1977. Compartmentalization of decay in trees. U.S. Dep. Agric. Agric. Inf. Bull. 405. 73 pp.

1785. Shigo, A. L., and Marx, H. G. 1979. Tree decay. An expanded concept. U.S. Dep. Agric. Agric. Inf. Bull. 419. 73 pp.

1786. Shigo, A. L., and Sharon, E. M. 1970. Mapping columns of discolored and decayed tissues in sugar maple, *Acer saccharum*. Phytopathology 60: 232–237.

1787. Shigo, A. L., and Shortle, W. C. 1979. Compartmentalization of discolored wood in heartwood of red oak. Phytopathology 69:710–711.

1788. Shigo, A., and Tippett, J. T. 1981. Compartmentalization of American elm tissues infected by *Ceratocystis ulmi*. Plant Dis. 65:715–718.

1789. Shigo, A. L., and Tippett, J. T. 1981. Compartmentalization of decayed wood associated with *Armillaria mellea* in several tree species. U.S. For. Serv. Res. Pap. NE-488. 20 pp.

1790. Shoemaker, R. A. 1964. Conidial states of some *Botryosphaeria* species on *Vitis* and *Quercus*. Can. J. Bot. 42:1297–1301.

1791. Shoemaker, R. A. 1965. Revision of some *Dimeriella* and *Dimerosporium* parasites of conifers. Can. J. Bot. 43:631–639.

1792. Shortle, W. C. 1979. Compartmentalization of decay in red maple and hybrid poplar trees. Phytopathology 69:410–413.

1793. Shurtleff, W., and Aoyagi, A. 1977. The book of kudzu. Autumn Press, Brookline, MA. 102 pp.

1794. Siccama, T. G., Weir, G., and Wallace, K. 1976. Ice damage in a mixed hardwood forest in Connecticut in relation to *Vitis* infestation. Bull. Torrey Bot. Club 103:180–183.

1795. Siggers, P. V. 1944. The brown spot needle blight of pine seedlings. U.S. Dep. Agric. Tech. Bull. 870. 36 pp.

1796. Sijam, K., Karr, A. L., and Goodman, R. N. 1983. Comparison of the extracellular polysaccharides produced by *Erwinia amylovora* in apple tissue and culture medium. Physiol. Plant Pathol. 22:221–231.

1797. Sikorowski, P. P., and Roth, L. F. 1962. *Elytroderma* mycelium in the phloem of ponderosa pine. Phytopathology 52:332–336.

1798. Simmons, E. G. 1981. *Alternaria* themes and variations. Mycotaxon 13:17–34.

1799. Simmons, E. G. 1982. Alternaria themes and variations (7–10). Mycotaxon 14:17–43.

1800. Simms, H. R. 1967. On the ecology of *Herpotrichia nigra*. Mycologia 59:902–909.

1801. Simons, M. D., Rothman, P. J., and Michel, L. J. 1979. Pathogenicity of *Puccinia coronata* from buckthorn and from oats adjacent to and distant from buckthorn. Phytopathology 69:156–158.

1802. Sinclair, W. A. 1964. Root- and butt-rot of conifers caused by *Fomes annosus*, with special reference to inoculum dispersal and control of the disease in New York. Cornell Univ. Agric. Exp. Stn. Mem. 391. 54 pp.

1803. Sinclair, W. A. 1967. Decline of hardwoods: possible causes. Proc. Int. Shade Tree Conf. 42:17–32.

1804. Sinclair, W. A., and Campana, R. J., eds. 1978. Dutch elm disease. Perspectives after 60 years. Cornell Univ. Agric. Exp. Stn. Search Agric. 8, No. 5. 52 pp.

1805. Sinclair, W. A., Cowles, D. P., and Hee, S. M. 1975. Fusarium root rot of Douglas-fir seedlings: suppression by soil fumigation, fertility management, and inoculation with spores of the fungal symbiont *Laccaria laccata*. For. Sci. 21:390–399.

1806. Sinclair, W. A., and Hudler G. W. 1980. Tree and shrub pathogens new or noteworthy in New York State. Plant Dis. 64:590–592.

1807. Sinclair, W. A., Smith, K. L., and Larsen, A. O. 1981. Verticillium wilt of maples: symptoms related to movement of the pathogen in stems. Phytopathology 71:340–345.

1808. Sinclair, W. A., and Stone, E. L. 1974. Boron toxicity to pines subject to home laundry waste water. Arborist's News 39:71–72.

1809. Sinclair, W. A., Stone, E. L., and Scheer, C. F. Jr. 1975. Toxicity to hemlocks grown in arsenic-contaminated soil previously used for potato production. HortScience 10:35–36.

1810. Sinden, J. L., DeVay, J. E., and Backman, P. A. 1971. Properties of syringomycin, a wide spectrum antibiotic and phytotoxin produced by *Pseudomonas syringae*, and its role in the bacterial canker disease of peach trees. Physiol. Plant Pathol. 1:199–213.

1811. Singh, P. 1978. Broom rusts of balsam fir and black spruce in Newfoundland. Eur. J. For. Pathol. 8:25–36.

1812. Singh, S. J., and Heather, W. A. 1982. Temperature sensitivity of qualitative race-cultivar interactions in *Melampsora medusae* Thüm. and *Populus* species. Eur. J. For. Pathol. 12:123–127.

1813. Singh, S. J., and Heather, W. A. 1982. Temperature-light sensitivity of infection types expressed by cultivars of *Populus deltoides* Marsh. to races of *Melampsora medusae* Thüm. Eur. J. For. Pathol. 12:327–331.

1814. Sinha, M. K., Singh, R., and Jeyarajan, R. 1970. Graphiola leaf spot on date palm (*Phoenix dactylifera*): susceptibility of date varieties and effect on chlorophyll content. Plant Dis. Rep. 54:617–619.

1815. Sinha, R. C., and Chikowski, L. N. 1980. Transmission and morphological features of mycoplasmalike bodies associated with peach X-disease. Can. J. Plant Pathol. 2:119–124.

1816. Sivanesan, A. 1977. The taxonomy and pathology of *Venturia* species. Bibl. Mycol. 59. 139 pp.

1817. Skelly, J. M. 1964. The nature and occurrence of an annual canker of

Acer saccharum Marsh. in Pennsylvania. M.S. thesis. Penn. State. Univ. 92 pp.

1818. Skerman, V. D. B., McGowan, V., and Sneath, P. H. A. 1980. Approved lists of bacterial names. Int. J. Syst. Bacteriol. 30:225–420.

1819. Skilling, D. D. 1977. The development of a more virulent strain of *Scleroderris lagerbergii* in New York State. Eur. J. For. Pathol. 7:297–302.

1820. Skilling, D., Kienzler, M., and Haynes, E. 1984. Distribution of serological strains of *Gremmeniella abietina* in eastern North America. Plant Dis. 68:937–938.

1821. Skilling, D. D., and Nicholls, T. H. 1974. Brown spot needle disease—biology and control in Scotch pine plantations. U.S. For. Serv. Res. Pap. NC-109. 19 pp.

1822. Skilling, D. D., and Nicholls, T. H. 1975. The development of *Lophodermium pinastri* in conifer nurseries and plantations in North America. Eur. J. For. Pathol. 5:193–197.

1823. Skilling, D. D., O'Brien, J. T., and Beil, J. A. 1979. Scleroderris canker of northern conifers. U.S. For. Serv. For. Insect Dis. Leafl. 130. 7 pp.

1824. Slagg, C. M., and Wright, E. 1943. Diplodia blight in coniferous seedbeds. Phytopathology 33:390–393.

1825. Sleeth, B., and Bidwell, C. B. 1937. *Polyporus hispidus* and a canker of oaks. J. For. 35:778–785.

1826. Smeltzer, D. L. K., and French, D. W. 1981. Factors affecting spread of *Cronartium comptoniae* on the sweetfern host. Can. J. For. Res. 11:400–408.

1827. Smerlis, E. 1962. Taxonomy and morphology of *Potebniamyces balsamicola* sp. nov. associated with a twig and branch blight of balsam fir in Quebec. Can. J. Bot. 40:351–359.

1828. Smerlis, E. 1967. Pathogenicity of *Phacidium abietis*. Plant Dis. Rep. 51:678–679.

1829. Smerlis, E. 1970. Pathogenicity of *Aleurodiscus amorphus*. Can. For. Serv. Bi-mon. Res. Notes 26:18.

1830. Smerlis, E. 1971. Pathogenicity tests of *Valsa* species occurring on conifers in Québec. Phytoprotection 52:28–31.

1831. Smerlis, E. 1973. Pathogenicity tests of some discomycetes occurring on conifers. Can. J. For. Res. 3:7–16.

1832. Smerlis, E., and Saint-Laurent, M. 1966. Pathogenicity of *Lophophacidium hyperboreum* Lagerberg. Plant Dis. Rep. 50:356–357.

1833. Smidt, M., and Kosuge, T. 1978. The role of indole-3–acetic acid accumulation by methyl tryptophan-resistant mutants of *Pseudomonas syringae* in gall formation on oleanders. Physiol. Plant Pathol. 13:203–214.

1834. Smith, C. O. 1928. Oleander bacteriosis in California. Phytopathology 18:503–518.

1835. Smith, C. O. 1934. Inoculations showing the wide host range of *Botryosphaeria ribis*. J. Agric. Res. 49:467–476.

1836. Smith, D. H., Lewis, F. H., and Wainwright, S. H. 1970. Epidemiology of the black knot disease of plums. Phytopathology 60:1441–1444.

1837. Smith, E. F. 1888. A date palm fungus (*Graphiola phoenicis* Poit.) Bot. Gaz. 13:211–213.

1838. Smith, E. M. 1975. Tree stress from salts and herbicides. J. Arboric. 1:201–205.

1839. Smith, E. M. 1981. The flowering crabapple—a tree for all seasons. J. Arboric. 7:89–95.

1840. Smith, E. M., and Treaster, S. A. 1981. Preventing habitual iron chlorosis of woody landscape plants. Ohio Agric. Res. Devel. Cent. Res. Circ. 263:26–29.

1841. Smith, G. C., and Brennan, E. G. 1984. Response of honeylocust cultivars to air pollution stress in an urban environment. J. Arboric. 10:289–293.

1842. Smith, H. C., and MacDonald, W. L., eds. 1982. Proceedings of the USDA Forest Service American Chestnut Cooperators' Meeting. West Virginia University Books, Morgantown. 229 pp.

1843. Smith, L. D. 1983. Major nutrient influence on *Verticillium dahliae* infections of *Acer saccharum*. J. Arboric. 9:277–281.

1844. Smith, L. D., and Neely, D. 1979. Relative susceptibility of tree species to *Verticillium dahliae*. Plant Dis. Rep. 63:328–332.

1845. Smith, R. B. 1973. Factors affecting dispersal of dwarf mistletoe seeds from an overstory western hemlock tree. Northwest Sci. 47:9–19.

1846. Smith, R. B., and Wass, E. F. 1979. Infection trials with three dwarf mistletoe species within and beyond their known ranges in British Columbia. Can. J. Plant Pathol. 1:47–57.

1847. Smith, R. S. Jr. 1966. Effect of diurnal temperature fluctuations on the charcoal root disease of *Pinus lambertiana*. Phytopathology 56:61–64.

1848. Smith, R. S. Jr. 1967. Verticicladiella root disease of pines. Phytopathology 57:935–938.

1849. Smith, R. S. Jr. 1973. Sirococcus tip dieback of *Pinus* spp. in California forest nurseries. Plant Dis. Rep. 57:69–73.

1850. Smith, R. S. Jr., and Bega, R. V. 1964. *Macrophomina phaseoli* in the forest tree nurseries of California. Plant Dis. Rep. 48:206.

1851. Smith, R. S. Jr., and Graham, D. 1975. Black stain root disease of conifers. U.S. For. Serv. For. Pest Leafl. 145. 4 pp.

1852. Smith, W. H. 1981. Air pollution and forests. Interactions between air contaminants and forest ecosystems. Springer-Verlag, New York. 379 pp.

1853. Snetsinger, R., and Himelick, E. B. 1957. Observations on witches'-broom of hackberry. Plant Dis. Rep. 41:541–544.

1854. Snow, G. A., Beland, J. W., and Czabator, F. J. 1974. Formosan sweetgum susceptible to North American *Endothia gyrosa*. Phytopathology 64:602–605.

1855. Snyder, E. B., and Derr, H. J. 1972. Breeding longleaf pines for resistance to brown spot needle blight. Phytopathology 62:325–329.

1856. Snydor, T. D., and Kuhns, L. 1976. Visual symptoms of copper toxicity on woody ornamentals. Weeds Trees Turf 15(7):58–59.

1857. Sobers, E. K. 1964. Cercospora diseases of *Ligustrum*. Proc. Fla. State Hortic. Soc. 77:486–489.

1858. Sobers, E. K. 1967. The perfect stage of *Coniothyrium concentricum* on leaves of *Yucca aloifolia*. Phytopathology 57:234–235.

1859. Sobers, E. K., and Alfieri, S. A. Jr. 1972. Species of *Cylindrocladium* and their hosts in Florida and Georgia. Fla. State Hortic. Soc. Proc. 85:366–369.

1860. Sobers, E. K., and Seymour, C. P. 1967. *Cylindrocladium floridanum* sp. n. associated with decline of peach trees in Florida. Phytopathology 57:389–393.

1861. Sommer, N. F. 1955. Sunburn predisposes walnut trees to branch wilt. Phytopathology 45:607–613.

1862. Southey, J. F., ed. 1978. Plant nematology. G.B. Minist. Agric. Fish. & Food GD1. 440 pp.

1863. Spaulding, P. 1922. Investigations of the white-pine blister rust. U.S. Dep. Agric. Bull. 957. 100 pp.

1864. Spaulding, P., and Bratton, A. W. 1946. Decay following glaze storm damage in woodlands of central New York. J. For. 44:515–519.

1865. Spaulding, P., and Hansbrough, J. R. 1932. *Cronartium comptoniae*, the sweetfern blister rust of pitch pines. U.S. Dep. Agric. Tech. Bull. 217. 21 pp.

1866. Spaulding, P., and MacAloney, H. J. 1931. A study of organic factors concerned in the decadence of birch on cut-over lands in northern New England. J. For. 29:1134–1149.

1867. Spencer, D. M., ed. 1978. The powdery mildews. Academic Press, New York. 565 pp.

1868. Spencer, D. M., ed. 1981. The downy mildews. Academic Press, New York. 636 pp.

1869. Spielman, L. J. 1983. Taxonomy and biology of *Valsa* species on hardwoods in North America, with special reference to species on maples. Ph.D. thesis, Cornell Univ., Ithaca, NY. 175 pp.

1870. Spielman, L. J. 1985. A monograph of *Valsa* on hardwoods in North America. Can. J. Bot. 63:1355–1378.

1871. Spiers, A. G. 1983. Host range and pathogenicity studies of *Marssonina brunnea* to poplars. Eur. J. For. Pathol. 13:181–196.

1872. Spiers, A. G. 1983. Host range and pathogenicity studies of *Marssonina castagnei* to poplars. Eur. J. For. Pathol. 13:218–227.

1873. Spiers, A. G. 1984. Comparative studies of host specificity and symptoms exhibited by poplars infected with *Marssonina brunnea*, *Marssonina castagnei*, and *Marssonina populi*. Eur. J. For. Pathol. 14:202–218.

1874. Spiers, A. G., and Hopcroft, D. H. 1983. Ultrastructural study of the pathogenesis of *Marssonina* species to poplars. Eur. J. For. Pathol. 13:414–427.

1875. Spooner, D. M. 1983. The northern range of eastern mistletoe, *Phoradendron serotinum* (Viscaceae), and its status in Ohio. Bull. Torrey Bot. Club 110:489–493.

1876. Spotts, R. A. 1980. Infection of grape by *Guignardia bidwellii*—factors affecting lesion development, conidial dispersal, and conidial populations on leaves. Phytopathology 70:252–255.

1877. Spotts, R. A., Covey, R. P., and Chen, P. M. 1981. Effect of low temperature on survival of apple buds infected with the powdery mildew fungus. HortScience 16:781–783.

1878. Sprague, R., and Heald, F. D. 1927. A witches' broom of the service berry. Trans. Am. Microscop. Soc. 46:219–247.

1879. Sproston, T. Jr., and Scott, W. W. 1954. *Valsa leucostomoides*, the cause of decay and discoloration in tapped sugar maples. Phytopathology 44:12–13.

1880. Squillace, A. E., Dinus, R. J., Hollis, C. A., and Schmidt, R. A. 1978. Relation of oak abundance, seed source, and temperature to geographic patterns of fusiform rust incidence. U.S. For. Serv. Res. Pap. SE-186. 20 pp.

1881. Srivastava, L. M., and Esau, K. 1961. Relation of dwarfmistletoe (*Arceuthobium*) to the xylem tissue of conifers. I. Anatomy of parasite sinkers and their connection with host xylem. Am. J. Bot. 48:159–167.

1882. Stack, R. W., and Ash, C. L. 1979. Botryodiplodia canker of American elm. (Abstr.) J. Arboric. 5:163–164.

1883. Staley, J. M. 1964. A new *Lophodermium* on ponderosa pine. Mycologia 56:757–762.

1884. Staley, J. M. 1965. Decline and mortality of red and scarlet oaks. For. Sci. 11:2–17.

1885. Staley, J. M. 1979. *Lophodermella cerina*, a pathogen of pine foliage. Abstr. 628. 9th Int. Cong. Plant Prot., Washington, DC.

1886. Staley, J. M., and Hawksworth, F. G. 1967. *Bifusella crepidiformis* on Englemann spruce. Plant Dis. Rep. 51:791–792.

1887. Stathis, P. D., and Plakidas, A. G. 1958. Anthracnose of azaleas. Phytopathology 48:256–260.

1888. Stathis, P. D., and Plakidas, A. G. 1959. Entomosporium leaf spot of *Photinia glabra* and *Photinia serrulata*. Phytopathology 49:361–365.

1889. Steekelenburg, N. A. M. van. 1972. Verwelkingziekte in clematis. Bedrijfsontwikkeling 3:855–857.

1890. Steinmetz, H. F., and Hilborn, M. T. 1937. A histological evaluation of low temperature injury to apple trees. Maine Agric. Exp. Stn. Bull. 388. 32 pp.

1891. Stephan, B. R. 1980. Prüfung von Douglasien-Herkünften auf Resistenz gegen *Rhabdocline pseudotsugae* in Infektionsversuchen. Eur. J. For. Pathol. 10:152–161.

1892. Stephan, B. R., and Hyun, S. K. 1983. Studies on the specialization of *Cronartium ribicola* and its differentiation on the alternate hosts *Ribes* and *Pedicularis*. Z. Pflanzenkr. Pflanzenschutz 90:670–678.

1893. Stephan, B. R., and Millar, C. S., compilers. 1975. *Lophodermium* on pines. Proc. 5th Eur. Colloq. For. Pathol., Schmalenbeck, April 1975. Mitt. Bundesanstalt für Forst- und Holzwirtschaft 108:1–201.

1894. Stephens, G. R., Turner, N. C., and De Roo, H. C. 1972. Some effects of gypsy moth (*Porthetria dispar* L.) and elm spanworm (*Ennomos subsignarius* Hbn.) on water balance and growth of deciduous forest trees. For. Sci. 18:326–330.

1895. Stermer, B. A., Scheffer, R. P., and Hart, J. H. 1984. Isolation of toxins of *Hypoxylon mammatum* and demonstration of some toxin effects on selected clones of *Populus tremuloides*. Phytopathology 74:654–658.

1896. Stern, A. C., ed. 1976. Air pollution. 3rd ed. Vol. 2. The effects of air pollution. Academic Press, New York. 684 pp.

1897. Stevens, N. E. 1926. Two species of *Physalospora* on citrus and other hosts. Mycologia 18:206–217.

1898. Stevens, N. E. 1933. Life history and synonomy of *Physalospora glandicola*. Mycologia 25:504–508.

1899. Stevens, N. E. 1933. Two apple black rot fungi in the United States. Mycologia 25:536–548.

1900. Stevens, N. E. 1936. Two species of *Physalospora* in England. Mycologia 28:331–336.

1901. Stewart, F. C. 1917. Witches-brooms on hickory trees. Phytopathology 7:185–187.

1902. Stewart, V. B. 1916. The leaf blotch disease of horse-chestnut. Phytopathology 6:5–19.

1903. Steyaert, R. L. 1949. Contribution à l'étude monographique de *Pestalotia* de Not. et *Monochaetia* Sacc. (*Truncatella* gen. nov. et *Pestalotiopsis* gen. nov.). Bull. Jard. Bot. de l'Etat Brux. 19:285–354.

1904. Steyaert, R. 1967. Considérations générales sur le genre *Ganoderma* et plus spécialement sur les espèces européennes. Bull. Soc. R. Bot. Belg. 100:189–211.

1905. Steyaert, R. 1967. Les *Ganoderma* palmicoles. Bull. Jard. Bot. Natl. Belg. 37:465–492.

1906. Steyaert. R. 1971. Species of *Ganoderma* and related genera mainly of the Bogor and Leiden herbaria. Persoonia 7:55–118.

1907. Stayaert, R. 1975. The concept and circumscription of *Ganoderma tornatum*. Trans. Br. Mycol. Soc. 65:451–467.

1908. Stillwell, M. A. 1964. The fungus associated with woodwasps occurring in beech in New Brunswick. Can. J. Bot. 42:495–496.

1909. Stipes, R. J., and Campana, R. J., eds. 1981. Compendium of elm diseases. American Phytopathological Society, St. Paul, MN. 96 pp.

1910. Stipes, R. J., and Phipps, P. M. 1971. A species of *Endothia* associated with a canker disease of pin oak (*Quercus palustris*) in Virginia. Plant Dis. Rep. 55:467–469.

1911. Stone, E. L. Jr. 1952. An unusual type of frost injury in pine. J. For. 50:560–561.

1912. Stone, E. L. Jr. 1953. Magnesium deficiency of some northeastern pines. Proc. Soil Sci. Soc. Am. 17:297–300.

1913. Stone, E. L. 1968. Microelement nutrition of forest trees: a review. Pages 132–175 in: Forest fertilization: theory and practice. Tennessee Valley Authority, Muscle Shoals, AL. 306 pp.

1914. Stone, E. L., and Baird, G. 1956. Boron level and boron toxicity in red and white pine. J. For. 54:11–12.

1915. Stone, E. L., and Cornwell, S. 1968. Basal bud burls in *Betula populifolia*. For. Sci. 14:64–65.

1916. Stone, E. L., and Greweling, T. 1971. Arsenic toxicity in red pine and the persistence of arsenic in nursery soils. Tree Planters Notes 22(1):5–7.

1917. Stone, E. L., Morrow, R. R., and Welch, D. S. 1954. A malady of red pine on poorly drained sites. J. For. 52:104–114.

1918. Stowell, E. A., and Backus, M. P. 1966–1967. Morphology and cytology of *Diplocarpon maculatum* on *Crataegus*. I. The *Entomosporium* stage. II. Initiation and development of the apothecium. Mycologia 58:949–960; 59:623–636.

1919. Streets, R. B., and Bloss, H. E. 1973. Phymatotrichum root rot. Monogr. 8. American Phytopathological Society, St. Paul, MN. 38 pp.

1920. Strider, D. L. 1976. Increased prevalence of powdery mildew of azalea and rhododendron in North Carolina. Plant Dis. Rep. 60:149–151.

1921. Strider, D. L., and Jones, R. K. 1978. Rust of fuchsia in North Carolina. Plant Dis. Rep. 62:745–746.

1922. Strobel, N. E., Hussey, R. S., and Roncadori, R. W. 1982. Interaction of vesicular-arbuscular mycorrhizal fungi, Meloidogyne incognita, and soil fertility on peach. Phytopathology 72:690–694.

1923. Strouts, R. G. 1973. Canker of cypresses caused by Coryneum cardinale Wag. in Britain. Eur. J. For. Pathol. 3:13–24.

1924. Struckmeyer, B. E., Kuntz, J. E., and Riker, A. J. 1958. Histology of certain oaks infected with the oak wilt fungus. Phytopathology 48:556–561.

1925. Struckmeyer, B. E., and Riker, A. J. 1951. Wound-periderm formation in white-pine trees resistant to blister rust. Phytopathology 41:276–281.

1926. Stuntz, D. E., and Seliskar, C. E. 1943. A stem canker of dogwood and madrona. Mycologia 35:207–221.

1927. Subirats, F. J., and Self, R. L. 1971. A new petal blight of camellias incited by Pestalotia sp. Plant Dis. Rep. 55:697–700.

1928. Sucoff, E. 1975. Effect of deicing salts on woody vegetation along Minnesota roads. Minn. Agric. Exp. Stn. Tech. Bull. 303. 49 pp.

1929. Sucoff, E., and Hong, S. G. 1976. Effect of NaCl on cold hardiness of Malus spp. and Syringa vulgaris. Can. J. Bot. 54:2816–2819.

1930. Sucoff, E., Hong, S. G., and Wood, A. 1976. NaCl and twig dieback along highways and cold hardiness of highway versus garden twigs. Can. J. Bot. 54:2268–2274.

1931. Surico, G., Comai, L., and Kosuge, T. 1984. Pathogenicity of strains of Pseudomonas syringae pv. savastanoi and their indoleacetic acid-deficient mutants on olive and oleander. Phytopathology 74:490–493.

1932. Surico, G., Iacobellis, N. S., and Sisto, A. 1985. Studies on the role of indole-3–acetic acid and cytokinins in the formation of knots on olive and oleander plants by Pseudomonas syringae pv. savastanoi. Physiol. Plant Pathol. 26:309–320.

1933. Sutherland, J. R. 1977. Corky root disease of Douglas-fir seedlings: pathogenicity of the nematode Xiphinema bakeri alone and in combination with the fungus Cylindrocarpon destructans. Can. J. For. Res. 7:41–46.

1934. Sutherland, J. R., Hopkinson, S. J., and Farris, S. H. 1984. Inland spruce cone rust, Chrysomyxa pirolata, in Pyrola asarifolia and cones of Picea glauca and morphology of the spore stages. Can. J. Bot. 62:2441–2447.

1935. Sutherland, J. R., Lock, W., and Farris, S. H. 1981. Sirococcus blight: a seed-borne disease of container-grown spruce seedlings in coastal British Columbia forest nurseries. Can. J. Bot. 59:559–562.

1936. Sutherland, J. R., and Van Eerden, E. 1980. Diseases and insect pests in British Columbia forest nurseries. B.C. Minist. For./Can. For. Serv. Joint Rep. 12. 55 pp.

1937. Šutić, D., and Dowson, W. J. 1963. The reactions of olive, oleander, and ash cross-inoculated with some strains and forms of Pseudomonas syringae (Smith) Stevens. Phytopathol. Z. 46:305–314.

1938. Sutton, B. C. 1969. Forest microfungi. III. The heterogeneity of Pestalotia de Not. Section Sexloculatae Klebahn sensu Guba. Can. J. Bot. 47:2083–2094.

1939. Sutton, B. C. 1971. Coelomycetes. IV. The genus Harknessia, and similar fungi on Eucalyptus. Commonw. Mycol. Inst. Mycol. Pap. 123. 46 pp.

1940. Sutton, B. C. 1977. Coelomycetes. VI. Nomenclature of generic names proposed for coelomycetes. Commonw. Mycol. Inst. Mycol. Pap. 141. 253 pp.

1941. Sutton, B. C. 1980. The Coelomycetes. Commonwealth Mycological Institute, Kew, Surrey, England. 696 pp.

1942. Sutton, B. C., and Lawrence, J. J. 1969. Black rib of willows in Manitoba and Saskatchewan. Plant Dis. Rep. 53:101–102.

1943. Sutton, T. B. 1981. Production and dispersal of ascospores and conidia by Physalospora obtusa and Botryosphaeria dothidea in apple orchards. Phytopathology 71:584–589.

1944. Sutton, T. B., Jones, A. L., and Nelson, L. A. 1976. Factors affecting dispersal of conidia of the apple scab fungus. Phytopathology 66:1313–1317.

1945. Sutton, T. B., and Shane, W. W. 1983. Epidemiology of the perfect stage of Glomerella cingulata on apples. Phytopathology 73:1179–1183.

1946. Suzuki, K., and Kiyohara, T. 1978. Influence of water stress on development of pine wilting disease caused by Bursaphelenchus lignicolus. Eur. J. For. Pathol. 8:97–107.

1947. Swai, I. S., and Hindal, D. F. 1981. Selective medium for recovering Verticicladiella procera from soils and symptomatic white pines. Plant Dis. 65:963–965.

1948. Swart, H. J. 1973. The fungus causing cypress canker. Trans. Br. Mycol. Soc. 61:71–82.

1949. Sweet, J. B. 1980. Fruit tree virus infections of woody exotic and indigenous plants in Britain. Acta Phytopathol. Acad. Sci. Hung. 15:231–238.

1950. Sweet, J. B., and Barbara, D. J. 1979. A yellow mosaic disease of horse chestnut (Aesculus spp.) caused by apple mosaic virus. Ann. Appl. Biol. 92:335–341.

1951. Swingle, D. B., and Morris, H. E. 1918. Plum pocket and leaf gall of Americana plums. Mont. Agric. Exp. Stn. Bull. 123:167–188.

1952. Swingle, R. U. 1938. A phloem necrosis of elm. Phytopathology 28:757–759.

1953. Swingle, R. U., and Bretz, T. W. 1950. Zonate canker, a virus disease of American elm. Phytopathology 40:1018–1022.

1954. Sylvestre-Guinot, G. 1981. Etude de l'émission des ascospores du Lachnellula willkommii (Hartig) Dennis dans l'Est de la France. Eur. J. For. Pathol. 11:275–283.

1955. Sylvia, D. M., and Sinclair, W. A. 1983. Phenolic compounds and resistance to fungal pathogens induced in primary roots of Douglas-fir seedlings by the ectomycorrhizal fungus Laccaria laccata. Phytopathology 73:390–397.

1956. Szirmai, J. 1972. An Acer virus disease in maple trees planted in avenues. Acta Phytopathol. Acad. Sci. Hung. 7:197–207.

1957. Sztejnberg, A., Azaizia, H., and Chet, I. 1983. The possible role of phenolic compounds in resistance of horticultural crops to Dematophora necatrix Hartig. Phytopathol. Z. 107:318–376.

1958. Sztejnberg, A., and Madar, Z. 1980. Host range of Dematophora necatrix, the cause of white root rot disease in fruit trees. Plant Dis. 64:662–664.

1959. Sztejnberg, A., Madar, Z., and Chet, I. 1980. Induction and quantification of microsclerotia in Rosellinia necatrix. Phytopathology 70:525–527.

1960. Tainter, F. H. 1973. Development of Cronartium comandrae in Comandra umbellata. Can. J. Bot. 51:1369–1372.

1961. Tainter, F. H., and French, D. W. 1971. The role of wound periderm in the resistance of eastern larch and jack pine to dwarf mistletoe. Can. J. Bot. 49:501–504.

1962. Tainter, F. H., and Gubler, W. D. 1973. Natural biological control of oak wilt in Arkansas. Phytopathology 63:1027–1034.

1963. Tainter, F. H., and Gubler, W. D. 1974. Effect of invasion by Hypoxylon and other microorganisms on carbohydrate reserves of oak-wilted trees. For. Sci. 20:337–342.

1964. Tainter, F. H., and Ham, D. L. 1983. The survival of Ceratocystis fagacearum in South Carolina. Eur. J. For. Pathol. 13:102–109.

1965. Tainter, F. H., Williams, T. M., and Cody, J. B. 1983. Drought as a cause of oak decline and death on the South Carolina coast. Plant Dis. 67:195–197.

1966. Takai, S. 1980. Relationship of the production of the toxin, cerato-ulmin, to synnemata formation, pathogenicity, mycelial habit, and growth of Ceratocystis ulmi isolates. Can. J. Bot. 58:658–662.

1967. Takai, S., and Hiratsuka, Y. 1984. Scanning electron microscope observations of internal symptoms of white elm following Ceratocystis ulmi infection and cerato-ulmin treatment. Can. J. Bot. 62:1365–1371.

1968. Takai, S., Richards, W. C., and Stevenson, K. J. 1983. Evidence for the involvement of cerato-ulmin, the Ceratocystis ulmi toxin, in the development of Dutch elm disease. Physiol. Plant Pathol. 23:275–280.

1969. Tao, D., Li, P. H., Carter, J. V., and Ostry, M. E. 1984. Relationship of environmental stress and Cytospora chrysosperma to spring dieback of poplar shoots. For. Sci. 30:645–651.

1970. Taris, B., and Avenard, J.-C. 1969. Comparaison de l'évolution au printemps des attaques de deux Taphrinales, Taphrina aurea (Pers.) Fr. et Taphrina deformans (Berk.) Tul., et se développant sur les jeunes bouquets foliaires de Populus et de Prunus persica. C. R. Hebd. Séances Acad. Sci., Paris D 268:3078–3081.

1971. Tarjan, A. C., and Howard, F. L. 1954. Detrimental effects of copper sprays to Norway maple in Rhode Island. Plant Dis. Rep. 38:58.

1972. Tate, R. L. 1980. Detection, description, and treatment of girdling roots on urban Norway maple trees. J. Arboric. 6:168.

1973. Tate, R. L. 1981. Characteristics of girdling roots on urban Norway maples. J. Arboric. 7:268–270.

1974. Taubenhaus, J. J., and Ezekiel, W. N. 1936. A rating of plants with reference to their relative resistance or susceptibility to Phymatotrichum root rot. Tex. Agric. Exp. Stn. Bull. 527. 52 pp.

1975. Tayal, M. S., Sharma, S. M., and Agarwal, M. L. 1981. Studies on the polyphenols, proteins, chlorophylls, IAA-oxidase, and amylases of normal and false smut infected leaves of Phoenix sylvestris. Indian Phytopathol. 34:337–339.

1976. Taylor, G. S. 1983. Cryptosporiopsis canker of Acer rubrum: some relationships among host, pathogen, and vector. Plant Dis. 67:984–986.

1977. Taylor, G. S., and Moore, R. E. B. 1979. A canker of red maples associated with oviposition by the narrow-winged tree cricket. Phytopathology 69:236–239.

1978. Tehon, L. R., and Daniels, E. 1925. Notes on the parasitic fungi of Illinois—II. Mycologia 17:240–249.

1979. Tehon, L. R., and Jacks, W. R. 1933. Smooth patch, a bark lesion of white oak. J. For. 31:430–433.

1980. Tekauz, A., and Patrick, Z. A. 1974. The role of twig infections on the incidence of perennial canker of peach. Phytopathology 64:683–688.

1981. Terashita, T. 1963. Studies on the diseases of Acacia dealbata. III.

Taxonomic opinion on the anthracnose fungus. Bull. Gov. For. Exp. Stn., Tokyo No. 155. 22 pp.

1982. Terashita, T. 1973. Studies of an anthracnose fungus on broad leaved trees in Japan, with special reference to the latency of the fungus. Bull. Gov. For. Exp. Stn., Tokyo No. 252. 85 pp.

1983. Têtu-Bernier, P., Allen, E., and Hiratsuka, Y. 1983. Bibliography of western gall rust. Can. For. Serv. Inf. Rep. NOR-X-250. 10 pp.

1984. Thielges, B. A., and Adams, J. C. 1975. Genetic variation and heritability of Melampsora leaf rust resistance in eastern cottonwood. For. Sci. 21:278–282.

1985. Thies, W. G. 1983. Determination of growth reduction in Douglas-fir infected by Phellinus weirii. For. Sci. 29:305–315.

1986. Thies, W. G., and Patton, R. F. 1970. The biology and control of Cylindrocladium scoparium in Wisconsin forest tree nurseries. Phytopathology 60:1662–1668.

1987. Thirumalachar, M. J., Rao, D. V. S., and Ravindranath, V. 1950. Telia of the rust on cultivated figs. Curr. Sci. 19:27–28.

1988. Thomas, B. J. 1981. Studies on rose mosaic disease in field-grown roses produced in the United Kingdom. Ann. Appl. Biol. 98:419–429.

1989. Thomas, B. J. 1982. The effect of Prunus necrotic ringspot virus on field-grown roses. Ann. Appl. Biol. 100:129–134.

1990. Thomas, B. J. 1984. Rose mosaic disease: symptoms induced in roses by graft inoculation with both prunus necrotic ringspot and apple mosaic viruses. Plant Pathol. 33:155–160.

1991. Thomas, C. S., Hart, J. H., and Cress, C. E. 1984. Severity of Endocronartium harknessii in two provenance stands of Pinus ponderosa in Michigan. Plant Dis. 68:681–683.

1992. Thomas, D. L., and Donselman, H. M. 1979. Mycoplasmalike bodies and phloem degeneration associated with declining Pandanus in Florida. Plant Dis. Rep. 63:911–916.

1993. Thomas, G. P. 1958. Studies in forest pathology. XVIII. The occurrence of the Indian paint fungus, Echinodontium tinctorium E. & E., in British Columbia. Can. Dep. Agric. Publ. 1041. 30 pp.

1994. Thomas, H. E. 1933. The quince-rust disease caused by Gymnosporangium germinale. Phytopathology 23:546–553.

1995. Thomas, H. E. 1934. Studies on Armillaria mellea (Vahl) Quel., infection, parasitism, and host resistance. J. Agric. Res. 48:187–218.

1996. Thomas, H. E., and Mills, W. D. 1929. Three rust diseases of the apple. Cornell Univ. Agric. Exp. Stn. Mem. 123. 21 pp.

1997. Thomas, L. K. Jr. 1980. The impact of three exotic plant species on a Potomac island. U.S. Natl. Park Serv. Sci. Monogr. Ser. No. 13. 179 pp.

1998. Thompson, E. O. 1939. Morphological differences in Taphrina caerulescens upon different species of Quercus. Univ. Kans. Sci. Bull. 26:357–366.

1999. Thompson, G. E. 1939. A leaf blight of Populus tacamahaca Mill. caused by an undescribed species of Linospora. Can. J. Res. C17:232–238.

2000. Thompson, G. E. 1941. Leaf-spot diseases of poplars caused by Septoria musiva and S. populicola. Phytopathology 31:241–254.

2001. Thompson, G. E. 1963. Decay of oaks caused by Hypoxylon atropunctatum. Plant Dis. Rep. 47:202–205.

2002. Thompson, R. L., ed. 1971. The ecology and management of the red-cockaded woodpecker. U.S. Dep. Interior Bur. Sport Fish. & Wildl. 188 pp.

2003. Thomson, A. J., Alfaro, R. I., Bloomberg, W. J., and Smith, R. B. 1985. Impact of dwarf mistletoe on the growth of western hemlock trees having different patterns of suppression and release. Can. J. For. Res. 15:665–668.

2004. Thomson, S. V., Schroth, M. N., Moller, W. J., and Reil, W. O. 1982. A forecasting model for fire blight of pear. Plant Dis. 66:576–579.

2005. Thomson, V. E., and Mahall, B. E. 1983. Host specificity by a mistletoe, Phoradendron villosum (Nutt.) Nutt. subsp. villosum, on three oak species in California. Bot. Gaz. 144:124–131.

2006. Tiangco, E. S., and Varney, E. H. 1970. A vein chlorosis or yellow-net disease of Forsythia caused by tobacco ringspot virus. (Abstr.) Phytopathology 60:579.

2007. Tiedemann, G., Bauch, J., and Bock, E. 1977. Occurrence and significance of bacteria in living trees of Populus nigra L. Eur. J. For. Pathol. 7:364–374.

2008. Timmer, L. W., Garnsey, S. M., Grimm, G. R., El-Gholl, N. E., and Schoulties, C. L. 1979. Wilt and dieback of Mexican lime caused by Fusarium oxysporum. Phytopathology 69:730–734.

2009. Timonin, M. I., and Self, R. L. 1955. Cylindrocladium scoparium Morgan on azaleas and other ornamentals. Plant Dis. Rep. 39:860–863.

2010. Tinnin, R. O., and Knutson, D. M. 1980. Growth characteristics of the brooms on Douglas-fir caused by Arceuthobium douglasii. For. Sci. 26:149–158.

2011. Tinus, R. W. 1981. Root system configuration is important to long tree life. Plants Landscape 4(1):1–5.

2012. Tippett, J. T., and Shigo, A. L. 1981. Barrier zone formation: a mechanism of tree defense against vascular pathogens. IAWA Bull. n.s. 2:163–168.

2013. Tippett, J. T., and Shigo, A. L. 1981. Barriers to decay in conifer roots. Eur. J. For. Pathol. 11:51–59.

2014. Tisserat, N., and Kuntz, J. E. 1983. Dispersal gradients of conidia of the butternut canker fungus in a forest during rain. Can. J. For. Res. 13:1139–1144.

2015. Tisserat, N., and Kuntz, J. E. 1983. Longevity of conidia of Sirococcus clavigignenti-juglandacearum in a simulated airborne state. Phytopathology 73:1628–1631.

2016. Tisserat, N., and Kuntz, J. E. 1984. Butternut canker: development on individual trees and increase within a plantation. Plant Dis. 68:613–616.

2017. Tobiessen, P., and Buchsbaum, S. 1976. Ash dieback and drought. Can. J. Bot. 54:543–545.

2018. Todhunter, M. N., and Beineke, W. F. 1984. Effect of anthracnose on growth of grafted black walnut. Plant Dis. 68:203–204.

2019. Tomerlin, J. R., and Jones, A. L. 1983. Effect of temperature and relative humidity on the latent period of Venturia inaequalis in apple leaves. Phytopathology 73:51–54.

2020. Toole, E. R. 1949. Fusarium wilt of staghorn sumac. Phytopathology 39:754–759.

2021. Toole, E. R. 1952. Two races of Fusarium oxysporum f. pernicosum causing wilt of Albizzia spp. Phytopathology 42:694.

2022. Toole, E. R. 1954. Rot and cankers on oak and honeylocust caused by Poria spiculosa. J. For. 52:941–942.

2023. Toole, E. R. 1955. Polyporus hispidus on southern bottomland oaks. Phytopathology 45:177–180.

2024. Toole, E. R. 1959. Decay after fire injury to southern bottom-land hardwoods. U.S. Dep. Agric. Tech. Bull. 1189. 25 pp.

2025. Toole, E. R. 1963. Cottonwood canker caused by Fusarium solani. Plant Dis. Rep. 47:1032–1035.

2026. Toole, E. R. 1966. Stem canker of red oaks caused by Fusarium solani. Plant Dis. Rep. 50:160–161.

2027. Toole, E. R. 1966. Root rot caused by Polyporus lucidus. Plant Dis. Rep. 50:945–946.

2028. Torrey, J. G., et al. 1983. International Conference on the Biology of Frankia. Can. J. Bot. 61:2765–2966.

2029. Touliatos, P., and Roth, E. 1971. Hurricanes and trees: ten lessons from Camille. J. For. 69:285–289.

2030. Trappe, J. M., and Fogel, R. D. 1977. Ecosystematic functions of ecto-mycorrhizae. Pages 205–214 in: The belowground ecosystem: a synthesis of plant-associated processes. J. K. Marshall, ed. Colo. State Univ., Range Sci. Dep., Sci. Ser. No. 26. 351 pp.

2031. Trappe, J. M., Stahly, E. A., Benson, N. R., and Duff, D. M. 1973. Mycorrhizal deficiency of apple trees in high arsenic soils. HortScience 8:52–53.

2032. Treshow, M. 1971. Fluorides as air pollutants affecting plants. Annu. Rev. Phytopathol. 9:21–44.

2033. Treshow, M., Anderson, F. K., and Harner, F. 1967. Responses of Douglas-fir to elevated atmospheric fluorides. For. Sci. 13:114–120.

2034. Trimble, G. R. Jr., and Tryon, E. H. 1974. Grapevines a serious obstacle to timber production on good hardwood sites in Appalachia. Nor. Logger 23(5):22, 23, 44.

2035. Trione, E. J. 1959. The pathology of Phytophthora lateralis on native Chamaecyparis lawsoniana. Phytopathology 49:306–310.

2036. Tripepi, R. R., and Mitchell, C. A. 1984. Metabolic response of river birch and European birch roots to hypoxia. Plant Physiol. 76:31–35.

2037. Trolinger, J. C., Elliott, E. S., and Young, R. J. 1978. Host range of Cristulariella pyramidalis. Plant Dis. Rep. 62:710–714.

2038. True, R. P. 1938. Gall development on Pinus sylvestris attacked by the Woodgate Peridermium and morphology of the parasite. Phytopathology 28:24–49.

2039. True, R. P., Barnett, H. L., Dorsey, C. K., and Leach, J. G. 1960. Oak wilt in West Virginia. West Virginia Agric. Exp. Stn. Bull. 448T. 119 pp.

2040. True, R. P., and Tryon, E. H. 1956. Oak stem cankers initiated in the drought year 1953. Phytopathology 46:617–622.

2041. True, R. P., Tryon, E. H., and King, J. F. 1955. Cankers and decays of birch associated with two Poria species. J. For. 53:412–415.

2042. Tryon, E. H. 1971. Frost damage to tree species related to time of budbreak. Proc. West Virginia Acad. Sci. 43:1–8.

2043. Tryon, E. H., and True, R. P. 1952. Blister-shake of yellow poplar. West Virginia Agric. Exp. Stn. Bull. 350T. 15 pp.

2044. Tsoumis, G. 1965. Structural deformities in an epidemic tumor of white spruce, Picea glauca. Can. J. Bot. 43:176–181.

2045. Tsuneda, A., Hiratsuka, Y, and Maruyama, P. J. 1980. Hyperparasitism of Scytalidium uredinicola on western gall rust, Endocronartium harknessii. Can. J. Bot. 58:1154–1159.

2046. Tsuneda, I., and Kennedy, L. L. 1980. Basidiospore germination and substrate preference in Fomes fomentarius and Fomitopsis cajaderi. Mycologia 72:204–208.

2047. Tucker, C. M., and Milbrath, J. A. 1942. Root rot of Chamaecyparis caused by a species of Phytophthora. Mycologia 34:94–103.

2048. Tucker, D. P. H., Lee, R. F., Timmer, L. W., Albrigo, L. G., and Brlansky, R. H. 1984. Experimental transmission of citrus blight. Plant Dis. 68:979–980.

2049. Turner, P. D. 1965. Infection of oil palms by *Ganoderma*. Phytopathology 55:937.

2050. Tuttle, M. A., and Gotlieb, A. R. 1985. Histology of Delicious/Malling Merton 106 apple trees affected by apple union necrosis and decline. Phytopathology 75:342–347.

2051. Uchida, J. Y., Aragaki, M., and Yoshimura, M. A. 1984. Alternaria leaf spots of *Brassaia actinophylla*, *Dizygotheca elegantissima*, and *Tupidanthus calyptratus*. Plant Dis. 68:447–449.

2052. Umbach, D. M., and Davis, D. D. 1984. Severity and frequency of SO$_2$-induced leaf necrosis on seedlings of 57 tree species. For. Sci. 30:587–596.

2053. Unger, L. S. 1972. Common needle diseases of spruce in British Columbia. Can. For. Serv. Pac. For. Res. Cent. For. Insect Dis. Surv. Pest Leafl. 39. 7 pp.

2054. Upadhyay, H. P. 1981. A monograph of *Ceratocystis* and *Ceratocystiopsis*. University of Georgia Press, Athens. 176 pp.

2055. Uphof, J. C. T. 1931. *Tillandsia usneoides* als Pflanzenschädling. Z. Pflanzenkr. 41:593–607.

2056. Valentine, F. A., Manion, P. D., and Moore, K. E. 1976. Genetic control of resistance to *Hypoxylon* infection and canker development in *Populus tremuloides*. Pages 132–146 in: Proc. 12th Lake States For. Tree Impr. Conf., U.S. For. Serv. Gen. Tech. Rep. NC-26. 206 pp.

2057. Van Arsdel, E. P. 1961. Growing white pine in the Lake States to avoid blister rust. U.S. For. Serv. Lake States For. Exp. Stn. Stn. Pap. 92. 11 pp.

2058. Van Arsdel, E. P. 1967. The nocturnal diffusion and transport of spores. Phytopathology 57:1221–1229.

2059. Van Arsdel, E. P. 1972. Some cankers on oaks in Texas. Plant Dis. Rep. 56:300–304.

2060. Van Arsdel, E. P. 1979. Symptoms and conditions of environmental tree disease. Weeds Trees Turf 18(6):16–20, 23, 26, 28–29.

2061. Van Arsdel, E. P. 1980. Managing trees to reduce damage from low-level saline irrigation. Weeds Trees Turf 19(6):26–28, 61.

2062. Van Arsdel, E. P., and Chitzanidis, A. 1970. Life cycle and spread of ash rust in Texas. (Abstr.) Phytopathology 60:1317.

2063. Van Arsdel, E. P., Riker, A. J., and Patton, R. F. 1956. Effects of temperature and moisture on the spread of white pine blister rust. Phytopathology 46:307–318.

2064. Van der Kamp, B. J., Gokhale, A. A., and Smith, R. S. 1979. Decay resistance owing to near-anaerobic conditions in black cottonwood wetwood. Can. J. For. Res. 9:39–44.

2065. Van Dyke, C. G., and Amerson, H. V. 1976. Interactions of *Puccinia sparganioides* with smooth cordgrass (*Spartina alterniflora*). Plant Dis. Rep. 60:670–674.

2066. Van Hook, J. M., and Busteed, R. C. 1935. Anthracnose of *Betula nigra*. Proc. Indiana Acad. Sci. 44:81.

2067. Van Sickle, G. A. 1969. Occurrence of *Cronartium comptoniae* in the Maritime Provinces. Plant Dis. Rep. 53:369–371.

2068. Van Sickle, G. A. 1973. A survey of production losses due to witches' broom of blueberry in the Maritime Provinces. Plant Dis. Rep. 57:608–611.

2069. Van Sickle, G. A. 1973. A quantitative survey for needle rust of balsam fir. Plant Dis. Rep. 57:765–766.

2070. Van Sickle, G. A. 1974. Growth loss caused by a needle rust (*Pucciniastrum goeppertianum*) of balsam fir. Can. J. For. Res. 4:138–140.

2071. Van Sickle, G. A. 1974. Nectria canker: a problem on black locust in New Brunswick. Plant Dis. Rep. 58:872–874.

2072. Van Sickle, G. A. 1975. Basidiospore production and infection of balsam fir by a needle rust, *Pucciniastrum goeppertianum*. Can. J. Bot. 53:8–17.

2073. Van Sickle, G. A. 1977. Seasonal periodicity in the discharge of *Pucciniastrum geoppertianum* basidiospores. Can. J. Bot. 55:745–751.

2074. Van Warmelo, K. T., and Sutton, B. C. 1981. Coelomycetes VII. *Stegonsporium*. Commonw. Mycol. Inst. Mycol. Pap. 145. 45 pp.

2075. Vassey, W. E., Gould, C. J., and Ryan, G. F. 1977. Disease resistant pyracantha for the Pacific Northwest. Ornamentals Northwest 1977–78 (Dec–Jan):4–6.

2076. Verrall, A. F., and May, C. 1937. A new species of *Dothiorella* causing die-back of elm. Mycologia 29:321–324.

2077. Vlamis, J., and Raabe, R. D. 1985. Copper deficiency of manzanita grown in a bark-sand mixture. HortScience 20:61–62.

2078. Voorhees, R. K. 1942. Life history and taxonomy of the fungus *Physalospora rhodina*. Fla. Agric. Exp. Stn. Bull. 371. 91 pp.

2079. Wadsworth, F. H. 1943. Lightning damage in ponderosa pine stands of northern Arizona. J. For. 41:684–685.

2080. Wagener, W. W. 1925. Mistletoe in the lower bole of incense cedar. Phytopathology 15:614–616.

2081. Wagener, W. W. 1939. The canker of *Cupressus* induced by *Coryneum cardinale* n. sp. J. Agric. Res. 58:1–46.

2082. Wagener, W. W. 1949. Top dying of conifers from sudden cold. J. For. 47:49–53.

2083. Wagener, W. W. 1957. The limitation of two leafy mistletoes of the genus *Phoradendron* by low temperatures. Ecology 38:142–145.

2084. Wagener, W. W. 1958. Infection tests with two rusts of Jeffrey pine. Plant Dis. Rep. 42:888–892.

2085. Wagener, W. W., and Davidson, R. W. 1954. Heart rots in living trees. Bot. Rev. 20:61–134.

2086. Wagener, W. W., and Mielke, J. L. 1961. A staining-fungus root disease of ponderosa, Jeffrey, and pinyon pines. Plant Dis. Rep. 45:831–835.

2087. Wainhouse, D. 1980. Dispersal of first instar larvae of the felted beech scale, *Cryptococcus fagisuga*. J. Appl. Ecol. 17:523–532.

2088. Wainwright, S. H., and Lewis, F. H. 1970. Developmental morphology of the black knot pathogen on plum. Phytopathology 60:1238–1244.

2089. Walker, J. T. 1971. *Dothiorella*, associated with cankers on Katsura-tree, causes wood discoloration in seedling trees. Plant Dis. Rep. 55:893–895.

2090. Walker, L. C. 1956. Foliage symptoms as indicators of potassium-deficient soils. For. Sci. 2:113–120.

2091. Wall, R. E., and Kuntz, J. E. 1964. Water-soluble substances in dead branches of aspen (*Populus tremuloides* Michx.) and their effects on *Fomes igniarius*. Can. J. Bot. 42:969–977.

2092. Wall, R. E., and Magasi, L. P. 1976. Environmental factors affecting Sirococcus shoot blight of black spruce. Can. J. For. Res. 6:448–452.

2093. Walla, J. A., and Riffle, J. W. 1981. *Fomes fraxinophilus* on green ash in North Dakota windbreaks. Plant Dis. 65:669–670.

2094. Walla, J. A., and Stack, R. W. 1980. Dip treatment for control of blackstem on *Populus* cuttings. Plant Dis. 64:1092–1095.

2095. Wallis, G. W. 1976. Growth characteristics of *Phellinus* (*Poria*) *weirii* in soil and on root and other surfaces. Can. J. For. Res. 6:229–232.

2096. Wallis, G. W., and Bloomberg, W. J. 1981. Estimating the total extent of *Phellinus weirii* root rot centers using above- and below-ground disease indicators. Can. J. For. Res. 11:827–830.

2097. Wallis, G. W., and Reynolds, G. 1965. The initiation and spread of *Poria weirii* root rot of Douglas-fir. Can. J. Bot. 43:1–9.

2098. Walter, J. M. 1946. Canker stain of planetrees. U.S. Dep. Agric. Circ. 742. 12 pp.

2099. Walter, J. M., Rex, E. G., and Schreiber, R. 1952. The rate of progress and destructiveness of canker stain of planetrees. Phytopathology 42:236–239.

2100. Walters, J. W., Hinds, T. E., Johnson, D. W., and Beatty, J. 1982. Effects of partial cutting on diseases, mortality, and regeneration of Rocky Mountain aspen stands. U.S. For. Serv. Res. Pap. RM-240. 12 pp.

2101. Wang, C.-G., Blanchette, R. A., Jackson, W. A., and Palmer, M. A. 1985. Differences in conidial morphology among isolates of *Sphaeropsis sapinea*. Plant Dis. 69:838–841.

2102. Ward, J. C., and Pong, W. Y. 1980. Wetwood in trees: a timber resource problem. U.S. For. Serv. Gen. Tech. Rep. PNW-112. 56 pp.

2103. Ware, G. H. 1982. Decline in oaks associated with urbanization. Pages 61–64 in: Urban and surburban trees: pest problems, needs, prospects, and solutions. B. O. Parks, F. A. Fear, M. T. Lambur, and G. A. Simmons, eds. Mich. State Univ. 253 pp.

2104. Wargo, P. M. 1977. *Armillariella mellea* and *Agrilus bilineatus* and mortality of defoliated oak trees. For. Sci. 23:485–492.

2105. Wargo, P. M. 1981. Defoliation and secondary-action organism attack: with emphasis on *Armillaria mellea*. J. Arboric. 7:64–69.

2106. Wargo, P. M., and Houston, D. R. 1974. Infection of defoliated sugar maple trees by *Armillaria mellea*. Phytopathology 64:817–822.

2107. Wargo, P. M., Parker, J., and Houston, D. R. 1972. Starch content in roots of defoliated sugar maple. For. Sci. 18:203–204.

2108. Wargo, P. M., and Shaw, C. G. III. 1985. Armillaria root rot: the puzzle is being solved. Plant Dis. 69:826–832.

2109. Watanabe, T., Smith, R. S. Jr., and Snyder, W. C. 1970. Populations of *Macrophomina phaseoli* in soil as affected by fumigation and cropping. Phytopathology 60:1717–1719.

2110. Waterman, A. M. 1943. *Diplodia pinea*, the cause of a disease of hard pines. Phytopathology 33:1018–1031.

2111. Waterman, A. M. 1947. *Rhizosphaera kalkhoffii* associated with a needle cast of *Picea pungens*. Phytopathology 37:507–511.

2112. Waterman, A. M. 1954. Septoria canker of poplars in the United States. U.S. Dep. Agric. Circ. 947. 24 pp.

2113. Waterman, A. M. 1955. The relation of *Valsa kunzei* to cankers on conifers. Phytopathology 45:686–692.

2114. Waterman, A. M. 1957. Canker and dieback of poplars caused by *Dothichiza populea*. For. Sci. 3:175–183.

2115. Waterman, A. M., and Marshall, R. P. 1947. A new species of *Cristulariella* associated with a leaf spot of maple. Mycologia 39:690–698.

2116. Waters, C. W. 1962. Significance of life history studies of *Elytroderma deformans*. For. Sci. 8:250–254.

2117. Waterworth, H. E., and Lawson, R. H. 1973. Purification, electron microscopy, and serology of the dogwood ringspot strain of cherry leafroll virus. Phytopathology 63:141–146.

2118. Waterworth, H. E., Lawson, R. H., and Monroe, R. L. 1976. Purification and properties of Hibiscus chlorotic ringspot virus. Phytopathology 66:570–575.

2119. Waterworth, H. E., and Povish, W. R. 1972. Tobacco ringspot virus from naturally infected dogwood, autumn crocus, and forsythia. Plant Dis. Rep. 56:336–337.

2120. Waterworth, H. E., and Povish, W. R. 1977. A yellow leafspot disease of *Ilex crenata* caused by tobacco ringspot virus. Plant Dis. Rep. 61:104–105.

2121. Watling, R., Kile, G. A., and Gregory, N. M. 1982. The genus *Armillaria*—nomenclature, typification, the identity of *Armillaria mellea*, and species differentiation. Trans. Br. Mycol. Soc. 78:271–285.

2122. Watson, A. J. 1941. Studies of *Botryosphaeria ribis* on *Cercis* and *Benzoin*. Plant Dis. Rep. 25:29–31.

2123. Waxman, S. 1975. Witches'-brooms sources of new and interesting dwarf forms of *Picea*, *Pinus*, and *Tsuga* species. Acta Hortic. 54:25–32.

2124. Wean, R. E. 1937. The parasitism of *Polyporus schweinitzii* on seedling *Pinus strobus*. Phytopathology 27:1124–1142.

2125. Weaver, D. J. 1974. Effect of root injury on the invasion of peach roots by *Clitocybe tabescens*. Mycopathol. Mycol. Appl. 52:313–317.

2126. Weaver, D. J. 1978. Interaction of *Pseudomonas syringae* and freezing in bacterial canker on excised peach twigs. Phytopathology 68:1460–1463.

2127. Weaver, D. J. 1979. Role of conidia of *Botryosphaeria dothidea* in the natural spread of peach tree gummosis. Phytopathology 69:330–334.

2128. Webber, J. 1981. A natural biological control of Dutch elm disease. Nature 292:449–451.

2129. Webber, J. F., and Brasier, C. M. 1984. The transmission of Dutch elm disease: a study of the processes involved. Pages 271–306 in: Invertebrate-microbial interactions. J. M. Anderson, A. D. M. Rayner, and D. W. H. Walton, eds. Cambridge University Press. 349 pp.

2130. Weber, G. F. 1957. Cold injury to young pine trees. Plant Dis. Rep. 41:494–495.

2131. Weber, G. F., and Roberts, D. A. 1951. Silky threadblight of *Elaeagnus pungens* caused by *Rhizoctonia ramicola* n. sp. Phytopathology 41:615–621.

2132. Webster, R. K., and Butler, E. E. 1967. A morphological and biological concept of the species *Ceratocystis fimbriata*. Can. J. Bot. 45:1457–1468.

2133. Wehlburg, C., and Cox, R. S. 1966. Rhizoctonia leaf blight of azalea. Plant Dis. Rep. 50:354–355.

2134. Wehmeyer, L. E. 1933. The genus *Diaporthe* Nitschke and its segregates. University of Michigan Press, Ann Arbor. 349 pp.

2135. Wehmeyer, L. E. 1941. A revision of *Melanconis*, *Pseudovalsa*, *Prosthecium*, and *Titania*. University of Michigan Press, Ann Arbor. 161 pp.

2136. Weidensaul, T. C., Leben, C., and Ellett, C. W. 1977. Reducing decay losses in hardwood forests and farm woodlots. Ohio State Univ. Coop. Exten. Serv. Bull. 629. 14 pp.

2137. Weidensaul, T. C., and Wood, F. A. 1973. Sources of species of *Fusarium* in northern hardwood forests. Phytopathology 63:367–371.

2138. Weimer, J. L. 1917. The origin and development of the galls produced by two cedar rust fungi. Am. J. Bot. 4:241–251.

2139. Weimer, J. L. 1917. Three cedar rust fungi, their life histories, and the diseases they produce. Cornell Univ. Agric. Exp. Stn. Bull. 390:509–549.

2140. Weingartner, D. P., and Klos, E. J. 1975. Etiology and symptomatology of canker and dieback diseases on highbush blueberries caused by *Godronia* (*Fusicoccum*) *cassandrae* and *Diaporthe* (*Phomopsis*) *vaccinii*. Phytopathology 65:105–110.

2141. Weir, J. R. 1916. *Keithia thujina*, the cause of a serious leaf disease of the western red cedar. Phytopathology 6:360–363.

2142. Weir, J. R. 1921. Note on *Cenangium abietis* (Pers.) Rehm on *Pinus ponderosa* Laws. Phytopathology 11:166–170.

2143. Weir, J. R., and Hubert, E. E. 1918. Notes on the overwintering of forest tree rusts. Phytopathology 8:55–59.

2144. Weiss, F., and Smith, F. F. 1940. A flower-spot disease of cultivated azaleas. U.S. Dep. Agric. Circ. 556. 28 pp.

2145. Welch, B. L., and Martin, N. E. 1974. Invasion mechanisms of *Cronartium ribicola* in *Pinus monticola* bark. Phytopathology 64:1541–1546.

2146. Welch. D. S. 1934. The range and importance of Nectria canker on hardwoods in the Northeast. J. For. 32:997–1002.

2147. Wells, B. W. 1939. A new forest climax: the salt spray climax of Smith Island, N.C. Bull. Torrey Bot. Club 66:629–634.

2148. Wells, O. O., and Dinus, R. J. 1978. Early infection as a predictor of mortality associated with fusiform rust of southern pines. J. For. 76:8–12.

2149. Wene, E. G., and Schoeneweiss, D. F. 1980. Localized freezing predisposition to Botryosphaeria canker in differentially frozen woody stems. Can. J. Bot. 58:1455–1458.

2150. Wenner, J. J. 1914. A contribution to the morphology and life history of *Pestalozzia funerea* Desm. Phytopathology 4:375–383.

2151. Wensley, R. N. 1964. Occurrence and pathogenicity of *Valsa* (*Cytospora*) species and other fungi associated with peach canker in southern Ontario. Can. J. Bot. 42:841–857.

2152. Werner, A., and Siwecki, R. 1978. Histological studies of infection processes by *Dothichiza populea* Sacc. et Briard in susceptible and resistant poplar clones. Eur. J. For. Pathol. 8:217–226.

2153. West, E. 1933. Powdery mildew of crape myrtle caused by *Erysiphe lagerstroemiae* n. sp. Phytopathology 23:814–819.

2154. Wester, H. V., Davidson, R. W., and Fowler, M. E. 1950. Cankers of linden and redbud. Plant Dis. Rep. 34:219–223.

2155. Wester, H. V., and Jylkka, E. W. 1959. Elm scorch, graft transmissible virus of American elm. Plant Dis. Rep. 43:519.

2156. Wester, H. V., and Jylkka, E. W. 1963. High incidence of Dutch elm disease in American elms weakened by elm scorch associated with breeding attacks of *Scolytus multistriatus*. Plant Dis. Rep. 47:545–547.

2157. Westing, A. H. 1966. Sugar maple decline: an evaluation. J. Econ. Bot. 20:196–212.

2158. Westing, A. H. 1969. Plants and salt in the roadside environment. Phytopathology 59:1174–1179.

2159. Westwood, M. N., and Cameron, H. R. 1978. Environment-induced remission of pear decline symptoms. Plant Dis. Rep. 62:176–179.

2160. Whetzel, H. H. 1945. A synopsis of the genera and species of Sclerotiniaceae, a family of stromatic inoperculate Discomycetes. Mycologia 37:648–714.

2161. Whigham, D. 1984. The influence of vines on the growth of *Liquidambar styraciflua* L. (sweetgum). Can. J. For. Res. 14:37–39.

2162. Whitcomb, C. E. 1986. Solving the iron chlorosis problem. J. Arboric. 12:44–48.

2163. Whitcomb, R. F., and Tully, J. G., eds. 1979. The mycoplasmas. Vol. III. Plant and insect mycoplasmas. Academic Press, New York. 351 pp.

2164. White, B. L., and Merrill, W. 1969. Pathological anatomy of *Abies balsamea* infected with *Melampsorella caryophyllacearum*. Phytopathology 59:1238–1242.

2165. White, J. H. 1919. On the biology of *Fomes applanatus* (Pers.) Wallr. Trans. R. Can. Inst. 12:133–174.

2166. White, O. E. 1948. Fasciation. Bot. Rev. 14:319–358.

2167. White, P. M. 1972. Plant tolerance for standing water: an assessment. Cornell Plantations 28:50–52.

2168. White, P. R., and Millington, W. F. 1954. The structure and development of a woody tumor affecting *Picea glauca*. Am. J. Bot. 41:353–361.

2169. White, P. R., Tsoumis, G., and Hyland, F. 1967. Some seasonal aspects of the growth of tumors on the white spruce, *Picea glauca*. Can. J. Bot. 45:2229–2232.

2170. White, R. P. 1930. Pathogenicity of *Pestalotia* spp. N.J. Agric. Exp. Stn. Annu. Rep. 43 (1929–30):264–268.

2171. White, R. P. 1930. Pathogenicity of *Pestalotia* spp. on rhododendron. Phytopathology 20:85–91.

2172. White, R. P. 1937. Rhododendron wilt and root rot. N.J. Agric. Exp. Stn. Bull. 615. 32 pp.

2173. White, R. P., and McCullock, L. 1934. A bacterial disease of *Hedera helix*. J. Agric. Res. 48:807–815.

2174. Whiteside, J. O. 1970. Etiology and epidemiology of citrus greasy spot. Phytopathology 60:1409–1414.

2175. Whiteside, J. O. 1971. Some factors affecting the occurrence and development of foot rot on citrus trees. Phytopathology 61:1233–1238.

2176. Whiteside, J. O. 1972. Histopathology of citrus greasy spot and identification of the causal fungus. Phytopathology 62:260–263.

2177. Whiteside, J. O. 1974. Environmental factors affecting infection of citrus leaves by *Mycosphaerella citri*. Phytopathology 64:115–120.

2178. Whiteside, J. O. 1981. Aberrant behavior of *Mycosphaerella citri* on freeze-killed citrus leaf tissue and its taxonomic and epidemiologic implications. Phytopathology 71:1108–1110.

2179. Whitney, H. E., and Johnson, W. C. 1984. Ice storms and forest succession in southwestern Virginia. Bull. Torrey Bot. Club 111:429–437.

2180. Whitney, R. D. 1962. Studies in forest pathology. XXIV. *Polyporus tomentosus* Fr. as a major factor in stand-opening disease of white spruce. Can. J. Bot. 40:1631–1658.

2181. Whitney, R. D. 1966. Germination and inoculation tests with basidiospores of *Polyporus tomentosus*. Can. J. Bot. 44:1333–1343.

2182. Whitney, R. D. 1977. *Polyporus tomentosus* root rot of conifers. Can. For. Serv. For. Tech. Rep. 18. 11 pp.

2183. Whitney, R. D., and Bohaychuk, W. P. 1976. Pathogenicity of *Polyporus tomentosus* and *P. tomentosus* var. *circinatus* on seedlings of 11 conifer species. Can. J. For. Res. 6:129–131.

2184. Wick, R. L., and Moore, L. D. 1983. Histopathology of root disease incited by *Thielaviopsis basicola* in *Ilex crenata*. Phytopathology 73:561–564.

2185. Wicker, E. F. 1965. A Phomopsis canker on western larch. Plant Dis. Rep. 49:102–105.

2186. Wicker, E. F. 1981. Natural control of white pine blister rust by *Tuberculina maxima*. Phytopathology 71:997–1000.

2187. Wicker, E. F., Laurent, T. H., and Israelson, S. 1978. Sirococcus shoot blight damage to western hemlock regeneration at Thomas Bay, Alaska. U.S. For. Serv. Res. Pap. INT-198. 11 pp.

2188. Wicker, E. F., and Wells, J. M. 1983. Intensification and lateral spread of *Arceuthobium laricis* in a young stand of western larch with stocking control. Can. J. For. Res. 13:314–319.

2189. Wicker, E. F., and Yakota, S. 1976. On the *Cronartium* stem rust(s) of five-needle pines in Japan. Ann. Phytopathol. Soc. Jap. 42:187–191.

2190. Widin, K. D., and Schipper, A. L. Jr. 1980. Epidemiology of *Melampsora medusae* leaf rust of poplars in the north central United States. Can. J. For. Res. 10:257–263.

2191. Widin, K. D., and Schipper, A. L. Jr. 1981. Effect of *Melampsora medusae* leaf rust infections on yield of hybrid poplars in the north-central United States. Eur. J. For. Pathol. 11:438–448.

2192. Wiens, D. 1964. Revision of the acataphyllous species of *Phoradendron*. Brittonia 16:11–54.

2193. Wiersum, L. K., and Harmanny, K. 1983. Changes in the water-permeability of roots of some trees during drought stress and recovery, as related to problems of growth in urban environment. Plant Soil 75:443–448.

2194. Wikström, C. 1976. Occurrence of *Phellinus tremulae* (Bond.) Bond. and Borisov as a primary parasite in *Populus tremula* L. Eur. J. For. Pathol. 6:321–328.

2195. Wikström, C., and Unestam, T. 1976. Decay pattern of *Phellinus tremulae* (Bond.) Bond. et Borisov in *Populus tremula* L. Eur. J. For. Pathol. 6:291–301.

2196. Wilhelm, S., and Taylor, J. B. 1965. Control of Verticillium wilt of olive through natural recovery and resistance. Phytopathology 55:310–316.

2197. Wilhour, R. G. 1970. The influence of ozone on white ash (*Fraxinus americana* L.). Penn. State Univ., Cent. Air Environ. Stud. Publ. 188-71. 86 pp.

2198. Wilkins, W. H. 1936, 1938. Studies in the genus *Ustulina* with special reference to parasitism. II. A disease of the common lime (*Tilia vulgaris* Hayne) caused by *Ustulina*. III. Spores—germination and infection. Trans. Br. Mycol. Soc. 20:133–156; 22:47–93.

2199. Wilkinson, T. G., and Barnes, R. L. 1973. Effects of ozone on $^{14}CO_2$ fixation patterns in pine. Can. J. Bot. 51:1573–1578.

2200. Wilks, D. S., Gersper, P. L., and Cobb, F. W. Jr. 1985. Association of soil moisture with spread of *Ceratocystis wageneri* in ponderosa pine disease centers. Plant Dis. 69:206–208.

2201. Williams, E. B., and Kuc, J. 1969. Resistance in *Malus* to *Venturia inaequalis*. Annu. Rev. Phytopathol. 7:223–246.

2202. Wills, W. H., and Lambe, R. C. 1978. Pathogenicity of *Thielaviopsis basicola* from Japanese holly (*Ilex crenata*) to some other host plants. Plant Dis. Rep. 62:1102–1106.

2203. Wilson, C. L. 1962. Brooming and galling of shagbark hickory in Arkansas. Plant Dis. Rep. 46:448–450.

2204. Wilson, C. L. 1963. Wilting of persimmon caused by *Cephalosporium diospyri*. Phytopathology 53:1402–1406.

2205. Wilson, C. L. 1965. Consideration of the use of persimmon wilt as a silvicide for weed persimmons. Plant Dis. Rep. 49:789–791.

2206. Wilson, E. E. 1928. Studies of the ascigerous stage of *Venturia inaequalis* (Cke.) Wint. in relation to certain factors of the environment. Phytopathology 18:375–418.

2207. Wilson, E. E. 1935. The olive knot disease: its inception, development, and control. Hilgardia 9:233–264.

2208. Wilson, E. E. 1947. The branch wilt of Persian walnut trees and its cause. Hilgardia 17:413–436.

2209. Wilson, E. E. 1965. Pathological histogenesis in oleander tumors induced by *Pseudomonas savastanoi*. Phytopathology 55:1244–1249.

2210. Wilson, E. E., and Magie, A. R. 1964. Systemic invasion of the host plant by the tumor-inducing bacterium *Pseudomonas savastanoi*. Phytopathology 54:576–579.

2211. Wilson, E. E., and Ogawa, J. M. 1979. Fungal, bacterial, and certain nonparasitic diseases of fruit and nut crops in California. Univ. Calif., Div. Agric. Sci., Berkeley. 190 pp.

2212. Wilson, E. E., Zeitoun, F. M., and Fredrickson, D. L. 1967. Bacterial phloem canker, a new disease of Persian walnut trees. Phytopathology 57:618–621.

2213. Wilson, M., and Henderson, D. M. 1966. British rust fungi. Cambridge University Press, Cambridge. 324 pp.

2214. Wingfield, M. J. 1983. Association of *Verticicladiella procera* and *Leptographium terebrantis* with insects in the Lake States. Can. J. For. Res. 13:1238–1245.

2215. Wingfield, M. J., and Blanchette, R. A. 1982. Association of pine wood nematode with stressed trees in Minnesota, Iowa, and Wisconsin. Plant Dis. 66:934–937.

2216. Wingfield, M. J., and Blanchette, R. A. 1983. The pine-wood nematode, *Bursaphelenchus xylophilus*, in Minnesota and Wisconsin: insect associates and transmission studies. Can. J. For. Res. 13:1068–1076.

2217. Wingfield, M. J., Blanchette, R. A., and Kondo, E. 1983. Comparison of the pine wood nematode, *Bursaphelenchus xylophilus*, from pine and balsam fir. Eur. J. For. Pathol. 13:360–370.

2218. Wingfield, M. J., and Knox-Davies, P. S. 1980. Association of *Diplodia pinea* with a root disease of pines in South Africa. Plant Dis. 64:221–223.

2219. Winkler, A. J., ed. 1949. Pierce's disease investigations. Hilgardia 19:207–264.

2220. Wisniewski, M., Bogle, A. L., and Wilson, C. L. 1984. Histopathology of canker development on peach trees after inoculation with *Cytospora leucostoma*. Can. J. Bot. 62:2804–2813.

2221. Witcosky, J. J., and Hansen, E. M. 1985. Root-colonizing insects recovered from Douglas-fir in various stages of decline due to black-stain root disease. Phytopathology 75:399–402.

2222. Wolf, F. A. 1912. Some fungous diseases of the prickly pear, *Opuntia lindheimeri* Engelm. Ann. Mycol. 10:113–134.

2223. Wolf, F. A. 1912. A new *Gnomonia* on hickory leaves. Ann. Mycol. 10:488–491.

2224. Wolf, F. A. 1926. The perfect stage of the fungus which causes melanose of citrus. J. Agric. Res. 33:621–625.

2225. Wolf, F. A. 1929. The relationship of *Microstroma juglandis* (Bereng.) Sacc. J. Elisha Mitchell Sci. Soc. 45:130–136.

2226. Wolf, F. A. 1930. A parasitic alga, *Cephaleuros virescens* Kunze, on citrus and certain other plants. J. Elisha Mitchell Sci. Soc. 45:187–205.

2227. Wolf, F. A. 1938. Life histories of two leaf-inhabiting fungi on sycamore. Mycologia 30:54–63.

2228. Wolf, F. A. 1939. Leafspot of ash and *Phyllosticta viridis*. Mycologia 31:258–266.

2229. Wolf, F. A., and Barbour, W. J. 1941. Brown-spot needle disease of pines. Phytopathology 31:61–74.

2230. Wolf, F. A., and Cavaliere, A. R. 1965. Two new species of leafblight fungi on *Kalmia latifolia*. Mycologia 57:576–582.

2231. Wolf, F. A., and Davidson, R. W. 1941. Life cycle of *Piggotia fraxini*, causing leaf disease of ash. Mycologia 33:526–539.

2232. Wolf, F. T., and Wolf, F. A. 1939. A study of *Botryosphaeria ribis* on willow. Mycologia 31:217–227.

2233. Wolf, F. T., and Wolf, F. A. 1952. Pathology of *Camellia* leaves infected by *Exobasidium camelliae* var. *gracilis* Shirai. Phytopathology 42:147–149.

2234. Wolswinkel, P., Ammerlaan, A., and Peters, H. F. C. 1984. Phloem unloading of amino acids at the site of attachment of *Cuscuta europaea*. Plant Physiol. 75:13–20.

2235. Woo, J. Y., and Martin, N. E. 1981. Scanning electron microscopy of *Cronartium ribicola* infecting *Ribes* leaves. Eur. J. For. Pathol. 11:7–15.

2236. Woo, J. Y., and Partridge, A. D. 1969. The life history and cytology of *Rhytisma punctatum* on bigleaf maple. Mycologia 61:1085–1095.

2237. Wood, F. A., and French, D. W. 1963. *Ceratocystis fimbriata*, the cause of a stem canker of quaking aspen. For. Sci. 9:232–235.

2238. Wood, F. A., and French, D. W. 1965. *Hypoxylon* canker of aspen: seasonal development of cankers and ascospore ejection in winter. Phytopathology 55:771–774.

2239. Woolhouse, H. W. 1983. Toxicity and tolerance in the responses of plants to metals. Pages 245–300 in: Physiological plant ecology. III. Responses to the chemical and biological environment. O. L. Lange, P.S. Nobel, C. B. Osmond, and H. Ziegler, eds. Springer-Verlag, Berlin. 799 pp.

2240. Working Group on International Cooperation in Forest Disease Research, Section 24, Forest Protection, International Union of Forestry Research Organizations. 1963. Internationally dangerous forest tree diseases. U.S. Dep. Agric. Misc. Publ. 939. 122 pp.

2241. Worley, R. E., Littrell, R. L., and Dutcher, J. D. 1980. A comparison of tree trunk injection and implantation of zinc capsules for correction of zinc deficiency. J. Arboric. 6:253–257.

2242. Wormald, H. 1954. The brown rot diseases of fruit trees. G.B. Minist. Agric. Fish. Tech. Bull. No. 3. 113 pp.

2243. Worrall, J. J., and Parmeter, J. R. Jr. 1982. Formation and properties of wetwood in white fir. Phytopathology 72:1209–1212.

2244. Worrall, J. J., and Parmeter, J. R. Jr. 1983. Inhibition of wood-decay fungi by wetwood of white fir. Phytopathology 73:1140–1145.

2245. Worrall, J. J., Parmeter, J. R. Jr., and Cobb, F. W. Jr. 1983. Host specialization of *Heterobasidion annosum*. Phytopathology 73:304–307.

2246. Wright, E. 1942. *Cytospora abietis*, the cause of a canker of true firs in California and Nevada. J. Agric. Res. 65:143–153.

2247. Wright, E. 1957. Cytospora canker of Rocky Mountain Douglas-fir. Plant Dis. Rep. 41:811–813.

2248. Wright, E., and Wells, H. R. 1948. Tests on the adaptability of trees and shrubs to shelterbelt planting on certain Phymatotrichum root rot infested soils of Oklahoma and Texas. J. For. 46:256–262.

2249. Wutscher, H. K., Cohen, M., and Young, R. H. 1977. Zinc and water-soluble phenolics levels in the wood for the diagnosis of citrus blight. Plant Dis. Rep. 61:572–576.

2250. Wyman, D. 1939. Salt water injury of woody plants resulting from the hurricane of September 21, 1938. Arnold Arbor. Bull. Popular Inf. Ser. 4, Vol. 7:45–51.

2251. Wyss, U. 1982. Virus-transmitting nematodes: feeding behaviour and effect on root cells. Plant Dis. 66:639–644.

2252. Yakota, S. 1975. Scleroderris canker of Todo-fir in Hokkaido, northern Japan. III. Dormant infection of the causal fungus. IV. An analysis of climatic data associated with the outbreak. Eur. J. For. Pathol. 5:7–12, 13–21.

2253. Yakota, S., Uozumi, T., and Matsuzaki, S. 1974. Scleroderris canker of

Todo-fir in Hokkaido, northern Japan. I. Present status of damage and features of infected plantations. II. Physiological and pathological characteristics of the causal fungus. Eur. J. For. Pathol. 4:65–74, 155–166.

2254. Yang, Y.-S., Skelly, J. M., and Chevone, B. I. 1982. Clonal response of eastern white pine to low doses of O_3, SO_2, and NO_2, singly and in combination. Can. J. For. Res. 12:803–808.

2255. Yarwood, C. E. 1957. Powdery mildews. Bot. Rev. 23:235–300.

2256. Yarwood, C. E. 1981. The occurrence of *Chalara elegans*. Mycologia 73:524–530.

2257. Yarwood, C. E., and Gardner, M. W. 1972. Powdery mildews favored by man. Plant Dis. Rep. 56:852–855.

2258. Yde-Andersen, A. 1979. Host spectrum, host morphology, and geographic distribution of larch canker, *Lachnellula willkommii*. A literature review. Eur. J. For. Pathol. 9:211–219.

2259. Yde-Anderson, A. 1979. Disease symptoms, taxonomy, and morphology of *Lachnellula willkommii*. A literature review. Eur. J. For. Pathol. 9:220–228.

2260. Yde-Anderson, A. 1979. *Lachnellula willkommii*—canker formation and the role of microflora. A literature review. Eur. J. For. Pathol. 9:347–355.

2261. Yde-Andersen, A. 1980. Infection process and the influence of frost damage in *Lachnellula willkommii*. A literature review. Eur. J. For. Pathol. 10:28–36.

2262. Yeoman, M. M., Kilpatrick, D. C., Miedzybrodzka, M. B., and Gould, A. R. 1978. Cellular interactions during graft formation in plants: a recognition phenomenon? Symp. Soc. Exp. Biol. 32:139–160.

2263. York, H. H., Wean, R. E., and Childs, T. W. 1936. Some results of investigations on *Polyporus schweinitzii* Fr. Science 84:160–161.

2264. Young, D. J., and Alcorn, S. M. 1984. Latent infection of *Euphorbia lathyris* and weeds by *Macrophomina phaseolina* and propagule populations in Arizona field soil. Plant Dis. 68:587–589.

2265. Young, J. M., Dye, D. W., Bradbury, J. F., Panagopoulos, C. G., and Robbs, C. F. 1978. A proposed nomenclature and classification for plant pathogenic bacteria. N.Z. J. Agric. Res. 21:153–177.

2266. Zabel, R. A. 1976. Basidiocarp development in *Inonotus obliquus* and its inhibition by stem treatments. For. Sci. 22:431–437.

2267. Zalasky, H. 1964. The histopathology of *Macrophoma tumefaciens* infections in black poplar. Can. J. Bot. 42:385–391.

2268. Zalasky, H. 1964. Nomenclature and description of *Diplodia tumefaciens* (Shear) Zalasky [≡*Macrophoma tumefaciens* Shear apud Hubert]. Can. J. Bot. 42:1049–1055.

2269. Zalasky, H. 1965. Process of *Ceratocystis fimbriata* infection in aspen. Can. J. Bot. 43:1157–1162.

2270. Zalasky, H. 1968. Penetration and initial establishment of *Nectria galligena* in aspen and peachleaf willow. Can. J. Bot. 46:57–60.

2271. Zalasky, H. 1975. Chimeras, hyperplasia, and hypoplasia in frost burls induced by low temperature. Can. J. Bot. 53:1888–1898.

2272. Zalasky, H. 1975. Low-temperature-induced cankers and burls in test conifers and hardwoods. Can. J. Bot. 53:2526–2535.

2273. Zalasky, H. 1976. Xylem in galls of lodgepole pine caused by western gall rust, *Endocronartium harknessii*. Can. J. Bot. 54:1586–1590.

2274. Zalasky, H. 1978. Stem and leaf spot infections caused by *Septoria musiva* and *S. populicola* on poplar seedlings. Phytoprotection 59:43–50.

2275. Zalasky, H. 1980. Lodgepole pine (*Pinus contorta* Dougl. var. *latifolia* Engelm.) shoot abnormalities from frost injury. Can. For. Serv. Bi-mon. Res. Notes 36:21–22.

2276. Zeikus, J. G., and Ward, J. C. 1974. Methane formation in living trees: a microbial origin. Science 184:1181–1183.

2277. Zentmyer, G. A. 1980. *Phytophthora cinnamomi* and the diseases it causes. Monogr. 10. American Phytopathological Society, St. Paul, MN. 96 pp.

2278. Ziller, W. G. 1955–1970. Studies of western tree rusts. II. *Melampsora occidentalis* and *M. albertensis*, two needle rusts of Douglas-fir. V. The rusts of hemlock and fir caused by *Melampsora epitea*. VI. The aecial host ranges of *Melampsora albertensis*, *M. medusae*, and *M. occidentalis*. VII. Inoculation experiments with pine stem rusts (*Cronartium* and *Endocronartium*). VIII. Inoculation experiments with conifer needle rusts (Melampsoraceae). Can. J. Bot. 33:177–188; 37:109–119; 43:217–230; 48:1313–1319; 48:1471–1476.

2279. Zheng, R. 1985. Genera of the Erysiphaceae. Mycotaxon 22:209–263.

2280. Ziller, W. G. 1968. Studies of hypodermataceous needle diseases. I. *Isthmiella quadrispora* sp. nov. causing needle blight of alpine fir. Can. J. Bot. 46:1377–1381.

2281. Ziller, W. G. 1974. The tree rusts of western Canada. Can. For. Serv. Publ. 1329. 272 pp.

2282. Ziller, W. G., and Funk, A. 1973. Studies of hypodermataceous needle diseases. III. The association of *Sarcotrochila macrospora* n. sp. and *Hemiphacidium longisporum* n. sp. with pine needle cast caused by *Davisomycella ampla* and *Lophodermella concolor*. Can. J. Bot. 51:1959–1963.

2283. Zoeten, G. A. de, Lauritis, J. A., and Mircetich, S. M. 1982. Cytopathology and properties of cherry leaf roll virus associated with walnut blackline disease. Phytopathology 72:1261–1265.

2284. Zwet, T. van der, and Keil, H. L. 1979. Fire blight. A bacterial disease of rosaceous plants. U.S. Dep. Agric. Agric. Handb. 510. 200 pp.

Many diseases and disorders are indexed under the name of the plant genus or the plant species but not under both. To locate all information about a disease or symptom, search under both the genus and species names. A page number in **boldface type** indicates that a relevant illustration appears on the page or on the facing page.

Abelia (abelia): *Meloidogyne* 298, root knot nematode 298

Abies (fir): *Aleurodiscus amorphus* 168, *Antennatula pinophila* 30, *Apostrasseria* 144, *Armillaria* 308, black mold 30, black stain root disease 368, blight 60, 138, *Botrytis cinerea* 60, brown felt blight 50, butt crack 308, butt rot 318–322, 328, canker 144, 168, 194, 350, *Cerrena unicolor* 358, *Cylindrocladium* 294, *Cytospora* 194, *Diaporthe eres* 144, dieback 144, 194, dwarf mistletoe 426, 432, *Echinodontium tinctorium* 348, *Epipolaeum abietis* 30, *Fomitopsis pinicola* 348, *Ganoderma applanatum* 334, *G. oregonense* 330, herbicide tolerance 460, *Herpotrichia juniperi* 50, *Inonotus* 318–320, *I. dryadeus* 328, *I. hispidus* 356, *Lachnellula* 236, *Laetiporus sulfureus* 346, *Lophodermium* 32, *Macrophomina phaseolina* 292, *Melampsora* 256–258, *Melampsorella caryophyllacearum* 264, *Milesina* 264, needle cast 32, 40, *Nothophacidium phyllophilum* 52, ozone tolerance 466, *Phacidium abietis* 52, *P. balsamicola* 144, *Phaeocryptopus nudus* 40, *Phaeolus schweinitzii* 322, *Phellinus pini* 350, *P. weirii* 324, *Phomopsis juniperovora* 138, *Pucciniastrum epilobii* 260, *P. goeppertianum* 262, *P. pustulatum* 262, root rot 292–294, 308, 322–324, rust 258–264, *Sarcotrochila balsameae* 52, *Schizophyllum commune* 358, *Sclerophoma phyllophila* 138, snow blight 52, sooty mold 30, sulfur dioxide tolerance 468, *Uredinopsis* 264, *Valsa* 194, *Verticicladiella procera* 370, *V. wageneri* 368, wetwood 382, witches'-broom 264, wood rot 318–324, 328–330, 334, 346–350, 356–358

A. alba (silver fir): brown felt blight 50, *Herpotrichia juniperi* 50, needle cast 42, *Rhizosphaera kalkhoffii* 42, *Sphaeropsis sapinea* 136, tip blight 136

A. amabilis (Pacific silver fir): brown felt blight 50, butt rot 318, 322, canker 194, dwarf mistletoe 426, 432, *Echinodontium tinctorium* 348, *Herpotrichia juniperi* 50, *Inonotus* 318, *Melampsorella caryophyllacearum* 264, *Phacidium abietis* 52, *Phaeolus schweinitzii* 322, *Phellinus pini* 350, *P. weirii* 324, *Pucciniastrum epilobii* 260, *P. goeppertianum* 262, root rot 318, 322–324, rust 260–264, snow blight 52, *Valsa* 194, witches'-broom 264, wood rot 318, 322–324, 348–350

A. balsamea (balsam fir): *Apostrasseria* 144, *Ascocalyx abietina* 232, brown felt blight 50, butt rot 318, 322, canker **144**, **194**–196, 236, *Cenangium* 230, *Cylindrocladium* 294, *Cytospora* **194**–196, *Diaporthe eres* 144, dieback **144,**

230, 236, dwarf mistletoe 434, flood tolerance 486, *Herpotrichia juniperi* 50, *Inonotus* 318, *Isthmiella faullii* 42, *Lachnellula* 236, *Leucostoma kunzei* 196, *Lophophacidium hyperboreum* 52, *Melampsorella caryophyllacearum* **264,** needle cast 42, nutrient deficiency **472,** *Phacidium abietis* 52, *Phaeolus schweinitzii* 322, *Phellinus pini* 350, *Phomopsis* **144,** pine wood nematode 380, *Pucciniastrum epilobii* 260, *P. goeppertianum* 262, root rot 294, 318, 322, rust 260–**264,** salt tolerance 454, snow blight 52, *Uredinopsis* **264,** *Valsa abietis* 194, witches'-broom **264,** wood rot 318, 322, 350

A. concolor (white fir): air pollution damage 462, black stain root disease 368, brown felt blight 50, canker 194, drought injury 478, dwarf mistletoe 426, 432, *Fomitopsis pinicola* 348, graft incompatibility 492, herbicide injury **456,** *Herpotrichia juniperi* 50, *Heterobasidion annosum* 314–316, *Inonotus dryadeus* 328, *Macrophomina phaseolina* 292, *M. occidentalis* 258, *Melampsorella caryophyllacearum* 264, mistletoe **424,** *Phacidium abietis* 52, *Phaeolus schweinitzii* 322, *Phellinus pini* 350, *P. weirii* 324, *Phoradendron bolleanum* **424,** *Pucciniastrum epilobii* 260, *P. goeppertianum* 262, root rot 292, 314, 322–324, 328, rust 256–264, salt tolerance 454, shoot blight 134, *Sirococcus conigenus* 134, snow blight 52, *Valsa abietis* 194, *Verticicladiella wageneri* 368, wetwood 382, witches'-broom 264, wood rot 314–316, 322–324, 328, 348–350

A. fraseri (Fraser fir): canker 236, *Cylindrocladium* 294, dieback **236,** *Lachnellula agassizii* **236,** *Macrophomina phaseolina* 292, *Phaeolus schweinitzii* 322, *Phytophthora cinnamomi* **288**–290, root rot **288**–294, 322, sulfur dioxide tolerance 468, wood rot 322

A. grandis (grand or lowland white fir): black mold **30,** black stain root disease 368, brown felt blight 50, butt rot 318–322, **328,** canker 194, canker rot **350,** drought injury 478, dwarf mistletoe 426, 432–434, *Echinodontium tinctorium* 348, *Epipolaeum abietis* **30,** fluoride tolerance 470, *Fomitopsis pinicola* 348, frost injury **478,** *Herpotrichia juniperi* 50, *Inonotus dryadeus* **328,** *I. hispidus* 356, *I. tomentosus* 318–**320,** *Lophodermium decorum* 32, *Melampsora medusae* 256, *Melampsorella caryophyllacearum* 264, needle cast 32, *Phacidium abietis* 52, *Phaeolus schweinitzii* 322, *Phellinus pini* **350,** *Pucciniastrum epilobii* **260,** *P. goeppertianum* 262, root rot 318, **320,** 322, 328, rust 256, **260**–264, snow blight 52, *Valsa* 194, *Verticicladiella wageneri* 368, wetwood 382, witches'-broom 264, wood rot 318–322, **328,** 348–**350,** 356

A. lasiocarpa (alpine or subalpine fir): brown felt blight 50–**52,** butt rot 318, 322, canker 194, dwarf mistletoe 426, 432–434, *Echinodontium tinctorium*

348, hail injury **488,** *Herpotrichia juniperi* 50–**52,** *Inonotus* 318, *Isthmiella quadrispora* 42, *Melampsorella caryophyllacearum* 264, needle cast 42, *Phacidium abietis* **52,** *Phaeolus schweinitzii* 322, *Phellinus weirii* 324, *Pucciniastrum epilobii* 260, *P. goeppertianum* 262, root rot 318, 322–324, rust 260–**264,** snow blight **52,** *Valsa abietis* 194, witches'-broom 264, wood rot 318, 322–324, 348

A. lasiocarpa var. *arizonica* (cork fir): *Arceuthobium douglasii* 432, dwarf mistletoe 432

A. magnifica (California red fir): *Arceuthobium abietinum* 426, brown felt blight 50, butt rot 322, canker 194, dwarf mistletoe 426, *Echinodontium tinctorium* 348, *Fomitopsis pinicola* 348, *Herpotrichia juniperi* 50, *Macrophomina phaseolina* 292, *Melampsorella caryophyllacearum* **264,** *Phaeolus schweinitzii* 322, *Phellinus weirii* 324, *Pucciniastrum goeppertianum* 262, root rot 292, 322–324, rust 262–**264,** *Valsa abietis* 194, witches'-broom **264,** wood rot 322–324, 348

A. magnifica var. *shastensis* (Shasta fir): *Phellinus pini* 350, root injury from waterlogged soil 486, wood rot 350

A. procera (noble fir): *Arceuthobium abietinum* 426, *A. tsugense* 426, 432, brown felt blight 50, butt rot 318, canker 194, dieback 136, 194, dwarf mistletoe 426, 432, *Herpotrichia juniperi* 50, *Inonotus* 318, *Melampsorella caryophyllacearum* 42, *Phaeolus schweinitzii* 322, *Phellinus pini* 350, *Pucciniastrum epilobii* 260, *P. goeppertianum* 262, *Rhizosphaera kalkhoffii* 42, root rot 318, 322, rust 260–264, *Sphaeropsis sapinea* 136, twig blight 136, *Valsa abietis* 194, witches'-broom 264, wood rot 318, 322, 350

A. sachalinensis (Sakhalin or Todo fir): *Ascocalyx abietina* 232, canker 232

A. sibirica (Siberian fir): needle cast 42, *Rhizosphaera kalkhoffii* 42

Acacia (acacia): anthracnose 122–124, cassytha 436, *Coniothyrium* 82, *Erysiphe polygoni* 16, *Ganoderma applanatum* 334, *Glomerella cingulata* 122–124, powdery mildew 16, *Schizophyllum commune* 358, *Thyronectria austro-americana* 204, twig dieback 82

A. auriculiformis (acacia): algal leaf spot 28, *Cephaleuros virescens* 28

A. baileyana (Cootamundra wattle): *Oxyporus latemarginatus* 328, wood rot 328

A. cyanophila (beach acacia or orange wattle): anthracnose **124,** *Colletotrichum gloeosporioides* 124, *Glomerella cingulata* 124

A. farnesiana (sweet acacia): *Ganoderma lucidum* 332, root rot 332

A. koa (koa): *Phaeolus schweinitzii* 322, root rot 322

A. longifolia (Sydney golden wattle): *Oxyporus latemarginatus* 328, wood rot 328

A. pendula (weeping myall): *Phymatotrichum omnivorum* 292, root rot 292

A. pubescens (hairy wattle): graft incompatibility 492

Acanthopanax sieboldianus: *Alternaria panax* 90, leaf blight 90

Acer (maple): *Alternaria* 90, anthracnose 104, bark rot 168, bleeding canker 286, *Botryosphaeria dothidea* 172, *B. obtusa* 176, butt rot 304, 334, canker 144, 166, 176, 186, 202, 210–212, 220, 286, canker rot 356–358, *Cerrena unicolor* 358, *Climacodon septentrionalis* 344, compartmentalization of wound-associated discolored and decayed wood 336–338, *Cristulariella* 62, *Cryptodiaporthe hystrix* 104, *C. salicella* 186, decline 444, *Diaporthe* 144, *D. eres* 144, dieback 144, 172, 176, 210, *Discella acerina* 104, *Discula* 104, *Eutypella parasitica* 220, *Fomes fomentarius* 346, *Fomitopsis pinicola* 348, *Ganoderma applanatum* 334, *G. lucidum* 332, girdling roots 488, *Globifomes graveolens* 346, *Hendersonia* 82, *Heterobasidion annosum* 314, *Hypoxylon* 226, *H. atropunctatum* 224, *H. deustum* 304, *H. mammatum* 222, *Inonotus glomeratus* 356, *Kabatiella apocrypta* 104, *Laetiporus sulfureus* 346, leaf blight 62, 104, leaf blister 22, leaf spot 54, 76, 90, 104, lightning injury 490, *Microsphaera aceris* 16, mistletoe 422, *Nectria cinnabarina* 210, *N. galligena* 212, nematodes 302, *Oxyporus populinus* 344, ozone tolerance 466, *Perenniporia medulla-panis* 168, *Phellinus igniarius* 342, *Phomopsis* 144, 146, *P. dubia* 144, *P. mali* 144, *Phoradendron serotinum* 422, *Phyllactinia guttata* 16, *Phytophthora cactorum* 166, 286, *Polyporus squamosus* 344, powdery mildew 16, *Pratylenchus* 302, *Rhytisma* 54, salt tolerance 454, *Schizophyllum commune* 358, sulfur dioxide tolerance 468, *Taphrina* 22, tar spot 54, *Valsa* 202, Verticillium wilt 374, wetwood 382, winter sunscald 484, wood decay 168, 304, 314, 334–338, 342–348, 356–358, *Xiphinema* 302

A. campestre (hedge maple): *Didymosporina aceris* 104, fluoride tolerance 470, leaf spot 54, 76, 104, *Phyllosticta minima* 76, *Rhytisma acerinum* 54, salt tolerance 454, sulfur dioxide tolerance 468, tar spot 54

A. circinatum (vine maple): *Cristulariella depraedens* 62, *Hypoxylon mediterraneum* 224, leaf blight 62, *Rhytisma punctatum* 54, tar spot 54

A. ginnala (Amur maple): herbicide tolerance 460, leaf spot 76, *Phyllosticta minima* 76, *Pseudomonas syringae* pv. *syringae* 160, *Rhytisma acerinum* 54, salt tolerance 454, tar spot 54, tobacco ringspot virus 414

A. glabrum (Rocky Mountain maple): butt rot 328, canker 198, *Cristulariella depraedens* 62, dieback 198, *Inonotus dryadeus* 328, leaf blight 62, *Polyporus squamosus* 344, *Rhytisma* 54, root rot 328, sulfur dioxide tolerance 468, tar spot 54, *Valsa sordida* 198, wood rot 328, 344

547

A. macrophyllum (bigleaf maple): butt rot 304, 328, canker **212,** *Didymosporina aceris* 104, herbicide injury **458,** herbicide tolerance 460, *Hypoxylon deustum* 304, *H. mammatum* 222, *Inonotus dryadeus* 328, leaf spot **54,** 104, *Nectria galligena* **212,** *Polyporus squamosus* 344, *Rhytisma* **54,** root rot 304, 328, tar spot **54,** wood rot 304, 328, 344
A. negundo (box elder): anthracnose 104, butt rot 304, canker 202, **220,** climbing bittersweet **438,** *Coniothyrium* 82, *Cristulariella moricola* **62,** *Didymosporina aceris* 104, dieback 202, *Eutypella parasitica* **220,** flood tolerance 486, fluoride tolerance 470, *Hendersonia* 82, herbicide injury 456–**458,** herbicide tolerance 460, *Hypoxylon deustum* 304, ice glaze injury 490, *Inonotus glomeratus* 356, *Kabatiella apocrypta* 104, leaf blight **62,** 104, leaf spot **76,** 104, maple mosaic virus 494, *Oxyporus populinus* 344, *Phyllosticta negundinis* **76,** *Polyporus squamosus* 344, *Rhytisma* 54, root rot 304, tar spot 54, tobacco ringspot virus 414, twig dieback 82, *Valsa ambiens* 202, wetwood 382, wood rot 304, 344, 356
A. nigrum, see *A. saccharum*
A. palmatum (Japanese maple): anthracnose **104,** *Cristulariella depraedens* 62, *Kabatiella apocrypta* 104, leaf blight 62, **104,** leaf spot **76,** 160, *Phyllosticta minima* **76,** *Pseudomonas syringae* pv. *syringae* 160, tip blight 160
A. pensylvanicum (striped maple): anthracnose 104, *Cristulariella depraedens* 62, *Discula* 104, frost susceptibility 482, leaf blight 62, 104, *Rhytisma punctatum* **54,** tar spot **54**
A. platanoides (Norway maple): adventitious roots **490,** adventitious shoots **494,** anthracnose 104, *Armillaria* 308, bleeding canker 286, burl **494,** canker 198, **202, 212**–214, **220,** 286, chimera 492, *Cristulariella* 62, *Didymosporina aceris* **104,** dieback 198, **202,** *Eutypella parasitica* **220,** flood tolerance 486, fluoride tolerance 470, freeze injury **484,** *Fusarium solani* 214, *Ganoderma lucidum* 332, girdling roots 488, hail injury **488,** herbicide tolerance 460, ice injury **488**–490, *Kabatiella apocrypta* 104, leaf blight 62, 104, leaf scorch **454,** leaf spot **104,** 160, lightning injury **490,** *Nectria galligena* **212,** ozone tolerance 466, pesticide injury 460, *Phytophthora cactorum* 286, *Pseudomonas syringae* pv. *syringae* 160, *Rhytisma acerinum* 54, root rot **308,** 332, salt damage **454,** tar spot 54, tip blight 160, *Valsa ambiens* **202,** *V. sordida* 198, Verticillium wilt **374**
A. pseudoplatanus (sycamore maple): bleeding canker 286, canker **220,** 286, *Cristulariella* 62, *Eutypella parasitica* **220,** leaf blight 62, leaf spot 54, 76, maple mosaic virus 494, ozone tolerance 466, *Phyllosticta minima* 76, *Phytophthora cactorum* 286, *Rhytisma* 54, salt tolerance 454, tar spot 54
A. rubrum (red maple): anthracnose 104, bacterial leaf scorch 384, bleeding canker 286, butt rot 304, canker 166, **202,** 214, **220,** 236, 286, canker-rot **358,** *Ceratocystis coerulescens* 362, *Cerrena unicolor* **358,** compartmentalization of wound-associated discolored and decayed wood **336–338,** *Cristulariella* 62, *Cryphonectria parasitica* 186, *Cryptosporiopsis* 166, **202,** dieback **202, 488,** *Eutypella parasitica* **220,** flood tolerance 486, frost susceptibility 482, *Fusarium solani* 214, gall 146, *Ganoderma* 332, girdling roots **488,** herbicide tolerance 460, *Hypoxylon deustum* 304, ice glaze injury 490, *Kabatiella apocrypta* 104, leaf blight 62, 104, leaf blotch **96,** leaf spot 76, 160, leaf scorch 386, nutrient deficiency **474,** *Oxyporus latemarginatus* 328, *O. populinus* 344, ozone tolerance 466, *Phomopsis* 146, *Phyllosticta minima* 76, *Phytophthora cactorum* 286, *Pseudomonas syringae* pv. *syringae* 160, *Rhytisma* **54,** root rot 304, 328, 332, salt damage **452**–454, sapstreak **362,** stem crack **334,** *Strumella coryneoidea* 236, tar spot **54,** tip blight 160, *Urnula craterium* 236, *Valsa* **202,** *Venturia acerina* **96,** Verticillium wilt **374,** wood rot 328, 332, **334, 336,** 344, 358
A. saccharinum (silver maple): anthracnose 104, *Armillaria tabescens* 312, bleeding canker 286, canker 192, 202, 214, 220, 286, *Cristulariella moricola* **62,** dieback 202, *Endothia gyrosa* 192, *Eutypella parasitica* 220, flood tolerance 486, fluoride tolerance 470, *Fusarium solani* 214, *Ganoderma lucidum* 332, herbicide injury **460,** herbicide tolerance 460, *Hericium erinaceus* 344, ice glaze injury 490, *Kabatiella apocrypta* 104, leaf blight **62,** 104, leaf blotch 96, leaf spot **76,** mistletoe 422, ozone injury 466, *Phoradendron serotinum* 422, *Phyllosticta minima* **76,** *Phytophthora cactorum* 286, *Rhytisma* 54, root rot 312, 332, salt tolerance 454, tar spot 54, *Valsa* 202, *Venturia acerina* 96, wood rot 312, 332, 344
A. saccharum (sugar maple), including subsp. *nigrum* (black maple): *Aleurodiscus oakesii* 168, anthracnose **104,** *Armillaria* **308,** 444, *Bjerkandera adusta* 346, bleeding canker 286, *Botryosphaeria obtusa* 444, burl **494,** butt rot 304, **334,** canker 144, 198, **202, 210**–214, **220,** 286, **344,** canker-rot 356–**358,** *Cerrena unicolor* **358, 444,** *Climacodon septentrionalis* **344,** compartmentalization of wound-associated discolored and decayed wood **338,** *Cristulariella* 62, decline **444,** *Dendrothele acerina* 168, *Diaporthe dubia* 144, dieback 144, **202,** 210, **362, 444,** *Discula* **104,** *Eutypella parasitica* **220,** flood tolerance 486, frost susceptibility 482, *Fusarium solani* 214, gall 146, 494, *Ganoderma applanatum* **334, 494,** *G. lucidum* 332, 444, hail injury **488,** herbicide injury 456, herbicide tolerance 460, *Hericium erinaceus* 344, *Hypoxylon deustum* 304, ice glaze injury 490, *Inonotus glomeratus* 356, *Kabatiella apocrypta* 104, leaf blight 62, **104,** leaf blotch 96, leaf scorch 478, leaf spot **76, 104,** *Nectria cinnabarina* **210,** 444, *N. coccinea* 442, *N. galligena* 212, **344,** nematode 302, *Oxyporus latemarginatus* 328, *O. populinus* **344,** ozone tolerance 466, *Phomopsis dubia* 144, *Phyllosticta minima* **76,** *Phytophthora cactorum* 286, *Polyporus squamosus* 344, *Rhytisma* 54, root injury from waterlogged soil **486,** root rot 304–**308,** 328, 332, salt damage 452–454, sapstreak **362,** sapwood stain **362,** *Schizophyllum commune* 358, slime flux **382,** smooth patch 168, *Stegonosporium* 444, tar spot 54, tobacco mosaic virus 414, *Trametes versicolor* 342, *Valsa* 444, *V. ambiens* **202,** *V. sordida* 198, *Venturia acerina* 96, *Verticillium dahliae* **374,** 444, 454, Verticillium wilt **374,** wood rot 304, 328, 332–334, **338,** 342–344, 356–**358,** *Xiphinema americanum* 302, *Xylaria* 306
A. spicatum (mountain maple): anthracnose 104, *Cristulariella depraedens* 62, *Kabatiella apocrypta* 104, leaf blight 62, 104, leaf blotch 96, leaf spot 76, *Phyllosticta minima* 76, *Rhytisma* 54, tar spot 54, *Venturia acerina* 96
A. tataricum (Tatarian maplè): anthracnose 104, *Kabatiella apocrypta* 104, leaf blight 104, leaf spot 76, *Phyllosticta minima* 76, salt tolerance 454
Aceria snetsingeri **20**
Acoelorrhaphe wrightii (Everglades, paurotis, or saw cabbage palm): false smut 92, *Ganoderma lucidum* 332, *Graphiola phoenicis* 92, leaf spot 92, root rot 332, *Stigmina palmivora* 92
Acremonium diospyri **372**
actinorhizae **502**
adventitious shoots and roots **494**
Aeglopsis chevalieri: greasy spot 72, *Mycosphaerella citri* 72
Aesculus (buckeye, horse-chestnut): bleeding canker 286, *Botryosphaeria dothidea* 172, *B. obtusa* 176, butt rot 304, 334, canker 172, 176, 286, dieback 172, 176, *Ganoderma applanatum* 334, *Guignardia aesculi* 76, 80, herbicide tolerance 460, *Heterobasidion annosum* 314, *Hypoxylon deustum* 304, leaf blotch 76, 80, lightning injury 490, mistletoe 422, *Oxyporus populinus* 344, *Phyllosticta sphaeropsoidea* 76, 80, *Phytophthora cactorum* 286, root rot 304, Verticillium wilt 376, wood rot 304, 314, 334, 344
A. californica (California buckeye): bacterial leaf scorch **384,** *Guignardia aesculi* 80, leaf blotch 80, mistletoe 422, *Phoradendron villosum* 422, *Phyllactinia guttata* 16, powdery mildew 16, *Taphrina aesculi* 22, yellow leaf blister 22
A. ×carnea (red horse-chestnut): apple mosaic virus 412, canker 144, *Diaporthe ambigua* 144, dieback 144, *Guignardia aesculi* 80, leaf blotch 80, *Phomopsis ambigua* 144, salt tolerance 454
A. glabra (Ohio buckeye): *Guignardia aesculi* 80, leaf blotch 80, ozone tolerance 466
A. hippocastanum (horse-chestnut): anthracnose 122–124, apple mosaic virus 412, canker **166,** 210–212, *Cerrena unicolor* 358, dieback 210, drought injury 478, *Glomerella cingulata* 122–124, *Guignardia aesculi* 76, **80,** herbicide tolerance 460, leaf blotch 76, **80,** *Nectria cinnabarina* 210, *N. galligena* 212, *Oxyporus populinus* 344, *Phyllosticta sphaeropsoidea* 76, 80, *Polyporus squamosus* 344, salt tolerance 452–454, *Schizophyllum commune* 358, sulfur dioxide tolerance 468, wood decay 344, 358
A. octandra (yellow buckeye): flood tolerance 486, *Guignardia aesculi* 80, leaf blotch 80, *Polyporus squamosus* 344, sulfur dioxide tolerance 468, wood decay 344
A. pavia (red buckeye): *Guignardia aesculi* 80, leaf blotch 80
A. turbinata (Japanese horse-chestnut): *Guignardia aesculi* 80, leaf blotch 80
Aethaloderma 30
Agave (agave, century plant): anthracnose 122–124, *Coniothyrium concentrica* 82, *Glomerella cingulata* 122–124, leaf spot 82, *Leptosphaeria obtusispora* 82, root rot 296, salt tolerance 454, *Sclerotium rolfsii* 296, southern blight 296
Agrobacterium pseudotsugae 154
A. radiobacter 156
A. rhizogenes 156
A. rubi 156
A. tumefaciens **156**
Ailanthus altissima (tree-of-heaven): *Botryosphaeria dothidea* 172, *B. obtusa* 176, *B. rhodina* 180, butt rot 334, canker 172, 176, 180, 210, 214, *Cerrena unicolor* 358, *Coniothyrium* 82, dieback 172, 176, 180, 210, fluoride tolerance 470, *Fusarium lateritium* 214, *Ganoderma applanatum* 334, herbicide tolerance 460, ice glaze injury 490, ozone tolerance 466, *Nectria cinnabarina* 210, *Phymatotrichum omnivorum* 292, root rot 292, salt tolerance 454, *Schizophyllum commune* 358, sulfur dioxide tolerance 468, twig dieback 82, Verticillium wilt 376, wood rot 334, 358
Aiphanes lindeniana (palm): lethal yellowing 390
air pollution damage **462–470**
Ajuga reptans (bugleweed): root rot **296,** *Sclerotium rolfsii* **296,** southern blight **296**
Albizia julibrissin (mimosa): alcoholic flux **166,** *Botryosphaeria dothidea* 172, *B. rhodina* 180, canker 172, 180, 210, 214, *Coniothyrium* 82, dieback 172, 180, 204, 210, *Fusarium lateritium* 214, *F. oxysporum* f. sp. *pernicosum* 166, **378,** Fusarium wilt **378,** *Ganoderma* 332, herbicide tolerance 460, *Oxyporus latemarginatus* 328, ozone tolerance 466, *Nectria cinnabarina* 210, root rot 328, 332, salt tolerance 454, *Schizophyllum commune* 358, *Thyronectria austro-americana* 204, twig dieback 82, wilt **378,** wood rot 328, 332, 358
A. lebbek (lebbek): gall 150, *Nectriella pironii* 150
A. procera: Fusarium wilt 378
alcoholic flux **166**
alder, see *Alnus*
black or European: *A. glutinosa*
Italian: *A. cordata*
mountain or thinleaf: *A. tenuifolia*
red: *A. oregona*
smooth or speckled: *A. rugosa*
Aleurites fordii (tung-oil tree): anthracnose 122–124, *Armillaria tabescens* 312, *Botryosphaeria dothidea* 172, *B. rhodina* 180, canker 180, *Cristulariella moricola* 62, dieback 122, 172, 180, *Glomerella cingulata* 122–124, leaf spot 62, *Schizophyllum commune* 358, root rot 296, 312, *Sclerotium rolfsii* 296, southern blight 296, wood rot 312, 358
Aleurodiscus 168
A. oakesii **168**
algal leaf spot **28**
algal pathogens **28**
Allagoptera arenaria (seashore palm): lethal yellowing 390
almond, see *Prunus, P. dulcis*
desert: *P. fasciculata*
flowering: *P. triloba*
Russian dwarf: *P. tenella*
Alnus (alder): *Botryosphaeria dothidea* 172, *B. obtusa* 176, butt rot 304, 334, canker 172, 176, 186, 210–212, *Cerrena unicolor* 358, *Cristulariella moricola* 62, *Cryptodiaporthe salicella* 186, dieback 172, 176, 210, *Fomes fomentarius* 346, *Fomitopsis pinicola* 348, *Ganoderma applanatum* 334, *Heterobasidion annosum* 314, *Hypoxylon deustum* 304, *H. mammatum* 222, *Inonotus tomentosus* 318, leaf spot 62, *Melampsoridium* 254, *Microsphaera penicillata* 16, *Nectria cinnabarina* 210, *N. galligena* 212, nitrogen-fixing nodules **502,** *Oxyporus latemarginatus* 328, ozone tolerance 466, *Phellinus igniarius* 342, *Phyllactinia guttata* 16, powdery mildew 16, root rot 304, rust 254, *Schizophyllum commune* 358, *Scorias spongiosa* **30,** sooty mold **30,** *Stereum gausapatum* 344, *Taphrina* 22, wood rot 304, 314, 318, 328, 334, 342–348, 358
A. cordata (Italian alder): salt tolerance 454
A. glutinosa (black or European alder): fluoride tolerance 470, ozone tolerance 466, salt tolerance 454, sulfur dioxide tolerance 468
A. oregona (red alder): blight 60, *Botrytis cinerea* 60, herbicide tolerance 460, *Hypoxylon mediterraneum* 224, *Inonotus obliquus* 352, *Nectria ditissima* 442

A. rugosa (smooth or speckled alder): flood tolerance 486, frost susceptibility 482, salt tolerance 454

A. tenuifolia (mountain or thinleaf alder): sulfur dioxide tolerance 468

Alternaria 90

A. panax 90

Amanita muscaria 502

A. pantherina 502

Amazonia 30

Amelanchier (serviceberry, shadbush): Apiosporina collinsii 20, black witches'-broom 20, blight 126, 162, Botryosphaeria obtusa 176, canker 198, 202, 210–212, Cerrena unicolor 358, Cladosporium 20, Cytospora 198, dieback 176, 198, 202, 244, Diplocarpon mespili 64, Erwinia amylovora 162, Erysiphe polygoni 16, fire blight 162, fluoride tolerance 470, frost susceptibility 482, Gymnosporangium 242–248, G. nidus-avis 244, Hendersonia 82, herbicide tolerance 460, Heterobasidion annosum 314, leaf spot 64, 244, Leucostoma 198, Monilinia amelanchieris 126, Nectria cinnabarina 210, N. galligena 212, powdery mildew 16, rust 242, 244–248, Schizophyllum commune 358, sulfur dioxide tolerance 468, Taphrina amelanchieri 22, Valsa ambiens 202, Verticillium wilt 376, witches'-broom 22

A. alnifolia (western serviceberry): ozone tolerance 466, sulfur dioxide tolerance 468

A. arborea (downy serviceberry): Cristulariella moricola 62, leaf spot 62

A. canadensis: Apiosporina collinsii 20, black witches'-broom 20, Oxyporus latemarginatus 328, wood rot 328

A. laevis (Allegheny serviceberry): salt tolerance 454

A. utahensis (Utah serviceberry): sulfur dioxide tolerance 468

ammonia injury 470

Amorpha (false indigo): Botryosphaeria obtusa 176

Ampelopsis (ampelopsis): Botryosphaeria obtusa 176, dieback 176, 210, downy mildew 12, Guignardia bidwellii 80, leaf spot 80, Nectria cinnabarina 210, as a pest 438, Plasmopara viticola 12, powdery mildew 14, Uncinula necator 14

A. arborea (pepper vine): fastidious xylem-inhabiting bacteria 384

Amsonia: Coleosporium apocynaceum 268, rust 268

anatto, see Bixa orellana

Ancistrocactus (hook cactus): Pseudomonas fluorescens 158, soft rot 158

Andromeda (bog rosemary): Exobasidium vaccinii 26

andromeda, Japanese, see Pieris japonica

antelope bush, see Purshia

Antennatula pinophila 30

anthracnose 68, 104–122

Aphelandra squarrosa (zebra plant): gall 148, Nectriella pironii 148

Apiognomonia errabunda 104–106, 110–112

A. quercina 110

A. veneta 110–112

Apioplagiostoma populi 122

Apiosporina collinsii 20

A. morbosa 152

Apostrasseria 144

apple, see Malus
common: M. pumila, M. sylvestris
plum-leaved: M. prunifolia

apple flat limb disease 412

apple mosaic virus 410–412, 448

apricot, see Prunus armeniaca

arabis mosaic virus 404, 414–416

Aralia (aralia): Alternaria panax 90, leaf blight 90

A. spinosa (Hercules club): Botryosphaeria dothidea 172, dieback 172, Oxyporus latemarginatus 328

aralia
false, see Dizogtheca elegantissima
geranium-leaf: Polyscias guilfoylei
Ming: P. fruticosa

Araucaria: Botryosphaeria dothidea 172, dieback 172

A. araucana (monkey puzzle): Pestalotiopsis funerea 128, salt tolerance 454

A. bidwillii (bunya-bunya): anthracnose 122–124, Glomerella cingulata 122–124, Pestalotiopsis funerea 128, Schizophyllum commune 358, wood rot 358

A. cunninghamii (hoop pine): Sphaeropsis sapinea 136, tip blight 136

A. heterophylla (Norfolk Island pine): Botryosphaeria rhodina 180, dieback 180, salt tolerance 454

arborvitae, see Platycladus, Thuja
eastern: T. occidentalis
giant: T. plicata
Japanese: T. standishii
Oriental: P. orientalis

Arbutus menziesii (madrone): Aleurodiscus diffisus 168, bleeding canker 286, Botryosphaeria dothidea 172, Coniothyrium 82, dieback 168, 172, Exobasidium vaccinii 26, Hendersonula toruloidea 168, herbicide tolerance 460, Heterobasidion annosum 314, leaf spot 82, Phellinus igniarius 342, Phytophthora cactorum 286, Rhytisma arbuti 54, root rot 314, smooth patch 188, tar spot 54, wood rot 314, 342

Arceuthobium 426–434

A. americanum 426–430

A. apachecum 426, 432–434

A. campylopodum 428, 432–434

A. douglasii 426, 430–432

A. gillii 426, 432–434

A. laricis 426, 432–434

A. occidentale 426–428, 432

A. pusillum 426, 434

A. tsugense 426, 430–432

A. vaginatum subsp. cryptopodum 426, 432–434

Arctostaphylos (bearberry, manzanita): bud burl 494, Coniothyrium 82, Exobasidium 26, Heterobasidion annosum 314, leaf gall 26, leaf spot 82, mistletoe 422, Phellinus igniarius 342, root rot 314, wood rot 342

A. uva-ursi (common bearberry, kinnikinnik): Chrysomyxa arctostaphyli 266, Exobasidium 26, rust 266, sulfur dioxide tolerance 468

Arecastrum romanzoffianum (queen palm): Botryosphaeria dothidea 172, false smut 92, Ganoderma zonatum 330, Graphiola phoenicis 92, leaf spot 92, lethal yellowing 390, nutrient deficiency 472, root rot 330, Stigmina palmivora 92

Arenga (arenga palm): false smut 92, Graphiola phoenicis 92

A. engleri (Englers palm): lethal yellowing 390

Arikuryroba schizophylla (Arikury palm): Ganoderma zonatum 330, lethal yellowing 390, root rot 330

Ariocarpus (living-rock cactus): Pseudomonas fluorescens 158, soft rot 158

Armillaria 308–312, 314–318, 322, 326, 444, 448

A. mellea 308–310

A. tabescens 312, 318

Aronia (chokeberry): Diplocarpon mespili 64, Erwinia amylovora 162, fire blight 162, Gymnosporangium clavariiforme 248, G. clavipes 246, leaf spot 64, rust 246–248

arrowwood, see Viburnum dentatum

Artemisia (sage brush): Botryosphaeria obtusa 176, dieback 168, 176, Perenniporia medulla-panis 168, Verticillium wilt 376, wood decay 168

artichoke, Jerusalem, see Helianthus

Articularia quercina 118

Aruncus dioicus (goatsbeard):

Cristulariella depraedens 62, Erwinia amylovora 162, fire blight 162, leaf blight 62

Ascocalyx abietina 230–232

Ascochyta cornicola 78

A. piniperda 134

A. syringae 130

ash, see Fraxinus, Sorbus, Zanthoxylum
Arizona or Modesto: F. velutina var. glabra
Berlandier: F. berlandierana
Biltmore: F. biltmoreana
black: F. nigra
blue: F. quadrangulata
Carolina or water: F. caroliniana
European: F. excelsior
flowering: F. ornus
green: F. pennsylvanica var. lanceolata
Himalayan: F. floribunda
'Moraine': F. holotricha cv. 'Moraine'
mountain: Sorbus
American: S. americana
European: S. aucuparia
Greene: S. scopulina
Sitka: S. sitchensis
Oregon: F. latifolia
prickly: Zanthoxylum
pumpkin: F. profunda
red: F. pennsylvanica
velvet: F. velutina
white: F. americana

Asimina triloba (pawpaw): Cristulariella moricola 62, flood tolerance 486, leaf spot 62, root rot 296, Sclerotium rolfsii 296, southern blight 296

aspen, see Populus
bigtooth or large-tooth: P. grandidentata
Chinese: P. adenopoda
European: P. tremula
quaking or trembling: P. tremuloides

Aster (aster): Coleosporium asterum 268, rust 268

aster, China, see Callistephus

Asteridiella 30

Asterina 30

Asteroma caryae 108

A. inconspicuum 108

A. tiliae 78

A. ulmeum 108

Asteromella 70

A. fraxini 70

Astrophytum (star cactus): Pseudomonas fluorescens 158, soft rot 158

Athelia rolfsii 296

Athyrium (lady fern): rust 264, Uredinopsis 264

Atriplex (quail brush, rabbitbrush, sage brush, salt bush, salt sage): Botryosphaeria dothidea 172, dieback 172, salt tolerance 454, Verticillium wilt 374

A. hortensis (orach): Pseudomonas syringae pv. syringae 160

Atropellis 230, 234

A. piniphila 234

A. tingens 234

Aucuba (aucuba): Alternaria 90, leaf spot 90

A. japonica (Japanese aucuba): anthracnose 122–124, dieback 124, Glomerella cingulata 122–124, root rot 296, Sclerotium rolfsii 296, southern blight 296

Aureobasidium apocryptum 104

A. pullulans 30, 138

Australian-pine, see Casuarina, C. equisetifolia

autumn browning 498

Avena sativa (oats): Puccinia coronata 250, rust 250

avens, see Geum
mountain: Dryas

Avignon berry, see Rhamnus infectoria

avocado, see Persea americana

azalea, see Menziesia, Rhododendron
Hinodegiri, Hiryu, or Kirishima: R. obtusum
hoary: R. canescens
Indian or macranthum: R. indicum

Korean: R. yedoense var. poukhanense
mock: Menziesia
snow: R. mucronatum
torch: R. kaempferi
western: R. occidentalis

Baccharis (baccharis): Botryosphaeria dothidea 172, B. obtusa 176, B. rhodina 180, dieback 172, 176, 180, salt tolerance 454

B. halmifolia (groundsel tree): salt tolerance 454

Bacillus megaterium 382

bacteria 154–166, 382–386
fastidious xylem-inhabiting bacteria 384–386

Bacterium pseudotsugae 154

bald cypress, see Taxodium distichum

ball moss, see Tillandsia recurvata

bamboo, see Bambusa, Nandina, Sasa
heavenly or sacred: Nandina

Bambusa (bamboo): Schizophyllum commune 358

banana, see Musa acuminata

banana shrub, see Michelia figo

barberry, see Berberis
Japanese: B. thunbergii

bark formation and wound healing 496

bark rot 168

basswood, see Tilia americana

Bauhinia (bauhinia, orchid tree): anthracnose 122–124, gall 148, Glomerella cingulata 122–124, Nectriella pironii 148

B. purpurea (orchid tree): Botrytis blight 60, Cristulariella moricola 62, leaf spot 62, salt tolerance 454, see also B. variegata

B. variegata (orchid tree): Cristulariella moricola 62, fruit spot 62, gall 148, Nectriella pironii 148, see also B. purpurea

bay, see Magnolia, Persea
bull: M. grandiflora
red: P. borbonia
sweet: M. virginiana

bayberry, see Myrica pensylvanica

bead tree, see Melia

bean
coral, see Erythrina herbacea
white: Sorbus aria

bean tree, see Laburnum

bearberry, see Arctostaphylos
common: A. uva-ursi

beautyberry, see Callicarpa, C. americana

beautybush, see Kolkwitzia amabilis

beech, see Carpinus, Fagus
American: F. grandifolia
blue: C. caroliniana
European: F. sylvatica

Berberis (barberry): Botryosphaeria dothidea 172, Dematophora necatrix 306, dieback 172, 210, herbicide tolerance 460, Meloidogyne 298, Nectria cinnabarina 210, Phyllactinia guttata 16, powdery mildew 16, root knot nematode 298, root rot 306, Verticillium wilt 376

B. atrocarpa: Cumminsiella mirabilissima 250, rust 250

B. thunbergii (Japanese barberry): flood tolerance 486, salt tolerance 454, sulfur dioxide tolerance 468

Betula (birch): anthracnose 114, 120, Armillaria 308, Botryosphaeria dothidea 172, B. obtusa 176, bronze birch borer 448, bud burl 494, butt crack 308, butt rot 332, canker 212, 352, canker rot 352, Climacodon septentionalis 344, cold tolerance 480, dieback 172, 176, 210, Discula betulina 114, Fomes fomentarius 346, Fomitopsis pinicola 348, Ganoderma applanatum 334, herbicide tolerance 460, Heterobasidion annosum 314, Hypoxylon mammatum 222, H. punctulatum 226, Inonotus obliquus 352, I. tomentosus 318, leaf blotch 114, 120, lightning injury 490, Marssonina betulae 120, Melampsoridium betulinum 254, Microsphaera or-

Betula (cont.)
nata 16, mistletoe 422, Nectria cinnabarina 210, N. galligena 212, 352, Oxyporus populinus 344, Phellinus igniarius 342, Phoradendron serotinum 422, Phyllactinia guttata 16, Piptoporus betulinus 346, Polyporus squamosus 344, powdery mildew 16, root rot 308, rust 254, salt tolerance 454, Schizophyllum commune 358, Scorias spongiosa 30, Stereum gausapatum 344, sulfur dioxide tolerance 468, Taphrina carnea 22, wood rot 308, 314, 318, 334, 342–348, 358

B. alleghaniensis (yellow birch): Aleurodiscus oakesii 168, apple mosaic virus 412, 448, twig blight 144, Armillaria 308, 448, butt rot 304, canker 144, 210, canker rot 342, 352, Ceratocystis coerulescens 362, Cerrena unicolor 358, decline 448, Diaporthe alleghaniensis 142–144, dieback 144, 448, Fomes fomentarius 346, frost susceptibility 482, Ganoderma lucidum 332, Hericium erinaceus 344, Hypoxylon deustum 304, ice glaze injury 490, Inonotus hispidus 356, I. obliquus 352, leaf and shoot blight 144, Melampsoridium betulinum 254, Nectria coccinea 442, N. galligena 210, 352, Oxyporus latemarginatus 328, Phellinus laevigatus 342, Phomopsis 144, Piptoporus betulinus 346, root rot 304, 308, 332, rust 254, salt tolerance 454, smooth patch 168, sterile conk 352, winter sunscald 484, wood rot 304, 308, 328, 344–346, 352, 356–358

B. ermanii: Melampsoridium betulinum 254, rust 254

B. glandulosa (bog or dwarf birch): Melampsoridium betulinum 254, rust 254

B. lenta (black or sweet birch): bleeding canker 286, canker 212, 286, frost susceptibility 482, Inonotus obliquus 352, Melampsoridium betulinum 254, Nectria galligena 212, Phytophthora cactorum 286, Piptoporus betulinus 346, prune dwarf virus 408, Prunus necrotic ringspot virus 408, rust 254, salt tolerance 454, wood rot 346, 352

B. nana: Melampsoridium betulinum 254, rust 254

B. nigra (river birch): anthracnose 114, Discula betulina 114, flood tolerance 486, leaf blotch 114, ozone tolerance 466, salt tolerance 454

B. occidentalis (water birch): Melampsoridium betulinum 254, rust 254

B. papyrifera (paper birch): anthracnose 114, apple mosaic virus 412, 448, butt rot 304, Cerrena unicolor 358, decline 448, dieback 200, 448, Discula betulina 114, flood tolerance 486, Hypoxylon deustum 304, Inonotus glomeratus 356, I. obliquus 352, leaf spot 114, 120, Marssonina betulae 120, Melampsoridium betulinum 254, nutrient deficiency 474, pesticide injury 460, Phaeolus schweinitzii 322, Piptoporus betulinus 346, prune dwarf virus 408, root rot 304, 322, rust 254, salt tolerance 454, sulfur dioxide injury 468, Valsa sordida 200, wetwood 382, wood decay 304, 322, 346, 352, 356–358

B. pendula (European white birch): anthracnose 122–124, Botryosphaeria dothidea 172–174, 484, canker 172–174, 484, dieback 174, drought stress 174, fasciation 492, flood tolerance 486, fluoride tolerance 470, Glomerella cingulata 122–124, ice glaze injury 490, leaf spot 120, Marssonina betulae 120, Melampsoridium betulinum 254, ozone tolerance 466, pesticide injury 460, Piptoporus betulinus 346, prune dwarf virus 408, rust

254, salt tolerance 454, wood decay 346, see also B. pubescens

B. platyphylla var. japonica (Japanese white birch): Melampsoridium betulinum 254, rust 254

B. populifolia (gray birch): anthracnose 114, bud burl 494, Discula betulina 114, flood tolerance 486, ice glaze injury 490, Inonotus obliquus 352, leaf spot 114, Melampsoridium betulinum 254, Piptoporus betulinus 346, rust 254, salt tolerance 454, wood decay 346

B. pubescens (European white birch): Melampsoridium betulinum 254, prune dwarf virus 408, rust 254, see also B. pendula

B. pumila (swamp birch): Melampsoridium betulinum 254, rust 254

Bignonia (cross vine): Botryosphaeria obtusa 176, dieback 176

bigtree, see Sequoiadendron giganteum

bilberry, see Vaccinium
bog: V. uliginosum
dwarf: V. caespitosum

birch, see Betula
black or sweet: B. lenta
bog or dwarf: B. glandulosa
European white: B. pendula, B. pubescens
gray: B. populifolia
Japanese white: B. platyphylla var. japonica
paper: B. papyrifera
river: B. nigra
swamp: B. pumila
water: B. occidentalis
yellow: B. alleghaniensis

bittersweet, see Celastrus
climbing: C. scandens
Oriental: C. orbicularis

Bixa orellana (anatto): algal leaf spot 28, Cephaleuros virescens 28

Bjerkandera adusta 344–346

blackberry, see Rubus

black knot 152

blackline disease 404, 492

black locust, see Robinia pseudoacacia

black mildew 30

black stain root disease 368

blackthorn, see Prunus spinosa

bladdernut, see Staphylea

blight 44, 52, 58–62, 84, 86, 90, 94, 100, 102, 130, 134–138, 160–164, 188, 206, 214, 296

blossom blast 160

bluebell, see Campanula

blueberry, see Vaccinium
box: V. ovatum
highbush: V. corymbosum
lowbush: V. angustifolium, V. myrtilloides

blue stain 350, 380

Blumeriella jaapii 66

Borassus aethiopum (palm): leaf spot 92, Stigmina palmivora 92

B. flabellifer (Palmyra palm): lethal yellowing 390

bottlebrush, see Callistemon
weeping: C. viminalis

Botryodiplodia 172, 180, 450
B. gallae 178
B. hypodermia 182, 208
B. theobromae 180, 360
Botryosphaeria aesculi 80
B. berengiana 172
B. bidwellii 80
B. dothidea 172–174, 176, 284, 484
B. melanops 178
B. obtusa 172, 176, 444
B. quercuum 172, 178
B. rhodina 172, 176, 180, 360, 484
B. ribis 172
B. stevensii 176
Botryotinia fuckeliana 60
Botrytis cinerea 60, 90
Bougainvillea (bougainvillea): anthracnose 122–124, Glomerella cingulata 122–124

box elder, see Acer negundo

boxwood, see Buxus
common: B. sempervirens

Brahea (rock palm): Botryosphaeria dothidea 172

brambles, see Rubus

Brasiliomyces trina 14

Brassaia actinophylla (schefflera): Alternaria panax 90, anthracnose 122–124, edema 486, Glomerella cingulata 122–124, leaf blight 90, Meloidogyne 298, root knot nematode 298, root rot 296, Sclerotium rolfsii 296, southern blight 296

bridal-wreath, see Spiraea ×vanhouttei

broom, see Cytisus, Genista

Broussonetia: Cerotelium fici 238, Phymatotrichum omnivorum 292, root rot 292, rust 238

B. papyrifera (paper mulberry): canker 210, 214, dieback 210, Fusarium solani 214, Nectria cinnabarina 210

brown felt blight of conifers 50–52

Brunchorstia pinea 232

Bucida buceras (black olive): Cristulariella moricola 62, leaf spot 62

buckeye, see Aesculus
California: A. californica
Ohio: A. glabra
red: A. pavia
yellow: A. octandra

Buckleya: Cronartium appalachianum 270, rust 270

buckthorn, see Rhamnus
alder: R. frangula
alder-leaved: R. alnifolia
California: R. californica
Carolina: R. caroliniana
European: R. cathartica
lance-leaved: R. lanceolata
rock: R. saxatalis

buffalo berry, see Shepherdia argentia, S. canadensis

bugleweed, see Ajuga reptans

bunya-bunya, see Araucaria bidwillii

burl 494

burning bush, see Euonymus

Bursaphelenchus lignicolus 380
B. xylophilus 380

butternut, see Juglans cinerea

buttonbush, see Cephalanthus

buttonwood, silver, see Conocarpus erectus

Butia capitata (jelly palm): false smut 92, Ganoderma zonatum 330, Graphiola phoenicis 92, leaf spot 92, root rot 330, Stigmina palmivora 92

Buxus (boxwood): dieback 210, Ganoderma lucidum 332, Meloidogyne 298, Nectria cinnabarina 210, Phyllactinia guttata 16, powdery mildew 16, Pratylenchus 302, nematode 298–302, root rot 332, salt tolerance 454, Verticillium wilt 374

B. sempervirens (common boxwood): dieback 206, Pseudonectria rouselliana 206, sulfur dioxide tolerance 468, Volutella buxi 206, winter injury 476

cacao, see Theobroma cacao

cacti: canker 72, rot 72, 82, 158, viral diseases 418

cactus
California barrel, see Ferocactus acanthodes
hatchet: Pelecyphora
hook: Ancistrocactus
living-rock: Ariocarpus
pin cushion: Mammillaria
prickly pear: Opuntia
star: Astrophytum
see also Cereus, Echinocactus, Pediocactus

Caesalpinia (dwarf poinciana): Botryosphaeria dothidea 172, dieback 172

calamondin, see ×Citrofortunella mitis

California laurel, see Umbellularia californica

Callicarpa (beautyberry): Botryosphaeria

obtusa 176, Coniothyrium 82, dieback 82, 176, 210, Nectria cinnabarina 210

C. americana (beautyberry): fastidious xylem-inhabiting bacteria 384, salt tolerance 454

Callistemon (bottlebrush): Cylindrocladium 294, gall 148, Nectriella pironii 148, root rot 294, salt tolerance 454

C. viminalis (weeping bottlebrush): gall 150, Sphaeropsis tumefaciens 150, witches'-broom 150

Callistephus (China aster): Coleosporium asterum 268, rust 268

Callitris preisii (Rottnest Island pine): Armillaria tabescens 312, root rot 312

Calluna vulgaris (heather): Armillaria 308, Heterobasidion annosum 314, root rot 308, wood rot 308, 314

Calocedrus decurrens (incense cedar): brown felt blight 50, canker 170, drought injury 478, Fomitopsis pinicola 348, Gymnosporangium libocedri 244–246, Herpotrichia juniperi 50, Heterobasidion annosum 314, Lophodermium juniperi 30, mistletoe 424, ozone tolerance 466, Pestalotiopsis funerea 128, Phaeolus schweinitzii 322, Phellinus pini 350, P. weirii 324, Phoradendron juniperinum 424, root rot 314, 322–324, rust 244–246, Seiridium cardinale 170, witches'-broom 244, wood decay 314, 322–324, 348–350

Calonectria 294

Calycanthus occidentalis (calycanthus, sweet shrub): Botryosphaeria obtusa 176, dieback 176, Phyllactinia guttata 14, powdery mildew 14

Calyptospora goeppertiana 260–262

Camellia (camellia): blight 60, 128, Botryosphaeria dothidea 172, Botrytis cinerea 60, dieback 172, edema 486, fluoride tolerance 470, Hendersonia 82

C. japonica (common camellia): anthracnose 122–124, Armillaria tabescens 312, blight 128, Cephaleuros virescens 28, Chalara elegans 294, Ciborinia camelliae 58, dieback 122–124, Exobasidium camelliae 26, flower blight 58, Glomerella cingulata 122–124, leaf and flower variegation (viral) 418, Pestalotiopsis 128, root rot 294, 312

C. oleifera (tea-oil plant): anthracnose 122–124, Glomerella cingulata 122–124

C. reticulata: anthracnose 122–124, Glomerella cingulata 122–124

C. saluensis: anthracnose 122–124, Glomerella cingulata 122–124

C. sasanqua (sasanqua camellia): anthracnose 122–124, canker 122, Ciborinia camelliae 58, dieback 122, edema 486, Exobasidium camelliae 26, flower blight 58, Glomerella cingulata 122–124, leaf and flower variegation (viral) 418, leaf gall 26

C. sinensis (tea): algal leaf spot 28, anthracnose 122–124, blight 128, butt rot 304, Cephaleuros parasiticus 28, Glomerella cingulata 122–124, Hypoxylon deustum 304, Pestalotiopsis maculans 128, root rot 304

Campanula (bluebell): Coleosporium campanulae 268, rust 268

camphor tree, see Cinnamomum camphora

canker 74, 102, 112, 116, 122, 132, 140–144, 164–236, 272–284, 342–362, 442–448

canker-rot 342–358

canker-stain 360

cape leadwort, see Plumbago capensis

Capnodium 30

Caragana arborescens (Siberian pea tree): Hendersonia 82, salt tolerance 454, sulfur dioxide tolerance 468

Carica papaya (papaya): Alternaria 90, anthracnose 122–124, Botryosphaeria

Carica papaya (cont.)
dothidea 172, *B. rhodina* 180, dieback 172, 180, fruit rot 90, *Glomerella cingulata* 122–124, leaf spot 90, root rot 296, *Sclerotium rolfsii* 296, southern blight 296, witches'-broom 396
Carissa grandiflora (Natal plum): blight **94,** *Botryosphaeria dothidea* 172, *B. obtusa* 176, dieback 172, 176, gall **150,** *Pseudomonas syringae* pv. *syringae* 160, *Rhizoctonia solani* **94,** *Sphaeropsis tumefaciens* **150**
Carnegiea gigantea (saguaro): cortical rot 82, *Erwinia carnegiana* **158,** *Hendersonia cerei* 82, saguaro virus 418, soft rot **158**
carob, see *Ceratonia*
Carpinus (hornbeam): anthracnose 116, *Botryosphaeria obtusa* 176, butt rot 334, canker 212, *Cerrena unicolor* 358, dieback 176, *Ganoderma applanatum* 334, *Gnomoniella carpinea* 116, *Heterobasidion annosum* 314, *Hypoxylon mammatum* 222, *H. tinctor* 226, *Microsphaera ellisii* 16, *Nectria galligena* 212, *Phellinus igniarius* 342, *Phyllactinia guttata* 16, powdery mildew 16, *Schizophyllum commune* 358, *Sphaerognomonia carpinea* 116, *Stereum gausapatum* 344, wood decay 314, 334, 342–344, 358
C. betulus (European hornbeam): anthracnose 116, fluoride tolerance 470, *Ganoderma lucidum* 332, *Gnomoniella carpinea* 116, *Monostichella robergei* 116, ozone tolerance 466, root rot 332, salt tolerance 454, *Sphaerognomonia carpinea* 116, sulfur dioxide tolerance 468
C. caroliniana (American hornbeam, blue beech): *Aleurodiscus oakesii* 168, anthracnose **116,** canker 214, flood tolerance 486, *Fusarium lateritium* 214, *Gnomoniella carpinea* **116,** *Hypoxylon atropunctatum* 224, *Monostichella robergei* 116, salt tolerance 454, smooth patch 168, *Sphaerognomonia carpinea* 116
C. japonica (Japanese hornbeam): anthracnose 116, *Sphaerognomonia carpinea* 116
carrotwood, see *Cupaniopsis*
Carya (hickory): *Botryosphaeria dothidea* 172, *B. obtusa* 176, *B. quercuum* 178, canker 212, *Cerrena unicolor* 358, *Climacodon septentrionalis* 344, cold tolerance 480, *Coniothyrium* 82, *Dendrothele candida* 168, dieback 172, 176–178, 210, drought injury 476, *Fomes fomentarius* 346, *Fomitopsis pinicola* 348, frost susceptibility 482, gall 146, *Ganoderma* 332, *G. applanatum* 334, *Globifomes graveolens* 346, *Gnomonia* 108, *Hendersonia* 82, *Hericium erinaceus* 344, herbicide tolerance 460, *Hypoxylon atropunctatum* 224, *H. punctulatum* 226, ice glaze injury 490, *Inonotus hispidus* 356, leaf spot 82, 108, *Microsphaera caryae* 16, mistletoe 422, *Nectria cinnabarina* 210, *N. galligena* 212, *Oxyporus populinus* 344, *Phellinus igniarius* 342, *P. spiculosus* 354, *Phomopsis* 146, *Phyllactinia guttata* 16, powdery mildew 16, root rot 306, 332, salt tolerance 454, *Schizophyllum commune* 358, smooth patch 168, witches'-broom 118, wood decay 332–334, 342–348, 354–358, *Xylaria* 306
C. aquatica (water hickory): bunch disease 398, downy leaf spot 118, flood tolerance 486, *Hericium erinaceus* 344, *Microstroma juglandis* 118, wood decay 344
C. cordiformis (bitternut hickory): bunch disease 398, *Gnomonia caryae* 108, leaf spot 108, 118, *Microstroma juglandis* 118
C. glabra (pignut hickory): canker 236, *Dendrothele strumosa* 168, flood tolerance 486, gall **146,** *Gnomonia caryae* 108, leaf spot 108, 118, *Melanconis juglandis* 132, *Microstroma juglandis* 118, *Phellinus spiculosus* 354, *Phomopsis* **146,** smooth patch 168, *Strumella coryneoidea* 236, *Urnula craterium* 236
C. illinoinensis (pecan): anthracnose 122–124, *Botryosphaeria dothidea* 172, *B. rhodina* 180, bunch disease **398,** canker 212, *Cristulariella moricola* 62, *Dendrothele candida* 168, dieback 172, 180, flood tolerance 486, *Glomerella cingulata* 122–124, *Gnomonia* 108, leaf spot 62, 108, 118, *Microsphaera caryae* 16, *Microstroma juglandis* 118, mistletoe 422, mycoplasmalike organism **398,** *Nectria galligena* 212, nematode 302, nutrient deficiency 474, nut rot 172, *Phoradendron serotinum* 422, powdery mildew 16, *Radopholus similis* 302, salt tolerance 454, smooth patch 168, witches'-broom **398**
C. laciniosa (shellbark hickory): sulfur dioxide tolerance 468
C. ovata (shagbark hickory): bunch disease 398, canker 236, *Cristulariella moricola* **62,** *Cryphonectria parasitica* 186, flood tolerance 486, *Gnomonia caryae* 108, leaf spot **62, 108,** 118, *Microstroma juglandis* 118, mistletoe 422, *Oxyporus latemarginatus* 328, *Phellinus spiculosus* 354, salt tolerance 454, *Strumella coryneoidea* 236, *Urnula craterium* 236
C. pallida (sand hickory): downy leaf spot 118, *Microstroma juglandis* 118
C. tomentosa (mockernut hickory): bunch disease 398, flood tolerance 486, *Gnomonia caryae* 108, leaf spot 108, 118, *Melanconis juglandis* 132, *Microstroma juglandis* 118, *Phellinus spiculosus* 354
Caryota mitis (cluster fishtail palm): *Botryosphaeria dothidea* 172, leaf spot 92, lethal yellowing 390, *Stigmina palmivora* 92
C. urens (wine palm): leaf spot 92, *Stigmina palmivora* 92
cascara, see *Rhamnus purshiana*
cassandra, see *Chamaedaphne calyculata*
cassava, see *Manihot esculenta*
Cassia (senna): *Botryosphaeria dothidea* 172, dieback 172
C. armata (senna): dodder **436**
Cassiope (cassiope): *Exobasidium vaccinii* 26
Cassytha filiformis (cassytha, dodder laurel, woe vine) **436**
Castanea (chestnut, chinkapin): bark rot 168, blight 186, *Botryosphaeria dothidea* 172, *B. obtusa* 176, *B. rhodina* 180, canker 186, *Cerrena unicolor* 358, *Ceratocystis fagacearum* 364, *Ciborinia candolleana* 56, *Cryphonectria parasitica* 186–188, dieback 172, 176, 180, drought injury 476, *Endothia parasitica* 186, *Fomitopsis pinicola* 348, *Ganoderma applanatum* 334, *Heterobasidion annosum* 314, *Hypoxylon punctulatum* 226, *Laetiporus sulfureus* 346, leaf blight 56, *Microsphaera americana* 16, mistletoe 422, oak wilt 364, *Perenniporia medullapanis* 168, *Phellinus igniarius* 342, *Phoradendron serotinum* 422, *Phyllactinia guttata* 16, *Phytophthora cinnamomi* 288, powdery mildew 16, root rot 288, *Stereum gausapatum* 344, wood rot 168, 314, 334, 342–348, 358
C. alnifolia (bush chinkapin): blight 186, canker 186, *Cryphonectria parasitica* 186
C. crenata (Japanese chestnut): blight 186, canker 186, *Cryphonectria parasitica* 186
C. dentata (American chestnut): blight 186–**188,** *Botryosphaeria quercuum* 178, canker **188,** 192, 236, *Cryphonectria parasitica* 186–**188,** dieback 178, *Endothia gyrosa* 192, *Hericium erinaceus* 344, *Strumella coryneoidea* 236, *Urnula craterium* 236, wood rot 344
C. mollissima (Chinese chestnut): anthracnose 122–124, blight **186**–188, canker **186**–188, *Ceratocystis fagacearum* 364, *Cryphonectria parasitica* **186**–188, *Glomerella cingulata* 122–124, oak wilt 364, *Trametes versicolor* 342, wood decay 342
C. pumila (Allegheny chinkapin): blight 186, canker 186, *Cryphonectria parasitica* 186
C. sativa (Spanish chestnut): blight 188, canker 188, *Cryphonectria parasitica* 188, dieback 168, *Hendersonula toruloidea* 168, viruslike symptoms 418
Castanopsis (chinkapin): *Ceratocystis fagacearum* 364, *Microsphaera americana* 16, oak wilt 364, powdery mildew 16
C. chrysophylla (giant chinkapin): *Cryphonectria parasitica* 186, *Ganoderma lucidum* 332, root rot 332
Castilleja (paintbrush): *Cronartium arizonicum* 280, *C. coleosporioides* 270, 278, *C. ribicola* 272, *Peridermium* 270, 280–282, rust 270, 278–282
Casuarina (Australian-pine): *Armillaria tabescens* 312, *Ganoderma tornatum* 330, root rot 312, 330
C. equisetifolia (Australian-pine, horsetail casuarina, horsetail tree): actinorhizae 502, *Armillaria tabescens* 312, ball moss **504,** freeze damage **340,** nitrogen-fixing nodules 502, root rot 312, slime flux **382,** stem crack **340**
Catalpa (catalpa): anthracnose 122–124, *Alternaria* 90, *Botryosphaeria dothidea* 172, *B. rhodina* 180, *Chalara elegans* 294, dieback 172, 180, 210, *Glomerella cingulata* 122–124, herbicide tolerance 460, ice glaze injury 490, leaf spot 90, *Microsphaera elevata* 16, *Nectria cinnabarina* 210, ozone tolerance 466, *Phyllactinia guttata* 16, *Phymatotrichum omnivorum* 292, powdery mildew 16, root rot 292–296, *Sclerotium rolfsii* 296, sooty mold 30, southern blight 296, sulfur dioxide tolerance 468, winter sunscald 484
C. speciosa (western catalpa): *Alternaria* 90, *Chalara elegans* 294, *Cristulariella moricola* 62, leaf spot 62, 90, *Phymatotrichum omnivorum* 292, root rot 292–296, salt tolerance 454, *Schizophyllum commune* 358, *Sclerotium rolfsii* 296, southern blight 296, *Trametes versicolor* 342, wood rot 342, 358
Catharanthus roseus (periwinkle): dodder **436**
Ceanothus (ceanothus, redroot): actinorhizae 502, *Botryosphaeria dothidea* 172, *B. obtusa* 176, dieback 172, 176, *Microsphaera ceanothi* 16, *M. sydowiana* 16, nitrogen-fixing nodules 502, powdery mildew 16, Verticillium wilt 374–376
C. sanguineus (redstem ceanothus): sulfur dioxide tolerance 468
C. velutinus (snowbrush): sulfur dioxide tolerance 468
cedar
  Alaska or Alaska yellow, see *Chamaecyparis nootkatensis*
  Atlantic white: *Chamaecyparis thyoides*
  Atlas: *Cedrus atlantica*
  deodar: *Cedrus deodara*
  eastern red: *Juniperus virginiana*
  incense: *Calocedrus decurrens*
  Japanese: *Cryptomeria japonica*
  northern white: *Thuja occidentalis*
  red: *Juniperus*
  salt: *Tamarix*
  southern red: *J. silicicola*
  stinking: *Torreya taxifolia*
  western red: *Thuja plicata*
Cedrus (cedar): *Cylindrocladium* 294, *Heterobasidion annosum* 314, *Pestalotiopsis funerea* 128, root rot 294, 314
C. atlantica (Atlas cedar): herbicide injury **456,** pine wood nematode 380, salt tolerance 454
C. deodara (deodar or deodar cedar): pine wood nematode 380
Celastrus orbicularis (Oriental bittersweet): flood tolerance 486
C. scandens (climbing bittersweet): anthracnose 122–124, blight **120,** *Botryosphaeria dothidea* 172, *B. obtusa* 176, dieback 172, 176, *Glomerella cingulata* 122–124, *Marssonina thomasiana* **120,** as a pest **438**
Celtis (hackberry, sugarberry): *Botryosphaeria obtusa* 176, dieback 176, *Cerrena unicolor* 358, *Hendersonia* 82
C. laevigata (sugarberry): butt rot **304,** *Cristulariella moricola* 62, downy mildew 12, eriophyid mite 20, flood tolerance 486, *Ganoderma lucidum* 332, *Hypoxylon deustum* **304,** leaf spot 62, mistletoe 422, *Phoradendron* 422, *Pleochaeta polychaeta* 14, powdery mildew 14, 18–20, *Pseudoperonospora celtidis* 12, root rot **304,** 332, *Sphaerotheca phytoptophila* 18–20, *Uncinula polychaeta* 14, witches'-broom 20
C. occidentalis (hackberry): downy mildew 12, eriophyid mite **20,** flood tolerance 486, herbicide injury 458, herbicide tolerance 460, ice glaze injury 490, *Laetiporus sulfureus* 346, mistletoe 422, nematode 302, *Oxyporus latemarginatus* 328, *Phoradendron serotinum* 422, *Polyporus squamosus* 344, powdery mildew 18–**20,** *Pseudoperonospora celtidis* 12, salt tolerance 454, *Sphaerotheca phytoptophila* 18–**20,** witches'-broom **20,** wood decay 328, 344–346, *Xiphinema americanum* 302
C. reticulata (netleaf hackberry): *Oxyporus latemarginatus* 328, wood rot 328
Cenangium 230
C. ferruginosum **230**
C. singulare 228
ceniza, see *Leucophyllum frutescens*
Cenococcum geophilum **500**
century plant, see *Agave*
Cephalanthus (buttonbush): *Botryosphaeria dothidea* 172, *B. obtusa* 176, *Coniothyrium* 82, dieback 172, 176, flood tolerance 486, leaf spot 82, *Microsphaera semitosta* 16, *Phyllactinia guttata* 16, powdery mildew 16, salt tolerance 454
Cephaleuros parasiticus 28
C. virescens **28**
Cephalosporium diospyri 372
Cephalotaxus harringtonia (Harrington plum yew): blight 138, *Phomopsis juniperovora* 138
Ceratocystis coerulescens **362**
C. fagacearum **364,** 372
C. fimbriata **360–362**
C. fimbriata f. sp. *platani* **360**
C. ulmi **366**
C. wageneri 310, **368**
Ceratoides (winter fat): Verticillium wilt 376
Ceratonia (carob): *Botryosphaeria dothidea* 172, dieback 172, Verticillium wilt 376
Cercidiphyllum japonicum (katsura tree): *Botryosphaeria dothidea* 172, *Cylindrocladium* 294, dieback 172, root rot 294, sulfur dioxide tolerance 468

Cercidium (paloverde): mistletoe 422, *Phoradendron californicum* 422

Cercis canadensis (redbud): anthracnose **114**, Botrytis blight **60**, *Botryosphaeria dothidea* **172,** *B. obtusa* 176, canker **172,** 212, decline **406,** dieback **172,** 176, downy mildew 12, flood tolerance 486, fluoride tolerance 470, *Ganoderma* 332, herbicide injury 456, herbicide tolerance 460, *Kabatiella* **114,** *Nectria galligena* 212, ozone tolerance 466, *Plasmopara cercidis* 12, root rot 332, *Schizophyllum commune* 358, sulfur dioxide tolerance 468, viruslike symptoms **406,** wetwood 382, wood rot 358

C. chinensis (Chinese redbud): ozone tolerance 466

Cercoseptoria 70

Cercospora 70, **86–88**

C. citri-grisea 72

C. concentrica 70

C. epicoccoides **86**

C. fuliginosa **88**

C. handelii **88**

C. kalmiae **88**

C. ligustri 88

C. liquidambaris **86–88**

C. nandinae **86**–88

C. pittospori **88**

C. platanicola 76–78

C. rosicola 76–78

C. sequoiae **86**

Cercosporella 70

Cereus (cactus): *Fusarium dimerum* var. *violaceum* 72, Fusarium rot 72

cereus, night-blooming, see *Hylocereus undatus*

Cerocarpus (mountain mahogany): actinorhizae 502, nitrogen-fixing nodules 502

Cerotelium fici **238**

Cerrena unicolor **358, 444,** 484

Chaenomeles japonica (flowering quince): apple mosaic virus 412, blight 126, 162, *Botryosphaeria dothidea* 172, *B. obtusa* 176, dieback 172, 176, 210, *Diplocarpon mespili* 64, *Erwinia amylovora* 162, *Gymnosporangium* 244, *G. clavipes* 246, *G. nidus-avis* 242, leaf spot 64, *Monilinia* 126, mosaic **412,** *Nectria cinnabarina* 210, rust 242–246, salt tolerance 454, sulfur dioxide tolerance 068, viruslike symptoms **412**

Chalara 360, 364

C. elegans **294**

Chamaebatiaria millefolium (fernbush): mistletoe 424, *Phoradendron juniperinum* 424

Chamaecyparis (cedar, cypress, false cypress): *Botryosphaeria dothidea* 172, *B. obtusa* 176, dieback 172, 176, *Fomitopsis pinicola* 348, *Gymnosporangium nootkatense* 240, *Heterobasidion annosum* 314, *Pestalotiopsis funerea* 128, *Phellinus pini* 350, root rot 314, rust 240, *Schizophyllum commune* 358, wood decay 314, 348–350

C. funebris (mourning cypress): blight 138, *Phomopsis juniperovora* 138

C. lawsoniana (Lawson cypress): blight 138, canker 170, *Phaeolus schweinitzii* 322, *Phomopsis juniperovora* 138, *Phytophthora cinnamomi* 288, *P. lateralis* **288,** root rot **288,** 322, *Seiridium cardinale* 170, *Sphaeropsis sapinea* 136, sulfur dioxide tolerance 468, tip blight 136

C. nootkatensis (Nootka cypress, Alaska or Alaska yellow cedar): blight 138, brown felt blight 50, canker 170, 194, *Herpotrichia juniperi* 50, *Kabatina juniperi* 138, *Seiridium cardinale* 170, sulfur dioxide tolerance 468, *Valsa abietis* 194

C. obtusa (Hinoki cypress): blight 128, 138, *Pestalotiopsis foedans* 128, *Phomopsis juniperovora* 138

C. pisifera (Sawara cypress): blight 138,

Cercospora sequoiae 86, needle blight 86, *Phomopsis juniperovora* 138, salt tolerance 454, sulfur dioxide tolerance 468

C. thyoides (Atlantic or southern white cedar): blight 138, *Dendrothele nivosa* 168, *Didymascella chamaecyparisii* 44, flood tolerance 486, *Gymnosporangium* 248, *Hendersonia* 82, *Phaeolus schweinitzii* 322, *Phomopsis juniperovora* 138, root rot 322, rust 248, salt tolerance 454, smooth patch 168

Chamaedaphne (leatherleaf): *Botryosphaeria obtusa* 176, dieback 176, *Exobasidium vaccinii* 26

C. calyculata (cassandra, leatherleaf): *Chrysomyxa* 266, rust 266

cherry, see *Cornus, Prunus*
   bird: *P. padus*
   bitter: *P. emarginata*
   black: *P. serotina*
   chokecherry: *P. virginiana*
   cornelian: *C. mas*
   European dwarf: *P. fruticosa*
   Higan or weeping: *P. subhirtella*
   holly-leaved: *P. ilicifolia*
   Japanese flowering: *P. serrulata, P. yedoensis*
   'Kwanzan': *P. serrulata* cv. 'Kwanzan'
   mahaleb: *P. mahaleb*
   mazzard or sweet: *P. avium*
   Nanking: *P. tomentosa*
   pin: *P. pensylvanica*
   sand: *P. pumila*
   Sargent: *P. sargentii*
   sour: *P. cerasus*
   western sand: *P. besseyi*

cherry leaf roll virus **404**–406, 416

cherry rasp leaf virus **408**

chestnut, see *Castanea*
   American: *C. dentata*
   Chinese: *C. mollissima*
   Japanese: *C. crenata*
   Spanish: *C. sativa*
   see also *Aesculus*

chimera **492**

chinaberry, see *Melia azedarach*

chinkapin, see *Castanea, Castanopsis*
   Allegheny: *Castanea pumila*
   bush: *Castanea alnifolia*
   giant: *Castanopsis chrysophylla*

Chionanthus (fringe tree): *Botryosphaeria dothidea* 172, dieback 172, gall 154, *Microsphaera syringae* 16, *Phyllactinia guttata* 16, powdery mildew 16, *Pseudomonas syringae* pv. *savastanoi* 154

Choisya ternata (Mexican orange): dieback 210, *Nectria cinnabarina* 210

chokeberry, see *Aronia*

chokecherry, see *Prunus virginiana*

cholla, see *Opuntia fulgida*

Chondroplea populea 184

Christmasberry, see *Heteromeles arbutifolia*

Chrysalidocarpus (cabada palm): leaf spot 92, *Stigmina palmivora* 92

C. cabadae (cabada palm): lethal yellowing 390

C. lutescens (areca, Madagascar, or yellow palm): *Ganoderma zonatum* 330, leaf spot 92, root rot 330, *Stigmina palmivora* 92

Chrysomyxa **266**

C. arctostaphyli **266**

C. ledi 266

C. ledicola **266**

C. weirii 266

Chrysothamnus (rabbitbrush): Verticillium wilt 376

Ciborinia 56–58

C. camelliae **58**

C. whetzelii 56–58

Cinnamomum (camphor tree, cinnamon tree): anthracnose 122–124, *Glomerella cingulata* 122–124, *Microsphaera cinnamomi* 16, *Phytophthora*

cinnamomi 288, powdery mildew 16, root rot 288, Verticillium wilt 376

C. camphora (camphor tree): anthracnose 122–124, *Armillaria tabescens* 312, *Botryosphaeria dothidea* 172, *B. rhodina* 180, dieback 172, 180, *Glomerella cingulata* 122–124, mistletoe 422, *Phoradendron serotinum* 422, root rot 312, *Schizophyllum commune* 358, wood rot 358

cinquefoil, see *Potentilla*
   shrubby: *P. fruticosa*

Cistus (rock-rose): Verticillium wilt 376

×*Citrofortunella mitis* (calamondin): gall 150, *Sphaeropsis tumefaciens* 150

Citrus (citrus): *Alternaria citri* 90, anthracnose 122–**124,** blight 60, 384, *Botryosphaeria dothidea* 172, *B. obtusa* 176, *Botrytis cinerea* 60, butt rot 334, canker 214, 334, *Capnodium* **30,** *Cephaleuros virescens* 28, *Chalara elegans* 294, collar rot 286, *Diaporthe citri* 142, dieback 172, 176, exocortis 400, fastidious xylem-inhabiting bacteria 384, fruit rot 90, *Fusarium lateritium* 214, *F. oxysporum* f. sp. *citri* 378, Fusarium wilt 378, *Ganoderma* 332, *G. applanatum* 334, *Glomerella cingulata* 122–124, greasy spot 72, leaf spot 28, 90, melanose 142, *Microsphaeropsis olivacea* 82, *Mycosphaerella citri* 72, nutrient deficiency 474, *Oxyporus latemarginatus* 328, ozone tolerance 466, *Phymatotrichum omnivorum* 292, *Phytophthora citrophthora* 286, *P. parasitica* 286, root rot 292–296, 308, 328–330, salt tolerance 454, *Schizophyllum commune* 358, *Sclerotium rolfsii* 296, sooty mold **30,** southern blight 296, sulfur dioxide tolerance 468, viroid 400, wood decay 332–334, young tree decline 384, 450

C. aurantiifolia (Key or Mexican lime): anthracnose 122–124, blossom blast 160, cassytha 436, dieback 168, Fusarium wilt 378, gall 150, *Glomerella cingulata* 122–124, *Hendersonula toruloidea* 168, *Pseudomonas syringae* pv. *syringae* 160, *Sphaeropsis tumefaciens* 150

C. aurantium (sour orange): cassytha 436

C. limon (lemon): anthracnose 122–**124,** *Armillaria tabescens* 312, Botrytis blight 60, dieback 168, fluoride tolerance 470, gall 150, *Glomerella cingulata* 122–**124,** *Hendersonula toruloidea* 168, ozone tolerance 466, root rot 312, *Sphaeropsis tumefaciens* 150

C. maxima (shaddock): cassytha 436

C. ×paradisi (grapefruit): anthracnose 122–124, bleeding canker 286, blight 384, **450,** blossom blight 160, cassytha 436, collar rot **286,** decline 384, **450,** *Diaporthe citri* **142,** dieback 168, fluoride tolerance 470, *Glomerella cingulata* 122–124, greasy spot **72,** *Hendersonula toruloidea* 168, melanose **142,** *Mycosphaerella citri* **72,** *Phytophthora parasitica* **286,** *Spiroplasma citri* 398, stubborn disease 398, young tree decline 384, **450**

C. reticulata (mandarin orange, tangerine): collar rot **286,** dieback 168, fluoride tolerance 470, fruit rot 160, *Hendersonula toruloidea* 168, leaf blight 160, *Phytophthora parasitica* **286,** *Pseudomonas syringae* pv. *syringae* 160

C. sinensis (sweet orange): black pit 160, blossom blight 160, cassytha 436, decline **302,** fluoride tolerance 470, gall 150, greasy spot **72,** *Mycosphaerella citri* **72,** nematode **302,** *Pseudomonas syringae* pv. *syringae* 160, *Radopholus similis* **302,** *Sphaeropsis tumefaciens* 150, *Spiroplasma citri* 398, stubborn disease 398

C. ×tangelo (tangelo): *Spiroplasma citri* 398, stubborn disease 398

Cladosporium 20

C. herbarum 30

C. humile 96

Cladrastis lutea (yellowwood): *Botryosphaeria dothidea* 172, dieback 172, flood tolerance 486, herbicide tolerance 460, *Oxyporus latemarginatus* 328, root rot **306,** *Xylaria polymorpha* **306**

Clarkia: *Pucciniastrum pustulatum* 262, rust 262

Clematis (clematis): anthracnose 122–124, *Botryosphaeria obtusa* 176, *Coniothyrium clematidis-rectae* 82, dieback 176, *Glomerella cingulata* 122–124, *Hendersonia* 82, wilt 82

Clerodendrum bungei (glory-bower): gall 148, *Nectriella pironii* 148

Clethra (summer sweet): *Botryosphaeria obtusa* 176, dieback 176

Climacodon septentrionalis **344**

Clitocybe tabescens 312

Clostridium 382

clover, yellow owl's, see *Orthocarpus luteus*

Coccoloba uvifera (sea grape): *Alternaria* 90, cassytha 436, *Cephaleuros virescens* 28, *Ganoderma applanatum* 334, leaf spot 28, 90

C. diversifolia (pigeon plum): algal leaf spot 28, *Cephaleuros virescens* 28

Coccomyces 66

Coccothrinax argentata (silver palm): leaf spot 92, *Stigmina palmivora* 92

Cocos nucifera (coconut palm): *Botryosphaeria dothidea* 172, *B. rhodina* 180, *Ganoderma* 330, leaf spot 92, lethal yellowing **390,** root rot 330, *Stigmina palmivora* 92

Codiaeum variegatum (croton): anthracnose 122–124, gall **148,** *Glomerella cingulata* 122–124, *Nectriella pironii* **148**

Coffea (coffee): anthracnose 122–124, *Ceratocystis fimbriata* 362, *Glomerella cingulata* 122–124

coffee, see *Coffea*
   wild: *Psychotria nervosa*

coffeeberry, see *Rhamnus californica*

Coleosporium **268**

C. asterum **268**

C. pacificum **268**

Colletotrichum 124

C. gloeosporioides 102, 122–**124**

C. griseum 124

C. limetticolum 124

comandra, northern, see *Geocaulon livida*

Comandra umbellata (bastard toadflax, comandra): *Cronartium comandrae* 270, 278, rust 270, 278

compartmentalization of wound-associated discolored and decayed wood **336–338,** 496

Comptonia peregrina (sweetfern): actinorhizae 502, *Botryosphaeria dothidea* 172, *Cronartium comptoniae* **270,** 276, dieback 172, *Gymnosporangium ellisii* 240, 248, nitrogen-fixing nodules 502, rust 240, 248, **270,** 276

coniferous seedlings: *Cylindrocladium* 294, *Fusarium* 296, *Macrophomina phaseolina* 294, nematode 302, root rot 292–296, *Xiphinema* 302

Coniothyrium **82**

C. fuckelii **82**

Conocarpus erectus var. *sericeus* (silver buttonwood): *Cylindrocladium* 294, root rot 294

Conoplea globosa 236

Coreopsis (coriopsis): *Coleosporium inconspicuum* 268, rust 268

Coriolus versicolor 342

cordgrass, see *Spartina*
   smooth: *S. alternifolia*

cork tree, see *Phellodendron*
   Amur: *P. amurense*

Cornus (dogwood): *Botryosphaeria dothidea* 172, *B. obtusa* 176, *B. rhodina* 180, canker 172, 212, 284, *Cerrena*

Cornus (cont.)
unicolor 358, collar rot 284, dieback 172, 176, 180, 210, drought injury 476, fluoride tolerance 470, herbicide tolerance 460, Heterobasidion annosum 314, Microsphaera pulchra 16, mistletoe 422, Nectria cinnabarina 210, N. galligena 212, Oxyporus populinus 344, Phellinus igniarius 342, Phoradendron serotinum 422, Phyllactinia guttata 16, Phytophthora cactorum 284, powdery mildew 16, Pseudomonas syringae pv. syringae 160, sulfur dioxide tolerance 468, tip blight 160, wetwood 382, wood rot 314, 342, 358
C. alba (Tatarian dogwood): salt tolerance 454
C. alternifolia (alternate-leaved dogwood): Hendersonia 82
C. amomum (silky dogwood): Cristulariella moricola 62, leaf spot 62, salt tolerance 454, witches'-broom 396
C. florida (flowering dogwood): anthracnose 122–124, Armillaria tabescens 312, Ascochyta cornicola 78, Botryosphaeria dothidea 174, Botrytis blight 60, canker 284–286, cherry leafroll virus 404, collar rot 284–286, Cristulariella moricola 62, decline 404, dieback 174, flood tolerance 486, frost susceptibility 482, Glomerella cingulata 122–124, Hypoxylon mediterraneum 224, ice glaze injury 490, leaf spot 62, 74, 78, Meloidogyne 298, ozone tolerance 466, Phyllosticta cornicola 78, Phytophthora cactorum 284–286, root knot nematode 298, root rot 312, salt tolerance 454, Septoria 74, tobacco ringspot virus 414, tomato ringspot virus 414, witches'-broom 396
C. kousa (kousa dogwood): leaf spot 78, Phyllosticta cornicola 78
C. mas (cornelian cherry): flood tolerance 486, fluoride tolerance 470, salt tolerance 454
C. nuttallii (Pacific dogwood): canker 284–286, collar rot 284–286, leaf spot 74, Phytophora cactorum 284–286, Septoria cornicola 74
C. racemosa (gray or panicled dogwood): flood tolerance 486, ozone tolerance 466, salt tolerance 454, tobacco ringspot virus 414, witches'-broom 396
C. sanguinea (blood-twig dogwood): salt tolerance 454
C. sericea (red-osier dogwood): Botryosphaeria dothidea 174, canker 174, dieback 174, see also C. stolonifera
C. stolonifera (red-osier dogwood): Botryosphaeria dothidea 484, canker 484, cold tolerance 480, flood tolerance 486, leaf spot 74, Septoria cornicola 74, tobacco ringspot virus 414, witches'-broom 396, see also C. sericea
Corticium rolfsii 296
Corylus (filbert, hazel, hazelnut): Botryosphaeria dothidea 172, B. obtusa 176, Botrytis blight 60, canker 212, dieback 172, 176, Heterobasidion annosum 314, Hypoxylon cohaerens 226, leaf blister 22, leaf spot 78, Nectria galligena 212, Phellinus igniarius 342, Phyllactinia guttata 16, Phyllosticta coryli 78, powdery mildew 16, Schizophyllum commune 358, sulfur dioxide tolerance 468, Taphrina coryli 22, wood decay 314, 342, 358
C. americana (American filbert, American hazelnut): anthracnose 114, leaf spot 78, 114, Microsphaera ellisii 16, Monostichella coryli 114, Phyllosticta coryli 78, powdery mildew 16, salt tolerance 454
C. avellana (European filbert): anthracnose 114, apple mosaic virus 412, Corynebacterium humiferum 382, fluoride tolerance 470, leaf spot 78, 114, Monostichella coryli 114, Phyllosticta

coryli 78, Tulare apple mosaic virus 412
C. cornuta (beaked filbert, beaked hazelnut): anthracnose 114, leaf spot 78, 114, Microsphaera hommae 16, Monostichella coryli 114, Phyllosticta coryli 78, powdery mildew 16, salt tolerance 454
Coryneum cardinale 170
Corypha elata (Gebang palm): lethal yellowing 390
Cotinus coggyria (smoke tree): Botryosphaeria rhodina 180, dieback 180, nematode 302, Pratylenchus 302, sulfur dioxide tolerance 468, Verticillium wilt 376
Cotoneaster (cotoneaster): Botryosphaeria dothidea 172, B. obtusa 176, canker 214, 286, Cephaleuros virescens 28, dieback 172, 176, 210, Diplocarpon mespili 64, Erwinia amylovora 162–164, fire blight 162–164, fruit rot 284–286, Fusarium lateritium 214, Gymnosporangium clavipes 246, G. confusum 244, leaf spot 64, Nectria cinnabarina 210, nematode 302, Phytophthora cactorum 284–286, powdery mildew 14, Pratylenchus 302, rust 244–246, salt tolerance 454
C. affinis: apple scab 96
C. divaricata (spreading cotoneaster): ozone tolerance 466, sulfur dioxide tolerance 468, winter injury 476
C. horizontalis (rock cotoneaster): dieback 284, fruit rot 284, ozone tolerance 466, Phytophthora cactorum 284
C. integerrimus: apple scab 96
C. salicifolia: Dematophora root rot 306
cotton, see Gossypium hirsutum
cottonwood, see Populus
black: P. trichocarpa
eastern: P. deltoides
Fremont: P. fremontii
Great Plains: P. sargentii
lanceleaf: P. ×acuminata
narrow leaved or yellow: P. angustifolia
Rio Grande: P. fremontii var. wislizenii
swamp: P. heterophylla
Cowania (cliff rose): Erwinia amylovora 162, fire blight 162
cowpea, see Vigna sinensis
cow wheat, see Melampyrum lineare
crabapple, see Malus
American: M. coronaria
prairie: M. ioensis
showy: M. floribunda
Siberian: M. baccata
southern wild: M. angustifolia
Toringo: M. sieboldii
cracks in bark or wood 338–340, 482–484
cranberry, see Vaccinium
mountain: V. vitis-idaea
cranberry bush, see Viburnum
American: V. trilobum
European: V. opulus
crape myrtle, see Lagerstroemia indica
Crataegomespilus: Erwinia amylovora 162, fire blight 162
Crataegus (hawthorn): anthracnose 122–124, blight 60, 126, Botryosphaeria dothidea 172, B. obtusa 176, Botrytis cinerea 60, canker 212, 248, dieback 172, 176, 248, Diplocarpon mespili 64, Erwinia amylovora 162, fire blight 162, Ganoderma applanatum 334, Glomerella cingulata 122–124, Gymnosporangium 242–244, 248, G. clavipes 246–248, G. juniperi-virginianae 240, Hendersonia 82, herbicide tolerance 460, Heterobasidion annosum 314, leaf spot 64, 82, 248, Monilinia johnsonii 126, Nectria galligena 212, ozone tolerance 466, Perenniporia fraxinophila 344, Phyllactinia guttata 16, Podosphaera clandestina 14–16, P. tridactyla 14, powdery mildew 14–16, rust 240–244, 246–248, salt tolerance 454, sul-

fur dioxide tolerance 468, tobacco ringspot virus 414, wood rot 314, 334, 344
C. columbiana (Columbia hawthorn): sulfur dioxide tolerance 468
C. crus-galli (cockspur thorn): Diplocarpon mespili 64, Gymnosporangium globosum 242, leaf spot 64, rust 242, salt tolerance 454
C douglasii (black hawthorn): sulfur dioxide tolerance 468
C. flava (yellow-fruited thorn): Gymnosporangium globosum 242, rust 242
C. intricata: Gymnosporangium globosum 242, rust 242
C. laevigata (English hawthorn): canker 248, dieback 248, Diplocarpon mespili 64, Gymnosporangium clavipes 248, G. confusum 244, 248, G. globosum 242, leaf spot 64, 248, rust 242–244, 246–248, see also C. monogyna
C. ×lavallei (Lavall's hawthorn): flood tolerance 486
C. mollis (hawthorn): Gymnosporangium globosum 248, rust 248
C. monogyna (English hawthorn): Gymnosporangium clavipes 248, rust 248, sulfur dioxide tolerance 468, see also C. laevigata
C. phaenopyrum (Washington hawthorn): Diplocarpon mespili 64, flood tolerance 486, leaf spot 64, sulfur dioxide tolerance 468
C. pruinosa: Gymnosporangium globosum 242, rust 242
creambush, see Holodiscus discolor
creosote bush, see Larrea
Cristulariella depraedens 62
C. moricola 62
C. pyramidalis 62
Cronartium 270–280, 282
C. appalachianum 270
C. coleosporioides 270, 276, 278–282
C. comandrae 270, 278
C. comptoniae 270, 276, 350
C. conigenum 270, 280
C. fusiforme 274
C. kamtschaticum 272
C. occidentale 270
C. quercuum 270, 274, 282
C. quercuum f. sp. fusiforme 274
C. quercuum f. sp. virginianae 274
C. ribicola 270–272
C. strobilinum 270, 280
cross vine, see Bignonia
croton, see Codiaeum variegatum
crowberry, see Empetrum
black: E. nigrum
Cryphonectria 186–190, 450
C. cubensis 190
C. parasitica 186–188
Cryptococcus fagisuga 440
Cryptodiaporthe hystrix 104
C. populea 184
C. salicella 184–186
C. salicina 186
Cryptomeria japonica (Japanese cedar): blight 138, Botryosphaeria dothidea 172, canker 170, Cercospora sequoiae 86, dieback 172, Heterobasidion annosum 314, needle blight 86, nematode 302, Pestalotiopsis funerea 128, Phomopsis juniperovora 138, Pratylenchus 302, root rot 296, 314, salt tolerance 454, Sclerotium rolfsii 296, Seiridium cardinale 170, southern blight 296, sulfur dioxide tolerance 468
Cryptosphaeria populina 218, 338
Cryptosporiopsis 202
cucumber tree, see Magnolia acuminata
Cumminsiella mirabilissima 250
C. sanguinea 250
Cunninghamia lanceolata (China fir): anthracnose 122–124, Glomerella cingulata 122–124
Cupaniopsis (carrotwood): Verticillium wilt 376
Cuphea (false heather): gall 148, Nectriella pironii 148

×Cupressocyparis leylandii (Leyland cypress): canker 170, Pestalotiopsis funerea 128, Seiridium cardinale 170, tip dieback 128
Cupressus (cypress): blight 138, Botryosphaeria obtusa 176, dieback 176, Heterobasidion annosum 314, mistletoe 424, Pestalotiopsis funerea 128, Phoradendron juniperinum 424, root rot 314, Sclerophoma pithyophila 138
C. arizonica (Arizona cypress): blight 86, 138, canker 170, Cercospora sequoiae 86, Gymnosporangium cupressi 248, mistletoe 424, Phomopsis juniperovora 138, rust 248, Seiridium cardinale 170
C. bakeri (Modoc cypress): blight 138, canker 170, Phomopsis juniperovora 138, Seiridium cardinale 170
C. forbesii (tecate cypress): canker 170, Seiridium cardinale 170
C. glabra (smooth-barked Arizona cypress): blight 138, Phomopsis juniperovora 138
C. goveniana (Gowen cypress): blight 138, canker 170, Phomopsis juniperovora 138, Seiridium cardinale 170
C. lusitanica (Portuguese cypress): blight 86, 138, canker 170, Cercospora sequoiae 86, Gymnosporangium clavipes 246, Phomopsis juniperovora 138, rust 246, Seiridium cardinale 170, Sphaeropsis sapinea 136, tip blight 136
C. macnabiana (MacNab cypress): canker 170, Cercospora sequoiae 86, needle blight 86, Seiridium cardinale 170
C. macrocarpa (Monterey cypress): blight 60, 86, 136–138, Botrytis cinerea 60, canker 170, Cercospora sequoiae 86, Phomopsis juniperovora 138, Seiridium cardinale 170, Sphaeropsis sapinea 136
C. pygmaea (Mendocino cypress): canker 170, Seiridium cardinale 170
C. sempervirens (Italian cypress): blight 60, 86, 136, 138, Botrytis cinerea 60, canker 170, Cercospora sequoiae 86, Phomopsis juniperovora 138, Seiridium cardinale 170, Sphaeropsis sapinea 136
C. torulosa (Bhutan cypress): blight 138, Phomopsis juniperovora 138
currant, see Ribes
alpine: R. alpinum
black: R. americanum, R. hudsonianum, R. nigrum
blood: R. sanguineum
blue: R. bracteosum
golden: R. aureum
red: R. rubrum, R. sativum, R. triste
skunk: R. glandulosum
swamp: R. lacustre
Cuscuta (dodder) 436
C. ceanothi 436
Cycas revoluta (Sago palm): Alternaria 90, anthracnose 122–124, Glomerella cingulata 122–124, leaf spot 90
Cyclaneusma minus 38
C. niveum 38
Cydonia oblonga (common quince): anthracnose 122–124, Botryosphaeria dothidea 172, B. obtusa 176, canker 212, Dematophora necatrix 306, Diaporthe ambigua 144, dieback 172, 176, Diplocarpon mespili 64, Erwinia amylovora 162, fire blight 162, fruit rot 126, 144, Glomerella cingulata 122–124, Gymnosporangium 242–244, 248, G. clavipes 246, Hendersonia 82, leaf spot 64, 82, Microsphaeropsis olivacea 82, Monilinia 126, Nectria galligena 212, Phomopsis ambigua 144, Phyllactinia guttata 16, Phymatotrichum omnivorum 292, Podosphaera clandestina 16, powdery mildew 16, root rot 292, 296, 306, rust 242–248, salt tolerance 454, Sclerotium rolfsii 296, sulfur dioxide tolerance 468

Cylindrocarpon 212, 442
C. mali 212
Cylindrocladiella parva **294**
Cylindrocladium 294
C. scoparium **294**
Cylindrosporella caryae 108
C. ulmea 108
Cylindrosporium **70**
C. fraxini 70
C. padi 66
Cyphomandra betacea (tree tomato): gall 148, Nectriella pironii 148
cypress, see Chamaecyparis, ×Cupressocyparis, Cupressus, Taxodium
　Alaska, Alaska yellow, or Nootka: Chamaecyparis nootkatensis
　Arizona: Cupressus arizonica
　bald: T. distichum
　Bhutan: Cupressus torulosa
　false: Chamaecyparis
　Gowen: Cupressus goveniana
　Hinoki: Chamaecyparis obtusa
　Italian: Cupressus sempervirens
　Lawson: Chamaecyparis lawsoniana
　Leyland: ×Cupressocyparis leylandii
　MacNab: Cupressus macnabiana
　Mendocino: Cupressus pygmaea
　Modoc: Cupressus bakeri
　Monterey: Cupressus macrocarpa
　mourning: Chamaecyparis funebris
　Portuguese: Cupressus lusitanica
　Sawara: Chamaecyparis pisifera
　smooth-barked Arizona: Cupressus glabra
　tecate: Cupressus forbesii
Cystopteris (bladder fern): rust 264, Uredinopsis 264
Cytisus (broom): Botryosphaeria obtusa 176, dieback 176, Microsphaeropsis olivacea 82
Cytophoma pruinosa **446**
Cytospora **194–202**, 214, 374
C. abietis 194
C. chrysosperma 74, 200
C. cincta 198
C. kunzei 196
C. leucostoma 198
C. nivea 200
Cytosporina 220

Dacrydium: Pestalotiopsis funerea 128
Daedalea unicolor 358
Daphne (daphne): Chalara elegans 294, fasciation **492**, root rot 294–296, Sclerotium rolfsii 296, southern blight 296, tobacco ringspot virus 414, Verticillium wilt 376
D. cneorum (garland flower): Chalara elegans 294, root rot 294
D. mezurum: tomato ringspot virus 414
Dasyscypha willkommii 236
decay of wood in roots, trunk, and limbs **304–358**
decline **440–450,** see also entries for individual plants
Delonix (royal poinciana): Schizophyllum commune 358
Dematophora necatrix **306**
Dendrothele 168
deodar, see Cedrus deodara
Deutzia (deutzia): Meloidogyne 298, root knot nematode 298
devilwood, see Osmanthus
Diaporthe 140–144, 148
D. alleghaniensis 142–144
D. cinerascens 142–144
D. citri **142**
D. cubensis 190
D. eleagni 140
D. eres 142–**144**
D. gardeniae **148**
Dibotryon morbosum 152
Dichotomanthes: Erwinia amylovora 162, fire blight 162
Dictosperma album (princess palm): lethal yellowing 390
Didymascella 44
D. thujina **44**
Didymosporina aceris **104**

dieback **122, 128, 132, 142–144, 172–186, 208–210, 214, 224–226, 230–232, 444–450**
Diervilla (bush honeysuckle): tobacco ringspot virus 414
Dimerosporium 30
Diospyros (persimmon): anthracnose 122–124, Botryosphaeria dothidea 172, B. obtusa 176, B. rhodina 180, Diaporthe diospyri 144, dieback 144, 172, 176, 180, flood tolerance 486, Ganoderma 332, Glomerella cingulata 122–124, Hericium erinaceus 344, Heterobasidion annosum 314, Microsphaeropsis olivacea 82, mistletoe 422, nematode 302, Phomopsis diospyri 144, Phoradendron serotinum 422, Radopholus similis 302, root rot 332, Schizophyllum commune 358, Verticillium wilt 376, wood rot 314, 344, 358
D. kaki (Japanese persimmon): Acremonium diospyri 372, persimmon wilt 372
D. texana (black or Texas persimmon): Acremonium diospyri 372, Alternaria 90, fruit rot 90, persimmon wilt 372
D. virginiana (common persimmon): Acremonium diospyri **372**, Cercospora fuliginosa **88**, flood tolerance 486, ice glaze injury 490, leaf spot **88**, persimmon wilt **372**, Podosphaera clandestina 16, powdery mildew 16, Schizophyllum commune 372
Diplocarpon maculatum 64
D. mespili **64**
D. rosae **66**
Diplodia 172, 178
D. longispora 178
D. natalensis 180
D. pinea 136
D. tumefaciens 150
Diplodina microsperma 186
D. salicis 186
Discella acerina 104
D. carbonacea 186
D. salicis 186
Discosporium populeum 184
Discula **104–106, 110–114,** 122
D. betulina **114**
D. platani 112
D. quercina 110
D. umbrinella 110
Distichlis spicata (marsh grass): Puccinia sparganioides 252, rust 252
Dizogtheca elegantissima (false aralia): Alternaria panax 90, leaf blight 90
Docynia: Erwinia amylovora 162, fire blight 162
dodder, see Cuscuta
dodder laurel, see Cassytha filiformis
Dodonea (hop seed bush): Verticillium wilt 376
dog urine damage **452**
dogwood, see Cornus
　alternate-leaved: C. alternifolia
　blood-twig: C. sanguinea
　flowering: C. florida
　gray or panicled: C. racemosa
　kousa: C. kousa
　Pacific: C. nuttallii
　red-osier: C. sericea, C. stolonifera
　silky: C. amomum
　Tatarian: C. alba
Dombeya: gall 148, Nectriella pironii 148
Dothichiza populea 184
Dothiorella 172–174
D. quercina **178**
D. ulmi 372
Dothistroma 70
D. pini 48
D. septospora 48
Douglas-fir, see Pseudotsuga menziesii
　big-cone: P. macrocarpa
downy mildew **486**
Dracena (dracena): gall 148, Nectriella pironii 148
Drepanopeziza populi-albae 120
D. populorum 120
D. punctiformis 120
drought damage 224, **340,** 450, **476–478**

Dryas (mountain avens): Erwinia amylovora 162, fire blight 162
Dryopteris (wood fern): Milesina 264, rust 264, Uredinopsis 264
Dutch elm disease **366,** 386
dwarf mistletoe **426–434**

Echinocactus (cactus, visnaga): Fusarium rot 72, Pseudomonas fluorescens 158, soft rot 158
Echinodontium tinctorium **348**
edema **486**
Elaeagnus (autumn or Russian olive, elaeagnus): actinorhizae 502, Armillaria tabescens 312, Botryosphaeria dothidea 172, B. obtusa 176, B. rhodina 180, canker 180, dieback 172, 176, 180, 208, 210, nematode 302, nitrogen-fixing nodules 502, root rot 312, salt tolerance 454, Verticillium wilt 376, Xiphinema americanum 302
E. angustifolia (Russian olive): bacterial wetwood 392, Botryosphaeria rhodina **180,** canker **140,** 198, **208,** dieback **180,** 198, **208–210,** fluoride tolerance 470, Hendersonia 82, herbicide tolerance 460, Leucostoma 198, Nectria cinnabarina 210, nematode 302, Phomopsis arnoldiae **140,** root rot 296, Sclerotium rolfsii 296, southern blight 296, Tubercularia ulmea **208,** twig canker 82, Valsa sordida 200, Xiphinema americanum 302
E. commutata (silverberry): canker 140, ozone tolerance 466, Phomopsis arnoldiae 140, Phymatotrichum omnivorum 292, Puccinia coronata 250, root rot 292, rust 250
E. multiflora (cherry elaeagnus): canker 140, Phomopsis arnoldiae 140
E. pungens (thorny elaeagnus): Rhizoctonia blight 94
Elaeis guineensis (oil palm): butt rot 304, **330,** Ganoderma **330–332,** Hypoxylon deustum 304, root rot 296, 304, **330,** Sclerotium rolfsii 296, southern blight 296
elder, see Acer, Sambucus
　blue: S. caerulea
　box: A. negundo
　American: S. canadensis
　European: S. nigra
　European red: S. racemosa
　southern: S. simpsonii
Elephantopus (elephant's-foot): Coleosporium vernoniae 268, rust 268
elm, see Planera, Ulmus
　American or white: U. americana
　Belgian: U. ×hollandica cv. 'Belgica'
　cedar: U. crassifolia
　Chinese: U. parvifolia
　cork or rock: U. thomasii
　Dutch: U. ×hollandica
　English: U. procera
　European field or smooth-leaf: U. carpinifolia
　European white or Russian: U. laevis
　Japanese: U. davidiana var. japonica
　red or slippery: U. rubra
　Scots or wych: U. glabra
　September: U. serotina
　Siberian: U. pumila
　water: P. aquatica
　winged: U. alata
elm mottle virus 406
elm phloem necrosis 388
elm yellows **388**
Elytroderma deformans **36**
E. torres-juanii 36
Empetrum (crowberry): Heterobasidion annosum 314
E. nigrum (black crowberry): Chrysomyxa 266, Exobasidium empetri 26, rust 266
Encoelia pruinosa **228**
Endocronartium harknessii 270, **278,** 280, **282**
Endothia 450
E. gyrosa **192**
E. parasitica 188

Enterobacter agglomerans 382
E. cloacae 382
Entomosporium mespili 64
Epilobium (fireweed): Pucciniastrum pustulatum 260, rust 260
epiphytes **504**
Epipolaeum abietis **30**
Erica (heath): Verticillium wilt 376
Erigeron (daisy fleabane): Coleosporium asterum 268, rust 268
Erigonum (erigonum, umbrella plant): Verticillium wilt 376
Eriobotrya japonica (loquat): anthracnose 122–124, Armillaria tabescens 312, Botryosphaeria dothidea 172, B. obtusa 176, B. rhodina 180, dieback 172, 176, 180, Diplocarpon mespili 64, Erwinia amylovora 162–**164,** fire blight 162–**164,** Glomerella cingulata 122–124, leaf spot 64, root rot 296, 312, scab **98,** Sclerotium rolfsii 296, southern blight 296, Spilocaea pyracanthae **98**
Eriophyes celtis **20**
Erwinia amylovora **162–164**
E. carnegiana 158
E. carotovora 158
E. herbicola 382, 482
E. nigrifluens 166
E. nimipressuralis 382
E. rubrifaciens 166
Erysiphe liriodendri 16
E. polygoni **14–16**
E. trina 14
Erythrina herbacea (coral bean): Rhizoctonia blight 94
Euantennaria 30
Eucalyptus (eucalyptus, gum): Botryosphaeria dothidea 172, B. rhodina 180, Botrytis blight 60, bud burl 494, canker 190, Cryphonectria cubensis 190, Cylindrocladium 294, dieback 172, 180, Ganoderma applanatum 334, Hendersonia eucalypticola 82, Laetiporus sulfureus 346, leaf spot 82, lignotuber 494, Microsphaeropsis olivacea 82, Phytophthora cinnamomi 288, powdery mildew 16, root rot 288, 294, 330, Schizophyllum commune 358, Sphaerotheca pannosa 16, wood rot 334, 346
E. camaldulensis (Murray red gum): canker 190, Cryphonectria cubensis 190
E. cinerea (silver dollar tree): Cercospora epicoccoides **86–88,** leaf spot 86–88
E. citriodora (lemon-scented gum): canker 190, Cryphonectria cubensis 190
E. deglupta: canker 190, Cryphonectria cubensis 190
E. diversicolor (karri gum): canker 192, Endothia gyrosa 192
E. globulus (blue or Tasmanian blue gum): Cercospora epicoccoides 86, Hendersonia eucalypticola **82,** leaf spot **82,** 86–88, Phaeolus schweinitzii 322, root rot 322
E. grandis (rose gum): canker **190,** Cryphonectria cubensis **190**
E. maculata (spotted gum): canker 190, Cryphonectria cubensis 190
E. microcorys (tallowwood): canker 190, Cryphonectria cubensis 190
E. paniculata (gray ironbark): canker 190, Cryphonectria cubensis 190
E. propinqua: canker 190, Cryphonectria cubensis 190
E. robusta (swamp mahogany): canker 190, Cryphonectria cubensis 190
E. saligna (Sydney blue gum): canker 190, Cryphonectria cubensis 190
E. tereticornis (forest red gum): canker 190, Cryphonectria cubensis 190
E. torelliana: Cryphonectria cubensis 190
E. urophylla: Cryphonectria cubensis 190
Eugenia (eugenia): Botryosphaeria dothidea 172, dieback 172
Euonymus (euonymus): dieback 168, 210, Hendersonula toruloidea 168, Microsphaera pusilla 16, Nectria cinnabarina 210, powdery mildew 16

E. alata (winged euonymus, winged spindletree): *Botryosphaeria dothidea* 484, canker 214, 484, dieback 208, *Fusarium lateritium* 214, salt tolerance 454, *Tubercularia ulmea* 208
E. atropurpurea (burning bush): anthracnose 120, *Marssonina thomasiana* 120
E. fortunei (evergreen euonymus, winter creeper): *Agrobacterium tumefaciens* **156,** anthracnose 122–124, crown gall **156,** flood tolerance 486, freeze injury **482,** *Glomerella cingulata* 122–124, salt tolerance 454, tomato ringspot virus 414
E. japonica (Japanese spindletree): anthracnose 122–**124,** fasciation 400, *Glomerella cingulata* 122–**124,** *Microsphaera euonymi-japonici* **14**–16, *Phymatotrichum omnivorum* 292, powdery mildew **14**–16, rhabdovirus 400, root rot 292
E. kiautschovica: tomato ringspot virus 414
Euphorbia lathyris (caper spurge): *Macrophomina phaseolina* 292, root rot 292
E. longan (longan): *Cephaleuros virescens* **28**
E. pulcherrima (poinsettia): *Alternaria* 90, *Botryosphaeria dothidea* 172, Botrytis blight 60, *Chalara elegans* 294, dieback 172, gall 148, leaf spot 90, *Nectriella pironii* 148, ozone tolerance 466, root rot 294
Europhium trinacriforme 448
Eutypella parasitica **220**
Exobasidium **26**
E. camelliae **26**
E. vaccinii **26**
E. vaccinii-uliginosi **26**
Exochorda (pearlbush): *Erwinia amylovora* 162, fire blight 162
Exosporium palmivorum 92

Fabraea maculata 64
Fagus (beech): *Bjerkandera adusta* 346, butt rot 304, 334, canker 212, 440–442, *Cerrena unicolor* 358, *Climacodon septentrionalis* 344, *Coniothrium* 82, *Dendrothele acerina* 168, dieback 210, *Fomes fomentarius* 346, *Fomitopsis pinicola* 348, *Ganoderma applanatum* 334, *Globifomes graveolens* 346, *Hericium erinaceus* 344, *Heterobasidion annosum* 314, *Hypoxylon* 222–226, *H. cohaerens* 226, *H. deustum* 304, *Inonotus hispidus* 356, *Laetiporus sulfureus* 346, leaf spot 82, lightning injury 490, *Microsphaera erineophila* 16, mistletoe 422, *Nectria cinnabarina* 210, *N. galligena* 212, *Oxyporus latemarginatus* 328, *O. populinus* 344, *Perenniporia medulla-panis* 168, *Phyllactinia guttata* 16, *Polyporus squamosus* 344, powdery mildew 16, root rot 304–306, *Schizophyllum commune* 358, *Scorias spongiosa* 30, smooth patch 168, wood rot 168, 314, 328, 334, 344–348, 356–358, *Xylaria* 306
F. grandifolia (American beech, beech): beech bark disease **440–442,** bleeding canker 286, *Botryosphaeria quercuum* 178, butt rot 304, **334,** canker 192, 236, 286, **342, 440–442,** *Ceratocystis coerulescens* 362, compartmentalization of wound-associated discolored and decayed wood **336,** dieback 178, *Endothia gyrosa* 192, flood tolerance 486, frost susceptibility 482, *Ganoderma applanatum* **334,** herbicide tolerance 460, *Hypoxylon* 222–226, *H. cohaerens* **226,** *H. deustum* 304, ice glaze injury 490, *Inonotus glomeratus* **356,** *I. obliquus* 352, *Nectria coccinea* var. *faginata* 440–**442,** *Phellinus igniarius* **342,** *Phytophthora cactorum* 286, root rot 304, salt tolerance 454, sterile conk **356,** *Strumella coryneoidea* 236, sulfur dioxide tolerance 468, *Urnula craterium* 236, winter sunscald 484, wood decay 304, 334–**336, 342,** 352, **356**
F. sylvatica (European beech): beech bark disease 440–442, bleeding canker 286, butt rot **304,** canker **192,** 286, *Endothia gyrosa* **192,** fluoride tolerance 470, *Ganoderma lucidum* 332, *Hypoxylon cohaerens* 226, *H. deustum* **304,** *Nectria coccinea* 442, *N. ditissima* 442, ozone tolerance 466, *Phytophthora cactorum* 286, root and butt rot **304**–306, 332, salt tolerance 454, sulfur dioxide tolerance 468
fall browning **498**
farkleberry, see *Vaccinium, V. arboreum*
fasciation **492**
Fatsia japonica (Japanese fatsia): gall 148, *Nectriella pironii* 148
Feijoa sellowiana (pineapple guava): Rhizoctonia blight 94
Fendlera (fendlera): *Gymnosporangium speciosum* 240, 246, rust 240, 246
fern
bladder, see *Cystopteris*
bracken: *Pteridium*
flowering: *Osmunda*
lady: *Athyrium*
oak: *Gymnocarpium*
ostrich: *Matteuccia*
polypody: *Polypodium*
sensitive: *Onoclea sensibilis*
sweetfern: *Comptonia peregrina*
wood: *Dryopteris*
fernbush, see *Chamaebatiaria millefolium*
Ferocactus acanthodes (California barrel cactus): cactus virus X 418
Ficus (fig): anthracnose 122–124, *Botryosphaeria dothidea* 172, *B. obtusa* 176, *B. rhodina* 180, dieback 172, 176, 180, gall 148, *Glomerella cingulata* 122–124, *Meloidogyne* 298, *Microsphaeropsis olivacea* 82, *Nectriella pironii* 148, *Ophiodothella* 84, root knot nematode 298, *Schizophyllum commune* 358, tar spot 84
F. aurea (strangler fig): *Cerotelium fici* 238, gall 148, *Nectriella pironii* 148, *Ophiodothella* 84, rust 238, tar spot 84
F. benjamina (weeping fig): canker **142**–144, *Diaporthe cinerascens* **142**–144, dieback 144, gall 148, *Nectriella pironii* 148, *Phomopsis cinerascens* 144, Verticillium wilt 376
F. carica (common fig): canker 144, 214, *Cerotelium fici* **238,** *Cylindrocladium* 294, *Dematophora necatrix* 306, *Diaporthe cinerascens* 144, dieback 144, 168, 210, *Fusarium lateritium* 214, gall 148, *Hendersonula toruloidea* 168, *Nectria cinnabarina* 210, *Nectriella pironii* 148, nematode 302, *Phymatotrichum omnivorum* 292, *Pratylenchus* 302, root rot 292–296, 306, rust **238,** *Sclerotium rolfsii* 296, southern blight 296, *Xiphinema index* 302
F. elastica (Indian rubber tree, rubber plant): *Alternaria* 90, anthracnose 122–**124,** *Armillaria tabescens* 312, Botrytis blight 60, canker **124,** *Glomerella cingulata* 122–**124,** leaf spot 90, root rot 312
fig, see *Ficus*
common: *F. carica*
strangler: *F. aurea*
weeping: *F. benjamina*
filbert, see *Corylus*
American: *C. americana*
beaked: *C. cornuta*
European: *C. avellana*
fir, see *Abies, Cunninghamia, Pseudotsuga*
alpine or subalpine: *A. lasiocarpa*
balsam: *A. balsamea*
California red: *A. magnifica*
China: *Cunninghamia lanceolata*
cork: *A. lasiocarpa* var. *arizonica*
Douglas-: *Pseudotsuga*
Fraser: *A. fraseri*
grand or lowland white: *A. grandis*
noble: *A. procera*
Pacific silver: *A. amabilis*
Sakhalin or Todo: *A. sachalinensis*
Shasta: *A. magnifica* var. *shastensis*
Siberian: *A. siberica*
silver: *A. alba*
white: *A. concolor*
fire blight **162–164**
fire scar **336**
firethorn, see *Pyracantha, P. coccinea* 'Lalandei': *P. coccinea* cv. 'Lalandei'
fireweed, see *Epilobium*
Firmiana (phoenix tree): *Botryosphaeria dothidea* 172, dieback 172
flannelbush, see *Fremontodendron*
fleabane, daisy, see *Erigeron*
flood damage **486**
flower blight **58–60,** 160
fluoride air pollutant injury **470**
flux: alcoholic or white **378,** 382, brown **382**
Fomes annosus 314
F. applanatus 334
F. connatus 344
F. everhartii 346
F. fomentarius **344**
F. fraxinophilus 344
F. igniarius 342
F. pini 350
F. pinicola 348
F. rimosus 344
Fomitopsis pinicola **348**
Forestiera (desert olive, forestiera): *Microsphaera neomexicana* 16, *M. syringae* 16, powdery mildew 16
F. acuminata (swamp privet): *Coleosporium minutum* 268, flood tolerance 486, gall 154, *Pseudomonas syringae* pv. *savastanoi* 154, *Puccinia sparganioides* 252, rust 252, 268
F. neomexicana (desert olive): gall 154, *Pseudomonas syringae* pv. *savastanoi* 154
F. segregata (Florida privet): *Puccinia sparganioides* 252, rust 252
Forsythia (forsythia): *Alternaria* 90, arabis mosaic virus 416, flood toleranca 486, gall **146**–148, 154, herbicide tolerance 460, leaf spot 90, *Meliodogyne hapla* **298,** *Nectriella pironii* 148, nematode **298**–302, *Phomopsis* **146,** *Pratylenchus* 302, *Pseudomonas syringae* pv. *savastanoi* 154, pv. *syringae* 160, root knot **298,** salt tolerance 454, tobacco ringspot virus 414–**416,** yellow net **416**
F. ×intermedia: elm mottle virus 406, ozone tolerance 466, salt tolerance 454, sulfur dioxide tolerance 468
F. viridissima: sulfur dioxide tolerance 468
Fortunella (kumquat): *Alternaria* 90, fruit rot 90, greasy spot 72, *Mycosphaerella citri* 72
Fragaria (strawberry): *Erwinia amylovora* 162, fire blight 162
frangipani, see *Plumeria*
Frankia **502**
Franklinia (franklinia): algal leaf spot 28, *Cephaleuros virescens* 28
Fraxinus (ash): anthracnose 106, bark rot 168, *Botryosphaeria dothidea* 172, *B. obtusa* 176, canker 212, 214, 446, *Cerrena unicolor* 358, decline 446, *Dendrothele* 168, dieback 172, 176, 200, 210, *Discula* 106, drought injury 476, *Fusarium* 214, *Ganoderma applanatum* 334, *Hendersonia* 82, herbicide injury 456, herbicide tolerance 460, *Heterobasidion annosum* 314, *Laetiporus sulfureus* 346, leaf spot 70, lightning injury 490, *Microsphaera fraxini* 16, mistletoe 422, *Mycosphaerella* 70, *Nectria cinnabarina* 210, *N. galligena* 212, *Oxyporus populinus* 344, ozone injury 466, *Perenniporia fraxinophila* 344, *P. medulla-panis* 168, *Phellinus igniarius* 342, *Phyllactinia guttata* 16, powdery mildew 16, *Puccinia sparganioides* 252, rust 252, *Schizophyllum commune* 358, smooth patch 168, *Valsa* 200–202, Verticillium wilt 376, wood rot 168, 314, 334, 342–346, 358
F. americana (white ash): anthracnose **106,** arabis mosaic virus 414, ash yellows **394, 484,** blight **106, 252,** butt rot 304, **334,** canker **252, 446, 484,** *Cristulariella moricola* **62,** *Cytophoma pruinosa* **446,** decline 414, 446, dieback **394, 446,** *Discula* **106,** drought damage 446, flood tolerance 486, freeze damage **394, 484,** frost susceptibility 482, *Fusicoccum* 446, gall 154, *Ganoderma lucidum* 332, herbicide tolerance 460, *Hypoxylon deustum* 304, ice glaze injury 490, leaf spot **62, 70,** mycoplasmalike organism **394,** *Mycosphaerella fraxinicola* 70, nematode 302, ozone injury 446, *Perenniporia fraxinophila* 344, *Pseudomonas syringae* pv. *savastanoi* 154, *Puccinia sparganioides* **252–254,** root rot 304, rust **252–254,** salt tolerance 454, sulfur dioxide tolerance 468, tobacco mosaic virus **414,** tobacco ringspot virus **414,** tomato ringspot virus 414, witches'-broom **394, 484,** wood rot 304, 332, 344, *Xiphinema americanum* 302
F. angustifolia: mycoplasmalike organism 394, ash yellows 394
F. berlandierana (Berlandier ash): *Puccinia sparganioides* 252, rust 252, witches'-broom 394
F. biltmoreana (Biltmore ash): leaf spot 70, *Mycosphaerella fraxinicola* 70
F. bungeana: mycoplasmalike organism 394, ash yellows 394
F. caroliniana (Carolina or water ash): *Puccinia sparganioides* 252, rust 252
F. excelsior (European ash): arabis mosaic virus 414, canker 154, fluoride tolerance 470, gall 154, mycoplasmalike organism 394, ozone tolerance 466, *Pseudomonas syringae* pv. *savastanoi* 154, salt tolerance 454, sulfur dioxide tolerance 468, tobacco mosaic virus 414, yellows 394
F. floribunda (Himalayan ash): *Pseudomonas syringae* pv. *syringae* 160
F. holotricha cv. 'Moraine' ('Moraine' ash): tobacco mosaic virus 414, tobacco ringspot virus 414
F. latifolia (Oregon ash): *Cylindrosporium* **70,** herbicide tolerance 460, leaf spot **70,** mycoplasmalike organism 394, *Mycosphaerella fraxinicola* **70,** *Perenniporia fraxinophila* 344, *Puccinia sparganioides* 252, rust 252, wood rot 344, yellows 394
F. nigra (black ash): anthracnose 106, *Discula* 106, ice glaze injury 490, *Inonotus hispidus* 356, leaf spot 70, *Mycosphaerella fraxinicola* 70, mycoplasmalike organism 394, *Oxyporus latemarginatus* 328, *Perenniporia fraxinophila* 344, *Puccinia sparganioides* 252, rust 252, salt tolerance 454, wood rot 328, 344, yellows 394
F. ornus (flowering ash): gall 154, mycoplasmalike organism 394, ozone tolerance 466, *Pseudomonas syringae* pv. *savastanoi* 154, pv. *syringae* 160, yellows 394
F. pennsylvanica (red ash): anthracnose 106, ash yellows 394, decline 446, dieback 446, *Discula* 106, fluoride tolerance 470, leaf spot 70, mycoplasmalike organism 394, *Mycosphaerella fraxinicola* 70, sulfur dioxide tolerance 468
F. pennsylvanica var. *lanceolata* (green ash): anthracnose 106, ash yellows 394, *Botryosphaeria quercuum* 178, decline 446, *Discula* 106, flood tolerance 486, *Ganoderma lucidum* 332, herbicide tolerance 460, leaf spot 70, mycoplasmalike organism 394, *Mycosphaerella fraxinicola* 70, nematode

F. pennsylvanica (cont.)
302, ozone injury 466, *Perenniporia fraxinophila* **344,** *Puccinia sparganioides* 252, root rot 332, rust 252, salt tolerance 454, sulfur dioxide tolerance 468, tobacco mosaic virus 414, tobacco ringspot virus 414, wood decay 332, **344,** *Xiphinema americanum* 302

F. potamophila: mycoplasmalike organism 394, yellows 394

F. profunda (pumpkin ash): flood tolerance 486, mycoplasmalike organism 394, ozone tolerance 466, *Puccinia sparganioides* 252, rust 252

F. quadrangulata (blue ash): *Botryosphaeria dothidea* **172,** canker **172,** dieback **172,** herbicide tolerance 460, leaf spot 70, *Mycosphaerella fraxinicola* 70, mycoplasmalike organism 394, ozone tolerance 466, *Perenniporia fraxinophila* **344,** *Puccinia sparganioides* 252, rust 252, wood decay 344, yellows 344

F. velutina (velvet ash): anthracnose 106, *Discula* 106, gall 154, *Perenniporia fraxinophila* 344, *Pseudomonas syringae* pv. *savastanoi* 154, *Puccinia sparganioides* 252, rust 252, wood decay 344

F. velutina var. *glabra* (Arizona or Modesto ash): anthracnose **106,** *Discula* **106,** fluoride tolerance 470, leaf blight **106,** salt tolerance 454

freeze damage **214, 340, 394,** 446, 454, **476–484**

Fremontodendron (flannelbush): Verticillium wilt 376

fringe tree, see *Chionanthus*

Fuchsia (fuchsia): *Botryosphaeria dothidea* 172, Botrytis blight 60, dieback 172, ozone tolerance 466, *Pucciniastrum pustulatum* 260–**262,** rust 260–**262**

Fumago vagans 30

Fusarium 72, 214–216, 292, 296, 378

F. dimerum var. *violaceum* **72**

F. lateritium **214**

F. lateritium f. sp. *pini* 216

F. moniliforme var. *subglutinans* **216**

F. oxysporum **296, 378**

F. oxysporum f. sp. *pernicosum* **378**

F. solani 214, 296

Fusicladium pyrorum 98

Fusicoccum 446

Gaillardia (gaillardia): *Coleosporium pacificum* 268, rust 268

gale, sweet, see *Myrica gale*

gall **26, 146–156, 274, 282, 298, 422, 494**

Ganoderma **330–334,** 444, 450

G. applanatum 330–**334**

G. brownii 332

G. curtisii 332

G. lucidum **332,** 444

G. oregonense **332**

G. tornatum 330

G. tsugae **332**

G. zonatum **330**

Gardenia (gardenia): *Alternaria* 90, *Erysiphe polygoni* 14–16, leaf spot 90, powdery mildew 14–16

G. jasminoides (cape jasmine, common gardenia): algal leafspot 28, canker **148,** *Cephaleuros virescens* 28, *Diaporthe gardeniae* 148, dieback 148, gall **148,** *Meloidogyne* **298,** root knot nematode **298,** sooty mold **30**

garland flower, see *Daphne cneorum*

Gaultheria (salal, wintergreen): *Exobasidium vaccinii* 26, *Microsphaera vaccinii* 16, powdery mildew 16

G. hispidula (creeping snowberry): *Chrysomyxa* 266, rust 266

G. shallon (salal): *Coniothyrium* **82,** leaf spot **82,** *Pestalopezia brunneo-pruinosa* 128, *Pestalotiopsis gibbosa* 128

Gaussia attenuata (Puerto Rican gaussia): lethal yellowing 390

gayfeather, see *Liatris*

Gaylussacia (huckleberry): *Exobasidium vaccinii-uliginosi* 26, flyspeck leaf spot 84, *Ophiodothella vaccinii* 84

Gelsemium (yellow-jessamine): *Botryosphaeria obtusa* 176, *B. rhodina* 180, dieback 176, 180

G. sempervirens (Carolina jessamine): gall 148, *Nectriella pironii* 148

Genista (broom): *Botryosphaeria obtusa* 176, dieback 176

Geocaulon livida (northern comandra): *Cronartium comandrae* 270, 278, rust 270, 278

Geum (avens): *Erwinia amylovora* 162, fire blight 162

Gibberella baccata 214

G. fujikurai var. *subglutinans* 216

Ginkgo biloba (ginkgo): anthracnose 122–124, *Glomerella cingulata* 122–124, nematode 302, *Oxyporus populinus* 344, ozone tolerance 466, *Pestalotiopsis funerea* 128, *Phymatotrichum omnivorum* 292, *Pratylenchus* 302, root rot 292, salt tolerance 454, sulfur dioxide tolerance 468

girdling roots **488**

Gleditsia aquatica (water locust): flood tolerance 486, *Linospora gleditsiae* 84, *Plagiosphaeria gleditschiae* 84, tar spot 84

G. japonica (Japanese honey locust): canker 204, *Thyronectria austro-americana* 204, wilt 204

G. macrantha (honey locust): *Ganoderma lucidum* 332, root rot 332

G. triacanthos (honey locust): anthracnose 122–124, bark rot 168, *Botryosphaeria dothidea* 172, *B. obtusa* 176, *B. rhodina* 180, butt rot 334, canker **204, 210,** compartmentalization of wounds **338,** dieback 172, 176, 180, flood tolerance 486, *Ganoderma applanatum* 334, *G. curtisii* 332, *G. lucidum* 332, *Glomerella cingulata* 122–124, *Hendersonia* 82, herbicide injury **458,** herbicide tolerance 460, ice glaze injury 490, *Inonotus hispidus* 356, *Laetiporus sulfureus* 346, *Linospora gleditsiae* 84, *Microsphaera ravenelii* 16, mistletoe 422, *Nectria cinnabarina* **210,** *Oxyporus latemarginatus* 328, *O. populinus* 344, ozone tolerance 466, *Perenniporia medulla-panis* 168, *Phellinus igniarius* 342, *P. spiculosus* 354, *Phoradendron serotinum* 422, *Phymatotrichum omnivorum* 292, *Plagiosphaeria gleditschiae* 84, powdery mildew 16, root rot 292, 306, **332,** salt tolerance 452–454, *Schizophyllum commune* 358, sulfur dioxide tolerance 468, tar spot 84, *Thyronectria austro-americana* **204,** witches'-broom 396, wood decay 168, 328, 332–334, 342–346, 354–358, *Xylaria mali* 306

Gloeosporium 104, 108–116, 122–124

G. acaciae 124

G. aridum 106

G. betulicola 114

G. caryae 108

G. coryli 114

G. elasticae 124

G. fructigenum 124

G. inconspicuum 108

G. lunatum 12

G. nervisequum 110–112

G. platani 112

G. quercinum 110

G. quercuum 110

G. robergei 116

G. ulmeum 108

G. ulmicolum 108

Glomerella acaciae 124

G. cingulata 102, **122–124**

G. miyabeana **102**

Glomopsis lonicerae 94

glory-bower, see *Clerodendrum bungei*

Gmelina arborea: *Ceratocystis fimbriata* 362

Gnomonia caryae **108**

G. dispora 108

G. errabunda 110

G. leptostyla **118**

G. nerviseda 108

G. pecanae 108

G. platani 112

G. quercina 110

G. veneta 110–112

Gnomoniella carpinea **116**

goatsbeard, see *Aruncus dioicus*

Godetia: *Pucciniastrum pustulatum* 262, rust 262

golden-chain tree, see *Laburnum anagyroides*

golden-rain tree, see *Koelreuteria*

goldenrod, see *Solidago*

Gonatorrhodiella highlei 442

gooseberry, see *Ribes*
common: *R. grossularia*
prickly: *R. cynosbati*

gorse, see *Ulex*

Gossypium hirsutum (cotton): *Phymatotrichum omnivorum* **292,** root rot **292**

graft union incompatibility **404, 492**

grape, see *Coccoloba, Mahonia, Vitis*
European bunch: *V. vinifera*
fox: *V. labrusca*
muscadine: *V. rotundifolia*
Oregon: *M. aquifolium*
sea: *C. uvifera*

grapefruit, see *Citrus, C. ×paradisi*

Graphiola congesta 92

G. phoenicis **92**

G. thaxteri 92

grass, marsh, see *Distichlis spicata, Spartina*

graybeard, see *Tillandsia usneoides*

greenbrier, see *Smilax*

green ring mottle virus **408**

Gremmeniella abietina 232

Grevillea (spider flower): *Botryosphaeria dothidea* 172, dieback 172, *Schizophyllum commune* 358

G. robusta (silk oak): *Botryosphaeria rhodina* 180, dieback 180, *Phymatotrichum omnivorum* 292, root rot 292

Grindelia (gumweed): *Coleosporium asterum* 268, rust 268

groundsel tree, see *Baccharis halmifolia*

grouseberry, see *Vaccinium scoparium*

Grovesinia pyramidalis 62

guava, see *Psidium guajava*
pineapple: *Feijoa sellowiana*

guayule, see *Parthenium*

Guignardia aesculi 76, **80**

G. bidwellii **80**

gum, see *Eucalyptus, Liquidambar, Nyssa*
black or sour: *N. sylvatica*
blue or Tasmanian blue: *E. globulus*
forest red: *E. tereticornis*
karri: *E. diversicolor*
lemon-scented: *E. citriodora*
Murray red: *E. camaldulensis*
rose: *E. grandis*
sour: *N. sylvatica*
spotted: *E. maculata*
sweet: *L. stryaciflua*
Formosan: *L. formosana*
Sydney blue: *E. saligna*
tupelo: *N. aquatica*

gumweed, see *Grindelia*

Gymnocarpium (oak fern): *Uredinopsis* 264, rust 264

Gymnocladus dioica (Kentucky coffeetree): *Ganoderma lucidum* 332, herbicide tolerance 460, ozone tolerance 466, root rot 332, Verticillium wilt 376

Gymnosporangium **240–248**

G. bermudianum 240

G. bethelii 248

G. biseptatum 248

G. clavariiforme 244, 248

G. clavipes 240–**242,** 244–**248**

G. confusum **242,** 244, **246**

G. connersii 242

G. cornutum 248

G. cupressi 248

G. effusum 248

G. ellisii 240, 248

G. fuscum 248

G. gaeumannii 240

G. globosum 240–242, **246–248**

G. juniperi-virginianae **240**–242

G. kernianum 248

G. libocedri **244–246**

G. nidus-avis 242–**244,** 248

G. nootktense 240

G. speciosum 240, **244–246**

G. tremelloides 248

Gyrostroma austro-americana 204

hackberry, see *Celtis occidentalis*
netleaf: *C. reticulata*

hail injury **488**

hairy root 156

Halesia (silverbell): *Hypoxylon punctulatum* 226, *Oxyporus populinus* 344, *Phaeolus schweinitzii* 322, root rot 322, wood rot 322, **344**

Hamamelis virginiana (witch hazel): *Botryosphaeria obtusa* 176, dieback 176, frost susceptibility 482, *Hendersonia* 82, leaf spot **76,** *Phyllosticta hamamelidis* **80,** sulfur dioxide tolerance 468

Haplopappus: *Coleosporium asterum* 268, rust 268

hawthorn, see *Crataegus, Raphiolepis*
black: *C. douglasii*
cockspur: *C. crus-galli*
Columbia: *C. columbiana*
English: *C. laevigata, C. monogyna*
India- or Indian: *R. indica*
Lavall's: *C. ×lavallei*
Washington: *C. phaenopyrum*
Yedda: *R. umbellata*
yellow-fruited: *C. flava*

hazel or hazelnut, see *Corylus*
American: *C. americana*
beaked: *C. cornuta*

hazel, witch, see *Hamamelis virginiana*

heartnut, see *Juglans ailantifolia*

heath, see *Erica*

heather, see *Calluna vulgaris*
false: *Cuphea*
mountain: *Phyllodoce*

heat injury **336,** 478–**480**

Hebe (hebe): Fusarium wilt 378, leaf spot 74, *Septoria exotica* 74, Verticillium wilt 376

H. speciosa (hebe): leaf spot **74,** *Septoria exotica* **74**

Hedera helix (English ivy): *Alternaria* 90, anthracnose 122–124, blight **158,** *Botryosphaeria obtusa* 176, dieback 176, gall 148, *Glomerella cingulata* 122–124, leaf spot 90, *Nectriella pironii* 148, as a pest **438,** sulfur dioxide tolerance 468, *Xanthomonas campestris* pv. *hederae* **158**

Helianthus (Jerusalem artichoke): *Coleosporium helianthi* 268, rust 268

Hemiphacidium convexum 52

H. longisporum 38, 52

H. planum 52

Hemizonia: *Coleosporium pacificum* 268, rust 268

hemlock, see *Tsuga*
Carolina: *T. caroliniana*
eastern: *T. canadensis*
mountain: *T. mertensiana*
western: *T. heterophylla*

Hendersonia 38, 82

H. eucalypticola 82

Hendersonula toruloidea **168,** 484

herbicide injury **456–460**

Hercules club, see *Aralia spinosa*

Hericium erinaceus 344

Herpobasidium deformans 94

Herpotrichia coulteri 50

H. juniperi **50–52**

H. nigra 50

Heterobasidion annosum 310, **314–316,** 338

Heteromeles arbutifolia (Christmas berry, toyon): *Botryosphaeria dothidea* 172, dieback 172, *Diplocarpon mespili* **64,** *Erwinia amylovora* 162, fire blight 162,

Heteromeles arbutifolia (cont.)
leaf spot **64,** scab **98,** *Spilocaea pyracanthae* **98**
Heterotheca: Coleosporium asterum 268, rust 268
Hevea brasiliensis (rubber tree, Para rubber tree): anthracnose 122–124, butt rot 304, *Ceratocystis fimbriata* 362, dieback 168, *Glomerella cingulata* 122–124, *Hendersonula toruloidea* 168, *Hypoxylon deustum* 304, root rot 304
Hibiscus (hibiscus, mallow): *Alternaria* **90,** *Armillaria tabescens* **312,** *Botryosphaeria dothidea* 172, *B. obtusa* 176, *B. rhodina* 180, canker 214, *Coniothyrium* 82, dieback 82, 172, 176, 180, edema 486, *Fusarium lateritium* 214, leaf spot **90,** 160, *Pseudomonas syringae* pv. *syringae* 160, root rot **312,** Verticillium wilt 376
H. rosa-sinensis (Chinese hibiscus): *Armillaria tabescens* 312, hibiscus chlorotic mottle virus 410, molybdenum deficiency 474, rhabdovirus 400, root rot 312, vein yellowing 400, viruslike symptoms **410**
H. syriacus (rose-of-sharon): Botrytis blight 60, canker 214, *Fusarium lateritium* 214, tobacco ringspot virus 414
hickory, see *Carya*
bitternut: *C. cordiformis*
mockernut: *C. tomentosa*
pignut: *C. glabra*
sand: *C. pallida*
shagbark: *C. ovata*
shellbark: *C. laciniosa*
water: *C. aquatica*
Higginsia 66
hills-of-snow, see *Hydrangea arborescens*
holly, see *Leea, Ilex, Nemopanthus*
American: *I. opaca*
Chinese: *I. cornuta*
dahoon: *I. cassine*
deciduous: *I. decidua*
English: *I. aquifolium*
Japanese: *I. crenata*
mountain: *Nemopanthus*
West Indian: *Leea*
Holodiscus discolor (creambush, oceanspray): *Erwinia amylovora* 162, fire blight 162, sulfur dioxide tolerance 468
Holozonia: Coleosporium pacificum 268, rust 268
honey locust, see *Gleditsia macrantha, G. triacanthos*
Japanese: *G. japonica*
honeysuckle, see *Diervilla, Lonicera*
bush: *Diervilla*
European fly: *L. xylosteum*
Japanese: *L. japonica*
Morrow: *L. morrowii*
Tatarian: *L. tatarica*
Utah: *L. utahensis*
Zabel's: *L. korolkowii* var. *zabelii*
hop hornbeam, American, see *Ostrya virginiana*
hops, see *Humulus lupus*
hop seed bush, see *Dodonea*
hornbeam, see *Carpinus, Ostrya*
American: *C. caroliniana*
European: *C. betulus*
hop or American hop: *O. virginiana*
Japanese: *C. japonica*
horse-chestnut, see *Aesculus, A. hippocastanum*
Japanese: *A. turbinata*
red: *A. ×carnea*
horsetail casuarina, horsetail tree, see *Casuarina equisetifolia*
Hovenia (Japanese raisin tree): *Ganoderma applanatum* 334
Howea belmoreana (Belmore sentry palm): lethal yellowing 390
huckleberry, see *Gaylussacia, Vaccinium*
California or evergreen: *V. ovatum*
mountain or thinleaf: *V. membranaceum*
tall red: *V. parvifolium*

Humulus lupus (hops): Prunus necrotic ringspot virus 408
Hydnum septentrionale 344
Hydrangea (hydrangea): Botrytis blight 60, dieback 210, *Erysiphe polygoni* 14–**16,** gall 148, *Hendersonia* 82, *Nectria cinnabarina* 210, *Nectriella pironii* 148, powdery mildew 14–**16**
H. anomala (climbing hydrangea): *Cristulariella moricola* 62, leaf spot 62
H. arborescens (hills-of-snow, wild hydrangea): gall 148, *Nectriella pironii* 148, *Pucciniastrum hydrangae* 260, root rot 296, rust 260, *Sclerotium rolfsii* 296, southern blight 296
H. macrophylla (French hydrangea): anthracnose 122–124, *Glomerella cingulata* 122–124, tobacco ringspot virus 414, tomato ringspot virus 414
H. paniculata (panicled hydrangea): *Pucciniastrum hydrangae* 260, rust 260, sulfur dioxide tolerance 468
Hylocereus undatus (night-blooming cereus): cactus virus X 418
Hyphoderma baculorubrense 168
Hyophorbe verschaffeltii (spindle palm): lethal yellowing 390
Hypoderma lethale 34
Hypodermella laricis 44
Hypoxylon **222–226,** 450, 484
H. atropunctatum **224,** 478
H. cohaerens **226**
H. deustum **304**
H. mammatum **222,** 338
H. mediterraneum **224**
H. necatrix 306
H. pruinatum 222
H. punctulatum **226**
H. tinctor **226**

ice injury **488**–490
Ilex (holly): *Alternaria* 90, blight 294, *Botryosphaeria dothidea* 172, *B. obtusa* 176, *B. rhodina* 180, butt rot 304, canker 212, *Cephaleuros virescens* 28, *Cerrena unicolor* 358, *Coniothyrium* 82, *Cylindrocladium avesiculatum* 294, dieback 172, 176, 180, *Ganoderma applanatum* 334, *Hypoxylon deustum* 304, leaf spot 28, 82, 90, *Microsphaera nemopanthis* 16, *Nectria galligena* 212, pesticide injury 460, *Phyllactinia guttata* 16, powdery mildew 16, *Rhytisma prini* 54, root rot 304, *Schizophyllum commune* 358, tar spot 54
I. aquifolium (English holly): canker 144, *Diaporthe crustosa* 144, fluoride tolerance 470, freeze injury **482,** *Phomopsis crustosa* 144, sulfur dioxide tolerance 468
I. cassine (dahoon holly): gall **150,** *Sphaeropsis tumefaciens* **150,** witches'-broom **150**
I. cornuta (Chinese holly): anthracnose 122–124, *Glomerella cingulata* 122–124, salt tolerance 454
I. crenata (Japanese holly): *Chalara elegans* **294,** *Meloidogyne* 298, ozone tolerance 466, Rhizoctonia blight 94, root knot nematode 298, root rot **294,** tobacco ringspot virus 414
I. decidua (deciduous holly): flood tolerance 486
I. opaca (American holly): Botrytis blight 60, canker 192, 214, *Chrysomyxa ilicina* 266, *Endothia gyrosa* 192, flood tolerance 486, *Fusarium solani* 214, gall 150, ozone tolerance 466, rust 266, salt tolerance 454, *Sphaeropsis tumefaciens* 150
I. vomitoria (yaupon): salt tolerance 452–454
India-hawthorn, see *Raphiolepis indica*
indigo
false, see *Amorpha*
green: *Rhamnus davurica*
Inonotus andersonii 356
I. circinatus 318–320
I. dryadeus 328

I. glomeratus 356
I. hispidus 354–356
I. obliquus 212, **352**
I. tomentosus 318–**320**
insecticide injury **460**
insects, as vectors, see vectors
insect damage in relation to disease:
aphids 38, bark beetles 310, 368, 370, 424, 484, borers (beetles) 198, 214, 222, 310, 372, 448–450, 484, borers (moth larvae) 72, 158, 198, 216, 280, cicada 222, defoliators 450, midges 72, 216, scales 178, 230, 450, weevils 214, 370
Ipomoea (morning-glory): *Coleosporium ipomoeae* 268, rust 268
Irenopsis 30
ironbark, gray, see *Eucalyptus paniculata*
ironweed, see *Vernonia*
Irpex lacteus 358
Isariopsis 70
Isthmiella crepidiformis **42**
I. faullii 42
I. quadrispora 42
ivy
Boston, see *Parthenocissus tricuspidata*
English: *Hedera helix*
poison: *Rhus radicans*
Ixora (ixora): *Alternaria* 90, *Coniothyrium* **82,** leaf blotch **82,** leaf spot 90, *Pestalotiopsis* **128**

Jacaranda (jacaranda): *Schizophyllum commune* 358
Japanese fatsia, see *Fatsia japonica*
Japanese pagoda tree, see *Sophora japonica*
Japanese raisin tree, see *Hovenia*
jasmine, see *Gardenia, Jasminum, Murraya*
cape: *G. jasminoides*
orange: *M. paniculata*
primrose: *J. mesnyi*
winter: *J. nudiflorum*
Jasminum (jasmine): *Botryosphaeria obtusa* 176, canker 148, dieback 176, *Nectriella pironii* 148, root rot 296, *Sclerotium rolfsii* 296, southern blight 296, Verticillium wilt 376
J. mesnyi (primrose jasmine): gall 154, *Pseudomonas syringae* pv. *savastanoi* 154, pv. *syringae* 160
J. nudiflorum (winter jasmine): gall 146, *Phomopsis* 146, tobacco ringspot virus 414
Jerusalem thorn, see *Parkinsonia aculeata*
jessamine, see *Gelsemium*
Carolina: *G. sempervirens*
jetbead, see *Rhodotypos*
Juglans (butternut, walnut): anthracnose 118, *Botryosphaeria dothidea* 172, *B. obtusa* 176, canker 132, 286, collar rot 286, *Dematophora necatrix* 306, dieback 132, 172, 176, *Discula* 110, *Gnomonia leptostyla* 118, herbicide injury 456, *Inonotus hispidus* 356, *Laetiporus sulfureus* 346, leaf spot 118, *Microsphaera juglandis-nigrae* 16, *Microstroma juglandis* 118, nematode 302, *Phellinus igniarius* 342, *Phyllactinia guttata* 16, *Phytophthora cactorum* 286, powdery mildew 16, *Pratylenchus* 302, root rot 306, *Schizophyllum commune* 358, *Sirococcus clavigignenti-juglandacearum* 132, wetwood 382, white mold 118, witches'-broom 118, wood rot 342, 346, 356–358
Juglans cv. 'Paradox' ('Paradox' walnut): blackline disease 404, cherry leafroll virus 404
J. ailantifolia (Japanese walnut): anthracnose 118, blight 160, bunch disease 396, dieback 132, *Gnomonia leptostyla* 118, leaf spot 118, *Melanconis juglandis* 132, *Xanthomonas campestris* pv. *juglandis* 160
J. ailantifolia var. *cordiformis* (heartnut): anthracnose 118, bunch disease 396,

canker 132, *Gnomonia leptostyla* 118, leaf spot 118, *Sirococcus clavigignenti-juglandacearum* 132
J. arizonica (Arizona walnut), see *J. major*
J. californica (California black walnut): *Agrobacterium tumefaciens* 156, anthracnose 118, crown gall 156, *Gnomonia leptostyla* 118, leaf spot 118, *Microstroma juglandis* 118
J. cinerea (butternut): *Agrobacterium tumefaciens* **156,** anthracnose 118, bunch disease **396,** canker **132,** 212, *Cristulariella moricola* 62, crown gall **156,** dieback **132,** 210, frost susceptibility 482, *Ganoderma applanatum* 334, *Gnomonia leptostyla* 118, ice glaze injury 490, *Laetiporus sulfureus* 346, leaf spot 62, 118, *Melanconis juglandis* **132,** *Microstroma juglandis* 118, *Nectria cinnabarina* 210, *N. galligena* 212, *Phellinus igniarius* **342,** root rot 306, *Sirococcus clavigignenti-juglandacearum* 132, witches'-broom **396,** wood decay 334, **342,** *Xylaria* 306
J. hindsii (Hinds or Northern California black walnut): *Agrobacterium tumefaciens* 156, anthracnose 118, blackline disease 404, blight 160, cherry leafroll virus 404, crown gall 156, *Dematophora necatrix* 306, dieback 168, *Gnomonia leptostyla* 118, *Hendersonula toruloidea* 168, leaf spot 118, *Microstroma juglandis* 118, root rot 306, *Xanthomonas campestris* pv. *juglandis* 160
J. major (Arizona walnut): anthracnose 118, *Gnomonia leptostyla* 118, leaf spot 118, *Microstroma juglandis* 118, *Phellinus weirianus* **346,** wood decay 346
J. mandshurica (Manchurian walnut): blossom blast 160, bunch disease 396, *Pseudomonas syringae* pv. *syringae* 160
J. microcarpa (little walnut): anthracnose 118, bunch disease **396,** dieback **132,** *Gnomonia leptostyla* 118, leaf spot 118, *Melanconis juglandis* **132,** *Microstroma juglandis* 118, witches'-broom **396**
J. nigra (black or eastern black walnut): anthracnose **118,** bunch disease 396, burl 494, canker 132, 140, **212–214,** chimera **492,** *Cristulariella moricola* 62, *Cylindrocladium* 294, flood tolerance 486, fluoride tolerance 470, frost susceptibility 482, *Fusarium lateritium* 214, *Gnomonia leptostyla* 118, herbicide tolerance 460, ice glaze injury 490, leaf spot 62, **118,** lightning injury 490, *Microstroma juglandis* **118,** *Nectria galligena* **212,** *Oxyporus latemarginatus* 328, ozone tolerance 466, *Phomopsis arnoldiae* 140, root rot 294–296, 306, salt tolerance 454, *Sclerotium rolfsii* 296, *Sirococcus clavigignenti-juglandacearum* 132, southern blight 296, sulfur dioxide tolerance 468, white mold **118,** *Xylaria* 306
J. regia (English or Persian walnut): *Agrobacterium tumefaciens* 156, anthracnose 118, Armillaria root rot 404, blackline disease **404,** 492, blight **160,** blossom blast 160, bunch disease 396, canker 144, **166,** 214, cherry leafroll virus **404,** crown gall 156, *Diaporthe ambigua* 144, dieback 132, 144, 168, *Erwinia nigrifluens* 166, *E. rubrifaciens* **166,** fluoride tolerance 470, fruit rot **160,** *Fusarium lateritium* 214, *Gnomonia leptostyla* 118, *Hendersonula toruloidea* 168, leaf spot 118, *Melanconis juglandis* 132, *Microstroma juglandis* 118, *Oxyporus latemarginatus* 328, ozone tolerance 466, *Phomopsis ambigua* 144, *Pseudomonas syringae* pv. *syringae* 160, root rot 404, salt tolerance 454, sulfur dioxide tolerance 468, wood rot 328,

J. regia (cont.)
Xanthomonas campestris pv. juglandis **160**
J. sinensis: anthracnose 118, Gnomonia leptostyla 118, leaf spot 118
juniper, see Juniperus
alligator: J. deppeana
Ashe: J. ashei
cherrystone or one-seed: J. monosperma
Chinese: J. chinensis
common: J. communis
creeping: J. horizontalis
needle: J. rigida
Pfitzer: J. chinensis cv. 'Pfitzerana'
red-berry: J. pinchotii
Rocky Mountain: J. scopulorum
savin: J. sabina
Sierra: J. occidentalis
shore: J. conferta
Utah: J. osteosperma
Juniperus (juniper, red cedar): Alternaria 90, blight 60, **86,** 90–92, **138,** Botryosphaeria dothidea 172, B. obtusa 176, Botrytis cinerea 60, cassytha 436, Cerrena unicolor 358, Coniothyrium fuckelii 82, Dendrothele nivosa 168, Didymascella tetraspora 44, dieback 172, 176, fluoride tolerance 470, gall 240–244, Gymnosporangium 240–248, herbicide tolerance 460, Heterobasidion annosum 314, Lophodermium juniperi **32,** Metacapnodium juniperi 30, Microsphaeropsis olivacea 82, mistletoe 424, nematode 302, Pestalotiopsis funerea 128, Phellinus pini 350, Phomopsis juniperovora **138,** Phoradendron 424, Pratylenchus 302, root rot 314, rust 240–248, salt tolerance 454, Schizophyllum commune 358, shoot blight 82, smooth patch 168, sooty mold 30, Stigmina 92, sulfur dioxide tolerance 468, wood rot 314, 350, 358
J. ashei (Ashe juniper): blight 138, drought damage **476,** Phomopsis juniperovora 138
J. chinensis (Chinese juniper): blight 138, canker 170, flood tolerance 486, gall 240, Gymnosporangium juniperi-virginianae 240, Phomopsis juniperovora 138, rust 240, Seiridium cardinale 170, sulfur dioxide tolerance 468
J. communis (common juniper): blight 138, canker 194, Coniothyrium fuckelii 82, dieback 194, gall 240, Gymnosporangium 248, G. clavariiforme 244, 248, G. clavipes 246, G. gaeumannii 240, G. globosum 240–242, Phomopsis juniperovora 138, rust 240–248, salt tolerance 454, shoot blight 82, Valsa 194
J. conferta (shore juniper): blight 138, Phomopsis juniperovora 138
J. deppeana (alligator juniper): gall **244,** Gymnosporangium kernianum 248, G. speciosum **244**–246, Perenniporia fraxinophila 344, rust **244**–248, Uredo apacheca 246, wood rot 344
J. formosana: blight 138, Phomopsis juniperovora 138
J. horizontalis (creeping juniper): blight 138, canker 194, dieback 194, gall 240, Gymnosporangium 242, G. clavipes 246, G. juniperi-virginianae 240, Phomopsis juniperovora 138, rust 240–242, 246, Valsa friesii 194
J. monosperma (cherrystone or one-seed juniper): blight 86, 138, Cercospora sequoiae 86, Gymnosporangium speciosum 246, mistletoe 424, Phomopsis juniperovora 138, Phoradendron juniperinum 424, rust 246
J. occidentalis (Sierra juniper): canker 170, Gymnosporangium 248, G. speciosum 246, mistletoe **424,** ozone tolerance 466, Phoradendron juniperinum **424,** rust 246–248, Seiridium cardinale 170
J. osteosperma (Utah juniper): blight 138,

gall 240, Gymnosporangium juniperi-virginianae 240, G. kernianum 248, G. speciosum 246, mistletoe 424, Phomopsis juniperovora 138, Phoradendron juniperinum 424, rust 240, 246–248
J. pinchotii (red-berry juniper): gall 240, Gymnosporangium juniperi-virginianae 240, rust 240
J. rigida (needle juniper): blight 138, Phomopsis juniperovora 138
J. sabina (savin juniper): blight 138, canker 170, Gymnosporangium clavipes 246, G. confusum **242,** G. fuscum 248, Phomopsis juniperovora 138, rust **242,** 246–248, Seiridium cardinale 170, sulfur dioxide tolerance 468
J. scopulorum (Rocky mountain juniper): blight 86, 138, Cercospora sequoiae 86, gall **240**–242, Gymnosporangium bethelii 248, G. clavipes 246, G. juniperi-virginianae **240,** G. nidus-avis **242,** mistletoe **424,** Phomopsis juniperovora 138, Phoradendron juniperinum **424,** rust 240–**242,** 246–248, sulfur dioxide tolerance 468, witches'-broom **242**
J. silicicola (southern red cedar): blight 86, 138, Cercospora sequoiae 86, gall 240–242, Gymnosporangium 240–242, Phomopsis juniperovora 138, rust 240–242
J. squamata: blight 138, Phomopsis juniperovora 138
J. virginiana (eastern red cedar): blight **86,** 138, canker 170, Cercospora sequoiae **86,** Dendrothele nivosa 168, flood tolerance 486, gall 240–242, Gymnosporangium **240**–242, G. clavipes **242,** 246–248, G. effusum 248, G. nidus-avis 244, herbicide tolerance 460, ice glaze injury 490, nematode 302, Perenniporia fraxinophila 344, Phomopsis juniperovora 138, rust **240**–242, 246–248, salt tolerance 454, Seiridium cardinale 170, smooth patch 168, sulfur dioxide tolerance 468, wood rot 344, Xiphinema americanum 302

Kabatiella **114**
K. apocrypta **104**
Kabatina juniperi **138**
K. thujae 138
Kageneckia oblonga: Erwinia amylovora 162, fire blight 162, scab 98, Spilocaea pyracanthae 98
Kalmia (American laurel): Botryosphaeria dothidea 172, bud burl 494, dieback 172, Exobasidium vaccinii 26, fluoride tolerance 470, Heterobasidion annosum 314, lignotuber 494, Microsphaera vaccinii 16, Pestalotiopsis 128, powdery mildew 16
K. latifolia (American or mountain laurel): Cercospora kalmiae **88,** Diaporthe kalmiae 144, flower blight 58, Hendersonia 82, leaf and twig blight 144, leaf spot **88,** necrotic ringspot 416, Ovulinia azaleae 58, ozone tolerance 466, Phomopsis kalmiae 144, twig blight 144
Kaskaskia gleditsiae 204
katsura tree, see Cercidiphyllum japonicum
Keithia thujina 44
Kentucky coffeetree, see Gymnocladus dioica
Kerria japonica (Japanese rose): Diaporthe japonica 144, dieback 144, 210, Erwinia amylovora 162, fire blight 162, Nectria cinnabarina 210, Phomopsis japonica 144
kinnikinnik, see Arctostaphylos uva-ursi
Klebsiella oxytoca 382
koa, see Acacia koa
Koelreutera (golden-rain tree): Botryosphaeria obtusa 176, dieback 176, 210, Nectria cinnabarina 210, ozone tolerance 466, Verticillium wilt 376

Kolkwitzia amabilis (beautybush): salt tolerance 454
Kretzschmaria clavus 304
K. deusta 304
kudzu, see Pueraria lobata
kumquat, see Fortunella
Kutilakesa pironii 148

Labrador tea, see Ledum, L. groenlandica
smooth: L. glandulosum
Labridella 128
Laburnum anagyroides (bean tree, golden-chain tree): Botryosphaeria dothidea 172, canker **214,** dieback 172, Fusarium lateritium **214,** herbicide injury 458, Microsphaeropsis olivacea 82, Pseudomonas syringae pv. syringae 160, rhabdovirus 400, sulfur dioxide tolerance 468, tip blight 160, veinal chlorosis 400
Laccaria **500–502**
Lachnellula 236
L. agassizii 236
L. willkommii **236**
Laetiporus sulfureus **346**
Lagerstroemia indica (crape myrtle): Botryosphaeria obtusa 176, dieback 176, Erysiphe lagerstroemiae **18,** Phyllactinia guttata 16, powdery mildew 14–**18,** Rhizoctonia blight 94, salt tolerance 454
L. parviflora: Erysiphe lagerstroemiae 18, powdery mildew 18
Lagophylla: Coleosporium pacificum 268, rust 268
Lansinum domesticum: gall 148, Nectriella pironii 148
larch, see Larix, Pseudolarix
Dahurian: L. gmelinii
Dunkeld: L. ×eurolepis
eastern: L. laricina
European: L. decidua
golden: P. kaempferi
Japanese: L. kaempferi
Lyall: L. lyallii
Siberian: L. sibirica
western: L. occidentalis
Larix (larch): blight 60, 138, Botryosphaeria obtusa 176, Botrytis cinerea 60, canker 236, dieback 176, Fomitopsis pinicola 348, Heterobasidion annosum 314, Inonotus tomentosus 318, Lachnellula willkommii 236, Laetiporus sulfureus 346, Melampsora 258, M. medusae 256, Melampsoridium betulinum 254, ozone tolerance 466, Phellinus pini 350, root rot 314, 318, rust 254–258, salt tolerance 454, Schizophyllum commune 358, Sclerophoma pithyophila 138, sulfur dioxide tolerance 468, wood rot 314, 318, 346–350, 358
L. decidua (European larch): blight 138, canker 194, 236, dieback 210, Lachnellula willkommii 236, Leucostoma kunzei 194, Melampsora medusae 256, M. occidentalis 258, Melampsoridium betulinum 254, Meria laricis 44, Mycosphaerella pini 48, Nectria cinnabarina 210, needle blight 44, ozone tolerance 466, Phomopsis juniperovora 138, rust 254–258, salt tolerance 454, shoot blight 134, Sirococcus conigenus 134, sulfur dioxide tolerance 468, Valsa 194
L. ×eurolepis (Dunkeld larch): canker 236, Lachnellula willkommii 236, Meria laricis 44, needle blight 44
L. gmelinii (Dahurian larch): canker 236, Lachnellula willkommii 236, Meria laricis 44, needle blight 44
L. kaempferi (Japanese larch): butt rot **322,** canker 194, 236, Lachnellula willkommii 236, Leucostoma kunzei 194, Melampsora medusae 256, M. occidentalis 258, Meria laricis 44, needle blight 44, ozone tolerance 466, Phaeolus schweinitzii **322,** root rot 322, rust 256–258, salt tolerance 454, sulfur dioxide tolerance 468, Valsa abietis 194

L. laricina (eastern larch, tamarack): Arceuthobium pusillum **434,** Ascocalyx abietina 232, butt rot 322, canker 194, **236,** dieback 210, dwarf mistletoe 426, **434,** ice glaze injury 490, Inonotus tomentosus 318, Lachnellula laricis 236, L. occidentalis 236, L. willkommii **236,** Leucostoma kunzei 194, Melampsora medusae 256, Melampsoridium betulinum 254, Meria laricis 44, Nectria cinnabarina 210, needle blight 44, Phaeolus schweinitzii 322, pine wood nematode 380, root rot 322, rust 254–256, salt tolerance 454, shoot blight 134, 136, Sirococcus conigenus 134, Sphaeropsis sapinea 136, sulfur dioxide tolerance 468, Valsa 194
L. lyallii (Lyall larch): Melampsora medusae 256, M. occidentalis 258, rust 256–258
L. occidentalis (western larch): Arceuthobium laricis **432**–434, butt rot 322, canker 236, drought injury 478, dwarf mistletoe 426, **432**–434, fluoride tolerance 470, Inonotus tomentosus 318, Lachnellula willkommii 236, Melampsora medusae 256, M. occidentalis 258, Meria laricis **44,** needle blight **44,** Phaeolus schweinitzii 322, Phellinus weirii **326,** root rot 318, 322, **326,** rust 256–258, sulfur dioxide tolerance 468, Valsa abietis 194
L. sibirica (Siberian larch): canker 236, Lachnellula willkommii 236, Meria laricis 44, needle blight 44
Larrea (creosote bush): Verticillium wilt 376
Lasiodiplodia 172
L. theobromae 180
Latania (Latan palm): lethal yellowing 390
laurel, see Kalmia, Laurus, Prunus, Umbellularia
American or mountain: K. latifolia
California: U. californica
cherry: P. laurocerasus, P. caroliniana
mountain: K. latifolia
Portugal: P. lusitanica
Laurus (laurel): Heterobasidion annosum 314
laurustinus, see Viburnum tinus
Lavandula (lavender): Botrytis blight 60
lead tree, see Leucaena
leaf blight **32–52, 56, 60–62, 80, 84, 90, 94, 110–112, 122–124, 128–130**
leaf blister diseases **22–26**
leaf curl **22**
leaf scorch **384–386, 452–470**
leaf spot **22–28, 34, 38–40, 46–48, 54, 62–70, 74–92, 96, 100, 104–110, 114, 118–120, 158, 176**
leatherleaf, see Chamaedaphne
lebbek, see Albizia lebbek
Lecanosticta 70
L. acicola 46
Ledum (Labrador tea): Exobasidium 26
L. glandulosum (smooth Labrador tea): Chrysomyxa 266, rust 266
L. groenlandica (Labrador tea): Chrysomyxa 266, rust 266
L. palustre (crystal tea): Chrysomyxa 266, rust 266
Leea coccinea (West Indian holly): gall 148, Nectriella coccinea 148
Leiophyllum (box sandmyrtle): Exobasidium vaccinii 26
Lembosia quercina 30
lemon, see Citrus limon
Lepteutypa cupressi 170
Leptodothiorella 76, 80
Leptographium 370, 448
L. terebrantis 368
Leptosphaeria 82, 128
L. coniothyrium 22
L. obtusispora 82
Leptothyrium 108
Leucaena (lead tree): Botryosphaeria dothidea 172, B. obtusa 176, dieback 172, 176
Leucocytospora 194, 198–200

L. kunzei 196
Leucophyllum frutescens (ceniza, Texas sage): gall 148, *Nectriella pironii* 148
Leucostoma 82, 126, 194
L. cincta **198**
L. kunzei **196**
L. niveum **200**
L. persoonii 194, **198**
Leucothoë (leucothoë): *Cylindrocladium* 294, *Exobasidium vaccinii* 26, root rot 294
Leveillula taurica 14
Liatris: Coleosporium laciniariae 268, rust 268
Libertella 218–220
Libocedrus, see *Calocedrus*
lichens **28, 504–506**
lightning injury **490**
Ligustrum (privet): *Alternaria* 90, anthracnose 122–124, *Botryosphaeria dothidea* 172, *B. obtusa* 176, cherry leafroll virus 404, *Dematophora necatrix* 306, dieback 172, 176, 210, edema 486, fluoride tolerance 470, gall 146, *Glomerella cingulata* 122–124, herbicide tolerance 460, *Heterobasidion annosum* 314, leaf spot 86–90, *Microsphaera fraxini* 16, *M. syringae* 16, *Nectria cinnabarina* 210, nematode 302, *Oxyporus latemarginatus* 328, *Phomopsis* 146, *Phymatotrichum omnivorum* 292, powdery mildew 16, *Pratylenchus* 302, *Pseudocercospora ligustri* 86–88, root rot 292, 306, salt tolerance 454, sulfur dioxide tolerance 468, *Verticillium* wilt 376
L. amurense (Amur privet): *Armillaria tabescens* 312, flood tolerance 486, leaf spot 88, ozone tolerance 466, *Pseudocercospora ligustri* 86–88, root rot 312, salt tolerance 454
L. japonicum (wax-leaf privet): *Cercospora* 86–88, gall 148, 154, leaf spot 86–88, *Nectriella pironii* 148, *Pestalotiopsis* **128,** prunus necrotic ringspot virus 408, *Pseudomonas syringae* pv. *savastanoi* 154
L. lucidum (glossy privet): *Cercospora* 86–88, gall 148, leaf spot 88, *Nectriella pironii* 148
L. obtusifolium var. *regelianum* (Regel's privet): flood tolerance 486
L. ovalifolium (California privet): gall 154, leaf spot 88, *Pseudocercospora ligustri* 86–88, *Pseudomonas syringae* pv. *savastanoi* 154, sulfur dioxide tolerance 468
L. sinense (Chinese privet): copper deficiency **474,** gall 148, *Nectriella pironii* 148
L.. vulgare (common privet): *Agrobacterium tumefaciens* **156,** anthracnose 122–124, crown gall **156,** flood tolerance 486, gall **156,** *Ganoderma lucidum* 332, *Glomerella cingulata* 122–124, leaf spot 88, ozone tolerance 466, *Pseudocercospora ligustri* 86–88, root rot 332, sulfur dioxide tolerance 468
lilac, see *Syringa*
Chinese: *S. ×chinensis*
common: *S. vulgaris*
Hungarian: *S. josikaea*
Japanese tree: *S. reticulata*
Persian: *S. ×persica*
lime, key or Mexican, see *Citrus aurantiifolia*
linden, see *Tilia*
American: *T. americana*
Crimean: *T. euchlora*
large-leaved European: *T. platyphyllos*
littleleaf or small-leaved European: *T. cordata*
silver: *T. petiolaris*
Lindera benzoin (spicebush): anthracnose 122–124, *Botryosphaeria dothidea* 172, *B. obtusa* 176, *Cristulariella moricola* 62, dieback 172, 176, *Glomerella cingulata* 122–124, *Hendersonia* 82, leaf spot 62
Linospora gleditsiae 84

L. tetraspora **84**
Liquidambar formosana (Formosan sweet gum): canker 192, 212, *Cercospora liquidambaris* 88, *Endothia gyrosa* 192, *Heterobasidion annosum* 314, leaf spot 88, *Nectria galligena* 212
L. stryaciflua (sweetgum): bleeding canker 172, 286, *Botryosphaeria dothidea* 172–174, 484, *B. obtusa* 176, *B. rhodina* 180, butt rot 334, canker 166, 172–174, 192, 214, 286, 484, *Cercospora liquidambaris* **86–88**, *Cerrena unicolor* 358, *Cylindrocladium* 294, dieback 172–176, 180, drought stress 174, *Endothia gyrosa* 192, flood tolerance 486, fluoride tolerance 470, *Fusarium solani* 214, *Ganoderma applanatum* 334, *G. lucidum* 332, *Globifomes graveolens* 346, herbicide tolerance 460, *Inonotus hispidus* 356, leaf spot **86–88**, mistletoe 422, nematode 302, *Oxyporus latemarginatus* 328, *O. populinus* 344, ozone tolerance 466, *Phoradendron serotinum* 422, *Phytophthora cactorum* 286, *Pratylenchus* 302, root rot 294, 332, salt damage **452–454**, *Schizophyllum commune* 358, sulfur dioxide tolerance 468, wood rot 332–334, 344–346, 356–358
Liriodendron tulipifera (tulip tree, yellow poplar): anthracnose 122–124, bark rot 168, *Botryosphaeria dothidea* 172, *B. obtusa* 176, butt rot 304, 334, canker **142–144**, 212–214, 286, *Ceratocystis coerulescens* 362, *Cerrena unicolor* 358, *Cristulariella moricola* 62, *Cylindrocladium scoparium* **294,** *Diaporthe eres* 144, dieback 172, 176, *Erysiphe liriodendri* 16, flood tolerance 486, frost injury **484,** frost susceptibility 482, *Fusarium solani* 214, *Ganoderma applanatum* 334, *Glomerella cingulata* 122–124, herbicide tolerance 460, *Hericium erinaceus* 344, *Hypoxylon deustum* 304, ice glaze injury 490, interveinal spots **476,** kudzu **438,** *Laetiporus sulfureus* 346, leaf spot 62, lightning injury 490, mistletoe 422, *Nectria magnoliae* 212, nematode 302, *Oxyporus latemarginatus* 328, *O. populinus* 344, ozone injury 466, *Perenniporia medulla-panis* 168, *Phomopsis* **142–144,** *Phyllactinia guttata* 16, *Phytophthora cactorum* 286, *Polyporus squamosus* 344, powdery mildew 16, *Pratylenchus* 302, *Rhytisma liriodendri* 54, root rot **294,** 304, salt tolerance 454, sapstreak 362, *Schizophyllum commune* 358, sulfur dioxide tolerance 468, tar spot 54, *Verticillium* wilt 374–376, wetwood 382, wood rot 168, 304, 328, 344–346, 358
Lirula macrospora 42
Lithocarpus densiflorus (tanbark oak, tanoak): *Ceratocystis fagacearum* 364, *Inonotus andersonii* 356, oak wilt 364, powdery mildew 18, *Sphaerotheca lanestris* 18, tobacco ringspot virus 414
Livistona chinensis (Chinese fan palm): lethal yellowing 390
locust, see *Gleditsia, Robinia*
black: *R. pseudoacacia*
honey: *G. macrantha, G. triacanthos*
Japanese honey: *G. japonica*
New Mexico: *R. neomexicana*
water: *G. aquatica*
longan, see *Euphorbia longan*
Lonicera (honeysuckle): blight **94,** *Botryosphaeria obtusa* 176, *Chalara elegans* 294, dieback 176, 210, *Ganoderma applanatum* 334, *Herpobasidium deformans* **94,** *Heterobasidion annosum* 314, *Microsphaera caprifoliacearum* 16, *M. lonicerae* 16–18, *Nectria cinnabarina* 210, powdery mildew 16–18, root rot 294
L. japonica (Japanese honeysuckle): as a pest 438
L. korolkowii var. *zabelii* (Zabel's honeysuckle): ozone tolerance 466, salt tolerance 454
L. morrowii (Morrow honeysuckle): flood tolerance 486, sulfur dioxide tolerance 468
L. tatarica (Tatarian honeysuckle): flood tolerance 486, salt tolerance 454, sulfur dioxide tolerance 468
L. utahensis (Utah honeysuckle): ash deposit 14
L. xylosteum (European fly honeysuckle): salt tolerance 454
Lophodermella 38, 82
L. concolor **38**
Lophodermium 32, 82
L. juniperi **32**
L. piceae 42
L. pinastri **32**
L. seditiosum **32–34**
Lophophacidium hyperboreum 52
loquat, see *Eriobotrya japonica*
lousewort, see *Pedicularis bracteosa*
Lyonia (lyonia): *Botryosphaeria obtusa* 176, dieback 176
Lysiloma bahamensis (Bahama lysiloma): gall 148, *Nectriella pironii* 148
Lysimachia: Coleosporium campanulae 268, rust 268

Macadamia (macadamia): *Botryosphaeria dothidea* 172, *B. rhodina* 180, dieback 172, 180, *Kretzschmaria clavus* 304, *Microsphaeropsis olivacea* 82, root rot 304
Machaeranthera: Coleosporium asterum 268, rust 268
Maclura pomifera (osage orange): *Botryosphaeria obtusa* 176, Botrytis blight 60, *Cerotelium fici* 238, dieback 176, flood tolerance 486, mistletoe 422, *Phoradendron* 422, rust 238, salt tolerance 454, *Schizophyllum commune* 358, *Verticillium* wilt 376
Macrophoma tumefaciens 150
Macrophomina phaseolina **292**
Madia (tarweed): *Coleosporium pacificum* 268, rust 268
madrone, see *Arbutus menziesii*
magnesium deficiency 472–**474**
Magnolia (bay, magnolia): *Alternaria* 90, *Botryosphaeria dothidea* 172, *B. obtusa* 176, canker 212, *Cerrena unicolor* 358, dieback 172, 176, *Fomitopsis pinicola* 348, *Ganoderma applanatum* 334, *G. lucidum* 332, *Hendersonia* 82, leaf spot 76, 82, 90, *Microsphaera magnifica* 16, *Nectria galligena* 212, nematode 302, ozone tolerance 466, *Phyllactinia guttata* 16, *Phyllosticta cookei* 78, powdery mildew 16, *Pratylenchus* 302, root rot 332, salt tolerance 454, *Schizophyllum commune* 358, *Tylenchorynchus* 302, *Verticillium* wilt 376, wetwood 382, wood rot 332–334, 348
M. acuminata (cucumber tree): frost susceptibility 482, ice glaze injury 490, ozone tolerance 466
M. fraseri (Fraser magnolia, mountain magnolia, ear-leaved umbrella tree): canker 212, *Cristulariella moricola* 62, frost susceptibility 482, leaf spot 62, 76, *Nectria magnoliae* 212, *Phyllosticta magnoliae* 78
M. grandiflora (bull bay, southern magnolia): algal leaf spot **28,** anthracnose 122–124, *Armillaria tabescens* 312, *Cephaleuros* **28,** *Coniothyrium fuckelii* **82,** *Cylindrocladium* 294, flood tolerance 486, *Glomerella cingulata* 122–124, leaf cast **496,** leaf spot **28, 78, 82,** *Microsphaeropsis olivacea* 82, *Phyllosticta magnoliae* **78,** root rot 294, 312, yellowing **498**
M. heptapeta (yulan magnolia): cucumber mosaic virus 418
M. macrophylla (great-leaved magnolia): mosaic 418
M. ×soulangiana (saucer magnolia):

cucumber mosaic virus 418, flood tolerance 486, leaf spot 78, 160, ozone tolerance 466, *Phyllosticta magnoliae* 78, *Pseudomonas syringae* pv. *syringae* 160
M. stellata (star magnolia): Botrytis blight **60**
M. tripetala (umbrella tree): *Cristulariella moricola* 62, ice glaze injury 490, leaf spot 62
M. virginiana (sweet bay): *Ciborinia gracilipes* 56, flood tolerance 486, leaf blight 56, leaf spot 78, *Oxyporus latemarginatus* 328, *Phyllosticta glauca* 78, *P. magnoliae* 78
mahogany
mountain, see *Cerocarpus*
swamp: *Eucalyptus robusta*
see also *Swietenia*
Mahonia (mahonia): *Cumminsiella mirabilissima* 250, gall 148, *Nectriella pironii* 148, *Puccinia* 250, rust 250
M. aquifolium (Oregon grape): *Cumminsiella mirabilissima* **250,** fluoride tolerance 470, rust **250,** sulfur dioxide tolerance 468, winter injury **476**
M. bealei (mahonia): *Cylindrocladium ellipticum* 294, leaf blight 294
M. nervosa (Cascades mahonia): *Cumminsiella mirabilissima* 250, rust 250
M. pinnata (cluster mahonia): *Cumminsiella mirabilissima* 250, rust 250
M. repens (creeping mahonia): *Cumminsiella mirabilissima* 250, rust 250
mallow, see *Hibiscus*
Malus (apple, crabapple): *Alternaria* 90, anthracnose 122–124, apple scab **96,** black rot **176,** blight 160–164, blossom blast 160, *Botryosphaeria dothidea* 172, *B. obtusa* **176,** *B. rhodina* 180, butt rot 334, canker 82, 144, 162–164, **176,** 198, 202, 342, *Climacodon septentrionalis* 344, *Coniothyrium fuckelii* 82, *Diaporthe* 144, dieback 172, **176,** 180, 198, 202, 210, *Diplocarpon mespili* 64, *Erwinia amylovora* 162–164, fire blight 162–164, flood tolerance 486, fluoride tolerance 470, *Fomes fomentarius* 346, *Fomitopsis pinicola* 348, fruit rot 90, 144, 212, *Ganoderma applanatum* 334, *Glomerella cingulata* 122–124, *Gymnosporangium* 242, 248, *G. clavipes* 246, *G. libocedri* 244, hail injury 488, *Hendersonia* 82, herbicide tolerance 460, *Heterobasidion annosum* 316, leaf spot 64, 82, 90, **176,** 242–248, mycoplasmalike organism 18, *Nectria cinnabarina* 210, *N. galligena* 212, *Oxyporus populinus* 344, ozone tolerance 466, *Phellinus igniarius* 342, *Phomopsis* 144, *Phyllactinia guttata* 16, *Podosphaera leucotricha* 14, 18, powdery mildew 14–18, *Pseudomonas syringae* pv. *syringae* 160, rust 242–248, salt damage 454, *Trametes versicolor* 342, *Valsa ambiens* 202, *Venturia asperata* 96, *V. inaequalis* **96,** wood rot 316, 342–348
M. angustifolia (southern wild crab): apple scab 96, *Venturia inaequalis* 96
M. ×arnoldiana: apple scab 96, *Venturia inaequalis* 96
M. baccata (Siberian crab): apple scab 96, *Erwinia amylovora* **162,** fire blight **162,** ozone tolerance 466, *Venturia inaequalis* 96
M. brevipes: apple scab 96, *Venturia inaequalis* 96
M. coronaria (American crab): apple scab 96, *Venturia inaequalis* 96
M. florentina: apple scab 96, *Venturia inaequalis* 96
M. floribunda (showy crabapple): apple mosaic virus 412
M. glaucescens: apple scab 96, *Venturia inaequalis* 96
M. ioensis (prairie crabapple): apple mosaic virus 412, apple scab 96, *Venturia inaequalis* 96

M. ×micromalus: apple scab 96, *Venturia inaequalis* 96
M. ×platycarpa: apple scab 96, *Venturia inaequalis* 96
M. prunifolia (plum-leaved apple): apple scab 96, *Venturia inaequalis* 96
M. pumila (common apple): *Agrobacterium* 156, apple mosaic virus 412, *Armillaria tabescens* 312, blight **162–164,** boron deficiency 472, *Botryosphaeria dothidea* 172, canker 144, **162–164,** 172, 176, 212–214, **358,** *Ceratocystis fagacearum* 364, cold tolerance 480, collar rot **286,** *Cristulariella moricola* 62, crown gall 156, *Cylindrocladium* 294, *Dematophora necatrix* **306,** dieback 168, 172, 176, *Diaporthe* 144, *Erwinia amylovora* **162–164,** fire blight **162–164,** flat limb disease **412,** fruit rot 126, 144, 172, 212, *Fusarium lateritium* 214, gall 156, *Ganoderma lucidum* 332, graft union necrosis 414, hairy root 156, *Hendersonula toruloidea* 168, herbicide injury **460,** herbicide tolerance 460, ice glaze injury 490, leaf spot 62, *Monilinia* 126, mycoplasmalike organism 18, *Nectria galligena* 212, nematode **302,** *Oxyporus latemarginatus* 328, ozone injury **466,** *Phomopsis* 144, *Phymatotrichum omnivorum* 292, *Phytophthora cactorum* 286, *P. megasperma* **286,** *Podosphaera clandestina* 16, *P. leucotricha* **18,** powdery mildew 16–**18,** *Pratylenchus penetrans* **302,** proliferation 18, root rot 292–**296, 306,** 312, 332, scab 96, *Schizophyllum commune* **358,** *Sclerotium rolfsii* **296,** southern blight **296,** sulfur dioxide tolerance 468, tobacco mosaic virus 414, tobacco necrosis virus 402, tobacco ringspot virus 414, tomato ringspot virus 414, *Trametes versicolor* 342, *Venturia inaequalis* **96,** witches'-broom 396, wood rot 312, 328, 332, 342, **358,** *Xiphinema americanum* **302,** *Xylaria* 306, see also *M. sylvestris*
M. ×purpurea: apple scab 96, *Venturia inaequalis* 96
M. ×scheideckeri: apple scab 96, *Venturia inaequalis* 96
M. sieboldii (Toringo crab): apple scab 96, *Venturia inaequalis* 96
M. sylvestris (common apple): apple scab 96, *Venturia inaequalis* 96, see also *M. pumila*
Malvaviscus arboreus var. *mexicanus* (Turk's-cap waxmallow): *Armillaria tabescens* 312, root rot 312
Mammillaria (pin cushion cactus): black rot 72, *Fusarium* rot 72, *Pseudomonas fluorescens* 158, soft rot 158
manganese deficiency **472–474**
Mangifera indica (mango): anthracnose 122–124, *Botryosphaeria dothidea* 172, *B. rhodina* 180, *Dendrothele candida* 168, dieback 168, 172, *Glomerella cingulata* 122–124, *Hendersonula toruloidea* 168, nematode 302, *Radopholus similis* 302, *Schizophyllum commune* 358, smooth patch 168
mango, see *Mangifera indica*
mangrove, see *Rhizophora mangle*
Manihot esculenta (cassava): anthracnose 122–124, *Botryosphaeria dothidea* 172, dieback 172, *Glomerella cingulata* 122–124
manzanita, see *Arctostaphylos*
maple, see *Acer*
 Amur: *A. ginnala*
 bigleaf: *A. macrophyllum*
 black: *A. saccharum* subsp. *nigrum*
 hedge: *A. campestre*
 Japanese: *A. palmatum*
 mountain: *A. spicatum*
 Norway: *A. platanoides*
 red: *A. rubrum*

Rocky Mountain: *A. glabrum*
 silver: *A. saccharinum*
 striped: *A. pensylvanicum*
 sugar: *A. saccharum*
 sycamore: *A. pseudoplatanus*
 Tatarian: *A. tataricum*
 vine: *A. circinatum*
maple mosaic virus 494
marigold, see *Tagetes*
Marssoniella juglandis 118
Marssonina 66, 70, 120
M. balsamiferae 120
M. betulae **120**
M. brunnea **120**
M. castagnei **120**
M. fraxini 70
M. juglandis 118
M. populi 120
M. rosae 66
M. thomasiana **120**
Massaria platani 82
Matteuccia (ostrich fern): rust 264, *Uredinopsis* 264
medlar, see *Mespilus germanica*
Melampsora 184, **256–258**
M. abietis-canadensis **258**
M. epitea **256**
M. farlowii 256
M. medusae **256**
M. occidentalis **256–258**
Melampsorella caryophyllacearum **264**
Melampsoridium 254
M. betulinum **254**
Melampyrum lineare (cow wheat): *Cronartium coleosporioides* 270, 278, rust 270, 278
Melanconis juglandis **128**
Melanconium oblongum 128
Melia (bead tree): *Botryosphaeria rhodina* 180
M. azedarach (chinaberry): anthracnose 122–124, *Botryosphaeria dothidea* 172, *B. obtusa* 176, canker 148, 214, dieback 168, 172, 176, 210, *Fusarium lateritium* 214, gall 148, *Glomerella cingulata* 122–124, *Hendersonula toruloidea* 168, mistletoe 422, *Nectria cinnabarina* 210, *Nectriella pironii* 148, *Oxyporus latemarginatus* 328, *Phoradendron serotinum* 422, *Phyllactinia guttata* 16, *Phymatotrichum omnivorum* 292, powdery mildew 16, root rot 292, wood rot 328
Meliola palmicola **30**
Meloidogyne **298–300**
M. incognita 298, 378
M. hapla **298**
M. javanica 298, 378
Menziesia (mock azalea): *Exobasidium vaccinii* 26, *Rhytisma arbuti* 54, tar spot 54
Meria laricis **44**
Mespilus germanica (medlar): *Diplocarpon mespili* 64, *Erwinia amylovora* 162, fire blight 162, fruit rot 126, *Gymnosporangium clavipes* 246, *G. confusum* 244, *G. globosum* 242, leaf spot 64, 242–246, *Monilinia fructigena* 126, *Podosphaera clandestina* 16, powdery mildew 16, rust 242–246
mesquite, see *Prosopis juliflora*
Metacapnodium juniperi 30
Metasequoia glyptostroboides (dawn redwood): *Botryosphaeria dothidea* 172, dieback 172, ozone tolerance 466, salt tolerance 454, sulfur dioxide tolerance 468
Methanobacter arbophilicum 382
Michelia figo (banana shrub): algal leaf spot 28, *Cephaleuros virescens* 28
Microsphaera **14,** 16, **18**
Microsphaeropsis olivacea 82
Microstroma album 118
M. juglandis **118**
mildew, downy **12,** powdery **14–20**
Milesina 264
mimosa, see *Albizzia julibrissin*
mistletoe **420–424,** dwarf **426–434**
mites, eriophyid **20**

MLO **388–398**
mock orange, see *Philadelphus*
 Lewis: *P. lewisii*
 sweet: *P. coronarius*
Moneses uniflora (single-delight): *Chrysomyxa* 266, rust 266
Monilinia **126**
M. fructicola **126**
M. laxa **126**
monkey puzzle, see *Araucaria araucana*
Monochaetia unicornis 170
Monostichella coryli **114**
M. robergei 116
Morenoella quercina 30
Morfea 30
morning-glory, see *Ipomoea*
Morus (mulberry): bacterial leaf scorch 386, *Botryosphaeria dothidea* 172, *B. obtusa* 176, butt rot 334, canker 212–214, dieback 172, 176, 210, *Fusarium lateritium* 214, *Ganoderma applanatum* 334, herbicide tolerance 460, *Inonotus hispidus* 356, *Microsphaeropsis olivacea* 82, *Nectria cinnabarina* 210, *N. galligena* 212, *Schizophyllum commune* 358, wetwood 382
M. alba (white mulberry): canker 214, dieback **168,** *Fusarium lateritium* 214, *Hendersonula toruloidea* **168,** ozone tolerance 466, *Phymatotrichum omnivorum* 292, root rot 292, 296, salt tolerance 454, *Sclerotium rolfsii* 296, southern blight 296, sulfur dioxide tolerance 468
M. microphylla (Texas mulberry): sulfur dioxide tolerance 468
M. rubra (red mulberry): canker **214,** *Cerotelium fici* 238, *Cytospora* 214, dieback **214,** flood tolerance 486, fluoride tolerance 470, freeze damage **214,** *Fusarium* 214, *Nectria cinnabarina* 214, rust 238
moss
 ball or bunch, see*Tillandsia recurvata*
 Spanish: *T. usneoides*
mountain ash, see *Sorbus*
 European: *S. aucuparia*
 Greene: *S. scopulina*
 Sitka: *S. sitchensis*
mulberry, see *Broussonetria, Morus*
 paper: *B. papyrifera*
 red: *M. rubra*
 Texas: *M. microphylla*
 white: *M. alba*
Murraya paniculata (orange jasmine): greasy spot 72, *Mycosphaerella citri* 72
Musa acuminata (banana): burrowing nematode 302, *Radopholus similis* 302
myall, weeping, see *Acacia pendula*
mycoplasma 388
mycoplasmalike organism **388–398,** 446
mycorrhizae 486, **500–502**
Mycosphaerella 70
M. aucupariae 78
M. citri **72**
M. dearnessii **46**
M. effigurata 70
M. fraxinicola **70**
M. opuntiae **72**
M. pini **48**
M. platanifolia 76–78
M. populicola 74
M. populorum **74**
M. rosicola 76–**78**
M. stigmina-platani 92
M. yuccae 70
Myrica cerifera (wax myrtle): *Botryosphaeria dothidea* 172, *B. obtusa* 176, *Cronartium comptoniae* 276, dieback 172, 176, *Gymnosporangium ellisii* 240, rust 240, 276, salt tolerance 454
M. faya (candleberry myrtle): algal leaf spot 28, *Cephaleuros virescens* 28
M. gale (sweet gale): *Cronartium comptoniae* 270, 276, *Gymnosporangium ellisii* 240, 248, rust 240–248, 270, 276
M. pensylvanica (bayberry): actinorhizae 502, *Gymnosporangium ellisii* 240,

248, nitrogen-fixing nodules 502, rust 240, 248, salt tolerance 454
myrtle, see *Lagerstroemia, Myrica, Myrtus, Vinca*
 candleberry: *Myrica faya*
 crape: *L. indica*
 Greek: *Myrtus communis*
 ground: *V. minor*
 wax: *Myrica cerifera*
Myrtus communis (Greek myrtle): fasciation **492**

Naemacyclus minor 38
N. niveus 38
Nandina (nandina, sacred bamboo): Verticillium wilt 376
N. domestica (heavenly bamboo, nandina): anthracnose 122–124, *Cercospora nandinae* **86–88,** *Glomerella cingulata* 122–124, leaf spot **86–88,** *Phymatotrichum omnivorum* 292, root rot 292
Nannorrhops ritchiana (Mazari palm): *Botryosphaeria obtusa* 176, lethal yellowing 390
nannyberry, see *Viburnum lentago*
nectarine, see *Prunus persica*
Nectria 208–212
N. cinnabarina **208–210,** 374, 444, 484
N. coccinea 212, 442
N. coccinea var. *faginata* 440–**442**
N. ditissima 212, 442
N. galligena **212, 344, 352,** 362, 440–442
N. haematococca 214
N. magnoliae 212
Nectriella pironii 148
Nectriodium ferrugineum 442
Nemopanthus (mountain holly): *Microsphaera nemopanthis* 16, powdery mildew 16
Neodypsis decaryi: lethal yellowing 390
Neopeckia coulteri 50
Nerium oleander (oleander): *Alternaria* 90, anthracnose 122–124, *Botryosphaeria obtusa* 176, dieback 176, dodder **436,** gall 150, **154,** *Glomerella cingulata* 122–124, leaf spot 90, olive knot **154,** *Pseudomonas syringae* pv. *savastanoi* **154,** pv. *syringae* 160, salt tolerance 454, *Schizophyllum commune* 358, *Sphaeropsis tumefaciens* 150
ninebark, see *Physocarpus, P. opulifolius*
 mallow: *P. malvaceus*
nitrogen-fixing nodules **502**
Nothophacidium phyllophilum 52
nutrient deficiencies **472–474**
Nyssa (gum, tupelo): *Botryosphaeria dothidea* 172, *Cerrena unicolor* 358, dieback 172, *Heterobasidion annosum* 316, salt tolerance 454, *Schizophyllum commune* 358, Verticillium wilt 376
N. aquatica (tupelo gum, water tupelo): flood tolerance 486, *Oxyporus latemarginatus* 328, wood rot 328
N. sylvatica (black or sour gum), including *N sylvatica* var. *biflora* (swamp tupelo): canker 212–214, 236, flood tolerance 486, *Fusarium solani* 214, *Ganoderma lucidum* 332, herbicide tolerance 460, *Hericeum erinaceus* 344, *Hypoxylon tinctor* 226, ice glaze injury 490, mistletoe **420–422,** *Nectria galligena* 212, *Oxyporus latemarginatus* 328, *O. populinus* 344, ozone tolerance 466, *Phoradendron serotinum* **420–422,** root rot 332, *Strumella coryneoidea* 236, sulfur dioxide tolerance 468, *Urnula craterium* 236, wood rot 328, 332, 344

oak, see *Grevillea, Lithocarpus, Quercus, Rhus*
  Arizona white: *Q. arizonica*
  bear or scrub: *Q. ilicifolia*
  black: *Q. velutina*
  blackjack: *Q. marilandica*
  blue: *Q. douglasii*
  bluejack: *Q. incana*
  bur: *Q. macrocarpa*
  California black or Kellog: *Q. kelloggii*
  California live or coast live: *Q. agrifolia*
  California scrub: *Q. dumosa*
  canyon live: *Q. chrysolepis*
  Chapman: *Q. chapmanii*
  cherrybark: *Q. falcata* var. *pagodifolia*
  chestnut: *Q. prinus*
  chinkapin: *Q. prinoides, Q. muehlenbergii*
  cork: *Q. suber*
  Dunn: *Q. dunnii*
  Durand: *Q. durandii*
  Durmast or sessile: *Q. petraea*
  Emory: *Q. emoryi*
  Engelmann or mesa: *Q. engelmannii*
  English: *Q. robur*
  Gambel: *Q. gambelii*
  Garry or Oregon: *Q. garryana*
  gray: *Q. grisea*
  holly: *Q. ilex*
  interior live: *Q. wislizenii*
  laurel: *Q. laurifolia*
  live or southern live: *Q. virginiana*
  Mexican blue: *Q. oblongifolia*
  Mexican live: *Q. obtusata*
  myrtle: *Q. myrtifolia*
  netleaf: *Q. rugosa*
  northern pin: *Q. ellipsoidalis*
  northern red: *Q. rubra*
  Nuttall: *Q. nuttallii*
  overcup: *Q. lyrata*
  pin: *Q. palustris*
  poison: *R. toxicodendron*
  post: *Q. stellata*
  running: *Q. pumila*
  sand live: *Q. virginiana* var. *maritima*
  scarlet: *Q. coccinea*
  shingle: *Q. imbricaria*
  Shumard: *Q. shumardii*
  silk: *G. robusta*
  silverleaf: *Q. hypoleuoides*
  southern red: *Q. falcata*
  swamp chestnut: *Q. michauxii*
  swamp white: *Q. bicolor*
  tanbark: *L. densiflorus*
  Texas red: *Q. texana*
  turkey: *Q. cerris, Q. laevis*
  valley: *Q. lobata*
  water: *Q. nigra*
  white: *Q. alba*
  willow: *Q. phellos*
  yellow chestnut: *Q. muehlenbergii*
oak decline **450**
oak wilt **364**, 386
oats, see *Avena sativa*
ocean-spray, see *Holodiscus discolor*
Oidiopsis 14
Oidium 14
Olea (olive): *Botryosphaeria dothidea* 172, dieback 172, *Schizophyllum commune* 358, Verticillium wilt 376
O. europaea (common olive): nematode 302, olive knot **154**, *Pratylenchus* 302, *Pseudomonas syringae* pv. *savastanoi* **154,** root rot 296, *Sclerotium rolfsii* 296, southern blight 296
oleander, see *Nerium oleander*
olive
  American, see *Osmanthus americanus*
  autumn or Russian: *Elaeagnus angustifolia*
  black: *Bucida buceras*
  common: *Olea europaea*
  desert: *Forestiera neomexicana*
  fragrant or sweet: *Osmanthus fragrans*
  holly: *Osmanthus heterophyllus*
Olpidium brassicae 402
Onoclea sensibilis (sensitive fern): rust 264, *Uredinopsis* 264

Ophiodothella fici 84
O. floridana 84
O. vaccinii **84**
Opuntia (prickly pear, prickly pear cactus): anthracnose 122–124, black rot **72**, *Botryosphaeria obtusa* 176, *B. rhodina* 180, cactus virus X 418, cladode spot **72**, *Erwinia* 158, *Fusarium dimerum* var. *violaceum* **72,** *Glomerella cingulata* 122–124, Hendersonia opuntiae 82, *Mycosphaerella opuntiae* **72**, *Phyllosticta concava* **72**, rot **72**, 158, 358, Sammons' Opuntia virus **418**, *Schizophyllum commune* 358, scorch 82, sun scald 82
O. fulgida (cholla): *Erwinia* 158, soft rot 158
orach, see *Atriplex hortensis*
orange
  Mandarin, see *Citrus reticulata*
  Mexican: *Choisya ternata*
  mock: *Philadelphus*
    Lewis: *P. lewisii*
    sweet: *P. coronarius*
  osage: *Maclura pomifera*
  sour: *Citrus aurantium*
  sweet: *Citrus sinensis*
  trifoliate: *Poncirus trifoliata*
orchid tree, see *Bauhinia purpurea, B. variegata*
Orthocarpus luteus (yellow owl's clover): *Cronartium coleosporioides* 270, 278, rust 270, 278
osage orange, see *Maclura pomifera*
osier, purple, see *Salix purpurea*
Osmanthus (devilwood, osmanthus): salt tolerance 454, Verticillium wilt 376
O. americanus (American olive): gall 154, *Pseudomonas syringae* pv. *savastanoi* 154
O. fragrans (fragrant or sweet olive): anthracnose 122–124, *Armillaria tabescens* 312, gall 148, 154, *Glomerella cingulata* 122–124, *Nectrielha pironii* 148, *Pseudomonas syringae* pv. *savastanoi* 154, root rot 312
O. heterophyllus (holly olive, holly osmanthus): *Dematophora necatrix* 306, gall 154, *Pseudomonas syringae* pv. *savastanoi* 154, root rot 306
Osmunda (flowering fern): rust 264, *Uredinopsis* 264
Osteomeles: Botryosphaeria obtusa 176, dieback 176, *Erwinia amylovora* 162, fire blight 162
Ostrya virginiana (American hop hornbeam): *Aleurodiscus oakesii* 168, anthracnose **116**, bark rot 168, *Botryosphaeria dothidea* 172, *B. obtusa* 176, burl **494**, canker **116,** 212, 236, *Cerrena unicolor* 358, *Dendrothele griseo-cana* 168, dieback 172, 176, flood tolerance 486, *Gnomoniella carpinea* 116, ice glaze injury 490, *Inonotus obliquus* 352, *Melampsoridium carpini* 254, *Microsphaera ellisii* 16, *Monostichella robergei* 116, *Nectria galligena* 212, *Oxyporus populinus* 344, *Perenniporia medulla-panis* 168, *Phellinus igniarius* 342, powdery mildew 16, rust 254, salt tolerance 454, smooth patch 168, *Strumella coryneoidea* 236, *Taphrina virginica* 22, *Urnula craterium* 236, wood decay 168, 342–344, 358
Ovulariopsis 14
Ovulinia azaleae **58**
Oxydendrum arboreum (sourwood): *Botryosphaeria obtusa* 176, canker 212, *Cristulariella moricola* 62, dieback 176, drought injury 476, flood tolerance 486, leaf spot 62, *Nectria galligena* 212, salt tolerance 454
Oxyporus latemarginatus **328**
O. populinus **344**
ozone damage 446, **462–466**

Pachysandra (pachysandra): ozone tolerance 466

P. terminalis (Japanese pachysandra): blight **206,** *Pseudonectria pachysandricola* **206,** *Volutella pachysandricola* 206
Paeonia (peony): Verticillium wilt 376
P. suffruticosa (tree peony): *Chalara elegans* 294, root rot 294
paintbrush, see *Castilleja*
palm
  areca, Madagascar, or yellow, see *Chrysalidocarpus lutescens*
  arenga: *Arenga*
  Arikury: *Arikuryroba schizophylla*
  Belmore sentry: *Howea belmoreana*
  cabada: *Chrysalidocarpus, C. cabadae*
  cabbage: *Sabal palmetto*
  Chinese fan: *Livistona chinensis*
  Christmas or Manila: *Veitchia merrillii*
  cluster fishtail: *Caryota mitis*
  coconut: *Cocos nucifera*
  date: *Phoenix*
    Canary Island: *P. canariensis*
    cliff: *P. rupicola*
    common: *P. dactylifera*
    pygmy or Roebelen: *P. roebelenii*
    Senegal: *P. reclinata*
    Sylvester or wild: *P. sylvestris*
  Englers: *Arenga engleri*
  Everglades, paurotis, or saw cabbage: *Acoelorrhaphe wrightii*
  Fiji fan: *Pritchardia pacifica*
  Gebang: *Corypha elata*
  Hildebrand's: *Ravenea hildebrandtii*
  jelly or pindo: *Butia capitata*
  key: *Thrinax morrisii*
  Kona: *Pritchardia affinis*
  lady: *Rhapis excelsa*
  Latan: *Latania*
  MacArthur: *Ptychosperma macarthurii*
  Mazari: *Nannorrhops ritchiana*
  Montgomery's: *Veitchia montgomeryana*
  oil: *Elaeis guineensis*
  Palmyra: *Borassus flabellifera*
  princess: *Dictosperma album*
  queen: *Arecastrum romanzoffianum*
  rock: *Brahea*
  royal: *Roystonea*
    Cuban: *R. regia*
    Florida: *R. elata*
  Sago: *Cycas revoluta*
  seashore: *Allagoptera arenaria*
  silver: *Coccothrinax argentata*
  solitaire: *Ptychosperma elegans*
  spindle: *Hyophorbe verschaffeltii*
  thread: *Washingtonia robusta*
  Thurston: *Pritchardia thurstonii*
  Washington: *Washingtonia*
  windmill: *Trachycarpus fortunei*
  wine: *Caryota urens*
Palmae (palms): *Alternaria* 90, anthracnose 122–124, false smut 92, *Ganoderma* **330,** *Glomerella cingulata* 122–124, *Graphiola phoenicis* 92, leaf spot 90–92, *Leptosphaeria* 128, lethal yellowing **390,** *Pestalotiopsis palmarum* 128, root and butt rot **330,** salt tolerance 454, *Schizophyllum commune* 358, *Stigmina palmivora* 92, wood rot 330, 358
palmetto, see *Sabal, Serenoa*
  dwarf: *Sabal minor*
  saw or scrub: *Serenoa repens*
paloverde, see *Cercidium*
Pandanus utilis (common screw pine): anthracnose 122–124, *Botryosphaeria dothidea* 172, *Glomerella cingulata* 122–124, lethal yellowing 390
papaya, see *Carica papaya*
Parkinsonia aculeata (Jerusalem thorn): gall 148, *Nectriella pironii* 148
Parthenium (guayule): *Coleosporium helianthi* 268, rust 268, Verticillium wilt 376
Parthenocissus quinquefolia (Virginia creeper, woodbine): bacterial leaf scorch 384, *Botryosphaeria obtusa* 176, canker 82, *Coniothyrium fuckelii* 82, *Cristulariella moricola* 62, dieback

176, downy mildew 12, fluoride tolerance 470, *Guignardia bidwellii* 80, leaf spot 62, 80, ozone tolerance 466, as a pest 438, *Plasmopara viticola* 12, powdery mildew 14, sulfur dioxide tolerance 468, *Uncinula necator* 14
P. tricuspidata (Boston ivy): downy mildew 12, *Guignardia bidwellii* **80,** leaf spot **80,** *Plasmopara viticola* 12
Paulownia (paulownia): *Botryosphaeria obtusa* 176, canker 144, *Diaporthe eres* 144, dieback 144, 176, fluoride tolerance 470, ozone tolerance 466, *Phyllactinia guttata* 16, powdery mildew 16, *Schizophyllum commune* 358, wood rot 358
pawpaw, see *Asimina triloba*
peach, see *Prunus persica*
pear, see *Opuntia, Pyrus*
  callery: *P. calleryana*
  Chinese or sand: *P. pyrifolia*
  common or European: *P. communis*
  evergreen: *P. kawakamii*
  Oriental: *P. betulifolia*
  prickly: *Opuntia*
  ussury: *P. ussuriensis*
pearlbush, see *Exochorda*
pea tree, Siberian, see *Caragana arborescens*
pecan, see *Carya illinoinensis*
Pedicularis bracteosa (lousewort): *Cronartium coleosporioides* 270, 278, *C. kamtschaticum* 272, rust 270–272, 278
Pediocactus (cactus): *Pseudomonas fluorescens* 158, soft rot 158
Pelecyphora (hatchet cactus): *Pseudomonas fluorescens* 158, soft rot 158
Peniophora gigantea 316
peony, see *Paeonia*
  tree: *P. suffruticosa*
Peperomia obtusifolia (baby rubber plant, peperomia): Rhizoctonia blight **94**
pepper, black, see *Piper nigrum*
pepper tree, see *Schinus*
  Brazilian: *S. terebinthefolius*
  California: *S. molle*
pepper vine, see *Ampelopsis arborea*
Peraphyllum: Erwinia amylovora 162, fire blight 162
Perenniporia fraxinophila **344**
P. medulla-panis 168
P. phloiophila **168**
periderm formation **496**
Peridermium 260, 270, 280–282
P. filamentosum 270, 280–282
P. peckii 260
periwinkle, see *Catharanthus roseus*
Peronospora rubi 12
P. sparsa **12**
Persea americana (avocado): *Alternaria* 90, anthracnose 122–124, *Botryosphaeria dothidea* 172, *B. obtusa* 176, *B. rhodina* 180, collar rot 286, *Cristulariella moricola* 62, *Dematophora necatrix* 306, dieback 172, 176, fruit rot 90, fruit spot 160, *Glomerella cingulata* 122–124, leaf spot 62, nematode 302, ozone tolerance 466, *Phytophthora cactorum* 286, *P. cinnamomi* 288, *Pratylenchus* 302, *Pseudomonas lauraceum* 158, *P. syringae* pv. *syringae* 160, *Radopholus similis* 302, root rot 288, 296, 306, salt tolerance 454, *Schizophyllum commune* 358, *Sclerotium rolfsii* 296, southern blight 296, sun blotch 400, tobacco mosaic virus 414, Verticillium wilt 376, viroid 400
P. borbonia (red bay): algal leaf spot **28,** *Cephaleuros virescens* **28,** flood tolerance 486, salt tolerance 454
persimmon, see *Diospyros*
  black or Texas: *D. texana*
  common: *D. virginiana*
  Japanese: *D. kaki*
Pestalopezia 128
Pestalotia 128
Pestalotiopsis **128**
P. funerea **128**

P. guepinii 128
P. maculans 128
pesticides, phytotoxicity of, **456–460**
Phacidiopycnis 144
Phacidium 52, 144
P. abietis **52**
Phaeocryptopus gaeumannii **40**
P. nudus 40
P. pinastri 40
Phaeolus schweinitzii **322**
Phellinus chrysoloma 148
P. everhartii **346**
P. igniarius **342**
P. pini **350**
P. robiniae **344**
P. spiculosus **354–356**
P. tremulae **342**
P. weirianus **346**
P. weirii 310, 314, **324–326**
Phellodendron (cork tree): *Verticillium* wilt 376
P. amurense (Amur cork tree): herbicide tolerance 460
Phialophora parasitica 372
Philadelphus (mock orange): fluoride tolerance 470, *Gymnosporangium speciosum* 240, *Microsphaeropsis olivacea* 82, *Phyllactinia guttata* 16, powdery mildew 16, rust 240, salt tolerance 454
P. coronarius (sweet mock orange): blight 160, canker **144**, *Diaporthe eres* **144**, dieback **144**, flood tolerance 486, ozone tolerance 466, *Phomopsis oblonga* 144, *Pseudomonas syringae* pv. *syringae* 160, sulfur dioxide tolerance 468
P. lewisii (Lewis mock orange): sulfur dioxide tolerance 468
P. ×virginalis: sulfur dioxide tolerance 468
Phlebiopsis gigantea 316
Phloeosporella padi 66
Phoenix (date palm): *Botryosphaeria dothidea* 172, false smut 92, *Graphiola phoenicis* 92, leaf spot 92, lethal yellowing 390, root rot 330, *Stigmina palmivora* 92
P. canariensis (Canary Island date palm): false smut **92**, Fusarium wilt 378, *Ganoderma zonatum* 330, *Graphiola phoenicis* **92**, leaf spot **92**, lethal yellowing **390**, root rot 330, *Stigmina palmivora* **92**
P. dactylifera (common date palm): false smut 92, *Graphiola phoenicis* 92, leaf spot 92, lethal yellowing 390, *Stigmina palmivora* 92
P. loureirii: leaf spot 92, *Stigmina palmivora* 92
P. reclinata (Senegal date palm): false smut 92, *Ganoderma zonatum* 330, *Graphiola phoenicis* 92, lethal yellowing 390, root rot 330, *Stigmina palmivora* 92
P. roebelenii (pygmy date or Roebelen palm): false smut 92, *Graphiola phoenicis* 92, leaf spot 92, *Stigmina palmivora* 92
P. rupicola (cliff date palm): leaf spot 92, *Stigmina palmivora* 92
P. sylvestris (Sylvester or wild date palm): lethal yellowing 390
phoenix tree, see *Firmiana*
Phoma 146
Phomopsis **138–148**
P. arnoldiae **140**
P. citri 142
P. elaeagni 140
P. gardeniae 148
P. juniperovora **138**
P. macrospora **142–144**
P. scabra 180
Phoradendron (mistletoe or American mistletoe) **420–424**
P. bolleanum subsp. *pauciflorum* **424**
P. californicum **422**
P. capitellatum 424
P. coryae **422**
P. densum 424
P. flavescens 422

P. hawksworthii 424
P. juniperinum **424**
P. macrophyllum 422
P. rubrum 422
P. serotinum 420–**422**
P. tomentosum **420**–422
P. villosum **422**
Photinia (photinia): *Botryosphaeria dothidea* 172, dieback 172, *Diplocarpon mespili* **64**, *Erwinia amylovora* 162, fire blight 162, *Gymnosporangium clavipes* 246, leaf spot **64**, 246, powdery mildew 14–16, rust 246, *Sphaerotheca pannosa* 14–16
P. glabra (Japanese photinia): *Armillaria tabescens* 312, root rot 312
Phragmidium 238
P. mucronatum **238**
Phragmocapnias 30
Phyllactinia corylea 14–16
P. guttata 14–16
Phyllodoce (mountain heather): brown felt blight 50, *Exobasidium vacciniiuliginosi* 26, *Herpotrichia juniperi* 50
Phyllosticta 70–**72, 76–78,** 130
P. ampelicida 80
P. concava **72**
P. cookei 78
P. cornicola **78**
P. coryli **78**
P. glauca 78
P. globigera 78
P. hamamelidis **76**
P. magnoliae **78**
P. minima **76**
P. negundinis **76**
P. platani 76–**78**
P. rosicola 76–**78**
P. sorbi **78**
P. sphaeropsoidea 76, 80
P. syringae 130
P. tiliae 78
P. viridis **70**
Phymatotrichopsis omnivora 292
Phymatotrichum omnivorum **292**
Physalospora glandicola 178
P. miyabeana 102
P. rhodina 180
Physocarpus (ninebark): *Botryosphaeria obtusa* 176, dieback 176, *Erwinia amylovora* 162, fire blight 162, *Heterobasidion annosum* 316, *Phyllactinia guttata* 16, powdery mildew 16–**18,** salt tolerance 454, *Sphaerotheca* **18,** witches'-broom **18**
P. malvaceus (mallow ninebark): sulfur dioxide tolerance 468
P. opulifolius (ninebark): flood tolerance 486
Phytophthora **128, 284–290,** 444
P. cactorum **284**–286
P. cambivora 286
P. cinnamomi 284–286, **288–290,** 318
P. citricola 284–286
P. citrophthora 284–286
P. cryptogea 284–286
P. drechsleri 286
P. gonapodyides 284
P. heveae 284
P. inflata **284**
P. lateralis 284, **288**
P. megasperma 284–**286**
P. parasitica 284–**286**
P. syringae 286
Picea (spruce): *Arceuthobium* 426, 430–434, *Botrytis* blight 60, brown felt blight 50, butt rot 314, 318–322, 334, canker 194–196, *Cenangium* 230, *Chrysomyxa* 266, *C. arctostaphyli* 266, *Cylindrocladium* 294, *Cytospora kunzei* 196, dwarf mistletoe 426, 434, *Fomitopsis pinicola* 348, *Ganoderma applanatum* 334, *Herpotrichia coulteri* 50, *H. juniperi* 50, *Heterobasidion annosum* 316, *Hypoxylon mammatum* 222, *Inonotus tomentosus* 318–320, *Laetiporus sulfureus* 346, *Leucostoma kunzei* 196, lightning injury 490, *Lophodermium piceae* 32, needle cast 32, 42, nematode 302, *Pestalotiopsis*

funerea 128, *Phaeolus schweinitzii* 322, *Phellinus pini* 350, *Pratylenchus* 302, *Rhizosphaera kalkhoffii* 42, root rot 294, 306, 314, 318–322, rust 266, *Schizophyllum commune* 358, *Sclerophoma pithyophila* 138, *Sirococcus conigenus* 134, *Sphaeropsis sapinea* 136, tip blight 60, 134–136, *Tylenchorynchus* 302, *Valsa* 194, witches'-broom 266, wood rot 316–322, 334, 346–350, 358, *Xylaria* 306
P. abies (Norway spruce): *Ascocalyx abietina* 232, brown felt blight 50, butt rot 316–318, 322, canker 194–196, *Cenangium* 230, *Chrysomyxa* 266, *Cylindrocladium* 294, *Cytospora kunzei* 196, dieback 232, dwarf mistletoe 434, flood tolerance 486, herbicide injury 456, *Herpotrichia juniperi* 50, *Heterobasidion annosum* 316, ice glaze injury 490, *Inonotus tomentosus* 318, *Cytospora kunzei* 196, *Leucostoma kunzei* 196, *Lophophacidium hyperboreum* 52, needle cast 42, nutrient deficiency 474, ozone tolerance 466, *Phacidium abietis* 52, *Phaeolus schweinitzii* 322, *Rhizosphaera kalkhoffii* 42, root rot 294, 316–318, 322, rust 266, salt tolerance 454, *Sarcotrochila piniperda* 52, *Sirococcus conigenus* 134, snow blight 52, *Sphaeropsis sapinea* 136, sulfur dioxide tolerance 468, tip blight 134–136, *Valsa* 194, witches'-broom 266, wood rot 316–318, 322
P. breweriana (Brewer spruce): *Arceuthobium abietinum* 426, 432, dwarf mistletoe 426, 432, *Lophodermium crassum* 32
P. engelmannii (Engelmann spruce): *Arceuthobium* 426, 430–434, brown felt blight **50,** butt rot 318, 322, canker 196, *Cenangium* 230, *Chrysomyxa* 266, *Cytospora kunzei* 196, dwarf mistletoe 426, 430–434, *Echinodontium tinctorium* 348, *Herpotrichia juniperi* **50,** *Inonotus dryadeus* 328, *I. tomentosus* 318, *Isthmiella crepidiformis* 42, *Leucostoma kunzei* 196, *Lophophacidium hyperboreum* 52, *Macrophomina phaseolina* 292, needle cast 42, *Phaeolus schweinitzii* 322, *Phellinus weirii* 324, *Rhizosphaera kalkhoffii* 42, root rot 292, 318, 322–324, 328, rust 266, *Sirococcus conigenus* 134, snow blight 52, sulfur dioxide tolerance 468, tip blight 134, witches'-broom 266, wood rot 318, 322, 328, 348
P. glauca (white spruce): *Arceuthobium* 426, 430, 434, *A. pusillum* 434, *Ascocalyx abietina* 232, butt rot 318, **320,** 322, canker 194–196, *Cenangium* 230, *Chrysomyxa* 266, *Cylindrocladium* 294, *Cytospora kunzei* 196, dieback 230–232, drought injury 478, dwarf mistletoe 426, 430, 434, *Echinodontium tinctorium* 348, flood tolerance 486, fluoride tolerance 470, *Inonotus tomentosus* 318–**320,** *Isthmiella crepidiformis* 42, *Leucostoma kunzei* 196, lichens **504,** *Lophophacidium hyperboreum* 52, needle cast **42,** ozone tolerance 466, *Phacidium abietis* 52, *Phaeolus schweinitzii* 322, pine wood nematode 380, *Rhizosphaera kalkhoffii* **42,** root rot 294, 318, **320,** 322, rust 266, salt tolerance 454, *Sarcotrochila piniperda* 52, *Sirococcus conigenus* 134, snow blight 52, *Sphaeropsis sapinea* 136, sulfur dioxide tolerance 468, tip blight 134–136, *Valsa* 194, winter browning 478, witches'-broom 266, wood rot 318, **320**–322, 348
P. mariana (black spruce): *Arceuthobium pusillum* **434,** *Ascocalyx abietina* 232, brown felt blight 50, butt rot 318, 322, canker 194–196, *Cenangium* 230, *Chrysomyxa* 266, *C. ledicola* **266,** *Cylindrocladium* 294, *Cytospora kunzei* 196, dieback 230–232, dwarf mis-

tletoe 426, **434,** flood tolerance 486, *Herpotrichia juniperi* 50, *Inonotus tomentosus* 318, *Isthmiella crepidiformis* **42,** *Leucostoma kunzei* 196, *Lophophacidium hyperboreum* 52, needle cast **42,** *Phaeolus schweinitzii* 322, *Phellinus pini* 350, *Rhizosphaera kalkhoffii* 42, root rot 294, 318, 322, rust **266,** *Sarcotrochila piniperda* 52, *Sirococcus conigenus* 134, snow blight 52, tip blight 134, *Valsa* 194, witches'-broom 266, wood rot 318, 322, 350
P. omorika (Serbian spruce): needle cast 42, *Rhizosphaera kalkhoffii* 42
P. orientalis (Oriental spruce): canker 196, *Cytospora kunzei* 196, *Leucostoma kunzei* 196, needle cast 42, *Rhizosphaera kalkhoffii* 42
P. pungens (blue or Colorado blue spruce): *Arceuthobium* 426, 430–434, brown felt blight 50, butt rot 318, canker 194–**196,** *Chrysomyxa* 266, *C. arctostaphyli* **266,** *Cytospora kunzei* **196,** dwarf mistletoe 426, 430–434, flood tolerance 486, fluoride tolerance 470, freeze injury **480,** herbicide injury **456,** herbicide tolerance 460, *Herpotrichia juniperi* 50, *Inonotus tomentosus* 318, *Leucostoma kunzei* **196,** *Lophophacidium hyperboreum* 52, needle cast **42,** nematode 302, 380, ozone tolerance 466, *Phacidium abietis* 52, pine wood nematode 380, *Rhizosphaera kalkhoffii* **42,** root rot 318, rust **266,** salt tolerance 454, *Sirococcus conigenus* **134,** snow blight 52, *Sphaeropsis sapinea* 136, sulfur dioxide tolerance 468, tip blight **134**–136, *Valsa abietis* 194, *Verticicladiella procera* 370, winter browning 478, witches'-broom **266,** wood rot 318, *Xiphinema americanum* 302
P. rubens (red spruce): air pollution damage 462–464, *Aleurodiscus canadensis* 168, butt rot 318, 322, canker 194–196, *Chrysomyxa* 266, *C. ledicola* **266,** *Cytospora kunzei* 196, dwarf mistletoe 426, 434, *Fomitopsis pinicola* **348,** ice glaze injury 490, *Inonotus tomentosus* 318, *Leucostoma kunzei* 196, *Lophophacidium hyperboreum* 52, *Phacidium abietis* 52, *Phaeolus schweinitzii* 322, root rot 318, 322, rust **266,** *Sarcotrochila piniperda* 52, snow blight 52, *Sirococcus conigenus* 134, tip blight 134, twig blight 168, *Valsa* 194, winter browning 478, witches'-broom 266, wood decay 318, 322, **348**
P. schrenkiana: needle cast 42, *Rhizosphaera kalkhoffii* 42
P. sitchensis (Sitka spruce): *Arceuthobium* 432, brown felt blight 50, butt rot 316–318, 322, *Chrysomyxa* 266, dwarf mistletoe 426, 434, flood tolerance 486, *Ganoderma oregonense* 332, *Herpotrichia juniperi* 50, *Heterobasidion annosum* 316, *Inonotus dryadeus* 328, *I. tomentosus* 318, *Melampsora medusae* 256, *M. occidentalis* 258, *Mycosphaerella pini* 48, needle blight 48, needle cast 42, *Phaeolus schweinitzii* 322, *Phellinus weirii* 324, *Rhizosphaera kalkhoffii* 42, root rot 316–318, 322–324, 328, rust 256–258, 266, *Sirococcus conigenus* 134, tip blight 134, witches'-broom 266, wood rot 316–318, 322–324, 328
Pieris japonica (Japanese andromeda): *Botryosphaeria dothidea* 172, dieback 172, 284, nematode 302, ozone tolerance 466, *Pestalotiopsis* 128, *Phytophthora* 284, *P. cinnamomi* **290,** root rot 290, *Tylenchorynchus* 302, wilt **290**
pigeon plum, see *Coccoloba diversifolia*
Piggotia coryli 114
P. fraxini 70
Pimenta (pimento): *Ceratocystis fimbriata* 362

pimento, see *Pimenta*
pine
  Aleppo, see *Pinus halepensis*
  Apache: *P. engelmannii*
  Australian: *Casuarina, C. equisetifolia*
  Austrian or black: *P. nigra*
  Benguet: *P. insularis*
  Bishop: *P. muricata*
  bristlecone: *P. aristata*
  Canary Island: *P. canariensis*
  cedar or spruce: *P. glabra*
  Chihuahua: *P. chihuahuana*
  Chinese: *P. tabuliformis*
  chir: *P. roxburghii*
  cluster: *P. pinaster*
  Coulter: *P. coulteri*
  Cuban: *P. caribaea*
  digger: *P. sabiniana*
  eastern white: *P. strobus*
  foxtail: *P. balfouriana*
  Himalayan white: *P. wallichiana*
  hoop: *Araucaria cunninghamii*
  Italian stone: *P. pinea*
  jack: *P. banksiana*
  Japanese black: *P. thunbergiana*
  Japanese red: *P. densiflora*
  Japanese white: *P. parviflora*
  Jeffrey: *P. jeffreyi*
  knobcone: *P. attenuata*
  limber: *P. flexilis*
  loblolly: *P. taeda*
  lodgepole: *P. contorta* var. *latifolia*
  longleaf: *P. palustris*
  Macedonian: *P. peuce*
  Mexican pinyon or Mexican stone: *P. cembroides*
  Mexican yellow: *P. patula*
  Monterey: *P. radiata*
  mountain or mugo: *P. mugo*
  Norfolk Island: *Araucaria heterophylla*
  pinyon: *P. edulis*
  pitch: *P. rigida*
  pond: *P. serotina*
  ponderosa: *P. ponderosa*
  red: *P. resinosa*
  Rottnest Island: *Callitris preisii*
  rough-barked Mexican: *P. montezumae*
  sand: *P. clausa*
  Scots: *P. sylvestris*
  screw, common: *Pandanus utilis*
  shore: *Pinus contorta* var. *contorta*
  shortleaf: *P. echinata*
  Sierra lodgepole: *P. contorta* var. *murrayana*
  singleleaf pinyon: *P. monophylla*
  slash: *P. elliottii* var. *elliottii*
  South Florida slash: *P. elliottii* var. *densa*
  southwestern white: *P. strobiformis*
  sugar: *P. lambertiana*
  Swiss stone: *P. cembra*
  table mountain: *P. pungens*
  Torrey: *P. torreyana*
  twisted-leaf: *P. teocote*
  umbrella: *Sciadopitys verticillata*
  Virginia: *P. virginiana*
  western white: *P. monticola*
  whitebark: *P. albicaulis*
pine wilt **370, 380**
pine wood nematode **380**
Pinus (pine, pinyon): *Arceuthobium* 426–434, *Ascocalyx abietina* 232, *Atropellis* 234, *Botryosphaeria dothidea* 172, *B. obtusa* 176, blister rusts 270–282, brown felt blight 50, butt rot 322, canker 194–196, 216, 230–234, 272–278, cassytha 436, *Coleosporium* 268, *Cronartium* 270–280, dieback 172, 176, 194–196, 216, 230–234, *Fomitopsis pinicola* 348, frost injury 482, *Fusarium moniliforme* var. *subglutinans* 216, *Ganoderma* 332, *G. applanatum* 334, *Gremmeniella abietina* 232, *Hemiphacidium longisporum* 52, *Hendersonia* 82, herbicide tolerance 460, *Herpotrichia coulteri* 50, *Heterobasidion annosum* 314–316, *Inonotus circinatus* 318–320, *I. tomentosus* 318–320, *I. hispidus* 356, *Laetiporus sulfureus* 346, lightning injury 490, littleleaf 290, *Lophodermella* 38, 82, *Lophodermium* 32, *Microsphaeropsis olivacea* 82, needle blight 34–36, 46–48, 128, needle cast 32–38, nematode 302, nutrient deficiencies 472–474, *Peridermium* 280–282, *Pestalotiopsis* 128, *Phaeocryptopus pinastri* 40, *Phaeolus schweinitzii* 322, *Phellinus pini* 350, *Phytophthora cinnamomi* 288–290, pitch canker 216, *Ploioderma* 34, *Pratylenchus* 302, root rot 288–290, 314–326, rust 268–282, *Sarcotrochila macrospora* 52, *Schizophyllum commune* 358, *Scleroderris lagerbergii* 232, *Sclerophoma pithyophila* 138, *Scorias spongiosa* 30, sooty mold 30, *Tylenchorynchus* 302, *Valsa* 194, wood rot 312–322, 332–334, 346–350, 358
  P. albicaulis (whitebark pine): *Arceuthobium* 426, 430–434, *Atropellis piniphila* 234, blister rust **272,** brown felt blight 50–**52,** butt rot 322, canker 234, **272,** *Cronartium ribicola* **272,** dwarf mistletoe 426, 430–434, *Herpotrichia* 50, *Lophodermella arcuata* 38, *Phaeolus schweinitzii* 322, root rot 322, *Sirococcus conigenus* 134, tip blight 134
  P. aristata (bristlecone pine): *Arceuthobium* 426, 430–432, blister rust 272, canker 272, *Cronartium ribicola* 272, dwarf mistletoe 426, 430–432, ozone tolerance 466
  P. attenuata (knobcone pine): black stain root disease 368, butt rot 322, canker 278, *Cronartium coleosporioides* 278, *C. comandrae* 278, dwarf mistletoe 426, 430, *Elytroderma deformans* 36, *Endocronartium harknessii* 282, gall 282, *Lophodermella morbida* 38, *Mycosphaerella dearnessii* 46, *M. pini* 48, needle blight 36, 46–48, ozone tolerance 466, *Phaeolus schweinitzii* 322, root rot 322, rust 278, 282, *Verticicladiella wageneri* 368
  P. balfouriana (foxtail pine): blister rust 272, brown felt blight 50, *Herpotrichia coulteri* 50
  P. banksiana (jack pine): *Arceuthobium* 426–428, 434, *Ascocalyx abietina* **232,** *Atropellis* 234, Botrytis blight 60, brown felt blight 50, butt rot 322, canker **232**–234, 276, *Cenangium* 230, *Coleosporium* 268, *Cronartium coleosporioides* 278, *C. comandrae* 278, *C. comptoniae* 276, *C. quercuum* 274, *Cylindrocladium* 294, dieback 194, 210, 230–234, dwarf mistletoe 426–428, 434, *Elytroderma deformans* 36, *Endocronartium harknessii* 282, flood tolerance 486, gall 274, 282, *Herpotrichia coulteri* 50, *Inonotus* 318, *Lophodermella concolor* 38, *Melampsora medusae* 256, *Mycosphaerella dearnessii* 46, *Nectria cinnabarina* 210, needle blight 36, 46, ozone tolerance 466, *Phacidium infestans* 52, *Phaeolus schweinitzii* 322, *Phellinus pini* 350, *Phomopsis juniperovora* 138, pine wood nematode 380, root rot 294, 318, 322, rust 256, 268, 274–278, 282, salt tolerance 454, *Sirococcus conigenus* 134, snow blight 52, sulfur dioxide tolerance 468, tip blight 134, *Valsa* 194, *Verticicladiella procera* 370, wood decay 318, 322, 350
  P. canariensis (Canary Island pine): *Endocronartium harknessii* 282, gall rust 282, *Mycosphaerella pini* 48, needle blight 48, *Sphaeropsis sapinea* 136, tip blight 136
  P. caribaea (Cuban pine): anthracnose 122–124, cone rust 280, *Cronartium conigenum* 280, *C. quercuum* f. sp. *fusiforme* 274, *Cyclaneusma minus* 38, fusiform rust 274, *Glomerella cingulata* 122–124, *Mycosphaerella dearnessii* 46, *M. pini* 48, needle blight 34, 46–48, needle cast 38, *Ploioderma lethale* 34, *Sphaeropsis sapinea* 136, sulfur dioxide tolerance 468, tip blight 136
  P. cembra (Swiss stone pine): *Cenangium* 230, dieback 230, pine wood nematode 380, *Sphaeropsis sapinea* 136, sulfur dioxide tolerance 468, tip blight 136
  P. cembroides (Mexican pinyon, Mexican stone pine): *Arceuthobium divaricatum* 426, *Cenangium* 230, dieback 230, dwarf mistletoe 426
  P. chihuahuana (Chihuahua pine): *Arceuthobium gillii* **432,** *Coleosporium ipomoeae* 268, cone rust **280,** *Cronartium conigenum* **280,** dwarf mistletoe 426, **432**–434, rust 268, **280**
  P. clausa (sand pine): *Armillaria tabescens* **312,** *Atropellis tingens* 234, butt rot 322, canker 216, 234, *Cronartium quercuum* 274, dieback 216, 234, *Fusarium moniliforme* var. *subglutinans* 216, gall 274, heart rot **350,** *Inonotus circinatus* **318–320,** *Phellinus pini* **350,** pine wood nematode 380, needle blight 34, *Phaeolus schweinitzii* 322, pitch canker 216, *Ploioderma lethale* 34, root rot **312, 318, 320,** 322, rust 274, *Verticicladiella procera* 370, wood decay **318, 320,** 322, 350
  P. contorta, including varieties *contorta* (shore pine), *latifolia* (lodgepole pine), and *murrayana* (Sierra lodgepole pine): *Arceuthobium* 426–434, *A. americanum* **426**–430, *Ascocalyx abietina* 232, *Atropellis* **234,** black stain root disease 368, brown felt blight 50, butt rot 322, canker 232–**234,** 276–**278, 282,** *Cenangium* 230, *Coleosporium asterum* 268, *Cronartium coleosporioides* **278,** *C. comandrae* 278, *C. comptoniae* 276, *Cyclaneusma minus* 38, dieback 194, 230–234, drought injury 478, dwarf mistletoe **426**–434, *Elytroderma deformans* 36, *Endocronartium harknessii* **282,** fluoride tolerance 470, *Herpotrichia* 50, *Inonotus* 318, *Lophodermella* **38,** *Melampsora medusae* 256, *M. occidentalis* 258, *Mycosphaerella dearnessii* 46, *M. pini* 48, needle blight 36, 46–48, needle cast **38,** ozone tolerance 466, *Phaeolus schweinitzii* 322, *Phellinus weirii* 324–326, pine wood nematode 380, root rot 318, 322–326, rust 256–258, 268, 276–**278, 282,** *Sirococcus conigenus* 134, sulfur dioxide tolerance 468, tip blight 134, *Valsa* 194, *Verticicladiella penicillata* **368,** *V. wageneri* 368, winter browning 478, wood rot 318, 322–326
  P. cooperi: Arceuthobium gillii 434, cone rust 280, *Cronartium conigenum* 280, dwarf mistletoe 434, limb rust 280, *Peridermium* 280
  P. coulteri (Coulter pine): *Arceuthobium* 426, 430–432, canker 276, *Coleosporium pacificum* 268, *Cronartium coleosporioides* 278, *C. comptoniae* 276, dwarf mistletoe 426, 430–432, *Elytroderma deformans* 36, *Endocronartium harknessii* 282, gall 282, needle blight 36, ozone tolerance 466, rust 268, 276–278, 282, *Sirococcus conigenus* 134, *Sphaeropsis sapinea* 136, tip blight 134–136
  P. densiflora (Japanese red pine): *Atropellis tingens* 234, canker **230,** 234, 276, *Cenangium ferruginosum* **230,** *Cronartium comptoniae* 276, *C. quercuum* 274, dieback **230,** 234, gall 274, *Mycosphaerella pini* 48, needle blight 48, needle cast 42, *Rhizosphaera kalkhoffii* 42, pine wood nematode 380, rust 274–276
  P. durangensis: cone rust 280, *Cronartium conigenum* 280, limb rust 280, *Peridermium* 280
  P. echinata (shortleaf pine): air pollution damage 462, *Atropellis* 234, butt rot 322, canker 216, 234, 276, *Cenangium* 230, *Coleosporium* 268, *Cronartium comandrae* 278, *C. comptoniae* 276, *C. quercuum* 274, dieback 216, 234, *Elytroderma deformans* 36, flood tolerance 486, *Fusarium moniliforme* var. *subglutinans* 216, gall 274, herbicide tolerance 460, *Inonotus circinatus* 318, littleleaf **290,** *Mycosphaerella dearnessii* 46, needle blight 34–36, 46, ozone tolerance 466, *Phaeolus schweinitzii* 322, *Phytophthora cinnamomi* **290,** pine wood nematode 380, pitch canker 216, *Ploioderma lethale* 34, root rot **290,** 318, rust 268, 274–278, *Verticicladiella procera* 370
  P. edulis (pinyon pine): *Arceuthobium divaricatum* 426, black stain root disease 368, butt rot 322, *Coleosporium* 268, *Cronartium occidentale* 270, dwarf mistletoe 426, *Elytroderma deformans* 36, *Hemiphacidium planum* 52, needle blight 36, *Phaeolus schweinitzii* 322, root rot 322, rust 268–270, snow blight 52, *Sphaeropsis sapinea* 136, sulfur dioxide tolerance 468, tip blight 136, *Verticicladiella wageneri* 368
  P. eldarica: comandra rust 278
  P. elliottii, including varieties *elliottii* (slash pine) and *densa* (South Florida slash pine): *Atropellis tingens* 234, butt rot 322, canker **216,** 234, **274,** *Cenangium* 230, *Coleosporium apocynaceum* 268, compartmentalization of wound-associated decayed wood **336,** cone rust 270, **280,** *Cronartium quercuum* f. sp. *fusiforme* **274,** *C. strobilinum* 270, **280,** dieback **216,** 234, fire scar **336,** flood tolerance 486, fluoride air pollutant injury **470,** *Fusarium moniliforme* var. *subglutinans* **216,** gall **274,** *Heterobasidion annosum* 314, *Inonotus circinatus* 318–320, *Lophodermella cerina* 38, *Mycosphaerella dearnessii* **46,** *M. pini* 48, needle blight 34, **46**–48, nematode 302, 380, nutrient deficiency **472,** *Phaeolus schweinitzii* 322, pine wood nematode 380, pitch canker **216,** *Ploioderma* **34,** *Radopholus similis* 302, root rot 314, 318–322, rust 268–270, **274, 280,** salt tolerance 454, Spanish moss **504,** *Sphaeropsis sapinea* 136, sulfur dioxide injury **468,** tip blight 136, *Verticicladiella procera* 370
  P. engelmannii (Apache pine): *Arceuthobium vaginatum* 426, 432, brown felt blight 50, cone rust 280, *Cronartium conigenum* 280, dwarf mistletoe 426, 432, *Herpotrichia coulteri* 50, limb rust 280, *Peridermium* 280
  P. flexilis (limber pine): *Arceuthobium* 426–432, blister rust 272, brown felt blight 50, butt rot 322, canker 272, *Cenangium* 230, *Coleosporium crowellii* 268, *Cronartium ribicola* 272, *Cyclaneusma minus* 38, dwarf mistletoe 426–432, *Hemiphacidium planum* 52, *Herpotrichia coulteri* 50, needle cast 38, *Phaeolus schweinitzii* 322, root rot 322, rust 268, 272, snow blight 52, sulfur dioxide tolerance 468
  P. glabra (cedar pine, spruce pine): *Coleosporium minutum* 268, *Cronartium comandrae* 278, *C. quercuum* 274, gall 274, *Mycosphaerella dearnessii* 46, needle blight 34, 46, *Ploioderma lethale* 34, rust 268, 274, 278
  P. halepensis (Aleppo pine): *Arceuthobium* 430–432, dwarf mistletoe 430–432, *Endocronartium harknessii* **282,** gall rust **282,** *Lophodermium seditiosum* 32, *Mycosphaerella dearnessii* 46, *M. pini* 48, needle blight 46–48, needle cast 32, pine wood nematode 380, *Sirococcus conigenus* 134, *Sphaeropsis sapinea* 136, tip blight 134–136
  P. hartwegii: cone rust 280, *Cronartium conigenum* 280
  P. insularis (Benguet pine): *Sphaeropsis sapinea* 136, tip blight 136

P. jeffreyi (Jeffrey pine): *Arceuthobium campylopodum* 430, black stain root disease 368, brown felt blight 50, butt rot 322, canker 276–278, *Cenangium* 230, *Coleosporium pacificum* 268, *Cronartium coleosporioides* 278–280, *C. comandrae* 278, *C. comptoniae* 276, *Cyclaneusma minus* 38, dieback 230, dwarf mistletoe 430, *Elytroderma deformans* 36, *Endocronartium harknessii* 282, gall 282, *Herpotrichia coulteri* 50, limb rust 280, *Macrophomina phaseolina* 292, *Mycosphaerella pini* 48, needle blight 36, 48, needle cast 38, ozone damage 464, ozone tolerance 466, *Peridermium filamentosum* 280, *Phaeolus schweinitzii* 322, root rot 292, 322, rust 268, 276–282, *Sirococcus conigenus* 134, sulfur dioxide tolerance 468, tip blight 134, *Verticicladiella wageneri* 368

P. lambertiana (sugar pine): *Arceuthobium californicum* 426, *Atropellis pinicola* 234, blister rust 272, black stain root disease 368, brown felt blight 50, butt rot 322, canker 234, 272, *Cenangium* 230, *Cronartium ribicola* 272, dieback 230, dwarf mistletoe 426, *Herpotrichia coulteri* 50, *Lophodermella arcuata* 38, *Melampsora medusae* 256, *M. occidentalis* 258, ozone tolerance 466, *Phaeolus schweinitzii* 322, root rot 322, rust 256–258, 272, *Sirococcus conigenus* 134, tip blight 134, *Verticicladiella wageneri* 368

P. lawsonii: cone rust 280, *Cronartium conigenum* 280

P. lumholtzii: Arceuthobium gillii 434, cone rust 280, *Cronartium conigenum* 280, dwarf mistletoe 434

P. michoacana: cone rust 280, *Cronartium conigenum* 280, limb rust 280, *Peridermium* 280

P. monophylla (singleleaf pinyon): *Arceuthobium divaricatum* 426, black stain root disease **368,** butt rot 322, *Coleosporium jonesii* 268, *Cronartium occidentale* 270, dwarf mistletoe 426, ozone tolerance 466, *Phaeolus schweinitzii* 322, root rot 322, rust 268–270, *Verticicladiella wageneri* **368**

P. montezumae (rough barked Mexican pine): limb rust 280, *Lophodermium seditiosum* 32, *Peridermium* 280

P. monticola (western white pine): *Arceuthobium* 426, 432–434, *Armillaria* 448, *Atropellis* 234, black stain root disease 368, blister rust 272, blue stain **350,** brown felt blight 50, butt rot 322, canker 234, 272, **448,** *Cenangium* 230, *Cronartium ribicola* 272, decline **448,** dieback 230, 234, drought injury 478, dwarf mistletoe 426, 432–434, *Europhium trinacriforme* 448, heart rot **350,** *Herpotrichia* 50, *Inonotus circinatus* 318, *I. tomentosus* **320,** *Leptographium* 448, *Lophodermella arcuata* 38, *Lophodermium durilabrum* 32, *Melampsora occidentalis* 258, *Mycosphaerella dearnessii* 46, *M. pini* 48, needle blight 46–48, needle droop 478, *Phaeolus schweinitzii* 322, *Phellinus pini* **350,** *P. weirii* **324**–326, root rot 318–**324,** 326, rust 258, sulfur dioxide tolerance 468, *Verticicladiella wageneri* 368, wood rot 318–**320,** 322, **324**–326, **350**

P. mugo (mountain or mugo pine): brown felt blight 50, butt rot 322, canker 276, *Cenangium* 230, *Coleosporium asterum* 268, *Cronartium comandrae* 278, *C. comptoniae* 276, *C. quercuum* 274, *Cyclaneusma minus* 38, *Cylindrocladium* 294, *Endocronartium harknessii* 282, fluoride air pollutant injury **470,** gall 274, 282, *Herpotrichia* 50, mycorrhizae **500,** *Mycosphaerella dearnessii* **46,** *M. pini* 48, needle blight **46**–48, needle cast 38, 42, *Phaeolus schweinitzii* 322, pine wood nematode 380, *Rhizosphaera kalkhoffii* 42, root rot 294, 322, rust 268, 274–278, 282, salt tolerance 454, *Sphaeropsis sapinea* **136,** sulfur dioxide tolerance 468, tip blight **136**

P. muricata (Bishop pine): *Arceuthobium occidentale* 426, *Cronartium comptoniae* 276, dwarf mistletoe 426, *Endocronartium harknessii* 282, gall 282, *Mycosphaerella pini* 48, needle blight 48, rust 276, 282, *Sphaeropsis sapinea* 136

P. nigra (Austrian or black pine): air pollution damage 464, *Armillaria* **310,** *Ascocalyx abietina* 232, *Atropellis* **234,** brown felt blight 50, canker 232–**234,** *Cenangium ferruginosum* **230,** *Coleosporium asterum* 268, *Cronartium comptoniae* 276, *C. quercuum* 274, *Cyclaneusma minus* 38, *Cylindrocladium* 294, dieback **230–234,** dwarf mistletoe 432, *Endocronartium harknessii* 282, gall 274, 282, *Herpotrichia coulteri* 50, *Inonotus* 318, *Lophodermium seditiosum* 32, *Mycosphaerella dearnessii* 46, *M. pini* **48,** needle blight 46–**48,** needle cast 32, 38, 42, ozone tolerance 466, pine wood nematode 380, *Ploioderma lethale* **34,** root rot 294, **310,** 318, rust 268, 274–276, salt damage **452–454,** *Sphaeropsis sapinea* **136,** sulfur dioxide tolerance 468, tip blight **136,** *Verticicladiella procera* 370, winter fleck **462**

P. oocarpa: cone rust 280, *Cronartium conigenum* 280

P. palustris (longleaf pine): butt rot 322, canker 216, *Coleosporium* 268, cone rust 270, 280, *Cronartium quercuum* f. sp. *fusiforme* 274, *C. strobilinum* 270, 280, dieback 216, *Fusarium moniliforme* var. *subglutinans* 216, lightning injury **490,** *Mycosphaerella dearnessii* 46, needle blight **46,** *Phaeolus schweinitzii* 322, pine wood nematode 380, pitch canker 216, root rot 322, rust 268–270, 274, salt tolerance 454, wood rot 322

P. parviflora (Japanese white pine): ozone tolerance 466

P. patula (Mexican yellow pine): cone rust 280, *Cronartium conigenum* 280, *Mycosphaerella pini* 48, needle blight 48, *Sphaeropsis sapinea* 136, tip blight 136

P. peuce (Macedonian pine): sulfur dioxide tolerance 468

P. pinaster (cluster pine): *Atropellis tingens* 234, canker 234, *Cenangium* 230, *Cronartium comptoniae* 276, dieback 230, 234, *Endocronartium harknessii* 282, gall 282, *Mycosphaerella dearnessii* 46, *M. pini* 48, needle blight 46–48, rust 276, 282, *Sphaeropsis sapinea* 136, tip blight 136

P. pinea (Italian stone pine): *Arceuthobium occidentale* 432, dwarf mistletoe 432, *Mycosphaerella dearnessii* 46, needle blight 46, salt tolerance 454, *Sphaeropsis sapinea* 136, tip blight 136

P. ponderosa (ponderosa pine): air pollution damage **462**–466, *Arceuthobium* 426–434, *A. americanum* **426,** *A. campylopodum* **428,** *A. vaginatum* subsp. *cryptopodum* **432,** *Armillaria* **310**–312, *Ascocalyx abietina* 232, *Atropellis* 234, black stain root disease 368, Botrytis blight 60, brown felt blight 50, butt rot 322, canker 232, **278,** *Cenangium* 230, *Coleosporium asterum* 268, *Cronartium arizonicum* 280, *C. coleosporioides* 278–280, *C. comandrae* 278, *C. comptoniae* 276, *C. quercuum* 274, decline **462,** dieback 230–232, **282,** drought injury 478, dwarf mistletoe **426, 428,** 430, **432,** 434, *Elytroderma deformans* **36,** *Endocronartium harknessii* 282, fluoride tolerance 470, gall 274, **282,** *Hemiphacidium planum* 52, *Herpotrichia coulteri* 50, *Heterobasidion annosum* 314–316, *Inonotus* 318, *Lophodermella* 38, *Lophodermium ponderosae* 32, *Macrophomina phaseolina* 292, *Melampsora medusae* 256, *M. occidentalis* 258, *Mycosphaerella dearnessii* 46, *M. pini* 48, needle blight **36,** 46–48, needle droop 478, *Phaeolus schweinitzii* 322, pine wood nematode 380, root rot 292, 312–318, 322, rust 256–258, 268, 274, 278–**280,** salt tolerance 454, *Sirococcus conigenus* 134, snow blight 52, sooty mold **30,** *Sphaeropsis sapinea* 136, sulfur dioxide tolerance 468, tip blight 134–136, *Verticicladiella wageneri* 368, winter browning 478

P. pseudostrobus: cone rust 280, *Cronartium conigenum* 280

P. pungens (table mountain pine): *Atropellis tingens* 234, butt rot 322, canker 216, 234, *Cenangium* 230, *Coleosporium asterum* 268, *Cronartium comandrae* 278, *C. comptoniae* 276, *C. quercuum* 274, dieback 216, 234, *Fusarium moniliforme* var. *subglutinans* 216, gall 274, needle blight 34, *Phaeolus schweinitzii* 322, pitch canker 216, *Ploioderma lethale* 34, root rot 322, rust 268, 274–278, *Sphaeropsis sapinea* 136, tip blight 136

P. radiata (Monterey pine): butt rot 322, *Cenangium* 230, *Coleosporium pacificum* **268,** *Cronartium comptoniae* 276, *Cyclaneusma minus* 38, dieback 160, dwarf mistletoe 426, *Endocronartium harknessii* 282, *Fusarium moniliforme* var. *subglutinans* 216, gall 282, *Inonotus* 318, *Melampsora medusae* 256, *M. occidentalis* 258, *Mycosphaerella dearnessii* 46, *M. pini* 48, needle blight 34, 46–48, needle cast 38, ozone tolerance 466, *Phaeolus schweinitzii* 322, pine wood nematode 380, *Ploioderma pedatum* 34, *Pseudomonas syringae* pv. *syringae* **160,** root rot 318, 322, rust 256–258, **268,** 276, 282, *Sphaeropsis sapinea* 136, tip blight 136

P. resinosa (red pine): *Arceuthobium pusillum* 434, *Ascocalyx abietina* **232,** *Atropellis tingens* 234, Botrytis blight 60, butt rot 322, canker 194–196, **232**–234, *Cenangium* 230, *Coleosporium* 268, *C. asterum* **268,** *Cronartium comptoniae* 276, *C. quercuum* 274, *Cylindrocladium* 294, *Cytospora kunzei* 196, dieback **194**–196, 230, **232,** 234, drought injury **194,** 478, dwarf mistletoe 434, flood tolerance 486, frost injury **478,** gall 274, *Heterobasidion annosum* **314–316,** ice glaze injury 490, *Inonotus* 318, *Leucostoma kunzei* 196, lichen **506,** lightning injury **490,** *Lophodermium seditiosum* 32, *Melampsora medusae* 256, mycorrhizae **500,** *Mycosphaerella dearnessii* 46, *M. pini* 48, needle blight 46–48, needle browning **496,** needle cast 32, needle droop **478,** ozone tolerance 466, *Phacidium infestans* 52, *Phaeolus schweinitzii* 322, pine wood nematode 380, root rot 294, **314–316,** 318, 322, rust 256, 268, 274–276, salt tolerance 454, *Sirococcus conigenus* **134,** snow blight 52, *Sphaeropsis sapinea* **136,** sulfur dioxide tolerance 468, tip blight **134–136,** *Valsa* **194,** *Verticicladiella procera* 370

P. rigida (pitch pine): *Ascocalyx abietina* 232, *Atropellis tingens* 234, canker 216, 232–234, **276**–278, *Cenangium* 230, *Coleosporium* 268, *Cronartium comandrae* 278, *C. comptoniae* **276,** *C. quercuum* 274, dieback 216, 232–234, *Fusarium moniliforme* var. *subglutinans* 216, gall 274, *Hemiphacidium convexum* 52, *Inonotus* 318, *Mycosphaerella dearnessii* 46, needle blight 34, 46, ozone tolerance 466, pine wood nematode 380, pitch canker 216, *Ploioderma lethale* 34, root rot 318, rust 268, 274, **276,** 278, salt tolerance 454, snow blight 52, sulfur dioxide tolerance 468

P. roxburghii (chir pine): *Sphaeropsis sapinea* 136, tip blight 136

P. sabiniana (digger pine): *Arceuthobium occidentale* **428,** 432, butt rot 322, *Cenangium* 230, *Cyclaneusma minus* 38, dieback 230, dwarf mistletoe 426–**428,** *Endocronartium harknessii* 282, gall rust 282, needle cast 38, ozone tolerance 466, *Phaeolus schweinitzii* 322, root rot 322, rust 282, *Sphaeropsis sapinea* 136, tip blight 136, wood rot 322

P. serotina (pond pine): *Atropellis tingens* 234, butt rot 322, canker 234, *Cronartium comandrae* 278, *C. quercuum* 274, dieback 234, gall 274, *Mycosphaerella dearnessii* 46, needle blight 34, 46, *Phaeolus schweinitzii* 322, *Ploioderma lethale* 34, root rot 322, rust 274, 278, *Sphaeropsis sapinea* 136, tip blight 136

P. ×*sondereggii* (longleaf × loblolly pine hybrid): *Mycosphaerella dearnessii* 46, needle blight 46

P. strobiformis (southwestern white pine): *Arceuthobium apachecum* 426, **432**–434, blister rust 272, canker 272, *Cronartium ribicola* 272, dwarf mistletoe 426, **432**–434, ozone tolerance 466

P. strobus (eastern white pine): air pollution damage 462–**464,** *Arceuthobium pusillum* 434, *Ascocalyx abietina* 232, *Atropellis* 234, bark formation **496,** black stain root disease 368, blister rust **272,** boron toxicity **454,** brown felt blight 50, butt rot 322, canker 196, **216,** 230–234, **272, 370,** *Cenangium ferruginosum* 230, *Cronartium ribicola* **272,** *Cyclaneusma minus* 38, *Cylindrocladium* 294, *Cytospora kunzei* 196, dieback 194–196, 230–236, drought injury 478, dwarf mistletoe 434, flood tolerance 486, fluoride tolerance 470, freeze damage 478, *Fusarium moniliforme* var. *subglutinans* **216,** heat injury **478,** *Hemiphacidium planum* 52, herbicide injury **456,** *Herpotrichia coulteri* 50, *Heterobasidion annosum* **316,** ice glaze injury 490, *Inonotus* 318, *Lachnellula agassizii* 236, *Leucostoma kunzei* 196, lightning injury **490,** *Lophophacidium hyperboreum* 52, *Mycosphaerella dearnessii* 46, needle blight 46, **464,** needle browning **454,** needle cast 32, 38, 42, needle droop 478, nutrient deficiency **472**–474, ozone tolerance 466, *Phacidium abietis* 52, *P. infestans* 52, *Phaeolus schweinitzii* 322, *Phellinus pini* 350, pine wood nematode 380, pitch canker **216,** *Rhizosphaera kalkhoffii* 42, root lesions **370,** root rot 294, **316**–318, 322, salt damage **452**–454, snow blight 52, *Sphaeropsis sapinea* 136, sulfur dioxide injury **468,** sun scald **478,** tip blight 136, *Valsa* 194, *Verticicladiella procera* **370,** *V. wageneri* 368, wilt **370,** winter fleck **464,** witches'-broom **494,** wood rot **316**–318, 322, 350, wound healing in bark **496**

P. sylvestris (Scots pine): *Arceuthobium* 430–434, *Ascocalyx abietina* **232,** *Atropellis* 234, Botrytis blight 60, butt rot 322, canker **232**–234, *Cenangium* 230, *Coleosporium* 268, *Cronartium comandrae* 278, *C. comptoniae* **276,** *C. quercuum* 274, *Cyclaneusma minus* **38,** *Cylindrocladium* 294, dieback 194,

P. sylvestris (cont.)
230, **232,** 234, dwarf mistletoe 430, 434, *Endocronartium harknessii* **282,** fluoride tolerance 470, *Fusarium moniliforme* var. *subglutinans* 216, gall 274, **282,** herbicide injury **456,** ice glaze injury 490, *Inonotus* 318, *Lophodermella cerina* 38, *Lophodermium pinastri* 32, *L. seditiosum* **32,** *Melampsora medusae* 256, *Mycosphaerella dearnessii* 46, *M. pini* 48, needle blight 46–48, needle cast **32, 38,** nutrient deficiency **472**–474, ozone tolerance 466, *Phacidium infestans* 52, *Phaeolus schweinitzii* 322, pine wood nematode **380,** root rot 294, 318, 322, rust 256, 268, 274, **276,** 278, 282, salt tolerance 454, *Sirococcus conigenus* 134, snow blight 52, *Sphaeropsis sapinea* 136, sulfur dioxide tolerance 468, tip blight 134–136, *Valsa* 194, *Verticicladiella procera* 370, witches'-broom **282, 494**

P. tabuliformis (Chinese pine): *Cenangium* 230, *Cronartium comptoniae* 276, rust 276

P. taeda (loblolly pine): *Armillaria tabescens* 312, *Atropellis* 234, butt rot 322, canker 216, 234, **274,** *Cenangium* 230, *Coleosporium* **268,** *Cronartium comandrae* 278, *C. comptoniae* 276, *C. quercuum* **274,** dieback 216, 234, flood tolerance 486, fluoride tolerance 470, *Fusarium moniliforme* var. *subglutinans* 216, gall 274, *Heterobasidion annosum* 314, *Inonotus circinatus* 318, littleleaf **290,** *Lophodermella cerina* 38, *Mycosphaerella dearnessii* 46, *M. pini* 48, needle blight 34, 46–48, nematode 302, 380, nutrient deficiency **472,** ozone damage 466, *Phaeolus schweinitzii* 322, *Phytophthora cinnamomi* **290,** pine wood nematode 380, pitch canker 216, *Ploioderma lethale* 34, *Radopholus similis* 302, root rot **290,** 312–314, 318, 322, rust **268, 274**–278, salt tolerance 454, *Sphaeropsis sapinea* 136, tip blight 136, *Verticicladiella procera* 370

P. teocote (twisted-leaf pine): cone rust 280, *Cronartium conigenum* 280

P. thunbergiana (Japanese black pine): *Arceuthobium occidentale* 432, blue stain **380,** canker **230,** *Cenangium ferruginosum* **230,** *Coleosporium asterum* 268, dieback **230,** dwarf mistletoe 432, *Endocronartium harknessii* 282, freeze injury **478–480,** gall 282, herbicide injury **456,** *Lophodermium pinastri* **32,** *Mycosphaerella dearnessii* 46, *M. pini* 48, needle blight 34, 46–48, needle cast **32**–34, 42, ozone tolerance 466, pine wood nematode **380,** *Ploioderma lethale* 34, *Rhizosphaera kalkhoffii* 42, rust 268, 282, salt tolerance **452**–454, sulfur dioxide tolerance 468

P. torreyana (Torrey pine): ozone tolerance 466

P. virginiana (Virginia pine): *Atropellis* 234, butt rot 322, canker 216, 234, *Cenangium* 230, *Coleosporium* 268, *Cronartium appalachianum* 270, *C. comandrae* 278, *C. comptoniae* 276, *C. quercuum* **274,** dieback 216, 234, flood tolerance 486, *Fusarium moniliforme* var. *subglutinans* 216, gall **274,** *Lophodermium seditiosum* 32, *Mycosphaerella dearnessii* 46, needle blight 34, 46, needle cast 32, ozone tolerance 466, *Phaeolus schweinitzii* 322, pine wood nematode 380, pitch canker 216, *Ploioderma lethale* 34, root rot 322, rust 268–270, **274**–278, salt tolerance 454, *Verticicladiella procera* 370

P. wallichiana (Himalayan white pine): canker 196, *Cytospora kunzei* 196, dieback 196, *Leucostoma kunzei* 196, needle cast 42, *Rhizosphaera kalkhoffii* 42, *Sphaeropsis sapinea* 136, tip blight 136

pinyon, see *Pinus cembroides, P. edulis, P. monophylla*

Piper nigrum (black pepper): nematode 302, *Radopholus similis* 302

Piptoporus betulinus **346**

Pisolithus tinctorius **500**

pistachio, see *Pistacia*

Pistacia chinensis (pistachio): *Botryosphaeria dothidea* 172, Botrytis blight 60, dieback 172, herbicide injury **460,** *Schizophyllum commune* 358, Verticillium wilt **376**

Pittosporum (pittosporum): *Alternaria tenuissima* 90, *Armillaria tabescens* 312, *Botryosphaeria rhodina* 180, *Cercospora pittospori* **88,** dieback 180, leaf spot **88**–90, rhabdovirus 400, root rot 312, salt tolerance 454, vein chlorosis 400

P. tobira (Japanese pittosporum): gall 148, *Nectriella pironii* 148, *Phymatotrichum omnivorum* 292, Rhizoctonia blight 94, root rot 292, 296, *Sclerotium rolfsii* 296, southern blight 296

Plagiosphaeria gleditsiae 84

Planera aquatica (water elm): *Ganoderma lucidum* 330, root rot 330

plane tree, see *Platanus*
London: *P. ×acerifolia*
Oriental: *P. orientalis*

Plasmopara cercidis 12

P. ribicola 12

P. viburni **12**

P. viticola **12**

Platanus (plane tree, sycamore): anthracnose 112, *Apiognomonia veneta* 112, *Botryosphaeria dothidea* 172, *B. obtusa* 172, butt rot 304, 334, canker 226, 360, *Ceratocystis fimbriata* 360–362, dieback 172, 176, 226, 360, *Ganoderma applanatum* 334, *Hendersonia desmazierii* 82, *Hypoxylon atropunctatum* 224, *H. deustum* 304, *H. mammatum* 222, *H. tinctor* 226, *Massaria platani* 82, *Microsphaera platani* 16, mistletoe 422, *Oxyporus populinus* 344, *Phyllactinia guttata* 16, powdery mildew 16, root rot 304, salt tolerance 454, *Schizophyllum commune* 358, twig dieback 82, wood decay 344

P. ×acerifolia (London plane): anthracnose **112,** *Apiognomonia veneta* **112,** bark formation **496,** canker 226, 360, *Ceratocystis fimbriata* 360–362, *Cristulariella moricola* 62, dieback 226, drought damage 340, fluoride tolerance 470, herbicide injury **460,** *Hypoxylon tinctor* 226, leaf spot 62, ozone tolerance 466, salt tolerance 454, stem crack **340,** sulfur dioxide tolerance 468, *Trametes versicolor* 342, winter damage **476,** wood rot 342

P. occidentalis (eastern sycamore, sycamore): anthracnose 110–**112,** *Apiognomonia veneta* 110–**112,** bacterial leaf scorch 180, 384–**386,** *Botryosphaeria rhodina* **180,** canker **112,** 144, **226, 360,** canker-stain **360,** *Ceratocystis fimbriata* **360**–362, *Cercospora platanicola* 76–78, *Cristulariella moricola* 62, *Diaporthe scabra* 144, decline 360, dieback 144, 180, **226,** 360, **386,** *Discula* 110–112, drought stress 180, fastidious xylem-inhabiting bacteria 384–**386,** flood tolerance 486, fluoride tolerance 470, frost susceptibility 482, herbicide tolerance 460, *Hericium erinaceus* 344, *Hypoxylon tinctor* **226,** ice glaze injury 490, leaf spot 62, 76–**78,** 92, lightning injury 490, mistletoe 422, *Mycosphaerella platanifolia* 76–78, *M. stigmina-platani* 92, *Oxyporus latemarginatus* 328, ozone injury 466, *Perenniporia fraxinophila* 344, *Phomopsis scabra* 144, *Phyllosticta platani* 76–**78,** root rot 328, salt tolerance 454, sapwood stain 360, sulfur dioxide tolerance 468, *Stigmina platani* 92, wetwood 382, wood rot 328, 344, *Xylemella fastidiosum* 384–**386**

P. orientalis (Oriental plane): anthracnose 112, *Apiognomonia veneta* 112, *Ceratocystis fimbriata* 360

P. racemosa (California sycamore): anthracnose 112, *Apiognomonia veneta* 112, mistletoe 422, ozone tolerance 466, *Phoradendron macrophyllum* 422

P. wrightii (Arizona sycamore): anthracnose 112, *Apiognomonia veneta* 112

Platycladus orientalis (Oriental arborvitae): *Armillaria tabescens* 312, canker 170, *Cercospora sequoiae* 86, *Didymascella thujina* 44, leaf blight 44, 86, 138, *Phomopsis juniperovora* 138, root rot 312, *Seiridium cardinale* 170

Plectosphaerella 72

Pleochaeta polychaeta 14

Ploioderma 34

P. lethale **34**

plum, see *Carissa, Coccoloba, Prunus, Syzygium*
American or wild: *P. americana*
beach: *P. maritima*
bullace or Damson: *P. insititia*
Canada: *P. nigra*
cherry or myrobalan: *P. cerasifera*
chickasaw: *P. angustifolia*
common plum or prune: *P. domestica*
flowering: *P. ×blireiana, P. cerasifera*
Jambolan or Java: *Syzygium cumini*
Japanese: *P. salicina*
Natal: *Carissa grandiflora*
pigeon: *Coccoloba diversifolia*
Sierra: *P. subcordata*
wild goose: *P. hortulana, P. munsoniana*

Plumbago capensis (cape leadwort): gall 148, *Nectriella pironii* 148

Plumeria (frangipani): *Botryosphaeria dothidea* 172, dieback 172

Podocarpus macrophyllus (southern yew): salt tolerance 454

Podosphaera clandestina 14–**16**

P. leucotricha 14, **18**

P. oxycanthae 14

P. tridactyla 14

poinciana
dwarf, see *Caesalpinia*
royal: *Delonix*

poinsettia, see *Euphorbia pulcherrima*

Pollaccia 100

P. saliciperda 102

Polypodium (polypody fern): *Milesina* 264, rust 264

Polyporus adustus 144

P. betulinus 346

P. circinatus 318–320

P. curtisii 332

P. dryadeus 328

P. glomeratus 356

P. graveolens 144

P. hispidus 356

P. lucidus 332

P. oregonense 332

P. schweinitzii 322

P. squamosus **344**

P. sulfureus 346

P. tomentosus 318–320

P. tsugae 332

P. tulipiferae 358

P. versicolor 342

Polyscias (aralia): gall 148, *Nectriella pironii* 148

P. fruticosa (Ming aralia): *Alternaria panax* 90, leaf blight 90

P. guilfoylei (geranium-leaf aralia): *Alternaria panax* **90,** leaf blight **90**

pomegranate, see *Punica granatum*

Poncirus (trifoliate orange): *Botryosphaeria obtusa* 176, *B. rhodina* 180, burrowing nematode 302, dieback 176, 180, *Radopholus similis* 302, root rot 306, *Xylaria* 306

poplar, see *Liriodendron, Populus*
balsam: *P. balsamifera*
black: *P. nigra*
Carolina: *P. ×canadensis*
gray: *P. canescens*
Lombardy: *P. nigra* cv. 'Italica'
Simon: *P. simonii*
white: *P. alba*
yellow: *Liriodendron tulipifera*

poplar mosaic virus **400**

poplar potyvirus **402**

Populus (aspen, cottonwood, poplar): *Agrobacterium tumefaciens* 156, *Alternaria* **90,** bark rot 168, *Botryosphaeria dothidea* 172, *B. obtusa* 176, bronzing **402,** butt rot 334, canker **74,** 142–144, 184, 200, 212–214, 218, 222, *Ceratocystis fimbriata* 362, *Cerrena unicolor* 358, *Climacodon septentrionalis* 344, crown gall 156, *Cryptodiaporthe populea* 184, *C. salicella* 186, *Cryptosphaeria populina* 218, dieback 142–144, 172, 176, 184, 200, *Diplodia tumefaciens* 150, gall 156, *Fomes fomentarius* 346, *Fomitopsis pinicola* 348, *Ganoderma applanatum* 334, *Heterobasidion annosum* 314, *Hypoxylon mammatum* 222, *Inonotus glomeratus* 356, leaf blight **90,** leaf blister 22–24, leaf spot **74,** *Leucostoma niveum* 200, lightning injury 490, *Melampsora* 256–258, mistletoe 422, *Mycosphaerella* **74,** *Nectria galligena* 212, *Oxyporus populinus* 344, ozone tolerance 466, *Perenniporia medullapanis* 168, *Phellinus tremulae* 342, *Phomopsis macrospora* 142–144, *Phoradendron macrophyllum* **422,** *Phymatotrichum omnivorum* 292, powdery mildew 14, root rot 292, rough bark disease 150, rust 256–258, salt tolerance 454, *Schizophyllum commune* 358, *Taphrina* 22–24, *Uncinula* 14, *Valsa sordida* 200, virus **400–402,** wetwood 382, wood rot 168, 314, 334, 342–348, 358

Populus cv. 'Northwest': vein-associated yellowing **402,** virus symptoms **402**

Populus cv. 'Robusta': canker 144, 184, *Cryptodiaporthe populea* 184, dieback 144, 184, mosaic **402,** *Phomopsis macrospora* 144, vein-associated mottling **402,** virus symptoms **402**

P. ×acuminata (lanceleaf cottonwood): canker 184, 200, dieback 184, 200, *Cryptodiaporthe populea* 184, *Melampsora occidentalis* 258, rust 258, *Valsa sordida* 200

P. adenopoda (Chinese aspen): canker 222, *Hypoxylon mammatum* 222

P. alba (white poplar): blight 100, **122,** *Apioplagiostoma populi* **122,** canker 184, **200,** 222, *Cryptodiaporthe populea* 184, dieback 168, 184, **200,** *Drepanopeziza populi-albae* 120, *Hendersonula toruloidea* 168, *Hypoxylon mammatum* 222, leaf spot **120,** *Leucostoma niveum* **200,** *Marssonina castagnei* **120,** *Melampsora* 258, rust 258, salt tolerance 454, *Schizophyllum commune* 358, *Valsa sordida* 200, *Venturia tremulae* 100, wood rot 358

P. angustifolia (narrow-leaved or yellow cottonwood): canker 200, 218, *Cryptosphaeria populina* 218, dieback 200, leaf spot 74, *Melampsora medusae* 256, *M. occidentalis* 258, *Mycosphaerella* 74, rust 256–258, sulfur dioxide tolerance 468, *Valsa sordida* 200

P. balsamifera (balsam poplar): blight **100,** canker 184, 200, 218, 222, 228, *Ciborinia whetzelii* 56, *Cryptodiaporthe populea* 184, *Cryptosphaeria populina* 218, dieback 184, 200, *Encoelia pruinosa* 228, flood tolerance 486, fluoride tolerance 470, herbicide tolerance 460, *Hypoxylon mammatum* 222, leaf blight 56, **84,** leaf spot **74,** 120, *Leucostoma niveum* 200, *Linospora*

P. balsamifera (cont.)
tetraspora **84,** *Marssonina* 120, *Melampsora* 258, *M. medusae* 256, *Mycosphaerella* **74,** *Oxyporus latemarginatus* 328, poplar mosaic 400, rhabdovirus **400,** rust 256–258, salt tolerance 454, sulfur dioxide tolerance 468, veinal chlorosis and necrosis **400,** *Valsa sordida* 200, *Venturia populina* **100**
P. ×*berolinensis:* canker 184, *Cryptodiaporthe populea* 184
P. ×*canadensis* (Carolina poplar): canker 144, 184, *Cryptodiaporthe populea* 184, dieback 144, fluoride tolerance 470, leaf spot 120, *Marssonina* 120, *Melampsora occidentalis* 258, *Phomopsis macrospora* 144, *Pseudomonas syringae* pv. *syringae* 160, rust 258, sulfur dioxide tolerance 468
P. canescens (gray poplar): leaf spot 120, *Marssonina castagnei* 120, salt tolerance 454, shoot blight 100, *Venturia tremulae* 100
P. cathayana: canker 184, *Cryptodiaporthe populea* 184
P. caudina: canker 144, dieback 144, *Phomopsis macrospora* 142–144
P. deltoides (eastern cottonwood): blight 102, butt rot **334,** canker **74, 142**–144, 184, **200,** 214, 218, chlorotic spots **400,** *Ciborinia whetzelii* 56, *Cryptodiaporthe populea* 184, *Cryptosphaeria populina* 218, dieback **142**–144, **200,** flood tolerance 486, fluoride air pollutant injury **470,** *Fusarium solani* 214, *Ganoderma applanatum* **334,** herbicide tolerance 460, *Hypoxylon mammatum* 222, ice glaze injury 490, leaf blight 56, leaf spot 74, **120,** *Marssonina brunnea* **120,** *M. populi* 120, *Melampsora abietis-canadensis* 258, *M. medusae* **256,** *Mycosphaerella* **74,** nematode 302, *Oxyporus latemarginatus* **328,** ozone injury 466, *Phomopsis macrospora* **142**–144, poplar mosaic virus **400,** poplar potyvirus **402,** root rot **328,** rust **256**–258, salt tolerance 452–454, sulfur dioxide tolerance 468, *Valsa sordida* **200,** veinal yellowing and necrosis **400–402,** *Venturia tremulae* 100, wetwood **382,** wood rot **328, 334,** *Xiphinema americanum* 302
P. ×*euramericana* (European-American hybrid poplars): leaf spot 120, *Marssonina brunnea* 120
P. fremontii (Fremont cottonwood), including var. *wislizenii* (Rio Grande cottonwood): canker 184, *Cryptodiaporthe populea* 184, dieback 168, *Hendersonula toruloidea* 168, leaf blister 22–24, leaf spot 120, *Marssonina brunnea* 120, *Melampsora medusae* 256, *M. occidentalis* 258, mistletoe 422, *Phoradendron macrophyllum* 422, rust 256–258, *Taphrina populi-salicis* 22–24
P. grandidentata (bigtooth or large-tooth aspen): *Apioplagiostoma populi* 122, blight 100, 122, canker 184, 200, 218, 222, 228, *Ciborinia whetzelii* 56, *Cryptodiaporthe populea* 184, *Cryptosphaeria populina* 218, decline 402, dieback 184, 200, *Encoelia pruinosa* 228, *Hypoxylon mammatum* 222, ice glaze injury 490, leaf blight 56, leaf spot 120, *Leucostoma niveum* 200, *Marssonina* 120, *Melampsora abietis-canadensis* 258, *M. medusae* 256, *Oxyporus latemarginatus* 328, poplar potyvirus 402, rust 256–258, salt tolerance 454, *Valsa sordida* 200, *Venturia tremulae* 100, wood rot 328
P. heterophylla (swamp cottonwood): *Melampsora abietis-canadensis* 258, rust 258
P. laurifolia: canker 184, *Cryptodiaporthe populea* 184
P. macdougalii: canker 200, *Valsa sordida* 200
P. maximowiczii: canker 144, 184, *Cryptodiaporthe populea* 184, dieback 144, *Phomopsis macrospora* 144
P. nigra (black poplar), including cv. 'Italica' (Lombardy poplar): *Armillaria tabescens* 312, blight 100, canker 160, **184,** 200, 214, 218, *Ciborinia whetzelii* 56, *Cryptodiaporthe populea* **184,** *Cryptosphaeria populina* 218, *Dematophora necatrix* 306, dieback 168, **184,** 200, fluoride tolerance 470, *Fusarium solani* 214, *Hendersonula toruloidea* 168, leaf blight 56, leaf spot 120, *Marssonina* 120, *Melampsora medusae* 256, poplar mosaic 400, *Pseudomonas syringae* pv. *syringae* 160, root rot 306, 312, rust 256, salt tolerance 454, sulfur dioxide tolerance 468, *Valsa sordida* 200, *Venturia populina* 100, wetwood 382
P. ×*petrowskiana:* canker 184, *Cryptodiaporthe populea* 184
P. sargentii (Great Plains cottonwood): canker 184, 200, *Cryptodiaporthe populea* 184, dieback 200, herbicide tolerance 460, leaf spot 120, *Marssonina populi* 120, *Melampsora abietis-canadensis* 258, *M. medusae* 256, rust 256–258, salt tolerance 454, *Valsa sordida* 200
P. simonii (Simon poplar): canker 184, *Cryptodiaporthe populea* 184, leaf spot 120, *Marssonina populi* 120
P. tremula (European aspen): blight 100, canker 184, 222, *Cryptodiaporthe populea* 184, *Hypoxylon mammatum* 222, leaf spot 120, *Marssonina brunnea* 120, poplar mosaic 400, *Venturia tremulae* 100
P. tremuloides (quaking or trembling aspen): anthracnose 122–124, *Apioplagiostoma populi* 122, blight **100, 122,** 160, butt rot 334, canker 184, **200,** 214, **218, 222, 228, 362,** *Ceratocystis fimbriata* 360–**362,** *Ciborinia whetzelii* **56,** cold tolerance 480, *Cryptodiaporthe populea* 184, *Cryptosphaeria populina* **218,** decline 402, dieback **200,** *Encoelia pruinosa* **228,** flood tolerance 486, fluoride tolerance 470, *Fomes fomentarius* 346, *Fomitopsis pinicola* 348, *Fusarium lateritium* 214, *Ganoderma applanatum* 334, *Glomerella cingulata* 122–124, herbicide tolerance 460, *Hypoxylon mammatum* **222**–224, ice glaze injury 490, leaf blight **56,** leaf spot 120, *Leucostoma niveum* 200, *Marssonina* 120, *Melampsora medusae* 256–258, *Oxyporus populinus* 344, ozone tolerance 466, *Phellinus tremulae* **342,** *Polyporus squamosus* 344, poplar mosaic 400, poplar potyvirus 402, *Pseudomonas syringae* pv. *syringae* 160, root rot 334, rust 256–258, salt tolerance 454, sulfur dioxide tolerance 468, tobacco necrosis virus 402, tobacco ringspot virus 414, *Valsa sordida* **200,** vein mottling **402,** *Venturia tremulae* **100,** wood rot 334, **342,** 344–348
P. trichocarpa (black cottonwood): blight **100,** canker 184, 214, 218, 228, *Ceratocystis fimbriata* 362, *Cryptodiaporthe populea* 184, *Cryptosphaeria populina* 218, *Encoelia pruinosa* 228, *Fusarium lateritium* 214, *Hypoxylon mammatum* 222, *H. mediterraneum* 224, *Inonotus glomeratus* 356, *I. obliquus* 352, leaf blight 84, leaf blister **24,** leaf spot 74, 120, *Linospora tetraspora* 84, *Marssonina populi* 120, *Melampsora occidentalis* 256–258, *Mycosphaerella* 74, *Oxyporus latemarginatus* 328, *Polyporus squamosus* 344, poplar mosaic 400, *Pseudomonas syringae* pv. *syringae* 160, rust 256, sulfur dioxide tolerance 468, *Taphrina populi-salicis* **24,** *Venturia populina* **100,** wood decay 328, 344, 352, 356
P. tristis: canker 184, *Cryptodiaporthe populea* 184
P. wilsonii: canker 200, *Valsa sordida* 200
Poria ambigua 328
P. andersonii 356
P. latemarginata 328
P. obliqua 352
P. spiculosa 354
P. weirii 324–326
potassium deficiency 472–**474**
Potebniamyces 144
Potentilla (cinquefoil, potentilla): *Erwinia amylovora* 162, fire blight 162, tobacco ringspot virus 414
P. fruticosa (shrubby cinquefoil): salt tolerance 454
powdery mildew **14–18**
Pratylenchus **301–302**
P. penetrans **301–302**
prickly pear, see *Opuntia*
Prinsepia: Erwinia amylovora 162, fire blight 162
Pritchardia affinis (Kona palm): lethal yellowing 390
P. pacifica (Fiji fan palm): lethal yellowing 390
P. thurstonii (Thurston palm): lethal yellowing 390
privet, see *Forestiera, Ligustrum*
Amur: *L. amurense*
California: *L. ovalifolium*
Chinese: *L. sinense*
common: *L. vulgare*
glossy: *L. lucidum*
Florida: *F. segregata*
Regel's: *L. obtusifolium* var. *regelianum*
swamp: *F. acuminata*
wax-leaf: *L. japonicum*
Prosopis juliflora (mesquite): bacterial wetwood 382, *Ganoderma lucidum* 332, *Leveillula taurica* 14, mistletoe **422,** *Phellinus badius* 346, *Phoradendron californicum* **422,** powdery mildew 14, root rot 332, salt tolerance 454, wood decay 332, 346
prune, see *Prunus domestica*
prune dwarf virus **408**
Prunus (stone fruits: almond, apricot, cherry, nectarine, peach, plum, prune): *Agrobacterium tumefaciens* 156, *Alternaria* 90, *Apiosporina morbosa* **152,** *Armillaria* 308, bark rot 168, black knot **152,** bladder plum 22, blight 60, 126, *Blumeriella jaapii* 66, *Botryosphaeria dothidea* 172, *B. obtusa* 176, *B. quercuum* 178, *Botrytis cinerea* 60, butt rot 334, canker 144, 172, 198, 212, *Cerrena unicolor* 358, collar rot 286, crown gall 156, *Cylindrocladium* 294, *Dematophora necatrix* 306, *Diaporthe perniciosa* 144, dieback 144, 172, 176, 198, *Erwinia amylovora* 162, fire blight 162, *Fomes fomentarius* 346, *Fomitopsis pinicola* 348, fruit rot 90, 126, gall 152, 156, *Ganoderma applanatum* 334, *G. brownii* 332, graft incompatibility **492,** herbicide tolerance 460, *Heterobasidion annosum* 314, *Laetiporus sulfureus* 346, leaf blister 24, leaf curl 22, leaf spot 66, 90, *Leucostoma* 198, mistletoe 422, *Nectria galligena* 212, nematode 302, *Perenniporia medulla-panis* 168, *Phaeolus schweinitzii* 322, *Phellinus igniarius* 342, *Phomopsis* 144, *Phoradendron serotinum* 422, *Phyllactinia guttata* 16, *Phytophthora* 286, plum pockets 22, *Podosphaera clandestina* 14, *P. tridactyla* 14, powdery mildew 14–16, *Pratylenchus* 302, root rot 294, 306–308, 322, *Schizophyllum commune* 358, *Taphrina* 22–24, tobacco mosaic virus 414, tobacco ringspot virus 414, tomato ringspot virus 414, *Valsa ambiens* 198, Verticillium wilt 376, witches'-broom 24, wood rot 168, 314, 322, 334, 342, 346–348, 358
P. americana (American or wild plum): *Apiosporina morbosa* 152, black kjot 152, *Blumeriella jaapii* 66, canker 198, dieback 198, leaf spot 66, *Leucostoma* 198, nematode 302, *Taphrina* 22, *Xiphinema americanum* 302
P. angustifolia (chickasaw plum): anthracnose 122–124, blight 126, brown rot 126, *Glomerella cingulata* 122–124, *Monilinia* 126, *Taphrina* 22
P. armeniaca (apricot): *Alternaria* 90, *Apiosporina morbosa* 152, black knot 152, blight **126,** 160, brown rot **126,** canker 160, 198, *Ceratocystis fimbriata* 362, *Cylindrocladium* 294, *Dematophora necatrix* 306, dieback 168, 198, fluoride tolerance 470, fruit rot 90, green ring mottle virus 408, *Hendersonula toruloidea* 168, leaf curl 22–24, leaf spot 90, *Leucostoma* 198, *Monilinia laxa* **126,** *Phymatotrichum omnivorum* 292, powdery mildew 16, *Pseudomonas syringae* pv. *syringae* 160, root rot 292–294, 306, salt tolerance 454, *Schizophyllum commune* 358, *Sphaerotheca pannosa* 16, sulfur dioxide tolerance 468, *Taphrina deformans* 22, *T. wiesneri* 24, witches'-broom 24
P. avium (mazzard or sweet cherry): *Agrobacterium tumefaciens* **156,** *Armillaria tabescens* 312, *Apiosporina morbosa* 152, black knot 152, blight 160, *Blumeriella jaapii* 66, blight **126,** 160, brown rot **126,** canker **160,** 198, cherry leafroll virus 404, cherry rasp leaf virus **408,** crown gall **156,** dieback 198, fluoride tolerance 470, green ring mottle virus 408, leaf spot 66, *Leucostoma* 198, *Monilinia laxa* **126,** ozone tolerance 466, *Podosphaera clandestina* 16, powdery mildew 16, prune dwarf virus 408, prunus necrotic ringspot virus **408,** *Pseudomonas syringae* pv. *syringae* **160,** root rot 312, rusty mottle virus **408,** stem pitting 414, sulfur dioxide tolerance 468, *Taphrina wiesneri* 24, witches'-broom 24, X-disease 392
P. besseyi (western sand cherry): *Apiosporina morbosa* 152, black knot 152, *Blumeriella jaapii* 66, leaf spot 66
P. ×*blireiana* (flowering plum): salt tolerance 454, see also *P. cerasifera*
P. caroliniana (cherry laurel): *Blumeriella jaapii* 66, leaf spot 66, see also *P. laurocerasus*
P. cerasifera (cherry, flowering, or myrobalan plum): *Apiosporina morbosa* 152, black knot 152, fluoride tolerance 470, prune dwarf virus 408, prunus necrotic ringspot virus 408, salt tolerance 454, sulfur dioxide tolerance 468
P. cerasus (sour cherry): anthracnose 122–124, *Apiosporina morbosa* 152, black knot 152, blight 126, *Blumeriella jaapii* **66,** brown rot 126, canker 198, **358,** *Cerrena unicolor* **358,** cherry rasp leaf virus 408, dieback 198, *Glomerella cingulata* 122–124, green ring mottle virus **408,** leaf curl 24, leaf spot **66,** *Leucostoma* 198, *Monilinia* 126, *Podosphaera clandestina* 16, powdery mildew 16, prune dwarf virus **408,** *Pseudomonas syringae* pv. *syringae* 160, sour cherry yellows **408,** stem pitting 414, sulfur dioxide tolerance 468, *Taphrina wiesneri* 24, Verticillium wilt **376,** witches'-broom 24, X-disease 392
P. domestica (common plum, prune): *Apiosporina morbosa* 152, black knot 152, blight 126, *Blumeriella jaapii* 66, brown line disease 414, brown rot **126,** canker 198, *Ceratocystis fimbriata* 362, dieback 168, 198, *Hendersonula toruloidea* 168, leaf spot 66, *Leucostoma* 198, *Monilinia fructicola* **126,** *Podosphaera clandestina* 16, powdery mildew 14–16, prunus necrotic ringspot virus 408, *Pseudomonas syringae* pv. *syringae* 160, *Sphaerotheca pannosa* 16, sulfur dioxide tolerance 468,

P. domestica (cont.)
Taphrina communis 22, *T. pruni* 22, tobacco mosaic virus 414, tomato ringspot virus 414
P. dulcis (almond): *Agrobacterium tumefaciens* 156, apple mosaic virus 412, bacterial leaf scorch **384,** blight 126, 160, *Blumeriella jaapii* 66, brown rot 126, canker 160, *Ceratocystis fimbriata* 362, crown gall 156, *Dematophora necatrix* 306, dieback 168, fastidious xylem-inhabiting bacteria **384,** *Hendersonia rubi* 82, *Hendersonula toruloidea* 168, leaf blight 82, leaf curl 22, leaf scorch **454,** leaf spot 66, *Monilinia* 126, nematode 302, *Oxyporus latemarginatus* 328, *Phymatotrichum omnivorum* 292, *Pratylenchus* 302, Prunus necrotic ringspot virus 408, *Pseudomonas syringae* pv. *syringae* 160, root rot 292, 306, 328, salt injury **454,** *Taphrina deformans* 22, *Xylemella fastidiosum* **384**
P. emarginata (bitter cherry): *Apiosporina morbosa* 152, black knot 152, *Blumeriella jaapii* 66, leaf spot 66, *Phyllactinia guttata* 16, powdery mildew 16, sulfur dioxide tolerance 468
P. fasciculata (desert almond): prunus necrotic ringspot virus 408
P. fruticosa (European dwarf cherry): *Taphrina wiesneri* 24
P. hortulana (wild goose plum): *Taphrina communis* 22, *T. pruni* 22
P. ilicifolia (holly-leaved cherry): leaf curl 24, *Podosphaera clandestina* **16,** powdery mildew **16,** *Taphrina wiesneri* 24
P. insititia (bullace or Damson plum): *Apiosporina morbosa* 152, black knot 152, canker 198, dieback 198, *Ganoderma lucidum* 330, *Leucostoma* 198, prune dwarf virus 408, root rot 332
P. laurocerasus (cherry laurel): salt tolerance 454, see also *P. caroliniana*
P. lusitanica (Portugal laurel): winter injury **476**
P. mahaleb (mahaleb cherry): Apiosporina morbosa 152, black knot 152, *Blumeriella jaapii* 66, leaf spot 66, *Oxyporus latemarginatus* 328, root rot 306, 328, X-disease 392, *Xylaria mali* 306
P. maritima (beach plum): *Apiosporina morbosa* 152, black knot 152, leaf curl 22, salt tolerance 454, *Taphrina* 22
P. munsoniana (wild goose plum): leaf curl 22, *Taphrina* 22
P. nigra (Canada plum): *Apiosporina morbosa* 152, black knot 152, bladder plum **22,** *Blumeriella jaapii* 66, leaf spot 66, *Taphrina communis* **22,** *T. pruni* 22
P. padus (bird cherry): *Apiosporina morbosa* 152, black knot 152
P. pensylvanica (pin cherry): *Apiosporina morbosa* 152, black knot 152, *Blumeriella jaapii* 66, canker 198, dieback 152, 198, frost susceptibility 482, leaf curl 24, leaf spot 66, *Leucostoma* 198, lichens **504,** prunus necrotic ringspot virus 408, *Taphrina wiesneri* **24,** witches'-broom **24**
P. persica (nectarine, peach): *Agrobacterium tumefaciens* 156, anthracnose 122–124, *Apiosporina morbosa* 152, *Armillaria tabescens* 312, black knot 152, blight 126, 160, *Botryosphaeria dothidea* 172, *B. obtusa* 176, *B. rhodina* 180, brown rot 126, canker 126, 144, 160, **176,** 180, **198,** 214, *Ceratocystis fimbriata* 362, *Cristulariella moricola* 62, crown gall 156, *Cylindrocladium* 294, *Dematophora necatrix* 306, *Diaporthe perniciosa* 144, dieback 126, 144, 172, 176, 180, **198,** flood tolerance 486, fluoride tolerance 470, freeze injury **482,** gall 152, 156, *Ganoderma lucidum* 332, *Fusarium lateritium* 214, *Glomerella cingulata* 122–124, graft union necrosis 414, green ring mottle virus 408, gummosis 172, **176,** 180, **198,** herbicide tolerance 460, leaf curl **22,** leaf spot 62, 92, *Leucostoma* **198,** *Meloidogyne* 298, *Monilinia* **126,** mycoplasmalike organism **392,** nematode 298–302, *Oxyporus latemarginatus* **328,** *O. populinus* 344, ozone tolerance 466, *Phomopsis mali* 144, *Podosphaera clandestina* 16, powdery mildew 14–16, *Pratylenchus* 302, *Pseudomonas syringae* pv. *syringae* 160, *Radopholus similis* 302, root knot 298, root rot 294–296, 306, 312, **328,** 332, *Schizophyllum commune* 358, *Sclerotium rolfsii* 296, southern blight 296, *Sphaerotheca pannosa* 14–16, *Stigmina carpophila* 92, sulfur dioxide tolerance 468, *Taphrina deformans* **22,** stem pitting 414, tomato ringspot virus 414, X-disease **392,** yellow bud mosaic 414, yellow leafroll **392,** witches'-broom 396, wood decay 312, **328,** 344
P. pseudocerasus: leaf curl 24, *Taphrina wiesneri* 24
P. pumila (sand cherry): *Apiosporina morbosa* 152, black knot 152
P. salicina (Japanese plum): *Apiosporina morbosa* 152, bladder plum 22, black knot 152, *Blumeriella jaapii* 66, leaf spot 66, powdery mildew 16, *Sphaerotheca pannosa* 16, *Taphrina* 22, X-disease 392
P. sargentii (Sargent cherry): salt tolerance 454
P. serotina (black cherry): *Apiosporina morbosa* **152,** black knot **152,** *Blumeriella jaapii* 66, burl 494, canker 198, 214, *Diaporthe pruni* 144, dieback 144, 198, flood tolerance 486, frost susceptibility 482, *Fusarium solani* 214, *Ganoderma lucidum* 332, ice glaze injury 490, *Laetiporus sulfureus* **346,** leaf spot 66, *Leucostoma* 198, *Monilinia seaveri* 126, nutrient deficiency 472–**474,** ozone tolerance 466, *Phomopsis pruni* 144, salt damage **452**–454, sulfur dioxide tolerance 468, *Taphrina farlowii* 22, *Trametes versicolor* 342, twig blight 126, 144, wood decay 332, 342, **346**
P. serrulata (Japanese flowering cherry): apple mosaic virus 412, blight **126,** *Blumeriella jaapii* 66, canker **198,** dieback **198,** fasciation **492,** fluoride tolerance 470, freeze injury **482,** green ring mottle virus 408, leaf curl 24, leaf spot 66, *Leucostoma* **198,** *Monilinia laxa* **126,** salt tolerance 454, sulfur dioxide tolerance 468, *Taphrina wiesneri* 24, witches'-broom 24, see also *P. yedoensis*
P. spinosa (blackthorn): *Apiosporina morbosa* 152, black knot 152, canker 198, dieback 198, *Leucostoma* 198
P. subcordata (Sierra plum): *Apiosporina morbosa* 152, black knot 152
P. subhirtella (Higan or weeping cherry): *Cristulariella moricola* 62, flood tolerance 486, leaf spot 62, salt tolerance 454
P. tenella (Russian dwarf almond): *Blumeriella jaapii* 66, leaf spot 66
P. tomentosa (Nanking cherry): *Apiosporina morbosa* 152, black knot 152
P. triloba (flowering almond): *Apiosporina morbosa* 152, apple mosaic virus 412, black knot 152
P. virginiana (chokecherry): *Apiosporina morbosa* **152,** black knot **152,** blight 126, *Blumeriella jaapii* **66,** canker 198, decline **392,** dieback **152,** 198, fluoride tolerance 470, gall **152,** herbicide tolerance 460, leaf spot **66,** *Leucostoma* 198, *Monilinia demissa* 126, mycoplasmalike organism **392,** *Podosphaera clandestina* 16, powdery mildew 16, reddening **392,** salt tolerance 454, sulfur dioxide tolerance 468, viruslike symptoms **408,** X-disease **392**
P. yedoensis (Japanese flowering cherry): *Taphrina wiesneri* 24, see also *P. serrulata*
Prunus necrotic ringspot virus **408–410**
Pseudocercospora ligustri **86**–88
Pseudolarix kaempferi (golden larch): canker 236, *Lachnellula willkommii* 236
Pseudomonas fluorescens 158, 382, 482
P. lauraceum **158**
P. savastanoi 154
P. syringae 154, 160, 164, 482
P. syringae pv. *savastanoi* **154**
P. syringae pv. *syringae* **160,** 164
Pseudonectria pachysandricola **206**
P. rousselliana 206
Pseudoperonospora celtidis 12
Pseudotsuga macrocarpa (big-cone Douglas-fir): needle cast 40, ozone tolerance 466, *Rhabdocline* 40
P. menziesii (Douglas-fir): *Arceuthobium* 430–432, *A. douglasii* 426, **430,** black stain root disease **368,** blight 60, 134, *Botryosphaeria dothidea* 172, Botrytis blight 60, brown felt blight 50, butt rot 322, **324–326,** canker 144, 172, 194–196, *Cenangium* 230, corky root 302, *Cylindrocladium* 294, *Cytospora kunzei* 196, *Diaporthe lokoyae* 144, dieback 144, 172, drought injury 478, dwarf mistletoe 426, **430**–432, *Echinodontium tinctorium* 348, fluoride tolerance 470, *Fomes fomentarius* 346, *Fomitopsis pinicola* 348, freeze injury **480,** *Fusarium oxysporum* **296,** gall **154,** *Ganoderma applanatum* 334, *G. oregonense* 332, herbicide tolerance 460, *Hericium erinaceus* 344, *Herpotrichia juniperi* 50, *Heterobasidion annosum* 314, ice glaze injury 490, *Inonotus tomentosus* 318, *Laetiporus sulfureus* 346, *Leucostoma kunzei* 196, lichens **506,** *Macrophomina phaseolina* 292, *Melampsora medusae* 256, *M. occidentalis* **258,** mycorrhizae **500,** *Mycosphaerella pini* 48, needle blight 48, needle cast **40,** nematode 302, ozone tolerance 466, *Pestalotiopsis funerea* 128, *Phacidium abietis* 52, *P. pini-cembrae* 52, *Phaeocryptopus gaeumannii* **40,** *Phellinus pini* 350, *P. weirii* **324–326,** *Phomopsis* 144, *P. juniperovora* 138, pine wood nematode 380, *Pratylenchus* 302, *Rhabdocline* **40,** root rot 292–**296,** 314, 318, 322, **324, 326,** rust 256–**258,** salt tolerance 454, *Schizophyllum commune* 358, *Sclerophoma pithyophila* 138, *Sirococcus conigenus* 134, snow blight 52, *Sphaeropsis sapinea* 136, sulfur dioxide tolerance 468, tip blight 134–136, *Valsa* 194, *Verticicladiella procera* 370, *V. wageneri* **368,** winter browning 478, wood rot 314, 318, 322, **324, 326,** 334, 344–350, 358, *Xiphinema bakeri* 302
Psidium guajava (guava): anthracnose 122–124, *Armillaria tabescens* 312, *Botryosphaeria dothidea* 172, *Cephaleuros virescens* 28, dieback 168, 172, *Glomerella cingulata* 122–124, *Hendersonula toruloidea* 168, leaf spot 28, root rot 312
Psychotria nervosa (wild coffee): gall 148, *Nectriella pironii* 148
Pteridium (bracken fern): rust 264, *Uredinopsis* 264
Pterocarya stenoptera (wingnut): blackline disease 404, *Botryosphaeria dothidea* 172, blackline 172
Ptychosperma elegans (solitaire palm): lethal yellowing 330
P. macarthurii (Macarthur palm): lethal yellowing 330
Puccinia **250–254**
P. brachypodii 250
P. caricina **250**
P. coronata **250**
P. graminis 250
P. koeleriae 250
P. peridermiospora 252
P. sparganioides **252–254**
Pucciniastrum **260–264**
P. epilobii **260**–262
P. goeppertianum 260–**262**
P. hydrangae **260**
P. pustulatum 260–**262**
P. vaccinii 260
Pueraria lobata (kudzu): *Botryosphaeria dothidea* 172, *B. obtusa* 176, dieback 172, 176, as pest **438**
Punica granatum (pomegranate): *Botryosphaeria dothidea* 172, dieback 172, powdery mildew 16, *Sphaerotheca pannosa* 16
Purshia (antelope bush): actinorhizae 502, nitrogen-fixing nodules 502
Pyracantha (firethorn): anthracnose 122–124, *Botryosphaeria dothidea* 172, *B. obtusa* 176, dieback 172, 176, *Erwinia amylovora* 162, fire blight 162, fluoride tolerance 470, fruit rot **284**–286, *Fusarium oxysporum* f. sp. *pyracanthae* 378, *Glomerella cingulata* 122–124, *Phytophthora cactorum* **284**–286, salt tolerance 454, scab 96–**98,** *Spilocaea pyracanthae* **98,** wilt 378
P. atalantioides: scab 98, *Spilocaea pyracanthae* 98
P. coccinea (firethorn): *Diplocarpon mespili* 64, *Erwinia amylovora* **164,** fire blight **164,** leaf spot 64, scab **98,** *Spilocaea pyracanthae* **98,** sulfur dioxide tolerance 468
P. coccinea cv. 'Lalandei' ('Lalandei' firethorn): ozone tolerance 466
P. crenulata: scab 98, *Spilocaea pyracanthae* 98
P. fortuneana: scab 98, *Spilocaea pyracanthae* 98
P. rogersiana: apple mosaic virus 412, powdery mildew 14, scab 98, *Spilocaea pyracanthae* 98
Pyrola (shinleaf): *Chrysomyxa pirolata* 266, rust 266
Pyrus (pear): *Agrobacterium rhizogenes* 156, *Alternaria* 90, anthracnose 122–124, blight 60, 126, 160–164, *Botryosphaeria dothidea* 172, *B. obtusa* 176, *B. rhodina* 180, *Botrytis cinerea* 60, canker 144, 162–164, *Cerrena unicolor* 358, *Coniothyrium* 82, *Diaporthe* 144, dieback 162–164, 172, 176, 198, 210, *Diplocarpon mespili* 64, *Erwinia amylovora* 162–164, fire blight 162–164, fruit rot 90, *Ganoderma applanatum* 334, *Glomerella cingulata* 122–124, *Gymnosporangium* 242–248, hairy root 156, *Hendersonia* 82, *Heterobasidion annosum* 314, *Hypoxylon mammatum* 222, leaf spot 64, 82, 90, 242–248, *Leucostoma* 198, *Monilinia* 126, *Nectria cinnabarina* 210, *Phellinus igniarius* 342, *Phomopsis* 144, *Phyllactinia guttata* 16, *Phymatotrichum omnivorum* 292, *Podosphaera clandestina* 16, powdery mildew 16, root rot 292, rust 242–248, salt tolerance 454, scab 96–**98,** *Schizophyllum commune* 358, wood rot 314, 334, 342, 358
P. betulifolia (Oriental pear): flood tolerance 486, pear decline 398
P. calleryana (callery pear): *Diplocarpon mespili* 64, flood tolerance 486, *Gymnosporangium globosum* **246,** leaf spot 64, **246,** ozone tolerance 466, pear decline 398, *Pseudomonas syringae* pv. *syringae* 160, rust **246,** salt tolerance 456, tip blight 160
P. communis (common or European pear): blight **160–164,** canker 144, 212, *Diaporthe* 144, *Dematophora necatrix* 306, *Erwinia amylovora* **162–164,** fire blight **162–164,** flood tolerance 486, fluoride tolerance 470, fruit rot 160, 284–286, herbicide tolerance 460, ice glaze injury 490, myco-

P. communis (cont.)
plasmalike organism **398,** *Nectria galligena* 212, *Oxyporus latemarginatus* 328, ozone tolerance 466, pear decline **398,** *Phomopsis* 144, *Phytophthora cactorum* 284–286, *Pseudomonas syringae* pv. *syringae* 160, root rot 306, 328, scab **98,** sulfur dioxide tolerance 468, tobacco mosaic virus 414, tobacco necrosis virus 402, *Venturia pirina* **98,** wood rot **306,** 328, *Xylaria* 306, *X. polymorpha* **306**
P. kawakamii (evergreen pear): salt tolerance 454
P. pyrifolia (Chinese or sand pear): *Armillaria tabescens* 312, pear decline 398, root rot 306, 312, *Xylaria* 306
P. sàtiva: scab 98, *Venturia pirina* 98
P. syriaca: scab 98, *Venturia pirina* 98
P. ussuriensis (ussury pear): pear decline 398, root rot 306, *Xylaria* 306

quail brush, see *Atriplex*
Quercus (oak): anthracnose 110, *Apiognomonia quercina* 110, *Armillaria mellea* 308, 450, *Articularia quercina* 118, *Bjerkandera adusta* 346, black mold 30, bleeding canker 166, 286, borers 450, *Botryodiplodia* 450, *B. gallae* 178, *Botryosphaeria dothidea* 172, *B. melanops* 178, *B. obtusa* 176–178, *B. quercuum* 178, *B. rhodina* 180, butt rot 334, *Brasiliomyces trina* 14, canker 166, 178, 212, 236, canker-rot 354–356, *Ceratocystis fagacearum* 364, *Cerrena unicolor* 358, *Climacodon septentrionalis* 344, *Cronartium* 270, 274, 280, *Cryphonectria* 450, decline 450, defoliating insects 450, *Dendrothele* 168, dieback 172, 176–180, 210, 224, 450, *Diplodia longispora* 178, drought damage 450, *Endothia* 450, *E. gyrosa* 192, *Erysiphe trina* 14, *Fomitopsis pinicola* 348, *Ganoderma* 332, 450, *G. applanatum* 334, *Globifomes graveolens* 346, *Hericium erinaceus* 344, *Heterobasidion annosum* 314, *Hypoxylon* 450, *H. atropunctatum* 224, *H. cohaerens* 226, *H. mammatum* 222, *H. punctulatum* 226, *Inonotus dryadeus* 328, *I. hispidus* 356, kudzu **438,** *Laetiporus sulfureus* 346, leaf blight 110, leaf blister 22, *Lembosia quercina* 30, lichens **506,** lightning injury 490, *Microsphaera* 16, *Microstroma album* 118, mistletoe **420**–422, *Nectria cinnabarina* 210, *N. galligena* 212, oak wilt 364, *Oxyporus populinus* 344, *Phellinus everhartii* 346, *P. igniarius* 342, *P. spiculosus* **354,** *Phoradendron* 420–422, *P. serotinum* **420,** *Phyllactinia guttata* 16, *Phytophthora* 286, *P. cactorum* 286, powdery mildew **14**–16, root rot 306–310, 328, rust 270, 274, 280, salt damage **452**–454, scale insects 450, *Schizophyllum commune* 358, *Sphaeropsis* 450, *Sphaerotheca lanestris* 14, *Stereum gausapatum* 344, *Strumella coryneoidea* 236, *Taphrina caerulescens* 22, tobacco mosaic virus 414, *Urnula craterium* 236, wetwood 382, witches'-broom 118, wood rot 308, 328, 332, 342–348, 354–358, *Xylaria* 306
Q. aegilops: tobacco mosaic virus 418
Q. agrifolia (California live or coast live oak): anthracnose 110, *Apiognomonia quercina* 110, bleeding canker 286, *Brasiliomyces trina* **14,** canker 192, 286, dieback 178, 224, *Diplodia* 178, drought stress 178, *Endothia gyrosa* 192, *Hypoxylon atropunctatum* 224, *Inonotus andersonii* 356, mistletoe 422, *Phytophthora cactorum* 286, powdery mildew **14, 18,** *Sphaerotheca lanestris* 14, **18,** tobacco mosaic virus 418, witches'-broom **18,** wood rot 356
Q. alba (white oak): air pollution damage 462, *Aleurodiscus oakesii* 168, anthracnose **110,** *Apiognomonia quercina* **110,** *Botryosphaeria quercuum* 178, butt rot 304, canker 178, 236, *Ceratocystis fagacearum* 364, *Cronartium quercuum* 274, *C. strobilinum* 280, *Cryphonectria parasitica* 186, dieback 178, **224,** flood tolerance 486, frost susceptibility 482, gall 146, herbicide tolerance 460, *Hypoxylon atropunctatum* **224,** *H. deustum* 304, *H. mediterraneum* 224, ice glaze injury 490, *Inonotus dryadeus* 328, *I. hispidus* 356, leaf blight **110,** oak wilt 364, *Oxyporus latemarginatus* 328, ozone tolerance 466, *Phomopsis* 146, *Pseudomonas syringae* pv. *syringae* 160, root rot 304, 328, rust 274, 280, salt tolerance 454, smooth patch 168, *Strumella coryneoidea* 236, sulfur dioxide tolerance 468, tobacco mosaic virus 418, *Urnula craterium* 236, wood rot 328, 356
Q. arizonica (Arizona white oak): *Aleurodiscus oakesii* **168,** *Cronartium conigenum* 280, *Inonotus andersonii* **356,** *Phellinus everhartii* **346,** rust 280, smooth patch **168,** wood rot **346, 356**
Q. bicolor (swamp white oak): anthracnose 110, *Apiognomonia quercina* 110, *Botryosphaeria quercuum* 178, canker 178, 236, *Cronartium strobilinum* 280, dieback 178, 224, *Hypoxylon mediterraneum* 224, *Oxyporus latemarginatus* 328, rust 280, salt tolerance 454, *Strumella coryneoidea* 236, *Urnula craterium* 236, wood rot 328
Q. cerris (turkey oak): *Inonotus hispidus* 356, viruslike symptoms 418, see also *Q. laevis*
Q. chapmanii (Chapman oak): *Cronartium strobilinum* 280, rust 280
Q. chrysolepis (canyon or canyon live oak): *Cronartium conigenum* 280, leaf scorch 478, powdery mildew 18, rust 280, *Sphaerotheca lanestris* 18
Q. coccinea (scarlet oak): anthracnose 110, *Apiognomonia quercina* 110, bacterial leaf scorch 386, *Botryosphaeria quercuum* 178, butt rot 304, canker 178, 236, *Ceratocystis fagacearum* 364, *Cronartium quercuum* 274, *Cryphonectria parasitica* 186, dieback 178, 224, 478, drought injury 478, fastidious xylem-inhabiting bacteria 386, *Hypoxylon deustum* 304, *H. mediterraneum* 224, ice glaze injury 490, *Inonotus andersonii* 356, *I. dryadeus* 328, oak wilt 364, ozone tolerance 466, root rot 304, 328, rust 274, salt tolerance 454, *Strumella coryneoidea* 236, sulfur dioxide tolerance 468, *Urnula craterium* 236, wood rot 304, 328, 356
Q. douglasii (blue oak): canker-rot **356,** dieback **356,** *Inonotus andersonii* **356,** mistletoe 422, *Phoradendron villosum* 422, tobacco mosaic virus 418, wood rot **356**
Q. dumosa (California scrub oak): *Cronartium quercuum* 274, rust 274
Q. dunnii (Dunn oak): *Cronartium conigenum* 280, rust 280
Q. durandii (Durand oak): anthracnose 110, *Apiognomonia quercina* 110
Q. ellipsoidalis (northern pin oak): anthracnose 110, *Apiognomonia quercina* 110, *Botryosphaeria quercuum* 178, canker 178, *Ceratocystis fagacearum* 364, *Cronartium quercuum* 274, dieback 178, *Inonotus dryadeus* 328, oak wilt 364, root rot 328, rust 274, wood rot 328
Q. emoryi (Emory oak): *Cronartium conigenum* 280, *Inonotus andersonii* 356, mistletoe **422,** *Phoradendron coryae* **422,** rust 280, wood rot 356
Q. engelmannii (Engelmann or mesa oak): tobacco mosaic virus 418
Q. falcata (southern red oak): anthracnose 110, *Apiognomonia quercina* 110, canker 192, *Ceratocystis fagacearum* 364, *Cronartium quercuum* 274, dieback 224, 478, drought injury 478, *Endothia gyrosa* 192, flood tolerance 486, *Globifomes graveolens* **346,** *Hypoxylon atropunctatum* 224, *Inonotus dryadeus* 328, *I. hispidus* 356, oak wilt 364, root rot 328, rust 274, salt tolerance 454, wood rot 328, **346,** 356
Q. falcata var. *pagodifolia* (cherrybark oak): anthracnose 110, *Apiognomonia quercina* 110, *Cronartium quercuum* 274, *Cylindrocladium* 294, flood tolerance 486, *Inonotus hispidus* 356, *Oxyporus latemarginatus* 328, *Phellinus spiculosus* 354, root rot 294, rust 274, wood rot 354–356
Q. gambelii (Gambel oak): *Hypoxylon mediterraneum* 224, *Inonotus andersonii* 356, ozone tolerance 466, sulfur dioxide tolerance 468, wood rot 356
Q. garryana (Garry or Oregon oak): *Aleurodiscus oakesii* 168, anthracnose 110, *Apiognomonia quercina* 110, canker 236, *Cronartium quercuum* 274, *Dendrothele candida* 168, *Hericium erinaceus* 344, *Hypoxylon mediterraneum* 224, *Inonotus andersonii* 356, *I. dryadeus* 328, *I. glomeratus* 356, *I. hispidus* 356, mistletoe **422,** *Oxyporus latemarginatus* 328, *Phaeolus schweinitzii* 322, *Phoradendron villosum* **422,** root rot 322, 328, rust 274, smooth patch 168, *Strumella coryneoidea* 236, *Urnula craterium* 236, wood rot 322, 328, 344, 356
Q. grisea (gray oak): *Cronartium conigenum* 280, rust 280
Q. hypoleucoides (silverleaf oak): *Cronartium conigenum* **270,** rust **270,** 280
Q. ilex (holly oak): powdery mildew 18, *Sphaerotheca lanestris* 18, tobacco mosaic virus 418
Q. ilicifolia (bear or scrub oak): canker 236, salt tolerance 452–454, *Strumella coryneoidea* 236, *Urnula craterium* 236
Q. imbricaria (shingle oak): *Botryosphaeria quercuum* 178, canker 178, *Ceratocystis fagacearum* 364, *Cronartium quercuum* 274, dieback 178, oak wilt 364, *Oxyporus latemarginatus* 328, ozone tolerance 466, rust 274, wood rot 328
Q. incana (bluejack oak): *Cronartium quercuum* 274, *C. strobilinum* 280, rust 274, 280
Q. kelloggii (California black or Kellogg oak): anthracnose 110, *Apiognomonia quercina* 110, herbicide tolerance 460, leaf scorch 478, *Phyllactinia guttata* **14,** powdery mildew **14,** 18, *Sphaerotheca lanestris* 18, tobacco mosaic virus 418
Q. laevis (turkey oak): *Armillaria tabescens* 312, *Ceratocystis fagacearum* 364, *Cronartium quercuum* 274, *C. strobilinum* 280, *Inonotus hispidus* 356, oak wilt 364, root rot 312, rust 274, 280, wood rot 312, 356, see also *Q. cerris*
Q. laurifolia (laurel oak): anthracnose 110, *Apiognomonia quercina* 110, *Armillaria tabescens* 312, canker-rot **356,** *Cronartium quercuum* 274, *C. strobilinum* 280, dieback 224, 478, drought injury 478, flood tolerance 486, *Hypoxylon atropunctatum* 224, *Inonotus glomeratus* 356, *I. hispidus* **356,** leaf spot 110, powdery mildew 18, root rot 312, rust 274, 280, salt tolerance 454, *Sphaerotheca lanestris* 18, wood rot 312, **356**
Q. lobata (valley oak): *Inonotus andersonii* 356, mistletoe 422, *Phoradendron villosum* 422, wood rot 356
Q. lyrata (overcup oak): canker 236, *Ceratocystis fagacearum* 364, dieback 224, flood tolerance 486, *Hypoxylon mediterraneum* 224, oak wilt 364, *Strumella coryneoidea* 236, *Urnula craterium* 236
Q. macrocarpa (bur oak): *Aleurodiscus oakesii* 168, anthracnose 110, *Apiognomonia quercina* 110, *Botryosphaeria quercuum* 178, canker 178, 236, *Ceratocystis fagacearum* 364, *Cronartium quercuum* 274, *C. strobilinum* 280, *Dendrothele griseo-cana* 168, dieback 178, flood tolerance 486, gall 146, *Hypoxylon mediterraneum* 224, oak wilt 364, ozone tolerance 466, *Phomopsis* 146, rust 274, 280, salt tolerance 454, smooth patch 168, *Strumella coryneoidea* 236, sulfur dioxide tolerance 468, *Urnula craterium* 236
Q. marilandica (blackjack oak): anthracnose 110, *Apiognomonia quercina* 110, *Botryosphaeria quercuum* 178, canker 178, 236, *Ceratocystis fagacearum* 364, *Cronartium quercuum* 274, dieback 178, 224, flood tolerance 486, *Hypoxylon atropunctatum* 224, *Inonotus andersonii* 356, leaf spot **110,** oak wilt 364, *Phellinus spiculosus* 354, rust 274, salt tolerance 454, *Strumella coryneoidea* 236, *Urnula craterium* 236, viruslike symptoms 418, wood rot 354–356
Q. michauxii (swamp chestnut oak): flood tolerance 486, *Oxyporus latemarginatus* 328, wood rot 328
Q. muehlenbergii (chinkapin oak, yellow chestnut oak): *Botryosphaeria quercuum* 178, canker 178, dieback 178, flood tolerance 486
Q. myrtifolia (myrtle oak): *Cronartium strobilinum* 280, rust 280, viruslike symptoms **418**
Q. nigra (water oak): anthracnose 110, *Apiognomonia quercina* 110, *Armillaria tabescens* 312, *Botryosphaeria quercuum* 178, canker 178, **192,** 214, 236, canker-rot **354,** *Ceratocystis fagacearum* 364, *Cronartium quercuum* 274, *C. strobilinum* 280, dieback 178, **224,** 478, drought injury 478, *Endothia gyrosa* **192,** flood tolerance 486, *Fusarium solani* 214, *Globifomes graveolens* **346,** herbicide tolerance 460, *Hypoxylon atropunctatum* **224,** *H. mediterraneum* 224, *Inonotus andersonii* 356, *I. dryadeus* 328, *I. hispidus* 356, leaf spot 110, mistletoe 422, oak wilt 364, *Oxyporus latemarginatus* 328, *Phellinus spiculosus* **354,** *Phoradendron serotinum* 422, powdery mildew 18, root rot 312, 328, rust 274, 280, salt tolerance 454, *Sphaerotheca lanestris* 18, *Stereum gausapatum* **344,** *Strumella coryneoidea* 236, *Taphrina caerulescens* 22, *Urnula craterium* 236, wood rot 312, **344–346, 354–356**
Q. nuttallii (Nuttall oak): canker 214, canker-rot **354,** flood tolerance 486, *Fusarium solani* 214, *Hericium erinaceus* **344,** *Inonotus hispidus* 356, *Oxyporus latemarginatus* 328, *Phellinus spiculosus* **354,** punk knot **354,** wood decay 328, **344, 354**–356
Q. oblongifolia (Mexican blue oak): *Cronartium conigenum* 280, *Inonotus andersonii* 356, mistletoe **422,** *Phoradendron coryae* **422,** rust 280, wood rot 356
Q. obtusata (Mexican live oak): tobacco mosaic virus 418
Q. palustris (pin oak): anthracnose 110, *Apiognomonia quercina* 110, bacterial leaf scorch 386, bleeding canker 286, canker **192,** 236, 286, *Ceratocystis fagacearum* 364, *Cronartium quercuum* 274, dieback **192,** *Endothia gyrosa* **192,** fastidious xylem-inhabiting bacteria 386, flood tolerance 486, *Globifomes graveolens* 346, herbicide injury 456, herbicide tolerance 460, *Inonotus dryadeus* 328, nutrient defi-

Q. palustris (cont.)
ciency **474,** oak wilt 364, ozone tolerance 466, *Phytophthora cactorum* 286, root rot 328, rust 274, salt tolerance 454, *Strumella coryneoidea* 236, sulfur dioxide tolerance 468, *Urnula craterium* 236, Verticillium wilt 376, wood rot 328, 346
Q. petraea (Durmast or sessile oak): sulfur dioxide tolerance 468, tobacco mosaic virus 418
Q. phellos (willow oak): anthracnose 110, *Apiognomonia quercina* 110, *Armillaria tabescens* 312, canker 192, 214, canker-rot **354–356,** *Cronartium quercuum* 274, dieback 478, drought injury 478, *Endothia gyrosa* 192, flood tolerance 486, *Fusarium solani* 214, *Inonotus dryadeus* **328,** *I. hispidus* **356,** leaf blister **22,** leaf spot 110, *Phellinus spiculosus* **354,** powdery mildew 18, punk knot **354,** root rot 312, **328,** rust 274, *Sphaerotheca lanestris* 18, *Taphrina caerulescens* **22,** tobacco mosaic virus 418, wood decay 312, **328, 354–356**
Q. prinoides (chinkapin oak): canker 236, *Cronartium quercuum* 274, rust 274, *Strumella coryneoidea* 236, *Urnula craterium* 236, see also *Q. muehlenbergii*
Q. prinus (chestnut oak): adventitious shoots **494,** air pollution damage 462, anthracnose 110, *Apiognomonia quercina* 110, *Armillaria* **308,** *Botryosphaeria quercuum* 178, canker 178, 236, *Cronartium quercuum* 274, *C. strobilinum* 280, dieback 178, **494,** frost susceptibility 482, gall 146, ice glaze injury 490, *Inonotus dryadeus* 328, *Phomopsis* 146, root rot **308,** 328, rust 274, 280, *Strumella coryneoidea* 236, sulfur dioxide tolerance 468, *Urnula craterium* 236, wood decay **308,** 328
Q. pumila (running oak): *Cronartium strobilinum* 280, rust 280
Q. robur (English oak): anthracnose 110, *Apiognomonia quercina* 110, canker 192, *Endothia gyrosa* 192, fluoride tolerance 470, *Hypoxylon mediterraneum* 224, ozone tolerance 466, salt tolerance 454, sulfur dioxide tolerance 468
Q. rubra (northern red oak): anthracnose **110,** *Apiognomonia quercina* **110,** *Armillaria tabescens* 312, bacterial leaf scorch 384–**386,** bleeding canker 286, *Botryosphaeria quercuum* **178,** butt rot 304, canker **178,** 192, 214, **236,** 286, *Ceratocystis fagacearum* **364,** *Ciborinia candolleana* 56, *Cronartium quercuum* 274, dieback **178,** 478, drought injury 478, *Endothia gyrosa* 192, fastidious xylem-inhabiting bacteria 384–**386,** flood tolerance 486, freeze damage **340,** frost susceptibility 482, *Fusarium solani* 214, gall 146, *Ganoderma lucidum* **332,** *Globifomes graveolens* 346, herbicide injury **458,** herbicide tolerance 460, *Hypoxylon deustum* 304, *H. mediterraneum* 224, ice glaze injury 490, *Inonotus andersonii* 356, *I. dryadeus* 328, iron deficiency **474,** leaf blight 56, leaf blister **22,** lightning injury **490,** oak wilt **364,** ozone tolerance 466, *Phellinus spiculosus* 354, *Phomopsis* 146, *Phytophthora cactorum* 286, root rot 304, 312, 328, **332,** rust 274, salt tolerance 454, *Sphaerotheca lanestris* 14, stem crack **340,** *Strumella coryneoidea* **236,** sulfur dioxide tolerance 468, *Taphrina caerulescens* **22,** *Trametes versicolor* **342,** twig blight **110,** *Urnula craterium* **236,** winter sunscald 484, wood rot 304, 312, 328, **332, 342,** 354–356, *Xylemella fastidiosum* 384–**386**
Q. rugosa (netleaf oak): *Cronartium conigenum* 280, rust 280
Q. shumardii (Shumard oak): anthracnose 110, *Apiognomonia quercina* 110, *Botryosphaeria quercuum* 178, canker 178, *Cronartium quercuum* **270,** 274, dieback 178, flood tolerance 486, leaf spot 110, ozone tolerance 466, rust **270,** 274
Q. stellata (post oak): *Aleurodiscus oakesii* 168, anthracnose 110, *Apiognomonia quercina* 110, *Armillaria tabescens* 312, *Botryosphaeria quercuum* 178, canker 178, 192, 236, *Ceratocystis fagacearum* 364, *Cronartium quercuum* 274, *C. strobilinum* 280, *Cryphonectria parasitica* 186, dieback 178, 224, *Endothia gyrosa* 192, flood tolerance 486, gall 146, *Hypoxylon atropunctatum* 224, *Inonotus dryadeus* 328, oak wilt 364, *Phomopsis* 146, root rot 312, 328, rust 274, 280, salt tolerance 454, smooth patch 168, *Strumella coryneoidea* 236, *Urnula craterium* 236, wood rot 312, 328
Q. suber (cork oak): canker 192, *Endothia gyrosa* 192, *Hypoxylon mediterraneum* 224, *Phymatotrichum omnivorum* 292, root rot 292
Q. texana (Texas red oak): *Ceratocystis fagacearum* 364, dieback 224, *Hypoxylon atropunctatum* 224, *H. mediterraneum* **224,** *Inonotus dryadeus* 328, oak wilt 364, root rot 328, wood rot 328
Q. velutina (black oak): anthracnose 110, *Apiognomonia quercina* 110, bleeding canker **166,** *Botryosphaeria quercuum* 178, canker **166,** 178, 192, 236, *Ceratocystis fagacearum* **364,** *Cronartium quercuum* 274, decline **450,** 478, dieback 178, **224, 450,** drought injury 478, *Endothia gyrosa* 192, gall 146, herbicide tolerance 460, *Hypoxylon atropunctatum* **224,** ice glaze injury 490, *Inonotus andersonii* 356, *I. dryadeus* 328, *I. hispidus* 356, leaf blister **22,** oak wilt **364,** ozone tolerance 466, *Phomopsis* 146, root rot 328, rust 274, stem crack **340,** *Taphrina caerulescens* **22,** viruslike symptoms 418, wood decay 328, 356, wound **338**
Q. virginiana (live or southern live oak): anthracnose 110, *Apiognomonia quercina* 110, bark rot **168,** bleeding canker 286, *Botryosphaeria quercuum* 178, *B. rhodina* 180, canker 178, **186,** 192, 286, *Cephalosporium diospyri* 372, *Ceratocystis fagacearum* **364,** 372, *Cronartium quercuum* 274, *C. strobilinum* 280, *Cryphonectria parasitica* **186,** decline 372, **450,** *Dendrothele acerina* 168, dieback 178–180, 224, **450,** *Endothia gyrosa* 192, herbicide tolerance 460, *Hyphoderma baculorubrense* 168, *Hypoxylon atropunctatum* 224, mistletoe **420,** oak wilt 180, **364,** 372, *Perenniporia phloiophila* **168,** *Phialophora parasitica* 372, *Phoradendron tomentosum* **420,** *Phytophthora cactorum* 286, rust 274, 280, salt tolerance 452–454, smooth patch 168, sulfur dioxide tolerance 468
Q. virginiana var. *maritima* (sand live oak): cassytha **436**
Q. wislizenii (interior live oak): leaf scorch 478, *Pseudomonas syringae* pv. *syringae* 160
quince
common, see *Cydonia oblonga*
flowering: *Chaenomeles japonica*

rabbitbrush, see *Atriplex, Chrysothamnus*
Radopholus similis (burrowing nematode) **302**
raisin tree, Japanese, see *Hovenia*
Raphiolepis indica (India- or Indian hawthorn): *Diplocarpon mespili* **64,** *Erwinia amylovora* 162, fire blight 162, leaf spot **64,** Verticillium wilt 376
R. umbellata (Yedda hawthorn): *Diplocarpon mespili* 64, leaf spot 64
rattle, yellow, see *Rhinanthus crista-galli*
Ravenea hildebrandtii (Hildebrand's palm): lethal yellowing 390
redberry, see *Rhamnus crocea*
redbud, see *Cercis canadensis*
Chinese: *C. chinensis*
redroot, see *Ceanothus*
redwood, see *Metasequoia, Sequoia, Sequoiadendron*
coast: *Sequoia sempervirens*
dawn: *M. glyptostroboides*
giant: *Sequoiadendron giganteum*
Rhabdocline pseudotsugae 40
R. weirii **40**
rhabdovirus **400**
Rhamnus (buckthorn): *Botryosphaeria dothidea* 172, *B. obtusa* 176, *Cerrena unicolor* 358, dieback 172, 176, 210, *Heterobasidion annosum* 314, *Nectria cinnabarina* 210, *Phellinus igniarius* 342, *Puccinia coronata* 250, rust 250, *Schizophyllum commune* 358, wood decay 314, 342, 358
R. alnifolia (alder-leaved buckthorn): *Puccinia coronata* 250, rust 250
R. californica (California buckthorn, coffeeberry): *Dematophora necatrix* 306, *Phyllactinia guttata* 16, powdery mildew 16, *Puccinia coronata* 250, root rot 306, rust 250
R. caroliniana (Carolina buckthorn): *Puccinia coronata* 250, rust 250
R. cathartica (common or European buckthorn): *Puccinia coronata* **250,** rust **250,** salt tolerance 454, sulfur dioxide tolerance 468
R. crocea (redberry): *Puccinia coronata* 250, rust 250
R. davurica (green indigo): *Puccinia coronata* 250, rust 250
R. frangula (alder buckthorn): canker 212, 484, dieback 208, *Nectria galligena* 212, *Puccinia coronata* 250, rust 250, salt tolerance 454, sulfur dioxide tolerance 468, *Tubercularia ulmea* 208, 484
R. infectoria (Avignon berry): *Puccinia coronata* 250, rust 250
R. lanceolata (lance-leaved buckthorn): *Puccinia coronata* 250, rust 250
R. purshiana (cascara): *Puccinia coronata* 250, rust 250
R. saxatalis (rock buckthorn): *Puccinia coronata* 250, rust 250
Rhapis (palm): leaf spot 92, *Stigmina palmivora* 92
R. excelsa (lady palm): leaf spot 92, *Stigmina palmivora* 92
Rhinanthus crista-galli (yellow rattle): *Cronartium coleosporioides* 270, rust 270
Rhizobium **502**
Rhizoctonia ramicola 94
R. solani **94**
Rhizophora mangle (mangrove): cassytha 436
Rhizosphaera kalkhoffii **42**
Rhododendron (azalea, rhododendron): *Agrobacterium tumefaciens* **156,** *Alternaria* 90, anthracnose 122–124, blight **60,** 126, **294,** *Botryosphaeria dothidea* 172–**174,** *B. obtusa* 176, *B. rhodina* 180, Botrytis blight **60,** bud burl 494, *Cephaleuros virescens* 28, *Cercospora handelii* **88,** *Chrysomyxa* 266, crown gall **156,** *Cylindrocladiella parva* **294,** *Cylindrocladium* 294, *Diaporthe eres* 144, dieback 144, 172, 176, 180, **284,** *Exobasidium* 26, flower blight 58, flower spot 90, fluoride tolerance 470, gall 26, **146, 156,** *Glomerella cingulata* 122–124, *Hendersonia* 82, herbicide tolerance 460, *Heterobasidion annosum* 314, iron deficiency **474,** leaf blight **174,** leaf spot 74, 82, **88, 142,** lignotuber 494, *Meloidogyne* 298, *Microsphaera* 16, *Monilinia azaleae* 126, mosaic **416,** necrotic ringspot 416, nematode 298, *Ovulinia azaleae* 58, *Pestalopezia rhododendri* 128, *Pestalotiopsis* **128,** *Phomopsis* **142, 146,** *P. rhododendri* 142, *Phytophthora* **284,** *P. cinnamomi* 288–**290,** *P. lateralis* 288, powdery mildew 16, *Pucciniastrum vaccinii* 260, Rhizoctonia blight 94, root knot 298, root rot 284, 288–**290, 294**–296, rust 260, 266, salt tolerance 454, *Sclerotium rolfsii* 296, *Septoria azaleae* 74, southern blight 296, wilt **290, 294,** winter injury **476,** wood rot 314
R. canescens (hoary azalea): herbicide tolerance 456, 460
R. caroliniana (Carolina rhododendron): flower blight 58, *Ovulinia azaleae* 58, ozone tolerance 466
R. catawbiense (catawba rhododendron): *Cercospora handelii* 88, *Chrysomyxa* 266, flower blight **58,** leaf spot 88, *Ovulinia azaleae* **58,** ozone tolerance 466, rust 266, sulfur dioxide tolerance 468
R. indicum (Indian or macranthum azalea): *Armillaria tabescens* 312, *Cercospora handelii* 88, flower blight **58,** leaf spot **74,** 88, *Ovulinia azaleae* **58,** root rot 312, *Septoria azaleae* **74**
R. kaempferi (torch azalea): ozone tolerance 466
R. lapponica (Lapland rosebay): *Chrysomyxa* 266, rust 266
R. macrophyllum (West Coast rhododendron): *Exobasidium vaccinii-uliginosi* 26, leaf spot 142, *Phomopsis ericaceana* 142, red leaf disease 26
R. maximum (rosebay): herbicide tolerance 460
R. minus (Piedmont rhododendron): *Chrysomyxa* 266, rust 266
R. mucronatum (snow azalea): flower blight 58, leaf spot 74, *Ovulinia azaleae* 58, ozone tolerance 466, *Septoria azaleae* 74
R. obtusum (Hinodegiri, Hiryu, or Kirishima azalea): *Cercospora handelii* 88, *Exobasidium vaccinii* **26,** leaf gall **26,** leaf spot 74, 88, ozone tolerance 466, *Septoria azaleae* 74
R. occidentalis (western azalea): herbicide tolerance 460
R. periclymenoides (azalea): *Exobasidium vaccinii* **26,** leaf gall **26**
R. ponticum (rhododendron): *Cercospora handelii* 88, leaf spot 88
R. yedoense var. *poukhanense* (Korean azalea): ozone tolerance 466
Rhodoleia (rhodoleia): *Botryosphaeria dothidea* 172, dieback 172
Rhodotypos (jetbead): *Erwinia amylovora* 162, fire blight 162
Rhus (sumac): *Botryosphaeria dothidea* 172, *B. rhodina* 180, canker 212, dieback 172, 180, *Fusarium oxysporum* f. sp. *callistephi* 378, f. sp. *rhois* 378, herbicide tolerance 460, *Heterobasidion annosum* 314, *Nectria galligena* 212, ozone tolerance 466, *Perenniporia medulla-panis* 168, *Phyllactinia guttata* 16, powdery mildew 16, *Schizophyllum commune* 358, *Verticillium* 376, wilt 376–378, wood rot 168, 314, 358
R. aromatica (fragrant sumac): Verticillium wilt **376**
R. glabra (smooth sumac): fluoride tolerance 470, salt tolerance 454, sulfur dioxide tolerance 468
R. radicans (poison ivy): as pest 438
R. toxicodendron (poison oak): *Cristulariella moricola* 62, leaf spot 62
R. typhina (staghorn sumac): *Cryphonectria parasitica* 186, fluoride tolerance 470, salt tolerance 454, sulfur dioxide tolerance 468
Rhytisma **54**
Ribes (currant, gooseberry): *Alternaria* 90, anthracnose 122–124, *Botryosphaeria dothidea* 172, *B. obtusa* 176, cane blight 82, *Coleosporium jonesii* 268, *Coniothyrium fuckelii* 82, *Cronartium occidentale* 270, *C. ribicola* **270**–

Ribes (cont.)
272, Dematophora necatrix 306, dieback 172, 176, 210, downy mildew 12, fluoride tolerance 470, Glomerella cingulata 122–124, Hendersonia 82, leaf spot 90, Melampsora ribesii-purpureae 258, Microsphaera 16, Nectria cinnabarina 210, Phyllatinia guttata 16, Plasmopara ribicola 12, powdery mildew 14–16, Puccinia caricina 250, root rot 306, rust 250, 258, 268, **270**–272, tobacco mosaic virus 414, Verticillium wilt 376, Xylaria 306

R. alpinum (alpine currant): Botrytis blight 60, Puccinia caricina 250, rust 250, salt tolerance 454, sulfur dioxide tolerance 468

R. americanum (black currant): Puccinia caricina 250, rust 250, see also R. nigrum

R. aureum (golden currant): Puccinia caricina 250, rust 250

R. bracteosum (blue currant): Puccinia caricina 250, Cronartium ribicola 272, rust 250, 272

R. cynosbati (prickly gooseberry): Cronartium ribicola 272, rust 272

R. glandulosum (skunk currant): Puccinia caricina **250**, rust **250**

R. grossularia (common gooseberry): Puccinia caricina 250, rust 250

R. hirtellum (gooseberry): Puccinia caricina 250, rust 250

R. hudsonianum (black currant): Puccinia caricina 250, rust 250, see also R. nigrum

R. inebrians (currant): Coleosporium jonesii 268, rust 268

R. inerme: Cronartium ribicola 272, rust 272

R. lacustre (swamp currant): Puccinia caricina 250, rust 250

R. nigrum (black currant): Cronartium ribicola 272, rust 272, see also R. americanum

R. petiolare: Cronartium ribicola 272, rust 272

R. prostratum: Cronartium ribicola 272, rust 272

R. roezlii: Cronartium ribicola 272, rust 272

R. rotundifolium: Cronartium ribicola 272, rust 272

R. rubrum (red currant): sulfur dioxide tolerance 468, see also R. sativum

R. sanguineum (blood currant): Puccinia caricina 250, rust 250

R. sativum (red currant): Puccinia caricina 250, rust 250, see also R. rubrum

R. triste (red currant): Puccinia caricina 250, rust 250, see also R. rubrum

Robinia neomexicana (New Mexico locust): Phellinus robiniae 344, wood decay 344

R. pseudoacacia (black locust): Alternaria 90, bark rot 168, Botryosphaeria dothidea 172, B. obtusa 176, butt rot 334, canker 144, 212–214, Cerrena unicolor 358, Chalara elegans 294, Diaporthe oncostoma 144, dieback 144, 172, 176, 210, drought injury 476, Erysiphe polygoni 14–16, flood tolerance 486, fluoride tolerance 470, frost susceptibility 482, Fusarium 214, Ganoderma applanatum 334, G. lucidum 332, herbicide tolerance 460, ice glaze injury 490, Laetiporus sulfureus 346, leaf blight 90, mistletoe 422, Nectria cinnabarina 210, N. galligena 212, nematode 302, nitrogen-fixing nodules **502**, Oxyporus latemarginatus 328, ozone tolerance 466, Perenniporia medulla-panis 168, Phellinus robiniae **344**, Phomopsis oncostoma 144, Phymatotrichum omnivorum 292, powdery mildew 14–16, root nodules **502**, root rot 292–294, 306, 332, salt tolerance 452–454, Schizophyllum commune 358, sulfur

dioxide tolerance 468, Tylenchorynchus 302, Verticillium wilt 376, witches'-broom 396, wood rot 168, 328, 332–334, **344**–346, 358, Xylaria 306

rock-rose, see Cistus
root knot nematode 298–300, 376
root nodules **502**
root rot **286–296, 304–332**
shoestring **308–310**
Rosa (rose): Agrobacterium tumefaciens **156**, Alternaria alternata 90, anthracnose 122–124, apple mosaic virus **410**, arabis mosaic virus 410, Armillaria tabescens 312, black spot **66**, Botryosphaeria dothidea 172, B. obtusa 176, B. rhodina 180, Botrytis blight **60**, canker 82, **144**, Cercospora rosicola 76–78, Coniothyrium 82, crown gall **156**, Cylindrocladium 294, Diaporthe eres 144, dieback 172, 176, 180, 202, **210**, Diplocarpon rosae **66**, downy mildew 12, Erwinia amylovora 162, fire blight 162, fluoride tolerance 470, gall **156**, Glomerella cingulata 122–124, hairy root 156, Hendersonia 82, herbicide tolerance 460, Heterobasidion annosum 314, leaf spot **12**, 76–**78**, Leptosphaeria coniothyrium 82, Meloidogyne 298, Mycosphaerella rosicola 76–78, Nectria cinnabarina **210**, nematode 298–302, Peronospora sparsa **12**, petal blight 90, Phomopsis **144**, Phragmidium 238, Phyllosticta rosicola 76–**78**, Phymatotrichum omnivorum 292, powdery mildew 14–**16**, Pratylenchus 302, prunus necrotic ringspot virus 408–**410**, root knot 298, root rot 292–296, 312, rose mosaic **410**, rose ring pattern 410, rust **238**, salt tolerance 454, Sclerotium rolfsii 296, southern blight 296, Sphaerotheca pannosa 14–**16**, tobacco ringspot virus 414, tobacco streak virus 410, tomato ringspot virus 414, Tylenchorynchus 302, Valsa ambiens 202, Verticillium wilt 376

R. floribunda: Phragmidium mucronatum **238**, rust **238**

R. multiflora (multiflora rose): Prunus necrotic ringspot virus 410

R. odorata (tea rose): Pseudomonas syringae pv. syringae 160

R. rugosa (Turkestan rose): salt tolerance 454

R. setigera (prairie rose): tobacco streak virus 410

rose, see Cowania, Kerria, Rosa
cliff: see Cowania
Japanese: K. japonica
multiflora: R. multiflora
prairie: R. setigera
tea: R. odorata
Turkestan: R. rugosa
rosebay, see Rhododendron maximum
Lapland: R. lapponica
Rosellinia necatrix 306
rosemary, bog, see Andromeda
rose-of-sharon, see Hibiscus syriacus
rot (root, trunk, wood) **286–296, 304–358**
Roystonea elata (Florida royal palm): false smut 92, Graphiola phoenicis 92, leaf spot 92, Stigmina palmivora 92

R. regia (Cuban royal palm): leaf spot 92, Stigmina palmivora 92
rubber plant
baby, see Peperomia obtusifolia
Indian: Ficus elastica
rubber tree, Para, see Hevea brasiliensis
Rubus (blackberry, brambles): Agrobacterium 156, Botryosphaeria dothidea 172, B. obtusa 176, crown gall 156, Dematophora necatrix 306, dieback 172, 176, 210, downy mildew 12, Erwinia amylovora 162, fire blight 162, gall 156, hairy root 156, Heterobasidion annosum 314, Nectria cinnabarina 210, Peronospora rubi 12, root rot 306,

tobacco ringspot virus 414, tomato ringspot virus 414, Verticillium wilt 376

R. procerus (blackberry): fastidious xylem-inhabiting bacteria 384
rust **238–282**
rusty mottle virus **408**
rye, see Secale cereale

Sabal (palmetto): leaf spot 92, salt tolerance 454, Schizophyllum commune 358, Stigmina palmivora 92

S. minor (dwarf palmetto): false smut 92, Graphiola phoenicis 92

S. palmetto (cabbage palm): false smut 92, Ganoderma applanatum 334, G. zonatum 330, Graphiola phoenicis 92, leaf spot 92, root rot 330, Stigmina palmivora 92
sage, Texas, see Leucophyllum frutescens
sage brush, see Artemisia, Atriplex
saguaro, see Carnegiea gigantea
salal, see Gaultheria shallon
Salix (willow): Agrobacterium tumefaciens 156, Aleurodiscus oakesii 168, blight **102**, Botryosphaeria dothidea 172, B. obtusa 176, B. rhodina 180, butt rot 334, canker **102**, 184–186, 198–**200**, 212–214, Cerrena unicolor 358, Ciborinia wisconsinensis 56, cold tolerance 480, crown gall 156, Cryptodiaporthe salicella 184–186, dieback 172, 176, 180, 184–186, **200**–202, 210, fluoride tolerance 470, Fomes fomentarius 346, Fomitopsis pinicola 348, frost susceptibility 482, Fusarium lateritium 214, gall 148, 156, Ganoderma applanatum 334, G. lucidum 332, Glomerella miyabeana **102**, Hendersonia 82, herbicide tolerance 460, Heterobasidion annosum 314, Hypoxylon mammatum 222, H. mediterraneum 224, ice glaze injury 490, Inonotus glomeratus 356, I. hispidus 356, leaf blight 56, Melampsora 258, M. epitea 256–**258**, Meloidogyne 298, mistletoe 422, mycoplasmalike organism **396**, Nectria cinnabarina 210, N. galligena 212, Nectriella pironii 148, nematode 298–302, Oxyporus populinus 344, Perenniporia fraxinophila 344, Phellinus igniarius 342, Phoradendron serotinum 422, Phyllactinia guttata 16, Phymatotrichum omnivorum 292, powdery mildew 14–16, Pratylenchus 302, Rhytisma salicinum 54, root knot 298, root rot 292, 332, rust **256**–258, scab **102**, Schizophyllum commune 358, smooth patch 168, tar spot 54, Trametes versicolor 342, Uncinula 14, Valsa ambiens 202, V. sordida **200**, Venturia saliciperda **102**, witches'-broom **396**, wood rot 314, 332–334, 342–348, 356–358

S. alba (white willow) including varieties calva (cricket-bat willow) and vitellina (golden willow): blight 102, canker 184, 198–**200**, Cryptodiaporthe populea 184, C. salicella 184, dieback 184, 200, flood tolerance 486, Glomerella miyabeana 102, Leucostoma 198–200, salt tolerance 454, scab 102, Valsa sordida 200, Venturia saliciperda 102

S. amygdaloides (peach-leaved willow): blight 102, canker 200, dieback 200, Glomerella miyabeana 102, scab 102, Valsa sordida 200, Venturia saliciperda 102

S. appendiculata: canker 200, Valsa sordida 200

S. aurita: blight 102, canker 200, dieback 200, Glomerella miyabeana 102, Leucostoma niveum 200, scab 102, Valsa sordida 200, Venturia saliciperda 102

S. babylonica (weeping willow): adventitious roots **494**, Armillaria **308**, bleeding canker 286, blight 102, canker 184, 200, 286, Cryptodiaporthe salicella 184, dieback 184, 200, Glome-

rella miyabeana 102, ozone tolerance 466, Phytophthora cactorum 286, root rot 308, scab 102, sulfur dioxide tolerance 468, Valsa sordida 200, Venturia saliciperda 102

S. bebbiana (bebb willow): blight 102, canker 184, Cryptodiaporthe salicella 186, dieback 184, Glomerella miyabeana 102, scab 102, Venturia saliciperda 102

S. ×blanda (Niobe willow): blight 102, Glomerella miyabeana 102, scab 102, Venturia saliciperda 102

S. capraea (goat willow): Bjerkandera adusta 346, blight 102, canker 184–**186, 200**, Cryptodiaporthe salicella 184–**186**, dieback 184–**186, 200**, Glomerella miyabeana 102, salt tolerance 454, scab 102, sulfur dioxide tolerance 468, Valsa sordida **200**, Venturia saliciperda 102, wood decay **346**

S. cordata (heart-leaved willow): blight 102, canker 186, Cryptodiaporthe salicella 184, Glomerella miyabeana 102, scab 102, Venturia saliciperda 102

S. daphnoides: canker 200, 222, Hypoxylon mammatum 222, Leucostoma niveum 200, Valsa sordida 200

S. discolor (pussy willow): blight 102, Botryosphaeria dothidea 174, canker 184, Cryptodiaporthe salicella 184, dieback 174, 184, Glomerella miyabeana 102, flood tolerance 486, scab 102, Venturia saliciperda 102

S. fragilis (crack willow): blight 102, canker 184, Cryptodiaporthe salicella 184, Glomerella miyabeana 102, salt tolerance 454, scab 102, sulfur dioxide tolerance 468, Venturia saliciperda 102

S. glauca: canker 200, Leucostoma niveum 200

S. hookerana (Hooker willow): canker 184–186, Cryptodiaporthe salicella 184–186

S. interior (sandbar willow): canker 184, Cryptodiaporthe salicella 184, flood tolerance 486

S. laevigata (polished or red willow): leaf blister 24, Taphrina populi-salicis 24

S. lucida (shining willow): blight 102, canker 200, dieback 200, Glomerella miyabeana 102, scab 102, Valsa sordida 200, Venturia saliciperda 102

S. ×mollissima: blight 102, Glomerella miyabeana 102, scab 102, Venturia saliciperda 102

S. nigra (black willow): blight 102, canker 184, 200, Cryptodiaporthe salicella 184, dieback 184, 200, flood tolerance 486, Glomerella miyabeana 102, Inonotus andersonii 356, mycoplasmalike organism 396, Polyporus squamosus 344, Oxyporus latemarginatus 328, salt tolerance 454, scab 102, sulfur dioxide tolerance 468, tobacco ringspot virus 414, Trametes versicolor 342, Valsa sordida 200, Venturia saliciperda 102, witches'-broom 396, wood decay 328, 342–344, 356

S. nigricans: blight 102, canker 200, dieback 200, Glomerella miyabeana 102, Leucostoma niveum 200, scab 102, Valsa sordida 200, Venturia saliciperda 102

S. pentandra (bay-leaved or laurel willow): blight 102, canker 200, dieback 200, Glomerella miyabeana 102, salt tolerance 454, scab 102, sulfur dioxide tolerance 468, Valsa sordida 200, Venturia saliciperda 102

S. purpurea (purple osier, purple willow): blight **102**, canker 200, dieback 200, Glomerella miyabeana 102, Leucostoma niveum 200, salt tolerance 454, scab **102**, sulfur dioxide tolerance 468, Trametes versicolor 342, Valsa sordida 200, Venturia saliciperda **102**, wood decay **342**

S. repens (creeping willow): canker 200, dieback 200, Leucostoma niveum 200, Valsa sordida 200

S. rigida: witches'-broom 396

S. scoulerana (Scouler willow): canker 184–186, Cryptodiaporthe salicella 184–186, Hypoxylon mammatum 222

S. sericea (silky willow): blight 102, Glomerella miyabeana 102, scab 102, Venturia saliciperda 102

S. sitchensis (Sitka willow): blight 102, canker 198, dieback 198, Glomerella miyabeana 102, scab 102, Venturia saliciperda 102

S. triandra: blight 102, Glomerella miyabeana 102, scab 102, Venturia saliciperda 102

S. viminalis (basket or osier willow): blight 102, canker 200, dieback 200, Glomerella miyabeana 102, Leucostoma niveum 200, salt tolerance 454, scab 102, Valsa sordida 200, Venturia saliciperda 102

salt bush, salt sage, see Atriplex
salt damage 452–454

Sambucus (elder, elderberry): Botryosphaeria dothidea 172, B. obtusa 176, Coniothyrium 82, dieback 172, 176, 200, 210, herbicide tolerance 460, leaf spot 82, Nectria cinnabarina 210, Oxyporus populinus 344, Phellinus igniarius 342, Phyllactinia guttata 16, powdery mildew 16, rhabdovirus 400, salt tolerance 454, Valsa sordida 200, vein chlorosis 400, wood rot 342–344

S. caerulea (blue elder): Oxyporus latemarginatus 328, sulfur dioxide tolerance 468, wood rot 328

S. canadensis (American elder): cherry leafroll virus 404, fastidious xylem-inhabiting bacteria 384, Schizophyllum commune 358, tobacco mosaic virus 414, tobacco ringspot virus 414, tomato ringspot virus 414, Verticillium wilt 376

S. nigra (European elder): cherry leafroll virus 404, fluoride tolerance 470, ozone tolerance 466, sulfur dioxide tolerance 468

S. racemosa (European red elder): cherry leafroll virus 404, fluoride tolerance 470, sulfur dioxide tolerance 468

S. simpsonii (southern elderberry): gall 148, Nectriella pironii 148

sandmyrtle, box, see Leiophyllum

Sapindus (soapberry): anthracnose 122–124, Glomerella cingulata 122–124

Sapium sebiferum (Chinese tallow tree): Alternaria 90, Botryosphaeria dothidea 172, dieback 172, iron deficiency 474, leaf spot 90, Phymatotrichum omnivorum 292, root rot 292, salt tolerance 454

sapstreak 362

Sarcotrochila 52

Sasa (bamboo): iron deficiency 472

Sassafras albidum (sassafras): anthracnose 122–124, Botryosphaeria obtusa 176, B. rhodina 180, canker 212, Cristulariella moricola 62, dieback 176, 180, flood tolerance 486, frost susceptibility 482, Ganoderma lucidum 332, Glomerella cingulata 122–124, ice glaze injury 490, leaf spot 62, mistletoe 422, mycoplasmalike organism 396, Nectria galligena 212, Phellinus igniarius 342, Phoradendron serotinum 422, Phyllactinia guttata 16, powdery mildew 16, root rot 306, 332, salt tolerance 454, sulfur dioxide tolerance 468, Verticillium wilt 376, witches'-broom 396, wood decay 332, 342, Xylaria polymorpha 306

Saxifraga: Melampsora arctica 258, rust 258

scab 68, 96–98, 102

schefflera, see Brassaia actinophylla
dwarf: Schefflera arboricola

Schefflera arboricola (dwarf schefflera): Alternaria panax 90, leaf blight 90

Schinus (pepper tree): Armillaria tabescens 312, Botryosphaeria dothidea 172, dieback 172, Ganoderma applanatum 334, G. lucidum 332, Inonotus hispidus 356, Laetiporus sulfureus 346, Phymatotrichum omnivorum 292, root rot 292, 312, 332, Schizophyllum commune 358, Verticillium wilt 376, wood decay 312, 332–334, 346, 356–358

S. molle (California pepper tree): burl 494

S. terebinthefolius (Brazilian pepper tree): gall 150, Sphaeropsis tumefaciens 150

Schizophyllum commune 358, 372, 484

Sciadopitys verticillata (umbrella pine): salt tolerance 454

Scirrhia acicola 46
S. pini 48

Scleroderris lagerbergii 232
Sclerophoma pithyophila 138
Sclerotinia azaleae 58
S. bifrons 56
S. camelliae 58
S. confundens 56
S. fuckeliana 60

Sclerotium bataticola 292
S. rolfsii 296

Scorias spongiosa 30
screw pine, common, see Pandanus utilis
Scytalidium 168
S. uredinicola 282

sea grape, see Coccoloba uvifera
Secale cereale (rye): Puccinia coronata 250, rust 250

Seiridium 128
S. cardinale 170, 484
S. cupressi 170
S. unicornis 170

senna, see Cassia, C. armata
Septoria 70, 74
S. azaleae 74
S. cornicola 74
S. corni-maris 74
S. exotica 74
S. floridae 74
S. musiva 74
S. populicola 74

Sequoia (redwood, sequoia): Fomitopsis pinicola 348, Pestalotiopsis funerea 128, Schizophyllum commune 358, wood rot 348, 358

S. sempervirens (coast redwood): Botrytis blight 60, burl 494, Cercospora sequoiae 86, Heterobasidion annosum 314, needle blight 86, ozone tolerance 466, root rot 314

Sequoiadendron giganteum (bigtree, giant sequoia): Botryosphaeria dothidea 172, Botrytis blight 60, butt rot 322, Cercospora sequoiae 86, dieback 172, Heterobasidion annosum 314, needle blight 86, ozone tolerance 466, Phaeolus schweinitzii 322, root rot 314

Serenoa repens (saw or scrub palmetto): black mildew 30, Ganoderma zonatum 330, Meliola palmicola 30, root rot 330, salt tolerance 454

serviceberry, see Amelanchier
Allegheny: A. laevis
downy: A. arborea
Utah: A. utahensis
western: A. alnifolia

service tree, see Sorbus domestica
wild: S. torminalis

shadbush, see Amelanchier
shaddock, see Citrus maxima

Shepherdia argentia (buffalo berry): actinorhizae 502, nitrogen-fixing nodules 502, salt tolerance 454, sulfur dioxide tolerance 468

S. canadensis (buffalo berry): Puccinia coronata 250, rust 250

shinleaf, see Pyrola
silk oak, see Grevillea robusta
Silphium: Coleosporium helianthi 268, rust 268

silverbell, see Halesia
silverberry, see Elaeagnus commutata
silver dollar tree, see Eucalyptus cinerea

single-delight, see Moneses uniflora
Sirococcus clavigignenti-juglandacearum 132
S. conigenus 134
S. strobilinus 134

Sistotrema brinkmannii 292
Skimmia (skimmia): tobacco ringspot virus 414

slime flux 382
Smilax (greenbrier): anthracnose 122–124, Glomerella cingulata 122–124

smog damage 462
smoke tree, see Cotinus coggyria
smooth patch 168
snowberry: see Gaultheria, Symphoricarpos
creeping: G. hispidula
mountain: S. oreophilus

snow blight 52
snow brush, see Ceanothus velutinus
snow mold 52
soapberry, see Sapindus
Solidago (goldenrod): Coleosporium asterum 268, C. delicatulum 268, rust 268

sooty mold 30
Sophora japonica (Japanese pagoda tree): canker 214, dieback 210, Fusarium lateritium 214, herbicide injury 458, Nectria cinnabarina 210, ozone tolerance 466, salt tolerance 454, sulfur dioxide tolerance 468, Verticillium wilt 376

Sorbaria (false spirea): Erwinia amylovora 162, fire blight 162

Sorbus (mountain ash): Alternaria 90, Botryosphaeria dothidea 172, B. obtusa 176, canker 144, 212, Diaporthe perniciosa 144, dieback 144, 172, 176, 200, 210, Diplocarpon mespili 64, Erwinia amylovora 162, fire blight 162, Gymnosporangium 242–248, Heterobasidion annosum 314, Hypoxylon mammatum 222, leaf spot 64, 78, 90, Nectria cinnabarina 210, N. galligena 212, Phellinus igniarius 342, Phomopsis mali 144, Phyllactinia guttata 16, Phyllosticta 78, Podosphaera clandestina 16, powdery mildew 16, root rot 242–248, salt tolerance 454, Schizophyllum commune 358, Valsa sordida 200, wood rot 314, 342, 358

S. americana (American mountain ash): anthracnose 122–124, canker 214, Fusarium lateritium 214, Glomerella cingulata 122–124, leaf spot 78, Phyllosticta sorbi 78

S. aria (white bean): apple scab 96, Venturia inaequalis 96

S. aucuparia (European mountain ash): apple scab 96, Botryosphaeria dothidea 174, 484, canker 162, 484, dieback 174, Erwinia amylovora 162, fire blight 162, flood tolerance 486, fluoride tolerance 470, ozone tolerance 466, salt tolerance 454, sulfur dioxide tolerance 468, Venturia inaequalis 96

S. domestica (service tree): apple scab 96, Venturia inaequalis 96

S. scopulina (Greene mountain ash): sulfur dioxide tolerance 468

S. sitchensis (Sitka mountain ash): canker 198, dieback 198, Leucostoma 198, sulfur dioxide tolerance 468

S. torminalis (wild service tree): apple scab 96, Venturia inaequalis 96

sourwood, see Oxydendrum arboreum
southern blight 296
Spanish bayonet, Spanish dagger, see Yucca, Y. gloriosa

Spanish moss, see Tillandsia usneoides
Spartina (cordgrass, marsh grass): Puccinia sparganioides 252–254

S. alterniflora (smooth cordgrass): Puccinia sparganioides 252–254, rust 252–254

Specularia: Coleosporium campanulae 268, rust 268

Sphaeropsis 136, 150, 172, 176–178, 450

S. ellisii 136
S. malorum 176
S. quercina 178
S. sapinea 136, 484
S. tumefaciens 150

Sphaerotheca 14–20
S. humuli 16
S. lanestris 14, 18, 418
S. pannosa 14–16
S. phytoptophila 20

spicebush, see Lindera benzoin
spider flower, see Grevillea
Spilocaea pomi 96
S. pyracanthae 98

spindletree, see Euonymus
Japanese: E. japonica
winged: E. alata

Spiniger meineckellus 316
Spiraea (spirea): Botryosphaeria obtusa 176, dieback 176, 210, Erwinia amylovora 162, fire blight 162, Heterobasidion annosum 314, Nectria cinnabarina 210, ozone tolerance 466, Podosphaera clandestina 16, powdery mildew 16, sulfur dioxide tolerance 468, Verticillium wilt 376

S. ×bumalda (Bumalda spirea): salt tolerance 454

S. ×vanhouttei (bridal-wreath): fluoride tolerance 470, ozone tolerance 466, salt tolerance 454, sulfur dioxide tolerance 468

spirea, false, see Sorbaria
Spiroplasma citri 392, 398
spot anthracnose 68
spruce, see Picea
black: P. mariana
blue or Colorado blue: P. pungens
Brewer: P. breweriana
Engelmann: P. engelmannii
Norway: P. abies
Oriental: P. orientalis
red: P. rubens
Serbian: P. omorika
Sitka: P. sitchensis
white: P. glauca

spurge, caper, see Euphorbia lathyris
Stagonospora 82
Staphylea (bladdernut): Hendersonia 82
Steccherinum septentrionale 344
Stegonosporium 444
S. pyriforme 444
Stegophora ulmea 110
Stereum gausapatum 344
Stigmina 92
S. concentrica 70
S. palmivora 92
stone fruits, see Prunus
Stranvaesia davidiana (stranvaesia): Diplocarpon mespili 64, Erwinia amylovora 162, fire blight 162, leaf spot 64

strawberry, see Fragaria
Strigula 28
Strumella coryneoidea 236
sugarberry, see Celtis laevigata
Suillus 502
sulfur dioxide injury 468
sumac, see Rhus
fragrant: R. aromatica
smooth: R. glabra
staghorn: R. typhina

summer sweet, see Clethra
sun scald 478, 482
sweetfern, see Comptonia peregrina
sweet gale, see Myrica gale
sweet gum, see Liquidambar, L. stryaciflua
Formosan: L. formosana

sweetleaf, see Symplocos
sweet shrub, see Calycanthus occidentalis

Swietenia (mahogany): Alternaria 90, canker 212, leaf spot 90, mistletoe 422, Nectria galligena 212, Phoradendron rubrum 422

sycamore, see Platanus
Arizona: P. wrightii
California: P. racemosa
eastern: P. occidentalis

Sydowia polyspora 138
Symphoricarpos (snowberry): *Alternaria* 90, *Botryosphaeria obtusa* 176, dieback 176, fruit rot 90, *Microsphaera symphoricarpi* 16, ozone tolerance 466, *Podosphaera clandestina* 16, powdery mildew 16, salt tolerance 454
S. albus: anthracnose 122–124, blight 94, *Glomerella cingulata* 122–124, *Herpobasidium deformans* 94, *Podosphaera clandestina* **16,** powdery mildew **16,** sulfur dioxide tolerance 468
S. oreophilus (mountain snowberry): sulfur dioxide tolerance 468
Symplocos (sweetleaf): bud gall 26, *Exobasidium symploci* 26
Syringa (lilac): *Ascochyta syringae* **130,** blight 60, **130,** 160, *Botryosphaeria obtusa* 176, *Botrytis cinerea* 60, cherry leafroll virus 404, *Cylindrocladium* 294, dieback 176, fluoride tolerance 470, *Heterobasidion annosum* 314, *Meloidogyne* 298, mycoplasmalike organism 396, root knot nematode 298, *Phymatotrichum omnivorum* 292, root rot 292–294, salt damage 454, *Trametes versicolor* 342, Verticillium wilt 376, witches'-broom 396, wood rot 314, 342
S. ×*chinensis* (Chinese lilac): blight 160, *Pseudomonas syringae* pv. *syringae* 160
S. josikaea (Hungarian lilac): mycoplasmalike organism 396, witches'-broom **396**
S. ×*persica* (Persian lilac): blight 160, *Pseudomonas syringae* pv. *syringae* 160
S. reflexa hybrid: witches'-broom 396
S. reticulata (Japanese tree lilac): blight 160, herbicide injury **460,** mycoplasmalike organism 396, ozone tolerance 466, *Pseudomonas syringae* pv. *syringae* 160, salt tolerance 454, sulfur dioxide tolerance 468, witches'-broom 396
S. sweginzowii: mycoplasmalike organism 396, witches'-broom 396
S. villosa hybrid: mycoplasmalike organism 396, witches'-broom 396
S. vulgaris (common lilac): anthracnose 122–124, blight **160,** elm mottle virus 406, freeze injury **484,** *Glomerella cingulata* 122–124, herbicide tolerance 460, *Microsphaera syringae* **14**–16, mycoplasmalike organism 396, *Oxyporus latemarginatus* 328, ozone tolerance 466, powdery mildew **14**–16, *Pseudomonas syringae* pv. *syringae* **160,** salt damage 454, sulfur dioxide tolerance 468, *Trametes versicolor* 342, witches'-broom 396, wood rot 342
Syzygium cumini (jambolan, Java plum): algal leaf spot 28, *Cephaleuros virescens* 28

Tagetes (marigold): *Coleosporium pacificum* 268, rust 268
tallow tree, Chinese, see *Sapium sebiferum*
tallowwood, see *Eucalyptus microcorys*
tamarack, see *Larix laricina*
tamarisk, see *Tamarix*
Tamarix (salt cedar, tamarisk): *Laetiporus sulfureus* 346, salt tolerance 454, *Schizophyllum commune* 358, wood decay 346, 358
tangelo, see *Citrus* ×*tangelo*
tangerine, see *Citrus reticulata*
tanoak, see *Lithocarpus densiflorus*
Taphrina **22–24**
T. caerulescens **22**–24
T. communis **22**
T. deformans **22**
T. populi-salicis **24**
T. wiesneri 22–**24**
tar spot **54**
tarweed, see *Madia*
Taxodium distichum (bald cypress): *Cercospora sequoiae* 86, flood tolerance 486, herbicide tolerance 460, leaf blight 86, ozone tolerance 466, salt tolerance 454, *Schizophyllum commune* 358, sulfur dioxide tolerance 468, wood rot 358
Taxus (yew): *Botryosphaeria dothidea* 172, dieback 128, 172, edema 486, herbicide tolerance 460, *Heterobasidion annosum* 314, *Pestalotiopsis funerea* 128, *Phellinus pini* 350, root rot 314, salt tolerance 454, wood rot 350
T. baccata (English yew): *Phomopsis juniperovora* 138, sulfur dioxide tolerance 468
T. brevifolia (Pacific yew): brown felt blight 50, *Herpotrichia* 50, *Phacidium dearnessii* 52, *P. taxicola* 52, snow blight 52, sulfur dioxide tolerance 468
T. canadensis (Canadian yew): *Phacidium taxicola* 52, snow blight 52
T. cuspidata (Japanese yew): edema **486,** flood tolerance 486, fluoride tolerance 470, freeze injury **480,** herbicide tolerance 460, waterlogged soil **486**
tea, see *Camellia sinensis, Ledum*
crystal: *L. palustre*
Labrador: *L. groenlandica*
smooth Labrador: *L. glandulosum*
tea oil plant, see *Camellia oleifera*
Tectona grandis (teak): *Botryosphaeria obtusa* 176, butt rot 304, *Hypoxylon deustum* 304, root rot 304
Thanatephorus cucumeris **94**
Thekopsora hydrangeae 260
T. vaccinii 260
Theobroma cacao (cacao): *Ceratocystis fimbriata* 362
Thielaviopsis basicola 294
thorn, see *Crataegus, Parkinsonia*
cockspur: *C. crus-galli*
Jerusalem: *P. aculeata*
yellow-fruited: *C. flava*
Thrinax morrisii (key palm): leaf spot 92, *Stigmina palmivora* 92
Thuja (arborvitae, cedar): *Alternaria* 90, blight 60, 138, *Botryosphaeria obtusa* 176, *Botrytis cinerea* 60, dieback 128, 176, *Ganoderma applanatum* 334, *Heterobasidion annosum* 314, *Pestalotiopsis funerea* 128, root rot 314, *Schizophyllum commune* 358, *Sclerophoma pithyophila* 138, twig dieback 90
T. occidentalis (arborvitae, eastern arborvitae, northern white cedar): *Armillaria tabescens* 312, autumn browning **498,** blight 128, 138, *Dendrothele nivosa* 168, *Didymascella thujina* 44, dog urine injury **452,** flood tolerance 486, fluoride tolerance 470, herbicide tolerance 460, ice glaze injury 490, leaf blight 44, ozone tolerance 466, *Pestalotiopsis funerea* **128,** *Phacidium infestans* 52, *Phaeolus schweinitzii* 322, *Phellinus pini* 350, *Phomopsis juniperovora* 138, root rot 312, 322, salt tolerance 454, smooth patch 168, snow blight 52, sulfur dioxide tolerance 468, tip blight **128,** *Valsa abietis* 194, wood rot 312, 322, 350
T. plicata (giant arborvitae, western red cedar): blight 138, brown felt blight 50, canker 170, 196, *Didymascella thujina* **44,** dieback 82, *Fomitopsis pinicola* 348, *Hendersonia thyoides* 82, *Herpotrichia juniperi* 50, *Inonotus tomentosus* 318, leaf blight **44,** *Leucostoma kunzei* 196, *Pestalotiopsis foedans* 128, *Phacidium sherwoodiae* 52, *Phaeolus schweinitzii* 322, *Phellinus pini* 350, *P. weirii* 324, *Phomopsis juniperovora* 138, root rot 318, 322–326, *Seiridium cardinale* 170, sulfur dioxide tolerance 468, wetwood 382, wood rot 318, 322–324, 348–350
T. standishii (Japanese arborvitae): *Didymascella thujina* 44, leaf blight 44
Thyronectria austro-americana **204,** 484
Tilia (basswood, linden): anthracnose 122–124, *Aureobasidium pullulans* 30, *Botryosphaeria* 172–176, 180, butt rot 304, canker 172–174, *Cerrena unicolor* 358, *Cladosporium herbarum* 30, *Dendrothele griseo-cana* 168, dieback 172–176, 180, 210, flood tolerance 486, *Fumago vagans* 30, *Glomerella cingulata* 122–124, herbicide injury 460, *Hypoxylon deustum* 304, mistletoe 422, *Nectria cinnabarina* 210, *Phellinus igniarius* 342, *Phoradendron serotinum* 422, *Phyllactinia guttata* 16, powdery mildew 16, root rot 304, smooth patch 168, sooty mold 30, wood rot 342
T. americana (American linden, basswood): *Asteroma tiliae* **78,** bark rot 168, butt rot 304, 334, canker **212,** 236, *Climacodon septentrionalis* 344, *Cristulariella moricola* 162, drought injury 476, fluoride tolerance 470, frost susceptibility 482, *Ganoderma applanatum* 334, herbicide tolerance 460, *Hypoxylon atropunctatum* 224, *H. deustum* 304, ice glaze injury 490, leaf spot 62, **78,** 160, *Nectria galligena* **212,** ozone tolerance 466, *Perenniporia medulla-panis* 168, *Phyllosticta tiliae* 78, *Polyporus squamosus* 344, *Pseudomonas syringae* pv. *syringae* 160, root rot 304, salt tolerance 454, *Strumella coryneoidea* 236, sulfur dioxide tolerance 468, *Urnula craterium* 236, wood decay 168, 334, 344
T. cordata (littleleaf linden, small-leaved European linden): bud burl 494, fluoride tolerance 470, ozone tolerance 466, salt tolerance 454, *Schizophyllum commune* 358, sulfur dioxide tolerance 468, *Trametes versicolor* 342, wood decay 168, 342, 358
T. euchlora (Crimean linden): ozone tolerance 466
T. petiolaris (silver linden): ozone tolerance 466
T. platyphyllos (large-leaved European linden): ozone tolerance 466, salt tolerance 454, sulfur dioxide tolerance 468
Tillandsia recurvata (ball moss or bunch moss) **504**
T. usneoides (graybeard, Spanish moss) **504**
toad flax, bastard, see *Comandra umbellata*
tobacco mosaic virus **414,** 418, 446
tobacco necrosis virus 402
tobacco ringspot virus **414–416,** 446
tomato ringspot virus 414, 446
tomato, tree, see *Cyphomandra betacea*
Torreya taxifolia (stinking cedar): *Alternaria* 90, leaf spot 90
toyon, see *Heteromeles arbutifolia*
Trachycarpus fortunei (windmill palm): lethal yellowing 390
tree-of-heaven, see *Ailanthus altissima*
Trichomerium 30
Truncatella hartigii 128
Tsuga (hemlock): *Arceuthobium* 426, 430–432, blight 60, 134, 138, Botrytis blight 60, *Cerrena unicolor* 358, dwarf mistletoe 426, 430–432, *Fomitopsis pinicola* 348, *Ganoderma tsugae* 332, *Heterobasidion annosum* 314, *Inonotus glomeratus* 356, *Laetiporus sulfureus* 346, lightning injury 490, *Melampsora* 256–258, *Pestalotiopsis funerea* 128, *Phaeolus schweinitzii* 322, *Phellinus pini* 350, *Pucciniastrum* 260, root rot 314, 322, rust 256–260, *Schizophyllum commune* 358, *Sclerophoma pithyophila* 138, *Sirococcus conigenus* 134, wood rot 314, 322, 332, 346–350, 356–358
T. canadensis (eastern hemlock): arsenic toxicity **454,** canker **194**–196, *Cytospora* **194,** *Didymascella tsugae* 44, flood tolerance 486, *Ganoderma tsugae* **332,** heat injury **480,** herbicide tolerance 460, ice glaze injury 490, *Inonotus tomentosus* 318, *Korfia tsugae* 52, *Leucostoma kunzei* 196, *Melampsora abietis-canadensis* 256–**258,** *M. farlowii* 256–258, needle browning **454,** ozone tolerance 466, *Phacidium abietis* 52, *Phaeolus schweinitzii* 322, *Pucciniastrum hydrangae* **260,** *P. vaccinii* 260, root rot 318, 322, rust 256, **258, 260,** salt tolerance 454, snow blight 52, sulfur dioxide tolerance 468, *Valsa* 194, winter browning 478, wood rot 318, 322, 332
T. caroliniana (Carolina hemlock): *Melampsora abietis-canadensis* 258, *M. farlowii* 256–258, *Pucciniastrum hydrangae* 260, rust 256–260
T. heterophylla (western hemlock): *Apostrasseria* 144, *Arceuthobium tsugense* 426, **430**–432, canker 144, *Diaporthe lokoyae* 144, dieback 144, dwarf mistletoe 426, **430**–432, *Echinodontium tinctorium* **348,** *Fomitopsis pinicola* 348, *Ganoderma applanatum* 334, *G. oregonense* **332,** *Inonotus dryadeus* 328, *I. tomentosus* 318, *Phaeolus schweinitzii* 322, *Phellinus weirii* 324, *Pucciniastrum vaccinii* 260, *Sirococcus conigenus* 134, root rot 318, 322–328, **332,** rust 260, sulfur dioxide tolerance 468, wood rot 318, 322–324, 328, **332**–334, **348**
T. mertensiana (mountain hemlock): *Arceuthobium tsugense* 426, **430**–432, black stain root disease 368, brown felt blight 50, dwarf mistletoe 426, **430**–432, *Echinodontium tinctorium* 348, *Fomitopsis pinicola* **348,** Herpotrichia juniperi 50, Inonotus dryadeus 328, *Melampsora medusae* 256, *Phaeolus schweinitzii* 322, *Phellinus weirii* 324, *Pucciniastrum vaccinii* 260, root rot 322–324, 328, rust 256, 260, *Verticicladiella wageneri* 368, wood rot 322–324, 328, **348**
Tubercularia ulmea 182, **208**–210, 484
T. vulgaris 208–210
Tuberculina maxima 272, 278
tulip tree, see *Liriodendron tulipifera*
tung-oil tree, see *Aleurites fordii*
tupelo, see *Nyssa*
swamp: *N. sylvatica* var. *biflora*
water: *N. aquatica*
Tupidanthus calyptratus: Alternaria panax 90, leaf blight 90
Tylenchorynchus **300–302**

Ulex (gorse): *Heterobasidion annosum* 314
Ulmus (elm): *Armillaria tabescens* 312, bark rot 168, *Botryodiplodia hypodermia* 182, *Botryosphaeria* 172–176, butt rot 334, canker 144, 182, 212–214, *Cephalosporium* 372, *Ceratocystis ulmi* 366, *Cerrena unicolor* 358, *Coniothyrium* 82, *Dendrothele griseo-cana* 168, *Diaporthe eres* 144, dieback 144, 172–176, 182, 210, 382, *Dothiorella ulmi* 372, Dutch elm disease 366, *Fusarium solani* 214, *Ganoderma applanatum* 334, *Heterobasidion annosum* 314, *Hypoxylon mammatum* 222, *H. punctulatum* 226, leaf blister 22, leaf spot 82, 108, lightning injury 490, *Microsphaera neglecta* 16, mistletoe 422, *Nectria cinnabarina* 210, *N. galligena* 212, nematode 302, *Oxyporus populinus* 344, ozone tolerance 466, *Perenniporia medulla-panis* 168, *Phellinus igniarius* 342, *Phoradendron serotinum* 422, *Phyllactinia guttata* 16, *Phymatotrichum omnivorum* 292, powdery mildew 16, root rot 292, 312, *Schizophyllum commune* 358, smooth patch 168, *Taphrina ulmi* 22, Verticillium wilt 376, wetwood 382, wood decay 168, 314, 334, 342–346, 358, *Xiphinema americanum* 302

U. alata (winged elm): *Ceratocystis ulmi* 366, Dutch elm disease 366, elm yellows **388**, flood tolerance 486, *Ganoderma lucidum* 332, herbicide tolerance 460, leaf spot 108, mycoplasmalike organism 388, root rot 332, *Stegophora ulmea* 108, wood rot 332

U. americana (American or white elm): *Aleurodiscus oakesii* 168, *Armillaria* **308**, bacterial leaf scorch 384–**386,** *Bjerkandera adusta* **344,** *Botryodiplodia hypodermia* 182, butt rot 304, canker 182, 192, **284,** *Ceratocystis ulmi* 366, *Chalara elegans* 294, cherry leafroll virus 404–**406,** dieback 182, **386,** *Discula* 110, Dutch elm disease **366,** 386, elm mosaic 404–**406,** elm yellows **388,** *Endothia gyrosa* 192, flood tolerance 486, fluoride tolerance 470, gall 148, *Ganoderma* 332, girdling roots 488, herbicide tolerance 460, *Hypoxylon deustum* 304, *H. mediterraneum* 224, ice glaze injury 490, *Inonotus dryadeus* 328, leaf spot **108,** lightning injury **490,** mistletoe 422, *Nectriella pironii* 148, *Perenniporia fraxinophila* 344, *Phoradendron serotinum* 422, *Phytophthora inflata* **284,** *Polyporus squamosus* **344,** root rot 294, 304–**308,** 328, 332, salt tolerance 454, smooth patch 168, *Stegophora ulmea* **108,** sulfur dioxide tolerance 468, tobacco ringspot virus 414, wetwood **382,** wood rot 304, 328, 332, **344,** *Xylaria mali* 306, *Xylemella fastidiosum* 384–**386**

U. carpinifolia (European field or smooth-leaf elm): *Botryodiplodia hypodermia* 182, canker 182, *Ceratocystis ulmi* 366, dieback 182, Dutch elm disease 366, elm mottle virus 406, leaf spot 108, *Stegophora ulmea* 108, sulfur dioxide tolerance 468

U. crassifolia (cedar elm): *Ceratocystis ulmi* 366, Dutch elm disease 366, elm yellows 388, flood tolerance 486, leaf spot 108, *Stegophora ulmea* 108

U. davidiana var. *japonica* (Japanese elm): *Ceratocystis ulmi* 366, Dutch elm disease 366, leaf spot 108, *Stegophora ulmea* 108

U. glabra (Scots or wych elm): bacterial leaf scorch 386, *Botryodiplodia hypodermia* 182, canker 182, *Ceratocystis ulmi* 366, dieback 182, Dutch elm disease 366, elm mottle virus 406, leaf spot 108, salt tolerance 454, *Stegophora ulmea* 108, sulfur dioxide tolerance 468

U. ×*hollandica* (Dutch elm), including cv. 'Belgica' (Belgian elm): leaf spot 108, *Stegophora ulmea* 108

U. laciniata: leaf spot 108, *Stegophora ulmea* 108

U. laevis (European white or Russian elm): *Ceratocystis ulmi* 366, Dutch elm disease 366, leaf spot **108,** *Stegophora ulmea* **108**

U. parvifolia (Chinese elm): *Alternaria* 90, *Botryodiplodia hypodermia* **182,** canker **182,** *Ceratocystis ulmi* 366, *Cristulariella moricola* 62, dieback **182,** Dutch elm disease 366, fluoride tolerance 470, leaf spot 62, 90, **108,** mycoplasmalike organism 396, ozone tolerance 466, *Stegophora ulmea* **108,** sulfur dioxide tolerance 468, witches'-broom 396

U. procera (English elm): *Botryodiplodia hypodermia* 182, canker 182, *Ceratocystis ulmi* 366, dieback 182, Dutch elm disease 366, leaf spot 108, *Stegophora ulmea* 108

U. pumila (Siberian elm): bacterial leaf scorch 386, *Botryodiplodia hypodermia* 182, canker 182, *Ceratocystis ulmi* 366, dieback 182, **208,** Dutch elm disease 366, flood tolerance 486, herbicide injury **458,** leaf spot 108, salt tolerance 454, *Stegophora ulmea* 108, *Tubercularia ulmea* **208,** wetwood **382**

U. rubra (red or slippery elm): canker 284, *Ceratocystis ulmi* 366, Dutch elm disease 366, elm yellows **388, 396,** flood tolerance 486, frost susceptibility 482, leaf spot 108, *Phytophthora inflata* 284, *Stegophora ulmea* 108, witches'-broom **396**

U. serotina (September elm): *Ceratocystis ulmi* 366, Dutch elm disease 366, elm yellows 388, ice glaze injury 490, leaf spot 108, *Stegophora ulmea* 108

U. thomasii (cork or rock elm): *Ceratocystis ulmi* 366, Dutch elm disease 366, *Hypoxylon mediterraneum* 224, leaf spot 108, *Stegophora ulmea* 108

Umbellularia californica (California laurel): anthracnose 122–124, canker 212, dieback 210, *Fomes fomentarius* 346, *Ganoderma applanatum* 334, *G. brownii* 332, *Glomerella cingulata* 122–124, leaf spot **158,** *Nectria cinnabarina* 212, *N. galligena* 212, *Pseudomonas lauraceum* **158,** *Schizophyllum commune* 358, *Trametes versicolor* 342, wood rot 330, 334, 342, 358

umbrella plant, see *Erigonum*

umbrella tree, see *Magnolia tripetala* ear-leaved: *M. fraseri*

Uncinula 14

Uredinopsis **264**

Uredo apacheca 246

Urnula craterium **236**

Ustulina deusta 304

U. vulgaris 304

Vaccinium (bilberry, blueberry, cranberry, farkleberry, huckleberry, whortleberry): *Alternaria* 90, anthracnose 122–124, *Botryosphaeria* 172–176, Botrytis blight 60, *Cristulariella moricola* 62, dieback 172, *Exobasidium* 26, fluoride tolerance 470, *Glomerella cingulata* 122–124, *Heterobasidion annosum* 314, leaf blight 90, leaf blister 26, leaf spot 26, 62, 84, *Microsphaera vaccinii* 16, *Monilinia vaccinii-corymbosae* 126, mummy berry disease 126, *Ophiodothella vaccinii* 84, powdery mildew 16, *Pucciniastrum goeppertianum* 262, *P. vaccinii* 260, red leaf disease 26, rust 260–262, *Schizophyllum commune* 358, swollen shoots 26, tobacco ringspot virus 414, witches'-broom 26, **262,** 396, wood rot 314

V. angustifolium (lowbush blueberry): *Pucciniastrum goeppertianum* 262, rust 262, sulfur dioxide tolerance 468, witches'-broom 262, 396, see also *V. myrtilloides*

V. arboreum (farkleberry): fly speck leaf spot 84, *Ophiodothella vaccinii* 84

V. caespitosum (dwarf bilberry): *Exobasidium dimorphosporum* 26, *E. vaccinii-uliginosi* 26, red leaf disease 26, swollen shoots 26

V. corymbosum (highbush blueberry): canker 144, *Diaporthe vaccinii* 144, dieback 144, gall 146, *Phomopsis* 146, *P. vaccinii* 144

V. membranaceum (mountain or thinleaf huckleberry): *Exobasidium vaccinii-uliginosi* 26, *Pucciniastrum goeppertianum* 262, red leaf disease 26, rust 262, witches'-broom 262

V. myrtilloides (lowbush blueberry): *Pucciniastrum goeppertianum* 262, rust 262, witches'-broom 262, 396, see also *V. angustifolium*

V. ovalifolium (blue whortleberry): *Pucciniastrum goeppertianum* 262, rust 262, witches'-broom 262

V. ovatum (box blueberry, California or evergreen huckleberry): *Exobasidium vaccinii* 26, *E. vaccinii-uliginosi* 26, *Pucciniastrum goeppertianum* **262,** red leaf disease 26, red leaf spot **26,** rust **262,** witches'-broom 262

V. parvifolium (tall red huckleberry):

Chrysomyxa 266, *Pucciniastrum goeppertianum* 262, rust 262, 266, witches'-broom 262

V. scoparium (grouseberry): *Pucciniastrum goeppertianum* 262, rust 262, witches'-broom 262

V. uliginosum (bog bilberry): *Exobasidium vaccinii-uliginosi* 26, *Podosphaera clandestina* 16, powdery mildew 16, *Pucciniastrum goeppertianum* 262, red leaf disease 26, swollen shoots 26, rust 262, witches'-broom 26, 262

V. vitis-idaea (mountain cranberry): *Exobasidium vaccinii-uliginosi* 26, *Pucciniastrum goeppertianum* 262, red leaf disease 26, rust 262, witches'-broom 262

Valsa 184, **194–202,** 444

V. abietis 194

V. ambiens 198–**202**

V. ceratosperma 194, 202

V. cincta 198

V. friesii 194

V. kunzei 16

V. leucostoma 198

V. leucostomoides 202

V. nivea 200

V. pini 194

V. sordida **200**

vectors, fungal: *Olpidium brassicae* 402

vectors, insect

bees 164

beetles

bark (Scolytidae) 364–370

long-horn borers (Cerambycidae) **380**

rove (Staphylinidae) 362

sap (Nitidulidae) 360–364

weevils (Curculionidae) 216, 368–370

flias 164

leafhoppers (Cicadellidae) 384–388, 392–398

planthoppers (Cixiidae) 390

spittlebugs (Cercopidae) 384–386

vectors, nematode: *Xiphinema* 302, 404–408

Veitchia merrillii (Christmas or Manila palm): leaf spot 92, lethal yellowing **390,** *Stigmina palmivora* 92

V. montgomeryana (Montgomery's palm): lethal yellowing 390

Venturia acerina **96**

V. asperata 96

V. chlorospora 102

V. inaequalis **96**

V. macularis 100

V. pirina **98**

V. populina 100

V. saliciperda **102**

V. tremulae **100**

V. rhamni 96

Vernonia (ironweed): *Coleosporium vernoniae* 268, rust 268

Veronica (veronica): leaf spot 74, *Septoria exotica* 74

Verticicladiella abietina 368

V. penicillata **368**

V. procera 318, 368–**370**

V. serpens 368

V. wageneri 314, 324, **368**

Verticillium **374–376,** 444, 454, 472

V. albo-atrum 374–376

V. dahliae **374–376,** 444, 454

Verticillium wilt **374–376**

Viburnum (viburnum): *Botryosphaeria* 172–176, 180, Botrytis blight 60, *Coleosporium viburni* 82, *Coniothyrium viburni* 82, *Cristulariella moricola* 62, *Dematophora necatrix* 306, dieback 172–176, 180, 210, downy mildew 12, *Hendersonia* 82, *Inonotus glomeratus* 356, leaf spot 62, 82, *Microsphaera sparsa* 16, *Nectria cinnabarina* 210, ozone tolerance 466, *Plasmopara viburni* 12, powdery mildew 16, root rot 306, rust 268, Verticillium wilt 376, wood rot 356

V. carlesii (Korean spice viburnum): ozone tolerance 466

V. dentatum (arrowwood): downy mildew 12, flood tolerance 486, *Plasmopara viburni* 12

V. lantana (wayfaring tree): salt tolerance 454, sulfur dioxide tolerance 468

V. lentago (nannyberry): flood tolerance 486

V. opulus (European cranberry bush): downy mildew **12,** gall 146, leaf blight **12,** *Phomopsis* 146, *Plasmopara viburni* **12,** salt tolerance 454, sulfur dioxide tolerance 468

V. tinus (laurustinus): *Armillaria tabescens* 312, root rot 312

V. trilobum (American cranberry bush): flood tolerance 486, salt tolerance 454

Vigna sinensis (cowpea): 400–402, 412

Vinca minor (ground myrtle): *Alternaria* 90, *Botryosphaeria obtusa* 176, dieback 176, leaf spot 90

vines, as pests **438**

Virginia creeper, see *Parthenocissus quinquefolia*

viroid 400

virus **400–418**

alfalfa mosaic 416

apple mosaic **410–412**

arabis mosaic 404, 410, 414–416

broad bean wilt 404

cactus virus 2 418

cactus virus X 418

cherry leaf roll **404–406,** 416

cherry rasp leaf 408

cucumber mosaic 404, 416–418

elm mosaic 406

elm mottle 406

green ring mottle 408

hibiscus chlorotic ringspot 410

poplar mosaic 400

poplar potyvirus **402**

prune dwarf 408

Prunus necrotic ringspot **408–410**

raspberry ringspot 416

rhabdovirus **400**

rusty mottle 408

saguaro 418

Sammons' Opuntia **418**

tobacco black ring 416

tobacco mosaic **414,** 418

tobacco necrosis 402

tobacco rattle 416

tobacco ringspot 404, **414–416**

tobacco streak 410

tomato ringspot 404, 414

zygocactus 418

Viscum album (mistletoe): 420–422

visagna, see *Echinocactus*

Vitis (grapevine): *Alternaria* 90, anthracnose 122–124, *Armillaria tabescens* 312, berry rot 60, 80, 90, *Botryosphaeria* 172–180, Botrytis blight 60, *Coniothyrium* 82, *Cristulariella moricola* 62, *Dematophora necatrix* 306, *Dendrothele griseo-cana* 168, dieback 82, 168, 172–180, downy mildew **12,** fastidious xylem-inhabiting bacteria 384, fluoride tolerance 470, *Glomerella cingulata* 122–124, *Guignardia bidwellii* 80, *Hendersonia* 82, *Hendersonula toruloidea* 168, herbicide injury 456–458, herbicide tolerance 460, leaf spot **12,** 62, 90, *Microsphaeropsis olivacea* 82, nematode 302, ozone tolerance 466, as pest 438, *Phymatotrichum omnivorum* 292, Pierce's disease 384, *Plasmopara viticola* **12,** powdery mildew 14, *Pratylenchus* 302, root rot 292, 306, 312, *Schizophyllum commune* 358, sulfur dioxide tolerance 468, tobacco mosaic virus 414, tobacco ringspot virus 414, tomato ringspot virus 414, *Uncinula necator* 14, Verticillium wilt 376, wood rot 358, *Xylemella fastidiosum* 384

V. labrusca (fox grape): *Endothia gyrosa* 192, *Hypoxylon mediterraneum* 224

V. rotundifolia (muscadine grape): fastidious xylem-inhabiting bacteria 384, Pierce's disease 384, *Xylemella fastidiosum* 384

V. vinifera (European bunch grape): fastidious xylem-inhabiting bacteria **384**, nematode 302, Pierce's disease **384**, root rot 296, *Sclerotium rolfsii* 296, southern blight 296, *Xiphinema index* 302, *Xylemella fastidiosum* **384**
Volutella buxi 206
V. pachysandricola 206

walnut, see *Juglans*
    Arizona: *J. arizonica, J. major*
    black or eastern black: *J. nigra*
    California black: *J. californica*
    English or Persian: *J. regia*
    Hinds or Northern California black: *J. hindsii*
    Japanese: *J. ailantifolia*
    little: *J. microcarpa*
    Manchurian: *J. mandshurica*
    Paradox: *Juglans* cv. 'Paradox'
Washingtonia (Washington palms): *Botryosphaeria dothidea* 172, dieback 172, leaf spot 92, lethal yellowing 390, *Stigmina palmivora* 92
W. robusta (thread palm): *Ganoderma zonatum* 330, leaf spot 92, root rot 330, *Stigmina palmivora* 92
waterlogged soil **486**
water stress **476**
wattle, see *Acacia*
    Cootamundra: *A. baileyana*
    hairy: *A. pubescens*
    orange: *A. cyanophila*
    Sydney golden: *A. longifolia*

waxmallow, Turk's-cap, see *Malvaviscus arboreus* var. *mexicanus*
wax myrtle, see *Myrica cerifera*
wayfaring tree, see *Viburnum lantana*
Weigela (weigela): *Meloidogyne* 298, root knot nematode 298, sulfur dioxide tolerance 468, *Verticillium* wilt 376
wetwood, bacterial **382**
wheat, cow, see *Melampyrum*
whortleberry, blue, see *Vaccinium ovalifolium*
willow, see *Salix*
    basket or osier: *S. viminalis*
    bay-leaved or laurel: *S. pentandra*
    bebb: *S. bebbiana*
    black: *S. nigra*
    creeping: *S. repens*
    cricket-bat: *S. alba* var. *calva*
    crack: *S. fragilis*
    goat: *S. capraea*
    golden: *S. alba* var. *vitellina*
    heart-leaved: *S. cordata*
    Hooker: *S. hookerana*
    Niobe: *S.* ×*blanda*
    peach-leaved: *S. amygdaloides*
    polished or red: *S. laevigata*
    purple: *S. purpurea*
    pussy: *S. discolor*
    sandbar: *S. interior*
    Scouler: *S. scoulerana*
    shining: *S. lucida*
    silky: *S. sericea*
    Sitka: *S. sitchensis*
    weeping: *S. babylonica*
    white: *S. alba*

wilt diseases **290, 362–380**
wingnut, see *Pterocarya stenoptera*
winter fat, see *Ceratoides*
winter creeper, see *Euonymus fortunei*
wintergreen, see *Gaultheria*
winter injury **476–484**
winter sunscald **482–484**
Wisteria (wisteria): *Alternaria* 90, *Botryosphaeria obtusa* 176, dieback 176, herbicide tolerance 460, leaf spot 90, as a pest 438
W. sinensis (Chinese wisteria): tobacco mosaic virus 414
witch hazel, see *Corylus, Hamamelis virginiana*
witches'-broom **18–20, 24, 36, 150, 242–244, 262–266, 394–398, 426–434, 494**
woe vine, see *Cassytha filiformis*
woodbine, see *Parthenocissus quinquefolia*

Xanthomonas campestris pv. *hederae* **158**
X. campestris pv. *juglandis* **160**
X. hederae 158
X. juglandis 160
Xiphinema **301**–302, 404–408
X. americanum **301–302**, 406–408
X. diversicaudatum 404
X. index 302
Xylaria mali 306
X. polymorpha **306**
Xylococculus betulae **440**– 442
Xylemella fastidiosum **384–386**

yaupon, see *Ilex vomitoria*
Yedda hawthorn, see *Raphiolepis umbellata*
yellow-jessamine, see *Gelsemium*
yellow poplar, see *Liriodendron tulipifera*
yellow rattle, see *Rhinanthus crista-galli*
yellowwood, see *Cladrastis lutea*
yew, see *Cephalotaxus, Podocarpus, Taxus*
    Canadian: *T. canadensis*
    English: *T. baccata*
    Harrington plum: *C. harringtoniana*
    Japanese: *T. cuspidata*
    Pacific: *T. brevifolia*
    southern: *P. macrophyllus*
Yucca (Spanish bayonet, Spanish dagger, yucca): *Alternaria* 90, *Botryosphaeria obtusa* 176, *Coniothyrium concentrica* 82, leaf spot 70, 82, 90, *Leptosphaeria obtusispora* 82, *Mycosphaerella yuccae* 70, salt tolerance 454, *Schizophyllum commune* 358
Y. gloriosa (Spanish dagger): *Cercospora concentrica* 70, leaf spot **70**, *Mycosphaerella yuccae* **70**, *Stigmina concentrica* 70

Zanthoxylum (prickly ash): *Botryosphaeria rhodina* 180
zebra plant, see *Aphelandra squarrosa*
Zelkova serrata (Japanese zelkova): dieback 210, leaf spot 108, *Nectria cinnabarina* 210, ozone tolerance 466, *Stegophora ulmea* 108

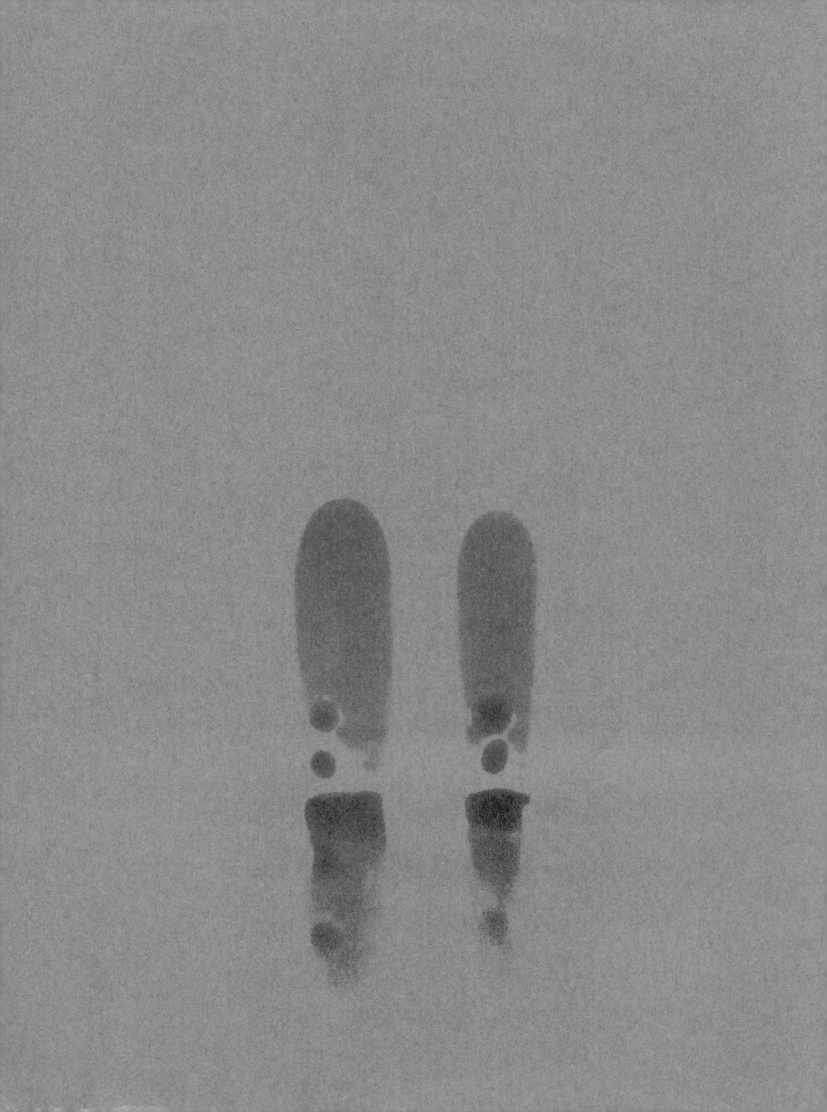

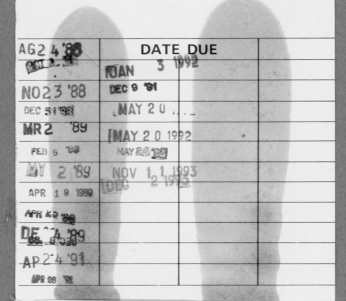